PHYSICS

Third Edition

R. Terrance Egolf
Rachel Santopietro

bju press
Greenville, South Carolina

PHYSICS
Third Edition

R. Terrance Egolf, CDR, USN (Retired)
Rachel Santopietro

Consultants
Eugene Chaffin, PhD
William Lovegrove, PhD
Richard Seeley, MS

Bible Integration
Bryan Smith, PhD
Brian Collins, MDiv

Project Editors
Donald Congdon, MA
Philip Johnston

Cover Design
Nathan Hutcheon

Page Design
BlackDot Group
Elly Kalagayan
Dan VanLeeuwen

Photo Acquisition
Joyce Landis
Rita Mitchell
Holly Nelson
Susan Perry

Text Permissions
Abigail Crockett
Anne Smith

Technical Consultant
Patricia Tirado, Dzign Associates

Illustrators
BlackDot Group
Hattie Bopp
John Cunningham
Aaron Dickey
R. Terrance Egolf
Preston Gravely
John Roberts
David Schuppert
Heather Stanley

Project Managers
Victor Ludlum, MS
Franklin Hall, MA

Second Edition by R. Terrance Egolf and Linda Shumate. First Edition by Rosemary A. Lasell and Paul Wilt

Text acknowledgments appear on page 740

Photograph credits appear on pages 741–743

Greenville, South Carolina 29614

ISBN 978-1-62856-206-4

15 14 13 12 11 10 9 8 7 6 5 4 3 2 1

I Sing the Mighty Power of God

I sing the mighty pow'r of God,
that made the mountains rise;
That spread the flowing seas abroad,
and built the lofty skies.
I sing the wisdom that ordained
the sun to rule the day;
The moon shines full at His command,
and all the stars obey.

I sing the goodness of the Lord,
that filled the earth with food;
He formed the creatures with His Word,
and then pronounced them good.
Lord, how Thy wonders are displayed,
where'er I turn my eye:
If I survey the ground I tread,
or gaze upon the sky!

There's not a plant or flow'r below,
but makes Thy glories known;
And clouds arise, and tempests blow,
by order from Thy throne;
While all that borrows life from Thee
is ever in Thy care,
And ev'rywhere that man can be,
Thou, God, art present there.

Isaac Watts

CONTENTS

Features of This Book

This book is designed just for you. Physics can be intimidating to many students, so we included many features and tools that will help you master this subject. Flip through the next few pages to see some of the helps that you will use this year.

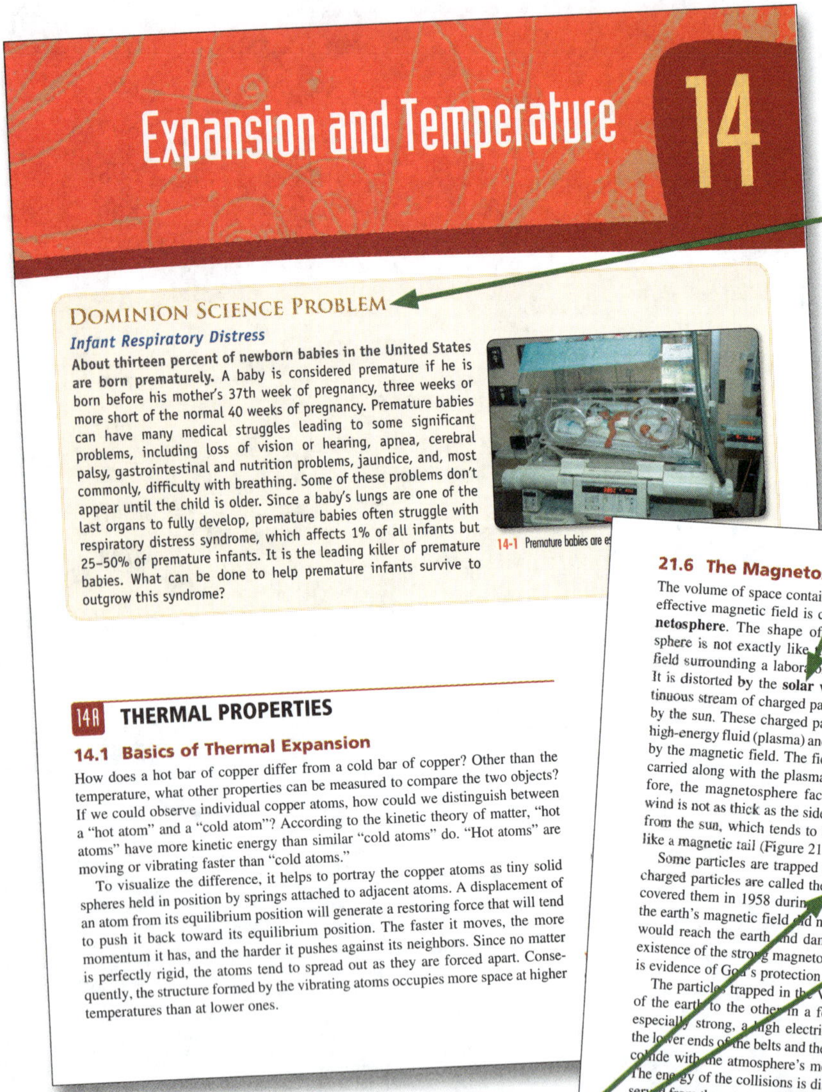

Expansion and Temperature 14

DOMINION SCIENCE PROBLEM

Infant Respiratory Distress

About thirteen percent of newborn babies in the United States are born prematurely. A baby is considered premature if he is born before his mother's 37th week of pregnancy, three weeks or more short of the normal 40 weeks of pregnancy. Premature babies can have many medical struggles leading to some significant problems, including loss of vision or hearing, apnea, cerebral palsy, gastrointestinal and nutrition problems, jaundice, and, most commonly, difficulty with breathing. Some of these problems don't appear until the child is older. Since a baby's lungs are one of the last organs to fully develop, premature babies often struggle with respiratory distress syndrome, which affects 1% of all infants but 25–50% of premature infants. It is the leading killer of premature babies. What can be done to help premature infants survive to outgrow this syndrome?

14-1 Premature babies are es

14A THERMAL PROPERTIES

14.1 Basics of Thermal Expansion

How does a hot bar of copper differ from a cold bar of copper? Other than the temperature, what other properties can be measured to compare the two objects? If we could observe individual copper atoms, how could we distinguish between a "hot atom" and a "cold atom"? According to the kinetic theory of matter, "hot atoms" have more kinetic energy than similar "cold atoms" do. "Hot atoms" are moving or vibrating faster than "cold atoms."

To visualize the difference, it helps to portray the copper atoms as tiny solid spheres held in position by springs attached to adjacent atoms. A displacement of an atom from its equilibrium position will generate a restoring force that will tend to push it back toward its equilibrium position. The faster it moves, the more momentum it has, and the harder it pushes against its neighbors. Since no matter is perfectly rigid, the atoms tend to spread out as they are forced apart. Consequently, the structure formed by the vibrating atoms occupies more space at higher temperatures than at lower ones.

Real-world challenges help you understand why the chapter's content is important.

Boldfaced terms are glossary terms you need to know.

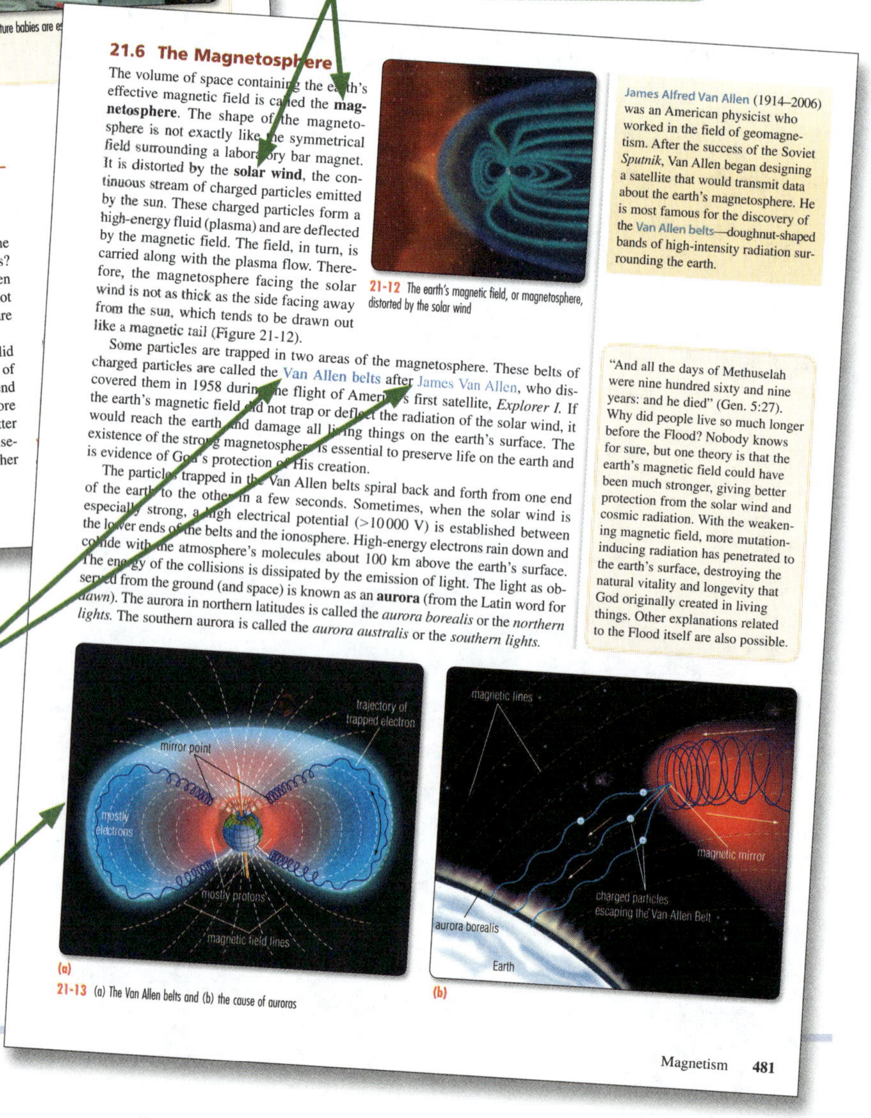

21.6 The Magnetosphere

The volume of space containing the earth's effective magnetic field is called the **magnetosphere**. The shape of the magnetosphere is not exactly like the symmetrical field surrounding a laboratory bar magnet. It is distorted by the **solar wind**, the continuous stream of charged particles emitted by the sun. These charged particles form a high-energy fluid (plasma) and are deflected by the magnetic field. The field, in turn, is carried along with the plasma flow. Therefore, the magnetosphere facing the solar wind is not as thick as the side facing away from the sun, which tends to be drawn out like a magnetic tail (Figure 21-12).

21-12 The earth's magnetic field, or magnetosphere, distorted by the solar wind

Some particles are trapped in two areas of the magnetosphere. These belts of charged particles are called the Van Allen belts after James Van Allen, who discovered them in 1958 during the flight of America's first satellite, *Explorer I*. If the earth's magnetic field did not trap or deflect the radiation of the solar wind, it would reach the earth and damage all living things on the earth's surface. The existence of the strong magnetosphere is essential to preserve life on the earth and is evidence of God's protection of His creation.

The particles trapped in the Van Allen belts spiral back and forth from one end of the earth to the other in a few seconds. Sometimes, when the solar wind is especially strong, a high electrical potential (>10 000 V) is established between the lower ends of the belts and the ionosphere. High-energy electrons rain down and collide with the atmosphere's molecules about 100 km above the earth's surface. The energy of the collisions is dissipated by the emission of light. The light as observed from the ground (and space) is known as an **aurora** (from the Latin word for *dawn*). The aurora in northern latitudes is called the *aurora borealis* or the *northern lights*. The southern aurora is called the *aurora australis* or the *southern lights*.

James Alfred Van Allen (1914–2006) was an American physicist who worked in the field of geomagnetism. After the success of the Soviet *Sputnik*, Van Allen began designing a satellite that would transmit data about the earth's magnetosphere. He is most famous for the discovery of the Van Allen belts—doughnut-shaped bands of high-intensity radiation surrounding the earth.

"And all the days of Methuselah were nine hundred sixty and nine years: and he died" (Gen. 5:27). Why did people live so much longer before the Flood? Nobody knows for sure, but one theory is that the earth's magnetic field could have been much stronger, giving better protection from the solar wind and cosmic radiation. With the weakening magnetic field, more mutation-inducing radiation has penetrated to the earth's surface, destroying the natural vitality and longevity that God originally created in living things. Other explanations related to the Flood itself are also possible.

(a) (b)

21-13 (a) The Van Allen belts and (b) the cause of auroras

Magnetism 481

Blue text links to margin notes which provide extra information or biographical details.

Labeled diagrams help to clarify important concepts.

More complicated equations have been devised to describe more gases over wider ranges of temperatures and pressures. However, the simplest equation, the ideal gas equation, is still the most useful approximation for the behavior of gases under typical conditions.

14.16 Using the Gas Laws to Solve Problems

Like Charles and Gay-Lussac, Forrest Bird was a scientist whose inspiration for his scientific pursuits derived from his passion for aviation. Bird was the son of a World War I pilot. He met Orville Wright when he was young and received his pilot's license before he could drive. Later, when flying for the military during World War II, he became motivated to improve the clumsy, inefficient breathing respirators available to pilots for high-altitude missions. After successfully modifying a captured German respirator, he became interested in the general principles of breathing and artificial respirators. As an experienced pilot, he visualized human respiration as air flowing in and out of the lungs due to the same forces that cause lift in aircraft.

Because of his successful work in designing high-altitude respirators for pilots, the Air Force sent Forrest Bird to medical school to study human respiration, and he completed several residencies. His first successful medical respirator was assembled from an aviation respirator mask and had a manual control knob used to artificially inflate the lungs of a man with emphysema. This device was quickly improved, and Bird began marketing it in 1955 as the first low-cost, mass-produced, portable respirator—the Mark 7. Its ease of use and portability vastly improved the quality of life of patients, such as polio victims, who formerly were confined to iron lungs (see Figure 14-21).

In the late 1960s, Bird experimented with regulating pressure in the cabins of jets. An Air Force doctor convinced him to develop a ventilator for infants. In Bird's own words, "I thought I could take a baby's lungs and treat it like the cabin of the Learjet," though this process would have to be much gentler. In 1969, Bird developed the Baby Bird respirator. After its initial success, it quickly gained wide acceptance, and infant mortality due to respiratory distress syndrome was reduced from 70% to 10%.

Ventilators for both adults and infants use the same principles. In the most basic sense, positive pressure respirators deliver a compressed air/oxygen mixture to a patient to assist or take the place of inhalation, increasing the pressure in the patient's lungs as the volume increases. The patient naturally exhales as the elasticity of the lungs and chest muscles compress the gas, exhausting it through the breathing passages. This process repeats several times per minute, based on a mechanical timer or in response to natural breath demands. Modern infant ventilators have several modes of operation, depending on the needs of the patient.

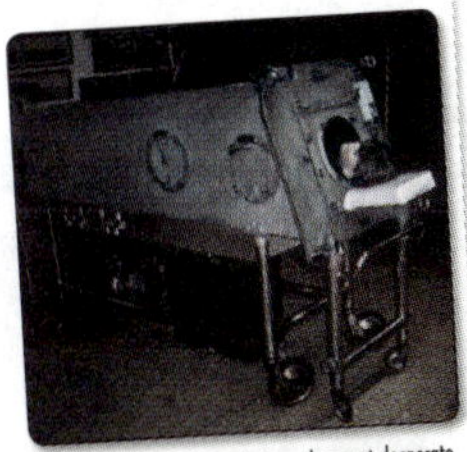

14-21 Before Bird's respirators, the most desperate respiratory patients used an iron lung.

Loving One's Neighbor Through Inventions and New Technologies

Out of desperation, Forrest Bird used an experimental version of his Intrapulmonary Percussive Ventilator because his wife was about to die from an advanced case of emphysema. The invention immediately cleared her lungs of secretions, improved her breathing, and added five more years to her life. Suggest other examples where scientific and technological developments could fulfill the second great commandment.

14C Section Review

1. What are the characteristics of an ideal gas?
2. Describe in your own words Charles's law. Under what conditions is this principle true?
3. Describe in your own words Boyle's law. Under what conditions is this principle true?
4. What does the ideal gas law take into account that the other gas laws assume to be constant?
5. Which of the gas laws applies most directly to the function of a positive-pressure ventilator?

Application of physics to the real-world challenge from the chapter introduction

Bible Integration boxes often contain thought-provoking questions.

Photos add interest to the text.

Problem-Solving Strategy margin boxes give helpful tips for attacking physics problems.

Example problems illustrate how these strategies work.

where θ_B is the Brewster's angle of incidence that produces the greatest polarization, and n_r and n_i are the refraction indices of the media containing the refracted and the incident rays, respectively. Thus, you can easily calculate the Brewster's angle of incidence that produces the greatest polarization in the reflected light beam.

EXAMPLE 25-5

Reflecting on Polarization: Brewster's Angle

What is Brewster's angle for an air-water interface? Use the indices of refraction from Table 24-1 on page 552.

Solution:

You find that $n_{air} = 1.0003$ and $n_{water} = 1.333$. Use Equation 25.6:

$$\tan\theta_B = \frac{n_r}{n_i} = \frac{n_{water}}{n_{air}}$$

$$\theta_B = \tan^{-1}\left(\frac{n_{water}}{n_{air}}\right) = \tan^{-1}\left(\frac{1.333}{1.0003}\right)$$

$$\theta_B \cong 53.\underline{1}1^\circ \cong 53.1^\circ$$

When the sun is about 37° above the horizon (90° − 53.1°), its rays reflecting off the surface of water are horizontally polarized.

25-25 Polaroid sunglasses contain a dichroic material whose vertical transmission axis blocks the horizontally polarized light reflected from the lake.

Problem-Solving Strategy 25.6

Remember that Brewster's angle is the incident angle of the incoming ray of light. The 90° angle is measured between the reflected ray and the refracted ray.

25.13 Polarization by Double Refraction

A second method of polarizing light uses a property of many mineral crystals called *double refraction* (or *birefringence*), which was first observed by Erasmus Bartholinus in 1669. It was so named because the indices of refraction are different through the different crystal planes. Some common examples of crystals with this property are aragonite, calcite, ice, mica, quartz, selenite, sugar, and topaz. In such a crystal, a refracted ray is split into two rays—the *ordinary* (*O*) ray and the *extraordinary* (*E*) ray. The ordinary ray follows Snell's law of refraction exactly. The extraordinary ray does not refract according to Snell's law but will be refracted at an angle dependent upon the microscopic structure of the crystal. It does emerge parallel to the ordinary ray, however. Both rays are polarized so that their planes of polarization are at right angles to each other.

25-26 Double refraction in Iceland spar calcite

A *Nicol prism*, invented in 1828 by William Nicol, is made from a calcite crystal that has been cut and cemented together again with Canada balsam, an adhesive often used in optical applications. The calcite produces an *O*-ray and an *E*-ray, but the *O*-ray is reflected by the Canada balsam so that the *E*-ray continues alone. This device produces polarized light. Nicol prisms may be used to identify unknown minerals. In this application, two Nicol prisms are placed in the path of an unpolarized beam of light with a thin section of the material of interest between them. The polarized light from the first prism enters the sample and is rotated by its crystal lattice. The amount of rotation is detected by rotating the second prism

Erasmus Bartholinus (1625–98) was a Danish mathematician who wrote the first scientific description of the polarization of light in a crystal of Iceland spar calcite.

25-27 Minerals in a thin section of rock sample can be identified with a polarizing microscope by using crossed Nicol prisms.

William Nicol (1768–1851) was a Scottish physicist.

Wave Optics 591

Facets of Physics put the spotlight on fascinating people, technologies, or concepts in physics.

FACETS of PHYSICS

Optical Testing

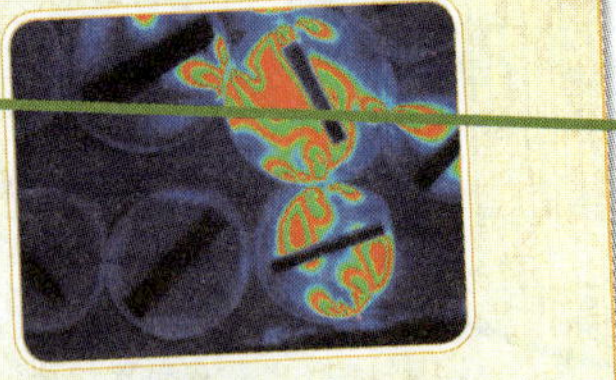

When designing buildings, bridges, aircraft, ships, and so forth, engineers employ a multitude of mathematical formulas to calculate the loads and the resulting stresses and strains on each member of their design. Analyzing loads is a difficult procedure. How can the engineer estimate the maximum weight of trucks, buses, and cars that will ever be on a bridge at any one time? To make sure that a design will do what it is supposed to do, engineers design in a "safety factor." A safety factor of three, for example, means that the design will handle loads three times as heavy as the analysis assumes as the maximum anticipated load.

With complex designs there are serial loads; that is, the load on one member may depend on the loads on other connecting members. This relationship often makes precise calculations of stresses and strains impossible. In the past, it was customary to increase the safety factor to compensate for possible imprecision, but this also increased the construction cost. Recent advances in photoelastic plastics and holography are helping engineers to accurately analyze complex stress patterns and increase the precision of their calculations.

Photoelastic plastics are created with one special property. Under normal conditions they reflect or transmit light evenly in all directions, but under stress they doubly refract light. Polarizers can be positioned to block out all light except the second refracted beam that follows a slightly different path. The engineer places a small-scale plastic model of his design in the light path between two crossed polarizers. At first, the unstressed model will appear black. But as the engineer applies different forces on the model, the stressed areas will appear as light regions. The intensity or color of the light is proportional to the strain in the material. This information can help the engineer to verify or improve his design.

Another new aid in analyzing stress is the hologram. First, an engineer makes a hologram of the machine part or model. After the hologram is completed, he returns it to the laboratory where it was exposed. He then illuminates both the hologram and the object. When the engineer superimposes the hologram on an image of light reflected directly from the object, the image is uniform. However, when the object is stressed, an interference fringe pattern forms in areas of strain. The advantage of this testing method is that the full-sized object made out of the actual materials can be tested. The engineer can experiment with many complex and shifting stresses.

As industrial materials and designs continue to be improved, photoelasticity and holography will be great helps in testing designs and keeping construction costs to a minimum.

The review questions fall into several categories, sometimes identified by bullets, abbreviations, or asterisks. Questions without bullets are recall questions designed to assess your basic knowledge. Questions with a blue bullet (✪) require you to apply your knowledge and often include mathematical calculations. A red bullet (✪) indicates that the topic will require outside reading and research. A "DS" or "DM" signals that a question pertains to material covered in the dominion science or dominion modeling chapter opener topic.

25C Section Review

1. Describe the difference between polarized and unpolarized light sources.
2. What happens to the different light waves that reflect off the surface of glass?
3. How can you tell if a crystalline substance causes double refraction?
4. What happens to the energy in the light waves that are *not* transmitted through a Polaroid filter?
5. At what angle should you align the transmission axis of a Polaroid filter in order to screen out the majority of direct sunlight reflected from a lake's surface?
6. What is the principal cause for the red color of sunsets?
7. ✪ What is Brewster's angle for an air-flint glass interface? The incident light ray is in air. See Table 24-1 for indices of refraction.

25C Objectives

After completing this section, I can

✓ describe the different polarization states of light.

✓ explain dichroism and polarization from Malus's law.

✓ explain how to determine the index of refraction of a material using Brewster's angle of incidence.

✓ discuss birefringence and explain how … mat…

✓ des… duc… tive…

✓ ex…

✓ sh…

Section Objectives summarize the information you should know and skills you should have acquired.

Chapter Review

In Terms of Physics

interference	574	Rayleigh angle (θ_R)	588
constructive interference	574	polarization state	589
destructive interference	574	unpolarized	589
superposition principle	574	polarized	589
interferometer	579	transmission axis	590
diffraction	584	dichroic	590
collimator	587	Malus's law	590
resolving power	588	Brewster's angle (θ_B)	590
Airy discs	588	Brewster's law	590

Complete list of all boldfaced terms with page numbers

Problem-Solving Strategies

25.1 (page 577) The symbol *n* in the interference equation is equal to two quantities: (1) the difference between path lengths (in whole wavelengths of the light) that produce the maximum and (2) the order number of the same maximum.

25.2 (page 577) When analyzing the *minima*, be sure to use the order number (*n*) of the next-higher maximum in the interference formula.

25.3 (page 586) When you determine SDs associated with diffraction grating ruling densities, the number of lines is *counted* and therefore does not affect SDs in calculations. Its reciprocal is a distance, however, so assume that grating spacing has three SDs.

25.4 (page 588) Remember that the Rayleigh angle (θ_R) is given in radians.

25.5 (page 590) Brewster's angle (θ_B) may be given in degrees or radians.

25.6 (page 591) Remember that Brewster's angle is the incident angle of the incoming ray of light. The 90° angle is measured between the reflected ray and the refracted ray.

Problem-Solving Strategies for the whole chapter

Review Questions

1. What is wave interference?
2. Two pulses of equal amplitude but completely out of phase move toward each other. Will they annihilate each other? Explain.
3. The two pulses in each figure travel toward each other. Will they interfere constructively or destructively?

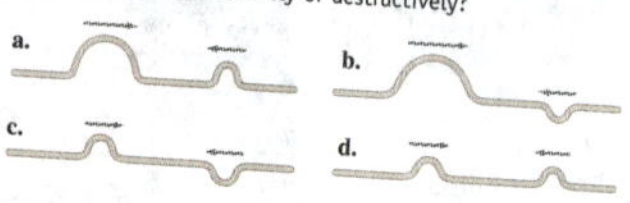

4. Why did Young's experiment demonstrate that light acts as a wave?
5. What does an interference pattern from two close slits look like?
6. The figure below shows a double-slit experiment. Is point *P* dark or bright? Why?

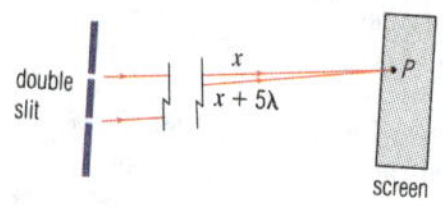

7. **a.** What factors can affect the distance between adjacent interference maxima in a double-slit experiment?
   b. Discuss the effect of each factor.
8. Describe briefly why gasoline spilled on water looks multicolored.
9. Referring to Question 8, why does a thick or opaque film not produce the same multicolored effect?
10. What are some characteristics of laser light?
11. An interferometer was instrumental in refining the precision of which physical constant?
12. What is one common use for diode lasers?
13. What is diffraction?
14. Why did Newton *not* notice light's diffraction?
15. Name three kinds of diffraction.
16. What can diffraction produce at the center of the shadow of a small spherical object?
17. Which kind of diffraction produces a distinct color spectrum?
18. What limits the ability of an optical system to resolve two visual objects that are separated by a small angle?
19. What is polarized light?
20. Name two ways to polarize light.
21. Discuss how you could determine the theoretical index of refraction of an opaque material by using reflected polarization. (For example, the index of refraction for gold is 0.47.)

594 Chapter 25

Chapter Review Questions, supported by diagrams where necessary

How to Use This Book

Though mastering physics may seem a daunting task, you can use this textbook to help you succeed. The following suggestions outline a strategy for staying focused as you assemble the principles and facts necessary to understand and use physics.

1. **Get the big picture.** Use the unit, chapter, and section divisions and headings to help you get oriented. Consider outlining the major ideas of the chapter as you study the material.
2. **Read with attention.** Take the time to pause while reading the text to look at the photos, diagrams, tables, and margin boxes. These features can help you make connections between ideas or visualize hard-to-grasp concepts. Try to think of everyday applications of the ideas you explore.
3. **Learn the highlighted formulas.** Don't just memorize them. Try to understand the relationships between the quantities represented in the formulas. Look for proportionalities—is one quantity directly proportional or inversely proportional to another? What happens to the other quantities when one changes? Create a list of the formulas that you frequently use. Identify the symbolic quantities in each formula. List any restrictions to using the formula. Refer to your list as you complete the homework problems.
4. **Focus on the Dominion Science Problems and Models.** These exercises reveal how science is the best way to understand and exercise dominion over God's creation. Take time to do extra research on the topic or to gain a deeper understanding of the mathematical relationships found in the models.
5. **Prepare for the chapter test.** Review objectives, section and chapter review questions, terms lists, and example problems to prepare for quizzes and tests. Use all of the tools this book provides, including the helpful appendices, glossary, and index.

UNIT 1

A Framework

The science called physics began in the dim ages before the time of Christ, when natural philosophers tried to make sense of the world around them. Studying the history of science in general and physics in particular will reveal that real strides in scientific investigation did not occur until phenomena and processes were observed critically and described mathematically.

Every course of study must have a beginning. Just as a day begins with a sunrise and a building must start with a foundation, you must establish a firm basis for the subject of physics. Unit 1 will provide you with a biblical philosophical framework that is necessary to appreciate the beauty, complexity, and orderliness of the physical universe. The methods of real science will be discussed, as well as its limitations. The wide-open fields of study and employment available to the student of physics will also be presented. The mathematics needed in this course are not exotic, but you may encounter unfamiliar applications of math you may have already studied, so you will study the math concepts that are essential to introductory physics.

A trapped miner understands the immense value of human life.

Foundations of Physics 1

Half-way through their evening shift at the Quecreek Mine, nine Pennsylvania coal miners accidentally breached an abandoned, water-filled mine tunnel that wasn't on the mining map. Water burst from the coal face, inundating them with approximately 60 million gallons of water. The miners, now blocked from the mine's surface in a tunnel flooded with chilly water, fought to survive without light, without heat, and without escape in a four-foot air pocket 240 feet underground.

Rescue operations commenced immediately. The rescuers began by boring a narrow hole into the tunnel air pocket in order to ascertain the miners' location and condition. But how did they manage to aim a six-inch drill hole at a tunnel hundreds of feet below ground?

1-1 The rescue of the miners involved drilling several holes where rescuers thought the miners were trapped.

1A WHY STUDY PHYSICS?

1.1 Physics to the Rescue

The lives of the nine miners trapped on July 24, 2002, in the flooded coal mine were saved due in large part to a special application of physics. Rescuers used several high-accuracy Global Positioning System (GPS) instruments and a correct map of the mine tunnel to determine which drilling position was most likely just above the trapped miners. A few feet of error in any direction would have caused the bore to miss the tunnel or, even worse, to damage a tunnel support and collapse the tunnel. And time was crucial to safeguarding the miners' lives.

After several anxious hours of drilling on the first day, the six-inch drill bit broke through the tunnel roof. The miners, relieved and excited by their imminent rescue, pounded on the drill shaft with their hammers, signaling that they were alive. Three days later after several setbacks, the rescuers pulled the miners to safety. While rescue workers cheered, friends, family, and people around the country rejoiced to see a tragedy turned into a triumph.

Few, however, thought to ask *why*. Why are we so moved by human suffering? Why does it seem right to devote immense resources to research and technology in order to help other humans? Why were these nine miners worth the effort required to rescue them?

1-2 A creationary scientist's conception of the appearance of the earth and universe during the first day of Creation

1.2 God, Man, and Physics

The answer to these questions has its beginnings in Genesis 1:1. In an instant, God started time, defined space, spoke matter into existence, and caused electromagnetic energy—which includes light—to radiate throughout the universe, forever

1-3 Though the new creation was immense, Adam and Eve were its centerpieces.

1-4 One of the constellation of 24 GPS satellites used for navigation and surveying

banishing absolute darkness from the world. Such unimaginable changes involving immense quantities of matter over intergalactic distances occurred according to God's spoken Word. The entire span of creation, from the billions of intricately spiraled galaxies in the vastness of space to the structures of elementary particles in the nucleons of atoms, celebrates and magnifies its Creator. For God created the universe to bring Himself glory.

Yet the new and colossal creation was complete only after God placed man in the middle of it. The first man and woman, Adam and Eve, were not like anything else He had made. They were created in the **image of God** (Gen. 1:26–27). This means that we, as humans, are like God in our abilities to love, create, reason, and discern between right and wrong. We are finite mirrors of His infinite glory. And because we bear God's image, we are precious in His sight. This is the reason that the nine miners in Pennsylvania were worth rescuing.

God not only created us in His image but also gave us a job to do. God directed us to have dominion over all the creatures of the land, sea, and air, as well as over the entire earth. We call this commandment the **Creation Mandate** (Gen. 1:26, 28). We are to exercise wise, informed, and good stewardship of the earth and its resources.

This command is written on the hearts and minds of all humans everywhere. For God's speech is not just true; it is also powerful. When He said, "Let the earth bring forth grass" (Gen. 1:11), His words brought abundant life to a lifeless earth. So also when God said, "Replenish the earth, and subdue it" (v. 28), He brought to life in the human heart a longing for dominion. We all desire to manage the world around us and maximize its usefulness in order to enrich our lives. This is the reason humans mine coal for fuel and develop advanced navigational technologies like GPS.

But not all dominion is good dominion—God never condones the reckless and wasteful exploitation of nature. Proper dominion has two objectives which are best summarized in the two great commandments declared by Jesus. The first commandment is "Thou shalt love the Lord thy God with all thy heart, and with all thy soul, and with all thy mind, and with all thy strength" (Mark 12:30). The second commandment is "Thou shalt love thy neighbor as thyself" (v. 31). As we exercise dominion, we must keep foremost in our minds our total devotion to God and also our love for our fellow image-bearers.

What does all of this have to do with physics? To exercise dominion we must learn about the world around us. Today, physics pervades the technologies that make our existence safer, healthier, and more convenient. It is also through physics that we have come to understand the design, structure, and workings of the material universe, and thus find even more reasons to praise God for His marvelous handiwork.

The first and second commandments are to love God and to love others. Jesus said that on these two laws hang all of God's other commands. (Matt. 22:40)

1-5 The Fall affected more than humans: all of nature came under the Curse.

1.3 Physics in a Fallen World

When Adam introduced sin into the human race, God punished him with a curse that affects all of creation (Gen. 3:14–19; Rom. 8:22). As God decreed, humans, though possessing immortal souls, became mortal, as did all animal life. Death reigned in the world. Subtle but important changes to the laws upholding the universe may have also occurred, as the material universe began its deterioration. Living things aged, diseases appeared, some animals became carnivores, and varieties of plants became weeds, some bearing thorns and brambles.

After the Fall, the physical world and the rest of the universe began to deteriorate. We do not know, but we can infer that nuclear decay began or dramatically increased, changing elements into other kinds of elements. Corrosion and other adverse chemical reactions also may have started. It is possible that the earth's crust became unstable and geologic tectonic activity began. In the heavens, stars exploded into supernovas, and in our own solar system there seems to have been a past period of intense meteoric bombardment that cratered the planets and moons. No longer is God's creation "very good." It is marred, with only part of its former glory remaining.

1-6 This object is the Crab Nebula (M1), the remnant of a star that exploded as a supernova in AD 1054.

Although the curse on the material universe is tragic, the most grievous aspects of the curse affected God's image-bearers themselves. As humans, we all inherited Adam's sinfulness, and it is present throughout our body and soul. This pervasive twisting of every person's being is called *human depravity*. Because we are now corrupted and fallen from our created state, we do not think or love as we should.

The Bible clearly teaches that the fallen mind is deeply flawed (1 Cor. 2:14; Eph. 4:17–18). This flaw is caused, not by a lack of intelligence, but by wrong loves. Since the Fall, human nature leads us to love ourselves supremely, and on our own we cannot bring ourselves to love God or others as we should. This misdirected love drives us to think about God, the world, and ourselves in twisted ways. We sin in rejecting evidences of God's handiwork and in attributing the marvels of creation to things other than Him. The apostle Paul stated that in his day people attributed God's work to a multitude of gods (Rom. 1:23) modeled on man and animals. Similarly, people today often attribute what God has done to a number of impersonal laws of nature. The details are different, but the result is the same. Humans—ancient and modern—attempt to understand the world without thinking about the one true God.

1-7 The Apostle Paul reasoned with the mockers and doubters on Mars Hill in Athens (Acts 17:18–32).

We cannot escape the fact that we live in a fallen world. In addition to the aging and deterioration of the physical world, unexpected disasters like earthquakes, tsunamis, and even the flooding of mines occur regularly. But what is even more tragic is that humans tend to blame God for their sufferings. Some show their rejection of God by openly denouncing Him. Others condemn Him by asserting that He is irrelevant to the problems of human existence. Faced with pain and suffering, some people attempt to cope by reasoning in the following way: "If there is a God and He permits disaster, calamity, and death without doing anything to prevent them from happening, then He must be a monster. So maybe He's not paying attention to everything—and maybe He is not even 'out there.' Isn't it better to say that God is not there than to say that He's a monster?" All such reasoning ignores the Bible's own explanation for human suffering. Moments after the Fall, God decreed, "Cursed is

1-8 Collapsed buildings in China's south-central Sichuan province, after the May 12, 2008, earthquake. More than 69 000 people died, more than 374 000 were injured, and more than 18 000 people are still missing. Such events are evidence for the fact that we live in a fallen world.

Following the 2004 Indonesian earthquake and tsunami, Richard Dawkins, a prominent British atheist and biology professor, wrote in a newspaper column, "It is psychologically possible to derive comfort from sincere belief in a nonexistent illusion, but—silly me—I thought believers might be disillusioned with an omnipotent being who had just drowned 125,000 innocent people (or an omniscient one who failed to warn them). Of course, if you can derive comfort from such a monster, I would not wish to deprive you."

Dawkins, Richard, *The Guardian*, 31 December 2004

the ground *for thy sake*" (Gen. 3:17; emphasis added). We are to blame. We have rebelled against God's dominion, and, as a consequence, He allows the world to rebel against our dominion. God is not a monster; God is just.

1-9 Aging and death are just a few of the Fall's consequences.

1.4 God's Work of Redemption

We live in a world of disobedience, suffering, and death, but God has not abandoned us. Even while pronouncing a curse on the serpent, God promised in Genesis 3:15 that He would redeem His world: "I will put enmity between thee [the serpent] and the woman, and between thy seed and her seed; it shall bruise thy head, and thou shalt bruise his heel." The story of Scripture is the story of how God is restoring His people to their original stature and calling. The center of this story is the triumph of Jesus Christ—God's own Son—over demons, sin, and death. In this triumph Jesus has dealt the kingdom of Satan a fatal blow and delivered His people from the kingdom of darkness, placing them into His own kingdom of light (Col. 1:13). These members of the "seed of the woman" will forever enjoy the blessings of the kingdom of God.

As we examine the story of Scripture, we learn that God's redemption is physical as well as spiritual. One of the ways that God restores the material world is through human beings who obey the Creation Mandate. King Solomon is a prime example; the vast wisdom that God gave him included knowledge of plants, animals, and natural processes. "[Solomon] spake of trees, from the cedar tree that is in Lebanon even unto the hyssop that springeth out of the wall: he spake also of beasts, and of fowl, and of creeping things, and of fishes" (1 Kings 4:33). Redemption is restoration, and here in these verses we read that Solomon accomplished—to some degree—the work of the Creation Mandate. Yet these scientific endeavors did not prevent Solomon from emphasizing the spiritual aspects of God's redemption. In fact, his scientific efforts may have made those spiritual aspects more profound. The next verse states that great people from near and far came to hear Solomon's wisdom—a wisdom that concerned God's kingdom as well as trees, birds, and fish.

1-10 Scientists are restored to the tasks of exercising dominion.

Jesus's life and ministry also demonstrate that redemption is physical as well as spiritual. He spent His days preaching the gospel of the kingdom and engaging in works of mercy as He healed the sick and cast out demons (Matt. 4:23–24). These works of mercy were not ploys for generating a crowd. They were Jesus's way of demonstrating that the gospel He preached was true. The kingdom had indeed come, and God was showing mercy to sinners by offering them freedom from sin and its consequences, including demon possession, injury, discouragement, and disease.

Christians are followers of this Jesus. He has called Christians to go into the entire world and preach His gospel (Matt. 28:18–20). If we are serious about imitating our Lord, we will take seriously our obligation to exercise dominion in a way that glorifies God and shows love to our fellow humans.

1-11 Saving human lives is one of the most important goals of exercising dominion. Here, one of the trapped miners arrives at the mine's surface.

1.5 Redemption and Physics

Physics is an amazing tool for exercising benevolent dominion. The miners' rescue mentioned earlier is just one example of the power of physics to save lives. But some believe that Christians should not give priority to studying science, even if it has the potential to help others. Yet if a Christian cannot get excited about saving lives, who can? Christians are the ones who believe that each person possesses intrinsic worth because he bears God's image. Christians also believe that the Son of God has commanded His people not only to proclaim the gospel but also to do as the Good Samaritan did. The same Lord who said, "Go ye therefore, and teach all nations" (Matt. 28:19), also said, "Go, and do thou likewise" (Luke 10:37).

By studying physics for the purpose of exercising merciful, God-glorifying dominion, we begin to redeem physics itself. As long as physicists believe that science is independent of and superior to belief in God, physics will remain a "high thing that exalteth itself against the knowledge of God" that must be brought "to the obedience of Christ" (2 Cor. 10:5). We accomplish this goal, not by constructing a totally new academic subject, but by placing physics in its proper context.

Viewed from the perspective of a Christian worldview, physics is a tool that people use to declare the glory of the almighty Creator and to help their fellow image-bearers live well in God's presence. You should approach this course as one step toward making physics what it ought to be. Even consider devoting your life to making a difference for God in the world of physics.

See Subsection 1.7 for a discussion of worldviews and their significance to the function of science.

1A Section Review

1. What single characteristic of humans makes them unique? How does this affect their value or worth?
2. Read Genesis 1:26–30. What do we call the commandment to humans given in these verses? Discuss in your own words how people can obey this commandment.
3. How do the commandments of Mark 12:30–31 relate to the commandment given in Genesis 1:28? What does physics have to do with these commandments?
4. What is the most significant negative consequence of the Fall when it comes to doing physics?
5. Why is it much harder to have dominion over a fallen creation?
6. How can a Christian scientist participate in God's present redeeming of the world?
7. What is physics if physicists deny God's involvement or even His existence? Give a Bible verse that expresses this idea.
8. (True or False) There is little historical or biblical evidence that humans conducted detailed studies of nature prior to about 300 BC (the time of the Greeks).

1A Objectives

After completing this section, I can

✓ state why human life is valuable.

✓ identify the purpose for doing physics.

✓ relate some of the consequences of God's curse after the Fall.

✓ discuss why doing science is often difficult.

✓ state how science properly done is part of God's plan of redemption.

1B WHAT IS PHYSICS?

1.6 The Nature of Science

Science can refer to any body of knowledge gained from observing the natural world (for instance, the science of entomology) as well as to the activities involved in acquiring scientific knowledge. Within the last century, the word *science* has also gained an impersonal and vague institutional meaning similar to the meaning of *government*. For example, some people think that *science* will find a cure for diabetes some day.

Mere observation of nature is not sufficient to qualify an activity as science however. By observing the relationships of living things, philosophers claim that they can discern the purpose of life. A painter intently observes the texture of bark on a

1-12 Though this artist is intently observing nature, she is not doing science.

A **phenomenon** is anything that the mind can perceive, whether real or imaginary. In physics, however, a phenomenon is anything perceptible by the senses or by scientific instruments.

The goal of model-making is **workability**. If a model does not work, it is discarded for a more useful model.

1-13 Wind tunnel tests produce a kind of model that aids physicists in understanding the natural world.

Dominion modeling is the theorizing about and analyzing of natural phenomena for the purpose of exercising dominion (Gen.1:28). It involves essentially all practical and theoretical sciences.

Examples of Scientific Models

Can you think of some examples of how models are used in science? Consider the fields of astronomy, biology, chemistry, geology, meteorology, and physics.

beech tree (*Fagus grandifolia*) to accurately represent it in a painting. Yet no one mistakes either of these activities—though both involve the careful scrutiny of nature—for natural science.

Furthermore, defining science as the study of nature does not mean that science deals only with things that are not artificial. Science can concern itself with any observable object or phenomenon—whether naturally occurring or man-made. And while observation involves the human senses, it may be enhanced by scientific instruments, the tools people invent to aid and augment their senses.

Science encompasses the knowledge acquired by its activities. The body of scientific information is incomplete, and what we have discovered through science is not unchanging, settled truth. Instead of providing us with absolute facts, science contributes to our construction of models of the world. **Models** are simplified and thus incomplete representations of reality. Their most important feature is their workability, how well they explain or describe a particular feature of reality; a model is useful only if it consistently explains observations and reliably predicts new ones. A model's representation of natural phenomena may be graphical, as in computer images of electron orbits, or physical, as in laboratory pendulums or models placed into a wind tunnel. However, models are more often **conceptual models**, the mental constructions that enable scientists to organize observations into understandable ideas. Scientists often express their more precise conceptual models as mathematical relationships or even as numerical computer models. Through dominion modeling, Christians can use models of reality to help them exercise dominion over God's world.

Thus, model-making is essential to science, and to physics in particular. Again, models are not—and have no claim to be—complete, entirely accurate, or even absolutely true like the Bible is true. Models are constructed for particular, limited purposes by imperfect people who are subject to ordinary human limitations (see the facet on pages 12–13). Yet good models help us to make sense of the world by providing useful answers to the questions they are designed to answer.

Although many people have tried to define science, it defies a simple definition. The more one attempts to develop a complete definition, the more the definition leaves out valid aspects of science. Based on the discussion in Section 1A, the following is a working definition of science:

> Science *is the collection of observations, inferences, and models produced through a systematic study of nature for the purpose of enabling humans to exercise good and wise dominion over God's world. The word* science *is also used to refer to the methods that produce the observations, inferences, and models.*

This definition accomplishes several things. It identifies science's "what" and "how," all-encompassing terms applicable to nearly every area of science. It also declares science's "why"—that science's purpose is more than satisfying human curiosity. True science is the logical and necessary result of human obedience to God's command to become stewards of His creation and to have dominion over the earth (Gen. 1:28).

1-14 Worldviews are like colored sunglasses. The world appears differently depending on the lenses through which it is viewed.

1.7 Science and Faith

Though it establishes science's purpose, the Bible is not a physics text or a scientific reference book. Nor would we want it to be! For physics and science change, but God's Word is the unchanging basis for the Christian **worldview**. Like a pair of eyeglasses, a person's worldview affects everything he looks at. A worldview is often defined as a perspective from which a person observes and understands the world—physics included. Secular physicists view the world through humanistic lenses that bring certain things into focus. Viewing the world through biblical lenses delivers quite a different scene.

One's worldview gives rise to the paradigms by which we understand various aspects of the world, our culture and society, and ourselves (for a definition and discussion of *paradigm*, see the facet on pages 8–9). Today, nearly all scientists work from within a *naturalistic paradigm*. They assume each physical phenomenon has only a natural cause and reject the existence, or at least the relevance, of anything supernatural. Thus, naturalistic science is essentially atheistic, even though individual scientists may believe in a god.

Naturalistic science concludes that physics is incompatible with the Bible. One major incompatibility is **uniformitarianism**, naturalism's assumption that the present is the key to the past. Uniformitarianism makes **inferences** about conditions in the past based on those observed today. It assumes that not only past processes but also the conditions under which the processes occurred were uniformly the same as they are today. This assumption supports the theory of biological evolution by encouraging high estimates of the ages of the universe and the earth. The major weaknesses of uniformitarianism are its lack of a sufficient cause to account for the beginning of all things and its lack of evidence for constructive evolutionary processes in today's world. Uniformitarianism relies on faith instead of evidence, for the uniformitarian paradigm arises from a belief system that rejects God as the Creator and a supernatural origin of the material universe.

Uniformitarianism is the belief that the present is the key to the past, especially regarding geology.

An **inference** is a logical conclusion obtained from premises that are known or assumed to be true.

In contrast, a very small number of scientists throughout the world hold to a completely different scientific paradigm. They believe that the Bible is the Word of God, and they therefore believe that all scientific activity must be done from within the **Christian worldview**. The Christian worldview assumes three things mentioned earlier:

1. God created the world for His glory.
2. This world is fallen.
3. God is redeeming this world to Himself.

1-15 The tenets of the Christian worldview

Christians find in the Bible, not huge incompatibilities with physics, but rather the best reasons to study physics. Science works because God designed the world according to regular and predictable underlying principles. For example, He promised that the cycle of seasons will not cease "while the earth remains" (Gen. 8:22). The world's regularity allows scientists to repeat experiments under the same conditions and obtain the same results each time. This principle of *uniformity* is essential to true science. But the Christian worldview distinguishes the principle of uniformity from uniformitarianism: the Bible records Creation, the Fall, and the Flood—universal changes in the earth's processes that uniformitarianism denies.

Like uniformitarianism, the *biblical paradigm* is also held by faith. However, the biblical paradigm has several significant advantages over uniformitarianism with regard to scientific work. It presents a more than adequate cause for the beginning of all things—the omnipotent Creator God. In addition, it provides a history of the earth that accounts for shortened life spans, present geological features, the extinction of species, and the diversity of people groups. The biblical paradigm also establishes the basis for engaging in science as obedience to the two great commandments (Matt. 22:37–39).

1-16 Formed over millions of years or in forty days and forty nights? Either inference depends on one's presuppositions.

A biblical paradigm for the sciences arises from the Christian worldview. It assumes that Christ is the foundation for every human activity because He is the source of all things and upholds all things by His powerful word (Heb. 1:3). It also assumes that the Bible is God's Word and thus accurate and reliable in its descriptions. This means that as you study physics, you should continually evaluate the facts and theories from the perspective of God's Word. You will not find laws in physics that contradict the Bible because God established the laws or ordinances of nature. But you may discover that physicists infer causes for phenomena without acknowledging the role of the Creator or without accepting the history of the earth as the Bible records it. In these cases, your knowledge of both the Bible and physics can lead you to different inferences that, instead, honor the Creator.

1.8 Structure of Science

A **continuum** is an unbroken sequence or expanse.

Scientific knowledge is constructed through an interlocking pattern of *theories*, *hypotheses*, and *laws*. Some present these terms as a **continuum** of increasingly affirmed scientific knowledge, but each is actually a distinct kind of knowledge that has a different purpose.

FACETS of PHYSICS

Thomas Samuel Kuhn: Philosopher of Science

Scientific knowledge continually improves over time to come closer to the truth, right? Not according to Thomas Kuhn, arguably the most influential American philosopher of science in the second half of the twentieth century.

Thomas Samuel Kuhn was born in 1922 in Cincinnati, Ohio. He received his higher education in physics at Harvard University. In an unusual twist, after his science education he served as a member of the history or philosophy faculties of prominent American universities: Princeton, the University of California at Berkeley, MIT, and his alma mater, Harvard. For most of his professional life, Kuhn wrote about and taught philosophy of science. His most influential book is titled *The Structure of Scientific Revolutions,* published in 1962. In it Kuhn directly tackles the positivist claim that science inevitably progresses toward truth. **Positivism** is the humanistic philosophy that came out of the European Enlightenment period, which began in the mid-1700s. In Positivism man is the measure of all things, and human reason, informed by science, is the only reliable path to truth. Since the Enlightenment, Western society has equated science with truth. Whatever does not conform to scientific fact is not true. Especially troubling is that some Christians fall prey to this way of thinking, even when it contradicts what the Bible says.

Kuhn dismantles Positivism by making two main points. First, he asserts that ***normal science serves the reigning paradigms.*** By "**paradigm**" Kuhn means "accepted examples of actual scientific practice—examples which include law, theory, application, and instrumentation together—[which] provide models from which spring particular coherent traditions of scientific research" (p. 10). Some familiar examples of paradigms are the heliocentric theory, the atomic theory, and Newtonian physics.

Thomas S. Kuhn, PhD (1922–1996)

Later, in his "Postscript," Kuhn gives a more detailed definition. He calls a paradigm a "disciplinary matrix" composed of (1) *symbolic generalizations* (like $F = ma$), (2) *beliefs in particular models*, (3) *values* (for example, he says that scientific questions must be askable and solutions must be

A scientific **theory** may be a general or specific model that attempts to account for a set of related observations or phenomena. It explains many different observations and inferences. For example, the *kinetic-molecular theory* is an overarching explanation for the structure, behavior, and properties of matter.

As a framework within the scientist's paradigm, a theory can also be a way of explaining events in the unobservable past. Such *origins theories* allow scientists to make complex inferences based on present-day observations. For example, the *plate tectonics theory* attempts to explain the origin, arrangement, and past and present movements of the large segments of earth's crust called tectonic plates.

When a scientist chooses to investigate a scientific question, he suggests a **hypothesis** as a temporary explanation for a phenomenon. A hypothesis serves as a starting point for further investigation and is structured so that it can be tested. Hypotheses are never suggested in a knowledge vacuum; scientists always rely on their knowledge of existing theories and the governing scientific paradigm to construct their hypotheses. An example of a modern hypothesis is *anthropogenic global*

1-17 Astrophysicists have suggested several theories to explain why the planet Uranus rotates "on its side" compared to the other planets in the solar system.

findable), and (4) representative examples or *exemplars* ("concrete problem-solutions that students encounter from the start of their scientific education" [p. 187]). The authoritarian role of paradigms is well-expressed in the following statement by Kuhn: "To desert the paradigm is to cease practicing the science it defines" (p. 34).

By "**normal science**" Kuhn means research based on these four things (especially the exemplars). The goal of normal science is to fulfill the promise of the exemplars, "by increasing the extent of the match between [the] facts [that the paradigm displays] and the paradigm's predictions, and by further articulation of the paradigm itself" (p. 24). He refers to this activity as the "mop-up work" that is always left behind after an old paradigm is discarded. And he states that this work is a mammoth task. He says that normal science is an "attempt to force nature into the preformed and relatively inflexible box that the paradigm supplies. No part of the aim of normal science is to call forth new sorts of phenomena; indeed those that will not fit the box are often not seen at all" (p. 24).

Examples of normal science include the search for new elementary particles in the nucleus, which falls within the paradigm of the atomic theory, and developing methods of gravitationally assisting the flight of space probes, which falls within the paradigms of universal gravitation and the heliocentric theory.

Kuhn refers to the day-to-day research of normal science as "puzzle-solving." He calls it that because, unlike resolving problems such as Mid-East violence, normal science research always knows the solution in advance. Investigation is aimed at confirming and more clearly describing the paradigm. One such puzzle was determining the actual speed of light.

The significance of reckoning with the idea that paradigms dominate normal science is that science is driven by faith. Consider these words by David Naugle: "Normal science forces nature to fit the dimensions of the reigning scientific paradigm. Like a pair of tinted glasses, paradigms color everything scientists see. . . . paradigms are like the umpires at a baseball game: they control all the action on the (scientific) field, but no one pays much attention to them" (*Worldview*, p. 200).

The second way in which Kuhn attacks Positivism is by asserting that ***scientific revolutions are paradigm shifts***. Here he argues against the "development-by-accumulation" view of the history of science (p. 2). The relationship between previous and subsequent theories is not that the former logically flow into the latter. Instead, the latter are revolutionary changes in the way the world is seen—paradigm shifts. The process, claims Kuhn, begins when scientists engaged in the puzzle-solving activities of normal science find anomalies. These anomalies are places in research where nature refuses to fit into the box provided by the paradigm. In time, science (or a particular field within science) comes to a state of crisis. That crisis leads to a revolution through which a new candidate for the paradigm gains prominence and is eventually confirmed.

Kuhn contends that these revolutions are destructive to the old theory and that the two (the old and the new theories) will always be incompatible. In particular, one cannot claim to derive the older paradigm from the newer or claim that the older paradigm is a special case of the newer. Consider this sentence from Kuhn: "To make the transition to Einstein's universe, the whole conceptual web whose strands are space, time, matter, force, and so on, had to be shifted and laid down again on nature whole" (p. 149). In this movement from one paradigm to another, Kuhn does not see a movement toward truth—he sees only movement. For Kuhn, ***science is not about truth but rather about a particular view of reality determined by the reigning paradigm.***

(References: *The Structure of Scientific Revolutions,* Thomas S. Kuhn. The University of Chicago Press, Chicago. 1996.
Worldview, David K. Naugle. William B. Eerdmans Publishing Company. 2002.)

warming, or global warming caused by human activities. This hypothesis is very controversial, and there is significant evidence both for and against it.

A scientific **law** models or describes an observed relationship among several often unrelated phenomena or properties under specified conditions. It is usually a simple statement expressed as a mathematical equation. For instance, Newton's second law of motion is commonly expressed as

$$F = ma.$$

Notice that none of the three quantities in the equation is in any way similar to the other. The first is a force, a push or pull; the second is mass, the quantity of matter in an object; and the third is acceleration, the rate of change of speed. Yet when combined in the second-law equation, these quantities yield a numerical and dimensional equivalence. The main restriction on this law is that the object must be moving at speeds much less than the speed of light.

The distinction between laws and theories is that *laws describe* what happens but *theories explain* why it happens. Scientists use theories, hypotheses, and laws within well-understood paradigms to carry out their work in a conventional way.

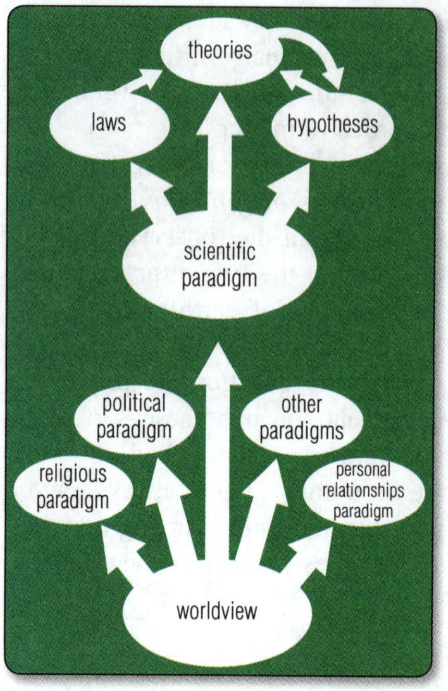

1-18 Hypotheses, theories, and laws exist and are formulated within a governing paradigm.

Dimension in this context means the measuring unit of each quantity.

Laws describe; theories explain.

1.9 Kinds of Science

There are two broad kinds of science that are defined by scientists' methods and approaches. The first and most common type is **empirical science**, which depends on present-day observations or experiments. The great majority of science being done today falls into this category. Empirical science is also called **operational science** because it provides workable or operational explanations of how the world functions. When a physicist smashes the nuclei of gold atoms in a high-energy particle accelerator to discover new elementary particles, he is doing operational science. Operational science is responsible for the advances in technology that revolutionized communications, data management, and medicine.

We noted earlier that the main purposes of science are to study God's creation in order to return glory to Him and to learn how to use God's creation for the benefit of our fellow humans. These are our two main objectives for exercising dominion over the earth. For this reason, we refer to operational science as **dominion science**. Dominion science is the practical science done to advance our knowledge of life, the earth, and the universe.

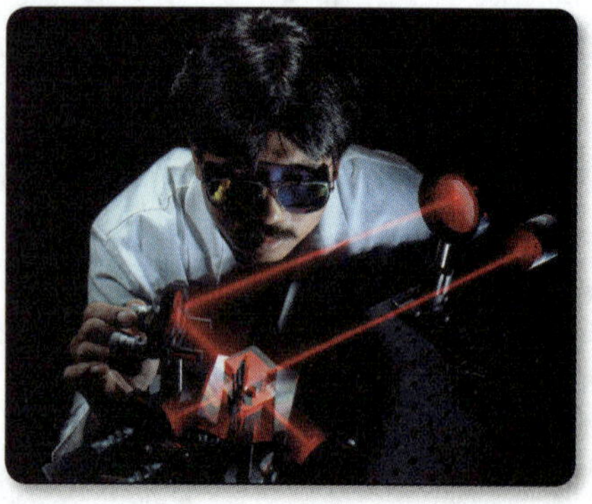

1-19 Scientific observations of present-day phenomena are the essence of operational science.

The second category of science creates hypotheses and develops theories and laws pertaining to events in the unobservable past. Humans are naturally curious about their origins, the beginnings of the world, and the past processes that produced what we see today. When scientists apply the methods used in operational science to problems originating in the unobservable past, they are doing historical science. Though scientists make observations in the present, they cannot reproduce the past events or processes that produced the data we see today. Thus, when constructing hypotheses or theories (explanations) in historical science, scientists must make inferences based upon certain assumptions. Often

Historical science is not empirical science (that is, science conducted from direct or recorded observations). Its conclusions are therefore more speculative and should be held with less confidence than those drawn from empirical science (see the facet on pages 12–13).

these assumptions, or **presuppositions**, are derived from the scientist's scientific paradigm, which, in turn, is based on the worldview within which he is working. Depending on the paradigms on which they are based, vastly different historical scientific models can be constructed from the same data.

1-20 Any event that occurred in the unobservable past, such as the Genesis Flood, can be a subject of historical science.

1.10 Physics as a Science

Physics encompasses everything from the tiny elementary particles that form the nuclei of atoms to the immense clusters of galaxies that contain the universe's billions of stars. Physics reveals the variety and wonderful orderliness of God's creation—a creation so orderly that it can often be described by relatively simple principles or mathematical formulas. In one sense, physics is important because it satisfies our God-given curiosity about nature by providing workable models of the world. Hence men and women, seeing that God does "great things and unsearchable; marvellous things without number" (Job 5:9), have dedicated their lives to unraveling the rich mysteries of creation.

1-21 If atoms collide hard enough, they break up into their elementary particles.

As with any science, physics includes elements of mathematics, reasoning, philosophy, and communication. *Physics* comes from the Latin word *physica*, which is derived from the Greek *fusika*, meaning the study or science of natural things. Until the mid-1800s, physics was called *natural philosophy*, but this phrase fell out of favor because professionals desired to separate the natural sciences from purely philosophical disciplines.

Physics can be subdivided into two general groups of knowledge based on the time in history when they emerged as distinct disciplines. *Classical physics* was developed during the period from 1600 to 1900, and *modern physics* encompasses all discoveries (apart from those of classical physics) from 1900 to the present. However, this division is somewhat arbitrary, and the periods partly overlap between 1890 and 1920.

1.11 Classical Physics

Classical physics began in the seventeenth century with the overturning of Aristotelian mechanical concepts. Isaac Newton, building on the discoveries of Kepler and Galileo before him, described how objects moved (*kinematics*) and developed laws that explained the causes of motion (*dynamics*). Newton's combination of kinematics and dynamics was the basis for the body of the most comprehensive scientific laws in history. They forever changed mankind's view of nature.

Mechanics is the study of how and why motion occurs.

Kinematics and dynamics are together called mechanics, the first body of foundational physical principles. Mechanical theory describes how forces affect motion and describes motion in lines, arcs, and circles. Closely related to motion are the concepts of mechanical energy and momentum. Another related area of mechanics is *statics*, which deals with the distribution of forces within unmoving and rigid bodies. Newton incorporated many principles of statics into his development of dynamics.

Portrait of Sir Isaac Newton after an original painting by Sir Godfrey Kneller, by Goldschmit, 1847, (Detail), Academie des Sciences, Paris, France/Archives Charmet/The Bridgeman Art Library

1-22 Galileo and Isaac Newton were key figures who began the revolution that resulted in classical physics.

The second body of foundational physical theory is electromagnetism. The fascinating effects of electrical charges and magnetism have been known since the time of the ancient Greeks, and throughout the eighteenth and early nineteenth centuries people investigated electricity and magnetism as separate phenomena.

Electromagnetism is the study of electricity and magnetism.

Thermodynamics is the study of thermal energy's relationship with and its conversion into other forms of energy.

1-23 One of James Watt's practical steam engines

Benjamin Franklin assigned the terms *positive* and *negative* to electrical charges. Charles Coulomb discovered the rule governing the forces between charges. Christian Oersted showed the relationship between electrical current and magnetism, and Michael Faraday demonstrated the opposite phenomenon in which alternating magnetic fields produce electrical currents. In 1864, James Clerk Maxwell discovered that electricity and magnetism were actually two manifestations of the same phenomenon, electromagnetism. Electromagnetism successfully predicted the existence of electromagnetic waves, which, when eventually discovered, were also found to compose visible light.

The science of **thermodynamics**, the third body of fundamental physical principles, came into existence as scientists attempted to understand the use of steam as a power source. Steam had one of its first practical applications in James Watt's steam engine, one of the most significant developments of the Industrial Revolution. Sadi Carnot, a young French engineer, developed an understanding of engine efficiency based on the caloric theory of heat. James Prescott Joule studied the transmission of mechanical energy and heat transfer. Others developed the statistical kinetic theory of particle motion and the concept of entropy. These efforts eventually led to the establishment of the laws of thermodynamics.

FACETS of PHYSICS

Limitations of Science

During each phase of scientific inquiry, scientists make personal choices. Science, therefore, is only as trustworthy as the men who participate in it. Using a scientific methodology does not guarantee that an explanation of nature is objective, nor can all the questions that mankind faces be answered by science. Science has four major limitations.

First, empirical science is limited to phenomena that can be observed. Instruments can increase man's ability to observe by extending the range of his natural senses. They can also give him sensory capabilities with which he was not created. Instruments can measure only present events or present effects caused by past events (e.g., radiocarbon dating a human fossil or determining the age of the Grand Canyon). No instrument can ever directly observe past events. Neither do instruments interpret the data that they measure. Data that an instrument measures is the same whether it is collected by a creationist or by an evolutionist. However, interpretations of this data sometimes differ widely.

Second, because man's knowledge is incomplete, science changes. Humans were created with this built-in limitation; only God has complete knowledge about everything. In addition, the models of the world that people create can quickly be replaced by different models that seem to explain observations better. In this course, you will learn about several well-established theories that were later replaced with more usable theories consistent with a new scientific paradigm (see the facet on pages 8–9). The only sure truths are found in God's Word, which is settled forever in heaven (Ps. 119:89). Scientific "truths" established by human reasoning and observation are tentative and are abandoned when they are proved wrong or when they cease to be useful. But the Bible, penned through the inspiration of an omniscient God, can never be proved wrong and is always useful.

Instruments that detect phenomena outside the range of human senses: [above], the Chandra X-ray Observatory and [below], the SNO neutrino detector

1.12 Modern Physics

The age of **modern physics** dawned late in the nineteenth century at a time when some physicists had concluded that nothing important was left to be discovered. They believed that the overarching principles of classical physics—Newtonian mechanics, electromagnetism, and thermodynamics—were all there was to know about the physical world and that only a few details remained to be worked out.

But a series of discoveries in the late 1800s made scientists realize that the universe was a lot more complex than they had thought. Radioactivity and penetrating x-rays revealed new and unanticipated aspects of both electromagnetic radiation and matter. The discovery of the electron demonstrated that the atom could be broken down into smaller parts. A few years later, scientists discovered the atomic nucleus and the orderly arrangement of electrons around it. Scientists then found that energy moves in packets called *quanta*. The discovery of the photoelectric effect led to the development of the photon model of electromagnetic energy. Many of these new discoveries were collected into a broad new area of physics called *quantum mechanics* that has provided fertile ground for research for many decades.

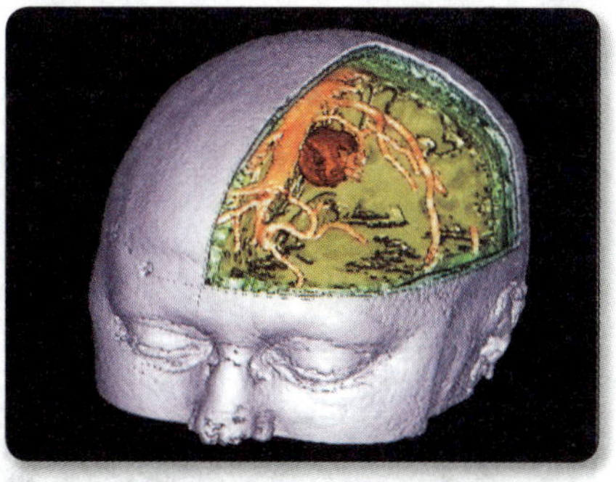

1-24 Scientific discoveries lead to practical technological applications, such as 3D magnetic resonance images (MRI).

Quantum mechanics is the study of the structure and behavior of molecules, atoms, and subatomic particles.

Third, science has difficulty defining and explaining the causes of natural phenomena. Many basic but abstract natural concepts and processes can be defined only through operational definitions. **Operational definitions** prescribe specific measurable tests necessary for a definition to apply. For example, the operational definition of *matter* is anything that has measurable mass and volume with the necessary measurements specified. The problems with such definitions are that they often do not include all the possible meanings of a term and that they involve words that are circularly related to the term being defined (e.g., *mass* and *matter*).

Neither can scientific study explain why natural phenomena occur. For instance, scientists can describe the characteristics of gravity and how masses interact, but it is unlikely that scientists will ever be able to explain what ultimately causes masses to attract each other or why they are not repelled instead. The Bible reveals that God's Word holds matter together: "And he is before all things, and by him all things consist" (Col. 1:17), and "who being the brightness of his glory, and the express image of his person, and upholding all things by the word of his power" (Heb. 1:3).

Earth and its moon form a gravitational system due to the proximity of their masses.

Fourth, questions about man's purpose and condition are beyond the realm of true science. This deserves special emphasis. For example, science cannot solve questions of

- morality, because God gives principles of conduct by direct revelation—"For we walk by faith, not by sight" (2 Cor. 5:7).
- human sociology, because man's nature is both physical and spiritual and also marred by sin (Jer. 17:9).
- origins, because no one was present at Creation to record the events—"Through faith we understand that the worlds were framed by the word of God, so that things which are seen were not made of things which do appear" (Heb. 11:3).

Science cannot solve questions of human morality or sociology.

The Creation-evolution debate is an example of a stormy controversy that cannot be solved through science. The origin of life cannot be repeated or observed. Furthermore, evolution and Creation have philosophical ramifications on morality and sociology, so conflict between the competing systems is to be expected. Followers of both systems are heavily biased by their respective beliefs, and this influences how they interpret their observations of natural phenomena. Both paradigms are based on faith, not on science, but the objects of faith are different. Evolutionists trust in blind chance and natural selection, or in an evolutionary principle said to be present in all matter. But creationists trust in God and His Word.

There is not an "old" physics and a "new" physics. Neither is classical physics out of date and no longer useful. In fact, the principles scientists use to send space probes to other planets and design car airbags are based on classical physics.

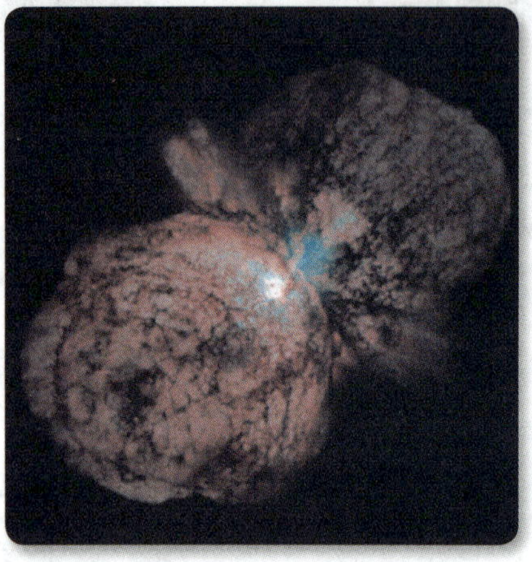

1-25 Nebulous bubbles of dust and gases surround the remnant star of Eta Carinae after it became a supernova in 1843.

In the midst of these dizzying discoveries, Albert Einstein's theorizing revealed the limitations of the centuries-old and well-established Newtonian mechanics and provided a new way of understanding the universe. These principles are summarized in his *special* and *general theories of relativity*. Continuing Einstein's and his contemporary physicists' efforts, theoretical physicists today still devote much effort to finding a "law of everything" that melds the fundamental forces of nature, quantum mechanics, and relativity into a Grand Unification Theory. Nevertheless, this goal has proved to be tantalizingly elusive.

1.13 What Physicists Study

We have analyzed physics historically, but what exactly do physicists do today? The many fields and occupations that involve physics could fill quite a long list, and many of these overlap with other branches of science. Some prominent examples include the following fields of study.

Astrophysics—combines astronomy and physics. Astrophysicists are interested in the movements and structures of celestial bodies, from meteoroids to galaxies. They may also calculate the trajectories of natural and man-made space objects.

Geophysics—combines geology and physics. Geophysicists study the interior, dynamics, and crustal structure of the earth, including its oceans and atmosphere. They are interested in both the cause of geomagnetism and its effects as it extends into the near-earth regions of space. Some geophysicists study the structure of the other planets and moons in our solar system.

Biophysics—combines biology and physics. Biophysicists are interested in the mechanics of living organisms, especially humans. They design noninvasive detectors, such as MRI and CAT scans, to produce images of the interior of the body. Biophysicists are also involved in developing radiation protection for nuclear workers as well as using radiation therapy to treat diseases.

1-26 Geophysicists have discovered the complex nature of the magnetic field that protects the earth from solar wind.

1-27 Biophysicists design "smart" artificial limbs.

Atmospheric physics—is applied in meteorology. Meteorologists work to develop accurate models of the atmosphere in order to improve weather forecasting and to make long-term predictions regarding global climatology. Meteorologists also work to understand the formations, motions, and effects of destructive weather conditions such as tornadoes and hurricanes.

Nuclear and High Energy Physics (HEP)—investigates the ultimate structure and particles of matter. Using particle accelerators, nuclear physicists study the nature of the nuclei of atoms and the numerous elementary particles that make up the nucleus. Nuclear physicists also develop more efficient nuclear fuel arrays for power and propulsion reactors. One of the main objectives of high-energy particle physics is to develop fusion as a widely available, economical energy source. Another objective is to experimentally identify important elementary particles whose existence is only theoretically suspected.

1-28 A computer-generated model of a hurricane system

1-29 A commercial nuclear power plant

Applied Physics—develops new and practical technologies. Physicists work in numerous capacities as they develop new products and applications that utilize physical phenomena. Many are, or work closely with, mechanical, electrical, and civil engineers. A sampling of such activities includes researching electro-optical communications, developing new composite materials, building more crash-proof and fuel-efficient cars, designing faster microprocessors, improving satellite communications, and designing aircraft, space vehicles, ships, and submarines. Physicists are involved with designing anything from pacemaker batteries and hearing aids to nuclear reactors and space surveillance radar systems.

Students who graduate with a college degree in physics can work in countries all over the world and seek employment in almost any sector of the economy. A quick Internet search reveals a huge selection of work and educational opportunities in physics and related fields. However, no matter how important finding a job may be, you should remember that the most important reason for learning about the physical universe is to give glory to God. The ministry of a preacher or missionary, or even a parent-teacher in a homeschool, can be greatly enhanced by a sound knowledge of physics. If you focus on the eternal purpose of your studies, God will meet your temporal needs. "But seek ye first the kingdom of God, and his righteousness; and all these things shall be added unto you" (Matt. 6:33).

1-30 One of the U.S. Navy's newest submarines, the USS North Carolina (SSN 777), is a product of applied physics.

1B Section Review

1. List three characteristics most people associate with science or doing science.
2. Why would merely looking at an insect not be considered science? What other activities might need to occur in order to qualify an action as science?
3. What are the end products of true science? How does this kind of knowledge differ from the unambiguous knowledge acquired from the Bible?
4. What are the main attributes of science as defined in this section? How is science related to the Creation Mandate?
5. What are the three tenets of the Christian worldview? How would this worldview affect a physicist's assumptions regarding the origin and operations of matter and energy?
6. Why is the principle of uniformity in natural phenomena not contrary to biblical science? How and why is this principle perverted in naturalistic science?

*7. According to the American philosopher Thomas Kuhn, what is *normal science*? Why does he call the routine activities of normal science "puzzle solving"?

8. Discuss the differences between a hypothesis, a theory, and a law.
9. Discuss the similarities and differences between operational science and historical science. What is another name for operational science that takes into account the Creation Mandate?
10. What are the three broad areas of classical physics?
11. State at least three areas of theoretical science included in modern physics.
12. (True or False) Physicists use particle accelerators to study the ultimate structure of the nucleus. Those that infer from these studies the nature of matter just moments after the big bang work from within a naturalistic worldview of historical science.

1B Objectives

After completing this section, I can

- ✓ state an operational definition for science in my own words.
- ✓ explain how a Christian can use science to exercise dominion over God's world.
- ✓ describe a Christian worldview.
- ✓ describe uniformitarianism.
- ✓ compare and contrast the two broad categories of science.
- ✓ describe classical and modern physics.
- ✓ list at least three possible occupations of physicists.

1-31 Questions in science can occur in any place and under any circumstance.

A **hypothesis** (plural, hypotheses) is a scientist's initial explanation that accounts for at least some observations of a natural phenomenon.

Tentative means to be hesitant or that something is not completely worked out.

A **bias** is an inclination or leaning based on personal experience and other influences in one's life.

As we learn more about microbiology, we realize that even the so-called simple forms of life are astonishingly complex. In microorganisms, indeed in all cells, we frequently find irreducible complexity—numerous symbiotic functions that cannot exist without the simultaneous existence of all of the others. Gradual evolution, the result of small changes over many millennia, cannot explain such relationships.

1C HOW DO PHYSICISTS WORK?

1.14 Introducing Scientific Methodology

Since Galileo, investigations in science have matured. Scientists have come to agree that scientific investigation must meet certain standards in order for its conclusions to be acceptable and useful. Scientists establish these standards to eliminate the possibility that wishful thinking or faulty reasoning take the place of accurate observation and logical analysis. Christian physicists in particular should seek to be as accurate and truthful as possible in order to protect their testimonies, help other people, and ultimately give glory to God.

Acceptable scientific standards result in a **scientific methodology**, commonly called the **scientific method**. While a *method* implies a particular sequence of distinct steps, this is rarely the way a scientist works. Rather, scientists follow whatever *methodology* ensures that the applicable scientific standards are met. The following subsections introduce the main components of scientific methodology.

1.15 Scientific Questions

Virtually all scientific investigations start with recognizing a problem that the scientist then expresses as a What, Where, How, Why, or even a When question. Questions occur when the existing theories or hypotheses fail to provide a ready explanation for an observation. For most scientists doing normal science, these questions continually arise as they proceed with their work. As they discover new facts and collect data, new questions arise. However, scientific questions can occur in response to many other circumstances, such as the following:

- An unexpected observation, such as noting that a new material absorbs all wavelengths of light that shine on it
- A pattern noted in numerical or physical data
- A prediction made by a theory
- A need to solve a technological problem, such as developing a cheaper and faster microprocessor
- A desire to develop a more accurate and less invasive medical diagnostic method

There is nothing mysterious about how a scientist arrives at scientific questions. The greater challenge for the scientist is to choose which question to investigate in order to find an answer.

1.16 Hypotheses

In response to a scientific problem, a scientist states a hypothesis, a tentative best guess at an explanation for the observation. It is a starting point for conducting an investigation and provides the direction for finding the solution. It immediately allows for collecting certain kinds of data while logically eliminating the consideration of other kinds of data. Note that a scientist's underlying biases influence his forming of a hypothesis. Creationary scientists have a different set of presuppositions than do naturalistic scientists; hence it is not surprising that these two groups of people often form different hypotheses to explain the same phenomenon or data.

In empirical science, a good hypothesis has several important qualities. First, the hypothesis must be reasonable. It should not suggest, for instance, that a scientific principle depends on one's geographic coordinates or on what day of the week it is.

Second, the hypothesis must be testable. Scientists test hypotheses by making predictions from them that additional observations and experiments can either prove or disprove. A hypothesis that cannot be tested is not empirically scientific, even though it may be true. No hypothesis regarding the origin of life is scientific because it cannot be tested. However, Special Creation as described in the Bible is more reasonable than other origins hypotheses because the probability of life coming into existence spontaneously is vanishingly small.

Third, the hypothesis should not contradict established principles. This qualification may be disregarded if the new hypothesis *explains observations better than the older theory*. For instance, Einstein's general theory of relativity explains anomalies in Mercury's orbit around the sun better than did the standard application of Newton's law of universal gravitation. The hypothesis may also contradict established principles if it *restricts the older theory to special cases*. Einstein's special theory of relativity restricts the applicability of Newton's laws of motion to objects traveling at speeds much less than the speed of light.

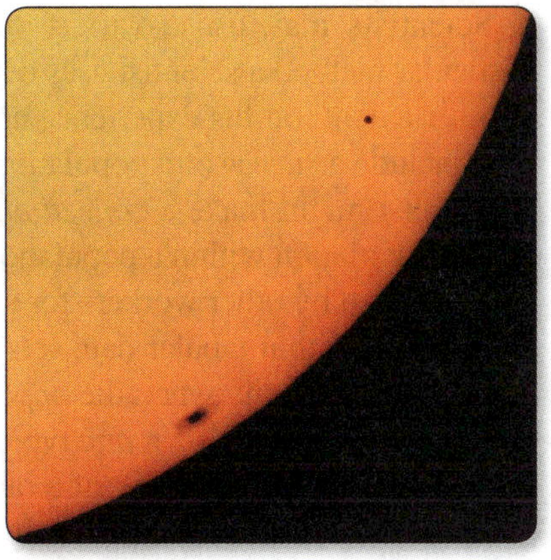

1-32 Mercury's orbit does not obey Newton's laws of motion. Einstein's theories provide a better description of Mercury's behavior.

Fourth, the hypothesis must explain all current observations and predict new ones. This constitutes the goals of normal science: to relate known data to newly discovered facts according to the current scientific paradigm and to point the way for future research. The kinetic-molecular theory of heat explains all the known facts about heat. It also predicts that heat involves a form of energy that is interchangeable with other forms of energy. For example, a certain amount of electrical energy should always produce the same amount of thermal energy. The prominent nineteenth century inventor James Joule later confirmed this prediction through a series of experiments. Nevertheless, even a hypothesis that makes workable predictions at one point in time may fail under different conditions at another time.

Fifth, the hypothesis should be as simple as possible, though this is not a hard-and-fast rule. Our present knowledge of God's creation indicates that the significant processes in the universe can often be expressed by simple rules. It is also true that some complex processes cannot be reduced to simple rules because God is not restricted by any man-made principle. Nevertheless, when two hypotheses are equally good, scientists typically choose the simpler one.

Keeping Things Simple

The goal of keeping a hypothesis as simple as possible comes from the principle known as Occam's Razor, which is attributed to a medieval English philosopher, William of Occam. When making a theory or hypothesis, one should not make more assumptions or set more conditions than absolutely necessary. The razor alludes to trimming away a theory's unnecessary conditions. Doing so allows the hypothesis to be as simple as possible and minimizes the potential for inconsistencies and other logical problems later on.

1.17 Testing Hypotheses: Observations

Before a scientist begins testing his hypothesis, he reviews other scientists' work that is related to the problem he is investigating. This is done to prevent him from unnecessarily duplicating other scientists' work (unless this is intended), and it helps him understand the limits of what is known about his problem.

Testing a hypothesis is done by collecting information through observation. **Observations** are made by seeing, hearing, smelling, feeling, or tasting. You see the blue-white light of the star Sirius (Alpha Canis Majoris) in the constellation Canis Major; you hear the characteristic call of a hermit thrush (*Catharus guttatus*); you smell the pungent skunk cabbage (*Symplocarpus fœtidus*); you feel the bristles on the underside of an earthworm (*Lumbricus terrestris*); and you taste the bitterness of the limonoid compounds in orange juice.

Scientific **observations** are the raw data, measurements, and descriptions of phenomena that are used to support or refute scientific hypotheses, theories, or laws.

1-33 You observe Sirius with your sense of sight.

Such observations produce data. **Data** may be *qualitative data*, such as the descriptive observations mentioned in the previous paragraph, or *quantitative data*, which is obtained from measurements using calibrated instruments. Though both kinds of data may be collected when testing a hypothesis, quantitative data is considered more **objective** because it is less subject to a scientist's observational biases.

To be **objective** means to be impartial, completely separate from personal bias.

1-34 This geologist samples the joints (fractures) in a rock that reveal the direction of the forces that shaped it.

Scientists may use a variety of observational methods to collect data. The particular method a scientist uses depends mainly on what he is trying to test and, to a lesser extent, on his experience and personal preferences. Observational methods may include a *survey* of a population of organisms or objects in their natural setting. A scientist might make a *collection*, representative samples of items to identify the limits of variation within a population. Some research lends itself to *mining data* that was recorded by other workers for a different purpose. Data mining can often be accomplished within tabular data sets by using a computer program. Sometimes data can be obtained only when the *opportunity* presents itself. Examples of such occurrences include observing a rare type of supernova, analyzing the action of a tsunami, or investigating the outbreak of a rare disease.

The most familiar and probably the most common test for a hypothesis is the *experiment*. In an experiment, a scientist attempts to demonstrate a cause-effect relationship between two or more phenomena. In a *controlled experiment*, a scientist artificially controls one or more variables to determine how they affect the single experimental variable that he is investigating. Controls are important for determining cause-effect relationships. Another advantage of the controlled experiment is that the same or different scientists can repeat the test exactly.

Finally, on rare occasions a scientist might resort to informed *trial and error*, in which a series of potential solutions to a problem are tested and eliminated. This process can be inefficient but has resulted in the discovery of some important technologies.

Before a scientist collects data to test a hypothesis or extend the limits of a theory, he must determine the minimum amount of data required to make his results statistically significant. A scientist can draw only tentative conclusions about the subject of his investigation if, for instance, he records only a few data points during an experiment.

1.18 Sharing Scientific Knowledge

1-35 Physicists must report their findings to receive credit for their discoveries.

A scientist's confirmed hypothesis or new discovery doesn't benefit anyone if he doesn't tell anyone about it. More importantly to the scientist, if he doesn't publish his findings, he will not be given credit for his discovery. It is possible that another scientist may publish his own findings on the same subject first. Thus, reporting the results of scientific research is an integral part of scientific methodology because it serves the interests of the general public, other scientists, and the researcher himself.

After collecting and analyzing the data, a scientist produces a public report of his scientific research. He writes a comprehensive paper describing the scientific problem and the data collection methods, including, if applicable, detailed descriptions of the experiments and equipment. He then carefully discusses his conclusions and the reasoning behind them. The report usually concludes with suggestions for further research.

If the investigator is a junior scientist working under a senior researcher, he will normally submit his report to his superior, who will then edit the report and suggest improvements. When a scientist is ready to publish his research, he submits his paper to one or more scientific journals that publish research in his field. Often, the journal editor will request from the scientist a list of other scientists who are qualified to review the report before it is published. This part of the publication process is called the **peer review** of the work. Good scientific methodology requires that all research be peer-reviewed prior to its publication to give some assurance that the work is valid and adheres to the highest scientific standards. Scientists who critique a work provide comments and suggestions to clarify the work's methods or to revise its conclusions. Reviewers can recommend that a report be rejected because it is fundamentally flawed in some way. This occurs if the scientist uses invalid reasoning or if his conclusions are inconsistent with the data.

After successfully completing the peer review process and making the suggested corrections, the scientist publishes his report. The work may be published in hard-copy journals or distributed as bound copies to research libraries. In today's digital world, the report can also be published in online versions of the journal and archived in Web-based libraries.

Though publishing scientific works should be a straightforward component of scientific methodology, this is, unfortunately, not always the case. Often the subject of the research or the reputation of the researcher determines which journals, if any, actually publish the work. This is particularly true when one considers how difficult it is for creationary scientists to publish their works. If a scientist is known to believe in biblical creationism—even if his research has nothing to do with the question of origins—he can find it very difficult to get his work published in the more prominent journals whose strong anti-creationism and anti-Christian biases are well-known. There are essentially no mainstream science journals that publish creation-science research or even non-historical (i.e. empirical) science research by known creationary scientists. To be published in these journals, any research that involves historical science *has* to be done within the deep-time, evolutionary, Bible-rejecting, naturalistic paradigm. Fortunately, there are now at least five journals devoted to creation science. The Internet provides an increasingly effective means for creationary scientists to deliver their discoveries to the public.

1-36 *Science*, *Nature*, and *PNAS* are three important general scientific journals.

1.19 Dominion Science Models and Problems

A Christian should see doing physics as part of his attempt to live out the Creation Mandate of Genesis 1:28. We learn about the world around us so that we can manage it for the glory of God and for the benefit of our fellow humans. To remind you of the connection between Genesis 1:28 and physics, each chapter in this textbook begins with an introduction titled "Dominion Modeling" or "Dominion Science Problem."

A dominion modeling topic supplies you with real-world data. You are asked to graph the data to discover an optimal mathematical relationship or create some other model that helps you understand the data better. You will often be asked in review questions to discuss the advantages and limitations of such methods. The dominion science problems ask you to solve real-world problems using what you will learn in the chapter. You will often reconsider the problem as you progress through the material. When the solution to the problem is revealed later in the chapter, you will be asked to discuss matters related to the solution. By focusing on these kinds of problems and models, you will see how physics is done in the real world. You will also see how physicists can do science for God's glory by obeying His most basic commands, to exercise dominion and love others.

Scientific Papers and Paradigms

How does the difficulty experienced by creationary scientists to get their papers published relate to Kuhn's limits of doing normal science? (See the facet on pages 8–9.)

1-37 Three scientific journals that publish research in biblical creationism

1C Section Review

1. Discuss the difference between the "scientific method" and a scientific methodology.
2. What is the purpose of a scientific hypothesis? How does a person's worldview affect the kinds of hypotheses he formulates?
3. Summarize the properties of a good hypothesis.
4. Astrophysicists observe a large uncharted asteroid smash into the moon, creating a crater 100 km in diameter. What kind of data collection would their research entail?
5. Why is peer review important? What is a serious weakness of this process, especially regarding work in historical science?
6. (True or False) Controlled experiments are artificial situations under regulated conditions established by scientists in order to study cause-effect relationships.

1C Objectives

After completing this section, I can

✓ summarize the process of doing science.

✓ describe the characteristics of a good hypothesis.

✓ explain how scientists communicate their findings.

Chapter Review

In Terms of Physics

image of God	2	normal science	9
Creation Mandate	2	law	10
first commandment	2	dimensional	10
second commandment	2	empirical science	10
science	5	operational science	10
phenomenon	6	dominion science	10
model	6	historical science	10
workability	6	presupposition	11
conceptual model	6	classical physics	11
dominion modeling	6	mechanics	11
worldview	6	electromagnetism	11
uniformitarianism	7	thermodynamics	12
inference	7	modern physics	13
Christian worldview	7	operational definition	13
continuum	8	scientific methodology	16
Positivism	8	scientific method	16
paradigm	8	observation	17
theory	9	data	17
hypothesis	9	peer review	18

Review Questions

1. What aspect of humans makes their lives worth saving?

2. How has human scientific reasoning been affected by the Fall? How does this affect physics?

3. How can physics be one way in which God is redeeming the world to Himself?

4. What is the most important characteristic of a scientific model? What does this indicate about the essential purpose of scientific work?

5. Is the underlying issue in the Creation-evolution controversy based on one side's doing true science and the other's engaging in pseudoscience, or is there something more fundamental involved? Discuss your answer.

6. How does a person hold to either naturalistic uniformitarianism or Special Creation? How empirically scientific are these paradigms?

7. Discuss the similarities and differences between a theory and a hypothesis.

8. What seems to be the main problem with making inferences when doing historical science?

9. Discuss the importance of physics as a science.

10. What key discoveries marked the transition from classical physics to modern physics?

11. Within what branch of physics would a physicist be working when doing the following?

- **a.** Analyzing the movement of and the forces within the San Andreas Fault
- **b.** Researching a gliding suit that allows a parachutist to fly as much as 120 miles from his jump to his landing
- **c.** Developing a method to use gravitational assist to propel a deep space probe into the outer limits of the solar system
- **d.** Using a particle collider to investigate the composition of the quarks that make up the nucleons of an atom

12. Why do scientists use a scientific methodology?

13. When does a scientific question occur? Discuss why this is a normal part of exercising dominion.

14. Give an example of a hypothesis that is true but untestable. Give an example of a hypothesis that may or may not be true, but is untestable.

15. A theoretical meteorologist develops a computer program that calculates hurricane motion vectors from weather satellite images. He then enters that data into another program that uses satellite photos of the storm to make predictions based on a past storm's motion. He then compares the storm's predicted track to its actual movement. What kind of data-collection method is he using?

16. How do the editors of scientific journals control the kind of science research that gets published? How can this situation control the impressions of other scientists and the general public?

True or False (17–30)

17. Humans and animals were made in the image of God.

18. The world we see today is only slightly changed from the original creation.

19. One of the ways that God redeems the physical world is through humans who exercise dominion over creation.

20. For Christians to use physics, they do not have to create an entirely new subject, but instead use physics for its intended purpose.

21. The purpose of science is to discover ever more accurate truths about nature.

22. All sciences require a belief system that assumes the universe is orderly and predictable.

23. A scientific law is a thoroughly proven scientific theory.

24. Historical science relies on empirical data from past processes to explain the evidence we see today.

25. Electromagnetism emerged as a distinct aspect of physics less than 150 years ago.

26. Rather than superseding classical physics, modern physics expands physics into areas that cannot be studied using the earlier techniques.

27. A physicist who incorporates string theory in his hypothesis about the structure of three-dimensional space is working in applied physics.

28. A physicist asks how he can more precisely measure the distances between atoms in a crystalline material. This is a valid scientific question.

29. A working hypothesis that makes accurate predictions may later fail in its prediction of a phenomenon.

30. The Internet now makes it possible for scientists to publish their findings immediately, bypassing the peer-review step of scientific methodology without compromising their scientific integrity.

✪31. Read Thomas Kuhn's *The Structure of Scientific Revolutions* and write a summary of his essay. How does his view of the history of science differ from the one generally depicted in the media?

✪32. Read selections from the following books: pages 58–67 of *Rocks of Ages* by Steven Jay Gould; pages 54–61 of *The God Delusion* by Richard Dawkins, entitled "NOMA"; and pages 243–278 of *Science and Grace* by Morris and Petcher. Write a paper contrasting the three different views of God and science—Science as God, Science and God (NOMA), and the God of Science.

This photo is of Jupiter's moon Io and its parent planet. Observations of Io from the earth provided data for the first useful measurement of the speed of light.

Measurement 2

Dominion Modeling

Measuring the Speed of Light

When you think of the fastest speed possible, you probably think of the speed of light. But can you imagine trying to measure the speed of light? Early Greek philosophers like Aristotle and early scientists like Kepler, Hooke, and Descartes believed that the speed of light was infinite. Others like Roger Bacon and Francis Bacon (no relation) thought it had a very large but finite value. Galileo did one of the first experiments to measure the speed of light. He had two observers positioned on hilltops less than a mile apart signal each other with lanterns. He didn't discover anything conclusive, though he suggested that the speed of light, if finite, was significantly greater than the speed of sound.

In 1676, a Danish astronomer named Ole Rømer observed something strange about the way Jupiter's moon Io disappeared in its orbit behind the planet. When the earth was nearest to Jupiter, the predicted time of disappearance was very close to the observed time. But when the earth was on the far side of its orbit from Jupiter, the observed time was many minutes later. Rømer was the first person to recognize that the time differences were due to the delay caused by light's having to travel the extra distance to the earth.

2A DIMENSIONS OF PHYSICS

2.1 Introduction

Four centuries before Christ, the Greeks reasoned from their presuppositions regarding the nature of the world to unravel its mysteries. Their commonsense solutions were accepted for nearly two thousand years. But when scientists during the Renaissance stopped trusting traditional "science" and began testing their observations with new instruments and improved mathematical systems, they discovered that God's universe was far more orderly than they had thought, often in unexpected ways.

The essence of physics is to measure the observable world and describe the principles that underlie everything in creation. Physical theoretical principles are usually expressed or defined by mathematical *formulas*. As instruments and measurements improve, scientists must constantly refine their theories to fit the data. To ensure that he has the best possible data, a physicist must know what to measure, how to measure it, how to record his measurements accurately, and how to use the measurements to produce mathematically meaningful results.

2-1 Ancient measuring instruments included balance scales, knotted cords, and hourglasses.

A **formula** is an equation that orders the relationship between given physical quantities.

A **metric system** is any measuring system in which different-sized units are related to each other by multiples of 10.

The **SI** is the standard metric system used by scientists worldwide.

Dimensions in physics can have a geometric meaning or refer to a *kind* of measured quantity.

2.2 The SI

Early in history, each nation maintained its own system of currency, weights, and measures. With the rapid growth of science during the eighteenth and nineteenth centuries, nations recognized the need for worldwide standardized units. France took the standardization initiative in 1799 by establishing a decimal metric system consisting of the kilogram and the meter. The metric concept gradually gained wide acceptance among scientists.

Throughout the nineteenth and twentieth centuries, researchers sought to identify new standard units of measure and to more accurately define the existing ones. In 1960, the current metric system was renamed the *Système International d'Unités*, from which we get the abbreviation SI. It consists of seven base or fundamental units and many other units derived from the base units. As new fields of science are established, new derived units continue to be added to the SI.

2.3 Meanings of Dimension

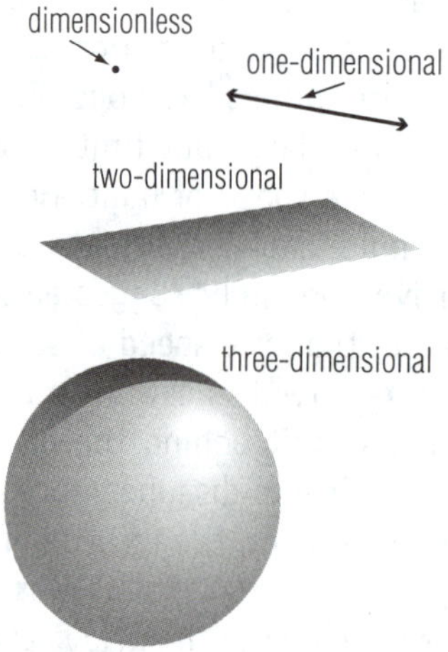

2-2 A point has no dimensions. A line exists in one dimension, a plane in two, and a sphere in three dimensions.

The word ***dimension*** has two meanings when used in physics. The first refers to the number of spatial coordinates required to describe an object. A point is dimensionless, a line occupies one dimension, and a plane exists in two dimensions. A volume of space or a physical object can exist only in three dimensions. The image on a TV screen is two-dimensional, but a horse (*Equus caballus*) is three-dimensional.

The second meaning of *dimension* refers to a kind of *measurable physical quantity*. Examples include length, time, mass, resistivity, hardness, and momentum. Each dimension is described by a specific type of unit or combination of units. For example, length is measured only with units of length (meter, foot, etc.), and density is limited to mass per unit volume (kg/m^3 or lbm/ft^3).

2.4 Fundamental Dimensions

The universe consists of three principal dimensions: **space**, **time**, and **matter**. These are intricately bound together. One cannot exist alone without the others. In fact, everywhere we look, we see this kind of simultaneous unity and diversity. Many have wondered why the world is like this. Christians say that the answer lies in the triune God. He is one Being existing in three Persons. We see unity and diversity everywhere because everything comes from a God Who is one and three at the same time.

2-3 The vast universe contains billions of galaxies such as Galaxy NGC 1512.

2.5 Space and Length

Space is the apparently limitless three-dimensional volume within which all matter exists and all physical phenomena occur. Matter must exist and be present in order for us to perceive the size of space. If you looked into a space that contained no matter, you would see nothing. You would not know if you were looking into an atom-sized or a galaxy-sized space! The physical separation of entities in space (distance) or the size of an object (which itself occupies space) are measured in units of length.

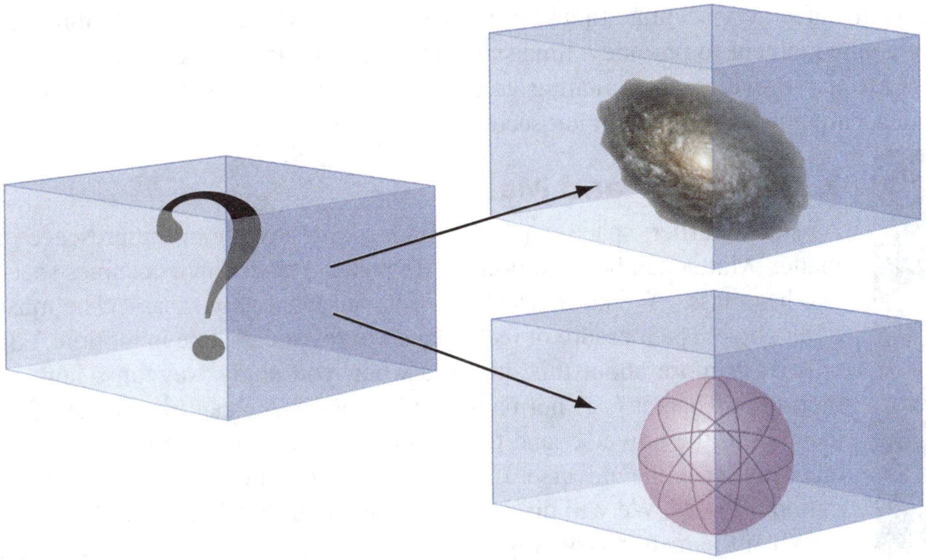

2-4 How large is a given space? The presence of matter fixes the size of a space.

2-5 A platinum-iridium (Pt-Ir) alloy meter standard

The metric unit of length, the **meter (m)**, was originally defined as one-ten-millionth (10^{-7}) of the distance between the earth's north geographic pole and the equator, as measured along the meridian (line of longitude) passing through Paris. Because such a distance could not be reproduced in a laboratory, much less measured accurately, scientists fabricated a corrosion-proof platinum-iridium bar bearing two scribe marks one meter apart. This *standard meter* was retained in the International Bureau of Weights and Measures at Sèvres, France, under stringent environmental conditions that prevented expansion and contraction due to temperature changes.

The SI unit of length is the **meter (m)**. A meter is the distance traversed by light in a specified time interval.

Platinum-iridium alloy is a precious metal that is highly corrosion resistant.

Even well-protected physical objects can be affected by environmental variables, so throughout the twentieth century scientists worked to abandon all references to dimensional standards based on man-made artifacts. At the same time, such objects became less useful as increased accuracy and precision were required. Scientists preferred instead to use extremely accurate and stable natural phenomena that could be measured in any properly equipped laboratory. In 1983, the meter was given its present definition in the SI as the distance traversed by light in a vacuum in *exactly* 1/299 792 458 second.

2.6 Time

Have you ever tried to define the concept of time? Think about it. Everyone *knows* what time is, but try to define it! One attempt at an operational definition could be, *"Time is a nonphysical continuum that orders the sequence of events and phenomena."* (A *continuum* is any expansive entity without gaps or breaks.) A moment of time exists in the indeterminable future, is experienced in the instantaneous present, and immediately moves into the past, where its significance can be understood only by one's reflecting upon past events. Notice the seamless, three-in-one character of time—again revealing the triune nature of its Creator. Time is best defined by specifying the method of measurement that clearly differentiates it from all other physical phenomena.

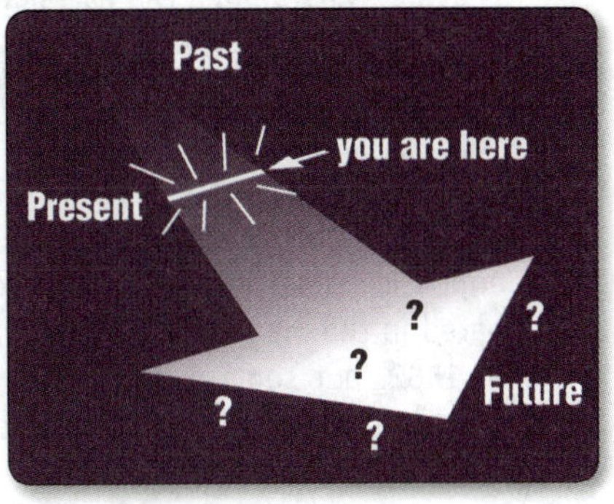

2-6 Our position in time is always the present. Our progress in time is always forward. Only in our memory can we experience the past.

The SI unit of time is the second (s). A second is the time interval of a specified number of vibrations of gaseous cesium-133 atoms resonating within special instruments.

The SI unit of time is the second (s). The second was originally defined as 1/86400 of a mean solar day (24 h × 60 min/h × 60 s/min = 86400 s). Unfortunately, the rotational period of the earth is not stable enough to define the second to the precision required for modern scientific purposes. In the mid-1950s, scientists invented very accurate timekeeping devices that used the resonant vibrations of a pure element to provide a time standard. A second is now defined as exactly 9 192 631 770 cycles of oscillating gaseous cesium-133 atoms. "Atomic clocks" are accurate to within three microseconds (μs) per year.

2-7 A modern cesium-beam time standard maintained by the National Institute of Standards and Technology

2.7 Matter and Mass

As stated earlier, space would be dimensionless without the presence of matter. Matter can be operationally defined as "that which occupies space and has mass." We know what space is, but what about mass? The **mass** of an object is a measure of its tendency to resist a change in motion. You will learn more about this attribute when you study Newton's laws of motion in Chapter 6. A normal object's mass is constant regardless of its location in the universe, and that mass has gravitational attraction for all other matter in the universe. Large amounts of matter produce large gravitational forces. We will discuss gravity in detail in Chapter 7.

When the metric system was established late in the eighteenth century, an attempt was made to define other physical properties that relate to the

FACETS of PHYSICS

Just a Second

Stop for a moment (a second?) and look around you. How many displays of time do you see? Your watch, the clock on the wall, the clock displayed in the corner of your computer desktop, your cell phone. Today, more so than at any other time in history, we order our lives by the clock. Busy personal schedules, manufacturing assembly lines, power generation, international financial transactions, GPS navigation, the Internet, and communications all require accurate and reliable timekeeping.

In the dim beginnings of history, humans observed celestial markers to determine the passage of time. In the first chapter of the book of Genesis, the Bible records that a day was defined as one complete rotation of the earth. Weeks were defined by the Creation Week, months were defined by the lunar cycle, and years were marked by the position of the sun against the heavens. After the Flood, determining an accurate annual calendar was a long and difficult process that culminated in the Gregorian calendar. This calendar was established in 1582, but some countries did not fully adopt it until the twentieth century due to political or religious reasons.

Accurately measuring smaller increments of time proved even more challenging. Early scientists, particularly astronomers, recognized the need to define smaller standard time units—first the hour, then the minute and second. They started with various devices such as calibrated candles, water clocks, and hour glasses. Eventually large, delicate, stationary clock-like instruments located in observatories were constructed. Public time was provided by ponderous clock-tower mechanisms; many were correct literally only twice a day. Some clocks had only one hand. There was little pressing need for accurate portable timepieces until the latter part of the sixteenth century. The necessity for accurately knowing the time came from what may seem like an unexpected quarter—ship navigation.

Far-flung merchant enterprises during the Age of Discovery required accurate ocean navigation to ensure the safe arrival of passengers and cargo in the shortest possible time. Knowing the ship's latitude was easily determined from observing the sun's angle above the horizon. Determining a vessel's longitude (its position east or west) depended on knowing the time at a geographic reference point, such as the port of departure, and subtracting that time from local time indicated by the sun's east-west position. The poor accuracy of shipboard clocks meter. For instance, volume could be defined by cubing a linear metric measurement. A unit of mass could then be defined as the mass of a specified metric volume. The mass of one cubic decimeter (1000 cm^3 or 1 L) of pure water was first used to define the **kilogram (kg)**.

2-8 Even an asteroid can produce enough gravity to have its own moon. Here, the 1.5 km diameter rock called Dactyl orbits the asteroid 243 Ida.

The standard **kilogram (kg)** is the mass of a cylinder manufactured from a precious alloy. It is the only SI standard based on a manufactured object.

Scientists soon recognized that even the relatively large volume of one liter could not be measured directly with the desired precision, so scientists manufactured a 1 kg cylinder of Pt-Ir alloy (called the *standard kilogram*) to replace the water standard. Its dimensions and the alloy mix of 90% platinum and 10% iridium are reproducible to a much higher level of precision than is possible with the water standard. The standard kilogram is also retained at the International Bureau of Weights and Measures. It is the *only* SI standard that is still based on a manufactured object. Even this may eventually change. Scientists are currently investigating various ways to redefine the kilogram. One method defines a kilogram as the mass that can be levitated in a precisely measured magnetic field of a specified strength. Another definition, which seems to have more promise for adoption, involves

made longitude determination very inaccurate, resulting in position errors of hundreds of miles. Many ships and the lives of their crews were lost due to such navigational errors.

After decades of failure, the first successful marine *chronometer* was built by an English carpenter and self-taught clock maker, John Harrison, in 1736. The problem of finding longitude at sea was solved, and with further refinements of his design, the manufacturing of precision timepieces and small portable watches became an established technology.

All clocks contain a component that subdivides time into regular intervals. Early clocks relied upon weight-driven pendulums that were subject to inaccuracies due to friction, temperature, and other pendulum errors. Later clocks and chronometers used spring-driven rotational escapements and free pendulums.

The advent of electronics early in the twentieth century was the motivation for the discovery of a more accurate and precise timekeeper. A clock containing a quartz crystal oscillator was invented at the Bell Laboratories in 1927. The periodic vibrations of the crystal subdivided time into millions of parts per second.

Throughout this developmental period, however, the second continued to be defined by scientists as 1/86 400 of a mean solar day. Although the invention of the quartz clock proved that the earth's rotation was irregular, quartz crystals could not be manufactured with sufficient uniformity to become a replacement time reference.

The development of the atomic clock in the 1950s provided the stable and reliable time standard scientists required. Since 1967, the second has been defined as the time required for gaseous cesium-133 atoms to resonate exactly 9 192 631 770

The first practical marine chronometer used to find longitude at sea, John Harrison's H-4

times inside of a specially designed instrument. These extremely accurate and precise atomic clocks, which are installed in laboratories across the globe, provide the reference time signals for today's modern technologies.

Excuse me, do you have the time?

2-9 One of the silicon spheres that may be used to redefine the kilogram

counting the number of atoms in a perfectly-machined sphere of pure silicon-28 having a mass of 1 kg. The spheres have been fabricated in Australia. When physicists finish counting, the number of atoms of silicon, not their mass, will define the kilogram.

The seven fundamental units of the SI are listed in Table 2-1. Note that there is a difference between the name of a physical dimension and its unit of measure. There is also a difference between the unit symbol and the mathematical symbol used to represent the dimensional quantity in physics formulas. The meter, second, and kilogram (and combinations of these units) will be the only units of concern for the chapters dealing with motion and mechanics in Unit 2. The other SI units will be introduced later as needed.

Table 2-1

Fundamental SI Notation

Dimension	Fundamental SI Unit	Unit Symbol	Formula Symbol
length	meter	m	*d, r, s, l*
time	second	s	*t*
mass	kilogram	kg	*m*
thermodynamic temperature	kelvin	K	*T*
amount of substance	mole	mol	*n*
electric current	ampere	A	*I*
luminous intensity	candela	cd	I_L

Units indicate kinds of dimensions in data, and *formula symbols* represent dimensional quantities in formulas.

SI derived units are obtained from combinations of SI base units or other approved or derived units.

The SI unit prefixes are derived from Greek, Latin, Spanish, and Scandinavian roots descriptive of the corresponding powers of ten.

2.8 SI Derived Units

Scientists deal with many more dimensional quantities than those listed in Table 2-1. For example, there is no fundamental unit for area. However, area can be represented by the square of length, l^2, which has the SI unit m^2. Similarly, speed is represented by the dimension of distance per unit time (d/t) with a unit of m/s. Such dimensions and their associated units are derived from the SI fundamental units and are called SI derived units. Some derived units have been given special names and symbols. Examples include the SI unit for force, the *newton* (N) ($1\text{ N} = 1\text{ kg}\cdot\text{m/s}^2$), and the SI unit for electrical charge, the *coulomb* (C) ($1\text{ C} = 1\text{ A}\cdot\text{s}$).

Some SI derived units are obtained from other derived units. For example, the SI unit for work and energy, the *joule* (J), is defined as

$$1\text{ J} = 1\text{ N}\cdot\text{m}.$$

2.9 Unit Conversions and Analysis

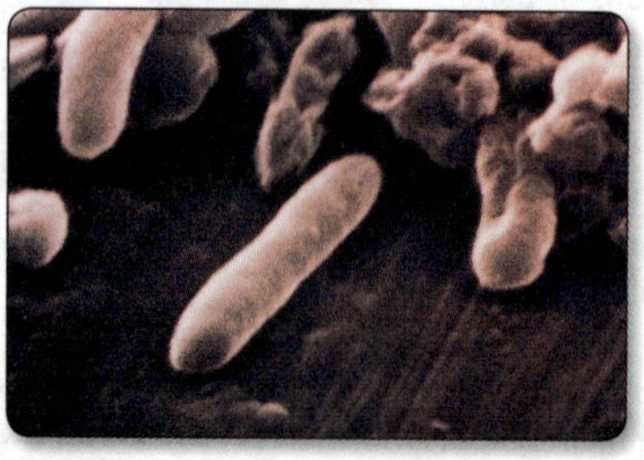

2-10 The size of a bacterium is measured in micrometers (μm).

Is the size of a virus more appropriately reported as 0.000 000 5 m or 500 nm? The basic SI units and their derived units often are not conveniently sized for use in every conceivable measurement. The founders of the metric system recognized that a method was needed to provide larger and smaller increments of the base units that would be convenient for any measurements that might be required.

Prefixes were developed that, when combined with a given *base unit,* indicate the size of the new unit. Table 2-2 lists the named metric system prefixes, the prefix symbols, and the factors by which the base unit is multiplied. Note that, for consistency, the gram is the *base* unit of mass. However, the SI *fundamental* mass unit is the kilogram. The prefixes that we will use most frequently are *kilo-, centi-,* and *milli-*.

It is often necessary in science to perform calculations with similar dimensions that have different-sized units or non-SI units. All similar dimensions should be converted to the same units before you calculate. You accomplish *unit conversion* by multiplying the quantity by a **conversion factor**, a fraction that contains both the given and the desired units. In order to preserve the size of the original measurement, the conversion factor must be *equivalent to one*—its numerator and denominator must be equal-sized quantities. To find the correct ratio, we start with an equation relating the two units.

For example,

$$1\ \text{m} = 100\ \text{cm}.$$

The factor required to convert *from* meters *to* centimeters is found by dividing both sides by 1 m.

$$\frac{\cancel{1\ \text{m}}}{\cancel{1\ \text{m}}} = \frac{100\ \text{cm}}{1\ \text{m}}$$

$$1 = \frac{100\ \text{cm}}{1\ \text{m}}$$

Note that the unit we are converting to is on top. If we needed to convert from centimeters to meters, the conversion factor would be $^{1\,\text{m}}/_{100\,\text{cm}}$. Both ratios are equal to 1 because the numerator and denominator are equivalent.

EXAMPLE 2-1

Finding a Conversion Factor

Convert 22.4 L to mL.

Solution:

$$1\ \text{L} = 1000\ \text{mL} \Rightarrow \frac{\cancel{1\ \text{L}}}{\cancel{1\ \text{L}}} = \frac{1000\ \text{mL}}{1\ \text{L}} \begin{matrix} \leftarrow \text{desired unit} \\ \leftarrow \text{given unit} \end{matrix}$$

$$22.4\ \text{L} = \frac{22.4\ \cancel{\text{L}}}{1} \times \frac{1000\ \text{mL}}{1\ \cancel{\text{L}}} = 22\,400\ \text{mL}$$

When several simultaneous unit conversions are required, such as when converting mi/h to m/s, we multiply the appropriate conversion factors together in any order, taking advantage of the commutative property of multiplication. We use a mathematical aid called the *unit analysis bridge* to help with conversion factor unit cancellation and to organize the solutions to physics problems.

Look at Figure 2-11. The horizontal line corresponds to the fraction bar (representing the *division operation*) in any fractional numbers involved in a calculation. The vertical lines correspond to *multiplication signs* between factors in the calculation. Identical units that appear on opposite sides of the horizontal line may be canceled as usual. To see how this process works, study Example 2-2 below.

After cancellation, the result is obtained by multiplying and dividing the remaining numbers and units. The key advantage of the bridge method is the convenient

TABLE 2-2

SI Unit Modifiers

Prefix	Symbol	Factor
exa-	E-	$\times10^{18}$
peta-	P-	$\times10^{15}$
tera-	T-	$\times10^{12}$
giga-	G-	$\times10^{9}$
mega-	M-	$\times10^{6}$
kilo-	k-	$\times10^{3}$
hecto-	h-	$\times10^{2}$
deka-	da-	$\times10^{1}$
(base)	—	$\times10^{0}$
deci-	d-	$\times10^{-1}$
centi-	c-	$\times10^{-2}$
milli-	m-	$\times10^{-3}$
micro-	μ-	$\times10^{-6}$
nano-	n-	$\times10^{-9}$
pico-	p-	$\times10^{-12}$
femto-	f-	$\times10^{-15}$
atto-	a-	$\times10^{-18}$

Problem-Solving Strategy 2.1

In calculations, the prefix of the SI unit of a measurement may be replaced by the exponential multiplier of the base unit.

The product of any number and 1 is the number.

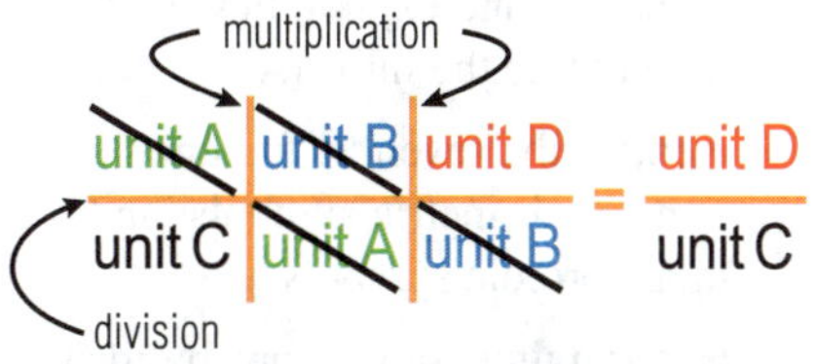

2-11 The unit analysis bridge

Problem-Solving Strategy 2.2

In a conversion factor, the unit you are converting *to* is in the same position relative to the fraction bar as the unit you are converting *from* in the given quantity.

Problem-Solving Strategy 2.3

The first factor in the unit analysis bridge should be the given data. The remaining data and conversion factors may be inserted in any order to cancel the given units and substitute the required units in their place.

arrangement of factors, which eases the cancellation of like units. When solving physics problems using the bridge utility, we can often check the setup of the solution by analyzing the units that remain after cancellation to verify that the resulting dimensional units are the ones required for the solution. This process is called *dimensional analysis*, or more commonly, **unit analysis**. Valid physics equations must have the same dimensions on both sides of the equal sign. Dimensional inequality is a guarantee that the solution will be incorrect. Of course, the equation can have equivalent dimensions and still have an incorrect arithmetic solution. Dimensional analysis is only a check on the setup of the solution.

$$\text{volume} = \text{volume}$$

$$V_{\text{sphere}}\,(\text{in m}^3) = \tfrac{4}{3}\pi r^3\ (\text{in m}^3)$$

2-12 A physics equation must have the same dimensions on both sides of the equation.

EXAMPLE 2-2

Using the Bridge: Unit Analysis

Convert 120 km/h to meters per second.

Solution:

Standard notation:

$$120\ \frac{\cancel{\text{km}}}{\cancel{\text{h}}} \times \frac{1000\ \text{m}}{\cancel{\text{km}}} \times \frac{1\ \cancel{\text{h}}}{60\ \cancel{\text{min}}} \times \frac{1\ \cancel{\text{min}}}{60\ \text{s}} = 33\ \frac{\text{m}}{\text{s}}$$

Bridge utility:

120 $\cancel{\text{km}}$	1000 m	1 $\cancel{\text{h}}$	1 $\cancel{\text{min}}$	= 33 m
$\cancel{\text{h}}$	$\cancel{\text{km}}$	60 $\cancel{\text{min}}$	60 s	s

given data   required units

Problem-Solving Strategy 2.4

The division operation is the inverse of multiplication. To indicate division by a quantity using the bridge utility, express the quantity and its unit as a fraction, invert the quantity *and its unit,* then write it as one of the factors in the bridge.

2A Objectives

After completing this section I can

✓explain the need for a standardized measurement system.

✓state the three fundamental dimensions and explain how they help define the universe.

✓identify the seven SI fudamental units, their metric symbols, and their formula symbols.

✓perform unit conversions by identifying and using appropriate conversion factors.

2A Section Review

1. Why did scientists seek to develop a standard metric system of measurement in the eighteenth and nineteenth centuries?
2. What are three kinds of dimensional quantities required to perceive the universe?
3. What is the current definition of the meter?
4. List the seven fundamental dimensional units of the SI, their unit symbols, and their variable symbols. Explain the difference between the unit symbol and the corresponding formula symbol.
5. What is an SI derived unit? Give an example.
6. Discuss three disadvantages of using a manufactured object as the standard meter bar.
7. Write an operational definition of time that includes a way of measuring it.
8. Discuss the two meanings of the term "dimension" in physics.
9. Convert the following quantities to the indicated units using the unit analysis bridge utility.
   a. 6.5 ft to m
   b. 3.25 km to m
   c. 325 m/s to km/h
   d. 1040 kg/m^3 to g/cm^3
10. Convert the speed of light to astronomical units (ua) per year.

2B PRINCIPLES OF MEASUREMENT

2.10 Instruments

God designed the human senses to provide all essential information for daily life. Early in history, many linear units of measure were based on the dimensions of the human body (e.g., "pace," "cubit," "span," and "hand"). As society and technology advanced, more accurate methods for measurement were required. Using their God-given intellectual capabilities, men invented the necessary tools to extend their senses and refine their perception. These tools, collectively called *instruments*, permitted more exact measurements.

2-13 It is likely that the workers had to use a variety of measuring instruments for the construction of the ark.

2-14 Lasers are finding practical uses as range finders and level references at building construction sites.

Instruments are invaluable to modern scientific research because they permit the accurate and precise measurement of quantities that unaided human perception can only estimate or that are completely outside the range of human senses. For example, lengths and distances can be measured with rulers as well as with lasers, radar, and sonar. Astronomical distances can be measured from stellar red shifts obtained with telescopic spectrometers, and atomic distances can be measured using instruments such as scanning tunneling electron microscopes and x-ray diffractometers. Besides measuring length, time, and mass, instruments have been developed to measure force, speed, temperature, light intensity, magnetism, electric fields, frequency, and a host of other physical properties.

All modern instruments have at least one limitation—they are all man-made. No matter how accurately an instrument can measure and how skillful the user, the tiny variations in dimensions and quality during its construction and the limitations of human perception guarantee that there will always be some error in the measurement.

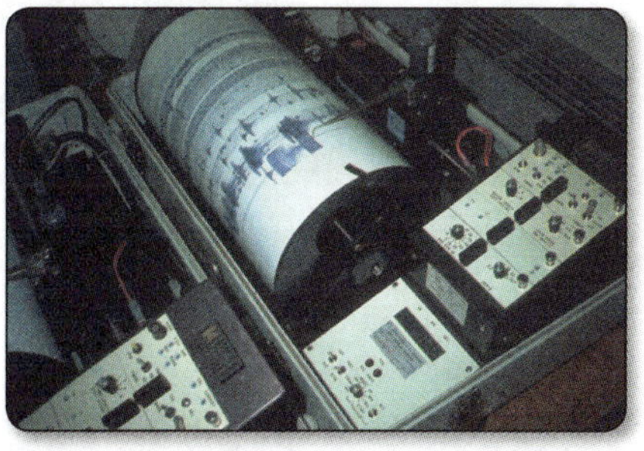

2-15 A seismograph recording an earthquake

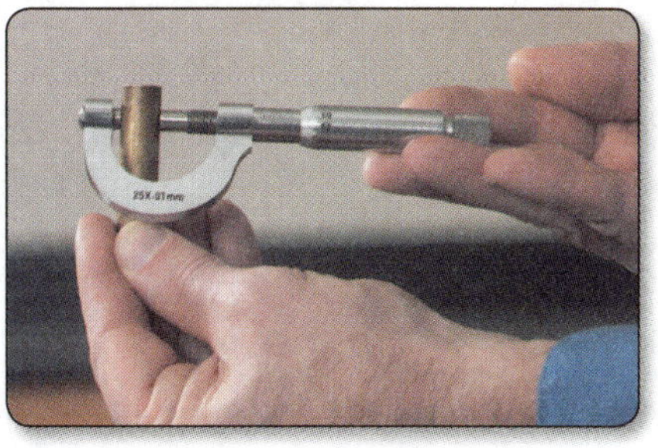

2-16 A micrometer caliper permits high-precision measurements of small lengths.

Small permissible variations in the accuracy of manufactured parts are called *manufacturing tolerances.*

2.11 Accuracy

How can you be sure when you measure the length of a pencil with a ruler that the observed length is the actual length? The question is answered by a qualitative term called **accuracy**. Whenever we make a measurement, we compare the object being measured to the graduated scale of an instrument. The instrument is accurate if it produces measurements very close to the actual value.

2-17 A student performs a zero calibration of a mechanical laboratory balance.

The accuracy of an instrument depends on the quality of its original design and construction *and* on how well it is maintained. Most scientific instruments can be adjusted so that the error of their measurements is kept within an acceptable limit. Periodic *calibration checks* are performed that compare an instrument's output to a highly accurate *prime standard* to verify that the instrument is accurate at one or more points on its scale. Zeroing a laboratory balance is a simple instrument calibration.

Accuracy is a *qualitative* evaluation of how close a measurement is to the actual or accepted value.

A **calibration check** of an instrument compares an indicated measurement with the corresponding accepted value.

The **actual value** is a theoretical concept indicating the result of measuring with a perfect instrument. An **accepted value** is the agreed-upon value established for a particular physical quantity by an authoritative source.

Error = observed − accepted
Absolute error =
|observed − accepted|

Percent error relates the measurement error as a percentage of the accepted value.

Importance of Accuracy

Christians should highly value accuracy in measuring because God does. God told the Israelites, "A . . . just measure shalt thou have" (Deut. 25:15). Why do you think accurate measurement is so important to God?

Obviously, the accuracy of a measurement also reflects the skill of the operator. The contribution of human error to a measurement can be reduced by averaging the results of several measurements that are taken either by the same or by different persons.

To make a *quantitative* assessment of the accuracy of measured data, we must compare the measurement to the *actual value* or *accepted value* and determine the **error**. Error is expressed as the simple difference of the observed and accepted values. The sign of the result indicates whether the observed value is larger (+) or smaller (−) than the accepted value. The *magnitude* of the error, or *absolute error*, is the absolute value of the difference.

Knowing the error of measured data is not always as useful as knowing how large the error is compared to the accepted value. In this case, we want to find the relative or **percent error**, which gives the measurement error as a percentage of the accepted value.

$$\text{percent error} = \frac{\text{observed value} - \text{accepted value}}{\text{accepted value}} \times 100\% \qquad (2.1)$$

You can also find the *absolute percent error* by substituting the absolute error in the numerator of Equation 2.1.

EXAMPLE 2-3

To Measure Is to Err: Percent Error

A student measured the density of a copper cylinder to be 8.83 g/cm^3. The accepted value for the density of copper metal is 8.94 g/cm^3. (a) What is the error of the student's measurement? (b) What is the percent error?

Solution:

a. $\text{Error} = \text{measured} - \text{accepted} = 8.83\ g/cm^3 - 8.94\ g/cm^3$
$\text{Error} = -0.11\ g/cm^3$

b. $$\text{percent error} = \frac{\text{measured} - \text{accepted}}{\text{accepted}} \times 100\%$$

$$\text{percent error} = \frac{8.83\ g/cm^3 - 8.94\ g/cm^3}{8.94\ g/cm^3} \times 100\%$$

$\text{percent error} = -1.23\%$

The negative error and percent error indicate that the measured density was less than the accepted density.

2.12 Precision

Precision is a *qualitative* evaluation of how exactly a measurement can be made.

Modern handheld calculators can display calculation results to 11 or more digits. Is it appropriate to report your homework problem results to this many figures? The answer is obviously no, but why not? The reason rests on the **precision** of the measured data given in the problem.

Precision is a *qualitative* term that describes the exactness of a number or measured data. Some physical quantities are known exactly because of the way in which they are defined. The speed of light and the length of a meter are two such quantities. Other numbers can be counted. When we count the number of people in the classroom, we know the number *exactly.* There are other mathematical quantities that are not exact, but we may express them to any degree of exactness we desire by specifying the number of decimal places to include. These are called irrational numbers. Their values are exact only when we write them using special symbols reserved for irrational numbers, such as the square root bracket ($\sqrt{\ }$), the Greek letter π (ratio of a circle's circumference to its diameter, C/d), and the letter e (base of natural logarithms). When an irrational number is written as a decimal, the number of decimal places indicates the precision of the irrational number. *Irrational numbers have potentially unlimited precision.*

Measured data does not have unlimited precision. When we make a measurement with a mechanical instrument, we read the major and minor scale division marks of the instrument. The digits in the measurement based on the scale marks are known for certain. But measurements almost never fall exactly on an instrument scale mark, so we must estimate the fraction of the distance between the marks. **When you use a mechanical *metric* instrument (one with scale subdivisions based on tenths), measurements should be estimated to the nearest $\frac{1}{10}$ of the smallest *decimal* increment.** For most measurements made in a high school laboratory, this *estimated digit* is the last (right-most) digit in the measurement. With finer graduations, more certain digits are possible, and the measured data potentially has greater precision (see Figure 2-18). The maximum precision of an instrument is limited by its finest scale subdivisions.

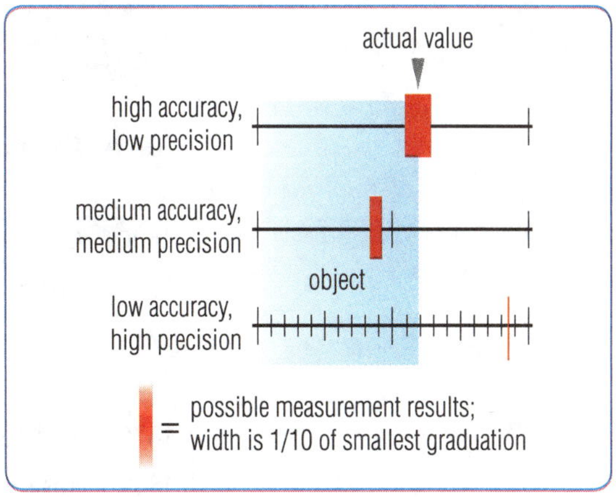

2-18 The finer the scale markings, the more precise the measurement. Measurement accuracy depends on instrument calibration as well as on the fineness of scale graduation.

The last digit that has any significance in a measurement is estimated. (See the discussion of significant digits in the next section.) The human act of measurement introduces some uncertainty into every measurement. We will assume for this course that the uncertainty of any measurement is plus-or-minus half of the place value of the estimated digit. For a meter stick whose smallest subdivision is the millimeter (0.001 m), the uncertainty is ±0.5 mm or ±0.0005 m. The uncertainty of most measurements you will make will be much smaller than half the place value of the estimated digit. In actual research, uncertainty is determined by a statistical process.

2-19 A measurement rarely falls exactly on a scale marking. The position of the measurement on the scale must be estimated.

2.13 Modeling the Speed of Light

Evaluating Ole Rømer's observations of Io and his data depended on some developing science and technology. In 1675, Christiaan Huygens had developed an accurate clock that could measure seconds. Giovanni Cassini had formulated a table of predicted sightings of Io, and he had just recently calculated the diameter of the earth's orbit. All of these pieces of information were necessary to measure the speed of light with any accuracy.

Since then, the speed of light has been determined many times by more sophisticated methods. In 1983, this quantity was defined as a constant (c). The speed of light appears in many scientific formulas, like Einstein's famous equation, $E = mc^2$. Table 2-3 on the next page lists the data for some of the more significant determinations of c using a variety of techniques. The uncertainties of the measurements, when known, are based on the method by which the speed of light was determined.

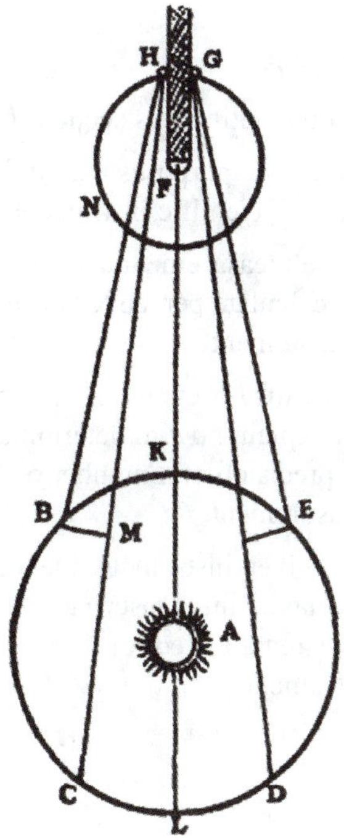

2-20 The diagram from Christiaan Huygens's paper describing the analysis of Rømer's experimental data

2B Section Review

1. Discuss why all measured data contains some error.
2. Describe one method for quantifying accuracy.
3. To what decimal place will you report measurements taken with the following instruments?
   a. a graduated cylinder with 1 mL subdivisions
   b. a metric ruler with 1 mm subdivisions
   c. a laboratory balance with 0.1 g subdivisions
   d. a thermometer with 2 °C subdivisions

Table 2-3

Historical Speed of Light Measurements

Date	Experimenter	Method	Speed (*c*) (km/s)	Uncertainty (± km/s)
1600	Galileo	Lanterns	"fast"	N/A
1676	Rømer/Huygens	Jovian moon occultation	215 000	unknown
1729	Bradley	Aberration of Light	301 000	unknown
1849	Fizeau	Spinning slotted wheel	315 300	unknown
1879	Michelson	Rotating mirror	299 910	50
1889	Rosa	Capacities method	300 000	800
1890	Comstock	18 star pairs	300 560	170
1898	Perot/Fabry	Electrometric method	299 730	300
1906	Rosa/Dorsey	Electrical method	299 803	10
1923	Mercier	Electrical measurements	299 700	30
1926	Michelson/Birge	Interferometer	299 798	15
1928	Karolus/Mittelstaedt	Kerr Cell shutter	299 778	10
1932	Pease/Pearson	Interferometer	299 774	10
1947	Essen/Gordon-Smith	Microwave cavity	299 792	3
1958	Froome	Interferometer	299 792.50	0.1
1973	Evenson	Laser method	299 792.467	0.0011
1978	Woods/Shotton/Rowley	Laser method	299 792.4588	0.0002
1983	International	Definition	299 792.458	0

2B Objectives

After completing this section, I can

✓ discuss the purposes and limitations of scientific instruments.

✓ define measurement accuracy and calculate percent error in a measurement.

✓ define measurement precision and explain how to determine the precision of a number or measurement.

✓ for a given instrument, identify the maximum measuring precision allowed for that instrument.

4. Which volume is reported with greater precision, 87.0 mL or 87.04 mL? Why?

✪5. Discuss some possible reasons that God did not create man with extended senses such as ultrasonic hearing or infrared vision.

✪6. You and your lab partner determined the density of a sample of pure mercury to be 13.52 g/cm^3. The accepted value for the density of mercury is 13.60 g/cm^3. Determine (a) the error, (b) the absolute error, and (c) the percent error of your result to two decimal places.

DM✪7. Using Table 2-3, create a scatter plot of the year versus the measurements of the speed of light for the years 1879 to 1983. Include error bars for the uncertainties of these measurements. For assistance in creating graphs from tabular data, trendlines, and error bars, see Appendix G.

DM 8. Refer to the graph you created in Question 7. What general trends do you see in both the value and accuracy of the speed of light data with time?

DM 9. Reread the facet on Thomas Kuhn in Chapter 1 (pages 8–9). How does this graph confirm his explanation of the purpose of "normal science"?

FACETS of PHYSICS

Fine-Tuning Measurements—Vernier Scales

How sharp are your eyes? Could you measure the size of a grain of sand (dimensions on the order of 0.1 mm) if you had a standard metric ruler? Probably, if you have average eyesight. Now, do you think you could measure the size of a large biological cell (around 0.001 mm) using the same instrument? You probably could not even see the cell. However, machinists who build precision mechanical equipment must often make measurements that are precise to hundredths or even thousandths of a millimeter.

A

B

Scientists also need to estimate measurements to much finer tolerances than is possible using simple visual estimation. Instruments that measure to such extreme precision use a specially calibrated auxiliary scale called a **vernier scale**. The term *vernier* (vur′nē-êr) is used in honor of the seventeenth-century French engineer, architect, and surveyor Pierre Vernier, who developed an instrument to more accurately measure angles for land surveys. Modern vernier scales are based on his invention. They are commonly used in micrometers, calipers, analytical balances, fine focusers for optical instruments, electrical potentiometers, and other precision measuring devices.

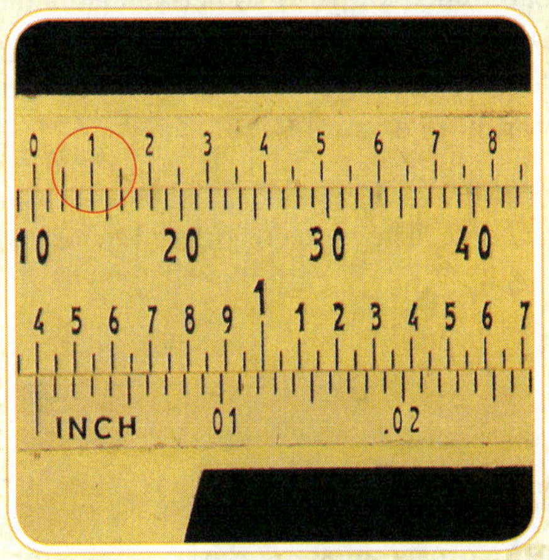

C The reading is 10.1 mm.

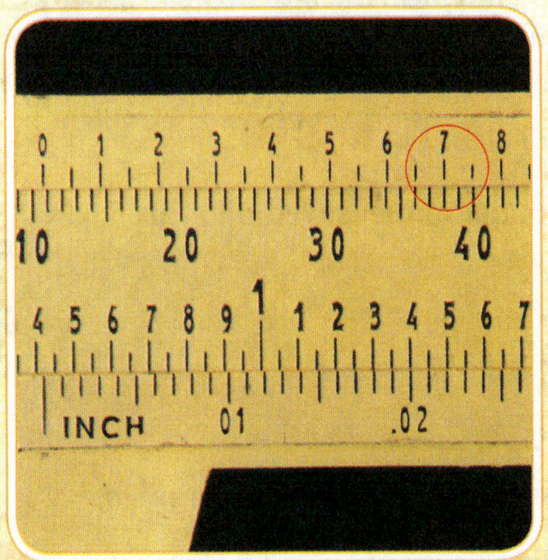

D The reading is 10.7 mm.

Vernier scales permit estimating the last significant digit one or more decimal places farther to the right than would be possible without the vernier. A vernier instrument contains two scales that lie alongside each other. One scale is fixed and calibrated in standard units such as millimeters or hundredths of an inch. We will call this the *standard scale*. The other scale that actually indicates the measurement, called the *vernier*, slides alongside the standard scale. It is subdivided in a similar fashion as the standard scale with one difference—10 subdivisions on the vernier scale exactly span one less than an integer multiple of 10 subdivisions on the standard scale. For example, 10 vernier subdivisions could span either 9 standard subdivisions or 49 standard subdivisions.

The number of vernier subdivisions determines the precision to which the instrument can be read. If there are 10 vernier subdivisions per unit, then the instrument can produce readings to $\frac{1}{10}$ (0.1) of the standard scale's smallest subdivision. If there are 50 vernier subdivisions, then the instrument can be read to $\frac{1}{50}$ (0.02) unit. Notice that with a vernier scale, it is possible to accurately estimate the last significant digit to a greater precision than just $\frac{1}{10}$ of the smallest standard metric scale calibration.

Figures C and D show how the vernier scale functions. The vernier scale scribe mark along the scale that lines up exactly with a standard scale subdivision mark is the vernier decimal reading. This decimal value is added to the value of the *smaller standard scale* subdivision nearest the index line (zero mark) of the vernier scale.

Vernier scales are just one of the tools scientists use to improve the precision with which they observe and measure the physical world. However, you should remember that any instrument is subject to errors built into it as well as to damage and wear. Man's observations in this universe will always be limited by the finite precision and accuracy of his instruments.

To be valid, a scientific experiment must be repeatable. This means that it yields the same results, within an acceptable range, each time it is repeated. Repeatability implies reliability. As a general rule, simple measuring procedures are more repeatable than complex ones.

Significant digits (SDs) in a measurement consist of all certain digits and the one estimated digit.

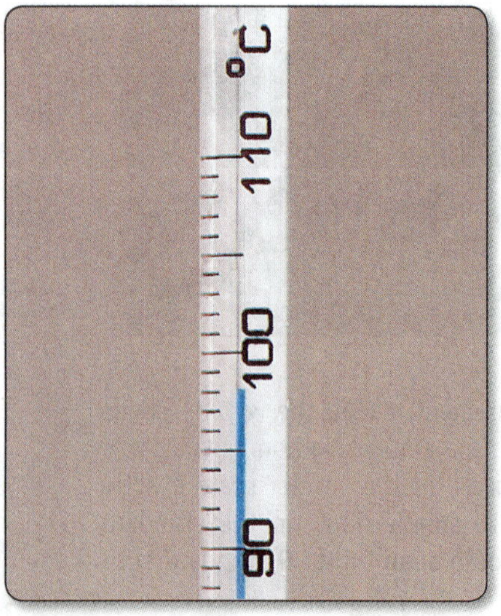

2-21 What is the temperature indicated by the thermometer? You should report 98.1 °C.

Problem-Solving Strategy 2.5

SD rules regarding zeros vary depending on whether the decimal point is written or is assumed due to placeholding zeros.

2C TRUTH IN MEASUREMENTS AND CALCULATIONS

2.14 Introducing Significant Digits

A scientist's integrity and professional reputation depend on the accuracy and lack of ambiguity in his observations and reports. If a scientist's observations and data cannot be reproduced by others using the same instruments and techniques, he loses the regard of his peers (along with the potential for future research funding). Scientists are careful to report their data only to the precision warranted by the instruments or methods used.

Precision is indicated by the kind and placement of digits reported in the data. These digits are called significant figures, or **significant digits (SDs)**. Only the digits that are certain based on the instrument scale calibrations and one estimated digit (in tenths of the smallest subdivision) may be reported as significant digits. Consequently, the last (right-most) significant digit is the estimated digit when recording measured data. Zeros are added as placeholders before or after the significant digits, as appropriate, to locate the decimal point.

2.15 Rules for Determining Significant Digits in Data

How can we tell which digits are significant when we read someone else's data? This is important to know, especially if a mathematical operation is to be performed using the data. Rules have been developed which make this process fairly easy. Each rule below is followed by one or more examples.

Rule 1—SDs apply only to measured data.

The following numbers are not *measured* data. They are considered to be *exact,* and therefore SDs are not applicable:

- Counted or pure numbers

  2 one dozen π $\sqrt{2}$

- Ratios that are *exactly* 1 by definition

  60 s/min
  100 cm/m

Rule 2—All nonzero digits in measured data are significant.

324.4 m	(4 SDs)
62 °C	(2 SDs)

Rule 3—All zeros between nonzero digits in measured data are significant.

100.3 mm	(4 SDs)
7.025 μm	(4 SDs)

Rule 4—For measured data *containing a decimal point:*

- **All zeros to the right of the last nonzero digit (trailing zeros) *are* significant.**

30.0 s	(3 SDs)
6.500 L	(4 SDs)

- **All zeros to the left of the first nonzero digit (leading zeros)** ***are not*** **significant.** They are merely placeholders for the decimal point.

0.040 m (Only last two digits are SDs.)

Rule 5—For measured data ***lacking*** **a decimal point:**

- **No trailing zeros are significant.**

1500 m (2 SDs)

We cannot tell if this measurement was estimated to the nearest 100 m, 10 m, or 1 m. Therefore, we can be assured only that at least the two nonzero digits are SDs. (SD Rule 2)

- **Scientific notation (see Appendix G) shows** ***only*** **significant digits in the decimal part of the expression.**

1.50×10^3 m (3 SDs)

This notation indicates that the measurement was estimated to the nearest 10 m.

- **In this textbook, a decimal point following the last zero indicates that the zero in the ones place** ***is*** **significant.**

1500. m (4 SDs)

This measurement was estimated to the nearest 1 m, so all four digits are SDs.

The position of the estimated digit in measured data containing trailing zeros is uncertain unless it has a decimal point.

Problem-Solving Strategy 2.6

Use scientific notation when trailing zeros would make the precision of measured data ambiguous.

Mathematical Operations with Measured Data

2.16 Calculator Precision

Why do we need to know about significant digits? Let's calculate the density of a stone with a mass of 5.85 g and a volume of 4.3 cm^3.

$$\text{density} = \frac{\text{mass}}{\text{volume}} = \frac{5.85 \text{ g}}{4.3 \text{ cm}^3} \cong 1.360\,465\,116 \text{ g/cm}^3$$

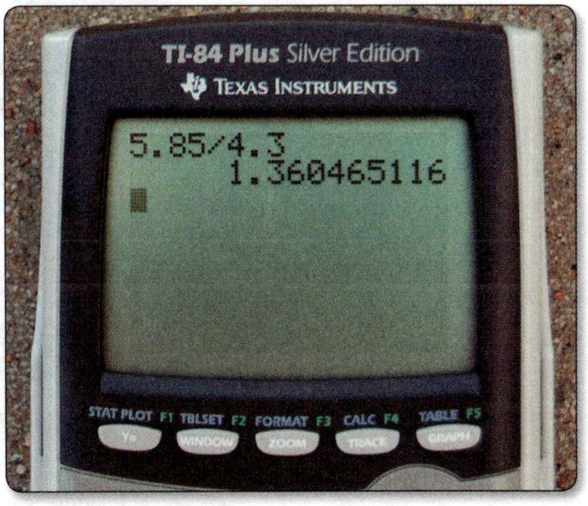

2-22 This calculator can display 11 digits in its answer.

The result as displayed indicates that the density calculation is precise to at least a billionth (10^{-9}) g/cm^3. However, the volume was known only to the nearest tenth (10^{-1}) cm^3 and the mass only to the nearest hundredth (10^{-2}) g. If we compare the precisions of the mass, the volume, and the calculated density, we find that the density displayed above has a far higher precision than the measured data. A scientifically valid calculation should have a precision that is about the same as the precisions of the contributing factors.

Scientists often perform calculations with measured data to derive other data, so they have established some additional rules in order to make sure that their results do not have more (or less) precision than the original data. The rules for mathematical operations that follow are convenient for high school work, but they do not give the scientifically acceptable precision in every case. Scientists doing research must determine the precision for each calculation based on approved statistical methods for a given kind of calculation and the kind of measured data used.

In this textbook, the symbol "≅" means that a truncated or rounded calculator result follows.

What if a town had a sign on its outskirts that read as follows:

Population	3,501
Altitude	2,955
Tumbleweeds	4,687
Total	11,143

Would there be any practical use for this total?

2.17 Math Rule 1—Adding and Subtracting Measured Data

A. Measured data must be the same kind of data and have the same units before they can be added or subtracted. (You cannot add apples and oranges!)

B. The sum or difference of measured data cannot have greater precision than the *least* precise measurement. Round results to the place value of the estimated digit in the least precise quantity.

Problem-Solving Strategy 2.7

When adding or subtracting numbers in a column, draw a vertical line to the right of the estimated digit in the *least* precise data. In the answer, round to the digit's place to the *left* of this line.

EXAMPLE 2-4

Adding SDs: Length

Add 1.54 m and 60 cm.

Solution:

First convert to common units.

$$\frac{1.54\ \cancel{\text{m}}}{} \Bigg| \frac{100\ \text{cm}}{1\ \cancel{\text{m}}} = 154\ \text{cm}$$

Add the lengths in a column.

$$\begin{array}{r} 15\underline{4}\ \text{cm} \\ +\underline{6}0\ \text{cm} \\ \hline 21\underline{4}\ \text{cm} \cong 2\underline{1}0\ \text{cm} \end{array}$$

Estimated digits are underlined. The sum must be rounded to the estimated place of the *least* precise number—to the nearest 10 cm. Review rounding rules in Appendix G.

In this textbook, the symbol "≅" also means that a calculated result has been rounded to the appropriate number of significant digits.

2.18 Math Rule 2—Multiplying and Dividing Measured Data

A. The product or quotient of measured data cannot have more SDs than the measurement with the fewest SDs. Round results as necessary. (The density that we calculated in Subsection 2.16 should be rounded to 2 SDs, or 1.4 g/cm^3, since the volume, 4.3 cm^3, has only 2 SDs.)

$$\begin{array}{r} 10.3\ \text{m} \\ \times\ 3.7\ \text{m} \\ \hline 7\,2\,1 \\ 3\,0\,9 \\ \hline 38.1\,1\ \text{m}^2 \cong 38\ \text{m}^2 \end{array}$$

Estimated digits or the results of operations with estimated digits are circled. As with the original measured data, the final answer can have only one estimated digit, so it is rounded as shown, which gives 2 SDs.

EXAMPLE 2-5

Multiplication of SDs: Area

Find the area of a rectangular solar cell array that is 10.3 m long and 3.7 m wide.

Solution:

$$A = l \times w = lw = (10.3\ \text{m})(3.7\ \text{m})$$

$$A = 3\underline{8}.11\ \text{m}^2$$

Since 3.7 m has only 2 SDs, only 2 SDs are allowed in the result.

$A \cong 38\ \text{m}^2$

Study the explanation provided in the margin for this rule.

B. The product or quotient of measured data and a pure number should not have more or less *precision* than the original measurement.

Examples:	7 × 1.75 cm = 12.25 cm	(*not* 12.3 cm)
	1.35 cm ÷ 7 ≅ 0.19 cm	(*not* 0.193 cm)

2.19 Math Rule 3—Compound Calculations

You should know how to efficiently use a modern scientific calculator. Multiple math operations can be completed simultaneously using nested parentheses and other functions. *Compound calculations* involve two or more simultaneous math operations.

A. If the operations in a compound calculation are all of the same kind (either addition/subtraction **or** multiplication/division), complete the operations simultaneously using standard order of operations before rounding to the correct significant digits (Math Rule 1). This practice eliminates excessive rounding errors (see Example 2-6).

EXAMPLE 2-6

Compound Calculations: Base-ic Math

The base for a street lamp is a concrete cylinder 0.85 m in diameter (d) and 1.5 m tall (h) (see Figure 2-23). Calculate the mass (m) of the lamp base in kilograms if the density of concrete is 2.83 g/cm^3.

Solution:

The mass of an object may be found from the product of its density (ρ) and its volume (V).

Given: $d = 0.85$ m; $r = d/2$; $h = 1.5$ m; $\rho = 2.83$ g/cm^3

Formulas: $V = \pi r^2 h$; $m = \rho V = \rho(\pi r^2 h)$

	ρ	pi	r^2	h	$m^3 \to cm^3$	$g \to kg$	
$m =$	$\frac{2.83\ \cancel{g}}{\cancel{cm^3}}$	π	$\frac{0.85\ \cancel{m} \times 0.85\ \cancel{m}}{2 \times 2}$	$1.5\ \cancel{m}$	$\frac{10^6\ \cancel{cm^3}}{1\ \cancel{m^3}}$	$\frac{1\ kg}{10^3\ \cancel{g}}$	$\cong 24\underline{0}8.82$ kg

Since this is a compound multiplication/division calculation, the result cannot have more SDs than the measured data with the fewest SDs (2). Therefore,

$$m \cong 2400 \text{ kg.} \qquad \text{(2 SDs allowed)}$$

B. If a solution to a problem requires the combination of both addition/subtraction and multiplication/division operations, rounding the intermediate solutions may introduce excess rounding errors into the final answer.

(1) For intermediate calculations, you should underline the estimated digit in the result and retain *at least* one extra digit beyond the estimated digit. Drop any remaining digits without rounding.

(2) Round the final calculation to the correct significant digits according to the applicable math rules, taking into account the underlined estimated digits in the intermediate answers. Carefully examine Example 2-7 on the next page to see how this works.

2.20 Math Rule 4—Angles and Trigonometry

A. Angles in the SI are measured in derived units called radians. A **radian (rad)** is the plane angle that subtends a circular arc equal in length to the radius of the circle (see Figure 2-24). There are 2π radians in 360 degrees. Standard degree units for angles (°) are not part of the SI, nor are they metric. In this textbook, angles may be expressed in either degrees or radians depending on the application. Angles measured with a protractor should be reported to the nearest 0.1 degree.

Compound calculations involve two or more simultaneous math operations performed on a calculator.

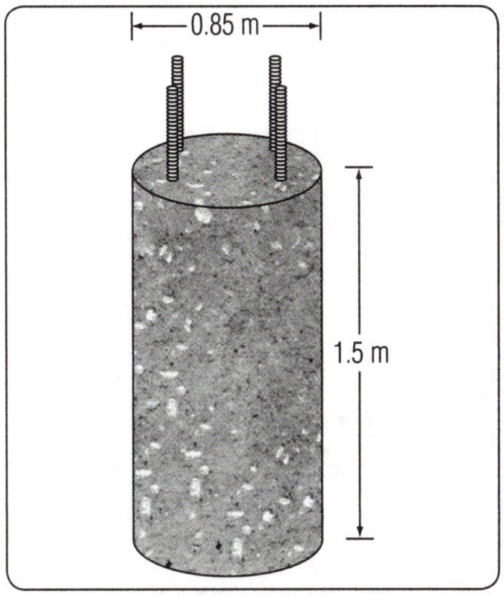

2-23 Concrete street lamp base

Note that the last two factors in the "bridge" convert the given units of volume and density to the required units of mass.

Problem-Solving Strategy 2.8

For problems requiring more than one calculation, do not round the intermediate answers, but note the last SD place for each answer as if you were. Round only the final answer, taking into account the SD placement in each of the intermediate answers.

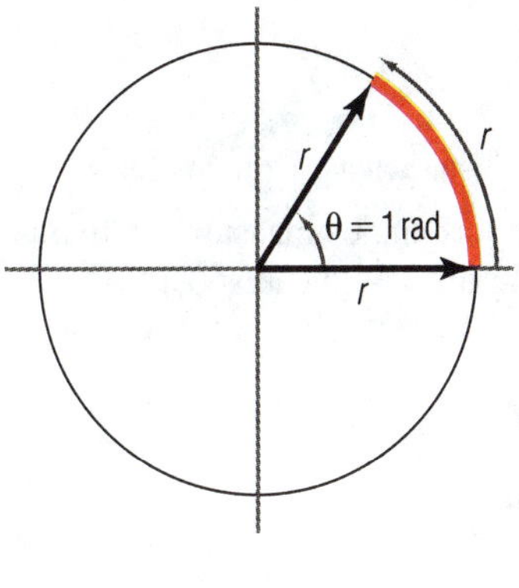

2-24 Definition of a radian

Special subscripts are often used to clearly identify variables and expressions in equations.

EXAMPLE 2-7

Compound Calculations: Sum of Areas

Three students are assigned the task of calculating the total floor area of the school's science lab. The first student finds that the area of the main lab floor (A_m) is 9.3 m by 7.6 m. Meanwhile, the second student measures the floor area of the chemical storage room (A_s) to be 3.35 m by 1.67 m. The third student determines that the closet floor area (A_c) is 93.5 cm by 127.6 cm.

What is the total floor area (A_{total}) in square meters?

Solution:

Given: $l_m = 9.3$ m, $w_m = 7.6$ m, $l_s = 3.35$ m, $w_s = 1.67$ m, $l_c = 127.6$ cm, $w_c = 93.5$ cm

The total area (A_{total}) is equal to the sum of all three areas, or $A_{total} = A_m + A_s + A_c$.

The area of any rectangular region is the product of length and width.

$$A_{total} = (lw)_m + (lw)_s + (lw)_c$$

Convert the closet measurements to meters.

$$l_c = \begin{array}{c|c} 127.6\ \cancel{\text{cm}} & 1\ \text{m} \\ \hline & 100\ \cancel{\text{cm}} \end{array} = 1.276\ \text{m}$$

$$w_c = \begin{array}{c|c} 93.5\ \cancel{\text{cm}} & 1\ \text{m} \\ \hline & 100\ \cancel{\text{cm}} \end{array} = 0.935\ \text{m}$$

$$A_{total} = (9.3\ \text{m} \times 7.6\ \text{m}) + (3.35\ \text{m} \times 1.67\ \text{m}) + (1.276\ \text{m} \times 0.935\ \text{m})$$

$$A_{total} = 7\underline{0}.6\ \text{m}^2 + 5.5\underline{9}4\ \text{m}^2 + 1.1\underline{9}3\ \text{m}^2$$

(Underlined digits show last SD according to Math Rule 2.)

$$\begin{array}{lr@{}l l} A_{total} = & 7\underline{0} & .6 & \text{m}^2 \\ & 5 & .5\underline{9}4 & \text{m}^2 \\ & 1 & .1\underline{9}3 & \text{m}^2 \\ \hline & 7\underline{7} & .387 & \text{m}^2 \cong 77\ \text{m}^2 \end{array}$$

(precise to 1 m^2, 2 SDs allowed)

B. The process for converting between degrees and radians is discussed in Appendix G but is summarized here.

degrees to radians:

$$\theta_{radians} = \theta_{degrees} \times \frac{\pi}{180°}$$

radians to degrees:

$$\theta_{degrees} = \theta_{radians} \times \frac{180°}{\pi}$$

These conversion factors are definitions, so the new angles retain the number of SDs of the original measurement.

C. Report angles resulting from trigonometric calculations to the lowest precision of any angles given in the problem. For example, if an angle given in the problem is stated in whole degrees, then any angles resulting from calculations should be expressed to the nearest whole degree.

Problem-Solving Strategy 2.9

Report angles measured with a protractor to the nearest 0.1°.

D. For simplicity, assume that trigonometric ratios for angles given in this textbook are pure numbers, so SD restrictions do not apply. Consequently, the SDs of trigonometric problem solutions will depend on the precision of the nonangular data given in the problem rather than the precision of the trigonometric ratios.

Sine, cosine, and tangent ratios are nonlinear functions of angles, so simple significant digit rules are not possible. Trigonometric ratios expressed to four decimal places will generally permit determining angles to within 0.1° precision.

2C Section Review

1. In each of the following quantities, which place value contains the estimated digit? How many significant digits are in each?

   a. 84 g **b.** 100 mL **c.** 3 **d.** 0.0020 m

2. Express the following measurements in scientific notation (see Appendix G if necessary). Be sure to include the units in your answers.

   a. 803 000 m **b.** 30.75 cm **c.** 0.15 mL **d.** 0.050 30 s

3. Which of the following measurements is the most precise? Explain.

   a. 7.40×10^3 cm **b.** 7400 cm **c.** 7400. cm

4. Using a calculator, find the decimal value of the following trigonometric expressions. Record your answers according to the rules in this section.

   a. sin 53° **b.** tan 28° **c.** cos $\pi/3$ **d.** sin π

5. Convert the following angles to the other angular form.

   a. 37° **b.** 125° **c.** $2\pi/3$ **d.** $\pi/8$

6. Using a flexible tape measure, you measure the circumference of the shaft of a machine bolt to be 5.98 cm. Calculate the bolt's diameter, adhering to the significant digit rules given in this section.

7. Report the answers for the following calculations using the correct number of SDs. Show your work, including the answers prior to rounding.

   a. 74.63 cm + 12.0 cm **c.** 110 m × 5 m

   b. 8.6 L − 667 mL **d.** 10.7 s ÷ 3

8. Find the unknown side and angles for the right triangle in the diagram. Report your answers with the correct SDs and include the appropriate units.

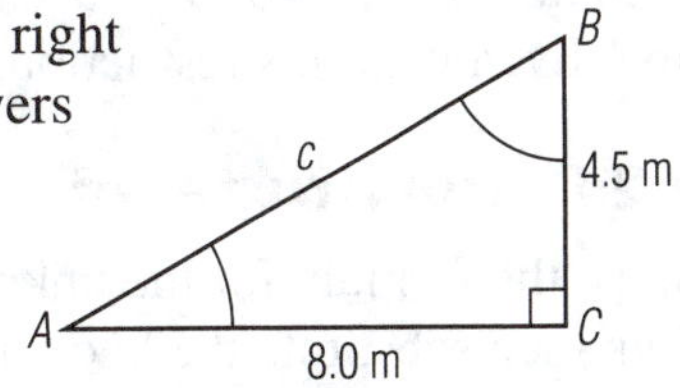

Problem-Solving Strategy 2.10

Consider trigonometric ratios of angles to be pure numbers for the purpose of determining the SDs of problems involving angles.

2C Objectives

After completing this section, I can

✓ write measured data using the allowed significant digits.

✓ express calculated answers using measured data with the correct number of significant digits.

✓ state the significant digit rules for measured angles, trigonometric ratios, and the results of nonalgebraic functions.

2D PROBLEM SOLVING

2.21 The Need for Problem Solving

To inquisitive scientists, questions about the structure or behavior of the universe are problems to be solved, and every problem has its own unique solution. You will be presented with a variety of problems throughout this course—not to make solving them more difficult, but to develop skills that will help you solve many problems that you will face. God made the world in such a way that humans must solve problems to live in obedience to Him. He made humans with the ability to solve these problems. Solutions to problems may require using one or more of the techniques that will be discussed in this section.

2-25 The fictional detective Sherlock Holmes was known for his uncanny ability to solve seemingly insurmountable problems using observation and deductive reasoning.

Problem-Solving Strategy 2.11

Read the problem carefully and make a sketch that accurately relates all known information. Note information that is extraneous.

The correct solution to a physics problem—or any problem, for that matter—is usually obtained by following an orderly process that leads eventually to the answer. This is not to say that you cannot occasionally stumble onto the answer, but it is not likely, and much effort can be wasted along the way. The four general steps that follow outline a process that can be applied to finding a solution to any physics problem.

2.22 Read the Exercise to Understand the Problem

This is the most important step, though students frequently rush through it. Read the problem statement completely through once to gain an overview of the problem and determine what you are being asked to find.

Read the problem a second time slowly and carefully to note what information has been given and to consider the principles relating to the problem. Identify the known and unknown variables and write down the data stated in the exercise. Also, watch for information that has nothing to do with solving the problem.

Make a simple sketch to help organize the information. It does not have to be an artistic Rembrandt, but neither should it be an abstract Picasso! The essential objects, surfaces, paths, vectors, values, and directions should be indicated in their correct relationships. Show the unknown using the correct formula symbol.

Problem-Solving Strategy 2.12

Select the appropriate method to solve the problem. Usually, you will be required to show a mathematical or graphical solution.

2.23 Determine the Method of Solution

Problem solutions may require physical constants, data from tables, conversion factors, or other information that is not expressed in the exercise.

Write down the formulas needed to solve the problem that relate the variables you identified in the first step. Sometimes you will have to perform one or more intermediate calculations to provide the missing information, so more than one formula may be required.

Consider the best tool for finding a solution. Some methods that will yield an answer include the use of the following:

algebra	analogy
estimation	reasoning
tables	guess and check
graphs	

In this course, you will usually have to show a mathematical or graphical solution. Even with this restriction, there may be several possible approaches.

2-26 Doing everything in an orderly fashion is a biblical principle that can be practiced every time you do homework (I Cor. 14:40).

2.24 Substitute and Solve

Solve the formula for the unknown quantity to isolate it on one side of the equation; then substitute the known values.

Write down each mathematical step on your paper before performing the calculations. Do not get into the habit of completing a solution on your calculator and writing down just the answer. Writing out the calculation first allows you to perform unit analysis, and it guides your mathematical operations. Later, if your answer is wrong, you can quickly locate your mistake. This step also permits your teacher to identify conceptual problems when grading your paper.

Work to make your solutions neat and orderly. The written work that you turn in for grading should be a methodical, step-wise process to the solution.

Problem-Solving Strategy 2.13

Solve the applicable formula(s) for the unknown quantity and substitute the known information. Show as much work as necessary so that another person can follow your reasoning.

2.25 Check Your Answer for Reasonableness

a. This step is the one most often missed by students. Reread the problem. Have you answered the right question? Does the result have the expected order of magnitude? Does the answer have the correct dimensional units? Performing these checks will provide additional assurance that the answer is correct (or at least reasonable).

b. Mentally make an estimation of the magnitude of the answer. You can usually estimate a quantity to at least some power of 10 that is within a factor of 2 or 3 of the correct calculator result. This kind of estimation is often called an *order of magnitude estimation.*

c. Be sure to simplify units, especially if there are derived units that stand for more complex combinations of SI units. For example, the newton (N) can replace $kg \cdot m/s^2$.

d. Express results in the correct number of significant digits. For mathematical solutions, you should carefully follow the rules discussed in Sections 2B and 2C for maintaining the correct number of SDs in your solution.

Order of magnitude indicates the nearest power of 10 that approximates a given quantity. 85000 ($\sim 10^5$) is three orders of magnitude larger than 43 ($\sim 10^2$).

Problem-Solving Strategy 2.14

Check the answer for reasonableness.

TABLE 2-4

Orders of Magnitude

Object	Size
Quasar, distance ($\sim 10^9$ ly)	$\sim 10^{24}$ m
Galaxy, diameter ($\sim 10^5$ ly)	$\sim 10^{21}$ m
Nearest star, distance (~ 4 ly)	$\sim 10^{17}$ m
Solar system, diameter	$\sim 10^{13}$ m
Earth's orbit, diameter	$\sim 10^{11}$ m
Moon's orbit, diameter	$\sim 10^{9}$ m
Earth's diameter	$\sim 10^{7}$ m
Large city	$\sim 10^{5}$ m
City park	$\sim 10^{2}$ m
Human	$\sim 10^{0}$ m
Insect	$\sim 10^{-2}$ m
Cell	$\sim 10^{-5}$ m
DNA molecule	$\sim 10^{-7}$ m
Inorganic molecule	$\sim 10^{-9}$ m
Atom	$\sim 10^{-10}$ m
Atomic nucleus	$\sim 10^{-15}$ m

EXAMPLE 2-8

Estimation

You calculate that a car traveling at 55 mi/h is moving 2.5 m/s. You should question this result since a person can easily run at that rate. To check your solution by estimation, you figure that, since there are about 2 km/mi, a car moving at 55 mi/h is traveling about 110 km/h, or about 10^5 m/h. There are 3600 seconds per hour, which is about 4×10^3 s/h. Mentally divide the distance by the time to find the speed. First, divide both numbers by 1000, leaving 100 divided by 4. The result is about 25 m/s. Since your estimate is ten times larger than the calculated result, you need to recheck your original work.

2D Section Review

1. When beginning a physics exercise, what should you consider doing in order to organize the information and visualize the circumstances?
2. In order to find the mass of a cubic meter of sea water, what information would you need to look up?
3. In most cases, which of the solution methods discussed in subsection 2.23 would be the least likely to produce an accurate answer?
4. How many orders of magnitude larger is 7250 m than 83 m?
5. Discuss several checks that you can perform to give greater assurance that your solution to the problem is correct.

For Questions 6–10: Firewood is purchased in units called cords. A cord of wood is a stack 4.00 ft high, 4.00 ft wide, and 8.00 ft long. The density of sugar maple wood is about 6.8×10^2 kg/m^3. What is the volume of a half-cord of sugar maple wood in cubic meters?

6. What quantity are you asked to find?
7. Make a sketch of the problem showing the required information.
8. Discuss any additional information that you must provide.
9. State any information that may be extraneous.
10. Solve the problem.

2D Objectives

After completing this section, I can

✓ explain why a problem-solving strategy is essential to my success as a physics student.

✓ use a problem-solving strategy to answer physics questions.

Chapter Review

In Terms of Physics

SI (International System of Units)	24	conversion factor	29
dimension	24	unit analysis	30
space	24	accuracy	31
time	24	error	32
matter	24	percent error	32
meter (m)	25	precision	32
second (s)	26	vernier scale	35
mass	26	significant digit (SD)	36
kilogram (kg)	27	radian (rad)	39
derived unit	28	order of magnitude	43

Problem-Solving Strategies

2.1 (page 29) In calculations, the prefix of the SI unit of a measurement may be replaced by the exponential multiplier of the base unit.

2.2 (page 29) In a conversion factor, the unit you are converting *to* is in the same position relative to the fraction bar as the unit you are converting *from* in the given quantity.

2.3 (page 30) The first factor in the unit analysis bridge should be the given data. The remaining data and conversion factors may be inserted in any order to cancel the given units and substitute the required units in their place.

2.4 (page 30) The division operation is the inverse of multiplication. To indicate division by a quantity using the bridge utility, express the quantity and its unit as a fraction, invert the quantity *and its unit,* then write it as one of the factors in the bridge.

2.5 (page 36) SD rules regarding zeros vary depending on whether the decimal point is written or is assumed due to place holding zeros.

2.6 (page 37) Use scientific notation when trailing zeros make the precision of measured data ambiguous.

2.7 (page 38) When adding or subtracting numbers in a column, draw a vertical line to the right of the estimated digit in the *least* precise data. In the answer, round to the digit's place to the *left* of this line.

2.8 (page 39) For problems requiring more than one calculation, do not round the intermediate answers, but note the last SD place for each answer as if you were. Round only the final answer, taking into account the SD placement in each of the intermediate answers.

2.9 (page 40) Report angles measured with a protractor to the nearest 0.1°.

2.10 (page 41) Consider trigonometric ratios of angles to be pure numbers for the purpose of determining the SDs of problems involving angles.

2.11 (page 42) Read the problem carefully and make a sketch that accurately relates all known information. Note information that is extraneous.

2.12 (page 42) Select the appropriate method to solve the problem. Usually, you will be required to show a mathematical or graphical solution.

2.13 (page 42) Solve the applicable formula(s) for the unknown quantity and substitute the known information. Show as much work as necessary so that another person can follow your reasoning.

2.14 (page 43) Check the answer for reasonableness.

Review Questions

1. Name the three essential dimensions of the universe.
2. Name two SI fundamental units that have exact definitions.
3. In typeset text, what is the difference between the abbreviation for meters and the formula symbol for mass?
4. Write the conversion factor needed to convert 10.052 kg to grams.
5. What factor should normally be the first term in the bridge utility when performing unit analysis?
6. List three instruments each for measuring time and distance, in increasing order of precision.
7. If the error for a classroom mass balance is −0.75 g, is the mass of a sample indicated by the balance more or less than the actual mass?
8. Which is more precise, $\sqrt{2}$ or 1.414 213 562? Explain your answer.
9. A Celsius thermometer is calibrated in 1° increments. To what decimal place should you report a temperature obtained with this thermometer?

DM**10.** If you did not create a graph of the speed of light data given in Table 2-3 on page 34, do so according to the instructions given in Section Review Question 8 on that page. Examine the graph and provide an explanation for the apparent change in the measured value of the speed of light with time.

11. State the number of significant digits in each of the following quantities:
a. 103 kg **b.** 1500. mL **c.** 1.607×10^{-2} A **d.** 2000 km **e.** 0.0034 °C **f.** 10 eggs **g.** 30 μm **h.** tan(70°)

12. Round the following quantities to 3 significant digits:
a. 23 432 cm **b.** 0.000 329 5 m **c.** 10.009 N **d.** 7659 kg

13. (True or False) The calculated density of an object having a mass of 23.80 g and a volume of 24.0 cm^3 is 0.992 g/cm^3. Justify your conclusion.

14. Give several examples of the kinds of additional information that may be required to solve physics exercises.

15. Discuss some reasons why you should write down all mathematical steps in a solution to a physics exercise.

16. By how many orders of magnitude do the following quantities differ?
a. 4500 ly vs. 8.6×10^6 ly
b. 8.32×10^6 L vs. 0.023 L
c. 160 kg vs. 2.0 g
d. 22 m vs. 12 755 km

17. Estimate the following:
a. The length of your shoe in centimeters.
b. The number of times your heart beats in one minute.
c. The area of your desk in square centimeters.
d. The distance from your school to your house in meters.

✪**18.** Convert the following measured data to the units indicated. Use scientific notation where appropriate.
a. 2.9 cm to m
b. 35.70 Hz to MHz
c. 235 mL to L
d. 2325 ft to km

✪**19.** Convert the density 4230 mg/mm^3 to kg/m^3. Use the bridge utility for unit conversion.

✪**20.** The accepted value for the mass of one electron is 9.11×10^{-31} kg. Two experimenters separately measured the mass of the electron three times each. The first experimenter obtained the following masses: 7.31×10^{-29} kg, 7.29×10^{-29} kg, and 7.32×10^{-29} kg. The other experimenter obtained 8.5×10^{-31} kg, 9.6×10^{-31} kg, and 9.1×10^{-31} kg. Which experimenter's results were more accurate? Which were more precise?

✪**21.** Perform the required operations on the following quantities. Express your answers using scientific notation, the correct number of SDs, and the correct units.
a. $(6.234\ \text{m})(4.91 \times 10^{-5}\ \text{N})$
b. $(1.17 \times 10^3\ \text{mL}) + (2.315 \times 10^2\ \text{mL})$
c. $(32\ \text{m})(9.45 \times 10^6\ \text{s}^{-1})$
d. $(3.14 \times 10^{23}\ \text{ly}) - (7.18 \times 10^{20}\ \text{ly})$
e. $(2.32 \times 10^{-4}\ \text{m})(4.331 \times 10^{-3}\ \text{m})$
f. $(7.913 \times 10^{-4}\ \text{kg}) \div (9.1 \times 10^{-7}\ \text{m})$
g. $(7.5812\ \text{g}) - (2.1 \times 10^{-1}\ \text{g})$
h. $(2.32 \times 10^{-4}\ \text{m}) \div (4.331 \times 10^{-3}\ \text{m})$
i. $(3.0 \times 10^8\ \text{m/s}) \div (3.661 \times 10^{10}\ \text{s}^{-1})$
j. $(9.81\ \text{m/s}^2)(7.3 \times 10^1\ \text{kg})$

✪**22.** Find the long side (hypotenuse) and the area of a flower bed in the shape of a right triangle that has legs measuring 12.5 m and 10. m.

✪**23.** A board for a shelf is 2.31 m long and 33.4 cm wide.
a. What is the area of the shelf in square meters?
b. What is the perimeter of the shelf in centimeters?
c. Which will reduce the area of the shelf more: (1) cutting 3.4 cm from its width, or (2) cutting 0.31 m from its length?

✪**24.** A rectangular swimming pool is 10.0 m wide and 15.0 m long. The water depth varies over the length of the pool in three 5.0 m zones. The zone under the diving board is 3.0 m deep. The shallow zone at the other end is 1.0 m deep. The zone in the middle has a sloping bottom changing from 3.0 m to 1.0 m deep.
a. What is the volume of the pool in cubic meters?
b. What is the volume of the pool in liters?
c. The density of water at pool temperature is 1.0 g/cm^3. What is the mass, in kilograms, of the entire volume of water in the pool?
d. If a swimmer is allotted 5.0 m^2 of surface swimming space, how many swimmers may be in the pool at one time?

✪**25.** Describe the current dimensional standard for each of the seven fundamental SI units. Use the National Institute of Standards and Technology (NIST) website or another authoritative source.

✪**26.** Evaluate the statement, "Numerical scientific data may contain only significant digits."

✪**27.** Write an essay discussing the history of the metric system. Include in your discussion the development of the definitions for the seven fundamental SI units.

✪**28.** Estimate the number of German shepherds in your state. List any assumptions you make and the sources of information that you used to make the assumptions.

DM✪**29.** Review the graph below, which contains 193 values for the speed of light measured by physicists between 1675 and 1983. How does this set of data agree with your explanation given in Question 13? What does this suggest about the need to consider all the data available when making scientific inferences?

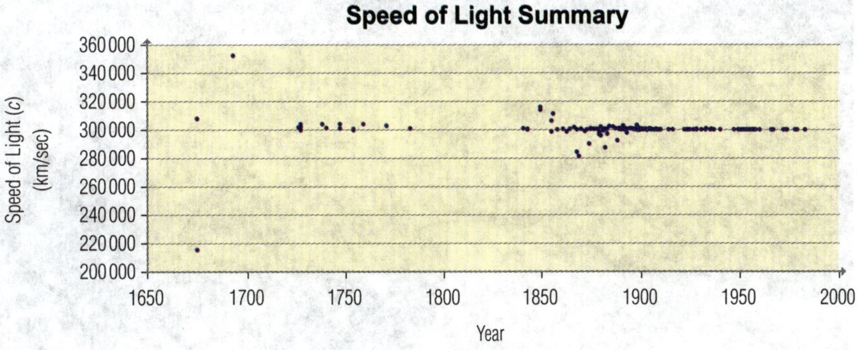

UNIT 2

Classical Mechanics

Classical mechanics had its beginnings with the work of Galileo in the sixteenth century. The height of the classical mechanics revolution came in the following century when Sir Isaac Newton discovered his three laws of motion. One of his greatest accomplishments was deducing the law of universal gravitation by analyzing the motion of the moon around the earth. For nearly 300 years, Newton's laws were the basis for understanding mechanical phenomena.

In Unit 2, you will learn how straight-line, two-dimensional, circular, and periodic motions are measured. You will delve into the nature of forces and how they can change motion and do work. You will also learn of the great mechanical conservation laws and the way various forms of energy are interchanged in order to conserve the total energy in the universe.

A US Navy F/A-18 Hornet fighter-bomber lands aboard an aircraft carrier. The pilot experiences significant negative acceleration in one dimension as the plane rapidly comes to a stop.

Motion in One Dimension 3

Dominion Modeling

Stopping Distance

You have probably heard the screeching of tires as car drivers apply their brakes to avoid a collision. Sometimes they are successful—sometimes not. Braking is a normal part of driving and is one of the most important skills to master when first learning to drive. You probably know that a car's stopping distance is related to its size and initial speed, the driver's reaction time, and the road conditions. Having an intuitive understanding of the stopping distance under existing conditions can make the difference between a potentially life-threatening collision and a safe stop. How can physics be used to analyze braking distance?

3A DESCRIBING MOTION

3.1 Introduction

Electrons moving through a tungsten light bulb filament generate light photons. These photons reflect off the letters on this page as your eyes scan across them and strike sensitive cells in your eyes. The retinas in your eyes generate charges that rush along the optic nerves to relay images to your brain. Meanwhile, the sunlight streaming in through the window gradually changes direction throughout the day as the earth spins around on its axis. The great wind and ocean current systems of the earth spiral in response to the planet's rotation. Each night, the moon waxes or wanes during its monthly orbit of the earth. The giant gaseous planets of our solar system revolve ponderously in their decades- or even centuries-long orbits about the sun. Distant galaxies retreat from us at hundreds and thousands of kilometers per second.

3-1 Motion is everywhere in the universe.

The universe is full of motion. The study of motion is called **mechanics**. Mechanics includes all kinds of motion, from simple one-dimensional motion along a straight line to complex three-dimensional aviation and space flight. This chapter investigates motion, speed and velocity, and acceleration.

3.2 What Is Motion?

Motion is the change of position with time. Kinematics is the description of *how* things move.

What is motion? Simply put, motion is a change of position. Scientists have divided mechanics into two branches: kinematics, to describe *how* objects move,

Dynamics is the study of the *causes* of motion.

and dynamics, to explain the *causes* of motion. Early Greek philosophers began their study of motion with the wrong question. They sought the causes of motion without understanding how things move in the first place. As a result, many of their ideas seem ridiculous to us today. *How* motion occurs (its description) must be known before its *causes* can be understood.

3.3 Mathematical Representations of Motion

If you were dropped into the middle of the ocean and you swam for an hour, how could you prove that you had changed position? Despite all your hard work, you might have swum in a circle. The only way to find out is to compare your first and last positions to a reference point, like an island. If your distance or direction from the island changed, then you could prove that you had moved.

To represent motion along a straight line, you identify the object's position on a number line. The number line is a one-dimensional coordinate system. Each position is compared to a reference point called the *origin*. By convention, positions to the *left* of the origin are assigned *negative* values, and positions to the *right* are assigned *positive* values. For example, an object 4 m to the left of the origin has a position of −4 m; an object 9 m to the right of the origin has a position of 9 m. The axis of the number line may be oriented in any direction, including vertically. *Up* is usually considered *positive*, and *down* is usually considered *negative*.

A variable with a subscript of 1 indicates an initial or earlier condition with respect to time. The symbol with a higher-numbered subscript indicates a later value relative to the initial value.

An object has moved if at one time its position is x_1 and at another time its position is x_2. Notice that each position is associated with a *time*: position x_1 occurs at time t_1 and position x_2 occurs at time t_2. The object's position at each time can be represented by an *ordered pair*, written (t_1, x_1) or (t_2, x_2).

The symbol s is the path distance traveled during a time interval. The path may be straight, curved, circular, or segmented.

Suppose you swim twenty-one lengths in a 50-meter pool. Because each length is 50 m, the distance you swim is 21 × 50 m = 1050 m. But notice that after swimming 1050 m, you end up only 50 m from the starting point (at the opposite end of the pool). The distance, or length, of the path (s) often differs from the net change in position. This change in position between two distinct points is called **displacement (d)**. In this example, s = 1050 m but **d** = 50 m forward from the starting point.

Scalar quantities contain one piece of information—a numerical value. Vector quantities contain two pieces of information—a numerical value and a direction. Magnitude is the absolute value of a numerical value and is always positive.

When a quantity contains only one piece of information, such as a distance of 1050 m or a temperature of −50 °C, it is a scalar quantity. A scalar value can be positive, negative, or zero. Scalars are symbolized by italicized letters, such as T for absolute temperature. When a quantity can be completely described only by stating both a magnitude and a *direction* (like 50 m forward, which is two pieces of information), the quantity is called a vector. Vectors are represented by bold roman characters, such as **d** for displacement. We will discuss vectors and their properties in more detail in Chapter 4. With motion along a straight line, the direction of a vector quantity is indicated by a positive or negative sign before the numerical magnitude of the vector. Forward motion is usually positive, and backward motion is usually negative.

In *one-dimensional motion*, the scalar symbol with the subscript of the coordinate axis (e.g., v_x) is identical to the more general vector notation for the same quantity (e.g., **v**).

Mathematical operations with vectors require special rules that you will learn in Chapter 4. To simplify vector operations in one-dimensional motion, we use a special notation that consists of the *scalar* symbol standing for the quantity with a subscript indicating the coordinate axis it is parallel to. For example, a displacement in the x-direction is symbolized by d_x. This notation represents *only* one-dimensional vectors, whereas the more general bold vector symbol can represent vectors in one, two, or three dimensions. The symbols d_x and **d** are equivalent if **d** is parallel to the x-axis.

In Figure 3-2 an object begins at position +1 m at time t_1. At time t_2, it reaches +4 m to the right of its original position. Then it moves left until it reaches +2 m

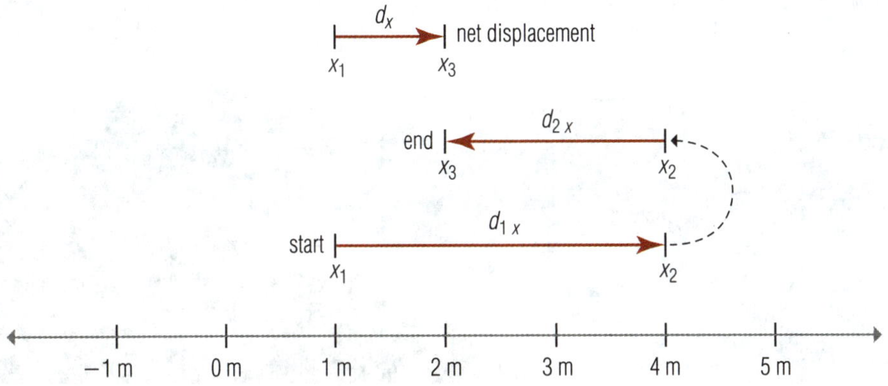

3-2 One-dimensional motion

at t_3. The object's displacement (d_x) for the interval t_1 to t_3 is its *net* change in position; that is,

$$d_x = d_{1\,x} + d_{2\,x}$$
$$d_x = (x_2 - x_1) + (x_3 - x_2) = x_3 - x_1$$
$$d_x = (+3 \text{ m}) + (-2 \text{ m})$$
$$d_x = +1 \text{ m, or 1 m to the right of the original position.}$$

The object has actually traveled over a path from x_1 to x_2 and back to x_3. The path distance is equal to the sum of the individual displacement magnitudes. Displacement magnitude is the absolute value of the difference of positions. This is true for linear motion in any number of dimensions.

$$s = |d_x| = |x_2 - x_1|$$

The path distance for the object in Figure 3-2 is determined as follows:

$$s = |d_{1\,x}| + |d_{2\,x}|$$
$$s = |(+4 \text{ m}) - (+1 \text{ m})| + |(+2 \text{ m}) - (+4 \text{ m})|$$
$$s = |+3 \text{ m}| + |-2 \text{ m}| = 3 \text{ m} + 2 \text{ m}$$
$$s = +5 \text{ m}$$

The object's *displacement* is +1 m, but it traveled a total *distance* of 5 m.

EXAMPLE 3-1

Picnic at the Lake: Displacement

A family leaves its campsite and travels by car to the picnic area on the north side of the lake. A sign at the campsite says that the picnic area is 2 km north by boat, but the children notice that the car's odometer advances by 10 km during the trip.

a. What is the car's displacement from the campsite when the family reaches the picnic area?
b. How far has the car traveled?
c. When the family returns to the campsite by the same road, what will be the car's displacement relative to its original starting point?
d. How far will the car have traveled?

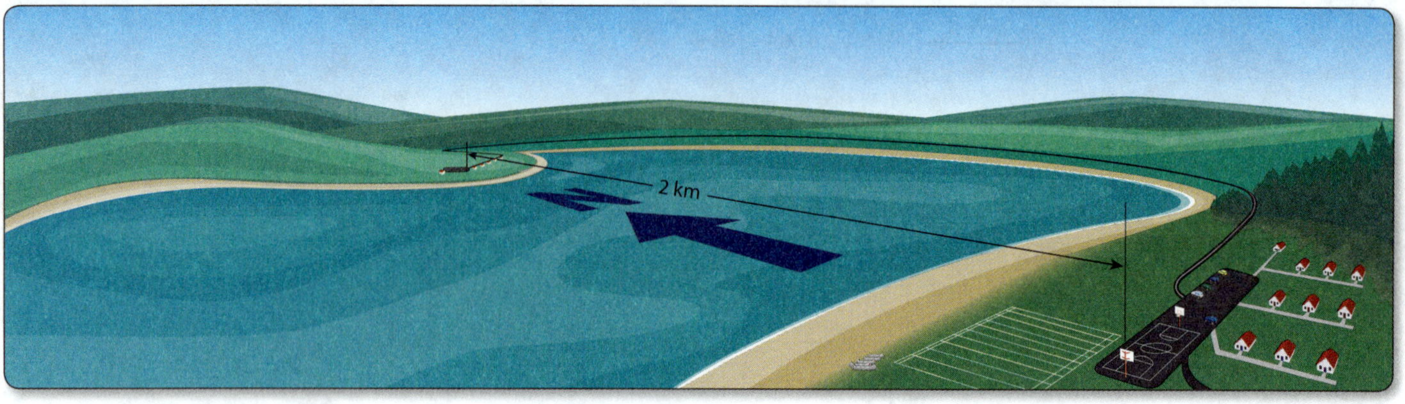

3-3 The distance across the lake is not the same as the distance traveled by car around the lake.

Solution:

Position an imaginary x-coordinate axis so that it passes through the initial and final positions, with north positive.

a. The car's total displacement is the straight-line distance and direction from its starting point to its final position. The car is 2 km north of its starting point.
b. The car has traveled 10 km, as its odometer shows.
c. When the car returns to the campsite, it will be at its starting position, so its total displacement is zero. For this answer, it is unnecessary to specify a direction since a zero displacement is the same in any direction.
d. The car has traveled a total of 10 km + 10 km, or 20 km.

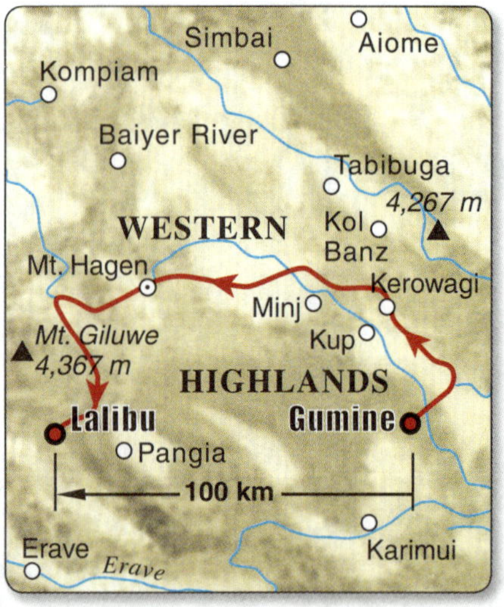

3-4 Map of the road between Gumine and Lalibu

EXAMPLE 3-2

Visiting Lalibu: Displacement in PNG

A map of Papua New Guinea (PNG) shows the two towns of Lalibu and Gumine separated by the mountains of the Western Highlands. Lalibu is 100 km due west of Gumine. A missionary visiting Lalibu from Gumine, traveling the only main road between the two towns, discovers that his odometer reading increases by 190 km. How far did he travel? What is his displacement for the trip?

Solution:

Draw the x-axis as an east-west line between Lalibu and Gumine, with its origin at Gumine. East is assumed to be positive. The missionary has traveled 190 km. However, he is now only 100 km from his starting point. His displacement is 100 km west ($d_x = -100$ km).

Problem-Solving Strategy 3.1

When graphing time-dependent data, always plot time on the horizontal axis and plot the dependent variable on the vertical axis.

3.4 Rate of Motion—Speed

You probably saw ordered pairs like (t, x) in algebra. In that class you plotted each pair as a point on a graph. You can also plot the (t, x) ordered pair on a graph called a **position-time graph**. This type of graph shows much useful information in a simple form.

For example, Figure 3-5a is the position-time graph of a train on an east-west track between Ames and Marshalltown, Iowa. The graph shows several (t, x) ordered pairs. From looking at this graph, you might think that the train jumped from point to point (city to city). Obviously this is impossible. The points are only

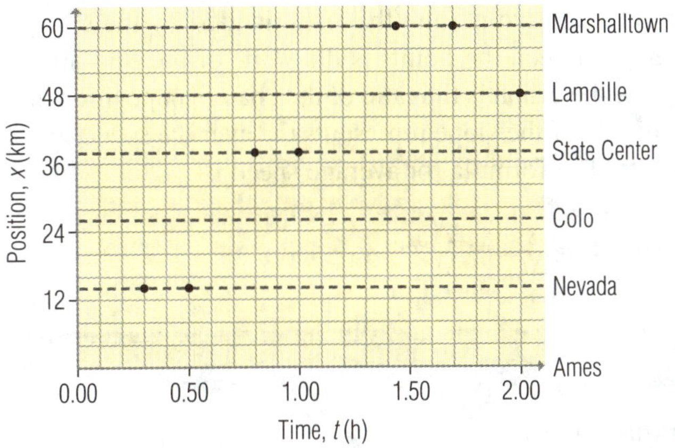

3-5a Plot of specific train positions and their times

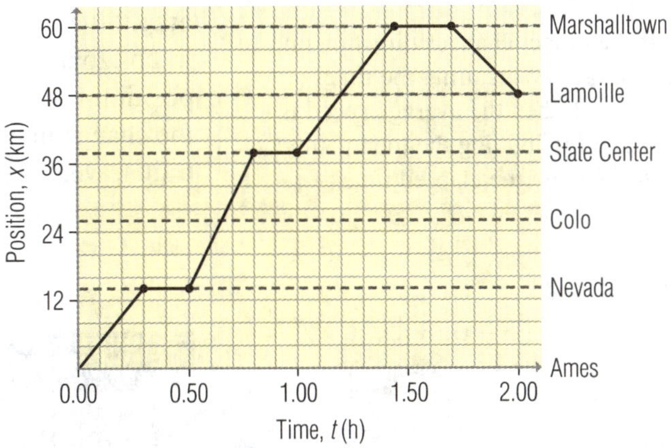

3-5b Continous plot of train position with time

the recorded positions of the train at certain times. If the train moved smoothly from point to point, you can draw lines connecting the points, as Figure 3-5b shows.

From Figure 3-5c you can find the displacement over a given time period. For example, suppose you want to know the displacement of the train from 0.60 h to 0.80 h. The graph shows that at $t_1 = 0.60$ h, the train's position is $x_1 = 22$ km east of its departure station. At $t_2 = 0.80$ h, its position is $x_2 = 38$ km east (State Center). The train's displacement during that period is the change in x during the interval. That is,

$$d_x = x_2 - x_1$$
$$d_x = (+38 \text{ km}) - (+22 \text{ km})$$
$$d_x = +16 \text{ km}.$$

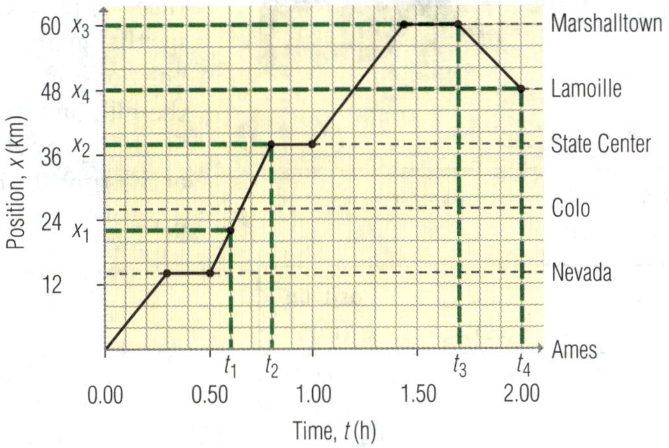

3-5c Time and position coordinate pairs for determining displacement

At 0.80 h, the train is 16 km farther east than it was at 0.60 h.

What if the displacement is negative? For instance, what is the displacement of the train from $t_3 = 1.70$ h to $t_4 = 2.00$ h? At t_3, $x_3 = 60$ km east (Marshalltown); at t_4, $x_4 = 48$ km east (Lamoille). The displacement is as follows:

$$d_x = x_4 - x_3$$
$$d_x = (+48 \text{ km}) - (+60 \text{ km})$$
$$d_x = -12 \text{ km}$$

In this answer, the negative displacement means that the train has moved backward (west) toward the origin.

A position-time graph also helps to determine the speed at which an object moves. The *average speed* ($\bar{v}$) of a train is the ratio of the distance traveled $|x_2 - x_1|$ over an interval of time $(t_2 - t_1)$. The absolute value sign is required because distance is always positive. If the train *does not change its direction,* its average speed is

$$\bar{v} = \frac{|x_2 - x_1|}{t_2 - t_1}.$$

The symbol Δ (Greek letter delta) conventionally means "change in," so Δx means "the change in position x" or $x_2 - x_1$. On a position-time graph, Δx is the difference between some position (x_2) and the initial position (x_1). The "change in time" (Δt, or $t_2 - t_1$), which is always positive, is the horizontal distance between the points on the timeline marking the beginning and the end of the time interval. In any linear

Problem-Solving Strategy 3.2

Whenever you must find the change of a quantity, always subtract the initial value from the final value.

The Greek letter "Δ" stands for the change in a quantity over a specified time interval. For example,

$$\Delta x = x_2 - x_1.$$

If the object changes direction one or more times during the time interval, then the average speed is found by dividing the total path distance by the elapsed time.

$$\bar{v} = \frac{s}{\Delta t},$$

where $s = |x_2 - x_1| + |x_3 - x_2| + \ldots + |x_n - x_{n-1}|$.

3-6a This skier must maximize her downhill speed in order to win.

Flag	Time (s)	Position (m)
Start	0.00	0.0
1	1.41	10.0
2	2.35	20.0
3	3.11	30.0
4	3.76	40.0
5	4.34	50.0

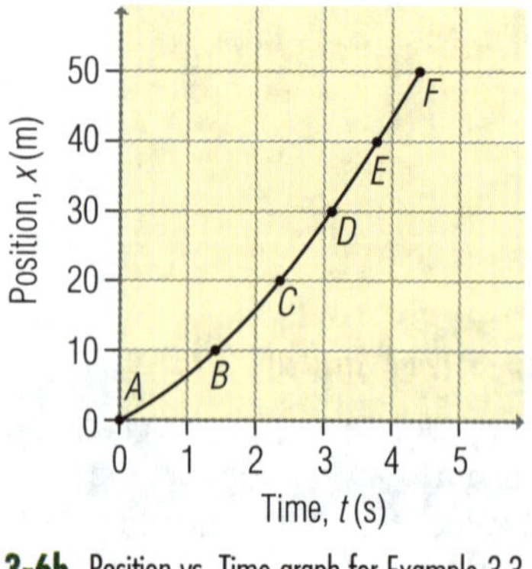

3-6b Position vs. Time graph for Example 3-3

A **tangent line** is a line that intersects a curve at a single point without crossing the curve.

The **limb** of a planet is the visible edge of the planetary disc seen against the background of space.

graph, the change in height between two points (the *rise*) divided by the change in horizontal distance (the *run*) between the points is the *slope* of the line. In the position-time graph, the slope of the graph is the ratio of the change in position and the change in time. The magnitude of the slope represents the average speed that a train travels between two points. The formula for average speed is

$$\bar{v} = \frac{s}{\Delta t} = \frac{|\Delta x|}{\Delta t}. \quad (3.1)$$

EXAMPLE 3-3

Downhill Skiing: Average Speed

A physics student, who also is a competition skier in Vermont, skis directly down a straight, 50.0 m long hillside that has a flag every 10.0 m. Each time she passes a flag, her coach records the time on a stopwatch. The table in the margin gives the time the skier passed each flag, and the position-time graph plots her motion. Her initial position is at the top of the hill, and the initial time is at the start of the descent. She passes the Start flag moving at 5.0 m/s. Find the skier's average speed (a) for the entire trip; (b) between flags 1 and 4; and (c) between flags 2 and 3.

Solution:

$$\bar{v} = \frac{|\Delta x|}{\Delta t}.$$

a. $$\bar{v}_{run} = \frac{|50.0 \text{ m} - 0.0 \text{ m}|}{4.34 \text{ s} - 0.00 \text{ s}} = \frac{50.0 \text{ m}}{4.34 \text{ s}} \cong 11.\underline{5}2 \cong 11.5 \text{ m/s}$$

b. $$\bar{v}_{1\text{–}4} = \frac{|40.0 \text{ m} - 10.0 \text{ m}|}{3.76 \text{ s} - 1.41 \text{ s}} = \frac{30.0 \text{ m}}{2.35 \text{ s}} \cong 12.\underline{7}6 \cong 12.8 \text{ m/s}$$

c. $$\bar{v}_{2\text{–}3} = \frac{|30.0 \text{ m} - 20.0 \text{ m}|}{3.11 \text{ s} - 2.35 \text{ s}} = \frac{10.0 \text{ m}}{0.76 \text{ s}} \cong 13.\underline{1}5 \cong 13.2 \text{ m/s}$$

3.5 Instantaneous Speed

As Example 3-3 shows, the average speed changes depending on the interval chosen. The skier's speed increases as her position changes from the top of the hill to the bottom. Suppose you want to know her speed at any one moment (her *instantaneous speed*). For example, what is her speed at the moment she passes the third flag (3.11 s, 30.0 m)? The slope of the position-time curve at that point is her speed. But how do you find a slope at one point on a nonlinear curve?

One method to estimate the slope is to construct a *tangent line* segment at a single point on the curve and measure the tangent's slope. If you could view the point on the graphed curve with a zoom-microscope, you would notice that as you increased the magnification power, the length of the curve segment visible in the eyepiece becomes shorter and the curve appears to become straighter. At maximum magnification, the curve is indistinguishable from the straight-line tangent at that point.

To help you understand this concept, consider the curve of the earth. Viewing it from the moon, you see that the *limb* of the earth is clearly circular; but as you approach the earth, the horizon becomes more sweeping and the curvature less distinct. By the time you touch down on land, the horizon appears to be flat. This small segment of the earth's 40 000 km circumference that you can see is essentially parallel to the tangent line at the point upon which you stand.

In a similar way, you can estimate the skier's instantaneous speed at the third flag by "zooming in" on the curve around the flag. A very rough estimate for her speed at this point can be made by finding the skier's average speed for the entire run. (This calculation was made in Example 3-3.) As depicted on the position-time graph (Figure 3-7a), your estimate would be the line connecting the top and the bottom points of the ski run. But the slope of this line is only roughly parallel to the tangent at the third flag (Point *D*).

To improve your estimate, try constructing a line between the endpoints of a shorter curve segment, for example, from Point *B* to Point *E*. This line is more nearly parallel to the curve at the third flag, so its slope provides a better estimate for the instantaneous speed at Point *D*. However, you can improve the estimate even more. Figure 3-7c shows a line through Points *C* and *E* that is nearly parallel to the curve at Point *D*. Thus its slope of 13.2 m/s is a good estimate of the slope of the tangent (and therefore the instantaneous speed) at the third flag. In summary, you estimate the instantaneous speed at a given point by calculating the average speed of a small interval containing that point—the smaller the interval, the better the estimate. This is the graphical basis of differential calculus.

The difference between average and instantaneous speed is important to automobile drivers, since the speed limit applies to instantaneous speed, not average speed. To maintain a particular speed on long trips (usually the speed limit), drivers often use cruise control. Although this device can maintain a certain average speed, it cannot guarantee that the automobile's instantaneous speed will always equal the speed for which it is set. Suppose, for instance, that a driver is traveling along a hilly road and the speed limit is 90 km/h (about 55 mph). The cruise control is set at that speed when the car begins to climb a hill. As the car climbs, it requires more fuel to maintain a speed of 90 km/h. When it reaches the top of the hill, the need for fuel abruptly decreases. If the cruise control cannot react quickly enough to reduce the fuel supply, the car speeds up and breaks the speed limit even though the cruise control was set at 90 km/h (Figure 3-8).

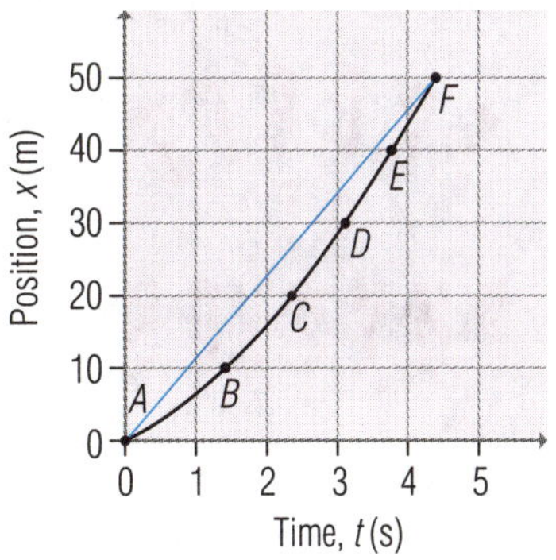

3-7a First estimate

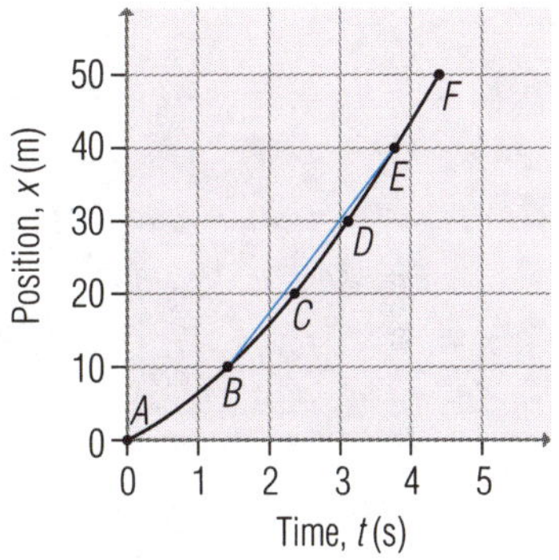

3-7b A better estimate

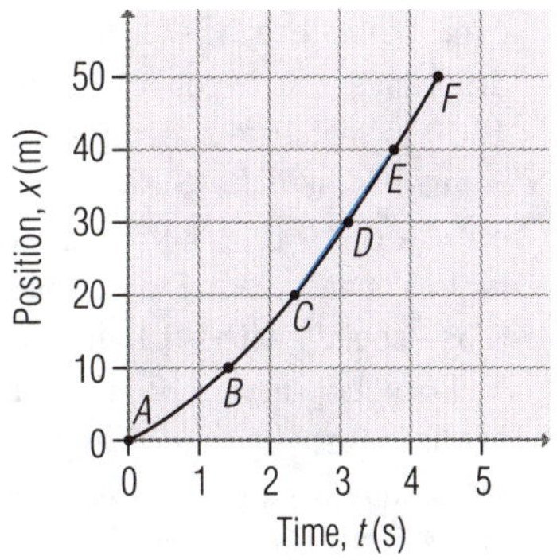

3-7c The best estimate

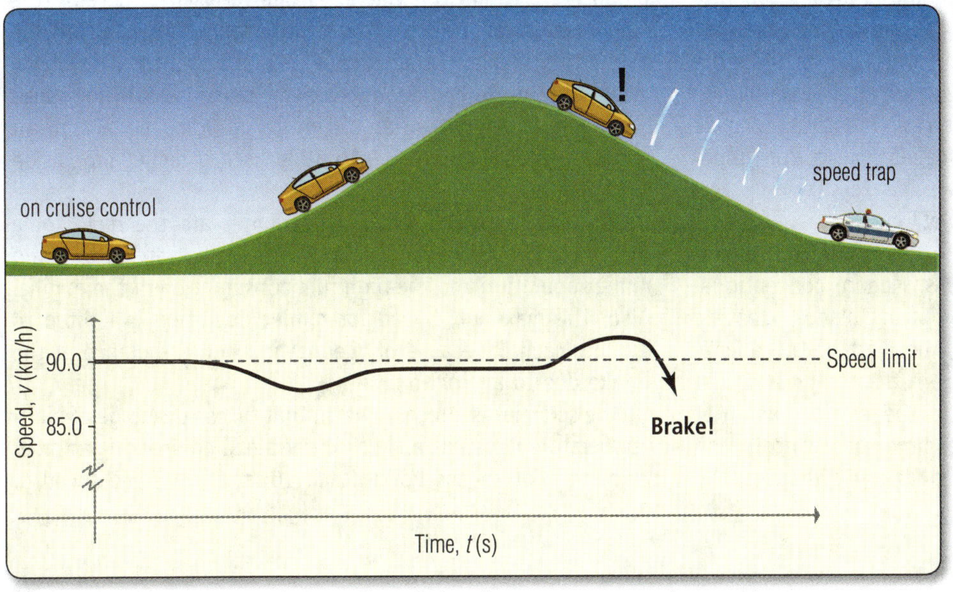

3-8 The diagram shows the car and terrain. The graph shows how instantaneous speed varies with time.

FACETS of PHYSICS

Hypersonic Jets
SCRAM!!!

The SR-71 *Blackbird* was the US Air Force's high-altitude, high-speed spy plane. It had special air-breathing engines that could operate beyond the Mach 3 limit for standard jet engines. These aircraft were taken out of military service in 1998.

How fast is fast? Scientists around the world are trying to redefine our definition of a quick trip. The fastest jet airplane ever built, the SR-71 *Blackbird* spy plane, was capable of 3.5 times the speed of sound (Mach 3.5) at altitudes over 100 000 feet. The speed record for any type of manned winged vehicle is held by the X-15 rocket-powered aircraft, at Mach 6.7. Rockets that launch space probes are capable of much higher speeds, but they cannot operate in the atmosphere at those speeds without being destroyed.

Rockets carry both fuel (usually liquid hydrogen) and an oxidizer (usually liquid oxygen), which allows them to operate independently of the atmosphere. In contrast, jet engines rely on atmospheric oxygen to burn their fuel and are therefore classified as "air breathing." NASA has been working to develop a vehicle equipped with an air-breathing engine capable of operating at very high speeds in the thin upper reaches of the earth's atmosphere. The key may be the ***scramjet*** *(supersonic combustion ramjet)* engine.

A little background is necessary to understand the differences among these propulsion systems. Modern conventional *jet* engines compress the air using rotating blades, creating high fuel-air pressure inside the engine. Part of the energy of the combustion gases is used to turn the compressor. The exhaust gases provide the thrust of the engine. As inlet air speed approaches Mach 3, the compressor's contribution to thrust practically disappears and actually reduces the engine's efficiency. At speeds above Mach 3, the compressor is irrelevant.

Scientists began researching *ramjet* technology during World War II for use in high-speed missiles. In a ramjet engine, the air is not compressed by any moving parts but simply by the speed of the vehicle forcing (ramming) the air into the engine intake, where the internal funnel-like geometry creates compression. Because it has no moving parts, a ramjet engine is lighter and mechanically simpler than a normal jet engine. However, a ramjet vehicle must have a booster propulsion system to accelerate it above Mach 3, because the compression required to operate the ramjet is not available below this speed. Though the vehicle is traveling at or above Mach 3, the airflow through the engine remains below the speed of sound (subsonic). The US military's *Talos* surface-to-air missile deployed from 1959 to 1979 used a ramjet propulsion system.

The biggest drawbacks of the ramjet are the need for a system to boost the vehicle to a high enough speed to begin ramjet operation and the high fuel consumption of the engine. Also, the maximum speed for ramjet operation is limited to around Mach 5 because of thermal-induced turbulence at the intake due to air friction.

The scramjet is a recent design that takes advantage of high-temperature materials and computer-aided engineering in order to operate at speeds up to Mach 10 (ten times the speed of

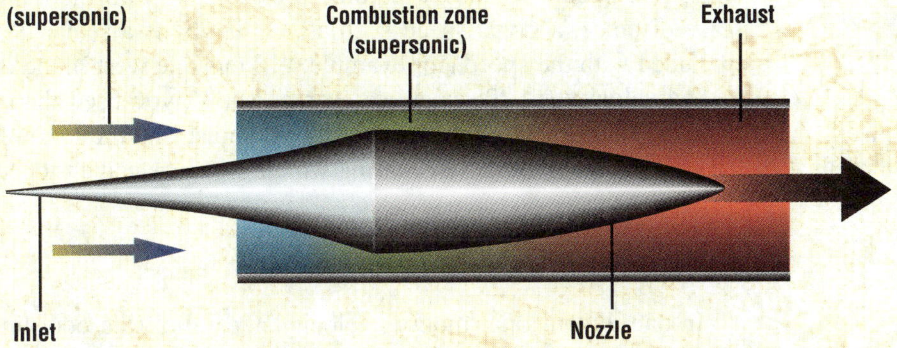

sound). It is called a supersonic ramjet because, in contrast to the older ramjets, the air continues flowing through the engine at hypersonic speeds (much greater than the speed of sound). Scramjets have some of the same drawbacks as ramjets, but if the program is successful, the payoffs could be tremendous.

Scramjets could replace current rocket-propulsion launch vehicles for small space payloads (such as communications satellites). Rocket propulsion is extremely expensive and inefficient, requiring a large, multistage rocket. Even the space shuttle, NASA's premier reusable launch vehicle, had to jettison its huge fuel tank, which then fell back to the earth and burned up.

Because it is theoretically more efficient and does not need to carry its oxidizer, a scramjet launcher achieving the same thrust would be lighter than a conventional launch vehicle. This would allow the payloads to increase while the launch-vehicle costs decrease.

Scramjets are also being considered for high-speed intercontinental commercial and passenger aircraft. Many improvements in metallurgy, fuels, and manufacturing processes must occur before this will become a reality.

The potentially great speed of a scramjet vehicle obviously has military applications. One application could be in land- or sea-based missiles used as quick interceptors of an enemy's ballistic missile.

This is an artist's rendering of NASA's X-43A scramjet prototype, which flew successfully at a speed of Mach 9.6 (7000 mi/h) on its last test flight on November 16, 2004.

3.6 Velocity

You learned earlier that a distance such as "15 km" is a scalar because a single piece of information completely describes it. But a displacement such as "3 km west" is a vector because it requires two pieces of information for its description. Like distance, a given speed such as "30 km/h" is a scalar. When a direction is included with the speed (for example, "10 m/s due west"), the quantity is a vector called **velocity (v)**. To calculate average speed, you used the total distance traveled. To find the *average velocity* ($\overline{\mathbf{v}}$), you must find the ratio of the displacement (**d**) to the time interval (Δt) during which the displacement occurred:

Average velocity is the displacement divided by the time interval:

$$\overline{\mathbf{v}} = \frac{\mathbf{d}}{\Delta t}$$

$$\overline{\mathbf{v}} = \frac{\mathbf{d}}{\Delta t} \quad (3.2)$$

In one-dimensional motion, velocity is parallel to a coordinate axis (just like displacement), so it can be denoted by a scalar symbol and a subscript. For example, the velocity vector parallel to the y-axis is denoted v_y.

What is the difference between $|\Delta x|$ in the average speed equation and **d** in the average vector velocity equation? The distance $|\Delta x|$ (or s) is always positive. Displacement (either **d** or d_x) may be positive or negative in one-dimensional motion and represents only the direct distance between two positions. The *magnitudes* of s and **d** will be equal if the object travels in only one direction during the interval Δt. Thus a car traveling east on a straight road might have an average speed of 50 km/h and also a velocity of 50 km/h due east. However, if the moving object changes direction during the time interval, its speed no longer equals the magnitude of the average velocity over the time interval. Consider, for instance, the earlier example of a campsite and a picnic area across the lake. Suppose the family car travels along the 10 km road at a speed of 50 km/h and took 12 min, or 0.2 h. Because the car's ultimate (net) change in position was only 2 km and not 10 km, the average velocity vector is 10 km/h due north.

Average velocity vector = 2 km north ÷ 0.2 h = 10 km/h north.

A position-time graph can show the velocity between positions. For example, turn to Figure 3-5b on page 53 and find the points on the graph where the slope of the line changes dramatically. The value of the slope (including the sign) between each pair of these points equals the average velocity for that interval. You can calculate velocities for the train by finding the slope of each of these four intervals:

$$\text{0 h to 0.30 h:}\quad \overline{v}_x = \frac{d_x}{\Delta t} = \frac{14\text{ km} - 0\text{ km}}{0.30\text{ h} - 0\text{ h}}$$

$$\overline{v}_x \cong +4\underline{6}.6\text{ km/h}$$

$$\text{0.50 h to 0.80 h:}\quad \overline{v}_x = \frac{d_x}{\Delta t} = \frac{38\text{ km} - 14\text{ km}}{0.80\text{ h} - 0.50\text{ h}}$$

$$\overline{v}_x = +80.\text{ km/h}$$

$$\text{1.00 h to 1.44 h:}\quad \overline{v}_x = \frac{d_x}{\Delta t} = \frac{60.\text{ km} - 38\text{ km}}{1.44\text{ h} - 1.00\text{ h}}$$

$$\overline{v}_x = +50.\text{ km/h}$$

$$\text{1.70 h to 2.00 h:}\quad \overline{v}_x = \frac{d_x}{\Delta t} = \frac{48.\text{ km} - 60\text{ km}}{2.00\text{ h} - 1.70\text{ h}}$$

$$\overline{v}_x = -40.\text{ km/h}$$

3-9 Trains often go straight for miles on tracks in the Midwest United States.

The average velocity for the last interval is negative because the train reverses direction. During the intervals 0.30–0.50 h, 0.80–1.00 h, and 1.44–1.70 h, the train has stopped at a station, and the initial and final positions are unchanged, so the train's speed is zero.

Figure 3-10 shows this data plotted on a **velocity-time graph**. On such a graph, lines above the horizontal zero velocity line show positive velocity (moving in the positive direction from the origin). Lines below this line are negative velocities (moving in the negative direction).

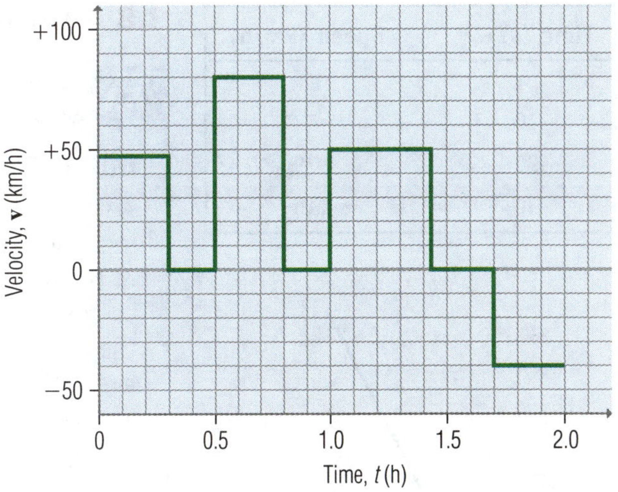

3-10 A velocity-time graph of the train between Ames and Lamoille stations, Iowa

3.7 Acceleration—Changing Velocity

The last problem was simple because it assumed that the train's velocity was uniform during each interval and that the velocity jumped instantly from one interval to the next (for instance, at 0.50 h the train instantly gained 80 km/h!). But you know from driving in a car that velocity changes frequently and smoothly—not in instantaneous jumps. This change in velocity with time is called **acceleration (a)**. *Average acceleration* ($\bar{\mathbf{a}}$) is the change in velocity ($\Delta\mathbf{v}$) divided by the time interval (Δt) of the velocity change:

$$\bar{\mathbf{a}} = \frac{\Delta\mathbf{v}}{\Delta t} \qquad (3.3)$$

The SI unit for acceleration is meters per second per second, usually written m/s^2.

The quantity $\Delta\mathbf{v}$ is a **difference vector**. You will learn more about difference vectors in Chapter 4. When an acceleration occurs, the initial and final velocities can have different magnitudes, different directions, or both. However, in one-dimensional motion, the initial and final velocities must be parallel to the line of motion (but may be in opposite directions). Therefore, $\Delta\mathbf{v}$ can be expressed in scalar-subscript notation as

$$\Delta v_x = v_{2x} - v_{1x}.$$

Recall that scalar-subscript notation is reserved *only* for vectors that lie parallel to the coordinate axis along which position is measured.

This statement means that the change in velocity in the x-direction is the difference of the initial and final velocities in the x-direction. The velocity difference points in the direction of the algebraic difference of the two velocities.

Acceleration is a vector of magnitude $\Delta v/\Delta t$ pointing in the same direction as $\Delta\mathbf{v}$. In one-dimensional motion, $\bar{a}_x$ may be positive, negative, or zero. A *positive* acceleration shows either an increase of speed in the positive direction or a decrease of speed in the negative direction. A *negative* acceleration shows a decrease of speed in the positive direction or an increase of speed in the negative direction. Zero acceleration means that velocity is constant.

You can find acceleration from a velocity-time graph. An object's average acceleration for an interval is the slope of the line joining the endpoints of the interval on the graph. Figure 3-11 is a hypothetical velocity-time graph of a car on a straight street. Note that speed varies nonuniformly with time. The average acceleration between times t_1 and t_2 is the slope of the dashed line. The *instantaneous acceleration* at a point is the slope of the tangent to the velocity-time curve at that point. Average acceleration does not have to equal the instantaneous acceleration from moment to moment.

Do not confuse a velocity-time curve with a position-time curve. The slope of a *position-time* curve is the *velocity*; the slope of a *velocity-time* curve is the *acceleration*.

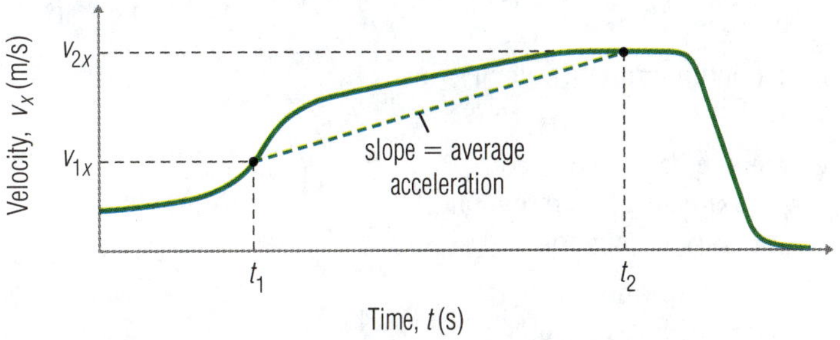

3-11 The slope of a line connecting two points on a velocity-time graph is the average acceleration between these two instants.

Time (s)	Speed (km/h)
0	5
10	42
20	79
30	116

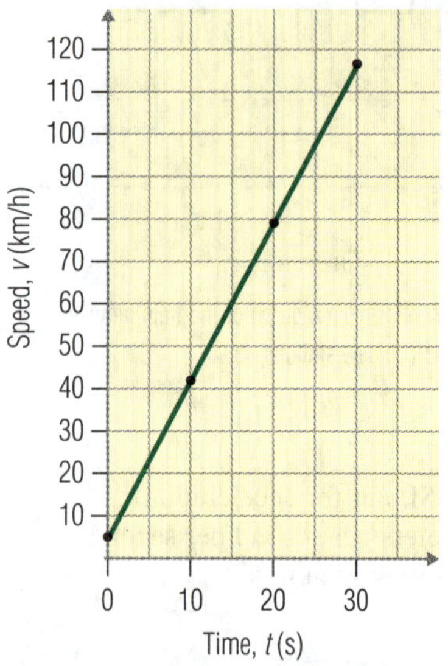

3-12 Velocity-time graph of the coasting car

Uniformly accelerated motion exists when the acceleration of an object is constant.

3A Objectives

After completing this section, I can

- ✓ define motion and describe one-dimensional motion mathematically.
- ✓ differentiate between distance and displacement.
- ✓ plot and evaluate motion on a position-time graph.
- ✓ approximate instantaneous speed on a position-time graph.
- ✓ differentiate between velocity and speed.
- ✓ plot and interpret a velocity-time graph.
- ✓ compute acceleration and approximate acceleration on a velocity-time graph.
- ✓ define uniformly accelerated motion and solve problems involving this concept.

3.8 Uniformly Accelerated Motion

EXAMPLE 3-4

Out of Gas: Uniform Acceleration

A man driving a car on a straight road runs out of fuel at the top of a long, straight hill. There is a gas station in the town in the valley at the bottom of the hill. The driver does not use his brakes while coasting in order to get as close as possible to the town. Assume that the car coasts freely down the hill to the level valley below. The driver notes his speedometer reading every 10 s. His readings are tabulated in the table, and Figure 3-12 shows a velocity-time graph constructed from the driver's data. What is the car's average acceleration for the trip down the hill?

Solution:

Assume that the car's coordinate axis is parallel to the slope of the hill and positive is downhill. The origin is placed at the top of the hill. Since the car's speed is in one direction, its speed can be plotted directly on a velocity-time graph. You can find the average acceleration by calculating the slope of the line joining the point of initial speed (0 s, 5 km/h) and the final speed (30 s, 116 km/h).

$$\bar{a}_x = \frac{\Delta v_x}{\Delta t} = \frac{(v_2 - v_1)_x}{\Delta t}$$

$$\bar{a}_x = \frac{116\text{ km/h} - 5\text{ km/h}}{30.\text{ s} - 0\text{ s}} = \frac{111\text{ km/h}}{30.\text{ s}} = \frac{3.7\text{ km}}{\text{h}\cdot\text{s}}$$

Notice that the units of hours and seconds are mixed in the answer. Convert the acceleration to the proper SI units (m/s^2) as follows:

$$\bar{a}_x = \frac{3.7\ \cancel{\text{km}}}{\cancel{\text{h}}\cdot\text{s}} \cdot \frac{1\ \cancel{\text{h}}}{3600\text{ s}} \cdot \frac{1000\text{ m}}{1\ \cancel{\text{km}}} \cong 1.0\underline{2}\text{ m/s}^2$$

$$\bar{a}_x \cong 1.0\text{ m/s}^2 \text{ downhill} \qquad \text{(2 SDs)}$$

In this example, the velocity-time curve is a straight line, so its slope is constant. The car's acceleration must also be constant. Therefore, the average acceleration ($\bar{a}_x$) is the same as the instantaneous acceleration (a_x). This kind of motion is called **uniformly accelerated motion**. We will study the significance of uniformly accelerated motion in the next section.

3A Section Review

1. Mathematical descriptions of change of position, rate of motion, and acceleration of objects fall within what area of physics?
2. Under what one condition may an object's displacement and the distance it has traveled between two positions be equal?
3. What is the difference between average speed and instantaneous speed?
4. State the kinematic quantity that is equivalent to the slope of
   a. a position-time graph.
   b. a velocity-time graph.
5. What is happening to an object experiencing negative acceleration?

6. Sketch and label the following graphs:

a. Velocity-time graph of a delivery truck traveling at a uniform 50 km/h.

b. Position-time graph of a fire ant (*Solenopsis invicta*) crawling the length of a 150 mm pencil during a 3-minute interval at a uniform rate.

c. Velocity-time graph of a freight yard switch engine initially moving at 8 km/h. During a 20 s interval the engine slows uniformly to a stop and then rests for 1 minute. During the next 20 s interval it uniformly speeds back up in the opposite direction to 8 km/h.

7. Sketch a position-time graph showing a continuously varying speed during a time interval. Using the sketch prove that, for at least one point during the interval, the instantaneous speed equals the average speed for the interval.

8. Assume that a yo-yo drops 1.0 m vertically from its owner's hand at a uniform speed of 0.80 m/s along its string. At the end of the string, it changes direction in 0.10 s and rises toward the owner's hand at a uniform speed of 0.60 m/s. Compute the average acceleration of the yo-yo at the bottom end of the string as it changes direction.

3B EQUATIONS OF MOTION

Many problems in physics involve approximately constant acceleration, represented as a straight portion of a velocity-time graph. The acceleration at any point along the straight-line segment is simply the slope of the line. Since this approximation of motion is common, we will develop equations to analyze it.

3.9 The First Equation of Motion

The acceleration formula Equation 3.3 can be rewritten as

$$\bar{a}_x = \frac{v_{2x} - v_{1x}}{\Delta t}.$$

What if you know the initial velocity and that the acceleration is constant but you want to find the final velocity? Solving for v_{2x},

$$\bar{a}_x \Delta t = v_{2x} - v_{1x}, \text{ or}$$

$$v_{2x} = v_{1x} + \bar{a}_x \Delta t. \tag{3.4}$$

This is the **first equation of motion**. This equation is also true for two or three dimensions using standard vector notation. However, for simplicity, we will retain the scalar-subscript form of vector notation for the equations of motion because we will always analyze motion with respect to only one coordinate axis at a time.

In algebra, you learned the working equation for a straight line:

$$y = mx + b,$$

where m is the line's slope and b is the point where the line crosses the y-axis (y-intercept). Similarly, the equation

$$v_{2x} = v_{1x} + \bar{a}_x \Delta t,$$

when rearranged as

$$v_{2x} = \bar{a}_x \Delta t + v_{1x},$$

The x-subscripts indicate that these vector quantities are all parallel to the same coordinate or position axis.

Problem-Solving Strategy 3.3

The first equation of motion is

$$v_{2x} = v_{1x} + \bar{a}_x \Delta t.$$

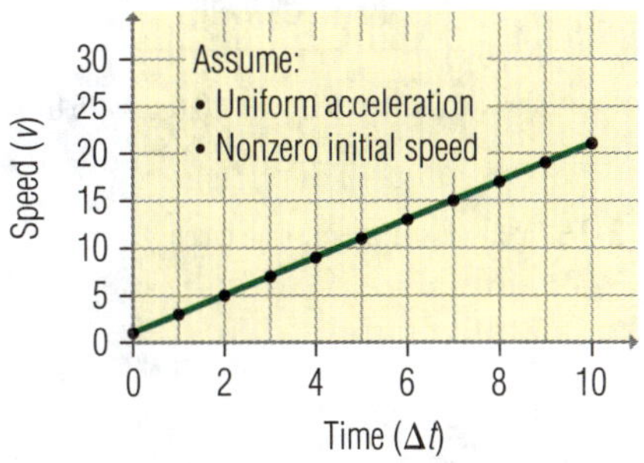

3-13 A graph of the velocity of a system experiencing a uniform acceleration during a time interval, Δt, according to the first equation of motion. The curve is a line intersecting the axis at the initial speed.

is the equation of a straight line on a velocity-time graph. Its slope is $\bar{a}_x$, and it crosses the velocity axis at $v_{1\,x}$. If you know the slope of a line and one point on the line, you can construct a linear graph. However, it is usually easier to use the information given in the problem to determine a second point on the velocity-time curve and then draw a straight line through the two points.

3-14 Launching an ASAT missile from an Air Force F-15 Eagle.

EXAMPLE 3-5

Antisatellite Missile: Graphing Velocity Versus Time

During the 1980s the United States developed a special missile system to destroy enemy spy satellites in low-earth orbits. The weapon was called the air-launched antisatellite missile, or air-launched ASAT missile. It was fired from an Air Force F-15 Eagle aircraft at a precise angle, speed, and altitude toward a region in the sky ahead of the target satellite. The missile destroyed the satellite by ramming into it at high speed. No explosives were used.

The missile launched itself when the aircraft was 11.6 km high at an angle of 65.0° to the ground (see Figure 3-14). The initial speed of the missile was 246 m/s when its engine fired. The average acceleration of the missile during the 15.0 s powered portion of its flight was 175 m/s^2.

Draw a velocity-time graph for the missile's flight while its engine was burning. Assume the missile flew a straight path for the entire 15.0 s interval.

Solution:

Choose the *x*-coordinate axis parallel to the missile's path, increasing upward. The origin is at the point of launch. Find the coordinates of two points on the line.

Given: $v_{1\,x} = +246$ m/s, $\bar{a}_x = +175$ m/s^2

Substitute the known information into the first equation of motion.

$$v_{2\,x} = v_{1\,x} + \bar{a}_x \Delta t$$

$$v_{2\,x} = (+246\ \text{m/s}) + (+175\ \text{m/s}^2)\Delta t$$

Find two points on the line:

1. For $t_1 = 0.0$ s, $v_{1\,x} = +246$ m/s; one point is (0.0 s, 246 m/s).
2. For $t_2 = 15.0$ s, $\Delta t = 15.0\ \text{s} - 0.0\ \text{s} = 15.0$ s.

$$v_{2\,x} = (+246\ \text{m/s}) + (+175\ \text{m/s}^{\cancel{2}})(15.0\ \cancel{\text{s}})$$

$$v_{2\,x} \cong +28\underline{7}1\ \text{m/s}$$

$$v_{2\,x} \cong 2870\ \text{m/s} \qquad \text{(3 SDs)}$$

The second point is (15.0 s, 2870 m/s).

Figure 3-15 is a velocity-time graph of the line through the two points.

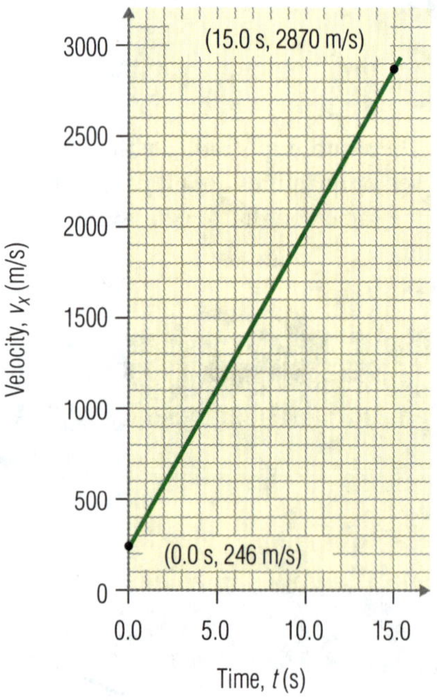

3-15 ASAT missile velocity-time graph

3.10 Determining Displacement Algebraically

In Example 3-5, it is possible to calculate how far the missile flew during the powered portion of its flight from the information given. The displacement of the missile is the distance it flew from the time the missile was released until its engine shut off. Knowing that

$$\bar{v}_x = d_x / \Delta t,$$

you can conclude that

$$d_x = \bar{v}_x \Delta t. \tag{3.5}$$

You already know the change in time (Δt), and the average velocity during an interval is

$$\bar{v}_x = \frac{v_{1\,x} + v_{2\,x}}{2}. \tag{3.6}$$

Problem-Solving Strategy 3.4

The average velocity during an interval is one-half the sum of the initial and final velocities.

$$\bar{v}_x = \frac{v_{1\,x} + v_{2\,x}}{2}$$

Substitute Equation 3.6 into Equation 3.5.

$$d_x = \left(\frac{v_{1\,x} + v_{2\,x}}{2}\right)\Delta t \tag{3.7}$$

$$d_x = {}^1\!/_2(v_{1\,x} + v_{2\,x})\Delta t$$

From Example 3-5, you know that the initial velocity of the missile is $v_{1\,x} = +246$ m/s, its final velocity is $v_{2\,x} = +2870$ m/s, and the powered flight time interval is $\Delta t = 15.0$ s. Substituting this information into Equation 3.7, you find that:

$$d_x = {}^1\!/_2[(+246\text{ m/s}) + (+2870\text{ m/s})](15.0\text{ s}) = +23\,\underline{3}70\text{ m}$$

$$d_x \cong +23.4\text{ km} \quad \text{(3 SDs)}$$

Therefore, the missile flies about 23.4 km (14.5 mi) powered by its rocket engine.

3.11 Determining Displacement Graphically

The area under the curve of a velocity-time graph is numerically equal to the displacement of the moving object. Figure 3-16 is the velocity-time graph of the ASAT missile. Because it is a straight line, the shaded area can be divided into a triangle and a rectangle. The area of the rectangle is base × width. For the rectangular area in the graph, the base $= \Delta t$, and the width $= v_{1\,x} - 0$ m/s, as shown. For any triangle, area is ½(base × height). Again, the base of the triangular area $= \Delta t$, and its height $= v_{2\,x} - v_{1\,x}$.

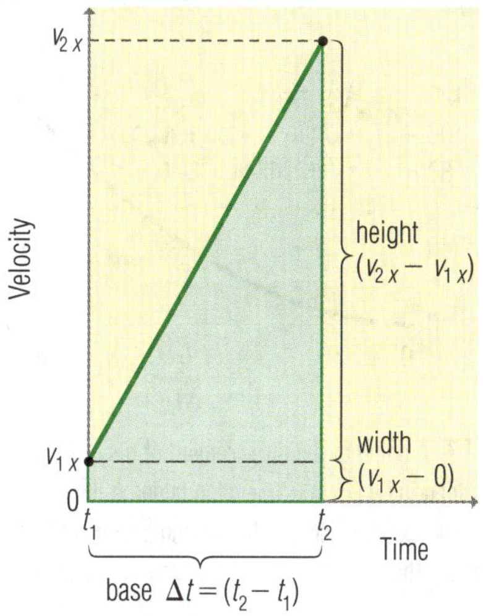

3-16 Using area on a velocity-time graph to estimate displacement

The total area of the region under the graph is the sum of the rectangular and triangular areas:

$$A_{\text{total}} = A_{\square} + A_{\triangle}$$

$$A_{\text{total}} = v_{1\,x}\Delta t + {}^1\!/_2(v_{2\,x} - v_{1\,x})\Delta t \tag{3.8}$$

Factoring and combining terms gives

$$A_{\text{total}} = {}^1\!/_2(v_{1\,x} + v_{2\,x})\Delta t.$$

Notice the similarity between this and Equation 3.7, which we have already used to calculate an answer of 23.4 km.

$$d_x = {}^1\!/_2(v_{1\,x} + v_{2\,x})\Delta t$$

Both methods yield the same answer.

The area of the region under a graph is often related in some way to the product of the variables that produce the graph. This is the basis for integral calculus.

3.12 The Second Equation of Motion

Sometimes it is not an object's displacement but its position that you want to find. For one-dimensional motion, the position is related to displacement by the equation

$$d_x = x_2 - x_1,$$

or rearranged,

$$x_2 = x_1 + d_x.$$

You have just seen in Equation 3.8 that

$$d_x = v_{1x}\Delta t + \tfrac{1}{2}(v_{2x} - v_{1x})\Delta t.$$

Recall that

$$\bar{a}_x = \frac{v_{2x} - v_{1x}}{\Delta t}, \text{ or}$$

$$\bar{a}_x \Delta t = v_{2x} - v_{1x}.$$

Substituting $\bar{a}_x \Delta t$ for $v_{2x} - v_{1x}$ in Equation 3.8 gives

$$d_x = v_{1x}\Delta t + \tfrac{1}{2}(\bar{a}_x \Delta t)\Delta t, \text{ or}$$

$$d_x = v_{1x}\Delta t + \tfrac{1}{2}\bar{a}_x(\Delta t)^2. \tag{3.9}$$

Substituting the definition of displacement for d_x in Equation 3.9 and solving for x, we obtain

$$x_2 = x_1 + v_{1x}\Delta t + \tfrac{1}{2}\bar{a}_x(\Delta t)^2. \tag{3.10}$$

Equations 3.9 and 3.10 are two forms of the **second equation of motion**. Observe that this formula is a quadratic equation in the form $y = c + bx + ax^2$, where the time interval (Δt) is the independent variable and the position (x_2) is the dependent variable. The average acceleration ($\bar{a}_x$), initial velocity (v_{1x}), and initial position (x_1) are constants. A plot of time-position ordered pairs using this equation will produce a parabolic curve.

Problem-Solving Strategy 3.5

The second equation of motion is

$$d_x = v_{1x}\Delta t + \tfrac{1}{2}\,\bar{a}_x\,(\Delta t)^2.$$

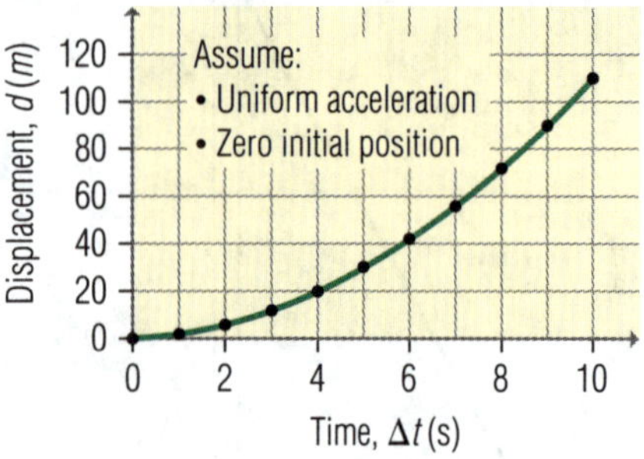

3-17 A graph of the displacement of a system experiencing uniform acceleration during a time interval, Δt, according to the second equation of motion. The curve describes a segment of a parabola.

3.13 The Third Equation of Motion

A third equation is useful when the time interval is not known or is irrelevant. Since

$$\bar{a}_x = \frac{v_{2x} - v_{1x}}{\Delta t}, \quad \text{then}$$

$$\Delta t = \frac{v_{2x} - v_{1x}}{\bar{a}_x}. \tag{3.11}$$

Also, you know from Equation 3.7 that

$$d_x = \tfrac{1}{2}(v_{1x} + v_{2x})\Delta t.$$

Now substitute Equation 3.11 for Δt above, and rearrange the terms to get

$$d_x = \frac{v_{2x} + v_{1x}}{2} \cdot \frac{v_{2x} - v_{1x}}{\bar{a}_x},$$

and simplify:

$$d_x = \frac{v_{2x}^2 - v_{1x}^2}{2\bar{a}_x} \tag{3.12}$$

$$x_2 = x_1 + \frac{v_{2x}^2 - v_{1x}^2}{2\bar{a}_x} \tag{3.13}$$

Equations 3.12 and 3.13 are two forms of the **third equation of motion**. You need to learn only the displacement form of the second and third equations of motion since the position form can be easily derived from the definition of displacement. The three equations of motion are helpful in solving most problems in *straight-line, constant acceleration* motion.

Problem-Solving Strategy 3.6

The third equation of motion is

$$d_x = \frac{v_{2x}^2 - v_{1x}^2}{2\bar{a}_x}.$$

	Second Equation	Third Equation
Displacement form	$d_x = v_{1x}\Delta t + \frac{1}{2}\bar{a}_x(\Delta t)^2$	$d_x = \dfrac{v_{2x}^2 - v_{1x}^2}{2\bar{a}}$
Position form	$x_2 = x_1 + v_{1x}\Delta t + \frac{1}{2}\bar{a}_x(\Delta t)^2$	$x_2 = x_1 + \dfrac{v_{2x}^2 - v_{1x}^2}{2\bar{a}_x}$

Problem-Solving Strategy 3.7

Use the *displacement form* of the equations of motion when only displacement is involved in a problem. Use the *position form* of the equations if position information is involved.

EXAMPLE 3-6

Stopping the Train: Negative Acceleration

A train is traveling at 26.8 m/s. If it brakes at a constant acceleration of -0.500 m/s^2, how far will the train continue to roll from the time it starts to brake until it comes to a complete stop?

Solution:

Assume the *x*-coordinate axis is oriented parallel to the train's motion and the train is moving in the positive direction. You want to find the *displacement* (d_x) for the unknown time interval (Δt) in which the train's velocity decreases from +26.8 m/s to 0.0 m/s.

Given: $\bar{a}_x = -0.500\ \text{m/s}^2$, $v_{1x} = +26.8\ \text{m/s}$, and $v_{2x} = 0.0\ \text{m/s}$.

More Difficult Solution:

First, solve for the time interval while the train was braking, using the first equation of motion:

$$v_{2x} = v_{1x} + \bar{a}_x\Delta t$$

$$0\ \text{m/s} = (+26.8\ \text{m/s}) + (-0.500\ \text{m/s}^2)\Delta t$$

$$\Delta t = \frac{0\ \cancel{\text{m/s}} - 26.8\ \cancel{\text{m/s}}}{-0.500\ \cancel{\text{m}}/\text{s}^{\cancel{2}}}$$

$$\Delta t = 53.6\ \text{s}$$

Now substitute Δt into the second equation of motion:

$$d_x = v_{1x}\Delta t + \tfrac{1}{2}\bar{a}_x(\Delta t)^2$$

$$d_x = (+26.8\ \text{m}/\cancel{\text{s}})(53.6\ \cancel{\text{s}}) + \tfrac{1}{2}(-0.500\ \text{m}/\cancel{\text{s}^2})(53.6\ \cancel{\text{s}})^2$$

$$d_x = 14\underline{3}6\ \text{m} - 71\underline{8}.2\ \text{m}$$

$$d_x \cong 7\underline{1}7\ \text{m} \cong 720\ \text{m} \qquad \text{(2 SDs)}$$

Easier Solution:

Substitute the known values into the third equation of motion:

$$d_x = \frac{v_{2x}^2 - v_{1x}^2}{2\bar{a}}$$

$$d_x = \frac{(0\ \text{m/s})^2 - (+26.8\ \text{m/s})^2}{2(-0.500\ \text{m/s}^2)}$$

$$d_x = \frac{-71\underline{8}.2\ \text{m}^{\cancel{2}}/\cancel{\text{s}^2}}{-1.000\ \cancel{\text{m}}/\cancel{\text{s}^2}} \cong 71\underline{8}.2\ \text{m}$$

$$d_x \cong 718\ \text{m} \qquad \text{(3 SDs)}$$

3-18 Long freight trains have a large mass and are hard to stop.

Both methods of solving this problem give essentially the same answer, but the second method is much more direct. It is almost always quicker to use one equation rather than two. Also, using two equations rather than one can introduce excessive rounding errors. This illustrates one way that a systematic approach to problem solving can save you time and improve precision.

3-19 Skydivers experience free fall for the initial portion of their descent.

Free fall is the vertical motion of an object under the influence of gravity alone.

Problem-Solving Strategy 3.8

Use the symbol g_y to represent the vector **g** when all motion is parallel to the y-axis. Indicate the magnitude of **g** using $|g|$, $|g_y|$, or $|\mathbf{g}|$.

3-20 World diagram for Example 3-7. Note that dropping rocks into deep canyons is dangerous and, in most parks, is illegal.

3.14 Free Fall

One common example of uniformly accelerated motion is **free fall**, the fall of an object under the influence of gravity alone with negligible air resistance. The magnitude of the acceleration due to gravity $|g|$ is almost constant near the earth's surface. We will assume that it is exactly 9.81 m/s². By convention, for problems involving gravitational acceleration, *up* is considered positive. Therefore, the gravitational acceleration vector at the earth's surface is

$$\mathbf{g} = g_y = -9.81 \text{ m/s}^2.$$

The subscript notation assumes that positive y is up.

The equations of motion can be adapted to problems involving free fall. The following equations are given only to show you that gravitational acceleration can be substituted directly for average acceleration in the equations of motion. Therefore, memorizing another set of equations is not necessary.

First equation: $$v_{2y} = v_{1y} + g_y\Delta t$$

Second equation: $$d_y = v_{1y}\Delta t + \tfrac{1}{2} g_y(\Delta t)^2$$

Third equation: $$d_y = \frac{v_{2y}^2 - v_{1y}^2}{2g_y}$$

EXAMPLE 3-7

Dropping a Stone: Free Fall

A physics student drops a stone from a scenic overlook on the rim of a canyon into the canyon below. (Note: *Dropped* means that the stone had zero initial velocity.) (a) Ignoring air resistance, what is its velocity after 4.0 s? (b) How far did the stone fall in that time?

Solution:

Given: $v_{1y} = 0$ m/s; $\Delta t = 4.0$ s; $\bar{a}_y = g_y = -9.81$ m/s²

Find v_{2y} when $t = 4.0$ s and d_y for the interval from $t_1 = 0$ s to $t_2 = 4.0$ s.

Assume that the y-coordinate system is oriented vertically, positive upward, and the origin is at the point where the stone is released.

To find v_{2y}, use the first equation of motion.

$$v_{2y} = v_{1y} + g_y\Delta t$$
$$v_{2y} = (0 \text{ m/s}) + (-9.81 \text{ m/s}^2)(4.0 \text{ s})$$
$$v_{2y} \cong -3\underline{9}.2 \text{ m/s} \cong -39 \text{ m/s}$$

The negative sign indicates that the stone is falling downward.

To find the vertical displacement, use the second equation of motion.

$$d_y = v_{1y}\Delta t + \tfrac{1}{2}g_y(\Delta t)^2$$
$$d_y = (0 \text{ m/s})(4.0 \text{ s}) + \tfrac{1}{2}(-9.81 \text{ m/s}^2)(4.0 \text{ s})^2$$
$$d_y \cong 0 \text{ m} - 7\underline{8}.4 \text{ m}$$
$$d_y \cong -78 \text{ m}$$

The negative sign of the displacement indicates that the stone moved in the downward (negative y) direction.

3.15 Modeling Stopping Distances Using Kinematics

Examining the motion of a vehicle that is braking to a stop is a practical application of kinematics. Without accounting for forces or energy, you can still learn how vehicle speed affects stopping distance.

Table 3-1 displays the data you will use to model the motion of a stopping car. The data is taken from the actual information used by traffic courts to evaluate the idealized performance of vehicles prior to a collision. The following assumptions are used:

- Skilled, alert driver
- Even, dry, clean, straight road surface
- Vehicle in good mechanical condition
- Four-wheel braking
- Same uniform deceleration for all speeds

TABLE 3-1

Stopping Distance Kinematic Data

Speed (mi/h)	Speed (m/s)	Driver Reaction Distance (0.75 s) (m)	Distance While Braking (m)	Total Stopping Distance (m)
10	4.5	3.4	1.5	4.9
15	6.7	5.0	3.7	8.7
20	8.9	6.7	6.4	13.1
25	11.2	8.4	9.8	18.1
30	13.4	10.1	14.3	24.4
35	15.6	11.7	19.2	30.9
40	17.9	13.4	25.0	38.4
45	20.1	15.1	31.7	46.8
50	22.3	16.8	39.0	55.8
55	24.6	18.4	47.2	65.7
60	26.8	20.1	56.4	76.5
65	29.1	21.8	66.1	87.9
70	31.3	23.5	76.8	100.3
75	33.5	25.1	88.1	113.2
80	35.8	26.8	100.0	126.8
90	40.2	30.2	129.5	159.7
100	44.7	33.5	156.7	190.2

3B Section Review

1. Write the first equation of motion. Under what condition(s) is this equation valid?
2. What quantities must be known in order to find the displacement of an object using the second equation of motion?
3. Which equation of motion should you use if you do not know the elapsed time of an object's motion, assuming the necessary quantities are known?
4. Discuss the differences, if any, between the one-dimensional quantities $\mathbf{d}$, $|\mathbf{d}|$, d, and d_x.

3B Objectives

After completing this section, I can

✓ compute the final velocity of a system using the first equation of motion when given the initial velocity and the uniform acceleration.

✓ compute the displacement of a system from a velocity-time graph.

✓ compute the displacement of a system using the second equation of motion.

✓ compute the displacement of a system using the third equation of motion when given only the initial and final velocities and the acceleration.

✓ solve problems involving uniform acceleration in the form of free fall.

✓ model the relationship between vehicle speed and stopping distance.

5. You throw a ball for your dog Procyon to fetch. "Pro" runs from right next to you to the ball lying 22 m away and brings it directly back to you. The round trip takes 11 s. You pat his head and give him a treat.

a. What was the net change in Procyon's position?
b. What was the total distance covered by the dog?
c. What was the dog's average speed?
d. What was the dog's average velocity for the entire interval?

6. The eccentric orbit of Pluto carries it as far as 7.376×10^9 km from the sun every 248 earth years. At this distance, how long, in hours, will it take a photon of light traveling at 3.00×10^8 m/s to travel between the sun and Pluto?

As an alternative to manually typing this data into a spreadsheet, your teacher may provide you with a file that you can import into a spreadsheet program.

DM **7.** Using a graphing calculator or a spreadsheet program with analytical graphing features, create the following three scatter plots of the car deceleration data (use the Initial Speed for the *x*-axis data):

a. Total stopping distance versus initial speed
b. Reaction time distance versus initial speed
c. Distance while braking versus initial speed

(*Note*: Review your calculator's manual and Appendix G for instructions on creating scatter plots.)

DM **8.** For each graph you created in Question 7, apply a best-fit curve using the built-in options of the calculator or program you are using. For example, you should try curves defined by linear, logarithmic, polynomial, or other functions. Look at the shape of the scatter plot and guess what kind of curve might best fit in order to reduce the number of trial-and-error attempts. At the same time, calculate the r^2 value for the curve fit. The curve that generates an r^2 value closest to 1 is the best-fit curve.

(*Note*: Review Appendix G for a brief discussion of the significance of r^2. It is not necessary that you know how this quantity is calculated using statistical methods.)

DM **9.** Using your graphing calculator or spreadsheet program, determine the equation for the best fit curve of the total stopping distance versus initial speed data set.

a. Calculate the total stopping distance for a car stopping from 80 mi/h. Explain any difference between the calculated distance and the stopping distance for 80 mi/h in Table 3-1.

b. An older production-model sports car can actually stop from 80 mi/h in 77.1 m. Give some reasons for the difference between the modeled braking distance and the actual braking distance. What does this say about the nature of models?

Chapter Review

In Terms of Physics

mechanics	49	magnitude	50
kinematics	49	vector	50
dynamics	50	position-time graph	52
displacement (**d**)	50	velocity (**v**)	58
scalar	50	velocity-time graph	59

acceleration (**a**)	59
difference vector	59
uniformly accelerated motion	60
first equation of motion	61
second equation of motion	64
third equation of motion	64
free fall	66

Problem-Solving Strategies

3.1 (page 52) When graphing time-dependent data, always plot time on the horizontal axis and plot the dependent variable on the vertical axis.

3.2 (page 53) Whenever you must find the change of a quantity, always subtract the initial value from the final value.

3.3 (page 61) **The first equation of motion is**

$$v_{2x} = v_{1x} + \bar{a}_x \Delta t.$$

3.4 (page 63) The average velocity during an interval is one-half the sum of the initial and final velocities.

$$\bar{v}_x = \frac{v_{1x} + v_{2x}}{2}$$

3.5 (page 64) **The second equation of motion is**

$$d_x = v_{1x}\Delta t + {}^1\!/_2\bar{a}_x(\Delta t)^2.$$

3.6 (page 64) **The third equation of motion is**

$$d_x = \frac{v_{2x}^2 - v_{1x}^2}{2\bar{a}_x}.$$

3.7 (page 65) Use the *displacement form* of the equations of motion when only displacement is involved in a problem. Use the *position form* of the equations if position information is involved.

3.8 (page 66) Use the symbol g_y to represent the vector **g** when all motion is parallel to the *y*-axis. Indicate the magnitude of **g** using $|g|$, $|g_y|$, or $|\mathbf{g}|$.

Review Questions

1. What does the study of kinematics describe?

2. What is the net change of position called?

3. How would you determine the average speed of a trip to Grandma's house?

4. How does increasing the number of data points on a position-time graph improve your ability to estimate instantaneous velocity at any point on the curve?

5. Describe the motion of an object if its velocity-time graph is horizontal.

6. What kind of kinematic information is provided by the slope of a velocity-time graph?

7. What is the general shape of a graph produced by
a. the first equation of motion?
b. the second equation of motion?

8. What is the area under a velocity-time graph, bounded by a specific time interval, numerically equal to?

9. a. Identify one common form of uniform acceleration.
b. Give an example of an object that would exhibit this kind of acceleration.

10. What vector has a magnitude equal to the instantaneous speed?

11. Which of the following quantities are *always* positive?

a. v	**c.** $|\bar{\mathbf{v}}|$	**e.** x	**g.** $|\mathbf{g}|$
b. d_x	**d.** $\bar{v}_y$	**f.** g_y	**h.** a

12. State which of the following are vector quantities, scalar quantities, or both.

a. distance traveled	**e.** acceleration
b. displacement	**f.** change in velocity
c. average velocity	**g.** change in position
d. instantaneous speed	**h.** change in time

True or False (13–19)

13. Only magnitudes that are positive are scalar quantities.

14. Average speed is never negative.

15. The slope of a line between two points on a velocity-time graph is the average velocity for that time interval.

16. Average speed is the total distance traveled divided by the time interval for the trip.

17. You must know the time interval (Δt) in order to use the third equation of motion.

18. The symbols g and **g** represent the same thing.

19. The average speed and the magnitude of the average velocity vector are equal only when total path distance and the magnitude of displacement are the same.

✪**20.** A railroad freight yard switch engine on a straight track travels 1000. m to the end of the track, then back 600. m, then forward 300. m.
a. What is the switch engine's final displacement from its starting position?
b. How far did it travel?

✪**21.** A sprinter in training runs along a straight 50.0 m track. At every 10.0 m is an observer with a stopwatch. The sprinter begins to run at the sound of the starting gun. The observers start their stopwatches at that same instant. Each observer stops his stopwatch as the runner passes him. After one trial run, the observers' stopwatches have the readings listed in the following table.
a. Construct a position-time graph for this sprint.
Calculate the sprinter's average speed for
b. the entire run.
c. the first 30.0 m.
d. the last 20.0 m.

Observer	Time (s)
start	0.00
1	1.10
2	2.20
3	3.30
4	4.10
5	4.90

The following graph represents the motion of a trolley along a straight track. Use the information presented here to answer Questions 22–27.

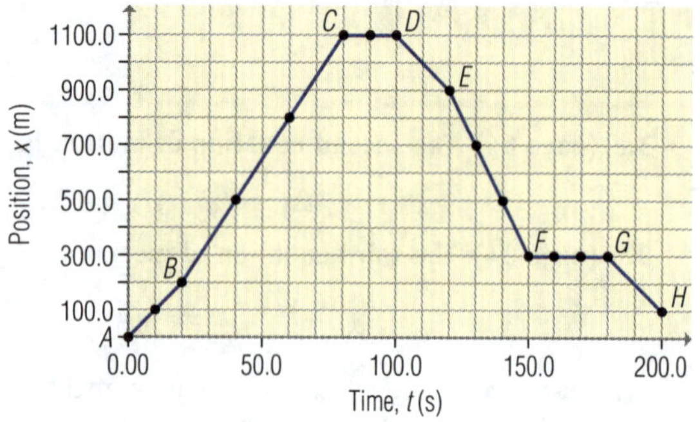

22. During which interval did the trolley move the farthest?

a. A to B **c.** E to F
b. B to C **d.** F to G

23. During which interval was the trolley's average speed the greatest?

a. A to B **c.** E to F
b. B to C **d.** F to G

24. Calculate the trolley's average speed for the interval in the answer to Question 23.

25. What is the trolley's speed at (90 s, 1100 m)?

26. During which interval(s) is the trolley's speed 15.0 m/s?

27. Using the position-time graph above, sketch a velocity-time graph for the trolley. Does the trolley undergo constant acceleration?

28. An auto manufacturer boasts that its newest sports car model "goes from 0 to 60 (mi/h) in 5.00 seconds!"

a. What is the car's acceleration (in m/s^2)?
b. How far does it travel as it accelerates from 0.0 mi/h to 60.0 mi/h (in meters)?

29. You are stuck on a two-lane road behind a truck traveling 64.4 km/h (40 mi/h). You pull into the oncoming lane to pass and immediately begin accelerating at 2.50 m/s^2. After passing the truck, 6.10 s have elapsed. You stop accelerating and pull back into your lane. How far down the road did you travel in the oncoming lane (in meters) during the time interval?

Use the velocity-time graph below to answer Questions 30–32.

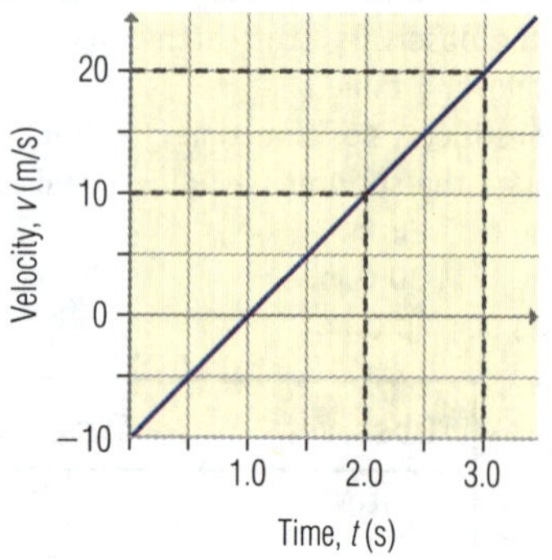

30. What happens to the object's direction of motion at 1.00 s?

31. Calculate

a. the average velocity from 0.50 s to 1.50 s.
b. the average acceleration from 0.50 s to 1.50 s.

32. What is the displacement

a. between 0.00 s and 1.00 s?
b. between 1.00 s and 2.00 s?
c. between 0.00 s and 3.00 s?

33. A science fiction author describes a space vehicle launching system consisting of a long linear track built at an angle into a mountain range on the moon. The space vehicle is mounted on a cradle that is accelerated by a magnetic field generated by the track at a constant 100. m/s^2.

a. How fast is the vehicle going after it has traveled the 2.88×10^4 m track?
b. How long does it take to reach this speed?

DM **34.** Refer to the three equations of motion to answer the following questions.

a. What condition is assumed to exist when using any of the three equations of motion?
b. Solve the third equation of motion for acceleration ($\overline{a}$).
c. Solve the first equation of motion for elapsed time (Δt). Substitute the result of Part b for a in this equation. Why is this action permissible? When analyzing braking distance problems, what is the value of the final speed, v_2? Simplify the resulting equation as much as possible.
d. Refer to the data for 55 mi/h from Table 3-1 on page 67. Calculate the car braking time from this initial speed using the equation you obtained in Part c.
e. Add the braking time from Part d to the assumed driver reaction time to obtain the total braking time from 55 mi/h.

DM **35. a.** Using the first equation of motion and the results from Question 34 Part d, determine the acceleration of a vehicle braking from 55 mi/h.
b. What is the sign of this acceleration? Explain its significance.
c. With the aid of a graphing calculator or spreadsheet program, create a velocity-time graph of a vehicle stopping from 55 mi/h. Use the first equation of motion. You may also plot several points on graph paper to create the graph.
d. What is the significance of the slope of this graph?

DM **36.** Why should Christians be concerned about modeling the stopping distance of vehicles?

37. You are given the following four position-time graphs:

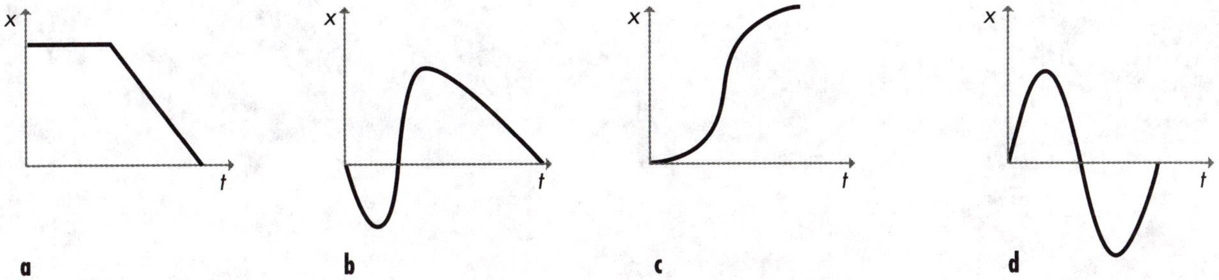

a b c d

Indicate which of the velocity-time graphs below corresponds best with each of the position-time graphs by pairing the appropriate letters.

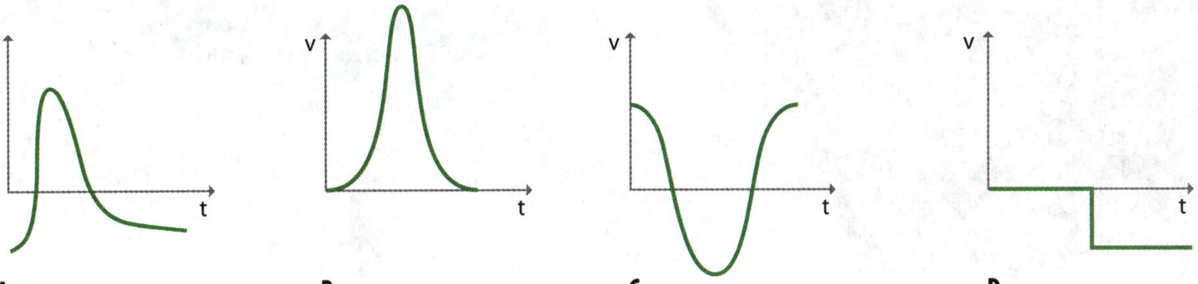

A B C D

38. Your average running speed is 8.0 m/s and your best friend's is 7.0 m/s. You both are 65 m from the front door of the school when the class bell rings, at which time you both sprint to get to class. How far are you in front of your friend when you reach the door of the school?

This NASA space shuttle, shortly after liftoff, had an increasing speed in a specific direction—straight up.

Vectors and Scalars

4

Dominion Science Problem

Accurate Hurricane Models

Hurricanes and cyclones caused almost 2 million deaths in the last two centuries. If you have never experienced one of these storms, consider yourself fortunate. Hurricanes are huge rotating storm systems in the earth's atmosphere that form over warm oceans in tropical or subtropical areas of the world. The winds in these storms exhibit complex motion in three dimensions. The air in a hurricane not only rotates but also spirals into the hurricane's eye from its periphery, rises vertically in the eye, and then spirals outward over the top of the storm. Wind speeds of over 320 km/h (200 mi/h) have been recorded for hurricanes that have made landfall in the United States.

But extreme winds are not the most dangerous aspect of a hurricane. Most deaths are caused by hurricane *storm surges* that inundate coastal areas with many meters of water, allowing huge ocean waves to assault human habitations far from the beach. A storm surge is the abnormally high sea level caused by the adverse combination of low air pressure, high winds piling water onshore, and a high tide. Often in the aftermath of a storm, disease and famine in the disaster area claim even more lives. How can we better model hurricanes to improve forecast accuracy and to more effectively evacuate people out of harm's way?

4-1 On May 2, 2008, cyclone Nargis smashed into the country of Myanmar (formerly known as Burma) in Southeast Asia. More than 2000 square miles were flooded, and an estimated 25 000 to 60 000 people were dead or missing.

4A PROPERTIES OF VECTORS AND SCALARS

4.1 Defining Vectors and Scalars

A manager gives a customer directions to her store: "Three blocks north and four blocks west of City Hall." The customer's trip can be represented by two arrows on a map, each having a specific length and direction.

An artillery officer knows that the muzzle velocity of a fired cannon projectile is 930 m/s. He must raise the barrel of his cannon to an elevation of 33° in order to hit a target. The initial speed of the projectile can be represented by a long arrow labeled 930 m/s at an angle of 33° above a horizontal line.

An engineer for a tire manufacturing company is designing a new, low-profile radial tire. In order to understand

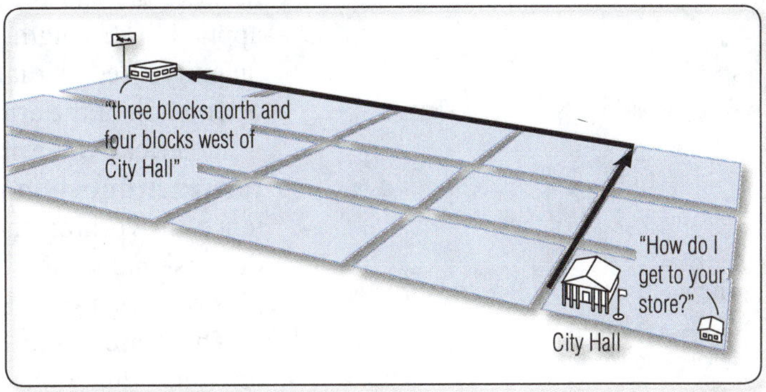

4-2 Each leg of the trip to the store is a vector.

4-3 The velocity vector describes both speed and direction.

The *magnitude* of a quantity is the absolute value of its numerical value. As such, magnitudes are always positive. It is important to remember that while all magnitudes of vectors are scalars, not all scalars are magnitudes of vectors.

The magnitude of **A** is written A or $|\mathbf{A}|$. The magnitude of a vector is also called its modulus. The magnitude of a vector may be indicated by a numerical label or graphically indicated by the relative length of the vector arrow.

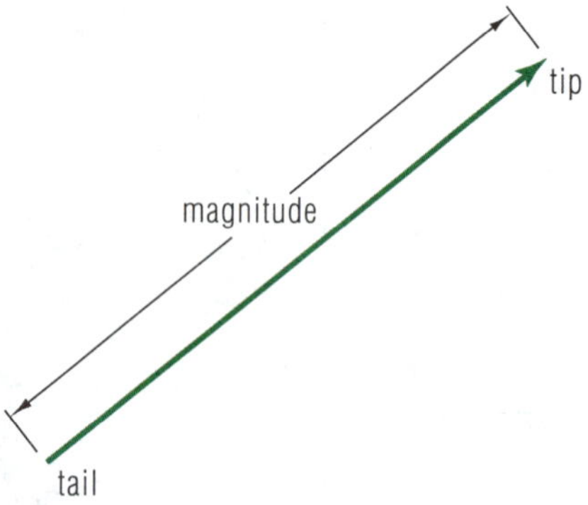

4-5 The essential parts of a vector diagram

The **vector angle (θ)** is the direction of the vector in relation to its **reference direction**, the direction of zero angle.

how the tire will respond to hard braking, he sketches a diagram of the various forces at work on the tire. He draws arrows representing the weight of the vehicle, the angular momentum of the wheel, braking torque, and the tire's friction with the road. Large quantities are represented by long arrows and small ones by short arrows. The diagram helps the engineer understand the dynamics needed to create a computer model of the tire design.

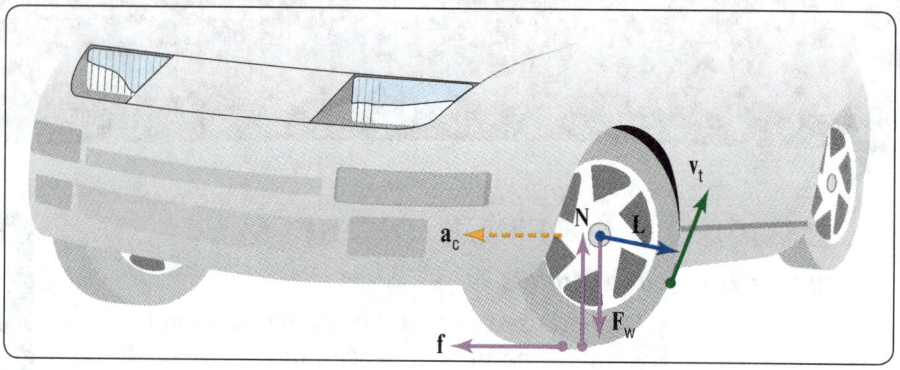

4-4 Each quantity in the diagram is represented by a vector.

In each case above, vectors are used to study quantities that have two pieces of information—a specific numerical value and a direction. Physicists use vectors to study dimensions such as displacement, velocity, force, acceleration, and magnetic fields. Vectors are drawn as arrows on *vector diagrams.*

Physicists also work with scalars. A scalar is a quantity that can be described completely by only a single numerical piece of information. Time, mass, volume, and temperature are all scalars because, unlike vectors, they have no direction associated with them. Some scalars are only positive, such as mass and time. Others, such as temperature, may be positive or negative, depending on the instant of measurement.

A vector symbol is printed with boldface roman type, **A**. Since writing boldface letters is difficult, you may use a regular letter with an arrow over it, $\vec{A}$. Upright lines around the vector symbol, $|\mathbf{A}|$, stand for its absolute value and specify its magnitude. As a shorthand notation for the magnitude, either the italicized letter or the absolute value sign will be used; that is,

$$A \equiv |\mathbf{A}| \quad \text{(handwritten, } A \text{ or } |\vec{A}|\text{)}.$$

The symbol ≡ means "is defined as."

4.2 Vector Angles

Vectors in diagrams provide important information. The tail of the arrow can indicate the starting point of motion or the point at which a force is physically applied. The length of the arrow is proportional to the magnitude of the vector quantity. Vector diagrams also include directional information indicated by the orientation of the arrow. The **vector angle**, labeled with the Greek letter θ, is given a numerical value in degrees or radians measured from a fixed direction. This **reference direction** is usually determined by convention or is selected for convenience in working the problem. This is especially important in two- and three-dimensional vectors. For one-dimensional vectors, direction can be indicated by a plus or minus sign, as stated in Chapter 3.

There are some *conventions* (generally agreed-upon rules) for establishing reference directions for angular measurement. This will be especially applicable to two-dimensional motion, which you will study in the next chapter. In the

Cartesian (two-dimensional) plane, the reference direction is the positive x-axis. Positive angles are measured counterclockwise and negative angles are measured clockwise around the origin. There is no limit to the size of geometric angles. An angle can be measured in either degrees or radians.

Map directions are always referenced to geographic north, which by convention is located at the *top* of the map or diagram. Angles referenced to true north are indicated by a capital "T" in place of the degree symbol. Map directions are usually measured clockwise from north. For example, 315T is northwest. When using north as the reference direction, angles are always positive and are reported only in degrees from 0T to 360T. This textbook will always give map directions using three digits (which is standard practice for navigational purposes). In this system, north is 000T, northeast is 045T, south is 180T, and so on. On the other hand, if a problem gives an angle with the degree symbol (e.g., 120°), then you know that a Cartesian reference direction is being used.

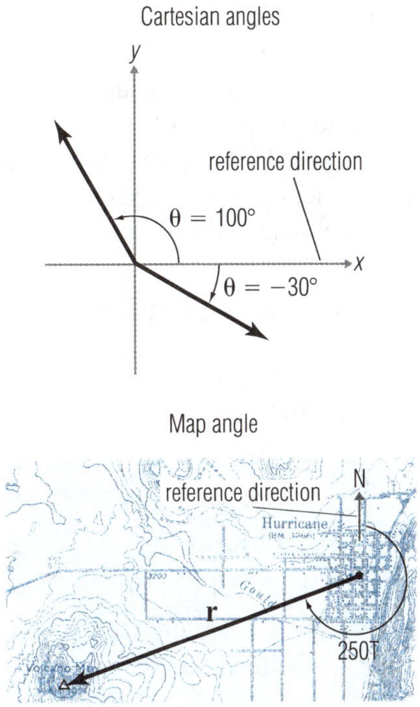

4-6 The method for measuring vector angles depends on the kind of angle.

4.3 Transporting Vectors

Drawing accurate vector diagrams requires a protractor and a ruler. For most physics problems, it is sufficient to sketch the approximate magnitude and orientation of each vector. Unless specifically requested, sketches don't have to be made exactly to scale.

4-7 This compass rose shows the cardinal and intercardinal directions, as well as degrees.

Vectors are equal if they have both the same magnitude and the same direction. Examine vectors **A** and **B** in Figure 4-8. Even though they are at different locations in the diagram, they have identical magnitudes and they point in the same direction, so **A** = **B**. Remember that vectors are models of the physical quantities that they represent. If necessary for the sake of a problem solution, it is perfectly permissible to transport a vector from one location to another in a diagram as long as you maintain its original length and orientation (see Figure 4-9).

A marine biologist is periodically monitoring the location of a submerged humpback whale (*Megaptera novaeangliae*) using a sonic transponder. He plots its position at three different times on a nautical chart. The whale's position with time is shown in Figure 4-10. The distance the whale moves between the first and second positions is represented by the length of the line segment between P_1 and P_2, written P_1P_2. The whale's *displacement* is indicated by an arrow drawn from P_1 to P_2 (labeled $\overrightarrow{P_1P_2}$). Displacement is a vector quantity because it specifies both distance and direction. Distance is a scalar because only the magnitude of the displacement, a single piece of information, is given.

Map direction angles referenced to true north are followed by the letter "T."

Modern directional compass scales are subdivided into 360°. The four principal compass directions, or *cardinal directions*, are represented by angles that are 90° apart: north is 000T, east is 090T, south is 180T, and west is 270T. *Intercardinal directions* (NE, SE, SW, and NW) are 45° from their associated cardinal directions.

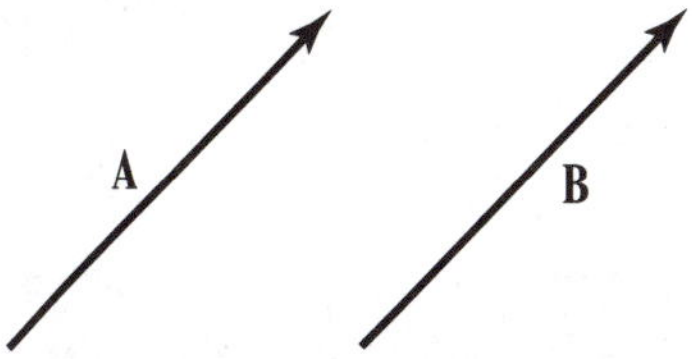

4-8 Equal vectors

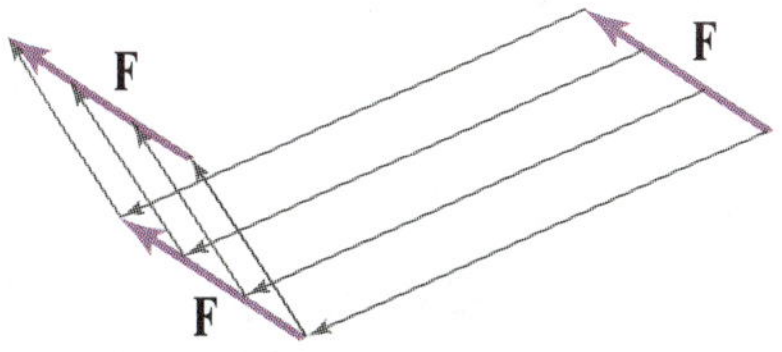

4-9 Vector transport

Problem-Solving Strategy 4.1

Vectors in a diagram can be transported from one point to another as long as their lengths and orientations remain unchanged.

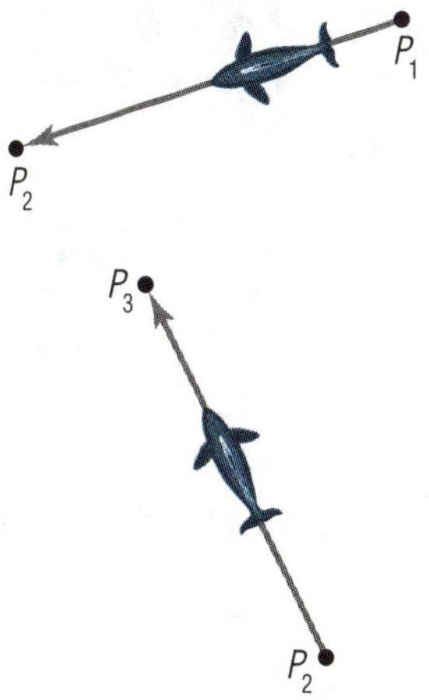

4-10 Positions and displacement of a whale

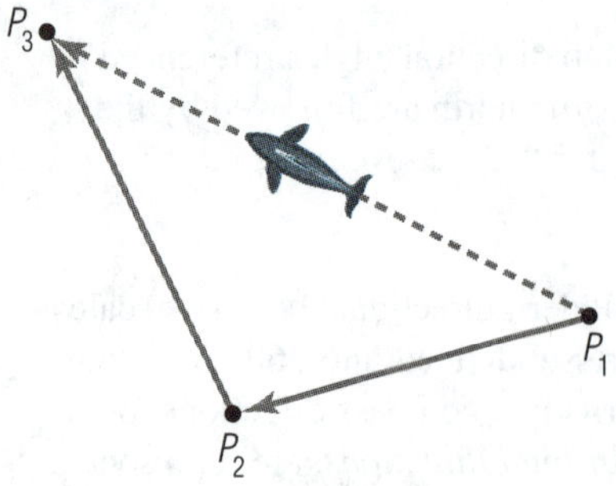

4-11 Positions and displacement of a whale

4A Objectives

After completing this section, I can

✓ differentiate between vectors and scalars.

✓ describe the properties of a vector.

✓ write the correct symbols for vector and scalar quantities.

✓ identify the vector angle according to the kind of reference direction specified.

If the whale moves on to P_3, the total distance that it traveled through the water from P_1 is the sum of the distances P_1P_2 and P_2P_3. Since the total distance is length alone, it is still just a scalar quantity. The whale's total displacement, however, is the vector from P_1 to P_3 (see Figure 4-11). The length of this vector is clearly different from the distance that the whale traveled. Vectors cannot be added by simple arithmetic. Fortunately, vector addition is not difficult to understand. You will learn this important skill in Section 4B.

4A Section Review

1. What is the key difference between scalar and vector quantities?
2. State whether the following quantities are scalars, vectors, or neither.

a. force	**e.** distance	**i.** mass
b. one-way sign	**f.** displacement	**j.** temperature
c. pressure	**g.** velocity	**k.** Wednesday
d. π (3.14 . . .)	**h.** speed	**l.** density

3. When dealing with vectors, what is the difference between **B** and B? Write another symbol that is equivalent to B.
4. Compare and contrast the properties of angles measured in a Cartesian coordinate system to those measured on a map.
5. Why can you transport a vector from one place to another in a vector diagram?

4B OPERATIONS WITH VECTORS: GEOMETRIC TECHNIQUES

When more than one quantity is involved in a problem, vectors often need to be added, subtracted, or multiplied. They may be multiplied and divided by scalars as well. Mathematicians have developed methods to simplify these operations and yield useful results. The geometric techniques of vector math can be done quickly with only a ruler and a pencil.

4.4 Adding Equal Vectors by Scalar Multiplication

The sum or difference of two or more vectors is called the **resultant**.

The simplest case of vector addition is adding a vector equal to itself. Graphically, this is accomplished by placing the tail of the second vector at the tip of the first. Their sum is the vector drawn from the tail of the first vector to the tip of the second. The vector sum, called the **resultant**, is twice as long as the original vector and is oriented in the same direction.

$$\mathbf{V} + \mathbf{V} = 2\mathbf{V}$$

The bold plus sign indicates vector addition. Note that this vector sum is also the product of a scalar number and the original vector. When a vector is multiplied by a positive scalar number, its length (magnitude) changes but its direction stays the same. The magnitude of vector 2**V** is $2V$.

In Figure 4-12, if **V** is a vector 6 units long, 2**V** would be a vector 12 units long pointing in the same direction as **V**. If the scalar multiplier is negative, the resulting vector points in the opposite direction from the original. Thus, −**V** (that is, $-1 \times$ **V**) is a vector 6 units long pointing in the opposite direction from **V**. Scalar coefficients may also be fractions, so $-\frac{1}{2}\mathbf{V}$ (or $-\mathbf{V}/2$) is a vector 3 units long oriented opposite to **V**.

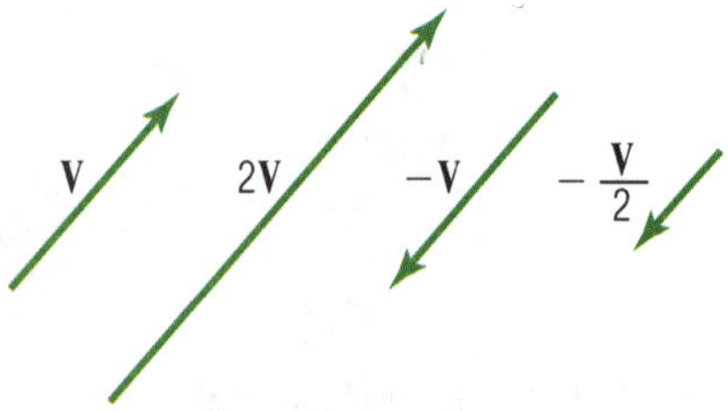

4-12 Scalar multiples of vector **V**

4.5 Adding Unequal Vectors Graphically

How are two different vectors added—for example, **A** + **B**? If they have different magnitudes but the same direction, then their resultant (**R**) is a vector whose magnitude is $R = A + B$, and it has the same direction as the original vectors. Again, the resultant vector extends from the tail of the first vector to the tip of the second. Observe that there is no difference in the resultant **R** between the operations **A** + **B** and **B** + **A** (see Figure 4-13). In other words, vector addition is *commutative.*

In most physics problems, however, vectors are usually unequal—they don't have the same magnitude and rarely point in the same direction. Let's look at a two-dimensional example. A student pilot flies 30 km north and then 40 km west on her solo flight, shown in Figure 4-14. The two flight legs are displacements that can be represented by vectors. To find the plane's net displacement from the airfield at the end of the second leg, you must sum the two leg vectors. Using the graphical method, each leg is drawn to scale on a sheet of paper or a map. Since vectors can be transported without changing their magnitude or direction, you can place the vectors head to tail in either order. As before, the sum of the two vectors is the resultant vector that stretches from the tail of one vector to the tip of the other.

The resultant is identical to the path the aircraft could have taken to fly directly from the airport to its position at the end of the second leg. You can find the magnitude of the resultant displacement vector graphically by measuring its length with a ruler and applying the map's scale factor. The airplane is found to be 50 km from the airfield. Notice that this is *not* the arithmetic sum of 30 km and 40 km. When the directions of the vectors are not identical, the magnitude of the resultant is not equal to the sum of the magnitudes of the two legs,

$$R \neq A + B.$$

The direction of the resultant is also different from either of the original vectors. You can measure the compass direction (vector angle) from the airport using a protractor. Measured clockwise from north using navigational angles, the aircraft bears 307T from the airport.

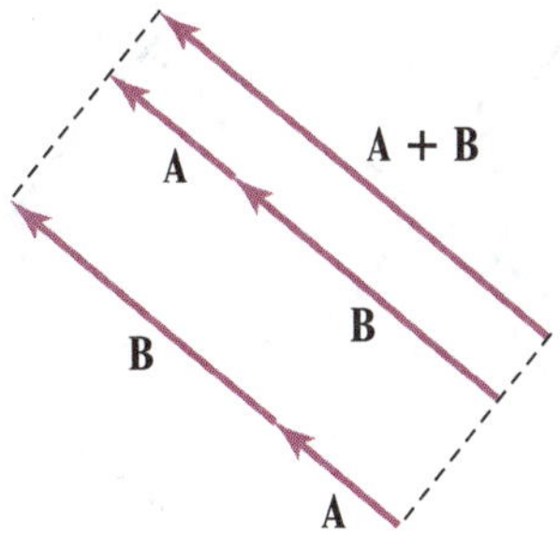

4-13 Commutative character of vector addition

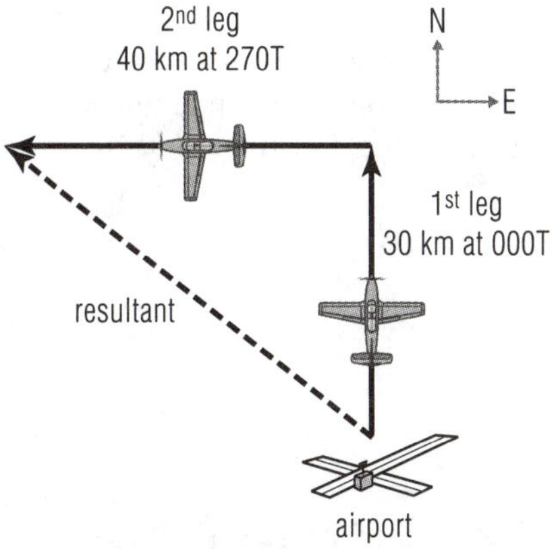

4-14 When adding vectors, the resultant stretches from the tail of one vector to the tip of the other.

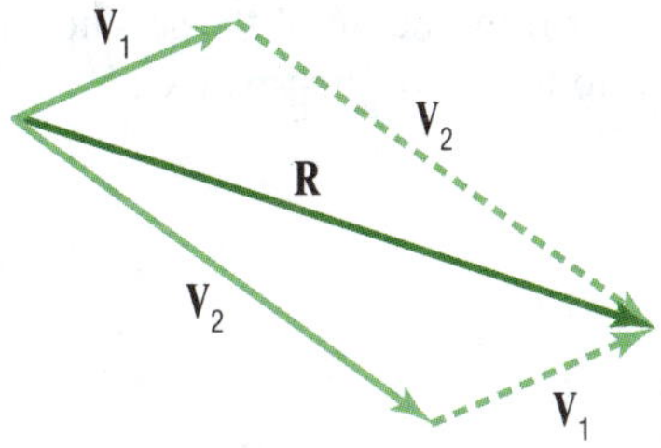

4-15 The resultant of two vectors can often be represented as the diagonal of a parallelogram with the sides formed by the two vectors placed tail to tip.

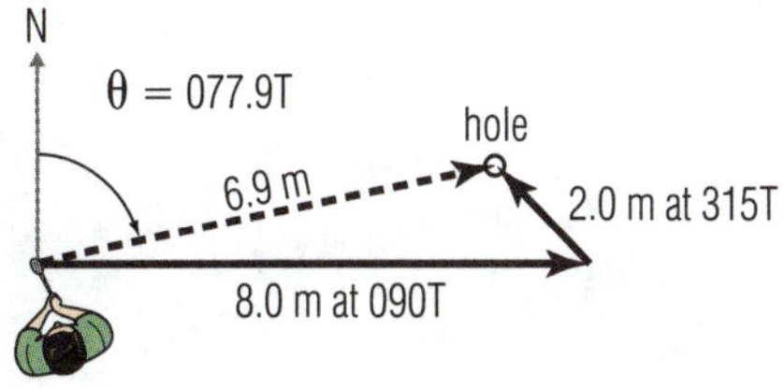

4-16 What would have been the right first putt?

EXAMPLE 4-1

Putting Around with Golf

A golfer is attempting to sink a putt. The ball rolls 8.0 m to the east and misses the hole. He putts again 2.0 m to the northwest and the ball finally drops into the cup. How far and in what direction should he have putted the ball in order to sink the shot on the first attempt?

Solution:

Refer to Figure 4-16. Draw vectors to scale that represent each putt and place them so that the tail of the second touches the head of the first (northwest is halfway between west and north, or 315T). Then draw the resultant vector from the tail of the first vector to the head of the second. According to the diagram scale, the resultant vector has a length of about 6.9 m. The angle as measured from north (the top of the diagram) using a protractor is 077.9T.

Note that when the resultant must be measured from a diagram, each vector must be drawn accurately to scale.

Three or more vectors may be added in a similar way. Place the tail of each vector at the tip of the previous one. Their resultant is the vector drawn from the

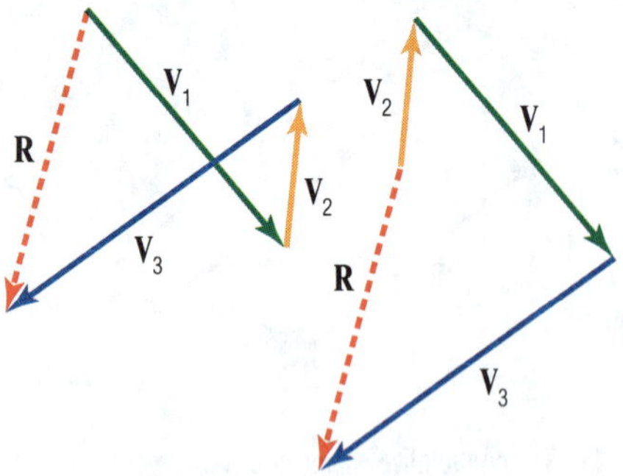

4-17 The resultant of a sum of vectors does not depend on the order of addition.

A vector's opposite is a vector with the same length pointing in the opposite direction.

tail of the first to the tip of the last. Because vector addition is commutative, the order of the vectors does not matter.

4.6 Vector Subtraction

In the expression **A−B**, the bold negative sign indicates vector subtraction. Recall from algebra that subtraction is equivalent to adding the opposite of a number—subtracting 3 is the same as adding −3. The vector expressions **A−B** and **A+(−B)** are equivalent. Remember that a vector's opposite has the same length but points in the opposite direction. Once you have found the opposite vector, vector addition is carried out to find the resultant. Note that, as with arithmetic subtraction, vector subtraction is *not* commutative; that is, **A−B** is not the same as **B−A** (see Figure 4-18).

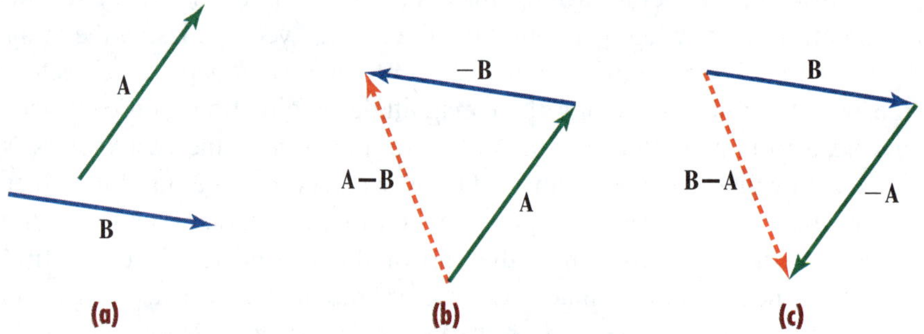

4-18 The difference of two vectors depends on the order of subtraction.

Problem-Solving Strategy 4.2

Quickly find the difference of two vectors (**A−B**) by placing them tail-to-tail and drawing the difference vector from the tip of the second (**B**) to the tip of the first (**A**).

A second method exists for easily finding the graphical difference of two vectors. Place the two vectors tail to tail, then draw the resultant vector from the tip of the second vector in the difference equation to the tip of the first.

A difference vector (e.g., **ΔV**) is the resultant of a vector subtraction operation (see Example 4-2). This method is useful for determining the *change* in a vector quantity. We saw a difference vector in Chapter 3 when we determined average acceleration, where

$$\bar{\mathbf{a}} = \frac{\Delta \mathbf{v}}{\Delta t}.$$

In this case, **Δv** is the difference vector of the velocity measured at two different times.

EXAMPLE 4-2

Vector Subtraction

Given the vectors **C** and **D**, find (**C−D**).

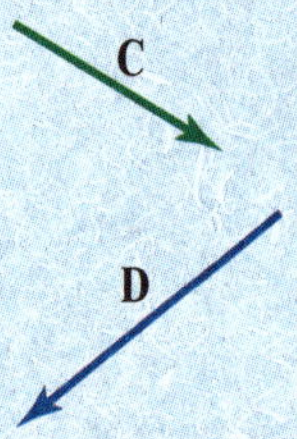

Solution—Two Methods:

Adding the opposite of the second to the first:

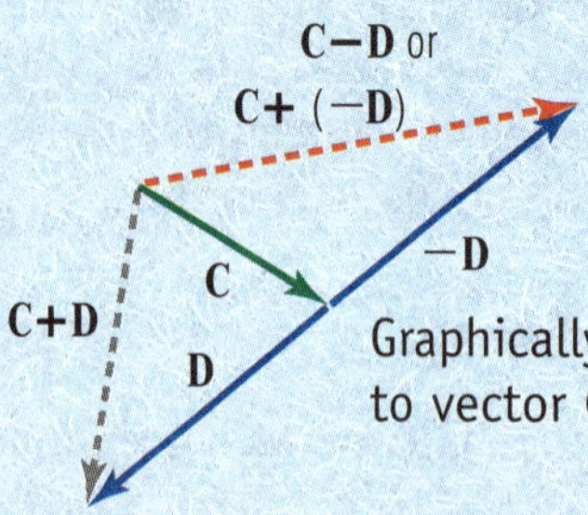

Graphically adding **−D** to vector **C**

Direct vector subtraction:

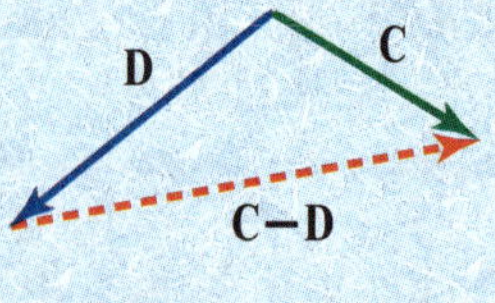

Note that the vector (**C−D**) is the same using either method.

4B Section Review

1. What is the sum of two or more vectors called?
2. Compare the vector 4**B** with the vector **B**.
3. Compare the vector **A** and the vector −**A**/3.

Copy vectors **A** and **B** onto your paper. Refer to them for Questions 4–6.

4. Graphically find the resultant of the following operations.

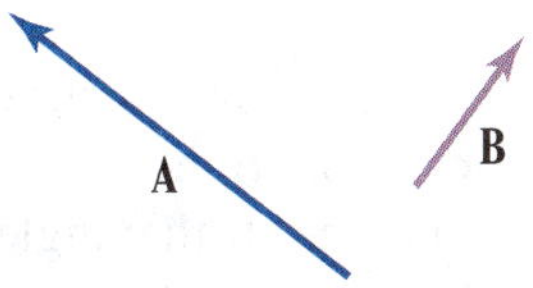

   a. **A**+**B**
   b. **B**+**A**
   c. **A**−**B**
   d. 2**B**+**A**

5. Describe the difference between the vector **A**−**B** and the vector **B**−**A**.
6. Determine the vector 2**B**−2**A** using the second method discussed in this section for finding the difference of two vectors. Describe how the vector 2**B**−2**A** compares to the vector **B**−**A**. Show how this illustrates the distributive property of equality.
7. You are given two vectors, **F** and **G**, so that $F = 1$ unit and $G = 2$ units. Sketch the vector sum **F**+**G** so that $|\mathbf{F}+\mathbf{G}| = 1$.

4B Objectives

After completing this section, I can

✓ add equal vectors by scalar multiplication.

✓ add unequal vectors graphically.

✓ subtract vectors graphically.

4C OPERATIONS WITH VECTORS: MATHEMATICAL TECHNIQUES

Graphical vector addition using sketches provides quick, convenient results but lacks precision and accuracy. Even carefully drawn vector diagrams are subject to drawing errors. For physics, mathematical techniques involving trigonometry allow you to perform vector addition with much greater accuracy.

4.7 Solving Triangles

In solving two-dimensional vectors, you will have to solve for unknown quantities in vector triangles using the mathematical tools taught in algebra and geometry. To help you find the measure of angles and lengths of sides in triangles, you must know the rudiments of trigonometry. We will review some of the more important trigonometric concepts here.

One useful principle of trigonometry is that two triangles are **similar triangles** when the three angles of one triangle equal the corresponding three angles of the other triangle. The angles, not the sides, determine similarity. Compare the triangles in Figure 4-19, which have different side lengths but equal corresponding angles. Upon closer inspection you should see that the lengths of the sides are indeed related: the ratios between corresponding sides of triangles A and B are all the same. In fact, all similar triangles have proportional sides. Physicists and mathematicians use this important relationship to solve vector triangle problems.

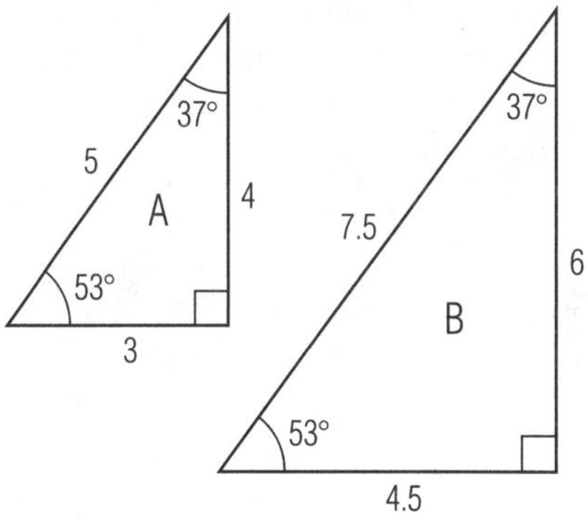

4-19 Similar triangles

The shorter sides of a right triangle are called the *legs* of the triangle. The *hypotenuse* is always opposite the right angle.

By far the most common triangle found in physics problems is the **right triangle**, which has one 90° angle. Right triangles possess several properties that make it easy to determine unknown sides and angles. Study Figure 4-20 carefully before you proceed further. The two acute angles, which are less than 90°, are labeled *A* and *B*. The measure of angle *A* is α and the measure of angle *B* is β. Side *a* is opposite acute angle *A*, while side *b* is opposite acute angle *B*. The hypotenuse, the line segment opposite the right angle *C*, is labeled *c*.

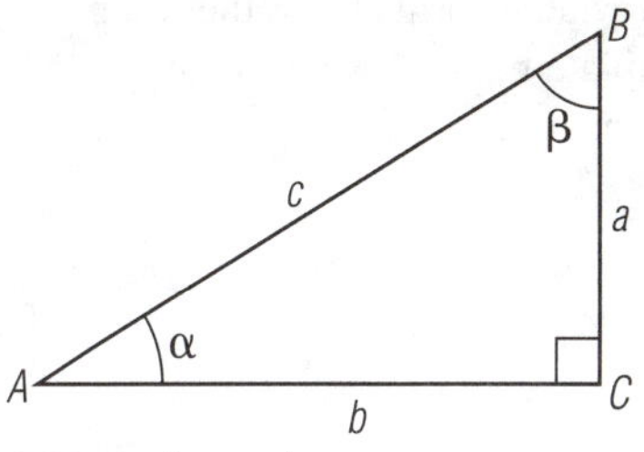

4-20 A right triangle

Because the sum of the interior angles of a triangle is always 180°, you can determine an unknown acute angle in a right triangle if you know the measure of the other acute angle. Similarly, if you know two of the sides of a right triangle, you can find the third according to the Pythagorean theorem.

$$a^2 + b^2 = c^2 \tag{4.1}$$

A significant benefit of using a right triangle is that one of the angles is *always* known—the right angle. As long as either two sides, or an acute angle and a side, are known, all the angles and all the sides in a right triangle can be determined mathematically. This can be accomplished by taking advantage of the proportionality of similar triangles. The values of ratios of corresponding sides in similar triangles are always the same. The ratios of the hypotenuse or sides that are either adjacent or opposite to a specified angle are given special names.

The **sine** of an acute angle is the ratio of the opposite side's length to the length of the hypotenuse. From Figure 4-20, you can identify the sines of angles α and β, which are written "sin α" and "sin β."

The angle measure, such as α, β, or θ, is called an *argument* of the trigonometric ratios sin, cos, and tan, and must be included. You should not say, "The sine is 0.866," but rather, for an angle α, "The sine of α is 0.866."

$$\sin\alpha = \frac{\text{opposite}}{\text{hypotenuse}} = \frac{a}{c}$$

$$\sin\beta = \frac{\text{opposite}}{\text{hypotenuse}} = \frac{b}{c}$$

The *adjacent side* of an acute angle in a right triangle is the shorter of the two sides that form the angle. The longest side is always the hypotenuse.

Similarly, the **cosine** of an angle is the ratio of the length of the angle's adjacent side to the length of the hypotenuse. The cosine of α is written "cos α."

$$\cos\alpha = \frac{\text{adjacent}}{\text{hypotenuse}} = \frac{b}{c}$$

$$\cos\beta = \frac{\text{adjacent}}{\text{hypotenuse}} = \frac{a}{c}$$

The **tangent** of an acute angle is the ratio of the length of the side opposite the angle to the length of its adjacent side. The tangent of α is written "tan α."

$$\tan\alpha = \frac{\text{opposite}}{\text{adjacent}} = \frac{a}{b}$$

$$\tan\beta = \frac{\text{opposite}}{\text{adjacent}} = \frac{b}{a}$$

When the names of trigonometric ratios are spoken, never use their abbreviated forms but rather their complete names—sine, cosine, and tangent.

The values of sine, cosine, and tangent for any angle are easily calculated using a scientific calculator.

EXAMPLE 4-3

Finding an Unknown in a Right Triangle

Study Figure 4-21. You are given one side, *a*, and the measure of one acute angle, β, in a right triangle. Solve for side *b* and the measure of acute angle α.

Solution:

1. You can find side b, using the tangent for angle β.

$$\tan\beta = \frac{\text{opposite}}{\text{adjacent}}$$

$$\tan 17° = \frac{b}{a} = \frac{b}{8.2}$$

$$b = 8.2 \tan 17°$$

$$b \cong 2.5$$

Whenever possible, perform all math operations involving trigonometric ratios simultaneously using the function keys on your calculator.

2. You can find the value of α, knowing that the angles of any triangle add up to 180°.

$$\alpha + \beta + 90° = 180°\text{, so}$$

$$\alpha = 180° - 90° - 17°$$

$$\alpha = 73°$$

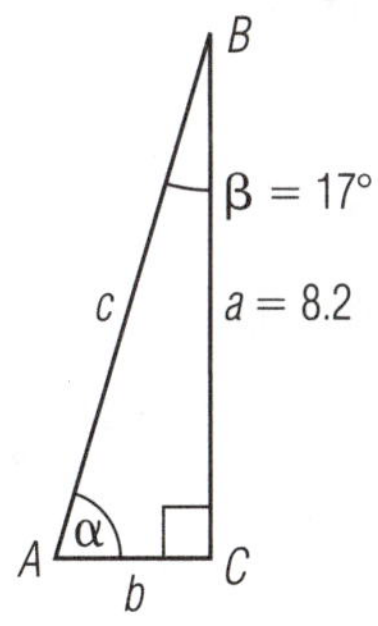

4-21 If you know one side and one acute angle in a right triangle, you can find the other unknowns.

Given only the sides of a right triangle, how can you find the measure of the two unknown acute angles? You could draw a scale model and measure the angles with a protractor, but this technique is not precise. Instead, it is better to use one of your calculator's inverse trigonometric functions (usually shown as a secondary function to the corresponding trig ratio keys—$\sin^{-1}$, $\cos^{-1}$, and $\tan^{-1}$). The argument for these operators is the ratio of the appropriate sides, and the output is the corresponding angle in the definition of the function.

Given the right triangle in Figure 4-22, suppose you want to find the measure of the angle whose opposite side is 7 units and whose adjacent side is 8 units.

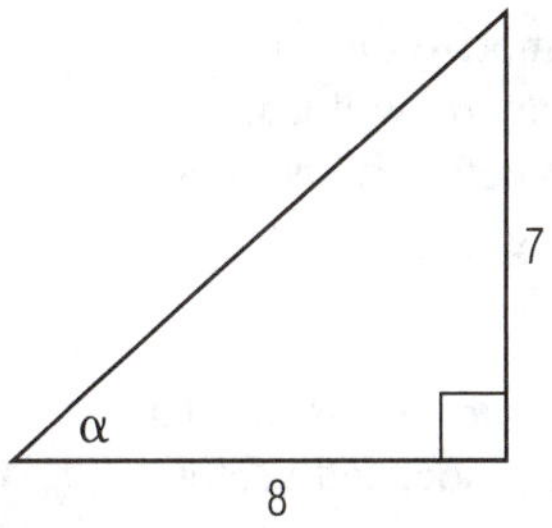

4-22 Finding α from its tangent ratio

$$\tan\alpha = \frac{\text{opposite}}{\text{adjacent}} = \frac{7}{8} = 0.8750$$

Using the inverse tangent function on your calculator gives you α directly:

$$\alpha = \tan^{-1}\left(\frac{7}{8}\right)$$

$$\alpha \cong 41.2°$$

The inverse tangent equation above may be read "Alpha is the angle whose tangent is 0.8750." The exponent -1 in an inverse trigonometric operator does not mean "reciprocal," so do not invert the expression.

Inverse trigonometric ratios are also known as arc-sine, arc-cosine, and arc-tangent.

4.8 Vector Components

Before we look at the mathematical methods for adding and subtracting vectors, we first need to develop the concept of **component vectors**. Every vector lying in a plane is the resultant of the sum of two special vectors called component vectors or **components**. To understand the relationship of components to the original vector, imagine two flashlights shining on a vector as in Figure 4-23. Note that the light beams are parallel to the vertical and horizontal axes of the vector's coordinate system. The shadows of the vector cast by the flashlights are projected onto each axis. These shadows represent the component vectors. In fact, components are often called *projections* of the vector onto the coordinate axes. Components are parallel to the coordinate axes and are therefore perpendicular to each other.

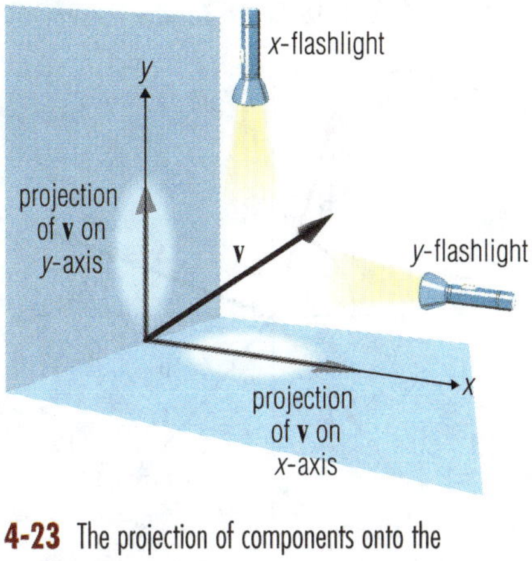

4-23 The projection of components onto the coordinate axes

Mathematically adding vectors involves combining their x-components and their y-components to obtain the components of the resultant vector. Quickly finding the components of a vector is therefore very important. This is where trigonometry comes into the picture. When a vector and its components are sketched together in a triangle, they form a *vector triangle.* The vector triangle for any given vector is a right triangle because the component vectors that form the legs (shorter sides) of the triangle are perpendicular to each other. As long as the vector's length and angular direction are known, you can find its components using trigonometry.

Vector components are labeled with the original vector's symbol and include a subscript of the axis to which the component is parallel. For example, the x-component of the displacement vector **d** is $\mathbf{d}_x$.

The angle used when determining a vector's components is usually the acute angle between the vector and its horizontal coordinate axis (the axis parallel to the horizontal component). We will call this the **reference angle (α)**. The reference angle may or may not be the same as the vector angle (θ) depending on the reference direction used for the problem. Regardless of the orientation of the original vector, the reference angle is always a positive acute angle within the vector triangle. When the reference angles for several vectors are required in a problem, you should label each angle with its vector name as a subscript (e.g., $\alpha_{\mathbf{A}}$, $\alpha_{\mathbf{B}}$).

The reference angle (α) of a vector is the acute angle between the vector and its horizontal component in the vector triangle. The relationships between reference angles and vector angles for the Cartesian and compass systems of angle measurements can be seen in Table 4-1. Though you don't need to memorize the table, with practice you will learn the relationship between vector and reference angles.

TABLE 4-1

Relationship Between θ and α

Quadrant	θ	α
I	$\theta = \alpha$	$\alpha = \theta$
II	$180° - \alpha$	$180° - \theta$
III	$180° + \alpha$	$\theta - 180°$
IV	$360° - \alpha$	$360° - \theta$
NE	$(090 - \alpha)$T	090T $- \theta$
SE	$(090 + \alpha)$T	$\theta -$ 090T
SW	$(270 - \alpha)$T	270T $- \theta$
NW	$(270 + \alpha)$T	$\theta -$ 270T

The scalar values of the component vectors are called **scalar components**. While the magnitude of a component vector is always positive, the scalar value may be positive, negative, or zero, depending on the orientation of the original vector. The sign indicates the direction of the component relative to its coordinate axis. Positive scalar components point up and to the right. Negative scalar components point down and to the left. Note that the one-dimensional vectors discussed in Chapter 3 had only one component—the vector itself.

A scalar component is the scalar value of the associated vector component.

When the component vectors of a given vector have been identified, we say that the vector has been *resolved into its components.* Completing the following steps will help you resolve vectors:

A vector is *resolved* into its components when its component vectors have been determined.

- *Select a convenient coordinate system.* Usually the coordinate axes will be vertical and horizontal, but other orientations are possible depending on the situation (see Figure 4-24).
- *Note the kind of directional system used*, Cartesian or map.
- *Determine the reference angle (α) for the vector.* Identify this angle and determine its magnitude from the geometry of the problem.
- *Find the magnitude of the horizontal scalar component*: $|V_x| = V \cos \alpha$, where V is the magnitude of the vector **V**.
- *Find the magnitude of the vertical scalar component*: $|V_y| = V \sin \alpha$.
- *Assign the correct signs to the scalar components based on their orientations to the coordinate axes.* The positive reference directions in the Cartesian system are the positive x- and y-axes. In the map system, north and east are positive. Always refer to the problem sketch when assigning scalar component signs.

Problem-Solving Strategy 4.3

Make it a habit to quickly identify the reference angles for all vectors in a problem. Resolving vector components then becomes a straightforward process.

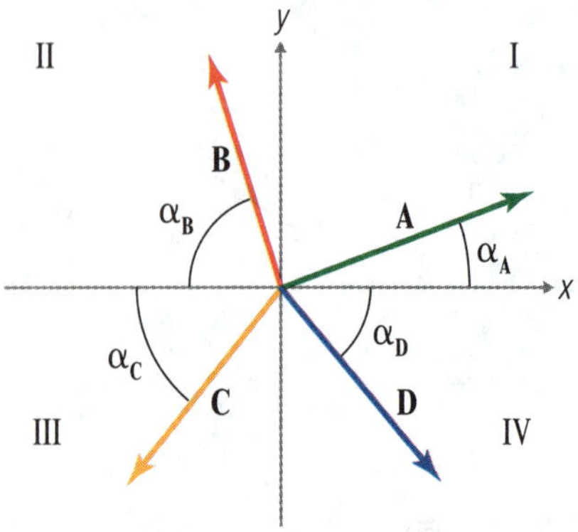

4-24 Reference angles of vectors in each quadrant

EXAMPLE 4-4

Resolving Vector Components

Vector **L** points at an angle of 37° above the *x*-axis of a graph and has a length of 15 units. Find the lengths of its components.

Solution:

Sketch vector **L** at the approximate vector angle given in the problem (see Figure 4-25). The vector forms the hypotenuse of a vector right triangle. The two unknown component vectors are the legs of the triangle.

The reference angle α is the acute angle between the horizontal side and the hypotenuse. Label the resulting right triangle. Once this is done, find the lengths of the two sides.

$$\sin\alpha = \frac{\text{opposite}}{\text{hypotenuse}}$$

Substituting the values for α and the hypotenuse gives

$$\sin 37° = \frac{|\mathbf{L}_y|}{15 \text{ units}}.$$

Solve for $|\mathbf{L}_y|$.

$$|\mathbf{L}_y| = (15 \text{ units})\sin 37°$$

$$|\mathbf{L}_y| \cong 9 \text{ units (same precision as original dimension)}$$

Similarly,

$$\cos\alpha = \frac{\text{adjacent}}{\text{hypotenuse}} \quad \text{or} \quad \cos 37° = \frac{|\mathbf{L}_x|}{15 \text{ units}}.$$

Solve for $|\mathbf{L}_x|$.

$$|\mathbf{L}_x| = (15 \text{ units})\cos 37°$$

$$|\mathbf{L}_x| \cong 12 \text{ units}$$

The original vector **L** can be thought of as the vector sum of vector $\mathbf{L}_x$ and $\mathbf{L}_y$.

$\mathbf{L} = \mathbf{L}_x + \mathbf{L}_y$, where

$\mathbf{L}_x$ = 12 units long pointing in the positive *x*-direction, and

$\mathbf{L}_y$ = 9 units long pointing in the positive *y*-direction.

Problem-Solving Strategy 4.4

When resolving a vector, always assign the signs of its scalar components based on the original vector's orientation in the problem sketch. Always start the solution to a problem with an accurate sketch.

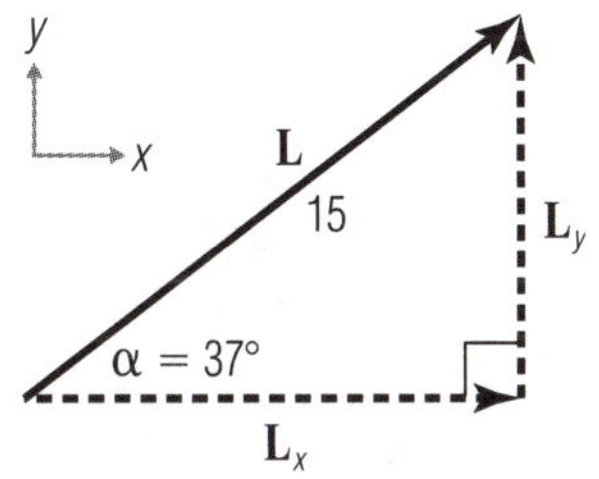

4-25 Find the vector components.

EXAMPLE 4-5

Resolving Displacement on a Slide

A little girl slides down a playground slide 4.0 m long. The slide forms a 40° angle with the horizontal. Determine the vertical and horizontal components of the girl's displacement from the top of the slide to the bottom.

Solution: Refer to Figure 4-26.

The girl's displacement, **d**, is the hypotenuse of the vector right triangle formed by the vertical and horizontal components, $\mathbf{d}_y$ and $\mathbf{d}_x$ respectively. From the problem diagram, we can see that d = 4.0 m and α = 40.°.

$|d_y| = d \sin\alpha = (4.0 \text{ m})\sin 40.° \cong 2.\underline{5}7$ m down

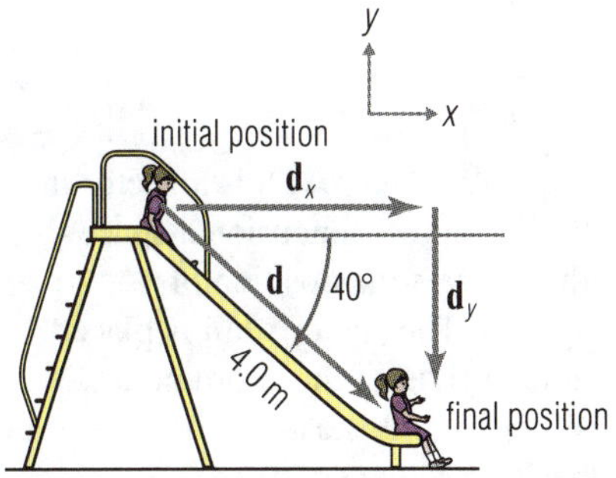

4-26 What are the components of the girl's displacement?

$d_y \cong -2.6$ m (down is negative y)

$|d_x| = d \cos \alpha = (4.0 \text{ m})\cos 40.° \cong 3.\underline{0}6$ m to the right

$d_x \cong +3.1$ m (right is positive x)

The girl's displacement components are 2.6 m down and 3.1 m to the right.

Vectors described using a map system of directions are resolved in the same way as vectors in a Cartesian coordinate system. The cardinal map directions must first be redefined as Cartesian axes. The east-west line becomes the x-axis and the north-south line the y-axis. Resolve the vectors as usual. Components may be labeled $\mathbf{V}_{\text{N-S}}$ and $\mathbf{V}_{\text{E-W}}$ to indicate that they are map vector components.

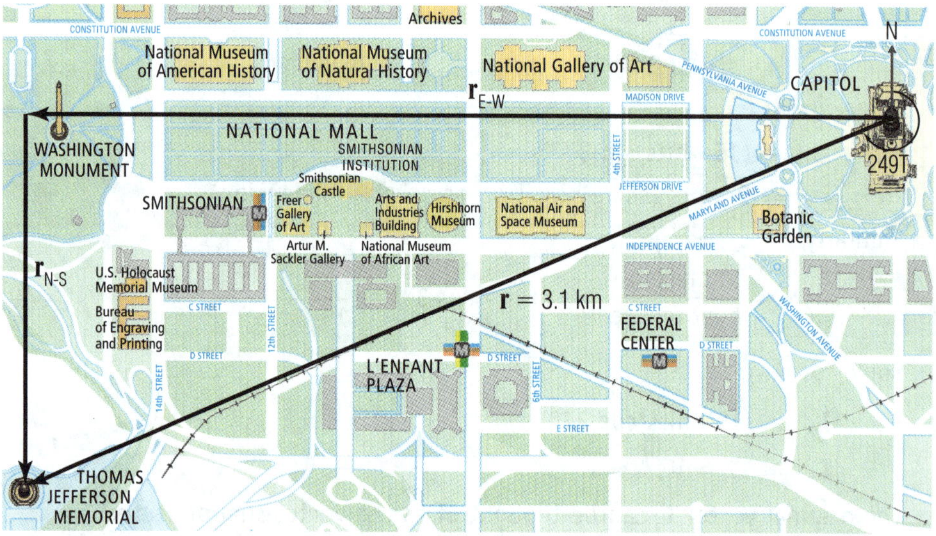

4-27 Map of the National Mall and its environs, Washington, D.C.

EXAMPLE 4-6

Where Is the Jefferson Memorial?

The Thomas Jefferson Memorial lies on a compass bearing 249T from the United States Capitol Building in Washington, D.C. Its distance from the Capitol is 3.1 km. Determine the components of the Jefferson Memorial's **position vector (r)** relative to the Capitol Building.

Solution: Refer to Figure 4-27.

The monument's position vector angle θ is 249° from north, or 249T. In order to find the vector components, the reference angle α must be determined.

Recall that α is the acute angle between the vector and the horizontal axis. For map problems, the horizontal axis is the east-west direction. Therefore, the reference angle in this problem is measured from the west direction to the position vector: $\alpha = 270\text{T} - 249\text{T} = 21°$. (Note that the difference of two directions is an *angle* and not a direction.)

A **position vector (r)** is a vector in two- or three-dimensional space that locates the position of a system. The vector's tail is placed at the origin of the coordinate system and its head is at the position of the system.

Now the east-west and north-south components may be easily found.

$|r_{N\text{-}S}| = r \sin\alpha = (3.1 \text{ km})\sin 21° \cong 1.\underline{11} \text{ km}$

$r_{N\text{-}S} \cong -1.1 \text{ km}$ (the north-south component points south)

$\mathbf{r}_{N\text{-}S} \cong 1.1 \text{ km south}$

$|r_{E\text{-}W}| = r \cos\alpha = (3.1 \text{ km})\cos 21° \cong 2.\underline{89} \text{ km}$

$r_{E\text{-}W} \cong -2.9 \text{ km}$ (the east-west component points west)

$\mathbf{r}_{E\text{-}W} \cong 2.9 \text{ km west}$

Component vectors may be summed to find the original vectors. This may seem to be an obvious statement, but you will need this skill to determine resultant vectors. First, let us look at a graphical example.

EXAMPLE 4-7

Obtaining Vectors from Their Components

Sketch the vectors that have the following components:

a. $K_x = 2$, $K_y = 5$ b. $P_x = 3$, $P_y = -1$

Solution:

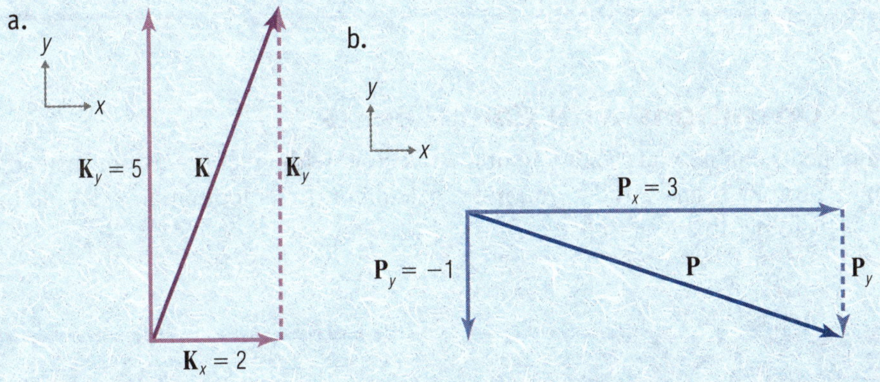

Three-dimensional vectors can be resolved into three components. In this case, the components are parallel to the x-, y-, and z-axes. By convention, the x- and y-axes lie in the horizontal plane and the z-axis is vertical. The following example shows how we obtain a three-dimensional vector from its components.

EXAMPLE 4-8

Rocket Launch

A physics student carries a model rocket from his car 200. m south along a sidewalk and 250. m west into a field to the launch site. He then launches the rocket to an altitude of 450. m. Find the rocket's distance from the car when it reaches its highest altitude.

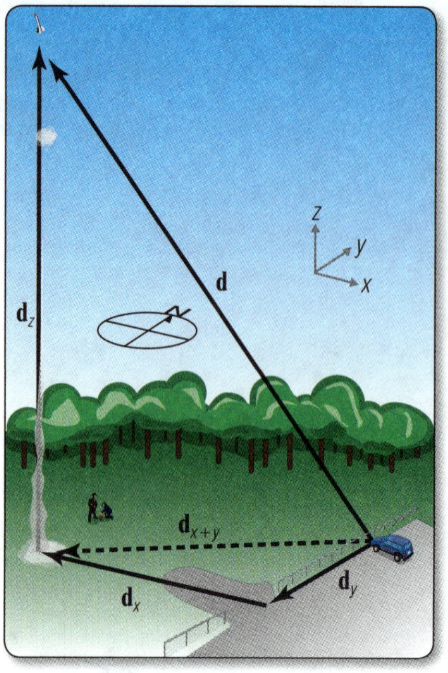

4-28 Displacement in three dimensions

A scalar component is to a component vector what the brand name "Boeing 747" is to the airplane itself. A vector is a physical phenomenon, and its vector symbol represents that physical entity. You cannot use a vector symbol in a strictly arithmetic operation any more than you can use the physical airplane in a sentence about the plane. Arithmetic and trigonometric operations work only with pure numbers. That is why you must convert components to equivalent scalar values before using them in nonvector equations.

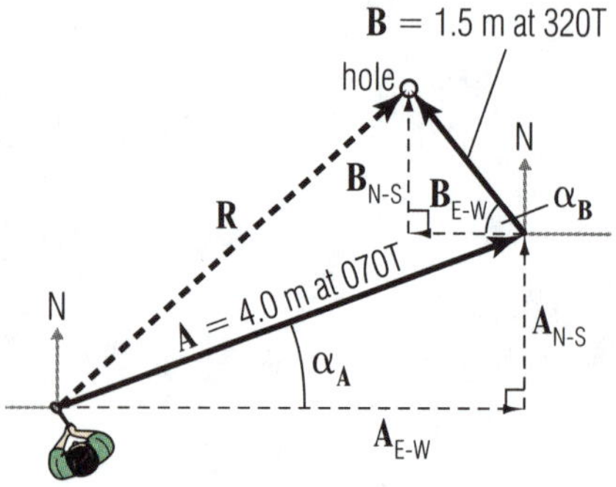

4-29 Find the resultant mathematically.

Solution:
If the car is considered the origin in a three-dimensional coordinate system, the *x*-axis represents the east-west direction, the *y*-axis represents the north-south direction, and the *z*-axis represents the vertical direction.

The final displacement is the vector sum of these three components:

$$\mathbf{d} = \mathbf{d}_x + \mathbf{d}_y + \mathbf{d}_z$$

The three scalar components are $d_x = -200.$ m, $d_y = -250.$ m, and $d_z = +450.$ m.

Examine Figure 4-28. The displacement of the rocket at its launch point on the ground is the vector sum of $\mathbf{d}_x$ and $\mathbf{d}_y$. The magnitude of this horizontal displacement is found using the Pythagorean theorem.

$$|\mathbf{d}_x + \mathbf{d}_y| = |\mathbf{d}_{x+y}| = \sqrt{d_x^2 + d_y^2} = \sqrt{(-200.\text{ m})^2 + (-250.\text{ m})^2}$$

$$|\mathbf{d}_{x+y}| \cong 32\underline{0}.1\text{ m}$$

Using this result as the base for the vertical displacement vector triangle, you can use the Pythagorean theorem again to find the magnitude of total displacement of the rocket, which is the distance you are looking for.

$$|\mathbf{d}_{x+y} + \mathbf{d}_z| = |\mathbf{d}| = \sqrt{d_{x+y}^2 + d_z^2} = \sqrt{(32\underline{0}.1\text{ m})^2 + (450.\text{ m})^2}$$

$$|\mathbf{d}| \cong 55\underline{2}.2\text{ m}$$

$$d \cong 552\text{ m}$$

4.9 Operations with Components

Components enable you to add or subtract vectors with great precision. Instead of using a protractor and a ruler, you use a calculator. Let's look at an example that allows us to put this tool to use.

EXAMPLE 4-9

Still Putting Around

Our impressive golfer attempts to sink a putt on the next hole. The ball rolls 4.0 m toward 070.0T and misses the hole (again). His second putt rolls 1.5 m toward 320.0T and drops into the cup. In what direction and how far should he have putted the ball to make the shot on the first attempt?

Solution: Refer to Figure 4-29.
Sketch the putt displacement vectors and label their magnitudes and vector angles. Develop the habit of sketching coordinate axes at the *tail* of each vector. Since the directions of the putts referred to compass directions, the axes should be north-south and east-west. Draw in the legs of the vector triangles, using each vector as a hypotenuse of its respective triangle. The legs must be parallel to the coordinate axes. Label the legs of the triangles to identify the components. The reference angles of the two displacement vectors are as follows:

$\alpha_A = 90.0T - 70.0T = 20.0°$

$\alpha_B = 320.0T - 270.0T = 50.0°$

The scalar components of **A** are $A_{N\text{-}S}$ and $A_{E\text{-}W}$:

$|A_{N\text{-}S}| = A \sin \alpha_{\mathbf{A}} = (4.0 \text{ m})\sin 20.0° \cong 1.\underline{3}6 \text{ m}$

$A_{N\text{-}S} \cong +1.\underline{3}6 \text{ m or } 1.\underline{3}6 \text{ m north}$

$|A_{E\text{-}W}| = A \cos \alpha_{\mathbf{A}} = (4.0 \text{ m})\cos 20.0° \cong 3.\underline{7}5 \text{ m}$

$A_{E\text{-}W} \cong +3.75 \text{ m or } 3.\underline{7}5 \text{ m east}$

The scalar components of **B** are $B_{N\text{-}S}$ and $B_{E\text{-}W}$:

$|B_{N\text{-}S}| = B \sin \alpha_{\mathbf{B}} = (1.5 \text{ m})\sin 50.0° \cong 1.\underline{1}4 \text{ m}$

$B_{N\text{-}S} \cong +1.\underline{1}4 \text{ m or } 1.\underline{1}4 \text{ m north}$

$|B_{E\text{-}W}| = B \cos \alpha_{\mathbf{B}} = (1.5 \text{ m})\cos 50.0° \cong 0.\underline{9}6 \text{ m}$

$B_{E\text{-}W} \cong -0.\underline{9}6 \text{ m or } 0.\underline{9}6 \text{ m west}$

The negative sign is used because the component $\mathbf{B}_{E\text{-}W}$ points west.

The resultant $\mathbf{R} = \mathbf{A} + \mathbf{B}$. But **R** also has components, so $\mathbf{R} = \mathbf{R}_{N\text{-}S} + \mathbf{R}_{E\text{-}W}$ (resultant vector components).

$\mathbf{R}_{N\text{-}S} = \mathbf{A}_{N\text{-}S} + \mathbf{B}_{N\text{-}S}$ and $\mathbf{R}_{E\text{-}W} = \mathbf{A}_{E\text{-}W} + \mathbf{B}_{E\text{-}W}$

$R_{N\text{-}S} = 1.\underline{3}6 \text{ m} + 1.\underline{1}4 \text{ m} = +2.\underline{5}0 \text{ m (or } 2.\underline{5}0 \text{ m north)}$

$R_{E\text{-}W} = 3.\underline{7}5 \text{ m} + (-0.\underline{9}6 \text{ m}) = +2.\underline{7}9 \text{ m (or } 2.\underline{7}9 \text{ m east)}$

From this information, the resultant that represents the putt that the golfer *wanted* to make can be found (see Figure 4-30).

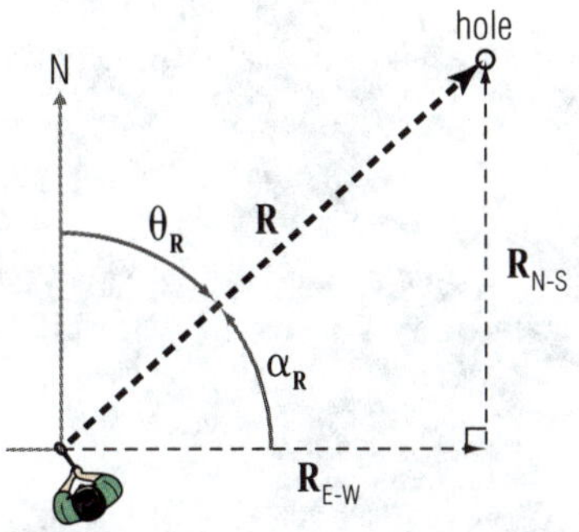

4-30 Adding resultant components

The magnitude of **R** is the distance of the putt. Use the resultant scalar components in the Pythagorean theorem:

$$R = \sqrt{R_{N\text{-}S}^2 + R_{E\text{-}W}^2} = \sqrt{(2.\underline{5}0 \text{ m})^2 + (2.\underline{7}9 \text{ m})^2}$$

$R \cong 3.\underline{7}4 \text{ m}$

$R \cong 3.7 \text{ m}$

Calculate the reference angle of the putt:

$$\alpha_{\mathbf{R}} = \tan^{-1}\left(\frac{\text{opposite}}{\text{adjacent}}\right)$$

$$\alpha_{\mathbf{R}} = \tan^{-1}\left(\frac{|R_{N\text{-}S}|}{|R_{E\text{-}W}|}\right) = \tan^{-1}\left(\frac{|+2.\underline{5}0 \cancel{\text{m}}|}{|+2.\underline{7}9 \cancel{\text{m}}|}\right)$$

Compute the reference angle to the same precision as the precision of any angles given in the problem.

$\alpha_{\mathbf{R}} \cong 41.\underline{8}6°$

$\alpha_{\mathbf{R}} \cong 41.9°$

The original putt directions were map directions given with respect to north. The direction of **R** is 41.9° north of east. To find the compass direction, subtract 41.9° from 090.0T:

$$\theta_{\mathbf{R}} = 048.1T$$

The required putt should have been 3.7 m in the direction of 048.1T.

Example 4-9 covers every step required for mathematically adding two vectors. Notice that you must convert vectors to scalar components in order to perform algebraic or trigonometric operations on them. At each step your solutions should show the calculations and the values of the variables.

Because vectors can describe displacements, velocities, accelerations, or forces, they can have a variety of different units. When you add or subtract two vectors, make sure that they have the same units.

The Good Samaritan and Saving Lives

Knowledge of the natural order can often enhance a person's attempts to show mercy toward others. The Good Samaritan's acts of kindness that Jesus commended actually required a working knowledge of the interaction of oil and wine. Because the Samaritan knew that wine and oil disinfect and protect wounds, he was able to effectively help a man in need.

By learning about the world around them, Christians are often better equipped to help others. As Christians learn how vectors can help save lives, they grow in their ability to show Christlike mercy. Even while learning how to mitigate the effects of hurricanes, Christians can hear the voice of their Lord: "Go, and do thou likewise" (Luke 10:37).

4.10 Using Vectors to Solve Problems

One of the most difficult problems with forecasting hurricane landfall is predicting the track of the storm. This uncertainty creates huge logistical problems for large areas threatened by the storm. People must be evacuated, buildings must be protected against damage, and disaster preparedness measures must be initiated.

How can meteorologists use vectors to improve hurricane track forecasting? For many years, images from US weather satellites have been used to observe the motions of hurricanes. The motion vector for the eye of the storm was used to predict the storm's track for a few hours after each photo was taken. But this information was not useful for long-term predictions. Analysts superimposed wind data, which was recorded by aircraft that flew through the hurricane itself, onto the satellite photos to better understand the storm's motions. But the data often arrived hours too late and had limited predictive usefulness.

In the 1990s, improvements in image processing algorithms allowed meteorologists to obtain motion vectors of all visible parts of the storm directly from satellite images. These atmospheric motion vectors (AMV) provide a much more comprehensive view of the storm motions in nearly real time. The next problem to solve involved using the AMV data in the storm track prediction computer models.

4-31 A weather satellite image of Hurricane Katrina with atmospheric motion vectors (AMV) superimposed

Since Hurricane Katrina, which devastated a large region in the southern United States in 2005, National Weather Service meteorologists have been experimenting with rapid data scans of satellite storm images—as frequently as 3 to 7 minutes apart—to acquire AMV data for numerical models of storm prediction. These computer models evaluate thousands of motion vectors and then calculate what forces must have been exerted on and by the storm in order to produce those vectors. The output of the models is a prediction of the direction and speed (another vector) that the storm will likely exhibit over a period of 24 to 120 hours. The rapid-scan experimental results have been very encouraging. As the accuracy of future hurricane track forecasts increases, evacuation efforts will begin earlier and become more effective. By better preparing for natural emergencies like hurricanes, we improve our ability to save human lives.

4C Objectives

After completing this section, I can

- ✓ determine the unknown parts of a right triangle by using trigonometric functions.
- ✓ identify the reference angle of a vector.
- ✓ resolve a vector into its scalar components.
- ✓ find the magnitude of a vector by adding its components.

4C Section Review

1. What property of similar triangles allows the development of trigonometric ratios for any acute angle in a right triangle?
2. What two geometric principles permit you to find an unknown angle or an unknown side in a right triangle?
3. Name (give the symbols for) the two-dimensional Cartesian component vectors of vector **F**.

DS4. What is an atmospheric motion vector (AMV)? What two pieces of information do AMVs contain?

DS5. How are AMVs used to protect human lives?

6. Solve the right triangles for the indicated sides or angles. Report angles to 0.1°.

a.

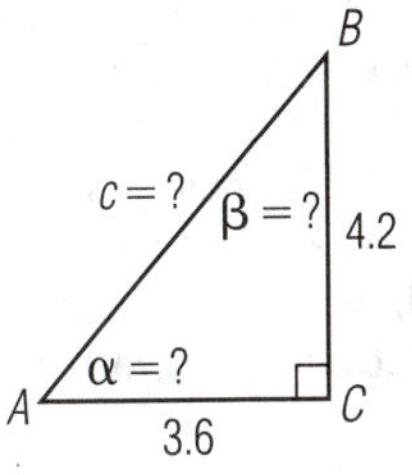

c.

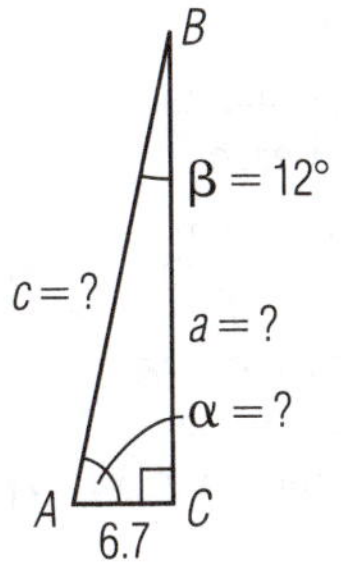

b.

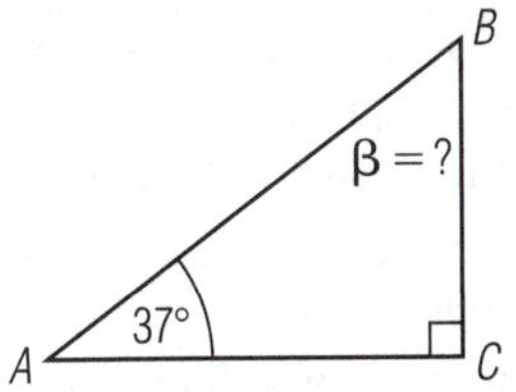

d.

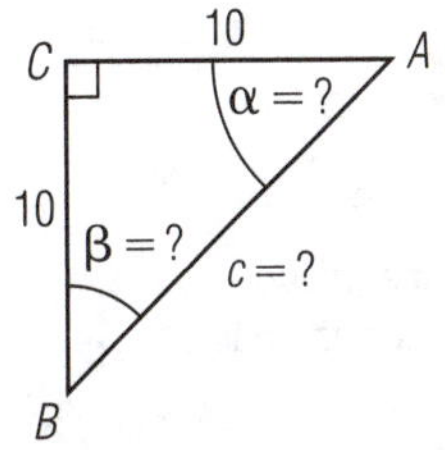

7. Determine the Cartesian vector components of a vector, **r**, that is 3.4 units long at an angle of 132°.

8. Using a ruler and a protractor, *graphically* add a vector that is 3.0 cm long at an angle of 230° to another vector that is 4.5 cm long at an angle of 0°.

9. *Mathematically* find $\mathbf{F}_1 + \mathbf{F}_2$ if $\mathbf{F}_1$ is 8.50 units at an angle of 30.0° and $\mathbf{F}_2$ is 10.25 units at an angle of 60.0°.

10. Find the difference vector $\mathbf{F}_1 - \mathbf{F}_2$ in Question 9 using mathematical techniques.

Chapter Review

In Terms of Physics

vector angle (θ)	74	tangent θ	80
reference direction	74	component vector	81
resultant	76	component (of a vector)	81
similar triangle	79	reference angle (α)	82
right triangle	79	scalar component	82
sine θ	80	position vector (**r**)	84
cosine θ	80		

Problem-Solving Strategies

4.1 (page 75) Vectors in a diagram can be transported from one point to another as long as their lengths and orientations remain unchanged.

4.2 (page 78) Quickly find the difference of two vectors (**A**−**B**) by placing them tail-to-tail and drawing the difference vector from the tip of the second (**B**) to the tip of the first (**A**).

4.3 (page 82) Make it a habit to quickly identify the reference angles for all vectors in a problem. Resolving vector components then becomes a straightforward process.

4.4 (page 83) When resolving a vector, always assign the signs of its scalar components based on the original vector's orientation in the problem sketch. Always start the solution to a problem with an accurate sketch.

Review Questions

1. What is the difference between a vector and a scalar?
2. State three physical quantities that can be represented by vectors. State three quantities that are scalars.
3. How will you show the difference between a vector symbol and its magnitude in your handwritten work? Give an example of each.
4. Write the compass directions for the four intercardinal points on a compass.
5. A model rocket is launched with an initial acceleration of 110 m/s^2 at an angle of 90.0° to the ground. Sketch a diagram of the rocket's initial acceleration vector (**a**) and appropriately label the sketch with all known information.
6. Draw vector **D**, shown in the diagram on your paper, and then draw the following scalar multiples of the vector.

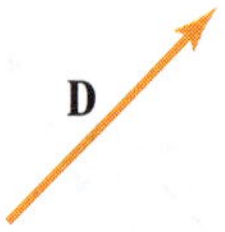

a. 4**D** **b.** $\frac{1}{3}$**D** **c.** −3**D** **d.** $-\frac{1}{2}$**D**

7. Consider two parallel vectors **M** and **N** that have magnitudes of 7 units and 5 units respectively. Show the directions that they must point relative to each other so that
   a. $|\mathbf{M}+\mathbf{N}|$ = 12 units.
   b. $|\mathbf{M}+\mathbf{N}|$ = 2 units.
   c. $|\mathbf{M}-\mathbf{N}|$ = 2 units.
   d. $|\mathbf{M}-\mathbf{N}|$ = 12 units.
8. Draw three vectors that, when added together graphically, give a zero resultant vector.
9. Given the components $\mathbf{T}_x$ and $\mathbf{T}_y$, write the formula you would use to find the magnitude of **T**.
10. Sketch the horizontal and vertical components of the following vectors and name them appropriately.

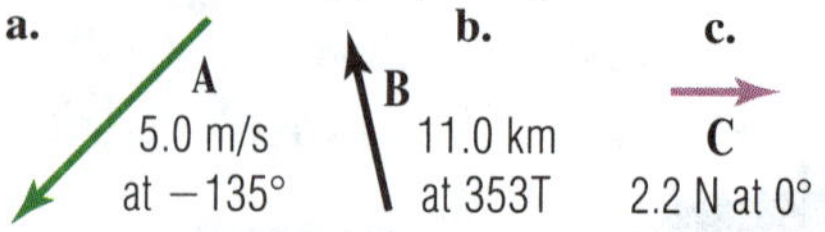

11. Write the names of the vertical and horizontal scalar components of the following vector quantities in symbols and indicate their signs:
    a. **v** = 1300 m/s at an angle of −25°
    b. **d** = 4.0 m bearing 250T
    c. **F** = 13 N at an angle of 78°
    d. **a** = 20 m/s^2 in a direction of 330T
12. Discuss the meanings of the symbols α and θ for specific angles in vector analysis. Under what conditions may they refer to the same angle?
13. When adding two vectors mathematically, why must you use the same coordinate system for finding the components of both vectors?
14. Discuss the difference between a scalar component and its magnitude.
15. A physics student is observing her horse (*Equus caballus*) in a rectangular field. Using herself as the origin, she notes the horse's initial and final position vectors ($\mathbf{r}_i$ and $\mathbf{r}_f$, respectively) over a 5-minute period. How could she indicate symbolically the total change of the horse's position during the time interval?

DS16. Standard atmospheric motion vector analysis was performed in the forecast of Hurricane Katrina's track and landfall. This analysis predicted a storm landfall within 24 km (15 mi) of the location of the actual event. Included in the prediction were estimates that some levees could be overtopped in New Orleans and that the potential for human suffering was great. Nearly ninety percent of the population of southeastern Louisiana was evacuated, yet 1464 people died. Besides accurate hurricane prediction, what other factors affect attempts to save lives?

✪17. Carefully draw the vector diagrams for each of the following situations and graphically determine the final displacement vector.
   a. A car drives 8.0 km west, then turns southwest for 3.0 km, and then drives south for an additional 4.0 km.
   b. An airplane flies 100.0 km north, then turns east and flies 100.0 km, then turns southwest and flies 141.4 km.

✪18. If the scalar components of **A** are A_x and A_y, write the formula for determining the magnitude of **A**.

✪19. Write the trigonometric formula for finding the reference angle of **A** in Question 18.

✪20. If $|D_x|$ = 3.5 cm and $|D_y|$ = 4.7 cm, calculate *D*. Show your work.

✪21. What are the *x*- and *y*-scalar components of the force vector **F**, which is 35.0 N oriented at 216.0°? Show your work.

✪22. In Parts a–d below, vectors have been written in terms of their scalar components. For each problem, sketch the reference coordinate system and the vector components in their correct orientation. Then sketch the resultant vector. Calculate the vector's magnitude, its reference angle (α), and its vector angle (θ). The following example problem shows what is expected in your written solutions.

EXAMPLE: A_x = 1.0 N; A_y = 2.0 N

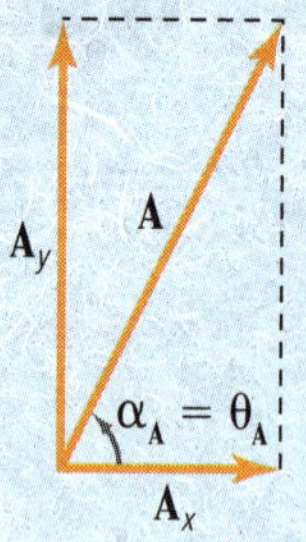

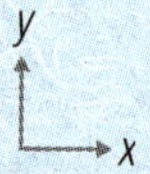

$$A = \sqrt{A_x^2 + A_y^2} = \sqrt{(1.0\text{ N})^2 + (2.0\text{ N})^2}$$

$$A \cong 2.\underline{2}3\text{ N} \cong 2.2\text{ N}$$

$$\alpha_A = \tan^{-1}\left(\frac{|A_y|}{|A_x|}\right) = \tan^{-1}\left(\frac{2.0\text{ N}}{1.0\text{ N}}\right) \cong 63.\underline{4}3° \cong 63.4°$$

A lies in Cartesian Quadrant I, so θ = α = 63.4°
A = 2.2 N at 63.4°

a. A_x = 3.0 m; A_y = −3.0 m
b. B_x = −2.00 N; B_y = 8.00 N
c. $C_{E\text{-}W}$ = −12.0 m/s; $C_{N\text{-}S}$ = −5.0 m/s
d. D_x = 0.0 m/s^2; D_y = 6.0 m/s^2

23. Resolve each vector into its component vectors.

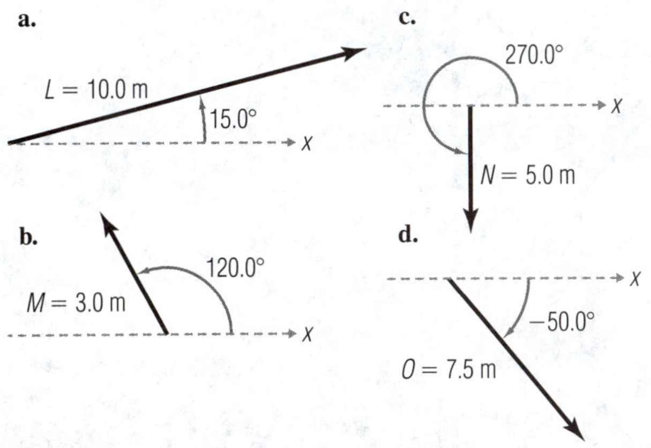

24. Use the components of **L**, **M**, **N**, and **O** in the previous problem to determine the following vectors:

a. **L + M**
b. **M + N**
c. **N + O**
d. **L − N**
e. **M − O**
f. **L + M + N + O**

25. A midfielder on the soccer team kicks the ball 20.0 m NW to his striker, who immediately kicks it 10.0 m NE into the goal. Describe the compass direction and distance that the midfielder would have had to kick the ball in order to place it in exactly the same point in the goal. Show your work.

26. Are the following properties true for vector addition? Support your answers with illustrative sketches. For each true relationship, give the name of the property of equality represented by the operation.

a. **A + B = B + A**
b. **A − B = B − A**
c. **(A + B) + C = A + (B + C)**
d. *n*(**A + B**) = *n***A** + *n***B**

27. A common housefly (*Musca domestica*) in a top corner of a 7.0 m × 10.0 m × 3.5 m room walks to the lowest opposite corner of the room.

a. What will be the magnitude of its net displacement after its epic trip?
b. If the fly walks only along the edges of the room, what is the shortest distance it must walk?
c. If it is not restricted to the edges, what is the distance of the most direct route the fly must walk to reach its destination?
d. Name the surfaces that the fly crosses in Part c.

These Formula race cars move in two dimensions on a flat track.

Motion in a Plane 5

Dominion Science Problem

Humanitarian Aid

Areas inhabited by people with the greatest needs are often the most inaccessible. Populated locations devastated and isolated by war or large-scale natural disasters quickly become uninhabitable. Homes and other shelters may be severely damaged or nonexistent. Utility networks may be destroyed or inoperable, resulting in the loss of electricity, water supplies, and sewage disposal systems. Roads and bridges may be impassable, cutting off the area from the outside, and airfields often become prime targets during wartime. The government's authority and services may dissolve, and the remaining local medical establishments may be overwhelmed with patients. In short, the whole fabric of civilized living unravels.

5-1 People in devastated areas need urgent care.

People surviving in such circumstances are in dire straits, urgently needing water, food, shelter, and medical care. The lack of transportation and communication amplifies these problems. How can humanitarian aid be delivered to such isolated, needy people?

5A KINEMATICS OF TWO-DIMENSIONAL MOTION

5.1 Position Vectors in Two Dimensions

Analyzing position and motion in two dimensions is only slightly more complicated than analyzing one-dimensional motion. Positions, displacements, velocities, and accelerations in two dimensions are all vector quantities that can be resolved into their components. You then can analyze the components using the equations of motion derived in Chapter 3.

Problem-Solving Strategy 5.1

The orientation of coordinate axes depends on the kind of motion being analyzed. When solving two-dimensional problems, it is usually best to align one of the axes parallel to the straight-line motion or to a force causing a change in motion.

In straight-line motion, you located an object by its position on a one-dimensional number line using coordinates. In two-dimensional motion, position is determined using the rectangular Cartesian coordinate system. Conventionally, the x-axis is horizontal and the y-axis is vertical, but any orientation and labeling is permissible as long as positions, directions, and angles are measured consistently within a given problem.

Position in two dimensions is determined by the position vector (**r**). The tail of the position vector is at the origin of the coordinate system, and the head is at the object's location. Figure 5-2a on page 94 shows the path of a fly on a window pane. At time t_1, the fly is at the tip of $\mathbf{r}_1$; at time t_2, the fly is at the tip of $\mathbf{r}_2$; at time t_3, the fly is at the tip of $\mathbf{r}_3$. The origin is at the lower left corner of the window.

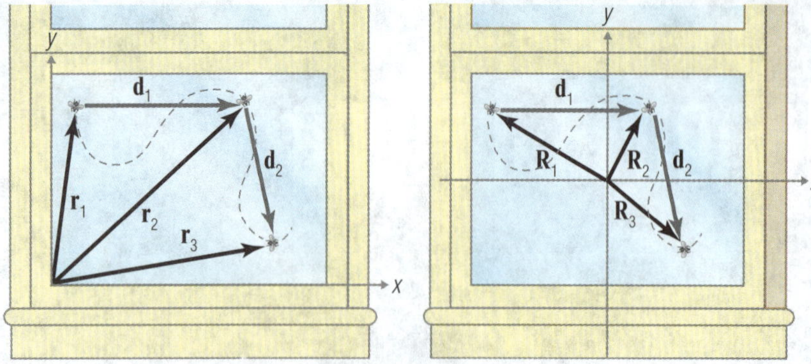

5-2 Position vectors showing the path of the fly on a window

Figure 5-2b shows the same path of the fly, but now the origin is at the center of the window. Although the positions of the fly are identical, the position vectors are different when measured from the different reference points. A **frame of reference** is simply the coordinate system within which motion is measured or observed. There are an infinite number of reference frames, and, as far as we can tell, there is no fixed "center of the universe" frame of reference from which all motion may be measured.

A **frame of reference** or **reference frame** is a coordinate system within which motion is measured. A frame of reference may be stationary, or, if it is moving, it can either have constant velocity or be accelerating.

Two-dimensional displacement:

$$\mathbf{d} = \mathbf{r}_2 - \mathbf{r}_1$$

Frame of Reference and the Christian Worldview

The concept of a frame of reference is important to the Christian worldview. The Bible is the Christian's frame of reference, the "coordinate system" from which all of life is to be viewed and evaluated.

5.2 Displacement in Two Dimensions

The displacement ($\mathbf{d}$) of an object is its change in position represented by the difference vector $\Delta\mathbf{r}$. The displacement for an interval of time is

$$\mathbf{d} = \Delta\mathbf{r} = \mathbf{r}_2 - \mathbf{r}_1,$$

where $\mathbf{r}_2$ is the position vector at the end of an interval and $\mathbf{r}_1$ is the initial position vector. For instance, the fly's displacement ($\mathbf{d}_1$) for the interval t_1 to t_2 in Figure 5-2 is

$$\mathbf{d}_1 = \mathbf{r}_2 - \mathbf{r}_1$$

for Figure 5-2a, or

$$\mathbf{d}_1 = \mathbf{R}_2 - \mathbf{R}_1$$

for Figure 5-2b.

The displacement is the same, no matter which reference frame is used.

EXAMPLE 5-1

Jogging Around

A beginning jogger starts from home and jogs four blocks west to corner *A*, turns and jogs three blocks south to corner *B*, and then turns east and jogs for another block before stopping to rest at point *R*. Using the jogger's home as the origin of the coordinate system, draw the jogger's complete path and the position vectors to the corners *A* and *B*, and position *R*.

a. What is the jogger's displacement from corner *A* to corner *B*?

b. What is his displacement from home to the resting spot?

c. What is the total distance that he jogged?

Use the unit "block" for all answers.

Solution: Refer to Figure 5-3.

a. $\mathbf{d}_2 = \mathbf{r}_B - \mathbf{r}_A$

$\mathbf{d}_2$ = 3 blocks south

b. $\mathbf{d}_R = \mathbf{r}_R - \mathbf{0} = \mathbf{r}_R$ (The **zero vector (0)** is any vector with zero magnitude. The position vector of the origin is **0**.)

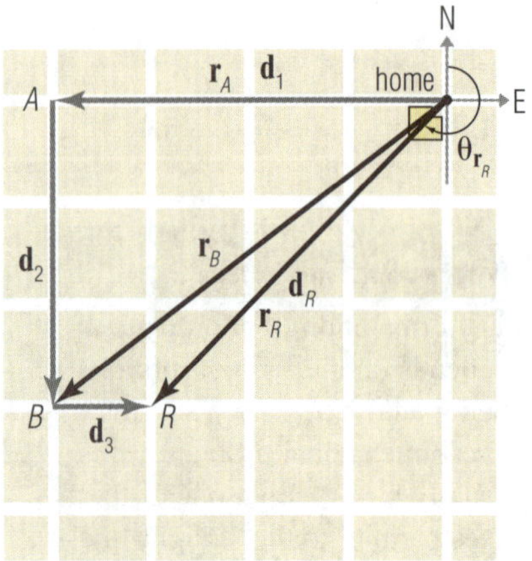

5-3 Position vectors of the jogger

The zero vector (0) is a vector with a magnitude of zero. It is required when determining displacements from the origin.

Determine the scalar components of r_R:

$r_{R_{E\text{-}W}} = -3$ blocks (west)

$r_{R_{N\text{-}S}} = -3$ blocks (south)

$$r_R = \sqrt{(-3 \text{ blocks})^2 + (-3 \text{ blocks})^2} = \sqrt{18 \text{ blocks}^2}$$

$$r_R = 3\sqrt{2} \text{ blocks} \cong \underline{4}.2 \text{ blocks}$$

$$\alpha_R = \tan^{-1}\left(\frac{|r_{R_{N\text{-}S}}|}{|r_{R_{E\text{-}W}}|}\right) = \tan^{-1}\left(\frac{|-3 \cancel{\text{blocks}}|}{|-3 \cancel{\text{blocks}}|}\right) = 45°$$

Since $\mathbf{r}_R$ is in the southwest quadrant, $\theta_R = 270\text{T} - \alpha_R$.

$\theta_R = 270\text{T} - 45° = 225\text{T}$ (southwest)

The rest position is about 4 blocks southwest.

c. Total distance jogged, d_{total}:

$d_{\text{total}} = d_1 + d_2 + d_3$

$d_{\text{total}} = (4 \text{ blocks} + 3 \text{ blocks} + 1 \text{ block})$

$d_{\text{total}} = 8$ blocks

5.3 Velocity and Speed in Two Dimensions

Average vector velocity and average speed between two positions in the Cartesian plane are determined just as they are in one dimension. Use vector notation when describing average vector velocity in two dimensions.

$$\text{Average Velocity:} \quad \overline{\mathbf{v}} = \frac{\Delta \mathbf{r}}{\Delta t} = \frac{\mathbf{d}}{\Delta t}$$

$$\text{Average Speed:} \quad \overline{v} = \frac{s}{\Delta t}$$

The *instantaneous velocity vector*, which is the velocity of an object at any given moment, always points in the direction that the object is moving at that instant in time. Figure 5-4 shows the instantaneous velocity of a car traveling around a curve. At time t_1 the vector points east, but at time t_2 the vector points north. In two- or three-dimensional motion, direction is indicated by the vector angles.

The instantaneous speed is equal to the magnitude of the instantaneous velocity, $v = |\mathbf{v}|$. For instance, in Figure 5-4, if $|\mathbf{v}_1|$ equals $|\mathbf{v}_2|$, and $|\mathbf{v}|$ remains constant through the curve, the speed of the car is constant. Depending on the path an object travels, the average speed ($\overline{v}$) and the magnitude of the average velocity ($\overline{\mathbf{v}}$) may be quite different. These quantities are equal only if $s = |\mathbf{d}|$.

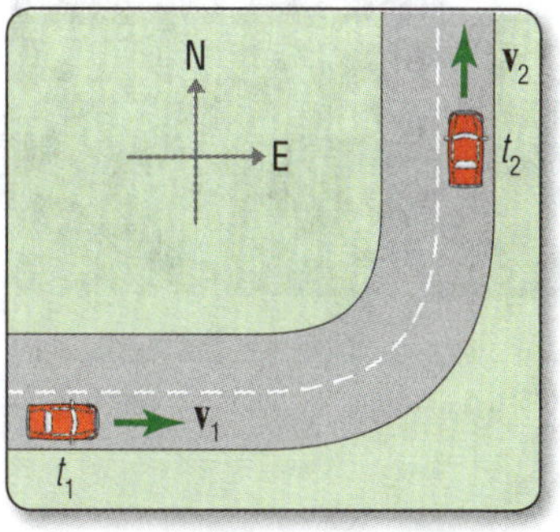

5-4 A car rounding a curve has a different instantaneous velocity at each point.

EXAMPLE 5-2

Swimming Across the River

A boy swims due north across a river that flows due east at a speed of 4.0 km/h (v_r). He swims at a speed of 3.0 km/h (v_b), heading directly across the river. His father is standing and watching from the north bank. What is the boy's velocity with respect to his father in km/h?

Solution:

Refer to Figure 5-5a. Orient the rectangular coordinate system east-west and north-south.

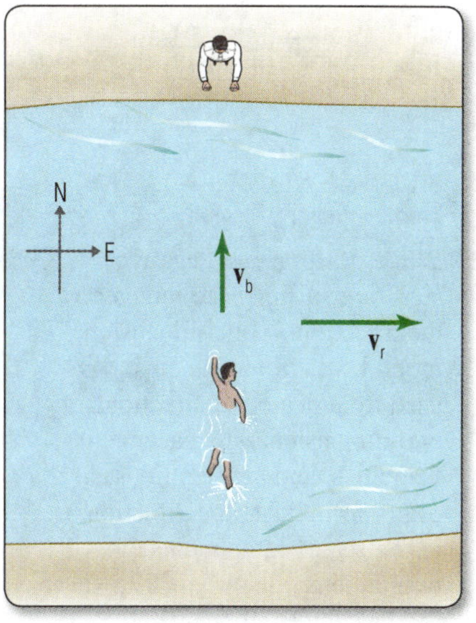

5-5a The boy swims across the river in the direction of his father.

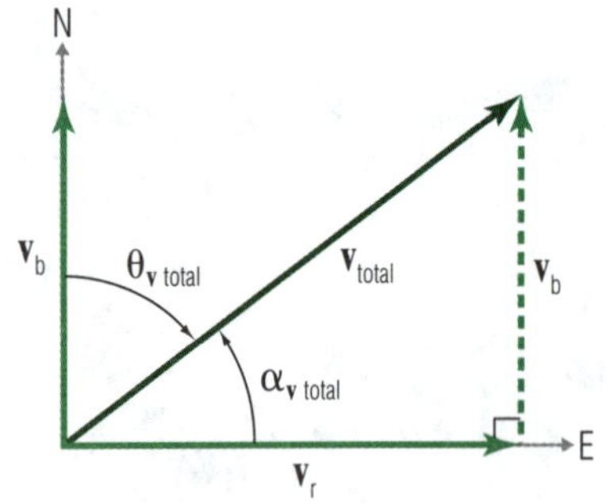

5-5b Vector diagram of the boy's component and resultant velocities

The boy's total velocity with respect to his father is the vector sum of his swimming velocity and the river's velocity:

$$\mathbf{v}_{total} = \mathbf{v}_b + \mathbf{v}_r$$

A vector diagram of the situation is shown in Figure 5-5b. Since $\mathbf{v}_b$ and $\mathbf{v}_r$ are parallel to the N-S and E-W axes respectively, they are the vector components of the boy's total velocity. Therefore, you can use scalar components to calculate the magnitude and direction of the total velocity vector.

Total speed:

$$v_{total} = \sqrt{(v_{E-W})^2 + (v_{N-S})^2} = \sqrt{v_r^2 + v_b^2}$$

$$v_{total} = \sqrt{(+4.0\ \text{km/h})^2 + (+3.0\ \text{km/h})^2}$$

$$v_{total} = 5.0\ \text{km/h}$$

Boy's direction:

$$\alpha_{\mathbf{v}_{total}} = \tan^{-1}\left(\frac{|v_b|}{|v_r|}\right) = \tan^{-1}\left(\frac{|+3.0\ \text{km/h}|}{|+4.0\ \text{km/h}|}\right)$$

$$\alpha_{\mathbf{v}_{total}} \cong 36.\underline{8}6° \cong 36.9°$$

The vector $\mathbf{v}_{total}$ is in the northeast quadrant, so $\theta_{\mathbf{v}_{total}} = 090\text{T} - \alpha_{\mathbf{v}_{total}}$.

$$\theta_{\mathbf{v}_{total}} = 090\text{T} - 36.9° = 053.1\text{T}$$

The father sees his son moving at 5.0 km/h in the direction 053.1T.

5-6 Acceleration in a straight line. The velocity of the golf ball changes in magnitude but not in direction.

5.4 Acceleration in Two Dimensions

A vector can change in any of three ways: magnitude only, direction only, or both magnitude and direction. Most real motion involves the third kind of change. Recall from Chapter 3 that acceleration is the rate of change of the velocity vector ($\Delta\mathbf{v}$). The **average acceleration** vector ($\bar{\mathbf{a}}$) is equal to the velocity difference divided by the time interval of the velocity change,

$$\bar{\mathbf{a}} = \frac{\mathbf{v}_2 - \mathbf{v}_1}{\Delta t} = \frac{\Delta\mathbf{v}}{\Delta t}. \tag{5.1}$$

Average Acceleration Vector:

$$\bar{\mathbf{a}} = \frac{\mathbf{v}_2 - \mathbf{v}_1}{\Delta t}$$

The direction of the average acceleration is always the same direction as the velocity difference vector, regardless of the direction of motion. For example, as a car slows to a stop at a traffic light, the magnitude of its final velocity is less than the magnitude of its initial velocity, so $\Delta\mathbf{v}$ points in the opposite direction from the car's motion. Therefore, as the car slows, $\bar{\mathbf{a}}$ points in the opposite direction from $\mathbf{v}$. When a moving object changes *direction*, its acceleration vector is always at an angle to its path. As with velocity, acceleration direction is indicated by a vector angle.

The *instantaneous acceleration* of an object is its acceleration at a particular moment. Its vector also points in the direction of the instantaneous velocity difference vector. In two-dimensional motion, determining the instantaneous change in velocity can be fairly complicated without the use of calculus. We will limit the discussion of acceleration to problems where either direction or speed vary, but not both. As with one-dimensional acceleration, if the acceleration is uniform, then the average and instantaneous accelerations are the same.

Problem-Solving Strategy 5.2

While positive and negative signs are sufficient to indicate the direction of one-dimensional motion, vector angles must be used to indicate the directions of two-dimensional displacements, velocities, and accelerations. Signs *do* indicate the directions of the scalar components of two-dimensional motion, however, because the components are one-dimensional vectors.

EXAMPLE 5-3

Taking the Curve

A car heading due east at 90.0 km/h enters a curve in the road. The curve ends with the car heading due north at 90.0 km/h. Determine the car's average acceleration vector ($\bar{\mathbf{a}}$) if the turn took 5.0 s.

Solution:

A diagram of the car on the curve is given in Figure 5-7a. Note the orientation of the initial and final velocity vectors.

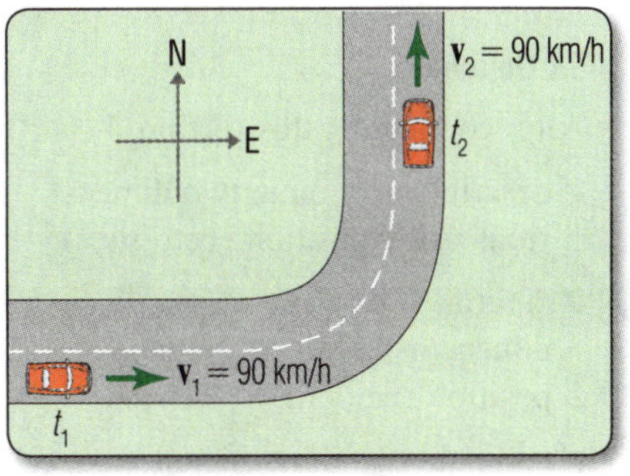

5-7a A car accelerates on a curve.

You know that $\mathbf{v}_1 = 90.0$ km/h east,

$\mathbf{v}_2 = 90.0$ km/h north, and $\Delta t = 5.0$ s.

Determine the velocity difference vector

$\Delta\mathbf{v} = \mathbf{v}_2 - \mathbf{v}_1 = \mathbf{v}_2 + (-\mathbf{v}_1)$

(see Figure 5-7b).

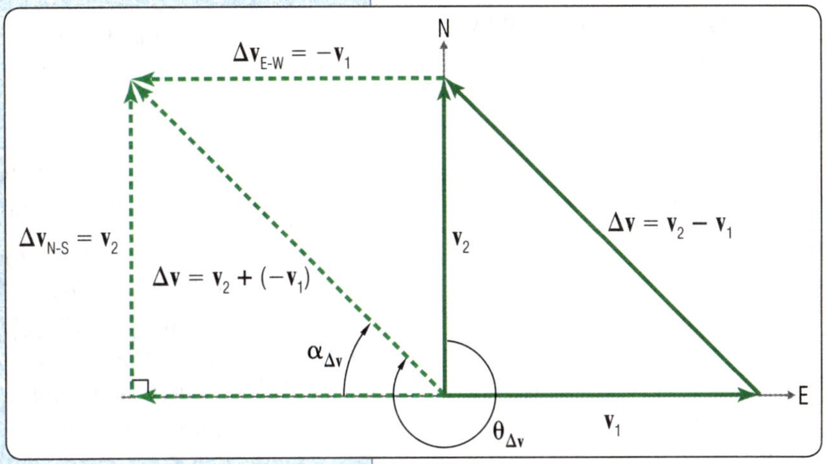

5-7b Vector diagram for the car on a curve

The scalar components of the velocity difference vector are as follows:

$\Delta v_{N\text{-}S} = v_{2_{N\text{-}S}} - v_{1_{N\text{-}S}}$

$\Delta v_{N\text{-}S} = (+90.0 \text{ km/h}) - (0.0 \text{ km/h})$

$\Delta v_{N\text{-}S} = +90.0 \text{ km/h}$

$\Delta v_{E\text{-}W} = v_{2_{E\text{-}W}} - v_{1_{E\text{-}W}}$

$\Delta v_{E\text{-}W} = (0.0 \text{ km/h}) - (+90.0 \text{ km/h})$

$\Delta v_{E\text{-}W} = -90.0 \text{ km/h}$

$$\Delta v = \sqrt{(\Delta v_{E\text{-}W})^2 + (\Delta v_{N\text{-}S})^2}$$

$$\Delta v = \sqrt{(-90.0 \text{ km/h})^2 + (+90.0 \text{ km/h})^2}$$

$$\Delta v \cong 127.\underline{2}7 \text{ km/h}$$

The magnitude of average acceleration is determined by substituting the known information into the magnitude form of Equation 5.1:

$$\bar{a} = \frac{\Delta v}{\Delta t} = \frac{127.\underline{2}7 \text{ km/h}}{5.0 \text{ s}} \left| \frac{1000 \text{ m}}{1 \text{ km}} \right| \frac{1 \text{ h}}{3600 \text{ s}}$$

$$\bar{a} \cong 7.\underline{0}7 \text{ m/s}^2$$

$$\bar{a} \cong 7.1 \text{ m/s}^2$$

Determine the orientation of $\bar{\mathbf{a}}$:

$$\alpha_{\Delta\mathbf{v}} = \tan^{-1}\left(\frac{|\Delta v_{N\text{-}S}|}{|\Delta v_{E\text{-}W}|}\right) = \tan^{-1}\left(\frac{|+90.0 \text{ km/h}|}{|-90.0 \text{ km/h}|}\right) = 45.0°$$

$\bar{\mathbf{a}}$ is in the northwest quadrant, so $\theta_{\bar{\mathbf{a}}} = 270.0\text{T} + \alpha_{\Delta\mathbf{v}}$

$\theta_{\bar{\mathbf{a}}} = 270.0\text{T} + 45.0° = 315.0\text{T}$ (northwest)

The average acceleration of the car during the 5.0 s interval is about 7.1 m/s² toward the northwest.

5A Objectives

After completing this section, I can

✓ describe position in two dimensions using position vectors.

✓ determine displacement in two dimensions as the difference of position vectors.

✓ compute two-dimensional velocity and speed.

✓ calculate two-dimensional acceleration as the difference of two velocities.

5A Section Review

1. In physics, two-dimensional kinematic position is most efficiently represented by what quantity?
2. Describe how displacement is calculated in two-dimensional motion. Include the formula you would use in your answer.
3. Discuss the difference between the quantities represented by the symbols $\mathbf{v}$, $\bar{\mathbf{v}}$, $|\bar{\mathbf{v}}|$, and $\bar{v}$.
4. How is the *direction* of the average acceleration vector, $\bar{\mathbf{a}}$, determined if you are given the initial and final velocities?
5. ✪ A visitor to Washington, D.C. walks directly from the White House along Pennsylvania Avenue to Constitution Avenue. The distance is 2.00 km in the direction 110.T. He turns left on Constitution Avenue and walks due east for an additional 0.87 km. When he stops, the dome of the US Capitol Building is 0.32 km due south of his position. He walks to the Capitol Building and enters the Rotunda under the dome. What is his displacement from the White House?
6. ✪ Famous quarterback Fleet O. Foote is attempting a quarterback draw by running up the middle toward his goal line at 15 ft/s (4.6 m/s). A linebacker hits him squarely, and their combined final velocity is 12 ft/s (3.7 m/s) at an angle of +120.° from the quarterback's original direction. The time it took to change direction at impact was 0.20 s. Choose an appropriate horizontal coordinate system and calculate the average acceleration vector of the quarterback during the tackle.
7. ✪ An athlete sprints a circular 400.0 m track in 50.0 s. Assume he runs at a constant speed.
   a. What is his speed over the entire course?
   b. What is the magnitude of his instantaneous velocity at any point along the circuit?
   c. If he starts facing west and runs counterclockwise, what is his velocity halfway around the circle?
   d. What is the average acceleration for one lap of the track?

5B PROJECTIONS

5.5 Projectiles

Flip a coin into the air—heads or tails? Fire a fast tennis ball deep into your opponent's left corner with your racquet—30, love! A competition high diver performs a perfect one-and-a-half gainer before making a clean entry into the water. Mt. Etna on the island of Sicily erupts, tossing fiery bombs of molten rock into high arcs. Each of these describes projectile motion.

A **projectile** is any flying object that is given an initial velocity and is then influenced only by external forces, such as gravity. In other words, a projectile can be propelled for only the first part of its flight, if at all. Projectiles include those objects usually associated with weapons, such as bullets, bombs, and missiles, as well as any other object that falls or is launched through air or space.

5-8 The lava bombs of a volcano are ballistic projectiles.

A projectile's path is called its **trajectory**. The unpowered portion of a projectile's flight is called a ballistic trajectory. A ballistic projectile is assumed to have only gravitational force acting on it. For simplicity's sake, we will disregard the effects of air resistance unless otherwise noted. This is a reasonable assumption for relatively slow-moving, small, dense objects.

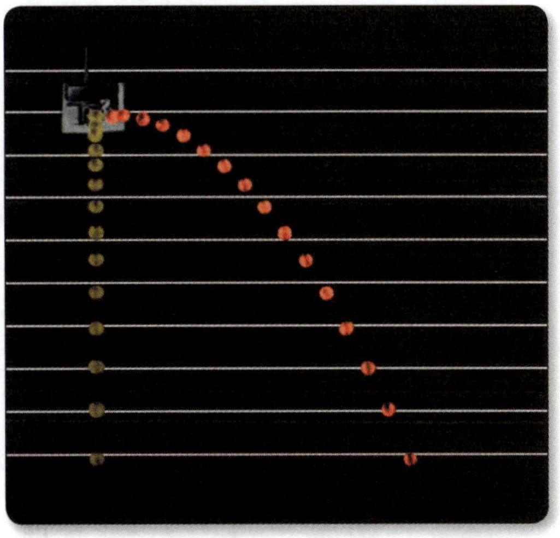

5-9 An experiment with golf balls demonstrates that horizontal motion and vertical motion are independent of each other.

5.6 Horizontal Projections

In Chapter 3 you studied the free fall of a dropped projectile. The path of motion was linear, and there was only one component of motion, parallel to the y-axis. A one-dimensional coordinate system was adequate for this kind of motion. However, if you kick a pebble off the top of a stairway, for example, you have another dimension to consider. As the pebble leaves the toe of your shoe, it is moving *horizontally*. In the total absence of gravity (ignoring air resistance), the pebble would continue at the same velocity indefinitely. However, because of the presence of gravity, the pebble also begins to fall *vertically*.

A **horizontal projection** is motion in which an object is initially propelled horizontally and then allowed to fall in a ballistic trajectory. The kinematics of the vertical and horizontal components of motion are completely separate, but they occur simultaneously to give a smooth, continuous path. Figure 5-9 compares the position of a dropped object with a second object projected horizontally at the same instant. You can see that at each time, the vertical positions of both objects in the stroboscopic photo are identical. This shows that the vertical component of motion does not depend on the horizontal component. The total velocity of a projectile at any time after launch is the vector sum of the horizontal and vertical velocity components.

The horizontal and vertical components of projectile kinematics (displacement, velocity, and acceleration) are treated as two separate one-dimensional problems to which we can apply the equations of motion learned in Chapter 3. Two quantities often of interest in horizontal projections are the time of flight, Δt, and the magnitude of the horizontal displacement, $|\mathbf{d}_x|$, called the **range (R)**.

Problem-Solving Strategy 5.3

Each component of motion can be analyzed completely independently from the other components using the equations of motion.

A **horizontal projection** is motion in which an object is initially propelled horizontally and then allowed to fall ballistically.

The **range (R)** of a projectile is the magnitude of the horizontal displacement of a projection.

The horizontal velocity of a projectile is assumed to be constant.

5.7 Projectile Motion: Horizontal Component

A projectile is ideally assumed to experience no horizontal acceleration. (This is not really a valid assumption because air resistance tends to slow projectiles significantly. However, for small, dense, slow-moving projectiles, it is a good approximation.) Zero acceleration in the horizontal direction means that the horizontal velocity is constant. The first equation of motion,

$$v_{2x} = v_{1x} + \bar{a}_x\Delta t,$$

becomes the horizontal scalar component equation

$$v_{2x} = v_{1x},$$

where v_{2x} and v_{1x} are the x-components of the projectile's velocity. Since the velocity's x-component is assumed to be constant, it will simply be called v_x.

The second equation of motion,

$$x_2 = x_1 + v_{1x}\Delta t + \tfrac{1}{2}\bar{a}_x(\Delta t)^2,$$

becomes the horizontal scalar component equation

$$x_2 = x_1 + v_x\Delta t,$$

where x_2 and x_1 are the x-components of the projectile's position vectors $\mathbf{r}_2$ and $\mathbf{r}_1$, respectively. This equation is often rearranged into its displacement form,

$$d_x = x_2 - x_1 = v_x\Delta t. \qquad (5.2)$$

The third equation of motion is meaningless when the acceleration is zero, since you would be dividing by zero.

5.8 Projectile Motion: Vertical Component

You learned in Chapter 4 that all objects experience a uniform downward acceleration due to gravity near the earth's surface. The scalar value of this acceleration is

$$g = -9.81 \text{ m/s}^2.$$

The initial vertical velocity of a horizontal projection is zero. The final vertical velocity of the projectile is due solely to the amount of time it has to fall. Choosing the positive direction to be upward and using the *y*-scalar components of vectors, the equations of motion take the forms listed below.

First Equation: $$v_{2y} = g_y\Delta t \quad (5.3)$$

Second Equation: $$d_y = \tfrac{1}{2}g_y(\Delta t)^2 \quad (5.4)$$

Third Equation: $$d_y = \frac{v_{2y}^2}{2g_y} \quad (5.5)$$

Problem-Solving Strategy 5.4

Do not learn a second set of motion equations for projectiles and free-fall problems. Use the equations in the forms presented in Chapter 3. Solve the equation for the unknown and then substitute the known scalar values, including their signs.

Problem-Solving Strategy 5.5

Remember when working with gravitational acceleration to substitute the *negative* scalar value of *g* for *a* in the equations of motion.

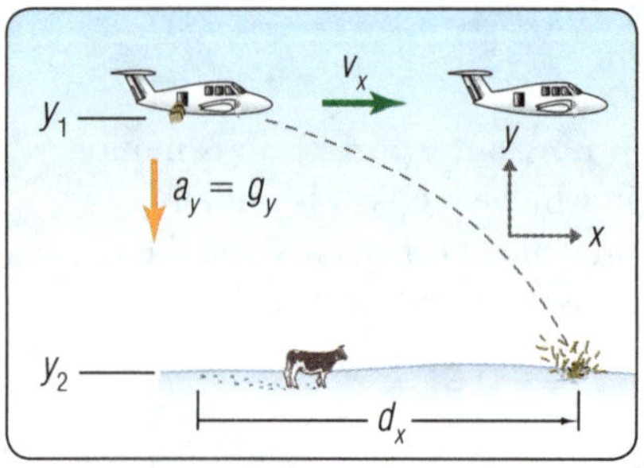

5-10 Aircraft dropping a bale of hay—disregard air resistance

EXAMPLE 5-4

Baling Out: A Horizontal Projection

A rancher in an airplane flying level at an altitude of 100. m and a speed of 190. km/h drops a bale of hay to hungry cattle stranded by a sudden snowstorm out on the range. If air resistance is negligible, how far in meters does the bale travel horizontally before it strikes the ground?

Solution:

Choose a Cartesian coordinate system with *y* positive upward and *x* positive in the direction of the plane's movement. Assume that the origin is at the position of the plane when the bale is released.

The horizontal distance that the bale traveled is the range, *R*, or the magnitude of the bale's horizontal displacement, $|d_x|$. You know the following information: $y_1 = 0$ m; $y_2 = -100.$ m (the ground); $v_{1y} = 0$ m/s; $v_{1x} = v_{2x} = v_x = +190.$ km/h; $g_y = -9.81$ m/s^2; $g_x = 0$ m/s.

From Equation 5.2, you know that

$$d_x = x_2 - x_1 = v_x\Delta t. \quad (1)$$

You know v_x, but what is Δt? The time that the bale takes to fall vertically to the ground is the time that it is able to travel horizontally, so its free-fall time must be calculated. Use the second equation of motion for this:

$$y_2 = y_1 + v_{1y}\Delta t + \tfrac{1}{2}g_y(\Delta t)^2 \quad (2)$$

The first two terms on the right side are zero, so Equation 2 reduces to

$$y_2 = \tfrac{1}{2}g_y(\Delta t)^2.$$

Solve for Δt:

$$\Delta t = \sqrt{\frac{2y_2}{g_y}} \quad (3)$$

Substitute the known values into Equation 3 to find the free-fall time:

$$\Delta t = \sqrt{\frac{2(-100.\ \cancel{\text{m}})}{-9.81\ \cancel{\text{m}}/\text{s}^2}}$$

$$\Delta t \cong 4.5\underline{15}\text{ s}$$

Substitute this time into Equation 1 to solve for range:

$$R = |d_x| = |v_x \Delta t| = |(190.\ \text{km/h})(4.5\underline{1}5\ \text{s})|$$

$$R = \frac{190.\ \cancel{\text{km}}}{\cancel{\text{h}}} \left| \frac{4.5\underline{1}5\ \cancel{\text{s}}}{} \right| \frac{1\ \cancel{\text{h}}}{3600\ \cancel{\text{s}}} \left| \frac{1000\ \text{m}}{1\ \cancel{\text{km}}} \right.$$

$$R \cong 23\underline{8}.2\ \text{m} \cong 238\ \text{m}$$

The bale travels horizontally about 238 m from the time it is dropped to the time it hits the ground.

Problem-Solving Strategy 5.6

You must convert all quantities having the same dimension to the same unit to take advantage of cancellation in the unit-analysis phase of problem solving.

5.9 Frames of Reference in Projections

To a person flying in the plane that dropped the hay in Example 5-4, it would look as if the hay were dropping straight down. A person on the ground, however, would say that the hay dropped in an arc. Their perceptions are different because they are comparing the motion to two different references. The person on the plane uses the plane as a reference with which to compare the movement of the hay bale. The hay is moving horizontally with the same speed as the plane, so it has zero horizontal velocity with respect to the plane. Since it does not move horizontally with respect to the plane, it appears to fall straight down due to the vertical component of its motion.

The person on the ground uses the ground as a reference with which to compare the bale's motion. The hay is moving horizontally with respect to the ground at a constant speed of 190. km/h. Therefore, during its fall it moves a considerable distance horizontally with respect to the ground. This is an example, as we discussed earlier, of the same motion being observed from different reference frames.

5-11 A B-2 bomber dropping free-fall bombs—projectiles

EXAMPLE 5-5

Fallen Rock Zone: A Horizontal Projection

A rock rolls down a talus slope and bounces horizontally off a ledge at the top of a vertical cliff. Its horizontal velocity as it becomes airborne is 10.0 m/s. It strikes the ground 30.0 m from the base of the cliff. (a) How high is the cliff? (b) What is the rock's speed just before it hits the ground?

Talus is the broken rock that accumulates at the bottom of cliffs. It is the result of mechanical weathering of the cliff.

Solution:

As usual, choose the coordinate system so that y is positive upward and x is positive away from the cliff. In this problem, let the origin be at ground level directly below the ledge (see Figure 5-12).

You know the following information: $d_x = +30.0$ m; $y_2 = 0$ m; $v_x = +10.0$ m/s; $g_y = -9.81$ m/s^2; and $g_x = 0$ m/s^2.

a. The vertical distance of fall determines the time of fall. The horizontal distance covered is a function of the same time interval and the horizontal velocity. Therefore, you can rearrange Equation 5.2 to find the time of fall.

$$\Delta t = \frac{d_x}{v_x}$$

$$\Delta t = \frac{30.0\ \cancel{\text{m}}}{10.0\ \cancel{\text{m}}/\text{s}}$$

$$\Delta t = 3.00\ \text{s}$$

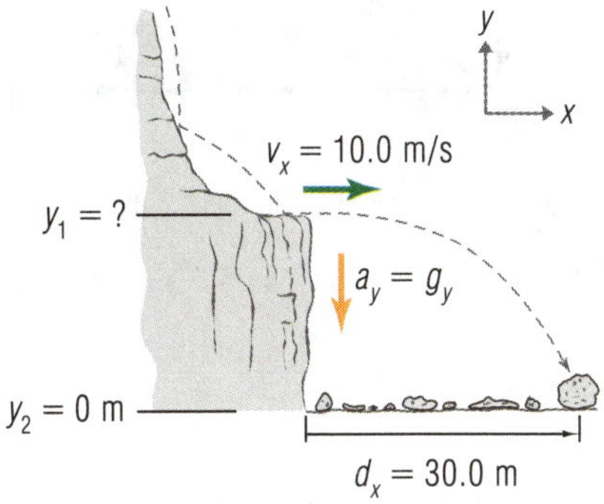

5-12 A rock falls from a cliff in a horizontal projection.

Substitute the time of fall into the second equation of motion in the vertical (Equation 5.4) to find the initial vertical position, y_1.

$$y_2 = y_1 + v_{1\,y}\Delta t + \tfrac{1}{2}g_y(\Delta t)^2$$

Both y_2 and $v_{1\,y}$ are zero, so the equation reduces as follows:

$$y_1 = -\tfrac{1}{2}g_y\,(\Delta t)^2$$

$$y_1 = -\tfrac{1}{2}(-9.81\ \text{m/s}^2)(3.00\ \text{s})^2$$

$$y_1 \cong +44.\underline{14}\ \text{m} \cong +44.1\ \text{m}$$

b. In order to find the total speed of the rock, you must find its horizontal and vertical velocity components and add them to find the resultant velocity for the instant it strikes the ground.

You already know the rock's horizontal velocity component, v_x. The vertical component can be calculated from the first equation of motion in the vertical (Equation 5.3) and the time of fall, Δt.

$$v_{2\,y} = v_{1\,y} + g_y\Delta t$$

$$v_{2\,y} = 0\ \text{m/s} + (-9.81\ \text{m/s}^2)(3.00\ \text{s})$$

$$v_{2\,y} \cong -29.\underline{4}3\ \text{m/s} \cong -29.4\ \text{m/s}$$

To determine the speed, the magnitude of the final velocity resultant, use the Pythagorean theorem.

$$v_2 = \sqrt{v_{2\,x}^2 + v_{2\,y}^2} = \sqrt{(+10.0\ \text{m/s})^2 + (-29.\underline{4}3\ \text{m/s})^2}$$

$$v_2 \cong 31.\underline{0}8\ \text{m/s} \cong 31.1\ \text{m/s}$$

The rock's speed just before it hits the ground is about 31.1 m/s.

5.10 Projection at an Angle

While some projectiles do begin with only horizontal velocity, most projectiles are launched at an angle. The horizontal and vertical accelerations are assumed to be the same as for horizontal projections, but the initial vertical velocity, $\mathbf{v}_{1\,y}$, is no longer zero. This difference makes analyzing the motion slightly more complicated.

For example, javelin throwers launch their javelins at an angle to the horizontal. Their objective is to throw the javelin as far, horizontally, as possible. In other words, they try to maximize the range of their throws.

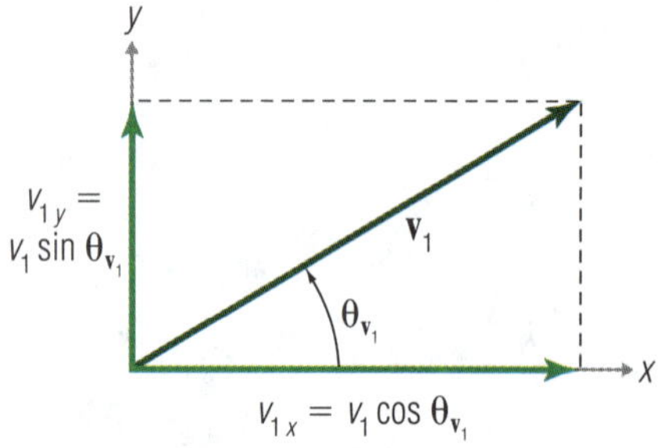

5-13 Resolving a velocity vector that is at an angle to the horizontal

Launching at an angle introduces an additional step—you must calculate the vector components of the initial velocity of the projectile in both the x and y directions. Figure 5-13 shows these vector components for a velocity vector making an angle θ with the positive x-axis. As the figure shows, the horizontal initial velocity component is

$$v_{1\,x} = v_1 \cos \theta_{\mathbf{v}_1},$$

while the vertical initial velocity component is

$$v_{1\,y} = v_1 \sin \theta_{\mathbf{v}_1}.$$

You must remember to substitute these expressions for the initial velocity terms in the equations of motion for the corresponding components. Once again, do not memorize a separate set of motion equations for projections at angles.

5.11 Graphing Projectile Motion

A projectile's vertical and horizontal displacement components can be easily calculated from the second or third equations of motion once the projectile's initial velocity components are identified. If displacement is measured from the launch point, the projectile's vertical and horizontal *position coordinates* can be determined from the position forms of the equations of motion. The trajectory of the projectile may be plotted on a graph from ordered pairs of vertical and horizontal position coordinates for various times after launch. Time-ordered coordinates are *ordered triples* (e.g., Δt, x, y).

Problem-Solving Strategy 5.7

When plotting the trajectory of a projectile, set up a table with the variables time, x-position, and y-position. Both x- and y-coordinates are time-dependent variables in a two-dimensional position plot.

EXAMPLE 5-6

Flight of an Arrow: Graphing Motion

An archer shoots an arrow at an angle of 45.0° to the horizontal with an initial speed of 41.6 m/s. (a) Ignoring air resistance, plot the vertical velocity of the arrow versus time every second for 6 s. (b) Then plot the height of the arrow versus range every second for 6 s.

Solution:

Choose a coordinate system with y positive upward and x positive to the right. Place the origin of the coordinate system at the point that the arrow leaves the bow string (see Figure 5-14).

a. The vertical velocity component is calculated from the first equation of motion.

$$v_{2y} = v_{1y} + g_y\Delta t$$

$$v_{2y} = v_1 \sin\theta + g_y\Delta t$$

$$v_{2y} = (41.6\text{ m/s})\sin 45.0° + (-9.81\text{ m/s}^2)\Delta t \qquad (1)$$

The vertical velocity component is calculated for each time interval from 0 to 6 seconds. The results are shown in the data table in the margin and plotted in Figure 5-15.

b. The position of the arrow is plotted using an ordered pair of x- and y-coordinates for each second of the flight. The x-coordinate is obtained from the second equation of motion in the horizontal. The y-coordinate is obtained from the second equation of motion in the vertical.

The x-coordinate:

$$x = x_1 + v_{1x}\Delta t + \tfrac{1}{2}a_x(\Delta t)^2$$

$$x = x_1 + (v_1\cos\theta)\Delta t + \tfrac{1}{2}g_x(\Delta t)^2$$

$$x = 0\text{ m} + (41.6\text{ m/s})(\cos 45.0°)\Delta t + \tfrac{1}{2}(0\text{ m/s}^2)(\Delta t)^2$$

$$x = (41.6\text{ m/s})(\cos 45.0°)\Delta t \qquad (2)$$

The y-coordinate:

$$y = y_1 + v_{1y}\Delta t + \tfrac{1}{2}g_y(\Delta t)^2$$

$$y = y_1 + (v_1\sin\theta)\Delta t + \tfrac{1}{2}g_y(\Delta t)^2$$

$$y = 0\text{ m} + (41.6\text{ m/s})(\sin 45.0°)\Delta t + \tfrac{1}{2}(-9.81\text{ m/s}^2)(\Delta t)^2$$

$$y = (41.6\text{ m/s})(\sin 45.0°)\Delta t + \tfrac{1}{2}(-9.81\text{ m/s}^2)(\Delta t)^2 \qquad (3)$$

The values of each coordinate are shown in the data table above. A smooth plot of these points produces the trajectory shown in Figure 5-16.

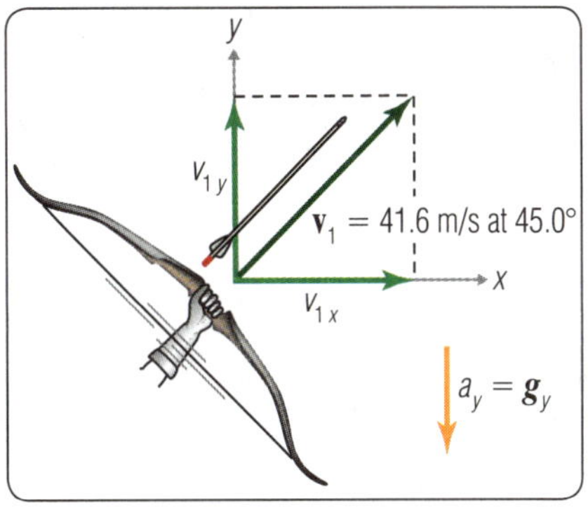

5-14 An archer shoots an arrow for maximum range.

Arrow Flight Data

Time (s)	v_y (m/s)	x (m)	y (m)
0	+29.4	0.0	0.0
1	+19.6	29.4	24.5
2	+9.8	58.8	39.2
3	0.0	88.2	44.1
4	−9.8	117.7	39.2
5	−19.6	147.1	24.5
6	−29.4	176.5	0.0

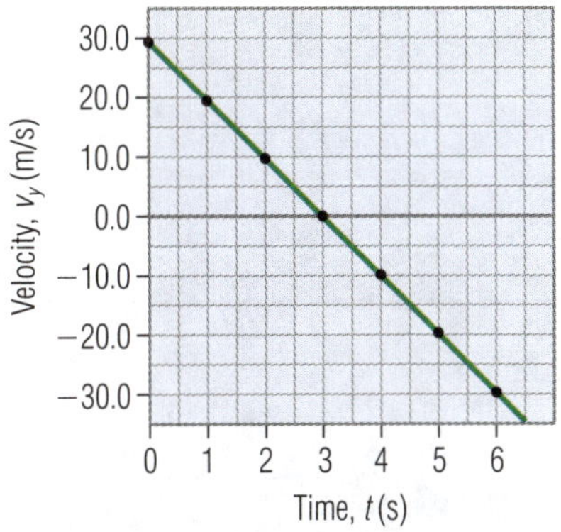

5-15 Arrow's vertical velocity versus time

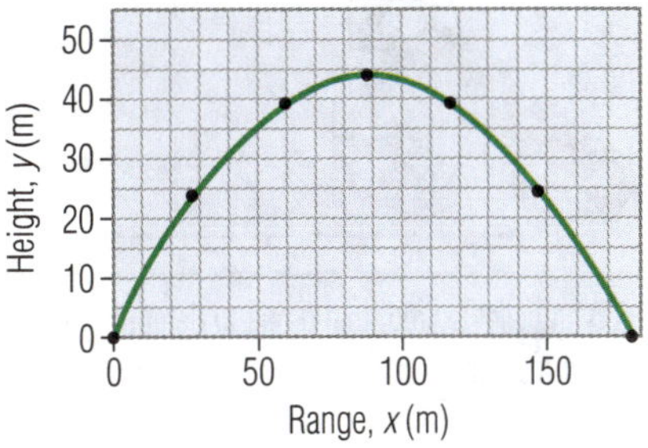

5-16 Arrow's height versus range

Notice in Example 5-6 that the arrow's vertical velocity is zero at 3 s and that the arrow reaches its maximum height at 3 s. In other words, the arrow stops going up and starts going down, which means its vertical motion must be stopped for an instant. In that same instant, the arrow reaches the peak of its flight. These facts apply to the flight of any projectile: at the peak of its trajectory, the vertical velocity is zero.

Ignoring air resistance and wind, the time it takes a projectile to go from a given height to its peak is the same time that it takes the projectile to fall from its peak to its original height. For example, 1 s before the peak of its flight, the arrow is at 39.2 m. Exactly 1 s after the peak, it has fallen back to a height of 39.2 m. In other words, the trajectory is symmetrical.

Another helpful fact about projectile flight is that the vertical speed is the same at corresponding heights. When the arrow rose to a height of 24.5 m, its vertical velocity was 19.6 m/s upward. When the arrow fell back to the height of 24.5 m, its vertical velocity was 19.6 m/s downward. The vertical speed was the same at the corresponding height.

What kind of curve is the graph of the ballistic trajectory in Figure 5-16? You can identify the category of the curve by the form of Equations 2 and 3 in the previous example. The independent variable, Δt, is squared in the second equation of motion. When the highest power of the independent variable is two, then the equation is called a *quadratic function* and its graph is a *parabola.*

5-17 The distance of a pass depends on time of flight and horizontal velocity.

5.12 Range Determination

In any projection, the range is determined by the time the projectile takes to fall to the ground (its time of flight) and its initial horizontal velocity. The time of flight is a function of the projectile's initial vertical velocity and its initial height above the ground. These factors are important regardless of whether you are throwing a winning pass in a football game or firing a cannon shell at a distant target.

EXAMPLE 5-7

Range of a Bullet: Projectile Motion

A target match rifle is clamped at a 30.0° angle to the horizontal so that the muzzle is 1.00 m above the ground. At t_1, when the bullet exits the barrel, it has a velocity of 370. m/s. (a) How long does it take the bullet to hit the ground? (b) What is the bullet's range?

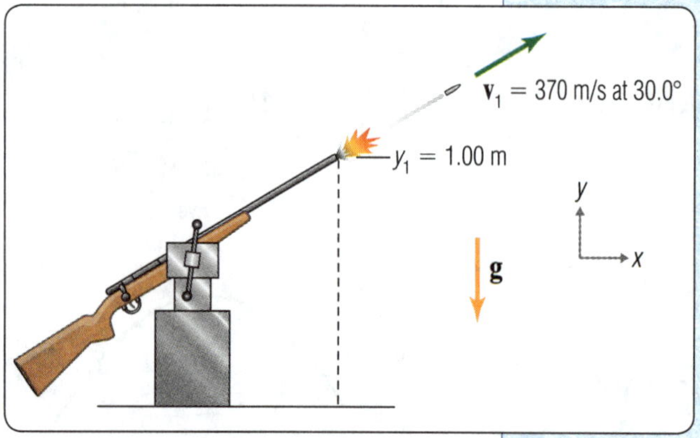

5-18 A bullet is fired from a rifle to determine its range at a specified angle.

Solution:

We select a Cartesian coordinate system oriented vertically and horizontally with x positive away from the rifle and y positive upward. Place the origin at ground level directly below the muzzle of the rifle.

a. As soon as the bullet exits the barrel, gravity accelerates the bullet downward. The time of flight is determined by how long the bullet takes to travel the vertical distance to the peak of the trajectory and then fall to the ground.

You know the following information: $\mathbf{v}_1$ = 370. m/s at 30.0°; y_1 = +1.00 m; y_2 = 0.00 m; and $g_y = -9.81\ \text{m/s}^2$.

You could work this problem in three steps by first calculating the time to the peak of the trajectory, then finding the height of the peak, and finally finding the time for the bullet to fall from the peak to the ground. This process requires using the first, third, and second equations of motion, in that order.

A more direct and elegant solution in one step involves calculating the elapsed time of the entire flight by solving the second equation of motion using the quadratic formula. Let's see how this works:
The standard form of a quadratic function is

$$f(x) = ax^2 + bx + c.$$

Since the final position of the bullet (y_2) is 0 m, you can rewrite the second equation of motion in the vertical in the form of a quadratic function and set it equal to zero:

$$y_2 = y_1 + v_{1y}\Delta t + \tfrac{1}{2}g_y(\Delta t)^2 = 0 \text{ m}$$

$$\tfrac{1}{2}g_y(\Delta t)^2 + v_{1y}\Delta t + y_1 = 0 \text{ m} \qquad (1)$$

Compared to the standard form, $\Delta t = x$, $\tfrac{1}{2}g_y = a$, $v_{1y} = b$, and $y_1 = c$.

The quadratic formula has the form

$$x = \frac{-b \pm \sqrt{b^2 - 4ac}}{2a}.$$

Solve Equation (1) for the unknown Δt using the quadratic formula:

$$\Delta t = \frac{-v_{1y} \pm \sqrt{v_{1y}^2 - 4(\tfrac{1}{2}g_y y_1)}}{2(\tfrac{1}{2}g_y)} = \frac{-v_{1y} \pm \sqrt{v_{1y}^2 - 2g_y y_1}}{g_y}$$

$$\Delta t = \frac{-v_1 \sin\theta \pm \sqrt{(v_1 \sin\theta)^2 - 2g_y y_1}}{g_y}$$

$$\Delta t = \frac{-(370.\text{ m/s})\sin 30.0° \pm \sqrt{[(370.\text{ m/s})\sin 30.0°]^2 - 2(-9.81\text{ m/s}^2)(+1.00\text{ m})}}{-9.81\text{ m/s}^2}$$

$\Delta t \cong 37.\underline{7}2$ s or $-0.005\,4\underline{0}4$ s

Time can only be positive, so the time of flight for the bullet is about 37.7 s.

b. The range (R) of the bullet is simply the product of the bullet's horizontal velocity and the time of flight:

$R = v_x\Delta t$

$R = (v_1 \cos\theta)\,\Delta t = (370.\text{ m/s})(\cos 30.0°)(37.\underline{7}2\text{ s})$

$R \cong 12\,\underline{0}86\text{ m} \cong 12\,100\text{ m}$

Problem-Solving Strategy 5.8

For any quadratic equation, $0 = ax^2 + bx + c$, you can solve for x using the formula

$$x = \frac{-b \pm \sqrt{b^2 - 4ac}}{2a}.$$

Problem-Solving Strategy 5.9

Use a quadratic solution of the second equation of motion to quickly obtain total time of flight for a projectile.

The range of the bullet calculated in Example 5-7 (more than 7.5 miles) is quite long compared to the typical range of a rifle bullet. This is because air resistance was ignored for this problem. Research has shown that a hunting rifle bullet can lose as much as 20% of its muzzle velocity in the first 100 m of flight, so the assumption is not realistic. Ignoring air resistance helps clarify the basic motion concepts without unnecessarily complicating the calculations.

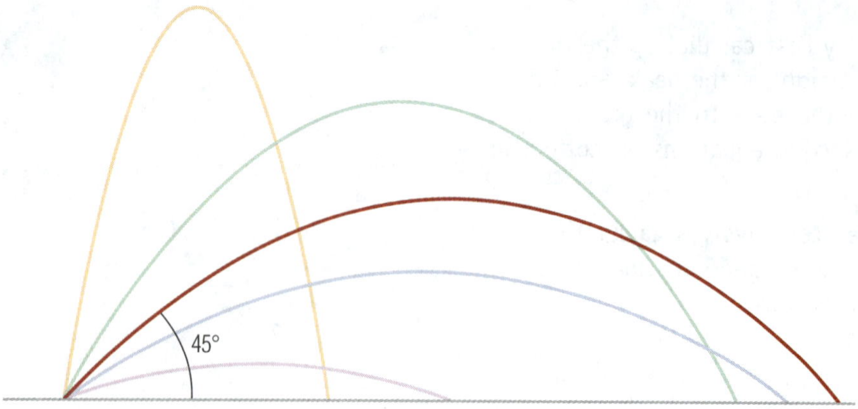

5-19 Range varies with launch angle for a fixed initial speed. Maximum range is obtained at an angle of 45°.

The maximum range that a projectile can attain for a given initial speed depends on its launch angle. You could determine this angle experimentally, or you could solve a series of problems with trial angles as in Example 5-7. Geometrically, the maximum theoretical range for a fixed launch speed is attained with a launch angle of 45°, assuming no air resistance. Real projectiles traveling at high speeds never reach their maximum theoretical range because of drag, but this angle still yields the maximum attainable range in the absence of wind or changes in air density. Range can also be increased by elevating the launch point. This is one reason armies try to capture and hold hills in battles.

5-20 Food and medical supplies being air dropped to refugees in Afghanistan

5.13 Using Projectile Motion to Solve Problems

Though warfare involves the use of ballistic projectiles like bombs and bullets, it can also use ballistic motion to provide humanitarian aid. As noted in the introduction, ground transportation is often completely disrupted in devastated regions. If there are no landing fields for winged aircraft, dropping aid packages from the air is the only option for providing assistance until ground access becomes available.

In past decades, many situations around the globe—whether war in Bosnia and Afghanistan, heavy rains in Somalia and Kenya, or a tsunami in the Indian ocean—have required aid from air drops. When large amounts of materials need to be delivered over a broad area, fixed-wing military cargo aircraft may be used. In one method of delivery, food and aid packages are bundled into refrigerator-sized reinforced cardboard boxes and rolled out of the backs of the planes at a relatively high altitude. This technique protects the aircraft from hostile ground fire and also allows the boxes to be torn apart by their speed through the air, scattering the packages over a large area. These individual aid packages follow a more-or-less ballistic path to the ground. If aid supplies are being delivered to an isolated village, the aircraft may pass at a low altitude over a clear area and drop the boxes from just a few tens of meters above the ground. The boxes cushion the fall and remain essentially intact, protecting their contents. Small parachutes are often used to stabilize the dropped materials. During the initial phases of the international response to the December 2004 tsunami in the Indian Ocean, Indian naval ships sent their helicopters into flooded coastal villages and dropped tents and food while hovering.

Though air-dropping supplies is not the most efficient method to provide aid to the victims of war and disaster, it often remains the only available solution during certain stages of emergency response because road travel may be hazardous or impossible. In addition, organizing an air drop may be much faster than organizing a truck convoy on the spur of the moment, especially if the air drops are planned as part of the overall combat operations. Saving lives and meeting human needs are important ways to love one's neighbor.

5B Section Review

1. Is a remote-controlled gas-powered model airplane a projectile? Explain your answer.
2. What kinematic quantities are assumed to be constant when analyzing horizontal projections?

3. At the same instant that a pebble is dropped from a tree house platform, another is kicked horizontally off the platform.
   a. Which pebble has the higher total velocity?
   b. Which one will hit the ground first?
4. Given a fixed initial projectile speed, v_1, what factors can be changed in order to maximize projectile range?
5. A marble resting near the edge of a 0.90 m high table is given an initial horizontal speed of 1.24 m/s. What will be its horizontal range from the table's edge when it strikes the floor?
DS 6. US Air Force C-17 Globemaster III aircraft were used to airdrop food to Afghan refugees in the early stages of the war against the Taliban in 2001. Calculate how far ahead of the drop zone a pilot would have needed to drop humanitarian aid packages if the delivery occurred at a speed of Mach 0.74 (263 m/s) and an altitude of 10 000 feet (3048 m) to avoid enemy fire. Assume the packages were released from their containers immediately and neglect air resistance.
7. The American bullfrog (*Rana catesbeiana*) can jump a distance of nearly 15 times its length! If a bullfrog starts on a horizontal log and leaps with a velocity of 4.40 m/s at an angle of 37.0° to the horizontal, what distance can it cover?
8. During the 1990 Olympics, an all-time record shot put was made—23.12 m (75 ft 10.2 in.) If the athlete released the shot at a height of 2.25 m with a velocity of 14.3 m/s at an angle of 35.0° to the ground, what was its vector velocity when it impacted the ground at the end of its flight?

5B Objectives

After completing this section, I can

- ✓ describe horizontal projections and the kinematic assumptions they involve.
- ✓ resolve the horizontal and vertical components of projectile motion.
- ✓ determine the velocity components of a system projected at an elevated angle.
- ✓ graph projectile motion.
- ✓ calculate the range of projectiles.
- ✓ explain how projectile motion can save lives.

FACETS of PHYSICS

Shot Put Release Angles

You have learned in this chapter that the range of a horizontal projection is maximized when the release angle is 45°. There are circumstances, however, in which this is not necessarily true.

The problem is biomechanics—how the muscles coordinate to produce the force that propels the shot on its trajectory. The classic relationship of range to projection angle assumes that the initial velocity is constant. However, in the shot put, the initial velocity is affected by other variables. The human body is structured to create more force in the horizontal direction than in the vertical direction. Even highly trained athletes can bench-press more weight than they can lift in a shoulder press. A larger release angle measured to the horizontal causes a greater percentage of the total force to come from the comparatively weaker vertically directed muscles. Thus, a larger release angle decreases the total force that the athlete can exert on the shot, resulting in a lower initial velocity and a shorter range of the shot.

The best release angle is different for each athlete, varying with his size, strength, and throwing technique, and may be anywhere from 26° to 38°.

Chapter Review

In Terms of Physics

Problem-Solving Strategies

5.1 (page 93) The orientation of coordinate axes depends on the kind of motion being analyzed. When solving two-dimensional problems, it is usually best to align one of the axes parallel to the straight-line motion or to a force causing a change in motion.

5.2 (page 96) While positive and negative signs are sufficient to indicate the direction of one-dimensional motion, vector angles must be used to indicate the directions of two-dimensional displacements, velocities, and accelerations. Signs *do* indicate the directions of the scalar components of two-dimensional motion, however, because the components are one-dimensional vectors.

5.3 (page 99) Each component of motion can be analyzed completely independently from the other components using the equations of motion.

5.4 (page 100) Do not learn a second set of motion equations for projectiles and free-fall problems. Use the equations in the forms presented in Chapter 3. Solve the equation for the unknown and then substitute the known scalar values, including their signs.

5.5 (page 100) Remember when working with gravitational acceleration to substitute the *negative* scalar value of g for a in the equations of motion.

5.6 (page 101) You must convert all quantities having the same dimension to the same unit to take advantage of cancellation in the unit-analysis phase of problem solving.

5.7 (page 103) When plotting the trajectory of a projectile, set up a table with the variables time, x-position, and y-position. Both x- and y-coordinates are time-dependent variables in a two-dimensional position plot.

5.8 (page 105) For any quadratic equation, $0 = ax^2 + bx + c$, you can solve for x using the formula

$$x = \frac{-b \pm \sqrt{b^2 - 4ac}}{2a}.$$

5.9 (page 105) Use a quadratic solution of the second equation of motion to quickly obtain total time of flight for a projectile.

Review Questions

1. An air traffic controller reports to the pilot of an inbound passenger aircraft the range and direction from his control tower to the plane. What kinematic quantity does this information represent?
2. A displacement in two-dimensional motion is the vector difference of what two quantities?
3. The direction of the average velocity vector ($\bar{\mathbf{v}}$) is the same as the direction of what related vector?
4. At what point does a rocket begin its ballistic trajectory?
5. What initial velocity component is always zero for a horizontal projection?
6. What simplifying assumption is made regarding the horizontal velocity of a projectile? Why is this not an accurate assumption?
7. **a.** What is the independent variable in a trajectory position graph?

   b. How are the x- and y-coordinates related to the independent variable?
8. What is the most direct method for determining the time of flight of a projectile, given its initial velocity and launch height?
9. At what angle to the ground should a batter hit the ball in order to maximize the range of the hit?
10. What kind of trajectory do droplets of water from a garden hose follow?
11. What is the magnitude of a projectile's vertical velocity at the peak of its trajectory?

    a. It equals the magnitude of the horizontal velocity.

    b. It is zero.

    c. It equals 9.81 m/s.

    d. None of the above.
12. A bullet is fired from a rifle and hits a target some distance away. Ignoring air resistance, compare the speed of the bullet upon impact with its original muzzle velocity

    a. when the target is the same height as the muzzle.

    b. when the target is 2 m higher than the muzzle.
13. Explain why a warplane dropping a bomb from a low altitude needs to turn away from its original path as soon as it releases the bomb.

True or False (14–23)

14. An acceleration is always a change in speed.
15. A position vector is constructed from an object's initial position to its final position.

16. Range is a vector.

17. In two-dimensional motion, the acceleration vector often points in a direction other than parallel to the direction of instantaneous velocity.

18. The horizontal and vertical components of an object's motion can be analyzed independently of each other.

19. A passenger on the Amtrak train *California Zephyr* and the driver of a car waiting for the train to pass through a crossing observe each other in the same reference frame.

20. The horizontal velocity of a projectile under the influence of only gravity is assumed to be constant.

21. The vertical speed of an object projected at a positive angle first increases and then decreases during its flight.

22. A flying spinning Frisbee is a ballistic projectile.

23. The graphs of quadratic functions are parabolas.

DS 24. The US aid effort during combat in Afghanistan was criticized by some humanitarian agencies because air drops are dangerous, costly, and inefficient. They argued that this kind of aid was simply for good media coverage. Even if these charges were valid, what are some reasons that air-dropping aid packages may have been the only alternative?

✪ 25. A motorized float in a Thanksgiving Day parade is traveling north at a speed of 16.0 km/h. Part of a "Pilgrim" woman's role in the thematic vignette is to walk across the float to call the men to dinner. She walks at a speed of 3.22 km/h east as the float passes the judge's booth, which is on the west side of the street. What is the woman's velocity with respect to the judge?

✪ 26. A car going 110. km/h southwest reaches a curve in the road. Assume that the curve is an arc of a circle. After rounding the curve, the car's velocity is 110. km/h west. The car takes 2.50 s to round the curve.

a. What is the car's acceleration?

b. What is the length of the car's path on the curve?

c. If the car instead maintained a steady speed of 100. km/h, how long would it take to round the curve?

DS ✪ 27. An aviation crew is flying an airplane to deliver supplies to a remote village. They intend to drop-deliver a cargo container attached to a parachute. The aircraft is flying at 165 km/h due east as the air crew drops the 100. kg crate. The parachute fails to open and the crate crashes to the ground 4.5 s after it was released.

a. What is the crate's vertical speed just before it hits the ground?

b. How high was the plane when the container was released?

c. The plane was 0.5 km west of a lake when it released the crate. Did the crate land in the lake?

✪ 28. An astronaut on the moon ($g_M = -1.63$ m/s^2) kicks a stone horizontally off a 45.0 m cliff with a speed of 2.50 m/s.

a. How far from the base of the cliff does the stone land?

b. How far from the base of the cliff would the stone have landed on the earth ($g_E = -9.81$ m/s^2, assuming no air resistance)?

✪ 29. A skydiver drops from an airplane that is 2000. m above the ground. An automatic timer opens his parachute exactly 10.0 s after he leaves the plane.

a. How far from the ground is he when his parachute is fully open, if it takes 1.0 s to open? (Assume no change in velocity until the chute is fully open.)

b. In the next 10.0 s, the skydiver's speed decreases uniformly to 10.0 m/s. How far does he fall during this time? (*Hint:* $d = \bar{v}\Delta t$)

c. After slowing, the skydiver's speed is constant until he lands on the ground. How long does the entire jump take?

✪ 30. A rifle's muzzle velocity is 400. m/s. The marksman's shoulder is 1.70 m high and the rifle is 1.50 m long.

a. What is the rifle's maximum possible range, neglecting air resistance?

b. Suppose the marksman fires at the broad side of a barn (height of the roof = 15.0 m) holding the rifle at a 10.0° angle above the horizontal. The ridge of the barn's roof is 100. m away from the end of the muzzle. Does the bullet strike the barn, or does it go over the barn?

c. What is the maximum initial vertical velocity component the bullet can have if it is to just graze the peak of the barn's roof at the midpoint of its trajectory? (Ignore the length of the rifle.)

d. At what angle would the marksman have to fire in order to have the initial vertical velocity component found in Part c?

✪ 31. A professional baseball player throws a ball as hard as he can at an angle of 30.0° above the horizontal. At the same instant, his 10-year-old son throws an identical ball directly upward at exactly half the speed of his father's pitch. Assume both balls are released at the same height. Which ball hits the ground first?

✪ 32. Two archers are trying to see who can obtain a greater range. The first archer shoots his arrow at a velocity of 65.0 m/s, 10.0° from the horizontal. The arrow is released 1.40 m above the ground. The other archer shoots his arrow from the same height at a velocity of 48.0 m/s, 45.0° from the horizontal. Who will win the contest?

✪ 33. A diver leaves the diving board, 10.0 m above the water, at a velocity of 12.0 m/s, 60.0° above the horizontal. If he takes 0.75 s to complete a somersault, how many full somersaults can the diver complete before hitting the water?

✪ 34. Draw a circle of 5.00 cm radius at the top of a piece of graph paper. The graphic scale of the drawing is 1 cm = 1 m. Assume a marble revolves at a uniform rate around the center of the circle once every 16.0 s. Directly below the circle, construct a velocity-time graph for the *x*-component of the marble's velocity vector. Choose at least 8 evenly spaced points around the circle to help define the velocity-time curve. Label the graph axes $\vec{v}_x$ and t. Below this graph, construct a velocity-time graph for the *y*-component of the marble's velocity vector using the same time-coordinate axis. Label the vertical axis $\vec{v}_y$.

a. Calculate the average scale speed of the marble.

b. What is the instantaneous speed of the marble at any time?

c. What is the average velocity for one complete revolution?

d. What kind of graphs do the two velocity components describe?

e. Discuss how the two graphs are related to each other. Include in your discussion answers to the following questions: When is one maximum compared to the other? When is one zero compared to the other? What is one curve doing when the other is decreasing?

FACETS of PHYSICS

Sir Isaac Newton (1642–1727)

Undoubtedly, the most influential person in the history of physics is Sir Isaac Newton. Besides formulating the law of universal gravitation and laying the foundation for modern dynamics, Newton experimented extensively with optics and shared in the discovery of the most profound mathematical system yet devised—calculus. This fundamental tool is used to describe the basic laws of science and engineering.

Three months before Newton was born, his father died. He was born prematurely and was so sickly that he was not expected to live past his first day. However, he survived and lived to the age of 85 years. As he grew up, he became so fascinated with books and experiments that he often forgot his chores. He later studied mathematics at Trinity College, Cambridge, a prestigious college in England. Though the college held to Aristotle's natural philosophy, Newton preferred the theories of more contemporary thinkers such as Galileo, Descartes, and Kepler. Even before he graduated, Newton had developed the general binomial theorem (a method of easily calculating the value of any simple binomial expression raised to any power) and had laid the foundation for what would become his system of calculus.

Shortly after receiving his degree in 1665, Newton returned home for two years because the Great Plague sweeping through England at that time threatened the university, resulting in its temporary closure. In his leisure time at the family farm, he improved his theory of light and color, worked out the details of calculus, and began to formulate his theory of gravitation.

Soon after he returned to Cambridge, at the young age of 26, Newton was elected to a professorship where he became a mathematics professor, a position that he held for much of his life. For several years he lectured on optics and investigated light spectra and refraction. From his studies, he concluded that refracting telescopes suffered from chromatic aberration, so he invented the first practical reflecting telescope, bringing him to the attention of the leading scientists of his time. In 1671, the Royal Society heard of Newton's telescope and elected him as a member. At that time, Newton published a paper about his theory of light and color. This paper drew Newton into a heated controversy with the Royal Society's instrumentalist, Robert Hooke, which deeply offended Newton, making him reluctant to publish more of his theories. Hooke and Newton remained bitter enemies until Hooke's death.

During the following years, Newton secretly matured his theory of gravity using differential calculus. In the early 1680s, a student named Edmund Halley asked him to analyze the problem of the orbits of comets. When Halley learned that Newton had already solved the problem and formulated a comprehensive theory of gravity, he insisted that Newton publish the work with Halley's own money. So in 1687 Newton's greatest work, *Philosophiæ Naturalis Principia Mathematica* (*The Mathematical Principles of Natural Philosophy*, called the *Principia* for short) was published. Here Newton set forth the theory of universal gravitation, the principles of dynamics, and many applications of these principles. He defined for the first time such concepts as force and mass. This work is the foundation of all modern mechanics. For ordinary objects, Newton's laws of mechanics have not been modified in the more than 300 years since the *Principia* was published.

The publication of the *Principia* established Newton as the foremost scientist in England. When Newton represented Cambridge at the famous convention that deposed the Catholic king James II, Newton came to the attention of politicians as well. As a result he was elected to Parliament in 1689 and served for two years. Later, Newton was employed at the mint to help redesign England's coins to make them more counterfeit-resistant. In 1699, he became Master of the Mint and eventually gave up his professorship at Cambridge to move to London. He gained such a reputation as a counterfeit expert that his name became a byword among counterfeiters, who saw many of their fellows hanged on Newton's account. Newton also moved the British currency from the silver standard to the gold standard. Many historians believe this change secured England's prosperity through the next century. Queen Anne knighted Newton in 1705 for his exceptional work at the Mint. He was the first scientist to be knighted.

During the latter years of the seventeenth century, Newton was busy with studies in alchemy. He was also very involved in religious studies, writing many tracts and papers, and defending the rationality and design of the Creation. Newton, unlike many contemporary scientists, did not separate science and religion. In observing the orbits of the planets, Newton wrote, "But it is not to be conceived that mere mechanical causes could give birth to so many regular motions, since comets range over all parts of the heavens in very eccentric orbits." He concluded, "This most beautiful system of the sun, planets, and comets could only proceed from the counsel and dominion of an intelligent and powerful Being."

Newton's work in calculus progressed throughout the period from 1677 to 1698, although he published almost nothing about it in fear of ridicule for his methods. On the other hand, the German mathematician Gottfried Wilhelm von Leibniz had been developing a similar system completely independently from Newton and had been publishing his work freely for decades. The two men clashed when the Royal Society published a report accusing Leibniz of plagiarism in 1711 (it was Newton who wrote the accusing summary statement in the report). Enmity between the two lasted until Leibniz's death in 1716. The controversy divided mathematicians in England from those on the Continent for many years. Though Leibniz's calculus notation and certain methods were later generally adopted by mathematicians everywhere, Newton's applications of calculus advanced the state of physics far beyond anything else in his time.

Like many scientists of his day, Newton had a wide range of interests and a critical distrust of unsound hypotheses. This attitude enabled him to contribute much to several areas of physics. He was both a good experimenter in his observations and a good theorist in his explanations. Nearly all great scientists after him, including Einstein and Bohr, gave tribute to his inspiring example as a scientist.

This astronaut's motions obey Newton's laws exactly because there are no frictional forces acting on him.

Dynamics 6

DOMINION SCIENCE PROBLEM

Aircraft Carrier Flight Operations

Because of the Fall, humans now exercise dominion over a broken world in which war is an ever-present threat. To exercise dominion responsibly, a nation must defend itself against those who wrongly exercise their dominion through aggressive warfare. Humans have used their considerable ingenuity to create weapons of war and defenses against those weapons. The methods and hardware of war evolved as new technologies made new modes of attack and defense possible. For many people, the ultimate weapon of war is the aircraft carrier, especially those of the US Navy.

6-1 The Nimitz-class aircraft carrier USS *Ronald Reagan* (CVN 76)

What are aircraft carriers? They are the behemoths of the Navy—all but one US carrier are the largest warships in the world—nuclear-powered, 24 stories tall, nearly 1100 feet long, with about 4.5 acres of flight deck, and carrying nearly 6000 men and women sailors as well as civilian contractors. They are virtually floating cities with an airport for a roof. Being mobile, they, with their 85 aircraft, can move to nearly any open ocean or coastal location in the world. This capability eliminates the need for airfields to be near a conflict in order to launch short-range tactical aircraft.

But even with the immense size of these ships, space is cramped. Modern fighter-bombers need 4500 ft to take off and more than 1500 ft to land unassisted. These are far longer distances than are available in the entire length of the ship! When combat flight operations occur, the flight deck is jammed with aircraft. How have naval architects provided the means to shorten these distances for shipboard operations?

The sole purpose of every combat ship of a navy is to carry out national policy under the direction of the national command authority. Aircraft carriers are very effective at conducting combat operations when needed in a time of war. However, they are even more effective at deterring war from occurring in the first place. This latter capability is also a part of exercising dominion and preserving life. American aircraft carriers have always been an implicit part of peace negotiations around the world.

6A THE HISTORY OF DYNAMICS

6.1 Early Greek Theories

The philosophers of Aristotle's day believed that objects could exhibit two types of motion: natural and unnatural. *Natural motion* was thought to be caused by some internal quality of an object that made it seek a certain "preferred" position without any application of force. For example, heated air rises, so its natural motion would be rising. Rocks fall; therefore, their natural motion would be falling. In this scheme of the universe, natural motions were thought to allow objects to reach their natural place, or place of rest. For heated air, the natural place was the heavens; for rocks, the

Greek natural philosophy attributed organic "inclination," or purposeful activity, to inanimate matter. In other words, their belief was that matter "knows" the kind of motion it should have. Any motion contrary to its natural inclination was considered unnatural motion.

natural place was the center of the earth. All motion other than natural motion was called *unnatural motion* and was thought to require applied force to be sustained.

Natural motions were divided into two subcategories, *terrestrial motions* (motions near the earth) and *celestial motions* (the motions of planets). This division of motions arose from the pagan Greeks' dualistic view of nature. The dualistic universe is divided into the celestial and the mundane; the divine and the mortal; the perfect and the imperfect. Celestial motions were thought to be purely circular. They represented the untouchable perfection of the celestial realm that was ruled by mathematics and pure reason. Terrestrial motion included all other forms of natural and unnatural motion. These two types of motion were considered to be completely separate and unrelated. This division helped to explain how heavenly bodies "automatically" moved in perfect circular paths.

According to Aristotle, an object's "heaviness" determined how "vigorously" it sought its natural place. A heavy object would move more vigorously toward the surface of the earth than a light object. Conversely, an object's vigorousness would measure its "heaviness." Were these conclusions supported by experience? It seemed so at the time.

Remember that Greek philosophy placed deductive reasoning on a qualitatively higher plane than experimentation. Even if experimental data contradicted a belief derived from reasoning, one could say that the experiment itself was faulty, since it was subject to the presupposed corruption of the material world.

6.2 Galileo's Investigations

Galileo chose a different perspective from which to study the problem of motion. His first task was to collect facts and establish an accurate *description of motion.* This description is known as kinematics. From this foundation, he inductively developed workable theories of dynamics (no scientist uses a strictly deductive approach in his work).

Suppose a feather, a tennis ball, a baseball, and a bowling ball were dropped from the top of a 30 m tower. How long would each object fall? Obviously, the bowling ball would hit the ground first, but why? If Aristotle's concept is correct, it is because heavier objects seek their natural positions near the center of the earth more vigorously than do lighter objects. Extending this logic, the speed of each object should be determined by its mass. The baseball, which has more than twice the mass of the tennis ball, should fall at least twice as fast. The bowling ball, about fifty times more massive, should fall fifty times as fast.

TABLE 6-1

Mass vs. Fall Time

Object	Mass (g)	Time (s)
Feather	14	15
Tennis Ball	57	3.0
Baseball	142	2.8
Bowling Ball	6804	2.6

Table 6-1 records the actual time required for each object to fall. Notice that the heavier objects fall at nearly the same rate. After a certain density is reached, the "heaviness" of the object has little effect. In this respect, Aristotle's theory of natural motions was shown to be wrong.

If a wind should blow during the experiment, the feather would be blown off course, and its falling time would be greatly increased. The falling times of the other objects would also be affected, but not as much as the feather's. Apparently, the presence of air has a great effect on the rate at which objects, especially light objects, fall.

6-2 The famous experiment on the moon proving Galileo's conjecture that all objects would fall at the same rate in the absence of air

Astronauts have performed experiments to test this conclusion. Colonel David Scott dropped a hammer and a feather in the vacuum at the moon's surface during the Apollo 15 mission in 1971. Without the presence of air, both objects fell at the same rate and hit the ground at the same time. Even though Galileo could not perform this experiment, he reached the same conclusion by considering the effects of air resistance on falling objects.

Aristotle believed that unnatural motions could only be maintained by the continuous application of force. Suppose you slide a large box of books from the foot of your bed into your closet. From experience, you know that when you stop pushing, the box stops almost immediately. This observation seems to show that Aristotle was correct.

Do other observations confirm this conclusion? Consider the flight of an arrow. The arrow begins to move when the bowstring applies a substantial amount of force. Yet the arrow still moves after it leaves the bow. According to Aristotle, some force must continue to push the arrow or it will stop and fall to the ground. Where can the force come from? Aristotle said that the air separates in front of the arrow, allowing the arrow to pass through easily. In the wake of the arrow, he said, the air snaps closed and provides the force to keep the arrow moving.

To disprove Aristotle's explanation, you would need to perform this experiment in a vacuum. If air provides the force to sustain the arrow's motion, the arrow should immediately cease moving if released in a vacuum. Instead, it continues on almost the same path as before. Since Aristotle was committed to his theory, he had to endow the air with special, unobservable properties to account for the arrow's motion. This kind of hypothesis, based on conjecture rather than observation, is called an *ad hoc* hypothesis. Aristotle's explanation was a direct consequence of the Greek philosophers' erroneous belief that inert matter had attributes similar to those of conscious organisms.

6-3 Disproving Aristotle's theory of the cause of an arrow's flight

The belief that inert matter exhibits purpose or intent is an example of an *ad hoc* hypothesis.

6.3 The Principle of Inertia

Galileo studied unnatural motion with an experiment similar to the following. Suppose that an unpolished metal cube rests on a wooden plane. If you give the cube a push, it soon stops moving. If the cube is polished and the plane is changed to oiled, polished metal, the cube travels a greater distance. Theoretically, if the cube and the plane were totally frictionless, the cube could slide forever without slowing down.

Now, suppose you arrange a frictionless, flexible, U-shaped strip of material as in Figure 6-4. This strip acts as a track for a frictionless ball. If you release the ball at the top of the track at the left end, the ball accelerates down the incline, rolls along the bottom of the curve, and then encounters a rising incline. The ball slows down as it rolls up the incline and stops when it reaches its original height (the reason this occurs will be considered in Chapter 9).

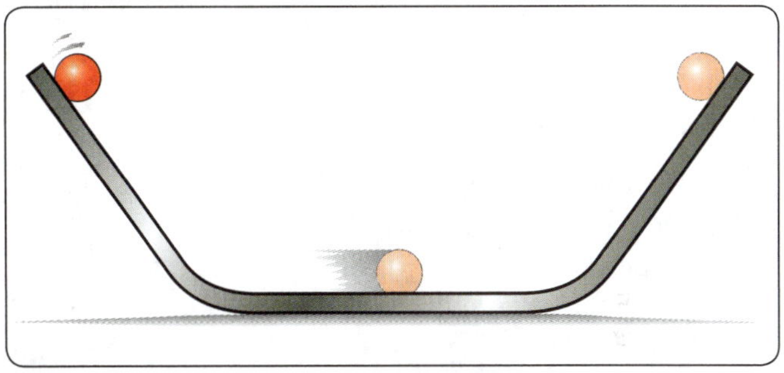

6-4 The motion of a ball on a frictionless, U-shaped track

What happens if the slope of the right end of the track is decreased and the experiment is performed again? The ball again decelerates up the right-hand incline, but its deceleration is less than before. It continues moving until it reaches its original height, although it has to move farther horizontally to reach that height. If the right-hand end of the track were bent to be completely horizontal, the ball would not experience any deceleration. It would naturally continue at a constant velocity for an indefinite amount of time and distance.

From such concepts, Galileo inductively asserted that the "natural" state of motion of an object could include "moving" as well as "resting." Galileo's **principle of inertia** states that an object will continue in its original state of motion unless some outside agent acts on it. A moving object does *not* require a continuous push to maintain a constant velocity. In fact, just the opposite is true! A push causes a *change* in an object's state of motion. This conclusion was truly amazing considering the limitations of the experimental apparatus of the time. The concept of inertia has served as the cornerstone of dynamics ever since Galileo.

Galileo's **principle of inertia** states that an object at rest tends to stay at rest and that an object in motion will continue in its current state of motion unless an outside agent acts on it.

6.4 Sir Isaac Newton

Newton carried forward the conceptual investigations of Galileo. By the time he was able to devote his considerable powers of deduction to the problem, much information about the movement of planets had been documented by the great astronomer Tycho Brahe, and Johannes Kepler had formulated his three laws of

6-5 Tycho Brahe (1546–1601) was a Danish astronomer who revolutionized astronomical observations. His discovery that comets were higher than the moon and that planets moved in noncircular orbits did much to complete the destruction of the dualistic belief in the perfection of the celestial spheres.

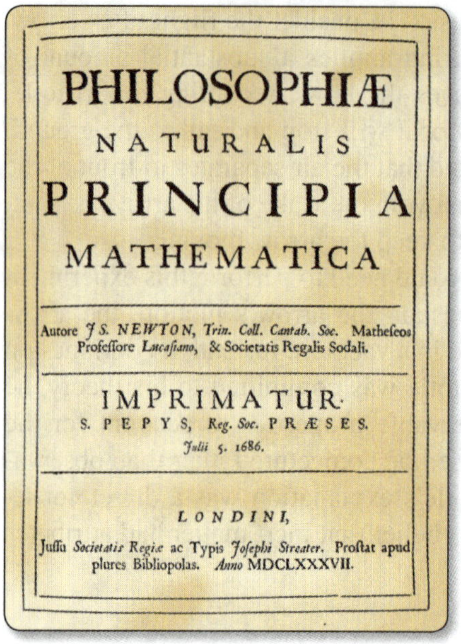

PHILOSOPHIÆ
NATURALIS
PRINCIPIA
MATHEMATICA

Autore *J S. NEWTON, Trin. Coll. Cantab. Soc.* Matheseos Professore *Lucasiano*, & Societatis Regalis Sodali.

IMPRIMATUR.
S. PEPYS, *Reg. Soc.* PRÆSES.
Julii 5. 1686.

LONDINI,
Jussu *Societatis Regiæ* ac Typis *Josephi Streater*. Prostat apud plures Bibliopolas. *Anno* MDCLXXXVII.

6-6 Newton's *Principia*

celestial mechanics (see Chapter 7). Newton gave credit to these and other scientists who had gone before him for preparing the way for his discoveries.

Newton's work grew from his study of gravitational attraction between astronomical bodies using differential calculus, which he had developed for the purpose. Eventually, he published his findings, which included his three laws of motion that defined the concepts of force (rather than an unspecified "agent"), mass, and momentum, in his famous work *Principia* (short for *Philosophiæ Naturalis Principia Mathematica*, or *Mathematical Principles of Natural Philosophy*). Newtonian dynamics has been validated repeatedly for ordinary objects over the three centuries since *Principia*. Only within the last century have scientists discovered that Newtonian physics does not hold true for particles that are extremely tiny (subatomic sized) and/or moving extremely fast (moving at a significant fraction of the speed of light).

6A Objectives

After completing this section, I can

- ✓ briefly summarize the historical development of the concepts of mass and inertia according to the Greeks, Galileo, and Newton.
- ✓ explain the paradigm responsible for each view of motion.
- ✓ identify Newton's great written work that established the principles of dynamics.

6A Section Review

1. According to Aristotle, how much faster should a 10 kg rock drop than a 1 kg rock? What was his explanation for the fact that rocks drop and smoke rises?
2. Which scientist first stated the principle of inertia?
3. When modern geologists using radiometric dating techniques find that the age obtained for a rock sample does not agree with other supposedly credible dating methods, they hypothesize reasons why the date is not in agreement with the expected age. What kind of explanations are these?
4. What did Galileo conclude was the "natural" state of motion for an object?
5. State the conditions under which Newton's laws do not accurately describe a system's motion.
6. Research the quotation attributed to Sir Isaac Newton in which he admits that his discoveries would not have been possible without the contributions of the great scientists who proceeded him.
7. Locate a copy of *Principia* in a library or bookstore, or research online. What interesting facts are associated with the original publication of this book?

6B FORCES

6.5 Summing Forces

Everyone intuitively understands that a push or a pull must be exerted in order for an object to be moved. "Pushes and pulls" belong to the class of physical phenomena called forces. All matter is held together by forces; matter exerts forces and has forces imposed on it. You may associate the word "force" with human activities and machines. Immense man-made forces are required to lift a space shuttle into orbit or to nudge an oil supertanker against a pier. However, these forces pale in comparison to the inconceivably powerful natural forces unleashed by a hurricane or the eruption of a solar flare (see Figure 6-7). You also consciously and unconsciously employ small forces in your daily life routines when you press the keys on a cell phone, breathe, swallow, walk, sit, pull a sweater on, or write.

A force is an agent that causes or opposes changes in motion of matter, or tends to do so.

6-7 A solar flare contains unimaginably large forces.

Forces have magnitudes and direction. They can easily be represented by vectors. Forces, for example, obey vector addition rules. Multiple forces can also act on an object simultaneously. In order to describe and predict how a force will affect an object's motion, we must know *where* the force is being applied. We will address that question later, but for now we will assume that the object on which forces are acting is a particle—with no physical size—and that all forces act on the same point. In later chapters dealing with circular motion and torques, we will consider what happens when forces do not act on the same point on an object.

Even though forces can be pushes or pulls, we will diagram forces as only pulls on the point of application of the force. This convention will help you avoid confusion when solving problems. The SI derived unit of force is the newton (N).

Problem-Solving Strategy 6.1

Always sketch forces as *pulls* in a force vector diagram.

The effects of simultaneous forces acting on an object are of primary interest in physics. Therefore, we will assign the vector sum of simultaneous forces the symbol

$$\Sigma \mathbf{F} \equiv \mathbf{F}_1 + \mathbf{F}_2 + \ldots + \mathbf{F}_n.$$

The Greek capital letter sigma (Σ) is the mathematical summation symbol that stands for the sum of all terms represented by the variable following it.

The symbol $\Sigma \mathbf{F}$ represents the sum of all simultaneous forces acting on an object. It is equivalent to the expression "net force."

6.6 Balanced and Unbalanced Force Sums

Forces do not always cause a change in motion. In Figure 6-8a, two people are pushing on a refrigerator equally hard but in opposite directions. As you would expect, no motion occurs. If the sum of all forces acting on an object is zero, then we say that the forces are balanced. Balanced **forces** cancel each other. The following mathematical expression is true for balanced forces:

When the sum of all simultaneous forces acting on an object equals zero, the forces are balanced.

$$\Sigma \mathbf{F} = 0 \Leftrightarrow \Sigma \mathbf{F}_x = 0 \text{ and } \Sigma \mathbf{F}_y = 0$$

The symbol ⇔ in mathematical statements means "if and only if."

In other words, when the force sum on an object is zero, the sums of the forces' x- and y-components are also zero.

In Figure 6-8b, one man has a cramp in one arm, so he cannot push as hard as the other man. The refrigerator moves toward the man who pushes less. **Unbalanced forces** change an object's state of motion.

A pair of balanced forces on an object is represented by two vectors with the same magnitude but opposite directions. When these vectors are summed, they add up to a zero resultant vector. If two forces act at less than 180° to each other, you can use either the geometric-addition method or the mathematical technique of vector resolution to determine the vector sum. Three or more forces acting on an object may sum to zero, but it is more likely that there will be a net resultant. The presence of unbalanced forces means that the force vectors do not cancel. Example 6-1 shows how unbalanced forces are used to move a ship in a desired direction.

6-8 Only *unbalanced* forces cause motion.

6-9 A Panama Canal mule engine

> **Problem-Solving Strategy 6.2**
>
> If one of a resultant's components is zero, then the resultant has the same magnitude and direction as the nonzero component.

EXAMPLE 6-1

Canal Mules: Summing Forces

Figure 6-10 shows a large ship entering the Gatún lock at the north end of the Panama Canal. The figure also shows two powerful "mule" engines running on railway tracks on both sides of the lock. They are attached to the ship with steel cables that allow them to pull the ship slowly into the lock basin as they move along the tracks.

If each mule engine exerts a pull of 1.50×10^5 N at an angle of 55.0° on either side to the front end (bow) of the ship, what is the magnitude and direction of the sum of the towing forces on the ship? Assume that the forces act at the same point on the centerline of the ship.

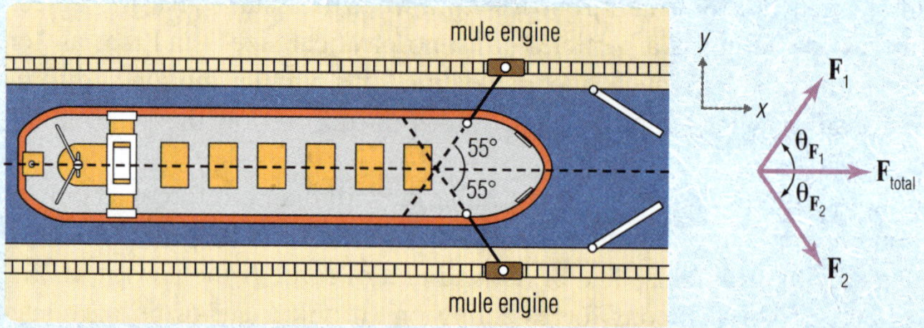

6-10 Vector diagram of mule engines towing a ship into a Panama Canal lock

Solution:

Choose a coordinate axis parallel to the length of the ship.

Given: $\mathbf{F}_1 = 1.50 \times 10^5$ N at $\theta_{\mathbf{F}_1} = +55.0°$;
$\mathbf{F}_2 = 1.50 \times 10^5$ N at $\theta_{\mathbf{F}_2} = -55.0°$

Find the total towing force ($\mathbf{F}_{\text{total}}$).

You know that the total towing force is the resultant of the vector sum of the two towing forces:

$$\mathbf{F}_{\text{total}} = \mathbf{F}_1 + \mathbf{F}_2$$

Every vector can be resolved into the sum of its components:

$$\mathbf{F}_{\text{total}} = \mathbf{F}_{\text{total}\,x} + \mathbf{F}_{\text{total}\,y}$$

The resultant components are the vector sums of the corresponding component vectors:

$$\mathbf{F}_{\text{total}\,x} = \mathbf{F}_{1x} + \mathbf{F}_{2x}$$
$$\mathbf{F}_{\text{total}\,y} = \mathbf{F}_{1y} + \mathbf{F}_{2y}$$

These equations are one-dimensional vector equations, so they may be expressed as scalar components in order to complete the algebra operations.

1. Identify reference angles: $\alpha_{\mathbf{F}_1} = \theta_{\mathbf{F}_1}$; $\alpha_{\mathbf{F}_2} = -\theta_{\mathbf{F}_2}$

2. Determine scalar components:

$F_{1x} = F_1 \cos \alpha_{\mathbf{F}_1} = (1.50 \times 10^5\ \text{N})\cos 55.0° \cong +0.86\underline{0}3 \times 10^5$ N (at 0°)

$F_{1y} = F_1 \sin \alpha_{\mathbf{F}_1} = (1.50 \times 10^5\ \text{N})\sin 55.0° \cong +1.2\underline{2}8 \times 10^5$ N (at 90°)

$F_{2x} = F_2 \cos \alpha_{\mathbf{F}_2} = (1.50 \times 10^5\ \text{N})\cos 55.0° \cong +0.86\underline{0}3 \times 10^5$ N (at 0°)

$F_{2y} = F_2 \sin \alpha_{\mathbf{F}_2} = (1.50 \times 10^5\ \text{N})\sin 55.0° \cong -1.2\underline{2}8 \times 10^5$ N (at 270°)

(signs of results are based on the geometry of the problem)

> Values of the force components in this example are expressed in scientific notation in terms of the largest exponential factors ($\times 10^5$ N). This allows us to add the components without further conversion.

3. Find the towing force sum components:

$F_{\text{total}\,x} = 2(+0.86\underline{0}3 \times 10^5\ \text{N}) \cong +1.72\underline{0}6 \times 10^5\ \text{N}$ (at 0°);

$F_{\text{total}\,y} = (+1.2\underline{2}8 \times 10^5\ \text{N}) + (-1.2\underline{2}8 \times 10^5\ \text{N}) = 0\ \text{N}$

4. The magnitude of the towing force is found by using the Pythagorean theorem:

$|\mathbf{F}_{\text{total}}| = F_{\text{total}} = \sqrt{F^2_{\text{total}\,x} + F^2_{\text{total}\,y}} = \sqrt{(+1.72\underline{0}6 \times 10^5\ \text{N})^2 + (0\ \text{N})^2}$

$F_{\text{total}} \cong 1.72\underline{0}6 \times 10^5\ \text{N}$

$F_{\text{total}} \cong 1.721 \times 10^5\ \text{N}$

The direction of $\mathbf{F}_{\text{total}}$ is the same as its x-component: 0°.

$\mathbf{F}_{\text{total}} = 1.721 \times 10^5$ N toward the bow (front) of the ship.

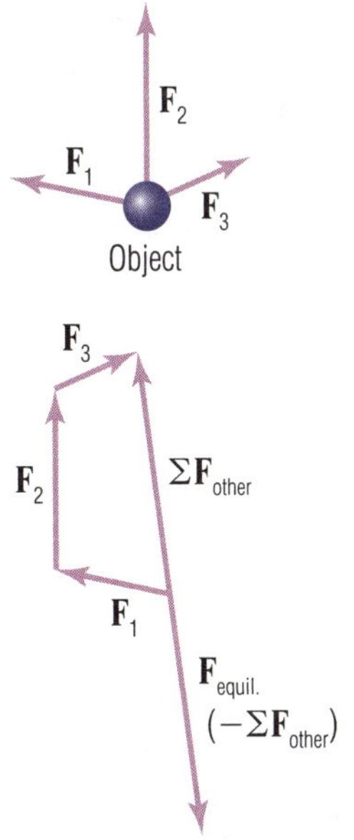

6-11 Resultant and equilibrant vectors

6.7 The Equilibrant Force

When a force sum is zero, then any one of the contributing forces exactly balances all of the others. A force that balances one or more other *concurrent* forces is called an **equilibrant force** (see Figure 6-11). The equilibrant of a set of unbalanced forces is a vector having the same magnitude as the vector sum of the other unbalanced forces but pointing in the opposite direction. In equation form, this means

$$\mathbf{F}_{\text{equil.}} = -\Sigma\mathbf{F}_{\text{other}},$$

where $\mathbf{F}_{\text{equil.}}$ is the equilibrant force and $\mathbf{F}_{\text{other}}$ are all the other unbalanced forces acting simultaneously on the object. You can use the equilibrant concept to determine an unknown force. If the sum of all forces on an object is zero, then any unknown force must be the equilibrant of all of the known forces.

The **weight ($\mathbf{F}_w$)** of an object is the force of gravity acting on it. This is a common force that must often be considered in mechanics problems. Example 6-2 demonstrates how weight can be an equilibrant force.

The term "concurrent" means "acting at a single point."

The **equilibrant** of a set of unbalanced forces is a vector having the same magnitude as the unbalanced force sum but pointing in the opposite direction.

The **weight (F_W)** of an object is the gravitational force on the object's mass.

EXAMPLE 6-2

Unknown Weight: Finding an Equilibrant

A shipyard has finished constructing a modular aluminum-and-steel superstructure for a new ship and is preparing to install it on the hull. Naval architects need to have a close approximation of the weight of the structure in order to calculate its effect on the buoyancy of the hull after it is installed.

In order to weigh the superstructure, two shipyard cranes lift and suspend it as the architects note the angle and lifting force exerted by the crane cables on the superstructure. The forces and angular information are provided in Figure 6-12. Determine the total lifting force. The equilibrant of the total lifting force is the unknown weight of the superstructure, $\mathbf{F}_w$.

Solution:

Choose a Cartesian coordinate system with axes horizontal and vertical. There are several ways to orient the coordinate system, but the least confusing is as shown in the figure. *Cable angles are measured from the vertical direction.*

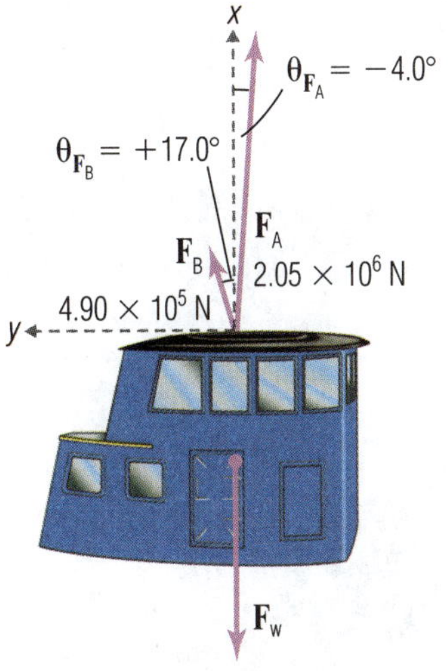

6-12 Force diagram for Example 6-2

From the figure, you know that $\mathbf{F}_A = 2.05 \times 10^6$ N at $-4.0°$ and $\mathbf{F}_B = 4.90 \times 10^5$ N at $+17.0°$.

Strategy: $\mathbf{F}_w$ is the equilibrant of the total lifting force:

$$\mathbf{F}_w = -\mathbf{F}_{lift}$$

The total lifting force is the vector sum of the two crane forces:

$$\mathbf{F}_{lift} = \mathbf{F}_A + \mathbf{F}_B$$

1. Determine the reference angles: The angles between the cables and the vertical are measured at the attachment point on the superstructure.

$\alpha_{\mathbf{F}_A} = 4.0°$; $\alpha_{\mathbf{F}_B} = 17.0°$

2. Determine the force sum components.

$|F_{Ax}| = (2.05 \times 10^6 \text{ N})\cos 4.0° \cong 2.0\underline{4}5 \times 10^6$ N (up)

$|F_{Ay}| = (2.05 \times 10^6 \text{ N})\sin 4.0° \cong 0.14\underline{3}0 \times 10^6 \text{ N} = 1.4\underline{3}0 \times 10^5$ N (right)

$|F_{Bx}| = (4.90 \times 10^5 \text{ N})\cos 17.0° \cong 4.6\underline{8}5 \times 10^5 \text{ N} = 0.46\underline{8}5 \times 10^6$ N (up)

$|F_{By}| = (4.90 \times 10^5 \text{ N})\sin 17.0° \cong 1.4\underline{3}2 \times 10^5$ N (left)

3. Determine the lifting force resultant components:

$F_{lift\,x} = F_{A\,x} + F_{B\,x} = (+2.0\underline{4}5 \times 10^6 \text{ N}) + (+0.46\underline{8}5 \times 10^6 \text{ N})$

$F_{lift\,x} \cong 2.5\underline{1}3 \times 10^6 \text{ N} \cong 2.51 \times 10^6$ N

$F_{lift\,y} = F_{A\,y} + F_{B\,y} = (+1.4\underline{3}0 \times 10^5 \text{ N}) + (-1.4\underline{3}2 \times 10^5 \text{ N})$

$F_{lift\,y} \cong -0.00\underline{2} \times 10^5 \text{ N} \cong 0$ N

4. The lifting force is $\mathbf{F}_{lift} = 2.51 \times 10^6$ N straight up since the horizontal component is essentially zero.

5. The superstructure's weight is the equilibrant to the sum of the lifting forces:

$$\mathbf{F}_w = -\mathbf{F}_{lift}$$

$$\mathbf{F}_w = -2.51 \times 10^6 \text{ N (straight down)}$$

6.8 Types of Forces

Gravitational Force

All forces can be classified as either fundamental forces or mechanical forces. The **fundamental forces** are inherently part of the atomic building blocks of matter. Scientists have identified four fundamental forces in the universe, from which all other forces are derived. Of the four, two were discovered as recently as the twentieth century.

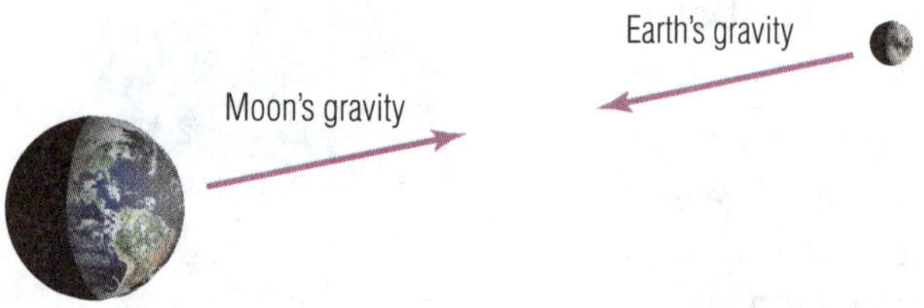

6-13 Both the earth and the moon attract each other through the gravitational force.

The first to be recognized and mathematically described (by Newton during the seventeenth century) was the **gravitational force**. This long-range force is proportional to the masses of the interacting objects and exerts its influence over theoretically infinite distances. It is the dominant force that organizes the structure of astronomical bodies, controls the motions of planets around the sun, and governs the celestial processes of the universe. All objects exert gravitational force on all other objects. You will study the effects of gravitational force in association with nearly all mechanics topics in Units 2 and 3, as well as its importance to modern physics in Unit 6.

6-14 A lightning strike is a large concentration of electromagnetic forces.

Electromagnetic Force

The existence of magnetic and electric forces had been known for millennia and studied to some extent in the seventeenth century, but these forces were not recognized as two components of the same fundamental force until the mid-nineteenth century when James Clerk Maxwell formulated the unifying theory of electromagnetism. The **electromagnetic force**, as it is now called, is also a long-range force. You have seen long-range electrostatic discharges in the form of lightning, and the *Voyager* space probes have detected the sun's magnetic field billions of miles from its source. The electromagnetic forces also act at very short distances. They are responsible for chemical bonding and other interactions among atoms and molecules. The electromagnetic force governs the fine structure of matter because it determines many of the physical and all of the chemical properties of substances. You will study electric fields in Chapter 19 and magnetism in Chapter 21. Combined electromagnetic theory will be studied in Chapter 22 and in Unit 6.

The four fundamental forces are
- Gravitational force
- Electromagnetic force
- Strong nuclear interaction force
- Weak nuclear interaction force

The noncontact forces are gravity and the electromagnetic forces. Gravity displays only attraction, while electromagnetic forces are both attractive and repulsive. No physical contact is required to transmit or exert these forces.

Strong and Weak Nuclear Forces

Early in the twentieth century, two short-range fundamental forces, or interactions, were discovered during the development of the quantum model of the atom. These forces act only over very short distances, less than a few femtometers (10^{-15} m), which is comparable to the diameter of protons and neutrons in the nuclei of atoms. Thus they are called nuclear forces. The **strong nuclear interaction force** binds together the particles forming the nucleus while the **weak nuclear interaction force** is involved in several modes of nuclear decay. You may study these two nuclear forces in advanced physics courses.

6.9 Classification of Forces

Prior to the last century, forces seemed to be easily classifiable into those that could act through space without physical contact and those that could be transmitted only through physical contact between objects.

Noncontact Forces

In one model, gravitational and electromagnetic forces are called action-at-a-distance forces to emphasize that no physical contact is necessary for transmission of the force. This model assumes that each kind of force is a property of the associated matter, which can somehow be exerted at a distance without requiring physical contact.

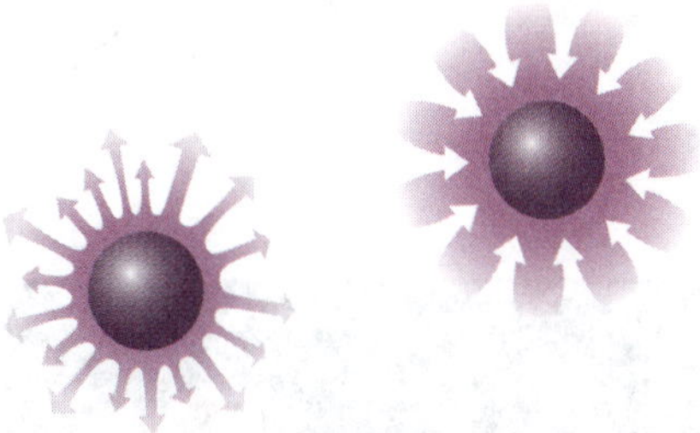

6-15 Action-at-a-distance forces are thought to emanate from certain objects and matter in general.

Another approach to explaining noncontact forces is the **field theory**. This theory proposes that the gravitational influence of matter and the attraction or repulsion of magnetic or charged particles are properties of the space surrounding the matter (the field). The matter exerts forces on other massive, magnetic, or charged particles when they are placed in their respective fields.

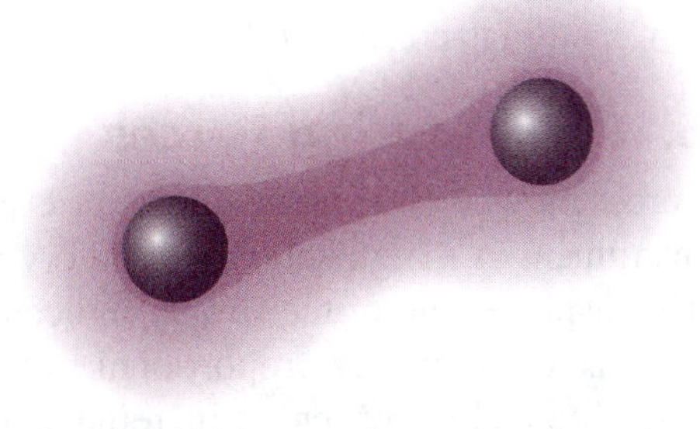

6-16 The field theory states that noncontact forces arise from the space surrounding certain objects and matter.

The development of quantum mechanics after World War II produced yet another model that tries to account for all of the fundamental forces. It suggests that **virtual particles** are exchanged between matter with like characteristics. The exchange of the virtual (nonmaterial) carrier particles supposedly creates the force. Virtual particles have been identified for the electromagnetic (photons) and nuclear

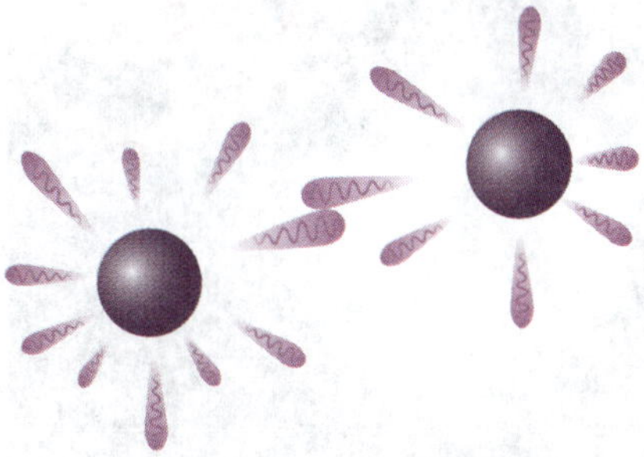

6-17 Particle exchange theories suggest that noncontact forces exist because matter exchanges nonmaterial particles associated with each kind of force. Here, photons are exchanged.

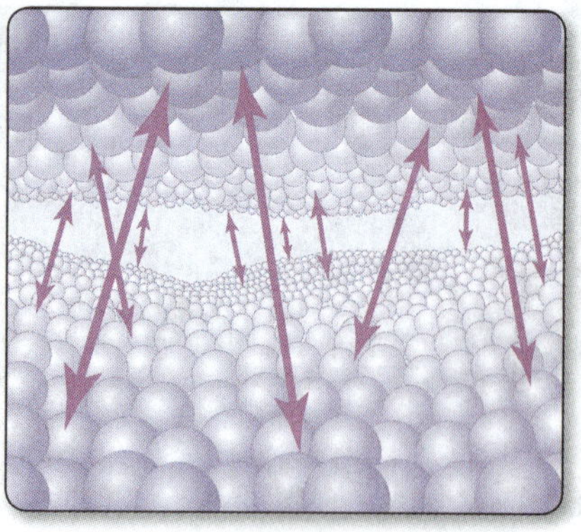

6-18 Friction begins with the electromagnetic attraction and irregularities of particles at the surfaces of objects.

Contact forces are derived from the interaction of electromagnetic forces among particles of matter. Contact forces require physical contact of objects in order to be transmitted or exerted.

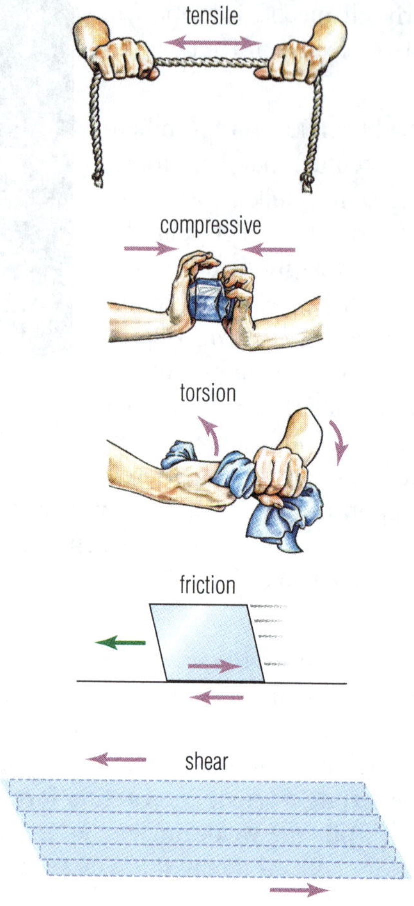

6-19 Mechanical or contact forces

forces (gluons), but no particle has yet been identified that carries the gravitational force (graviton?). You will study some aspects of these three force models in later chapters.

Contact Forces

As mentioned earlier, the electromagnetic force is responsible for both the structure of the atom and the chemical bonding between atoms. The electromagnetic forces in the electron orbital structures of atoms can either attract or repel other atoms. Most atoms have some tendency to attract and hold onto nearby atoms, but no two atoms can occupy the same space at the same time. When aggregates of atoms (any kind of matter) are forced together or pulled apart, the particles exert forces between each other. This gives rise to the concept of contact (or *mechanical*) forces (forces that are transmitted only by physical contact between objects).

Contact forces include *tension* or **tensile forces** (forces that tend to pull things apart), **compressive forces** (forces that tend to push things together or to crush), **torsion** (twisting forces), **friction** (forces that oppose motion between two objects in contact), and **shear forces** (forces that tend to cause layers within matter to slide past one another). Mechanical forces and the gravitational force will be the forces of primary concern in Units 2 and 3.

According to Scripture, the advent of forces in creation must have been during the very first instants of time (Gen. 1:1). All four fundamental forces appeared as matter was brought into existence—gravity from the matter itself; electromagnetic forces in the chemical bonds between the atoms of the matter; and the nuclear forces that held the atomic nuclei together. The material world continues to exist as a direct result of God's providence in sustaining these forces. "For by him [Jesus] were all things created, that are in heaven, and that are in earth. . . . and by him all things consist" (Col. 1:16–17).

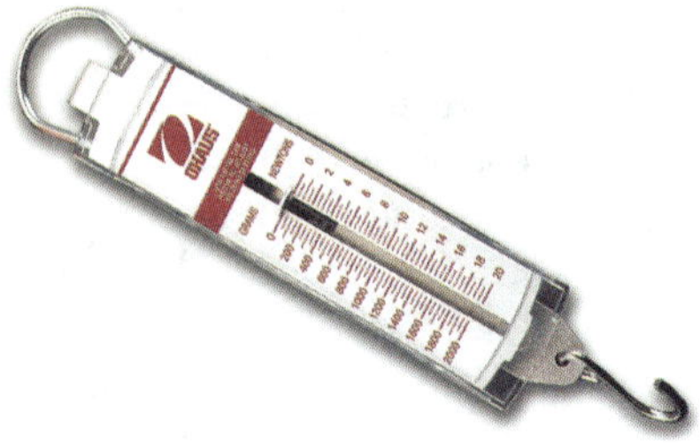

6-20 A typical laboratory spring scale

6-21 A pressure gauge indicates the force per unit area within a fluid.

6.10 Measuring Forces

Forces may be measured using a variety of instruments. A typical force-measuring instrument in a high-school laboratory is the *spring scale*, shown in Figure 6-20. The displacement of the spring, indicated by the change in position of the pointer from zero, is directly proportional to the force applied to the spring, so the scale is calibrated in appropriate increments to indicate the applied force.

Materials engineers use a type of force-sensing detector called a *load cell* to measure large forces applied to material samples. It consists of a crystalline substance, such as pure quartz, that produces an electrical signal when the crystal is squeezed by the applied force.

Forces may also be inferred using instruments such as *pressure gauges*, *ballistic pendulums*, and *accelerometers*. These and other instruments will be discussed later in this course.

In the classroom, a **force table**, shown in Figure 6-22, demonstrates the addition of force vectors. When the attachment ring is stationary directly above the center of the table, the applied forces are balanced. With this apparatus, the equilibrant force can be determined directly by finding the weight of the suspended mass generating the force.

6-22 A force table

6B Section Review

1. Give a definition of "force" that considers its potential to change motion.
2. Name and briefly describe five mechanical forces.
3. Write the symbol that represents the resultant of simultaneous concurrent forces.
4. a. What is the force called that exactly balances a nonzero force sum acting on an object?
   b. What is its magnitude and direction compared to the original force sum?
5. List the four fundamental forces and briefly describe each.
6. Three concurrent forces act simultaneously on a point: $\mathbf{F}_1 = 15.0$ N at 25.0°, $\mathbf{F}_2 = 22.0$ N at 73.0°, and $\mathbf{F}_3 = 8.0$ N at 265.0°. Determine the force sum, $\Sigma\mathbf{F}$.
7. Your dog Rover suddenly exerts a force of 200 N on his leash. What is the equilibrant of this force vector?
8. A force table is set up in a physics laboratory and three spring scales are hooked to the central ring and the notched perimeter of the table. The first scale reads 12.5 N at an angle of 0.0° and the second scale reads 17.3 N at an angle of −34.0°. If the three forces are balanced, what are the reading and angle of the third spring scale?

6B Objectives

After completing this section, I can

- ✓ write a vector equation that represents the sum of forces acting simultaneously on a system.
- ✓ sum forces using mathematical vector addition techniques.
- ✓ identify the equilibrant of a nonzero system of forces.
- ✓ identify the four fundamental forces in nature and describe their main properties.
- ✓ classify forces by their method of interaction with matter.
- ✓ identify five types of contact forces.
- ✓ describe ways to measure or detect forces.

6C NEWTON'S LAWS OF MOTION

6.11 Systems and Newton's Laws

Dynamics is the study of the effect of forces on the motion of objects. The central principles of dynamics have been summarized in a series of statements known as **Newton's laws of motion**. The term *laws* does not mean that these principles hold true in every case or that, like God's laws, they are immutable. Scientific laws are man's attempt to organize his observations of the universe in a sensible and predictable way. Consequently, they are imperfect and subject to change. A law describes a relationship among observable phenomena that has been experimentally validated. It is usually a mathematical expression that the majority of scientists accept as true for nearly all practical situations; any exceptions are usually well understood or else have not yet been discovered.

Before discussing Newton's laws, you must understand what they are used for. Physicists study the mechanics of **systems**. You may think of a system as a complex, interconnected collection of parts such as an electrical system, a plumbing system, or even the postal system. In physics, a system is whatever is inside an imaginary boundary chosen by the physicist. It may have one or more parts. We will usually study systems with only one or two objects to minimize the complexity of the problem. The "system" concept changes the way we define the word "object." An object may be a single, discrete particle, but it

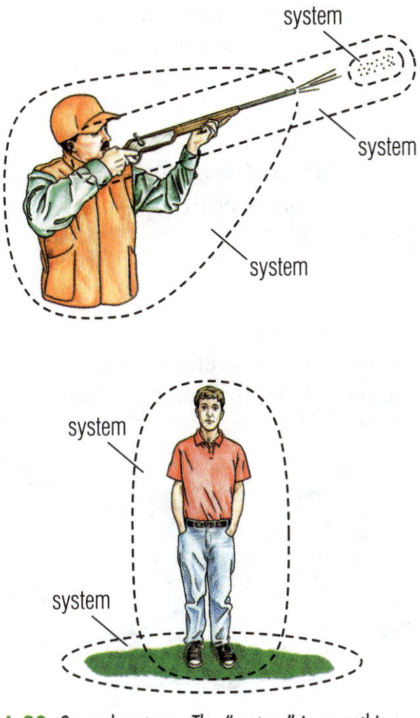

6-23 Several systems. The "system" is everything inside the boundary.

The word *system* has another meaning as well. In the phrase "system of forces" it refers to the set of related forces acting on an object. There can also be a system of forces acting on a system.

could also be a rock, a quantity of gas molecules, a flock of geese, a space probe, a planet, or even you.

Everything applicable to the problem and affecting the system exists within the system boundary. Everything outside the boundary (its **surroundings**) may be ignored if it doesn't somehow act on the system. Developing the skill to correctly identify "the system" in a problem will vastly simplify the solutions to physics exercises.

6.12 Newton's First Law of Motion

Newton's first law generalizes Galileo's principle of inertia. Newton wrote in *Principia*, *"Every body continues in its state of rest, or of uniform motion in a right line, unless it is compelled to change that state by forces impressed upon it."* A more modern version of this statement is: A system at rest will remain at rest, and a moving system will move continuously with a constant velocity (constant speed and direction) unless acted on by outside unbalanced forces.

The first law establishes two conditions:

- If all external forces on a system are balanced, then its velocity remains constant. That is, no acceleration occurs.
- If all forces acting on a system are not balanced, then a nonzero resultant force exists and the velocity changes, resulting in an acceleration.

Problem-Solving Strategy 6.3

After sketching a physics problem, determine the system in the problem, draw a circle around it, and identify all forces acting *on* the object(s) inside the circle.

Mathematically, the first law can be expressed as

$$\Sigma \mathbf{F} = 0 \Leftrightarrow \mathbf{a} = 0.$$

This statement means that the sum of all forces on a system is zero if and only if the acceleration of the system is zero.

An equivalent statement of the first law is

$$\Sigma \mathbf{F} \neq 0 \Leftrightarrow \mathbf{a} \neq 0.$$

Newton's first law defines the condition of balanced forces or zero force sum. If all forces are balanced, then a system's acceleration is zero. The converse is also true.

The first law defines the condition of a zero force sum on a system. However, zero total force does *not* mean that there are no forces on a system, only that their vector sum, or resultant, is zero.

Everyday experience sometimes seems to contradict the first law of motion. For example, if you do not continue to pedal a bike, it seems to slow down and stop "naturally." What causes this? From the first law you can predict that there is some unbalanced force that slows the object—friction.

The concept of *mass* was invented by scientists in order to quantify *inertia.*

Even when an unbalanced force is imposed on the system, the system tends to resist a change in motion. This tendency was identified by Galileo as inertia. The concept of mass was developed by scientists to quantify inertia.

An object is in mechanical equilibrium when the vector sum of all applied forces is zero.

When the vector sum of all forces on a system is zero, it is said to be in mechanical equilibrium. A system in equilibrium has a constant velocity. A system at rest (in a given reference frame) is in equilibrium with a constant velocity equal to zero. We will refer to mechanical equilibrium frequently in our study of mechanics.

Newton's first law predicts that without unbalanced forces, moving objects will travel in straight lines. When the young shepherd David defeated the Philistine hero Goliath, he propelled the pebble with a sling similar to the one shown in Figure 6-24. The principles of operation of a sling are simple, but putting them into practice is not. When the sling is whirled in a circle, the sling applies an unbalanced force to the stone that causes it to move in a circular path. When one of the thongs is released, the unbalanced force is removed from the pebble. It begins a straight-line trajectory in the direction of its instantaneous velocity at the moment of release.

6-24 The velocity of a pebble is continually changing until it is released.

6.13 Newton's Second Law of Motion

The second law is the most general of the three laws because it gives an operational working definition of the concept of force. It tells us how to measure a force, assuming that we can measure the mass and acceleration of a system. Before we state this law, let us consider an experiment that illustrates the relationship between mass, acceleration, and unbalanced forces.

Case 1: If you pull a 1 kg mass across a frictionless surface with *constant* force as measured with a spring scale, you observe a change in velocity (an acceleration) of the mass.

6-25a Case 1—Force causes a mass to accelerate.

Case 2: If you pull the same mass with a larger constant force you note a larger acceleration. If you repeat with many different forces, you begin to recognize a pattern.

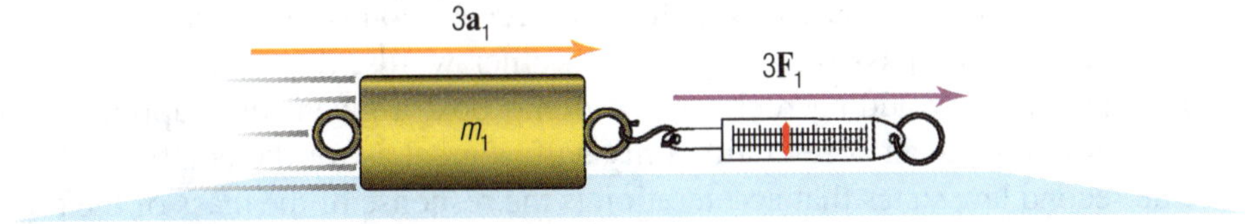

6-25b Case 2—Acceleration is proportional to the force.

Case 3: If you pull the same mass with two identical forces acting in *opposite* directions, you observe that no acceleration occurs since $\Sigma\mathbf{F} = 0$.

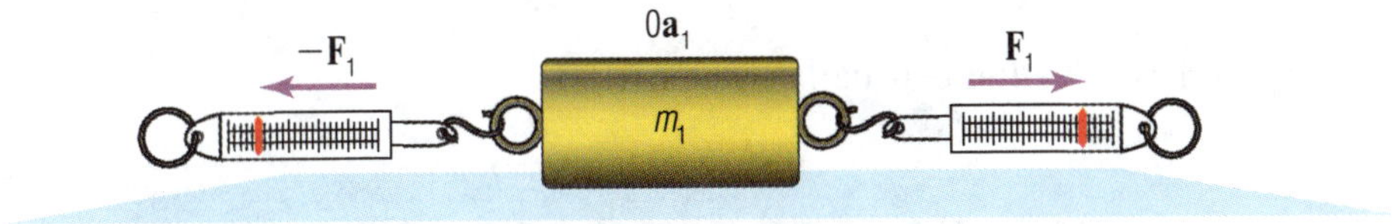

6-25c Case 3—Zero net force causes zero acceleration.

Case 4: If you pull the same mass with two identical forces in the *same* direction, you observe that the acceleration is twice as great as with one force.

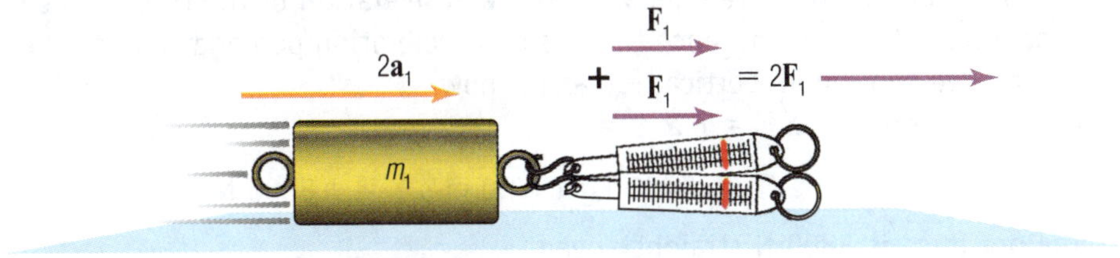

6-25d Case 4—Acceleration is *directly* proportional to force.

Case 5: If you pull a 2 kg mass with the same force as in Case 1, you note that the acceleration is half as great.

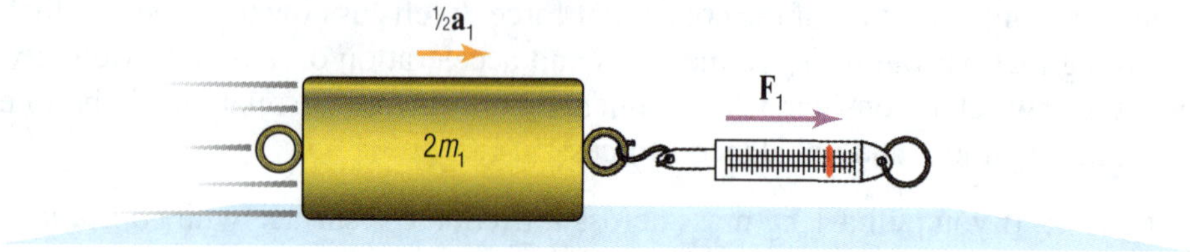

6-25e Case 5—Acceleration is inversely proportional to mass.

In the *Principia*, Newton wrote, *"The change in motion is proportional to the motive force impressed; and is made in the direction of the right line in which that force is impressed."*

In other words:

> The acceleration of a system is directly proportional to the sum of the forces (resultant force) acting on the system and is in the same direction as the resultant.

6-26 The forces needed to accelerate a Navy aircraft to takeoff speed are provided by a catapult.

In a mathematical form, the second law is written

$$\mathbf{a} = \frac{\Sigma \mathbf{F}}{m}. \qquad (6.1)$$

From Equation 6.1 you can see that the acceleration of a system is inversely proportional to the mass of the system. In other words, if you double the mass of a dump truck by adding rocks to the dump box, a given force applied by its engine will produce an acceleration only half as great as for the empty truck.

The second law states that acceleration is the response of the mass of the object to an unbalanced force. The force is the cause, and acceleration is the response of the mass. The SI unit of force, the newton (N), is formally defined as follows:

1 N is the *total force* that, when applied to exactly 1 kg, produces an acceleration of exactly 1 m/s^2. In SI base units,

$$1\ \text{N} = \frac{1\ \text{kg}\cdot\text{m}}{\text{s}^2} = 1\ \text{kg}\cdot\frac{1\ \text{m}}{\text{s}^2}.$$

> Newton's second law establishes that a system's acceleration is proportional to the net force imposed upon it and is inversely proportional to its mass:
>
> $$\mathbf{a} = \frac{\Sigma \mathbf{F}}{m}$$

The second law equation is usually written

$$\Sigma \mathbf{F} = m\mathbf{a}. \qquad (6.2)$$

EXAMPLE 6-3

Acceleration from Force: The Second Law

A 4400 kg truck accelerates straight ahead at 2.0 m/s^2. What is the net force on the truck?

Solution:

Choose a coordinate system parallel to the acceleration of the truck, which is parallel to the ground. Since there is no acceleration perpendicular to the ground, we can ignore vertical forces for now.

Applicable formula: $\Sigma \mathbf{F} = m\mathbf{a}$

$$\Sigma F_x = (4400\ \text{kg})(+2.0\ \text{m/s}^2) = +8800\ \text{N}$$

The net force is 8800 N straight ahead.

6-27 An accelerating truck

EXAMPLE 6-4

Great Shot: Quantifying Inertia—Mass

An athlete in the shot put event exerts a force with a magnitude of 200. N on the shot. If the magnitude of the shot's acceleration is 20. m/s^2 in the initial direction of its path, what is the mass of the shot? Assume that the athlete accelerates the shot in a straight line.

Solution:

Choose a coordinate axis parallel to the motion of the shot. Only motion along this axis will be considered.

$$F_x = ma_x$$

$$m = \frac{F_x}{a_x} = \frac{+200.\ \text{N}}{+20.\ \text{m/s}^2} = \frac{+200.\ \text{kg}\cdot\cancel{\text{m/s}^2}}{+20.\ \cancel{\text{m/s}^2}}$$

$$m = 10.\ \text{kg}$$

6-28 A shot-putter

The previous two examples were one-dimensional situations. However, the second law applies in two and three dimensions as well. The second law formula is a vector equation. Consequently, both sides can be resolved into corresponding component equations:

$$\Sigma F_x = ma_x$$

$$\Sigma F_y = ma_y$$

$$\Sigma F_z = ma_z \text{ (for three dimensions)}$$

Problem-Solving Strategy 6.4

To determine the net force vector acting on a system, multiply each component of acceleration by the mass to find the net force components. Sum the components to find the resultant force.

6.14 Newton's Third Law of Motion

Newton recognized the need for one additional postulate to complete the fundamental laws of dynamics. The existence of the third law can be demonstrated using two spring scales, as described below.

Each scale is a separate system. Hook the spring scales together and pull them in opposite directions. Both record the same force when the springs are slightly stretched and held stationary. The force read on the right-hand scale is being exerted by the left-hand scale and vice versa. This shows that there are two forces involved in each spring scale system: the force applied *by the scale* on something outside its system, and the force acting *on the scale* from outside the system.

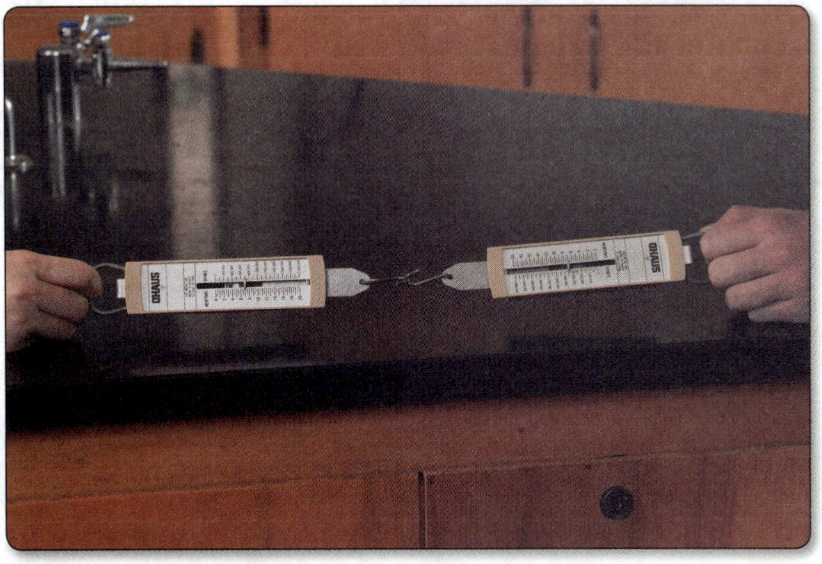

6-29 An action-reaction force pair

Newton's third law describes this balancing process. In his own words, *"To every action there is always opposed an equal reaction: or, the mutual actions of two bodies upon each other are always equal, and directed to contrary parts."* While the English translation of Newton's *Principia* used the words "action" and "reaction," these terms are not completely descriptive. Action-reaction force pairs can exist even if there is no visible action involved. To express this concept in more familiar terms:

Newton's third law reveals that every force is always accompanied by a second force acting in the opposite direction. It has been said, "You cannot touch without being touched."

> If system X exerts a force on system Y, then Y exerts a force of the same magnitude on X but in the opposite direction.

We denote the force of object X acting on Y by $\mathbf{F}_{X \to Y}$, where the arrow means "acts on." The other force in the action-reaction force pair is $\mathbf{F}_{Y \to X}$. In equation form, the third law can be expressed as

$$\mathbf{F}_{X \to Y} = -\mathbf{F}_{Y \to X}.$$

Forces have four properties as a consequence of the third law that you must keep in mind when solving physics problems:

- *All* forces occur in pairs.
- Each force in an action-reaction pair has the same magnitude.
- Each force acts in the opposite direction in line with the other force of the pair.
- Each force acts on a *different* system.

Students who don't understand the fourth property often have difficulties solving physics problems. According to the third law of motion, *every* force in nature is one of a pair of forces. Each force of an **action-reaction force pair** acts on a different system. When analyzing acceleration using the second law equation, you must first isolate the system of interest. Identify the action-reaction pairs, and include in your diagram only those forces that are acting *on* the system, not the forces that the system exerts on something else.

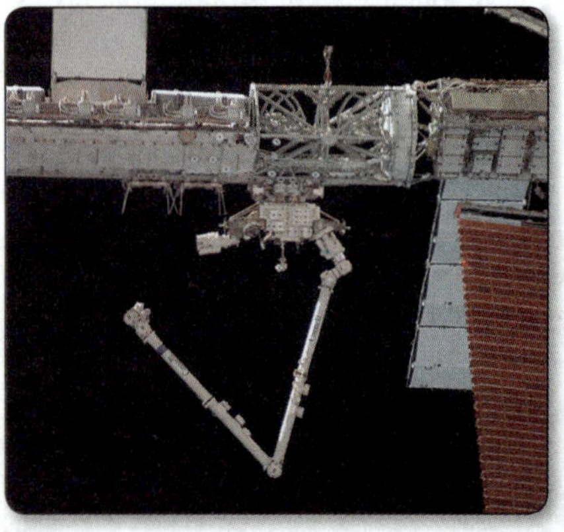

6-30a The Canadarm 2 mechanical arm serving at the International Space Station

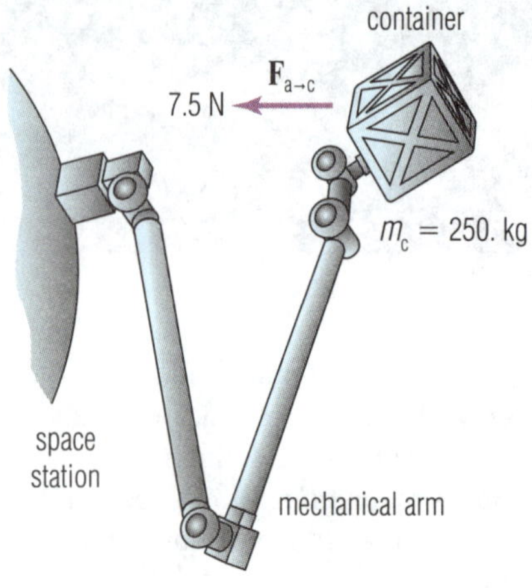

6-30b A manipulator arm moving a freight container

EXAMPLE 6-5

Cargo in Space: The Third Law

A remote-controlled robotic arm attached to the International Space Station is used for construction of the space station and the positioning of modular cargo containers in various external storage locations. While unloading a space shuttle, the arm exerts a force of 7.5 N on a 250. kg cargo container.

a. What force does the cargo container exert on the robotic arm?
b. What is the magnitude of the acceleration of the cargo container?
c. If the arm sustains this force for 5.0 seconds in a straight path, how fast will the container be moving?

Solution: See Figure 6-30b.

a. The force the manipulator arm exerts on the container is

$$F_{a \to c} = 7.5 \text{ N}.$$

Therefore, the reaction force component of the container on the arm that is in line with the arm's force is

$$F_{c \to a} = 7.5 \text{ N in the opposite direction.}$$

b. The system of interest is the cargo container. The only force acting on the container is the force exerted by the manipulator arm. (For now, we can ignore the force exerted by the earth's gravity, since it affects everything at the space station equally.)

$$F_{a \to c} = m_c a_c$$

$$a_c = \frac{F_{a \to c}}{m_c} = \frac{7.5\ \text{N}}{250.\ \text{kg}} = \frac{7.5\ \cancel{\text{kg}} \cdot \text{m/s}^2}{250.\ \cancel{\text{kg}}}$$

$$a_c = 0.030\ \text{m/s}^2$$

c. To find the container's speed after 5.0 seconds, use the first equation of motion,

$$v_2 = v_1 + a_c \Delta t.$$

Assume the container started at rest (in the space station's frame of reference), so $v_1 = 0$ m/s.

Substitute: $v_2 = 0\ \text{m/s} + (0.030\ \text{m/s}^2)(5.0\ \text{s})$

$$v_2 = 0.15\ \text{m/s}$$

Considering the large mass of the cargo container, the operator of the robotic arm must allow enough time to slow the container at its destination to avoid a damaging collision!

6.15 Using Forces to Solve Problems

Aircraft flight requires all kinds of forces for taking off and landing. During takeoff, lift is generated by the wings as they move through the air. The faster an aircraft can accelerate, the more quickly the lift can be generated to make the aircraft become airborne. When landing, the aircraft must rapidly decelerate after the wheels touch the ground. This is normally accomplished using friction through brakes and air resistance. On an aircraft carrier, these accelerations and decelerations must happen very quickly within a distance of a few hundred feet.

Modern aircraft carriers use some exceptional machines to assist aircraft with takeoff and landing. Currently, aircraft carriers are equipped with *steam catapults* to launch their fixed-wing aircraft. These are long, piston-like devices installed just below the flight deck. A sliding *shuttle* attached to the piston sits on top of the flight deck. The aircraft's front landing gear or the fuselage is attached to the shuttle with a tow bar or bridle. A *holdback* device is attached to the back side of the landing gear. The holdback allows the plane to run its engines at full throttle without moving. A *jet blast deflector door* is raised behind the aircraft during launch to shield people and other aircraft from its engine's high-velocity exhaust. At launch, steam from the reactor plant fills the catapult cylinders, the piston pulls the tow bar, breaking a specially-designed link in the holdback and hurling the aircraft down the deck at more than 3.5 gees. The plane becomes airborne as it leaves the flight deck.

Recovering an airplane on an aircraft carrier relies on an entirely different machine. The pilot flies the aircraft at the right angle, speed, and altitude to touch down at a point on the deck where the tail hook of his plane can snag one of four *arresting wires* placed across the landing area. The arresting wire is needed because the landing area is just too short for the aircraft to stop using its brakes alone. In fact, the pilot goes to full power as the aircraft touches down just in case the hook misses, so he has enough speed to take off and try to land again! As the arresting wire is pulled out from two hydraulic friction drums below deck, the plane rapidly decelerates from 130 knots to zero knots within a distance of less than 70 m (230 ft) by the fluid friction generated inside the drums.

6-31 An aircraft carrier catapult launch can generate more than 3.5 times the normal gravitational acceleration.

Aircraft carriers also increase the wind blowing down the deck by steering into the wind during launches and recoveries. By steaming at full speed, this technique can add more than 35 knots of wind, plus the wind speed itself, to the airspeed, creating the lift when the plane is catapulted.

6-32 A modern navy fighter-bomber recovering aboard an aircraft carrier

FACETS of PHYSICS

Flight and Newton's Third Law

At 412 775 kg (910 000 lb), the Boeing 747-400ER jumbo jet is still one of the largest commercial airliners in the world. In this chapter you learned that for every action there is an equal and opposite reaction (Newton's third law). If the reaction you desire is to get such a large plane into the air, the engines must generate an enormous amount of thrust.

While the jet is idling in line, waiting its turn to take off, the engine's turbines are rotating, generating an amount of thrust that is balanced by the force of the brakes on the wheels. If you have ever flown on a commercial jet, you remember the feeling as the jet rounds the end of the runway. You can hear the engines revving, and then you feel the explosive burst of power as the brakes are released. What is happening? As the engine exerts a great force on the exhaust gas by blowing it out the back nozzle, the exhaust gases exert a tremendous force on the plane, propelling it forward. When it is traveling fast enough to generate sufficient lift, the plane rises into the air. The 747-400 typically takes off at 290 km/h (180 mi/h), a speed attained by four engines that can generate up to 276 000 N (62 100 lb) of thrust each.

Thrust can be generated in two basic ways: by taking a large amount of air and increasing its speed a little, or by taking a smaller amount of air and increasing its speed greatly. The modern turbofan gas turbine engine combines both methods and is highly efficient.

The 747-400 lands at a speed of 260 km/h (160 mi/h), but how does the plane stop once it's on the ground? There are two systems deployed to assist the brakes in slowing the plane. First, spoilers raised on the top surface and flaps lowered on the trailing edge of the wing provide wind resistance. Second, thrust reversers enable the plane's engines to help slow the plane. After the plane touches down, you hear what may sound like the engines revving in reverse. Remember, however, that jet engines can't run backward. Instead, the thrust reversers mechanically deploy flaps into the exhaust to deflect the exhaust gases toward the front of the plane. This creates a strong net-braking force on the plane. The wing flaps, the brakes, and the thrust reversers all help slow the plane after it lands.

Applied physics, along with other disciplines, has been very important in the proliferation and development of aviation. To show you how far we have come, the 45 m economy section of a 747-400 is longer than the Wright brothers' first flight at Kitty Hawk, North Carolina!

6.16 Weight and Mass

Weight

The earth's matter mysteriously produces a measurable attraction on objects near the earth. This force keeps your feet planted firmly on the ground. Even more mysteriously, every object produces a gravitational force on every other object in the universe. Gravitational attraction between objects is related to the distance between them as well as their masses, so small and distant objects exert far less force than massive and near objects. We will investigate gravitational theory in greater depth in Chapter 7.

As we have discussed, the force of planetary gravitational attraction on an "average-sized" object is called the weight of the object, F_w. Weight is directly proportional to the object's mass. In equation form,

$$F_w \propto m, \text{ or}$$

$$F_w = km,$$

where k is a proportionality constant with units of N/kg. If we analyze the base units associated with k, $(\text{kg·m/s}^2)/\text{kg}$, we find that they simplify to m/s^2, which is the unit of acceleration. Substituting the symbol a for k into the weight equation gives $F_w = am$ or, after transposing the right side,

$$F_w = ma, \qquad (6.3)$$

which has a form similar to Newton's second law. However, rather than describing the response of a mass to a net force, this equation assigns a value to a particular force—the weight of the object. In Chapter 3 you learned that the acceleration due to gravity for freely falling objects is $\mathbf{g} = 9.81 \text{ m/s}^2$ straight down. The local acceleration due to gravity is the response of the mass of an object to the gravitational force acting on it (its weight).

It is common practice to substitute **g** for acceleration in the vector form of Equation 6.3:

$$\mathbf{F}_w = m\mathbf{g} \qquad (6.4)$$

Thus, a mass of 1 kg has a weight of 9.81 N and a mass of 10 kg has a weight of 98.1 N. Equation 6.4 is true no matter what the condition or orientation of the object. You can calculate the weight of an object using Equation 6.4 even if it is *not* falling. The weight of a rock of mass m resting on the ground is equal to $m|g|$. An object moving at "normal" speeds has the same weight as it does when stationary. (This is *not* the case for objects moving at a significant fraction of the speed of light—see Chapter 27.) For objects within a few thousand meters of the earth's surface, the proportionality between weight and mass is nearly constant over the entire planet.

Equation 6.4 is a vector equation. The magnitude of an object's weight vector is equal to $|m\mathbf{g}|$, and the weight vector points in the same direction as the gravity vector, which is normally considered to be straight down (toward the center of the earth). The scalar component form of Equation 6.4 is

$$F_{wy} = mg_y.$$

There are other units for weight besides the newton. The most familiar unit in the United States is the pound, which is equivalent to the English Engineering System (EES) unit of force, the **pound-force (lbf)**. One pound-force is equivalent to the weight, measured at the earth's surface, of a mass equal to one EES **pound-mass (lbm)**. None of the measurement units based on the EES commonly used in the United States are approved by NIST for use with SI units in scientific research. We will normally avoid using EES units in this textbook.

Problem-Solving Strategy 6.5

Calculate the weight of an object by using the formula $\mathbf{F}_w = m\mathbf{g}$. The direction of the weight vector is always in the same direction as **g**.

6-33 Sport utility vehicles are heavier than standard passenger cars.

> **Problem-Solving Strategy 6.6**
>
> When calculating the weight of an object, report only its magnitude unless you need its vector form for solving other portions of the problem.

EXAMPLE 6-6

A Massive Problem: Weight and the Second Law

What is the mass of an 1.80×10^4 N SUV?

Solution:

$$F_w = |m\mathbf{g}| = m|\mathbf{g}|$$

$$m = \frac{F_w}{|\mathbf{g}|} = \frac{1.80 \times 10^4 \text{ N}}{9.81 \text{ m/s}^2}$$

$$m \cong 18\underline{3}4 \text{ kg} \cong 1830 \text{ kg}$$

Mass is always a positive scalar quantity.

EXAMPLE 6-7

Weighing in on Mars: A different *g*

If Mars has a gravitational acceleration of 3.69 m/s^2, how much would the SUV in Example 6-6 weigh there?

Solution:

$$F_w = |m\mathbf{g}_{\text{Mars}}| = |(18\underline{3}4 \text{ kg})(-3.69 \text{ m/s}^2)|$$

$$F_w \cong 67\underline{6}7 \text{ N} \cong 6770 \text{ N}$$

Mass

The method for measuring mass should be familiar to you from other science courses. A typical mechanical high-school balance is a seesaw-like device that compares the force of gravity on an unknown mass to the force on known masses. A mass balance would not work in orbit where there is no apparent gravity. Depending on the type of balance, the known masses may be individual calibrated masses or, more commonly, sliders on a calibrated arm. When the known masses are adjusted in order to align the balance at the zero index mark, the sum of the known masses must equal the unknown mass.

Other kinds of instruments can measure mass. Electronic scales measure the weight of an object and digitally display the corresponding mass based on Equation 6.3.

Trace amounts of unknown substances in a sample of matter can be identified from their characteristic molecular masses using a mass spectrometer. This instrument accelerates charged atoms or molecules using a known force in a vacuum chamber to determine their masses.

Some of the force-measuring instruments mentioned in Section 6B can be configured to measure mass as well. Several mass-measuring instruments are pictured in Figure 6-34.

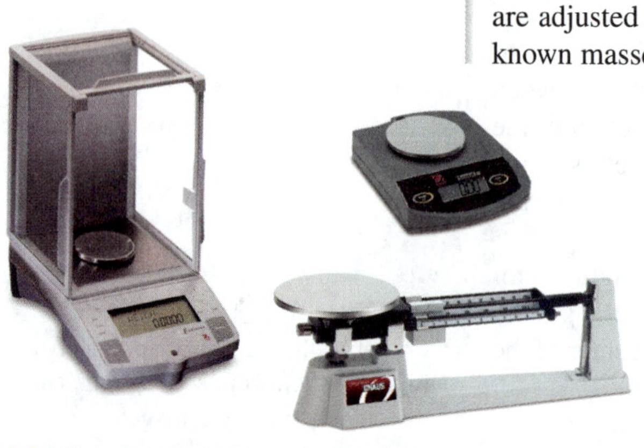

6-34 Various mass-measuring instruments

6C Section Review

1. Batter up! With his feet firmly planted on the ground, the batter twists his body during the swing and exerts a force on the bat as it strikes the ball, sending the ball deep into the outfield. Sketch a diagram that defines the "system" for the purpose of describing the ball's motion at

the instant it is hit. Include in your sketch any vector quantities that you think are acting *on* the system.

2. What can you conclude about the forces acting on an airplane if it is maintaining constant altitude, constant speed, and constant direction?
3. Does the force you use to roll a bowling ball down the lane depend on the ball's acceleration, or does the ball's acceleration depend on the force that you exert? Explain your answer.
4. A physics student exerts a force of 625 N straight down while standing on the ground. What other force must exist according to Newton's third law of motion? Describe the magnitude and the orientation of the force and identify the system upon which it acts.
5. (DS) a. Calculate the acceleration of a 16 600 kg Navy fighter-bomber launched by a catapult on an aircraft carrier if it accelerates from 0 km/h to 266 km/h in 2.0 seconds.
   b. Calculate the net force of the catapult on the aircraft.
   c. Why is it important for physicists and engineers to know this kind of information?
6. Compute your weight in newtons. (1 lbm = 0.454 kg)
7. The gravitational acceleration at the surface of the moon is 1.62 m/s^2. If you can barely lift 195 lbf on the earth, what would be the maximum weight that you could lift on the moon in newtons?
8. A 0.170 kg puck rests on frictionless ice. Two hockey players strike the puck simultaneously. One force is 350. N exerted at 020.0T, and the other force is 600. N exerted at 065.0T. What is the magnitude and direction of the puck's acceleration at the instant that it is struck?

6C Objectives

After completing this section, I can

- ✓ define *system* as used by physicists.
- ✓ state Newton's first, second, and third laws in my own words.
- ✓ explain why the first law defines mechanical equilibrium.
- ✓ perform dynamics calculations using Newton's second law.
- ✓ identify the forces and systems subject to Newton's third law for a given dynamics problem.
- ✓ calculate the weight or mass of a system as needed using Newton's second law.

Chapter Review

In Terms of Physics

Problem-Solving Strategies

6.1 (page 117) Always sketch forces as *pulls* in a force vector diagram.

6.2 (page 118) If one of a resultant's components is zero, then the resultant has the same magnitude and direction as the nonzero component.

6.3 (page 124) After sketching a physics problem, determine the system in the problem, draw a circle around it, and identify all forces acting *on* the object(s) inside the circle.

6.4 (page 127) To determine the net force vector acting on a system, multiply each component of acceleration by the mass to find the net force components. Sum the components to find the resultant force.

6.5 (page 131) Calculate the weight of an object by using the formula $\mathbf{F}_w = m\mathbf{g}$. The direction of the weight vector is always in the same direction as $\mathbf{g}$.

6.6 (page 132) When calculating the weight of an object, report only its magnitude unless you need its vector form for solving other portions of the problem.

Review Questions

1. What common phenomenon likely accounts for the Greek natural philosophers' observation that objects naturally tend to slow down and that heavier objects fall faster than lighter objects?
2. What did Galileo conclude about the natural tendency of any moving object?
3. Which of the following statements would Newton disagree with? Why?
   a. An object's motion changes in response to an unbalanced force.
   b. For every action, there is an equal and opposite reaction.
   c. A force is always needed to maintain a constant velocity.
   d. The acceleration of an object is proportional to the net force acting on it and is inversely proportional to its mass.
4. Galileo referred to an "agent" that causes a change in the state of motion of an object. What is this agent?
5. At what point in the Creation event did the fundamental forces likely appear?
6. What force causes the transmission of light?
7. Which model of noncontact force suggests that the space around an object is involved in conveying the force to another object?
8. The sum of all forces on an object is called
   a. the weight vector.
   b. the resultant force.
   c. the equilibrant force.
   d. the reaction force.
9. What is the minimum number of forces that must act on a system in order to have
   a. an unbalanced force on the system?
   b. a balanced force on the system?
10. What do you know about a system of unknown forces if it has a nonzero equilibrant?
11. The diagram shows a raccoon (*Procyon lotor*) resting on a log.
    a. If you need to find the forces acting on the raccoon, what is the system?
    b. What is the system if you must find the forces on the log?
    c. Identify the system if you must determine the force exerted by both the raccoon and the log together.

12. A hockey puck slides across the ice with a constant velocity. Does this mean that there is no force acting on the puck? Explain your answer.
13. According to Newton's second law, doubling a net force on a system of constant mass will have what result?
14. What characteristic of the acceleration vector ($\mathbf{a}$) of a system is identical to the force sum vector ($\Sigma \mathbf{F}$) acting on the system?

DS 15. When a navy jet lands on an aircraft carrier, identify the action and reaction forces at the tailhook of the aircraft.

16. A hot-air balloon is floating at a constant altitude of 3000 m.
    a. Against what is the total weight of the balloon, basket, passengers, and hot air acting, according to Newton's third law of motion?
    b. Describe the reaction force to this combined weight.
17. What is the unit of weight in the SI? in the EES?
18. Do all mass-measuring devices involve a balance of some sort? Explain your answer.

True or False (19–28)

19. Sir Isaac Newton originated the concept of inertia.
20. Forces are really only pulls applied by one system on another.
21. All forces on an object cause a change in the object's motion.
22. Every fundamental force is at work in a flower vase sitting on a shelf.
23. It is possible for a scientific law to *not* hold true in every case where it is tested.
24. All three laws of motion apply in every instance where a force is exerted on a system.
25. A heavier object has a larger gravitational force acting on it than a lighter object does. Therefore, according to Newton's second law, it will experience a greater acceleration toward the ground than the lighter.
26. The "action" force never acts on the same system as the "reaction" force.
27. The value of g on the moon is smaller than on the earth because objects have smaller masses on the moon than on the earth.
28. Astronauts in the International Space Station can use a standard laboratory balance when finding the mass of an object.

29. Three cattlemen are handling an unruly calf (*Bos taurus*). All three have ropes around the calf's neck. One man is directly in front of the calf, exerting a force of 120. N. A second man is at a right angle to the calf on its right, pulling with a force of 110. N. The third man is on the calf's left and ahead of it at a 30.0° angle from the rope of the first man, pulling with a force of 116 N. If the calf does not move, what forces is it exerting?
30. A 6.67×10^{-2} kg leopard frog (*Rana pipiens*) jumps into the air with an acceleration of 0.300 m/s^2 straight up. What is the net force on the frog as it jumps?
31. Three men are trying to push a 3.08×10^3 kg car up an icy hill. The driver pushes with a force of 430. N straight up the hill. Each of the other men pushes on a rear corner of the car with a force of 708 N at an angle of 45.0° from the direction the driver is pushing. The other forces on the car can be represented by a force of 800. N straight downhill. What is the magnitude of the acceleration of the car and in which direction?

32. A 25.0 kg bicycle is traveling at a speed of 10.0 m/s.
   a. If the bike comes to rest in 2.0 s, what is the magnitude of its acceleration?
   b. What net force is needed to provide this acceleration?
   c. How far will the bike travel in the 2.0 s?

DS 33. a. Calculate the deceleration of a 24 500 kg aircraft landing on an aircraft carrier with the aid of an arresting hook if it slows from 241 km/h to 0 km/h in 2.0 seconds within a distance of 70.1 m.
   b. What is the net force of the arresting wire on the aircraft?

34. The earth's mass is $m_E = 5.97 \times 10^{24}$ kg. An object of mass m_o falls from a height of 100. m, where it was at rest.
   a. How long does the object fall?
   b. Write the symbol for the force that the earth exerts on the object.
   c. Write the symbol for the force that the object exerts on the earth.
   d. What is the formula for determining the acceleration of the earth toward the object? Use the symbols from Parts b and c.
   e. What is the value of m_o if the earth moves 1.00×10^{-15} cm toward the object during the object's fall toward the earth?

35. The earth exerts a force of 1.00 N on an object in free fall. What is the object's mass?

36. What is the weight of a 5 g mass in newtons? (Assume that the mass is at rest at the surface of the earth.)

37. A mass m_1 has an acceleration a_1 when it is acted upon by a net force of magnitude F. If a second mass, $m_2 = 3m_1$, is acted upon by the same force F, what is its acceleration compared to a_1?

38. A mass of 2.50 kg hangs from a string.
   a. What force does the mass exert on the string?
   b. What force does the string exert on the mass?

39. A hunting bow imparts an acceleration of 5300. m/s² to a 29.3 g arrow at an angle of 15.0° to the horizontal. What is the horizontal component of the force exerted by the bow on the arrow?

40. A raindrop's terminal velocity (the velocity at which the falling drop stops accelerating due to air resistance) is $\mathbf{v} = -1.0$ m/s. If the force of air resistance can be modeled by the equation $\mathbf{F}_r = -b\mathbf{v}$, and $b = 4.9 \times 10^{-3}$ kg/s, what is the raindrop's mass?

41. You are test driving a brand new 1350 kg BMW sports car. You find that the car accelerates from 0 to 96.0 km/h in 5.8 s along a level stretch of interstate (speed limit 60 mi/h). Assuming that the acceleration is constant, solve the following problems.
   a. Calculate the magnitude of the net force that produces the acceleration.
   b. Calculate the distance the car covers during the period of acceleration.
   c. If the road is inclined at an angle of 7.0° to the horizontal but the car can still attain 96 km/h in 5.8 s, is the magnitude of the net force the same as in Part a? How does it compare? Explain.

42. A 1.50×10^3 kg car is pulling a 600. kg trailer. Leaving a traffic stop, the car accelerates at 0.200 m/s².
   a. What is the magnitude of the total force on the trailer?
   b. What is the magnitude of the total force on the car?
   c. What is the magnitude of the total force on the system consisting of both car and trailer?

43. A soccer ball of mass 0.380 kg bounces off the head of the sweeper. The incoming speed is 14.0 m/s and the outgoing speed is 17.0 m/s in the opposite direction. The ball is in contact with the player's head for 0.080 s.
   a. Determine the magnitude of the ball's average acceleration during the moment of the play.
   b. What is the magnitude of the force provided by the player's head to cause this acceleration?

44. Research and write a brief report about the discovery of calculus. Include a discussion of the contention between Newton and Leibniz.

Aristotle

The ancient Greeks spent much time trying to explain their surroundings, but their philosophical approach led to some unusual conclusions. For example, at one time there were two schools of thought about motion. One school said that all things are in continual motion; the other said continual motion is impossible. However, some philosophers disagreed with both schools. Among them was Aristotle (384–322 BC).

Aristotle used his reason to explain motion. He began with what he thought was obvious—heavy objects fall, and the heavier they are, the faster they fall. He reasoned from this fact that every object moves to its natural place and stops. The heavier an object is, the more vigorously it seeks its resting place. Therefore, a falling object's speed is proportional to its weight. Since "everybody knew" that heavy objects fall faster than light objects, Aristotle thought it unnecessary to experiment to prove his theory.

Aristotle applied his reason to a wide variety of other subjects. In logic, he invented the deductive syllogism, which we still use today to test hypotheses and theories. His writings have had a profound influence on Western thought in fields as diverse as psychology, ethics, speech, poetry, logic, physics, metaphysics, astronomy, meteorology, and biology. Usually he used his "common-sense" philosophy, but he was not completely opposed to experimentation for his biological treatises. In light of his philosophical approach and broad range of topics, it is not surprising that a few of his ideas were incorrect. What is surprising is the insight and wisdom he possessed in spite of the limitations of his day.

The rings of Saturn are composed of innumerable particles of dust and ice moving silently in great, nearly circular paths around the planet.

Circular Motion 7

Dominion Modeling

Saturn's Moons

Saturn, the ringed planet, is giving up its secrets. In 1997, the Cassini-Huygens mission, a cooperative venture of NASA, the European Space Agency (ESA), and the Italian Space Agency (ASI), was launched to investigate Saturn and its moons. Since the mission's arrival in the Saturnian system in 2004, scientists have uncovered some amazing things. Prior to the mission, Saturn had only eighteen known moons. However, information from many sources has revealed a stunning sixty moons in the Saturnian system, including five tiny ones detected by the Cassini-Huygens probe.

This mission has investigated Saturn and its moons, peering through the nearly opaque atmosphere of Titan, Saturn's largest moon, and landing a probe on that strange moon's dark surface. It has made several close encounters with the moon Enceladus, which, encrusted with frozen water, is one of the shiniest objects in the solar system. The moons vary in size from tiny Anthe, which is approximately 2 km (1 mi) in diameter, to huge Titan, with a diameter of 5151 km (3201 mi), which is slightly broader than the continental United States. The time the moons take to orbit Saturn also varies widely, from half a day to nearly four years! The average radii of the orbits of these moons range from 134 000 km to nearly 25 million km. Is there any way we can model the arrangement of Saturn's moons according to the principles of circular motion?

7-1 Saturn, with its magnificent rings

Enceladus: (en SEL uh dus)

Funding Faulty Science?

The Cassini-Huygens mission received funding from institutions and lots of attention from scientists because both Enceladus and Titan have conditions which scientists believe favor the evolution of life. How are the priorities for scientific research determined by scientists' and politicians' presuppositions?

7.1 Introduction

In the preceding chapters, we examined the properties of linear motion and simple two-dimensional projections that occur when a system is uniformly accelerated. Many other forms of motion exist in the "real world." One kind of motion that is extremely common and very important today is *rotational* or *circular* motion. Life in the modern world would be inconceivable without circular motion. We would not have wheels, printing press rollers, computer hard drives, compact disc players, steam turbines, electric motors, gas engines, or ferris wheels, to name just a few things.

7-2 The Galilean moons of Jupiter also clearly demonstrate the mechanics of celestial circular motion.

Certain natural systems approximate circular motion. Ancient Greeks believed that the celestial motions of the heavenly bodies followed "perfect" circles. Kepler showed that the celestial motions were elliptical instead. However, most planetary orbital motion is nearly circular to a first approximation. Planets also spin on their axes so that a point on a planet's surface describes a circular path. Humans have mimicked celestial motions with artificial satellites such as the International Space Station and the GPS satellites.

We will begin our study of circular motion by describing how systems move in a circular path—the kinematics of circular motion. Then, as we did for linear motion, we will investigate some of the causes of circular motion, including gravity.

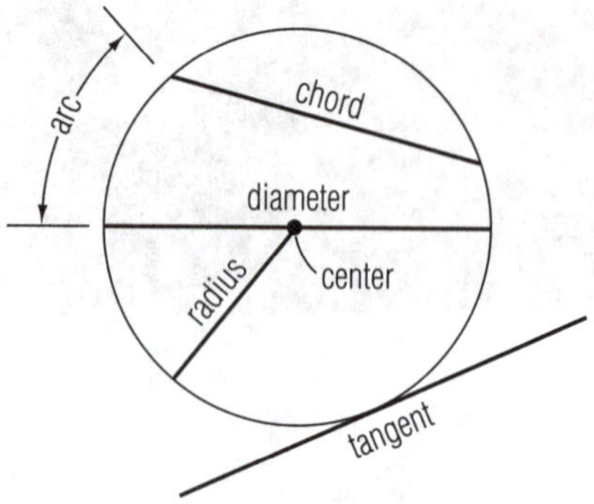

7-3 Geometric parts of a circle

7A CIRCULAR MOTION

7.2 Defining Circular Motion

Circular motion occurs when an object moves in a two-dimensional circular path. There are two general categories of circular motion. If an object rotates about an axis that passes through the object itself, then we say the object is spinning or has **spin**. If the object circles an axis that does *not* pass through the object itself, then we say that it *revolves* about an axis in **orbital motion**. The earth experiences both spin and orbital motion. Orbital motion is the more general case, but the formulas that will be developed can be applied to points on spinning objects as well.

You may want to stop for a moment here to review some terms used for circles. A *chord* is any straight line that connects two points on a circle. A *tangent* is a line that touches the circle in only one point and does not lie in the interior of the circle. An *arc* is a curved line segment that is part of a circle (see Figure 7-3).

7.3 Establishing Position in Circular Motion

By definition, the positions of an object experiencing circular motion lie in a plane. What is the best method for describing the position in circular motion? Physicists use position vectors for noncircular two-dimensional motion. How do position vectors work for circular motion?

A child is playing on a merry-go-round on a playground. He sits still while the merry-go-round moves at a constant speed. Since the merry-go-round rotates, the child travels in a circle. How would you describe his position during this motion?

You should begin by choosing a coordinate system to use as a position reference. Figure 7-4 shows three possible coordinate systems to use. Which is simplest? Figures 7-4a and b (and any coordinate system in which the origin is not centered on the rotational axis of the circle) give position vectors of different lengths and angles for different points in the child's path. The coordinate system in Figure 7-4c, however, gives position vectors with the same length for every point of the child's path. This is the simplest coordinate system for analyzing circular motion.

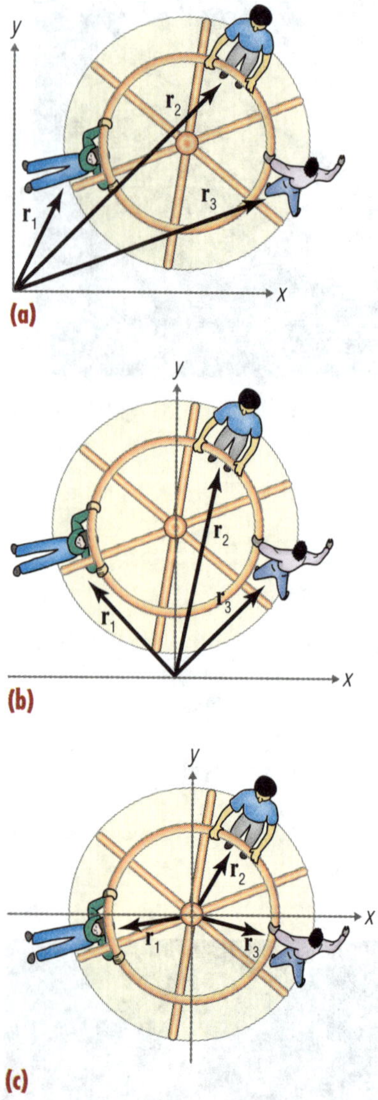

7-4 Three possible coordinate systems for describing circular motion. Figure (c) shows the system that is simplest to use.

The position vector of a rotating object at any time t_n is $\mathbf{r}_n$. The magnitude of $\mathbf{r}$ is just the radius of the circular path, r. The position of the radial vector is uniquely described by its angle, θ. Therefore, the position of an object in a circular path can be given by the ordered pair (r, θ). These are known as *polar coordinates.*

In Chapter 2 we introduced the SI unit for angular measurement—the *radian* (rad). Recall the definition of a radian:

> One radian is equal to the central angle of a circle that subtends an arc of the circle's circumference whose length is equal to the length of the radius of the circle.

There are exactly 2π rad in one complete circle. If necessary, review the discussion in Chapter 2 addressing degree-radian conversions and other properties of radian angle measurement.

The kinematics of circular motion are analogous to those of linear motion. Where motion in straight-line kinematics is measured in units of length, such as the meter, change of position in circular motion is measured in angular units. An angle θ is the angular position of the position vector from a reference axis or reference direction and is analogous to the linear position x on a number line. It can have either a positive or negative value.

An angular position cannot be considered a position vector in the same way that a number line coordinate can represent a one-dimensional position vector. Although angles and angular changes have magnitude and direction, they cannot be represented by straight arrows, and they do not obey all of the rules for vector operations. Consequently, sums and differences of angles are scalar and not vector quantities.

7.4 Speed and Velocity in Circular Motion

The symbol ω represents the time-rate of change of angular position, or the angular speed. The angular speed can be described by the angular change divided by time, or $\Delta\theta/\Delta t$. This is a scalar quantity. In common usage, angular speed may be expressed as the number of rotations per unit time. Each rotation is called a *revolution.* Angular speed thus can have units of *revolutions per unit time* (e.g., revolutions per minute [rpm], such as indicated by an engine tachometer). Rotational rate may have decimal values, such as 3.4 rpm. If angular speed is constant, then the rotating object is experiencing **uniform circular motion**.

As with the symbol for any scalar quantity, the Greek letter omega (ω) is not bold when representing angular speed.

In science, angular speed is expressed in *radians per second* because radians and seconds are the SI units for angle and time, respectively. When an object experiences uniform circular motion, its angular speed indicates the number of radians that a point in a rotating object passes through each second. Example 7-1 demonstrates a simple conversion from revolutions per minute to SI units.

Angular speed is the time-rate of change of angular position:

$$\omega = \frac{\Delta\theta}{\Delta t} = \frac{\theta_2 - \theta_1}{\Delta t}$$

It may be expressed in revolutions per unit time (e.g., rpm) or radians per second (s^{-1}).

EXAMPLE 7-1

Angular Speed Conversion

Convert 45 rpm to rad/s.

Solution:

Use the unit analysis bridge utility. The first term in the bridge is the given quantity.

$$\frac{45\ \cancel{\text{rev}}}{\cancel{\text{min}}}\left|\frac{1\ \cancel{\text{min}}}{60\ \text{s}}\right|\frac{2\pi\ \text{rad}}{\cancel{\text{rev}}} \cong 4.7\ \text{rad/s or } 4.7\ \text{s}^{-1}$$

The radian is a dimensionless SI unit, so "rad" is often omitted from angular speed units. Therefore, "rad/s" may appear as just "s^{-1}."

The velocity vector of a particle in circular motion is tangent to the circular path of the particle. This velocity is called **tangential velocity ($\mathbf{v}_t$)**. The tangential velocity vector is perpendicular to the radius of the circular path and, thus, to the position vector of the rotating object.

The magnitude of tangential velocity is *tangential speed,* $v_t = |\mathbf{v}_t|$. Tangential speed is equal to the length of arc traveled along the circular path in one second, or

$$v_t = \frac{\overset{\frown}{l}}{\Delta t}.$$

Arc length ($\overset{\frown}{l}$) is equal to the radius of the circle times the change of angle, $r \times \Delta\theta$, where $\Delta\theta$ is in radians. Therefore, average tangential speed, $\bar{v}_t$, is determined by

The tangential velocity of a rotating object is perpendicular to the object's position vector:

$$\mathbf{v}_t \perp \mathbf{r}$$

Problem-Solving Strategy 7.1

Tangential speed can be calculated by the formula $v_t = r\omega$ as well as by $v_t = r\Delta\theta/\Delta t$.

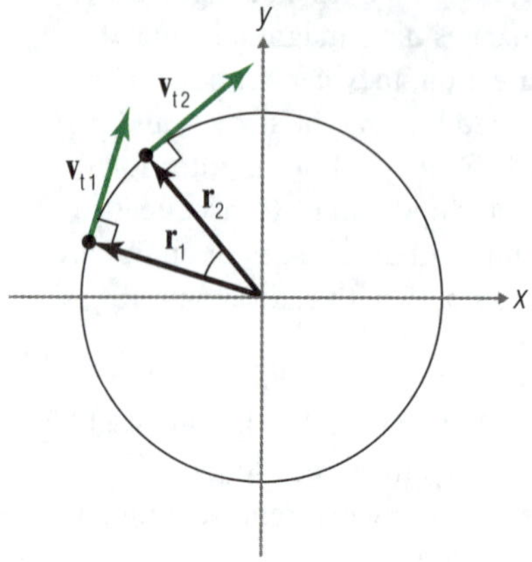

7-5a Velocity vectors for $\mathbf{r}_1$ and $\mathbf{r}_2$

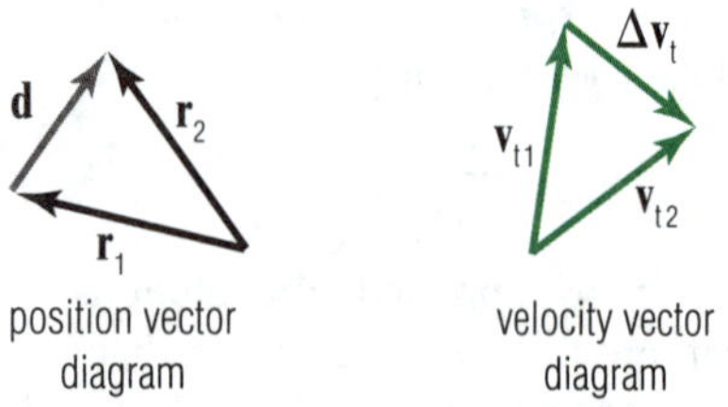

7-5b Adding vectors to form a triangle

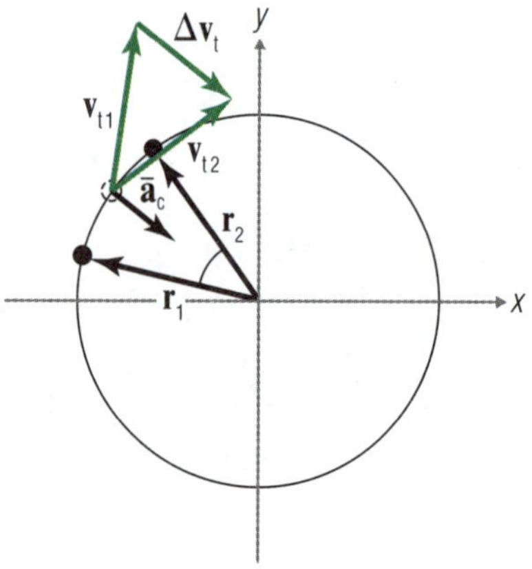

7-5c Centripetal acceleration

Centripetal is derived from the Latin words for "center-seeking."

$$\bar{v}_t = \frac{r\Delta\theta}{\Delta t}. \tag{7.1}$$

Tangential speed has linear motion units of m/s.

7.5 Acceleration in Circular Motion

Referring back to the merry-go-round, in uniform circular motion the child's speed is constant, but his tangential velocity vector continually changes direction. Thus, the child experiences an acceleration. The question is, what is the magnitude and the direction of his acceleration? You can find his average acceleration for an interval just as you did in earlier chapters:

$$\bar{\mathbf{a}} = \frac{\Delta\mathbf{v}}{\Delta t} \tag{7.2}$$

Figure 7-5a shows the velocity vectors for positions $\mathbf{r}_1$ and $\mathbf{r}_2$. Figure 7-5b shows the displacement vector triangle and velocity difference vector triangle as well. Since $\mathbf{v}_{t\,1} \perp \mathbf{r}_1$, $\mathbf{v}_{t\,2} \perp \mathbf{r}_2$, $v_{t\,1} = v_{t\,2} = v_t$, and $r_1 = r_2 = r$, the two triangles in Figure 7-5b are similar. Therefore, the ratios of the corresponding sides are equal:

$$\frac{\Delta v_t}{d} = \frac{v_t}{r}$$

Multiplying both sides by d gives

$$\Delta v_t = \frac{v_t d}{r}.$$

Substituting this expression for Δv in Equation 7.2, you get

$$\bar{a} = \frac{v_t d/r}{\Delta t} = \frac{v_t d}{r\Delta t}, \text{ so}$$

$$\bar{a} = \frac{v_t}{r} \cdot \frac{d}{\Delta t}.$$

But, for very small time intervals, $d/\Delta t$ is just v_t. So

$$\bar{a} = \frac{v_t}{r} \cdot v_t, \text{ or}$$

$$\bar{a} = \frac{v_t^2}{r}. \tag{7.3}$$

Equation 7.3 is an expression for the *magnitude* of the acceleration experienced by the child rotating on the merry-go-round in uniform circular motion.

For very small time intervals (Δt), the tangential velocity at the beginning of the interval, $\mathbf{v}_{t\,1}$, is almost parallel to the velocity at the end of the interval, $\mathbf{v}_{t\,2}$. The difference vector $\Delta\mathbf{v}_t$ is essentially perpendicular to both $\mathbf{v}_{t\,1}$ and $\mathbf{v}_{t\,2}$ and points in the direction of $-\mathbf{r}$. Since $\bar{\mathbf{a}} = \Delta\mathbf{v}/\Delta t$, $\bar{\mathbf{a}}$ also points in the direction of $-\mathbf{r}$. That is, the child's instantaneous acceleration vector always points toward the center of the child's circular path. Such acceleration is called centripetal **acceleration ($\mathbf{a}_c$)**.

During uniform circular motion, the magnitude of the child's centripetal acceleration is constant at

$$a_c = \frac{v_t^2}{r}\ \text{m/s}^2.$$

In other words, the magnitude of centripetal acceleration is directly proportional to the square of the tangential speed and inversely proportional to the radius of the circular motion. The direction of $\mathbf{a}_c$ is opposite to the child's position vector direction, toward the center of the circle. This is true for all circular motion at constant radius and speed.

EXAMPLE 7-2

Slings and Stones: Centripetal Acceleration

A boy in a corn field is attempting to hit a can on a fence post with a 30.0 g stone thrown from a sling. As he swings the sling horizontally over his head in a circle, the stone has a tangential speed of 25.0 m/s. The sling is 1.20 m long from hand to stone. What is the magnitude of the centripetal acceleration on the stone?

Solution:

Note that Equation 7.3 does not involve the mass of the revolving object, so the mass of the stone is irrelevant.

Substitute the known values into Equation 7.3:

$$a_c = \frac{v_t^2}{r} = \frac{(25.0\ \text{m/s})^2}{1.20\ \text{m}}$$

$$a_c \cong 52\underline{0}.8\ \text{m/s}^2 \cong 521\ \text{m/s}^2$$

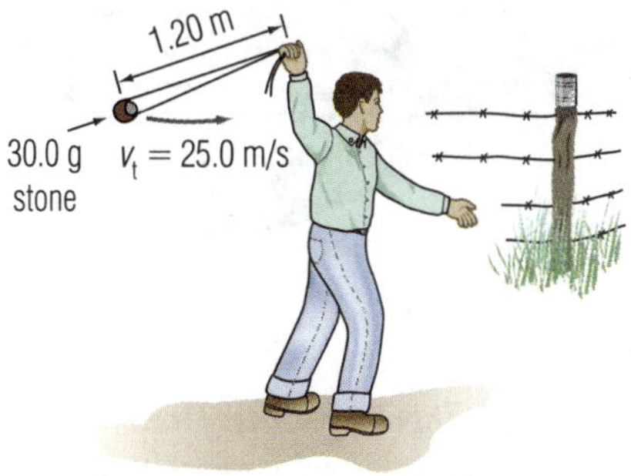

7-6 Slinging stones in residential areas is not recommended!

We can also use the angular speed to determine centripetal acceleration. First, we must express the tangential speed in terms of angular speed.

$$v_t = r\omega \tag{7.4}$$

Unit analysis of Equation 7.4:

$$\frac{\text{m}}{\text{s}} \Rightarrow \text{m} \cdot \frac{\langle\text{rad}\rangle}{\text{s}} = \frac{\text{m}}{\text{s}}$$

(Radians are dimensionless in terms of SI base units.)

If we substitute Equation 7.4 for v_t into Equation 7.3 we obtain

$$a_c = \frac{(r\omega)^2}{r} = \frac{r^{\not{2}}\omega^2}{\not{r}}$$

$$a_c = r\omega^2. \tag{7.5}$$

Equation 7.5 can be converted directly into a vector equation for centripetal acceleration:

$$\mathbf{a}_c = -\mathbf{r}\omega^2 \tag{7.6}$$

Equation 7.6 is the vector equation for centripetal acceleration. Equation 7.3 is a non-vector equation that provides only the magnitude of the centripetal acceleration.

Note that the negative sign indicates that the centripetal acceleration is oriented opposite to the direction of the position vector, **r**, which is consistent with the empirical observations on the merry-go-round.

7.6 Angular Velocity

Angular speed is the scalar measure of the number of radians through which the position vector of an object moves per unit time. It may also be considered the magnitude of a related vector quantity called **angular velocity (ω)**. A *uniform* angular velocity implies that the rate *and* direction of this aspect of circular motion are constant. How can a vector of circular motion be constant?

Here we introduce the **right-hand rule for circular motion**. Wrap the four fingers of your right hand around the axis in the direction of rotation of the object. Your extended thumb held at a right angle to the fingers will point in the direction of the angular velocity vector (see Figure 7-7). This is one of the few occasions in which you will be concerned about the third dimension, defined by

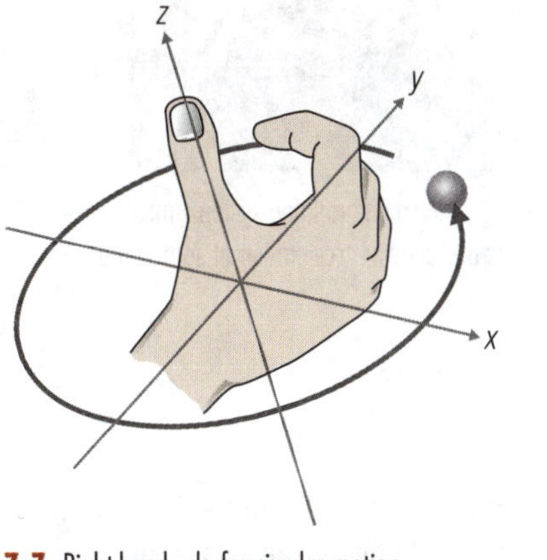

7-7 Right-hand rule for circular motion

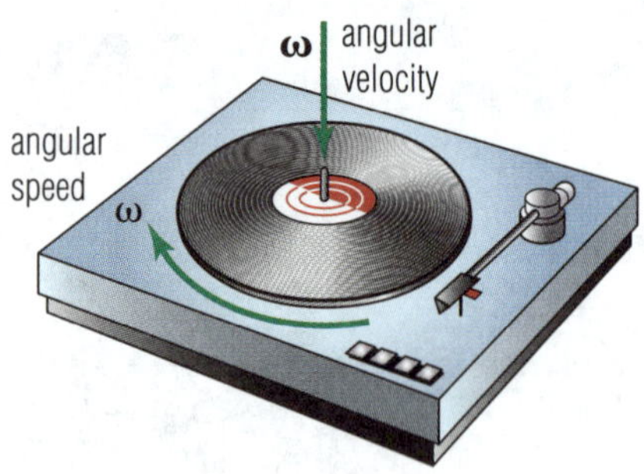

7-8 An LP turntable

Average angular velocity and instantaneous angular velocity are equal for uniform circular motion, unlike many other vector quantities of circular motion, such as tangential velocity and centripetal acceleration. Why is this so?

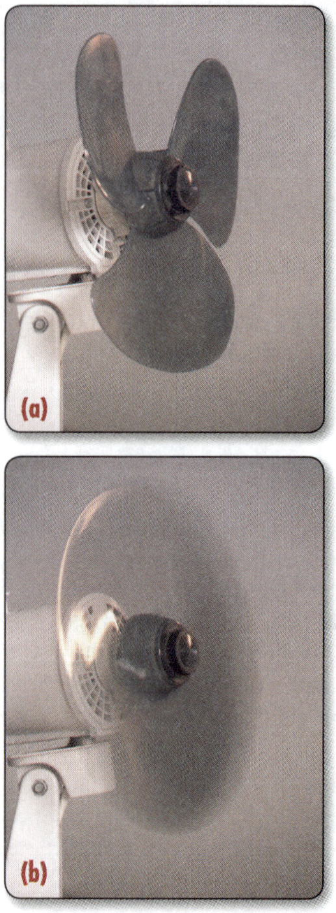

7-9 All man-made rotating systems must experience angular acceleration at some time.

Angular acceleration (α) is the time-rate of change of angular velocity.

the *z*-axis, which is parallel to the axis of rotation. The angular velocity vector points in the positive *z*-direction for counterclockwise rotation and in the negative *z*-direction for clockwise rotation. Similarly, positive scalar values of $\boldsymbol{\omega}$ indicate counterclockwise rotation, and negative values indicate clockwise rotation. The magnitude of $\boldsymbol{\omega}$ is indicated by arrow length, as is usual for vectors. The vector's orientation does not change unless the direction of circular motion changes.

EXAMPLE 7-3

An LP Record: Angular Velocity

Before the advent of cassette tapes, audio CDs, and MP3 players, long playing (LP) records were popular with music listeners. An LP rotates clockwise on a record player turntable at $33\frac{1}{3}$ rpm. Determine the magnitude and direction of the angular velocity vector of an LP record. Assume up is positive.

Solution:

The angular speed is $\omega = 33\frac{1}{3}$ rpm $= 33.333$ rpm and rotation is clockwise. Convert rpm to radians per second as follows:

$$\omega = \frac{33.333\ \text{rev}}{\text{min}} \left| \frac{1\ \text{min}}{60\ \text{s}} \right| \frac{2\pi\ \text{rad}}{1\ \text{rev}} \cong 3.490\ \text{rad/s}$$

$$\omega \cong 3.49\ \text{s}^{-1}$$

Applying the right-hand rule for circular motion to the record as it is revolving, the right thumb points downward. Therefore, angular velocity is

$$\boldsymbol{\omega} = -3.49\ \text{s}^{-1}.$$

7.7 Nonuniform Circular Motion

Uniform circular motion in rotating systems does not instantly come into existence. When you start the engine of your car, its rotating parts start at rest, and at some later time they are rotating at a constant angular speed as the engine is idling. A ferris wheel stops in order for passengers to get on or off, but it takes some time for it to get up to speed during the ride and then to slow back down to a stop again. Every man-made mechanical device that rotates must change its speed of rotation at some time or other. In the same manner, the daily rotation of the earth itself is slowing at a rate of about 2 milliseconds over the course of every hundred years or so. These examples demonstrate that **nonuniform circular motion** is a natural and common consequence of any circular motion.

We will not present the derivation of kinematic equations for nonuniform circular motion due to the level of the mathematics required. However, we can discuss some of the properties from an intuitive perspective and analyze specific examples. At the end of this section we will present some summary equations of circular kinematics that will look surprisingly familiar.

7.8 Angular Acceleration

Let us consider how the motion of a compact disc changes in a CD player when you press the "Play" button. The angular velocity of any point on the CD increases from zero to some nonzero value. The change of angular velocity is called angular acceleration (α). Average angular acceleration ($\bar{\boldsymbol{\alpha}}$) can be expressed in the same form as average linear acceleration,

$$\bar{\boldsymbol{\alpha}} = \frac{\Delta\boldsymbol{\omega}}{\Delta t} = \frac{\boldsymbol{\omega}_2 - \boldsymbol{\omega}_1}{\Delta t}. \tag{7.7}$$

Unit analysis for Equation 7.7:

$$\frac{\langle \text{rad} \rangle}{\text{s}^2} \Rightarrow \frac{\langle \text{rad} \rangle/\text{s}}{\text{s}} = \frac{\langle \text{rad} \rangle}{\text{s}^2} = \text{s}^{-2}$$

Units are radians per second squared, or s^{-2}.

The magnitude of $\boldsymbol{\alpha}$ is proportional to the difference of the angular velocities at the beginning and end of the time interval, Δt, and its direction is in the direction of the angular velocity difference vector, $\Delta\boldsymbol{\omega} = \boldsymbol{\omega}_2 - \boldsymbol{\omega}_1$. Since $\boldsymbol{\omega}$ is oriented along the axis of rotation (z-axis), then $\boldsymbol{\alpha}$ also coincides with the rotational axis. The sign of angular acceleration indicates the direction of change of angular velocity, just as for rectilinear (i.e., straight-line) motion.

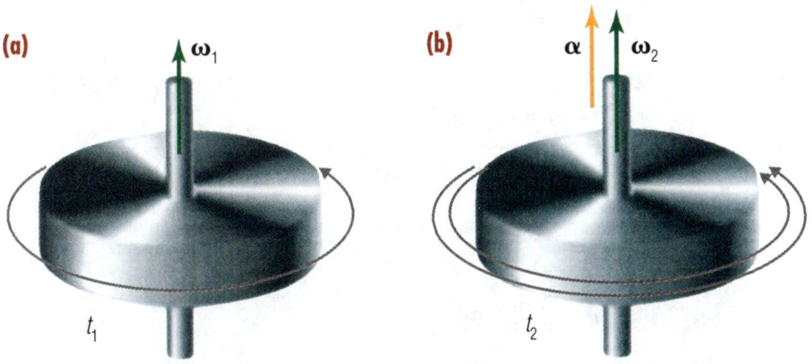

7-10 Angular acceleration is parallel to the rotational axis. It points in the same direction as the change of angular velocity.

7.9 Tangential Acceleration

We have already discussed how the direction of a revolving particle's tangential velocity changes in response to the centripetal acceleration of the particle in uniform circular motion. How do we describe the change of tangential speed as a revolving particle speeds up or slows down in nonuniform circular motion? The time-rate of change of the *magnitude* of tangential velocity is called **tangential acceleration (a_t)**. The following equation defines average tangential acceleration:

$$\bar{a}_t = \frac{\Delta v_t}{\Delta t}$$

Using Equation 7.4, we can replace Δv_t to obtain the following derivation:

$$\bar{a}_t = \frac{v_{t_2} - v_{t_1}}{\Delta t} = \frac{\omega r_2 - \omega_1 r}{\Delta t}$$

$$\bar{a}_t = \frac{(\omega_2 - \omega_1)r}{\Delta t} = \frac{\Delta\omega r}{\Delta t} = \frac{\Delta\omega}{\Delta t} \cdot r$$

$$\bar{a}_t = \bar{\alpha} r \tag{7.8}$$

Calculus shows us that as the time interval approaches zero, Equation 7.8 becomes the instantaneous tangential acceleration

$$a_t = \alpha r. \tag{7.9}$$

The direction of instantaneous tangential acceleration is, by definition, tangent to the circular path at the particle's position. If tangential speed is increasing, then tangential acceleration is in the same direction as rotation. If tangential speed is decreasing, tangential acceleration points in the opposite direction of rotation.

EXAMPLE 7-4

The Frisbee Factor: Acceleration in Nonuniform Circular Motion

You fling a Frisbee from rest to 360. rpm in 0.10 s. The disk is 23.0 cm (0.230 m) in diameter. Calculate (a) the angular acceleration and (b) the tangential acceleration of a point on the edge of the Frisbee.

Solution:

First, convert angular speed to SI units:

$$\frac{360.\ \text{rev}}{\text{min}} \cdot \frac{2\pi}{\text{rev}} \cdot \frac{1\ \text{min}}{60\ \text{s}} \cong 37.\underline{6}9\ \text{s}^{-1}$$

a. Since the Frisbee started from rest, $\Delta\omega = 37.\underline{6}9\ \text{s}^{-1}$ (the sign depends on whether you throw right- or left-handed).

Angular acceleration is calculated from Equation 7.7:

$$\bar{\alpha} = \frac{\Delta\omega}{\Delta t} = \frac{37.\underline{6}9\ \text{s}^{-1}}{0.10\ \text{s}}$$

$$\bar{\alpha} \cong 37\underline{6}.9\ \text{s}^{-2} \cong 377\ \text{s}^{-2}$$

b. The radius of the Frisbee is (0.230 m)/2. Using Equation 7.8 and the result from Part a, you get:

$$\bar{a}_t = \bar{\alpha}r = (37\underline{6}.9\ \text{s}^{-2})(0.230\ \text{m})/2$$

$$\bar{a}_t \cong 43.\underline{3}4\ \text{m/s}^2 \cong 43.3\ \text{m/s}^2$$

7.10 Equations of Circular Motion

Three equations of circular motion can be developed that are directly analogous to their counterparts for linear motion. In these equations, angular velocity ($\boldsymbol{\omega}$) is substituted for linear velocity, and angular acceleration ($\boldsymbol{\alpha}$) is substituted for linear acceleration. Recall that $\Delta\theta$ is analogous to linear displacement, *d*. As with the linear motion equations, these equations are valid only for conditions of uniform acceleration.

TABLE 7-1

Equations of Uniformly Accelerated Motion

Equation	Linear Motion	Circular Motion
First	$v_2 = v_1 + \bar{a}\Delta t$	$\omega_2 = \omega_1 + \bar{\alpha}\Delta t$
Second	$d = v_1\Delta t + \frac{1}{2}\bar{a}(\Delta t)^2$	$\Delta\theta = \omega_1\Delta t + \frac{1}{2}\bar{\alpha}(\Delta t)^2$
Third	$d = \frac{v_2^2 - v_1^2}{2\bar{a}}$	$\Delta\theta = \frac{\omega_2^2 - \omega_1^2}{2\bar{\alpha}}$

EXAMPLE 7-5

Motor Startup Time

An electric motor normally runs at 1800 rpm. Assuming a uniform angular acceleration of 64 s^{-2}, how many seconds will the motor require to go from stopped to full speed?

Solution:

Use the first equation of circular motion because it contains all of the known information and the unknown variable (Δt).

First, we must convert the common angular speed to SI units:

$$\omega = \frac{1800 \text{ rev.}}{\text{min}} \left| \frac{1 \text{ min}}{60 \text{ s}} \right| \frac{2\pi}{1 \text{ rev.}} \cong 1\underline{8}8 \text{ s}^{-1}$$

Rearrange the first equation of circular motion to solve for Δt:

$$\Delta t = \frac{\omega_2 - \omega_1}{\overline{\alpha}}$$

Substitute and solve for Δt.

$$\Delta t = \frac{1\underline{8}8 \text{ s}^{-1} - 0 \text{ s}^{-1}}{64 \text{ s}^{-2}} = \frac{188/\text{s}}{64/\text{s}^2}$$

$$\Delta t \cong 2.\underline{9}3 \text{ s} \cong 2.9 \text{ s}$$

7-11 Large AC motors may take several seconds to attain full speed when starting.

7A Section Review

1. Describe the coordinate system that is usually chosen for analyzing circular motion and state at least one advantage for this choice.
2. What position variable in circular motion is analogous to the number-line position variable in rectilinear motion?
3. What is the difference between uniform circular motion and nonuniform circular motion? Give an example of each.
4. A private aviation helicopter's main rotor blades rotate at approximately 450 rpm. What is the angular speed of the blades of the rotor in SI units?
5. The main rotor blades of the helicopter in Question 4 are 5.0 m long from the drive hub to tip. What is the tangential speed of the blades (a) at the drive hub, (b) 1.0 m from the hub, and (c) at the blade tip?
6. The rotor blades of a helicopter on the ground with its engine idling rotate at 48 rpm. If it takes 12 s for the pilot to increase rotor speed to 460 rpm when taking off, what is the angular acceleration on the rotor blades?
7. If it takes the earth 23 h 56 min 4 s to rotate once on its axis, how large an angle, in degrees and radians, will it rotate through in 1.00 hour?
8. What is the centripetal acceleration of a person standing on the equator of the earth?

7A Objectives

After completing this section, I can

- ✓ define circular motion.
- ✓ describe how position, speed, and velocity are measured in circular motion.
- ✓ compute the centripetal acceleration of a revolving object.
- ✓ show how the angular velocity vector is defined using the right-hand rule.
- ✓ calculate angular and tangential acceleration.
- ✓ compare and contrast the three equations of linear motion with the equations of circular motion.

A **central force** is any force that acts in a line between the centers of two objects or systems.

Centrifugal Force

Not to be confused with centripetal (*center-seeking*) force, **centrifugal** (*center-fleeing*) **force** is defined from either an accelerated or an inertial reference frame.

In an accelerated reference frame, like the one of a child on a merry-go-round, centrifugal force is a *pseudo-force*. It has no evident source and exists merely as a mathematical necessity to explain the child's tendency to accelerate away from the center of rotation due to the law of inertia.

In an inertial reference frame, like the puck-string system, centrifugal force is the third-law reaction force exerted by the rotating object *on the string*. It opposes the tension in the string that exerts the centripetal force to accelerate the object into circular motion. But our studies focus more on the forces acting *on the rotating object*, not the string, so for us centrifugal force is unimportant.

If the net force on the left side of a second-law equation is imposed on a rotating system with mass m and radius r, then the centripetal acceleration (v_t^2/r) is the *magnitude* of the response of the system.

7B DYNAMICS OF CIRCULAR MOTION

7.11 Centripetal Force

We saw in Section 7A that any object in uniform orbital motion about an axis experiences centripetal acceleration ($\mathbf{a}_c$) toward the center of rotation. We also know that acceleration occurs only when unbalanced forces act on a system, and the direction of the acceleration is in the direction of the net force on the system ($\Sigma\mathbf{F}$). The unbalanced force sum that produces this acceleration is called **centripetal force ($\mathbf{F}_c$)**. This is one of several types of center-seeking or *central forces*.

Let us first imagine that a puck-like object on a perfectly frictionless surface is connected by a string to a vertical rod rigidly imbedded in the surface so that the puck and string can freely rotate around the rod. The puck is set to rotating around the rod in orbital motion. We may assume that any forces acting parallel to the rod (the axis of rotation), and thus perpendicular to the surface of rotation, cancel out. Therefore, the net force acting on the rotating puck consists only of the force exerted by the string that causes centripetal acceleration.

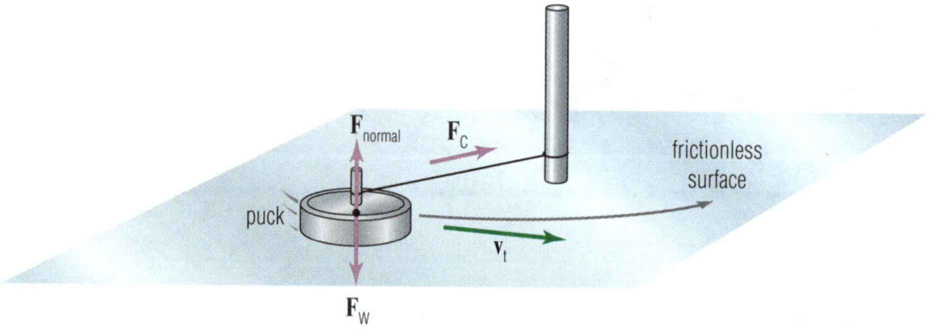

7-12 The string exerts a centripetal force on the puck.

Newton's second law is expressed in the equation

$$\Sigma\mathbf{F} = m\mathbf{a}.$$

From this, we can develop the second-law centripetal force equation by substituting the appropriate centripetal terms:

$$\mathbf{F}_c = m\mathbf{a}_c \tag{7.10}$$

Equation 7.10 would be more useful to find the magnitude of the centripetal force on the puck if we could use easily measurable parameters of the system described above. We can substitute Equation 7.3 from the previous section for a_c, which gives us the magnitude for $\mathbf{F}_c$:

$$F_c = ma_c = m\left(\frac{v_t^2}{r}\right)$$

$$F_c = \frac{mv_t^2}{r} \tag{7.11}$$

Equation 7.11 permits us to use easily measurable quantities to calculate the *magnitude* of the centripetal force that produces circular motion for a given object under a specific set of circumstances.

A string can exert centripetal force through tension, as we'll see in the following example. A planet may exert a centripetal force on a satellite through gravity (see Section 7C).

EXAMPLE 7-6

A Tethered Model Airplane: Centripetal Force

A model airplane enthusiast is flying his gas-powered plane in a circle tethered to a 10.0 m control line (see Figure 7-13a). The model has a mass of 0.90 kg and it takes 4.0 s to complete one circuit. (a) What is the magnitude of the force that the modeler must exert on the control line? (b) If he shortens the line to 5.0 m to land the airplane without reducing its flying speed, what force will he have to exert on the line?

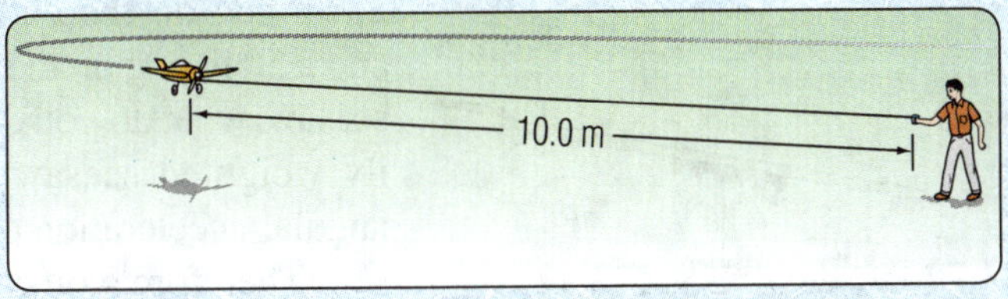

7-13a The modeler exerts a centripetal force through the control line to the model.

Solution:

a. You must find the tangential speed (v_t) before you can use Equation 7.11. You know that the distance of one circuit is the circumference of the circle $C = d = v_t\Delta t$. You can solve this equation for v_t after finding the circumference.

$$C = v_t\Delta t$$

$$v_t = \frac{C}{\Delta t} = \frac{2\pi r}{\Delta t} = \frac{2\pi(10.0 \text{ m})}{4.0 \text{ s}}$$

$$v_t \cong 1\underline{5}.7 \text{ m/s}$$

Substitute tangential speed into Equation 7.11 to solve for centripetal force as follows:

$$F_c = \frac{mv_t^2}{r} = \frac{(0.90 \text{ kg})(1\underline{5}.7 \text{ m})^2}{10.0 \text{ m}}$$

$$F_c \cong 2\underline{2}.1 \text{ N} \cong 22 \text{ N}$$

b. In this part, m and v_t are the same as above, but $r = 5.0$ m.

$$F_c = \frac{mv_t^2}{r} = \frac{(0.90 \text{ kg})(1\underline{5}.7 \text{ m})^2}{5.0 \text{ m}}$$

$$F_c \cong 4\underline{4}.3 \text{ N} \cong 44 \text{ N}$$

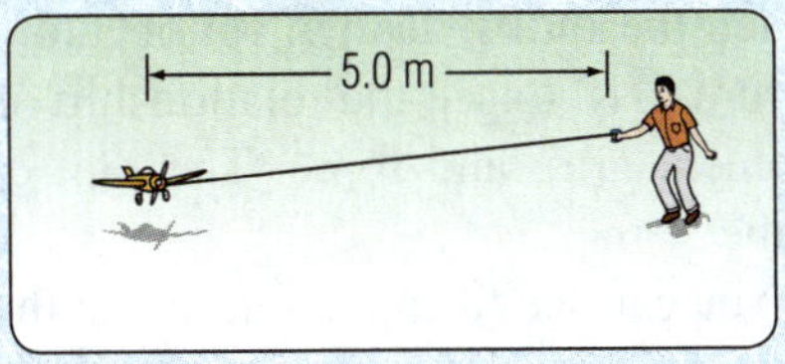

7-13b Landing the model plane

As expected, when the modeler shortened the line to half its length, he had to exert twice the force to maintain the same speed.

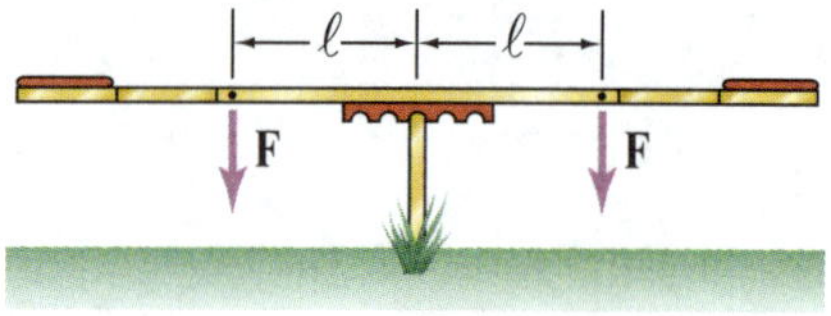

7-14 An empty seesaw balances at its center.

7.12 Torque

Accelerated *linear* motion results from a nonzero net force, but what causes accelerated *rotation*? To answer this question, let us perform some experiments with a playground seesaw.

How do you balance a seesaw? If it is empty, you can position the seesaw so that its center is the pivoting point. In this position the seesaw remains at rest. Since half of the seesaw is on either side of the pivot, each end is attracted to the earth by an equal force. When the pivot is not at the center of the seesaw, the longer side is attracted to the earth with a greater force, and the seesaw rotates.

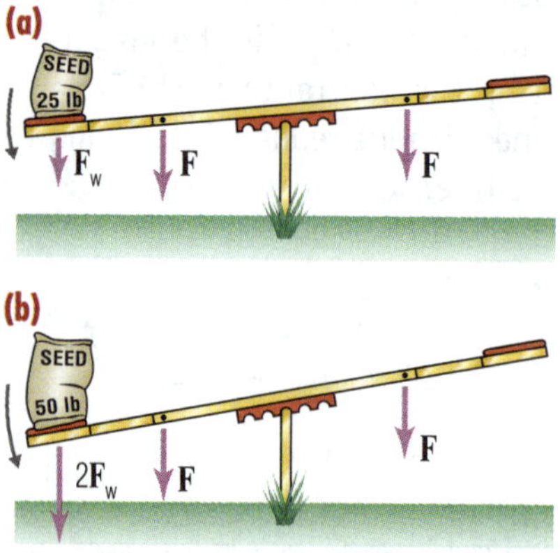

7-15 A greater weight causes greater angular acceleration.

Suppose you place a 25 lb bag of grass seed on one side of a balanced seesaw. What happens? The seesaw rotates, with the weighted side going down. If you place a 25 lb bag on the end of a balanced seesaw, and a 50 lb bag on the end of another identical seesaw, what do you observe? The more heavily weighted seesaw rotates with twice the angular acceleration of the less burdened seesaw. Therefore, you can conclude that angular acceleration is proportional to force.

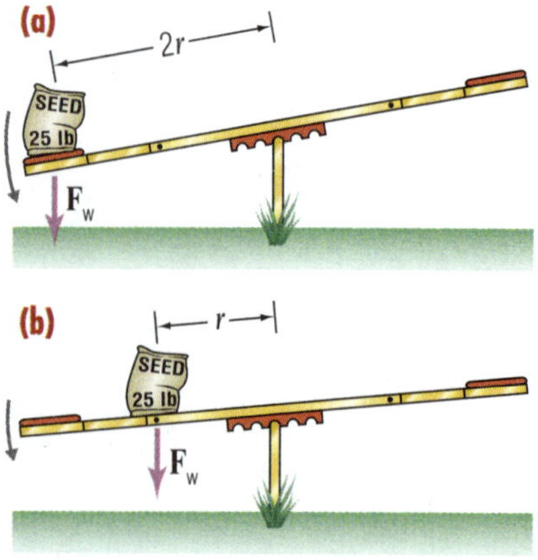

7-16 Angular acceleration increases with distance of the applied force from the pivot point.

Suppose identical 25 lb bags of seed are placed on two identical balanced seesaws. One bag is placed at an end of one seesaw, and the second bag is placed halfway between the pivot and an end of the other seesaw. You observe that the seesaw with the bag at the end rotates with twice the angular acceleration of the other seesaw. Therefore, angular acceleration is proportional to the force's position vector (**r**)—the distance from the pivot to the place that the force is applied.

You can see that angular acceleration is proportional to both the force and the force's position vector. Their product yields the vector quantity called **torque (τ)**. It is not possible to multiply two vectors together algebraically since they are not true numbers. However, the magnitude of a torque can be calculated by the formula

$$\tau = rF \sin \theta, \qquad (7.12)$$

where r is the magnitude of the position vector pointing from the center of rotation to the point at which the force is applied, F is the magnitude of the applied force, and θ is the smallest angle (<180°) between vectors **r** and **F** when they are positioned tail-to-tail. The quantity $r \sin \theta$ is called the **moment arm** (ℓ) of a torque. See Figure 7-18 to see the relationship of torque, moment arm, and force. The unit of torque is the N·m.

You can see from the figure that the magnitude of torque is the product of the component of the position vector that is perpendicular to the applied force (the moment arm) and the force. Maximum torque is obtained from a given force when it is perpendicular to its

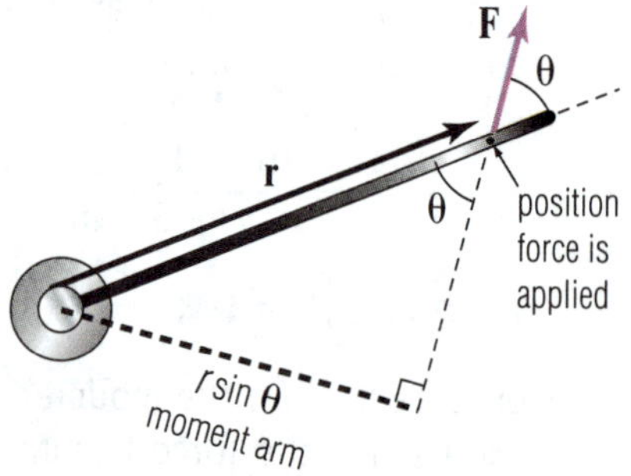

7-17 Force and its moment arm

A **torque** exists when a force is exerted on a system at a distance from the system's point of rotation.

The **moment arm** is often called a *lever arm* for simple lever problems.

Problem-Solving Strategy 7.2

The unit of torque is the N·m. Do not confuse this unit with the N·m that is a unit of energy and work which is equivalent to the joule (J).

position vector. If the applied force is parallel to its position vector, the torque is zero.

Torque has the same role in rotational motion that force has in linear motion. That is, *angular acceleration is produced by unbalanced torques.* When the sum of all torques on a system is zero, it will rotate either at a constant angular speed or not at all. The condition of zero net torques is called **rotational equilibrium**. Mathematically,

$$\Sigma \boldsymbol{\tau} = 0 \text{ N}\cdot\text{m}$$

is the condition for rotational equilibrium, just as Newton's first law established that

$$\Sigma \mathbf{F} = 0 \text{ N}$$

is the condition for translational equilibrium.

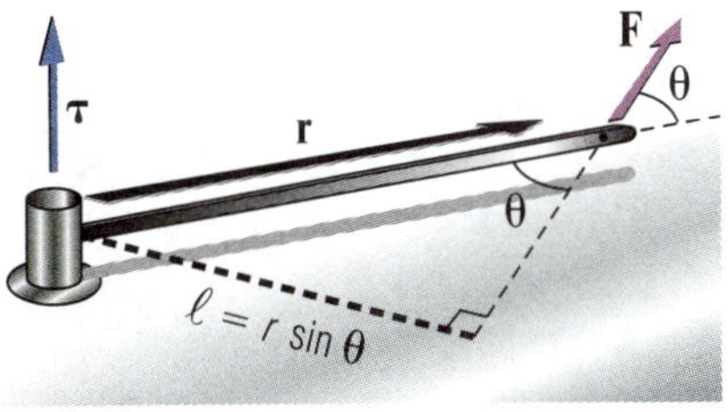

7-18 The geometry of a torque vector. Torque (τ) is perpendicular to both the lever arm and the applied force.

EXAMPLE 7-7

Balancing Act: Summing Torques

Two boys on a playground want to balance a seesaw. One boy weighs 240 N, and the other boy weighs 260 N. The seesaw is 3.6 m long and of negligible mass. Assuming that the boys sit at the ends of the seesaw, how far from the middle must they place the pivot point in order to balance?

Solution:

Assume that the seesaw is horizontal when balanced, which means the boys' position vectors are horizontal. The boys' vertical weight vectors are then perpendicular to their position vectors. Therefore, the torque magnitude formula reduces to

$$\tau = rF \sin 90° = rF.$$

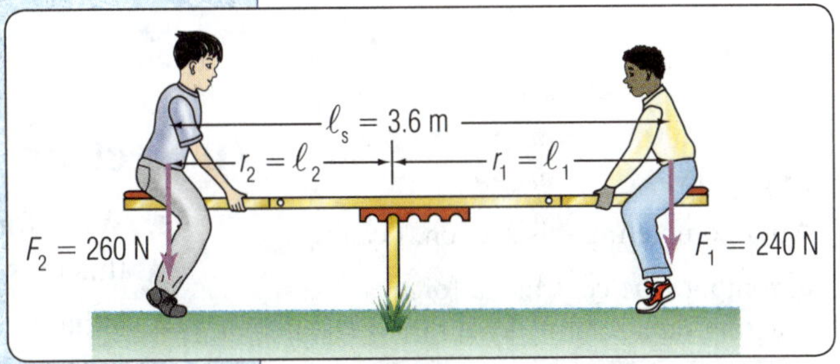

7-19 Balancing a seesaw when one child is heavier than the other

Let boy 1 be the lighter boy and boy 2 be the heavier boy.

$$F_1 = 240 \text{ N} \qquad r_1 = \ell_1$$

$$F_2 = 260 \text{ N} \qquad r_2 = \ell_2$$

$$\ell_1 + \ell_2 = \ell_s \text{ (length of seesaw)} \qquad (1)$$

Since the seesaw is in rotational equilibrium, the magnitudes of the torques must be equal:

$$\Sigma \tau = \tau_1 + \tau_2 = 0$$

$$\tau_1 = -\tau_2$$

$$|\tau_1| = |-\tau_2|$$

$$\ell_1 F_1 = \ell_2 F_2$$

This equation, called the **law of moments**, is true for any balanced seesaw. Rearranging,

$$\frac{F_1}{F_2} = \frac{\ell_2}{\ell_1}.$$

Multiplying both sides by ℓ_1,

$$\ell_1 \cdot \frac{F_1}{F_2} = \ell_2. \qquad (2)$$

Substituting Equation 2 for ℓ_2 into Equation 1 gives

$$\ell_1 + \frac{\ell_1 F_1}{F_2} = \ell_s.$$

Factoring out ℓ_1,

$$\left(1 + \frac{F_1}{F_2}\right)\ell_1 = \ell_s.$$

Now solve for ℓ_1:

$$\left(1 + \frac{240\text{ N}}{260\text{ N}}\right)\ell_1 = \left(\frac{260\text{ N} + 240\text{ N}}{260\text{ N}}\right) = 3.6\text{ m}$$

$$\ell_1 = (3.6\text{ m})\left(\frac{260\text{ N}}{260\text{ N} + 240\text{ N}}\right)$$

$$\ell_1 \cong 1.\underline{8}7\text{ m} \cong 1.9\text{ m}$$

The pivot point is about 1.9 m from the lighter boy. The midpoint of the seesaw is 3.6 m/2, or 1.8 m. Therefore, the pivot is about 0.1 m from the middle on the side of the heavier boy.

7B Objectives

After completing this section, I can

✓ compute the centripetal force acting on a rotating system.

✓ define torque and calculate the torques acting on a rotational system.

✓ state the conditions required for rotational equilibrium.

✓ state the law of moments and solve problems using this principle.

7B Section Review

1. What force causes centripetal acceleration? What is the name given to this class of forces?
2. What measurable properties of a revolving system determine the magnitude of centripetal force?
3. Under what conditions does rotational equilibrium exist?
4. What is a torque? Give an example.
5. Calculate the centripetal force that acts on a 35.0 kg child riding a merry-go-round at a distance of 2.2 m from the spin axis. The merry-go-round is spinning at 15 rpm.
6. A 73.0 kg father and his 45.0 kg son are sitting on opposite sides of a 4.00 m seesaw at the playground. The pivot point is fixed at the midpoint of the board. Where must the father position himself to horizontally balance the seesaw if his son sits at the end of the board?
7. An auto mechanic is struggling to turn a bolt in the engine compartment of a car. He applies a force of 180. N on a wrench 18.0 cm from the bolt. Due to space restrictions, he must exert the force at an angle of 30.0° from the end of the wrench. How much torque is he actually applying to the bolt?
8. What is the centripetal force acting on a 66.0 kg person standing on the equator of the earth?

7C UNIVERSAL GRAVITATION

Centripetal accelerations change the direction of the motions of objects from straight lines to curves. Different forces can cause centripetal acceleration, but perhaps the most important centripetal force is gravity. One of Newton's greatest achievements was to connect gravitation with other forces that cause circular motion. Newton's work was a breakthrough in a long line of studies of the motions of the planets.

7.13 Ancient Ideas

If you had not been taught that the earth orbits the sun, it would probably seem obvious to you that the earth is the unmoving center of the universe. Except during an earthquake, you do not feel the earth move. Any clear day or night you can watch the sun, moon, and stars travel around the earth. This logical idea is called the **geocentric** (or earth-centered) **theory**.

From earliest history men have tried to explain the motions of the planets. Priests in Mesopotamia learned to predict the motions of the planets. They often used this knowledge to deceive their kings with astrology. The ancient Greeks, including Aristotle, were among the first to use the predictions to try to explain the planets' motions. The earth was, naturally, the center of Aristotle's universe.

7-20 The geocentric cosmos

One man who disagreed with the geocentric theory was Aristarchus of Samos, who lived in the third century before Christ. He thought that the sun was the center of the universe and that all the planets, including the earth, revolved around it. His **heliocentric** (or sun-centered) **theory** seemed ridiculous to his contemporaries.

During Roman times a Greek scholar named Ptolemy modified Aristotle's theory to fit the actual motions of the planets. The planets do not travel in perfectly circular orbits around the earth, as the ancient astronomers could see even without telescopes. Among other complicated ideas, Ptolemy's theory included the idea that planets traveled in small circles called *epicycles* as they traveled around the earth in their principal or *deferent* orbits. Concepts like this made Ptolemy's theory almost correspond to the observations of the planets' motions.

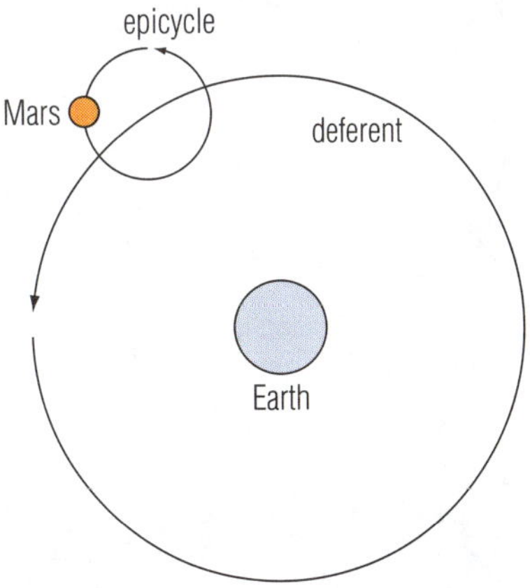

7-21 The Ptolemaic geocentric system needed epicycles to approximate the observed motions of the planets.

For centuries, Ptolemy's geocentric theory was accepted almost universally. Ptolemy's theory, however, did not quite agree with all astronomical observations. Late medieval astronomers tried to alter the theory to fit the observations. One astronomer, Nicolaus Copernicus, concluded that the geocentric idea itself was

Aristarchus of Samos (ca. 310–230 BC) was a Greek mathematician and astronomer who was the first proponent of the heliocentric theory. He also described the daily axial rotation of the earth.

Claudius Ptolemaeus (second century AD) was an Alexandrian astronomer and geographer.

Ptolemy believed that he needed to add epicycles to the geocentric theory in order to explain why some planets exhibited retrograde (backward) motion at certain points in their orbits.

Nicolaus Copernicus (1473–1543) was a Polish astronomer, mathematician, religious cleric, and government leader.

7-22 The original cover of Copernicus's paper describing his heliocentric theory

faulty. He found it simpler to explain the observations by saying that the planets orbit the sun in circular orbits. Copernicus's heliocentric theory was not much more accurate than Ptolemy's geocentric theory, but it was much simpler.

The Danish astronomer Tycho Brahe disagreed with both Copernicus and Ptolemy. His own theory placed the earth, unmoving, at the center of the universe with the sun orbiting the earth and all the other planets orbiting the sun. Tycho's greatest contribution to astronomy was his detailed, accurate observations of the heavens. Near the end of his life, Brahe hired Johannes Kepler, a mathematician, to interpret his observations.

7.14 Kepler's Laws

Kepler began the task of organizing Brahe's data with the idea that the observations were more reliable than any theory. He tried several theories but rejected them all—one because one of its predictions differed from Brahe's data by an angle of eight-sixtieths of a degree (about ¼ the width of a full moon). Kepler devised theories of his own about the planets' motions, which also did not work. Finally, he came to the conclusion that a presupposition of astronomical theory was wrong—the idea that the planets move at constant speeds. He united the idea of varying speed with the heliocentric theory to formulate **Kepler's laws** of planetary motion.

Kepler's second law states that the position vector of a planet travels through equal areas in equal times if the vector is drawn from the sun. In Figure 7-24a, area A equals area B, and the planet takes the same time to travel from position 1 to position 2 as it takes to travel from position 3 to position 4. The planet travels farther from 1 to 2, so it must be traveling faster in that interval.

If the planet's speed varies, yet its position vector travels through equal areas in equal times, the planet's orbit cannot be circular, and it must be farther from the sun when it is traveling more slowly. A planet in a circular orbit would always be the same distance from the sun. *Kepler's first law* states that the orbit of a planet is an ellipse with the sun at one focus, as in Figure 7-24b.

A New Point of View

The exchange of the geocentric for the heliocentric model of the solar system is probably the most significant example of a scientific paradigm shift. This was not merely a gradual change from a lesser to a greater truth, but a revolution in the way people viewed the heavens. Thomas Kuhn discusses in his book *The Copernican Revolution* how this change of view profoundly affected not only astronomy but also man's view of himself and of his importance in the universe.

7-23 Johannes Kepler (1571–1630) was a German mathematician who postulated the three laws of celestial mechanics.

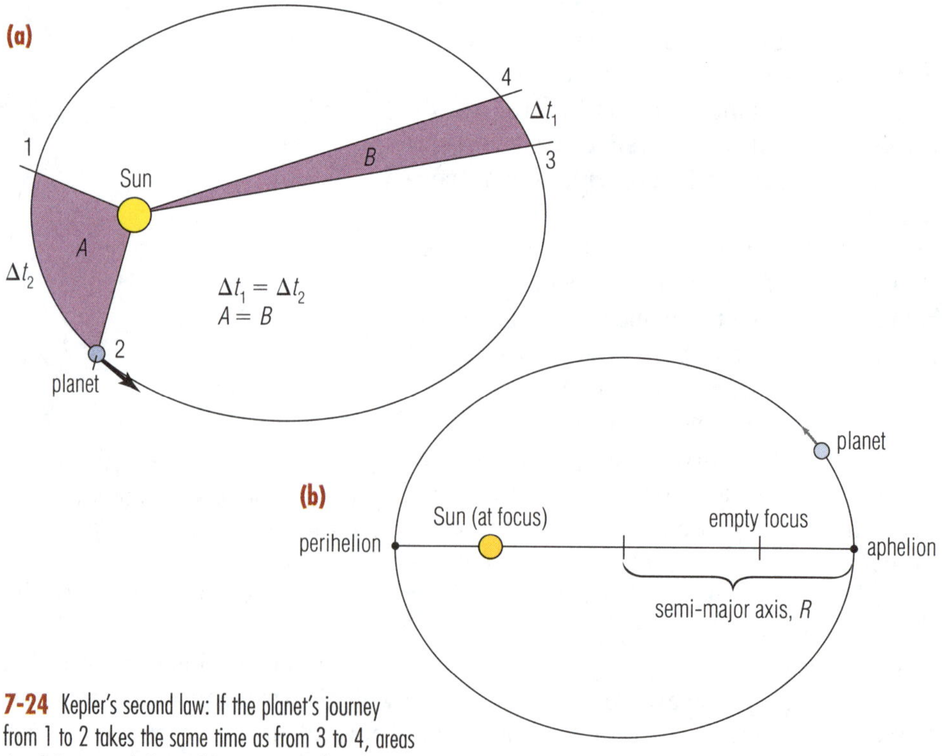

7-24 Kepler's second law: If the planet's journey from 1 to 2 takes the same time as from 3 to 4, areas *A* and *B* are equal. Figure (b) gives terms describing a planet's orbit.

Kepler's third law relates the size of each planet's orbit to the time it takes to complete one orbit. The size of the orbit is represented by its **semi-major axis (*R*)**, which is half of the distance between the planet's nearest position to the sun (its *perihelion*) and the planet's farthest position from the sun (its *aphelion*). The time the planet takes to complete an orbit is called its **period (*T*)**. Kepler found that for every planet in the solar system, the ratio of the square of the period and the cube of the semi-major axis is the same. In symbols,

$$\frac{T^2}{R^3} = K, \tag{7.13}$$

where K is called **Kepler's constant (*K*)**. Each orbital system has its own K. For instance, the moons of Jupiter have one K, while the planets in the solar system have a different K. Within each system, however, K is the same.

We can determine the value of K for solar planets by substituting the known values of the earth's orbit into Equation 7.13. The semi-major axis may be expressed in terms of **astronomical units (ua)**. One astronomical unit is the mean distance of the earth from the sun. The semi-major axis of the earth's orbit can be assumed to be 1 ua, and its period is, of course, one year. Kepler's constant for the earth (and any planet orbiting the sun) is then just

$$K_E = \frac{T_E^2}{R_E^3} = \frac{(1\ \text{y})^2}{(1\ \text{ua})^3} = 1\ \text{y}^2/\text{ua}^3.$$

If we rearrange Equation 7.13, we can then develop a simple method for determining the orbital period for any of the other planets in the solar system if we know their semi-major axes in ua:

$$\frac{T_{\text{planet}}^2}{R_{\text{planet}}^3} = \frac{T_E^2}{R_E^3}$$

$$\frac{T_{\text{planet}}^2}{R_{\text{planet}}^3} = K_E$$

$$T_{\text{planet}}^2 = R_{\text{planet}}^3 K_E$$

$$T_{\text{planet}}^2 = R_{\text{planet}}^3 (\text{y}^2/\text{ua}^3)$$

$$T_{\text{planet}} = \sqrt{R_{\text{planet}}^3 (\text{y}^2/\text{ua}^3)} \tag{7.14}$$

This equation is good only for planets, comets, asteroids, etc., orbiting the sun. It cannot be used for any planetary satellite or moon.

The **period (*T*)** of an object that repeats a specific motion, such as an orbit, is the length of time to complete one entire cycle. The units of T are time units. The SI unit for time is the second. In equations relating to planetary orbits, the period should be expressed in units of earth-years (y).

Problem-Solving Strategy 7.3

Remember that when using Kepler's constant $K = \text{y}^2/\text{ua}^3$ in his third law equation, R must be in terms of astronomical units and T must be in years.

EXAMPLE 7-8

A Period of Saturn: Kepler's Third Law

The semi-major axis of Saturn's orbit is 1.427×10^9 km. What is its orbital period in years?

Solution:

We are given the semi-major axis of Saturn's orbit, $R = 1.427 \times 10^9$ km, and we are asked to find Saturn's orbital period in (earth) years. We must first convert Saturn's semi-major axis dimension into astronomical units (ua) in order to use Equation 7.14.

$$R_{\text{Saturn}} = \frac{1.427 \times 10^9 \ \cancel{\text{km}}}{} \left| \frac{1 \text{ ua}}{1.496 \times 10^8 \ \cancel{\text{km}}} \right. \cong 9.53\underline{8}7 \text{ ua}$$

Substituting for R_{planet} in Equation 7.14:

$$T_{\text{Saturn}} = \sqrt{867.\underline{8}9 \text{ y}^2} \cong 29.4\underline{5}9 \text{ y}$$

$$T_{\text{Saturn}} \cong 29.46 \text{ y}$$

7-25 One of the few situations in which a person is almost completely free of the sensation of gravity

7-26 The orbital motion of the moon was the basis for Newton's unraveling the secret of gravitation.

7.15 Newton

Since the time of Creation, people have gazed at the sky and tried to make sense of the motions of the moving stars and planets, and how these relate to the great celestial lights, the sun and the moon. We, in modern times, have grown up with the notion that gravity controls the motions of celestial bodies, but this concept was completely unknown and unsuspected until the time of Isaac Newton in the seventeenth century.

Humans are subject to the force of gravity from the time of conception (although, in the womb, the buoyancy of amniotic fluid somewhat offsets the effects of gravity on the fetus). Whether we're conscious of it or not, the force of gravity is a part of our life on the earth until our last breaths are taken. Prior to Newton, it was impossible to imagine what life would be like without gravity. Even today, only astronauts in orbit and skydivers in free fall can experience the sensation of the absence of gravity.

Newton recognized that the force that causes an apple to fall to the ground was the same force that caused the moon to fall toward the earth. Whether an apple landed on his head or even contributed to conceptualizing his theory is a matter of conjecture. Through experimental observations and inspired guesses, he was able to extend this idea to the rest of the solar system.

Newton had the benefit of being able to refer to the observations and conclusions of the great astronomers Tycho Brahe, Johannes Kepler, and Edmond Halley. These men had compiled an immense amount of data that demonstrated conclusively that planets, moons, and comets moved in elliptical orbits. Newton, therefore, had a number of premises to work with as he began to consider the causes for orbital motion.

- He recognized that the moon accelerates toward the earth because of gravitational force but misses it because of its sideways orbital motion. Therefore, the gravity that causes the acceleration exists not only on the earth but also at least as far as the moon.

- His experiments revealed that the gravitational force between two objects depends on two physical properties—distance and mass.

Newton started with the formula for centripetal acceleration.

$$a_c = \frac{\bar{v}^2}{r}$$

Knowing the period, T, and the orbital radius, he calculated the centripetal acceleration caused by the mass of the earth both at the surface of the earth and at the moon's center of mass. Newton's genius lay in his recognizing the relationship between these centripetal accelerations and the radial distances at which these values were calculated. He found that though the radius of the earth (the distance between the earth's center of mass and the surface) was $\frac{1}{60}$ of the distance between the centers of mass of the earth and the moon, the centripetal acceleration at the moon was $\frac{1}{3600}$ of that at the surface of the earth $[\frac{1}{3600} = (\frac{1}{60})^2]$. From this he concluded that the gravitational force between two objects is proportional to the inverse square of the distance between them.

$$F_g \propto \frac{1}{r^2}$$

Next, he realized that the gravitational force between two masses must be dependent on both masses symmetrically so that interchanging their positions would not change the result. Mathematically, the force could depend on either the sum $(M + m)$ or the product (Mm) of the two masses, raised to some power n, so

7-27 Newton believed that the masses of both the earth and the moon were involved in the gravitational force between them.

$$F_g = k(M + m)^n \quad \text{or} \quad F_g = k(Mm)^n,$$

where k is an unknown proportionality constant. Since $F_g = ma_g$,

$$ma_g = k(M + m)^n \quad \text{or} \quad ma_g = k(Mm)^n.$$

Dividing both sides by m,

$$a_g = \frac{k(M + m)^n}{m} \quad \text{or} \quad a_g = \frac{k(Mm)^n}{m}.$$

Galileo had shown that the acceleration due to gravity alone of mass M on a mass m was independent of the smaller mass, m. Everything falls toward M with the same acceleration, ignoring air resistance. In the first formula for a_g, there is no exponent n that eliminates dependence on m. In the second formula for a_g, only an exponent of $n = 1$ eliminates a dependence on m. Therefore, Newton concluded that

$$a_g = \frac{k(Mm)^1}{m} = kM$$

$$F_g \propto Mm.$$

When he combined gravity's two dependencies on mass and distance, the resulting proportionality became

$$F_g \propto \frac{Mm}{r^2}.$$

Newton's **law of universal gravitation** states that the gravitational force between two objects is directly proportional to their two masses and inversely proportional to the distance between their centers of mass squared. The proportionality constant in Newton's law of universal gravitation is G.

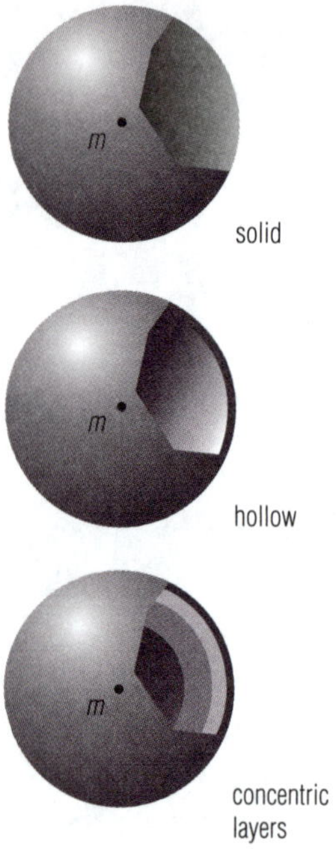

7-28 Solid spheres, hollow spheres, and concentrically layered spheres can be considered point masses.

Henry Cavendish (1731–1810) was an English chemist and physicist.

Newton's proportionality can be made an equation by simply including a proportionality constant, which conventionally is called G:

$$F_g = G\frac{Mm}{r^2} \tag{7.15}$$

The symbol G is called the **universal gravitational constant**. Do not confuse the constant G with the planetary acceleration due to gravity, g. The value of G was not known in Newton's time. It would not be determined until almost two centuries later. Equation 7.15 is called Newton's law of universal gravitation.

7.16 Properties of the Law of Universal Gravitation

The law of universal gravitation allows us to predict the gravitational force between two objects of given masses at a specified distance, but it does not indicate why gravitation exists, how it is propagated, or what gravitation is. In addition, the law is valid only for point-like masses. If the objects have appreciable size or are irregular in shape or composition, then the formula is only an approximation of the force between them unless they are very much farther apart than their physical sizes. Newton, using integral calculus that he developed for the purpose, showed that finite spheres with spherical mass distribution (not necessarily *uniform* mass distribution) could be considered point masses. For example, the mass of an empty spherical steel water tank (the kind that stores your town's water, perhaps) could be considered to be concentrated at the center of the sphere for the purposes of determining the force of gravity on it.

The law is not an independent law of motion. It merely describes the force of attraction of one object on the other. Gravity is always an attractive force, never repulsive. A repulsive gravitational force might be the result of a particle having a "negative mass." While particles with zero mass are known, no particle with a negative mass has ever been identified, nor is one believed to be possible.

7.17 Cavendish

The gravitational force on an object and its mass can be measured directly, but in order to calculate G, the mass of the earth had to be known as well. For nearly two centuries after Newton discovered the law of universal gravitation, the value of G remained a mystery. Late in the eighteenth century, the eccentric British physical scientist Henry Cavendish was encouraged by a fellow scientist to devise an experiment to determine the density of the earth. Then, when the density value was known, he could "weigh the earth" or more accurately, determine the mass of the earth. This experiment was the first step in determining the value of G.

Cavendish was the son of a wealthy family. He was very reclusive, avoiding contact with people, including his servants, most of his life. Biographies indicate that he was a meticulous observer who discovered and described the gas that was eventually called hydrogen, and he determined that water consists of two gases, thus closing the book on the Greek view that water was one of the four basic elements of the world.

7.18 The Cavendish Experiment

Cavendish's experiment measured the gravitational force exerted between known masses in his laboratory. The room-sized apparatus consisted of two small lead spheres mounted on the ends of a rod suspended by a quartz thread. This assembly was permitted to rotate freely on the thread, producing a torque in the thread as it twisted. This kind of apparatus is called a **torsion balance**. Another pair of much larger lead spheres was symmetrically fixed in such a way that the center of mass of each lay in the same plane with the centers of mass of the lead spheres mounted on the torsion balance. The torsion arm was suspended between the large masses so that as it rotated, the smaller masses would symmetrically approach the larger. See Figure 7-29 for a modern example of Cavendish's apparatus.

Cavendish determined beforehand how much torque was required to twist the quartz thread through various angles. By adjusting the distances between the centers of mass (r), he measured the deflection angles, which yielded a torque value. Knowing the physical dimensions of the apparatus permitted him to calculate the gravitational force exerted by the masses on each other. Knowing this gravitational force between the lead spheres enabled him to scale up his findings to calculate the density of the earth by knowing the gravitational force between an object and the earth.

Although Cavendish's data could have been used to establish the value of G in Newton's law of universal gravitation, Cavendish never calculated G. Almost 75 years passed before other scientists used Cavendish's results to determine this valuable constant. In modern SI units, Cavendish's data yields a value of $6.754 \times 10^{-11}\ \text{N}\cdot\text{m}^2/\text{kg}^2$. This is within 1% of the currently accepted value of $6.674 \times 10^{-11}\ \text{N}\cdot\text{m}^2/\text{kg}^2$.

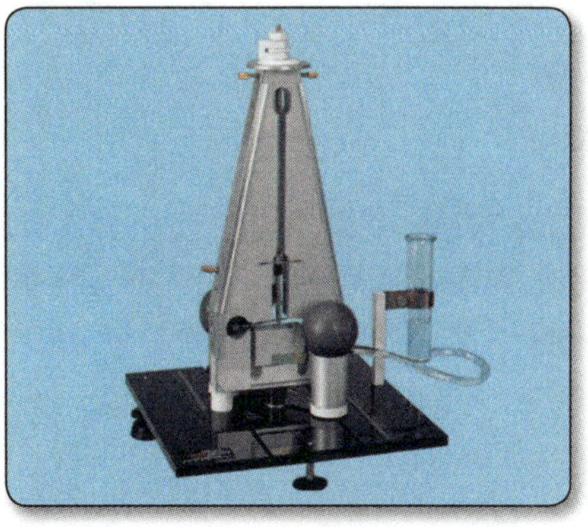

7-29 A modern Cavendish experiment apparatus

The modern value of G is $6.674 \times 10^{-11}\ \text{N}\cdot\text{m}^2/\text{kg}^2$.

7.19 Computing the Mass of Earth

Cavendish's experiment is often repeated by modern physics students and used to determine the value of G. Once one knows the value of G, he can estimate the mass of the earth. All he has to do is weigh an object of known mass (m) to find its gravitational attraction at the earth's surface (r_E), and then solve for M. For example, say he uses a 10.0 kg mass. Its weight ($F_w = F_g$) would be 98.1 N. The radius of the earth is approximately 6.37×10^6 m. Solving Equation 7.15 for M:

$$M = \frac{F_W r_E^2}{Gm}$$

$$M = \frac{(98.1\ \text{N})(6.37 \times 10^6\ \text{m})^2}{(6.674 \times 10^{-11}\ \text{N}\cdot\text{m}^2/\text{kg}^2)(10.0\ \text{kg})}$$

$$M \cong 5.9\underline{6}4 \times 10^{24}\ \text{kg}$$

From this mass, he can calculate the density of the earth (ρ_E), which is of importance to geologists and also to astronomers. The student assumes that the earth is a sphere and of uniform composition. Density is the mass per unit volume, so he has to calculate the volume of the earth in cubic meters.

$$V_E = \tfrac{4}{3}\pi r_E^3$$

$$V_E = \tfrac{4}{3}\pi(6.37 \times 10^6\ \text{m})^3$$

$$V_E \cong 1.08 \times 10^{21}\ \text{m}^3$$

The earth's density is calculated by:

$$\rho_E = \frac{M}{V_E}$$

$$\rho_E = \frac{5.9\underline{6}5 \times 10^{24}\ \text{kg}}{1.08 \times 10^{21}\ \text{m}^3}$$

$$\rho_E \cong 55\underline{2}3\ \text{kg/m}^3 \text{ or } 5.52\ \text{g/cm}^3$$

Thus, the average density of the earth is $5.52\ \text{g/cm}^3$.

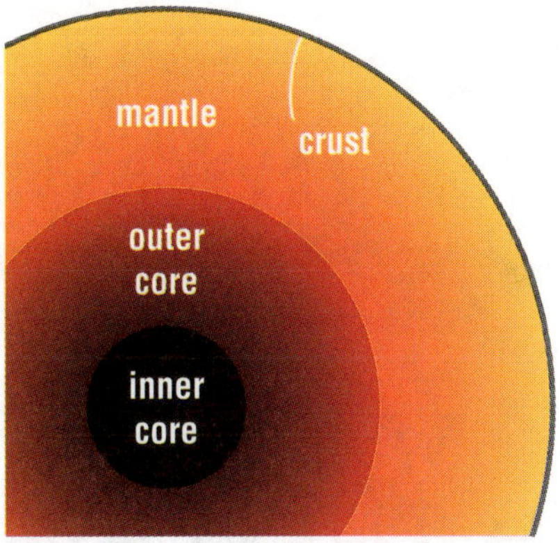

7-30 A determination of the average density of the earth reveals that the earth's composition is not uniform.

This result is interesting because the average density of rocks from the crust of the earth is only about 3 g/cm^3, which indicates that the interior of the earth must be much denser. Careful measurement of the speeds of seismic waves passing through the interior of the earth indicates that the density of the matter at the center of the earth is as high as 13–15 g/cm^3.

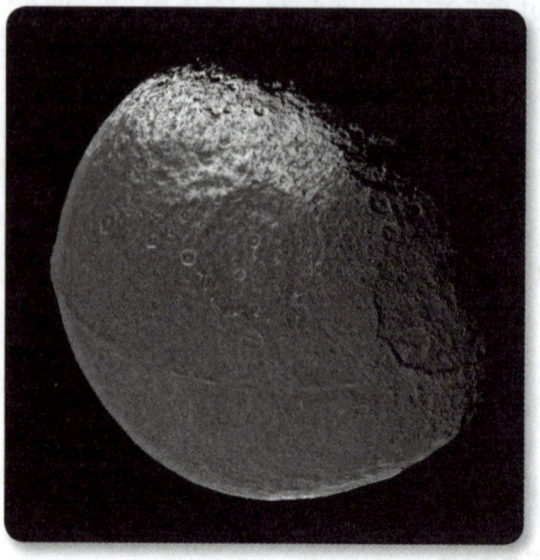

7-31 Iapetus is one of Saturn's many strange moons. One side is as black as coal, and the other is bright white. It also has an equatorial ridge, making it look like an unbroken walnut shell.

7.20 Modeling Saturn's Moons Using the Principles of Circular Motion

Saturn's moons are in orbit around the planet, so they should behave according to the principles taught in this chapter. But how well do real-world systems follow theory? To answer this question, you will evaluate several relationships from the data provided in Table 7-2. This table lists 20 of Saturn's 60+ named moons, arranged in order of their distance from Saturn.

Dominion and Saturn's Moons

How can investigating the properties of Saturn's moons be seen as part of exercising dominion?

TABLE 7-2

Saturn Moon Data

Name	Diameter (km)	Mass (kg)	Average Orbital Radius (km)	Orbital Period (d)
Pan	30	5×10^{15}	133 584	0.6
Atlas	31	1.1×10^{16}	137 670	0.6
Prometheus	86	2.7×10^{17}	139 380	0.6
Pandora	81	2.2×10^{17}	141 720	0.6
Epimetheus	113	5.6×10^{17}	151 422	0.7
Janus	179	2.01×10^{18}	151 472	0.7
Mimas	397	3.8×10^{19}	185 404	0.9
Methone	3	2×10^{13}	194 440	1.0
Enceladus	504	8.4×10^{19}	237 950	1.4
Calypso	21	8×10^{17}	294 619	1.9
Telesto	24	8×10^{17}	294 619	1.9
Tethys	1066	6.18×10^{20}	294 619	1.9
Dione	1123	1.05×10^{21}	377 396	2.7
Helene	33	8×10^{17}	377 396	2.7
Rhea	1529	2.49×10^{21}	527 108	4.5
Titan	5151	1.35×10^{23}	1 221 930	15.9
Hyperion	292	1.77×10^{19}	1 481 010	21.3
Iapetus	1472	1.88×10^{21}	3 560 820	79.3
Phoebe	220	4.0×10^{18}	12 869 700	545.1
Albiorix	32	2.1×10^{16}	16 266 700	774.6

7C Section Review

1. What are epicycles? Who proposed their existence?
2. Do the moons of Saturn have the same Kepler's constant as the moons of Uranus?
3. How does Cavendish's experiment disprove the idea that only planet-sized masses cause gravitational attraction?
4. DM ✪ Can you generalize the relationship between the diameters of Saturn's moons and their masses? Assume each moon is a solid sphere. How does the mass of a solid sphere relate to its volume, and what dimension is a sphere's volume calculated from? Create a scatter plot of the data and find a trend line that confirms your estimate.
5. DM ✪ Does the mass or size of a moon seem to determine its orbit's size or period? Considering what you have learned in this chapter, explain your answer. Create a scatter plot of average orbit radius versus moon diameter to confirm your answer.
6. ✪ If Venus's orbit has a semi-major axis of 0.723 ua, calculate its orbital period.
7. ✪ Calculate Kepler's constant for Io, a moon of Jupiter, if $T_{\text{Io}} = 1.769$ days and $R_{\text{Io}} = 421\,600$ km.
8. ✪ Calculate the gravitational force that exists between a dog with a mass of 53.0 kg and a cat with a mass of 3.5 kg if they are sitting on opposite sides of the room with their centers of mass 7.5 m apart.

7C Objectives

After completing this section, I can

- ✓ compare and contrast the geocentric and heliocentric models of the solar system.
- ✓ state Kepler's three laws of planetary motion.
- ✓ compute Kepler's constant for a given system of planets, moons, or satellites.
- ✓ outline Isaac Newton's discovery of the law of universal gravitation.
- ✓ summarize the Cavendish experiment that led to the determination of the universal gravitational constant.
- ✓ discuss a calculation of the average mass of the earth and the result's implications of the earth's structure.
- ✓ demonstrate through modeling how Saturn's system of moons confirms the principles of circular motion.

Chapter Review

In Terms of Physics

circular motion	138	torque ($\boldsymbol{\tau}$)	148
spin	138	moment arm (ℓ)	148
orbital motion	138	rotational equilibrium	149
angular speed (ω)	139	law of moments	149
uniform circular motion	139	geocentric theory	151
tangential velocity ($\mathbf{v}_t$)	139	heliocentric theory	151
centripetal acceleration ($\mathbf{a}_c$)	140	Kepler's laws	152
angular velocity ($\boldsymbol{\omega}$)	141	semi-major axis (R)	153
right-hand rule for circular motion	141	period (T)	153
nonuniform circular motion	142	Kepler's constant (K)	153
angular acceleration ($\boldsymbol{\alpha}$)	142	astronomical unit (ua)	153
tangential acceleration (a_t)	143	universal gravitational constant (G)	156
centripetal force ($\mathbf{F}_c$)	146	Newton's law of universal gravitation	156
centrifugal force	146	torsion balance	156

Problem-Solving Strategies

7.1 (page 140) Tangential speed can be calculated by the formula $v_t = r\omega$ as well as by $v_t = r\Delta\theta/\Delta t$.

7.2 (page 148) The unit of torque is the N·m. Do not confuse this unit with the N·m that is a unit of energy and work which is equivalent to the joule (J).

7.3 (page 153) Remember that when using Kepler's constant $K = \text{y}^2/\text{ua}^3$ in his third law equation, R must be in terms of astronomical units and T must be in years.

Review Questions

1. State at least one requirement for uniform circular motion.
2. When the origin of the coordinate system of an orbiting object is at the center of rotation, the object's position vector is parallel to what part of the circle?
3. Why does the complete coordinate system of a rotating object include the *z*-axis?
4. The quantity "60 rpm" is an example of what kinematic variable?
5. If the radius of an orbiting object increases and its tangential speed remains the same, what must happen to its angular speed? (This would occur if the space shuttle increased its orbital altitude without increasing its speed.)
6. What is the direction of the centripetal acceleration vector in circular motion?
7. What happens to the centripetal acceleration of an orbiting object if the radius of its orbit decreases while its tangential speed increases? (Assume its mass remains constant.)
8. Describe a simple method for determining the direction of the angular velocity vector of a rotating object.
9. What other vector quantity is parallel to angular acceleration? Describe what is happening to the object's motion when the angular acceleration vector points in the negative-*z* direction.
10. What condition must exist for the three equations of circular motion to hold true?
11. What force is responsible for centripetal acceleration?
12. If an object is experiencing uniform circular motion, is the centripetal force vector constant? Explain your answer.
13. If you swing a lead sinker on the end of a string horizontally over your head in a uniform motion, then suddenly shorten the string, what happens to centripetal force if you maintain the tangential speed constant?
14. If an object has unbalanced torques acting on it, what kind of mechanical condition exists?
15. What condition must exist for rotational equilibrium? To what principle of linear motion is this analogous?
16. According to historical records, who was the first to propose that the earth revolves around the sun?
17. Why did the Greek astronomer Ptolemy believe that epicycles were necessary in his geocentric model of the solar system?
18. What long-held belief about the motions of the planets did Kepler have to discard in order to develop his second law?
19. What is the shape of each planet's orbit around the sun?
20. At what part of a planet's orbit does it travel fastest?
21. What restriction applies to using Kepler's constant?
22. Name some of the other scientists who, through their earlier discoveries, contributed to the discovery of the law of universal gravitation by Newton. Which of the scientists mentioned in this section was a contemporary and friend of Newton?
23. How does gravitational attraction depend on the distances between two objects? From which reference points is this distance measured?
24. Why was it important to Cavendish that the weight of the lead spheres in his apparatus be accurately known when he calculated the earth's average density?
25. The universal gravitational constant can be determined only through experimentation. What are some enhancements to the basic Cavendish apparatus that might permit greater experimental precision compared to what was possible in the eighteenth century? Consider location as well as physical attributes.

True or False (26–35)

26. The tangential velocity of an orbiting object is always perpendicular to its radial position vector.
27. Angular speed is a scalar quantity. It includes both magnitude and the sign indicating its direction of motion.
28. The first and second equations of circular motion can be used with either vector or scalar forms of the variable quantities.
29. Centripetal acceleration is opposite to centripetal force.
30. The tangential speed of a revolving object will increase if its orbital radius increases and the centripetal force acting on it remains the same.
31. A torque vector is perpendicular to the plane containing the applied force and the moment arm.
32. Early heliocentric theories were not much more accurate than the best geocentric theory at predicting the positions of the planets in the heavens.
33. The earth is closer to the sun during winter in the northern hemisphere. This is why winter seems to drag out so long—the earth is moving more slowly at this point in its orbit.
34. A less massive falling object will not accelerate toward a planet as much as a more massive object.
35. Cavendish had to know the value of *G* before he could compute the earth's density.

36. ✪ A typical angular speed of an AC induction motor is 3600 rpm. Convert this speed to radians per second.
37. ✪ A fishing sinker on a string revolves in a circle of radius 1.00 m with a constant speed of 2.75 m/s. What is its centripetal acceleration?
38. ✪ A car drives around a curve with a 125 m radius. The car's maximum centripetal acceleration is 2.60 m/s^2.
    a. What is the car's maximum speed, in km/h, around the curve?
    b. If the car goes twice its maximum velocity for the curve and at the same centripetal acceleration as for Part a, what will be the radius of its path?
39. ✪ A June bug (*Phyllophaga corrosa*) is clinging to the rim of a 61.0 cm bicycle wheel that is rolling with an angular speed of 1.45 revolutions per second. What is the magnitude of the centripetal acceleration on the insect?
40. ✪ If the bicycle in Question 39 started at rest and took 5.0 s to attain the angular speed, what was the magnitude of the June bug's average angular acceleration?
41. ✪ If the idle speed of a compact disc in a multi-speed CD-ROM drive is 100 rpm, what is the angular acceleration if it takes 2.2 s for the disc to speed up to 2500 rpm?
42. ✪ A 92.2 g blue jay (*Cyanocitta cristata*) wheels through the trees in order to land at a backyard feeding station. If it flies with a constant speed of 24.0 km/h and makes a tight horizontal turn with a radius of 3.50 m, how much centripetal force must its wings exert to make the turn?

43. A 1.40 m thread is tied to a 50.0 g laboratory mass that a physics student is swinging in a horizontal circle around her head. If a force of 4.46 N will break the thread, what is the maximum angular speed at which she can revolve the mass without breaking the thread?

44. If a mechanic exerts 80.5 N·m while using an 18.0 cm wrench on a stuck bolt that does not budge, how much force are the sides of the bolt's head exerting on the mechanic's wrench if the head of the bolt is only 2.0 cm across?

45. Mercury orbits the sun with an average speed of 4.78×10^4 m/s. About how long does Mercury take to orbit the sun once?

46. What is the gravitational attraction between a dating couple whose centers of mass are 0.630 m apart if the boy weighs 715 N and the girl weighs 465 N?

DM **47.** Calculate Kepler's constant for the Saturnian system of moons. Express the constant in terms of km^3/d^2.

DM **48.** Create a scatter plot comparing the cube of each moon's orbital radius with the square of its period. (You may have to limit the range of data in order to display all the points on a single graph, depending on the method you use to create the plot.)

a. How well does the data compare to a linear fit of this plot?

b. The slope of the trendline curve is equal to what orbital quantity?

c. What do the results of this evaluation indicate about God's ordinances that Kepler described mathematically?

DM **49.** Verify the validity of the centripetal force equation (7.11) for the Saturnian orbital system by completing the following steps. Be sure to first convert all data to base SI units.

a. Select three of Saturn's moons, such as Janus, Tethys, and Iapetus, to evaluate. These three have a wide spread of orbital radii. Compute Saturn's gravitational force on each of these moons. The mass of Saturn is 5.685×10^{26} kg.

b. Determine the tangential speed of each of the moons by substituting the gravitational force from Part a above as the centripetal force in Equation 7.11.

c. Calculate the period of each moon's orbit as determined by its calculated tangential speed found in Part b.

d. Compare the calculated orbital period with the observed data. Suggest some reasons why this model does or does not agree with observation.

50. The diameter of a typical minivan tire is 64.8 cm. A stone is stuck in the tread of the right rear tire. What is the magnitude and direction of the stone's angular velocity vector if the van is traveling at 10.0 km/h?

51. The mean distance between the centers of the earth and the moon is 3.84×10^8 m.

a. At an instant in time, how far from the earth on a line connecting the centers of the earth and the moon would an object have to be in order to experience equal forces from the earth and the moon?

b. Why would this position not be a stable location between the two bodies?

c. After some research, discuss the locations around two gravitational bodies that permit a third object to experience a stable orbit.

52. Write a brief biography of Edmond Halley. Include in the paper a discussion of the influence of his works on Newton's gravitational theory.

53. Calculate the radius of a "lunasynchronous" orbit. The period of one lunar day, T_M, is equal to 27.3 days (2.36×10^6 s).

54. Prove that the semi-major axis of an ellipse is equal to the average distance of any point in an ellipse from both foci.

FACETS of PHYSICS

Roller Coasters

Why are some people addicted to roller coasters? Speed, near weightlessness in the vertical drops, and high acceleration all help produce the adrenaline rush that we associate with thrill rides. Extreme rides achieved new heights in the 1970s with the creation of the looping roller coaster. Those early versions seem tame compared to modern coasters, but the first loop was quite an innovation. If you look closely, you will notice that a loop on a roller coaster is not really a circle but rather an upside-down teardrop called a *clothoid loop*. Why is this necessary?

Start by considering a frictionless circular loop. As the coaster train enters the bottom of the loop and climbs to the top, its speed decreases. For the ride to be successful, the initial velocity must be great enough to ensure that the coaster will make it over the top of the loop. However, as the velocity increases, so does the centripetal force (the gee-force) experienced by the rider. At the velocity needed to complete the circular loop, the rider would experience a force equivalent to 6 gees, or six times the force of gravity, as he enters the bottom of the loop. People not accustomed to such forces start feeling uncomfortable when experiencing 3.5 gees and may start blacking out beyond 5 gees as the blood is forced out of the brain. Obviously, a circular loop would be neither safe nor popular.

The clothoid loop is a clever way of getting around the difficulty. The coaster train approaches the loop on a linear track and then enters a shallow rising curve with a regularly decreasing radius. Because the radius of curvature at the bottom of the loop is large, the centripetal acceleration exerted on the rider is lower. The upper section of the loop, ±65° from a vertical line bisecting the loop, is approximately a circle. The ride down the loop is the reverse of the ride up. In circular motion, when all other factors are kept constant, as the radius decreases, the tangential speed increases. Now we have two influences at work—the ascension of the loop tending to decrease the train's velocity, and the steadily decreasing radius of curvature, which tends to increase its velocity. The interplay of these two influences ensures that the train has enough energy to go over the top while maintaining accelerations that rarely exceed 3.7 gees. As the train falls and its speed increases, the curvature radius steadily increases so that the acceleration when the train goes through the bottom of the loop is acceptable.

Where should you sit on the roller coaster to get the best ride? Because the train is long, riders in the front, middle and back experience the ride differently. The ride is designed so that the rider traveling at the train's center of mass will have a smooth ride and, at most, feel weightless only briefly. The center of mass of the train is always traveling at the velocity designed for the loop at each point, and it will have the lowest velocity at the peak of the loop.

Imagine you are in the front of the train. As the train rises to the top of the loop, the front cars are pushed over the top faster than the design velocity and, since they are ahead of the center of mass, experience their lowest velocity not directly at the peak, but slightly past the peak. There they feel held back until the center of mass goes over the top and the train begins plunging down. The front cars experience a slight moment of "hang time." The back cars of the train experience their lowest velocity before they reach the peak of the loop. They hang back slightly on the way up. As the center of mass reaches the peak and accelerates, the last cars are whipped over the top. Since riders in the back are going faster than the design velocity for that point, they may feel forced into their seats.

Sailing ship rigging lines transmit mechanical forces, and pulleys change the direction of those forces.

Applying Newton's Laws 8

Dominion Science Problem

Head-On Collisions

Almost 37 600 drivers and passengers are killed in car accidents every year in the United States, averaging 1 death every 14 minutes. About 10% of these fatalities occur when a driver accidentally or purposefully veers into oncoming traffic and experiences a head-on collision. And these statistics don't even include those who are injured. Head injuries, back injuries, broken bones, whiplash, and other more extensive trauma can plague those who do survive head-on collisions. How can we prevent head-on collisions from taking place, especially on roads where cars travel at high speeds and collisions are more likely to be fatal?

8-1 The results of a head-on collision. There were no survivors.

8A SIMPLIFYING PROBLEMS

Newton's laws are fundamental to mechanics. By applying them systematically, you can solve almost any mechanical problem. But suppose you have a complex problem involving several interacting bodies. How do you apply Newton's laws to such a problem?

8.1 Free-Body Diagrams

A **free-body diagram** is a vector diagram that isolates a single object in a multi-object system in order to analyze the forces on it.

The first step in simplifying a complex problem is to isolate each body and consider it as a separate system. For example, Figure 8-2 is a real-world diagram showing three blocks connected by strings. A physics student is pulling upward on the string attached to block 1. To find the net force on block 1, draw a free-body diagram of that block. That is, isolate block 1 as the system of interest and determine all the external forces that act *directly* on the system.

What forces act directly on block 1? Gravity pulls down on the block. You can see that the hand does not act directly on the block but on the string, which then exerts the force on the block. Similarly, blocks 2 and 3 do not pull directly on block 1, but the string tied to the bottom of block 1 exerts a downward force on it. The forces of the strings on the block are called **tension (T)** forces because the strings *pull* on the block. These three forces—the force of the upper string, $\mathbf{T}_1$; the force of the lower string, $\mathbf{T}_2$; and the force of gravity (the block's weight), $\mathbf{F}_w$—are

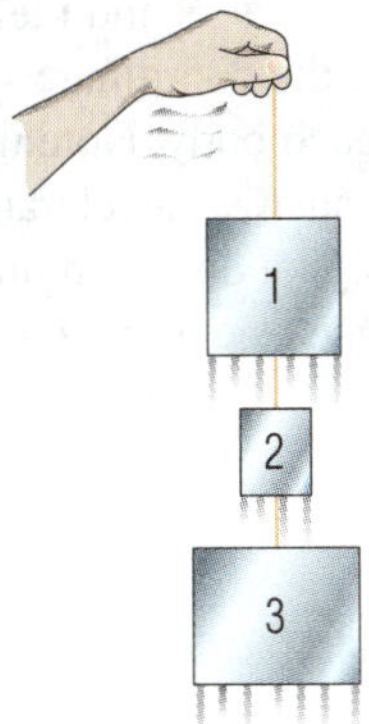

8-2 A multi-object system

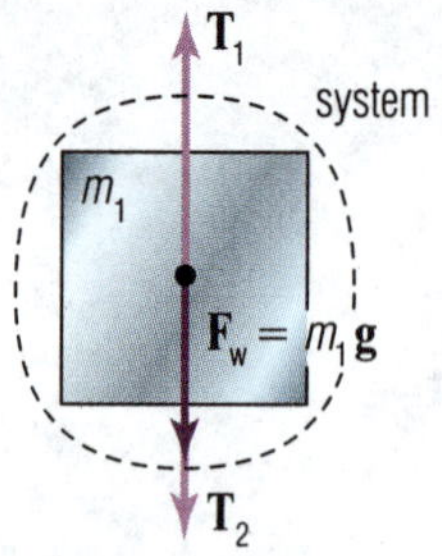

8-3 Free-body diagram of block 1

the only forces that act directly on block 1. Figure 8-3 shows the free-body vector diagram of block 1.

Keep in mind that a string connecting two objects exerts the same tension on both objects but in different directions.

8.2 Transmitting Forces in a Line: Ideal Strings

The strings attached to objects in this chapter are **ideal strings**. Physicists use the concept of the ideal string when studying multiple-body systems where the objects are not touching but are connected. Ideal strings have the following properties:

- They have no mass; therefore, they do not affect the acceleration of the system of objects.
- They do not stretch when exerting a force.
- They exert only pulling forces, not pushes. A pushing force would crumple the string.
- Forces are exerted only in line with the string.

Real strings and ropes approximate ideal strings as long as tension forces are limited to avoid stretching or breaking. For example, the force $\mathbf{T}_1$ on block 1 in Figure 8-3 has the same magnitude and direction as the force of the hand on the string. As long as the pull is continuous, an ideal string also holds objects at fixed distances. Therefore, all the objects connected by a string and pulled will have the same speed and acceleration. Although real strings have mass and do stretch somewhat, the mass and stretch are usually small enough to be ignored. So the assumption that a string is ideal is unlikely to introduce noticeable error into your calculations. You can assume that all strings are ideal unless otherwise stated.

8.3 Systems of Connected Objects

There are many approaches to analyzing the dynamics of a system of objects. If all the masses are known, it is generally easier to determine the motion of the system as a whole first and then analyze the forces on the individual objects. You need to identify what information is given in order to determine the approach to solving the problem. There is no "cookbook" method for solving mechanics problems.

Problem-Solving Strategy 8.1

Make accurate sketches of multi-body systems that include strings and pulleys. Indicate all known and unknown force vectors and their components in the coordinate system used for each body. Note unknown values such as acceleration, but do not assign a direction until after the analysis is completed.

Using free-body diagrams, you can modify the problem-solving method you learned in Chapter 2 to fit problems in mechanics. Most of the changes are additions to Step 1, where it was suggested that you sketch the problem. To apply Newton's laws, you should first draw a diagram of the entire arrangement. This is called the "world" diagram. Include in this diagram all connections (such as ideal strings) and an arrow showing the direction of motion if that is given. When you have drawn the overall arrangement, you may have to draw one or more free-body vector diagrams. Then select a coordinate system for each object. This does not necessarily have to be the same for all objects, especially if they are moving in different directions.

After you have drawn free-body diagrams, you can continue with the problem-solving method, applying Newton's laws to each diagram separately. When you have reached a solution, check it to see if it is reasonable. It should have the proper units and should agree with any given information. For example, if you find that the individual accelerations for two objects connected by a string are either of different magnitudes or in opposite directions, then you know that you have made a mistake.

EXAMPLE 8-1

Pulling Strings: Analyzing Systems of Objects

The masses of the blocks in Figure 8-4a are $m_1 = 5.0$ kg and $m_2 = 1.0$ kg. The hand pulls up on the string with a force of 120 N. What is the magnitude of the tension in the string between the blocks?

Solution:

Since the blocks are connected by an ideal string, you should first treat the two blocks as a single 6.0 kg system. The tension in the string connecting the two blocks is not shown because it is a force within the system and cannot affect the system's motion. Only forces from outside a system can affect the motion of the system as a whole.

The acceleration of block 2 equals the acceleration of the system. You can then determine the force on block 2 that will produce that acceleration of the block according to Newton's second law. That force is the tension in the string tied to block 2.

Choose a coordinate system with y vertical.

Calculate the acceleration of the two-block system:

$$\Sigma \mathbf{F}_{sys} = m_{sys}\mathbf{a} = \mathbf{F}_{hand} + \mathbf{F}_w$$

$$m_{sys}a_y = F_{hand\,y} + F_{wy}$$

$$a_y = \frac{F_{hand\,y} + m_{sys}g_y}{m_{sys}}$$

$$a_y = \frac{(+120\text{ N}) + (6.0\text{ kg})(-9.81\text{ m/s}^2)}{6.0\text{ kg}}$$

$$a_y \cong +1\underline{0}.1\text{m/s}^2\text{ (up)} \qquad (1)$$

Note that all the forces on the system are parallel to the y-axis.

Calculate the tension on block 2 using the acceleration from (1) and Newton's second law for block 2:

$$\Sigma \mathbf{F}_2 = m_2\mathbf{a} = \mathbf{T}_2 + \mathbf{F}_{w\,2}$$

$$\mathbf{T}_2 = m_2\mathbf{a} - \mathbf{F}_{w\,2}$$

$$T_{2\,y} = m_2a_y - F_{w2\,y} = m_2a_y - m_2g_y$$

$$T_{2\,y} = m_2(a_y - g_y) = (1.0\text{kg})[(+1\underline{0}.1\text{m/s}^2) - (-9.81\text{m/s}^2)]$$

$$T_{2\,y} \cong +1\underline{9}.9\text{ N} \cong +20.\text{ N (up)}$$

The tension in the string on block 2 is about 20 N straight up.

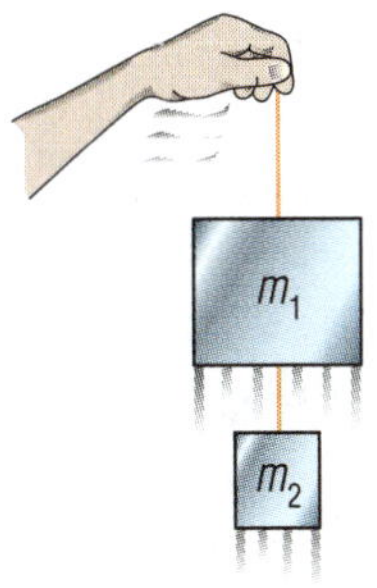

8-4a World diagram of the two-block system

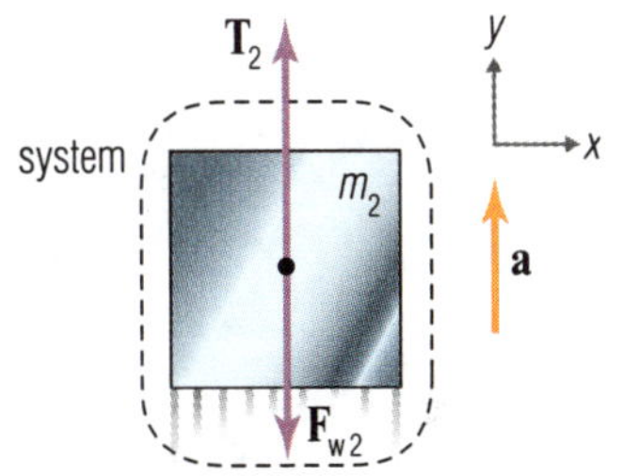

8-4b Free-body diagram of block 2

8A Section Review

1. Explain why a free-body diagram is essential to analyze the dynamics of an object.
2. Describe the properties of an ideal string. Under what conditions do ropes and real strings act like ideal strings?

8A Objectives

After completing this section, I can

- ✓ construct free-body diagrams of parts within a system of connected objects.
- ✓ identify the external forces acting on a system in a free-body diagram.
- ✓ state the properties of an ideal string.
- ✓ solve dynamics problems involving objects connected by strings.

3. Compare the tension exerted at one end of a straight string to the tension exerted at the other end of the string.

4. (True or False) The force that a system exerts on the surface that supports it is *not* included in a free-body diagram of the system.

5. (True or False) If a truck towing two trailers of equal mass in tandem accelerates away from a gas pump, the force exerted by the truck on the first trailer equals the force exerted on the second trailer by the first.

⊙6. Using a diagram, sketch the location, magnitude, and direction of all forces on a 1 kg brass sphere hanging from a string.

⊙7. A family is skating at an ice rink. The 58.2 kg mother is holding the hand of her 35.5 kg daughter. The father grabs his wife's free hand and pulls horizontally with a constant force of 100. N. Assume that the skates glide without friction on the ice and that the family's hands and arms approximate ideal strings. How much net force does the daughter experience?

⊙8. Two mountain climbers tied together with an "ideal rope" have slipped on an icy incline. Fortunately, the first climber was belayed to a piton in the ice. The climbers are hanging from their ropes on a frictionless 65.0° incline. The first climber and his gear have a mass of 93.0 kg. The second climber's total mass is 67.0 kg.

 a. What is the tension in the rope tied to the piton?

 b. What is the tension in the rope between the climbers?

8B TRANSMITTING MECHANICAL FORCES

8.4 Transmitting Forces Around a Corner: Ideal Pulleys

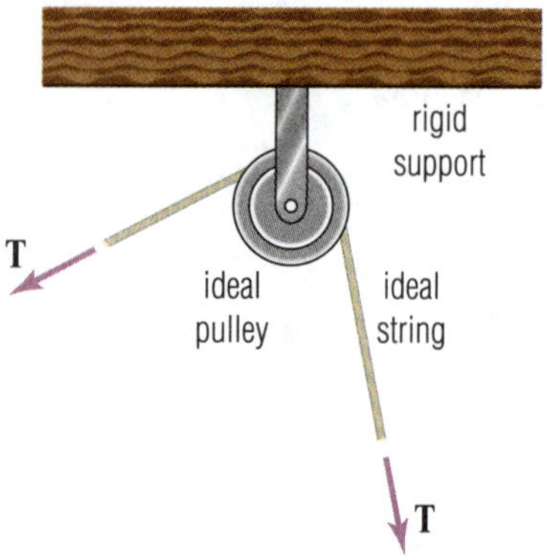

8-5 An ideal pulley changing the direction of a tension force

An ideal string (or a real one, for that matter) cannot exert a force in a direction other than in line with the string. However, one of the main reasons for using a string-like connection between two systems is to exert forces on systems that are not in line with one another. The direction of tension in a string can be changed by using a *pulley*. Physicists employ **ideal pulleys** when considering theoretical systems connected by ideal strings.

An ideal pulley has the following characteristics:

- It consists of a grooved wheel and an axle. The pulley can be mounted to a structure outside the system of interest or attached directly to the system.
- The axle is frictionless.
- The motion of the string around the wheel is frictionless.
- It changes the direction of the tension in the string without diminishing its magnitude.

Actual mechanical systems use pulleys to change the direction of tensional forces as well (e.g., flagpoles, sailboat rigging, etc.). If real systems did not use pulleys, but rather allowed a rope to wrap around an edge of a building or structural beam, for example, the rope would rapidly wear, and the force available at the end of the rope would be greatly reduced because of friction.

In Figure 8-6b, $\mathbf{F}_{wh}$ is the weight of the hanging block. The block exerts this force on the string, which in turn exerts a force **T′** on the block on the tabletop.

The reaction force of the tabletop block on the hanging block is **T**. The magnitude $T' = F_{\text{w h}}$, but the directions are not the same. The pulley changes the direction of the transmitted force. As you will see in the following example, adjusting the orientation of the coordinate system for each object in a connected system simplifies the analysis.

EXAMPLE 8-2

Getting the Hang of Pulleys: Changing the Direction of a Force

A 1.50 kg block (m_t) rests on a frictionless table. It is attached to a hanging block of mass $m_h = 1.00$ kg by a string passing over an ideal pulley mounted to the table's edge. What is the acceleration of the hanging block?

Solution:

This problem involves linear motion, but each of the elements of the system is moving in a different direction. The tabletop block is moving horizontally, and the hanging block is moving vertically. The tension of the string on each block and the blocks' accelerations have the same magnitudes, but their directions are different.

Choose a coordinate system for each block that maintains the same axis parallel to the motion of the block in its part of the system.

The *x*-axis for the tabletop block is horizontal and is pointing to the right. The *x*-axis for the hanging block is *vertical* and is pointing *downward.* This way, whatever direction the acceleration will be, it will have the correct direction relative to the block you are examining.

Perform free-body analysis of the hanging block using Newton's second law:

$$\Sigma \mathbf{F}_h = m_h \mathbf{a}$$

$$\mathbf{T} + \mathbf{F}_{wh} = m_h \mathbf{a}$$

or in scalar component notation,

$$T_x + m_h g_{h\,x} = m_h a_x. \quad (1)$$

(Note that $g_{h\,x}$ points in the *positive* *x*-direction due to the rotated reference frame of the hanging block. Therefore, $g_{h\,x} = -g$.)

Find T_x by analyzing the tabletop block. Since the block experiences no vertical motion, any vertical forces must cancel out, and you can work with the scalar components.

$$\Sigma F_x = m_t a_x = T'_x$$

$$T'_x = m_t a_x \quad (2)$$

The acceleration of the tabletop block (a_x) in this equation has the same magnitude and direction relative to the *x*-coordinate axis as a_x for the hanging block.

You know that $T_x = -T'_x$ because they are an action-reaction force pair transmitted by the ideal string. After replacing T_x in Equation 1, substitute Equation 2 into Equation 1 for T'_x to solve for a_x.

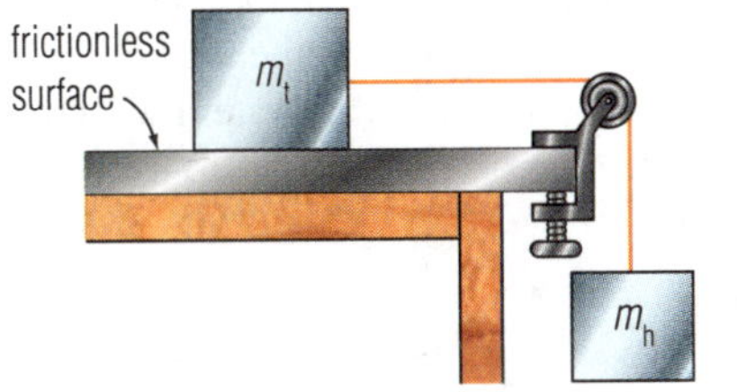

8-6a World diagram of the two blocks

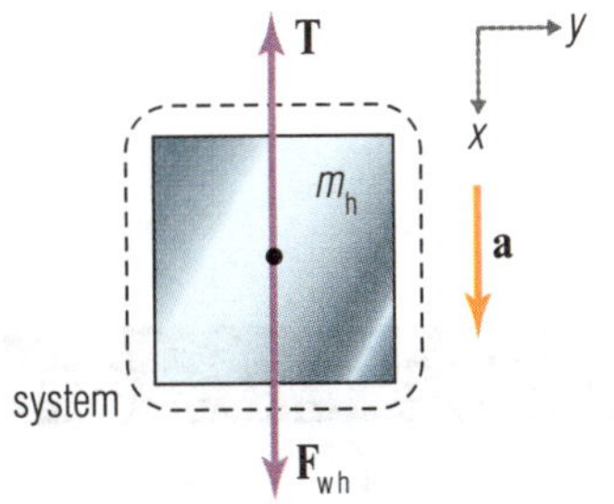

8-6b Forces on the hanging block

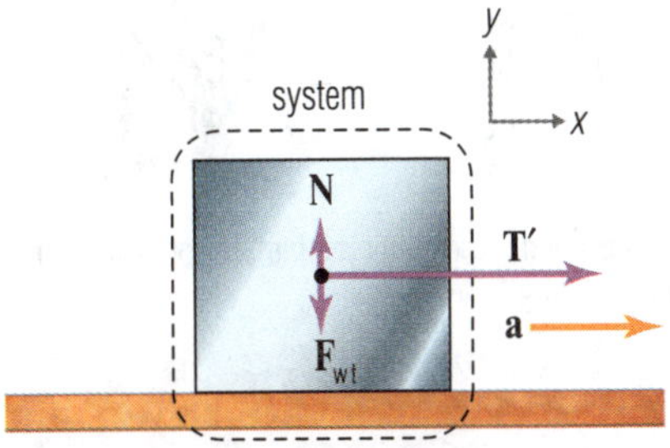

8-6c Forces on the tabletop block

Problem-Solving Strategy 8.2

The notation for tension vectors transmitted by strings connecting objects A and B is as follows: If the string exerts a tension **T** on object A, then it exerts a tension **T′** on object B. $|\mathbf{T}| = |\mathbf{T}'|$ and $\mathbf{T} = -\mathbf{T}'$ (opposite directions).

$$-T'_x + m_h g_{hx} = m_h a_x$$
$$-(m_t a_x) + m_h(-g) = m_h a_x$$
$$-m_h a_x - m_t a_x = m_h g$$
$$-a_x(m_h + m_t) = m_h g$$
$$a_x = \frac{m_h g}{-(m_h + m_t)} = -\frac{m_h}{(m_h + m_t)} \cdot g$$

Substituting the known values from the problem statement gives the following:

$$a_x = -\frac{m_h}{(m_h + m_t)} \cdot g$$
$$a_x = -\frac{1.00\ \text{kg}}{(1.00\ \text{kg} + 1.50\ \text{kg})} \cdot (-9.81\ \text{m/s}^2)$$
$$a_x \cong +3.9\underline{2}4\ \text{m/s}^2 \cong +3.92\ \text{m/s}^2$$

The hanging block is accelerating about 3.9 m/s^2 in the positive-*x* direction in its frame of reference. In the reference frame of the observer, the hanging block is accelerating *downward*.

Problem-Solving Strategy 8.3

Use different frames of reference to analyze separate objects in a system as long as the same coordinate axis in each reference frame is parallel to the motion of the system within the reference frame.

Another interesting problem involves two hanging masses on a string over a set of pulleys.

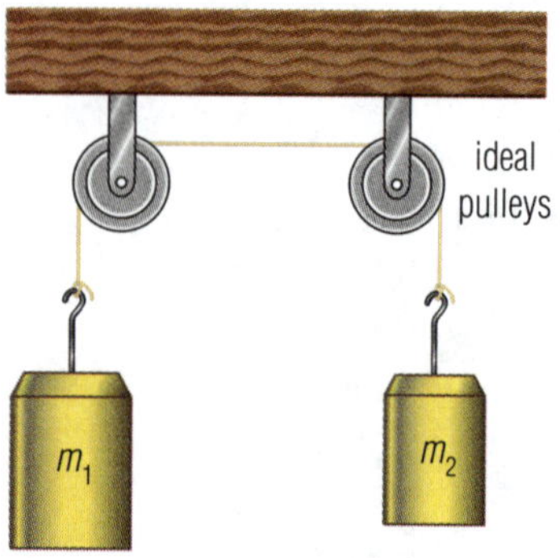

8-7a World diagram of the masses and pulleys in Example 8-3

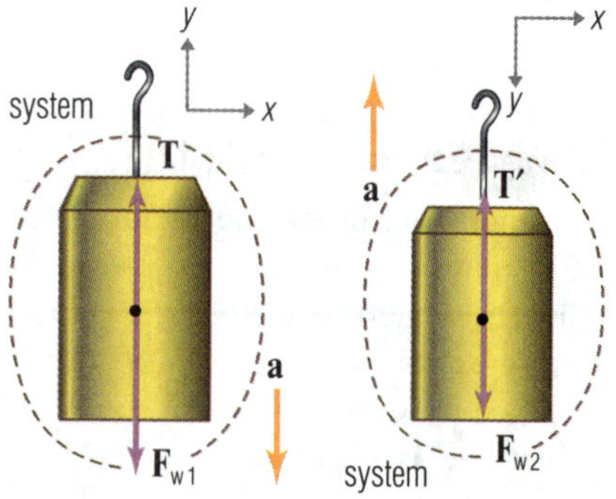

8-7b Free-body diagrams of mass 1 and mass 2

EXAMPLE 8-3

Balancing Act: Another Pulley Problem

Two lab masses, $m_1 = 1.50$ kg and $m_2 = 1.00$ kg, are connected by an ideal string that passes over two ideal pulleys (to keep the masses from interfering with each other). What is the net force on each mass?

Solution:

For mass 1, choose the coordinate axis parallel to the expected vertical motion with the *y*-axis positive upward. Choose the coordinate system for mass 2 to be the vertical with the *y*-axis positive *downward*. Whatever movement occurs in the system will be in the same direction relative to each mass's reference frame. Consequently, $\mathbf{a}_1 = \mathbf{a}_2 = \mathbf{a}$. The strategy for finding the force sum on each mass involves calculating system acceleration and then solving Newton's second law, $\Sigma\mathbf{F} = m\mathbf{a}$, for each mass.

Analyze the net force on each mass:

Mass 1:

$$\Sigma\mathbf{F}_1 = m_1\mathbf{a}$$
$$\mathbf{T} + \mathbf{F}_{w1} = m_1\mathbf{a}$$
$$T_y + m_1 g_{1y} = m_1 a_y$$
$$T_y = m_1 a_y - m_1 g_{1y} \quad (1)$$

Note that $g_{1y} = g$.

Mass 2:

$$\Sigma\mathbf{F}_2 = m_2\mathbf{a}$$
$$\mathbf{T}' + \mathbf{F}_{w2} = m_2\mathbf{a}$$
$$T'_y + m_2 g_{2y} = m_2 a_y$$
$$T'_y = m_2 a_{2y} - m_2 g_{2y} \quad (2)$$

Due to the rotated reference frame, $g_{2y} = -g$.

The magnitude of the tension in the string is the same throughout the string, but the tension vectors at the ends act on the masses in opposite directions relative to their associated frames of reference. Therefore, T_y and T'_y are an action-reaction force pair and $T_y = -T'_y$.

Combine Equations 1 and 2 and make appropriate substitutions:

$$m_1a_y - m_1g_{1\,y} = -T'_y$$

$$m_1a_y - m_1g_{1\,y} = -(m_2a_y - m_2g_{2y})$$

$$m_1a_y - m_1g = -[m_2a_y - m_2(-g)]$$

$$m_1a_y - m_1g = -m_2a_y - m_2g$$

$$m_1a_y + m_2a_y = m_1g - m_2g$$

Solve for a_y:

$$a_y(m_1 + m_2) = (m_1 - m_2)g$$

$$a_y = \frac{(m_1 - m_2)g}{(m_1 + m_2)}$$

$$a_y = \frac{(m_1 - m_2)}{(m_1 + m_2)} \cdot g$$

Substitute the known values from the problem statement:

$$a_y = \frac{(1.50\text{ kg} - 1.00\text{ kg})}{(1.50\text{ kg} + 1.00\text{ kg})} \cdot (-9.81\text{ m/s}^2)$$

$$a_y = -1.9\underline{6}2\text{ m/s}^2$$

The result indicates that mass 1 is dropping (negative-y direction in its reference frame) and mass 2 is rising (also negative-y direction in its reference frame), which is logical, since mass 1 weighs more.

The net force on mass 1 is

$$\Sigma F_{1\,y} = m_1a_y$$

$$\Sigma F_{1\,y} = (1.50\text{ kg})(-1.9\underline{6}2\text{ m/s}^2)$$

$$\Sigma F_{1\,y} \cong -2.9\underline{4}3\text{ N}$$

$$\Sigma F_{1\,y} \cong -2.94\text{ N (downward).}$$

The net force on mass 2 is

$$\Sigma F_{2\,y} = m_2a_y$$

$$\Sigma F_{2\,y} = (1.00\text{ kg})(-1.9\underline{6}2\text{ m/s}^2)$$

$$\Sigma F_{2\,y} \cong -1.9\underline{6}2\text{ N}$$

$$\Sigma F_{2\,y} \cong -1.96\text{ N (upward).}$$

8.5 Transmitting Forces on Inclines

It is important to realize that coordinate systems are *chosen.* As you have seen, the x-axis does not have to be horizontal, and the positive y-axis does not have to be opposite the direction of gravity. The only restriction is that the x-axis must be perpendicular to the y-axis. Sometimes the forces on an object will be neither horizontal nor vertical, but at an angle. In such cases it is often helpful to define the coordinate system so that one axis is in the same direction as most of the forces. For example, Figure 8-8a shows a block sitting on an incline. Two of the forces on the block are parallel to the incline, and one force is perpendicular to the incline. Therefore, a convenient choice of coordinates would place the x-axis along the incline and the y-axis perpendicular to it, as in the free-body diagram Figure 8-8b.

Figure 8-8b shows one force that is easy to overlook. That force is the **normal force (N)**, the force exerted by the incline on the block. This force is called the

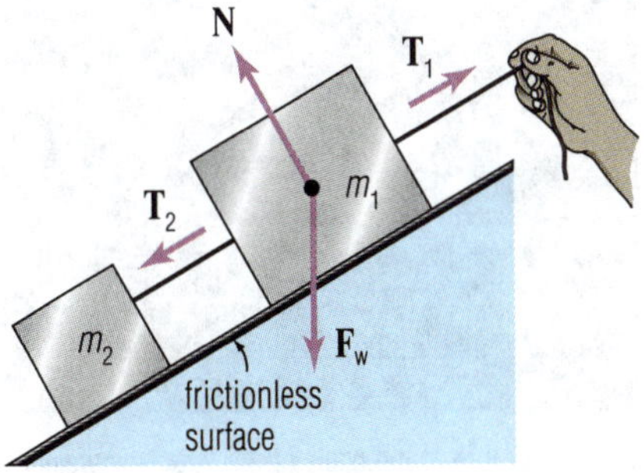

8-8a World diagram of a block on an incline

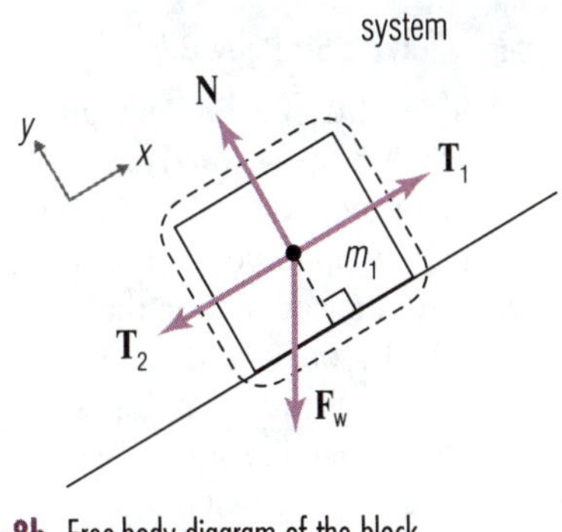

8-8b Free-body diagram of the block

normal force because the term *normal* in mathematics means "perpendicular to." The normal force exerted by a surface on an object in contact with it is *always* perpendicular to the surface.

To understand the normal force, examine an object resting on a flat table. The force of the earth's gravity on the object pulls it down. This is the object's weight ($\mathbf{F}_w$). Its weight causes the object to exert a force normal to the table that is equal in magnitude and direction to the object's weight. However, this latter force is *not* the weight of the object. Figure 8-9 illustrates the difference between these two forces, where $\mathbf{F}_{w\ norm}$ represents the normal component of the object's weight acting on the table. The table, according to Newton's third law, must exert an opposing force of equal magnitude on the object. This is the normal force (**N**). Therefore, the normal force can be expressed as the negative of the force exerted by the object's weight vector perpendicular to the table. If the *y*-axis is vertical,

$$\mathbf{N} = -\mathbf{F}_{w\ norm} = -\mathbf{F}_{w\,y}\,. \tag{8.1}$$

If the table is tilted as in Figure 8-10, the object's weight no longer presses fully on the table. Instead, the object exerts a force on the table equal to the magnitude of the component of its weight that is perpendicular to the incline, so the normal force decreases accordingly.

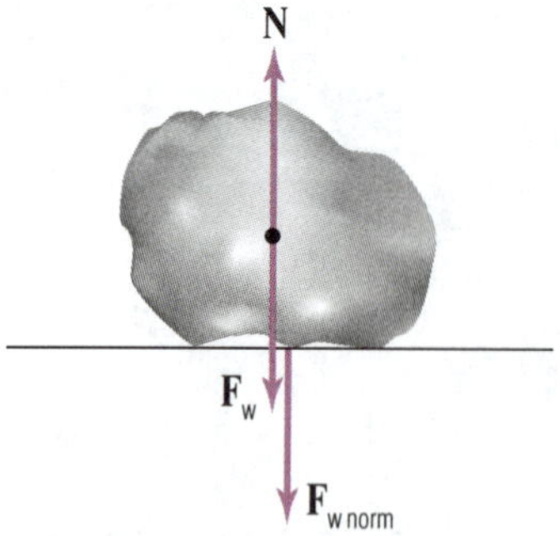

8-9 Relationship between normal force, weight, and the normal component of weight on a horizontal surface

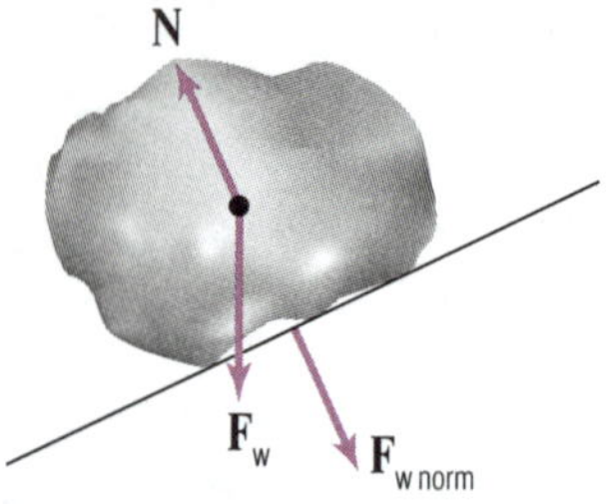

8-10 The relationship between normal force and weight on an incline

Problem-Solving Strategy 8.4

The vector of the force of gravity is always assumed to point straight down. However, the normal component of the object's weight varies with the incline of the surface.

EXAMPLE 8-4

The Normal Force on an Incline

A block rests on a 30.0° incline. If the block weighs 50. N, what is the normal force on the block?

Solution:

Choose a rectangular coordinate system so that the *x*-axis is parallel to the inclined surface and the *y*-axis is perpendicular to it. Figure 8-11b is the free-body diagram showing $\mathbf{F}_w$ resolved into a normal component, $\mathbf{F}_{w\,y}$, and a down-slope component, $\mathbf{F}_{w\,x}$. The magnitude of $\mathbf{F}_w$ is 50. N. A right triangle is formed with a hypotenuse of 50. N.

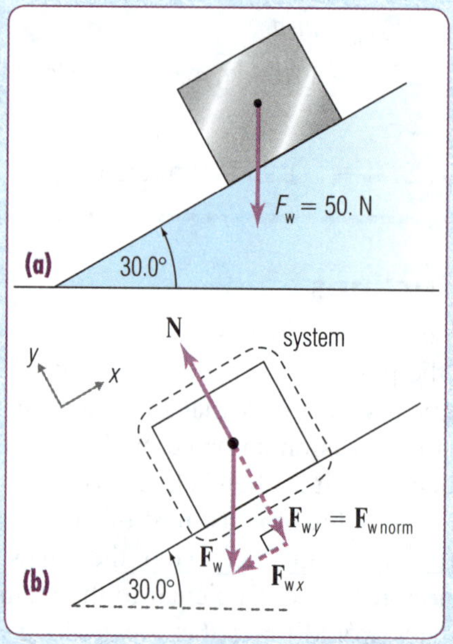

8-11 (a) World diagram and (b) free-body diagram of a block on an incline

The weight vector triangle consists of the block's weight and its two components. The angle between $\mathbf{F}_w$ and $\mathbf{F}_{w\,y}$ equals the incline angle, 30.0°.

Calculate the normal component of weight:

$|F_{w\,y}| = F_w \cos\theta = (50.\ \text{N})\cos 30.0°$

$|F_{w\,y}| \cong 4\underline{3}.3\ \text{N}$

$F_{w\,y} \cong -4\underline{3}.3\ \text{N}$ (negative-*y* direction)

Calculate the normal force:

$N_y = -F_{w\,y}$

$N_y \cong -(-4\underline{3}.3\ \text{N}) = +4\underline{3}.3\ \text{N}$

$N_y \cong +43\ \text{N}$

8-12 Because of the angle of the track, the weight of these cars is not perpendicular to the track.

Problem-Solving Strategy 8.5

For inclines, the conventional alignment for the positive *x*-axis is *up* the slope of the incline. This choice is completely up to you but is consistent with the convention that "up" is positive. If you should choose to have positive directions point down, be sure to change the sign of the negative scalar constant *g*.

8.6 Measured Weight

Weights are usually measured with spring scales or some form of electronic scale. The scale exerts an upward force that balances the object's weight. Usually a display indicator shows the amount of force needed to balance the object's weight. For a bathroom scale, this third-law reaction force is the normal force that the scale exerts on the object. In an *unaccelerated reference frame*, the normal force is equal in magnitude but opposite in direction to the weight of an object resting on a horizontal surface.

EXAMPLE 8-5

Weight: Normal Force Read from a Scale

A 25.0 kg crate rests on a package scale. The scale's display shows the magnitude of the normal force of the scale on the crate. The crate and the scale are at rest. What weight does the scale indicate?

Solution:

You want to find N, the magnitude of the normal force. Make a free-body diagram for the crate:

$$\Sigma \mathbf{F} = m\mathbf{a}$$

From Newton's first law, you know that the scale-crate system is in translational equilibrium, so $\mathbf{a} = 0$ and

$$\Sigma \mathbf{F} = 0 \text{ N}.$$

The force sum on the crate is

$$\mathbf{N} + \mathbf{F}_w = 0 \text{ N}.$$

Using y-components:

$$N_y + F_{wy} = 0 \text{ N}$$

$$N_y = -F_{wy} = -mg_y$$

$$N_y = -(25.0 \text{ kg})(-9.81 \text{ m/s}^2)$$

$$N_y \cong +24\underline{5}.2 \text{ N} \cong +245 \text{ N}$$

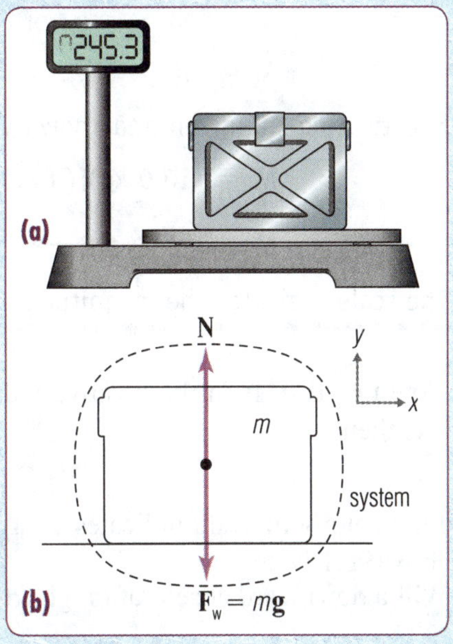

8-13 (a) World diagram of the crate; (b) Free-body diagram of the crate

The indicated weight is the magnitude of the normal force exerted by the scale, or about 245 N.

8.7 Apparent Weight

A spring scale is an accurate measure of weight when the scale and object are not accelerating. But what happens if they *are* accelerating?

EXAMPLE 8-6

Giving Weight a Lift: Accelerated Scales

A 10.0 kg box rests on a bathroom scale in an elevator. The elevator is accelerating upward from rest with a uniform acceleration of $+2.00 \text{ m/s}^2$. What weight does the scale indicate?

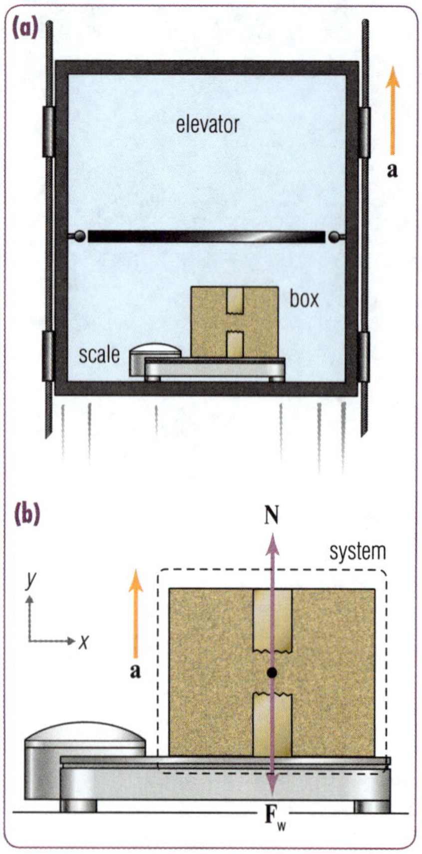

8-14 (a) World diagram of the box on a scale in an elevator; (b) Free-body diagram of the box

Solution:

The scale reading is the magnitude of the normal force (N) that the scale exerts on the box. The system is the box. The bathroom scale and the elevator are outside the system. You can find the normal force by identifying all of the forces acting on the box.

$$\Sigma \mathbf{F} = m\mathbf{a}$$

$$\mathbf{N} + \mathbf{F}_w = m\mathbf{a}$$

Using the y-components,

$$N_y + F_{w\,y} = ma_y. \qquad (1)$$

Solve Equation 1 for the vertical normal force component:

$$N_y = ma_y - F_{w\,y}$$

Since $F_{w\,y} = mg_y$, you now have

$$N_y = ma_y - mg_y, \text{ or}$$

$$N_y = m(a_y - g_y). \qquad (2)$$

Substituting the given and known values into Equation 2 gives you

$$N_y = (10.0\ \text{kg})[(+2.00\ \text{m/s}^2) - (-9.81\ \text{m/s}^2)]$$

$$N_y \cong +11\underline{8}.1\ \text{N} \cong +118\ \text{N}.$$

The scale indicates the magnitude of N_y, or about 118 N.

Note that if a_y is in the positive-y direction, as stated in the example problem above, then

$$|ma_y - mg_y| > |mg|,$$

and the bathroom scale indicates an **apparent weight** greater than the weight of the box (98.1 N).

Will a downward acceleration give the same result?

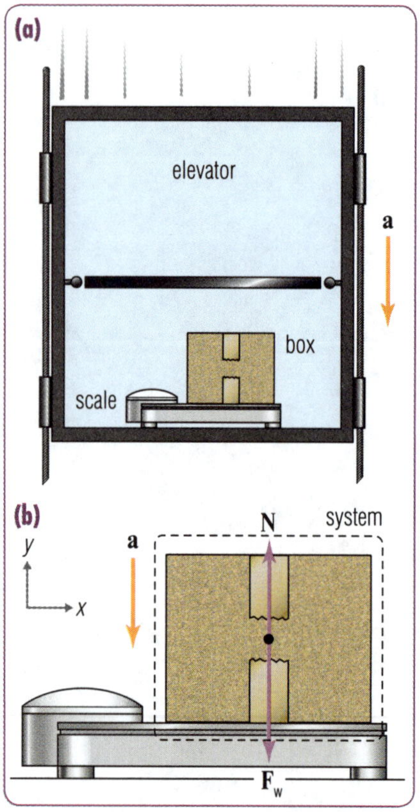

8-15 (a) World diagram of the box on a scale in the elevator; (b) Free-body diagram of the box

EXAMPLE 8-7

Weight Loss the Easy Way

The bathroom scale and the box from Example 8-6 are now accelerating downward at $-2.00\ \text{m/s}^2$. What weight does the bathroom scale indicate?

Solution:

Again, the scale gives the magnitude of **N**. The scalar values of Equation 2 in the previous example are not specified, so the magnitude of the normal force is still

$$N_y = m(a_y - g_y)$$

Substituting the values into this equation gives you

$$N_y = (10.0\ \text{kg})[(-2.00\ \text{m/s}^2) - (-9.81\ \text{m/s}^2)]$$

$$N_y = +78.1\ \text{N}.$$

The magnitude of the normal force is about 78 N. The box's apparent weight when the elevator is accelerating downward is less than its actual weight.

The last situation to be evaluated is the elevator dropping in free fall. Real elevators are designed to prevent this, but in this scenario its downward acceleration would be equal to **g**.

EXAMPLE 8-8

Weight in Free Fall

Suppose the elevator from the previous examples somehow breaks and falls freely down its shaft. What would be the indicated weight of the box?

Solution:

Even if the box, the scale, and the elevator are all in free fall, the normal force on the box is still

$$N_y = m(a_y - g_y).$$

But $a_y = g_y$, so

$$N_y = mg_y - mg_y, \text{ and}$$

$$N_y = 0 \text{ N}.$$

The scale exerts no force on the box. Since you measure weight by the reading of a scale, you might say that the box is *weightless.* Actually, the weight of the box has not vanished. It weighs no less than it would if it were stationary; the earth still exerts the same force of gravity on it. However, since both the box and the scale are in free fall, the scale can't exert a force on the box. You could say that the box is "normal force-less!"

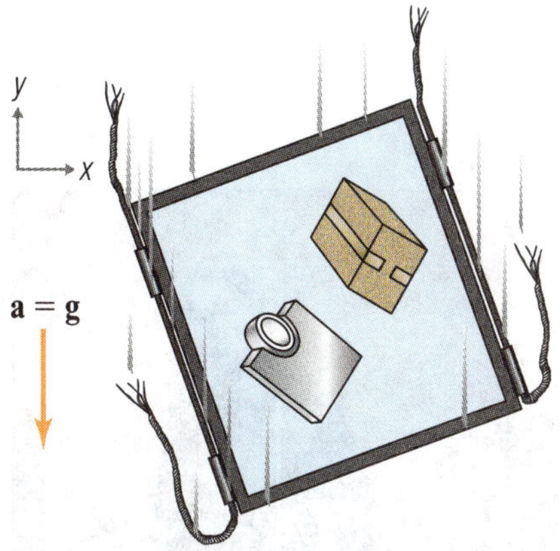

8-16 World diagram of the box, scale, and elevator in free fall

8-17 A "normal force-less" astronaut

You probably have heard that astronauts in space are weightless. Although the force of gravity lessens with increasing distance from the earth, astronauts are still influenced by gravity. However, their vehicle is in orbit, or free fall, around the earth. Since the spacecraft and its occupants fall at the same rate, the astronauts seem to float. Therefore, it only appears that they have no weight.

As you have seen, apparent weight equals real weight when the object being weighed is not accelerating. However, when the object is accelerating, its apparent weight is not the same as the gravitational force on the object.

8B Section Review

1. How can the direction of a tensional force be changed without diminishing the force?
2. Explain why assigning separate coordinate systems to different objects within a moving system sometimes simplifies the analysis of the system.
3. What force acting on a system is assumed to always act vertically, regardless of the orientation of the system?
4. (True or False) The normal force acting on a system is the third-law reaction force to the system's weight.
5. ✪ Two masses are suspended by a string placed over a pulley supported by the ceiling ($m_1 = 6.20$ kg and $m_2 = 7.70$ kg).
   a. Which object will fall?
   b. What will its acceleration be?

8B Objectives

After completing this section, I can

- ✓ solve dynamics problems involving pulleys that change the direction of tensional forces.
- ✓ assign appropriate coordinate systems to simplify the analysis of dynamical problems involving multipart systems.
- ✓ find the normal force exerted on objects resting on horizontal and inclined surfaces.
- ✓ determine the apparent weight of stationary and accelerated objects.
- ✓ explain why objects in orbit are not weightless.

6. A male African elephant (*Loxodonta africana*) has a mass of 5450 kg. Calculate the normal force that supports this critter on a horizontal surface.

7. The elephant in Question 6 jumps off of a 10 m cliff into a cool water hole, making an immense splash in the process. Calculate the acceleration of the earth toward the elephant if the earth's mass is 5.974×10^{24} kg.

8. Assume that an elevator carrying a bathroom scale supporting a 25.0 kg mass is rising at a uniform +4.00 m/s. It slows to a stop with a uniform acceleration of $-2.25\ m/s^2$. What weight does the bathroom scale display while the elevator is slowing?

8C FRICTION

8.8 Friction Defined

Have you ever wondered how it is possible for you to walk? According to Newton's first law, if there are no unbalanced external forces on a person, he cannot begin to move. Consider a woman standing in the middle of a driveway. What are the external forces on her? The force of gravity pulls down on her, and the normal force of the ground pushes up. Neither of these forces can propel her forward. The force exerted by her leg muscles is an *internal* force and cannot change her motion directly. How can she create an external force?

8-18 The normal force and the force of gravity on this woman cannot produce forward motion.

Newton's third law answers this question. As the woman attempts to walk, her leg muscles, through her feet, exert a force backward and down on the pavement. The pavement therefore exerts a force upward and forward on the foot, ultimately propelling the woman forward. The upward component of this reaction force is included in the normal force. The forward component of the reaction force is called **friction (f)**. Without friction, the woman could not move forward.

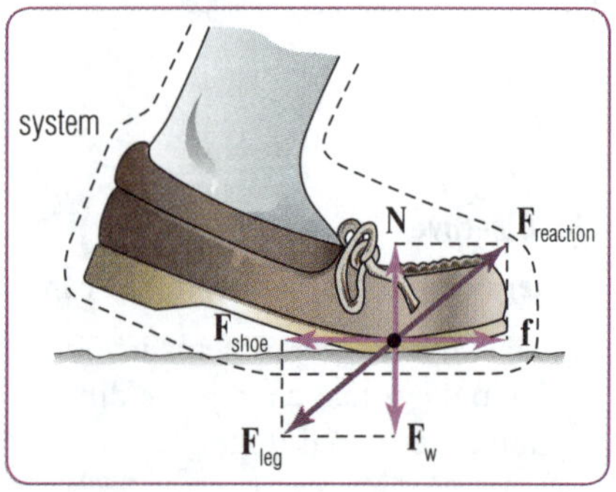

8-19 Friction provides the force needed for walking.

Ice is an example of a nearly frictionless surface. Suppose the woman were in the middle of a smoothly frozen, frictionless pond, instead of in a driveway. She could try to exert a backward force on the ice, but her foot would just slip backward. She could still exert a downward force on the ice, and the ice would push back with an equal force, but unless there were some frictional force she could not move forward.

Another kind of motion that is impossible without friction is rolling. As a tire rolls, it pushes backward on the road; consequently, the road pushes forward on the tire. This forward force of friction moves the car forward. When there is little friction (on an icy street, for example) and the car's tires spin, they exert little or no horizontal force on the ground. The car cannot move forward on level ground when friction is absent. Similarly, a wheel on a frictionless incline will not roll but will simply slide without rolling.

8-20 A large amount of friction is necessary to accelerate a dragster to high speeds.

The friction that makes walking, rolling, and similar motions possible is called **traction (f_t)**. Traction is also the name used for friction that prevents unwanted motion. For example, when a car rounds a curve, its tendency is to continue in a straight line. Traction provides the centripetal force to keep the car on the road. Sometimes traction alone would not be adequate to change the car's path. In such cases the curve is often banked so that a component of the normal force also points toward the center of the curve, adding to the centripetal acceleration caused by traction.

Friction does have positive aspects, but sometimes it gets in the way. Friction is, in general, a force between two surfaces that opposes the relative motion of the surfaces. Every time one surface slides across another—for example, a box sliding across a carpet—the sliding object exerts a force on the other surface in the direction of sliding. The surface therefore exerts a reaction force on the sliding object to oppose the sliding. Friction occurs because of the interaction of molecules of the two surfaces. The closer together they get, the stronger the interactions become. If you smooth two surfaces, at first you decrease friction by removing obstacles to the motion, just like removing speed bumps on the road allows a car to travel faster. However, when the surfaces become smooth enough for most of their atoms to touch, the atoms attract each other, increasing friction. Try sliding apart plates of glass that are stacked together.

8.9 Magnitude of Friction

What affects the magnitude of the friction force? Consider two boxes of the same size and material but different weights. Both are pulled across the same surface. Which experiences a greater frictional force? The heavier box does. Apparently, weight affects friction. Suppose a rectangular box is dragged across a surface, first on a small side, then on a large side. In which case is the frictional force greater? It is the same in both cases. The frictional force is not affected by the area of contact between the surfaces. Now suppose a box is pulled across a level surface and then across the same surface inclined to the horizontal. In which case is the force of friction greater? It is greater with the level surface.

One force that is related to weight and that decreases on an incline is the normal force. It seems reasonable to assume that the normal force and the frictional force are related. In fact, scientists have found experimentally that

$$f \propto N.$$

This proportion can be written as an equation by adding a proportionality constant:

$$f = \mu N \tag{8.2}$$

The constant, the Greek letter mu (μ), is called the coefficient of friction. There is a characteristic coefficient for each pair of materials in contact that must be found by experiment.

Problem-Solving Strategy 8.6

Only the absolute values of friction and normal force may be used in the friction formula. The formula is *not* a vector equation.

The coefficient of friction (μ) is the constant of proportionality between the magnitude of a system's normal force and the magnitude of friction. Because friction depends on the materials in contact, the frictional coefficient is unique for each pair of materials.

8.10 Kinetic Friction

The force of friction on an object depends not only on the object's weight and the material it is made of, but also on the object's state of motion. You may have noticed this already. For example, more force is required to start a heavy box moving than to keep it moving. Therefore, for each set of materials, there are two coefficients of friction: the coefficient of moving or kinetic friction (μ_k) and the coefficient of static friction (μ_s). Usually, $\mu_s > \mu_k$. For the magnitude of **kinetic friction ($\mathbf{f_k}$)**,

$$f_k = \mu_k N. \tag{8.3}$$

Kinetic friction has several properties that generally hold true, although there are exceptions. The kinetic frictional force

- is oriented parallel to the contact surface.
- opposes the motion of the system of interest.
- depends in some ways on the kinds of materials in contact and the condition of the surfaces. This property tends to be highly variable and dependent on other factors.
- is generally *independent* of the relative speed of the sliding surfaces as long as the characteristics of the surfaces remain unchanged (that is, no phase,

Problem-Solving Strategy 8.7

The *magnitudes* of variable vectors are represented by the italicized form of the vector symbol (f and N, for example). The *scalar component* symbols for friction and normal force will be f_x and N_y, respectively. These forces have only one nonzero component if they are parallel to the reference frame coordinate axes.

The constant of proportionality between the normal force on a sliding system and kinetic friction is called the *coefficient of kinetic friction*, μ_k.

structural, or chemical change occurs that might alter the surface material properties).

- is generally *independent* of the surface area of contact between two objects.
- is *directly proportional* to the normal force acting on the sliding object.

8.11 Static Friction

However, for **static friction ($\mathbf{f_s}$)** (friction between stationary objects), it is not necessarily true that $f_s = \mu_s N$. When two surfaces attempt to slide against each other due to a shearing force, friction will prevent them from sliding until the applied motivating force parallel to the surfaces exceeds the static friction. In other words, static friction can vary from zero up to a maximum value, $f_{s\,max}$. As a mathematical expression,

$$0 \leq f_s \leq f_{s\,max}.$$

As long as the applied force parallel to the surfaces is less than $f_{s\,max}$, static friction will exactly cancel out the applied force and the surfaces will not slip. Only when $F > f_{s\,max}$ will the surfaces begin to slide. The magnitude for *maximum* static friction between two materials in contact is found from

$$f_{s\,max} = \mu_s N. \quad (8.4)$$

Static friction also has properties that are generally true. Experimentation has shown that static friction

- can be any value between zero and a maximum value characteristic for the two materials in contact.
- is oriented parallel to the contact surface.
- opposes the motion of the system of interest.
- depends on the kinds of materials and condition of the contact surfaces.
- is normally independent of the contact surface area.

EXAMPLE 8-9

Where's the Rub?: Coefficient of Friction

A block of vulcanized rubber weighing 5.00 N rests on a horizontal sheet of clean glass. The rubber block has a screw eye of negligible weight inserted in one side. A spring scale is hooked into the screw eye and the assembly is dragged across the glass at a constant velocity (see Figure 8-21). The spring scale indicates a force of 5.35 N. What is the coefficient of kinetic friction for vulcanized rubber on glass?

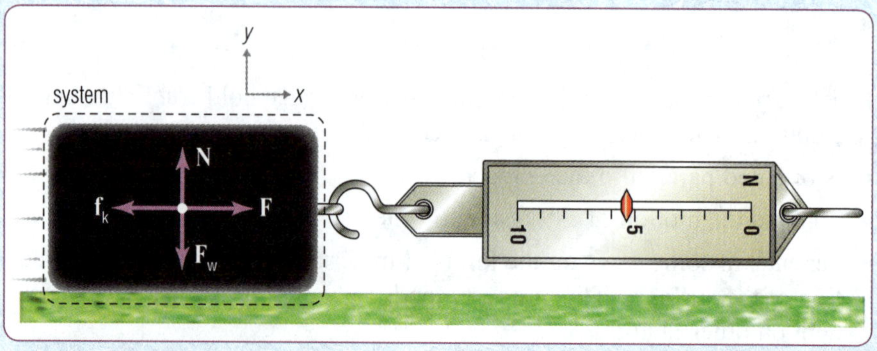

8-21 World view of the rubber block resting on a glass surface

Solution:

Choose a Cartesian coordinate system with the x-axis parallel to the glass surface and the y-axis perpendicular to the surface.

The velocity of the rubber block is constant, so $\mathbf{a} = 0\ \text{m/s}^2$. Therefore the net force on the block must be zero according to Newton's first law.

The force sum on the block is

$$\Sigma \mathbf{F} = \mathbf{F} + \mathbf{f}_k + \mathbf{N} + \mathbf{F}_w = 0\ \text{N}.$$

Analyze the y-components to find N.

$$\Sigma F_y = F_y + f_{k\,y} + N_y + F_{w\,y} = 0\ \text{N} \qquad (1)$$

In the vertical direction, F_y and $f_{k\,y}$ have zero components. Equation 1 reduces to

$$N_y + F_{w\,y} = 0\ \text{N}$$

$$N_y = -F_{w\,y}.$$

The magnitude of the normal force is just the magnitude of the weight of the block:

$$N = F_w$$

In the x-direction,

$$\Sigma F_x = F_x + f_{k\,x} + N_x + F_{w\,x} = 0\ \text{N}. \qquad (2)$$

The components N_x and $F_{w\,x}$ are zero, so Equation 2 reduces to

$$F_x + f_{k\,x} = 0\ \text{N, and}$$

$$f_{k\,x} = -F_x\,.$$

The magnitude of the kinetic friction force equals the magnitude of the pulling force on the block:

$$f_k = F$$

Solve Equation 8.3 for the coefficient of kinetic friction and substitute the known quantities:

$$\mu_k = \frac{f_k}{N} = \frac{F}{F_w} = \frac{5.35\ \cancel{\text{N}}}{5.00\ \cancel{\text{N}}}$$

$$\mu_k = 1.07 \qquad (\mu_k \text{ is dimensionless})$$

The coefficient of kinetic friction between rubber and glass is 1.07.

The next example shows how friction opposes the inception of motion and other forces causing motion.

EXAMPLE 8-10

Giving a Shove: Static Friction

A metal crate having a mass of 100. kg rests on the smooth concrete floor of a warehouse. The coefficients of friction for metal on concrete are $\mu_s = 0.94$ and $\mu_k = 0.83$. A worker is able to exert a horizontal force $F = +1250$ N. (a) Can he push the crate across the floor without assistance? In other words, is $F > f_{s\,max}$? (b) If so, what is the crate's acceleration?

Solution:

a. On a horizontal surface, the magnitude of the normal force on a system is just the magnitude of its weight, $N = F_w = |mg|$. The maximum static friction that can be generated by the crate on the concrete is as follows:

$$f_{s\,max} = \mu_s N = \mu_s |mg|$$

$$f_{s\,max} = (0.94)|(100.\text{ kg})(-9.81\text{ m/s}^2)|$$

$$f_{s\,max} \cong 9\underline{2}2\text{ N}$$

Since the force the man exerts is greater than the maximum static friction, the man can move the crate without help.

b. Once the crate begins to move, kinetic friction takes over. The horizontal second-law force sum on the crate is

$$\Sigma F_x = F_x + f_{k\,x} = ma_x. \quad (1)$$

Kinetic friction is determined from Equation 8.3:

$$f_k = \mu_k N = \mu_k |mg|$$

$$f_k = (0.83)|(100.\text{ kg})(-9.81\text{ m/s}^2)|$$

$$f_k \cong 8\underline{1}4\text{ N}$$

Substitute the known values into Equation 1, taking into account the relative directions of the force vectors:

$$F_x + f_{k\,x} = ma_x$$

$$(+1250\text{ N}) + (-8\underline{1}4\text{ N}) = (100.\text{ kg})a_x$$

$$a_x = \frac{+4\underline{3}6\text{ N}}{100.\text{ kg}}$$

$$a_x \cong +4.\underline{3}6\text{ m/s}^2 \cong +4.4\text{ m/s}^2$$

8C Section Review

1. What is friction? Discuss ways to reduce friction between two materials.
2. Write the formula for kinetic friction. Why can the formula *not* be expressed as a vector equation? (*Hint*: Consider the directions and magnitudes of the vector quantities in such an equation.)
3. Why is the coefficient of static friction usually larger than that of kinetic friction for the same materials?
4. Why is the static friction between two materials not constant?
5. ✪ A block with a mass of 9.00 kg is pulled at a constant speed across a horizontal tabletop with a spring scale. The scale reads 61.8 N. Calculate the coefficient of kinetic friction.
6. ✪ A warehouse worker is unloading a 240. kg crate from a truck using an "ideal rope" passing over a single ideal pulley above the truck. What is the minimum horizontal frictional force that the floor must exert on the worker to keep him from slipping while slowly lifting the crate if the rope between the pulley and the man makes an angle of 25.0° with the vertical?

8C Objectives

After completing this section, I can

- ✓ explain how friction and traction are essential for producing motion in walking and driving.
- ✓ compare and contrast the different types of friction.
- ✓ calculate the magnitudes of kinetic and static friction.
- ✓ discuss the factors that contribute to kinetic and static friction.
- ✓ calculate coefficients of friction.
- ✓ determine friction in dynamical problems.

8D MORE APPLICATIONS

8.12 Rolling Friction

Friction for rolling objects is not the same as sliding friction. When an object such as a wheel rolls, every point on the rim of the wheel is in motion relative to the surface on which it is rolling *except* the point where the wheel touches the surface. If the point of contact were completely frictionless, then the wheel would just slide along the road surface and no external torque could be applied to the wheel to roll it. In the real world, some friction always exists between the wheel and the road, causing the wheel to roll. The wheel experiences additional friction because it adheres to the road, tending to stick as it lifts away from the surface.

What is the magnitude of traction on a wheel of a vehicle that is propelled in some way? The applied force (F_{app}) can be either an external force such as a push or gravity, or it can be a force internal to the vehicle, such as a motor or engine. Such forces can be represented as a force vector on the wheel in the direction of application.

When a wheel turns, there is always some friction present as its axle rubs in its bearings. In a more complex vehicle, such as a car, there are many points in the car's *drive train* where friction reduces the force that the engine supplies to the wheels. The sum total of all these points of friction that retard the freedom of motion of the wheel, including the friction forces between the wheel and the surface over which it rolls, is called **rolling friction ($\mathbf{f_r}$)**.

The magnitude of the net propelling force that the wheel applies to the road surface (F_{prop}) is the difference between the magnitudes of the applied force and the rolling friction on the wheel.

$$F_{prop} = F_{app} - f_r$$

The propelling force can be represented diagrammatically by the sum of the external force applied to the wheel by the vehicle, pointing in the direction of motion, and the rolling friction force opposing the applied force, pointing opposite the direction of motion. Since

$$\mathbf{F}_{prop} = \mathbf{F}_{app} + \mathbf{f}_r,$$

the method shown in Figure 8-23 does not introduce errors, and this notation will be useful in later chapters.

The magnitude and direction of the traction force on the wheel depends on the source of the applied force and magnitude of rolling friction. If the applied force is a force external to the vehicle (such as gravity), then the traction vector is equal to just the rolling friction vector. If the applied force includes a torque on the axle of the wheel supplied by the engine, then the magnitude of the traction force is equal to that

8-22 Brakes in a car purposely increase rolling friction in order to slow the car when required.

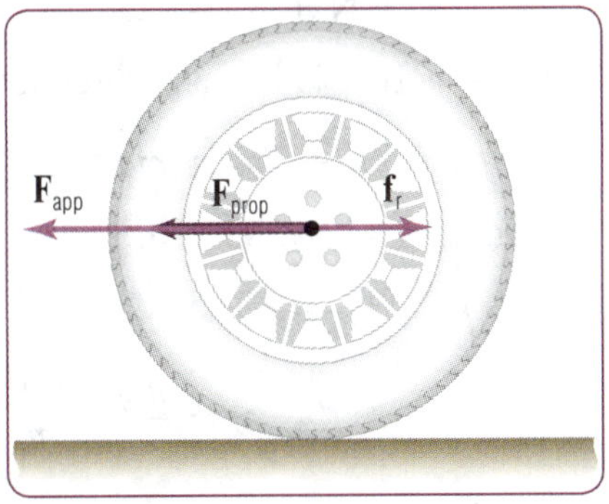

8-23 The net propelling force on the wheel of a vehicle

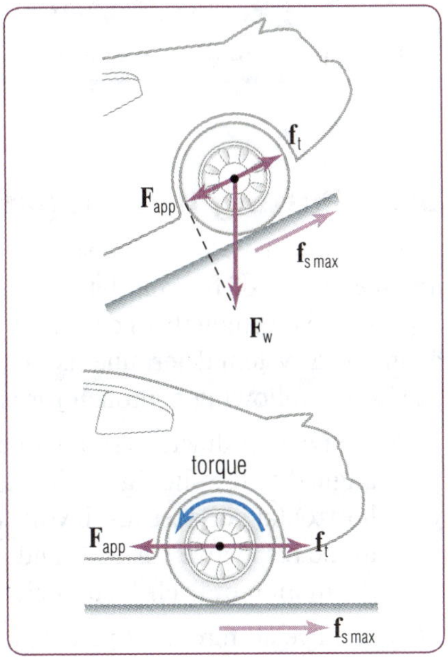

8-24 The magnitude and direction of the traction vector depends on the origin of the forces turning the wheel.

Inexperienced drivers in snow often fail to account for the different orientations of the traction vector when coasting and applying power to the wheels. A car idling as it coasts down an icy hill can begin to slide if the rolling friction exceeds the static friction available between the tires and the road. Adding a *little* power from the engine while descending the hill may counteract the rolling friction enough to prevent a skid by reducing the rolling friction to within the traction limits available.

of the propulsion force in the absence of any externally applied forces. The wheel will not slip as long as the magnitude of the traction force does not exceed the maximum static friction that can exist between the wheel and the road.

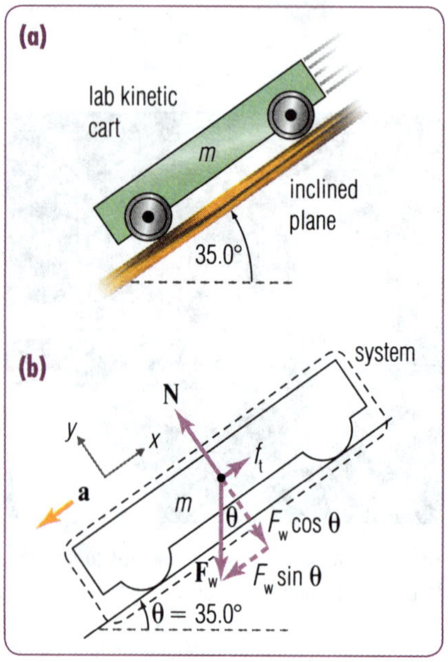

8-25 (a) World diagram of rolling cart on an incline; (b) Free-body diagram of the cart

EXAMPLE 8-11

On a Roll

A 2.0 kg wheeled laboratory cart rolls, without slipping, down an inclined plane that makes an angle of 35.0° with the horizontal. If the magnitude of rolling friction of the cart is 0.10 N, what is the cart's acceleration?

Solution:

The coordinate system is aligned as in Figure 8-25b.

The cart is propelled by gravity. To find the cart's acceleration, you have to know the net force acting on the cart:

$$\Sigma \mathbf{F} = m\mathbf{a}$$

$$\mathbf{f}_t + \mathbf{N} + \mathbf{F}_w = m\mathbf{a}$$

Since motion is in the x-direction, evaluate the x-components:

$$f_{tx} + N_x + F_{wx} = ma_x \qquad (1)$$

Since $f_{tx} = f_{rx}$, $N_x = 0$, and $F_{wx} = mg \sin\theta$:

$$f_{rx} + mg \sin\theta = ma_x$$

$$a_x = \frac{mg \sin\theta + f_{rx}}{m}$$

$$a_x = \frac{(2.0 \text{ kg})(-9.81 \text{ m/s}^2)\sin 35.0° + (+0.10 \text{ N})}{2.0 \text{ kg}}$$

$$a_x \cong -5.\underline{57} \text{ m/s}^2 \cong -5.6 \text{ m/s}^2$$

The cart's acceleration is about 5.6 m/s^2 down the incline.

8.13 Putting It All Together

Now that you have learned the process of free-body analysis, the property of inclines, and the effects of friction, let's look at several examples that combine all of these aspects. If you study them carefully, you will better appreciate how multiple forces acting on a system determine its motion.

Use the following approach for analyzing inclined-plane dynamics:

a. Assign coordinate systems to each system element so that the x-axis is aligned to the sliding surface and pointing *up* the slope. If other inclined or horizontal surfaces are involved, the x-axes should all be oriented parallel to the respective surfaces and point in the same general direction relative to the motion of each system element.

b. Resolve all forces acting on each element of the system into their components relative to the coordinate system for that system element.

c. Determine the maximum static friction possible for the two materials at the angle of incline.

d. Sum the nonfriction forces parallel to the sliding surface for the entire system and compare to the maximum static friction for the system to determine the dynamic state of the system—if the nonfriction forces are less than $f_{s\,max}$, the system remains at rest; if greater, the system is accelerating.

e. If the system is accelerating, calculate the kinetic friction.

f. Sum the x-component forces, including kinetic friction, to find acceleration according to Newton's second law.

EXAMPLE 8-12

Incline, Strings, and Friction

Steel blocks $m_1 = 2.50$ kg and $m_2 = 15.0$ kg rest on horizontal and inclined surfaces as shown in Figure 8-26a. The incline is 38.0° with the horizontal. The blocks are connected by an ideal string passing over an ideal pulley. For this system, $\mu_s = 0.61$ and $\mu_k = 0.50$. If the blocks are initially held at rest, (a) determine whether the blocks will move when released, and (b) if so, calculate the acceleration.

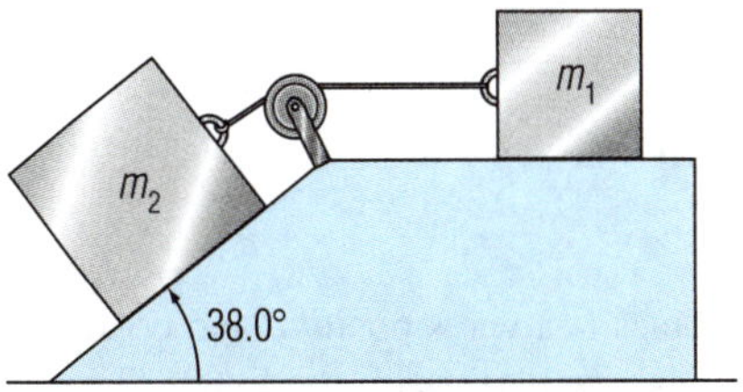

8-26a World diagram of blocks and planes for Example 8-12

Solution:

a. Assign coordinate systems to each block. The x- and y-axes will be aligned as in Figures 8-26b and c. Note that if the system does slide, the motion of each block will be parallel to the x-direction.

If the system accelerates, both blocks will have the same acceleration (**a**) since they are connected by an ideal string.

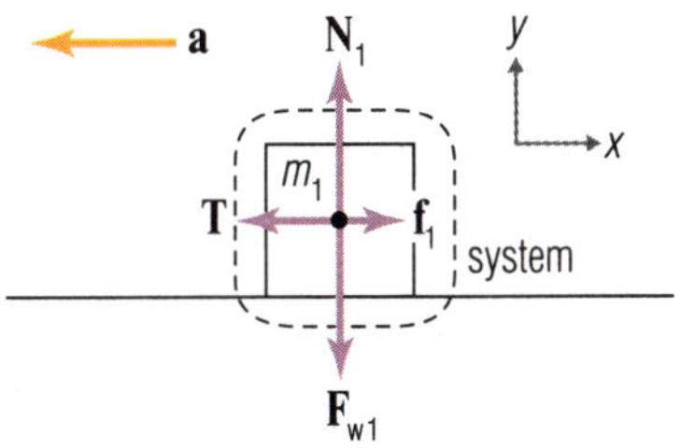

8-26b Free-body diagram for block 1

b. Define the force sums on each block.

Force sum on block 1:

$$\Sigma \mathbf{F}_1 = m_1\mathbf{a}$$

$$\mathbf{T} + \mathbf{f}_1 + \mathbf{N}_1 + \mathbf{F}_{w\,1} = m_1\mathbf{a}$$

$$\mathbf{T} + \mathbf{f}_1 + \mathbf{N}_1 + m_1\mathbf{g} = m_1\mathbf{a}$$

Force sum on block 2:

$$\Sigma \mathbf{F}_2 = m_2\mathbf{a}$$

$$\mathbf{T}' + \mathbf{f}_2 + \mathbf{N}_2 + \mathbf{F}_{w\,2} = m_2\mathbf{a}$$

$$\mathbf{T}' + \mathbf{f}_2 + \mathbf{N}_2 + m_2\mathbf{g} = m_2\mathbf{a}$$

c. Determine the magnitude of the normal forces acting on the blocks. The normal force is needed in order to calculate maximum theoretical static friction on each block. Rewrite the force sum equations using the y-scalar components.

$T_y + f_{1\,y} + N_{1\,y} + m_1g_{1\,y} = m_1a_y$

The components T_y, $f_{1\,y}$, and a_y are all zero, and $g_{1\,y} = g$. The equation reduces to

$N_{1\,y} + m_1g = 0$ N and

$$N_{1\,y} = -m_1g$$

$$N_1 = |-m_1g|$$

$$f_{s\,max\,1} = \mu_s N_1 = \mu_s|-m_1g|. \quad (1)$$

$T'_y + f_{2\,y} + N_{2\,y} + m_2g_{2\,y} = m_2a_y$

The components T'_y, $f_{2\,y}$, and a_y are all zero, and $g_{2\,y} = g \cos \theta$ because of the rotated reference frame. The equation reduces to

$$N_{2\,y} + m_2g \cos \theta = 0 \text{ N}$$

$$N_{2\,y} = -m_2g \cos \theta$$

$$N_2 = |-m_2g \cos \theta|$$

$$f_{s\,max\,2} = \mu_s N_2 = \mu_s|-m_2g \cos \theta|. \quad (2)$$

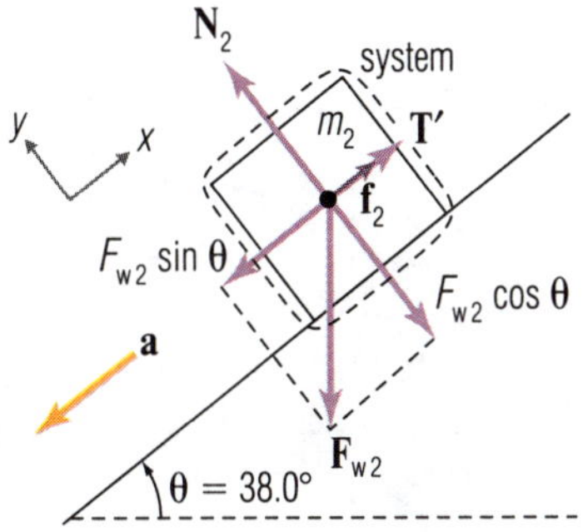

8-26c Free-body diagram for block 2

d. Evaluate the forces parallel to the sliding surface. The down-slope gravitational force on block 2 ($m_2g \sin \theta$) is the only force available to overcome the frictional forces ($f_{1\,x}$ and $f_{2\,x}$) and accelerate the system. If the x-weight component is less than the sum of the theoretical maximum static friction forces on blocks 1 and 2, then the blocks will not move.

Mass is always positive and cos θ is positive between 0° and 90°. Therefore, both terms may be factored from an absolute value expression without changing its value.

The total theoretical maximum static friction $f_{s\ max}$ for the system of blocks is

$$f_{s\ max} = f_{s\ max\ 1} + f_{s\ max\ 2}. \quad (3)$$

Substitute Equations 1 and 2 into Equation 3, and solve for $f_{s\ max}$:

$$f_{s\ max} = \mu_s|-m_1g| + \mu_s|-m_2g\cos\theta|$$

$$f_{s\ max} = \mu_s(|-g|\,m_1 + |-g|\,m_2\cos\theta)$$

$$f_{s\ max} = \mu_s|-g|\,(m_1 + m_2\cos\theta)$$

$$f_{s\ max} = (0.61)(9.81\ \text{m/s}^2)[2.5\ \text{kg} + (15.0\ \text{kg})\cos 38.0°]$$

$$f_{s\ max} = (0.61)(9.81\ \text{m/s}^2)(14.\underline{3}2\ \text{kg})$$

$$f_{s\ max} \cong 8\underline{5}.6\ \text{N}$$

The down-slope gravitational force on block 2:

$$F_{w2\,x} = m_2g\sin\theta$$

$$F_{w2\,x} = (15.00\ \text{kg})(-9.81\ \text{m/s}^2)\sin 38.0°$$

$$F_{w2\,x} = -90.\underline{5}9\ \text{N}$$

Comparing gravitational force to maximum system static friction, $|F_{w2\,x}| > f_{s\ max}$, so the blocks will accelerate.

e. Determine the *x*-scalar component force-sums on the blocks. These components determine the acceleration of the system:

$$T_x + f_{1\,x} + N_{1\,x} + m_1g_{1\,x} = m_1a_x$$

The components $N_{1\,x}$ and $g_{1\,x}$ are zero. The equation reduces to

$$T_x + f_{1\,x} = m_1a_x$$

$$T_x = m_1a_x - f_{1\,x}. \quad (4)$$

$$T'_x + f_{2\,x} + N_{2\,x} + m_2g_{2\,x} = m_2a_x$$

The component $N_{2\,x}$ is zero and $g_{2\,x} = g\sin\theta$ due to the rotated reference frame.

The equation reduces to

$$T'_x + f_{2\,x} + m_2g\sin\theta = m_2a_x. \quad (5)$$

The tension on each block transmitted by the ideal string is related by the equation

$$T'_x = -T_x.$$

Combine Equations 4 and 5 after making appropriate substitutions:

$$-T_x + f_{2\,x} + m_2g\sin\theta = m_2a_x \quad \text{(Equation 5)}$$

$$-(m_1a_x - f_{1\,x}) + f_{2\,x} + m_2g\sin\theta = m_2a_x \quad \text{(Substitute Equation 4 into 5)}$$

$$-m_1a_x + f_{1\,x} + f_{2\,x} + m_2g\sin\theta = m_2a_x \quad \text{(Remove parentheses)}$$

$$f_{1\,x} + f_{2\,x} + m_2g\sin\theta = m_1a_x + m_2a_x \quad \text{(Transpose term)}$$

$$f_{1\,x} + f_{2\,x} + m_2g\sin\theta = a_x(m_1 + m_2) \quad \text{(Factor)}$$

$$\frac{f_{1\,x} + f_{2\,x} + m_2g\sin\theta}{(m_1 + m_2)} = a_x \quad (6)$$

The friction terms in Equation 6 are the kinetic friction forces (f_k) on the blocks.

f. Calculate f_k for each block.

$$f_{1x} = f_{k1} = \mu_k N_1 = \mu_k|-m_1 g|$$

$$f_{k1} = (0.50)|-(2.50\text{ kg})(-9.81\text{ m/s}^2)|$$

$$f_{k1} \cong 1\underline{2}.2\text{ N}$$

$$f_{2x} = f_{k2} = \mu_k N_2 = \mu_k|-m_2 g\cos\theta|$$

$$f_{k2} = (0.50)|-(15.0\text{ kg})(-9.81\text{ m/s}^2)\cos 38.0°|$$

$$f_{k2} \cong 5\underline{7}.9\text{ N}$$

In which direction do the friction force vectors point? Using the same reasoning as in Part d, we can assume that the blocks will accelerate in the negative x-direction. Since friction opposes motion, the values of $\mathbf{f}_k$ will be positive.

g. Calculate the system acceleration. Substitute the friction values into Equation 6 and solve for acceleration.

$$a_x = \frac{(+1\underline{2}.2\text{ N}) + (+5\underline{7}.9\text{ N}) + (15.0\text{ kg})(-9.81\text{ m/s}^2)\sin 38.0°}{(2.50\text{ kg} + 15.0\text{ kg})}$$

$$a_x \cong -1.\underline{1}7\text{ m/s}^2 \cong -1.2\text{ m/s}^2$$

Whew! The blocks in Figure 8-26a will accelerate to the left at a rate of 1.2 m/s^2.

EXAMPLE 8-13

Skiing Downhill

A skier and her equipment have a total mass of 63.0 kg. She skis on wet snow down a straight slope that makes a 6.0° angle with the horizontal. She pushes herself off with an initial velocity of 1.00 m/s down the 800. m long slope and reaches the bottom 47.5 s later. (a) What is her average acceleration? (b) What is the coefficient of friction between her skis and the snow?

Solution:

You can use the second equation of motion to find $\bar{a}$, Newton's laws to find the net horizontal and vertical forces, and Equation 8.3 to find μ_k.

Assume that positive-x is *up* the slope.

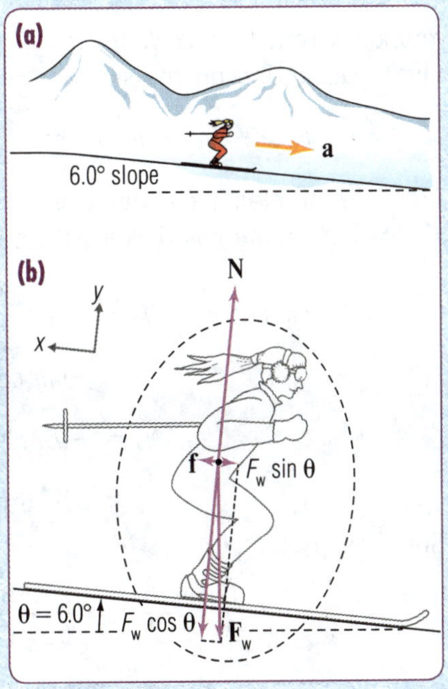

8-27 (a) World diagram of the skier; (b) Free-body diagram of the skier

a. Find $\bar{a}$:

$$d_x = v_{1x}\Delta t + \tfrac{1}{2}\bar{a}_x(\Delta t)^2$$

$$\tfrac{1}{2}\bar{a}_x(\Delta t)^2 = d_x - v_{1x}\Delta t$$

$$\bar{a}_x = \frac{2(d_x - v_{1x}\Delta t)}{(\Delta t)^2}$$

$$\bar{a}_x = \frac{2[-800.\text{ m} - (-1.00\text{ m/s})(47.5\text{ s})]}{(47.5\text{ s})^2}$$

$$\bar{a} \cong -0.66\underline{7}0\text{ m/s}^2$$

$$\bar{a} \cong 0.667\text{ m/s}^2\text{ (down the slope)}$$

b. Determine the coefficient of friction:

$$\Sigma\mathbf{F} = m\mathbf{a}$$

$$\mathbf{F}_w + \mathbf{N} + \mathbf{f} = m\mathbf{a}$$

Evaluate the x-component forces:

$$F_{wx} + N_x + f_{kx} = ma_x$$

Since $N_x = 0$ N and $F_{wx} = mg_x = mg\sin\theta$,

$$mg\sin\theta + f_{kx} = m\bar{a}_x$$

$$f_{kx} = m\bar{a}_x - mg\sin\theta$$

$$f_{kx} = m(\bar{a}_x - g\sin\theta)$$

$$f_{kx} = (63.0\text{ kg})[(-0.66\underline{7}0\text{ m/s}^2) - (-9.81\text{ m/s}^2)\sin 6.0°)]$$

$$f_{kx} \cong +22.\underline{5}8\text{ N (up-slope).}$$

You know that $f_k = \mu_k N$. To find N, you must evaluate the y-components of the forces acting on the skier:

$$F_{wy} + N_y + f_{ky} = ma_y$$

From the problem's geometry, you know that $f_{ky} = 0$ N, and $a_y = 0\text{ m/s}^2$. $F_{wy} = mg_y = mg\cos\theta$. Substitute these values to calculate N_y:

$$mg\cos\theta + N_y = 0\text{ N}$$

$$N_y = -mg\cos\theta$$

$$N_y = -(63.0\text{ kg})(-9.81\text{ m/s}^2)\cos 6.0°$$

$$N_y \cong 61\underline{4}.6\text{ N}$$

Solve for μ_k:

$$\mu_k = \frac{f_k}{N} = \frac{|f_{kx}|}{|N_y|} = \frac{22.\underline{5}8\text{ N}}{61\underline{4}.6\text{ N}}$$

$$\mu_k \cong 0.036\underline{7}3 \cong 0.0367$$

The coefficient of friction is a very slippery 0.0367.

8.14 Using Forces to Solve Problems

How could forces prevent head-on collisions? One way is to install a barrier that uses forces to prevent cars and trucks from crossing the road into oncoming traffic.

The familiar wedge-shaped concrete barriers in the medians of heavily traveled highways are called *Jersey barriers*, so named because they were first designed for New Jersey highways. Though apparently simple, their design is the result of a significant amount of testing and applied physics.

Their purpose is to prevent injury to the driver and damage to the vehicle by using only the minimum force necessary to prevent vehicles from crossing into oncoming traffic. To do this, the barriers must provide the right amount of force at the correct angle. Typically, the barriers stand about 1 m tall. They are made of white reinforced concrete and are anchored to the road. Their faces consist of a shallow-sloped lower section and a steeply pitched upper section (See Figure 8-28). Often they are fitted with reflectors to increase their visibility. Since the installation of the original barriers in the early 1960s, many designs based on this pattern have been adopted by various states and even by other countries.

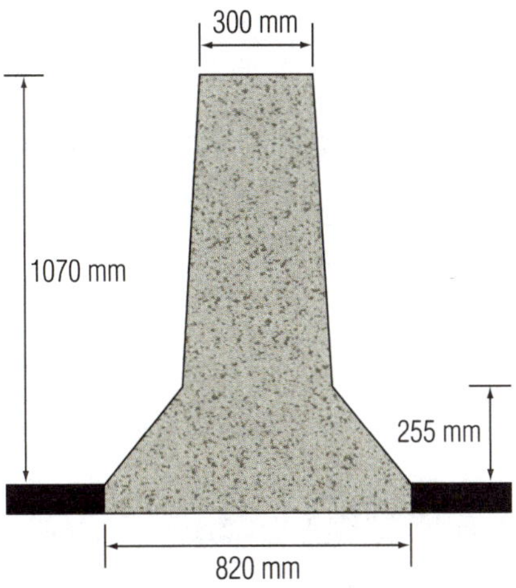

8-28 The dimensions of a standard Jersey barrier

Jersey barriers reduce accidents by forcing straying vehicles back onto the road in an upright position. If a car hits the barrier at a glancing angle, the wheels ride up onto the lower section with the shallow slope, thus turning the wheels and returning the car to the road with minimal damage. If impact occurs at a steeper angle, the car's front bumper will hit the steeply-angled face, lifting the vehicle upward and reducing friction between its wheels and the road. The front tire then descends to the barrier's lower section, which redirects the vehicle back onto the road. If the distance between the wheel and front bumper is small, as with a passenger car, then these processes are almost simultaneous.

The success of the Jersey barriers is indicated by the results—highways that had Jersey barriers installed experienced a significant reduction of head-on collisions and a *complete elimination of fatalities* due to cross-over collisions. After installing Jersey barriers in California (which has more miles of multilane highways than does any other state), there was a reduction of fatal collisions by 36% and of fatalities by 43%. Though not as fascinating as the more technological applications of physics, the design of these barriers is an instance of exercising dominion that achieves one of the best possible goals: preserving the lives of fellow human beings.

8D Section Review

8D Objectives

After completing this section, I can

✓ identify all the factors that contribute to rolling friction.

✓ draw a free-body diagram of a vehicle wheel that accounts for all the important forces acting on the wheel.

✓ solve complex dynamical problems involving multipart systems, strings, inclines, and friction.

1. Discuss the steps for dynamically analyzing systems of connected objects.
2. Under what condition will kinetic friction slow a sliding object?
3. What causes a wheel to roll over a surface?
4. For a system of mass *m* sliding on a straight inclined surface, on what does its acceleration ultimately depend?
5. A physics student is experimentally determining the coefficient of static friction for rubber on glass. She places a 150. g rubber stopper on a rectangular sheet of glass, and then she slowly raises one end of the glass. She notes that the angle of the incline at the instant the rubber begins to slip is 54.0°. What is the coefficient of static friction for rubber on glass?
6. A cute little 78 g Mongolian gerbil (*Meriones unguiculatus*) is placed on a smooth plastic ramp inclined at an angle of 40.0° to the horizontal. The coefficient of static friction between the gerbil and the ramp is 0.82. Describe the change in motion (if any) that occurs after the

animal is released. (Assume that the gerbil does not actively contribute to the motion!)

DS✪7. Draw a free-body diagram of a car's left front wheel as it strikes a Jersey barrier with a glancing blow. (Hint: Make the drawing from a bird's-eye view. Show the wheel, its axle, the steering pivot point, and the external forces acting on the wheel.)

Chapter Review

In Terms of Physics

free-body diagram	165	friction (**f**)	176
tension (**T**)	165	traction ($\mathbf{f_t}$)	176
ideal string	166	coefficient of friction (μ)	177
ideal pulley	168	kinetic friction ($\mathbf{f_k}$)	177
normal force (**N**)	171	static friction ($\mathbf{f_s}$)	178
apparent weight	174	rolling friction ($\mathbf{f_r}$)	181

Problem-Solving Strategies

8.1 (page 166) Make accurate sketches of multi-body systems that include strings and pulleys. Indicate all known and unknown force vectors and their components in the coordinate system used for each body. Note unknown values such as acceleration, but do not assign a direction until after the analysis is completed.

8.2 (page 169) The notation for tension vectors transmitted by strings connecting objects A and B is as follows: If the string exerts a tension **T** on object A, then it exerts a tension **T′** on object B. $|\mathbf{T}| = |\mathbf{T}'|$ and $\mathbf{T} = -\mathbf{T}'$ (opposite directions).

8.3 (page 170) Use different frames of reference to analyze separate objects in a system as long as the same coordinate axis in each reference frame is parallel to the motion of the system within the reference frame.

8.4 (page 172) The vector of the force of gravity is always assumed to point straight down. However, the normal component of the object's weight varies with the incline of the surface.

8.5 (page 172) For inclines, the conventional alignment for the positive *x*-axis is *up* the slope of the incline. This choice is completely up to you but is consistent with the convention that "up" is positive. If you should choose to have positive directions point down, be sure to change the sign of the negative scalar constant *g*.

8.6 (page 177) Only the absolute values of friction and normal force may be used in the friction formula. The formula is *not* a vector equation.

8.7 (page 177) The *magnitudes* of variable vectors are represented by the italicized form of the vector symbol (*f* and *N*, for example). The *scalar component* symbols for friction and normal force will be f_x and N_y, respectively. These forces have only one nonzero component if they are parallel to the reference frame coordinate axes.

Review Questions

1. What kind of diagram isolates a system so that the forces acting on the system can be analyzed?
2. A bucket is suspended in a well shaft by a rope, which is wound around a windlass. A boy turns the crank of the windlass to lower the bucket. Draw a free-body diagram for the bucket.
3. Why can't you "push" a rope or string?
4. Discuss why it is advisable to assign different coordinate systems to the connected parts of a system that do not move in the same line.
5. How can you change the direction of a force transmitted by a string?
6. What is *always* true about a normal force?
7. A family of three enters an elevator: a 75 kg father, a 55 kg mother, and a 40 kg son.
   a. Compare the weight of the father when the elevator is accelerating upward to his weight when it is at rest.
   b. Compare the normal force on the mother when the elevator is accelerating upward to when it is stationary.

c. If a difference exists in either Part a or b, is the difference the same magnitude for all three family members or is the change proportional to some quantity? Explain your answer.

8. What kind of reference frame does a passenger inside an airliner experience during the plane's takeoff roll? Would Newton's laws of motion appear to work normally during takeoff? Explain.

9. Explain why dragging a heavy trunk across the floor is harder if you pull horizontally than if you pull at an upward angle.

10. State two situations where kinetic or static friction is beneficial. State two situations where either kind of friction is undesirable.

11. Which of the following does friction generally depend on?
   a. surface area of contact
   b. the force pressing two surfaces together
   c. the speed of relative motion

12. What factors contribute most to kinetic friction?

13. Two wooden boxes of mass 500. g each are pulled across a wooden counter with a constant velocity of 10 cm/s. One box is a cube with faces of 64 cm^2 area. The other box is a rectangle, and the face contacting the counter has a surface area of 90 cm^2. Which box experiences a greater frictional force?

14. If $f_{s\ max}$ for an object resting on a horizontal surface is 10.0 N, what is the static friction acting on it if the sum of the *other* horizontal forces totals 5.5 N?

15. Which is usually greater for a pair of materials in contact, μ_s or μ_k? State at least one reason for your answer.

16. A concrete block rests on the bed of a pickup truck. The truck slowly accelerates so that the block does not slide relative to the bed. Name the force acting on the block that gives it a horizontal acceleration.

17. When you are rolling down an inclined sidewalk in your Radio Flyer wagon, what factors will determine the magnitude of your acceleration?

18. Compare the magnitudes of propelling force and traction when the tires on a car spin while trying to accelerate on a gravel surface.

DS 19. Describe the components of the force that a Jersey barrier exerts on a vehicle's wheel that impacts the barrier at a glancing angle. As the vehicle rides up on the barrier, what happens to the traction between the car and the road? Explain your answer.

True or False (20–29)

20. The total force on the last car in a railroad train is smaller than the total force on the first car when the train speeds up or slows down.

21. The normal force exerted by a surface on a system is *always* perpendicular to the surface.

22. Astronauts aren't weightless in orbit; they just lack a normal force. News reports err when they state that astronauts are weightless.

23. You wake up after a one-hour nap on an elevator. While yawning and stretching, you suddenly feel heavier. The elevator must be slowing to a stop.

24. According to Newton's third law of motion, when a skydiver jumps out of an airplane, the earth falls toward him with the same acceleration that he falls toward the earth.

25. A striped bass (*Morone saxatilis*) suspended from a hook at the end of a vertical fishing line experiences a normal force.

26. The formula for determining kinetic friction is *not* a vector equation.

27. The force of static friction on an object always has a magnitude of $f_s = \mu_s N$.

28. Static friction is an "intelligent force." Within certain limits, it "knows" how large it needs to be in order to exactly balance an imposed nonfriction force.

29. Rolling friction on a wheel is exerted only by the surface over which it rolls.

✪30. A sewing thread will break if the tension in it exceeds 5.40 N. What is the maximum unaccelerated mass that this thread can support without breaking?

✪31. A construction crew is roofing a house. They use a rope connected to an electric winch to pull up packages of shingles from the ground. The rope can have a maximum tension of only 500. N of force without breaking. Each package of shingles has a mass of 50.0 kg.
   a. Draw a free-body diagram of one package of shingles as it is being lifted by the winch.
   b. With what maximum acceleration can the crew raise the shingles?
   c. If the shingles must be lifted 8.00 m, how long will the lift take at maximum acceleration?

✪32. A 4.50 kg bowl of potato salad rests on a level picnic table.
   a. What is the normal force of the picnic table on the salad bowl?
   b. The picnic table tips so that it makes a 30.0° angle with the ground. Is the magnitude of the normal force greater than, less than, or equal to the value of the normal force when the table was level?

✪33. A frictionless playground slide makes an angle of 36.0° with the ground. It is 6.00 m long. A 22.5 kg child holds herself in place at the top of the inclined surface of the slide.
   a. Assume the girl is the system. Draw a free-body diagram of all of the forces acting on the girl-system.
   b. Find the scalar components of the force of gravity on the girl parallel to the slide and perpendicular to the slide.
   c. What normal force does the slide exert on her?
   d. What is the net force on the girl if she stops holding herself back?
   e. What is her resulting acceleration?
   f. How long does it take the girl to slide to the ground?
   g. What is her velocity at the end of the slide?

✪34. A man on a crowded elevator decides to stand on his newly purchased metric bathroom scale in order to provide more room. As the elevator accelerates downward at 0.200 m/s^2, the man looks down at the scale to see how well his one-day-old diet is coming. He sees that the scale reads 22.0 N less than he weighed the day before.
   a. If the man's mass is 110.0 kg, how much will the scale read when he steps on it in his house to show his wife his accomplishment?
   b. Will this be the same as the reading in the elevator?
   c. If not, how much weight did he really lose, if any?

35. In yet another elevator situation, a physics student receives permission from the building manager to perform an experiment. She suspends a 1.00 kg mass from the ceiling with a spring scale. She hooks an identical mass to the bottom of the first with a second spring scale. The elevator accelerates downward at 2.00 m/s^2.

a. Treat each mass as a system and sketch a free-body diagram for each during the interval of the acceleration.

b. Predict the reading on each of the spring scales. (Ignore the masses of the scales.)

36. A 3.00 kg mass and a 5.00 kg mass are attached to the ends of a string over a pulley. They are both held in place so that they hang 1.00 m below the bottom of the pulley. When they are released, how much time will elapse before the smaller mass hits the pulley?

37. A cook finds that in order to pull his 4.59 kg cast iron skillet across his steel stove top, he must apply more than 18.0 N of force. If he exerts more than 18.0 N of force on the skillet, it moves. If he exerts 18.0 N of force or less, it does not move.

a. What is the maximum value for the force of static friction?

b. What is the coefficient of static friction for cast iron on steel?

38. A physics teacher holds a 4.00 kg textbook against a vertical wall with a horizontal force of 180. N so that the book does not slide down the wall. The coefficient of static friction between the wall and the book is 0.280.

a. Determine the magnitude of the normal force of the wall on the book.

b. What is the maximum possible static friction force on the textbook?

c. If the teacher reduces his force on the textbook until it is just about to slip, what is the magnitude of the teacher's force at this point?

39. A 0.100 kg brass mass is placed on a horizontal steel surface. For brass on steel, μ_s is 0.510 and μ_k is 0.440.

a. If the mass moves with a constant speed of 0.500 m/s, what is the friction on the mass?

b. If the mass is not moving, what is the maximum theoretical static friction?

40. A car going 32.0 km/h on dry, level pavement jams on its brakes and skids to a stop. The car's mass is 1250 kg. The coefficient of kinetic friction for rubber on asphalt is $\mu_k = 0.400$.

a. What is the net force on the car?

b. What is the car's acceleration?

c. How far does the car travel before stopping?

41. Two ox teams are competing in an ox pull at a state fair. Each must move a loaded sled to remain in the competition. The load on the sled is increased after each team has had an opportunity to try to move it. Team 1 can exert a maximum force of 1.50×10^4 N; team 2 can exert a maximum force of 1.80×10^4 N. The coefficient of friction between the metal sled and the gravel surface is $\mu_s = 0.50$.

a. What is the mass of the greatest load that team 1 can move?

b. What is the mass of the greatest load that team 2 can move?

c. If mass is added in 500 kg increments, what is the mass of the first load that team 2 can pull but team 1 cannot?

42. A 9.50×10^4 kg locomotive is stopped on a track on a slope in the Rocky Mountains. The slope makes an angle of 15.0° with the horizontal. The locomotive's wheels are locked. If the coefficient of static friction between the locomotive's steel wheels and the steel track is 0.57, will the train slip?

DS **43.** Analyze the reduction of traction resulting from a collision with a Jersey barrier. Assume that a 1590 kg vehicle's weight is equally distributed on all four wheels when it is on the road surface. During the collision, the barrier exerts an upward force on the front left tire equal to half of the weight it normally supports. Assume the static coefficient of friction between the tires and the road is equal to 1.0. Calculate the remaining maximum traction as a percentage of the normal maximum traction.

44. You are approaching a level curve in the road in your Jaguar XKR. The curve has a uniform radius of 50.0 m. Assuming that the pavement is concrete, what is the maximum constant speed at which you can negotiate the turn without skidding?

45. A 75.0 kg professional dog walker in New York City is out airing two of his charges. All of a sudden, the large 54.5 kg rottweiler decides to head east while the smaller 29.5 kg basset hound heads west, exerting an 18.0 N pull. The human pulls west with 80.0 N in order to stop the rottweiler, but the whole group accelerates east at a uniform 0.30 m/s^2.

a. If the rottweiler's leash makes an angle of 30.0° with the ground and the basset hound's leash makes an angle of 40.0° with the ground, what is the magnitude of tension in each leash?

b. What is the normal force on the dog walker?

46. Prove that, for a system on an incline, the angle between its weight vector and the *y*-component of the weight vector is congruent to the incline angle. Use a formal two-column proof similar to the method discussed in *Geometry* published by BJU Press.

FACETS of PHYSICS

Artificial Gravity

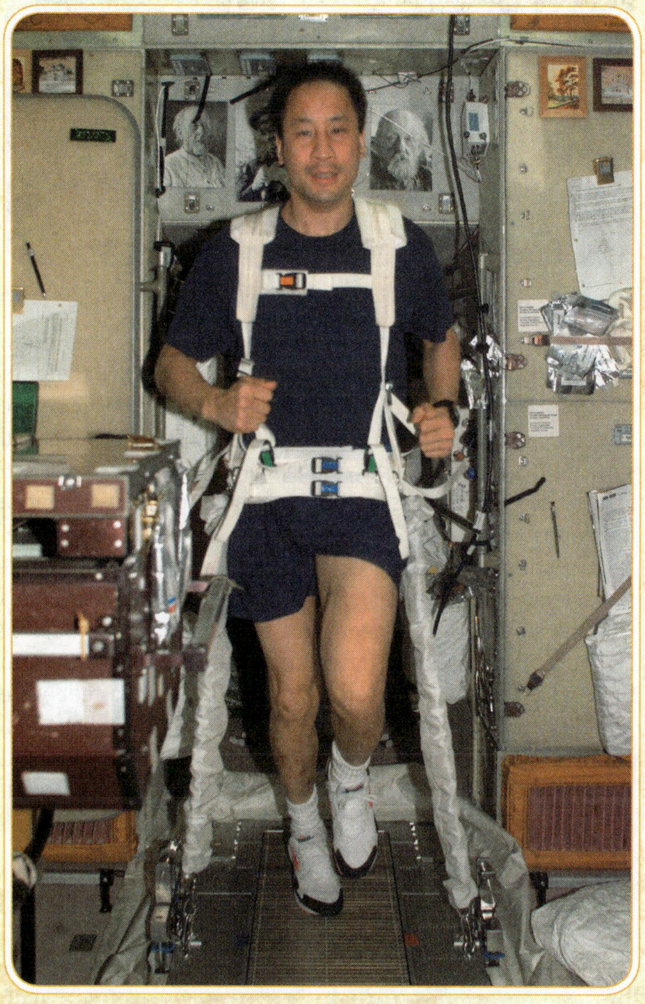

One of the reasons astronauts do so many science experiments in space is that they can study physical and biological processes without the effects of gravity, and the harvest of knowledge from these experiments shows great promise of being rich indeed. Why, then, are scientists interested in creating an artificial gravitational environment in space? There are two main reasons. One focuses on basic science research. By eliminating other variables, scientists can study how varying the simulated gravitational force affects biological and physical processes like plant growth and crystal formation. The International Space Station contains a centrifuge module designed to perform these kinds of experiments, which can generate simulated gravitational forces up to twice that of Earth. The other interest is in human physiology, whereby scientists try to find ways to keep astronauts healthy during long space flights. Our bodies are designed to operate in a "normal force" world, and in the free fall of space travel, astronauts experience a long list of physiological changes called *deconditioning*. These changes include immune system suppression, decreased blood volume and red blood cell count, loss of bone density, muscle atrophy (including the heart), as well as disorientation and space sickness caused by a lack of normal gravity-related sensory information.

If there is to be any hope of manned exploration to places farther than the moon, scientists have to know how to keep the astronauts healthy enough to withstand the flight and strong enough to perform the mission when they arrive. There are two approaches to the problem: counteractive measures and preventive measures. Both have limitations. Counteractive measures include exercise and diet. Astronauts have been exercising in space since the Gemini IV flight in 1965. While astronauts in the cramped quarters of the early space capsules exercised with stretchy bands, residents of the International Space Station spend several hours a day riding a stationary bike or running on a treadmill. This aerobic exercise strengthens the heart muscle as well as both upper and lower body muscle groups and slows the loss of conditioning.

Many scientists are trying to prevent the ill effects of space travel by providing some sort of artificial gravity. The only way to do this is through circular motion. The centripetal acceleration necessary for circular motion is experienced as a normal force that your body cannot distinguish from the earthly gravitational force. An astronaut can experience simulated gravity if he is spun in a person-sized centrifuge for a few hours a day with his head toward the center and his feet at the perimeter. There are a few problems with this approach, however. The centripetal force experienced by your body depends on its tangential speed. Since each part of the body is traveling at a different tangential speed, there will be a considerable difference in the "gravity" experienced between the head and the feet. This gravity gradient lessens the effectiveness of the treatment. Also, it normally takes astronauts several days of space flight for their bodies to adjust to the disorientation of the sensory nervous system. It is not clear whether they would be able to tolerate repeatedly switching back and forth between artificial gravity and weightlessness.

The other idea, which is still only theoretical, is to generate constant artificial gravity by spinning the entire space station. This could take the form of either a spinning doughnut (called a torus) or two units joined by a tether that spin around the center of mass of the system. This idea has its own problems, however. Artificial gravity would not feel the same or act quite the same as natural gravity. If you are in a torus-shaped

station, "down" is toward the rim. Thus, as you walked down the hallway, your head would be pointed toward the center of rotation. If you turned your head suddenly so that you were now looking toward the wall, the fluid in your ears would slosh to one side, causing vertigo similar to seasickness. Any kind of nodding or twisting of the head might cause balance problems.

It might be interesting to take an elevator ride from a docking station at the center of rotation to the outer edge of a rotating station. At the center the tangential speed would be zero, so there would be no normal force. You would have to float into the elevator and make sure that your feet were pointing "down"—at least where "down" would be when you arrived—and that you were facing in the direction of the spin. You will see why in a minute. Two things would happen as you traveled outwards: you would travel from the center to the rim, and you would accelerate from a tangential speed of zero at the center to the tangential speed of the rim. If the elevator had three stages in its outward journey—acceleration, constant velocity, and deceleration—what would you experience during the trip? First, because you would be weightless, as the elevator accelerated toward the edge, the ceiling would exert a force on you. While the elevator was traveling outward at constant velocity, the wall behind your back would be pushing on you, accelerating you to the local tangential speed. During the trip you would float down to the floor and start experiencing a force on your feet. As the elevator decelerated, the force on your feet would increase, and when you stopped, you would be able to walk out the door, having the same tangential speed as the hallway. You can see how important it would be to get yourself oriented properly before the ride. You wouldn't want to arrive upside down with your face pushed against the wall of the elevator!

The rotation of the station would create some other odd effects. Things that fall or are dropped wouldn't appear to fall straight down but would appear to curve in a direction opposite to the spin. If you held a ball four feet off of the floor, its tangential speed would be slower than that of your feet on the floor because the ball would be four feet closer to the center of rotation. If you dropped the ball, it wouldn't gain any tangential speed during the fall. Because the floor would be traveling faster than the ball, rather than landing at your feet it would land a short distance away. Raising the release point would increase the difference in tangential speed relative to the floor and thus would increase the amount of deflection. You can see that in a rotating space station any kind of ball-related sports would be difficult.

Mount Saint Helens erupting on May 18, 1980. A volcanic eruption can release vast quantities of energy in a short time.

Work and Energy

9

Dominion Science Problem

A Renewable Energy Resource

Coal, oil, and natural gas supply 85% of America's energy needs and 86% of the world's. Though these fossil fuels are the most plentiful sources of energy, demand for all kinds of energy is increasing rapidly, and as demand for energy increases, fuel costs increase across the globe. Most scientists believe that the world's supply of fossil fuels in the ground is fixed and that its rate of depletion is increasing. Many predict that oil reserves will be exhausted before the end of this century and coal reserves within just a few centuries. In addition to these limitations, fossil fuel combustion is a principal source of air pollution in most cities, affecting the health and lives of millions of people. In place of fossil fuels, what significant source of renewable energy can be developed domestically, has a low impact on the environment, and is economical to produce?

9-1 Fossil fuels are important sources of energy, but they are becoming more costly and may be exhausted in the not-too-distant future.

9.1 Introduction

On May 18, 1980, an explosion shattered the countryside in southwestern Washington. Mount Saint Helens erupted for the first time in more than one hundred years, hurling nearly 27 km^3 of solid rock and ice 4 km into the air. The volcano devastated more than 500 km^2 of the surrounding land. To cause the same amount of destruction using nuclear bombs, one Hiroshima-sized bomb would have to be set off every second for seven and one-half hours. The neighborhood of the volcano was changed from a life-filled region to a wasteland. This explosion is an example of the unrestrained release of energy.

Even explosions, however, can be contained and made to do useful work. A gasoline automobile engine, for example, is powered by explosions. Gas and

One description of energy is the ability to do work.

air flow into a cylinder where a piston compresses them into a small volume. When a spark ignites the gas, the mixture explodes, pushing the piston away from the spark plug. This push is the source of the force that turns the wheels of the car.

Both explosions illustrate one attribute of energy: **energy** has the ability to do work. That is, energy can move matter. The work may be useful, as in an engine, or destructive, as in a volcano's eruption. This chapter will discuss the relationship between work and energy.

9A WORK

9.2 Defining Work

The word *work* can mean many different things. It can mean exceptionally hard labor, an artistic achievement, a job, or any physical or mental effort. In physics, however, it means the energy required to move or change the state of a system. You can easily understand that it takes more energy to move a more massive object than it takes to move a less massive one. For example, you must use more energy to carry a 10 kg box 10 m than to carry a 5 kg box the same distance. Also, you use more energy to move the same object a longer distance. It takes more energy to move a box 10 m than to move the same box 5 m.

Considering these facts, it is not surprising that **work (*W*)** is defined as the product of the force component that is parallel to an object's motion and the distance that the object is moved. Mechanical work is done *by* a force *on* a system. It is important to identify the force for which work is to be calculated and to identify the system on which it acts. In symbols,

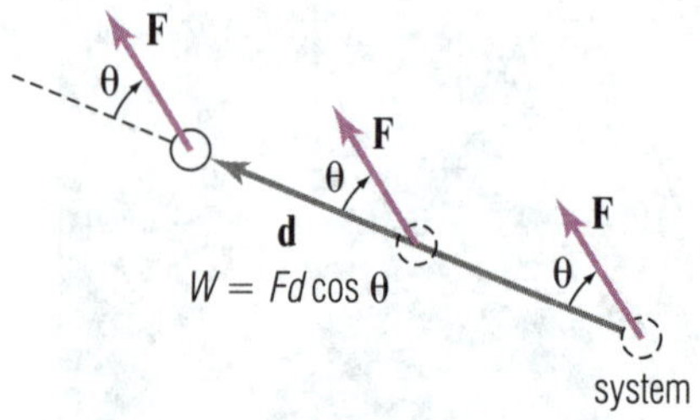

9-2 Relationship of force, displacement, and work

$$W \equiv Fd \cos \theta, \tag{9.1}$$

where W is the work done by a force $\mathbf{F}$ through a displacement $\mathbf{d}$. It is not necessary for the force causing the work to be parallel to the displacement. The angle θ is the smallest angle ($\leq 180°$) between the force and displacement vectors when they are placed tail-to-tail. Notice that work itself is a scalar and can be positive, negative, or zero. Work is the scalar product of the magnitude of the displacement (distance, d) and the magnitude of the force component that is parallel to the displacement ($F \cos \theta$).

The angle between the force on a system and the system's displacement determines the quality of work done. If $\theta < 90°$, Equation 9.1 is positive; therefore, positive work is done because the cosine of angles less than 90° is positive.

If $90° < \theta \leq 180°$, the cosine of the angle is negative, and work is negative. Negative work by the force occurs when the force acting on the system opposes the motion of the system.

If the angle between the force and displacement is 90° (they are perpendicular), there is zero work done by the force on the system because $\cos 90° = 0$.

Problem-Solving Strategy 9.1

To determine whether work by a force is positive or negative, find the force's vector component that is parallel to the displacement of the system. If it is in the same direction as the displacement, work is positive. If the force vector component is opposite to the direction of displacement, then work is negative.

The quality of work is determined by the angle between the applied force $\mathbf{F}$ and the direction of displacement $\mathbf{d}$:

If $0° \leq \theta < 90°$, work is positive.

If $\theta = 90°$, work is zero.

If $90° < \theta \leq 180°$, work is negative.

9.3 Units of Work

The units of work can be found from Equation 9.1. In the SI, F is measured in newtons and d is measured in meters.

$$W = (\text{newtons})(\text{meters}) = \text{N}\cdot\text{m}$$

The SI unit of work and energy, the newton-meter, has been given a special derived unit name, the **joule (J)**.

$$1\ \text{J} \equiv 1\ \text{N} \times 1\ \text{m}$$

The joule is utilized for all forms of energy and work. For mechanical work, the units N·m may be used. Be sure you do not confuse this unit with the unit of torque, also the N·m. Torque, unlike work and energy, is *never* expressed in joules.

9.4 Calculating Work

Any kind of force can do work. Contact forces as well as action-at-a-distance forces move objects and therefore do work. For example, a weightlifter exerts a contact force on a barbell that he does work on by lifting it, and the earth exerts the force of gravity on a falling rock that it does work on by pulling it down.

Although any kind of force *may* do work, not every force on an object *will* do work. For example, suppose a rock is at rest on a table. The earth exerts a gravitational force on the rock, but the rock does not move. How much work has the earth done on the rock?

$$W_g = F_g \times 0 \text{ m} = 0 \text{ J}$$

The earth does no work on the rock. Now suppose that the rock is pulled along the horizontal table by a string tied to it. The rock moves a distance of *d*. Now what is the work that the earth does on the rock?

$$W_g = F_{gx} d = F_w d \cos \theta$$

$$W_g = F_g d \cos 90°$$

$$W_g = F_g d \times 0 = 0 \text{ J}$$

The earth still does no work on the rock because the component of the force of gravity on the rock in the direction of the rock's motion is zero. The tension in the string, however, does positive work on the rock.

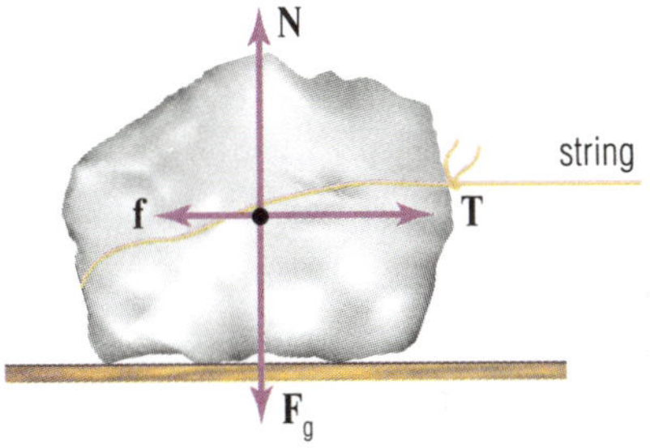

9-3 A rock pulled by a string and various forces that can do work on the rock

Problem-Solving Strategy 9.2

Gravity does positive work on a system when the system loses height, and it does negative work when the system gains height.

Problem-Solving Strategy 9.3

Whenever you must calculate the gravitational force on a system in order to compute work done by or against gravity, be sure to use the *magnitude* of gravitational force $|mg|$. The work formula uses the magnitudes of force and displacement, *not* their scalar values.

EXAMPLE 9-1

Calculating Work: Vertical Force

A 65.0 kg physics student jumps off a 2.00 m high masonry wall. How much work does the earth do on the student in his descent to the ground?

Solution:

$\mathbf{F}_g = \mathbf{F}_w$ and is in the same direction as $\mathbf{d}$; so

$$W = F_w d \cos 0° = F_w d \times 1 = F_w d$$

$$W = |mg|d$$

$$W = |(65.0 \text{ kg})(-9.81 \text{ m/s}^2)|(2.00 \text{ m})$$

$$W \cong 12\underline{7}5 \text{ kg·m}^2/\text{s}^2$$

$$W \cong 1280 \text{ N·m} = 1280 \text{ J}$$

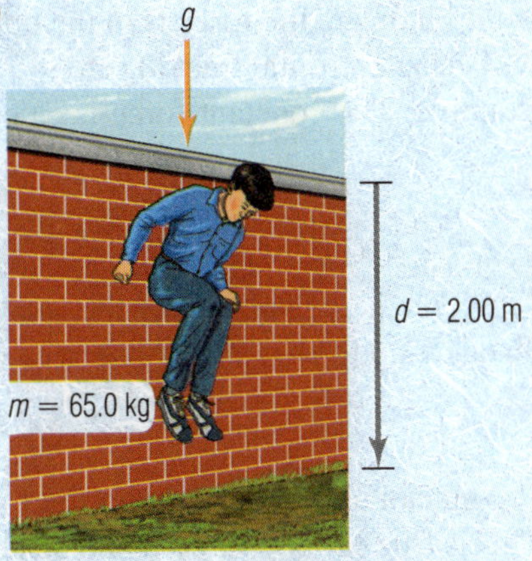

9-4 The earth doing work on a student

Recall that the component of the force *parallel* to the displacement is $F \cos \theta$.

EXAMPLE 9-2

Calculating Work: Force at an Angle to Displacement

A man pushes a lawn mower with a force of 30.0 N directed 60.0° below the horizontal. The lawn mower moves 20.0 m. How much work did the man do on the lawn mower?

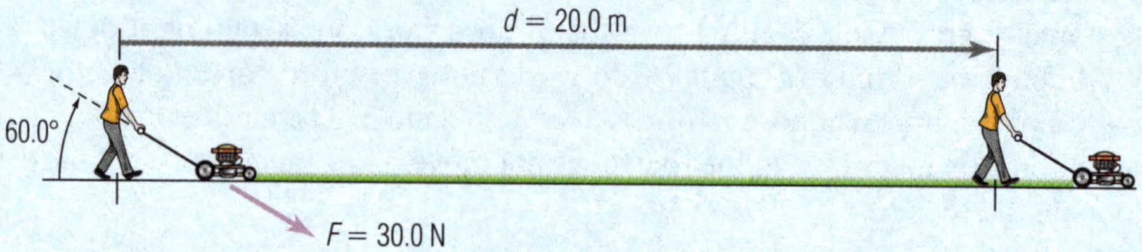

9-5 A man is doing work with the lawn mower and on the lawn mower.

Solution:

$$W = Fd \cos \theta$$

The angle between the force and the displacement vectors is 60.0°.

$$W = (30.0 \text{ N})(20.0 \text{ m})(\cos 60.0°)$$

$$W \cong 30\underline{0}.0 \text{ N·m} \cong 300. \text{ J}$$

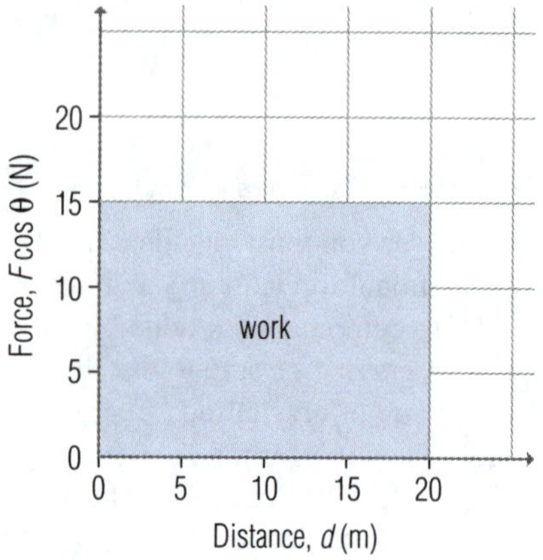

9-6 A force-distance graph for work done by a constant force

9.5 Determining Work Graphically: Constant Forces

Advanced methods of analysis often use graphs to show how two related variables determine a third. You have seen in previous chapters how the area under a curve approximates a quantity—finding displacement from the area under a velocity-time graph. Similarly, it is sometimes useful to plot the force doing work versus the displacement. As you may suspect, the area under such a graph approximates the work done on a system by the force.

We will first consider a simple case involving a constant force. The vertical axis represents the component of the force parallel to the displacement. Positive values indicate the force is in the same direction as the displacement, while negative values indicate the force is in the opposite direction. For constant forces, this graph will be a horizontal line. Figure 9-6 gives the *force-distance graph* for the previous example. The constant force is calculated as follows:

$$|F_x| = (30.0 \text{ N})(\cos 60.0°)$$
$$|F_x| = 15.0 \text{ N}$$

The force is the same after the mower has traveled 0 m, 5 m, 15 m, and any other distance up to $d = 20$ m. The area of the shaded block is

$$A = |F_x|d = W.$$

You can use the area under the force-distance curve to represent the work done. This is true for any force-distance curve. If you have trouble recalling the equation used to calculate work, you can use geometry to find the area under the force-distance curve. Remember, the physical size of the graphical area (e.g., in cm^2) is not important. It is the product of the dimensions represented by the horizontal and vertical axes that matters.

9.6 Determining Work Graphically: Variable Forces

A force-distance graph is especially helpful for a varying force. For example, the external force required to stretch most springs is proportional to the displacement that the spring has been stretched. That is,

$$\mathbf{F}_{ex} \propto \mathbf{d},$$

where $\mathbf{F}_{ex}$ is the external force and $\mathbf{d} = \Delta x = x_2 - x_1$. The position x_1 is the normal or relaxed length of the spring, called the **equilibrium position**. This observation is known as **Hooke's law**. It can be made into an equation by including a proportionality constant:

$$\mathbf{F}_{ex} = k\mathbf{d} \qquad (9.2)$$

The proportionality constant k is called the **spring constant**, which is nearly constant for real springs. In a spring system, the motion and the force are assumed to act along the same one-dimensional coordinate axis, with the origin placed at the spring's equilibrium position. Notice that as a force is applied to stretch or compress the spring, the displacement is in the same direction as the external force, so work done *on* a spring by an external force is always positive.

Robert Hooke (1635–1703) was an English scientist and inventor who pursued a wide range of interests, from microscopy to architecture. He is best known for first using the term "cell" in biology and for formulating Hooke's law in physics.

The SI units of the spring constant k are N/m.

As we have done with other idealized dynamics analyses, we will rely on ideal springs in discussions involving Hooke's law. Ideal springs have no mass, and the value of k for the spring is truly constant throughout its range of displacements.

An ideal spring has no mass, and its spring constant is truly constant through any displacement of the spring.

Many systems in nature act similarly to a spring that is displaced from an equilibrium position, particularly systems that tend to vibrate or oscillate. Any force that varies in proportion to the magnitude of displacement from an equilibrium position is called a Hooke's-law force.

Any force that is proportional to the displacement of a system from an equilibrium position is called a Hooke's-law force. Systems that generate such forces include vibrating strings and bells.

9.7 Work and Ideal Springs

How much work is done to stretch a spring from its equilibrium position by Δx? Plotting F_{ex} versus Δx on a graph, you get Figure 9-7. The shaded triangular area under the graph bounded by x_1 and x_2 is the work on the spring between these two points. The area of the shaded triangle is as follows:

$$A = \tfrac{1}{2}bh$$

$$A = W_{ex} = \tfrac{1}{2}(\Delta x)(k\Delta x)$$

$$W_{ex} = \tfrac{1}{2}k(\Delta x)^2$$

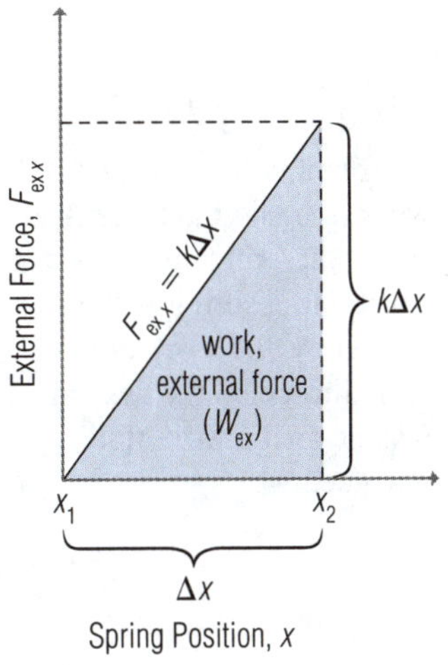

9-7 A force-distance graph for a force acting on a spring

Notice that the height of the triangle is the change in the x-coordinate times the slope of the line. In this case, the slope of the graph is the spring constant, k.

Figure 9-7 shows the work done *by* a force ($\mathbf{F}_{ex}$) *on* the spring. What about the force that a spring exerts? From Newton's third law, you know that the spring exerts a reaction force of

$$\mathbf{F}_s = -\mathbf{F}_{ex}.$$

If the spring is stretched by an external force, the force that the *spring* exerts is

$$\mathbf{F}_s = -k\mathbf{d}. \qquad (9.3)$$

Notice that the spring force is opposite to the direction of displacements from its equilibrium position caused by an external force. Therefore, the work done *by* a spring on a system connected to it is negative under these conditions. This equation holds true whether the spring is stretched or compressed from its equilibrium position.

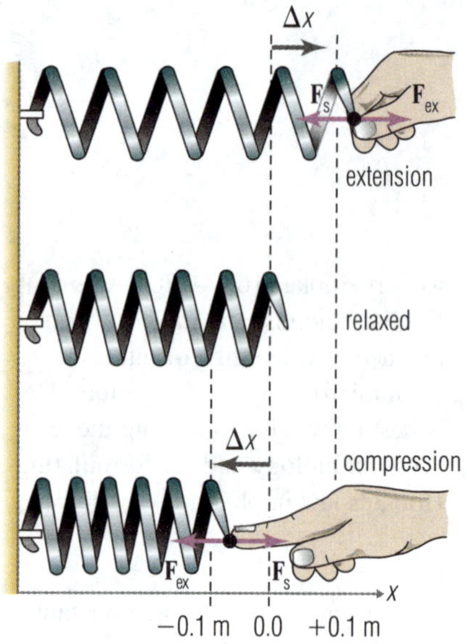

9-8 Comparison of forces on a stretched, relaxed, and compressed spring

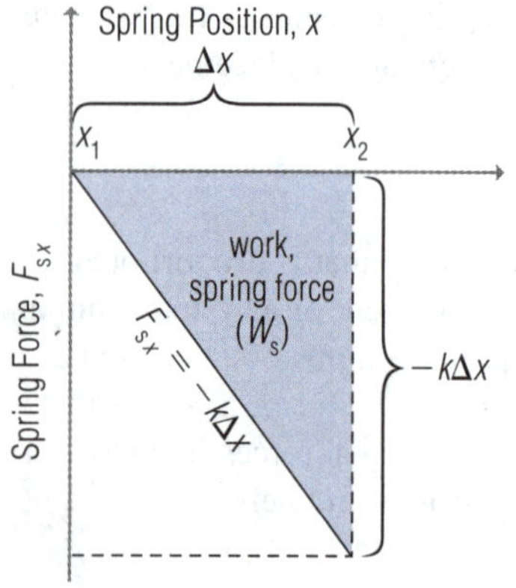

9-9 Force-distance graph for a spring force (Hooke's law)

Problem-Solving Strategy 9.4

You must take care that you calculate the work done by the force that is specified in the problem. If an object attached to a spring is the system of interest, the work done by an external force on the system is not the same as the work done by the spring force.

For example, if an ideal spring is attached to a wall on the left, as in Figure 9-8, and stretched to the right from $x_1 = 0$ m to $x_2 = 0.1$ m, the external force on the spring must be directed to the right. The spring exerts a force toward the left, as you can see from $F_{s\,x}$:

$$F_{s\,x} = -kd_x = -k(x_2 - x_1)$$

$$F_{s\,x} = -k(0.1 \text{ m} - 0 \text{ m}) = -0.1k \text{ N}$$

Since $F_{s\,x} < 0$, the spring force points to the left. If the spring is compressed to the left from $x_1 = 0$ m to $x_2 = -0.1$ m, the external force on the spring is toward the left. The spring, therefore, exerts a force to the right.

$$F_{s\,x} = -kd_x = -k(x_2 - x_1)$$

$$F_{s\,x} = -k(-0.1 \text{ m} - 0 \text{ m}) = +0.1k \text{ N}$$

How much work does an ideal spring do as it is stretched or compressed by an external force? Figure 9-9 gives the force-distance graph for a spring that is stretched. The shaded area represents the work done *by* the spring force.

$$W_s = A = \tfrac{1}{2}(\Delta x)(-k\Delta x)$$

$$W_s = -\tfrac{1}{2}k(\Delta x)^2 \qquad (9.4)$$

Notice that the negative work region is *below* the horizontal axis of the graph.

Equation 9.4 clearly shows that the work done by the spring on a system in this situation is negative. The spring force acts on the system in the direction opposite to the system's displacement.

EXAMPLE 9-3

Springing to Work: Work by a Variable Force

A spring with a spring constant $k = 0.50$ N/m is stretched from its equilibrium position of $x_1 = 0.00$ m to a position of $x_2 = +0.25$ m. (a) How much work is done on the spring by the external force (W_{ex})? (b) How much work does the spring do (W_s)?

Solution:

a. Work by the external force:

$$W_{ex} = \tfrac{1}{2}k(\Delta x)^2 = \tfrac{1}{2}(0.50 \text{ N/m})(+0.25 \text{ m} - 0.00 \text{ m})^2$$

$$W_{ex} \cong 0.015\underline{6}2 \text{ J} \cong 0.0156 \text{ J}$$

b. Work by the spring force:

$$W_s = -\tfrac{1}{2}k(\Delta x)^2 = -\tfrac{1}{2}(0.50 \text{ N/m})(+0.25 \text{ m} - 0.00 \text{ m})^2$$

$$W_s \cong -0.015\underline{6}2 \text{ J} \cong -0.0156 \text{ J}$$

9.8 Power

The amount of work required for a warehouse worker to carry a pallet of new coffeemakers one box at a time from the floor to a shelf 4 m above the floor is the same as for a forklift to raise the whole pallet at once. Neglecting the mass of the pallet, the only difference between the two instances is the amount of time needed to do the work. Up to now you have not been concerned with how long it takes to accomplish work, only that it occurs at all. Obviously, your employer would not be happy if your work ethic followed this principle! We use machines because they help us to do work faster and with less effort. We will discuss simple machines in Chapter 10.

The time rate of work done on a system by a force is called the **power (*P*)** of the force. If the total work done over an interval of time is considered, then the *average power* ($\overline{P}$) is computed by

$$\overline{P} = \frac{\text{work done by } \mathbf{F} \text{ during } \Delta t}{\Delta t} = \frac{W}{\Delta t}. \tag{9.5}$$

The rate at which a force does work at any instant is called the *instantaneous power*, or just power.

Customers inquiring about power in a car engine often ask how fast the car can go. They intuitively associate power with speed. This relationship does, in fact, exist. We can expand Equation 9.5 to include the work equation

$$\overline{P} = \frac{W}{\Delta t} = \frac{Fd\cos\theta}{\Delta t}. \tag{9.6}$$

But $d/\Delta t = \overline{v}$ (or just v for uniform motion). Equation 9.6 then becomes

$$P = Fv\cos\theta. \tag{9.7}$$

Power is a scalar quantity that is proportional to the force acting on a system and the speed of the system.

The unit of power is obtained from the dimensions of work per unit time, consisting of the SI units of joules per second. Because power is such a frequently encountered quantity in physics, a derived unit called the **watt (W)** was created in honor of James Watt (1736–1819), a Scottish engineer who was prominent in the development of the first practical steam engine.

$$1 \text{ watt} = 1 \text{ W} = 1 \text{ J/s}$$

The **average power** ($\overline{P}$) of a source of energy is the work accomplished during a time interval divided by the time interval:

$$\overline{P} = \frac{W}{\Delta t}$$

The SI unit of power is the **watt (W)**.
1 W = 1 J/s

EXAMPLE 9-4

Spring Power

How much power is needed to compress a 0.100 m spring with a spring constant $k = 1.25$ N/m to half its length (a) in 10.0 s? (b) in 5.0 s?

Solution:

Work by an external Hooke's law force is calculated from Equation 9.5.

a. Power for the 10.0 s compression:

$$\overline{P} = \frac{\frac{1}{2}k(\Delta x)^2}{\Delta t} = \frac{(1.25 \text{ N/m})(0.050 \text{ m} - 0.100 \text{ m})^2}{2(10.0 \text{ s})}$$

$$\overline{P} \cong 1.5\underline{6} \times 10^{-4} \text{ W} \cong 1.6 \times 10^{-4} \text{ W}$$

b. Power for the 5.0 s compression:

$$\overline{P} = \frac{(1.25 \text{ N/m})(0.050 \text{ m} - 0.100 \text{ m})^2}{2(5.0 \text{ s})}$$

$$\overline{P} \cong 3.1\underline{2} \times 10^{-4} \text{ W} \cong 3.1 \times 10^{-4} \text{ W}$$

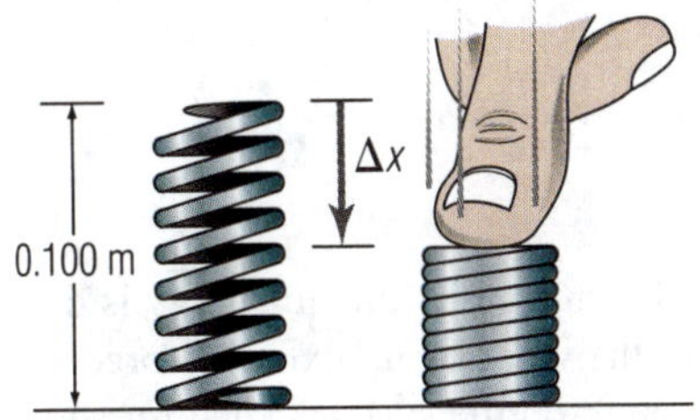

9-10 Compression of a spring

9A Objectives

After completing this section, I can

✓ define the physical concept of work.

✓ calculate work, given a force and displacement.

✓ describe ideal springs and the conventions used to measure displacement in a spring system.

✓ graphically determine work done by constant and Hooke's law forces.

✓ calculate mechanical power for a system.

9A Section Review

1. a. To what physical quantities is mechanical work proportional?
   b. What kind of quantity is mechanical work?
2. How can you tell whether positive or negative work has been done by a particular force on a system?
3. How can you find the spring constant from a force-displacement graph of an ideal spring?
4. a. To what physical quantities is mechanical power proportional?
   b. What kind of quantity is mechanical power?
5. If you slowly lift a 0.95 kg textbook 1.2 m vertically from your desk to a bookshelf, how much work have you done?
6. A 1.95 kg female great horned owl (*Bubo virginianus*) silently swoops directly from a branch 12.4 m high in a tree to capture an unfortunate field mouse (*Apodemus sylvaticus*) on the ground—a 24.0 m distance from the owl's perch. Calculate the work done by gravity on the owl during its downward flight.
7. A crane slowly lifts a 1140 kg load of steel girders vertically 28.0 m.
   a. How much work was done by the crane on the girders?
   b. How much work was done by gravity?
8. A 950. kg car is initially moving at 95.0 km/h. If the car is stopped in 6.50 s by a force of 3860 N opposite to the direction of motion, what is the stopping power of the car's brakes in watts?

9B ENERGY

9.9 Kinetic Energy

A force performs mechanical work on a system in order to move the system from one place to another. Motion implies that the system has a velocity, and changes of motion imply accelerations. Moving objects can do work when they strike something else. Just consider the damage a flying baseball can do to a windowpane! Evidently, a moving object possesses the ability to do work. This mechanical energy associated with motion is called **kinetic energy (*K*)**. Kinetic energy is a positive scalar quantity measured in joules in the SI.

The change in kinetic energy is due to the work by *all* external forces acting on the system. Such forces include propelling forces from a rocket engine or a baseball bat, frictional forces, and the force of gravity. You will see how each of these affects kinetic energy in this and later chapters.

The **work-energy theorem** states that the total work done on a system by all the external forces acting on it is equal to the change in the system's kinetic energy.

$$W_{total} = \Delta K, \text{ or}$$

$$W_{total} = K_2 - K_1 \quad (9.8)$$

From Newton's second law you know that

$$\Sigma \mathbf{F} = m\mathbf{a}.$$

The notation F_m is a shorthand way of saying "the force component parallel to the motion that caused the displacement."

Now the work equation can be rewritten as a first-law expression

$$W_{total} = \Sigma F_m d = (ma)d, \quad (9.9)$$

where ΣF_m is the sum of all force components parallel to the movement during the displacement, *d*. From the third equation of motion, you know that

$$d = \frac{v_2^2 - v_1^2}{2\bar{a}}.$$

Multiply both sides by $\bar{a}$.

$$\bar{a}d = \frac{v_2^2 - v_1^2}{2} = \tfrac{1}{2}(v_2^2 - v_1^2)$$

Substitute for $\bar{a}d$ in the right side of Equation 9.9 and distribute the terms.

$$W_{\text{total}} = m\bar{a}d = m\tfrac{1}{2}(v_2^2 - v_1^2)$$

$$W_{\text{total}} = \tfrac{1}{2}mv_2^2 - \tfrac{1}{2}mv_1^2 \qquad (9.10)$$

Comparing Equation 9.10 to Equation 9.8, it seems logical to define kinetic energy as

$$K \equiv \tfrac{1}{2}mv^2. \qquad (9.11)$$

Problem-Solving Strategy 9.5

Kinetic energy (*K*) can only be positive. For any given object, *K* varies as the square of its speed.

EXAMPLE 9-5

Take a Shot: Bullet Kinetic Energies

Calculate the kinetic energy in joules of two 150 grain (0.009 72 kg) 30-caliber rifle bullets propelled with different amounts of powder to produce different muzzle velocities. One bullet has a muzzle velocity of 703 m/s, while another has a muzzle velocity of 952 m/s.

Solution:

$$K = \tfrac{1}{2}mv^2$$

$$K_1 = \tfrac{1}{2}(0.009\,72\ \text{kg})(703\ \text{m/s})^2$$

$$K_1 \cong 24\underline{0}1\ \text{J} \cong 2.40 \times 10^3\ \text{J}$$

$$K_2 = \tfrac{1}{2}(0.009\,72\ \text{kg})(952\ \text{m/s})^2$$

$$K_2 \cong 44\underline{0}4\ \text{J} \cong 4.40 \times 10^3\ \text{J}$$

9-11 A bullet's high speed gives it a high kinetic energy.

Observe that the second bullet has almost twice the kinetic energy of the first bullet even though its muzzle velocity is less than a third greater.

EXAMPLE 9-6

Braking: Changing Kinetic Energy

A 1.00×10^4 kg loaded truck traveling at 20.0 m/s brakes until its speed is 10.0 m/s. (a) How much work was done on the truck to slow it? (b) If the truck takes 50.0 m to slow, what was the total braking force applied?

9-12 Braking a heavy truck requires a lot of work.

Solution:

a. The total work done on the truck to slow it can be calculated from its change of kinetic energy:

$$W_{total} = \Delta K = \tfrac{1}{2}mv_2^2 - \tfrac{1}{2}mv_1^2$$

$$W_{total} = \tfrac{1}{2}m(v_2^2 - v_1^2)$$

$$W_{total} = \tfrac{1}{2}(1.00 \times 10^4 \text{ kg})[(10.0 \text{ m/s})^2 - (20.0 \text{ m/s})^2]$$

$$W_{total} = -1.50 \times 10^6 \text{ J}$$

The work done on the truck is negative because the truck lost energy as it slowed.

b. By definition,

$$W_{total} = \Sigma F_m d.$$

Divide both sides by distance to find the net force on the truck:

$$\Sigma F_m = \frac{W_{total}}{d} = \frac{-1.50 \times 10^6 \text{ kg}\cdot\text{m}^{\not 2}/\text{s}^2}{50.0 \not\text{m}}$$

$$\Sigma F_m = -3.00 \times 10^4 \text{ N}$$

The net braking force is negative because it opposes the motion of the truck.

Recall that ΣF_m in this example is the product of the net force and cos θ. Since $\theta = 180°$ (ΣF_m opposes motion), $\cos 180° = -1$. Therefore, the value of ΣF_m is negative.

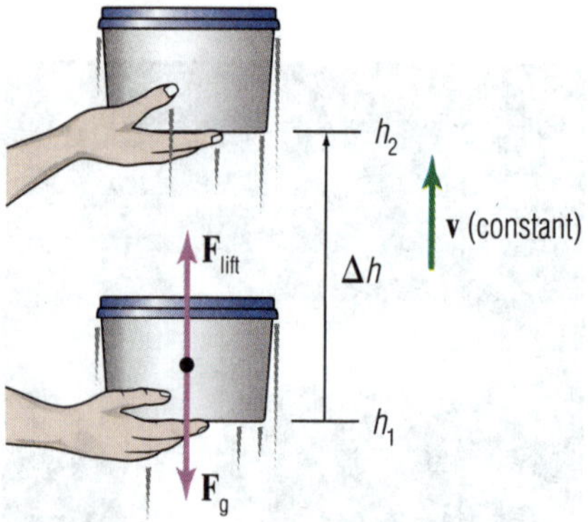

9-13 When a container is lifted with constant speed, the work done on the container does not change its kinetic energy.

9.10 Potential Energy

Not all work changes kinetic energy. Suppose you slowly lift a box from the floor at a constant speed to a height that will allow you to place it on a shelf. If the speed of the box is uniform throughout the lift, the kinetic energy of the box does not change. The work simply changes the position of the box. However, before you lifted the box, it had little potential for falling (unless there was a hole in the floor). Once on the shelf, the box has a greater potential for falling to the floor. If it were to fall, its kinetic energy would increase as gravity did work on the box. While it rests on the shelf, therefore, the box is said to have gravitational potential energy. **Potential energy (U)** is energy due to an object's condition or position relative to some reference point assumed to have zero potential energy. As with kinetic energy, the unit of potential energy is the joule.

There are other forms of potential energy besides gravitational potential energy. They include elastic potential energy, which comes from the forces between atoms, and electrical potential energy, which comes from the electrical forces between charges. Potential energy results from work done against a force. For example, the box was lifted to a shelf, increasing its gravitational potential energy by working against gravity. Similarly, a spring is stretched by working against the elastic (spring) force, thus increasing the elastic potential energy of the system to which it is connected.

Potential energy (U) is the energy associated with the position or condition of a system relative to a reference point of zero potential energy.

Problem-Solving Strategy 9.6

Different forms of potential energy are indicated by a subscript identifying the force to which the potential energy corresponds. For example, U_g is gravitational potential energy and U_s is elastic potential energy of a spring.

9.11 Conservative and Nonconservative Forces

Not every force has potential energy associated with it. Only forces that are **conservative forces** have their own potential energy. A force is a conservative force if either of the following two conditions is true:

- The net work done by the force on a system as it moves between any two points is independent of the path followed by the system.

Conservative forces can store work energy as some form of mechanical potential energy.

- The net work done by the force on a system that follows a closed path is zero. A system follows a closed path if it begins and ends at the same point in space.

Some examples of conservative forces are

- the gravitational force,
- any central force (a force acting between the centers of two systems),
- any Hooke's law force.

Conservative forces are so named because the energy expended when doing work against them is stored as potential energy and can be regained as kinetic energy.

Other forces do not have these attributes. If the net work done by a force on a system over a closed path is *not* zero, or if the work done by a force on a system between two points depends on the path taken, then the force is a nonconservative force. Examples of nonconservative forces include

- kinetic frictional force,
- internal resistance forces (e.g., spring internal resistance),
- fluid drag.

When work is done against nonconservative forces, the energy is not stored as potential energy but is converted into other forms of mechanically unusable energy, such as heat, sound, or light.

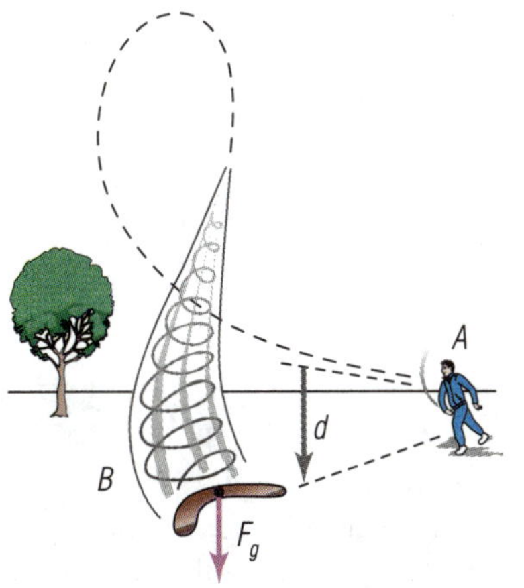

9-14 The work done on a system by a conservative force is independent of the path of the system.

Nonconservative forces convert work energy into mechanically unusable forms of energy.

9.12 Gravitational Potential Energy

All masses, from atoms to galaxies, exert gravitational forces on other masses. If the masses are separated, then the gravitational attraction tends to pull them together. Thus, two masses separated by some distance have the potential to be drawn together by the gravitational force exerted between them. The work required to move these masses apart against the force of gravity is equal to their *gravitational potential energy.*

Near the earth's surface, the force of gravity on an object is nearly constant at $\mathbf{F}_g = m\mathbf{g}$. The work done against gravity in lifting an object at constant speed goes entirely to increasing the object's potential energy. Consider an elevator of mass m that is lifted with constant speed from a height h_1 to a height h_2. Since the elevator's velocity is constant, the acceleration is zero, and there is no net force on the elevator. That is, the lifting force just balances the force of gravity, or the elevator's weight, neglecting frictional forces. The lifting force must therefore have the same magnitude, but opposite direction, as the force of gravity:

$$\mathbf{F}_{\text{lift}} = -\mathbf{F}_g = -m\mathbf{g}$$

$$F_{\text{lift}} = |mg| \quad \text{(upward)}$$

Problem-Solving Strategy 9.7

Recall that the scalar value of gravitational acceleration is $g = -9.81 \text{ m/s}^2$. Use the absolute value of mg when determining work involving weight or gravitational forces.

The work that the lift cables do against gravity in lifting the elevator is equal to the product of the force they exert on the elevator and the elevator's change in height.

$$W_{\text{lift}} = F_{\text{lift}}d = |mg|(h_2 - h_1)(\cos 0°)$$

$$W_{\text{lift}} = |mg|\Delta h$$

Since the force of gravity opposes the lifting force, their components along the direction of the elevator's motion are opposite. Therefore, the work that gravity does on the elevator as it rises is negative.

$$W_g = F_g d = |mg|(h_2 - h_1)(\cos 180°)$$

$$W_g = -|mg|\Delta h \quad (9.12)$$

The elevator's potential energy (U_g) changes by the amount of work the lift cables do on the elevator:

$$\Delta U_g = W_{\text{lift}} = |mg|\Delta h \qquad (9.13)$$

Does it make any difference whether the change in height begins at the earth's surface or at 10 m above the surface? No, as long as the object stays near enough to the earth's surface so that $F_g = |mg|$ there is no significant difference.

If you compare Equations 9.12 and 9.13, you will notice a fundamental relationship between the work done by a force and the potential energy associated with that force. In the case of the gravitational force,

$$W_g = -\Delta U_g\,. \qquad (9.14)$$

Equation 9.14 is true for any conservative force. In words, work must be done *against* a force in order to increase the potential energy of a system with respect to that force. Conversely, as potential energy of a system decreases with respect to a force, the force does positive work on the system.

Problem-Solving Strategy 9.8

Work by a conservative force on a system is always equal to the negative of the change of the system's potential energy with respect to that force.

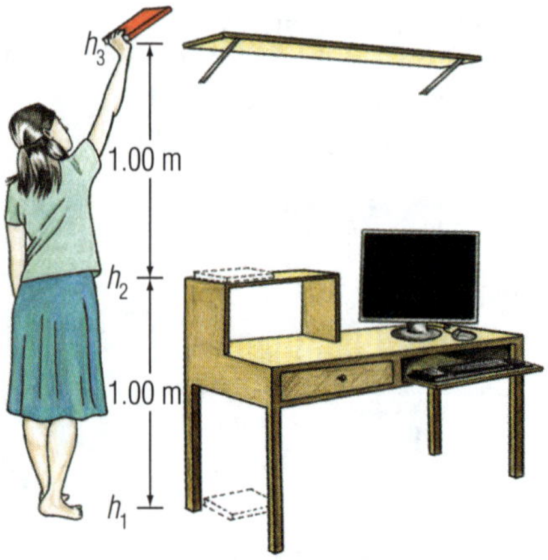

9-15 Lifting the book increases its gravitational potential energy.

EXAMPLE 9-7

Greater Potential: Potential Energy

A student lifts a 0.750 kg book from the floor to her 1.00 m high desk with a constant velocity. Later, she moves the book at a constant velocity to her bookshelf, 2.00 m above the floor. (a) How much work did she do in the first lift? (b) In the second? (c) What is the change in the book's gravitational potential energy for each lift?

Solution:

a. Work for the first lift:

$$W_1 = Fd_1 = |mg|\Delta h_1$$

$$W_1 = |mg|(h_2 - h_1)$$

$$W_1 = |(0.750\text{ kg})(-9.81\text{ m/s}^2)|(1.00\text{ m} - 0.00\text{ m})$$

$$W_1 \cong 7.3\underline{5}7\text{ J}$$

b. Work for the second lift:

$$W_2 = |mg|(h_3 - h_2)$$

$$W_2 = |(0.750\text{ kg})(-9.81\text{ m/s}^2)|(2.00\text{ m} - 1.00\text{ m})$$

$$W_2 \cong 7.3\underline{5}7\text{ J}$$

Notice that $W_1 = W_2$.

c. Since it is assumed that the lifts were made at constant speeds, the change of kinetic energy is zero. No work was expended in changing kinetic energy. All the work went into changing gravitational potential energy.

Therefore:

$$W_{\text{lift}} = \Delta U_g$$

$$\Delta U_{g\,1} = W_1 = 7.36\text{ J}$$

$$\Delta U_{g\,2} = W_2 = 7.36\text{ J}$$

For locations near the earth's surface, the *change* in potential energy is the same as long as the change in height is the same.

Notice in Part c of the previous example that the work done by the *lifting force* is compared to the change in potential energy with respect to the *gravitational force*. In this case, both are positive. The work of one force on a system often causes a change in potential energy with respect to another force also acting on the system. You must be careful when working such problems so that each force is correctly associated with its corresponding work or energy.

9.13 Relative Potential Energy

What is the gravitational potential energy of the book in Example 9-7? The gravitational potential energy of a system is measured by the amount of work gravity can do on it in the absence of any other restraining forces. Since the force of gravity on an object is constant as long as it is near the earth, the work gravity does varies only with the distance the object can fall. So we could say that

$$U_g = |mg|h, \qquad (9.15)$$

where h is the distance that the object can fall.

How far can an object fall? Figure 9-16 points out the difficulty in answering this question. A rock rests on the edge of a cliff. What is its potential energy? It depends on your *height reference point*—the point from which you measure h. If you believe the rock cannot fall at all because it is firmly wedged in at the top of the cliff, then $h = 0$ m and $U_g = 0$ J. If you expect that the rock could fall to the tree, then the tree is the height reference and the rock is at $h = +10$ m; so $U_g = 10|mg|$ J. If you think it's possible that the rock could fall to the surface of the water, then the water's surface is the height reference. The rock's height is $h = +40$ m and $U_g = 40|mg|$ J. If you allow the rock to fall to the bottom of the river, then its height is $h = +50$ m and $U_g = 50|mg|$ J. Which value of the potential energy is "correct"? Each of the values of potential energy is relative to the selected reference point. Since you need to know only the *change* in potential energy (ΔU_g), which reference point you choose does not matter. However, after choosing a reference point, you should use it consistently throughout the problem.

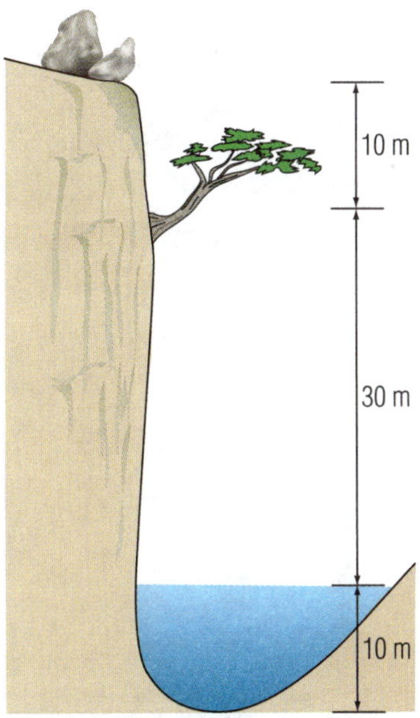

9-16 The potential energy of the rock depends on the point from which you measure *h*.

Problem-Solving Strategy 9.9

When solving gravitational potential energy problems, choose a height reference point that makes sense to you and simplifies the problem.

EXAMPLE 9-8

Loss of Potential Energy: Falling Objects

A 6.50 kg bowling ball falls from an observation deck to the roof of the building, rolls to the edge, falls off, and then makes a bull's-eye dive into an open manhole in the street below (see Figure 9-17). Calculate the ball's change in potential energy for the fall, using as height reference points (a) the roof, (b) the street, and (c) the bottom of the manhole.

Solution:

Choose a coordinate system as in Figure 9-17 with y positive upward. The origin will be at the heights specified in each part of the problem.

a. Origin at the roof; deck ($h_1 = +5.0$ m) to bottom of the manhole ($h_2 = -25.0$ m):

$$\Delta U_g = |mg|\Delta h = |mg|(h_2 - h_1)$$

$$\Delta U_g = |(6.50 \text{ kg})(-9.81 \text{ m/s}^2)|[(-25.0 \text{ m})-(+5.0 \text{ m})]$$

$$\Delta U_g \cong -19\underline{1}2 \text{ J}$$

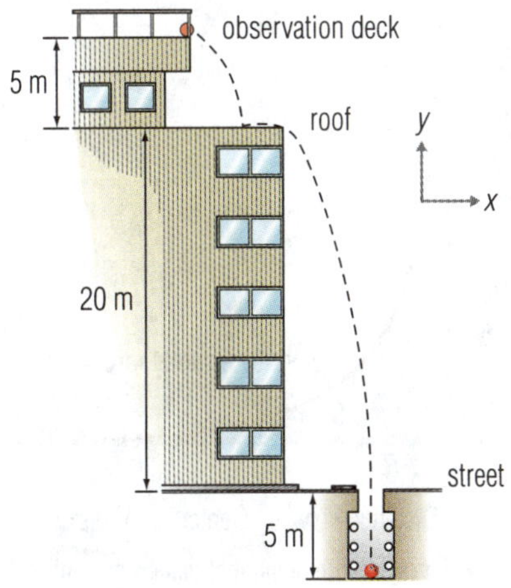

9-17 The ball's gravitational potential energy decreases as it falls.

b. Origin at the street; deck (h_1 = +25.0 m) to the bottom of the manhole (h_2 = −5.0 m):

$$\Delta U_g = |(6.50\ \text{kg})(-9.81\ \text{m/s}^2)|[(-5.0\ \text{m})-(+25.0\ \text{m})]$$

$$\Delta U_g \cong -191\underline{2}\ \text{J}$$

c. Origin at the bottom of the manhole; deck (h_1 = +30.0 m) to the bottom of the manhole (h_2 = 0.0 m):

$$\Delta U_g = |(6.50\ \text{kg})(-9.81\ \text{m/s}^2)|[(0.0\ \text{m})-(+30.0\ \text{m})]$$

$$\Delta U_g \cong -191\underline{2}\ \text{J}$$

The change of potential energy for the entire fall is the same regardless of the height reference point.

9-18 The gravitational potential for any object is $|g|h$ as long as the same reference point is selected.

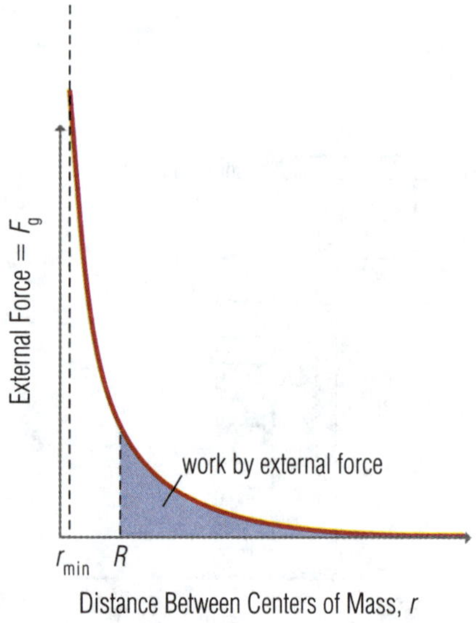

9-19 Force-distance graph that defines gravitational potential from a distance R

9.14 Gravitational Potential

Gravitational potential energy depends on the mass of an object and its position within the gravitational *field.* The strength of a **field** is measured by the force it exerts per unit mass. For locations near the surface of the earth, the *gravitational field strength* is simply 9.81 N/kg. So the gravitational force on an object, or its weight, is the product of the object's mass and the gravitational field strength, $F_w = m(9.81\ \text{N/kg}) = |mg|$. Similarly, the gravitational potential energy of the object is the product of its mass and its **gravitational potential, $U_g(r)$,** which is defined as its potential energy per kilogram at a specified distance r from a zero reference distance. For locations near the surface of the earth, $U_g(r)$ is simply $|g|h$ for any mass m.

But what if you want to calculate an object's potential energy at some location where gravitational field strength is not equivalent to $|g|$ (i.e., $F_w \neq |mg|$)? You can find the work done by any gravitational force by using a force-distance curve. You learned in Chapter 7 that the gravitational force between two objects can be calculated from Newton's law of universal gravitation,

$$F_g = G\frac{Mm}{r^2},$$

where r is the distance between the centers of the two masses. The force that one mass exerts on another is plotted versus distance in Figure 9-19.

Suppose that mass m starts at a distance R from mass M and moves away at a constant velocity. A force with the same magnitude as the gravity between the masses but with the opposite direction must move m. The work that this force does is equal to the shaded area in Figure 9-19. Calculus shows that the area is equal to

$$W_{ext} = G\frac{Mm}{R}.$$

Since the work was done at a constant velocity (ΔK = 0 J), it all became potential energy. The change in potential energy of m is therefore

$$\Delta U_g = G\frac{Mm}{R}. \qquad (9.16)$$

For gravitational potential far from the earth, as well as for potential energy near the earth, the reference point is arbitrary. However, in general gravitational problems it is customary to choose the potential energy at an infinite separation as the zero reference point. For any distance r between M and m,

$$\Delta U_g = U_{g\,\infty} - U_{g\,r}$$

$$\Delta U_g = 0 - U_{g\,r},$$

where $U_{g\,r}$ means "the gravitational potential energy at a separation of r."
Therefore, substituting for ΔU_g back into Equation 9.16,

$$-U_{g\,r} = G\frac{Mm}{r}$$

$$U_{g\,r} = -G\frac{Mm}{r}. \quad (9.17)$$

Since the constant G, the masses, and the distance are always positive, gravitational potential energy is negative for any measurable separation distance. However, increasing r decreases the absolute value of U_g, making it less negative, so potential energy is greater (less negative) the farther apart the two objects are.

Consequently, the **gravitational potential** (potential energy per unit mass) for any object of mass m at any distance r from mass M is simply

$$U_g(r) = -G\frac{M}{r}, \quad (9.18)$$

where M is the source of the gravitational field, and r is the distance from the center of mass of the source. The unit of gravitational potential is joules/kg.

9.15 Gravitational Work Function

How does the change in gravitational potential energy relate to the work that gravity does? Remember that the external force that pulls m away from M is equal in magnitude but opposite in direction to the gravitational force between the masses. So their components in the direction of motion are opposite. Therefore,

$$W_{ex} = F_{ex}d = -F_g d = -W_g.$$

The change in potential energy is

$$\Delta U_g = W_{ex} = -W_g$$

$$\Delta U_g = -W_g,$$

which is essentially the same as Equation 9.14. Recall that gravitational force is a conservative force, so the work and change of potential energy does not depend on what path the masses take as they move toward or away from each other.

Though gravitational potential energy with reference to an infinite distance from earth is negative, we are more interested in *changes* of potential energy due to changes in height. With increasing height, the change in gravitational potential energy is always positive.

Gravitational potential, $U_g(r)$, is the gravitational potential energy per unit mass at distance r from the source of the gravitational field. When referenced to an infinite distance,

$$U_g(r) = -G\frac{M}{r}.$$

When referenced to the surface of the earth,

$U_g(r) = U_g = |g|h$; where $h = r - r_E$.

Problem-Solving Strategy 9.10

When finding the change of potential energy from one finite distance to another between two objects, ΔU_g is just the difference of the final and initial gravitational potential energies at the two positions. Be sure that you use the correct value of r in the denominator of each term.

EXAMPLE 9-9

Rocket Science: Gravitational Work

How much work would it take to move a 4.50×10^4 kg rocket from the earth's surface to an infinite distance from the earth? The radius of the earth is 6.38×10^6 m; the mass of the earth is 5.97×10^{24} kg.

Solution:

Assuming that the starting position of the rocket is r_E, the earth's radius, the work necessary to move it to an infinite distance from the earth is

$$W = G\frac{Mm_r}{r_E},$$

where M is the mass of the earth and m_r is the mass of the rocket.

$$W = (6.67 \times 10^{-11}\ \text{N}\cdot\text{m}^2/\text{kg}^2)\left[\frac{(5.97 \times 10^{24}\ \text{kg})(4.50 \times 10^4\ \text{kg})}{6.38 \times 10^6\ \text{m}}\right]$$

$$W \cong 2.808 \times 10^{12}\ \text{J} \cong 2.81 \times 10^{12}\ \text{J}$$

9-20 The booster rockets provided most of the work necessary to lift the space shuttle into orbit around the earth.

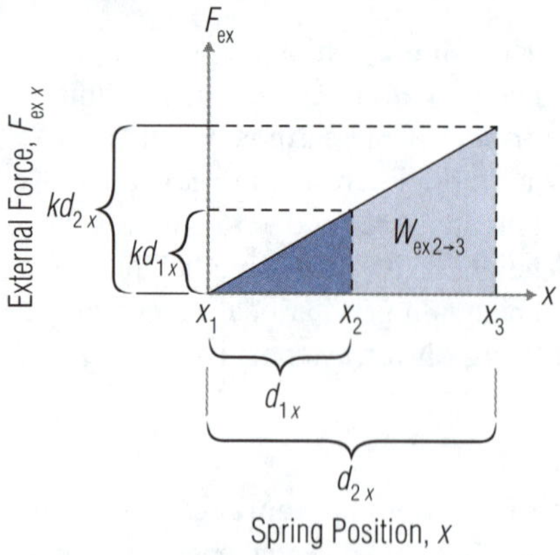

9-21 Work done on a spring between two positions

Recall that the spring displacement $d_x = \Delta x = x_2 - x_1$.

9.16 Elastic Potential Energy

Elastic forces in metals come from the tendency of a metal's atoms to remain at a fixed distance from one another. If the atoms are pushed closer together, they repel each other. If they are pulled farther apart, they attract one another. That is why a spring pulls or pushes back to its equilibrium position if it is stretched or compressed. External forces work against the spring force as the spring is stretched or compressed. Therefore, **elastic potential energy** can be produced with respect to spring-like (elastic) forces, which are conservative forces.

The work done by an external force (W_{ex}) *against* an elastic force to stretch or compress a spring at a constant velocity is equal to the change in the spring's elastic potential energy. The amount of work needed to stretch a spring from a length of x_2 to a length of x_3 is shown as the lightly shaded area in Figure 9-21. Both lengths are *measured from the spring's equilibrium position*, x_1. This area is equal to the area of the large triangle minus the area of the small triangle.

$$W_{ex\,2\to3} = \Delta U_s = \tfrac{1}{2}kd_{2x}^2 - \tfrac{1}{2}kd_{1x}^2$$

$$W_{ex\,2\to3} = \tfrac{1}{2}k(d_{2x}^2 - d_{1x}^2)$$

If the spring begins at its natural length ($x_1 = x_2$), then d_{1x} is zero, and

$$W_{ex} = \tfrac{1}{2}kd_{2x}^2.$$

Once again, we determine the work done by an external force on the spring by evaluating the change of potential energy with respect to the elastic force.

EXAMPLE 9-10

Change in Spring Potential Energy

A spring with a spring constant of 5.00 N/m is stretched from a length of 0.500 m to a length of 0.700 m. What is its change in potential energy if its rest length (x_1) is 0.450 m?

Solution:

The spring's initial position:

$$x_2 = 0.500\text{ m};\ d_{1x} = 0.500\text{ m} - 0.450\text{ m} = 0.050\text{ m}$$

The spring's final position:

$$x_3 = 0.700\text{ m};\ d_{2x} = 0.700\text{ m} - 0.450\text{ m} = 0.250\text{ m}$$

$$\Delta U_s = \tfrac{1}{2}k(d_{2x}^2 - d_{1x}^2)$$

$$\Delta U_s = \tfrac{1}{2}(5.00\text{ N/m})[(0.250\text{ m})^2 - (0.050\text{ m})^2]$$

$$\Delta U_s \cong 0.1\underline{5}0\text{ J} \cong 0.15\text{ J}$$

9B Section Review

1. a. What kind of mechanical energy is specifically related to a system's position or condition, irrespective of its motion?
   b. Is there more than one form of this energy? Explain.
2. Discuss why positive work by a force on a system reduces the potential energy of the system with respect to that force.

9B Objectives

After completing this section, I can

- ✓ differentiate between kinetic and potential energy.
- ✓ calculate the kinetic energy of a system.
- ✓ differentiate between conservative and nonconservative forces.
- ✓ compute the potential energy of conservative forces.
- ✓ define gravitational potential and compute work done against gravity.

3. What is the main restriction when using Equation 9.15 to determine relative gravitational potential energy?

4. A daredevil astronaut with a mass of m considers parachuting from orbit to the earth's surface wearing an asbestos space suit. How would the gravitational work on his body during the jump compare to the work done by gravity after he comes to his senses and rides a space shuttle back to the earth?

5. What is the kinetic energy of a 1.00 g hailstone falling at 8.50 m/s?

6. What is the total work required to stop an 85.0 kg football halfback moving at 4.50 m/s?

7. What is the change of the gravitational potential energy (with respect to the earth) of a 120. g meteoroid that falls from a distance of 3.80×10^5 km to the surface of the earth?

8. A spring with a spring constant of 4.50 N/m is compressed from its rest length of 0.600 m to a length of 0.300 m. What is the change in the potential energy of the spring?

9C TOTAL MECHANICAL ENERGY

9.17 Defining Total Mechanical Energy

All mechanical work on a system can be subdivided into the work done by conservative forces (W_{cf}) and the work done by nonconservative forces (W_{ncf}), such as kinetic friction and drag. We can determine their contributions to the change of system kinetic energy with the following equation:

$$W_{total} = W_{cf} + W_{ncf} = \Delta K$$

You know that $W_{cf} = -\Delta U$ because potential energy is associated only with conservative forces. Therefore,

$$-\Delta U + W_{ncf} = \Delta K.$$

Solve for the work done by nonconservative forces.

$$W_{ncf} = \Delta K + \Delta U$$

$$W_{ncf} = (K_2 - K_1) + (U_2 - U_1)$$

Rearranging terms,

$$W_{ncf} = (K_2 + U_2) - (K_1 + U_1), \text{ and}$$

$$W_{ncf} = \Delta(K + U).$$

The work by **nonconservative forces** on a system is equal to the change of the sum of kinetic and potential energies of the system.

Therefore, the work-energy theorem (see subsection 9.9) includes the principle that the work done by *nonconservative forces* is equal to the change of the system's total energy. The term $K + U$ has a special name in physics—**total mechanical energy (E)**. The definition of total mechanical energy is

$$E \equiv K + U. \quad (9.19)$$

Total mechanical energy (E) is the sum of a system's kinetic and potential energies:

$$E \equiv K + U.$$

You can conclude that the work accomplished by all nonconservative forces on a system during a certain process is equal to the change of total mechanical energy of a system:

$$W_{ncf} = \Delta E \quad (9.20)$$

9.18 Conservation of Mechanical Energy

This consequence of the work-energy theorem actually helps us to understand how energy exchanges during a process. If we assume that there are no nonconservative forces acting on a system, then the left side of Equation 9.20 becomes 0 J. We routinely do this when we say that no friction or drag occurs during motion. If there is no change of total mechanical energy, then change of kinetic energy must be equal but opposite to the change of potential energy:

$$\Delta E = \Delta K + \Delta U = 0 \text{ J}$$
$$\Delta K = -\Delta U$$

When only conservative forces are acting on a system, the increase of kinetic energy is equal to the loss of mechanical potential energy and vice versa. This is the principle of **conservation of mechanical energy**.

This special case of the work-energy theorem demonstrates the conservation of mechanical energy.

A good problem-solving strategy is to set the sum of the initial kinetic and potential energies of a system equal to the sum of their final values.

$$K_1 + U_1 = K_2 + U_2 \qquad (9.21)$$

If only conservative forces are involved, then Equation 9.21 will represent a true equality. Rearrange terms as necessary to solve for the unknown.

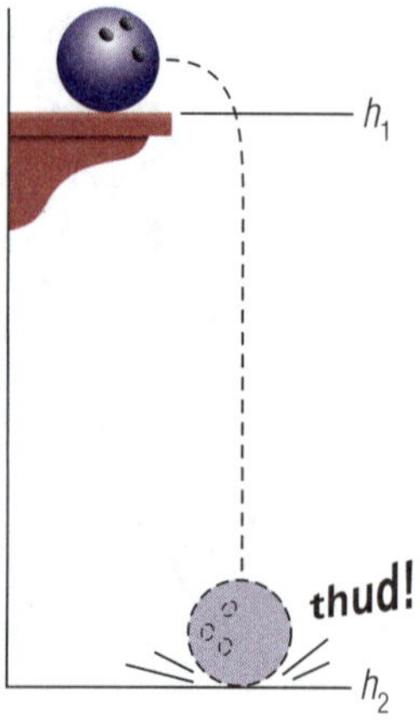

9-22 All the ball's gravitational potential energy is converted into kinetic energy during its fall.

EXAMPLE 9-11

Dropping the Ball: Conservation of Energy

A 5.45 kg bowling ball falls from rest on a shelf 2.0 m above the floor. What is its speed just as it strikes the floor? Assume that only conservative forces act on the ball as it falls.

Solution:

Assume that the height reference point is at floor level. Substitute the known values into Equation 9.21.

$$K_1 + U_1 = K_2 + U_2$$

$$\tfrac{1}{2}mv_1^2 + |mg|h_1 = \tfrac{1}{2}mv_2^2 + |mg|h_2$$

$$0 \text{ J} + |(5.45 \text{ kg})(-9.81 \text{ m/s}^2)|(2.0 \text{ m}) = \tfrac{1}{2}(5.45 \text{ kg})v_2^2 + 0 \text{ J}$$

$$1\underline{0}6 \text{ J} = \tfrac{1}{2}(5.45 \text{ kg})v_2^2$$

$$v_2 = \sqrt{\frac{2(1\underline{0}6 \text{ J})}{5.45 \text{ kg}}}$$

$$v_2 \cong 6.\underline{2}3 \text{ m/s} \cong 6.2 \text{ m/s}$$

The ball is falling at about 6.2 m/s when it hits the floor.

If a nonconservative force was involved in the process, then the two sides of Equation 9.21 will be unequal unless the nonconservative force is accounted for:

$$K_1 + U_1 = K_2 + U_2 + W_{ncf}$$

We will consider the conservation of mechanical energy in greater depth in Chapter 10.

9.19 Using Conservation of Mechanical Energy to Solve Problems

Potential energy is the energy of position in relation to a conservative force, such as weight produced by gravity. The height of an object gives it the potential to perform useful work. Rain- and melt-waters high in the mountains possess potential energy that turns into kinetic energy as they flow downhill. If you have ever waded in a fast-running stream, you realize that moving water has a lot of energy. People have been exploiting this natural energy source since the earliest times. In the past, water wheels provided the power for forges, sawmills, and mills for grinding grain. One of the reasons towns were located along rivers was to take advantage of this natural power source.

Though not commonly used today, water wheels are built where the elevation of the stream changes significantly. Usually a channel is constructed to divert some of the water flow to the wheel. To even out the water supply to the wheel, the diverted water from the stream often fills a pond behind a dam. When water power is needed, the dam's sluice gate is opened and flowing water turns the wheel. As the wheel turns, gears transmit the power to the machinery. After the water has done its work, it empties back into the stream at a lower elevation than the point at which it was diverted. During the trip from diversion to reentry, the water's potential energy is converted into kinetic energy. The difference in height between the water level of the supply and the discharge from the wheel is called the *head* of the water supply, which is proportional to the work that can be extracted from the falling water.

Today, we use the same principle to generate electricity using hydroelectric dams. The dam establishes a near-constant water level (the head). The water flows through a supply tube in the bottom of the dam, called the *penstock*, and into the casing of a water turbine. The water falls vertically through the turbine, adding more head to the water supply, and then flows out a discharge penstock to the downstream river. The turbine shaft spins under the flow of water and turns a large wheel-shaped electromagnet (the rotor) that spins within a stationary array of wire coils (the stator). As you will learn in later chapters, a magnet moving past a wire conductor induces an electrical potential in the wire. Notice that the energy of the water undergoes several transformations. The potential energy of the water is converted to kinetic energy as the water flows down the penstocks. Some of the kinetic energy of the water is transformed into the kinetic energy of the turbines and the generators, which then produce electrical energy.

Good and Wise Dominion?

Construction of the Three Gorges Dam is expected to displace nearly 2.5 million people when the valley is fully flooded, cause the extinction of several unique animal species, and flood 1300 archaeological and cultural sites. It also may have contributed to the 7.8 magnitude Sichuan Province earthquake in May, 2008.

At the same time, the project will reduce flooding of the Yangtze River and erosion of its banks downstream, and also substantially improve air quality in the region by eliminating the need for coal-fired power plants. And it has already increased upstream shipping by a factor of six through its system of locks.

Is this large-scale project an instance of exercising good and wise dominion? Is it the best means of exercising dominion in this situation?

Some dams, instead of relying simply on the incoming stream, re-create their own head of water by pumping water back up to the upper reservoir after it has done its work. These pumping stations actually expend more energy pumping the water back up than they gain as it flows down, but they are profitable because they exploit the law of supply and demand. Utilities may use electricity for pumping water upstream during times of low demand (e.g., at night) and generate electricity during times of peak demand when they can sell it for higher rates. Also, during times of drought, they are able to produce more electricity from the limited supply of water.

9-23 A view of the Three Gorges Dam on the Yangtze River in Yichang, central China's Hubei province

The largest hydroelectric facility in the world is the Three Gorges Dam, located in Hubei Province in China. It is an immense engineering project, creating a reservoir the length of Lake Superior (but having only slightly more than 1% of its surface area). The dam required more than 27 million cubic meters of concrete and enough steel to make 63 Eiffel Towers. The operational head for full power is 81 m but the dam can contain a head of 91 m. There will be 32 main generators that can produce 22.4 GW of electrical power when it becomes fully operational, supplying up to 3% of China's energy needs. Its turbine-generators are more than 90% efficient.

9C Objectives

After completing this section, I can
- ✓ define total mechanical energy.
- ✓ demonstrate how total mechanical energy is conserved in a system.
- ✓ compute the kinetic or potential energy of a system according to the rule of the conservation of mechanical energy.
- ✓ discuss how hydroelectric generators employ the conservation of total mechanical energy to provide an economical and nonpolluting source of electrical energy.

9C Section Review

1. What kinds of forces can change kinetic energy? Potential energy can be changed with respect to what kinds of forces?
2. What kind(s) of mechanical energy does a roller coaster have halfway down its first hill?
3. What kind(s) of energy relative to its rest position does a pendulum bob have as it hangs motionless on its string?
4. What kinds of forces could be at work if mechanical energy is not conserved in a process? Describe one such process.

DS 5. On what kind of energy conversions does hydropower depend?

DS✪ 6. Calculate the volume of water that must flow through a single Three Gorges Dam generator in 1 s assuming a head of 85 m, an efficiency of 90.%, and a capacity of 700. MW. (*Hints*: Assume all the potential energy available in the reservoir head is converted to kinetic energy in the turbine. Determine the mass flow rate of water through the turbine. Use the density of water to compute volume.)

Chapter Review

In Terms of Physics

energy	196
work (W)	196
joule (J)	196
equilibrium position	199
Hooke's law	199
spring constant (k)	199
ideal spring	199
Hooke's-law force	199
power (P)	201
watt (W)	201
kinetic energy (K)	202
work-energy theorem	202
potential energy (U)	204
conservative force	204
nonconservative force	205
field	208
gravitational potential ($U_g(r)$)	208
elastic potential energy	210
total mechanical energy (E)	211
conservation of mechanical energy	212

Problem-Solving Strategies

9.1 (page 196) To determine whether work by a force is positive or negative, find the force's vector component that is parallel to the displacement of the system. If it is in the same direction as the displacement, work is positive. If the force vector component is opposite to the direction of displacement, then work is negative.

9.2 (page 197) Gravity does positive work on a system when the system loses height, and it does negative work when the system gains height.

9.3 (page 197) Whenever you must calculate the gravitational force on a system in order to compute work done by or against gravity, be sure to use the *magnitude* of gravitational force $|mg|$. The work formula uses the magnitudes of force and displacement, *not* their scalar values.

9.4 (page 200) You must take care that you calculate the work done by the force that is specified in the problem. If an object attached to a spring is the system of interest, the work done by an external force on the system is not the same as the work done by the spring force.

9.5 (page 203) Kinetic energy (K) can only be positive. For any given object, K varies as the square of its speed.

9.6 (page 204) Different forms of potential energy are indicated by a subscript identifying the force to which the potential energy corresponds. For example, U_g is gravitational potential energy and U_s is elastic potential energy of a spring.

9.7 (page 205) Recall that the scalar value of gravitational acceleration is $g = -9.81\ m/s^2$. Use the absolute value of mg when determining work involving weight or gravitational forces.

9.8 (page 206) Work by a conservative force on a system is always equal to the negative of the change of the system's potential energy with respect to that force.

9.9 (page 207) When solving gravitational potential energy problems, choose a height reference point that makes sense to you and simplifies the problem.

9.10 (page 209) When finding the change of potential energy from one finite distance to another between two objects, ΔU_g is just the difference of the final and initial gravitational potential energies at the two positions. Be sure that you use the correct value of r in the denominator of each term.

Review Questions

1. What is the equation that defines work?

2. In the work equation, what is the significance of the cosine of the angle between the force and displacement vectors?

3. Which of the following is the unit for work?

a. N **b.** kg **c.** m **d.** J **e.** J/s

4. Explain how negative work on a system occurs.

5. Give an example of a force that acts on a system during a displacement but does no work.

6. What kind of graph can represent mechanical work? How is the graph used to do this?

7. What is the shape of the work region on the graph of a constant force versus displacement?

8. What is the shape of the work region on the graph of a Hooke's law force?

9. For a given object near the surface of the earth, what is the only variable factor that affects gravitational work?

10. Define *power.*

11. Kinetic energy is

a. energy of motion.

b. energy of position.

c. all of the energy that an object has.

d. independent of mass.

12. What equation expresses the work-energy theorem?

13. Define *mechanical potential energy* in your own words.

14. Explain, using an example, how positive work by a conservative force can result in a negative change of potential energy with respect to the same force.

15. What is the main restriction for using the equation $\Delta U_g = |mg|\Delta h$?

16. A brick falls a distance Δh, with a potential energy change of $\Delta U_g = |mg|\Delta h$. If no work is done on the earth-brick system, what is the change in the brick's kinetic energy, ΔK?

a. 0 J **b.** $-|mg|\Delta h$ **c.** $+|mg|\Delta h$ **d.** $-mv$

17. What is the sum of the kinetic energy and the potential energy of a system called?

True or False (18–28)

18. Work is a scalar quantity.

19. Every force acting on a system during a displacement does work on the system.

20. The graphical area under a force-displacement curve represents power.

21. Ideal springs follow Hooke's law only when stretched.

22. The force exerted by an ideal spring is a conservative force.

23. External forces that stretch or compress a spring do positive work on the spring.

24. The power delivered by a force can be increased by reducing the distance over which the force works in a given time.

25. The unit of power is J/s, or W.

26. The kinetic energy of a system is directly proportional to its mass and its speed.

27. Only conservative forces can change the kinetic energy of a system.

28. It doesn't matter what height you measure gravitational potential energy from as long as you are consistent within a given problem.

29. Calculate the work done by the earth on a 2.00 kg brick that falls from the roof of a 10.0 m building to the ground.

30. A 165 g hockey puck moves a distance of 1.00 m while the hockey stick is pushing it with a force of 200. N. How much work does the stick do on the puck?

31. A woman pushes on a baby stroller at an angle of 60° below the horizontal. If she does 400. J of work on the stroller while it moves 20.0 m, how much force does she exert?

32. An empty spring cart for cafeteria trays has a load of trays placed on it. The spring constant is 490 N/m, and the trays push the spring down 0.50 m.

a. What force do the trays exert on the spring?

b. Draw a tray force-distance diagram for the situation.

c. How much work did the trays do on the spring when they were loaded into the cart?

33. How much power is needed to compress a spring from its rest length of 0.150 m to half that in 10.0 s? The spring constant is 625 N/m.

34. A 165 g hockey puck resting on a level ice surface is struck by a stick with a force of 225 N. The stick is in contact with the puck over a straight-line distance of 0.840 m.

a. How much work does the stick do?

b. What is the change in the puck's kinetic energy?

c. What is the puck's kinetic energy before it is hit?

d. What is the puck's speed when the stick breaks contact?

35. An athlete running at a speed of 3.00 m/s decides to speed up on a long, flat stretch. After running 120. m, the runner is moving at a speed of 4.50 m/s. He has a mass of 75.0 kg.

a. What is his kinetic energy before he speeds up?

b. What is his kinetic energy after he speeds up?

c. How much work was done in speeding up?

d. What net force changed his speed?
e. From Newton's second law, what was his acceleration?
f. From the first equation of motion, how long did it take him to speed up?

36. A 1.50 kg model train initially traveling at 0.50 m/s accelerates to 1.00 m/s in 10. s.
a. What is the change in the train's kinetic energy?
b. How much power was expended to accelerate the train?

37. A 2.00×10^3 kg block of granite is lifted at constant velocity from the bottom of a quarry to the top. The quarry is 176 m deep.
a. How much work was done to the granite block?
b. What is the change in potential energy of the block?
c. If the lift took 70.0 s, how much power was expended?

38. A hydraulic lift raises at constant velocity a 975 kg car 3.00 m to the deck of a ship. What is the change in the car's potential energy?

39. A spring whose rest length is 35.0 cm and whose spring constant is 8.00 N/m hangs from a beam. The spring is slowly stretched, with no acceleration, to a length of 50.0 cm.
a. How much work was required to stretch the spring?
b. What is the change in the spring's elastic potential energy?

40. An engine stretches a spring at constant velocity 0.35 m from its rest position, doing 75 J of work.
a. What is the spring's change in elastic potential energy?
b. Determine its spring constant.

41. An ideal spring is attached to a 0.100 kg block on one end and to the wall on the other end. The block rests on a frictionless table. The spring is stretched at constant velocity until it is 2.00 cm longer than its equilibrium position, and then the spring and block are released. The spring constant is 8.00 N/m.
a. What is the work used to stretch the spring?
b. What is the potential energy of the block when the spring is stretched?
c. What is the change in the block's potential energy from the time it is released to the time it reaches the spring's equilibrium position?
d. What is the change in the block's kinetic energy from the time it is released to the time it reaches the spring's equilibrium position?
e. What is the block's velocity when it reaches the spring's equilibrium position after it is released?

42. A 0.165 kg ball is thrown upward with a velocity of 10.0 m/s.
a. What is the ball's initial kinetic energy as it is released?
b. At the peak of its flight, the ball's velocity is zero. What is the change in its kinetic energy from the time it is thrown to the time it reaches the peak of its flight?
c. What is the ball's change in potential energy from the time it is thrown to the time it reaches the peak of its flight, assuming that no work is done on the earth-ball system?
d. How high does the ball rise if it is released at 2.0 m above the ground?

43. A 2.00 kg brick falls from rest a distance of 1.00 m. Find ΔU_g, ΔK, and ΔE.

44. A 1.00×10^6 kg asteroid moves from an infinite distance from the sun to a distance of 1.75×10^{11} m from the center of the sun.
a. What is the force of the sun on the asteroid at 1.75×10^{11} m?
b. What is the change in the asteroid's potential energy?
c. How much work does the sun do on the asteroid?
d. If the asteroid returns to its original distance, what will be the change in its potential energy during its return?

45. A 1.00×10^4 kg space probe, after being launched from a space shuttle that is 1.25×10^7 m from the earth's center, travels to its orbit, an additional 2.00×10^7 m farther away from the earth.
a. What is the probe's potential energy with respect to the earth's surface at the time it is launched?
b. What is the probe's potential energy with respect to the earth's surface when it reaches its orbit?
c. How much work was done against gravity to move the probe to its orbit after its launch from the shuttle?
d. If the probe falls back to the earth's surface after attaining its orbit, how much potential energy will it lose?
e. How much work will the earth have done on the probe in such a fall?
f. By how much would the probe's kinetic energy change during such a fall (ignore air resistance in the atmosphere)?
g. If the probe has zero radial velocity before it starts its plunge, what is its speed as it reaches the earth's surface (ignore air resistance in the atmosphere)?

DS **46.** Currently, hydropower supplies only 19% of the world's energy needs, but accounts for 63% of the world's renewable energy. How could hydropower provide more energy for the world with minimal effects to the land, wildlife, and local populations?

Return to the earth from orbit depends on the fact that mechanical energy is *not* conserved as it is converted into nonmechanical forms, such as heat and light. This is an artist's conception of a replacement for the space shuttle as it prepares to reenter the earth's atmosphere.

Conservation of Energy 10

DOMINION SCIENCE PROBLEM

Elevator Safety

Since the beginning of civilization, people have lived together in cities. In early Western culture, Athens, Rome, Carthage, and Alexandria were all centers of human activity and industry. As towns and cities became more crowded, people began to build upwards. Stairways allowed for a few additional floors, but residents wanted to crowd even more living and working space into the limited ground area available. Vertical urban development required not only strong yet economical building materials, but also convenient access to the many floors above ground level.

Elevators appeared in historical references as long ago as the third century BC. The earliest elevators, like the ones that lifted gladiators up to the Coliseum's arena level, were powered by animals, waterwheels, or even people. The Industrial Revolution in the 1800s provided the strong, inexpensive building materials, such as steel and concrete, that vertical building required. Much taller buildings could be built, but walking up many floors was laborious, and impossible for the infirm. Elevators became more common, but they were notoriously unreliable. The risk of their catastrophic failure made people wary of using them. What could be done to make access to tall buildings safer and more convenient?

10-1 Manually propelled elevators were used to deliver animals to the main arena level of the Roman coliseums.

10A TOTAL MECHANICAL ENERGY

10.1 Conservation Laws

Scientists acknowledge that no truly creative processes can be observed today because of the existence of conservation laws. A **conservation law** states that something is *conserved*; that is, something remains constant under certain conditions. For example, you learned in the last chapter that a system's total mechanical energy remains constant as long as no nonconservative work is done on the system. Scientists have discovered that mass and electric charge, as well as energy, seem to be conserved in all processes except under the most extreme conditions.

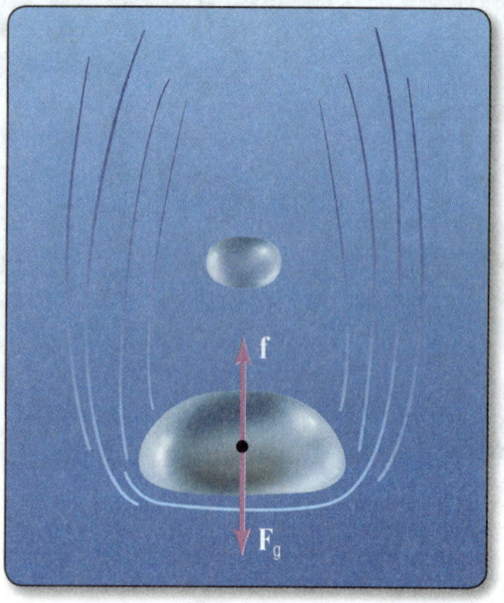

10-2 Forces on a falling raindrop at terminal velocity

10-3 The plasma gases generated by the extremely hot heat shield tiles of the space shuttle bathe the vehicle in a fiery glow during atmospheric reentry. The plasma glow is seen here through the flight deck windows.

10-4 Because there is no air resistance to slow them down, meteors strike the moon's surface while traveling at great speeds. As a result, they may have produced huge craters such as Clavius, shown here.

When Total Mechanical Energy *Is Not* Conserved

10.2 Air Resistance

Under what conditions is total mechanical energy not conserved? When work is done on the system by *nonconservative forces*, the total mechanical energy changes. This work is often done by frictional forces. One kind of frictional force is *air resistance*.

If you put your hand just outside the window of a moving car, you will feel air resistance. It is a reaction force. Your hand, moving with the car, pushes on air molecules, so the air molecules push back on your hand. If the car goes faster, you must exert more force to keep your hand in the air stream as the force of air resistance increases on your hand.

Consider a raindrop falling through the air. There are two forces on the drop: the force of gravity pulling it down and air resistance pushing upward against it. As the drop's velocity increases, the air resistance on it also increases. Eventually the drop's velocity is high enough that the air resistance balances the force of gravity. The drop stops accelerating and continues to fall with a constant velocity. This velocity is called the drop's **terminal velocity**. We are fortunate that air resistance prevents raindrops from accelerating throughout their fall. Otherwise, a raindrop falling from 6000 m would be traveling at nearly the speed of sound by the time it reached the earth's surface!

When the raindrop reaches its terminal velocity, its kinetic energy remains constant. However, since it falls, its potential energy decreases. What happens to the potential energy that the drop loses? It is changed to thermal energy and lost as heat into the air and within the raindrop. When an object falls through the air, its temperature increases.

In a short fall through the air, the rise in an object's temperature is not measurable. However, in a long or high-speed fall, this temperature increase is important. For example, when a spacecraft falls through the earth's atmosphere, its temperature may exceed 2500 °C. Therefore, all spacecraft intended to return to the earth must be protected by heat shields. The tragic loss of the space shuttle *Columbia* in 2003 is believed to be due to a breach in the integrity of its heat-shielding tile system.

Thermal energy is not included in an object's total *mechanical* energy. If potential or kinetic energy is changed to thermal energy, total mechanical energy is not conserved. Thus, for an object's fall through the air, total mechanical energy is not conserved. Spacecraft engineers use atmospheric drag to significantly slow the vehicle from orbital speed before it reaches an altitude in the atmosphere where other means are used to slow it to landing speed.

How can energy be conserved for a falling object? By removing air resistance. For example, on a planetary body that has little or no atmosphere, like the moon, there is no significant air resistance, and total mechanical energy is conserved during a fall.

10.3 Friction

Other frictional forces also change mechanical energy to thermal energy. For example, if you lightly push a wooden block across a table, it will probably stop before reaching the edge. Its kinetic energy changes to thermal energy. The temperatures of the block and the table rise slightly.

The increase in temperature because of friction is especially evident in vehicle brakes. Brakes use friction to stop a vehicle's wheels from spinning. After a car comes to a stop, its brakes are often hot. Overheated brakes can be a problem in loaded trucks or in any vehicle making a long descent down a mountain.

Friction can also change mechanical energy into sound (i.e., acoustic) energy. Some of the kinetic energy of a spinning wheel becomes the sound of squealing brakes. A creaking door results when friction at the door's hinges converts kinetic energy to sound. That is why lubricating a squeaking part, and thus decreasing friction, often stops the squeaking.

Not all frictional forces are obvious. A spring, for instance, has internal frictional forces that convert the spring's mechanical energy to thermal energy or sound. The internal forces of friction in the spring eventually stop its oscillations even if no external force acts on it. Most spring problems referred to in this textbook involve springs that have insignificant internal friction, another characteristic of an *ideal spring*. Ideal springs also have no mass, and they obey Hooke's law ($F = -k\Delta x$) perfectly. Ideal springs, like ideal strings, only approximate reality.

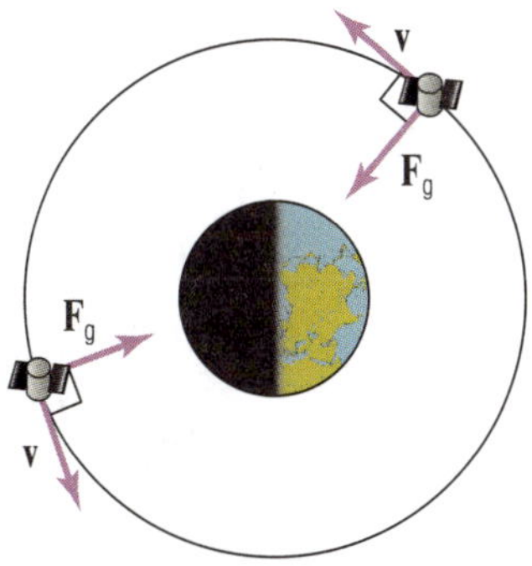

10-5 Because $\mathbf{v} \perp \mathbf{F}_g$, the instantaneous displacement is always perpendicular to the gravity vector as well. Therefore, the earth does no work on a satellite in a circular orbit.

When Total Mechanical Energy *Is* Conserved

10.4 Conservative Forces

We have discussed frictional forces that do not conserve total mechanical energy. What about *conservative forces*, forces that do conserve total mechanical energy?

A conservative force between two bodies always pushes or pulls along the line joining the centers of the bodies. You remember from Chapter 7 that this kind of force is called a **central force**. All conservative forces are central forces. One force of this kind is the force of gravity.

Consider a satellite orbiting the earth in a perfectly circular orbit. How much work does the earth do on the satellite? Figure 10-5 shows that the satellite moves in a direction perpendicular to the force of gravity. Therefore, the component of the gravitational force in the direction of the satellite's motion is zero. The earth does no work on the satellite as long as the satellite continues in a circular orbit.

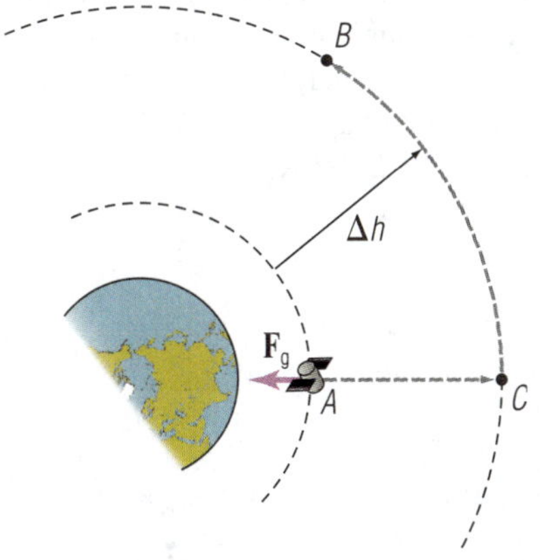

10-6a The work needed to move a satellite from A to C to B equals $F_g\Delta h$.

10.5 Path-Independence of Conservative Forces

What if the satellite moves from its orbit? Suppose it gains altitude from point A to point B. How much work must be done on the satellite against gravity? Consider the path followed by the satellite in Figure 10-6a. Assume that the satellite moves from A to C to B. To move it radially (without acceleration) from A to C (Δh), opposite the direction of the force of gravity, will require a force with the same magnitude as the force of gravity, but in the opposite direction ($-\mathbf{F}_g$). The work done on the satellite is

$$W_{AC} = F_g\Delta h.$$

To move the satellite from C to B, no gravitational work occurs, since the satellite moves along a constant-radius circular orbit. The total work in moving from A to C to B is

$$W_{AB} = W_{AC} = F_g\Delta h.$$

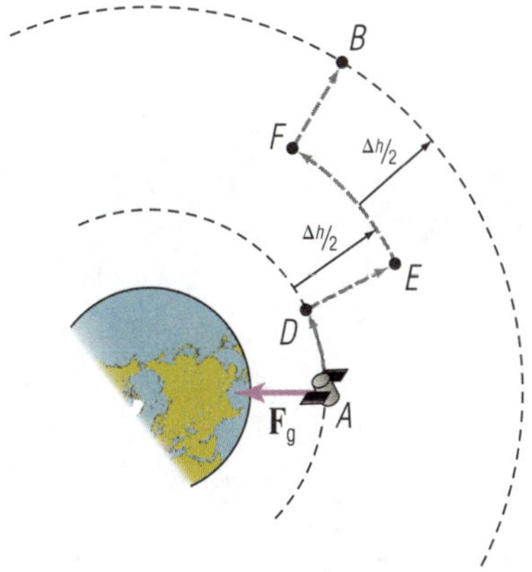

10-6b The work needed to move a satellite from A to B does not depend upon the path taken between A and B.

What if the satellite instead moves from A to D to E to F to B, as Figure 10-6b shows? The satellite moves from A to D and from E to F in circular arcs, so no gravitational work is done on the satellite. To move the satellite from D to E along a radial line, a force of $-\mathbf{F}_g$ is needed for a distance of $\Delta h/2$. The work done against gravity is

$$W_{DE} = F_g\frac{\Delta h}{2} = \tfrac{1}{2}F_g\Delta h.$$

A force of $-\mathbf{F}_g$ is again needed to move the satellite from F to B, a distance of $\Delta h/2$.

$$W_{FB} = F_g\frac{\Delta h}{2} = \tfrac{1}{2}F_g\Delta h$$

The total work done in moving the satellite from A to D to E to F to B is

$$W_{AB} = \tfrac{1}{2}F_g\Delta h + \tfrac{1}{2}F_g\Delta h = F_g\Delta h.$$

This is the same as the work done to move the satellite from A to C to B. In fact, any path from A to B can be thought of as a series of moves in the direction of $-\mathbf{F}_g$ and perpendicular to $-\mathbf{F}_g$. In each case, the work done against gravity is the same. This concept is called *path-independence*. For any conservative force, the work done on an object against the force depends only on the beginning and ending positions of the object. The work done is the same for all paths connecting those points.

Problem-Solving Strategy 10.1

You can simplify many problems by using the conservation of total mechanical energy. As you learned in the last chapter, $E = K + U$, and $K_1 + U_1 = K_2 + U_2$, or $\Delta K = -\Delta U$.

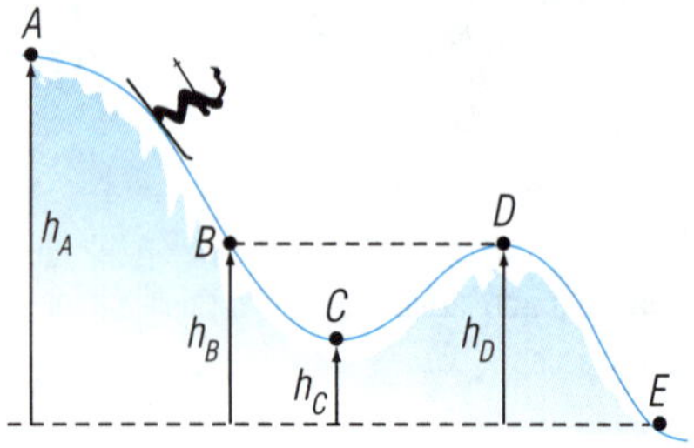

10-7 What is the skier's speed at point B?

EXAMPLE 10-1

The Ups and Downs of Skiing

The skier in Figure 10-7 begins at point A with a speed of 1.00 m/s. If h_A = 100.0 m and h_B = 50.0 m, what is the skier's speed at point B? His mass is 85.0 kg, and friction and air resistance are negligible.

Solution:

Assume that the height reference point is E. The strategy for this problem involves first finding the change in gravitational potential energy, ΔU_g. Knowing that $\Delta K = -\Delta U_g$ when no work is done on the skier-system by nonconservative forces, you can then solve for the final velocity using the definition of kinetic energy.

a. Find ΔU_g:

$$\Delta U_g = |mg|h_B - |mg|h_A = |mg|(h_B - h_A)$$

$$\Delta U_g = |(85.0\ \text{kg})(-9.81\ \text{m/s}^2)|(50.0\ \text{m} - 100.0\ \text{m})$$

$$\Delta U_g \cong -41\underline{6}90\ \text{J}$$

b. Find ΔK:

$$\Delta K = -\Delta U_g = -(-41\underline{6}90\ \text{J})$$

$$\Delta K = +41\underline{6}90\ \text{J}$$

c. Solve for v_B:

$$\Delta K = \tfrac{1}{2}mv_B^2 - \tfrac{1}{2}mv_A^2$$

$$\Delta K = \tfrac{1}{2}m(v_B^2 - v_A^2)$$

$$v_B^2 - v_A^2 = \frac{2(\Delta K)}{m}$$

$$v_B^2 = \frac{2(\Delta K)}{m} + v_A^2$$

$$v_B = \sqrt{\frac{2(\Delta K)}{m} + v_A^2} = \sqrt{\frac{2(+41\underline{6}90\ \text{J})}{85.0\ \text{kg}} + 1.00\ \text{m}^2/\text{s}^2}$$

$$v_B \cong 31.\underline{3}3\ \text{m/s} \cong 31.3\ \text{m/s}$$

The skier's speed at point B is about 31.3 m/s.

Suppose the skier in Figure 10-7 continues down the hill and over the hump. What happens to his energy? As he skis to point C, his potential energy changes to additional kinetic energy. Then at point C, he begins to go uphill. His kinetic energy decreases as his potential energy increases. At point D his potential energy is $|mg|h_D = |mg|h_B$. Since his total mechanical energy is conserved, his kinetic energy must be the same as it was at point B, and his speed is again v_B. As the skier continues over the hump and down the hill, his potential energy once again changes to kinetic energy. At the bottom of the hill, point E, his potential energy is zero, having changed completely to kinetic energy.

What is the effect of the hump on the skier's kinetic energy at E? Since his kinetic and potential energies are the same at D as at B, his speed, and thus his kinetic energy at E, is the same as it would be if there were no hump and he skied directly from A to B to E. This shows again that an object's energy change between two points is not affected by the path taken between those points.

10.6 Escape Speed

When total mechanical energy is conserved, potential energy can be changed to kinetic energy, and kinetic energy can be changed to potential energy. As an object falls from a shelf, its potential energy changes to kinetic energy. If it is tossed from the floor to the shelf, its kinetic energy changes to potential energy. Similarly, when a meteoroid is pulled from far away in space and impacts the earth's surface, its potential energy changes to kinetic energy; and when a spacecraft at the earth's surface is launched away from the earth, the kinetic energy delivered by its engines changes to potential energy. The **escape speed** is the minimum speed that an object of mass m requires to leave a larger object of mass M so that mass m cannot return due to gravitational attraction alone.

The **escape speed** is the minimum speed that an object of mass m requires to leave a larger object of mass M so that mass m cannot return due to gravitational attraction alone.

The zero-height reference point for the launch is the center of the earth. Assume that total mechanical energy does not change.

$$\Delta E = \Delta K + \Delta U_g = 0 \text{ J}$$

$$\Delta K = -\Delta U_g$$

Substitute the kinetic and gravitational potential energy formulas for both sides of the equation. Then solve for the escape speed, which is the theoretical initial velocity at the radius of the earth, v_R.

$$\tfrac{1}{2}mv_\infty^2 - \tfrac{1}{2}mv_R^2 = -G\frac{Mm}{R}$$

$$\tfrac{1}{2}\cancel{m}(v_\infty^2 - v_R^2) = -G\frac{M\cancel{m}}{R}$$

$$v_\infty^2 - v_R^2 = -2G\frac{M}{R}$$

$$-v_R^2 = -2G\frac{M}{R} - v_\infty^2$$

$$v_R^2 = 2G\frac{M}{R} + v_\infty^2$$

$$v_R = \sqrt{2G\frac{M}{R} + v_\infty^2}$$

In this case, M is the mass of the earth, m is the mass of the rocket, and R is the earth's radius. The minimum escape speed can be calculated when the speed of the spacecraft at an extreme distance is zero.

$$v_R = \sqrt{2G\frac{M}{R}} \qquad (10.1)$$

Problem-Solving Strategy 10.2

Escape speed from the surface of a planet does not depend on the mass of the launched object—it depends only on the mass and the radius of the planet.

10-8 Since the beginning of the space program, only a scant handful of spacecraft have attained sufficient speed to escape the gravitational tug of our sun. Shown here is an artist's rendering of one of the *Voyager* probes looking back from interstellar space toward the distant solar system with its sun.

Notice that the rocket's mass did not affect the calculation. The escape speed depends only upon the gravitational force of the planet and its radius. It is the same for all objects within that planet's gravitational field.

EXAMPLE 10-2

To Infinity . . . or Even Further: Escape Speed

A 1.00×10^4 kg rocket is launched at a speed of v_R from the surface of the earth, 6.38×10^6 m from the earth's center. It is intended to travel an infinite distance from the earth and arrive "there" with zero speed ($v_\infty = 0$ m/s). Assume that there is no air resistance. What is the smallest initial speed that will launch the rocket to achieve this objective?

Solution:

You know that the final speed is 0 m/s. Substitute the known values for *G*, *M*, and *R* to solve for initial speed.

$$v_R = \sqrt{2G\frac{M}{R}}$$

$$v_R = \sqrt{2(6.67 \times 10^{-11}\ \text{N}\cdot\text{m}^2/\text{kg}^2)(5.97 \times 10^{24}\ \text{kg})/6.38 \times 10^6\ \text{m}}$$

$$v_R \cong 11\underline{1}70\ \text{m/s} \cong 1.12 \times 10^4\ \text{m/s}$$

The rocket's minimum initial speed would have to be around 11.2 km/s!

10A Objectives

After completing this section, I can

- ✓ describe energy conservation in physics and provide examples of phenomena in which energy is conserved.
- ✓ state the conditions under which terminal velocity occurs.
- ✓ describe how friction can reduce or eliminate total mechanical energy in a system.
- ✓ define central forces and explain why they conserve total mechanical energy.
- ✓ solve problems that show that the action of conservative forces is path-independent.
- ✓ explain the derivation of the planetary escape speed formula and state the formula's assumptions.

10A Section Review

1. Why is the terminal velocity of a 1 g lead pellet higher than that of a 1 g piece of paper?
2. An energetic golden hamster (*Mesocricetus auratus*) is running at a constant speed on its frictionless exercise wheel (25.0 cm diameter, weighing 25.0 g). If the hamster is considered part of the exercise wheel system (his efforts are *internal* to the system), how much work is being done on the hamster-wheel system?
3. What is the main difference between a conservative force and a nonconservative one acting on a system?
4. A physics student removes his textbook from his locker, carries it to the classroom, lays it on the floor under his desk, picks it up at the end of class, and returns it to its place in his locker.
   a. What was the net change in gravitational potential energy on the book?
   b. What was the net work done by gravity on the book?
5. ✪ You drop a 100. g scoop of ice cream from a cone 1.20 m above the ground. Oops! Calculate the scoop's gravitational potential energy, its kinetic energy, and its total mechanical energy when it is 1.20 m, 0.80 m, 0.40 m, and 0.00 m from the ground.

6. A 6.80×10^4 kg space shuttle moving at 7.90×10^4 m/s reenters the earth's atmosphere 80.0 km above the earth's surface.

   a. What is the shuttle's total mechanical energy with respect to the earth's surface?

   b. It performs aerodynamic braking to slow sufficiently to land. Assuming landing speed is 346 km/h, how many joules of energy must be dissipated as thermal energy by drag with the atmosphere?

   c. If just 0.1% of the thermal energy generated by reentry is absorbed by the shuttle, what design feature keeps it from melting?

7. A 1.0 kg block sliding along a horizontal frictionless surface at 2.0 m/s encounters an ideal spring attached to a vertical wall. If the spring constant is 80. N/m, how far will the spring be compressed when the block is brought to rest?

8. Calculate the escape velocity from the earth's moon.

FACETS of PHYSICS

Conserving Energy

Every day we are bombarded by admonitions to "conserve energy." Appliances are proclaimed to be energy efficient, and power companies offer free advice on how to conserve energy. But, as we learned in this chapter, the energy put into a process is equal to the energy leaving the process—no matter what we do. Why, then, do we make a special effort to conserve it?

This seeming contradiction results from our definition of terms. Our statement that "energy into and out of a process is equal" means that the total amount of energy in the universe remains constant. However, not all energy is equally usable. When we use energy to do work in our homes or cars, it becomes less usable. For example, usable electrical energy can become lost thermal energy, such as what occurs in a light bulb.

Those who urge us to conserve energy really mean that we should try to keep energy in a usable form. While the total amount of energy in the universe is constant, the amount of usable energy always decreases. We can accelerate this decrease by using energy less efficiently, or we can slow it down by using energy wisely and more efficiently. Therefore, when you hear a plea to "conserve energy," remember that it simply means to use only as much energy as you need (Turn off those lights when you leave the room!), and when you use it, to use it efficiently.

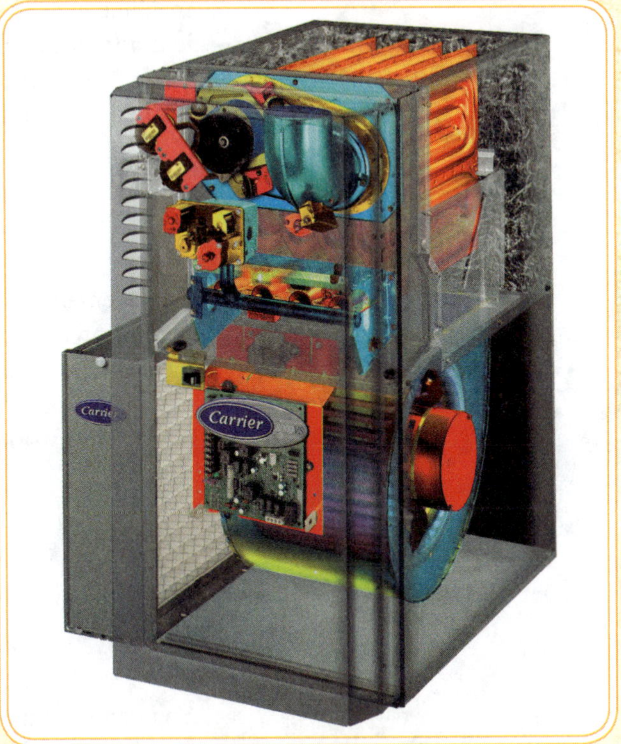

Some modern gas-fueled hot air furnaces are more than 96% efficient.

10B SIMPLE MACHINES

10.7 Why Machines?

A large stone rests in the middle of the front yard of your newly built house. Removing the stone will allow you to establish a beautiful level lawn. How will you remove it? You cannot simply lift such a large, sunken stone, even if you have several helpers. You may consider using dynamite to shatter the stone, but detonating explosives in a residential area is prohibited. You might think of hiring a crane to remove the rock, but that operation would be very expensive.

You would probably decide, finally, to remove the stone by using tools you already have or can borrow. To pull the rock out with a lawn tractor, you would first dig around it with a shovel and put a chain around it. On the side of the rock nearest the tractor, you could place boards to form a ramp under the stone to make it slide more easily from its bed. Opposite the ramp, neighbors with pry bars could loosen the rock as the tractor pulls. This strategy would probably remove the stone from its hole in your lawn.

10-9 Even complicated yard maintenance can be done with simple tools.

Machines and the Creation Mandate

Inventing machines is another way humans have obeyed the Creation Mandate. Machines better enable humans to exercise dominion over the world that God has placed under their stewardship.

Machines in physics are simple devices that change the magnitude or direction of a force.

In this illustration, several machines have been mentioned. Complex machines such as the crane and the tractor seem indispensable in modern life. However, simple machines, like the shovel, ramp, and pry bars, are not obsolete. Why are these called machines? A **machine** is a device that changes the magnitude or direction (or both) of an applied force. The force may be provided by a human or by a nonhuman source of energy like an electric motor. Although machines make some tasks easier by providing an advantage that would not otherwise be available, they do not create energy. In fact, because of friction, the amount of work a machine produces is less than the energy supplied to it.

10.8 Mechanical Advantage

Most simple machines amplify the effect of an applied force. How much does a machine amplify a force? To answer this question, we must define **ideal mechanical advantage (IMA)**. One definition of ideal mechanical advantage is the force the machine produces in the absence of friction, divided by the force that is applied to the machine. That is,

$$\text{IMA} = \frac{F_{\text{out}}}{F_{\text{in}}} \text{ (no friction).} \qquad (10.2)$$

Mechanical work is accomplished by a force component acting parallel to a displacement. In this discussion of simple machines, we will assume that input and output forces act in the same direction as the displacement of the machine where the force is exerted (e.g., $\theta = 0°$). We make this reasonable simplification to avoid having to resolve the forces into their components. We can then represent the mechanical work function as simply

$$W = Fd, \qquad (10.3)$$

since $\cos 0° = 1$.

For a real machine, the work done by the input force, W_{in}, equals the work done by the output force, W_{out}, plus the work done by friction, W_{f}:

$$W_{\text{in}} = W_{\text{out}} + W_{\text{f}}$$

You can see that friction reduces the output work in all real machines. However, for ideal (i.e., frictionless) machines, frictional work is zero and

$$W_{\text{in}} = W_{\text{out}} \text{ (no friction).} \qquad (10.4)$$

You can rewrite Equation 10.4 by substituting the definition of work from Equation 10.3,

$$W_{in} = F_{in}d_{in}, \text{ and } W_{out} = F_{out}d_{out}, \text{ so}$$

$$F_{in}d_{in} = F_{out}d_{out} \text{ (no friction).}$$

Rearranging,

$$\frac{F_{out}}{F_{in}} = \frac{d_{in}}{d_{out}} \text{ (no friction).} \quad (10.5)$$

The left side of Equation 10.4 is the definition of IMA; so

$$\text{IMA} = \frac{d_{in}}{d_{out}} \text{ (no friction).} \quad (10.6)$$

EXAMPLE 10-3

Moving Up: Ramps and Ideal Mechanical Advantage

A ramp is a simple machine called an inclined plane. A 250. kg piano must be moved to the second floor of an apartment building 3.00 m above the ground. The movers intend to use a 10.0 m frictionless ramp to move the piano. (a) What is the ideal mechanical advantage of the ramp? (b) What input force is needed to move the piano?

Solution:

When considering such a problem, you must ask "What am I basically trying to accomplish?" In this case, you are trying to move the weight of a massive 250. kg piano vertically from street level to an apartment 3.00 m up. These numbers are the "out" quantities in the IMA equations. The "in" quantities are the force exerted and distance covered by the workers.

a. Use Equation 10.6 to find the IMA:

$$\text{IMA} = \frac{d_{in}}{d_{out}} \text{ (no friction)}$$

$$\text{IMA} = \frac{10.0 \text{ m}}{3.00 \text{ m}}$$

$$\text{IMA} \cong 3.3\underline{3}\overline{3} \cong 3.33$$

Notice that IMA is dimensionless—the units always cancel out.

b. Now, solve for the input force or effort. Use Equation 10.2:

$$\text{IMA} = \frac{F_{out}}{F_{in}} \text{ (no friction).}$$

F_{out} is the weight of the piano. Rearrange to solve for F_{in}:

$$F_{in} = \frac{F_{out}}{\text{IMA}} = \frac{|mg|}{\text{IMA}} = \frac{|(250.\text{ kg})(-9.81\text{ m/s}^2)|}{3.3\underline{3}3}$$

$$F_{in} \cong 73\underline{5}.8 \text{ N} \cong 736 \text{ N}$$

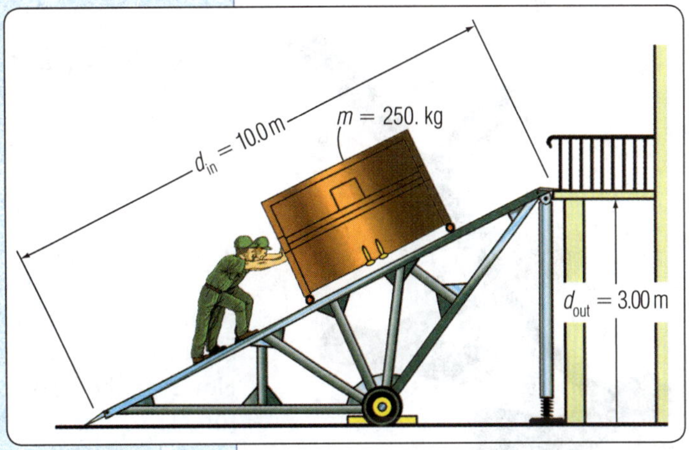

10-10 Simple machines make Herculean tasks easier.

Only 736 N is needed to push the piano up the ramp, compared to about 2450 N to lift the piano vertically. Using the ramp made the job much easier. However, notice that the piano traveled more than 3 m to reach the second floor. The movers "paid" for their easier job by moving the piano farther.

The word *ideal* in IMA implies that IMA is perfect and thus unrealistic. In real life, friction is always present. The **actual mechanical advantage (AMA)** of a machine is defined as the output force when friction is present, divided by the input force. The actual mechanical advantage is more easily measured than predicted.

We generally use a machine to make a job easier because the amount of force we either can or want to exert is less than the force needed to do the task without using the machine. Therefore, in most cases, we want the machine's IMA to be greater than 1. This means that the output force is greater than the input force and that, to accomplish the job, we have to move the load farther using the machine than we would if we moved the load directly. We will see how this works as we study various kinds of simple machines.

There are two basic classes of simple machines: inclined planes and levers. Although at first glance, shovels, pulleys, and so forth do not seem to fit into either category, closer examination shows that they do.

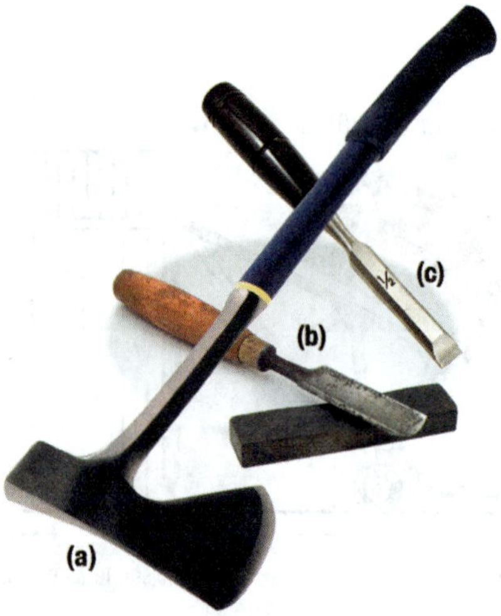

10-11 Types of inclined surfaces: (a) an axe (a wedge); (b) a gouge (a single inclined plane); (c) a chisel (also a single inclined plane)

10.9 Inclined Planes

Wedges, *screws*, and *ramps* are or contain **inclined planes**. Ramps were discussed above. A wedge is usually two inclined planes placed base-to-base. For example, an axe's cutting edge is a wedge. A wedge may also be a single inclined plane, such as a chisel or scalpel, that is sloped on only one side. Wedges are usually used as cutting or splitting tools.

A **screw** is a metal shaft surrounded by a helically coiled wedge. The wedge forms an inclined surface. It is much like a curving road going up a mountain. Few cars can produce enough force or traction to drive straight up a mountain. However, the slope of a winding road is smaller than that of a straight road; therefore, less force is needed to propel the car forward and up. Figure 10-12 shows the difference between the force needed to counter gravity on a straight road and the force needed on a winding road. Similarly, a screw uses a spiral inclined plane to make it easier to exert great force parallel to the shaft of the screw. With a climbing road, the closer the successive loops are, the less each loop slopes and the easier it is to move the car forward. Of course, the car must travel much farther

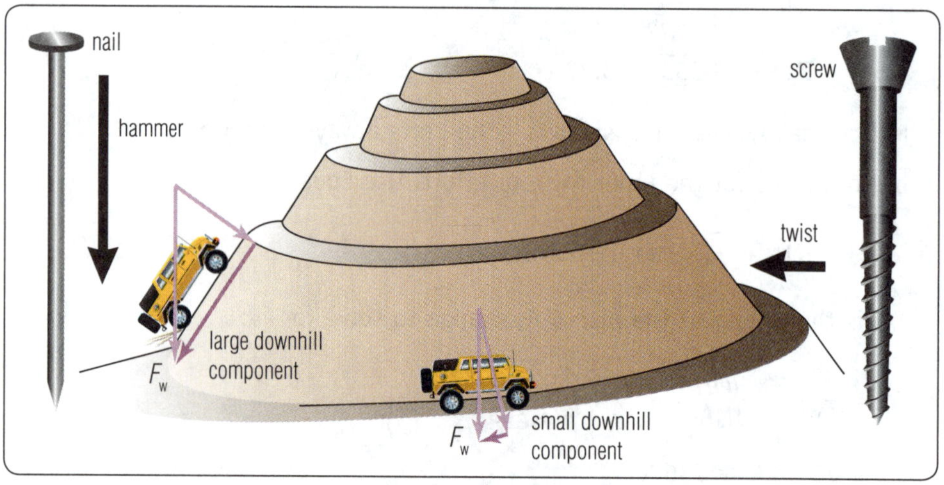

10-12 A screw contains a winding inclined plane. A winding road is easier to ascend than a road going straight up a mountain.

along the road to reach the mountaintop. Similarly, for a screw, the shorter the distance is between two successive threads (the **pitch**), the easier it is to turn the screw. However, a screw with a small pitch must be turned more times than a screw with a large pitch.

10.10 Levers

A **lever** is usually a rigid bar that turns around a pivot, the *fulcrum*. In use, an input or *effort force* (F_e) is applied at one end of the lever, called the *effort arm* (ℓ_e), and the output force (the *resistance* or *load*, F_r) is applied by the *resistance arm* (ℓ_r). You remember from Chapter 7 that a force applied perpendicular to a line passing through a pivot point is a torque. If the effort and resistance torques acting on the same pivot point are in equilibrium, according to the *law of moments* (see Section 7.13), the lever will be stationary and

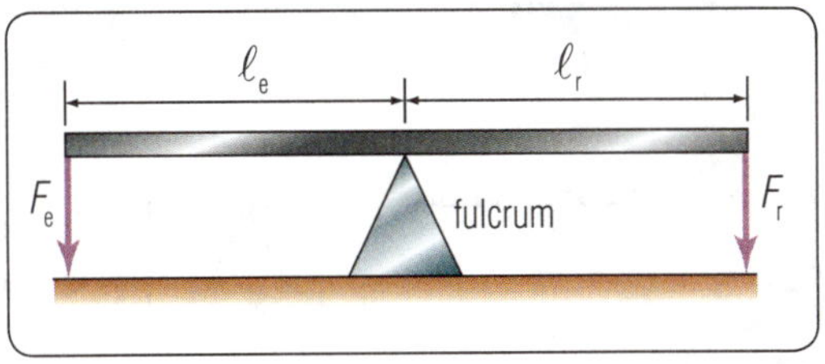

10-13 Parts of a simple lever system

$$F_e\ell_e = F_r\ell_r. \qquad (10.7)$$

When Equation 10.7 is true, there are no net torques on the lever and the lever just balances on the fulcrum. The ideal mechanical advantage of a lever system, then, is

$$\text{IMA} = \frac{\ell_e}{\ell_r},$$

while the actual mechanical advantage can be determined from the actual forces involved:

$$\text{AMA} = \frac{F_r}{F_e} \qquad (10.8)$$

Problem-Solving Strategy 10.3

Equation 10.8 can be used for determining the AMA of any system by substituting the actual effort and resistance forces into the equation.

There are three kinds of levers. In **first-class levers**, shown in Figure 10.14, the fulcrum is between the resistance and the effort forces. Some common first-class levers are seesaws, crowbars, and scissors. The IMA of a first-class lever varies depending on the relative lengths of the effort arm and the resistance arm. Since

The IMA of first-class levers can be greater than, less than, or equal to 1.

$$\text{IMA} = \frac{\ell_e}{\ell_r},$$

IMA > 1 if $\ell_e > \ell_r$; IMA $= 1$ if $\ell_e = \ell_r$; and IMA < 1 if $\ell_e < \ell_r$.

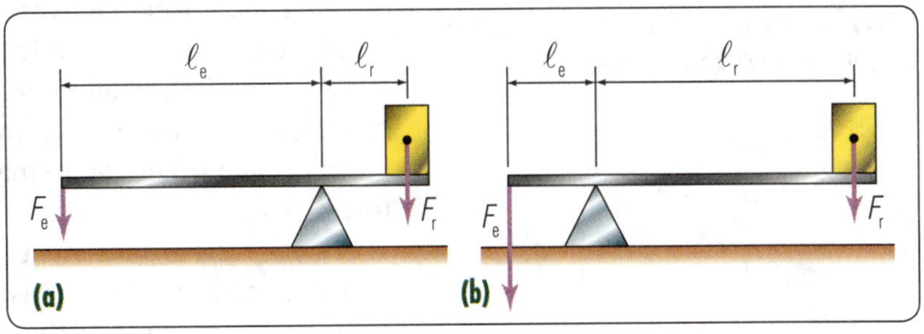

10-14 A first-class lever. (a) IMA > 1; (b) IMA < 1

It is generally desirable for a first-class lever IMA to be greater than 1. If the IMA were less than 1, you would have to exert more effort on the machine than it delivers to the resistance. To illustrate this, try cutting through a thick cord with a pair of scissors by placing the cord between the blades near the pivot point. Then try cutting the same cord using the tips of the blades. You must apply much greater effort to cut the cord using the tips, assuming you are able to do it at all.

In **second-class levers**, the resistance is between the fulcrum and the effort force. As you can see in Figure 10-15, the effort arm in a second-class lever system is *always* longer than the resistance arm, so the IMA is always greater than 1.

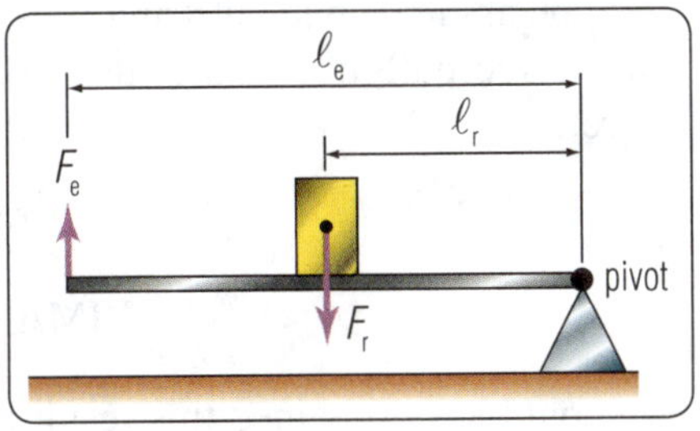

10-15 A second-class lever

The IMA of second-class levers is always greater than 1.

Some common second-class levers are hinged doors, wheeled tote bags, and nutcrackers.

Third-class levers always have an IMA less than 1.

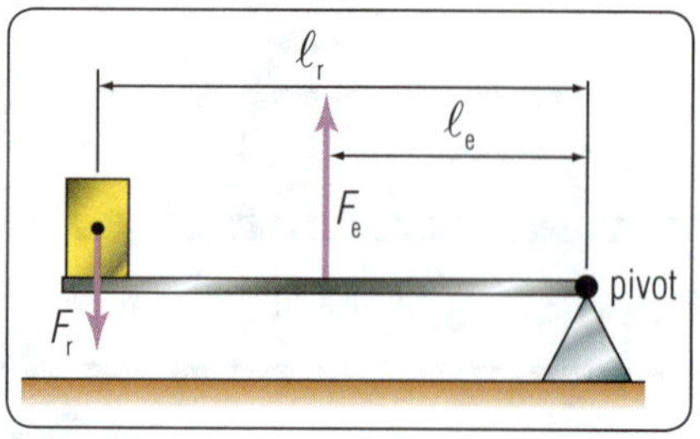

10-16 A third-class lever

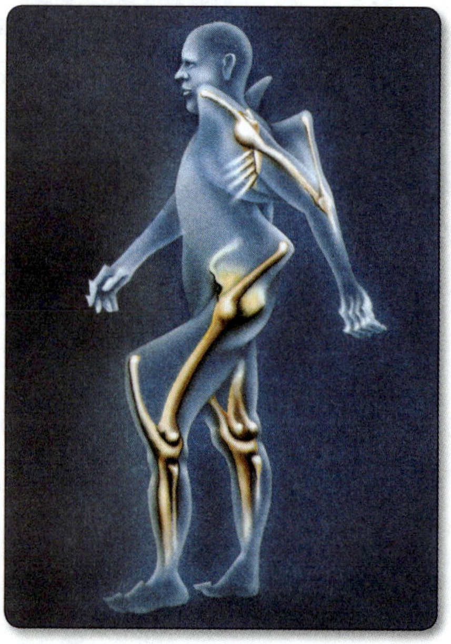

10-17 A human with first-class levers for appendages would be extremely ungainly and inefficient.

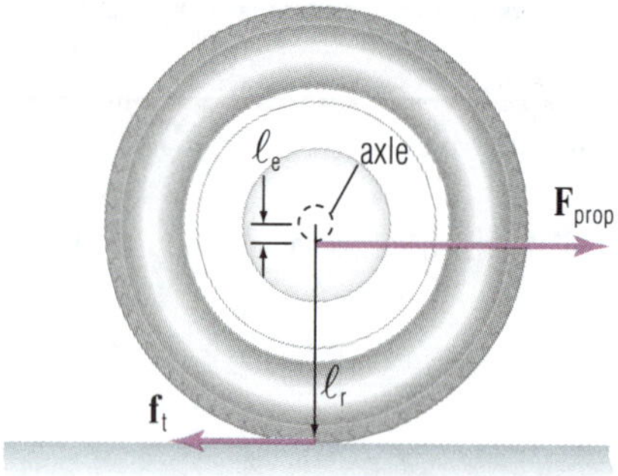

10-18 A vehicle's drive wheels act as a form of third-class lever.

A **pulley** is a modified first-class lever that can reposition itself indefinitely.

In **third-class levers**, the effort force is between the fulcrum and the resistance. As usual,

$$\text{IMA} = \frac{\ell_e}{\ell_r},$$

but here ℓ_e is always *less* than ℓ_r. Therefore, the IMA is always less than 1. A third-class lever diminishes input force. What good is it then?

A third-class lever moves a resistance a large distance, while the effort moves only a small distance. Brooms and tweezers are examples of common third-class levers. But the most common and important application was employed by our Creator in the appendages of every creature with a rigid endo- or exoskeleton. Nearly every major skeletal joint in your body is the fulcrum of a third-class lever. There is a simple biomechanical reason for this design. Muscles develop force at the subcellular level, and this force is applied to our limbs and appendages. Even with millions of cells contracting in unison, the total effort distance a muscle can generate is fairly small. In order to attain maximum possible movement, vertebrates and arthropods were designed with third-class levers in arms, legs, and nearly every other place where a muscle is attached to a rigid support.

10.11 Wheels, Pulleys, and Pulley Systems

A lever has one major limitation: it can move only so far while doing its job. At some point, the effort arm or the resistance arm can no longer move, either because of an obstruction or because the associated force can no longer be applied effectively. It would be ideal if the lever arm could be continually repositioned in order to keep the effort and resistance forces applied. This is exactly what wheels and pulleys do.

Wheels are modified levers. The fulcrum of a wheel-lever is the axle. A force applied at the edge of a steering wheel turns a load near its center. This is a second-class lever, and it makes the load easier to turn. Larger steering wheels, which have longer lever arms (radii), are easier to turn than smaller wheels. For an automobile drive wheel, a force delivered by the engine near the center of the wheel turns it against the force of static friction between the wheel and the road (the resistance force). This is yet another example of a third-class lever. It enables a force moving a short distance to move a load a large distance. Larger diameter wheels move a car farther than smaller wheels for the same angular distance that the force moves.

A *pulley* is a grooved wheel that turns on an axle. You learned how pulleys function in Chapter 8. The radius of the wheel is fixed, of course, but the radius forms both the effort arm and the resistance arm of this special lever. Effort and resistance forces are applied by a rope or string that lies in the groove as it passes over the pulley. The string enters and leaves the pulley tangentially, so the forces are perpendicular to the radii of the pulley. The pulley axle at the center of the radii acts as the fulcrum. Thus, a pulley is a kind of first-class lever.

If you assume an ideal string passes through the pulley, then for every meter the effort force pulls the string, the string also pulls the resistance load one meter. Therefore,

$$d_{in} = d_{out}, \text{ and}$$

$$\text{IMA} = \frac{d_{in}}{d_{out}} = 1.$$

You could have obtained this result by taking the ratio of the lengths of the effort and the resistance arms. Since they are the radii of the same circle, the IMA is 1.

Pulleys can be used in one of two ways. When a pulley is attached to a rigid support, it is called a *fixed pulley*. The advantage of a fixed pulley is that it changes the direction of an applied force. The applied force produces tension in the string that passes undiminished through the pulley. It is often easier to pull down on a rope, putting your weight into it, than to pull up on the load directly. Also with pulleys, you are not limited by your reach when lifting a load to a greater height.

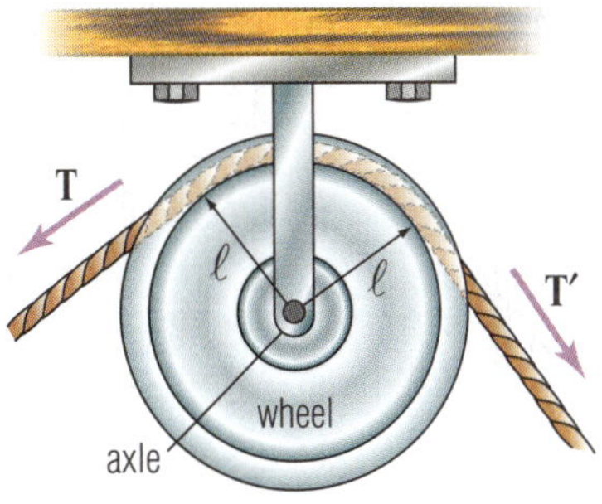

10-19 A simple fixed pulley

For a single fixed pulley, IMA = 1. The main advantage of a fixed pulley is that it changes the direction of the force in a string or rope, thereby allowing the user to pull downward and use his weight to lift an object.

10-20 A mining scoop shovel uses a fixed pulley to redirect its force.

A *movable pulley* works differently. The pulley is directly attached to the load that exerts the resistance force. One end of the string or rope is attached to a rigid support, and a force is applied to the other end. Figure 10-21b shows a free-body diagram for the pulley-load system in the absence of friction. The rope is assumed to be ideal. If all forces on the system in the diagram are in equilibrium, then

$$\Sigma \mathbf{F} = \mathbf{T} + \mathbf{T}' + m\mathbf{g} = 0 \text{ N}.$$

You can rewrite this expression using vertical scalar components. Since $\mathbf{T} = \mathbf{T}'$,

$$T_y + T_y = 2T_y = -mg_y$$

$$T_y = -\frac{mg_y}{2}.$$

In other words, each rope carries half the weight of the load. In Figure 10-21b, the resistance force (F_r) is the weight of the chest. The man holds only one end of the rope, so only one of the tension forces is the effort force, $T' = F_e$. Comparing the magnitudes of the forces, you find that

$$F_r = 2T = 2F_e.$$

The ideal mechanical advantage of the movable pulley is

$$\text{IMA} = \frac{F_r}{F_e} = \frac{2\cancel{F_e}}{\cancel{F_e}} = 2.$$

You can see then that a movable pulley doubles the effort force. However, it also doubles the distance over which the force must be applied. If the man in Figure 10-21a pulls the rope 1 m, the pulley and storage chest move only ½ m.

A system of fixed and movable pulleys is called a **block and tackle system**. Figure 10-22a shows a block and tackle system with two fixed and two movable pulleys. The effort force equals the uniform tension in the continuous rope that runs through the ideal block and tackle system. The upper fixed pulleys change only the

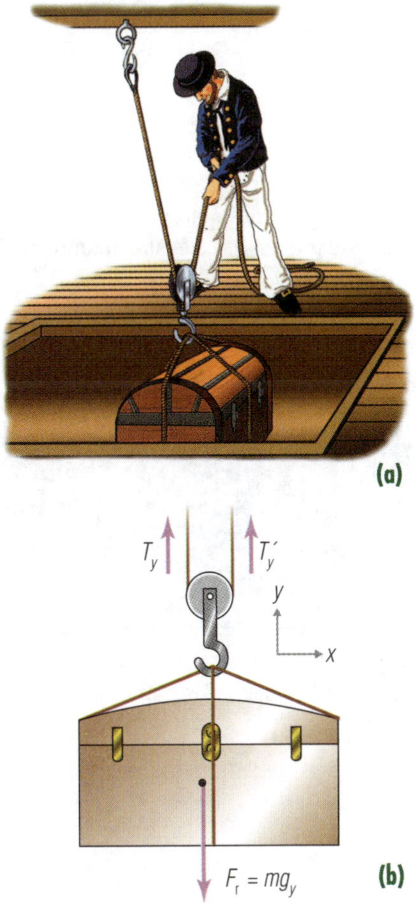

10-21 (a) Using a single movable pulley; (b) Free-body diagram of the chest

A movable pulley is attached to the load to be moved. Movable pulleys individually have an IMA = 2.

direction of the tensional force. The lower movable pulleys, as Figure 10-22b shows, double the effort force with each pulley. Let's see how this happens.

The net force on the load is the sum of its weight and the four tensional forces in the ropes running through the pulleys attached to the load. If these forces are in equilibrium, then

$$\Sigma \mathbf{F} = m\mathbf{g} + \mathbf{T} + \mathbf{T} + \mathbf{T} + \mathbf{T} = 0 \text{ N}.$$

Once again, the resistance force is the weight of the load and the effort force is just the tension in the rope. Therefore,

$$4\mathbf{T} = -m\mathbf{g}$$

$$4T = |mg|$$

$$4F_e = F_r.$$

The IMA is

$$\text{IMA} = \frac{F_r}{F_e} = \frac{4\cancel{F_e}}{\cancel{F_e}} = 4.$$

Notice that with four ropes pulling up on the load, the IMA is four; with two ropes pulling on the load, as for the movable pulley, the IMA is two; and with one rope pulling on the load, as for the fixed pulley, the IMA is one. This suggests that the ideal mechanical advantage of a pulley system is equal to the number of ropes supporting the load. This is true for any arrangement of pulleys.

10.12 Using Simple Machines to Solve Problems

In 1852, Elisha Otis solved the problem of developing a safe elevator. He worked for a company in New York City that manufactured beds in a multi-story building, a good place for an elevator. After working to develop a safe and reliable elevator for his employer, he unveiled his design for an elevator safety mechanism at the New York World's Fair. He staged a demonstration at the Crystal Palace in which he rode his elevator to a dizzying height and then had an assistant cut the supporting cable with an axe. The watching crowd gasped and then applauded as the safety mechanism activated and the elevator platform locked into place. Otis's invention ushered in the era of accessible skyscrapers.

Modern elevators depend on simple machines. In the standard traction elevator, one that is raised and lowered by cables, the cables are attached at one end to the passenger cab and at the other end to a counterweight. This system of cables runs up and over *sheaves*—essentially fixed pulleys—attached to an electric motor, usually through a series of gears. As the sheaves rotate, friction moves the cables.

If the supporting cables or lift mechanism fails, the passenger cab normally would plummet to the bottom of the elevator shaft, with potentially fatal results. However, a modern elevator has multiple safety systems, which also depend on simple machines, to prevent this from happening. Traction elevators are lifted by multiple, very strong cables, each able to support the elevator's weight on its own in case one or even several cables break. The failure of the lift motor presents a different kind of problem. A separate cable attached to the elevator cab runs through pulleys to a centrifugal governor installed in the elevator shaft mechanism. Within the governor are two flyweights. If the lifting motor fails or the main cables somehow become detached, the descending cab causes the governor to spin fast enough to throw out the flyweights by centrifugal force until they catch the ratchets in the outer casing of the governor. The governor engages a rack-and-pinion pull rod that pulls on cables that engage a friction brake on the cab, which brings it to a stop. Some elevators may also use an electromagnetic induction braking system. If all else fails, a shock absorber system at the bottom of the elevator shaft cushions the elevator's impact. Each of these mechanisms has a single purpose: to preserve human life.

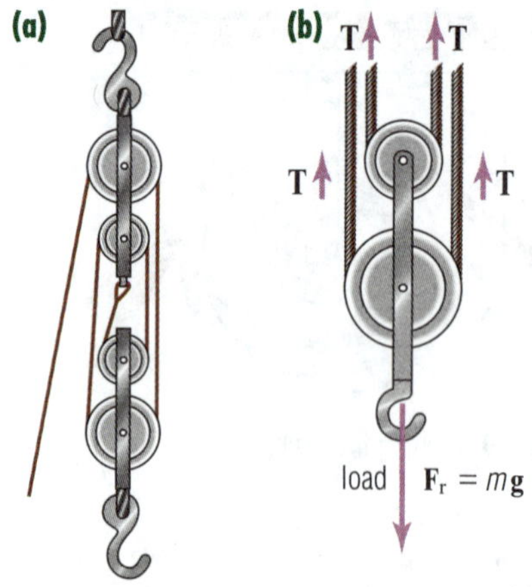

10-22 Block and tackle systems are used to multiply an available lifting force many times.

Problem-Solving Strategy 10.4

The IMA of a block and tackle system is equal to the number of ropes physically supporting the load or resistance force.

Elisha Graves Otis (1811–61), an American master mechanic, invented the elevator safety brake. He and his sons started a company that is the largest elevator manufacturer in the world today.

10-23 A modern elevator shaft showing the numerous cables associated with the lift and safety systems

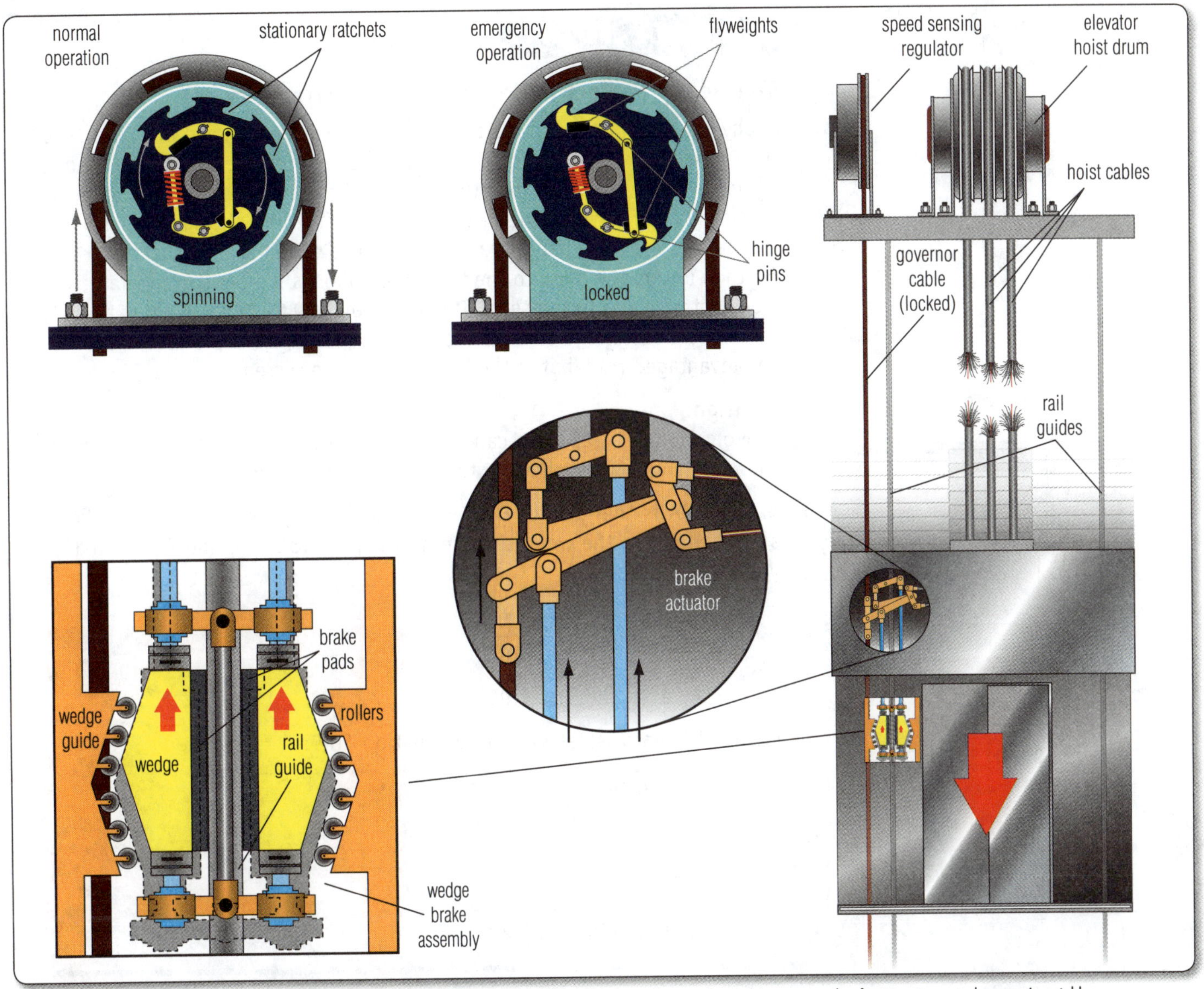

10-24 A modern elevator safety mechanism and its function. The development of the elevator safety mechanism is an example of using science to love one's neighbor.

10.13 Mechanical Efficiency

You saw earlier that an ideal simple machine conserves energy. No nonconservative forces act within an ideal simple machine. The work put into the machine equals the work output of the machine. No energy is lost. Real machines do not work this way. Friction between moving parts causes wear. Edges become dull. Axles squeak. Ropes stretch nonelastically. Parts flex and give off heat. The work output of a real machine is never equal to the work put in. It's always less.

Engineers (and consumers) are interested in getting as much energy out of a process as possible compared to the energy available at the beginning of the process. Since energy translates into work, they compare the output work to the work or energy put into a process to determine the **efficiency (η)** of the process. The efficiency is most easily determined by finding the actual mechanical advantage for the process, then dividing the AMA by the IMA.

$$\eta = \frac{\text{AMA}}{\text{IMA}} \times 100\% \qquad (10.9)$$

We will consider other methods for determining efficiency in later chapters.

EXAMPLE 10-4

The Nearly Perfect Pulley: Mechanical Efficiency

A physics student sets up the block and tackle system shown in Figure 10-25 and attaches a 10 N spring scale to the free end of the string. The load for his experiment is a 1 kg laboratory mass that is hooked to the bottom pulley. He slowly pulls down on the free end of the string until the mass is suspended by the tackle and is hanging motionless. The spring scale indicates 3.30 N. He then lowers the mass to the tabletop, unhooks it from the apparatus, and weighs it with the spring scale. The mass weighs 9.82 N. (a) What is the ideal mechanical advantage of the pulley system? (b) What is the actual mechanical advantage? (c) What is the efficiency of the system?

Solution:

a. From Figure 10-25, you can see that there are three strings supporting the mass and lower pulley, so the IMA for the system is 3.

b. The AMA is the ratio of F_r to F_e. The effort force is $F_e = 3.30$ N. The resistance force is the weight of the mass, $F_r = 9.82$ N. Using Equation 10.8, you can find the AMA:

$$\text{AMA} = \frac{F_r}{F_e} = \frac{9.82 \text{ N}}{3.30 \text{ N}}$$

$$\text{AMA} \cong 2.9\underline{7}5 \cong 2.98$$

c. The efficiency is calculated by Equation 10.9:

$$\eta = \frac{\text{AMA}}{\text{IMA}} \times 100\%$$

$$\eta = \frac{2.9\underline{7}5}{3} \times 100\%$$

$$\eta \cong 99.\underline{1}6\% \cong 99.2\%$$

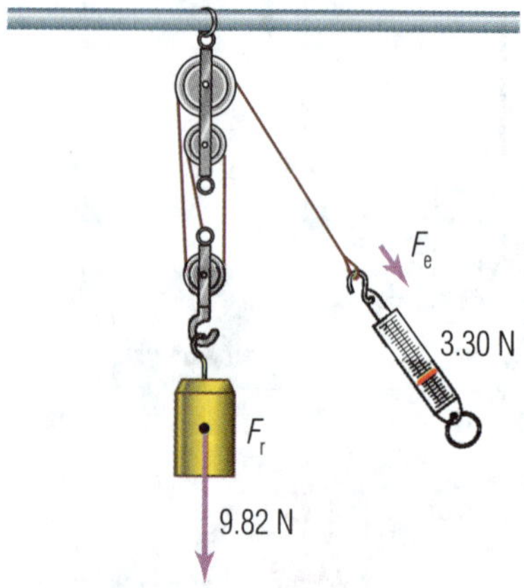

10-25 A block and tackle system with three supporting strings

Stationary pulley systems are nearly 100% efficient. Their efficiency decreases when the string runs through the pulleys because of the kinetic friction occurring at many points in the system.

Consumers are very interested in the efficiency of the products they buy. Fuel efficiency in cars and furnaces, electrical efficiency in air conditioners and laptop computers, and well-designed hand tools that provide maximum force with minimum effort are all examples of product efficiency. The federal government often mandates that manufacturers prominently post energy efficiency ratings on products prior to their sale.

10B Objectives

After completing this section, I can

- ✓ state the purpose of a simple machine.
- ✓ identify the parameters necessary to calculate the mechanical advantage of a simple machine.
- ✓ determine the mechanical advantages of inclined planes, levers, wheels, and pulley systems.
- ✓ recognize the three different kinds of lever systems.
- ✓ compute the efficiency of a simple machine.

10B Section Review

1. What force in all its forms do you ignore when determining ideal mechanical advantage?
2. How does the operator of a simple machine that magnifies force "pay" for this mechanical advantage?
3. DS Does the electric lift motor for a real elevator do more, less, or the same amount of work as would be necessary to move the elevator in the absence of friction? Explain your answer.

DS4. What is the ideal mechanical advantage of a lift cable sheave in a traction elevator?

5. What quantity would you calculate to determine the overall effect of friction and other nonconservative forces on a process?

✪6. A frictionless ramp forms a 20.0° angle with the horizontal. A 5.00 kg load rests on the ramp. Calculate the magnitude of the force that will hold the load stationary on the ramp, and then determine the ideal mechanical advantage of the ramp using this information.

✪7. A 75.0 kg worker must raise a 2.40×10^3 N shipping crate a few centimeters in order to remove an electrical extension cord that has become trapped under the crate. If he intends to use a 2.00 m pry bar and his entire weight to lift the crate, what is the maximum distance from the crate the fulcrum can be placed that will allow him to just barely move the crate? Assume that half the crate's weight bears on the crowbar.

✪8. What length of rope must a sailor pull in order to lift a boat 1.80 m from the ship's deck with the block and tackle shown in the diagram in the margin?

✪9. A block and tackle pulley system is 96.5% efficient when a load is being lifted. The effort force is 22.6 N. If the ideal mechanical advantage of the system is 6, how heavy is the load?

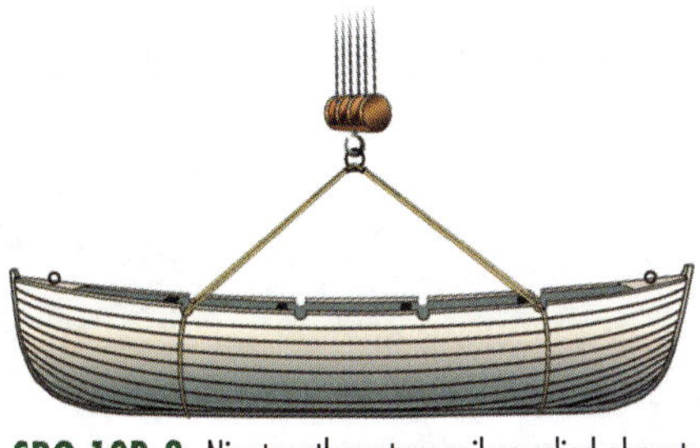

SRQ 10B-8 Nineteenth-century sailors relied almost exclusively on block and tackle rigging to operate their ships.

Chapter Review

In Terms of Physics

conservation law	219	screw	228
terminal velocity	220	pitch	229
central force	221	lever	229
escape speed	223	first-class lever	229
machine	226	second-class lever	229
ideal mechanical advantage (IMA)	226	third-class lever	230
actual mechanical advantage (AMA)	228	block and tackle system	231
inclined plane	228	efficiency (η)	233

Problem-Solving Strategies

10.1 (page 222) You can simplify many problems by using the conservation of total mechanical energy. As you learned in the last chapter, $E = K + U$, and $K_1 + U_1 = K_2 + U_2$, or $\Delta K = -\Delta U$.

10.2 (page 224) Escape speed from the surface of a planet does not depend on the mass of the launched object—it depends only on the mass and the radius of the planet.

10.3 (page 229) Equation 10.8 can be used for determining the AMA of any system by substituting the actual effort and resistance forces into the equation.

10.4 (page 232) The IMA of a block and tackle system is equal to the number of ropes physically supporting the load or resistance force.

Review Questions

1. What is a conservation law? How does the existence of conservation laws give evidence for a complete Creation?
2. How can you tell that nonconservative forces are working in a mechanical process?
3. A feather falls through the air. Is its total mechanical energy conserved? Why or why not?
4. An astronaut throws a rock upward while standing on the airless moon. After the rock leaves the astronaut's hand, is the rock's total mechanical energy conserved?
5. A speeding car screeches to a stop. What happened to its kinetic energy?

6. A mass is attached to a hanging spring that is located far from any source of thermal energy. The temperature of the spring increases as it oscillates. What happens to its total mechanical energy?

7. How much work is done by gravity on an airliner flying level at 30 000 ft between London and New York?

8. Three balls with identical masses are released from rest at the same height, *h*, as shown in the figure, and frictional forces are negligible. (Assume the balls slide and do not roll.)
   a. Which ball will be moving fastest at ground level: *A*, *B*, or *C*?
   b. Which will strike the ground first?

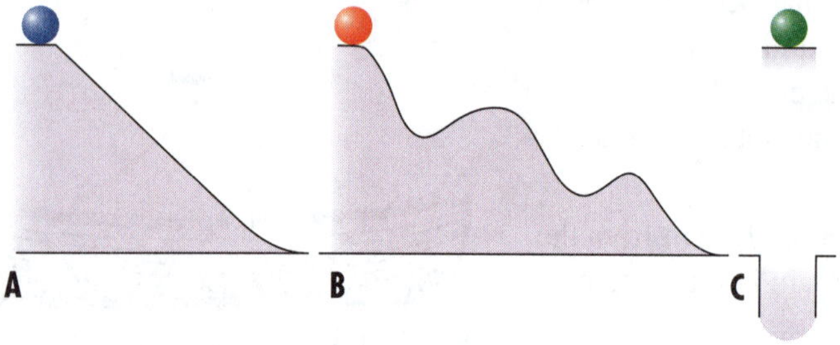

9. Explain why the escape speed from the earth's gravity for a baseball is the same as for an interplanetary rocket.

10. What is the principal difference between a simple machine and a complex machine?

11. Define *ideal mechanical advantage* (IMA). Why is ideal mechanical advantage not attainable in real systems?

12. Describe at least three different mathematical ratios that will yield the IMA for simple machines. Define each quantity the first time you use it in your answer.

13. Which one of the following is *not* a simple machine?
   a. bolt
   b. automobile engine
   c. broom
   d. baseball bat
   e. knife
   f. scissors
   g. stairway
   h. crowbar

14. What is the actual mechanical advantage of a machine whose output force is double its input force?

15. Why is it easier to measure actual mechanical advantage than it is to predict it?

16. Name the two basic kinds of simple machines.

17. Which one of the following does *not* contain an inclined plane?
   a. knife
   b. axe
   c. mountain road
   d. wood plane
   e. broom
   f. screw

18. Which has a greater mechanical advantage, a second-class lever or a third-class lever?

19. What is the ideal mechanical advantage of a movable pulley?

20. What is the ideal mechanical advantage of the block and tackle pictured here?

21. Describe the most direct way to calculate the mechanical efficiency of a machine.

True or False (22–31)

22. Gravity is a conservative force.

23. Astronauts beam to the earth TV images of spinning objects floating in the habitable compartments of a space station. Energy is conserved by spinning objects under these weightless conditions.

24. Some thermal energy is transferred to a falling raindrop due to its change of position.

25. Static friction can be both a conservative force and a nonconservative force.

26. The presence of an atmosphere increases the required escape speed of a vehicle launched from a planet's surface.

27. Machines with lower mechanical advantages require more work to do a certain job.

28. A wheel is a modified lever.

29. A screw with a large pitch has a small mechanical advantage.

30. First-class levers always have an ideal mechanical advantage of 1 or more.

31. The ideal mechanical advantage for any conceivable block and tackle system is equal to the number of ropes or strings physically supporting the load, not including the one exerting the effort force.

✪32. A 3.00 kg ball is thrown with a velocity of 10.0 m/s upward. If it is released at a height of 1.60 m, to what height will it rise before it stops?
   a. Solve the problem using the principles of the conservation of energy.
   b. Solve the problem using the equations of motion.

✪33. A frictionless surface is inclined 35.0° to a horizontal surface. A 1.40 kg ball rests on the incline at a height of 1.00 m above the horizontal surface. An identical ball is suspended 1.00 m above the horizontal surface. If both balls are released at the same instant, how long after release will each ball strike the horizontal surface?

✪34. A 1250 kg chunk of solid iron is all that is left of a speeding asteroid after it has skimmed the surface layers of the sun. It has sufficient radial speed to reach a theoretically infinite position from the sun.
   a. How much does the asteroid's gravitational potential energy change?
   b. If its radial speed at the sun is v_1 and its speed is 0 m/s at an infinite distance, what is v_1?

✪35. A 2.00 g penny is dropped from the top of a 110. m building. How far has the penny fallen when its speed reaches 45.0 m/s?

✪36. A ramp allows a small force exerted over 10 m to raise a load 1 m vertically. What is the ramp's ideal mechanical advantage?

✪37. A sailor exerts 330. N of force to raise a sail at a constant speed using a single fixed pulley. The sail and rope weigh 275 N.
   a. What is the ideal mechanical advantage of the pulley?
   b. What is the actual mechanical advantage?
   c. What is the efficiency of the pulley?
   d. What would you suggest could be done to improve the efficiency of the pulley?

✪38. A machine bolt requires 30 complete turns in order to thread it into a hole 6.00 cm deep. What is the pitch of the bolt?

DS✪39. Calculate the minimum work required for the electric lift motor of an elevator in the Sears Tower to lift a 2250 kg elevator cab 412 m to the building's Skydeck at an average speed of 8.0 m/s. What is the total mechanical energy of the elevator cab when it is halfway to the top? Calculate the average power produced by the electric motor if the lift to the Skydeck takes 45.0 s.

✪**40.** A 785 kg piano must be lifted 4.50 m to a performance stage in a large auditorium.

a. How much work must be done to move the piano?

b. If the piano is moved up a frictionless incline to the stage, how much work is done?

c. A block and tackle is used instead, and an effort force of 2670 N is applied to lift the piano at a constant speed. If 13.5 m of rope is pulled through the block and tackle system, what is its efficiency?

✪**41.** A spool is a cylinder mounted on a shaft around which a rope or string is wrapped multiple times. Two spools are mounted on the same shaft and turn as a single unit. One spool has a diameter of 0.300 m, while the smaller spool's diameter is 0.200 m. A string wrapped around the larger spool is tied to a 1.50 kg load. You pull on the string wrapped around the smaller spool, rotating the shaft and lifting the load attached to the larger spool. How much force must you exert on the string in order to lift the load at a constant speed? Ignore friction in the axle.

When landing, the space shuttle possesses a large amount of momentum, which has to be eliminated using wheel brakes and a drag chute. This landing rollout took more than a mile, while some took as many as two miles.

Momentum 11

Dominion Science Problem

Car Accident Injuries and Fatalities

On the way to school, traffic slows to a crawl. There's an accident ahead. Oh, no! Now you'll be late! We can be somewhat exasperated with the inconveniences that car accidents create for us, but we should be even more concerned about the injuries and fatalities they cause. As long as there are cars, there will be accidents. In the United States alone, there were almost 6.5 million car accidents in 2005, amounting to total losses of over $230 billion, injuring 2.9 million people, and killing over 40 000 people. What can engineers and scientists do to improve car design and built-in safety features to prevent fatalities and lessen the severity of injuries caused by car accidents?

11-1 Car accidents are often scenes of fatalities and serious injuries.

11A PRINCIPLES OF MOMENTUM

11.1 Introduction

"Seven . . . Six . . . Five . . . We have go for main engine start . . . Two . . . One . . . We have liftoff!" These words, spoken on April 12, 1981, marked the beginning of the first flight of the space shuttle *Columbia*. Thousands cheered as the *Columbia* blasted into space. Although the space shuttle was called a marvel of modern technology, it used the same principle as its predecessors, the early rockets, to get into space. That same principle also applies to such activities as playing marbles, target riflery, and bowling. The principle is the *conservation of momentum.*

11.2 Definition of Momentum

Momentum (p) is a measure of an object's inertia of motion. Isaac Newton called this property of an object its "quantity of motion." For example, an object with large momentum is harder to stop than an object with small momentum. What physical properties do you think affect momentum?

The first aspect of momentum should be obvious after a little thought. Which is easier to stop—a 10 000 kg truck moving at 50 km/h or a 0.05 kg tennis ball moving at the same speed? You can stop the tennis ball with one hand, but your whole body would not stop the truck! It seems clear that the mass of an object affects its momentum.

Which is easier to stop—a 0.01 kg bullet moving at 1000 m/s or the same bullet moving at a speed of 1 m/s? You can safely catch the slower bullet in your hand,

The unit of momentum is kg·m/s. It has no special derived unit in the SI.

Linear momentum of an object with mass m and velocity $\mathbf{v}$ is defined by the product $\mathbf{p} = m\mathbf{v}$. Momentum is a vector quantity and is always in the same direction as the object's velocity vector.

but the faster bullet would cause great damage before it stops. So, an object's velocity also affects its momentum.

It is not surprising, then, that momentum is defined as the product of mass and velocity:

$$\mathbf{p} = m\mathbf{v} \tag{11.1}$$

Linear momentum is a vector in the same direction as the velocity vector. In the SI, mass is measured in kilograms, and velocity is measured in meters per second. Consequently, the unit of momentum is kg·m/s. The unit for momentum does not have a special name.

EXAMPLE 11-1

11-2 Whose momentum is greater?

Law of Gross Tonnage: Momentum

Which has greater momentum, an 85.0 kg halfback running south at 9.00 m/s or a 135 kg defensive lineman running at 6.00 m/s in the same direction?

Solution:

Calculate the magnitude of momenta for each athlete using the scalar values in Equation 11.1:

$$p_x = mv_x$$

Halfback:

$$p_{\text{hb}} = (85.0\ \text{kg})(9.00\ \text{m/s})$$

$$p_{\text{hb}} = 765\ \text{kg·m/s}$$

Lineman:

$$p_{\text{dl}} = (135\ \text{kg})(6.00\ \text{m/s})$$

$$p_{\text{dl}} = 810.\ \text{kg·m/s}$$

The defensive lineman has greater momentum than the halfback.

11.3 Changing Momentum: Constant Mass

Since an object's mass is usually constant, a change in momentum is almost always a change in velocity—that is, an acceleration. Therefore,

$$\Delta\mathbf{p} = \Delta(m\mathbf{v}), \text{ and}$$

$$\Delta\mathbf{p} = m\Delta\mathbf{v}. \tag{11.2}$$

Dividing both sides by the time for the change (Δt) gives

$$\frac{\Delta\mathbf{p}}{\Delta t} = m\frac{\Delta\mathbf{v}}{\Delta t}. \text{ But}$$

$$\frac{\Delta\mathbf{v}}{\Delta t} = \bar{\mathbf{a}}, \text{ so}$$

$$\frac{\Delta\mathbf{p}}{\Delta t} = m\bar{\mathbf{a}}.$$

Problem-Solving Strategy 11.1

You will be introduced to many equations in this chapter, but you do not need to memorize them. So many conditions apply to using most of the equations that you probably could derive them from the principles you will learn before you could recall them or find them in your notes. Physics is not just about memorizing equations.

Remember, Newton's second law is

$$\Sigma \mathbf{F} = m\mathbf{a}, \text{ so}$$

$$\Sigma \mathbf{F} = \frac{\Delta \mathbf{p}}{\Delta t}. \tag{11.3}$$

In fact, Equation 11.3 was how Newton originally stated his second law. A change in momentum results from an unbalanced force.

The sum of the unbalanced external forces on a system is equal to the time-rate of change of the momentum of the system.

11.4 Impulse

An unbalanced force on an object changes its momentum. Does it make any difference how long the force acts? Consider a train starting from rest at a station. If its engine exerts a force to turn its wheels for 1 second, will its speed change significantly? Probably not. However, if its engine exerts a force for 1 minute, its speed, and therefore its momentum, will change substantially.

By rearranging Equation 11.3, it is easy to see the effect that the time has on the change in momentum:

$$\Delta \mathbf{p} = \Sigma \mathbf{F} \Delta t \tag{11.4}$$

The longer the force is applied (Δt), the more the momentum changes. The product of a force and the time the force is exerted is called the **impulse (I)** of the force.

$$\mathbf{I} = \Sigma \mathbf{F} \Delta t \tag{11.5}$$

By substituting Equation 11.5 into Equation 11.4, you can see that

$$\Delta \mathbf{p} = \mathbf{I}. \tag{11.6}$$

Although any force can be an impulse, we use the term "impulsive force" to refer to the forces on a system that are large compared to the other forces. When a ball is dropped to the ground, gravity is the impulsive force. At the instant that a bat hits a ball, however, the force the bat exerts on the ball is the impulsive force, since it is large compared to the force of gravity.

The following example shows the importance of impulse.

The change of momentum of a system is directly proportional to the force sum acting on it and the time interval during which the force was imposed.

The impulse of a net force on a system is equal to the change in momentum of the system.

$$\mathbf{I} = \Delta \mathbf{p}$$

The magnitude and direction of the impulse vector is the same as the momentum difference vector.

EXAMPLE 11-2

Landing Flat-Footed: Impulse

A 50.0 kg woman jumps to the ground from a wall that is 2.00 m high. At the instant she hits the ground, her velocity is 6.26 m/s downward. After she lands, her velocity is zero. (a) What is the change in her momentum? (b) If she hits the ground flat-footed with her knees locked, taking 0.100 s to stop, what is the force on her feet? (c) If she flexes her knees and lands gradually, taking 0.800 s to stop, what is the force on her feet?

Solution:

Assume all motion takes place in the vertical. Use the *y*-scalar components for the calculations.

a. Calculate her change in momentum. Use Equation 11.2:

$$\Delta p_y = m\Delta v_y$$

$$\Delta p_y = m(v_2 - v_1)_y = (50.0 \text{ kg})[0 \text{ m/s} - (-6.26 \text{ m/s})]$$

$$\Delta p_y = +313 \text{ kg·m/s (up)}$$

The magnitude of a decelerating force can be reduced by increasing the amount of time that a given change of momentum takes place. This is the principle behind automobile air bags.

b. Calculate the vertical force on her feet when she lands flat-footed. Use Equation 11.3:

$$\Sigma F_y = \frac{\Delta p_y}{\Delta t}$$

$$\Sigma F_y = \frac{+313\ \text{kg}\cdot\text{m/s}}{0.100\ \text{s}} = +3130\ \text{N} \qquad \text{(landing flat-footed)}$$

c. Calculate the vertical force when she lands with flexed legs:

$$\Sigma F_y = \frac{\Delta p_y}{\Delta t}$$

$$\Sigma F_y = \frac{+313\ \text{kg}\cdot\text{m/s}}{0.800\ \text{s}}$$

$$\Sigma F_y \cong +39\underline{1}.2\ \text{N} \cong 391\ \text{N} \qquad \text{(landing with flexed legs)}$$

The difference in force is substantial. Thus, by spreading out the time of a collision, the force on the colliders can be lessened.

FACETS of PHYSICS

The Ubiquitous Turbine

What do jets, hydroelectric plants, and windmills have in common? They all rely on turbines to operate. A *turbine* is a machine with blades or vanes that transforms a flowing fluid's energy into rotational motion, which may then be used to generate other forms of energy. The word comes from the Latin word *turbo*, which means "vortex."

The earliest record of the use of turbine-like machines is of water wheels used by the Greeks and Romans to grind corn during the first century BC. These simple water wheels were partially immersed in a river or stream so that the force of the moving water could turn the wheel. Windmills were another early application of turbine technology. Persian water-pumping windmills appeared in the ninth century. China also may have pioneered the use of windmills for grinding grain and pumping water for irrigation during the same period. Water wheel efficiency improved when people discovered that delivering water to the *top* of the wheel provided more power than that obtained by powering the wheel from the bottom. By the thirteenth century, these so-called *overshot* water wheels were becoming the dominant water power source in Europe.

The next major advance in turbine technology occurred in the early 1800s, when the French mining engineer Claude Burdin designed a rotary machine in which water flowed from the center outward. He also coined the word "turbine" in his description of his design. His student Benoit Fourneyron built the first practical water turbine and went on to build turbines producing tens of horsepower.

Wind is still used today to pump water for irrigation.

The first hydroelectric adaptation of the turbine occurred in 1880 to power commercial lighting in Grand Rapids, Michigan. With the advances in steam thermodynamics theory during the mid-1800s, steam turbines became a reality. The first steam turbine-powered electrical generator was placed in operation in 1884. Since that time, turbines have been widely

11.5 Conservation of Momentum

Firing a high-caliber rifle causes a powerful recoil. When a rifle fires, the exploding gunpowder exerts a force on the bullet, $\mathbf{F}_{r \to b}$, for a time Δt. The bullet, therefore, exerts a reaction force of $\mathbf{F}_{b \to r}$ on the rifle for the same time, Δt, according to Newton's third law. The impulse of the force on the bullet is

$$\mathbf{I}_b = \mathbf{F}_{r \to b}\Delta t.$$

The impulse of the bullet on the rifle is

$$\mathbf{I}_r = \mathbf{F}_{b \to r}\Delta t = -\mathbf{F}_{r \to b}\Delta t.$$

The change in the bullet's momentum is

$$\Delta \mathbf{p}_b = \mathbf{I}_b = \mathbf{F}_{r \to b}\Delta t.$$

The rifle's momentum also changes.

$$\Delta \mathbf{p}_r = \mathbf{I}_r = \mathbf{F}_{b \to r}\Delta t$$

$$\Delta \mathbf{p}_r = -\Delta \mathbf{p}_b$$

$$\Delta \mathbf{p}_r + \Delta \mathbf{p}_b = 0 \text{ kg·m/s}$$

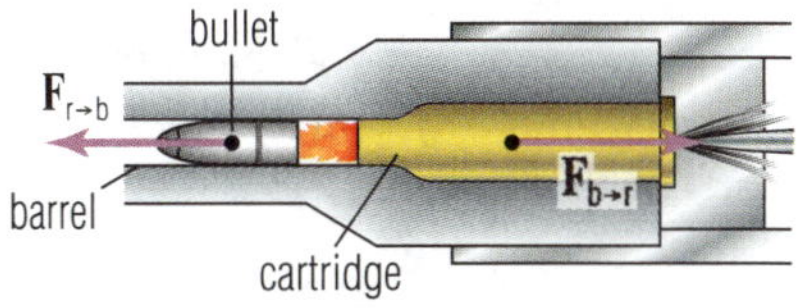

11-3 Reaction forces on a bullet and the rifle as the cartridge fires

Forces between parts inside the boundaries of a system are called *internal forces*. All forces exerted on a system that originate outside the boundaries of the system are *external forces*.

used for generating electricity and ship propulsion. Within the last century, efficient wind turbines have been built to harness the wind to produce electricity.

Gas turbines, largely developed during the first half of the twentieth century, use the energy produced by combustion to turn their blades. A fuel and air mixture is burned within the turbine, producing a rapidly expanding stream of hot gases. These gases turn the blades of the turbine, thereby producing rotational motion. Gas turbines are used in both propeller and jet aircraft as well as in a wide variety of other applications.

How do engineers design turbines for jets, electrical plants, nuclear power plants, and windmills? In each case they must consider three variables: the turbine principles they wish to apply, the density of the working fluid, and the speed at which the fluid flows. Steam turbines and water turbines differ in shape because their working fluids have different densities; steam turbines and wind turbines differ in shape because their working fluids operate at different speeds.

Turbines use one of two principles of physics to generate power: the reaction principle (using Newton's third law) or the impulse principle (using Newton's second law). In a *reaction turbine*, the working fluid exerts its weight or pressure to turn the turbine. The turbine converts the fluid's potential energy of high pressure into work. Consequently, the fluid's pressure decreases as it passes through the turbine. Ships often use reaction turbines for propulsion because of the relatively low speed of their propeller shafts. In an *impulse turbine*, high velocity fluid strikes the turbine blades. The fluid's change of momentum exerts an impulse on the blades, which provides the rotational force. The fluid's pressure remains relatively constant while passing through the turbine rotor, but its velocity drops as some of its energy is converted to work. Real-world turbines actually employ both principles, but one principle tends to dominate the overall picture. Wind generators are mainly reaction turbines, while hydroelectric turbines

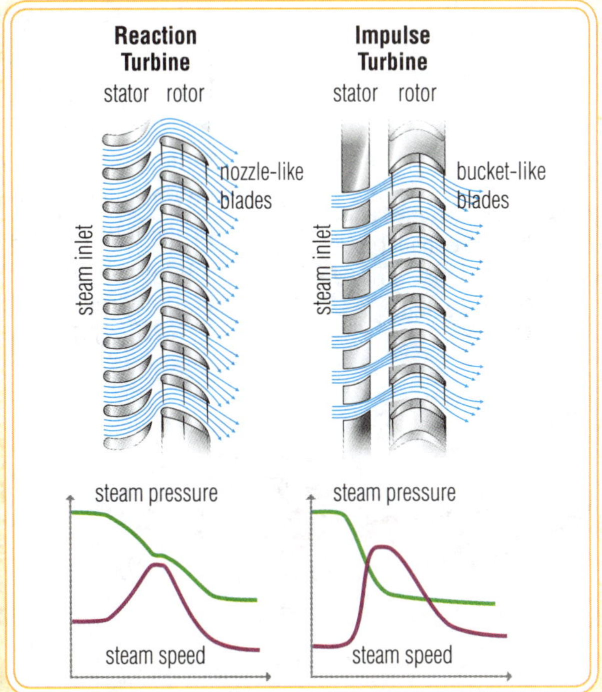

(a) Reaction turbines turn fluid potential energy into rotor kinetic energy, losing pressure in the process. (b) Impulse turbines convert change of momentum into an impulsive force that acts on the turbine rotor.

largely use the impulse principle. Gas turbines can be either reaction or impulse turbines since they come in a huge variety of forms that are used in many diverse applications.

Nearly every facet of life is affected and enhanced by turbine technology, whether the steam, water, or wind turbines that drive electrical generators, the gas turbines that propel passenger aircraft, or the marine turbines that turn merchant vessel and oil tanker propellers. Turbines are just another way in which technology is effectively employed for exercising dominion.

Recoil is the velocity given to one object within a system by other parts of the system.

The forces exerted by the exploding powder in the rifle shell come from *within* the system. Internal forces do not change the overall momentum of a system—only external forces do.

Momentum of a system is conserved when the net external force acting on the system is zero.

The change in the rifle's momentum is observed as recoil. However, if the rifle and the bullet together are considered the system, the overall momentum of the system does not change.

This discussion reveals the principle of conservation of momentum: when no unbalanced external forces act on a system, the momentum of the system is conserved (stays the same). Notice that the momentum of the bullet and the momentum of the rifle both changed. However, they changed equally in opposite directions so that the momentum change for the system was zero.

Suppose a hunter with a rifle is stranded at the middle of a frozen, frictionless lake. How can he propel himself to the edge of the lake? Because there is no friction between his boots and the ice, he cannot walk. Suppose he fires the gun. The bullet is given momentum in one direction. Since no outside force is assumed to act on the rifle-bullet system in the direction of motion, the rifle must gain momentum in the opposite direction to balance the bullet's gain in momentum. Of course, when the rifle moves, the man holding it must move too, since he is part of the rifle component of the system. Therefore, when the man fires the rifle, he moves in the direction opposite the bullet's direction. He will eventually reach the shore even if he fires only once. If he fires the gun repeatedly, he will reach the shore more quickly than if he fires only once.

11.6 Changing Momentum: Variable Mass

A space vehicle is in the same predicament as the hunter on frictionless ice—there is nothing against which it can push in order to move. The space vehicle's plight may be solved by a means similar to the hunter firing his gun. Instead of ejecting bullets, a chemical rocket exhausts burned rocket fuel, while an ion propulsion engine ejects charged particles. Such material is called *propellant mass*. As the propellant mass gains momentum in one direction, the spacecraft gains the same amount of momentum in the opposite direction. Therefore, engineers design the combustion chambers of space engines to rapidly eject as much mass as possible in order to develop maximum momentum.

The rocket nozzle shapes and accelerates the flow of exhaust gases to produce the maximum thrust for the mass of exhaust ejected.

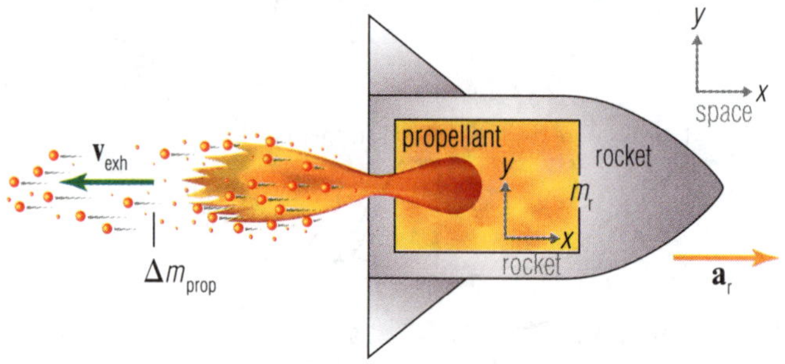

11-4 Conservation of momentum in a rocket-exhaust system

Let's examine how rocket propulsion in space works. Assume that the mass of propellant that is ejected as exhaust is Δm_{prop}. The momentum of this propellant exhaust is the product $\Delta m_{prop} v_{exh}$. The exhaust velocity is measured relative to the *rocket*. If we assume that there are no external forces acting on the rocket-propellant system, then the total change in momentum of the system must be zero. Therefore, the momentum of the rocket must change by an amount equal to the change of momentum of the propellant exhaust. The change of momentum of the rocket must then be

$$\Delta p_{r\,x} = -(\Delta m_{prop}) v_{exh\,x}.$$

The velocity of the exhaust is in the opposite direction to the momentum of the rocket.

If we divide both sides by the time (Δt) it takes to exhaust Δm_{prop} of propellant, we obtain the impulsive force of the exhaust on the rocket.

$$F_{exh\,x} = \frac{\Delta p_{r\,x}}{\Delta t} = -\left(\frac{\Delta m_{prop}}{\Delta t}\right) v_{exh\,x}$$

The term $\Delta m_{prop}/\Delta t$ is the rate of propellant mass flow, measured in kg/s. The impulsive force of the exhausted propellant, $F_{exh\,x}$, is called **thrust**. Thrust is the force that accelerates the rocket. Expressing this equation in the form of Newton's second law, we get

$$F_{exh\,x} = m_r a_{r\,x}, \text{ or}$$

$$-\left(\frac{\Delta m_{\text{prop}}}{\Delta t}\right)v_{\text{exh}\,x} = m_{\text{r}}a_{\text{r}\,x}\,, \qquad (11.7)$$

where m_r is the combined mass of the rocket and the remaining propellant within the rocket. Unlike previous examples, the spacecraft's mass continually changes as long as it ejects propellant mass in its exhaust. The left side of Equation 11.7 normally remains relatively constant until the propellant is gone. Therefore, on the right side of the equation, as the rocket's mass decreases with time, its acceleration must increase. This case of nonuniform acceleration requires more advanced mathematics to analyze the rocket's motion. You *cannot* use the three equations of motion for systems where mass and acceleration vary.

EXAMPLE 11-3

Rocket Science in Deep Space: Conservation of Momentum

A 5.10×10^5 kg space probe is at rest in an assumed fixed reference frame in deep space. In order to maneuver toward a comet nucleus, the probe fires its engine, which exhausts propellant at a mass flow rate of 1.00×10^4 kg/s with a speed of 700. m/s. What is the probe's initial acceleration the instant the engine fires?

Solution:

Assume that all motion is parallel to the x-coordinate axis with the probe moving in the positive direction. Use scalar components for your computations.

Solve Equation 11.7 for acceleration:

$$-\left(\frac{\Delta m_{\text{prop}}}{\Delta t}\right)v_{\text{exh}\,x} = m_{\text{r}}a_{\text{r}\,x}$$

$$a_{\text{r}\,x} = -\left(\frac{\Delta m_{\text{prop}}}{\Delta t}\right)\frac{v_{\text{exh}\,x}}{m_{\text{r}}}$$

Since the mass of the rocket has not changed appreciably, substitute the known quantities into this equation:

$$a_{\text{r}\,x} = -(1.00 \times 10^4\ \cancel{\text{kg}}/\text{s})\frac{(-700.\text{m/s})}{(5.10 \times 10^5\ \cancel{\text{kg}})}$$

$$a_{\text{r}\,x} \cong +13.\underline{7}2\ \text{m/s}^2 \cong +13.7\ \text{m/s}^2$$

The initial instantaneous acceleration of the probe is almost 14 m/s^2.

11A Section Review

1. Define momentum. What must a system be doing in order for momentum to exist?
2. Mathematically, what kind of physical quantity is momentum? How do you determine the direction of momentum?
3. When a dropped object falls toward the ground, is its change of momentum positive or negative? When it hits the ground, is its change of momentum positive or negative? Assume that up is positive.

11A Objectives

After completing this section, I can

- ✓ define linear momentum and state the properties of a system that contribute to momentum.
- ✓ compute linear momentum under varying conditions.
- ✓ identify impulsive forces and compute the impulse on a system.
- ✓ discuss the concept of conservation of momentum and the conditions under which it is true.
- ✓ describe how the conservation of momentum applies to rocket propulsion and thrust in space.

4. For a given change of momentum, will the force producing the impulse be larger or smaller if the change of momentum takes longer to occur?
5. Calculate the momentum of a 145 g baseball moving at 46.9 m/s at an angle of 30.0° to the horizontal.
6. A 4.32 kg rifle is clamped to a frictionless cart in a test firing range, pointing at the distant target. The rifle is fired electronically. The 9.72 g bullet slug exits the muzzle 0.001 80 s later with a speed of 835 m/s. What is the momentum of the bullet?
7. What is the total impulsive force of the burning propellant from the cartridge on the bullet slug in Question 6?
8. What is the recoil velocity of the rifle in Question 6?
9. An astronaut is working around the International Space Station (ISS), attached by a tether cord. The snap on the tether fails and he finds himself adrift, but stationary relative to the ISS. This astronaut was a former physics student and knows that if he can throw some objects away from himself in the right direction, he can propel himself back to the station. His tool kit has a 1.20 kg torque wrench, a 0.70 kg nut driver, and a 0.35 kg bolt. The astronaut, his suit, and all of his gear have an initial mass of 105.00 kg. If he throws away each of the items in his tool kit at 3.0 m/s, one at a time, what will be his final speed relative to the ISS?

11B COLLISIONS

11.7 Linear Collisions

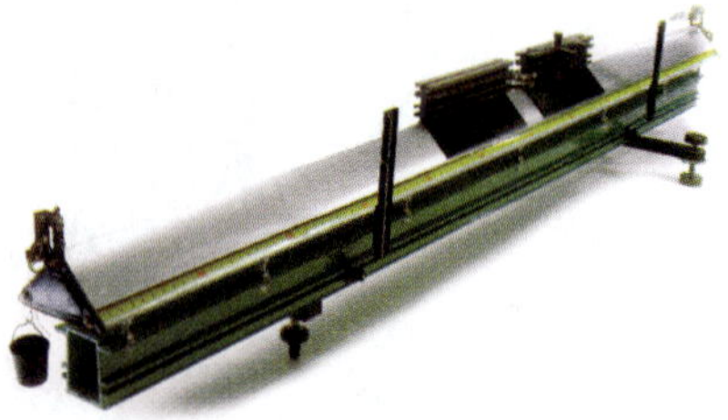

11-5 Gliders on an air track

In collision problems, the subscripts "bfr" and "aft" refer to properties or conditions before and after the collision, respectively.

Problem-Solving Strategy 11.2

Evaluate the effect of all external forces on the system as a whole. If there is no net external force, then the total change of momenta of all parts of the system is zero and momentum is conserved.

We have discussed the conservation of momentum when *internal* forces separate objects within a system. Now let us discuss the conservation of momentum of colliding bodies. In general, the same principle applies as long as we consider the colliding objects as parts of a single system—the momentum of the system is conserved unless an unbalanced *external* force acts on the system.

The air track shown in Figure 11-5 is a traditional physics laboratory apparatus that demonstrates conservation of momentum in collisions. The track is pierced by hundreds of holes through which air flows to form a nearly frictionless cushion of air to support Y-shaped gliders. The gliders move in a linear path along the straight track. Suppose glider 1 with mass m_1 on the left part of the track is given a velocity of $\mathbf{v}_1$ toward the right. Glider 2 with mass m_2 is at rest to the right of m_1. We will consider the system to include both m_1 and m_2. See Figure 11-6a on the next page. There are no external unbalanced forces on the system. Therefore, the total momentum of the system is conserved and

$$\Delta\mathbf{p}_{\text{total}} = 0 \text{ kg·m/s}.$$

What happens when m_1 collides with m_2? We know that the total momentum before the collision is

$$\mathbf{p}_{\text{total bfr}} = \mathbf{p}_{1\text{ bfr}} + \mathbf{p}_{2\text{ bfr}},$$

where the subscripts "1" and "2" identify the properties of gliders 1 and 2, respectively, and the subscript "bfr" identifies the properties *before* the collision. The total momentum of the system after the collision is

$$\mathbf{p}_{\text{total aft}} = \mathbf{p}_{1\text{ aft}} + \mathbf{p}_{2\text{ aft}},$$

where the subscript "aft" identifies the properties *after* the collision.

Since no external force is exerted on the system, the momentum of the system before and after the collision must be the same.

$$\mathbf{p}_{\text{total bfr}} = \mathbf{p}_{\text{total aft}}$$

If we expand these expressions, we can see how the individual momenta change but the overall system momentum is unchanged.

$$\mathbf{p}_{1\text{ bfr}} + \mathbf{p}_{2\text{ bfr}} = \mathbf{p}_{1\text{ aft}} + \mathbf{p}_{2\text{ aft}}$$

$$\mathbf{p}_{1\text{ bfr}} - \mathbf{p}_{1\text{ aft}} = \mathbf{p}_{2\text{ aft}} - \mathbf{p}_{2\text{ bfr}}$$

$$-(\mathbf{p}_{1\text{ aft}} - \mathbf{p}_{1\text{ bfr}}) = \mathbf{p}_{2\text{ aft}} - \mathbf{p}_{2\text{ bfr}}$$

$$-m_1(\mathbf{v}_{1\text{ aft}} - \mathbf{v}_{1\text{ bfr}}) = m_2(\mathbf{v}_{2\text{ aft}} - \mathbf{v}_{2\text{ bfr}})$$

$$m_1(\mathbf{v}_{1\text{ aft}} - \mathbf{v}_{1\text{ bfr}}) = -m_2(\mathbf{v}_{2\text{ aft}} - \mathbf{v}_{2\text{ bfr}})$$

This relationship can be summarized in the equation

$$\Delta\mathbf{p}_1 = -\Delta\mathbf{p}_2. \qquad (11.8)$$

Before the collision, m_2 is at rest; so $\mathbf{v}_{2\text{ bfr}} = 0$ m/s. Therefore,

$$m_1\mathbf{v}_{1\text{ aft}} - m_1\mathbf{v}_{1\text{ bfr}} = -m_2\mathbf{v}_{2\text{ aft}}. \qquad (11.9)$$

We do not know either $\mathbf{v}_{1\text{ aft}}$ or $\mathbf{v}_{2\text{ aft}}$. Since there are two unknowns in Equation 11.9, we cannot solve it. We need another equation involving the unknowns before we can find their values. This requires a closer look at the kind of collision that occurred.

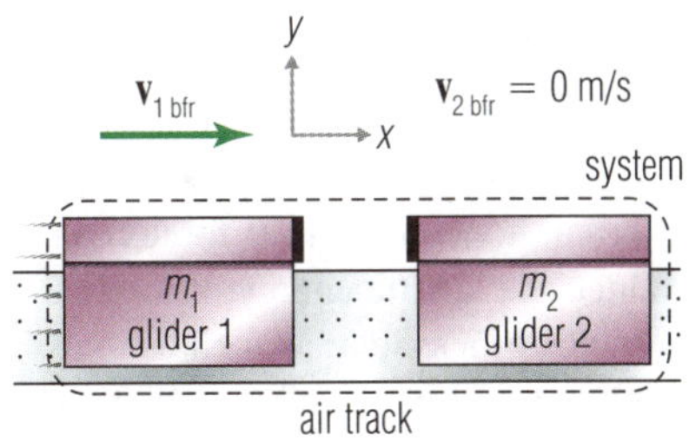

11-6a Before collision

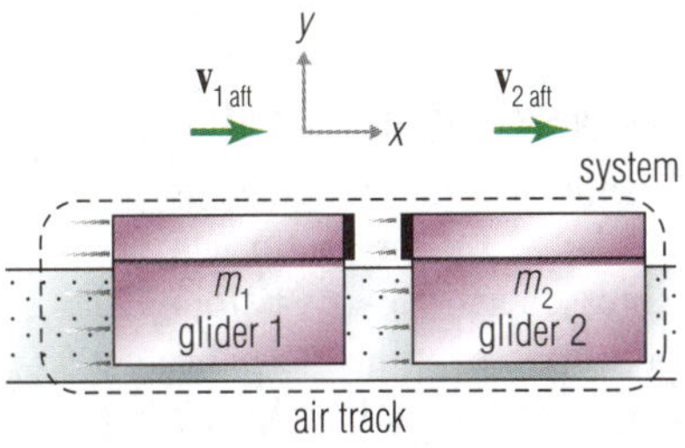

11-6b After collision

11.8 Collision Modes

Objects may collide in one of three ways or *modes*. In an elastic collision, only conservative forces act; therefore, total mechanical energy is conserved. This is a strictly theoretical mode of collision, but many real systems approach this condition, such as the collision of billiard balls rolling on a hard, smooth surface, or the collision of air molecules.

In a partially elastic collision, one or more of the colliding masses is deformed (changes its shape), but the masses rebound away from the collision. The material deformation converts some of the system's kinetic energy to work and elastic potential energy, and to thermal energy, sound, or other nonmechanical energy. Most real collisions are partially elastic. In an inelastic collision, the colliding masses stick together. The collision converts a large portion of the initial kinetic energy to work on the parts of the system; therefore, kinetic energy is not conserved. Examples of such a collision are a bullet imbedding itself in a tree and a scoop of ice cream falling to the floor.

Collisions can occur in three modes. Elastic collisions conserve both momentum and mechanical energy. Partially elastic collisions and inelastic collisions do not conserve mechanical energy because some is converted into nonmechanical energy (e.g., the colliding objects become deformed, produce sparks, or make noise). In inelastic collisions, objects stick together.

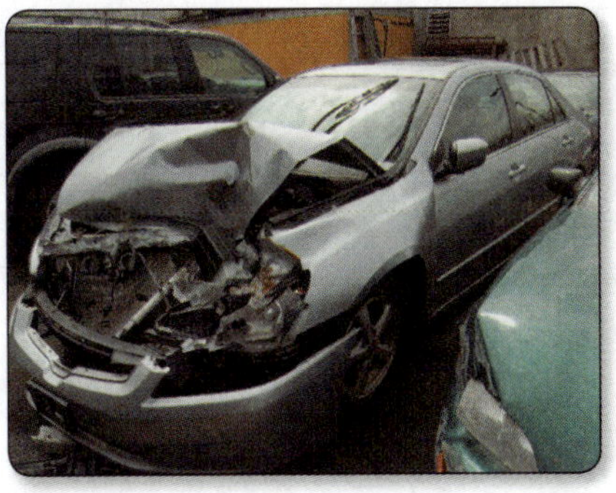

11-7 Car manufacturers design their vehicles so that collisions will be as inelastic as possible—dissipating the initial mechanical energy as nonmechanical energy to reduce injury to the occupants.

11.9 Elastic Collisions

Let's suppose that the collision between the gliders on the air track is elastic and x is positive to the right. Kinetic energy is conserved. Therefore, the following relationships are true:

$$K_{\text{total bfr}} = K_{\text{total aft}}, \text{ and}$$

$$\tfrac{1}{2}m_1v_{1\text{ bfr }x}^2 + \tfrac{1}{2}m_2v_{2\text{ bfr }x}^2 = \tfrac{1}{2}m_1v_{1\text{ aft }x}^2 + \tfrac{1}{2}m_2v_{2\text{ aft }x}^2$$

In this example, $v_{2\text{ bfr }x} = 0$ m/s; so

$$\tfrac{1}{2}m_1v_{1\text{ bfr }x}^2 = \tfrac{1}{2}m_1v_{1\text{ aft }x}^2 + \tfrac{1}{2}m_2v_{2\text{ aft }x}^2. \qquad (11.10)$$

Internal, nonconservative forces can change the kinetic energy of internal particles of a system as well as their momenta. While, ideally, momentum is conserved within a system, kinetic energy usually is not.

Problem-Solving Strategy 11.3

Quantities with multiple subscripts, such as $v_{1\text{ bfr }x}$, can be verbally expressed as "vee-one-before." The subscript x is written to indicate a scalar component but may be left unspoken if no confusion will result.

Earlier in the discussion we obtained Equation 11.9:

$$m_1\mathbf{v}_{1\text{ aft}} - m_1\mathbf{v}_{1\text{ bfr}} = -m_2\mathbf{v}_{2\text{ aft}}$$

Since the problem is one-dimensional, we can rewrite Equation 11.9 in scalar component notation as

$$m_1 v_{1\text{ aft }x} - m_1 v_{1\text{ bfr }x} = -m_2 v_{2\text{ aft }x}. \qquad (11.11)$$

Combining Equations 11.10 and 11.11, we can solve for $v_{1\text{ aft }x}$ and $v_{2\text{ aft }x}$. We find that

$$v_{1\text{ aft }x} = \left(\frac{m_1 - m_2}{m_1 + m_2}\right) v_{1\text{ bfr }x}\text{, and} \qquad (11.12)$$

$$v_{2\text{ aft }x} = \left(\frac{2m_1}{m_1 + m_2}\right) v_{1\text{ bfr }x}. \qquad (11.13)$$

These equations hold for any elastic collision where a moving mass m_1 collides with a resting mass m_2. If $v_{1\text{ aft }x}$ is negative, it simply means that m_1 rebounds from the collision with a speed $v_{1\text{ aft}}$.

11-8 Bumper cars undergo partially elastic collisions.

EXAMPLE 11-4

Tracking Momentum: An Elastic Collision

Two 50.0 g gliders are on an air track. One glider with m_1 is given a velocity of 2.00 cm/s to the right toward the other glider, which is stationary. What are the velocities of the gliders after the collision?

Solution:

Assume x is positive to the right. Use Equations 11.12 and 11.13 to solve for the final velocities of m_1 and m_2, respectively.

Velocity of glider 1 after collision:

$$v_{1\ \text{aft}\ x} = \left(\frac{m_1 - m_2}{m_1 + m_2}\right) v_{1\ \text{bfr}\ x}$$

$$v_{1\ \text{aft}\ x} = \left(\frac{50.0\ \text{g} - 50.0\ \text{g}}{50.0\ \text{g} + 50.0\ \text{g}}\right)(+2.00\ \text{cm/s})$$

$$v_{1\ \text{aft}\ x} = 0\ \text{cm/s}$$

Velocity of glider 2 after collision:

$$v_{2\ \text{aft}\ x} = \left(\frac{2m_1}{m_1 + m_2}\right) v_{1\ \text{bfr}\ x}$$

$$v_{2\ \text{aft}\ x} = \left(\frac{2(50.0\ \text{g})}{50.0\ \text{g} + 50.0\ \text{g}}\right)(+2.00\ \text{cm/s})$$

$$v_{2\ \text{aft}\ x} = +2.00\ \text{cm/s}$$

The moving glider stops, and the stationary glider moves off with the velocity that m_1 originally had. The momentum of each object has changed in the collision, but the total momentum of the system of objects has remained the same.

EXAMPLE 11-5

Irresistible Force: An Unequal Elastic Collision

An out-of-control 100.0 kg ice skater with a velocity of 10.0 m/s north careens toward a 10.0 kg stationary plastic drum that marks the edge of safe ice on the pond. What are the velocities and momentums of the skater and the drum after they collide elastically?

Solution:

Assume that y is positive north and that all motions are frictionless.
Ice skater's velocity after the collision:

$$v_{1\ \text{aft}\ y} = \left(\frac{m_1 - m_2}{m_1 + m_2}\right) v_{1\ \text{bfr}\ y}$$

$$v_{1\ \text{aft}\ y} = \left(\frac{100.0\ \text{kg} - 10.0\ \text{kg}}{100.0\ \text{kg} + 10.0\ \text{kg}}\right)(+10.0\ \text{m/s})$$

$$v_{1\ \text{aft}\ y} \cong +8.1\underline{8}1\ \text{m/s} \cong +8.18\ \text{m/s}$$

Drum's velocity after the collision:

$$v_{2\ \text{aft}\ y} = \left(\frac{2m_1}{m_1 + m_2}\right) v_{1\ \text{bfr}\ y}$$

$$v_{2\ \text{aft}\ y} = \left(\frac{2(100.0\ \text{kg})}{100.0\ \text{kg} + 10.0\ \text{kg}}\right)(+10.00\ \text{m/s})$$

$$v_{2\ \text{aft}\ y} \cong +18.\underline{1}8\ \text{m/s} \cong +18.2\ \text{m/s}$$

Total momentum before the collision:

$$p_{\text{total bfr }y} = p_{1\text{ bfr }y} + p_{2\text{ bfr }y} = m_1v_{1\text{ bfr }y} + m_2v_{2\text{ bfr }y}$$

$$p_{\text{total bfr }y} = (100.0\text{ kg})(+10.0\text{ m/s}) + (10.0\text{ kg})(0\text{ m/s})$$

$$p_{\text{total bfr }y} = +1.00 \times 10^3\text{ kg}\cdot\text{m/s}$$

Total momentum after the collision:

$$p_{\text{total aft }y} = p_{1\text{ aft }y} + p_{2\text{ aft }y} = m_1v_{1\text{ aft }y} + m_2v_{2\text{ aft }y}$$

$$p_{\text{total aft }y} = (100.0\text{ kg})(+8.1\underline{8}1\text{ m/s}) + (10.0\text{ kg})(+18.\underline{1}8\text{ m/s})$$

$$p_{\text{total aft }y} \cong +99\underline{9}.9\text{ kg}\cdot\text{m/s}$$

$$p_{\text{total aft }y} \cong +1000.\text{ kg}\cdot\text{m/s or } +1.000 \times 10^3\text{ kg}\cdot\text{m/s}$$

The moving man barely slows down, while the stationary drum shoots off at nearly twice the man's original velocity. This is true for elastic collisions between a very large moving mass and a relatively small stationary mass. Once again, the momenta of the two objects is very different after the collision, but the total momentum of the system is unchanged.

EXAMPLE 11-6

Exercise in Futility: Another Unequal Elastic Collision

A boy shoots a 0.0300 kg cork from a pop gun with a velocity of 5.00 m/s to the right toward a 6.80 kg stone placed on a fence post. What are the velocities of each after the elastic collision? Assume no friction.

Solution:

Assume that x is positive to the right.

The velocity of the cork after the collision:

$$v_{1\text{ aft }x} = \left(\frac{0.0300\text{ kg} - 6.80\text{ kg}}{0.0300\text{ kg} + 6.80\text{ kg}}\right)(+5.00\text{ m/s})$$

$$v_{1\text{ aft }x} \cong -4.9\underline{5}6\text{ m/s}$$

$$v_{1\text{ aft }x} \cong -4.96\text{ m/s (to the left)}$$

The velocity of the stone after the collision:

$$v_{2\text{ aft }x} = \left(\frac{2(0.0300\text{ kg})}{0.0300\text{ kg} + 6.80\text{ kg}}\right)(+5.00\text{ m/s})$$

$$v_{2\text{ aft }x} \cong +0.043\underline{9}2\text{ m/s}$$

$$v_{2\text{ aft }x} \cong +0.0439\text{ m/s (to the right)}$$

The cork bounces back with nearly its original speed, while the stone hardly budges to the right. This result is typical of an elastic collision of a small moving mass with a large stationary mass. If the stone had been immovable, attached to an extremely large mass, the cork would have bounced back with exactly its original speed. We assume that this is the case when we discuss an elastic collision with a solid wall.

11.10 Inelastic Collisions

We have discussed idealized elastic collisions, but what about the other extreme—idealized inelastic collisions—where the colliding objects stick completely together following the collision? Going back to the air track example, suppose that glider m_2 has tacky glue on the end facing glider m_1. When m_1 collides with it, the gliders stick together. Therefore, both gliders must have the same velocity after the collision:

$$v_{1\text{ aft }x} = v_{2\text{ aft }x} = v_{\text{aft }x}$$

Equation 11.9 becomes:

$$m_1 v_{1\text{ aft }x} - m_1 v_{1\text{ bfr }x} = -m_2 v_{2\text{ aft }x}$$

$$m_1 v_{\text{aft }x} - m_1 v_{1\text{ bfr }x} = -m_2 v_{\text{aft }x}$$

$$-m_1 v_{1\text{ bfr }x} = -m_1 v_{\text{aft }x} - m_2 v_{\text{aft }x}$$

$$m_1 v_{1\text{ bfr }x} = m_1 v_{\text{aft }x} + m_2 v_{\text{aft }x}$$

$$m_1 v_{1\text{ bfr }x} = (m_1 + m_2) v_{\text{aft }x} \quad (11.14)$$

Since we know all these quantities, except the final speed of the system, we can find $v_{\text{aft }x}$ without using a second equation.

EXAMPLE 11-7

Sticking Together: Inelastic Collisions

A 1.00 kg kinetic cart with a velocity of 1.00 m/s to the right collides with a stationary 2.00 kg cart. Both carts are equipped with mechanisms that firmly latch the carts together when they collide. (a) What is the velocity of the carts after the collision? (b) Is the system's kinetic energy conserved?

Solution:

Assume that x is positive to the right.

a. Solve for the final velocity of the system:

$$m_1 v_{1\text{ bfr }x} = (m_1 + m_2) v_{\text{aft }x}$$

$$v_{\text{aft }x} = \left(\frac{m_1}{m_1 + m_2}\right) v_{1\text{ bfr }x} = \left(\frac{1.00\text{ kg}}{1.00\text{ kg} + 2.00\text{ kg}}\right)(+1.00\text{ m/s})$$

$$v_{\text{aft }x} = +0.33\underline{3}\bar{3}\text{ m/s} \cong +0.333\text{ m/s}$$

b. Is kinetic energy conserved? If it is, then Equation 11.10 should hold true.

$$\tfrac{1}{2} m_1 v_{1\text{ bfr }x}^2 \stackrel{?}{=} \tfrac{1}{2} m_1 v_{\text{aft }x}^2 + \tfrac{1}{2} m_2 v_{\text{aft }x}^2$$

$$m_1 v_{1\text{ bfr }x}^2 \stackrel{?}{=} (m_1 + m_2) v_{\text{aft }x}^2$$

$$\left(\frac{m_1}{m_1 + m_2}\right) v_{1\text{ bfr }x}^2 \stackrel{?}{=} v_{\text{aft }x}^2$$

$$v_{1\text{ bfr }x} \sqrt{\frac{m_1}{m_1 + m_2}} \stackrel{?}{=} v_{\text{aft }x}$$

$$(+1.00\text{ m/s}) \sqrt{\frac{1.00\text{ kg}}{1.00\text{ kg} + 2.00\text{ kg}}} \stackrel{?}{=} +0.33\bar{3}\text{ m/s}$$

$$+0.5\underline{7}73\text{ m/s} \neq +0.33\underline{3}\bar{3}\text{ m/s}$$

Because the two speeds are not equal, kinetic energy is *not* conserved during an inelastic collision.

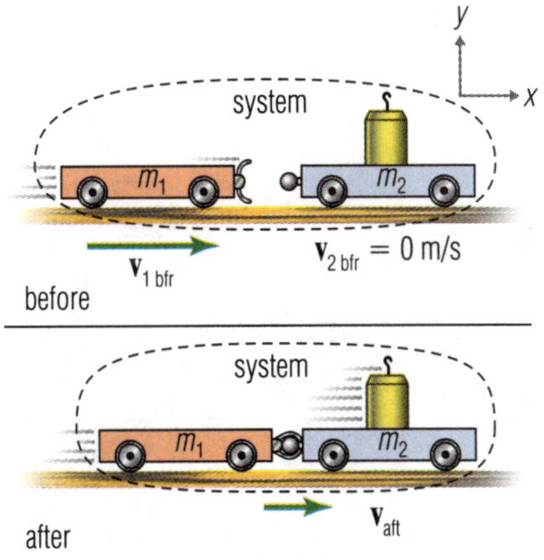

11-9 An inelastic collision

In an inelastic collision, kinetic energy is not conserved because some of the initial kinetic energy is converted to work to accelerate the stationary object and to decelerate the moving object to the final velocity of the combined system of objects. After absorbing the work, which is internal to the system, some of the energy is also manifested as other forms of energy, such as thermal energy, sound, or light.

11.11 Ballistic Pendulums

11-10 A ballistic pendulum

An experimental device that depends on completely inelastic collisions is called a **ballistic pendulum**. It is normally used to determine the energy of projectiles such as military and civilian bullets. Indirectly, this information can be used to find projectile velocity.

A ballistic pendulum consists of a uniform block of a dense material attached to one or more pendulum arms. Figure 11-10 shows a typical ballistic pendulum. The projectile to be tested impacts the pendulum and embeds itself in the block. The bullet converts essentially all of its kinetic energy into pendulum potential energy as the pendulum rises. An angle indicator records the maximum angle attained by the pendulum arm, and engineers are then able to calculate the kinetic energy and velocity of the projectile at impact.

EXAMPLE 11-8

On the Rise: A Ballistic Pendulum

A 10.0 g (0.010 kg) bullet strikes a stationary 2.00 kg ballistic pendulum. The pendulum rises a vertical distance of 12.0 cm (0.120 m). If the bullet remains embedded in the pendulum, what is the bullet's speed at impact?

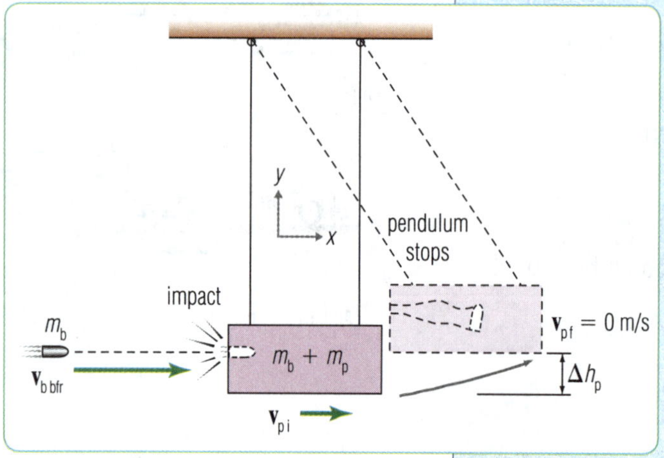

11-11 The bullet's kinetic energy is converted to the pendulum's potential energy.

Solution:

The pendulum-bullet system's total mechanical energy is conserved. When the pendulum pauses at the top of its swing, it has gained a potential energy of $\Delta U_g = m_p|g|\Delta h_p$, where Δh_p is the height it rises. At the maximum height, it has no kinetic energy. The pendulum's change in kinetic energy is as follows:

$$\Delta K_p = \tfrac{1}{2}m_p v_{p\,f}^2 - \tfrac{1}{2}m_p v_{p\,i}^2 = 0\text{ J} - \tfrac{1}{2}m_p v_{p\,i}^2$$

$$\Delta K_p = -\tfrac{1}{2}m_p v_{p\,i}^2$$

The velocity of the pendulum and bullet the instant after the bullet imbeds itself ($v_{p\,i}$) is found from the following:

$$\Delta K_p = -\Delta U_{g\,p}$$

$$-\tfrac{1}{2}\cancel{m_p}v_{p\,i}^2 = -\cancel{m_p}|g|\,\Delta h_p$$

$$v_{p\,i}^2 = 2|g|\Delta h_p$$

$$v_{p\,i} = \sqrt{2|g|\,\Delta h_p}$$

$$v_{p\,i} = \sqrt{2|-9.81\text{ m/s}^2|(+0.120\text{ m})}$$

$$v_{p\,i} \cong 1.5\underline{3}4\text{ m/s}$$

Initial pendulum velocity is to the right.
The velocity of the bullet just before impact can be determined from Equation 11.14.

$$m_b v_{b\ bfr\ x} = (m_b + m_p)v_{p\ i\ x}$$

$$v_{b\ bfr\ x} = \frac{m_b + m_p}{m_b} \cdot v_{p\ i\ x}$$

$$v_{b\ bfr\ x} = \frac{0.0100\ \text{kg} + 2.00\ \text{kg}}{0.0100\ \text{kg}} \cdot (+1.5\underline{3}4\ \text{m/s})$$

$$v_{b\ bfr\ x} \cong +30\underline{8}.3\ \text{m/s} \cong +308\ \text{m/s}$$

The bullet's speed just before impact is about 308 m/s.

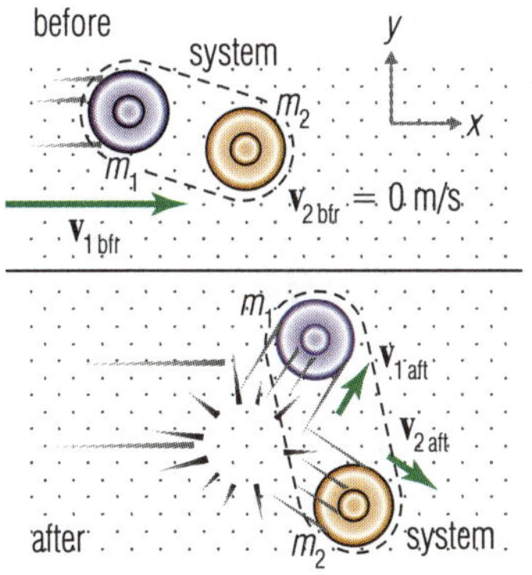

11-12a Pucks colliding on an air table

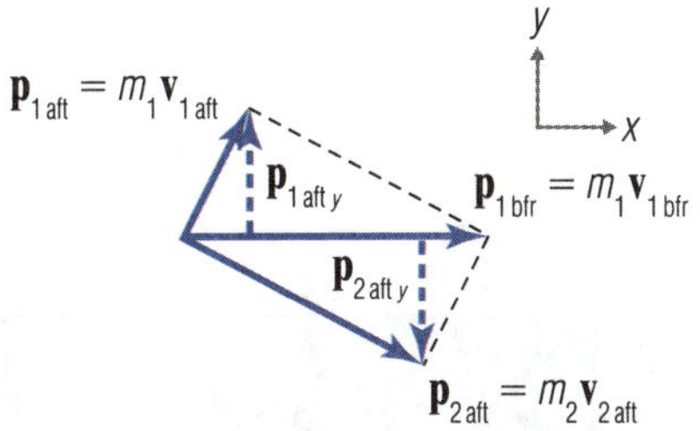

11-12b Resolving the momenta of two-dimensional collisions

11.12 Two-Dimensional Collisions

Most collisions occur in two or three dimensions. Momentum is conserved in two- and three-dimensional collisions as well as in one-dimensional collisions, but the equations are more complicated for more dimensions.

An excellent apparatus for demonstrating two-dimensional collisions is an air table, very similar to an air hockey table. Its surface has numerous holes through which air flows to suspend pucks of various masses and to permit nearly frictionless motion.

In Figure 11-12a, puck m_1 approaches the stationary puck m_2 from the left with velocity $\mathbf{v}_{1\ bfr}$. Puck 1 elastically strikes puck 2 off-center and caroms with a velocity $\mathbf{v}_{1\ aft}$. At the same time, m_2 rebounds with velocity $\mathbf{v}_{2\ aft}$. Momentum is conserved, and Equation 11.8 holds true.

$$\Delta\mathbf{p}_1 = -\Delta\mathbf{p}_2$$

$$m_1\mathbf{v}_{1\ aft} - m_1\mathbf{v}_{1\ bfr} = -(m_2\mathbf{v}_{2\ aft} - m_2\mathbf{v}_{2\ bfr}),\ \text{or}$$

$$m_1\mathbf{v}_{1\ aft} - m_1\mathbf{v}_{1\ bfr} = -m_2\mathbf{v}_{2\ aft},$$

since $\mathbf{v}_{2\ bfr} = 0$ m/s. Collecting terms for momenta before and after the collision,

$$-m_1\mathbf{v}_{1\ bfr} = -m_1\mathbf{v}_{1\ aft} - m_2\mathbf{v}_{2\ aft},\ \text{and}$$

$$m_1\mathbf{v}_{1\ bfr} = m_1\mathbf{v}_{1\ aft} + m_2\mathbf{v}_{2\ aft}.$$

In order for momentum to be conserved, the x-component of the momentum before collision must equal the sum of the x-components afterward. The same is true of the y-components. But $m_1\mathbf{v}_{1\ bfr}$ is only in the x-direction—it has no y-component—while $m_1\mathbf{v}_{1\ aft}$ and $m_2\mathbf{v}_{2\ aft}$ both have y-components. This difficulty is resolved by Figure 11-12b. The y-components of the two momenta after the collision exactly cancel each other. To find the final velocities of objects involved in a two-dimensional collision, obtain the components of their velocities from the x- and y-components of their momenta.

When two identical particle-like objects collide elastically, the post-collision angle between the objects' paths is exactly 90°. As the amount of deformation in one or both objects increases, and the collision becomes more inelastic, the angle becomes smaller. For completely inelastic collisions, the rebound angle is zero (see the illustrations in Figures 11-13a and b).

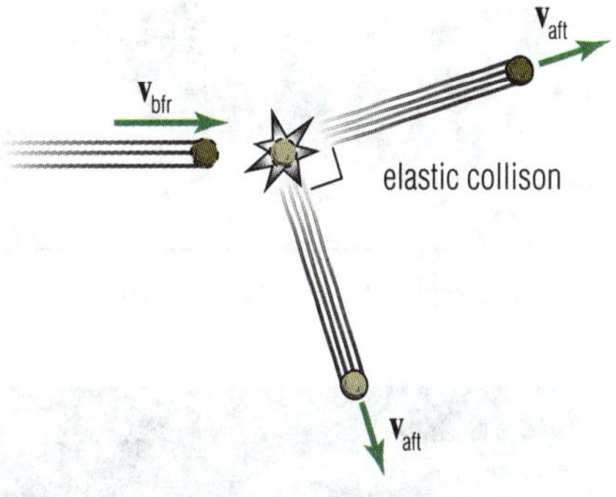

11-13a The rebound angle for an off-center, perfectly elastic collision is always 90°.

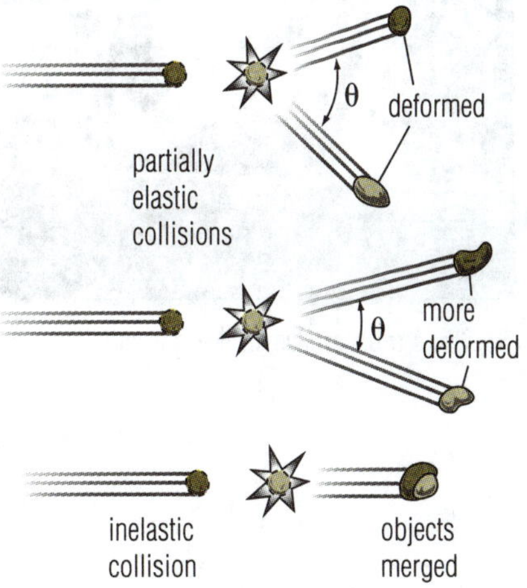

11-13b Rebound angles decrease as collisions become less elastic. The rebound angle is 0° for an inelastic collision.

11.13 Using Changes in Impulse and Momentum to Solve Problems

Christians have biblical reasons to design vehicles that will preserve and protect human life (Gen. 9:6). Injuries and fatalities in car accidents are reduced by reducing the impulsive forces on the passengers' bodies. If a person's body can be cushioned or parts of the car can compress to absorb energy, the magnitude of the impulse can be decreased by stretching out the time in which the change in momentum takes place. Intelligently designing cars to accomplish this requires that engineers understand the forces and times involved in a collision. But how can they make these kinds of measurements to guide technological development?

In order to design cars and safety features for maximum effect, engineers need feedback about the susceptibility of the human body to forces and impulses. Make seatbelts or airbags too stiff, and they can cause injuries. Make them too soft, and they may not provide adequate protection. This is where crash-test dummies enter the picture.

11-14 A crash-test dummy family portrait

11-15 A crash-test dummy during impact

Crash-test dummies were first used in the 1940s to test ejection seats in military fighter aircraft. Before this, engineers resorted to using cadavers. When the American automaker Ford started including seatbelts in its cars, auto engineers began using dummies that better resembled humans in weight, form, and reactions to external forces. Automaker General Motors has developed a whole family of crash-test dummies that represent different sizes and kinds of people, including adults, children, infants, and even pregnant women.

Different dummies are used for different types of tests, such as rear impacts, frontal impacts, side impacts, and roll-overs. Dummies used for these tests contain sensors in different places such as the chest, head, neck, back, legs, and pelvis. Some have specific designs to measure the possibilities of injury to the back and neck. These sensors are one of three kinds: load sensors, motion sensors, or accelerometers. Load sensors measure the force exerted on a certain part of the body. Motion sensors measure how much a part of the body, like the chest, is compressed or moved. Accelerometers measure how fast a certain part of the body changes speed. These sensors collect data and send it to specific input channels on a computer. Automated algorithms process the data to compute the likelihood of injury by comparing the impulse generated to the susceptibility of various joints and body parts to injury. Usually, this data can be synchronized with video and high-speed photographs of the test collision.

Thanks to collision tests using crash test dummies, many car parts—including airbags, seatbelts, padded dashboards, sturdy fenders, and collapsible steering columns—have been designed to provide people with the best possible protection during a car collision. Other aspects of the car, like the fuel system and the engine compartment, are being designed to "fail safely," so they do not add to the severity of the collision by starting a fire or allowing the engine to penetrate the passenger compartment.

11B Section Review

1. Under what conditions will momentum be completely conserved in a collision between two objects?
2. (True or False) If the only forces at work on a system are internal, momentum is conserved.
3. Under what conditions will kinetic energy be conserved during a collision?
4. Following an off-center elastic collision between two objects, what is the vector sum of the objects' momenta?

DS**5.** Describe the geometry and class of collisions evaluated in automobile crash tests.

DS**6.** What objectives do engineers and car manufacturers have when trying to decrease the impulse on passengers for any given collision?

7. An air track glider of mass $m_1 = 1.80$ kg with an initial velocity of 1.20 m/s to the right collides elastically with a stationary glider of mass $m_2 = 5.00$ kg. Calculate the speed and direction of the two gliders after the collision.

8. An air track glider ($m_1 = 1.80$ kg) moving to the right collides with a stationary glider (m_2). What must the mass of the second glider be in order for its velocity after the collision to be 1.2 times the first glider's initial velocity?

9. A 6.80 g bullet traveling at 1010 m/s to the right imbeds itself in a 3.00 kg cube of oak sitting on a fence post.

a. What is the velocity of the block-and-bullet mass after the impact?

b. How much mechanical energy was converted to internal work, thermal, and other forms of energy at impact?

10. A 0.250 kg air table puck traveling at 1.00 m/s to the right rebounds elastically off of a 2.00 kg stationary puck. The first puck rebounds with a velocity of 0.870 m/s at an angle of +60.° from the positive x-axis. What is the velocity of the second puck after the collision?

11B Objectives

After completing this section, I can

✓ describe the factors involved in an ideal one-dimensional (linear) collision between two objects.

✓ state the three kinds of ideal collisions and the factors that determine whether or not momentum and energy are conserved.

✓ solve various one-dimensional collision problems to demonstrate the conservation of momentum.

✓ describe a ballistic pendulum and its purpose, and determine the speed of impacting projectiles.

✓ solve simple two-dimensional collision problems by accounting for conservation of momentum.

✓ discuss how controlling impulsive forces on passengers can reduce injuries and fatalities during car collisions.

FACETS of PHYSICS

Gravity Assists

Ever since the first space flight, men have dreamed of exploring the farthest reaches of our solar system. This dream was considered impossible using typical chemical rocket launchers and propulsion systems. The spacecraft loses much of its initial velocity just escaping from earth orbit. If it is to explore the solar system, it has to use fuel in order to counteract the constant pull of the sun's gravity and regain an appropriate velocity. The amount of fuel that would have to be carried for such a mission is enormous. What great things could be accomplished if only we could get free energy!

The scope of exploration was greatly expanded when aeronautical researchers found a source of virtually free energy available to spacecraft if they can only escape earth orbit. What is this new technology? Well, it is not new. In fact, this force has been around since God formed the universe. It's not really even technology. Energy can be imparted to a spacecraft, increasing its speed, by the exploitation of a planet's gravity and angular momentum to "slingshot" the craft out into space as it passes by the planet. This technique is called a *gravity assist* (or flyby) and has two main advantages. First, because it does not use fuel for acceleration, the amount of fuel the craft must carry is greatly reduced, decreasing its weight. Second,

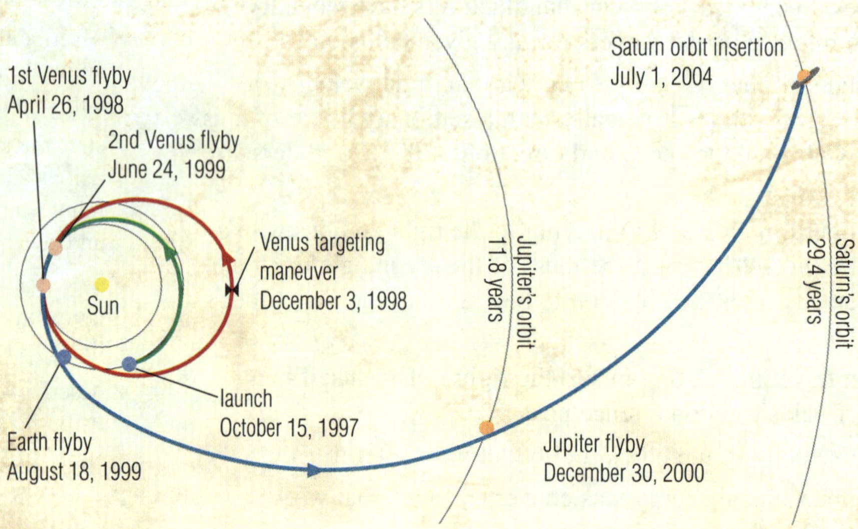

greater speed means it will take less time to reach the destination. The discovery of this method of trajectory design has made possible the launch of unmanned probes to planets that were considered unreachable by traditional methods.

We usually think of the launch vehicle sitting on the pad as a stationary object. That vehicle, though not in earth orbit, is already orbiting the sun along with the earth. When the rocket launches, it travels far enough away from the earth so that, although the earth's gravity is still present, the sun's gravity is the predominant influence, and the spacecraft begins to go into its own orbit around the sun. What happens when that spacecraft approaches a planet that is traveling in the same direction? The planet has a large *angular momentum* in its orbit around the sun. As the spacecraft comes within the planet's gravitational influence, it begins to fall toward the planet. However, since it is traveling so fast, it doesn't go into orbit around the planet but escapes, traveling in a different direction from that which it came in. Relative to the planet, the spacecraft's velocity is the same as before the interaction. Because the planet is traveling so much faster, however, while the spacecraft is "stuck" to the planet (under the gravitational influence), the spacecraft accelerates. There is an exchange of angular momentum between the planet and the spacecraft. The spacecraft gains angular momentum and the planet loses angular momentum. The planet is so massive that the loss doesn't appreciably affect its motion, but because the spacecraft's mass is so small, it can accelerate greatly. Relative to the sun, the spacecraft is going much faster when it leaves the planet than before it arrived. Gravity assists can also be used to decelerate a spacecraft by flying by in a direction opposite to the planet's orbit.

The trajectory designed for the mission must take into account not just the position of the destination planet, but the positions of the intermediate planets as well. This is why there is often only a certain time that a mission can be launched. The larger planets, because of their greater masses, are ideal for gravity assists. The trajectory of the Cassini-Huygens mission to Saturn (launched in 1997) included two flybys of Venus and one each of Earth and Jupiter. Notice that the spacecraft was first launched toward the sun even though its final destination was in the opposite direction. This trajectory imparted an acceleration which would have taken 68000 kg (75 tons) of fuel to achieve in a direct trajectory. The Cassini-Huygens mission arrived in the Saturnian system in July 2004, received a two-year mission extension in 2008, and may continue providing valuable data of the planet and its moons for years beyond 2010. Other current missions using gravity assist or gravity braking are the *Messenger* probe to Mercury (arriving at its final orbit around that planet in 2011) and the New Horizons mission to Pluto, arriving in 2015.

The *Cassini-Huygens* probe arriving at Saturn after a seven-year trip

11C CENTER OF MASS AND ANGULAR MOMENTUM

11.14 Explosions and Center of Mass

A fireworks rocket streaks into the sky. The initial burst throws out sprays of light. It isn't your imagination if it seems like the whole pattern of color continues to climb with the same direction and speed as the original rocket. If you do not look carefully, the parts of the explosion just seem to scatter randomly. However, if it were possible to plot the positions of all the particles of the rocket during the burst, you would find that their average positions would correspond to the original trajectory of the rocket's center of mass. **Center of mass** is an important and useful concept in physics for studying the behavior of systems containing two or more objects. The center of mass is the point in space where the mass of the entire system is assumed to be located for the purpose of determining the momentum of the system. Though conceptually easy to understand, actually determining its location requires calculus for all but the simplest systems. In general, the center of mass is proportionately closer to the parts of a multi-body system that have the greatest mass. Thus, in the moments following the rocket's burst, the most massive parts of an exploding fireworks rocket will be closest to the point of explosion (because they have the greatest inertia), and the least massive will be the farthest away.

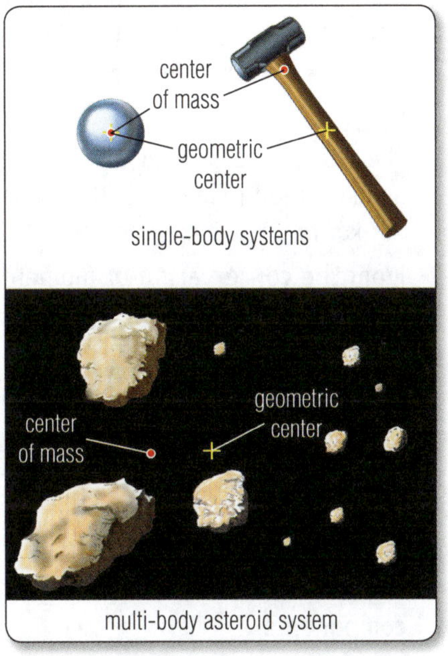

11-17 Centers of mass for single- and multi-body systems

If you could run the explosion in reverse, it would seem that all of the particles converged in one great simultaneous inelastic collision. This is a reasonable approximation of the situation if we ignore the mass of the explosive.

If we consider the entire collection of particles resulting from an explosion to be the system, then the most significant forces acting on the particles during the instant of the explosion are internal forces. Because internal explosive forces are usually very large compared to external forces such as gravity and air resistance, they will predominate when affecting the particle momenta at the instant of the explosion. Internal forces do not affect the total momentum of the system. Therefore, the sum of the momenta of all particles immediately after an explosion or disintegration is equal to the initial momentum of the object (assuming external forces are insignificant),

$$\mathbf{p}_{\text{sys bfr}} = \Sigma \mathbf{p}_{\text{i aft}},$$

where $\Sigma \mathbf{p}_i$ is the vector sum of the momenta of all of the individual particles.

Although momentum is conserved to a good approximation, the total kinetic energy of the particles is not. The forces that accelerate particles in explosions and other kinds of disintegrations come from chemical, mechanical, or other nonconservative sources.

11-16 To a first approximation, the center of mass of the glowing particles of an exploding fireworks rocket initially follows the same path that the rocket would have followed.

Momentum is a vector quantity that can be oriented in any direction in three dimensions. Each component of the momentum vector has its own vector equation.

The momentum of an object before an explosion is equal to the sum of the momenta of its parts after the explosion if external forces acting on the system are insignificant compared to the internal forces involved.

EXAMPLE 11-9

Breaking Up: Conservation of Momentum and Explosions

A 12.0 kg blob-like object is rising vertically at a speed of 20.0 m/s. At some point, it explodes apart two-dimensionally into three smaller parts. The 1.0 kg part continues straight up at 40.0 m/s. The 3.0 kg part flies away horizontally to the right at 10.0 m/s. Find the direction and speed of the remaining 8.0 kg piece.

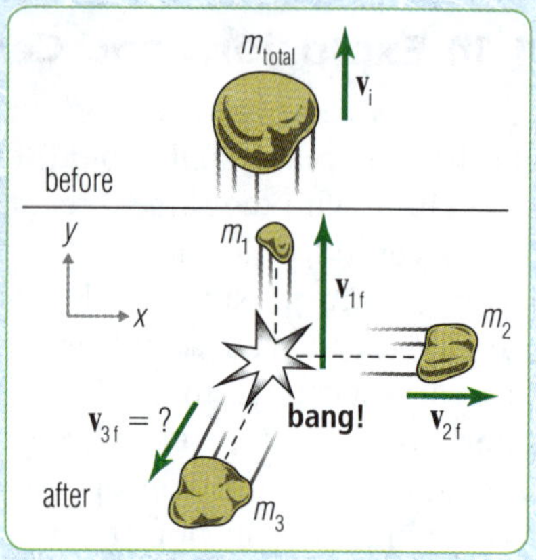

11-18 The explosion remnants

Solution:

The coordinate system and velocities are as indicated in Figure 11-18. Masses $m_1 =$ 1.0 kg, $m_2 =$ 3.0 kg, and $m_3 =$ 8.0 kg.

From the conservation of momentum, you know the following equation relationships are true:

$$\mathbf{p}_{\text{total i}} = \mathbf{p}_{\text{total f}}$$

$$m_{\text{total}}\mathbf{v}_{\text{i}} = m_1\mathbf{v}_{1\text{ f}} + m_2\mathbf{v}_{2\text{ f}} + m_3\mathbf{v}_{3\text{ f}} \qquad (1)$$

Rewrite Equation 1 in terms of its x- and y-components:

$$m_{\text{total}}v_{\text{i}\,x} = m_1v_{1\text{ f}\,x} + m_2v_{2\text{ f}\,x} + m_3v_{3\text{ f}\,x}, \text{ and}$$

$$m_{\text{total}}v_{\text{i}\,y} = m_1v_{1\text{ f}\,y} + m_2v_{2\text{ f}\,y} + m_3v_{3\text{ f}\,y}$$

Solve each of these equations for the unknown horizontal and vertical components.

x-component:

$$v_{3\text{ f}\,x} = \frac{m_{\text{total}}v_{\text{i}\,x} - m_1v_{1\text{ f}\,x} - m_2v_{2\text{ f}\,x}}{m_3}$$

$$v_{3\text{ f}\,x} = \frac{(12.0\text{ kg})(0\text{ m/s}) - (1.0\text{ kg})(0\text{ m/s}) - (3.0\text{ kg})(+10.\text{ m/s})}{8.0\text{ kg}}$$

$$v_{3\text{ f}\,x} = -3.\underline{7}5\text{ m/s}$$

y-component:

$$v_{3\text{ f}\,y} = \frac{m_{\text{total}}v_{\text{i}\,y} - m_1v_{1\text{ f}\,y} - m_2v_{2\text{ f}\,y}}{m_3}$$

$$v_{3\text{ f}\,y} = \frac{(12.0\text{ kg})(+20.\text{ m/s}) - (1.0\text{ kg})(+40.\text{ m/s}) - (3.0\text{ kg})(0\text{ m/s})}{8.0\text{ kg}}$$

$$v_{3\text{ f}\,y} = +25\text{ m/s}$$

Determine the velocity of the third piece:

$$|\mathbf{v}_{3\text{ f}}| = \sqrt{v_{3\text{ f}\,x}^2 + v_{3\text{ f}\,y}^2} = \sqrt{(-3.\underline{7}5\text{ m/s})^2 + (+25\text{ m/s})^2}$$

$$|\mathbf{v}_{3\text{ f}}| \cong 2\underline{5}.2\text{ m/s}$$

$$\alpha = \tan^{-1}\frac{|v_{3\text{ f}\,y}|}{|v_{3\text{ f}\,x}|} = \tan^{-1}\frac{|+25\text{ m/s}|}{|-3.\underline{7}5\text{ m/s}|}$$

$$\alpha \cong 81.5°$$

The third piece moves away at about 25 m/s at 81.5° above the horizontal to the left.

11.15 Angular Momentum

A mass being whirled around on a string or a spinning wheel has momentum, but its momentum is somewhat different from the momentum of a mass traveling in a straight line. Mass moving in a circle is said to have **angular momentum (L)**, which means momentum associated with circular motion.

Let's look at a dimensionless particle of mass m orbiting a point at a distance r with tangential speed v_t. The magnitude of the particle's instantaneous linear (tangential) momentum, p, is

$$p = mv_t.$$

Angular momentum is the product of the linear momentum and the radial distance of the particle from the axis of rotation. Therefore,

$$L = rp = mrv_t. \qquad (11.15)$$

The units of angular momentum are kg·m²/s. There are no derived SI units for angular momentum.

Recall that tangential speed is related to the radius of the circular path and the angular speed by the formula

$$v_t = r\omega.$$

So, angular momentum for a *single orbiting particle* is

$$L = mr^2\omega. \qquad (11.16)$$

Similar to linear momentum and external forces, angular momentum is conserved unless an external force is applied to the rotating system. For example, when a mass is whirled on a string of radius r with an angular speed of ω, if the string's length is reduced to half the radius, its tangential speed must double and its angular speed must increase as the square of the change in tangential speed. This is the same principle that a competition ice skater uses when performing a spin. His rotational speed is slower with his arms extended and much faster when the mass of his arms is held close to his body.

Assuming that no external forces act on the system:

$$\Delta L = m_2 r_2^2 \omega_2 - m_1 r_1^2 \omega_1 = 0 \text{ kg·m}^2\text{/s}$$

$$m_2 r_2^2 \omega_2 = m_1 r_1^2 \omega_1$$

Mass doesn't change, so $m_1 = m_2 = m$. The radius reduces to half, so

$$r_2 = \frac{r_1}{2}.$$

Substituting:

$$mr_2^2\omega_2 = mr_1^2\omega_1$$

$$\cancel{m}\left(\frac{r_1}{2}\right)^2\omega_2 = \cancel{m}r_1^2\omega_1$$

$$\frac{\cancel{r_1^2}}{4}\omega_2 = \cancel{r_1^2}\omega_1$$

$$\omega_2 = 4\omega_1$$

The angular speed of the mass quadruples.

11-19 The secret to a successful hammer throw is developing maximum angular momentum.

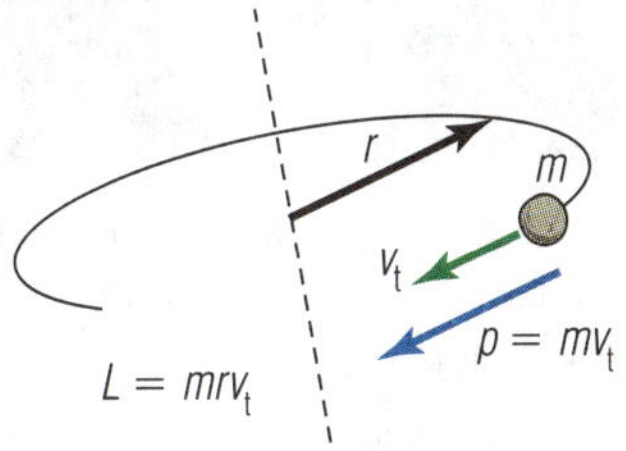

11-20 Derivation of Equation 11.15

11-21 The flywheel of a heavy machine stores angular momentum.

The magnitude of angular momentum of a revolving particle is equal to the product of the orbit radius and the linear momentum of the particle. The unit of angular momentum is kg·m²/s.

11-22 A skater draws his arms in to his axis of rotation to increase his rate of spin without changing his angular momentum.

Angular momentum of a particle is a vector quantity that is, technically, the product of two vectors—tangential momentum and the radial position vector. While its magnitude can be found from either Equation 11.15 or 11.16, its direction is found using the rotational right-hand rule (see Chapter 7). With the fingers of the right hand curled around the rotational axis in the direction of rotation, the extended thumb points in the direction of the angular momentum vector. Counterclockwise rotation has positive angular momentum.

11C Objectives

After completing this section, I can

✓ define the center of mass of a system.

✓ describe how an "ideal explosion" conserves a system's momentum.

✓ define angular momentum and identify the properties of a system that contribute to angular momentum.

11C Section Review

1. When analyzing the momentum of a system of particles resulting from an explosion, discuss at least two assumptions that are made to simplify the analysis.
2. If a trap shooter shatters a flying clay pigeon with a blast of shot from his shotgun, will the particles follow the original trajectory of the intact clay pigeon? Explain your answer.
3. What must occur for the angular momentum of a rigid spinning system to increase?
4. Six children are riding on the perimeter of a playground merry-go-round. If they all lean outward simultaneously, will the merry-go-round speed up, slow down, or continue spinning as before? Explain what happens.
5. ✪ Two identical air track gliders of mass 0.125 kg approach each other at equal speeds of 0.500 m/s. Assume that the two gliders are a system. Since the center of mass of a system of objects is the point at which all of the mass can be assumed to be concentrated, choose a convenient coordinate system and determine the location of the center of mass of the system for the following times:
   a. 0.5 s before the collision
   b. collision
   c. 0.5 s after the collision

   What is the momentum of the center of mass for this system?
6. ✪ Determine the speed of the center of mass of a system of the two gliders described in Question 5 if glider 1 is stationary and glider 2 approaches glider 1 at 0.500 m/s to the right. What is the momentum of the system in this case?

Chapter Review

In Terms of Physics

momentum (**p**)	239
impulse (**I**)	241
recoil	244
conservation of momentum	244
thrust	244
elastic collision	247
partially elastic collision	247
inelastic collision	247
ballistic pendulum	252
center of mass	257
angular momentum (**L**)	259

Problem-Solving Strategies

11.1 (page 240) You will be introduced to many equations in this chapter, but you do not need to memorize them. So many conditions apply to using most of the equations that you probably could derive them from the principles you will learn before you could recall them or find them in your notes. Physics is not just about memorizing equations.

11.2 (page 246) Evaluate the effect of all external forces on the system as a whole. If there is no net external force, then the total change of momenta of all parts of the system is zero and momentum is conserved.

11.3 (page 248) Quantities with multiple subscripts, such as $v_{1\,\mathrm{bfr}\,x}$, can be verbally expressed as "vee-one-before." The subscript x is written to indicate a scalar component but may be left unspoken if no confusion will result.

Review Questions

1. What is linear momentum? How does it differ from any other kind of momentum?

2. How do you determine the direction of the linear momentum vector?

3. What is the momentum of a 10 000 kg truck at rest?

4. Define *impulse*. For a constant net external force on a system, what determines the magnitude of the force's impulse?

5. How does a gymnast reduce or eliminate injury from a fall by using the principle of impulse?

6. At the instant it is struck, a baseball has the following forces acting on it: the bat's impact, the earth's gravitational force, the sun's gravitational force, atmospheric drag, and the pseudo-force due to the Coriolis effect. Which of these forces could be considered an impulsive force?

7. State the law of conservation of momentum.

8. What basic condition must exist for a system's momentum to remain unchanged?

9. Why is momentum *not* conserved when a rock is dropped?

10. A wad of soft clay is thrown against a wall and sticks in place. Is the clay's momentum conserved? Explain your answer.

11. For a given rocket engine at a fixed throttle setting, the force exerted by the engine is constant. Which rocket will achieve the higher final velocity with the same thrust setting—a rocket with a large propellant-to-payload ratio or one with a small ratio?

12. Which of the following are elastic collisions?

a. A piece of clay collides with a baseball and sticks to it.
b. A golf ball strikes an identical resting golf ball, which moves away with the original ball's speed while the original ball stops.
c. A polystyrene block strikes a table edge and bounces off, squashed.
d. A bullet strikes and lodges in a ballistic pendulum.
e. A steel ball bearing dropped on a steel plate rebounds with the same speed it had before the impact.

13. Which of the following is *not* conserved in an inelastic collision?

a. kinetic energy
b. momentum
c. mass
d. total mechanical energy

14. Under which situation will a moving object (m_1) completely transfer its mechanical energy to a stationary object (m_2) during a one-dimensional collision?

a. $m_1 << m_2$
b. $m_1 = m_2$
c. $m_1 >> m_2$
d. m_1 sticks to m_2

15. A physicist observes a subatomic particle passing through a special apparatus called a bubble chamber that makes the tracks of particles visible. It decays at one point into at least two particles that also leave tracks in the chamber. Assuming that the tracks in the figure all lie in the same plane, how can he conclude that there must have been at least one additional undetected particle resulting from this disintegration?

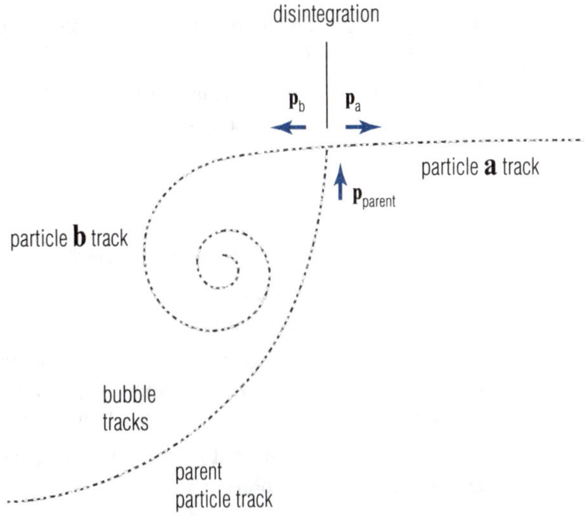

16. Using your knowledge of angular momentum, explain why aerial ropewalkers use long bars to maintain balance during their stunts.

17. A planet orbiting a star moves closer to the star. If no outside forces act on the planet-star system, what would have to happen to the planet's orbital speed?

18. A competition high diver often tucks in order to obtain as many revolutions as possible before entering the water. Explain why timing the point at which he straightens out is critical to successfully completing the dive. Does the diver actually stop rotating?

19. The engine of an Alaskan bush pilot's plane sputters and dies, and the pilot bails out with a parachute and a survival kit. He lands in the middle of a perfectly smooth frozen lake. There is no wind, no friction, and no wood on the lake. The survival kit contains two-weeks' supply of canned food, a can opener, a hatchet, and matches. If he stays in the middle of the lake, he will probably freeze to death before help comes. However, if he can reach the lake's wooded shore, he can survive the week needed to reach town. Using the principles of momentum conservation he learned in high school, how can he reach the lake's shore before he freezes to death?

20. A motorcyclist collides with a bumble bee (*Bombus griseocollis*) squarely in the middle of his helmet visor while driving 113 km/h (around 70 mi/h). Assuming the collision is elastic (*not* a good assumption), which receives the larger impulse?
   a. the motorcyclist
   b. the bumble bee
   c. neither—the impulses are equal

DS21. What kinds of design features can auto manufacturers include in their vehicles to reduce the severity of a collision impulse on passengers?

22. For a given speed of rotation, which property of the rotating system is more significant to its angular momentum: the mass of the system or the way the mass is distributed around the rotating axis? Explain.

True or False (23–32)

23. The object with the greatest velocity always has the greatest momentum.
24. Under normal circumstances, the momentum of an object changes because its velocity changes.
25. Any unbalanced external force changes an object's momentum.
26. A car that maintains a constant speed around a corner has constant momentum.
27. Gravity cannot be an impulse because it does not act suddenly like an explosive force or a baseball bat striking a ball.
28. An inelastic collision is a collision in which momentum is not conserved.
29. Most collisions are partially elastic as long as no external forces act on the system of colliding objects.
30. Ideally, the motion of the center of mass of an object is not affected if the object explodes.
31. Object 1 collides elastically with stationary object 2. Both move off in the same direction. From this scenario, you can conclude that object 1 had more mass than object 2.
32. The angular momentum vector points in the same direction as the motion of the rotating object.

33. What is the momentum of a 0.50 kg kitten running with a velocity of 1.00 m/s to the left?

34. Compare the momentum of a 1500. kg car traveling at 20. km/h north with the momentum of a 900. kg car traveling at 40. km/h north.

35. A 1.00 kg ball traveling at a velocity of 3.00 m/s west strikes a wall and rebounds with a velocity of 3.00 m/s east. The ball contacts the wall for 0.0100 s.
   a. What is the ball's momentum before it hits the wall?
   b. What is the ball's momentum after it hits the wall?
   c. What is the ball's change in momentum?
   d. What force must the wall exert on the ball in order to change its momentum?

36. A 0.400 kg cart traveling 0.0750 m/s to the right encounters an ideal spring and compresses it. In the next 0.125 s, the spring exerts an average force of 0.480 N to the left on the cart.
   a. What is the cart's momentum before it reaches the spring?
   b. What is the change in the cart's momentum as a result of the encounter?
   c. What is the cart's momentum after it leaves the spring?
   d. What is the cart's velocity after it leaves the spring?

37. A 975 kg car travels northeast at 8.00 m/s.
   a. What is its momentum?
   b. What impulse is needed to stop the car?
   c. What force is needed to stop the car in 10.0 s?
   d. How long would it take a 7.80×10^3 N force to stop the car?

38. A 10.0 g bullet leaves the muzzle of a 4.00 kg rifle with a velocity of 850. m/s east.
   a. What is the bullet's momentum?
   b. What is the rifle's momentum?
   c. What is the recoil velocity of the rifle?
   d. If the explosion took 0.002 00 s, how much force does the rifle exert on the rifleman's shoulder?

39. A 1.50×10^3 kg freight dolly travels north at a speed of 1.50 m/s. A 125 kg box drops vertically onto the dolly in an inelastic collision.
   a. What is the dolly's momentum before the box drops onto it?
   b. What is the momentum of the box and dolly together after the box lands on the dolly?
   c. Is this an example of conservation of momentum within a system? Explain your answer.

40. A 25.0 g golf ball traveling to the right with a speed of 10.0 m/s collides elastically with a stationary 145 g baseball. Calculate the velocity of each ball after the collision.

41. A 176 g billiard ball (m_1) collides elastically with an identical stationary billiard ball (m_2). With a speed of 1.50 m/s, m_2 moves off in the original direction of m_1.
   a. What was the initial speed of m_1 ($v_{1\ bfr}$)?
   b. What is the final speed of m_1 ($v_{1\ aft}$)?

42. A croquet ball with a mass of 0.369 kg moves to the left with a speed $v_{1\ bfr}$ of 11.9 m/s toward a stationary golf ball. Assuming that both balls are particles and they collide elastically, the golf ball moves off with a speed $v_{2\ aft}$ of 21.3 m/s in the direction of $\mathbf{v}_{1\ bfr}$.
   a. What is the mass of the golf ball?
   b. What is the final velocity of the croquet ball?

43. A 10.0 g bullet traveling at 850. m/s embeds itself in a stationary 3.60 kg block of wood resting on a frictionless surface. Calculate the final speed of the block of wood.

44. A 0.0150 kg bullet strikes a stationary 2.485 kg ballistic pendulum. The pendulum rises 0.200 m before stopping.
   a. How much does the pendulum-bullet system's potential energy change?
   b. How much does its kinetic energy change?
   c. What is the pendulum-bullet system's speed after the bullet strikes the pendulum?
   d. What is the bullet's original speed?

45. Is it possible for two objects within the same system to collide elastically and both be at rest after the collision? Explain your answer.

46. Research the theory and operation of spacecraft ion propulsion systems. Discuss the advantages and disadvantages of ion propulsion versus conventional combustion rocket engines.

47. If a nuclear particle at rest decays (splits apart) into two identical pieces, explain why the two parts *have* to rebound in exactly opposite directions if no other real or virtual particles are emitted.

The Egg Nebula (CRL-2688) is an unusual celestial object located approximately 3000 light-years from the earth. In this false-color image, the shells of ejected material are illuminated by strange narrow beams of light that highlight the wavelike structure of the nebula surrounding the hidden star.

Periodic Motion 12

Dominion Modeling

Mach Speed

BOOM! The windows rattle, and you hear a dull, deep, loud thud. What was that? When a jet accelerates past the speed of sound, it creates a large burst of acoustic energy called a *sonic boom*. Sonic booms occur because all objects that move through the air at high speeds create high pressure waves in front of them and low pressure waves behind them. Any wave moving through the air is limited by the speed of sound because of the laws of physics. So when a jet approaches the speed of sound, the cone of high-density pressure waves created in front of its nose move closer together and eventually merge into a single shock wave. The trailing edge of this cone intersects the ground well behind the aircraft. The sudden change of pressure at our ears is perceived as a burst of sound similar to an explosion. An aircraft that exceeds the speed of sound usually creates two shockwaves: one from the high-pressure region at the front of the plane and one from the low-pressure region behind the aircraft where the area of low pressure waves suddenly reverts to normal pressure. Depending on the hearer's distance from and orientation to the aircraft, one or two sonic booms may be heard.

12-1 Some aircraft can travel faster than the speed of sound.

Bullets, most missiles, and some jets can travel faster than the speed of sound in a *supersonic* or *hypersonic* condition. Because the speed of sound is not the same at all altitudes, engineers developed a measure of speed through air called **Mach (M) speed**, the ratio of the object's speed to the speed of sound at that altitude. In this chapter's dominion modeling exercise, you will investigate sound's wave properties and the factors that affect its speed.

There are four general speed *regimes*:
subsonic—$v <$ M1
transonic—M0.8 $< v <$ M1.2
supersonic—M1 $< v <$ M5
hypersonic—M5 $< v$

12.1 Introduction

Vibrations are all around us. Vibrating air carries sound to our ears. Vibrating electromagnetic fields provide light for our eyes. Vibrating atoms give a feeling of heat through our sense of touch. We depend on vibrations in our everyday life, yet many people are oblivious to them.

We also use vibrations to make our lives easier. Precisely controlled electronic vibrations in the form of radio waves speed communication. Microwave ovens generate electromagnetic vibrations that force water molecules in food to vibrate, heating the food quickly. Clocks depend on regular vibrations to show the passing of time. The repetitive motion of an engine's pistons produces the energy needed to move a car.

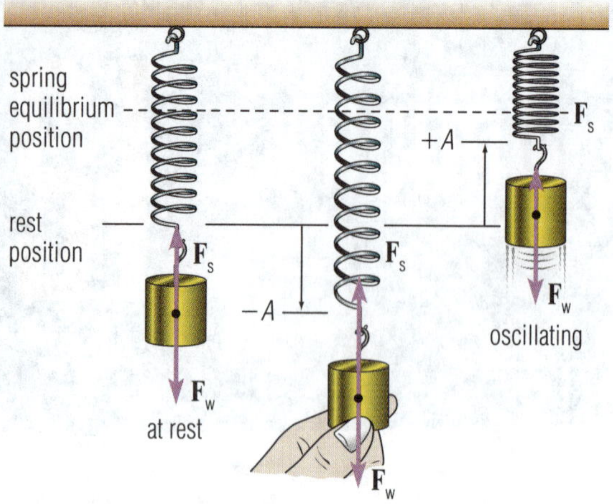

12-2 A laboratory mass oscillates on the end of a spring.

12A SIMPLE HARMONIC MOTION

12.2 Describing Periodic Motion

Vibrations are repetitive motions. For example, if you pluck a violin string, it will vibrate many times. Motion that repeats at a constant rate is called **periodic motion**.

Figure 12-2 shows a laboratory mass attached to an ideal spring hanging from a support. The oscillation of the hanging spring-mass system is an example of periodic motion. Initially, the mass is attached to the spring and allowed to hang at rest. At this position, the force of the spring pulling up on the mass equals the force of gravity pulling down on the mass. Since the mass is at rest, this position is called the *rest position*, or *equilibrium position*, of the system. Note that this is *not* the equilibrium position of the spring alone. When the mass is attached, the spring is stretched downward until an upward force equal to the weight of the mass is developed, according to Hooke's law, by the spring.

To begin the oscillations, the mass is pulled down to $y = -A$. Here the upward force of the spring is greater than the force of gravity on the mass, so when the mass is released, it is accelerated upward. When the mass reaches the system equilibrium position, the forces on it are again balanced, so at this instant it experiences no acceleration (Newton's first law). Velocity is maximum as the mass passes through the equilibrium position. After it rises above the equilibrium position, the spring force continues to decrease while gravity maintains a constant force on the mass. The net force on the mass increases in the downward direction and decelerates the mass until it comes to rest at $y = +A$. The net force then propels the mass downward. The mass again travels through its equilibrium position at maximum velocity, and the increasing spring force finally draws the mass to a temporary halt at $y = -A$. At this point the cycle starts over again.

The force sum on the mass is expressed in the equations $\Sigma\mathbf{F} = \mathbf{F}_W + \mathbf{F}_S$ and $\Sigma F_y = mg_y + (-k\Delta y)$, where the y-axis is vertical and positive upward.

The mass will continue moving up and down forever in the absence of frictional forces. However, in the real world frictional forces are always present. A spring may oscillate for a long time, but it will eventually stop. The effect of friction is called damping. See Section 12C for a discussion of damped oscillators.

Friction that opposes the restoring force in oscillating systems produces damping, the gradual reduction of the amplitude of oscillations.

One reason the mass keeps moving is that the net force on it always tends to return it to its equilibrium position. Below the equilibrium position, the upward pull of the spring is stronger than the downward pull of gravity. Above the equilibrium position the spring's pull weakens, and if the mass passes the spring's rest length, it even changes to a downward push. Thus the net force on the mass always tends to restore the mass to its equilibrium position. The force is therefore called a **restoring force ($\mathbf{F}_r$)**. Often two or more forces must be added to find the restoring force.

12.3 Characteristics of Simple Harmonic Motion

There are many kinds of periodic motion, but periodic motion that is controlled by a restoring force proportional to the system displacement from its equilibrium position is called **simple harmonic motion (SHM)**. The restoring force in simple harmonic motion is described by the formula

$$F_{r\,x} = -k\Delta x, \qquad (12.1)$$

Hooke's-law force may be summarized by the equation $\mathbf{F}_{Hooke} = -k\Delta x$. The force exerted by the spring is proportional to the displacement of the spring and opposite in direction to its displacement.

where k is a constant and Δx (or d_x) is the displacement of the system from the equilibrium position. Notice that Equation 12.1 is identical in form to Hooke's law, so a spring-mass system can approximate simple harmonic motion in the absence of friction. Many natural systems involving periodic motion approximate simple harmonic motion.

The following table summarizes the relationships during simple harmonic motion between restoring force, system velocity, and acceleration.

Table 12-1

Characteristics of an Oscillating System

Position	$x = +A$	$0 < x < +A$	$x = 0$	$-A < x < 0$	$x = -A$	$-A < x < 0$	$x = 0$	$0 < x < +A$	$x = +A$
Restoring Force	$F_{x\ max}\ (-)$	$\lvert F_x \rvert \downarrow$	$F_x = 0$	$\lvert F_x \rvert \uparrow$	$F_{x\ max}\ (+)$	$\lvert F_x \rvert \downarrow$	$F_x = 0$	$\lvert F_x \rvert \uparrow$	$F_{x\ max}\ (-)$
Acceleration	$a_{x\ max}\ (-)$	$\lvert a_x \rvert \downarrow$	$a_x = 0$	$\lvert a_x \rvert \uparrow$	$a_{x\ max}\ (+)$	$\lvert a_x \rvert \downarrow$	$a_x = 0$	$\lvert a_x \rvert \uparrow$	$a_{x\ max}\ (-)$
Velocity	$v_x = 0$	$\lvert v_x \rvert \uparrow$	$v_{x\ max}\ (-)$	$\lvert v_x \rvert \downarrow$	$v_x = 0$	$\lvert v_x \rvert \uparrow$	$v_{x\ max}\ (+)$	$\lvert v_x \rvert \downarrow$	$v_x = 0$

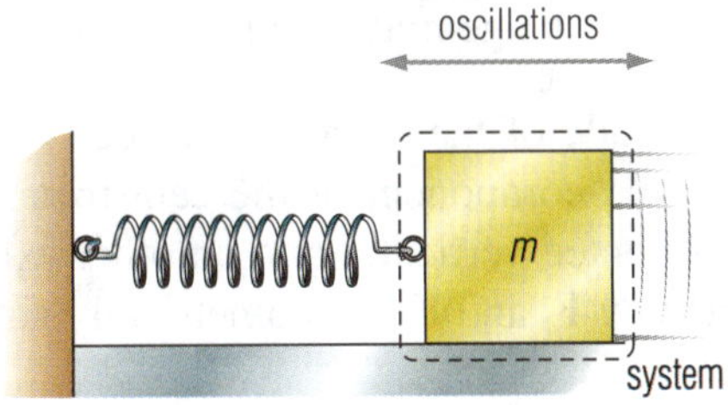

12-3 Diagram of a horizontal oscillating system on a frictionless surface

Simple harmonic motion is different from the straight-line or circular motions that you have studied. One characteristic of simple harmonic motion is that there is a maximum displacement from the equilibrium condition. The mass on the spring never goes farther from its equilibrium position than it was originally pulled, and a pendulum never rises above the height from which it was originally released. This maximum displacement is called the **amplitude** (***A***). An object undergoing simple harmonic motion goes from a distance A on one side of its equilibrium point to a distance A on the other side of its equilibrium point and then back again.

The mass in Figure 12-2 repeatedly goes through the same set of motions: from $-A$ to 0 to A to 0 to $-A$. This set of motions is called a **cycle**. The time that the mass takes to complete a cycle is constant. For example, if the mass moves from $-A$ to $+A$ to $-A$ in 1 s, the next time it moves from $-A$ to $+A$ to $-A$ will also take 1 s. This time is called the mass's *period.* You studied the concept of period in Chapter 7 in association with planetary orbits—another form of periodic motion. Sometimes you want to know how many cycles the mass will complete in a unit of time. This is called the mass's **frequency** (***f***) and is usually measured in cycles per second, or hertz (Hz). A hertz is defined as follows:

$$1 \text{ cycle/s} = \text{s}^{-1} = 1 \text{ Hz}$$

Since frequency is the number of cycles completed per unit of time, and period is the amount of time needed to complete a cycle, frequency and period are reciprocal quantities. In symbols,

$$f = \frac{1}{T} \text{ and } T = \frac{1}{f}.$$

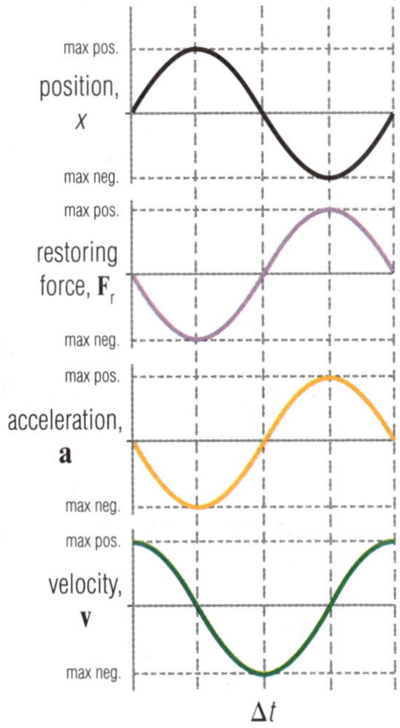

12-4 Graphs showing the relationships between position, restoring force, acceleration, and velocity of an oscillating system

The unit of frequency is cycles per second, or s^{-1}. It is also given the unit hertz (Hz).

Heinrich Rudolf Hertz (1857–94) was a German physicist whose experiments laid the foundation for the wireless telegraph and radio.

Problem-Solving Strategy 12.1

Remember that period and frequency are reciprocals of each other.

12.4 Reference Circle

Circular motion is similar to simple harmonic motion. Some aspects of an object's position in simple harmonic motion are related to position in circular motion. Figure 12-5 shows a ball rotating in a circle at a constant speed on the end of a thin rod. Light shines directly on the rotating ball from the left, casting the ball's shadow on the wall at the right. The shadow moves as the mass in Figure 12-2 moved, from a central "equilibrium" position down to a maximum "amplitude," then back up through the central position to an upward "amplitude," back down, and so on, as long as the ball is rotating. In fact, the motion of the ball's shadow and the motion of the spring-mass system can be synchronized. This helps to analyze the mass's motion. Assume that the mass of the ball equals the mass of the spring-mass system.

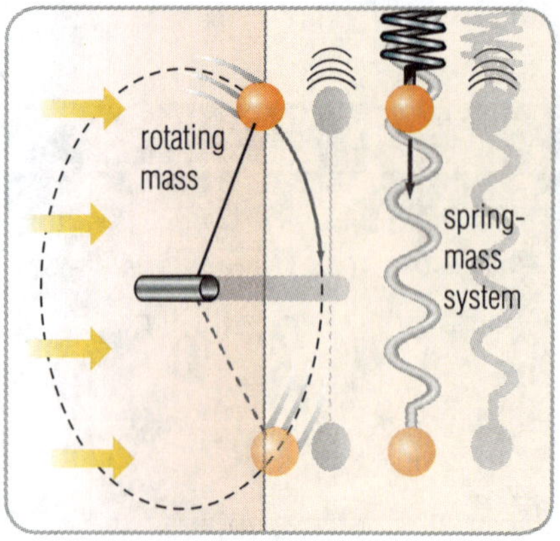

12-5 Circular motion and spring-mass system oscillations are mathematically similar.

When the motions of the ball's shadow and the mass are synchronized, the mass's period equals the ball's period. If the ball's speed around the circle is v_t and its orbital radius is r, then

$$T = \frac{\text{circumference}}{\text{tangential speed}} = \frac{2\pi r}{v_t}. \quad (12.2)$$

In Figure 12-6, the magnitude of the "restoring force" (F_r) that seems to act on the shadow is equal to the x-component of the centripetal force (F_c) on the ball (assuming the x-axis is oriented vertically and no other external forces act on the ball). The triangle formed by $\mathbf{F}_r$ and $\mathbf{F}_c$ is geometrically similar to that formed by Δx and r. Therefore,

$$\frac{F_r}{\Delta x} = \frac{F_c}{r}, \text{ or}$$

$$\frac{F_r}{F_c} = \frac{\Delta x}{r}. \quad (12.3)$$

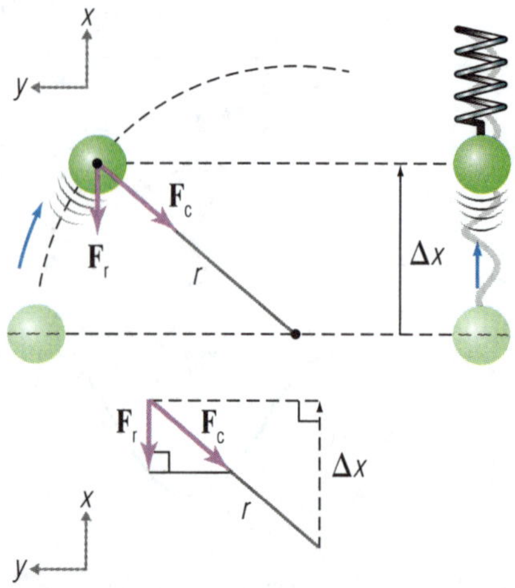

12-6 The geometry showing the similarity between circular motion and oscillations

But $F_r = k\Delta x$, and

$$F_c = ma_c = \frac{mv_t^2}{r}.$$

So, combining these formulas in Equation 12.3,

$$\frac{k\cancel{\Delta x}}{mv_t^2/r} = \frac{\cancel{\Delta x}}{r}, \text{ or}$$

$$\frac{kr}{mv_t^2} = \frac{1}{r}. \quad (12.4)$$

By rearranging Equation 12.4, you can see that

$$\frac{r^2}{v_t^2} = \frac{m}{k}, \text{ or}$$

$$\frac{r}{v_t} = \sqrt{\frac{m}{k}}. \quad (12.5)$$

Substitution of the right side of Equation 12.5 for r/v_t in Equation 12.2 gives

$$T = \frac{2\pi r}{v_t} = 2\pi\frac{r}{v_t}, \text{ or}$$

$$T = 2\pi\sqrt{\frac{m}{k}}. \quad (12.6)$$

Problem-Solving Strategy 12.2

A longer period means that the spring system moves slower and takes longer to complete one cycle. The larger the mass, the longer the period. The "stiffer" the spring, the shorter the period.

You can now calculate the period of the spring-mass system's motion without constructing a corresponding circular motion diagram. The period is proportional only to the square root of the oscillating system's mass, if the quantity k is considered a constant that characterizes the elastic property of the system.

EXAMPLE 12-1

SHM in a Spring-Mass System

A 0.190 kg mass is hung from an ideal spring with a spring constant $k = 76.0$ N/m. It is set in motion by displacing it 3.00 cm downward, then releasing it. (a) What is the period of its motion? (b) What is the frequency?

Solution:

a. Use Equation 12.6 to solve for period:

$$T = 2\pi\sqrt{\frac{m}{k}} = 2\pi\sqrt{\frac{0.190\text{ kg}}{76.0\text{ N/m}}}$$

$$T \cong 0.31\underline{4}1\sqrt{\frac{\cancel{\text{kg}}}{(\cancel{\text{kg}}\cdot\cancel{\text{m}})/(\cancel{\text{m}}\cdot\text{s}^2)}}\ \text{(unit cancellation)}$$

$$T \cong 0.314\text{ s}$$

b. Take the reciprocal of the period to find the frequency:

$$f = \frac{1}{T} = \frac{1}{0.31\underline{4}1\text{ s}}$$

$$f \cong 3.1\underline{8}3\text{ s}^{-1} \cong 3.18\text{ Hz}$$

Notice that the initial displacement, or amplitude, of the mass has no effect on period (T) or frequency (f).

12A Section Review

1. How does the equilibrium position of a mass hanging on a spring compare to the rest position of the spring alone?
2. What is the property of all oscillating systems that tends to return the system to its equilibrium position?
3. What differentiates simple harmonic motion from other periodic motion?
4. If the mass of a system in simple harmonic motion is increased, will the period of the system increase or decrease?
5. An ideal spring with a spring constant of 55.0 N/m is suspended from a support. The spring is 12.0 cm long. What is the stretched length of the spring if a 500. g laboratory mass is hung from the spring?
6. What is the period of the spring-mass system in Question 5?
7. What is the frequency of the spring-mass system in Question 5?
8. Convert the frequency of a sound at 261.6 cycles/s (middle C) to Hz.

12A Objectives

After completing this section, I can

- ✓ describe the motion of a system exhibiting periodic motion.
- ✓ discuss how periodic motion is caused by a restoring force and is opposed by damping.
- ✓ analyze simple harmonic motion (SHM) using proper nomenclature.
- ✓ calculate the period of an oscillating system exhibiting SHM.

Divination and Dominion

Dowsing for water is a form of divination. The Bible forcefully condemns divination (Lev. 19:26; Deut. 18:10). Both divination and science are attempts to learn about and manage the created world. Why is science mandated and divination forbidden?

12-7 Huygens's pendulum clock, based on Galileo's design

12B PERIODIC MOTION AND THE PENDULUM

12.5 Historical Overview

The origin of the **pendulum** is unclear. There is evidence in cave paintings that pendulums were used in some occult way to dowse for water. The pendulum was a mystery to the ancient Greeks because it defied the notion that a heavy weight would single-mindedly seek its natural place at the center of the earth. To have a weight on a rope continuously swing up and down in an arc did not fit with Aristotle's natural order of things.

It was not until the sixteenth century that the pendulum was first scientifically investigated by Galileo Galilei. Although educated in Aristotelian science, he concluded early in his life as a scientist that much of what he had learned was often contradicted by simple observation. For example, while still a student, he observed in the cathedral at Pisa, Italy, that the periods of swinging chandeliers having equal-length chains were independent of the amplitudes of their swings. This was an important discovery that eventually led to research in pendulum regulators for clocks.

The periods of both pendulums and spring-mass systems in simple harmonic motion are independent of the amplitudes of their initial displacements. In this section, we will investigate the properties of pendulums and how they approximate simple harmonic motion.

12.6 Description of Pendulum Motion

All the mass of an ideal pendulum is assumed to be centered at the end of the pendulum arm. This is a reasonable approximation as long as the mass of the arm is negligible compared to the mass at the end of the arm.

A simple or ideal pendulum consists of a mass suspended from an ideal string or a massless rod of length L, called the **pendulum arm**. All of the pendulum's mass (m) is assumed to reside at a point at a distance ℓ from the point of suspension. Therefore, for an ideal pendulum, $\ell = L$. In this section, the symbol ℓ specifically means the distance between the pendulum's pivot point and its *center of mass*. When a pendulum swings, its center of mass follows an arc of a circle with radius ℓ. The need to emphasize this point will be clear at the end of this section when we discuss physical (nonideal) pendulums.

It is important that we specify the time that we inventory the forces acting on the pendulum as the *instant of release*. As the pendulum begins to fall, other forces appear that affect the acceleration of the pendulum.

The forces acting on a pendulum at rest are its weight ($m\mathbf{g}$) and the tension in the pendulum arm ($\mathbf{T}_p$). When the pendulum is hanging still, these two forces are in equilibrium. This hanging position is the equilibrium position for a pendulum. Suppose an experimenter pulls the pendulum to the side so that the pendulum arm makes an angle θ with its equilibrium position. At the instant he releases it, the only forces acting on the mass are still just the weight of the mass and tension in the arm.

12.7 Pendulum Restoring Force

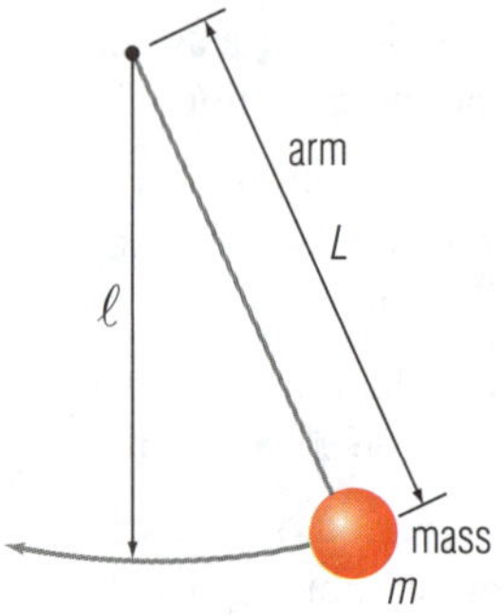

12-8 Parts of a simple pendulum

What is the restoring force that causes the pendulum to return to its original equilibrium position? The only freedom of movement that the pendulum mass has is tangential along a circular path of radius ℓ. What force acts in a tangential direction?

The weight of the pendulum mass is always directed vertically downward. Therefore, weight alone cannot move the mass back to its equilibrium position. If the string were cut, the pendulum could only move downward. The tension applied by the pendulum arm on the mass is oriented radially along the arm, so it cannot cause tangential motion by itself either. Only the sum of these two forces acting at an angle forms the restoring force that returns the pendulum to its equilibrium position.

At rest, the tension in the pendulum arm ($\mathbf{T}_p$) is due only to the weight exerted by the pendulum mass. We will designate this tension vector $\mathbf{T}'_w$. It is nothing more than the reaction force to the component of the weight of the mass acting in line with the pendulum arm ($\mathbf{T}_w$). The vector $\mathbf{T}_w$ is equal in magnitude and opposite in direction to $\mathbf{T}'_w$ according to the definition of third-law force pairs (see Figure 12-9). The magnitude of this component is

$$T_w = |mg| \cos\theta .$$

The other component of the pendulum's weight is perpendicular to $\mathbf{T}_w$. This is the component of weight tangent to the circular path. Its magnitude is calculated by

$$|mg| \sin\theta .$$

This is the force that produces the tangential acceleration responsible for the motion of the pendulum. We will call this vector sum the "restoring force" ($\mathbf{F}_r$) because it is the net force sum on the pendulum mass.

$$F_r = |mg| \sin\theta \qquad (12.7)$$

At the instant of release, the tangential or restoring force is equal to the vector sum of the pendulum tension vector $\mathbf{T}_p$ and the weight vector $m\mathbf{g}$. The vector equation defining the restoring force is

$$\mathbf{F}_r = \mathbf{T}_p + m\mathbf{g}.$$

We will see shortly that this force does not even approximate the simple harmonic motion restoring force except under certain conditions.

After it is released, the pendulum accelerates as it falls toward its equilibrium position. Several interesting things happen. As the pendulum's tangential speed increases, it is being accelerated toward the center of its circular path according to the principles of circular motion. The centripetal force ($\mathbf{F}_c$) that causes this centripetal acceleration is added to the tension in the pendulum arm supporting the weight of the mass. In equation form,

$$\mathbf{T}_p = \mathbf{T}'_w + \mathbf{F}_c,$$

where $T'_w = T_w = |mg| \cos\theta$ and $F_c = mv_t^2/r$. The vector sum of $\mathbf{T}_p$ and $m\mathbf{g}$ is no longer tangent to the pendulum path but tends to rotate toward its interior (see Figure 12-10).

Just as Newton's second law predicts, the acceleration of a pendulum is directly proportional to the net force acting on the pendulum mass. The tangential acceleration vector ($\mathbf{a}_t$) and the centripetal acceleration ($\mathbf{a}_c$) sum together to yield the **total acceleration ($\mathbf{a}_{total}$)** of the pendulum mass. The total acceleration vector points into the interior of the circular path ahead of the system's radial position vector if the system is accelerating and behind the position vector if the system is slowing. The force that causes this total acceleration is the pendulum's "restoring force."

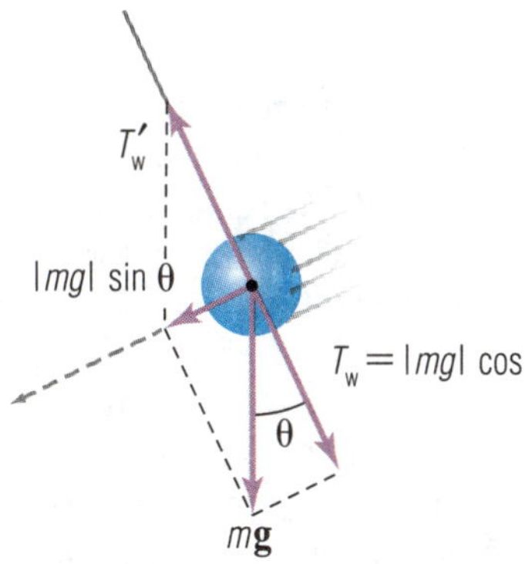

12-9 Forces on a pendulum

The "restoring force" in a pendulum is the vector sum of the pendulum's weight and the tension in the pendulum arm.

A pendulum experiencing large-amplitude swings does not follow the rules for simple harmonic motion.

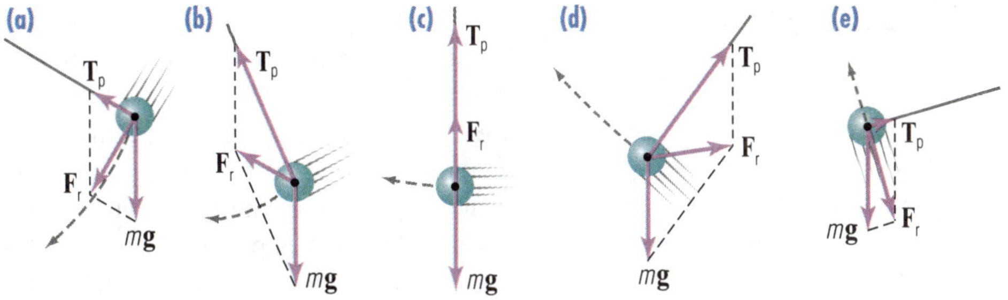

12-10 Relationships between forces and position of a pendulum swinging with a large amplitude

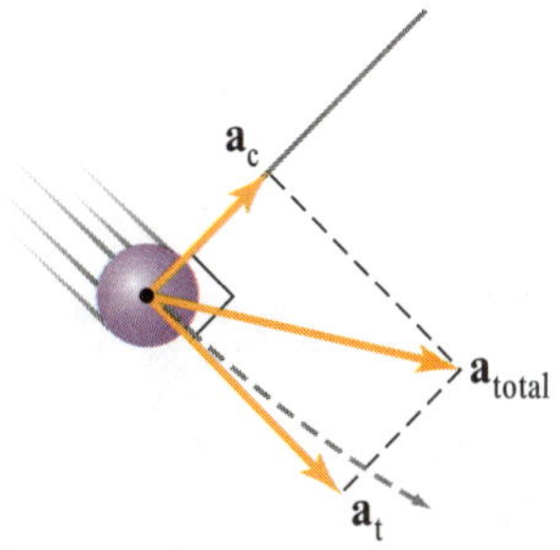

12-11 Components of total acceleration

In nonuniform circular motion, such as that of a pendulum, the vector sum of tangential and centripetal accelerations is its **total acceleration ($\mathbf{a}_{total}$)**.

In the inertial reference frame we are using to analyze the forces on a pendulum mass, the total reaction force to T_p *while the pendulum is moving* is called the centrifugal force on the pendulum. Since this is a force the pendulum mass exerts on the pendulum arm, it does not contribute to the forces acting on the mass to accelerate it in pendulum motion. Thus, it is irrelevant for this discussion.

Tangential acceleration decreases to zero as the pendulum reaches the bottom of its arc because the tangential component of gravitational acceleration disappears at the equilibrium position. As the pendulum rises in the second half of its swing, tangential acceleration begins to increase in the opposite direction to reduce tangential speed. The total acceleration vector points behind the pendulum, indicating that the vector sum of total tension and weight (the "restoring force") acts in the same direction. Examine Figures 12-10 and 12-11 to see how this interplay of forces and accelerations controls a pendulum's movement. It should be evident that the force restoring a pendulum to its equilibrium position is not proportional to the magnitude of the pendulum's displacement, nor does it act in the opposite direction to the displacement. So it seems that a pendulum's motion does not conform to the rules for simple harmonic motion.

You can see that as the pendulum's initial displacement (its amplitude) from its vertical equilibrium position increases, the tangential speed at the bottom of its swing increases. Centripetal force on an orbiting object is proportional to the *square* of its tangential speed, so the "restoring force" on the pendulum is significantly affected by its speed—much more so than by its displacement for large initial angles ($\theta > \pi/8$ rad, or more than 22.5°).

12.8 Small Amplitude Pendulum Motion

If we consider a pendulum starting its swing from a small displacement angle ($\theta < \pi/8$ rad from the vertical), then we can make some helpful approximations to analyze the periodic motion of a pendulum. With small initial displacement angles, the pendulum does not fall very far before it reaches its equilibrium position, nor does it acquire any appreciable tangential speed during its swing. Thus, centripetal force is negligible and tension force in the pendulum arm is essentially dependent only on the weight of the mass and the displacement angle. The restoring force acting on the pendulum is then equal to the vector sum of just the pendulum arm tension and the pendulum's weight. The restoring force vector is also essentially parallel to the displacement of the pendulum mass from equilibrium—a key requirement for simple harmonic motion.

The restoring force of a pendulum oscillator for small angles is dependent on the weight of the pendulum and the pendulum's displacement angle.

The magnitude of the pendulum's initial perpendicular displacement (d) is the length of the side opposite to the initial angle θ in the right triangle formed by the pendulum arm ℓ and the vertical. You can see that

$$d = \ell \sin \theta.$$

The semi-arc length of the pendulum's path ($\overset{\frown}{d}$)—half the pendulum swing—is

$$\overset{\frown}{d} = r\theta = \ell\theta,$$

where θ is in radians. However, for small angles, $\sin \theta \cong \theta$. Therefore,

$$d = \ell \sin \theta \cong \ell\theta.$$

But $\ell\theta = \overset{\frown}{d}$, so for small θ,

$$d \cong \overset{\frown}{d}.$$

The initial perpendicular displacement is nearly the same as the arc length of the pendulum's path from equilibrium. It follows that

$$\sin \theta = \frac{\overset{\frown}{d}}{\ell}.$$

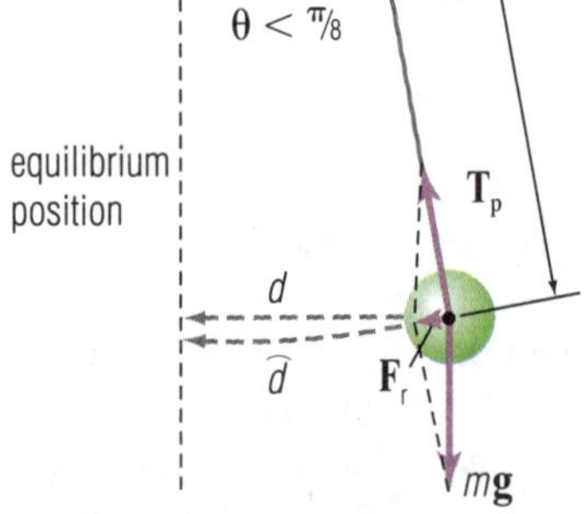

12-12 Forces on a pendulum with a small-amplitude swing

If you substitute this into the restoring force equation (Equation 12.7), you obtain the following:

$$F_r = |mg|\sin\theta$$

$$F_r = |mg|\cdot\frac{\overset{\frown}{d}}{\ell}$$

$$F_r = \frac{|mg|}{\ell}\cdot\overset{\frown}{d} \qquad (12.8)$$

For any given pendulum, the first factor in Equation 12.8 is a constant (k). We have previously shown that $d \cong \overset{\frown}{d}$, so Equation 12.8 has a form of the Hooke's-law force equation,

$$F = kd,$$

where $k = |mg|/\ell = m|g|/\ell$. If you substitute this value for k into Equation 12.6 (simple harmonic motion period formula), you obtain:

$$T = 2\pi\sqrt{\frac{m}{k}} \qquad \text{(Equation 12.6)}$$

$$T = 2\pi\sqrt{\frac{m}{m|g|/\ell}}$$

$$T = 2\pi\sqrt{\frac{\cancel{m}\ell}{\cancel{m}|g|}}$$

$$T = 2\pi\sqrt{\frac{\ell}{|g|}} \quad \text{(for small amplitudes)} \qquad (12.9)$$

Therefore, for small initial displacement angles, a pendulum's motion closely approximates simple harmonic motion. Longer pendulum arms produce longer periods of swing (pendulums appear to swing more slowly). With greater gravitational acceleration, the swing period is shorter (a pendulum swings more rapidly).

The pendulum's period is *not* related to the mass of the pendulum, for the same reason that a heavier mass does not fall with greater acceleration than a smaller mass. Although the larger mass weighs more and has a greater force acting on it, its larger inertia neutralizes the larger weight.

Problem-Solving Strategy 12.3

When calculating the period of a pendulum, be sure to use the *magnitude* of g. If you use the scalar value, the result will be an imaginary number.

For small initial displacements, the period of a pendulum is proportional to the square root of pendulum arm length and inversely proportional to the square root of gravitational acceleration.

EXAMPLE 12-2

Determining Gravitational Acceleration

A physics student discovers from his studies that he can determine the acceleration of gravity at his location using just a pendulum and a digital timer. He carefully measures the period of his 0.750 m pendulum 20 times and calculates an average period of 1.739 s. (a) What is gravitational acceleration at his location? (b) Based on what you know about the accepted value of g near the earth's surface, what could you deduce about the student's location relative to sea level?

Solution:

a. Rearrange Equation 12.9 to solve for the magnitude of g:

$$T = 2\pi\sqrt{\frac{\ell}{|g|}}$$

$$|g| = \frac{4\pi^2\ell}{T^2}$$

$$|g| = \frac{4\pi^2(0.750 \text{ m})}{(1.739 \text{ s})^2}$$

$$|g| \cong 9.7\underline{9}0 \text{ m/s}^2 \cong 9.79 \text{ m/s}^2$$

b. You know that gravitational acceleration decreases with distance from the center of the earth. Therefore, assuming his measurements are accurate, you can deduce that the student lives at a high elevation, perhaps in a mountainous region, since his value of $|g|$ is less than the accepted sea-level value of 9.81 m/s^2.

FACETS of PHYSICS

Foucault Pendulum

12.9 Physical Pendulums

Pendulums considered up to this point could be called "ideal pendulums." All of their mass is concentrated at a point, and the pendulum arm is assumed to be massless. Real systems that swing as a pendulum have their mass distributed to some extent along the length of the pendulum arm. In fact, any object that can swing from a pivot point acts like a pendulum. These systems are called **physical pendulums**. Determining the period of physical pendulums must take into account the distribution of mass and its distance from the center of rotation. This characteristic is constant for any given pendulum and is represented by a quantity called the moment of inertia (I) of the swinging mass. A table of moments of inertia for various-shaped objects is provided in Appendix F.

The period of a physical pendulum has been determined experimentally to be

$$T = 2\pi\left(\frac{\ell}{|g|}\right)^{1/2}\left(\frac{I}{m\ell^2}\right)^{1/2},$$

where the second factor containing the moment of inertia (I) for the physical pendulum "corrects" the basic period formula for the mass distribution of the physical pendulum. When this equation is simplified, we have

$$T = 2\pi\sqrt{\frac{I}{|mg|\ell}}. \qquad (12.10)$$

The moment of inertia (I) of an object quantifies the distribution of its mass around its rotational center.

Problem-Solving Strategy 12.4

The larger the moment of inertia for a physical pendulum, the farther its center of mass is from the center of rotation. Thus, larger moments of inertia correspond to longer pendulum oscillation periods.

Do you believe that the earth rotates once a day? If so, can you prove it? You cannot feel the rotation since the angular speed is about 0.0007 rpm. One strong proof that the earth rotates is the Foucault pendulum, an invention of the French physicist Jean Bernard Leon Foucault.

In 1851 Foucault hung a pendulum from the dome of the Pantheon in Paris. The pendulum consisted of a 28-kilogram iron ball on the end of an approximately 67.5 meter wire, which was attached to its support so that it could swing in any direction. The period of this extra-long pendulum was about 16.5 seconds—it would swing less than four times each minute.

Since the pendulum's only contact with the earth was its support, which exerted only a vertical force on it, the pendulum would not follow the rotation of the earth. If the earth truly rotated, it would move under Foucault's pendulum so that the pendulum's orientation would appear to change. Foucault set the pendulum in motion by pulling the bob to one side and releasing it. He then drew a line on the floor to mark the initial direction of the pendulum's swing. At first the pendulum seemed to continue to follow the line, but eventually a change became noticeable. After 9 hours, the pendulum was swinging perpendicular to the line. After nearly 32 hours, the pendulum's swing had rotated through a full circle. Since there were no unbalanced forces on the pendulum, Foucault proved that the earth must rotate.

You might wonder why the pendulum took 32 hours to complete one cycle. This effect is due to the location of the pendulum. At one of the earth's poles, the pendulum would take exactly 23.93 h to complete one full *precession* cycle as the earth rotates under the pendulum. This was actually done in 2001 at the South Pole. At the equator, the pendulum would not precess at all. However, at any in-between latitude, the period of precession is equal to 23.93 h divided by the sine of the latitude angle. So at the latitude of Paris (48.8N), the precession period should be

$$T_{\text{precession}} = \frac{23.93\text{ h}}{\sin 48.8°} = 31.8\text{ h}.$$

Not only is the precession period affected by pendulum location, but the *direction* of precession is affected as well. In northern latitudes, the precession is clockwise, while it is counterclockwise in the southern hemisphere. This is the phenomenon also responsible for the Coriolis effect, which determines the global cyclical wind and ocean current motions.

Christiaan Huygens (1629–95) was a prominent Dutch physicist and contemporary of Newton. He produced many discoveries in optics, astronomy, and physics, including pendulum theory, geometric optics, and the nature of light.

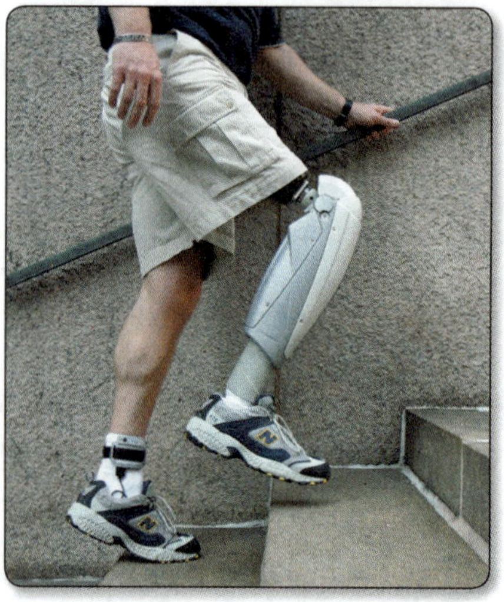

12-13 A smart prosthetic leg that works similarly to a natural leg

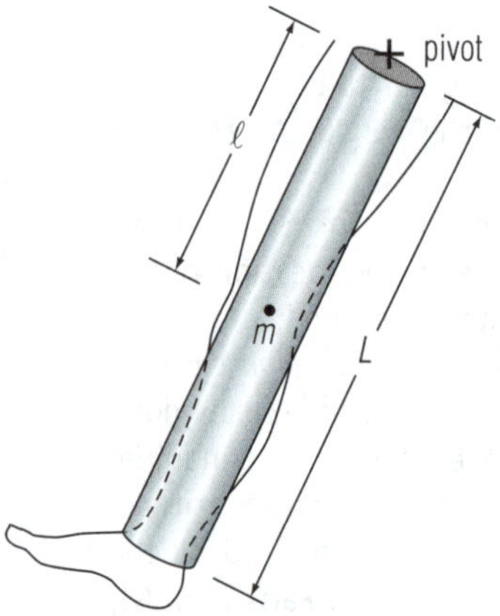

12-14 A schematic leg

Much research has gone into understanding the properties of physical pendulums. Beginning with Christiaan Huygens, the inventor of the first pendulum-regulated clock (see Figure 12-7 on page 270), scientists and inventors have attempted to improve the periodicity and accuracy of pendulums. The most notable of these efforts was the development of a chronometer that would withstand the rigors of the sea so that mariners could determine their longitude. Within the last century, pendulum-like devices were developed to control countless machines and activate safety systems, such as seatbelts and airbags in cars.

In order to understand the walking gait of people, you must know how to determine the natural period of the human leg. This knowledge directly affects the design of athletic footwear and of prosthetic limbs for amputees. The military's research into augmenting human limbs in "powered combat suits" also involves these principles. Learning about and using the principles of such pendulum-like systems are examples of exercising dominion over the world in ways that protect and show love to others.

EXAMPLE 12-3

Period of a Pendulum—Your Leg

What is the natural period of a human leg? This is an important consideration for "smart" prosthetic legs that adjust to the motion of the user.

Solution:

First, model an average human leg as a thin solid rod of length $L = 1.5$ m and mass $m = 16.0$ kg. From Appendix F, the moment of inertia for a thin solid rod is $I = mL^2/3$. The pendulum arm ℓ is the distance from the pivot point to the center of mass. Assume for this example that the mass is uniformly distributed throughout the leg; then $\ell = L/2$.

Use Equation 12.10 to determine the period of a human leg.

Substitute the known information into the formula and solve for T.

$$T = 2\pi\sqrt{\frac{I}{m|g|\ell}} = \sqrt{\frac{\cancel{m}L^{\cancel{2}}/3}{\cancel{m}|g|(\cancel{L}/2)}}$$

$$T = 2\pi\sqrt{\frac{2L}{3|g|}} = 2\pi\sqrt{\frac{2(1.5\text{ m})}{3|-9.81\text{ m/s}^2|}}$$

$$T \cong 2.\underline{0}0\text{ s} \cong 2.0\text{ s}$$

The natural period for an entire human leg is about 2 seconds.

12B Section Review

1. Why was the study of pendulums a problem to Greek natural philosophers?
2. a. When a pendulum is pulled to the side and then released, what forces are acting on the pendulum mass at the instant of release?
   b. What is the effect of the vector sum of these forces?

3. a. What additional force is acting on the pendulum at the bottom of its swing?

   b. Why does this force exist?

4. What limitation exists for considering pendulum motion as simple harmonic motion?

5. (True or False) The period of a pendulum is directly proportional to the length of the pendulum arm.

6. a. What forces must you consider when determining the "restoring force" that causes a pendulum to return to its equilibrium position?

   b. For large amplitude swings, how does the "restoring force" vary with the position of the pendulum arm?

✪7. a. What is the period of an ideal pendulum that has a mass of 0.250 kg attached to the end of a pendulum arm that is 0.650 m long? Assume that it is swinging at the earth's surface.

   b. If the same pendulum were taken to the moon, what would its period be?

✪8. What is the length of an ideal pendulum that swings with a period of 1.40 s near the earth's surface?

✪9. Find the period of a pendulum consisting of a thin solid rod weighing 8.50 kg that is 3.50 m long.

12B Objectives

After completing this section, I can

✓ briefly discuss the historical origins of pendulums.

✓ describe the basic structure of a pendulum.

✓ explain the motion of a pendulum, including the forces that cause its motion.

✓ state the conditions under which pendulum motion approximates simple harmonic motion.

✓ identify the factors that determine the period of a pendulum.

✓ calculate pendulum period.

✓ compare ideal pendulums with physical ones and calculate the period of a physical pendulum.

12C OSCILLATIONS IN THE REAL WORLD

12.10 Damped Oscillations

When an ideal spring-mass system is set to oscillating, it will theoretically continue its simple harmonic motion indefinitely in the absence of any frictional effects. If we hang a real mass from a vertical physical spring in a laboratory and set it to oscillating, in a relatively short time the amplitude of the oscillations is reduced to a slight bobbing that eventually stops. Resistance within the spring and the drag of the air on the mass produce forces that slow the motion of the oscillating mass. An oscillator that experiences such forces is called a **damped harmonic oscillator**. We can find such oscillators in everything from hand bells to diving boards.

12-15 Harmonic oscillators naturally experience damping.

A **damped harmonic oscillator** would experience simple harmonic motion in the absence of all frictional forces.

The mathematical modeling of damping is somewhat complex, and the derivation of the applicable equations is beyond the scope of this textbook. However, we can consider some of the characteristics of damped oscillators. Damping in an oscillator that is modeled by an ideal spring occurs because of frictional drag from the medium. The magnitude of the frictional force is approximately proportional to the velocity of the system through the medium. This can be written in equation form as

$$\mathbf{f}_x = -\beta \mathbf{v}_x,$$

where β is a friction proportionality constant. The friction force opposes the restoring spring force on the system. The net force sum acting on the system is then

$$\Sigma F_x = F_{\text{spring}\,x} + f_x, \text{ or}$$
$$\Sigma F_x = (-k\Delta x) + (-\beta v_x).$$

Rewriting the second equation in the form of Newton's second law, we have

$$-k\Delta x - \beta v_x = ma_x. \quad (12.11)$$

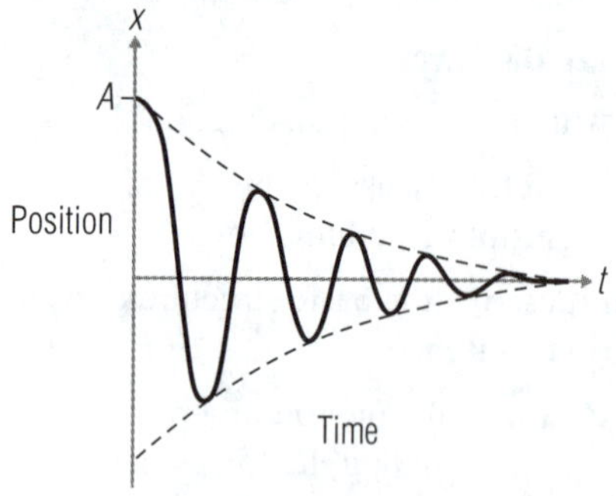

12-16 The amplitude of a damped oscillator gradually diminishes until motion stops.

In Equation 12.11, velocity, and thus the frictional damping force, is dependent on the system's displacement from equilibrium. Velocity is zero at the maximum system displacement and maximum when displacement is zero. When the system is initially released, the spring force accelerates the mass and there is little damping friction. But as system velocity increases, the damping force increases proportionately, so the mass does not accelerate as much as it otherwise would. A lower acceleration means that there is less speed to carry the system beyond the equilibrium point against the increasing restoring force of the spring on the other side. The displacement at the far side of the oscillation is less than the initial displacement. When the system begins to return, there is less initial acceleration because of the smaller displacement. As velocity increases toward the middle of the oscillation, frictional force again reduces the net force accelerating the mass. The net effect is that the amplitude for each half-cycle is smaller than the one before it. Eventually, the motion completely stops.

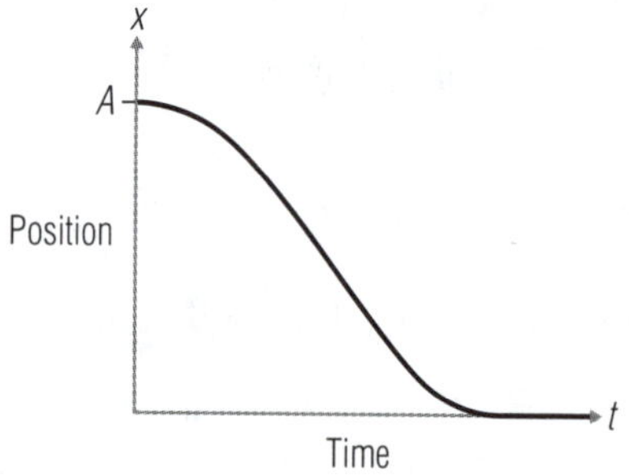

12-17 An overdamped oscillator does not complete even one cycle.

The medium surrounding the oscillator has a large effect on its behavior. Consider a mass hanging from a vertical spring suspended in a thick, viscous liquid such as heavy motor oil or a gooey syrup. If the mass is displaced and released, it slowly moves back to its equilibrium position and no further. In fact, the system is not even an oscillator since it cannot complete even a small fraction of a cycle before its motion is damped out. This motion is called an **overdamped oscillation**.

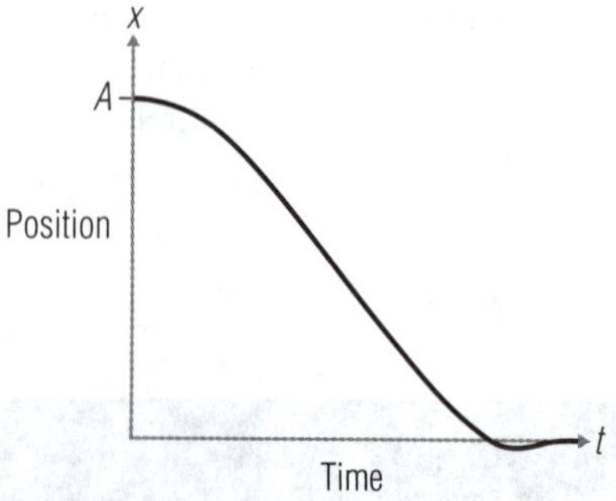

12-18 A critically damped oscillator barely overshoots the equilibrium point before damping stops all motion.

If we gradually thin the liquid by adding an appropriate solvent, we will eventually be able to get the system to just barely overshoot the equilibrium point before it comes to a stop—one truncated cycle, if you will. When the oscillator and medium viscosity are matched so that the mass hardly overshoots the equilibrium point, then the system is **critically damped**. If we continue to thin the liquid medium, then the system behaves like a damped oscillator as described earlier in this subsection.

12.11 Driven Oscillations and Resonance

Although all real vibrations are stopped by friction if left alone, they can be sustained by the application of an additional force to the system. This is known as *driving* the system. The force cannot be applied in a haphazard manner, however. To illustrate this, let us look at a child on a swing.

12-19 Pushing a child on a swing is most effective if the pushes are given at the natural frequency of the child-swing system.

A swing can be considered a simple pendulum. If you give the child a small push, the swing will begin to move in damped harmonic motion and with a small amplitude. The child will probably want to be pushed "faster." Changing tactics, if you perform five very rapid pushes in a short period of time, you note that the amplitude is still not very large and actually works against the swinging motion. What is the correct way to push the swing so that you can increase the amplitude without overexerting yourself and causing uncomfortable accelerations to the child?

You've learned from experience that giving a small push when the swing is at its maximum height (amplitude) produces the desired effect. This works because you are applying a force at the frequency of the swing-oscillator that is assisting the restoring force of the system. The frequency at which the force is most effective in increasing the swing amplitude is called the **natural oscillation frequency (f_0)**. As you push the child in this way, your pushes are in *phase* with the swing's oscillations. Engineers call such pushes in an oscillating system *pulses*.

> Objects tend to vibrate at a characteristic frequency called the **natural oscillation frequency (f_0)**, also called the *resonant frequency*.

If the added force exceeds the frictional damping forces and is in phase with the restoring force on the system, then the oscillator's amplitude will increase. When the oscillations increase in amplitude due to the force supplied at the natural frequency of the system, they are called **driven oscillations**. Driven oscillations can produce very large amplitude oscillations—a condition called **resonance**. Resonance sometimes has a good effect, such as the pure resonating tone in a musical instrument, but sometimes it causes damage.

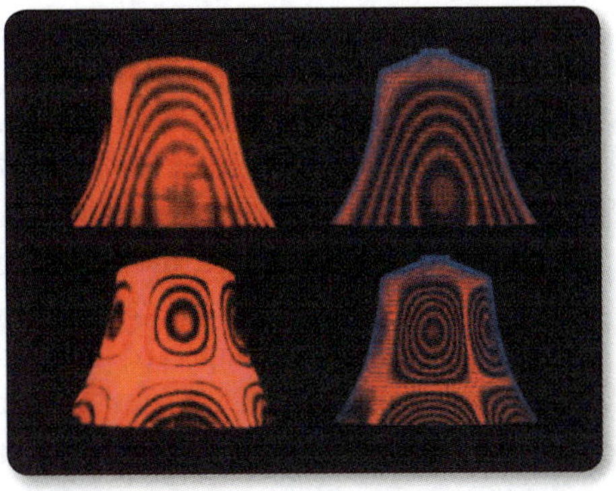

12-20 The resonating regions of the bell are not uniformly distributed over the bell's surface, as seen in these photos taken in polarized light.

The condition of **resonance** is indicated by large amplitude oscillations. These **driven oscillations** result from the periodic application of a force to the system.

One example of the damage resonance can cause is its effect on buildings during an earthquake. Earthquakes are low-frequency vibrations that travel through the earth. Tall buildings tend to suffer greater damage than shorter buildings because their natural frequencies are lower than those of smaller buildings. You can demonstrate this fact by clamping long and short rulers to a desktop and comparing their vibrations. The long ruler vibrates at a lower frequency than the shorter one does. The lower natural frequency of taller buildings is closer to the vibrational frequency of the ground during an earthquake, and resonant conditions are more easily induced, causing much greater damage.

Complex structures such as buildings and bridges have many natural frequencies. Engineers expend a lot of effort and use computer modeling extensively in order to ensure that their designs do not develop resonance under anticipated wind, road traffic vibration, or even earthquake conditions.

A driven oscillator has three forces acting on it:

- The restoring force, or Hooke's-law force;
- The damping resistance force that opposes the velocity of the system;
- The pulsed force that is applied in the same direction as the restoring force.

The applied force may occur as discrete pulses at intervals of T or $T/2$ seconds, but many natural sources of driving forces often vary sinusoidally with time.

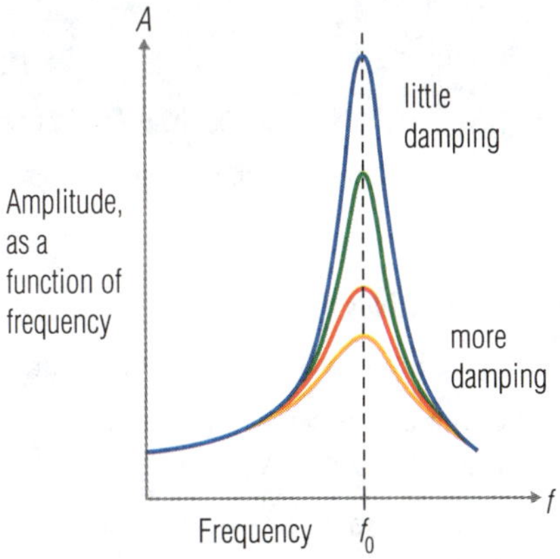

12-21 A graph of frequency versus amplitude showing how damping limits the peak amplitude of a resonating system

Natural sources of oscillation-driving forces include sound, water, and earthquake waves. Each of these can be modeled by a sinusoidal function.

Oscillation amplitude is a function of the frequency of the system. Maximum amplitude occurs when the frequency of the applied force matches the natural frequency of the oscillator. Theoretically, the maximum resonant amplitude is infinite in the absence of damping. However, in every real system there is always some damping, so the maximum amplitude is limited by the damping effects on the system or by the catastrophic failure of the system. These effects become more significant with increasing amplitude, which eventually limits the resonant amplitude to a maximum value.

One well-known example of the catastrophic potential of uncontrolled oscillation is the collapse of the Tacoma Narrows Bridge near Tacoma, Washington. This bridge, spanning a narrow section of Puget Sound, was often buffeted by strong winds blowing through the gorge. In the four short months it was open to traffic, the bridge often swayed in the breeze and regularly experienced transverse wave oscillation along its length, making a drive across the bridge a roller-coaster-like experience.

On November 7, 1940, a 42 mi/h wind blew through the Narrows. The aerodynamic properties of the bridge, combined with its elastic structural design, caused it to enter a condition known as *aeroelastic flutter*, a kind of self-created oscillation. Instead of gently oscillating in transverse waves along its length, the bridge began to oscillate torsionally, twisting back and forth at a frequency of 0.2 Hz. After 45 minutes of this motion, critical bridge components, fatigued beyond their limits, failed, and the bridge's main span crumbled and collapsed into the water.

The bridge was not destroyed by driven oscillations, which depend on the application of periodic external forces near the natural frequency of a system (i.e., simple resonance). Rather, the failure illustrates how poor design can lead to self-created conditions that can cause catastrophic failure with the application of a constant force. The collapse of the bridge had worldwide repercussions in bridge design, and suspension bridges are safer today because of this event. Learning from our mistakes is yet another example of dominion science.

12-22 The collapse of the Tacoma Narrows bridge

12C Objectives

After completing this section, I can

- ✓ compare and contrast ideal and real oscillating systems.
- ✓ discuss the effects of damping on real oscillating systems.
- ✓ explain how damping is classified according to system motion.
- ✓ discuss the factors that contribute to resonance, including driven oscillations.
- ✓ identify factors that limit resonance amplitude.

12C Section Review

1. What eventually happens to a real oscillating system without any outside assistance?
2. Assuming that an ideal spring restoring force acts on an oscillating system, what is the major source of damping force?
3. To which property of an oscillating system is the magnitude of the external damping force proportional?
4. What kind of condition exists when the damping force is so great that no overshoot of the equilibrium position occurs after the system is released?
5. The escapement system in a pendulum-regulated clock applies a small force to the pendulum arm every cycle. What are these momentary forces called, and what kind of oscillations are occurring?
6. At what frequency should force be applied to increase the amplitude of an oscillator?
7. During World War II, army field commanders noticed that floating pontoon bridges quickly fell apart when large numbers of troops marched across them in step. The marching cadence was very close to the natural frequency of the pontoon structures, producing damaging resonance in the bridges. Discuss one way the commanders could (and did) solve the problem without redesigning the bridges.

12D WAVES

12.12 Defining Waves

Simple harmonic motion applies most directly to oscillations of isolated objects, such as a mass attached to a spring or a bob on a string. **Waves**, on the other hand, are oscillations of extended bodies. For example, water waves are oscillations of a body of water. A wave disturbance is just one way energy can be transmitted between two points. A wave stores and carries energy by periodically converting the kinetic energy of particles (in the case of a mechanical wave) into potential energy and back again.

The material through which the waves travel is called the **medium**. An oscillation in the medium is called a **disturbance**. The disturbance usually begins at one point in the medium and travels from there. If you drop a pebble into a smooth pond, the water immediately around the pebble begins to move up and down, and soon the water at the edge of the pond is oscillating. The disturbance has traveled to the edge of the medium.

One thing to notice about a wave is that the *disturbance* travels; the *medium* does not move far. A cork floating in the pond mentioned above bobs up and down but stays in the same area as the disturbance caused by the pebble passes by. The disturbance traveled, but the water supporting the cork did not.

A wave is complicated. Suppose you tie one end of a rope to a tree and hold the other so that the rope is taut. Then you shake the end of the rope. In this example, the rope is the medium for the wave.

How can you describe the wave in the rope? A graph can help. Refer to Figure 12-23. Figure (a) shows the rope 0.5 s after it was shaken. The disturbance has pulled knot 1 down, but it has not yet reached knot 2. Figure (b) shows the rope 1 s after it was shaken. The disturbance has now passed knot 1 and has dropped knot 2. The shape of the disturbance has not changed, but the wave has moved. The graphs in figures (a) and (b) are called *waveform graphs.* A waveform graph is essentially a snapshot of the entire wave at an instant in time. Figures (c) and (d) show a different way of graphing the wave. Figure (c) is a graph of knot 1's vertical position versus time. Figure (d) is the graph for knot 2. These latter graphs are called *vibration graphs.* Vibration graphs show the position of a single point in the medium over a span of time.

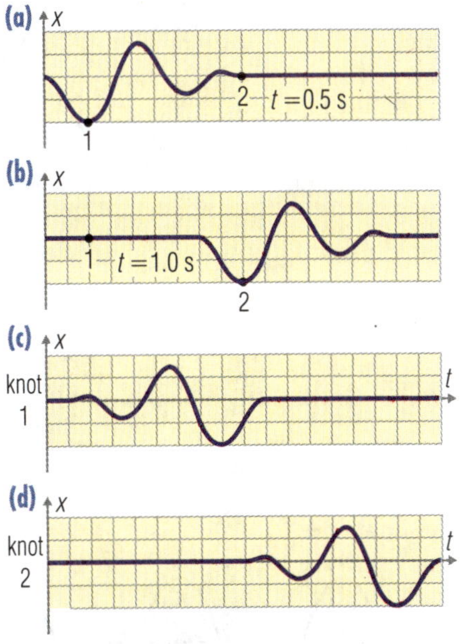

12-23 Waveform graphs for (a) $t = 0.5$ s and (b) $t = 1.0$ s. Figures (c) and (d) are vibration graphs for knots 1 and 2, respectively.

12.13 Types of Waves

There are two basic kinds of waves. A **longitudinal wave** is a disturbance that displaces the medium along its line of travel. A spring, for example, vibrates longitudinally when it stretches and compresses along its length. A **transverse wave** is a disturbance that displaces the medium perpendicular to its line of travel. For example, the wave along a snapped string is a transverse wave. Many waves have both longitudinal and transverse components.

Any physical medium can carry a longitudinal wave. Figure 12-24 shows an air-filled tube with a closed end. A hammer strikes one end of the tube, causing it to vibrate. The end of the tube pushes the air molecules next to it, and these, in turn, push other molecules. This sets up a longitudinal disturbance traveling down the tube.

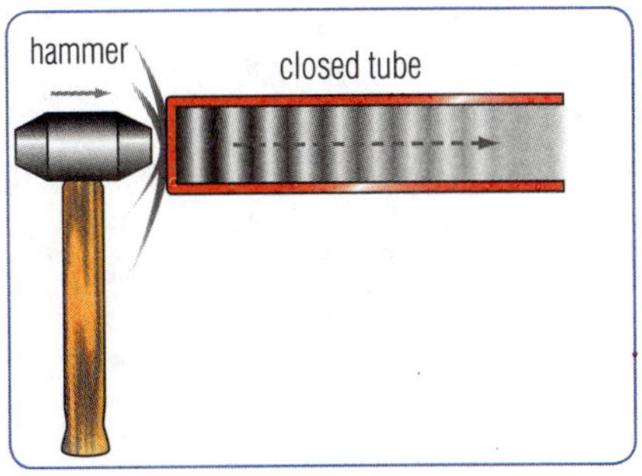

12-24 A longitudinal wave

As the wave travels, the air molecules in some regions are forced together, while the molecules in other regions are pushed apart. The regions where the molecules are pushed together have higher density and pressure than the average for the medium and are called **compression** zones. The regions where the molecules are spread apart have lower density and pressure and are called **rarefaction** zones. These zones form and move with the wave.

A longitudinal wave travels more quickly in a medium whose molecules are closer together (more dense). Therefore, longitudinal waves such as sound travel more rapidly through solids than through gases. For this reason, nineteenth-century Indians in the American West could put their ears to a railway track and detect approaching trains long before the trains could be heard through the air.

Some mechanical transverse waves, unlike longitudinal waves, cannot pass through gases or liquids. However, solids can transmit all forms of mechanical transverse waves. You will learn in later chapters that light is a transverse wave that does not require a medium for propagation.

An example of a wave with both transverse and longitudinal components is a water wave. The transverse and longitudinal forces combine to make a force that causes the particles in a wave to travel in a vertical circle. Some kinds of earthquake waves are also "combination" waves.

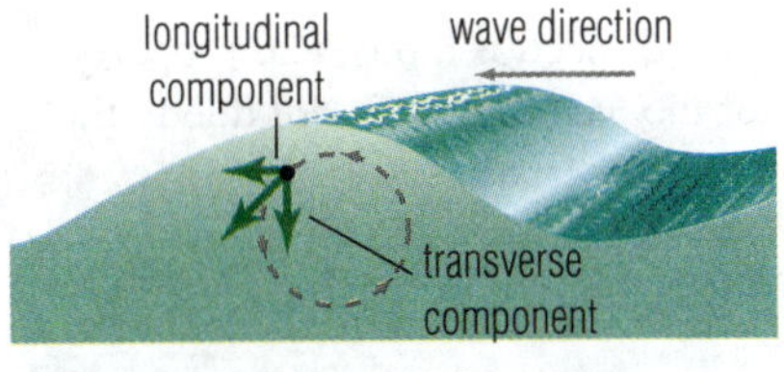

12-25 Velocity components of a particle in a water wave

12.14 Periodic Waves

When a disturbance regularly repeats, it produces a periodic wave. Periodic waves, like the waves resulting from a single disturbance, can carry information and energy from one place to another. Waves carry information the same way a code carries information. For example, if soldiers are told that one whistle means "move ahead," while two whistles means "stay put," then when they hear one whistle, they think "move ahead." Similarly, your ears and eyes regard waves as a code. When the cells in your eyes detect a light wave that vibrates 4.3×10^{14} times each second, they report "red."

Waves also carry energy. For example, tsunamis—huge water waves caused by earthquake shocks—carry an earthquake's energy far from the quake's origin. The 9.2 magnitude earthquake in December 2004 near the Indonesian island of Sumatra produced a series of tsunamis that swept across the Indian Ocean, killing nearly 230 000 people in eleven nations bordering that body of water. The waves traveled as far as Africa and were detectable in the Atlantic Ocean at Africa's Cape of Good Hope.

To describe waves, we use some of the same terms that are used to describe harmonic motion. For a transverse wave, the *amplitude*, A, is the greatest distance the wave displaces a particle from its average position. A longitudinal wave's physical amplitude is usually half the distance between the positions of maximum and minimum densities that it produces:

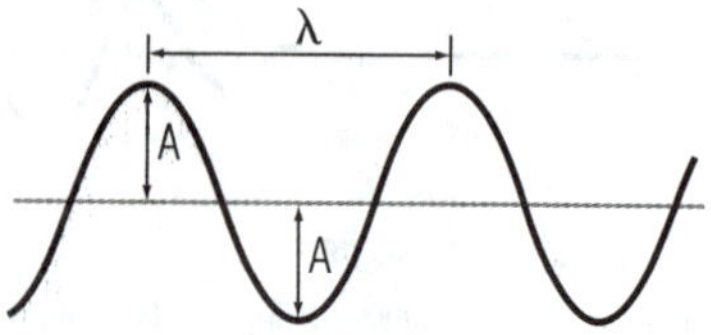

12-26 A wave can be described by its amplitude (A) and its wavelength (λ).

$$A = \tfrac{1}{2}(y_{\text{peak}} - y_{\text{trough}}) \text{ (transverse), or}$$

$$A = \tfrac{1}{2}(x_{\text{max}} - x_{\text{min}}) \text{ (longitudinal)}$$

One **wavelength (λ)** is the distance from one peak (or compression zone) to the next, or from one trough (or rarefaction zone) to the next. The wave completes one *cycle* as it moves through one wavelength. The wave's *frequency* is the number of cycles a wave completes per unit of time.

Each wave that we will study moves at a constant speed called the **wave speed (v)**. This is the speed of the *disturbance*, not of the medium. For example, suppose a traffic light turns red when a careless driver is not looking. The other cars stop, but this driver continues until he strikes the car in front of him. That car hits the next car, which hits the next, and soon the car at the front of the line is struck. It may be 1 s from the time the careless driver rams the last car to the time the first car, 20 m ahead, is struck. The wave has traveled at 20 m/1 s = 20 m/s, or 72 km/h. None of the cars approaches this speed.

For a periodic wave, this speed is related to wavelength and frequency. Since wavelength is the number of meters per cycle and frequency is the number of cycles per second, the product of wavelength and frequency is the number of meters the wave travels per second—the wave speed. In equation form,

$$\lambda f = v, \qquad (12.12)$$

where v is the wave's speed through the medium. By analyzing the units on the left side of Equation 12.12, you can see that the appropriate units for speed appear on the right:

$$\lambda f \Rightarrow \frac{\text{m}}{\cancel{\text{cycle}}} \cdot \frac{\cancel{\text{cycle}}}{\text{s}} = \frac{\text{m}}{\text{s}} \Rightarrow v$$

Wave speed is a physical property of the medium. If the medium is uniform, λ and f are inversely proportional.

Wave speed is a physical property of the *medium*—it is *not* due to the shape or frequency of the wave passing through the medium.

Problem-Solving Strategy 12.5

For a uniform medium, the wave speed is constant for all frequencies. Therefore, frequency and wavelength are inversely proportional. For example, if frequency doubles, then wavelength is halved.

EXAMPLE 12-4

Working With Waves: Determining Wavelength

A photon of red light has a speed of 3.00×10^8 m/s and a frequency of 4.30×10^{14} Hz. What is its wavelength?

Solution:

$$\lambda f = v$$

$$\lambda = \frac{v}{f} = \frac{3.00 \times 10^{8} \text{ m/s}}{4.30 \times 10^{14} \text{ s}^{-1}}$$

$$\lambda \cong 6.9\underline{7}6 \times 10^{-7} \text{ m} \cong 698 \text{ nm}$$

EXAMPLE 12-5

Analyzing Waves

Figure 12-27a shows a wave graph, and Figure 12-27b shows a vibration graph for the same wave. Find the wave's (a) amplitude (A), (b) wavelength (λ), (c) frequency (f), (d) period (T), and (e) speed (v).

Solution:

a. Amplitude:

$$A = \tfrac{1}{2}(d_{\text{peak}} - d_{\text{trough}}) = \tfrac{1}{2}[(+10 \text{ cm}) - (-10 \text{ cm})]$$

$$A = 10 \text{ cm}$$

b. Wavelength: $\lambda = 20$ cm (from Figure 12-27a)

c. Frequency:

$$f = \frac{n \text{ cycles}}{\Delta t} = \frac{1 \text{ cycle}}{2 \text{ s}} \text{ (from Figure 12-27b)}$$

$$f = 0.5 \text{ s}^{-1}$$

d. Period:

$$T = 1/f = \frac{1}{0.5 \text{ s}^{-1}}$$

$$T = 2 \text{ s}$$

e. Speed:

$$v = \lambda f = (0.2 \text{ m})(0.5 \text{ s}^{-1})$$

$$v = 0.1 \text{ m/s}$$

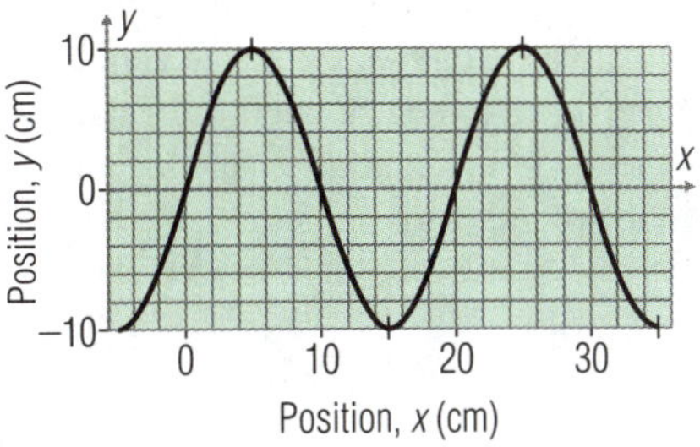

12-27a Wave graph

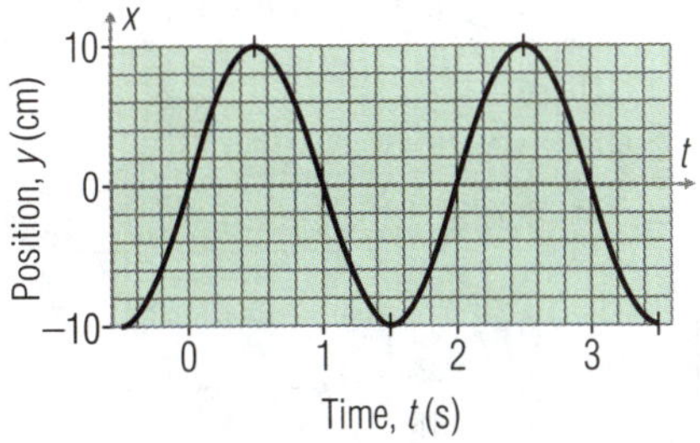

12-27b Vibration graph

12.15 Sound Waves

Sound waves are an important part of everyday life. Sound waves are longitudinal pressure waves that come from a vibrating body, such as a tuning fork, and are detected by your ears. When you strike a tuning fork, it vibrates, and the vibration alternately compresses and rarefies the air around it. This compression and rarefaction sets up a pressure wave that reaches your eardrum. Your eardrum vibrates in response to the wave and stimulates nerves in your inner ear through a mechanical linkage of tiny bones in your middle ear. Your brain interprets the neural impulses from this vibration as sound.

Because sound waves travel by compressing and rarefying their medium, they must have a material medium to travel through. Sound cannot travel through a vacuum. Like other longitudinal waves, sound travels most rapidly through solids and most slowly through gases. Thus, you can hear a sound sooner under water than in air, and even sooner if your head is in contact with a solid. You may have noticed that you hear sounds originating on your head bones differently from sounds only a short distance away. For example, if you scratch a pencil with your fingernail as you

TABLE 12-2

Speed of Sound in Various Media

Medium (at 25 °C)	Speed (m/s)
air	346
hydrogen	1310
glycerol	1904
water	1497
aluminum	6420
Pyrex glass	5640

The loudness of a sound (β) of intensity I_s is computed by the formula

$$\beta = (10\text{ dB})\log\left(\frac{I_s}{10^{-12}\text{ W/m}^2}\right).$$

Problem-Solving Strategy 12.6

The common logarithm (log) of a number is just the exponent of 10 that gives the argument x in log (x). The ratio of I_s over 10^{-12} W/m² yields the relative strength of the sound compared to a reference intensity. Taking the log of this ratio gives the magnitude of the ratio.

hold it a few inches from your ear, you can hardly hear the sound. If you hold the pencil between your teeth as you scratch it, you can hear the sound more clearly. Your solid skull conducts sound much better than the air next to your ear.

The sound you hear depends not only on the characteristics of the wave but also on the characteristics of your hearing. Three basic characteristics, influenced by both the sound wave and your hearing, describe what you hear. *Loudness* is the interpretation your hearing gives to the intensity of the wave. Sound **intensity (I_s)** is the amount of power transported by the wave per unit area and is measured in watts per square meter (W/m²). It is proportional to the square of the wave's amplitude.

The intensity of the sound does not directly equal perceived loudness. A sound must be ten times as intense as another in order to be perceived as twice as loud. The least intense sound a human ear can detect is about 10^{-12} W/m². A sound with an intensity of 10^{-11} W/m² seems only twice as loud as the least intense sound. Therefore, scientists have devised a scale of loudness that reflects this characteristic of our hearing. Loudness was originally measured in *bels*, where an increase of 1 bel doubles the loudness of the sound and increases the sound intensity by a factor of 10. Now scientists measure loudness in tenths of bels, or **decibels (dB)**. A loudness of 10 dB is barely audible to most people. A sound that is 20 dB is twice as loud as a 10 dB sound.

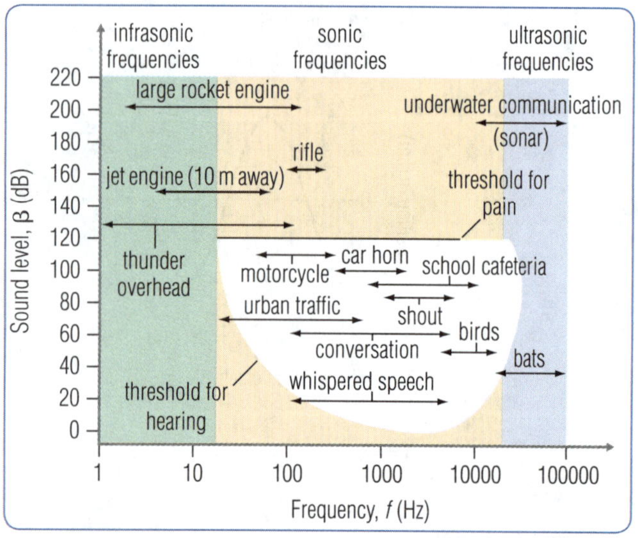

12-28 Comparison of sound levels for various sources of sound and the range of human hearing

A second characteristic of sound is its *frequency*, which you interpret as **pitch**, or how high or low the sound is. A sound with high frequency has a high pitch, and a sound with low frequency has a low pitch. A young person can detect sound frequencies from 20 Hz to 20 000 Hz. Some animals can detect sounds of higher and lower frequencies. Dogs, for instance, detect sounds from 15 Hz to 50 000 Hz, and bats and porpoises detect sounds with frequencies as high as 120 000 Hz.

Most sounds are actually combinations of waves of several frequencies. The particular combinations of frequencies can cause two sounds with the same pitch and loudness to sound different, thus giving sound a third characteristic—**quality**. For example, a trumpet playing middle C at 40 decibels sounds much different from an oboe playing middle C at 40 decibels. Actually, each instrument emits multiples of the frequency of middle C as well as the frequency itself. The frequency of the dominant note is called the **fundamental**; the multiples of this frequency are called **harmonics**. Each instrument produces different harmonics. The oboe tends to emit odd harmonics, while the trumpet emits both odd and even harmonics. Thus, the quality of the sounds from the instruments is different even when their pitch and loudness are the same.

The three characteristics of sound—loudness, pitch, and quality—affect the way sound is perceived. Sounds with high pitch tend to seem louder than sounds of low pitch. Also, the loudness of a sound affects its quality. You are less likely to hear the harmonics in a barely audible sound than in a sound that is comfortably loud. Therefore, you cannot fully describe one characteristic of sound without also describing the other characteristics.

12.16 Doppler Effect

So far, we have assumed that the observer and the wave source are not moving with respect to each other. Relative motion changes the observer's perception of the waves. Imagine that you are standing beside a long conveyor belt with milk bottles on it. Two bottles pass you each second. If you begin walking in the direction the bottles are moving, perhaps only one bottle will pass you each second. If you turn around and walk toward the source of the bottles, three bottles may pass you each second. The bottles are coming from the source at a constant rate, but the frequency of the bottles passing you depends on the relative velocity of you and the source of bottles. Similarly, the perceived frequency of any sound waves depends on the relative velocity of the observer and the sound source. This is called the **Doppler effect**.

The Doppler effect is responsible for the change in frequency that you hear as a car approaches, passes, and moves away from you. As the car approaches, you will hear a higher pitch than you would hear if there were no relative motion. As the car moves away, the pitch will drop to a lower-than-normal pitch. When the car, the source of the sound waves, is moving toward you, the crests of the waves are hitting your ear more frequently. This is perceived as a higher pitch. When the source of the sound waves is moving away from you, the crests of the waves hit your ear less frequently, and this is perceived as a lower pitch. Notice that this is a relative phenomenon. The actual sound emitted by the car does not change. We are so used to this phenomenon that we form an impression of how fast a car is going by the degree of the Doppler effect as it drives by.

Christian Andreas Doppler (1803–53) was an Austrian physicist who first described the effect of relative motion on perceived frequencies in the electromagnetic spectrum.

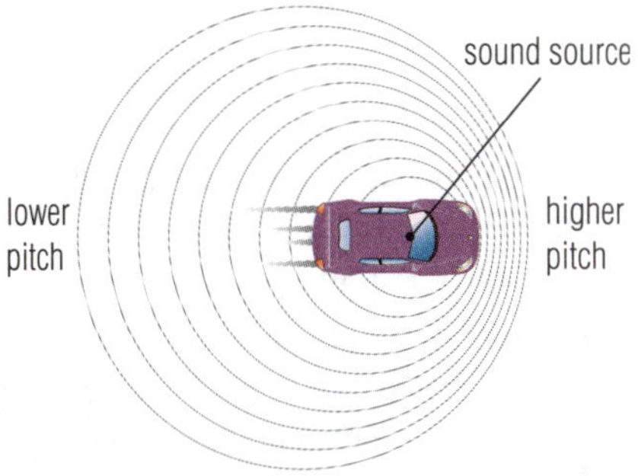

12-29 The Doppler effect

12.17 Modeling Mach Speed

As with any wave, the speed of sound is a function of the medium through which it travels. The density of the medium is one factor that controls the speed of sound through the medium. The density of a fluid depends on its composition, its pressure, and its temperature. Interestingly, with the atmosphere, only its composition and temperature seem to significantly affect the speed of sound through it. Since both temperature and composition of air change with altitude, the speed of sound in air also changes with altitude. Examine the data table below relating Mach speed, air temperature, air density, and altitude. Mach 1 is the speed of sound at a given altitude, temperature, and density.

Table 12-3

Atmospheric and Mach 1 (M1) Speed Data

Region	Altitude (km)	M1 Ground Speed (m/s)	Average Air Temperature (°C)	Air Density (kg/m³)	Air Pressure (hPa*)
Troposphere	0	340.3	15.0	1.2	1013
	2	332.5	2.0	1.0	800
	4	324.6	−11.0	0.82	620
	6	316.5	−24.0	0.66	470
	8	308.1	−37.0	0.53	360
	10	299.5	−50.0	0.41	260
Tropopause	12	295.1	−56.5	0.31	190
	14	295.1	−56.5	0.23	140
	16	295.1	−56.5	0.17	100
	18	295.1	−56.5	0.12	75
	20	295.1	−56.5	0.088	55
Stratosphere	22	296.4	−54.5	0.064	40
	24	297.7	−52.5	0.046	29
	26	299.1	−50.5	0.034	22
	28	300.4	−48.5	0.025	16
	30	301.7	−46.5	0.018	12
	32	303.0	−44.5	0.013	9
	34	306.5	−38.9	0.0096	7

*hPa is the abbreviation for hectopascals, or hundreds of pascals (N/m^2).

Suppressing Sonic Booms

In many residential and rural areas of the United States, military and commercial planes are prohibited from creating sonic booms. People and animals do not appreciate being subjected to large pulses of acoustic energy. How can these planes still operate at supersonic speeds without disturbing the local population? How is observing this rule an example of loving one's neighbor?

12D Objectives

After completing this section, I can

✓ differentiate waves from other kinds of periodic motion.

✓ describe a wave using proper nomenclature.

✓ classify waves by the way their medium is displaced during oscillations.

✓ calculate wave speed, frequency, or wavelength, given any two of the three parameters.

✓ describe sound waves and the properties of sound as interpreted by the sense of human hearing.

✓ define the Doppler effect and explain how it occurs.

✓ model the speed of sound in the atmosphere and identify the atmospheric factors that determine Mach speed.

12D Section Review

1. Explain how a physical wave moves through a medium. What actually moves?
2. Discuss the difference between waveform graphs and vibration graphs.
3. Describe longitudinal and transverse waves. Which category do sound waves fall under? Light waves?
4. Discuss the relationship between intensity and perceived loudness of sound waves.
5. a. Why does the pitch of a car's engine sound higher than normal as the car approaches you and then lower than normal as it moves away?
   b. What is this phenomenon called?
6. DM Explain why airlines and the military need to know how the speed of sound changes with altitude, especially in the vicinity of populated areas.
7. ✪ The speed of sound in air at 20 °C is 343 m/s. What is the wavelength of a 440. Hz sound wave?
8. ✪ What is the period of the sound wave in Question 7?
9. ✪ If a sound has an intensity of 10^{-7} W/m^2, how loud is the sound in dB?
10. DM✪ Create scatter plots using your graphing calculator or a graphing spreadsheet program for the following relationships. Plot altitude on the horizontal axis for each, since it is the independent variable in each relationship. If possible, create your plots on a single set of coordinate axes.
    a. Air pressure vs. altitude
    b. Air density vs. altitude
    c. Air temperature vs. altitude
    d. Sound speed vs. altitude
11. DM✪ Referring to the graphs created in Question 10, identify which of these four parameters are related based on the similarity of the shapes of their graphs. Verify your assessment by graphing them against each other and performing an r^2 fit evaluation according to the instructions found in Appendix G.

Chapter Review

In Terms of Physics

Mach (M) speed	265
periodic motion	266
damping	266
restoring force ($\mathbf{F}_r$)	266
simple harmonic motion (SHM)	266
amplitude (A)	267
cycle	267
frequency (f)	267
hertz (Hz)	267
pendulum	270
pendulum arm (L)	270
total acceleration ($\mathbf{a}_{total}$)	271
physical pendulum	275
moment of inertia (I)	275
damped harmonic oscillator	277
overdamped oscillation	278
critically damped oscillation	278
natural oscillation frequency (f_0)	278

driven oscillation	279	wavelength (λ)	282
resonance	279	wave speed (v)	282
wave	280	intensity (I_s)	284
medium	280	decibel (dB)	284
disturbance	280	pitch	284
longitudinal wave	281	quality	284
transverse wave	281	fundamental	284
compression	281	harmonic	284
rarefaction	281	Doppler effect	285

Problem-Solving Strategies

12.1 (page 267) Remember that period and frequency are reciprocals of each other.

12.2 (page 268) A longer period means that the spring system moves slower and takes longer to complete one cycle. The larger the mass, the longer the period. The "stiffer" the spring, the shorter the period.

12.3 (page 273) When calculating the period of a pendulum, be sure to use the *magnitude* of g. If you use the scalar value, the result will be an imaginary number.

12.4 (page 275) The larger the moment of inertia for a physical pendulum, the farther its center of mass is from the center of rotation. Thus, larger moments of inertia correspond to longer pendulum oscillation periods.

12.5 (page 282) For a uniform medium, the wave speed is constant for all frequencies. Therefore, frequency and wavelength are inversely proportional. For example, if frequency doubles, then wavelength is halved.

12.6 (page 284) The common logarithm (log) of a number is just the exponent of 10 that gives the argument x in log (x). The ratio of I_s over 10^{-12} W/m^2 yields the relative strength of the sound compared to a reference intensity. Taking the log of this ratio gives the magnitude of the ratio.

Review Questions

1. What is periodic motion?

2. What is the name for periodic motion that is controlled by a restoring force $F_{r\,x} = -k\Delta x$ and is not subject to friction or damping?

3. What is the net force on a simple harmonic oscillator at its equilibrium position?

4. Which of the following is *not* used to describe simple harmonic motion?
a. frequency, f
b. wavelength, λ
c. amplitude, A
d. period, T

5. Name two examples that approximate simple harmonic motion.

6. Mass m_1 undergoes simple harmonic motion on a spring. It has a period T_1. If it is replaced by a mass $m_2 = 2m_1$, is the new period double the old ($T_2 = 2T_1$)?

7. Under what conditions can a pendulum's motion be considered simple harmonic motion?

8. A clock contains a pendulum that swings from one side to the other in 1.00 s near the earth's surface. Assume that it exhibits simple harmonic motion. What must its period be?

9. What kind of friction naturally opposes periodic oscillations?

10. Which kind of damping would be most comfortable in a car shock absorber?
a. overdamped
b. critically damped
c. slightly damped

11. What is the name of the frequency at which energy is most efficiently added to an oscillating system?

12. What condition occurs when the energy added to a driven oscillator exceeds the natural damping of the system?

13. A cork floats in a smooth pond. A pebble is dropped next to it and creates a wave with a wave speed of 10.0 m/s. If the wave travels to the shore 1000 m away, how much closer to the shore will the cork be after 10 s?

14. What is the difference between a waveform graph and a vibration graph?

15. What are the two main forms of wave propagation?

16. Classify the following waves according to type (one of the two major types or a combination).
a. wave in an iron bar that is struck by a hammer parallel to its length
b. wave in a taut rope whose end is snapped
c. sound wave
d. wave of collisions in a traffic chain-reaction collision
e. waves from an earthquake
f. light wave

17. Why can you hear sounds sooner when your ear is placed to the ground than when you are standing up?

18. Which one of the following do waves *not* transmit?
a. information
b. matter
c. energy

19. Sound travels fastest in
a. gases.
b. liquids.
c. solids.

20. Which of the following determines a sound's pitch?
a. frequency
b. intensity
c. quality

21. Decibels measure a sound's
   a. intensity.
   b. loudness.
   c. pitch.
   d. quality.

22. On what does the quality of a sound depend?

DM23. Table 12-3 is divided into three atmospheric regions. What property of the atmosphere have scientists used to define these regions? Which region has the lowest Mach speed values? In which region could an aircraft fly the fastest without exceeding the speed of sound? What other considerations would make this region not the most efficient region to fly in?

True or False (24–35)

24. A laboratory mass hanging from an ideal spring can oscillate in simple harmonic motion.

25. The acceleration experienced by an oscillating system is directly proportional to the system's displacement from its equilibrium point.

26. Simple harmonic motion is analogous to uniform circular motion.

27. An oscillating system's maximum displacement from one extreme position to the other is its amplitude.

28. The position of an ideal vibrating system at time $t + T$ is the same as it was at time t.

29. A weaker spring (smaller k) implies a higher frequency of oscillation.

30. Ignoring air resistance and other forms of friction, there are only two forces that act on a pendulum at any point in its swing—its weight and the total tension in the pendulum arm.

31. Assuming an initial pendulum angle of $\pi/4$ when it is released, as the pendulum swings through the lowest position in its arc, the weight vector of the mass is larger because of centrifugal force.

32. A pendulum will probably swing more slowly at the top of a high mountain than at its base.

33. Damping tends to oppose the acceleration of a harmonic oscillator.

34. A physical pendulum is a damped oscillator.

35. In order to cause resonance to occur in a system with a higher natural frequency, you must apply a larger force than that required for a system with a lower natural frequency.

✪36. A mass is attached to the end of a spring. Then it is displaced 4.00 cm from its equilibrium position. In the next 0.500 s, the mass moves through its equilibrium position, stops 4.00 cm on the other side, then returns to its original displacement. It continues to vibrate at this rate. What are its amplitude, period, and frequency?

✪37. A 1.00 kg mass hangs from a spring. It is set in motion with a period of 6.28 s. What is the spring constant, k?

✪38. A mass on a spring with a spring constant of 5.00 N/m has a frequency of $f = 0.250$ Hz when undergoing simple harmonic motion.
   a. What is its period?
   b. What is the mass?

✪39. A spring stretches by 0.0500 m when a 2.55 kg mass is hung from it. The mass is then pulled down 0.100 m and released. It undergoes simple harmonic motion.
   a. What is the force that the mass exerts on the spring?
   b. What is the spring constant?
   c. What is the amplitude of the simple harmonic motion?
   d. What is the period of the mass's motion?
   e. What is the frequency of the mass's motion?

✪40. A spring-mass system oscillates with an amplitude of 0.100 m. The spring constant is 100. N/m. Take the potential energy of the system to be 0 J at the equilibrium position ($x = 0$ at equilibrium).
   a. What is the mass's spring potential energy at $x = -A$?
   b. What is the mass's total mechanical energy at $x = -A$?
   c. What is the mass's kinetic energy at $x = 0$?
   d. If the mass is 1.00 kg, what is its speed at $x = 0$?
   e. Find the period and frequency of the motion.

✪41. A pendulum has a period of 1.28 s on the earth's surface. What period will the same pendulum have on the surface of the moon ($|g_{moon}| = 1.62\ m/s^2$)?

✪42. What is the period of a 1.00 m long pendulum on Mars ($|g_{Mars}| = 3.71\ m/s^2$)?

✪43. If a given pendulum has a length ℓ and period T, what would be the length of a second pendulum if its period were $T/3$?

✪44. A blue light has a wavelength of 4.55×10^{-7} m and a speed of 3.00×10^8 m/s. What is its frequency?

✪45. Refer to the wave graph to answer the questions below.

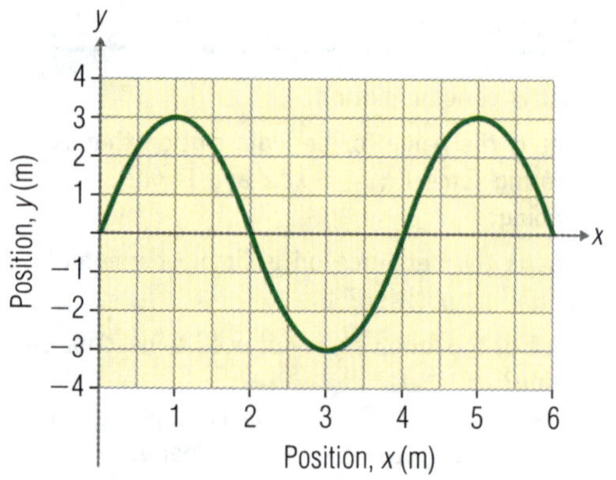

   a. What is the wavelength?
   b. What is the amplitude?
   c. The wave's speed is 64 m/s. What is its frequency?
   d. What is the wave's period?

✪46. A 5.50 kg laboratory mass rests on a surface with coefficients of friction $\mu_s = 0.30$ and $\mu_k = 0.25$. A horizontal spring is attached to the mass and is stretched 0.160 m. The mass remains motionless. However, if the spring is stretched 0.161 m, the mass slips. What is the spring constant?

✪47. A device is built to shoot a 150. g puck across a frictionless surface using a triggered spring launcher. The spring is compressed 8.0 cm and has a spring constant of 45 N/m. What is the initial acceleration of the puck when the spring trigger is released? Describe the puck's acceleration as a function of position throughout the entire launch. If you are familiar with calculus, describe the acceleration of the puck with respect to time. (*Hint*: A graph would be useful in your answer.)

DM✪**48.** Refer to the scatter plot of air temperature vs. Mach speed that you created for 12D Section Review Question 11, or create one if you have not already done so. Considering the variability of atmospheric conditions and the kinds of instruments a jet pilot would typically use, is this the most practical model for determining the Mach speed limits for his aircraft? If not, suggest a more practical relationship to plot.

DM✪**49.** The Aérospatiale-BAC Concorde, which began service in 1976, was the first supersonic passenger airplane to operate in Europe and the United States. Public outcry arose because of the sonic booms it created over populated areas, resulting in the significant curtailment of supersonic flights over land and restrictions on which airports Concorde could use. The last Concorde aircraft were decommissioned in 2003. Discuss the balance between the development of convenient technology and the necessity to protect public health and peace of mind.

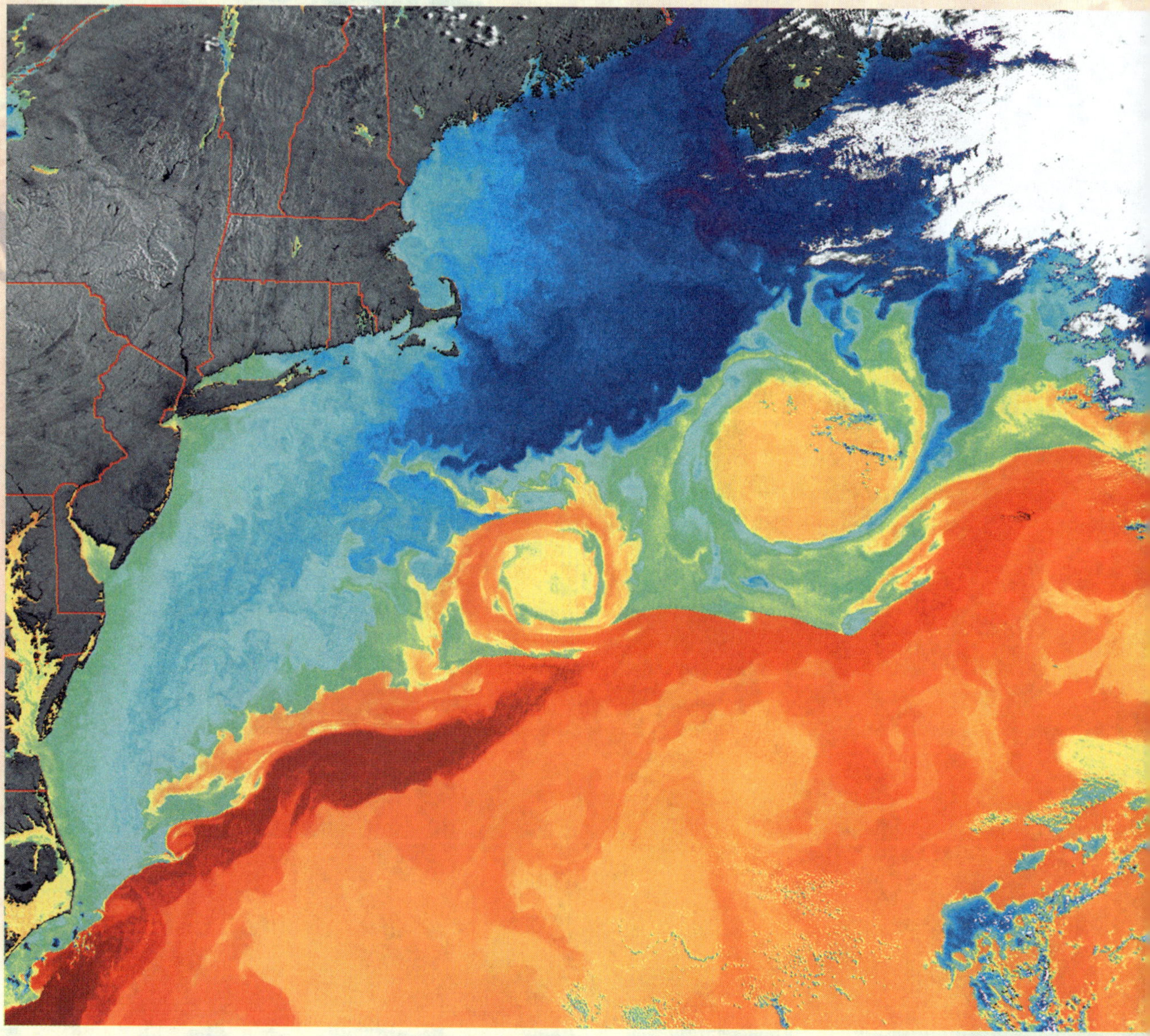

UNIT
3

Thermodynamics and Matter

The first faltering steps to harness the energy of steam in the eighteenth century provided the impetus for a totally new area of science—thermodynamics, one of the main areas of classical physics. Scientists soon discovered that thermodynamic principles underlie every process in which energy is liberated or absorbed, from a ship's snorting steam engine to the great heat engine of the earth's oceans. The laws of thermodynamics also provide the strongest evidence against a naturalistic origin to the universe.

In Unit 3, you will learn how the current model of matter became established as scientists gained a better understanding of thermodynamic principles. You will investigate the phases or states of matter and learn why phase changes take place. You will also learn about thermal energy and how it can produce useful work according to the laws of classical mechanics.

Most of the matter in the visible universe is in the bulk of stars, such as our sun. Therefore, most visible matter in the universe is in the plasma state.

Properties of Matter 13

Dominion Science Problem

Replacing Asbestos

Asbestos has been used for centuries—for Egyptian clothes, Persian embalming, Greek napkins, Chinese suits of armor, and Roman tablecloths. It was prized for its fire resistance, thermal insulation, and strength. But even in ancient times, people suspected that it might cause health problems. Pliny the Elder, a Roman scientist and historian, noticed that slaves working in asbestos mines were prone to lung diseases. He suggested that they wear some kind of facemask to avoid breathing asbestos fibers. But knowledge of the use and dangers of asbestos was almost lost during the Middle Ages.

13-1 Asbestos fibers used for insulation can become airborne when the insulation is damaged, producing a serious respiratory health hazard.

With the advent of steam locomotion during the Industrial Revolution, asbestos made a comeback. Asbestos was used for insulation in fireboxes, boilers, and refrigerators, and for pipes in trains, boats, automobiles, and even homes and home appliances. But there was a hidden danger. Modern research has revealed that airborne asbestos particles cause lung diseases, several types of cancer, and heart problems. In the 1970s, asbestos was banned from use in home insulation. But the need for insulation hasn't gone away. What could be used in the place of asbestos?

13A THEORIES OF MATTER

13.1 Introduction

The seemingly infinite ways that matter is manifested in nature can range from awe-inspiring to mundane. The continental mountain ranges, vast canyons, and river valleys give evidence of tremendous forces that operated in the past to move immense quantities of matter in their formation.

13-2 It is likely that cataclysmic forces during and after the Flood were responsible for building these mountains.

13-3 Vehicle Assembly Building, Kennedy Space Center

Man's efforts seem puny by comparison, but huge structures such as the Vehicle Assembly Building at the Kennedy Space Center, extraordinary water diversion and hydroelectric projects such as the Three Gorges Dam in China, and the Palm Island reclamation projects, which created the largest man-made islands in the world off the coast of Dubai, showcase man's considerable ability to manipulate all forms of matter for his purposes.

The advancement of science and technology depends on a thorough understanding of the nature of matter. Without this understanding, we would not be able to obey the Creation Mandate, given to Adam in the first chapters of the book of Genesis, to subdue the earth. Beginning with this chapter, you will investigate some of the properties of matter and see how the principles of classical mechanics determine those properties.

13.2 Nature of Matter

Mankind has been curious about the nature of matter since the dawning of historical times. Matter is the tangible stuff of the universe that surrounds us and is commonly separated into four states or phases: solids, liquids, gases, and plasmas.

Solids

Solids are normally considered to be fairly rigid, having definite shape and volume. Examples of solids are bowling balls, granite, and steel girders. Solids are usually thought to be massive and dense, but they also can be finely divided and have little density (dust, for example).

Liquids

Liquids have about the same density as the solid state and have a definite volume. However, liquids assume the shape of their containers even if they do not fill them. Liquids can also flow from one point to another under the influence of a force. Examples of pure liquids include water, mercury, and ethanol.

13-4 The three phases of water

Gases

Gases are characterized by low density (under normal circumstances), and they can flow just like the liquid phase. When a gas is placed in an evacuated container, it completely fills it. Gases can also be easily compressed and rarified. The atmosphere is composed of a mixture of gases.

Plasmas

The plasma phase exists when the particles of matter have enough kinetic energy that at least some of their electrons become stripped away. This occurs at very high temperatures. Physically, plasmas behave as a gas or liquid (they are fluid-like) and consist of a neutral mixture of electrons and positively charged particles. Plasmas exist in the sun and most stars, fluorescent light bulbs, and neon lights.

The first three states have been well understood for centuries. The fourth was not even suspected until late in the nineteenth century. The states of matter are difficult to define with any precision. Many common materials can have properties of two or more states, depending on conditions, and some materials in one state have properties normally associated with other states. For example, a finely ground solid material can partially fill its container and flow from one place to another like a liquid (e.g., the sand in an hourglass). Water may exist in any of three states, depending on temperature and pressure. It can even exist in all three states at once if conditions are right. Therefore, associating a state of matter with a particular material is not entirely correct, even though we tend to do it all the time.

In order to understand the reason for the states of matter, we will begin with a review of the background of the current theory of the structure of matter.

13-5 A fluidlike solid

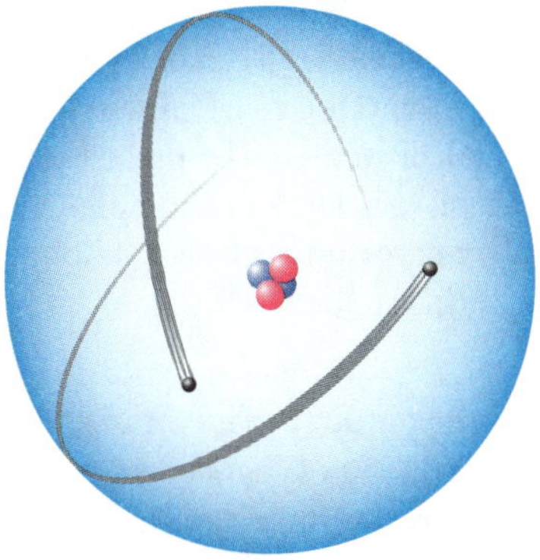

13-6 Model of a helium atom

John Dalton (1766–1844) was an English scientist whose atomic theory contributed to the development of a table of atomic weights and a mathematical basis for chemistry.

An **atom** is the smallest particle that exhibits the chemical properties of its element.

13.3 Particles of Matter

The basic chemical building blocks of matter are called **elements**. These are substances that cannot be broken down into simpler substances by ordinary chemical means. In the early nineteenth century, John Dalton noticed that elements always combine in definite ratios of masses. For example, for every gram of hydrogen in water there are always 8 g of oxygen. To explain this observation, Dalton suggested that matter is made up of small particles that he called **atoms**. All atoms of an element have the same mass, which is unique to that element. Thus, as a first guess, Dalton suggested that oxygen atoms are eight times as massive as hydrogen atoms. (Actually, oxygen atoms are sixteen times as massive as hydrogen atoms, but there are two hydrogen atoms for each oxygen atom in water.) He further theorized that atoms are never created, destroyed, or even changed. Dalton's atoms were hard, indivisible spheres. His theory was useful in correlating many observations.

Since Dalton's time, other researchers have discovered that atoms can be subdivided. The major components of an atom are *protons*, *neutrons*, and *electrons*. Positively charged protons and neutral particles called neutrons, which are similar in mass to protons, are bound together to form the atom's nucleus. Electrons—negatively charged entities about 1/1860 of the mass of a proton—occupy a region of space around the nucleus.

These major constituents of the atom are made of even smaller components called *elementary particles*, including *quarks*. What exactly these even more fundamental things are made of scientists do not fully understand. To say that they are physical entities composed of energy merely moves the mystery back one level to the question "What is energy?" To answer that question as we did in Chapter 9—"Energy is the ability to do work"—is inadequate here. The truth is that scientists do not really know what elementary particles are made from.

13-7 A computer-generated image of the spray of elementary particles from the destroyed nucleus of an atom in a particle accelerator

A pure sample of an element is composed of atoms that all have the same number of protons. For instance, every oxygen atom has eight protons, while every gold atom has seventy-nine protons. Atoms can exist independently, but they are usually found bound to one or more other atoms in order to establish a more advantageous arrangement of electrons around their nuclei. The group may consist of atoms of one element, or it may include atoms of two or more elements. Such groups are called **molecules** or **formula units**, depending on the kind of bonding between the atoms. If the atoms are of different elements, the combination is called a **chemical compound**. Atoms and groups of atoms can gain or lose electrons. The charge imbalance between the nuclei and the electrons results in charged particles called **ions**.

An **ion** is a charged particle consisting of a single atom or two or more atoms bonded together that have a mismatch in the total number of electrons compared to protons in their nuclei.

A **relative mass unit** is an arbitrarily chosen mass defined as one unit. Masses of atoms and molecules are then divided by this unit to obtain their masses in the relative unit.

The **atomic mass unit** is also known as the dalton (Da), especially in biochemical studies. The molecular masses of very large macromolecules are sometimes given in kilodaltons (kDa).

The **unified atomic mass unit (u)** is a relative unit of mass equal to exactly one-twelfth of the mass of a carbon-12 atom.

Avogadro's constant or **number** (N_A) is the number of particles in one mole of a substance. The **mole** is the SI unit for quantity. There are 6.022×10^{23} particles in a mole.

13-8 Each sample contains 1 mole of a substance.

Pressure is defined as force per unit area:

$$P \equiv \frac{F}{A}$$

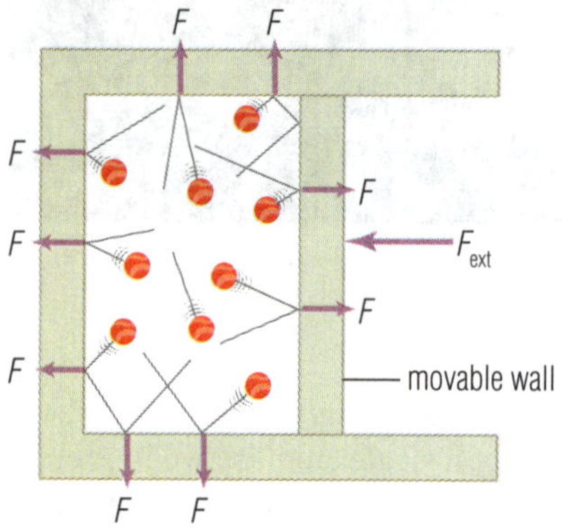

13-9a The total force the marbles exert on the sides of the container is a function of the number of marbles, their speed, and the frequency of collisions.

13.4 Atomic Mass

For physicists, one of the most important properties of matter is its mass. The mass of an object is a combination of the masses of the particles that it contains. While physicists measure the masses of objects in grams or kilograms, the extremely tiny masses of atoms and molecules make such units *very* inconvenient. For example, the mass of an iron atom is only 9.274×10^{-23} g.

Early in the 1800s, scientists recognized that it made more sense to compare the masses of atoms using a relative mass unit. John Dalton suggested using the mass of a hydrogen atom as this relative unit. Thus, an oxygen atom is about 16 times the mass of a hydrogen atom. Later, the relative unit was redefined as $\frac{1}{16}$ the mass of an oxygen atom. This unit came to be known as the **atomic mass unit (amu)**, which persisted until 1961, when the international community again redefined the **unified atomic mass unit (u)** as $\frac{1}{12}$ the mass of an unbound carbon-12 atom. The average mass of atoms of each chemical element in unified atomic mass units is provided in the periodic table of the elements in Appendix H.

Scientists sometimes need to know, at least approximately, the number of atoms or molecules in a sample of a substance. In fact, the mass of a pure sample can be used to "count" its particles. A **mole (mol)** is the amount of substance containing 6.022×10^{23} particles. This number has a special name, **Avogadro's constant** (N_A) or *Avogadro's number*. The mole is the SI unit for quantity.

You can use the mole to relate the atomic mass of a pure substance listed in the periodic table to the mass in grams of a sample of this same substance. The mass of a mole of substance can be found by expressing the sample's mass in atomic mass units and converting it to grams. For example, a carbon-12 atom has an atomic mass of 12.00 u. So Avogadro's number of carbon-12 atoms (i.e., 1 mole) will have a mass of 12.00 g.

13.5 An Application of the Kinetic Theory: Gas Pressure

One area of observation that Dalton's atomic model explains well is the behavior of gases. Using his model, scientists formed a good approximation of real gases that is known as the *ideal gas model.* You will study this model in greater detail in Chapter 14. This model assumes that the particles of a gas have no volume, interact with each other only through perfectly elastic collisions, and are in constant random motion. These assumptions make their behavior much easier to study and are usually good approximations of reality.

Imagine a box with low sides, one of which can be moved. A number of marbles rest on the bottom and are free to move in constant random motion. The marbles collide elastically with each other and with the sides of the container. In these two-dimensional collisions, the marbles exert impulsive forces on each other and on the sides of the container. The collision forces on the container are opposed by the rigid sides, except on the movable side of the box. The movable side will tend to be pushed outwards, enlarging the space in which the marbles are confined, unless the side is supported by an opposing force. The sum of the impulsive forces on the sides of the box at any instant divided by the length of the sides (the "area" of the side in this two-dimensional example) defines the *pressure* that the marbles exert on the container. Since their motion is random, the marbles collide with equal areas of the box's walls with about equal force and frequency; so they exert the same pressure on each wall of the container.

What factors affect the "pressure" that the marbles exert on the box? Whatever affects either the frequency of collisions with the box or the force of these collisions will affect the pressure. For a given average kinetic energy (which equates to average velocity if the marbles are identical), the number of marbles in the box affects the frequency of collision. Logically, the more marbles that are in the box, the more marbles there are to collide with the box, so collisions will be more frequent if their average speed is constant.

The "area" of the box sides also affects the collision frequency. If the box becomes smaller, the marbles will have less room to travel before striking a wall and will therefore collide with walls more frequently. The average speed of the marbles ($\overline{v}$) also affects the frequency of collision. If the marbles are moving faster, they will collide with the walls of the box more often. Combining these factors, the frequency of collision (f) is proportional to the number of marbles (N) and the average speed of the marbles and is inversely proportional to the area of the box (A). In symbols,

$$f \propto \frac{N \cdot \overline{v}}{A}. \tag{13.1}$$

The main factor affecting the force of the collisions is the average speed of the marbles. Remember from Chapter 11 that the force a marble exerts in a collision is equal to its change in momentum divided by the time required for the collision. If we assume that the average collision is perpendicular to the wall of the box, then the final momentum of the gas particle is equal to but opposite of its initial momentum before impact. Thus, the magnitude of the force of an average collision is

$$|F_x| = \left|\frac{m\Delta\overline{v}_x}{\Delta t}\right| = \left|\frac{m(\overline{v}_{\text{aft}\,x} - \overline{v}_{\text{bfr}\,x})}{\Delta t}\right|.$$

But, $\overline{v}_{\text{bfr}\,x} = -\overline{v}_{\text{aft}\,x}$, so

$$|F_x| = \left|\frac{m[\overline{v}_{\text{aft}\,x} - (-\overline{v}_{\text{aft}\,x})]}{\Delta t}\right|, \text{ and}$$

$$F = \left|\frac{2m\overline{v}_x}{\Delta t}\right| = \frac{2m\overline{v}}{\Delta t}. \tag{13.2}$$

The force of each collision is affected by the average speed of the marbles.

The "pressure" of the marbles on the sides of the box is proportional to both the frequency of collisions (Equation 13.1) and the force of each collision (Equation 13.2). In symbols:

$$P \propto f \cdot F$$

$$P \propto \frac{N\overline{v}}{A} \cdot \frac{2m\overline{v}}{\Delta t}$$

$$P \propto \frac{2Nm(\overline{v}^2)}{A\Delta t}$$

Therefore, gas pressure is proportional to $m(\overline{v}^2)$, which is proportional to the average kinetic energy of the gas particles.

According to the ideal gas model, gas molecules behave as the marbles in a box behave, but in three dimensions. The gas molecules exert pressure on their container by colliding with it, and this pressure is proportional to their average kinetic energy. Average kinetic energy of a gas is measured indirectly by a thermometer as temperature. Thus, adding energy to a gas increases the average kinetic energy

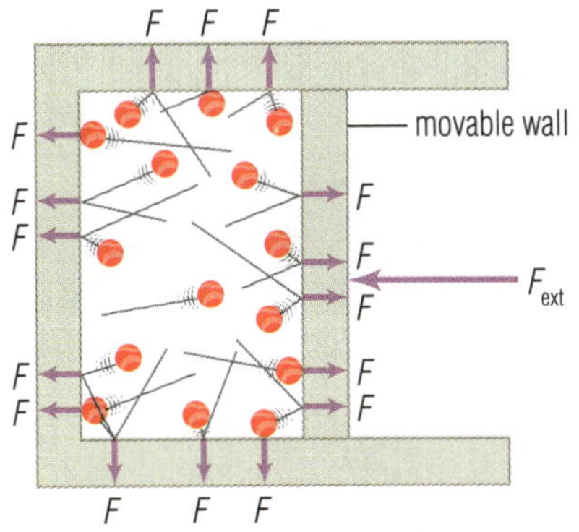

13-9b More marbles means more total force on the sides.

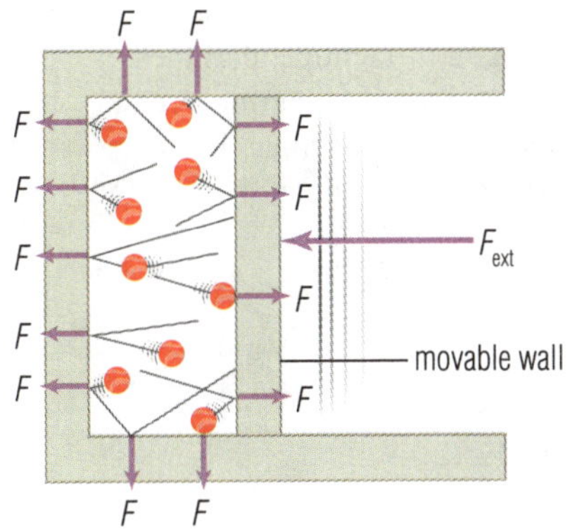

13-9c In a smaller box, the marbles do not travel as far before another collision occurs. The result is a higher frequency of collisions.

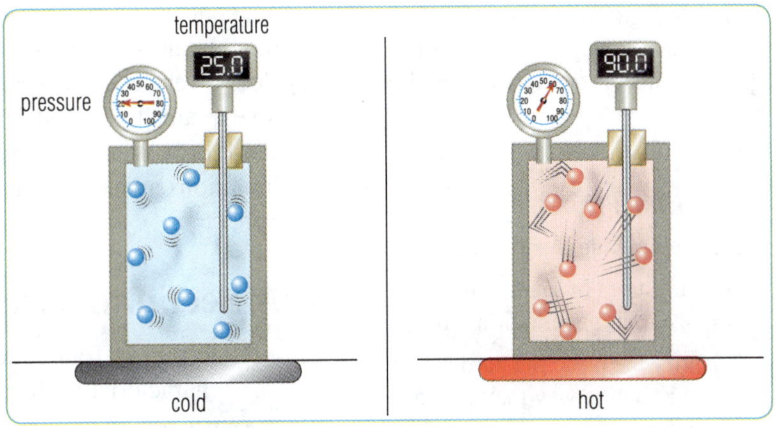

13-10 Pressure of a confined gas increases as its temperature increases.

of its particles, which appears as an increase in gas temperature. This topic will be discussed in more detail in Chapters 14 and 15.

13.6 Solids, Liquids, and the Kinetic Theory

The particles in solids are usually chemically bonded together or in such close association that their intermolecular forces are very strong, holding them rigidly in place. These particles can move only short distances from an equilibrium position, being pulled back and forth in a manner similar to simple harmonic motion. The velocity of the vibrations determines the particles' kinetic energy. Even though their range of motion is limited, solid particles may still possess a large amount of kinetic energy, indicated by a high temperature.

In liquids, the particles are still in close association but with far more mobility. Interparticle forces in liquids cause properties that we observe as cohesion and adhesion. **Cohesion** is the intermolecular attraction that similar particles in a liquid have for each other. **Adhesion** is the intermolecular attraction between particles of dissimilar materials. Since a substance's particles in the liquid state have far more kinetic energy than in the substance's solid phase, liquids generally exist at relatively higher temperatures at the same pressure.

Cohesion is the intermolecular bonding of identical particles. **Adhesion** is the intermolecular bonding of dissimilar substances through attraction at their surfaces.

13-11 The particles of solids and liquids are close together and exert significant forces on each other.

The concept of matter as a collection of numerous, extremely tiny particles in continuous motion forms the classical **kinetic theory of matter**. The kinetic theory does a very good job predicting the behavior of matter under many conditions. It also explains to a good first approximation how matter changes from one of the standard three states to another. As with any model, it is not perfect. We will note its limitations as we discuss various aspects of the properties of matter in this and later chapters.

The **kinetic theory of matter** is an interpretation of the physical properties of matter that relies strictly on classical Newtonian dynamics. It states that all matter consists of tiny particles in constant random motion.

13A Objectives

After completing this section, I can

- ✓ list the states of matter and summarize their general properties.
- ✓ identify the principle particles of matter and the way they group together.
- ✓ identify the factors that contribute to gas pressure according to the kinetic theory.
- ✓ discuss the general behavior of the particles in solids and liquids according to the kinetic theory of matter.

13A Section Review

1. Name the four states of matter and briefly describe their distinguishing characteristics.
2. What is the smallest particle of matter that determines the characteristics of an element?
3. What are the major components of atoms? Identify their charges (if any), relative masses, and their locations within the atom.
4. What are the names of the groups of atoms that determine the characteristics of their compounds?
5. What is the atomic mass unit? Why did scientists establish such a unit?
6. What theory is most effective in describing the properties of gases?

7. Discuss the factors that determine the pressure a gas will exert at any instant in time.
8. What is the relationship between the temperature of a gas and the kinetic energy of the gas particles?

13B STATES OF MATTER

13.7 Arrangement of Particles in Solids

Particles in a solid have little kinetic energy compared to the cohesive forces or chemical bonds between them. Therefore, the particles of a solid have little motion and are held in fixed arrangements. In most solids, this arrangement is a repeating pattern. Such solids are called **crystalline solids**. The smallest pattern of particles that is repeated throughout the solid, the *unit cell*, is composed of particles in a three-dimensional pattern. Figure 13-12 shows the geometric arrangement of ions in sodium chloride (NaCl). The distances between the atoms and the angles between planes of atoms are fixed. For sodium chloride, the angles at the corners of its unit cell are ninety degrees. Table salt is a good example of a crystalline compound. Most metals are also crystalline arrays of atoms.

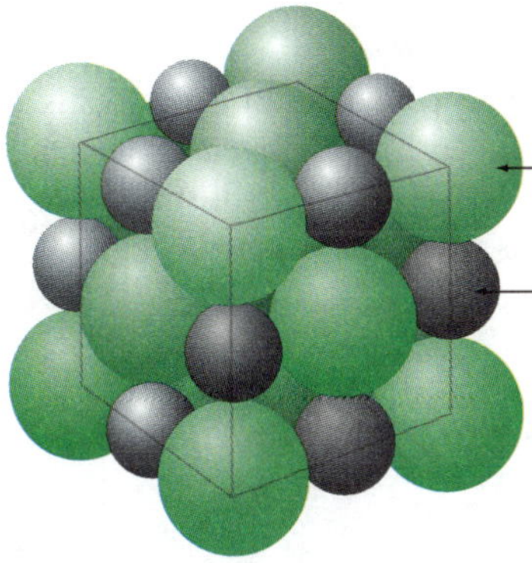

13-12 The face-centered cubic unit cell of sodium chloride—table salt

Solids whose molecules do not form repeating patterns are called **amorphous solids**. The atoms or molecules may be in any arrangement, although the particular arrangement a solid assumes does not change unless the solid is melted. Glass is probably the most familiar amorphous solid. Solid materials may also consist of combinations of crystalline and amorphous substances. These materials may be organized in recognizable patterns or structures that are very large compared to atomic sizes. These **heterogeneous solids** include important organic solids and semi-solid materials, such as wood and bone.

13.8 Solids Under Pressure: The Elastic Modulus

We usually assume that a solid moving in response to a force does not change its shape—it's perfectly rigid. As with many concepts in an introductory physics course, this ideal condition doesn't really exist. Solids *can* change shape, sometimes significantly, in response to certain forces.

When we discussed the transmission of forces by an ideal string, we assumed that the string did not stretch and tension was transmitted undiminished from end to end. A *tensile* force tends to pull particles of the string apart. Since atoms in a solid *can* move a limited amount, the particles will tend to spread out and the material will stretch.

A wire is structurally similar to a string. It can be considered a long, thin cylinder of solid metal. Suppose we apply a tension to the wire by pulling on both ends. How much will the wire stretch? The answer depends on the following three factors: stress, strain, and the wire's elastic modulus.

Stress (σ) (also called *tensile* or *normal stress*) is related to the tension force normal to the cross-sectional area of the wire. In this context, stress is defined as the force per unit area:

normal stress = perpendicular force ÷ area

$$\sigma = \frac{F_\perp}{A}$$

The symbol $F_\perp$ means the "force perpendicular to" something—the cross-sectional area in this case.

Strain (ε) (also called *tensile* or *linear strain*) is the amount that the wire stretches ($\Delta\ell$) divided by the initial length of the wire (ℓ_i):

linear strain = change in length ÷ initial length

$$\varepsilon = \frac{\Delta \ell}{\ell_i}$$

Strain is usually expressed as a simple decimal with no units or as a percent of the initial length.

Finally, if we take the ratio of the normal stress to the linear strain we obtain the wire's **elastic modulus (*E*)**:

$$\text{elastic modulus} = \frac{\text{normal stress}}{\text{linear strain}}$$

$$E = \frac{\sigma}{\varepsilon}$$

Young's Modulus

The elastic modulus of a material is also called *Young's modulus*. Its ratio is valid over the range of stresses and strains where the material obeys Hooke's law.

The SI units of the elastic modulus reduce to newtons per square meter, or N/m^2. These are the same units as pressure. The SI derived unit for pressure and related quantities is the pascal ($1\ Pa = 1\ N/m^2$). Elastic moduli (plural for modulus) are determined experimentally for many solids and tabulated in standard references. For brevity, the values of elastic moduli for most solids are usually listed in terms of GPa (10^9 Pa).

The elastic modulus is a measure of the material's stiffness, or resistance to change in shape. For example, it takes more force to change the shape of a strip of carbon steel (E = 203 GPa) than to proportionately change the shape of a strip of rubber (E = 0.0017 GPa). If a wire's elastic modulus, its cross-sectional area, its initial length, and the tension exerted on it are known, we can estimate its change in length.

Elastic moduli are determined under specified conditions. Variations in temperature, humidity, and moisture content can radically alter the elastic moduli.

$$\Delta \ell = \frac{F_\perp \cdot \ell_i}{AE} \tag{13.3}$$

EXAMPLE 13-1

Stressing: Strain and Elastic Modulus

Two men pull on the ends of a 10.00 m section of copper wire (E = 130 GPa) with a cross-sectional area of $2.50 \times 10^{-6}\ m^2$. They exert a force of 500. N. How long is the wire as they pull on it?

Solution:

You must find the total length of the wire as the men are pulling on it, or

$$L = \ell_i + \Delta \ell,$$

where L is the stretched length of the wire.

$$\Delta \ell = \frac{F_\perp \cdot \ell_i}{AE} = \frac{(500.\ \cancel{N})(10.00\ m)}{(2.50 \times 10^{-6}\ \cancel{m^2})(130 \times 10^9\ \cancel{N}/\cancel{m^2})}$$

$$\Delta \ell \cong 0.01\underline{5}3\ m$$

$$\Delta \ell \cong 0.015\ m$$

$$L = 10.0\underline{0}\ m + 0.01\underline{5}\ m \cong 10.02\ m$$

Problem-Solving Strategy 13.1

When calculating $\Delta \ell$ using the elastic modulus, E, remember that you are finding the *change* in length of the object, not the final length. You must add the change in length to the initial length to find the stretched (or compressed) length.

Mechanical forces can cause other kinds of deformations in solids. *Compressive forces* tend to crush or push the particles of matter together. *Shearing forces* tend to cause layers of particles within the solid to slide parallel to each other like deforming a deck of cards. Each of these forces has its own set of moduli. The compressive elastic modulus is usually the same as the tensile elastic modulus. Both elastic moduli together are called the Young's modulus after Thomas Young.

Thomas Young (1773–1829) was an English physician and physicist best known for Young's modulus and for his experiments with interference that proved the wave theory of light.

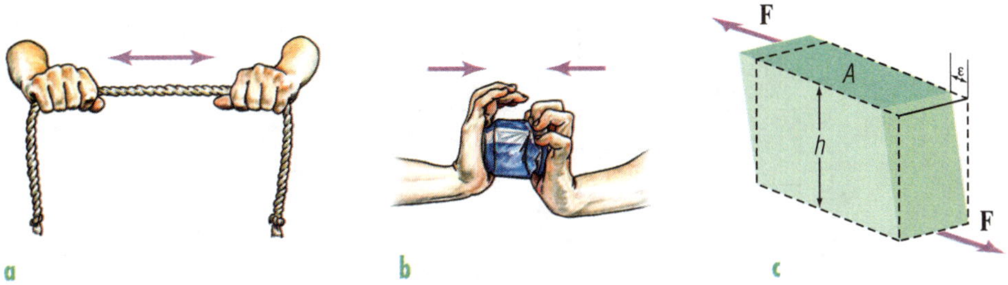

13-13 (a) Tensile, (b) compressive, and (c) shearing forces

Some materials, such as spider silk, exhibit *strain hardening* with increased stress. Silk's elastic modulus rapidly increases with stress until it breaks.

Shear stress and strain occur when the lines of action of forces are parallel but do not act through the same point on a system or object. Consider the rectangular prism (parallelepiped) in Figure 13-13c. The forces act on opposite surfaces in opposite directions. The shear stress equals the force exerted parallel to the surface, divided by the surface area. The shear strain is the ratio of deformation of the object parallel to the force, divided by the separation of the two surfaces of the object. The **shear modulus (*G*)** is the ratio of shear stress to shear strain:

$$G = \frac{\text{shear stress}}{\text{shear strain}}$$

The shear modulus has no special name. It usually has a smaller magnitude than the Young's modulus for a given material.

13.9 Stress-Strain Graph

If the tensile stress and corresponding strain are plotted together, a graph similar to Figure 13-14 on the next page is obtained. Each kind of material has a characteristic stress versus strain graph. Note that there are several key regions along the graph that are indicative of how the material responds to stresses. Where the graph is linear, the Young's modulus is the slope of the line and the solid acts as a perfectly elastic material (i.e., it obeys Hooke's law). The deformation is completely reversible. In other words, it will return to its original shape when the stress is removed. The maximum strain that can be experienced without permanent deformation is called the *proportional limit.* If the material is deformed beyond this limit, some permanent change in shape occurs, but it still behaves somewhat elastically. The limit of reversible deformation is called the *elastic limit.* Beyond the elastic limit, the material permanently deforms with little elastic character. Comparatively little stress is required beyond the elastic limit to cause a large change of shape. Eventually, after enough force is applied, the solid breaks.

A number of material properties determine the shape of the curve in Figure 13-14. The hardness, brittleness, and number of cycles of stress all contribute to the shape of the graph for that material. For example, some materials *work harden.* In other words, when stress is applied in a cyclic

Table 13-1

Approximate Elastic Moduli for Selected Materials

Material	Young's Modulus E (GPa)	Shear Modulus G (GPa)
Aluminum alloy	70	25
Bone		
tension	18	
compression	9	
Brass	100	36
Brick	15	
Copper (element)	130	42
Glass (Pyrex)	70	23
Gold (element)	78	
Iron (cast)	90	
Iron (element)	211	
Lead (element)	13	5.6
Lucite (PMMA)	3.0	
Nickel (element)	200	77
Nylon	2.8	
Pine		
tension	14	
compression	0.042	
Polyethylene	0.22	
Rubber	0.0017	
Silk (silk worm)	16	
Silk (spider, max.)	0.28	
Steel	203	84
Tungsten (element)	41.1	150

Note: The values given in Table 13-1 are generalized from various sources. They represent typical values rather than actual test results.

A **farrier** is a professional who cares for horse hooves and shoes horses.

way, the material becomes harder or more brittle. This has the effect of making the stress-deformation curve steeper and moving the curve closer to the vertical axis. Farriers use this property of iron to harden a horseshoe by repeatedly striking the cooling shoe with a hammer as they form it. Similarly, vibrations are responsible for causing cracks in the metal alloy framework of aircraft.

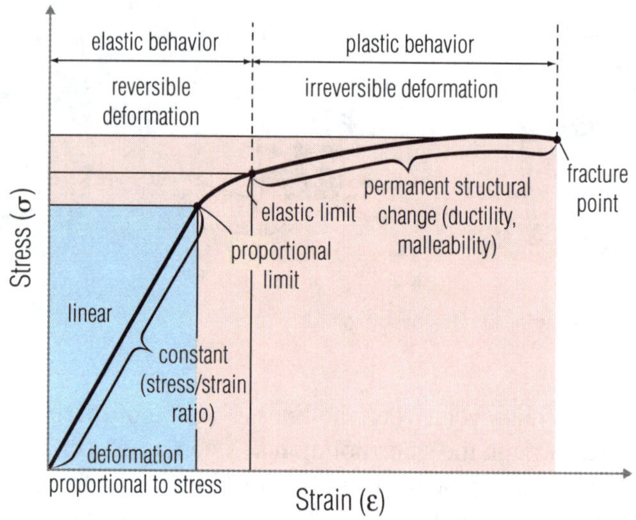

13-14 Graph of stress vs. strain

13.10 Transition Between the Solid and Liquid States

In most materials, the change from solid to liquid takes place at a predictable temperature called the **melting point**. At the melting point, the molecules of a solid gain enough kinetic energy to break out of their rigid arrangements and move more freely with respect to each other. The temperature at which a liquid becomes a solid is less predictable, but usually is the same as its solid form's melting point. When a liquid freezes, it gives up energy, so its molecules are constrained to stay in a rigid arrangement.

It *is* possible to cool a liquid to a temperature below its freezing point and maintain a liquid state. Such a liquid is said to be *supercooled.*

A solid absorbs energy when it melts. Molecules held in their solid phase positions require additional kinetic energy in order to break out of their rigid pattern. The absorption of energy by melting ice is the process that cools an iced drink. The ice simply absorbs thermal energy from the drink.

The melting point of a solid also depends on pressure. Most substances contract when they freeze. For these, higher pressure helps to force the molecules together and makes possible freezing at higher temperatures. There is an important exception to this general rule—water. Water expands when it freezes, so higher pressure hinders freezing. Water contracts when it melts, so higher pressure allows melting at lower temperatures. Simply adding pressure to ice at temperatures near but just below 0 °C can cause it to melt. This, however, requires *a lot* of pressure—approximately 120 atmospheres to lower the melting temperature by 1 °C.

The phrase "120 atmospheres" of pressure means 120 times the normal atmospheric pressure at sea level.

This property of water can be demonstrated with a block of ice, a wire, and two very heavy weights (see Figure 13-15). The apparatus should be at the same temperature as the ice in order to eliminate the effects of heat conduction along the wire from the surroundings. The block of ice must be supported. The weights are fastened to the ends of the wire and the wire is placed over the ice block. The small diameter of the wire coupled with the force exerted by the attached weights produces many atmospheres of pressure on the ice directly under the wire. If the setup is observed over a period of time, the wire will be seen to slowly pass vertically through the ice. The pressure exerted by the wire is sufficient to melt a thin layer

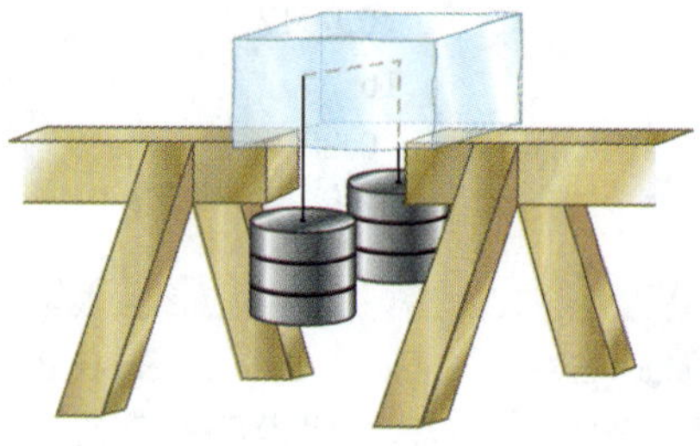

13-15 Experiment demonstrating regelation

of water molecules below the wire that then flow around it and freeze on the upper, low-pressure side of the wire. Eventually, the wire and attached weights will fall free of the ice block, which will remain intact. This process of melting ice under pressure and then refreezing it is called **regelation**. The movement of glaciers over the ground is lubricated, in part, with liquefied ice (water) from regelation due to the massive weight of the overlying glacier ice.

Other factors besides temperature and pressure can affect the melting and freezing points of liquids, including the presence of solutes. Investigating the freezing and melting points of chemical solutions is usually a topic covered in basic chemistry courses. In this course we will concern ourselves only with physical factors affecting phase changes.

13.11 Using Properties of Solids to Solve Problems

As previously mentioned, glass is an amorphous solid, one whose solid state has no specific arrangement of its molecules. Glass can be softened into a more fluid state, essentially melted, and drawn into long, thin fibers. You may have made glass fibers in the laboratory by heating glass rods or tubing, then stretching the molten glass. The arrangement of atoms in a very thin glass fiber is very similar to that found in its softened state.

People have known how to make glass fibers since ancient times. Phoenicians, Greeks, and medieval societies used glass fibers to decorate their pottery and glassware. But they were unable to draw glass thin enough to have a useful amount of flexibility for manufacturing other kinds of materials, such as cloth.

FACETS of PHYSICS

Ice Skating

Ice skating is a popular winter sport, especially in regions where ponds and lakes freeze in winter. Every skating enthusiast knows that this sport requires well-designed skates and a smooth ice surface. But not all skaters may be aware of the most important requirement for skating: the unique properties of water.

Unlike any other liquid, water expands just before it freezes. The ice takes up more space than the same mass of cold water, so the ice is less dense. Therefore, it floats and forms a smooth covering on the surface. If freezing water contracted and sank to the bottom, ice skaters would have a hard time skating (unless they wanted to wear scuba gear)!

For many years it was believed that the tendency of ice to melt under pressure (*regelation*) was the principle method for lubricating ice skates so that they glide easily. However, this belief ignores the fact that the average skater cannot generate more than about 6 atmospheres of pressure on the ice—even with a sharp blade—which is insufficient to cause regelation.

A more plausible explanation was first proposed by Michael Faraday in the nineteenth century. Ice naturally has a thin film of liquid water at the ice-air interface when the ice is at a temperature between −10 °C and the freezing point. It is this thin water film that provides the lubrication necessary for enjoyable skating. Because warmer ice is softer, the slickest ice is at a temperature of about −7 °C, where the layer of liquid water molecules is still thick enough to provide lubrication.

Through the centuries young people have enjoyed the sport of ice skating. That wonderful ability to glide gracefully (or maybe not so gracefully) across the glistening ice is possible only because of the unique properties of water.

Dale Kleist (1909-98) was inducted into the National Inventor's Hall of Fame in 2006 for his contribution to the invention of fiberglass.

Owens-Corning is the world's largest manufacturer of fiberglass and related products, including fiberglass-reinforced plastics and resin products.

13-16 Molten glass fibers being extruded from a perforated plate in a step of the fiberglass manufacturing process

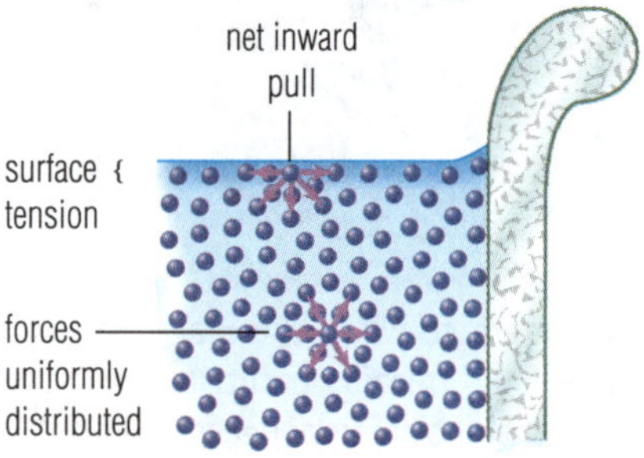

13-17 Cohesive forces in a liquid

In 1932, an engineering graduate student named Dale Kleist, who was working for the glass-blowing company Owens-Illinois, was given the assignment to find a way to fuse glass blocks for architectural use in order to weatherproof their joints. He decided to load a metal-spraying gun (the kind used to bronze baby shoes) with molten glass. It produced a spray of long, very fine glass fibers. The usefulness of his accidental discovery as an insulator was immediately recognized. The company later merged with Corning Glass, another company experimenting with glass fiber making, to begin making a product they called Fiberglas, which eventually became the generic term we know today—fiberglass. Fiberglass has many excellent properties for the purpose of insulation. As *glass wool*, it traps air in innumerable dead pockets that prevent the transfer of thermal energy by air movement through the material. It is also fireproof, so it can be used in most high-temperature applications. And it presents only a minor health hazard compared to asbestos.

The production of fiberglass usually begins with marbles of raw glass. The marbles are melted in a furnace and the liquid glass flows through a perforated platinum plate under the influence of gravity. If a long thread of glass is desired, the filament can be fastened to a winder to pull it as thin as 180-millionths of an inch, about a hundredth the width of a human hair. If a short thread of glass is desired, compressed air or steam under high pressure is blown on the glass fibers, causing an explosion of shorter glass fibers that fall on a conveyor belt below.

Fiberglass finds its way into many everyday applications, like house and other insulation applications, and is basically the same glass wool form originally developed by Owens-Corning. The fibers can be woven into mats of fabric having fire-resistant properties. When mixed with plastics or resins, it is called glass-reinforced plastic (GRP) and is used to produce many structural materials, such as those found in car bodies, boat hulls, fishing poles, shower stalls and tubs, and even whole buildings used as temporary residences for disaster victims.

13.12 Introduction to Fluids

Liquids and gases do not have fixed shapes because their particles (atoms, molecules, or ions) have too much kinetic energy to be rigidly held in place. This property allows them to assume the shape of their containers and to flow from one place to another under the influence of an impelling force. For this reason, they are both classified as *fluids*. Fluids have several common properties not shared with solids. Likewise, liquids have properties that distinguish them from gases.

13.13 Liquids and Surface Tension

Surface tension is a distinctly liquid property. Although liquid particles can slide freely over one another, cohesive forces between them are still significant. The strength of the cohesive forces depends on the liquid. Let's assume that the liquid is a molecular compound. Some nonpolar organic liquids have very low cohesive forces. Polar liquids, such as water, have much higher cohesive properties. Cohesion has little effect on the motion of molecules within the bulk of a liquid sample because the forces on each particle are balanced in all directions. However, molecules at the air-liquid interface (the surface) are not completely surrounded by symmetrical cohesive forces. The external forces on these particles are not balanced, and the net force on a surface molecule is inward toward the bulk of the liquid. Consequently, the layer of molecules at the surface of a liquid acts like a skin that resists penetration from outside or from within. This is why it is possible for small but dense objects such as pins, paper clips, and insects to float on the water's surface and why water droplets are nearly spherical. This phenomenon is called **surface tension**.

Figure 13-18 shows different ways in which surface tension is manifested. Figure (a) is a photograph of a small droplet of water. The surface tension is so strong compared to the weight of the drop that it tends to pull the water molecules together into a volume having the smallest possible surface area. The resulting geometric shape approximates a sphere. Surface tension also explains the characteristic concave meniscus of water (b), the convex meniscus of mercury in a tube (c), and why you can overfill a glass with water.

In some situations, adhesion to an adjoining surface attracts a liquid's surface molecules more than cohesion does. When adhesion is strong, the surface molecules are affected by this force. The net force, the sum of adhesion and cohesion, causes the liquid molecules to spread out over the other surface, even against gravity. The liquid can move against gravity until the weight of the rising liquid balances the adhesive force. This effect is called **capillarity**. The concave meniscus made by water in a test tube is a result of capillarity. Capillarity also permits liquids to flow into fibrous and porous materials. This property is what makes cloth and paper towels so absorbent. Ground water moves through porous rock by a combination of capillarity and gravity.

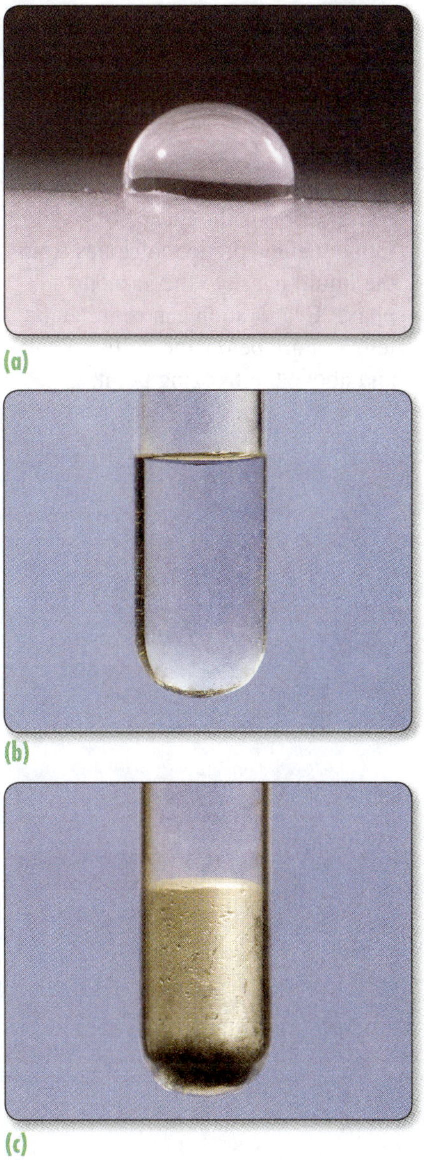

13-18 Examples of surface tension: (a) a droplet of water; (b) meniscus of water; (c) meniscus of liquid mercury

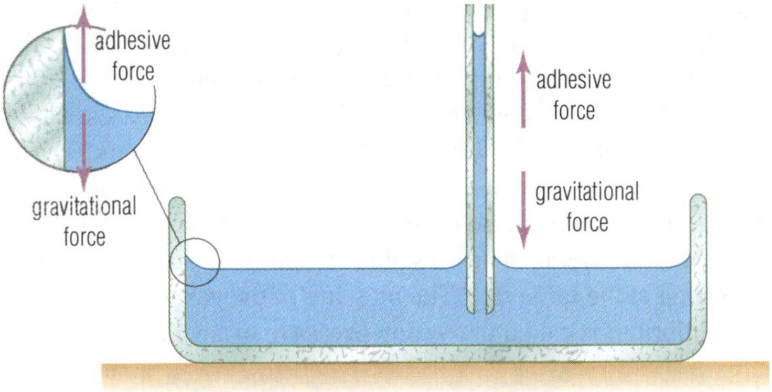

13-19 Adhesion causes a liquid to rise up a thin tube and to form a concave meniscus.

13.14 Vaporization and the Gaseous State

The gaseous phase is characterized by very energetic, widely separated particles. Under standard conditions, most gases are elements or molecular compounds. According to the kinetic model of matter, these particles collide with each other and boundary surfaces in nearly elastic collisions.

To a greater or lesser extent, nearly any substance can change states spontaneously from its solid or liquid phase to its gas phase. To do this, its particles must have sufficient kinetic energy to overcome the cohesive forces of the substance at its surface. This process is called **vaporization**. Vaporization can occur under almost any set of conditions. The rate of vaporization for a given substance is controlled by the temperature of the substance, the rate of heat transfer into the substance, and the concentration of the gas phase in contact with the substance.

Though vaporization is usually associated with liquids, many solids also experience vaporization, completely skipping the intermediate liquid phase. When particles in a solid acquire enough kinetic energy to escape the solid phase and enter the gaseous phase at temperatures below the melting point of the solid, the process is called **sublimation**. You may have observed sublimation if you stored ice cubes for a long time in your freezer. They shrink over time as the water molecules sublimate from the solid ice.

Boiling is a form of vaporization characterized by the rapid formation of vapor bubbles within a liquid. For a given liquid and a constant rate of heat transfer into

Boiling is rapid vaporization. When a liquid boils, its vapor pressure exceeds atmospheric pressure.

Evaporation is the escape of individual energetic molecules from the liquid phase to the gaseous phase. Evaporation can occur at any temperature below the boiling point and above the freezing point.

the liquid, the temperature at which boiling occurs is controlled by the total pressure on the liquid (imposed by the atmosphere or by the boundaries of a container). Since boiling of pure substances occurs at distinct temperatures, *boiling points* are tabulated as a property of pure substances under standard pressure, usually sea-level pressure (1 atm). Water, for example, boils at 100 °C at 1 atm.

Evaporation is vaporization of a liquid at temperatures below the boiling point and above the freezing point of the liquid for the prevailing pressure. This form of vaporization occurs only at the natural surface of a liquid. Evaporation is the primary mechanism for returning water from rivers, lakes, and the ocean to the atmosphere in the water cycle. Surface tension prevents liquid molecules with average or less-than-average kinetic energies from escaping from the body of liquid through the surface interface. How, then, do liquids evaporate without being heated?

The term *average* implies that some liquid molecules have less and some have more energy than the average kinetic energy of the molecules. When a molecule with more kinetic energy than average reaches the surface, the cohesive forces may not be enough to hold it there. Such molecules can escape from the liquid and enter the gaseous phase of energetic molecules above the liquid. As molecules from the liquid escape, they carry their kinetic energy away from the liquid. The average kinetic energy of the remaining molecules decreases, because, on the average, only less-energetic molecules remain. Since temperature is proportional to the average kinetic energy of the liquid, the liquid cools as it evaporates.

If the liquid is enclosed in a container, then the vapor, like all gases, exerts pressure on its boundaries, including the liquid. As more liquid molecules escape to the gaseous phase, the pressure increases until the number of gaseous molecules returning to the liquid phase equals those evaporating, and net vaporization stops. This is an example of *dynamic equilibrium*—a stable condition resulting from two opposing processes occurring at the same rate. The pressure of the gas when the closed system has reached equilibrium is called the **vapor pressure** of the liquid.

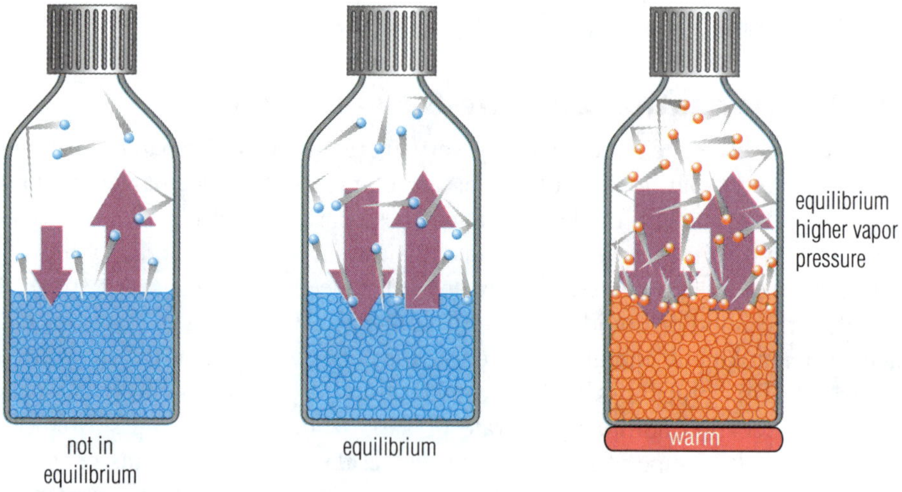

13-20 Vapor pressure is measured when a substance's gaseous phase is in equilibrium with its liquid phase in a closed container. Liquids in open containers generally do not reach equilibrium.

The equilibrium vapor pressure required to stop evaporation in a closed container increases with increasing liquid temperature. Thus, vapor pressure depends on the kind of liquid and its temperature. Comparing the tabulated vapor pressures of liquids at the same temperature gives a measure of their *volatility*. Liquids with low cohesive forces have high vapor pressures. We call such liquids *volatile*. If the cohesive forces are strong, then vapor pressure at a given temperature tends to be lower, and we call these liquids *nonvolatile*. A liquid in an open container will continue to evaporate, even if its temperature is in equilibrium with its surroundings, because of thermodynamic principles that you will study in Chapter 16.

13.15 Condensation and Solidification

If a liquid is in an open container, the pressure generated by its vapor is negligible unless there is a huge volume of liquid, such as that found in the earth's bodies of water. Since the volume of the atmosphere is so large, the pressure exerted by water vapor under these conditions is not likely to reach the equilibrium vapor pressure of water at daytime temperatures, so evaporation continues. However, at night, the air temperature may drop nearer to the equilibrium vapor pressure temperature of the water's surface. The water vapor molecules begin to stick together because they have insufficient kinetic energy to rebound during collisions. Eventually, visible droplets of water form in the phase change called **condensation**. In a natural setting, condensation is called dew and fog. Unlike boiling, condensation does not occur at a discrete temperature and pressure. It also depends on other factors, such as the concentration of the vaporized molecules and the availability of appropriate microscopic sites, called *condensation nuclei*, on which to condense.

Temperature is the dominant factor for controlling the solidification of a liquid, though pressure can affect this process under extreme conditions. This phase transition is called **freezing**. The temperature at which freezing occurs at standard pressure is called the *freezing point*. The freezing point and melting point are almost always the same temperature for a pure substance. From a kinetic model perspective, kinetic

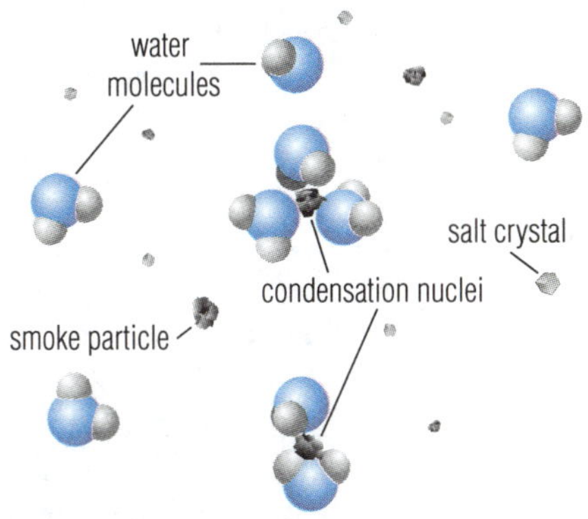

13-21 Condensation nuclei are necessary for the formation of fog and cloud droplets. (Water molecule size not to scale.)

FACETS of PHYSICS

Relative Humidity

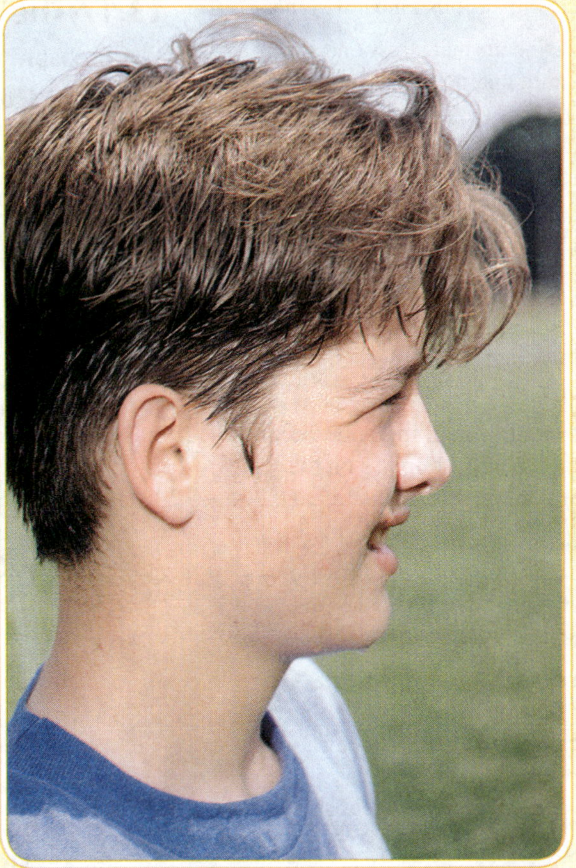

On a hot, damp summer day you might have complained to your friends about your discomfort: "It's not the heat—it's the humidity!" Perhaps the weatherman said that the humidity was 95% that day. Have you ever wondered, "95% of what?"

Absolute humidity is simply the amount of water vapor per unit volume of air. A desert in midsummer and a seaport on a cold, clammy winter day may have the same absolute humidity. The difference in comfort lies in the temperature.

Relative humidity is the ratio of the absolute humidity to the amount of water vapor that could be contained per unit volume in the air at that temperature with no condensation. This quantity is expressed as a percent. The relative humidity is what your body responds to. We need some humidity to live. However, the cooling mechanism of the human body depends on evaporation. When relative humidity is near 100%, little water or perspiration can evaporate. This blocks our cooling mechanism and makes us much warmer than we would be at the same temperature with lower relative humidity. So actually the cliché is true: It's not the heat—it's the *relative* humidity.

energy is removed from liquid particles through the transfer of thermal energy to the point that interparticle forces lock the particles rigidly into fixed positions, according to the type of solid (either crystalline patterns or a random, amorphous arrangement).

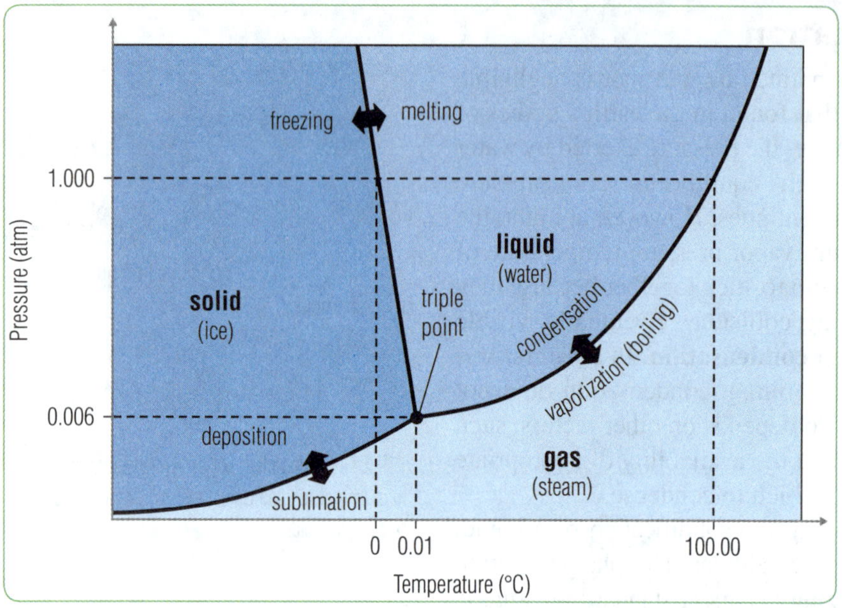

13-22 Phase diagram of pure water

Solidification, or freezing, does not have to occur between the liquid and solid states. Many gases can enter their solid phase spontaneously. When the gas temperature quickly drops below the freezing point, the vapor may skip the liquid phase entirely and change directly to the solid. This phase change is called **deposition**. Frost forms by deposition. As with condensation, deposition does not occur at a set temperature. The process depends on the temperature of the surface on which the gas deposits, the concentration of the gas phase, and the availability of freezing sites.

13.16 Phase Diagrams

Figure 13-22 is a special graph called a **phase diagram**. It shows the relationship of the phases of a substance compared to controlling factors such as pressure and temperature. Other factors such as solute concentration may also be plotted on a phase diagram. This phase diagram of pure water shows regions where water is a solid, liquid, and gas. The boundaries between these regions involve specific kinds of phase changes, depending on the direction of the change.

13.17 The Triple Point

You probably noticed the point on Figure 13-22 labeled **triple point**, where the three phase regions come together on the graph. This point represents the combination of temperature and pressure at which all three phases of water coexist. The triple point for water occurs at 0.01 °C and 0.006 atm (611.73 Pa). Other substances have triple points as well. You will learn the significance of this point when you study temperature scales in Chapter 14.

13B Objectives

After completing this section, I can

- ✓ describe the arrangement of particles in various kinds of solids.
- ✓ define normal stress and linear strain and how these quantities determine the elastic modulus of a solid.
- ✓ interpret a stress-strain graph and identify the conditions for various kinds of material deformation.
- ✓ describe various properties of fluids that depend on interactions of their particles with each other and other materials.
- ✓ explain how evaporation occurs, including its reliance on the vapor pressure of the liquid.
- ✓ discuss the properties of a gas and the mechanisms of the phase changes between liquids and gases, and between solids and gases.
- ✓ use the kinetic-molecular theory of matter to account for the various phase changes identified in a phase diagram.
- ✓ define the triple point of a substance.

13B Section Review

1. Name the different kinds of pure solids and discuss the differences between the arrangements of particles in each.
2. What is the significance of a material's elastic modulus? Would a larger or smaller elastic modulus be desirable for metal used in bridge supports?
3. To what is the slope of the linear portion of a stress-strain graph equal? Would the graph for a block of hard rubber be steeper or shallower in this region compared to the graph for a block of steel?
4. DS Solid Pyrex glass has an elastic modulus of 70 GPa. When the glass is heated to about 1400 °C, it begins to soften. What happens to its elastic modulus at this temperature? From a kinetic-molecular perspective, explain why this happens.
5. What property of liquids is *not* shared by gases? What kind of force is responsible for this property?

6. Is vapor pressure a property of the liquid phase or the gaseous phase? Explain.
7. Compare the volume occupied by a certain mass of liquid to the volume occupied by the same mass of gas under similar conditions.
8. What is the purpose of a phase diagram?
✪9. A section of a rubber band having a cross-sectional area of 3.14×10^{-6} m^2 is stretched 2.00 cm. If the original length of the rubber band is 6.00 cm, how much force is required to stretch it?
DS✪10. Glass fibers are comparatively strong when pulled parallel to their length, but they shatter without warning when the applied stress exceeds the elastic limit. What kind of stress is this? The stress-strain curve of a glass fiber is provided in the adjacent figure. Is a glass fiber ductile or brittle? How can you tell from the graph?

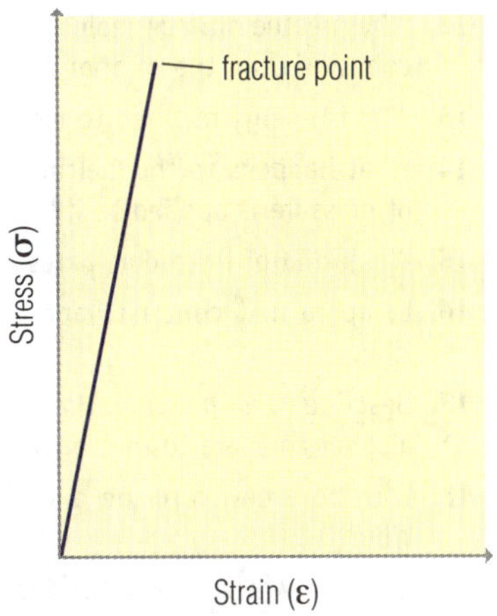

SRQ 13B-10 Stress-strain graph for glass fibers

Chapter Review

In Terms of Physics

element	295
atom	295
molecule	295
formula unit	295
chemical compound	295
ion	295
atomic mass unit (amu)	296
unified atomic mass unit (u)	296
mole (mol)	296
Avogadro's constant (N_A)	296
cohesion	298
adhesion	298
kinetic theory of matter	298
crystalline solid	299
amorphous solid	299
heterogeneous solid	299
stress (σ)	299
strain (ε)	299
elastic modulus (E)	300
Young's modulus	300
shear modulus (G)	301
melting point	302
regelation	303
surface tension	304
capillarity	305
vaporization	305
sublimation	305
boiling	305
evaporation	306
vapor pressure	306
condensation	307
freezing	307
deposition	308
phase diagram	308
triple point	308

Problem-Solving Strategies

13.1 (page 300) When calculating $\Delta \ell$ using the elastic modulus, E, remember that you are finding the *change* in length of the object, not the final length. You must add the change in length to the initial length to find the stretched (or compressed) length.

Review Questions

1. List and describe the four states of matter.
2. What are the chemically simplest substances called?
3. What did Dalton think atoms were?
4. How is the modern model of atoms different from Dalton's?
5. Silver atoms have an average mass of 107.9 u. What is the mass, in grams, of 6.022×10^{23} silver atoms?
6. Discuss the dependence of gas pressure on the number of gas particles, their kinetic energy, and the surface area of the container.
7. What happens to the pressure of a gas in a sealed container when its temperature rises?
8. What properties of matter are stated in the kinetic theory of matter?
9. What are the two common ways that the particles of pure solids can be arranged?
10. Which state of matter (solid, liquid, or gas) has the fastest-moving molecules at a given temperature?
11. Will a wire snap back to its original length if it is stretched past its elastic limit?

12. Why are the surface molecules of a liquid more affected by cohesion than the interior molecules are?
13. Why is energy required to melt a solid?
14. What happens to the melting temperature of ice when a lot of pressure is applied to it?
15. What kind of force dominates capillarity?
16. Compare and contrast vaporization, boiling, and evaporation.
17. Describe *vapor pressure*. How does the vapor pressure of ethyl alcohol compare to that of water at the same temperature?
18. Will more dew form on a cool night or on a warm night? Why?
19. Between which two states does sublimation occur?
20. According to a phase diagram, such as Figure 13-22, what phase tends to exist under conditions of low pressure and high temperature?

True or False (21–30)

21. Man has a biblical commandment to study and use matter in all of its forms.
22. Based on the descriptions of the four states of matter given in this chapter, plasma must be the most common state of matter in the visible universe.
23. The existence of gas pressure is explained well by the kinetic theory of matter.
24. Gas molecules exert pressure on a container by colliding with the container.
25. Strain is related to force, and stress is related to the material deformation caused by a force.
26. The elastic modulus has units of pressure.
27. A large elastic modulus implies that the material is "stretchy."
28. Some soft metals can be hardened by someone striking them repeatedly with hammer-like blows.

*29. An ice skater depends on regelation for smoothly sliding skates.

30. The adhesion of water molecules to each other results in surface tension.

✪31. How many grams of iron metal are contained in 1.00×10^{12} atoms of iron (atomic mass = 55.85 u)?

✪32. A student pushes down on a polystyrene cube with a force of 10.0 N. The 7.50 cm high cube is compressed to a height of 6.00 cm.
   a. What is the stress on the cube?
   b. What is the strain on the cube?
   c. What is the cube's Young's modulus?

✪33. A wooden two-by-four (3.81 cm $\times$ 8.89 cm) stud in an exterior wall of your house supports a load of 1340 N. Describe the kind of stress applied and determine its magnitude.

✪34. If the originally 2.36 m (93 in.) long stud in Question 33 shortens by 1.65 cm under load, what is the strain of the piece of lumber?

✪35. Assuming a stud experiences a normal stress of 3.00×10^5 N/m^2 and a strain of 0.0072, what is the modulus of elasticity for the wood?

✪36. A 65.0 kg physics student jumps stiff-legged from a 1.00 m wall vertically to a concrete sidewalk. He stops in 0.050 s as he lands. If the cross-sectional area of his femur (thigh bone) is 5.00 cm^2, calculate the stress on the bone as he lands on the pavement. (*Hint*: Calculate his landing speed and the resulting impulse of the force as he stops, and then calculate the stress on the bone. Also, consider that he has *two* legs.)

DS✪37. In the 1990s, concerns surfaced that glass wool in fiberglass home insulation was a possible carcinogen that could cause respiratory cancers. In 1994, the National Toxicology Program (NTP) inferred this from animal experimentation, but in 2001, after further study, the International Agency for Research on Cancer (IARC) concluded that these fears were groundless. What does this highlight about the development of technology and social responsibility?

At the same pressure, a given mass of hot air occupies more volume and is less dense than cool air. Thus, hot-air balloons can float in the surrounding cool air.

Expansion and Temperature 14

Dominion Science Problem

Infant Respiratory Distress

About thirteen percent of newborn babies in the United States are born prematurely. A baby is considered premature if he is born before his mother's 37th week of pregnancy, three weeks or more short of the normal 40 weeks of pregnancy. Premature babies can have many medical struggles leading to some significant problems, including loss of vision or hearing, apnea, cerebral palsy, gastrointestinal and nutrition problems, jaundice, and, most commonly, difficulty with breathing. Some of these problems don't appear until the child is older. Since a baby's lungs are one of the last organs to fully develop, premature babies often struggle with respiratory distress syndrome, which affects 1% of all infants but 25–50% of premature infants. It is the leading killer of premature babies. What can be done to help premature infants survive to outgrow this syndrome?

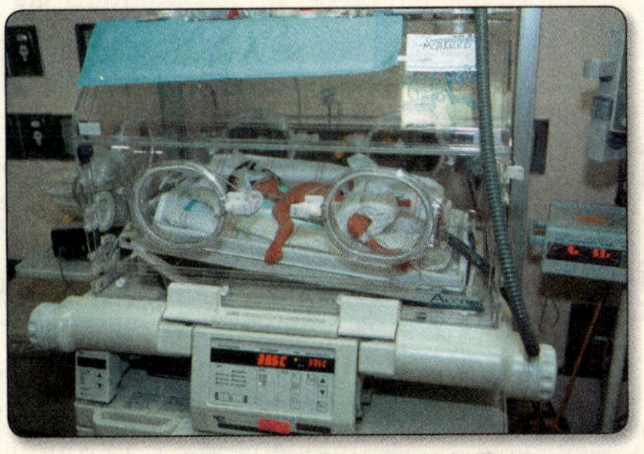

14-1 Premature babies are especially susceptible to respiratory distress.

14A THERMAL PROPERTIES

14.1 Basics of Thermal Expansion

How does a hot bar of copper differ from a cold bar of copper? Other than the temperature, what other properties can be measured to compare the two objects? If we could observe individual copper atoms, how could we distinguish between a "hot atom" and a "cold atom"? According to the kinetic theory of matter, "hot atoms" have more kinetic energy than similar "cold atoms" do. "Hot atoms" are moving or vibrating faster than "cold atoms."

To visualize the difference, it helps to portray the copper atoms as tiny solid spheres held in position by springs attached to adjacent atoms. A displacement of an atom from its equilibrium position will generate a restoring force that will tend to push it back toward its equilibrium position. The faster it moves, the more momentum it has, and the harder it pushes against its neighbors. Since no matter is perfectly rigid, the atoms tend to spread out as they are forced apart. Consequently, the structure formed by the vibrating atoms occupies more space at higher temperatures than at lower ones.

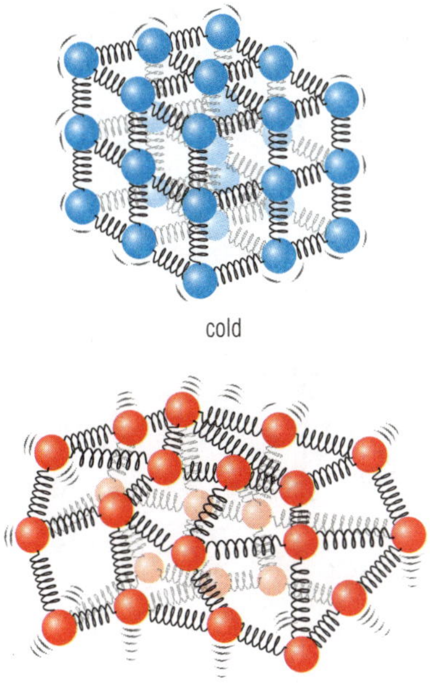

14-2 The atoms of solids can be considered as particles connected to each other with springs. The more kinetic energy they have, the harder they push on their neighbors.

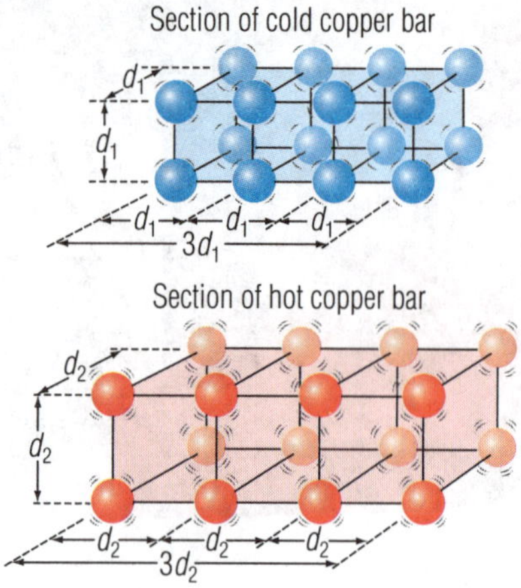

14-3 Thermal expansion—the more atoms in a row, the greater the expansion in that direction

Imagine a copper bar made up of sixteen atoms (Figure 14-3). When it is cold, its atoms vibrate with small amplitude and low velocity, so the lattice takes up little space. However, if the bar's temperature rises, each atom will vibrate faster and move farther from its equilibrium position. As a result, the lattice expands and occupies a greater volume, and the bar expands in every direction.

Since the interatomic spaces increase, the bar will expand more in the directions where more atoms are lined up in a row. In the illustration, the bar is two atoms high (one interatomic space), two atoms wide (one interatomic space), and four atoms long (three interatomic spaces). As the interatomic distance increases with temperature, the bar expands the same amount in height and width, but three times that amount in length. For this reason, when a wire is heated it expands much more in length than in diameter because its length is proportionately so much greater (more atoms in a line). This effect can be easily observed in the larger amount of sag experienced by an overhead high-voltage power line in the summer compared to its sag in the winter. However, a micrometer caliper would be required to detect the wire's seasonal change in diameter.

14-4 Thermal expansion can buckle railroad tracks on hot days.

The expansion of a heated material produces noticeable and sometimes undesirable effects. For example, railroad tracks that are laid too close end-to-end in cold weather will sometimes buckle in hot weather. Railroad builders try to leave a small gap between track sections for the track to expand, or they use alloys that expand minimally at higher temperatures. Even with these precautions, some tracks occasionally buckle on very hot days, causing hazardous conditions for trains. When designing mechanical systems, engineers normally make provision for thermal expansion and contraction in ways that do not interfere with a structure's function. Bridge road surfaces, for example, are usually connected by interlocking joints that prevent buckling when the bridge expands or contracts. The bridge's road-support structures themselves often rest on rollers. Even concrete sidewalks have fibrous expansion joint fillers or gaps between sections to prevent buckling.

14-5 Engineered solutions can prevent expansion-related damage in structures. This is an expansion joint in a bridge road surface.

Glassware is also affected by thermal expansion. Ordinary bottle glass expands significantly when it is heated. However, it is a poor heat conductor. If hot water is poured into a cold thick glass bottle, the inner glass surface quickly warms and expands. The middle and outer thickness of glass, however, is not heated as much and expands little. Because glass is brittle, this difference in expansion rates can cause the bottle to crack. Two ways to avoid this problem are to use thin glass so that the entire thickness heats up and expands almost simultaneously, or to use glass that expands very little. The second choice is more practical for normal use since thin glass breaks easily. Pyrex glass is a kind of strong glass that does not expand significantly when heated.

14.2 Linear Thermal Expansion of Solids

Suppose a 100 cm aluminum rod is uniformly heated along its entire length. Is there any way to predict how much it will expand? Physicists determined the answer to this question by experimentally measuring the change of length of a metal rod with a corresponding change of temperature and plotting the results on a graph. Since the change of length is caused by the change of temperature, temperature change is the independent variable and length is the dependent variable. Figure 14-6 is a typical plot of experimental data.

Mathematically, a straight line graph is described by the familiar equation

$$y = mx + b.$$

Similarly, the equation for the line in Figure 14-6 could be expressed as

$$\ell = m\Delta t + \ell_i, \qquad (14.1)$$

where ℓ is the length of the rod as a function of its temperature, ℓ_i is the rod's initial length at the reference temperature, and Δt is the difference from the reference temperature. The slope (m) is related to the initial length of the rod (the slope is steeper for a longer rod and less steep for a shorter rod) and includes a constant denoted α.

$$m = \alpha\ell_i$$

Replacing m in Equation 14.1 gives

$$\ell = (\alpha\ell_i)\Delta t + \ell_i.$$

The rod's length has changed by

$$\Delta\ell = \ell_f - \ell_i, \text{ so}$$

$$\Delta\ell = \alpha\ell_i\Delta t. \qquad (14.2)$$

This equation applies to the thermal expansion of any solid material. The constant α, called the **coefficient of linear expansion**, is different for each material. Table 14-1 gives the linear expansion coefficients for some common solids.

The units of the coefficient of linear expansion (α) are $°C^{-1}$ or K^{-1}. The initial length ℓ_i may be given in any unit of length, and $\Delta\ell$ will be expressed in the same unit. Of course, Δt must be expressed in the same temperature unit as that of α. The "length" of an expanding object may actually be its linear dimension in any direction. Since each dimension increases in proportion to its size, α is the same for any direction.

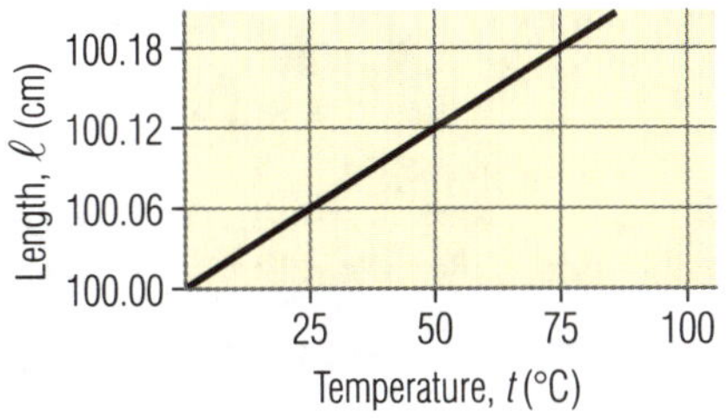

14-6 Thermal expansion graph

TABLE 14-1

Approximate Coefficients of Expansion for Selected Solids

Material	Linear Expansion (α) ($\times 10^{-6}$ °C^{-1})	Volume Expansion ($\beta = 3\alpha$) ($\times 10^{-6}$ °C^{-1})
Aluminum	23.1	69.3
Brass	20.3	60.9
Copper	16.5	49.5
Glass (ordinary)	8.7	26.1
Glass (Pyrex)	3.2	9.6
Gold	14.2	42.6
Ice	51	153
Iron	11.8	35.4
Lead	28.9	86.7
Pine wood (along grain)	5	15
Pine wood (across grain)	30	90
Silver	18.9	56.7
Steel (mild)	11.7	35.1
Steel (stainless)	17	51

EXAMPLE 14-1

A Bridge for Every Season: Thermal Expansion

A steel bridge is 200.00 m long on a winter day with a temperature of −15.0 °C. How many centimeters longer is it on a warm summer day with a temperature of 35.0 °C? ($\alpha_{steel} = 11.7 \times 10^{-6}$ °C^{-1})

Solution:

Compute the change in temperature:

$$\Delta t = 35.0 \text{ °C} - (-15.0 \text{ °C}) = +50.0 \text{ °C}$$

Problem-Solving Strategy 14.1

The thermal expansion equation can be simplified if the length of the object, ℓ_i, is measured at the reference temperature of 0.0 °C. Then the change of length is just

$$\Delta\ell = \alpha\ell_i t,$$

where ℓ_i is the reference length and t is the new Celsius temperature.

Compute the change of length:

$$\Delta \ell = \alpha \ell_i \Delta t$$

$$\Delta \ell = (11.7 \times 10^{-6}\ {}^{\circ}\mathrm{C}^{-1})(200.00\ \mathrm{m})(+50.0\ {}^{\circ}\mathrm{C})$$

$$\Delta \ell = 0.117\ \mathrm{m},\ \text{or}$$

$$\Delta \ell = 11.7\ \mathrm{cm}$$

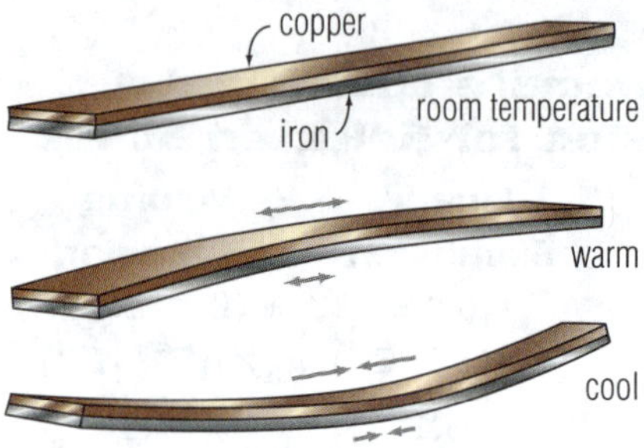

14-7 A bimetallic strip bends when heated or cooled.

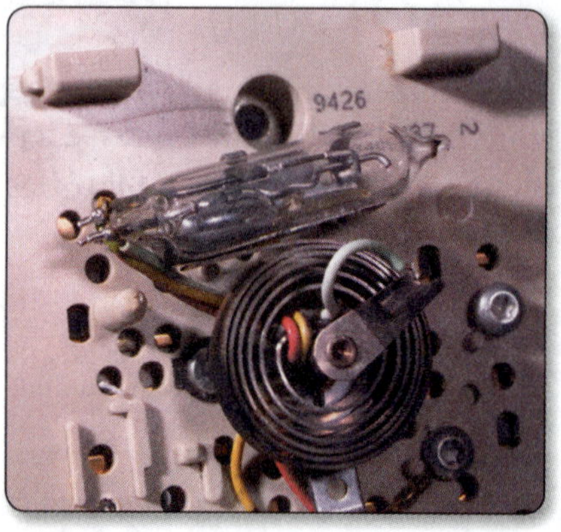

14-8 A common application for a bimetallic strip—a household thermostat

Different materials expand by different amounts with equal temperature change. This fact can be an advantage for some applications. For example, suppose strips of copper and iron metal are bonded along their entire lengths. When the resulting **bimetallic strip** is heated, the copper expands more than the iron does, and the strip curves with the iron toward the inside. Bimetallic strips are commonly used as the temperature-sensitive components of a thermostat. A long, thin strip of the two metals is fabricated in a spiral shape with the copper on the inside of the spiral to maximize the effect of the temperature change on the sensor. As temperature rises, the spiral expands, and as temperature drops, the spiral contracts. Movement of the end of the bimetallic strip controls an electrical contact that, in turn, operates the heating or cooling system.

14.3 Thermal Volume Expansion of Solids

What is the volume of a solid after it is heated? For regular solids such as cubes and spheres, we can predict new volumes by calculating new dimensions based on the linear expansion relationship and using these dimensions in the appropriate volume formula. But what if the solid has an irregular shape? A formula similar to the linear thermal expansion formula (Equation 14.2) has been developed that predicts the change in volume,

$$\Delta V = \beta V_i \Delta t, \tag{14.3}$$

where β is called the **coefficient of volume expansion** and is a constant for each material. For each material, $\beta = 3\alpha$. See Table 14-1 for a list of some representative coefficients of volumetric expansion.

EXAMPLE 14-2

Bulking Up: Volumetric Expansion

A solid sphere of ordinary glass has a volume of 7250.0 cm³ at 20.0 °C. Find its volume at 80.0 °C.

Solution:

Calculate the change of temperature:

$$\Delta t = 80.0\ {}^{\circ}\mathrm{C} - 20.0\ {}^{\circ}\mathrm{C} = +60.0\ {}^{\circ}\mathrm{C}$$

Compute the change of volume. Obtain the value of β from Table 14-1.

$$\Delta V = \beta V_i \Delta t$$

$$\Delta V = (26.1 \times 10^{-6}\ ^{\circ}\text{C}^{-1})(7250.0\ \text{cm}^3)(+60.0\ ^{\circ}\text{C})$$

$$\Delta V \cong 11.\underline{3}5\ \text{cm}^3$$

Compute the final volume:

$$V = V_i + \Delta V$$

$$V = 7250.0\ \text{cm}^3 + 11.\underline{3}5\ \text{cm}^3$$

$$V \cong 7261.\underline{3}5\ \text{cm}^3 \cong 7261.4\ \text{cm}^3$$

Problem-Solving Strategy 14.2

Be sure to use the appropriate coefficient of expansion for solid materials. Use α for linear expansion and β for volumetric expansion.

14.4 Thermal Expansion of Liquids

Liquids respond much more to temperature changes than do solids. Because the molecules in a liquid are bound less tightly to each other than in a solid, an increase in kinetic energy leads to a proportionately greater increase in volume in a liquid than it does in a solid. Thus, coefficients of volume expansion for liquids are generally greater than those for solids, as a comparison of Tables 14-1 and 14-2 shows. When measuring the volume of a liquid after heating, you can generally disregard any effects that expansion of the container might have on the accuracy of your results.

The volumetric coefficients of expansion of liquids tend to be as much as several orders of magnitude larger than those for solids because molecules in the liquid state are bound together much less firmly.

EXAMPLE 14-3

Gasoline Spill: A Consequence of Liquid Expansion

A car's owner tops off his car's 75.7 L stainless steel gas tank in the morning when the gas temperature is 10.0 °C. He drives a short distance to work. The car then stands in the sun until mid-afternoon, when the gas temperature is 23.9 °C. How much gas spills out of the tank's overflow vent because of thermal expansion? Calculate this (a) by ignoring the expansion of the tank, and then (b) by including this expansion. The volumetric expansion coefficient of the stainless steel tank is $\beta_t = 51 \times 10^{-6}\ ^{\circ}\text{C}^{-1}$.

Solution:

a. Calculate the volume of expansion of the gasoline alone (ΔV_g):

$$\Delta t = 23.9\ ^{\circ}\text{C} - 10.0\ ^{\circ}\text{C} = +13.9\ ^{\circ}\text{C}$$

$$\Delta V_g = \beta V_{g\,i} \Delta t$$

$$\Delta V_g = (960. \times 10^{-6}\ ^{\circ}\text{C}^{-1})(75.7\ \text{L})(+13.9\ ^{\circ}\text{C})$$

$$\Delta V_g \cong +1.0\underline{1}0\ \text{L}$$

Ignoring the expansion of the gas tank, slightly more than a liter of gasoline spills out of the tank.

b. Considering the expansion of the tank as well, the tank can hold ΔV_t more gas, and this additional volume reduces the gas spilled.

Determine the volumetric expansion of the tank:

$$\Delta V_t = \beta V_{t\,i} \Delta t$$

Problem-Solving Strategy 14.3

There is no linear coefficient of expansion for fluids because, unlike solids, fluids do not have fixed dimensions.

TABLE 14-2

Approximate Coefficients of Expansion (β) for Selected Liquids Near Room Temperature ($\times 10^{-6}\ ^{\circ}\text{C}^{-1}$)

Acetone	1490
Carbon disulfide	1150
Ethyl alcohol	1120
Gasoline	960
Glycerin	485
Mercury	183
Methyl alcohol	1134
Turpentine	900
Water	207

$$\Delta V_t = (51 \times 10^{-6}\ °C^{-1})(75.7\ L)(+13.9\ °C)$$

$$\Delta V_t \cong +0.05\underline{3}6\ L$$

The amount of gas that vents off in this case is:

$$\Delta V_g - \Delta V_t = 1.0\underline{1}0\ L - 0.05\underline{3}6\ L \cong 0.9\underline{5}6\ L$$

$$\Delta V_g - \Delta V_t \cong 0.96\ L$$

Slightly less than a liter of gas is spilled.

Example 14-3 illustrates why you should stop fueling your car when the automatic delivery nozzle shuts off. The nozzle setpoint is designed to leave expansion room in your tank so that gas does not spill out of the vent due to temperature changes. Spilling gas in this way is poor stewardship because it wastes money, can damage the pavement, and pollutes the ground. It also creates a fire hazard.

In physics, an "empty container" is one that is completely evacuated, having no particles of any kind inside when the gas is released into it.

14.5 Thermal Expansion of Gases

Gas particles are so far apart at normal pressures and temperatures, compared to solids, that they exert little or no force on each other. For this reason, gas particles in an evacuated space will continue to spread apart until they encounter the container boundaries, if any exist. In fact, the volume occupied by a given number of gas particles will expand to fill any empty (evacuated) container. The larger the container, the lower the pressure that will be exerted by the fixed number of particles. However, if the container is designed so that it can change volume in order to maintain a constant gas pressure, gases will then have temperature-dependent volume like liquids and solids. Most gases have volumetric expansion coefficients around $3700 \times 10^{-6}\ K^{-1}$. Table 14-3 lists the expansion coefficients of several common gases. These coefficients are quite similar because gas particles interact very little since they are so far apart. Therefore, one gas behaves much like another. You will study the response of gases to temperature and pressure changes in more detail in Section 14C.

Gases are far more compressible than either solids or liquids. Therefore, the coefficients of volumetric expansion for gases are all determined and reported at a constant standard pressure. Standard pressure is the average atmospheric pressure measured at mean sea level.

TABLE 14-3

Coefficient of Volume Expansion (β) for Selected Gases at Standard Pressure ($\times 10^{-6}\ K^{-1}$)

Air	3670
Carbon dioxide	3740
Helium	3665
Hydrogen	3660

14A Objectives

After completing this section, I can

- ✓ describe how thermal expansion occurs at the particle level.
- ✓ describe the consequences of thermal expansion in real-world situations.
- ✓ mathematically model linear and volumetric thermal expansion.
- ✓ compare the thermal expansion of solids, liquids, and gases.

14A Section Review

1. Briefly explain from the basis of the classical kinetic theory why most materials expand when heated.
2. Give some examples of engineered solutions for problems stemming from expansion and contraction of materials.
3. What three factors determine the amount of thermal expansion by a solid in a given dimension?
4. How is the volume expansion coefficient (β) related to the linear expansion coefficient (α) for solids?
5. Aluminum and copper are two metals that were commonly used for electric wiring in homes in the United States during the 1960s and 1970s. If a large electrical current is carried by the wire, its temperature can increase significantly. What is the difference in linear expansion experienced by a 1.00 m length of aluminum wire versus 1.00 m of copper wire for a +30.0 °C change in temperature?
6. How much will 1.000 cm^3 of mercury expand in volume for a +1.00 °C temperature change?

Facets of Physics

Ball and Ring

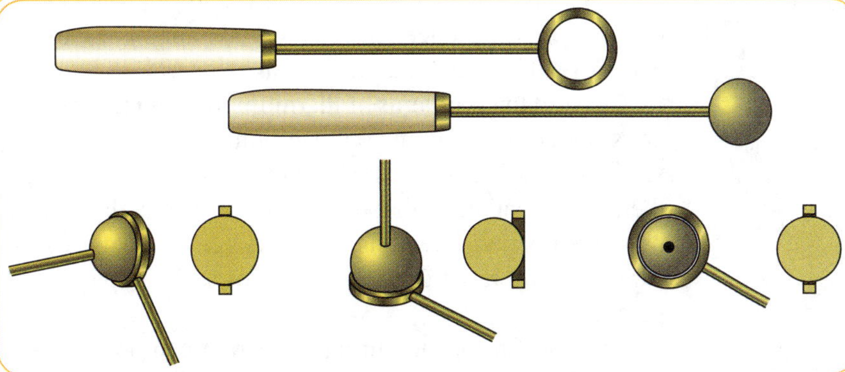

If you heat a metal washer, does the hole in the middle become larger or smaller? One way to find the answer is to use a ball-and-ring demonstrator. The ball and ring are made of the same metal so that they expand at the same rate. When they are both at room temperature, the ball just fits through the ring, but what happens if one or the other is heated?

If only the ball is heated, obviously it will no longer fit through the ring. If only the ring is heated, experiments show that the ball fits loosely through the ring. Therefore, the hole expands when it is heated, but by how much? You can find out by heating both the ball and the ring. Experiments show that a heated ball just fits through a heated ring. That is, the hole expands the same distance that the ball expands. A hole in a metal ring expands and contracts exactly as much as a disk of the same material and size would expand or contract.

14B MEASURING TEMPERATURE

14.6 Thermometers

Since materials expand in proportion to a temperature change, they provide a way of measuring temperature. Galileo used this property of expansion to make the first known **thermometer**. His *thermoscope*, as it was called, was based on the expansion of air and consisted of an air-filled glass bulb connected to a tube immersed in an open dish of water. When the temperature of the air in the glass bulb rose, it expanded, forcing the water down the tube. When the air temperature dropped, it contracted, allowing the water level to rise.

The drawback to the thermoscope is that it responds to changes in atmospheric pressure as well as temperature. When the pressure of the surrounding air is high, it forces the water up the tube even when the temperature has not changed. Thus, without

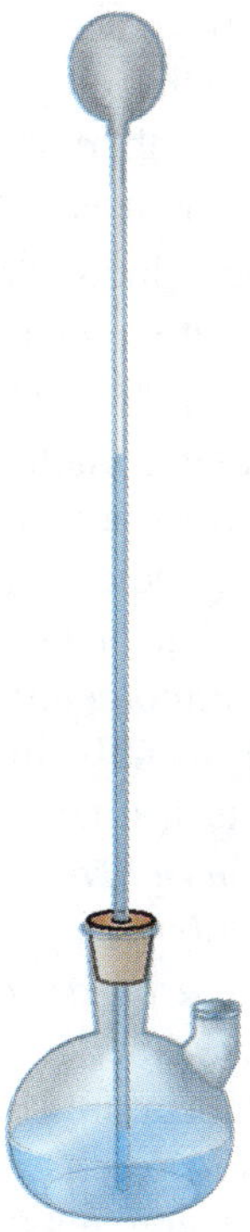

14-9 Galileo's thermoscope

A **thermometer** is any device that uses a thermometric property to indicate temperature. A *thermometric property* is any characteristic of a material or instrument that varies directly with temperature.

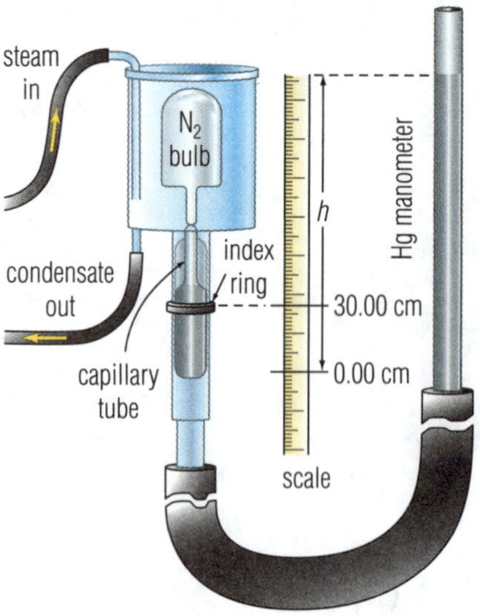

14-10 A gas thermometer

A **thermometric property** of a substance is any measurable property that changes in proportion to a corresponding change of temperature. Common thermometric properties are linear and volume expansion, electrical resistance, color, and electrical potential.

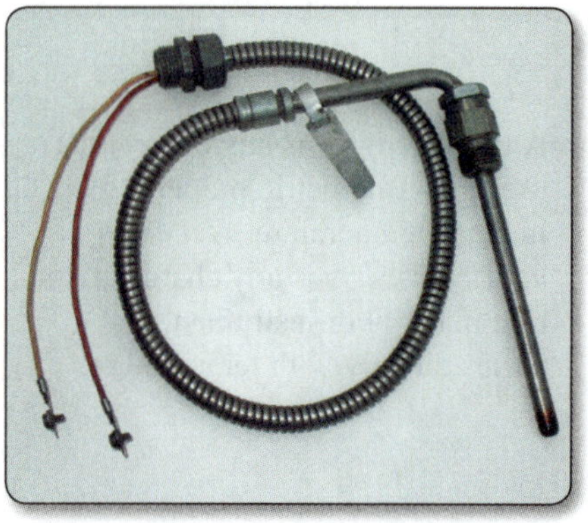

14-11 One kind of remote electrical temperature sensor is called an RTD (resistive temperature detector).

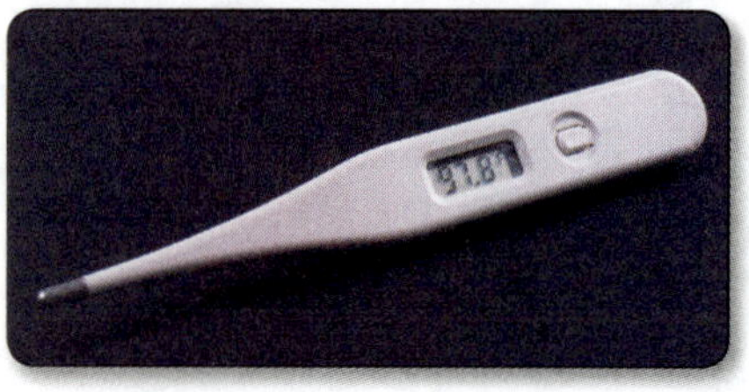

14-12 Digital thermometers are common household items for monitoring fever.

additional instrumentation, we can never be sure whether a thermoscope is showing a change in temperature or a change in pressure.

A gas thermometer can be a better tool for measuring temperature if the gas volume is maintained constant. This can be done by connecting the gas column with a column of mercury. When the gas thermometer is read, the mercury column position is adjusted so that the left-hand column height is returned to its original position, thus restoring the gas volume to the reference volume condition. The difference in height between the left- and right-hand mercury columns represents the pressure difference required to maintain a constant volume. The height difference is read on a calibrated scale as temperature.

To avoid the problems of thermoscopes and the inconvenience of gas thermometers, modern thermometers use liquids such as glycol (an alcohol) or mercury sealed in fine glass tubes under vacuum. The liquids expand so much more than their glass containers that the expansion of the glass usually has a negligible effect on the temperature reading. For very sensitive measurements, the thermometer is carefully calibrated to take into account the expansion of the glass at each temperature.

To be useful for measuring temperature, a liquid-filled thermometer must have several features:

- The thermometer must be calibrated so that the length of the column of liquid corresponds to a specific temperature.
- The thermometer must be sensitive enough that a small change in temperature produces a noticeable change in liquid level. This requirement is met by making the glass envelope sufficiently narrow so that a small change in the liquid's volume makes a large change in the column's length.
- The thermometer should respond quickly to changes in temperature. A liquid thermometer responds quickly if the glass of the bulb is thin enough.
- The mass of the thermometer must be small in relation to the bulk of the object whose temperature is being measured. If you were to attempt to measure the temperature of a few milliliters of a liquid with a standard laboratory thermometer, the thermal energy transferred between the mass of the thermometer and the liquid sample would materially affect the reading.
- The medium used in the thermometer must be appropriate for the range of temperatures being measured. Since mercury freezes at –39 °C, mercury thermometers cannot be used to measure temperatures at or below that point.

Liquid thermometers depend on the **thermometric property** of volumetric expansion in proportion to temperature change. Other thermometric properties are employed to indicate temperature. We have already mentioned how bimetallic strips can control heating and cooling systems. They are also used in large dial-type thermometers. Electrical resistance is another thermometric property of matter that usually increases with temperature. Some electronic temperature measuring instruments use special sensors called *resistive temperature detectors (RTD)* or *thermocouple junctions* to simply and reliably detect temperature at remote locations.

Mechanical thermometers present temperature directly to the user using the *analog* display mode. Analog displays vary continuously in a manner analogous to the way temperature changes. *Digital thermometers* use digital electronics (microprocessors) to process temperature inputs and to display changes of temperature in discrete increments. Their output precision is limited by the precision of the circuitry. These thermometers have become very popular for nearly all applications. Their main drawback is that they require electricity to work.

14.7 Temperature Scales

Since there are instruments to measure temperature, there must be units in which to measure it. A thermometer may have a column of liquid that rises to a particular height when it reaches a certain temperature, but what should that height be called? Historically, this decision was not settled until the nineteenth century. The issue concerned choosing appropriate reference points to establish a temperature scale.

The Fahrenheit Scale

Probably the most familiar temperature scale to people in the United States is the Fahrenheit scale, devised by Daniel Gabriel Fahrenheit. He initially made his zero point the lowest temperature that he could obtain with a water-ice-salt solution, and his other fixed point was the normal temperature of the human body. Following a suggestion from Newton, he set the temperature of the human body as 12° and used this as the upper reference temperature. These choices resulted in degrees that were too large to be practical, so he subdivided them into eighths, and renumbered, which then gave a human body temperature of 96°.

After recalibrating his thermometer in the smaller units, he discovered that pure water boiled near 212° and froze near 32°. There are 180° between these two points, so he decided to make these physical properties of pure water his new reference points. Once again he re-marked his scale, and, as a result, the average human body temperature, since it was no longer a fixed reference point, was measured as 98.6 °F. These freezing and boiling temperatures of pure water are valid at one atmosphere of pressure.

Daniel Gabriel Fahrenheit (1686–1736) was a German scientist who invented the mercury thermometer, established the Fahrenheit temperature scale, and discovered the effect of atmospheric pressure on the boiling points of liquids.

Reference standards for various kinds of measuring scales are also called *fiducial* points. The fiducial points for the Fahrenheit and Celsius scales are the freezing and boiling points of pure water.

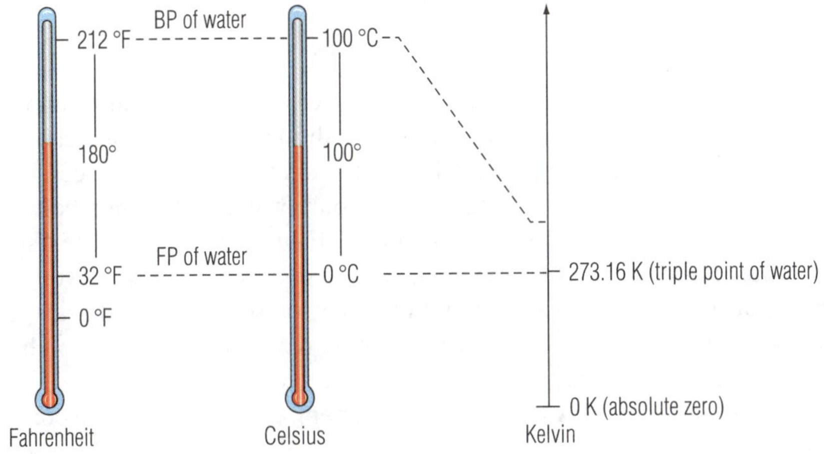

14-13 Comparison of three temperature scales showing fiducial points

The Celsius Scale

Anders Celsius also used the freezing and boiling points of water as reference points for calibrating his scale. However, Celsius, following the growing metric movement in Europe, chose to have 100 degrees between the freezing and boiling points of water. Consequently, the freezing point of pure water is 0 °C and the boiling point is 100 °C. This is the temperature scale used by most of the world other than the United States and is especially convenient for the scientific, technical, and engineering communities.

Anders Celsius (1701–44) was a famous Swedish astronomer who is most noted for developing the metric temperature scale for meteorological studies. He was part of the expedition that confirmed Newton's prediction that the earth was an oblate spheroid.

The Absolute or Kelvin Scale

Both the Fahrenheit and the Celsius scales have negative temperatures. For many years, no one knew how large the negative temperatures could get. For this reason, it was known that 60° was *not* twice the temperature of 30° in either temperature scale. What was needed was a single absolute reference temperature from which all others could be measured so that physical properties that depended on proportional temperatures could be correctly determined. After experimenting with the temperature-volume relationship of gases, scientists extrapolated the minimum possible theoretical temperature to be –273.15 °C. British physicist Lord Kelvin suggested in 1854 that this temperature be established as the **absolute zero** for measuring all temperatures. This temperature is not directly measurable because it cannot be reached. However, another single reference point was known and experimentally measurable—the triple point of water—where all three phases of water coexist. This point occurs at 0.01 °C and 611.73 Pa. So scientists define the triple point temperature to be *exactly* 273.16 Celsius degrees above the absolute zero point, which provides the *measurable* reference point for the absolute temperature scale.

The temperature unit on the absolute or Kelvin scale is called the kelvin (K) in honor of Lord Kelvin's many contributions to physical science. The kelvin is the same temperature interval as the Celsius degree on a thermometer. The kelvin unit is never used in conjunction with the word "degree," nor with the degree symbol. Therefore, when spoken, an absolute temperature is followed by the word "kelvins" rather than "degrees kelvin," and when written the unit symbol is just "K" and *not* "°K."

William Thomson, Lord Kelvin (1824–1907) was a prominent nineteenth-century Scottish mathematician and physicist who conducted investigations in many areas of science. He established the energy basis for thermodynamics and mechanics. He used thermodynamic principles to demonstrate that the earth could not be as old as evolutionists believed it had to be.

Absolute zero is the temperature of matter with no thermal energy. This is not achievable in the real world because there must be a colder location for the thermal energy to flow to in order to cool the substance to absolute zero.

The measurable fiducial point for the Kelvin scale is the triple point of water: +0.01 °C or 273.16 K.

14.8 Conversions Between Temperature Scales

Conversions between temperature scales are straightforward. The Celsius and Kelvin scales differ only by an added constant. Kelvin temperatures are *always* larger than Celsius, so converting from Celsius degrees to kelvins involves adding the constant:

$$T = t_C + 273.15$$

Converting from kelvins to Celsius is just the opposite:

$$t_C = T - 273.15$$

The symbol T represents an unknown absolute temperature. The symbol t represents an unknown Celsius or Fahrenheit temperature.

The constant is usually rounded to the decimal place of the given temperature before conversion. According to the significant digit addition rules, the result cannot have any more decimal places than the number with the fewest decimal places.

Between the freezing and boiling points of pure water, 180 Fahrenheit degrees correspond to 100 Celsius degrees. Therefore, Fahrenheit degrees are "smaller" than Celsius degrees. The two scales have different zero points as well (on the Celsius scale, 0 °C is the freezing point of water; on the Fahrenheit scale, 0 °F is 32° below the freezing point of water). For these reasons, the conversion between Celsius and Fahrenheit temperatures is a bit more complicated.

William Rankine (1820–72) was a contemporary and coworker of Lord Kelvin. He proposed an absolute temperature scale based on the Fahrenheit degree. The Rankine scale was eventually discarded with the adoption of the Kelvin scale by the SI as the standard absolute temperature scale.

The easiest and most flexible formula to remember for temperature conversion between Celsius and Fahrenheit is

$$t_{required} = k(t_{given} + 40°) - 40°,$$

where

- $t_{required}$ is the temperature in the scale that you need,
- t_{given} is the temperature in the given scale, and
- k is a factor that depends on the direction of the conversion.

The precision of the 40° terms using this method should match the given temperature data. The converted temperature should also have the same precision as the given temperature.

For conversion from Fahrenheit to Celsius, the factor k is $^5/_9$ (from the ratio $^{100°}/_{180°}$). You can remember that k is less than 1 because Celsius temperatures are lower than corresponding Fahrenheit temperatures. The factor $^5/_9$ gives smaller temperatures.

$$t_C = {}^5/_9\,(t_F + 40°) - 40°$$

Similarly, for conversions from Celsius to Fahrenheit, the factor k is $^9/_5$ (from the ratio $^{180°}/_{100°}$). Since Fahrenheit temperatures are higher than the corresponding Celsius temperatures, this factor gives higher temperatures.

The formulas for converting between Celsius and Fahrenheit may also be expressed using decimal factors:

$$t_F = 1.8(t_C + 40°) - 40°$$

$$t_C = 0.5\overline{5}\,(t_F + 40°) - 40°$$

$$t_F = \frac{9}{5}(t_C + 40°) - 40°$$

To convert from Kelvin to Fahrenheit, simply convert from Kelvin to Celsius and then to Fahrenheit.

EXAMPLE 14-4

Converting from Celsius to Kelvin

Convert 78 °C to kelvins.

Solution:

$$T = t + 273.15$$

$$T = 78\ °C + 273°$$

$$T = 351\ K$$

$$78\ °C = 351\ K$$

The given information is provided to the nearest degree, so the conversion constant is also rounded to the nearest degree.

EXAMPLE 14-5

Converting from Kelvin to Celsius

The boiling point for elemental oxygen is 90.2 K. What is this temperature in degrees Celsius?

Solution:

$$t = T - 273.15$$

$$t = 90.2\ K - 273.2°$$

$$t = -183.0\ °C$$

$$90.2\ K = -183.0\ °C$$

Note that the conversion constant was rounded to tenths of a kelvin before subtraction to match the decimal places of the given information.

EXAMPLE 14-6

Converting from Fahrenheit to Celsius

The temperature of water in your water heater at home is 131 °F. What is its Celsius temperature?

Solution:

$$t_C = \frac{5}{9}(t_F + 40°) - 40°$$

$$t_C = \frac{5}{9}(131\ °F + 40°) - 40°$$

$$t_C = 55\ °C$$

$$131\ °F = 55\ °C$$

Problem-Solving Strategy 14.4

It is recommended that you express temperatures mainly in the Celsius scale. The Celsius scale is more convenient, and you will become more familiar with the metric system as you use it.

14B Objectives

After completing this section, I can

✓ describe the construction and function of various thermometers.

✓ discuss the historical development of the three principal temperature scales and identify their fiducial points.

✓ convert temperatures between different temperature scales.

14B Section Review

1. What is a *thermometric* property? Give an example of one discussed in this section.
2. What was the key historical issue affecting the establishment of scientifically useful temperature scales?
3. What is the *measurable* reference temperature for the absolute temperature scale? What physical property establishes this temperature?
4. What two aspects complicate the conversion between Celsius and Fahrenheit temperatures?
5. Assume that 1.000 cm^3 of mercury expands 1.80×10^{-4} cm^3 for a +1.00 °C temperature change. If the degree scale marks on a thermometer are 1.00 mm apart, what must be the diameter of the cylindrical capillary tube in the thermometer to accommodate this volume change?
6. What is the average human body temperature in kelvins?
7. For a comfortable room temperature setting on a thermostat, 72 °F is often chosen. What is this temperature in Celsius?
8. Calculate absolute zero in degrees Fahrenheit.

Jacques Charles (1747–1823) was a French physicist who, after hearing of the Montgolfier brothers' airship balloon experiment, built his own balloon using hydrogen. His balloon stayed aloft for 45 minutes before landing in a field, where it was attacked and destroyed by superstitious farmers.

Joseph Louis Gay-Lussac (GAY le SAK) (1778–1850) was a French chemist and physicist credited with discovering the relationship between gas temperature and pressure. He conducted many experiments involving a gas balloon. During one such experiment, he reached a height of over 21 000 feet above sea level.

14C GAS LAWS

14.9 Ideal Gases

The gaseous phase of a substance has no definite volume or shape. In order to measure how much of a gas is present (how much mass or how many particles), you must establish a standard pressure and temperature. This, in turn, establishes a certain known volume for the gas under these standard conditions. As you study the interaction of a gas's temperature, pressure, and volume, you will use the idea of an **ideal gas** for the same reason that you used an ideal spring and other ideal concepts in earlier chapters—to make the basic properties of gases easier to understand.

An ideal gas has the following properties:

- Ideal gas particles are infinitesimally small—they have no volume.
- Ideal gas particles do not interact with each other except when they collide. Their collisions are perfectly elastic.

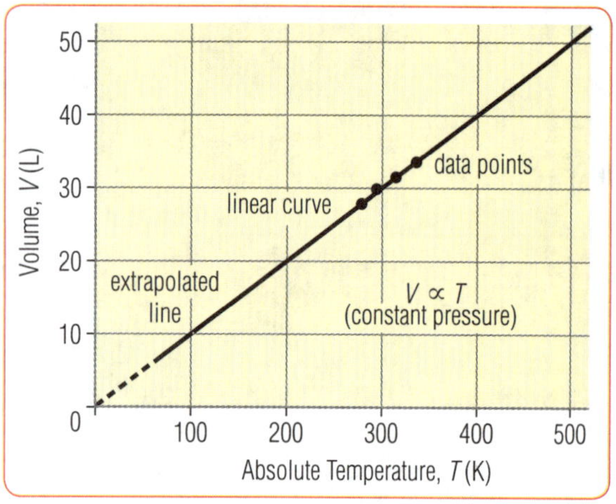

14-14 Graph showing the relationship between temperature and volume of an ideal gas in Charles's law

14.10 Charles's Law

In the late 1700s, Jacques Charles and Joseph Gay-Lussac studied the effect of temperature on the volume of a gas at constant pressure. Both men concluded that all gases expand at nearly the same rate, about 1/273 of their volume at 0 °C for each Celsius degree rise in temperature. Figure 14-14 shows this relationship. Extending the graph by extrapolation back to zero volume gives a temperature around –273 °C. More accurate experimentation in the nineteenth and twentieth centuries defined the theoretical minimum temperature to be –273.15 °C.

A gas theoretically has zero volume at absolute zero (0 K). (Actually, gases condense to the liquid phase above this temperature, so the line in Figure 14-14 does not apply for very low temperatures.) The graph represents a simple proportion:

$$V \propto T, \text{ or}$$

$$V = kT,$$

where k is an appropriate proportionality constant and T is absolute temperature measured in kelvins. This equation can be rearranged as

$$\frac{V}{T} = k.$$

The volume of a gas (at constant pressure) divided by its *Kelvin* temperature is a constant. If the temperature changes from T_1 to T_2,

$$\frac{V_1}{T_1} = k \text{ and } \frac{V_2}{T_2} = k, \text{ so}$$

$$\frac{V_1}{T_1} = \frac{V_2}{T_2}. \qquad (14.4)$$

Equation 14.4, called **Charles's law**, is true for any fixed quantity (mass) of an ideal gas *as long as Kelvin temperature is used* and pressure is the same when measuring volume and temperature.

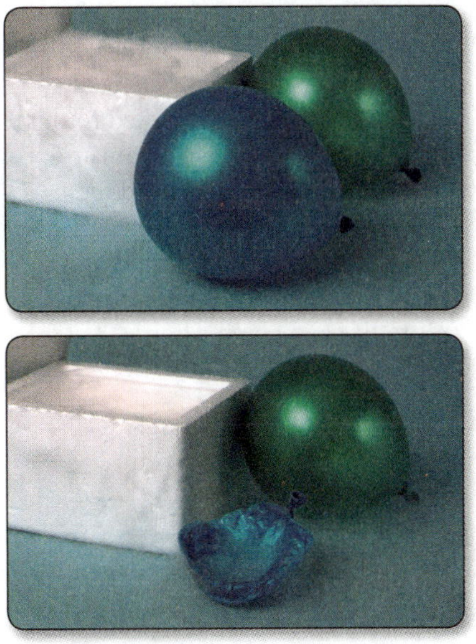

14-15 The volume of air in this balloon has contracted significantly with a reduction of absolute temperature after immersion in liquid nitrogen.

EXAMPLE 14-7

Using Charles's Law

What volume will a gas occupy at 277 °C if it occupies 0.500 L at 27 °C? Assume pressure remains constant.

Solution:

The "initial" conditions are V_1 = 0.500 L and T_1 = 27 °C + 273° = 300. K. The "final" condition is T_2 = 277 °C + 273° = 550. K, and V_2 is the unknown.

Solve Equation 14.4 for V_2 and substitute the known information to find the "final" volume:

$$\frac{V_1}{T_1} = \frac{V_2}{T_2}$$

$$V_2 = V_1\left(\frac{T_2}{T_1}\right)$$

$$V_2 = (0.500\text{ L})\left(\frac{550.\text{ K}}{300.\text{ K}}\right) \leftarrow (\text{"correction factor"} > 1)$$

$$V_2 \cong 0.91\underline{66}\text{ L} \cong 0.917\text{ L}$$

Notice that the appropriate ratio of the change in temperatures (>1) "corrects" the given volume to yield the unknown volume after the temperature change. Temperature increased; therefore, according to Charles's law, volume must increase.

Problem-Solving Strategy 14.5

When solving gas-law problems, the solution process can be greatly simplified by evaluating beforehand what effect the changing property has on the property of the unknown quantity, assuming the other variables are held constant. Multiply the given value by an appropriate ratio of the known variables to yield the unknown value.

When could you see Charles's law in effect? Try heating a bowl of water tightly wrapped with plastic food wrap in a microwave oven for several minutes. If the wrapping doesn't leak, the bowl will soon have a bulging plastic dome covering it. The water vapor and gases within heat up and expand. The plastic wrap maintains approximate atmospheric pressure in the trapped gases so you can see their change in volume with temperature.

Robert Boyle (1627–91) was an Irish-born English physicist and devout Christian, most noted for his critique of Aristotelian chemistry and alchemy, which opened the door to true chemistry. He also investigated the properties of gases and discovered the importance of air to life and combustion.

14.11 Boyle's Law

For Charles's law to be true, pressure must be constant. What if pressure is allowed to vary and temperature is held constant? Robert Boyle, the brilliant Christian physicist and chemist, investigated this possibility. To carry out his experiments, Boyle invented a J-shaped glass tube, called the Boyle's law tube. The short end of the J was melted shut, and the long end remained open. When Boyle poured mercury into the open end of the tube, he trapped some air in the closed end. He then measured this initial volume of air. By pouring more mercury into the tube and measuring the differences in the liquid metal's height between the long and short ends of the tube, he could determine the increase of pressure on the trapped air. He was careful to keep the temperature constant during the experiment. With each addition of mercury he measured the air's volume. He found that when the pressure was doubled, the air's volume was halved.

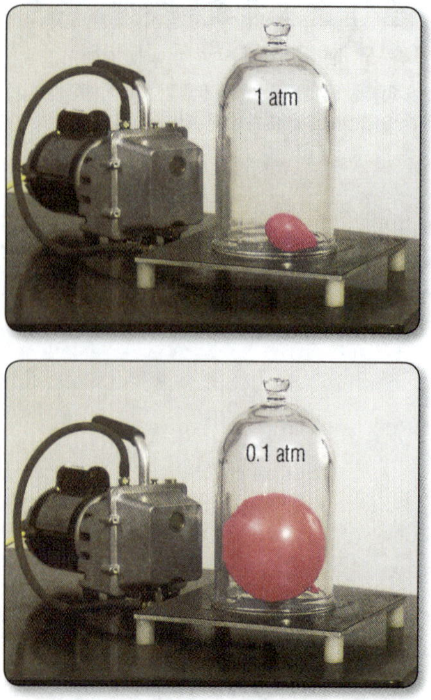

14-17 The volume of the balloon in differing air pressures illustrates Boyle's law.

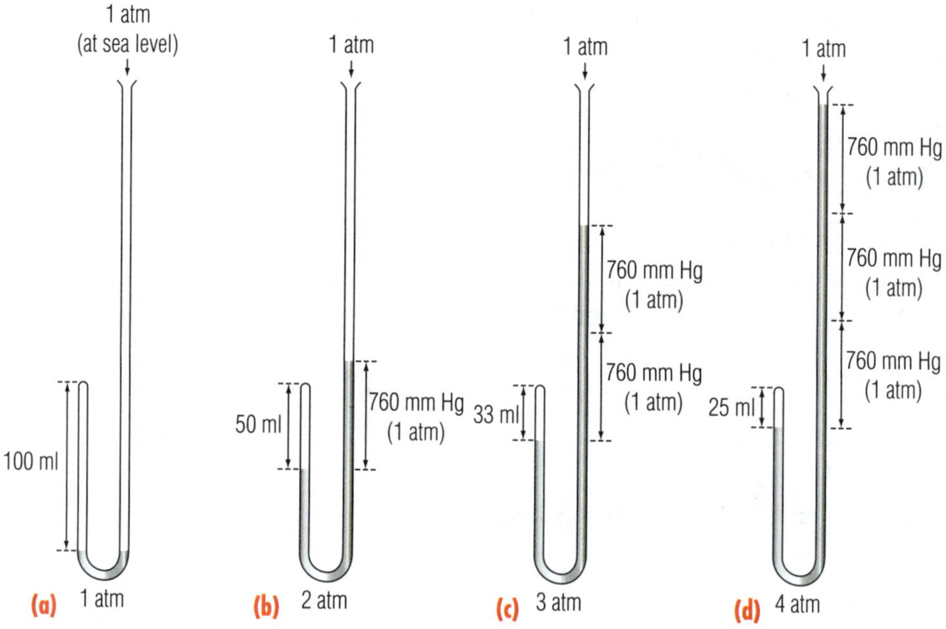

14-16 Boyle's law experiment. Note that the level of mercury in the open tube is measured relative to the level in the closed end.

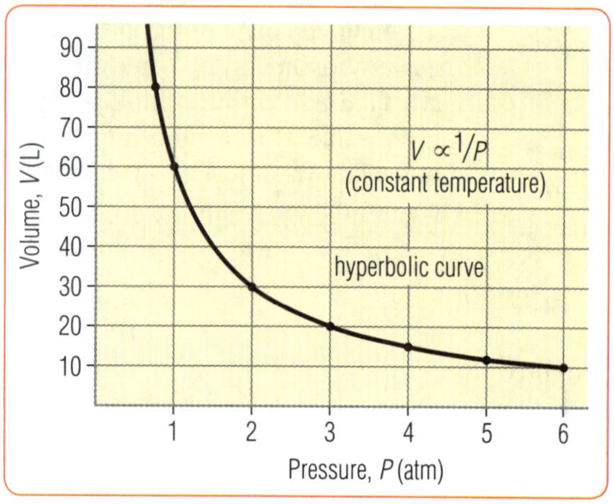

14-18 Graph showing the inverse relationship of pressure and volume according to Boyle's law

After experimenting thoroughly, Boyle established that the volume of a gas is inversely proportional to the pressure imposed on the gas as long as its temperature remains constant. Mathematically, this is equivalent to saying that the product of a gas's volume and pressure is constant. That is,

$$VP = k.$$

If the pressure and volume of a sample of gas change while temperature remains constant,

$$V_1P_1 = k, \text{ and } V_2P_2 = k, \text{ so}$$

$$V_1P_1 = V_2P_2. \quad (14.5)$$

Equation 14.5, called **Boyle's law**, is true for any fixed quantity of an ideal gas at constant temperature.

EXAMPLE 14-8

Using Boyle's Law

A chemist collects 2.50 L of oxygen from a chemical gas generator at a pressure of 800. torr. What volume will the oxygen occupy at a pressure of 760. torr? Assume that temperature remains the same.

Solution:

You know that pressure changes from 800. to 760. torr. According to the inverse relationship of Boyle's law, a pressure decrease means that the volume must be greater at the lower pressure if temperature remains constant. The "correction factor" for volume must be greater than 1. This is confirmed if we solve Boyle's law equation for the unknown V_2:

$$V_1P_1 = V_2P_2$$

$$V_2 = \frac{V_1P_1}{P_2} = V_1\left(\frac{P_1}{P_2}\right)$$

$$V_2 = (2.50\ \text{L})\left(\frac{800.\ \cancel{\text{torr}}}{760.\ \cancel{\text{torr}}}\right) \leftarrow (\text{"correction factor"} > 1)$$

$$V_2 \cong 2.6\underline{3}1\ \text{L} \cong 2.63\ \text{L}$$

Pressure decreased; therefore volume must increase.

The *torr* is a non-SI unit of pressure commonly used by chemists for routine laboratory work. 760. torr = 1 atm. The derivation of the torr will be discussed in Chapter 17.

Problem-Solving Strategy 14.6

It is not necessary to memorize the equations for Charles's or Boyle's laws if you can remember how gas volumes respond to changes of temperature and pressure. Use the appropriate "correction factor" and the given volume to obtain the required volume.

You probably have experienced a painful reminder of Boyle's law if you have rapidly descended or ascended a mountain or flown in an airplane when you had a head cold. If the air pressure change external to your ear cannot be equalized quickly enough, the volume of the air trapped in your middle ear changes, stretching your eardrum (tympanic membrane). The pain can usually be relieved by swallowing, chewing gum, or yawning to open the Eustachian tube to your middle ear.

14.12 Gay-Lussac's Law

Suppose the volume of a gas is held constant while the pressure and temperature are allowed to vary. What will happen? Gay-Lussac experimented to answer this question and found that the gas's pressure divided by its Kelvin temperature is constant as long as its volume is constant. That is,

$$\frac{P}{T} = k.$$

If the gas's pressure or temperature changes,

$$\frac{P_1}{T_1} = k \text{ and } \frac{P_2}{T_2} = k, \text{ so}$$

$$\frac{P_1}{T_1} = \frac{P_2}{T_2}. \qquad (14.6)$$

Equation 14.6, called **Gay-Lussac's law**, is true for any fixed quantity of an ideal gas at constant volume when T is expressed in kelvins.

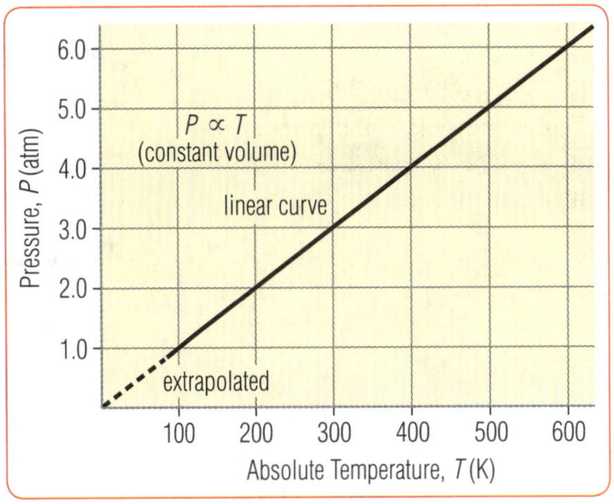

14-19 Graph showing the direct proportional relationship between temperature and pressure for a sample of an ideal gas

EXAMPLE 14-9

Using Gay-Lussac's Law

Air in a closed container is initially at a temperature of 27 °C and a pressure of 760. torr. What is the pressure of the air if it is cooled to −78 °C?

Solution:

The temperature decreases from 300. K to 195 K; therefore, according to Gay-Lussac's law, pressure must drop in a constant-volume container. The pressure "correction factor" must be less than 1.

$$\frac{P_1}{T_1} = \frac{P_2}{T_2}$$

$$P_2 = \frac{T_2 P_1}{T_1} = P_1\left(\frac{T_2}{T_1}\right)$$

$$P_2 = (760.\ \text{torr})\left(\frac{195\ \text{K}}{300.\ \text{K}}\right) \leftarrow (\text{"correction factor"} < 1)$$

$$P_2 = 494\ \text{torr}$$

The factor that "corrects" the initial pressure for the temperature change is less than 1.

Gay-Lussac's law can be observed in action when a soccer ball that is full and rigid on a warm summer day becomes soft and flaccid on a cold winter day.

14.13 The Combined Gas Law

The three gas laws that we have studied thus far had a constraint on one gas property—pressure, temperature, or volume. What if all three properties are permitted to change without any constraint? Can we still predict the value of an unknown property? The answer, of course, is yes. Ideal gases behave in such a way that if the temperature and pressure of a given quantity of gas change simultaneously, its final volume is the same as if temperature and pressure changed stepwise, one property at a time.

You could use two steps to calculate an answer when both temperature and pressure vary, but it is not necessary. The **combined gas law** equation shows that the same effect can be achieved in one mathematical operation:

The only constraint when using the combined gas law is that the quantity (number of particles) of gas must remain constant.

$$\frac{P_1 V_1}{T_1} = \frac{P_2 V_2}{T_2} \qquad (14.7)$$

Equation 14.7 is true as long as the quantity of gas is constant.

Any of the three gas laws can be derived from the combined gas law equation by holding the appropriate property constant. However, it should not be necessary to memorize all four (or even three) gas law equations if the effects of the basic laws are understood. Just use the appropriate "correction factors" on the required property to find the answer according to the respective gas laws.

EXAMPLE 14-10

The Combined Gas Law

A 1.00 L sample of gas initially at 27 °C and 760. torr is cooled to −123 °C at a pressure of 400. torr. What is its volume under the new conditions?

$$\frac{P_1V_1}{T_1} = \frac{P_2V_2}{T_2}$$

$$V_2 = V_1\left(\frac{T_2}{T_1}\right)\left(\frac{P_1}{P_2}\right)$$

given volume	temperature correction (< 1)	pressure correction (> 1)

$$V_2 = (1.00\ \text{L})\left(\frac{150.\ \text{K}}{300.\ \text{K}}\right)\left(\frac{760.\ \text{torr}}{400.\ \text{torr}}\right)$$

$$V_2 = 0.950\ \text{L}$$

Note that since temperature decreased, according to Charles's law, volume should decrease (temperature "correction" < 1). Pressure decreased; therefore, according to Boyle's law, volume should increase (pressure "correction" > 1).

14.14 The Ideal Gas Law

If Robert Boyle's experiments with a J-tube or similar device were performed at different temperatures, we would discover that the product of pressure and volume is directly proportional to the absolute temperature. In an equation,

$$PV = KT, \tag{14.8}$$

where K is the temperature-dependent proportionality constant.

If we take two samples of the gas having the same volume, temperature, and pressure and combine them so that their temperature and pressure are unchanged, the gas volume doubles. This means that K must double as well to maintain the equality of Equation 14.8. Therefore, K is directly proportional to the *amount* of gas, or the number of gas particles present. We can express this proportionality as

$$K = kN, \tag{14.9}$$

where N is the number of gas particles in the sample and k is the proportionality constant known as **Boltzmann's constant**. This constant has the value

$$k = 1.381 \times 10^{-23}\ \text{J/K}.$$

To avoid using the extremely large numbers associated with individual particles in a gas sample, we normally express the amount of a gas in moles (mol). Recall from Chapter 13 that a mole of any kind of particle (atoms, molecules, etc.) is Avogadro's number (N_A) of particles.

The number of particles is

$$N = nN_A,$$

where n is the number of moles of gas. Equation 14.9 can now be rewritten

$$K = nkN_A.$$

Ludwig Boltzmann (1844–1906) was an Austrian physicist and mathematician best known for his development of statistical mechanics, which states that the visible properties of matter are determined by the properties of the atoms and molecules of which it is made.

Avogadro's constant is $N_A = 6.022 \times 10^{23}$ particles/mol.

The product of Boltzmann's and Avogadro's constants, kN_A, is conveniently combined into one constant called the ideal or **universal gas constant (*R*)**:

$$R = kN_A = (1.38066 \times 10^{-23}\,\text{J/K})(6.022137 \times 10^{23}\,\text{mol}^{-1})$$

$$R \cong 8.31452\ \text{J/(K·mol)} \cong 8.315\ \text{J/(K·mol)} \qquad \text{(to four SDs)}$$

Equation 14.8 can be rewritten as

$$PV = nRT, \qquad (14.10)$$

which is known as the **ideal gas law**. Observe that rearranging Equation 14.10 to isolate the constant R places all of the possible gas variables in one ratio:

$$R = \frac{PV}{nT}$$

The ideal gas law predicts that this ratio should hold true for any pressure (pressure determines how close the particles are to each other). Real gases hold closely to this relationship up to several atmospheres of pressure.

The ideal gas constant as given thus far is expressed in SI units. In order to use it in this form, volume must be expressed in cubic meters, temperature in kelvins, and pressure in pascals. Since gas-law experiments may be conducted using other units, either the experimental units must be converted to SI units or the constant R must be converted to the units being used. Table 14-4 provides values of R in commonly used volume and pressure units. Note that temperature must *always* be expressed in kelvins regardless of the value of R used.

Table 14-4

Units for Use in Ideal Gas Law

V	P	T	n	R
m^3	N/m^2	K	mol	8.315 J/(K·mol)
m^3	Pa	K	mol	8.315 m^3·Pa/(K·mol)
L	atm	K	mol	0.08207 L·atm/(K·mol)
L	torr	K	mol	62.36 L·torr/(K·mol)

The ideal gas law is useful because we can determine one of a gas's four properties if the other three are known. In order to compare the properties of one gas to another, data for both must be obtained under the same conditions. The scientific community has defined **standard temperature and pressure (STP)** to be 273.15 K and 1.013×10^5 Pa respectively. Obviously, their equivalents in other unit systems may be used if the situation requires.

EXAMPLE 14-11

Using the Ideal Gas Law: Find Volume

Find the volume of 1.00 mole of hydrogen in liters at STP.

Solution:

Solve Equation 14.10 for volume and substitute the known values for the other quantities. Note that we may use either value of R that includes L in its unit. The other appropriate parameters must be used as well.

$$V = \frac{nRT}{P} = \frac{(1.00\ \cancel{\text{mol}})(0.08207\ \text{L}\cdot\cancel{\text{atm}}/\text{K}\cdot\cancel{\text{mol}})(273.15\ \cancel{\text{K}})}{1.00\ \cancel{\text{atm}}}$$

$$V \cong 22.\underline{4}1\ \text{L} \cong 22.4\ \text{L}$$

One mole of an ideal gas at STP occupies a volume of 22.4 L.

EXAMPLE 14-12

Using the Ideal Gas Law: Find Pressure

If 2.00 moles of ammonia occupy 50.0 L at 350. K, what is the pressure of the gas in pascals?

Solution:

In this case you will solve Equation 14.10 for pressure. Since you need to find pressure in pascals, $R = 8.315\ m^3{\cdot}Pa/K{\cdot}mol$.

In order to use this value of R, you must convert liters to cubic meters:

$$\frac{50.0\ \cancel{L}}{} \left| \frac{1\ m^3}{1 \times 10^3\ \cancel{L}} \right. = 0.0500\ m^3$$

Solve for pressure:

$$P = \frac{nRT}{V} = \frac{(2.00\ \cancel{mol})(8.315\ \cancel{m^3}{\cdot}Pa/\cancel{K}{\cdot}\cancel{mol})(350.\ \cancel{K})}{0.0500\ \cancel{m^3}}$$

$$P \cong 1.1\underline{6}4 \times 10^5\ Pa \cong 1.16 \times 10^5\ Pa$$

14.15 Real Gases

The ideal gas model assumes that particles of a gas have no volume and interact only through collisions. Of course, real gas particles *do* occupy space and attract each other. These properties can usually be ignored if the particles are so far apart that their volume is insignificant and their motion is so rapid that the forces between them have little effect. Thus, the ideal gas model is accurate for gases at high temperatures and low pressures.

At higher pressures or lower temperatures, a gas begins to deviate from the predictions of the ideal gas model. At first, volumes at higher pressures or lower temperatures are *less* than the ideal gas law predicts. The attraction between molecules reduces their impulsive collision forces with the container. As a result, the real gas cannot occupy as much volume at a given pressure as predicted by the ideal gas model. At extremely high pressures or extremely low temperatures, the gas molecules are so close together or have so little kinetic energy that their individual volumes are no longer insignificant compared to the volume of the gas sample as a whole. In this state, the gas begins to resist further compression, and the volume is *more* than the ideal gas law would predict for that pressure and temperature.

One of the first men to realize the limitations of the ideal gas model and to devise an experimentally corrected equation was Johannes van der Waals. His gas law is

$$\left(P + a \cdot \frac{n^2}{V^2}\right)(V - bn) = nRT.$$

The factors a and b are constant for each gas. The constant a represents a pressure correction for the force between gas molecules. The factor b is a volumetric correction based on the moles of molecules present. These factors must be found experimentally for each gas. This equation holds for most gases over a wide range of temperatures and pressures. Those gases that obey the van der Waals equation are called **van der Waals gases**.

Johannes Diederik van der Waals (1837–1923) was a Dutch physicist and professor known for his work on the continuity of the liquid and gaseous states as well as formulation of the van der Waals equation and other descriptive equations. He also studied electrolytic dissociation and the thermodynamic theory of capillarity. He was awarded the Nobel Prize in Physics in 1910.

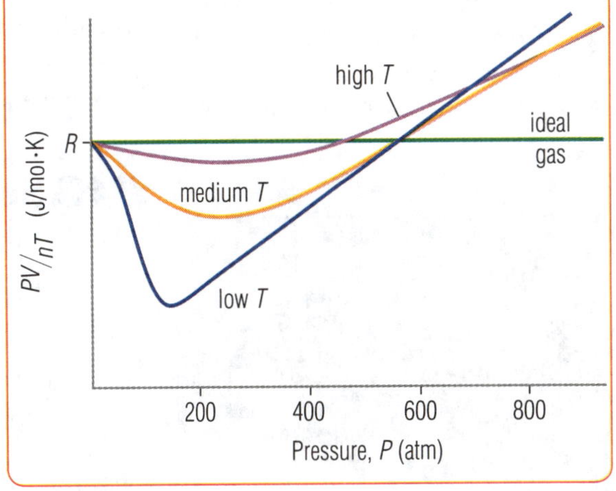

14-20 Comparison of the behavior of real gases at various temperatures and pressures to an ideal gas. Note that only at higher temperatures and lower pressures do real gases approximate ideal gases.

More complicated equations have been devised to describe more gases over wider ranges of temperatures and pressures. However, the simplest equation, the ideal gas equation, is still the most useful approximation for the behavior of gases under typical conditions.

14.16 Using the Gas Laws to Solve Problems

Like Charles and Gay-Lussac, Forrest Bird was a scientist whose inspiration for his scientific pursuits derived from his passion for aviation. Bird was the son of a World War I pilot. He met Orville Wright when he was young and received his pilot's license before he could drive. Later, when flying for the military during World War II, he became motivated to improve the clumsy, inefficient breathing respirators available to pilots for high-altitude missions. After successfully modifying a captured German respirator, he became interested in the general principles of breathing and artificial respirators. As an experienced pilot, he visualized human respiration as air flowing in and out of the lungs due to the same forces that cause lift in aircraft.

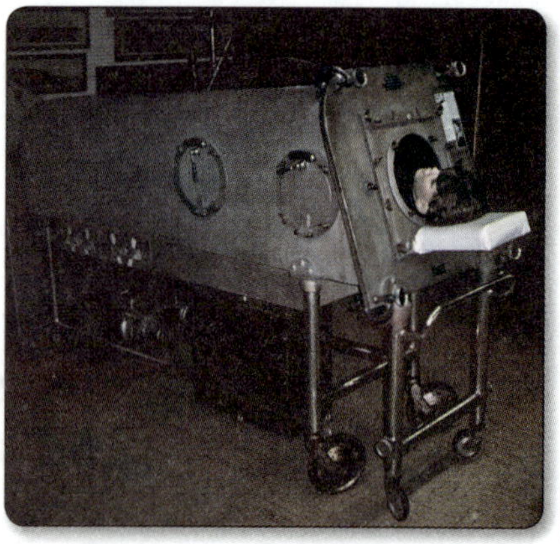

14-21 Before Bird's respirators, the most desperate respiratory patients used an iron lung.

Because of his successful work in designing high-altitude respirators for pilots, the Air Force sent Forrest Bird to medical school to study human respiration, and he completed several residencies. His first successful medical respirator was assembled from an aviation respirator mask and had a manual control knob used to artificially inflate the lungs of a man with emphysema. This device was quickly improved, and Bird began marketing it in 1955 as the first low-cost, mass-produced, portable respirator—the Mark 7. Its ease of use and portability vastly improved the quality of life of patients, such as polio victims, who formerly were confined to iron lungs (see Figure 14-21).

In the late 1960s, Bird experimented with regulating pressure in the cabins of jets. An Air Force doctor convinced him to develop a ventilator for infants. In Bird's own words, "I thought I could take a baby's lungs and treat it like the cabin of the Learjet," though this process would have to be much gentler. In 1969, Bird developed the Baby Bird respirator. After its initial success, it quickly gained wide acceptance, and infant mortality due to respiratory distress syndrome was reduced from 70% to 10%.

Loving One's Neighbor Through Inventions and New Technologies

Out of desperation, Forrest Bird used an experimental version of his Intrapulmonary Percussive Ventilator because his wife was about to die from an advanced case of emphysema. The invention immediately cleared her lungs of secretions, improved her breathing, and added five more years to her life. Suggest other examples where scientific and technological developments could fulfill the second great commandment.

Ventilators for both adults and infants use the same principles. In the most basic sense, positive pressure respirators deliver a compressed air/oxygen mixture to a patient to assist or take the place of inhalation, increasing the pressure in the patient's lungs as the volume increases. The patient naturally exhales as the elasticity of the lungs and chest muscles compress the gas, exhausting it through the breathing passages. This process repeats several times per minute, based on a mechanical timer or in response to natural breath demands. Modern infant ventilators have several modes of operation, depending on the needs of the patient.

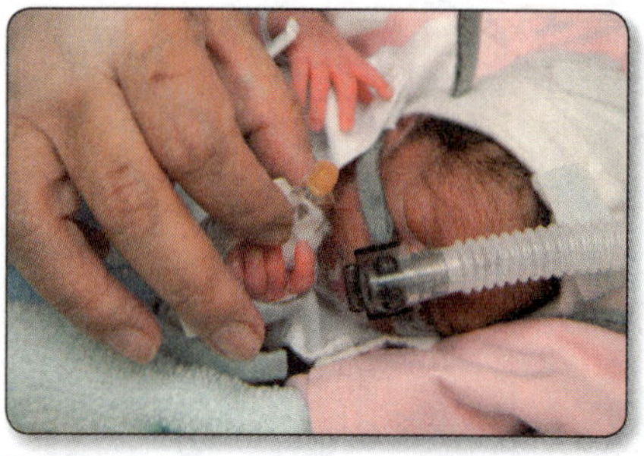

14-22 Modern respiratory ventilators are far less confining and simpler than their iron-lung predecessors.

14C Section Review

1. What are the characteristics of an ideal gas?
2. Describe in your own words Charles's law. Under what conditions is this principle true?
3. Describe in your own words Boyle's law. Under what conditions is this principle true?
4. What does the ideal gas law take into account that the other gas laws assume to be constant?
5. DS Which of the gas laws applies most directly to the function of a positive-pressure ventilator?

6. An ideally elastic balloon contains 0.650 L of air at 25.0 °C. What is the volume of the balloon after the sun warms it to 42.0 °C? Assume that pressure within the balloon remains constant.

7. A hot air balloon contains 120. m^3 of air while moored to the ground. Air pressure is 1.00 atm and temperature is 60.0 °C. The balloon launches and rises until the air pressure is 0.95 atm. What is the new volume of the hot air in the balloon, assuming that the hot air temperature remains constant and the amount of hot air does not change?

8. What is the new volume of hot air in the balloon in Question 7 if the hot air temperature decreases to 48.0 °C at the final altitude?

9. How many moles of an ideal gas would it take to occupy 1.00 L at STP?

14C Objectives

After completing this section, I can

- ✓ describe the properties of an ideal gas.
- ✓ state the relationships and restrictions involving pressure, temperature, volume, and quantity of gas for each of the gas laws.
- ✓ perform calculations using the gas laws.
- ✓ state the standard conditions for measuring gas volume, and state the molar volume of a gas under these conditions.
- ✓ differentiate between ideal and real gases and explain how real gases depart from the ideal gas model under various conditions.
- ✓ discuss how an understanding of gas laws was the basis for vastly improving the survival rate of premature infants.

Chapter Review

In Terms of Physics

coefficient of linear expansion (α)	315
bimetallic strip	316
coefficient of volume expansion (β)	316
thermometer	319
thermometric property	320
absolute zero	322
ideal gas	324
Charles's law	325
Boyle's law	326
Gay-Lussac's law	327
combined gas law	328
Boltzmann's constant (k)	329
universal (ideal) gas constant (R)	330
ideal gas law	330
standard temperature and pressure (STP)	330
van der Waals gas	331

Problem-Solving Strategies

14.1 (page 315) The thermal expansion equation can be simplified if the length of the object, ℓ_i, is measured at the reference temperature of 0.0 °C. Then the change of length is just $\Delta\ell = \alpha\ell_i t$, where ℓ_i is the reference length and t is the new Celsius temperature.

14.2 (page 317) Be sure to use the appropriate coefficient of expansion for solid materials. Use α for linear expansion and β for volumetric expansion.

14.3 (page 317) There is no linear coefficient of expansion for fluids because, unlike solids, fluids do not have fixed dimensions.

14.4 (page 323) It is recommended that you express temperatures mainly in the Celsius scale. The Celsius scale is more convenient, and you will become more familiar with the metric system as you use it.

14.5 (page 325) When solving gas-law problems, the solution process can be greatly simplified by evaluating beforehand what effect the changing property has on the property of the unknown quantity, assuming the other variables are held constant. Multiply the given value by an appropriate ratio of the known variables to yield the unknown value.

14.6 (page 327) It is not necessary to memorize the equations for Charles's or Boyle's laws if you can remember how gas volumes respond to changes of temperature and pressure. Use the appropriate "correction factor" and the given volume to obtain the required volume.

Review Questions

1. How are molecules of hot water and molecules of cold water different?
2. What factors cause the arrangement of particles in solid matter to require more space as the material becomes hotter?
3. What features are designed into a vehicle bridge to accommodate the expansion of materials as they are heated?
4. What two factors determine how much a metal rod will expand in length for a given temperature change?
5. Suppose you want to buy gold *by volume*. Would you get more for your money if you heat the gold or cool it before measuring it? (*Hint*: Temperature does not affect the gold's weight.)
6. Why is a thermoscope alone a poor instrument for measuring temperature?
7. How does a mercury or alcohol thermometer indicate temperature?
8. Which one of the following is *not* the freezing point of water?
   a. 32 °F
   b. 0 K
   c. 0 °C
   d. ~273 K
9. Which one of the following is the boiling point of water?
   a. 100 °F
   b. 273 K
   c. 100 °C
   d. 373 °C
10. Which temperature scale is based on a single fiducial point? What is the value of the fiducial point for this scale?
11. Which temperature unit represents the smallest change in temperature: the Celsius degree, the kelvin, or the Fahrenheit degree?
12. Under what conditions does $V_1/T_1 = V_2/T_2$ for a gas?
13. When using the gas laws, what temperature scale must be used?
14. For every gas law involving ideal gases, except the ideal gas law, what is assumed about the sample of gas?
15. Write the ideal gas law.
16. What is the product of Boltzmann's constant and Avogadro's constant called?

True or False (17–25)

17. Materials expand when heated because the atoms physically become larger, taking up more space.
18. The coefficient of volume expansion is the cube of the coefficient of linear expansion.
19. The coefficients of expansion for gases are very similar because gas particles are so far apart that they do not affect each other at normal temperatures and pressures except when they collide.
20. A thermometric property must be proportional to temperature.
21. It is best if a thermometer's mass is large compared to the mass of the material whose temperature it is measuring so that it will be more sensitive.
22. The accepted fiducial point of the absolute temperature scale is 0 K.
23. Charles's law states that the volume of an ideal gas is directly proportional to its absolute temperature, assuming pressure is held constant.
24. Boyle's law states that the volume of an ideal gas is proportional to the reciprocal of the gas's pressure, assuming temperature is held constant.
25. Real gases adhere closely to the ideal gas law at high pressures and low temperatures.

✪26. A bar of gold is 10.0 cm × 5.00 cm × 5.00 cm at room temperature, 25.0 °C. What is its change of volume at 50.0 °C?

✪27. An uncapped 1.00 L jar is filled to the brim with glycerin early in the morning, when the temperature is 10.0 °C. How much glycerin spills over the brim that day if the temperature reaches 30.0 °C? Ignore the jar's expansion.

✪28. At 100.0 °C, what will be the volume of mercury that has a volume of 1.00 cm^3 at 0.00 °C?

✪29. The pine mast of a New England charter sailing ship is 20.000 m tall from keel to peak on a frigid (−40.0 °C) winter day. The following summer, the mast expands to 20.007 m on a hot day. Assuming that the mast is at a uniform temperature during each measurement, what is the mast's temperature on the summer day? (The grain of the wood runs lengthwise in the mast.)

✪30. Convert the following temperatures to Fahrenheit.
   a. 25 °C
   b. −15 °C
   c. −40 °C
   d. 273 K
   e. 373 K

✪31. Convert the following temperatures to Celsius.
   a. 0 K
   b. 503 K
   c. 328 K
   d. 113 °F
   e. 59 °F

✪32. Convert the following temperatures to Kelvin.
   a. 12 °C
   b. 217 °C
   c. 122 °C
   d. −73 °C
   e. 104 °F

✪33. A gas occupies a volume of 2.00 L at 612 K and 760. torr. What volume does it occupy at 306 K and 760. torr?

✪34. A gas is confined to a 1.50 L expandable container at 300. K and 380. torr. What is its pressure if it is allowed to expand to a volume of 2.00 L at constant temperature?

✪35. A 2.5 L airtight, rigid container full of carbon dioxide at 325 K is heated to 525 K, where its pressure is 150 torr. What was its original pressure?

✪36. Find the volume in liters of 3.5 moles of oxygen at 20.0 °C under 760.0 torr.

✪37. A total of 1.75 L of acetylene gas is collected from a chemical generator at 735 torr and 35.0 °C. How many moles of acetylene were collected?

DS✪38. Some of the people Bird tested his inventions on were on the verge of death, and some of these people died, including several infants. Is this kind of experimentation morally acceptable?

✪39. Write a brief report on the development of the temperature scales used in scientific research. Discuss the current advances and standards for determining the key fiducial points.

✪40. Write a brief report on the work of Jacques Charles.

Count Rumford (1753–1814)

Scientists try to explain what they observe with theories. From time to time, new observations show that an old theory must be discarded and a new theory devised. A striking example of this process is Count Rumford's cannon-boring experiment.

Benjamin Thompson, who later became Count Rumford, was born to poor parents in Woburn, Massachusetts. His family could not afford to keep him in school. At the age of 19, Thompson improved his lot by marrying a rich widow from Rumford (now Concord), New Hampshire. When the American colonies were driven to war against England, Thompson sided with the British as a spy. But his sedition was discovered, and he had to flee to England. There his knowledge of colonial affairs won him a military position.

Following the war, Thompson's services were no longer needed in England. He had no hope of returning to America, where he was considered a traitor. Of the few choices remaining, he accepted a minor position in the Bavarian government; yet before Thompson went he persuaded George III to knight him by promising to work undercover in Germany. Thompson's skill in politics soon earned him a position as the minister of war and police. He introduced social reforms, promoted the potato plant, reorganized the outdated Bavarian army, and supervised the production of modern weapons. In 1792 the Bavarian government awarded him the title Count Rumford.

Along with everything else Thompson was doing, he found time to conduct scientific experiments. While watching munitions plants bore their brass cannons, he noticed that the searing heat never died down no matter how long the cannon was bored. His contemporaries believed that heat was a fluid locked in the metal's interstices and released by friction. But if that were the case, then the fluid would be used up in the process of boring. Thompson theorized that heat is not a fluid, but somehow related to motion.

Rumford investigated other aspects of heat, especially convection. He was the first to detect convection currents and the expansion of cooling water. Using this knowledge, he invented an efficient stove, a double boiler, and a drip coffee-maker. He also improved the common chimney and fireplace.

Rumford's plans to return to the United States were frustrated when President John Adams would not allow him to direct a military academy he had designed. So Rumford returned to England in 1798 and founded the Royal Institution of London, an organization to promote science among the common people. Its employees included Sir Humphry Davy and Michael Faraday. In 1804 Rumford left London for Paris, where he was warmly received by Napoleon Bonaparte. The count remained in Paris until his death ten years later.

James Joule
(1818–89)

A fundamental law of physics is the conservation of energy. It is surprising, therefore, that no one discovered this law (or even defined energy) until the nineteenth century. The major experiments that prove this law were first performed by James Prescott Joule.

A sickly child of wealthy parents, James had to be educated at home until he was fourteen. Always experimenting, young James had to settle for makeshift equipment from the kitchen until his father could build him a laboratory. When he went to Manchester University, he studied under the famous chemist John Dalton, who had formulated the theory that atoms are indivisible spheres.

Joule was especially interested in electricity and the heat that an electric current produces. His early experiments showed that the heat generated per unit time by electricity in a wire is proportional to the product of the wire's resistance and the square of the current ($P \propto RI^2$). This fact is now called Joule's law.

Whether the electricity was produced by chemical energy (in a voltaic cell) or by mechanical energy (in an electromagnetic generator he had invented), the resulting heat was always the same. Joule contrived an experiment to see if mechanical energy could produce heat directly without first becoming electricity. The success of his experiment proved that any form of energy, whether electrical, chemical, or mechanical, produces an equivalent amount of heat.

Joule went a step further and devised four experiments to determine the precise mechanical equivalent of heat. In his most famous experiment, a paddle wheel, driven by the mechanical energy in a falling weight, heated the water in a tub just enough to be measured by a sensitive thermometer. With more careful experiments he refined his results until, by 1850, he found that 772 foot-pounds of work always raise the temperature of one pound of water one Fahrenheit degree. He proved that heat transfer is not the flow of a fluid, as was popularly believed, but the transfer of a form of energy. He showed that this energy is conserved during work. His careful experiments paved the way for general acceptance of the conservation of energy.

Joule's work went unnoticed until William Thomson, Lord Kelvin, discovered it. Kelvin then asked Joule to collaborate with him in several experiments, the most famous of which demonstrated the "Joule-Thomson effect." This discovery, that the temperature of gas falls when it is allowed to expand, ushered in the huge refrigeration industry later in the century. With Thomson's help, Joule received acclaim as a scientist. In his last years, the British government granted him a pension in honor of his achievements.

The sun is the source of most of the thermal energy at the earth's surface.

Thermal Energy and Heat 15

DOMINION MODELING

Metals and Heat

Why are different types of metals used for different purposes? Specific metals and mixtures of metals, called alloys, are used for cookware, jet engines, steam boilers, catalytic converters, and many other applications. Metals are good conductors of heat and electricity and generally possess strength, ductility, and malleability. But why do different metals have different thermal properties? Can this difference be related to other measurable properties of metals?

15-1 Why are different metals used for different purposes?

15A THEORIES OF HEAT

15.1 Introduction

What makes one object hot and another cold? What exactly is heat? Scientists have studied these questions for centuries, and their conclusions have changed as they have learned more about how heated and unheated objects behave.

15.2 The Kinetic Theory

Modern scientists agree that what we call *heat* is really the movement of a form of energy called thermal energy. According to the kinetic theory of matter, all substances contain tiny, constantly moving particles. The particles in hot materials move faster than the particles in colder materials. **Thermal energy** is the sum of the kinetic energy of the random motion of the particles. The average kinetic energy of the particles is proportional to the temperature of the material. However, temperature alone does not indicate how much thermal energy is present. We will look into this aspect later. For now, let's review some important observations that give evidence for the kinetic-molecular theory of matter.

The kinetic theory is also called the kinetic-molecular theory of matter because the two concepts of tiny discrete particles in constant motion are inseparable.

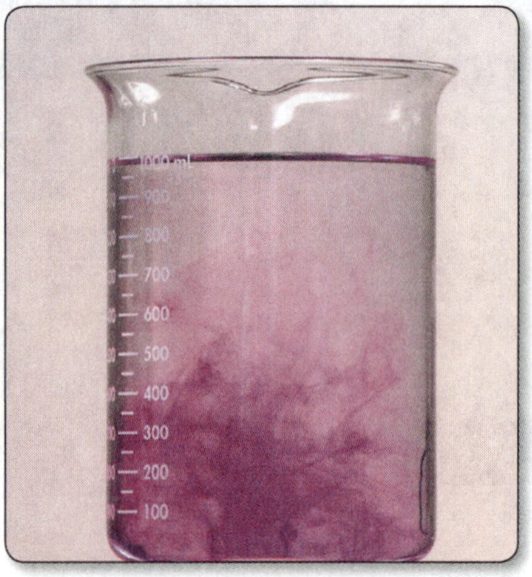

15-2 The spreading of this dye throughout the water illustrates the principle of diffusion.

15.3 Matter Can Be Subdivided

Suppose that you place a crystal of blue dye in a beaker of water and stir it until it is completely dissolved and all the water has become dark blue. The crystal has somehow divided itself and spread throughout the water in the beaker. If you pour the contents of the beaker into a much larger container of water and stir the solution again, the color becomes less intense but is still completely distributed throughout the larger container. The process may be repeated many times. The color becomes progressively fainter but is still present. The material from the crystal seems to keep dividing indefinitely. For this to be true, there must be a multitude of minuscule particles. These particles are called molecules, atoms, or ions.

15.4 Diffusion

You have seen evidence that particles exist. Now let's consider evidence that these particles are in motion. Suppose you drop a crystal of the same blue dye into a beaker of water *without* stirring. If you wait patiently, you will see that the crystal dissolves and that eventually the dye and the water completely mix (though this may take a very long time). This observation can be explained by assuming that the water molecules are in constant motion. Their motion distributes particles of the dye throughout the water. The spreading of a substance through particle motion alone is called **diffusion**.

Diffusion takes place much faster in gases than in liquids. This observation leads us to believe that gas molecules move faster than liquid molecules. For example, when a few drops of household ammonia are spilled from a bottle on one side of a room, the scent of ammonia can soon be smelled on the far side of the room. Judging by the speed at which the ammonia drifts, air molecules must move at high speeds, around 500 m/s or more at room temperature.

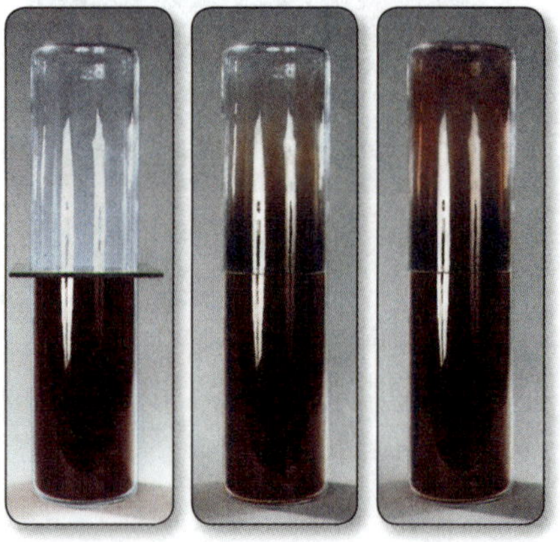

15-3 Diffusion of gases is clearly seen in the mixing of bromine vapor and air.

Scientists can visually demonstrate diffusion in gases using bromine vapor and air. Bromine vapor is brown and dense; it is about five and one-half times as dense as air. (It is also poisonous—that is why only qualified people should perform this demonstration.) Suppose that a container of air is placed upside down over a similar container of bromine vapor that has its open end facing up but covered by a glass plate (see Figure 15-3). When the glass plate is slid out from between the containers, the brown bromine vapor will start diffusing upward into the transparent air. Air will also move downward into the bromine vapor. This cannot be explained by gravity, because the lighter air goes downward, while the heavier bromine vapor moves upward. After a period of time, the two gases will be evenly mixed. The molecules of the air and the bromine must each have motion of their own to mix themselves so quickly and thoroughly. This process occurs even more rapidly at higher temperatures.

15.5 Brownian Motion

Robert Brown (1773–1858) was an English botanist who first described Brownian motion while studying pollen grains under a microscope.

In 1827 Robert Brown discovered more direct evidence that matter is in constant motion. Brown used a microscope to examine plant spores floating in water. He saw the spores jostling back and forth as if they were being struck repeatedly from different sides. Yet nothing was touching them except the water in which they were floating. The explanation for his observation is that water molecules are in constant random motion, and they collide with the spores. This effect, seen only in tiny particles, is called **Brownian motion**.

It is also possible to see Brownian motion in smoke particles. While looking through a microscope, introduce a small amount of smoke before the objective lens and illuminate it with a bright light. You will be able to clearly see the smoke particles jostling about in vibratory motion. This motion is caused by collisions with high-speed air molecules.

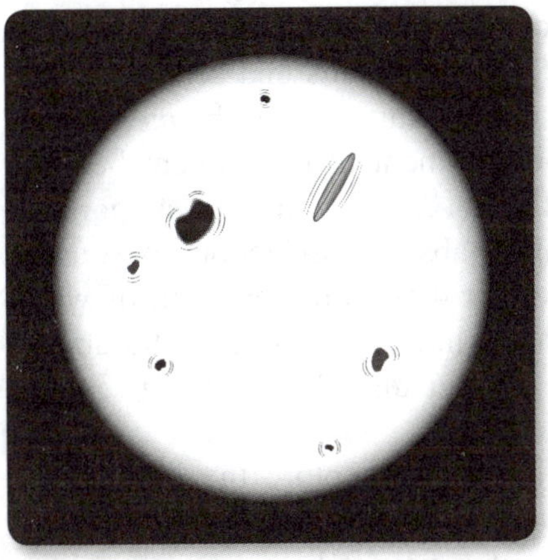

15-4a Brownian motion of microscopic particles suspended in a liquid or gas can be observed using a light microscope.

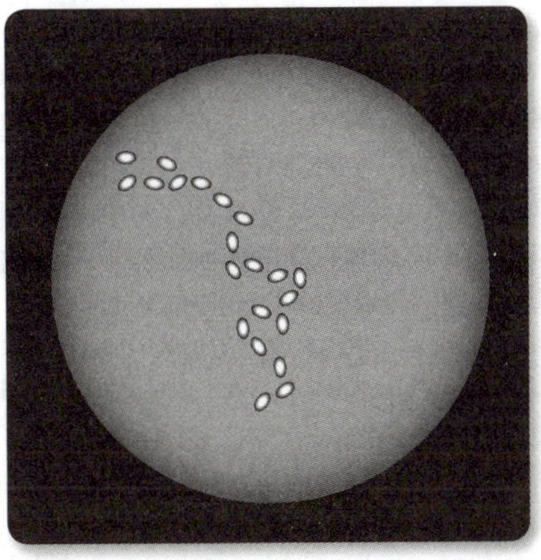

15-4b Time-lapse microphotography shows the position of an inanimate particle experiencing Brownian motion at given time intervals.

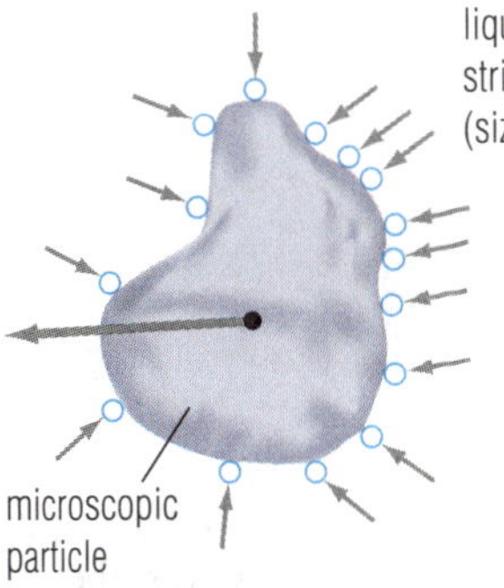

15-4c Brownian motion of microscopic particles is caused by the random, asymmetrical collisions of liquid or gas molecules against the particles.

15.6 The Caloric Theory

Many eighteenth-century scientists believed the **caloric theory**—the view that heat is a material fluid, called *caloric*, that flows from hot bodies to cold bodies. Earlier scientists, including Francis Bacon, Robert Boyle, and Robert Hooke, had previously advocated the rival kinetic theory—the theory that thermal energy is a property of an object arising from the motion of its particles. Still, such notable scientists as Joseph Black, who discovered carbon dioxide, and Antoine Lavoisier, the famous French chemist, enthusiastically endorsed the caloric theory.

A major blow to the caloric theory came with the famous cannon-boring experiment conducted by Count Rumford. According to the caloric theory, the amount of thermal energy produced in drilling a metal should be proportional to the volume of the material the drill removed. The caloric fluid was believed to reside between the most elementary particles of the metal and be released when the particles were separated.

However, Rumford observed no correlation between the volume of the material removed and the thermal energy produced. On the contrary, he observed that the greatest heat was produced when a dull drill removed only a little metal after a great amount of mechanical work was expended. An almost unlimited supply of thermal energy could be produced under such conditions. In a report to the Royal Society in 1798, he theorized that thermal energy was the result of motion in the elementary particles of the metal. Humphry Davy continued this line of attack on the caloric theory in the early 1800s.

In 1840 James Prescott Joule of Manchester, England, began a series of experiments designed to unite thermal and mechanical energy under a single conservation law. In a lecture that he delivered in 1847, Joule administered the *coup de grâce* to the faltering caloric theory. He showed that since caloric can be converted to kinetic or potential energy, it cannot possibly be a material substance. He explained the source of the heating in Rumford's experiment as a change from mechanical to thermal energy. The caloric theory quickly became a thing of the past. The paradigm shift from the caloric to the kinetic-molecular model of thermal energy and heat is one of the most profound scientific revolutions in history because it allowed the science of thermodynamics to develop rapidly.

15.7 The Mechanical Equivalent of Heat

The results of Joule's research, published in 1845, described a device in which falling weights turned a set of paddles in an enclosed drum of water. After conducting a series of careful experiments, he reported that it required an average

Joseph Black (1728–99) was a Scottish physician who conducted research in the thermal properties of matter and discovered carbon dioxide. Black also proposed the concepts of latent and specific heat. His work was foundational in the establishment of the science of chemistry.

Antoine Lavoisier (1743–94) was a brilliant French chemist often called the "Father of Modern Chemistry." He discovered the conservation of mass and identified oxygen as the active agent in combustion and respiration. He was beheaded in the French Revolution.

Humphry Davy (1778–1829) was an English chemist who specialized in discovering many new elements and investigating their properties. He discovered the law of combining proportions that supported Dalton's atomic theory of matter. He also conducted qualitative research in many areas of science.

of 819 foot-pounds-force (ft-lbf) of work to raise the temperature of a pound of water one degree Fahrenheit (he later refined his results and settled on 772 foot-pounds-force). Its equivalent in thermal energy units, called the *mechanical equivalent of heat*, varied somewhat from experiment to experiment because of frictional losses, but the values obtained were close enough to support the idea that the discrepancies would disappear in the absence of friction. In SI units, Joule's mechanical equivalent of heat was about 4.50 N·m per gram of water per degree Celsius. At the time, this amount of heat was assigned the unit of **calorie (cal)**, which was the standard unit of heat until late in the twentieth century.

The **calorie (cal)** was the unit of heat for many years. However, the joule is now the primary unit for both energy and heat in most scientific research. The calorie, or more specifically, the dietary Calorie (1 Calorie = 1 kcal or 1000 cal), is still in use for quantifying the energy content of foods.

Joule performed several different experiments to explore the idea of the mechanical equivalent of heat. His experiments did not agree exactly. For the equivalent of 1 cal, Joule reported 4.14 N·m of work for heating water with a perforated piston, 4.27 N·m for the temperature drop of rapidly expanding air, and 4.16 N·m for experiments involving water moving through narrow tubes. By 1850 he had settled on a value of 4.15 N·m for 1 cal. The currently accepted value of the mechanical equivalent of thermal energy is 4.186 N·m = 1 cal (at 15 °C). In honor of the fundamental work Joule accomplished in thermodynamics, the N·m was renamed the joule, the SI derived unit of energy, work, and heat.

15A Objectives

After completing this section, I can

- ✓ summarize the evidence for the kinetic-molecular model of matter.
- ✓ distinguish between diffusion and Brownian motion.
- ✓ discuss the basis of the caloric theory and its demise in the face of Joule's discovery of the mechanical equivalent of heat.
- ✓ compare the different units of heat and energy.

15A Section Review

1. According to the kinetic theory, describe how thermal energy is manifested in matter.
2. How is temperature related to thermal energy?
3. Discuss three evidences that support the kinetic-molecular theory of matter.
4. What observation produced a key piece of evidence that contradicted the caloric theory of heat and led to its eventual demise?
5. What logical argument proved that heat could not be a material fluid called *caloric*?
6. State the SI equivalent for 1 cal.
7. Ignoring losses to the surroundings, if 10.0 J of work is done stirring 2.0 L of water, how much thermal energy is transferred to the water?

15B THERMAL ENERGY AND MATTER

15.8 Heat Capacity

Recall from earlier chapters that momentum and velocity are related but different quantities. It is not always the fastest-moving object that has the greatest momentum. For example, a slow-moving truck may have more momentum than a speeding motorcycle because the truck has far more mass than the motorcycle. Momentum is the product of mass and velocity.

The **heat capacity (*C*)** of an object is the quantity of thermal energy required to raise the temperature of the *entire object* 1 degree Celsius. Units are J/°C. Heat capacity is a unique property of each object.

Heat (*Q*) is the quantity of thermal energy that is transferred to or from a system. It is not correct to say that an object contains a certain amount of heat.

Similarly, it is not always the hottest object that contains the greatest amount of thermal energy. There are two other factors involved here—the mass of the object and its ability to hold thermal energy. Both of these factors are included in the object's **heat capacity (*C*)**, which is the amount of thermal energy required to raise the temperature of the entire object one degree Celsius. The amount of thermal energy added to or taken from a system is called **heat (*Q*)**. The SI unit for heat capacity is joules per degree Celsius (J/°C). Heat capacity of an object is to thermal energy as mass is to momentum. Heat capacity may be calculated by the formula

$$C = \frac{Q_{\text{object}}}{\Delta t}, \tag{15.1}$$

where Δt is the change of temperature of the object,

$$\Delta t = t_{\text{final}} - t_{\text{initial}}.$$

Heat is analogous to mechanical work. Mechanical work on or by a system involves the transfer of mechanical energy to or from the system. Heat involves the movement of thermal energy. So, it is technically incorrect to say that a system *has* a certain amount of heat, just as it is improper to say that a system *has* a certain amount of work. Heat is not a property of a system as thermal energy is. It takes great discipline to use the two terms correctly.

Just as momentum is a function of a fixed property of an object (its mass) and a variable property (its speed), the thermal energy of an object is dependent on its heat capacity (a fixed property) and its temperature (a variable property).

Recall that lowercase t refers to Celsius temperature as well as time. This is another case in physics where a single symbol is used for more than one quantity. The context of the problem will indicate what dimensional property is meant.

15.9 Specific Heat

Mass is a property of a specific object, not of a kind of substance. Similarly, heat capacity is a property of an object. To compare the relative "heaviness" of one substance to another, scientists divide the mass of a sample of the substance by its volume, which yields the *specific density* of the material—its mass per unit volume. Similarly, to find the thermal energy capacity of a substance (not an object), scientists divide the heat capacity of an object by its mass to find its *specific heat capacity*—its heat capacity per gram. The **specific heat (c_{sp})** of a substance is the amount of thermal energy required to raise the temperature of 1 g of the substance one degree Celsius. It is properly expressed in SI units of joules per gram-degree Celsius (J/g·°C). Older textbooks sometimes refer to units of calories per gram-degree Celsius. Comparing the definitions of specific heat and calorie, we can see that the specific heat of water is, by definition, exactly 1 cal/g·°C (at 15 °C). In SI units, the specific heat of water is about 4.18 J/g·°C (near room temperature). The specific heats of some common substances are listed in Table 15-1 on page 344. Specific heat varies somewhat with temperature and pressure. However, the specific heats of most solids are nearly constant for temperatures between 20 °C and 100 °C and pressures near 1 atm.

Specific heat capacity, or just **specific heat (c_{sp})**, is the amount of heat required to raise the temperature of 1 g of a substance 1 °C.

The definition of the calorie, dependent as it is on the heat capacity of water (which varies with temperature), must be given with an associated temperature. Typically, the calorie is defined at 15 °C, although other temperatures are used as well.

How can we find the specific heat of a material? Using the definition of specific heat, if we add a known amount of thermal energy (Q) to an object of known mass (m) and note the corresponding temperature change (Δt) in degrees Celsius, then we can find the specific heat of the material from the equation

$$c_{sp} = \frac{Q}{m\Delta t}. \tag{15.2}$$

EXAMPLE 15-1

Taking Some Heat: Determining Heat Capacity

When 89.7 J of thermal energy is added to a 10.0 g aluminum block at 20.0 °C, the block's temperature rises to 30.0 °C. Compute (a) the specific heat of aluminum and (b) the heat capacity of the block.

Solution:

The system is the aluminum block.

a. Computing specific heat:

$$c_{sp\ Al} = \frac{Q}{m\Delta t} = \frac{89.7\ \text{J}}{(10.0\ \text{g})(10.0\ ^\circ\text{C})}$$

$$c_{sp\ Al} = 0.897\ \text{J/g}\cdot{}^\circ\text{C}$$

Equation 15.1 can be used to find the heat capacity of both objects and substances. However, to find the heat capacity of uniform substances, it is better to multiply the substance's specific heat by its mass. Example 15-1 takes this approach in Part b.

Table 15-1

Specific Heat of Selected Substances

Substance	c_{sp} (J/g·°C)
air (27 °C/300 K)	1.16
aluminum	0.897
benzene	1.75
brass	0.375
carbon (graphite)	0.709
chlorine	0.479
copper	0.385
ethyl alcohol	2.44
glass (crown)	0.670
gold	0.129
iron	0.449
lead	0.129
mercury	0.140
oil (olive)	1.79
oxygen	0.918
platinum	0.133
silver	0.235
tin	0.228
tungsten	0.132
water (solid/ice)	2.09
water (liquid)	4.18
water (gas/steam)	2.01
wood	1.70
zinc	0.388

The free electrons found in metals are unlocalized valence electrons involved in forming the bonds between metal atoms. We will discuss this further in Subsection 15.13.

The particles in liquids and gases not only oscillate, but they also can rotate and move along linear paths (translation). Molecules also exhibit complicated internal motions among their atoms. Thus, the kinetic energies associated with fluid particles tend to have a complex effect on their measurable properties.

b. Computing heat capacity:

$$C = mc_{sp\ Al}$$

$$C = (10.0\ g)(0.897\ J/g{\cdot}°C)$$

$$C = 8.97\ J/°C$$

The amount of heat that is needed to raise the temperature of an object a certain number of degrees is released when the object is cooled the same number of degrees. If, in the previous example, the aluminum block were cooled from 30 °C to 20 °C, the cooling process would release 89.7 J of heat.

15.10 Modeling Specific Heat

Compare the specific heats of silver and gold in Table 15-1. Both are metals and both have free electrons, which cause them to be good conductors of heat. Both metals naturally exist in the same kind of crystal lattice structure. When these metals are heated, atoms in the lattice can only vibrate in place, since metal bonds keep atoms from significantly changing position. So, if the two metals' particle arrangements and thermal motions are nearly identical, what accounts for the difference in the value of specific heat of these two metals? It must have to do with some difference in the metal atoms themselves. An element's measurable properties are mainly due to the mass and electron structure of its atoms. Since the electron structures of the two metals are similar, it seems reasonable that there must be some relationship between an element's mass (atomic weight) and its specific heat value. Look at the following table that gives the values of the specific heat and **molar masses** of several pure metals. Later, you will be asked to make a model relating these two quantities.

Table 15-2

Physical Data for Selected Metals

Metal	Molar Mass (g/mol)	Specific Heat (J/g·°C) at 25 °C
Mg	24.31	1.024
Al	26.98	0.897
Cr	52.00	0.450
Fe	55.85	0.449
Co	58.93	0.421
Cu	63.55	0.384
Zn	65.41	0.388
Ag	107.86	0.235
W	183.84	0.132
Au	196.97	0.129
Pb	207.21	0.127

Ref: *Handbook of Chemistry and Physics*, 86th ed. David R. Lide, Ed. CRC Press. 2005

15.11 Conservation of Thermal Energy

When an object gains heat, its surroundings lose the same amount of heat. Thermal energy is conserved if it is the only form of energy exchanged in a process. In this chapter we will assume that thermal energy is conserved, so the heat gained by a system is lost by its surroundings and vice versa. Symbolically,

$$Q_{\text{system}} = -Q_{\text{surroundings}}, \text{ or}$$
$$Q_{\text{system}} + Q_{\text{surroundings}} = 0 \text{ J}.$$

Such expressions are called *heat-balance equations*.

Problem-Solving Strategy 15.1

Heat-balance problems often result in slightly different calculated values of heat lost versus heat gained. This is due to the different precisions of the various measured quantities and the rounding conventions involved in the calculations. Differences of a few percent are acceptable for the purposes of this course.

EXAMPLE 15-2

Conservation of Thermal Energy: The Calorimeter

A 10.0 g aluminum block has an initial temperature of 19.7 °C. It is placed in an **adiabatic** vessel—one that allows no heat to enter or leave its contents—that contains 20.0 g of water at 31.1 °C. The two parts of the system come to equilibrium at a final temperature of 30.0 °C. Compare the heat gained by the metal block to the heat lost by the water.

Solution:

The system consists of the water and the aluminum block. It does *not* include the vessel for this problem. You are seeking the quantity of heat lost and gained by the respective parts of the system. Equation 15.2 must be rearranged to solve for Q:

$$Q = mc_{\text{sp}}\Delta t$$

Calculate the heat gained by the water:

$$Q_{H_2O} = (mc_{\text{sp}}\Delta t)_{H_2O} = [mc_{\text{sp}}(t_f - t_i)]_{H_2O}$$

$$Q_{H_2O} = (20.0 \text{ g})(4.18 \text{ J/g}\cdot{}^\circ\text{C})(30.0\ {}^\circ\text{C} - 31.1\ {}^\circ\text{C})$$

$$Q_{H_2O} \cong -9\underline{1}.9 \text{ J} \cong -92 \text{ J} \quad (1)$$

The heat "gained" by the water is negative, so thermal energy is actually lost from the water. There is a net flow of thermal energy out of the water and into the aluminum block.

Calculate the heat gained by the block:

$$Q_{Al} = (mc_{\text{sp}}\Delta t)_{Al} = [mc_{\text{sp}}(t_f - t_i)]_{Al}$$

$$Q_{Al} = (10.0 \text{ g})(0.897 \text{ J/g}\cdot{}^\circ\text{C})(30.0\ {}^\circ\text{C} - 19.7\ {}^\circ\text{C})$$

$$Q_{Al} \cong +92.\underline{3}9 \text{ J} \cong +92.4 \text{ J} \quad (2)$$

Notice that $Q_{Al} = -Q_{H_2O}$ or $+92 \text{ J} = -(-92 \text{ J})$ expressed to 2 SDs.

An **adiabatic** process is one that does not exchange thermal energy with its surroundings.

15-5a Components of a calorimeter

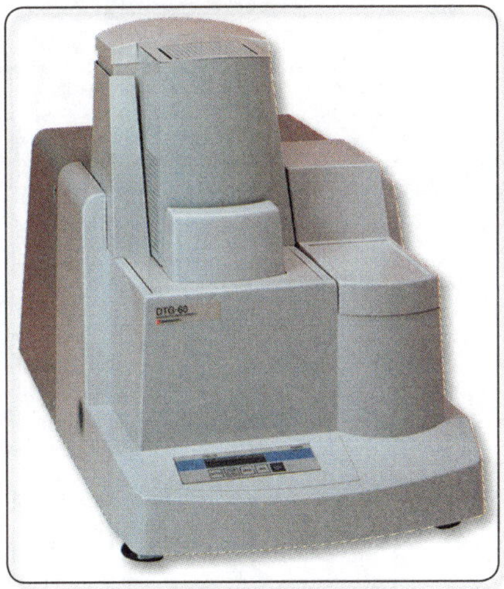

15-5b A research-grade calorimeter

In the previous example, an adiabatic container was specified. There are no truly adiabatic containers. Therefore, we must take into account the thermal energy lost or gained by the container and the surrounding air. A **calorimeter** is a container designed to minimize the exchange of thermal energy with its surroundings. It consists of an insulated outer cup, an inner cup, an insulated lid, and provision for a stirrer and a thermometer. Figure 15-5a is a photograph of a typical school laboratory calorimeter. Figure 15-5b is a research-quality calorimeter. It is possible to determine the heat capacity of a calorimeter by using the principle of the conservation of energy.

The title of Example 15-3 refers to Song of Solomon 2:15: "Take us the foxes, the little foxes, that spoil the vines"—the little foxes are an allegorical reference to the small iniquities in our lives that rob us of our joy in our Lord just as the calorimeter affects, if only in a small way, the thermal energy transferred between its contents.

EXAMPLE 15-3

Finding the Little Foxes: Calorimeter Heat Capacity

10.0 g of water at room temperature is placed into a calorimeter. The equilibrium temperature of the calorimeter and water is 20.0 °C. Then 10.0 g of water heated to 40.0 °C is added to the water in the calorimeter. The new equilibrium temperature is 27.0 °C. Find the heat capacity of the calorimeter.

Solution:

The system is the calorimeter, the cool water, and the additional hot water. You measure the water temperature and assume that the calorimeter's temperature is the same at equilibrium. The essential concept at work here is this: After the hot and cool water mix and you determine the heat gained and lost by the two quantities of water, any difference in heat *must* be due to the heat transferred to or from the calorimeter, based on the conservation of thermal energy.

$$Q_{hw} + Q_{cw} + Q_{cal} = 0\text{ J} \qquad (1)$$

Calculate the heat transfer of the cool water:

$$Q_{cw} = (mc_{sp}\Delta t)_{cw} = [mc_{sp}(t_f - t_i)]_{cw}$$

$$Q_{cw} = (10.0\text{ g})(4.18\text{ J/g}\cdot°\text{C})(27.0\text{ °C} - 20.0\text{ °C})$$

$$Q_{cw} \cong +2\underline{9}2\text{ J} \qquad (2)$$

Calculate the heat transfer of the hot water:

$$Q_{hw} = (mc_{sp}\Delta t)_{hw} = [mc_{sp}(t_f - t_i)]_{hw}$$

$$Q_{hw} = (10.0\text{ g})(4.18\text{ J/g}\cdot°\text{C})(27.0\text{ °C} - 40.0\text{ °C})$$

$$Q_{hw} \cong -54\underline{3}.4\text{ J} \qquad (3)$$

Substitute results (2) and (3) into Equation (1) and solve for the heat transfer of the calorimeter:

$$Q_{hw} + Q_{cw} + Q_{cal} = 0\text{ J}$$

$$Q_{cal} = -(Q_{hw} + Q_{cw})$$

$$Q_{cal} = -(-54\underline{3}.4\text{ J} + 2\underline{9}2\text{ J})$$

$$Q_{cal} \cong +2\underline{5}1 \qquad (4)$$

The calorimeter heat capacity is calculated from Equation 15.1:

$$C_{cal} = \frac{Q_{cal}}{\Delta t} = \frac{+2\underline{5}1\text{ J}}{(27.0\text{ °C} - 20.0\text{ °C})}$$

$$C_{cal} \cong 3\underline{5}.8\text{ J/°C} \cong 36\text{ J/°C}$$

Using a calorimeter with a known heat capacity, the heat capacity or specific heat of an unknown sample can be easily determined.

EXAMPLE 15-4

Experimentally Determining Specific Heat

A calorimeter with a heat capacity of 36.0 J/°C contains 20.0 g of water at 50.0 °C. A 10.0 g metal alloy sample is at 20.0 °C. When the sample is placed into the calorimeter, the final equilibrium temperature is 49.0 °C. What is the specific heat of the alloy?

Solution:

The system is the calorimeter, the alloy sample, and the hot water. Equation 15.2 permits you to calculate the specific heat of the alloy.

$$c_{sp\ alloy} = \frac{Q_{alloy}}{(m\Delta t)_{alloy}} \qquad (1)$$

You need to determine the heat transfer of the alloy sample when it is placed into the calorimeter in order to solve Equation (1). From the conservation of energy principle, you know that

$$Q_{hw} + Q_{alloy} + Q_{cal} = 0\ \text{J}. \qquad (2)$$

First, determine the heat lost by the hot water:

$$Q_{hw} = (mc_{sp}\Delta t)_{hw} = [mc_{sp}(t_f - t_i)]_{hw}$$

$$Q_{hw} = (20.0\ \text{g})(4.18\ \text{J/g·°C})(49.0\ \text{°C} - 50.0\ \text{°C})$$

$$Q_{hw} \cong -8\underline{3}.6\ \text{J}$$

Next, determine the heat lost by the calorimeter:

$$Q_{cal} = (C\Delta t)_{cal} = [C(t_f - t_i)]_{cal}$$

$$Q_{cal} = (36.0\ \text{J/°C})(49.0\ \text{°C} - 50.0\ \text{°C})$$

$$Q_{cal} \cong -3\underline{6}.0\ \text{J}$$

Solve Equation (2) for the heat gained by the alloy (Q_{alloy}):

$$Q_{alloy} = -(Q_{hw} + Q_{cal}) \qquad (3)$$

Substitute Equation (3) for Q_{alloy} into Equation (1) and determine $c_{sp\ alloy}$:

$$c_{sp\ alloy} = \frac{-(Q_{hw} + Q_{cal})}{(m\Delta t)_{alloy}} = \frac{-(Q_{hw} + Q_{cal})}{[m(t_f - t_i)]_{alloy}}$$

$$c_{sp\ alloy} = \frac{-[(-8\underline{3}.6\ \text{J}) + (-3\underline{6}.0\ \text{J})]}{(10.0\ \text{g})(49.0\ \text{°C} - 20.0\ \text{°C})}$$

$$c_{sp\ alloy} \cong 0.41\underline{2}\ \text{J/g·°C} \cong 0.41\ \text{J/g·°C}$$

Problem-Solving Strategy 15.2

Solving heat-balance problems requires finding the heat exchanged (both magnitude and sign) for every part of the system within the adiabatic boundary. The conservation of heat within an adiabatic boundary assures you that the sum of all heat transfers will be zero.

15-6 Engineers use calorimetric procedures in order to determine the rate of heat transfer and the efficiency of their power plants.

The procedure just described in this example is similar to many such heat balance calculations performed by scientists and engineers for a variety of reasons. For example, the latent heats required to bring about various phase transitions are determined using calorimetric procedures.

15.12 Heat and Phase Transitions

Every material has a characteristic specific heat capacity. You would expect that any thermal energy gained by an object would cause an increase in temperature, but this is not always the case.

Consider an ice cube in a freezer ice cube tray. Assume that we have inserted a thermocouple junction (an electrical temperature sensor) in the water before it freezes. When we are ready for the demonstration, we quickly remove the instrumented ice cube from the tray, connect it to a temperature display instrument, and immerse the cube in a beaker of ice and water. A laboratory burner is lighted under the beaker.

The initial temperature of the ice cube is around –10 °C, the temperature of the freezer. We record the temperature every ten seconds as the ice begins to warm. At first, the temperature rises rapidly. At 0 °C, however, we note that the cube's temperature stops rising. The ice cubes in the beaker, including our "wired" cube, are all shrinking in size as they melt. When the ice fragments are nearly gone, we observe that the temperature starts rising again, but more slowly than before. The water's temperature continues to rise at a steady rate until the water nears the boiling point (numerous small bubbles form and detach from the bottom of the beaker). Then the rate of temperature rise drops off, and the temperature becomes constant when the water reaches a rolling boil.

You may wonder why the temperature does not rise continuously during the entire period. The flame has been adding thermal energy to the beaker at the same rate the whole time. The answer lies in the water's changes of state. The thermal energy that the water absorbed without changing temperature was used to break the bonds between the water molecules in the ice. The same energy is released when the phase change is reversed. That is, the amount of thermal energy absorbed when ice melts is released when water freezes. Similarly, the amount of thermal energy absorbed when water vaporizes is released when steam condenses. This amount of thermal energy per unit mass is a constant for a given phase change of each substance. The **latent** (hidden) **heat of fusion** (**L_f**) of a substance is the amount of thermal energy required to melt 1 kg of the substance at its melting point. The **latent heat of vaporization** (**L_v**) is the amount of thermal energy required to vaporize 1 kg of a substance at its boiling point. The values of these quantities are determined at standard conditions so that they may be compared. Table 15-3 on page 349 lists some heats of fusion and vaporization for selected substances.

Heats of fusion and vaporization are determined by the following relationships:

$$L_f = \frac{Q_{melt}}{m} \quad (15.3)$$

$$L_v = \frac{Q_{boil}}{m} \quad (15.4)$$

Units of L_f and L_v are kJ/kg (or J/g).

Table 15-3

Heats of Fusion and Vaporization with Associated Phase Transition Temperatures for Selected Substances

Substance	Melting Point t_m (°C)	Heat of Fusion L_f (kJ/kg)	Boiling Point t_b (°C)	Heat of Vaporization L_v (kJ/kg)
Acetone	−94.8	98.0	56.1	501
Aluminum	660.3	397.0	2520	10 900
Carbon Dioxide	−56.6	205	—	—
Copper	1084.6	208.7	2563	4726
Ethanol	−114.1	78.4	78.3	837
Gold	1064.2	63.7	2857	1645
Iron	1538	247.3	—	—
Lead	327.5	23.0	1750	866
Mercury	−38.8	1.4	356.7	295
Nitrogen	−210.0	25.3	−195.8	198
Oxygen	−218.8	13.8	−183.0	213
Silver	961.8	104.7	2163	2323
Tungsten	3422	284.5	5900	4820
Water	0.0	333.5	100.0	2256

EXAMPLE 15-5

Determining Heat of Fusion: A Single Phase Change

When 10.0 g of ice is placed in a calorimeter with a heat capacity of 36.0 J/°C, the equilibrium temperature is 0.0 °C. Then 20.0 g of water at 50.0 °C is added, and the new equilibrium temperature is 5.2 °C. The ice has melted. Calculate the heat of fusion (L_f) of ice.

Solution:

The heat of fusion is calculated from Equation 15.3:

$$L_f = \frac{Q_{melt}}{m} \text{ or}$$

$$Q_{melt} = mL_f \tag{1}$$

Note which quantities are changing and which are constant. The ice melts, but its temperature does not change. The cold meltwater from the ice (subscript "cw") warms from 0.0 °C to 5.2 °C. The warm water (subscript "ww") cools from 50.0 °C to 5.2 °C. The calorimeter experiences the same temperature change as the cold water. Therefore, heat flows from the warm water into the ice to melt it and into the resulting cold water and the calorimeter.

From the conservation of energy in an adiabatic container you know that

$$Q_{cal} + Q_{cw} + Q_{ww} + Q_{melt} = 0 \text{ J.}$$

Solve for Q_{melt}:

$$Q_{melt} = -(Q_{cal} + Q_{cw} + Q_{ww})$$

Substitute Equation (1) for Q_{melt} and solve for L_f:

$$m_{ice}L_f = -(Q_{cal} + Q_{cw} + Q_{ww})$$

$$L_f = \frac{-(Q_{cal} + Q_{cw} + Q_{ww})}{m_{ice}} \qquad (2)$$

Now calculate each of the heat expressions in Equation (2):

Calorimeter: $Q_{cal} = (C\Delta t)_{cal} = (36.0\ \text{J/°C})(+5.2\ \text{°C})$

$Q_{cal} \cong +1\underline{8}7\ \text{J}$

Cold water: $Q_{cw} = (mc_{sp}\Delta t)_{cw} = (10.0\ \text{g})(4.18\ \text{J/g·°C})(+5.2\ \text{°C})$

$Q_{cw} \cong +2\underline{1}7\ \text{J}$

Warm water: $Q_{ww} = (mc_{sp}\Delta t)_{ww} = (20.0\ \text{g})(4.18\ \text{J/g·°C})(-44.8\ \text{°C})$

$Q_{ww} \cong -37\underline{4}5\ \text{J}$

Solve Equation (2) for L_f:

$$L_f = \frac{-[1\underline{8}7\ \text{J} + 2\underline{1}7\ \text{J} + (-37\underline{4}5\ \text{J})]}{10.0\ \text{g}}$$

$$L_f \cong 3\underline{3}4\ \text{J/g} \cong 330\ \text{J/g}$$

This result, obtained in a typical high school laboratory calorimeter, agrees reasonably well with the accepted value of 333.5 J/g for the heat of fusion of water.

EXAMPLE 15-6

From Ice to Steam: Thermal Energy Needed for Two Phase Changes

How much thermal energy is required to raise the temperature of 10.0 g of water from −10.0 °C to 110.0 °C?

Solution:

This problem involves five parts. We must determine the thermal energy changes for the following:

(1) Heating the ice from −10.0 °C to 0.0 °C (Q_{ice});

(2) Converting the ice to liquid water at 0.0 °C (Q_{melt});

(3) Heating the liquid water from 0.0 °C to its boiling point at 100.0 °C (Q_w);

(4) Converting the liquid water to gaseous water (steam) at 100.0 °C (Q_{boil});

(5) Heating the steam from 100.0 °C to 110.0 °C (Q_s).

$$Q_{total} = Q_{ice} + Q_{melt} + Q_w + Q_{boil} + Q_s \qquad (1)$$

Heats of fusion and vaporization can be obtained from Table 15-3. Specific heats can be obtained from Table 15-1.

Determine the heat transferred for each term on the right side of Equation (1):

Heat ice: $Q_{ice} = (mc_{sp}\Delta t)_{ice} = (10.0\ g)(2.09\ J/g{\cdot}°C)(+10.0\ °C)$

$Q_{ice} = +20\underline{9}.0\ J$

Melt ice: $Q_{melt} = m_{ice}L_f = (10.0\ g)(333.5\ J/g)$

$Q_{melt} = +33\underline{3}5\ J$

Heat water: $Q_w = (mc_{sp}\Delta t)_w = (10.0\ g)(4.18\ J/g{\cdot}°C)(+100.0\ °C)$

$Q_w = +41\underline{8}0\ J$

Boil water: $Q_{boil} = m_wL_v = (10.0\ g)(2256\ J/g)$

$Q_{boil} = +22\,\underline{5}60\ J$

Heat steam: $Q_s = (mc_{sp}\Delta t)_s = (10.0\ g)(2.01\ J/g{\cdot}°C)(+10.0\ °C)$

$Q_s = +20\underline{1}.0\ J$

Sum all of the heat quantities to find the total thermal energy transferred:

$Q_{total} = 20\underline{9}\ J + 33\underline{3}5\ J + 41\underline{8}0\ J + 22\,\underline{5}60\ J + 20\underline{1}\ J$

$Q_{total} \cong 30\,\underline{4}85\ J$

$Q_{total} \cong 30\,500\ J$ (Allowable precision is to nearest 100 J.)

Problem-Solving Strategy 15.3

Calculating the thermal energy required to change a solid substance to a vapor, or vice versa, can be broken down into five heat-transfer steps. The key to success is to use the proper specific heat value for the corresponding phase and the correct latent heat for each phase change. Be sure to observe proper significant digit rules when summing the final result.

15B Section Review

1. Discuss why, when evaluating heat processes involving water, it might be more convenient to use the calorie unit.
2. Define and clearly differentiate the terms temperature, heat, and thermal energy.
3. Discuss the difference between the heat capacity of a solid steel cylinder and its specific heat.
4. As thermal energy is added to a solid, why does its average temperature stop rising when it begins to melt?
5. What is the quantity of heat required to melt a gram of a substance at its melting point called? What do we call the quantity of heat per gram that must be removed to freeze the substance at its freezing point?
6. In a laboratory exercise, you are required to tentatively identify a 65.0 g sample of metal from Table 15-1 by its specific heat, using a calorimeter. The calorimeter and the 50.0 g of distilled water that it contains are both at a temperature of 24.0 °C. The initial temperature of the metal just before it is placed into the calorimeter is 98.5 °C. The final temperature of the calorimeter and its contents is 26.5 °C. The calorimeter's heat capacity is 39.5 J/°C. What metal is the sample most likely made of?
7. How much heat must be lost in order to completely condense 100.0 g of nitrogen at its condensation point?

15B Objectives

After completing this section, I can

✓ define the heat capacity of an object.

✓ compare specific heat capacity with heat capacity and explain why the former is more descriptive of a substance.

✓ using the principle of the conservation of thermal energy, determine heat capacities for various objects and materials.

✓ explain why solids have heats of fusion and liquids have heats of fusion and vaporization.

✓ perform heat balance calculations involving phase transitions.

8. Compare the heat loss required to condense 1.0 g of steam to water at 100.0 °C and the heat lost to cool 1.0 g of water at 100.0 °C by 1.0 °C. The specific heat of water at 100.0 °C is 4.22 J/g·°C. Which, gram for gram, is more damaging to the unprotected skin—water at its boiling point or condensing steam?

DM 9. Construct a scatter plot of specific heat versus molar mass for the pure metals listed in Table 15-2 on page 344. Then construct a best-fit curve through the data points. What kind of relationship, if any, exists?

DM 10. a. Construct a scatter plot of the metals' specific heats versus the reciprocals of their molar masses. Construct a best-fit curve through the data points. If possible, force the *y*-intercept to pass through zero. Display the trendline's equation and the r^2 value of the trendline fit. What relationship exists between the variables in this plot? Is this expected? Explain.

b. What are the units associated with the slope of the trendline found in Part a? What property of metals does this combination of units describe?

c. Divide the slope of the trendline and its units found in Part b by the universal gas constant $R = 8.315$ J/(K·mol) and round the result to 1 SD. The resulting number is called the *dimensionless heat capacity* (C^*), which is a property of nearly all metals.

Using the relationship found in Question 10c, the nineteenth-century French physicist Pierre Louis Dulong and chemist Alexis Thérèse Petit discovered that the specific heats of most crystalline solids can be found from the equation $c_{sp} = 3R/M$. This relationship is called the *Dulong-Petit law*.

15C MECHANISMS FOR HEAT TRANSFER

There are three mechanisms for the transfer of thermal energy from one system to another: conduction, convection, and radiation.

15.13 Conduction

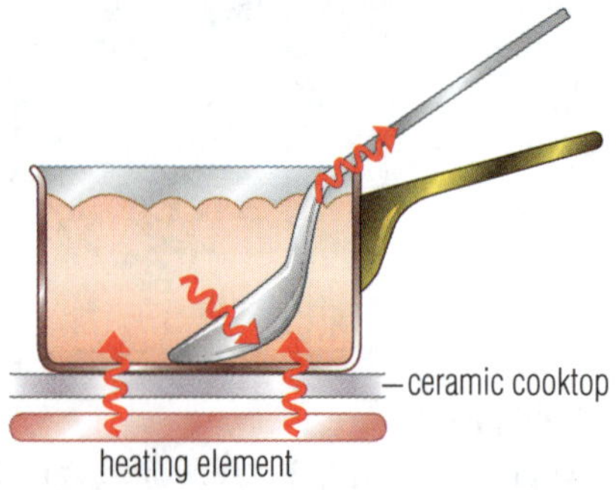

15-7a Conduction

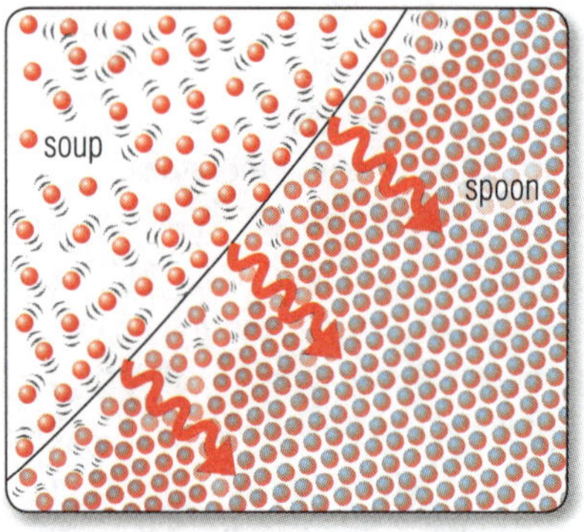

15-7b Thermal conduction takes place at the molecular level.

Hold a stainless steel spoon in a pot of boiling soup. You will begin to feel the spoon handle growing warm. Eventually, you will either have to drop the spoon or withdraw it from the soup to prevent being burned. This is an example of **conduction**. The hot soup particles accelerate the motion of the metal atoms in the bowl of the spoon. These rapidly vibrating molecules collide with the adjacent, slower-moving atoms within the metal, transmitting kinetic energy to the atoms farther up the handle. The rapid motion and associated kinetic energy is transmitted from atom to atom along the handle to your fingers.

Some materials conduct thermal energy easily, while others act as thermal insulators. For example, if you had used a wooden spoon to stir the soup, you could have held onto the spoon indefinitely. To understand the difference, we must return to the atomic model. Recall that the major subatomic particles are protons, neutrons, and electrons. The protons and neutrons are in the nucleus, while the electrons surround the nucleus. Some substances hold their electrons firmly. Other substances, especially metals, hold some of their valence electrons so loosely that it takes little energy to remove them. In fact, some electrons are so independent of their nuclei that they are called *free electrons*. These free electrons readily transmit kinetic energy between atoms throughout a substance. Substances with free electrons are therefore good thermal conductors. Substances lacking free electrons do not conduct thermal energy well, so they are thermal insulators.

Substances that contain free electrons, such as metals, generally are good conductors of heat. Substances that do not contain free electrons, such as most solid ionic and molecular compounds, are generally poor thermal conductors (good insulators).

15.14 Convection

Air is a poor thermal conductor. However, a hot water radiator surrounded by air can heat an entire room. This is possible because air transports thermal energy by convection. **Convection** moves thermal energy by physically moving material that has thermal energy. Gravity is the main force that causes this motion. The hot radiator transmits heat to the air mass directly around it through conduction and *radiation* (see Subsection 15.15). This warmer air expands, becoming less dense, and is displaced upward by the cooler, denser air around it. The cool air is soon heated and replaced with cooler air from the room. The cycle continues, setting up a convection current of rising and falling air in the room. Eventually, the warmed air cools, falls, and returns to the radiator to be heated again. In order to make most efficient use of this convection process, a well-designed house has its convective heat sources near the floor.

15-8 Convection

Water, as well as air, transmits thermal energy by convection. In bodies of water, the cooling process is especially interesting. Water is densest at 4 °C. A body of water—a lake, for example—cools from the top. When the top layer cools, it sinks and is replaced by warmer water from deeper in the lake. Since water at 4 °C is densest, no water can cool to below 4 °C until all the water in the lake has reached that temperature. When all the water is 4 °C, the surface layer cools to lower temperatures, and the lake freezes over from the top down. Since ice is less dense than liquid water at the same temperature, it floats. The net effect of this phenomenon is that freezing is delayed throughout the depth of the lake as long as possible. In deep lakes, even after the surface freezes, some unfrozen water exists under the ice all winter. This phenomenon, ordained by the Creator, preserves aquatic life through the winter months.

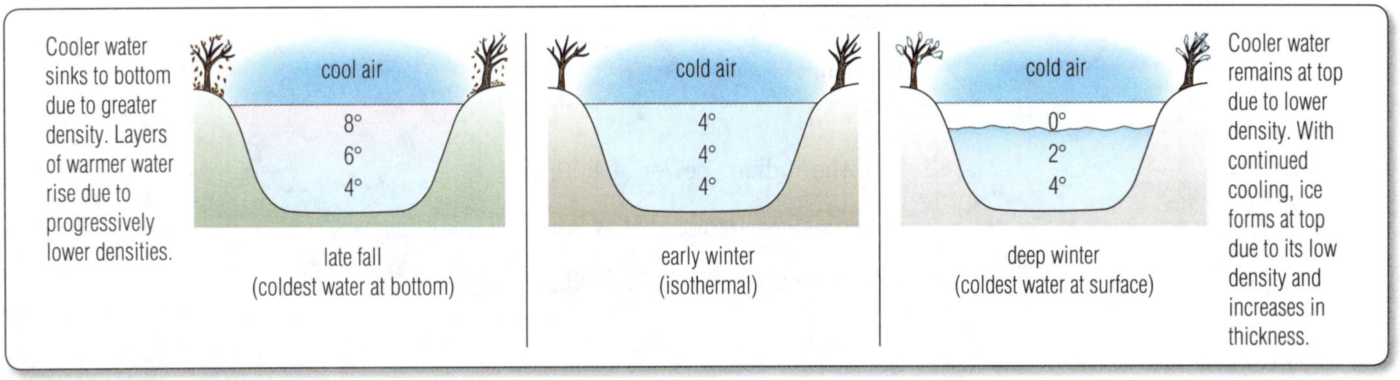

15-9a Surface-to-bottom temperature profile in a freshwater lake as it cools in winter

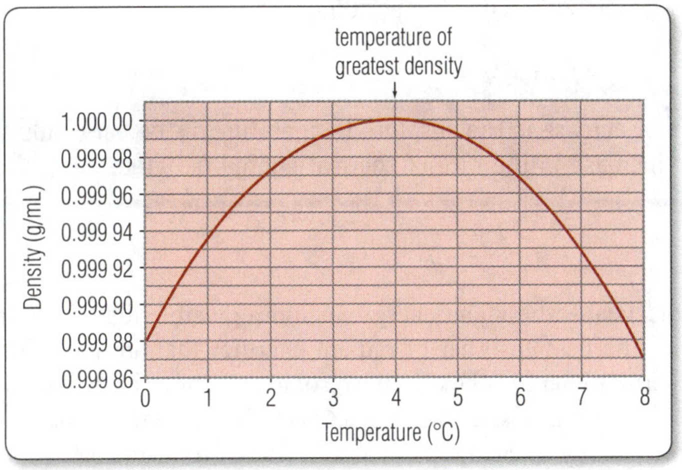

15-9b Density of water near its freezing point

Radiant energy, of which visible light is one form, is different from thermal energy, which can exist only in matter. Radiant energy is transformed into thermal energy when it is absorbed into matter. It is important to distinguish between these two forms of energy.

Josef Stefan (1835–93) was an Austrian physicist known for discovering the relationship between the radiant energy of an object and the fourth power of its absolute temperature (Stefan's law).

15.15 Radiation

The earth is warmed by thermal energy from the sun. However, since the density of the particles between the sun and the earth is so low, the sun's thermal energy cannot be transmitted by conduction or convection. The sun's electromagnetic energy reaches the earth by **radiation**, which is converted to thermal energy when it is absorbed by matter. The sun is not the only object that transfers thermal energy by radiation. Every object with a temperature greater than absolute zero transfers thermal energy in this way. "Cold" objects emit less radiation than "hot" objects. **Stefan's law** gives the correspondence between absolute temperature (T) and the power of its radiant energy (S),

$$S = \sigma T^4, \qquad (15.5)$$

where σ is a proportionality constant, called the **Stefan-Boltzmann constant** ($\sigma = 5.67 \times 10^{-8}$ W/(m^2·K^4)), and T is the Kelvin temperature of the object's surface. Because S is proportional to temperature to the fourth power, an object radiates sixteen times as much energy at a given absolute temperature as it radiates at half that temperature.

EXAMPLE 15-7

Thermal Energy and Radiation: Stefan's Law

What is the ratio of the radiant power of an iron bar at 600. K to its radiant power at 200. K?

Solution:

$$S = \sigma T^4$$

Find the radiant power at 600. K.

$$S_{600\text{ K}} = \sigma(600.\text{ K})^4 \text{ W/m}^2$$

Find the radiant power at 200. K.

$$S_{200\text{ K}} = \sigma(200.\text{ K})^4 \text{ W/m}^2$$

Find the ratio of the radiant energies.

$$\frac{S_{600\text{ K}}}{S_{200\text{ K}}} = \frac{\sigma(600.\text{ K})^4}{\sigma(200.\text{ K})^4} = \frac{\cancel{\sigma}(3)^4 \cdot \cancel{(200.\text{ K})^4}}{\cancel{\sigma}\cancel{(200.\text{ K})^4}}$$

$$\frac{S_{600\text{ K}}}{S_{200\text{ K}}} = (3)^4 = 81$$

A threefold increase in the absolute temperature of the bar yields an eighty-one-fold increase in the rate of thermal energy radiated.

Black objects absorb radiant energy, including visible light. Therefore, a black object is also the best absorber of radiant energies that are the most easily converted to thermal energy. (That is why solar collectors on rooftops are painted black.) For similar reasons, a black object is the best radiator. Since lighter objects do not absorb energy as easily, there is less energy to re-radiate. A perfect (ideal) radiator and absorber is called a **blackbody**.

15C Section Review

1. Which mechanism of heat transfer involves the physical movement of matter?
2. Which mechanism(s) of heat transfer require(s) a material medium?
3. Why is it *not* correct to say that the sun is radiating thermal energy?
4. What seems to be the mechanism for the rapid transmission of thermal energy *within* metals?
5. How is the radiant energy of an object related to its Kelvin temperature?
6. ✪ How much radiant power (in W/m^2) does a person's skin emit if its temperature is 32.0 °C?

15C Objectives

After completing this section, I can

✓ compare and contrast the three modes of heat transfer: conduction, convection, and radiation.

✓ give examples of each kind of heat transfer mode.

✓ explain why open bodies of water freeze from the top down.

✓ use Stefan's law to determine the radiant energy emission of an object.

Chapter Review

In Terms of Physics

Problem-Solving Strategies

15.1 (page 345) Heat-balance problems often result in slightly different calculated values of heat lost versus heat gained. This is due to the different precisions of the various measured quantities and the rounding conventions involved in the calculations. Differences of a few percent are acceptable for the purposes of this course.

15.2 (page 347) Solving heat-balance problems requires finding the heat exchanged (both magnitude and sign) for every part of the system within the adiabatic boundary. The conservation of heat within an adiabatic boundary assures you that the sum of all heat transfers will be zero.

15.3 (page 351) Calculating the thermal energy required to change a solid substance to a vapor, or vice versa, can be broken down into five heat-transfer steps. The key to success is to use the proper specific heat value for the corresponding phase and the correct latent heat for each phase change. Be sure to observe proper significant digit rules when summing the final result.

Review Questions

1. Give two evidences that matter is in constant motion.
2. What is the name of the discredited theory that heat is an invisible fluid?
3. Describe briefly one experiment that cast doubt on the above-mentioned theory of heat.
4. What theory states that thermal energy is energy due to the motion of particles of matter?
5. What is the difference between heat capacity and specific heat?
6. Discuss the factors that determine the heat capacity of an object.
7. What laboratory device can thermally isolate a process from its surroundings?
8. Briefly describe the steps that you would take to determine the specific heat of a small sample of a metal.
9. What assumption must you make in order to apply the principle of the conservation of thermal energy to a process?
10. Which contains more thermal energy, 1 g of liquid water at 100 °C or 1 g of steam at 100 °C? Explain your answer.
11. A glass of ice water rests on a table in a 20.0 °C room. From the time the ice begins to melt to the time the last bit of ice melts, how much does the temperature of the water change? Why?

12. You place a cold frying pan on a hot electric burner. A minute later, when you touch the pan's metal handle, you get burned. By what process did thermal energy travel from the burner to your hand?
13. How is thermal energy transferred through a vacuum?
14. If your home is heated by convection, where in a room should your source of thermal energy be located?
15. Is the rate that an object radiates energy twice as much at 2000 K as it is at 1000 K? If not, what is the relative rate at the higher temperature?

True or False (16–31)

16. Temperature alone does not determine the thermal energy of an object.
17. The dissolving of a sample of a solid substance is a key evidence for particle motion in matter.
18. Gas particles move faster than liquid particles.
19. According to the caloric theory, cold objects absorb caloric from hotter objects.
20. Joule appealed to the logic that energy cannot be converted to matter, and vice versa, by purely mechanical means, to discredit the caloric theory of heat.
21. The mechanical equivalence of heat is demonstrated where 1 N·m of work on a gram of water will raise its temperature 1 °C.
22. It is possible for two objects made of different materials to have the same heat capacity.
23. Speed is to momentum as temperature is to heat capacity.
24. The quantity of heat lost or gained by an object is inversely proportional to its temperature change.
25. Under normal conditions, the specific heat of most solids is nearly constant.
26. The term *adiabatic* means "constant thermal energy."
27. The thermal energy that must be lost in order to freeze a kilogram of a substance at its freezing point is its latent heat of fusion.
28. In order to transfer heat between two points by conduction, there must be a continuous path of matter between the points.
29. In Chemistry you learned that compounds of nonmetal elements usually form bonds where their valence electrons are closely shared between the atoms. You would expect such compounds to easily conduct heat.
30. Convection can take place only in fluids.
31. Only glowing objects radiate energy.

⭘32. A 20.0 g lump of chromium has a heat capacity of 9.03 J/°C. What is the specific heat of chromium?

⭘33. What is the heat capacity of a 100. g bar of silver?

⭘34. If a spool of copper wire has a heat capacity of 13.6 J/°C, what is the spool's mass?

⭘35. A 30.0 g gold coin and a 30.0 g silver coin sitting on a windowsill each receive 63.0 J of energy from the sun. If they were originally at the same temperature, which will be warmer after the energy is absorbed?

⭘36. A 425 g glass jar has a temperature of 20.0 °C. If $c_{sp\ glass} =$ 0.837 J/g·°C, what is the jar's temperature after it absorbs 3560 J of heat?

⭘37. A 50.0 g asbestos pad's temperature rises from 25.0 °C to 27.5 °C when 1.05×10^5 J of heat is absorbed.
   a. What is the pad's heat capacity?
   b. What is the specific heat of asbestos?

⭘38. A 10.0 g lump of lead is placed in boiling water (100.0 °C) until its temperature is constant. Then the lead is removed, quickly dried, and placed into 10.0 g of water at 20.0 °C in an adiabatic calorimeter. The final temperature of the lead and water is 22.4 °C.
   a. How much heat does the water gain?
   b. How much heat does the lead lose?
   c. What is the lump's heat capacity?
   d. What is the specific heat of lead?

⭘39. 25.0 g of boiling water is added to a calorimeter near room temperature (20.0 °C). The final temperature of the calorimeter and water is 81.5 °C.
   a. How much heat does the water lose?
   b. What is the calorimeter's heat capacity?

⭘40. How much heat does 1.00 g of steam/water release as it cools from 105.0 °C to 55.0 °C?

⭘41. A 10.0 g lump of metal at 77.6 °C is dropped into a calorimeter containing 10.0 g of ice and 10.0 g of water at 0.0 °C. 1.0 g of the ice melts before thermal equilibrium is reestablished.
   a. What is the final temperature of the mixture?
   b. How much heat does the ice that melts gain?
   c. How much heat do the water and calorimeter gain?
   d. What is the lump's heat capacity?
   e. What is the metal's specific heat?

⭘42. In order to double the rate of radiant energy output of a hot plate, how much hotter must the plate be? Express your answer to three significant digits.

DM⭘43. Predict the specific heat capacity (c_{sp}) of manganese (Mn) at 298 K (25 °C) using the value for its molar mass (*M*) (listed as atomic weight in the periodic table [Appendix H]), and the Dulong-Petit law (see Section Review Question 10 on page 352). If you did not complete that exercise, use the formula $c_{sp} = 3R/M$, where *R* is the universal gas constant, given in J/(K·mol). Calculate the percent error if the accepted specific heat capacity for Mn is 0.479 J/g·K.

⭘44. 50.0 g of liquid oxygen at its boiling point is sealed in a 50.0 g adiabatic aluminum calorimeter cup, initially at room temperature (25.0 °C). Is there enough thermal energy in the cup to vaporize the entire quantity of liquid oxygen at its boiling point? Assume pressure in the calorimeter remains constant. (*Hint*: Converting all temperatures to kelvins will make this problem easier to solve.)

DM⭘45. What is the significance of the Dulong-Petit law, which expresses the specific heats of metals in a crystalline state as a function of their atomic weight? How does this relate to the heat capacity of each atom of a metal? Do you think this could be generalized to apply to liquids and gases of compounds? Why or why not?

DM⭘46. The Dulong-Petit law is generally valid for crystalline solids near room temperature (25 °C). Does this law hold true for other temperatures? Your teacher can provide you specific heat capacity data for selected metals over a range of Kelvin temperatures. Using the techniques discussed in this chapter, evaluate the applicability of the Dulong-Petit law for these temperatures.

FACETS of PHYSICS

Nicolas Sadi Carnot (1796–1832)

Early steam engines were too inefficient to be practical, except in a few rare cases. The first engines were used in mines to pump out water, but the engines' massive coal-fired burners converted only 1% of the thermal energy into mechanical energy. After James Watt invented the first practical steam engine in 1769 (with a 2% efficiency), other men tried to make further improvements. Among them was Sadi Carnot, a French military engineer whose theories proved as useful to science as to engineering.

Nicolas Léonard Sadi Carnot was actively involved in the French government with the rest of his family. His father served on the five-man Directory that ruled France before Napoleon's reign, and Napoleon later appointed him minister of war. Years later, Carnot's nephew became president of France's Third Republic. Carnot himself was an engineer in the French army reserves most of his life.

Carnot became intrigued with the problem of designing good steam engines. Instead of becoming lost in the mechanical detail, as other engineers had done, Carnot reduced the operation of a steam engine to its bare essentials: a source of heat (the boiler), a receiver of unused heat (the condenser), and a working substance (water). He realized that the working substance does not matter in an ideal steam engine; what matters is the difference in temperature between the source of heat and the receiver. In 1824 he published his theories in a short, nontechnical paper entitled *Reflections on the Motive Power of Fire*.

Many details of the work do not follow modern thermodynamic theory because Carnot knew only of the caloric theory of heat. But the theorems that he deduced about the "Carnot cycle" were significant. Yet Carnot's discoveries were ignored until long after his death, when Lord Kelvin and Rudolf Clausius rediscovered his work. His ideas helped them to discover the second law of thermodynamics.

Although Carnot continued his scientific investigations after 1824, he never published his later findings. He died in relative obscurity in 1834 at the age of 36, when a cholera epidemic struck Paris. Most of his belongings and papers were buried with him to avoid spreading the disease. In 1878 his brother Hippolyte published a series of notes Carnot had compiled during his studies years earlier. These notes indicate that he had rejected the caloric theory after publishing his original *Reflections* paper and that he had suggested the concept of the mechanical energy equivalent of heat, though his value was much lower than that determined experimentally by Joule. Sadi Carnot's ground-breaking discoveries earned him the title of "Father of Thermodynamics," and his work is the basis for all cyclical heat engines ever developed.

Steam is a convenient medium for transforming thermal energy into other useful forms of energy. The efficient use of steam requires knowledge of thermodynamics.

Thermodynamic Laws 16

Dominion Science Problem

Miserable Heat

As the sun and the thermometer readings rise in the summer, we get more and more uncomfortable. But the hotter temperatures affect more than just our comfort. Hot air holds more humidity, and both of these factors can affect commerce and human health. Agriculture, meat processing, photography, restaurants, textiles, computers, manufacturing, libraries, and many industries rely on cooler temperatures and low humidity to function efficiently and to protect their customers and products. Diseases surface and spread faster in hot, humid weather. In recent decades, intense heat waves in the summertime have killed thousands of people in the United States. How can technology help us cope with this continuing problem?

16-1 Sometimes summer temperatures become miserably high.

16A THE ZEROTH AND FIRST LAWS

16.1 Introduction

So far you have studied two forms of energy—mechanical and thermal. Mechanical energy is a property of tangible objects. Two types of mechanical energy are potential energy and kinetic energy. Potential energy, one result of work, can be changed to kinetic energy—energy of motion. A motionless ball poised at the top of an inclined plane has potential energy. As the ball rolls down the inclined plane, it gains kinetic energy as it loses potential energy. When the ball reaches the bottom of the plane, the last of its gravitational potential energy disappears and it rolls along the flat surface with only kinetic energy. Total mechanical energy (E) is the sum of the kinetic and potential energies due to the motion and position of physical objects.

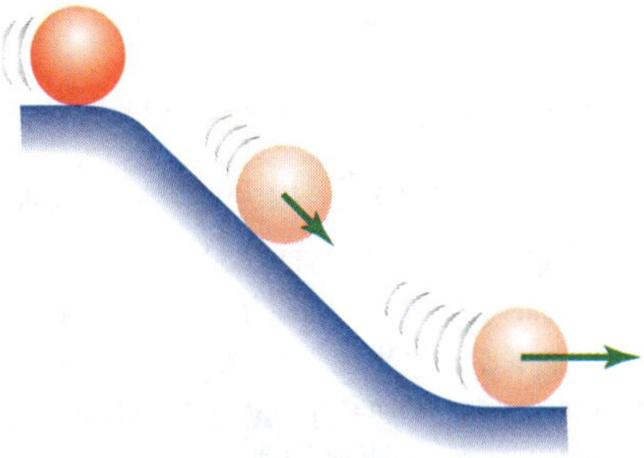

16-2 Mechanical potential energy can change to mechanical kinetic energy.

Once the appropriate reference points for determining the system's motion and position are established, the system's total mechanical energy can be known. Subsequently, only the changes of the system's total mechanical energy (ΔE) are important from a physics perspective.

Thermal energy is due to the rapid, random motion of the molecular, atomic, and subatomic particles of matter. Like mechanical energy, thermal energy can be subdivided into potential energy and kinetic energy. The particles of a substance are constantly moving; therefore, they have kinetic energy. Their average kinetic energy is proportional to the temperature of the substance.

The innumerable particles of a substance also have potential energies, but it is extremely complicated to determine the reference points to establish these energies. The sum total of the particle kinetic and potential energies is called **internal energy (U)**. As with total mechanical energy, once the internal energy of a system is established or assumed for a given state of the system, only the change in internal energy becomes important. The changes of internal energy (ΔU) are usually much easier to calculate than the internal energy states at the beginning and end of a process.

16.2 The Zeroth Law of Thermodynamics

Some materials conduct heat well, while other materials are good insulators. An ideal insulating wall through which no thermal energy can pass is called an *adiabatic* boundary. An ideal conductor of thermal energy is called a **diathermic** material. Although no completely adiabatic or diathermic materials exist, the concepts are useful for simplifying discussions.

As you have probably noticed, when you place a hot object in contact with a cold object, the cold object becomes hotter, and the hot object becomes cooler. In time, the two objects reach the same temperature. That is, they reach a state of **thermal equilibrium**. Thermal energy has flowed from the hot object to the cold object. As you learned in Chapter 15, the flow of thermal energy is called *heat*.

The zeroth law of thermodynamics states that if two systems are in thermal equilibrium with a third, then they are in thermal equilibrium with each other.

Figure 16-3 depicts what is called the zeroth law of thermodynamics. This law states that two systems that are in thermal equilibrium with a third system must be in thermal equilibrium with each other. The law received its name because it is more basic than either the first or second law but was formulated after the other laws already had been named.

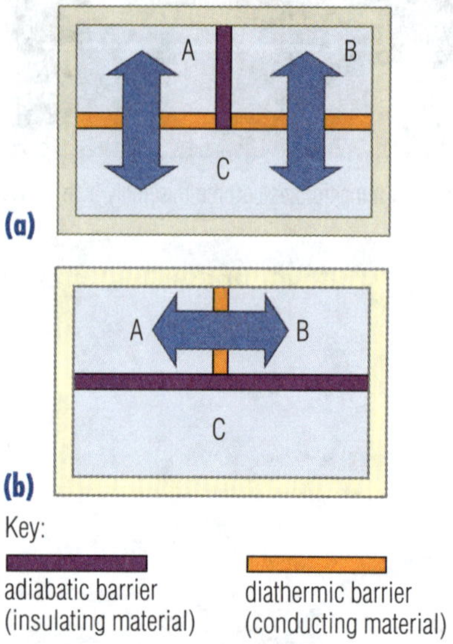

16-3 Three systems in thermal equilibrium demonstrating the zeroth law of thermodynamics. (a) If systems A and B are individually in thermal equilibrium with system C, then (b) they must be in thermal equilibrium with each other.

In (a), system A is separated from system B by an adiabatic barrier, and both are linked to system C by diathermic barriers. Although A and B cannot exchange thermal energy with each other, both can exchange thermal energy with C. In (b) the barriers have been switched so that A and B can exchange thermal energy with each other, but C is thermally insulated from both A and B. According to the zeroth law, if no net energy exchange occurs in (a), then none will occur in (b).

The principle becomes still more understandable when you view it in terms of temperature. In (a), systems A and B are at the same temperature as C. (You must, of course, assume that each system is itself at a uniform temperature.) System A must therefore be at the same temperature as system B. When the adiabatic barrier between them is replaced with a diathermic barrier, neither system will warm (that is, transfer thermal energy to) the other. Moreover, since all three systems are at the same temperature, no thermal energy can flow in any direction, regardless of how the barriers are placed.

16.3 The General Law of Conservation of Energy—The First Law

The symbol W_{ncf} represents the mechanical work done on a thermodynamic system by external nonconservative forces. This symbol is different from that used in earlier chapters because in studying thermodynamics, it is usually more important to analyze the work a system does on its surroundings (W).

Before progressing further, we need to gather together what we know about energy transfers in order to establish a fundamental or general energy conservation principle. Energy can be added to a system by either of two processes: (a) mechanical work (W_{ncf}) through the application of external nonconservative forces or (b) heat transfer (Q) through temperature differences with the system's surroundings. The energy transferred to the system by these means is seen as a change of the internal energy of the system (ΔU) or as a change in the system's total mechanical energy (ΔE) or both. The *general energy conservation law* can be written

$$Q + W_{ncf} = \Delta U + \Delta E. \tag{16.1}$$

When we discussed basic mechanics, we assumed that no heat transfers took place and no changes of internal energy occurred ($\Delta U = 0$ J). When discussing basic thermodynamics principles, we will make the assumption that the total mechanical energy of the system is constant ($\Delta E = 0$ J). Therefore, the only forces doing work on the system are nonconservative forces. Equation 16.1 can then be rewritten as

$$Q + W_{ncf} = \Delta U. \quad (16.2)$$

Rather than being concerned about the work done on a system, as we were in our study of mechanics, it is more useful in thermodynamics to evaluate the effects of a system on its surroundings; therefore the work notation in Equation 16.2 needs to be modified slightly. Recall from our study of Newton's laws that every force on a system is paired with an opposite force acting on another system. It follows that if a system's surroundings do mechanical work on it through the application of a nonconservative force, then the system must also do work on its surroundings via the reaction force. The work by the system on its surroundings is therefore the negative of the work on the system ($W_{ncf} = -W$). Equation 16.2 then becomes

$$Q - W = \Delta U, \text{ or}$$

$$Q = \Delta U + W. \quad (16.3)$$

Equation 16.3 is the mathematical statement of the first law of thermodynamics. The first law extends the principle of the conservation of total mechanical energy to internal energy as well. We will investigate the applications and limitations of the first law in the following discussion of thermodynamic processes.

The **surroundings** of a system are everything in the universe that is outside the system's boundaries.

The work done on a thermodynamic system is the negative of the work done on the system's surroundings. This is a direct consequence of Newton's third law of motion: every force is balanced by an equal but opposite force on a different system.

$$W = -W_{ncf}$$

The **first law of thermodynamics** states that the heat transferred to or from a system is equal to the sum of the change of the system's internal energy and the work the system does on its surroundings.

$$Q = \Delta U + W$$

16.4 Heat Engines

Thermal energy can be changed into mechanical energy. An engine called a **heat engine** can do mechanical work by absorbing and discharging heat. The simplest "machine" that converts thermal energy to work is an expanding gas. For a gas to expand usefully, it must be confined in an expandable container. One such container is a cylinder fitted with a gas-tight piston. Assuming that the piston is massless and frictionless simplifies calculations. As the gas expands, it is no longer in thermal equilibrium. It heats or cools unevenly. To avoid this problem, we allow the gas to expand in extremely minute steps, letting it return to thermal equilibrium between steps. Consequently, the gas expands without ever being far from thermal equilibrium. A process that proceeds in this way is called a *quasi-static process*.

For quasi-static processes, the gas pressure inside the cylinder is at all times in equilibrium with the external pressure. When the gas is compressed from a volume V_1 to a volume V_2 by an external pressure P, work is done on the gas. (You can verify that energy is added to the gas because it warms when it is compressed.)

The simplified formula for work is

$$W = Fd. \quad (16.4)$$

Pressure is force per unit area, so force exerted by the gas equals pressure times the cross-sectional area of the piston:

$$F = P_{gas}A$$

Substituting for force in Equation 16.4,

$$W = (P_{gas}A)d.$$

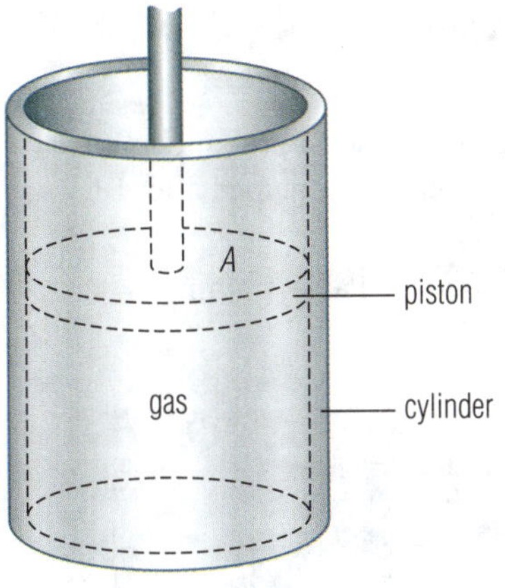

16-4 A simple heat engine

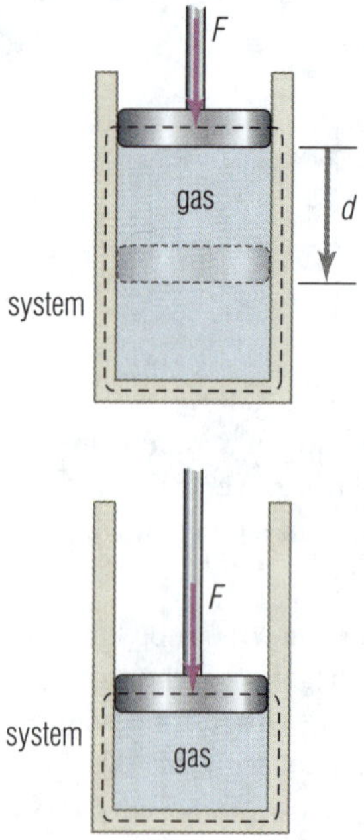

16-5 A gas is compressed by a pressure, *P*.

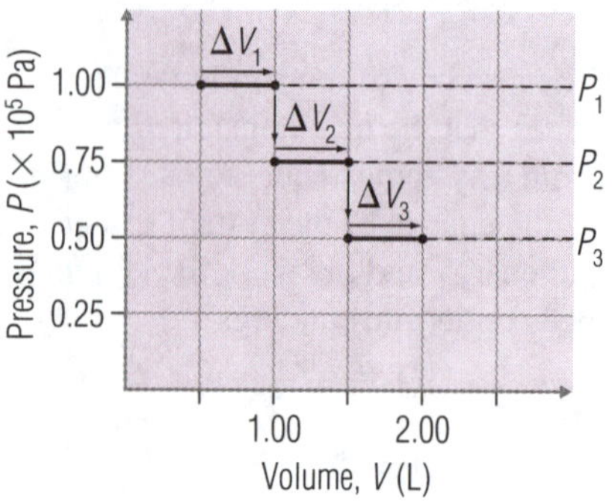

16-6 A *P-V* diagram of an expanding gas in a quasi-static process

Recall that the SI pressure unit is the **pascal (Pa)**, named after Blaise Pascal

$$1 \text{ Pa} = 1 \text{ N/m}^2$$

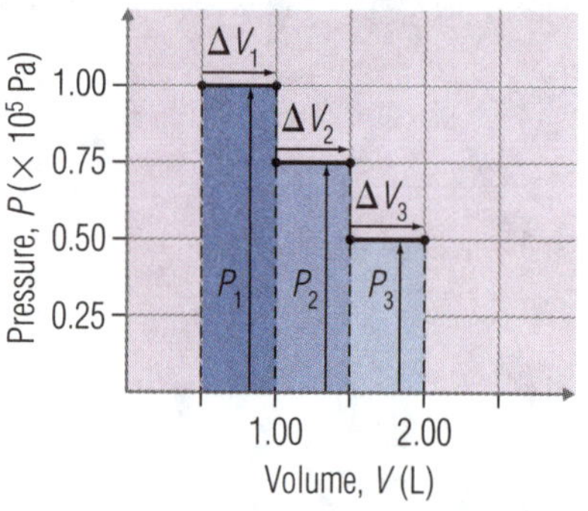

16-7 An area under a curve on a *P-V* diagram indicates the work done by a process between the two states represented by the endpoints of the curve.

But the cross-sectional area of the cylinder times the displacement distance of the piston is just the volume change (ΔV) of the gas in the cylinder, so the work done *by* the gas as it is compressed by a constant external pressure is

$$W = P_{\text{gas}}\Delta V.$$

The change of volume when a gas is compressed is negative. Therefore, the work done by the gas *on its surroundings* (see Subsection 16.3) is negative when it is compressed. This result makes sense, since the surroundings are losing energy and the gas is gaining energy.

When the gas expands against the applied pressure, the gas does *positive* work on its surroundings. (A gas that does work by expanding loses thermal energy, so it cools.) In general, for expansion or contraction against a constant pressure, the work done *by* the gas is

$$W = P(V_2 - V_1).$$

The pressure against the gas is not always constant. When the pressure varies during an expansion or a contraction, the work is the sum of the work done in each step. In a quasi-static process, you can assume that for a small volume change the pressure is constant, so

$$W_{\text{total}} = (P\Delta V)_1 + (P\Delta V)_2 + (P\Delta V)_3 + \ldots + (P\Delta V)_n. \qquad (16.5)$$

16.5 *P-V* Diagrams

Equation 16.5 becomes easier to solve if you plot the pressure versus the volume on a graph called a ***P-V* diagram**. For instance, in Figure 16-6 the gas is originally at 0.50 L and 1.00×10^5 Pa. It expands to 1.00 L ($\Delta V = 0.50$ L) at 1.00×10^5 Pa. Then, while the volume remains constant, the pressure changes to 0.75×10^5 Pa. The gas expands to 1.50 L at 0.75×10^5 Pa. The pressure is again lowered to 0.50×10^5 Pa. Finally, the gas expands to 2.00 L. The total work done by the gas is as follows:

$$W_{\text{total}} = (P\Delta V)_1 + (P\Delta V)_2 + (P\Delta V)_3$$

$$W_{\text{total}} = (1.00 \times 10^5 \text{ Pa})(+0.50 \text{ L}) + (0.75 \times 10^5 \text{ Pa})(+0.50 \text{ L}) + (0.50 \times 10^5 \text{ Pa})(+0.50 \text{ L})$$

$$W_{\text{total}} = 0.50 \times 10^5 \text{ Pa·L} + 0.3\underline{7}5 \times 10^5 \text{ Pa·L} + 0.25 \times 10^5 \text{ Pa·L}$$

$$W_{\text{total}} = 1.1\underline{2}5 \times 10^5 \text{ Pa·L}$$

$$W_{\text{total}} \cong 1.13 \times 10^5 \text{ Pa·L}$$

Since 1 Pa = 1 N/m^2, and 1 L = 0.001 m^3,

$$\frac{1.13 \times 10^5 \text{ N}}{\cancel{\text{m}^2}} \cdot \frac{1}{\cancel{\text{L}}} \cdot \frac{0.001 \cancel{\text{m}^3}}{1 \cancel{\text{L}}} = 113 \text{ N·m} = 113 \text{ J}.$$

Notice that each term in the work equation is equal to the block of the area under the corresponding step of the graph of the quasi-static process on the *P-V* diagram (Figure 16-7). That is, the area under the curve representing the process on a *P-V* diagram is equal to the absolute value of the work done by the gas during the process. This result is true for any confined gas process (expansion or contraction). The sign of the work depends on whether the gas gains or loses energy. If the gas expands, it does work on its surroundings, so the sign is positive. If the gas contracts or is compressed, the surroundings do work on it, so the sign is negative.

EXAMPLE 16-1

Work of an Expanding Gas

A gas originally at 1.00×10^5 Pa and 1.00 L expands until its volume is 2.00 L and its pressure is 0.50×10^5 Pa. How much work is done by the gas?

Solution:

The gas in the cylinder is the system. Work is determined relative to the system's surroundings.

The work by the gas is related to the numerical value of the entire shaded area of Figure 16-8. The easiest way to find the area is to calculate the area of the triangle above the dotted line and add it to the area of the rectangle below the dotted line.

The area of the triangle is

$$A_{\triangle} = \tfrac{1}{2}bh, \text{ where } b = |\Delta V| = +1.00 \text{ L and } h = |\Delta P| = +0.50 \times 10^5 \text{ Pa}$$

$$A_{\triangle} = \tfrac{1}{2}(1.00 \text{ L})(0.50 \times 10^5 \text{ Pa})$$

$$A_{\triangle} = 0.25 \times 10^5 \text{ Pa·L} = 25 \text{ N·m}$$

$$A_{\triangle} = 25 \text{ J.}$$

The rectangle's area is

$$A_{\square} = bh, \text{ where } b = |\Delta V| = +1.00 \text{ L and } h = P_2 = 0.50 \times 10^5 \text{ Pa}$$

$$A_{\square} = (1.00 \text{ L})(0.50 \times 10^5 \text{ Pa})$$

$$A_{\square} = 0.50 \times 10^5 \text{ Pa·L} = 50. \text{ N·m}$$

$$A_{\square} = 50. \text{ J.}$$

The total area is then

$$A_{total} = A_{\triangle} + A_{\square}$$

$$A_{total} = 25 \text{ J} + 50. \text{ J}$$

$$A_{total} = 75 \text{ J.}$$

The total area under the curve equals the absolute value of the work done, so

$$|W| = 75 \text{ J.}$$

Since the gas expanded, the work done by the gas is positive.

$$W = +75 \text{ J}$$

The gas does about 75 J of work on its surroundings as it expands.

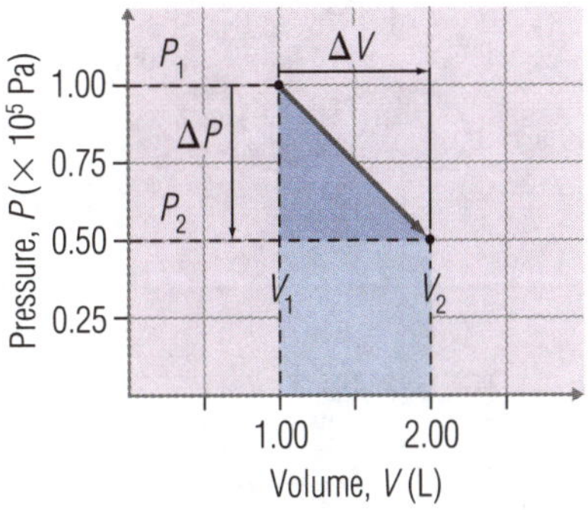

16-8 A graph showing the work done by an expanding gas for Example 16-1

The mathematical technique we use in Example 16-1 illustrates the great importance of *integral calculus* in physics. Since the pressure and volume shown in Figure 16-8 change in a linear (straight-line) fashion, ordinary algebra lets us calculate the area of the shaded space. However, many real-world processes aren't perfectly linear (see Figure 16-10). In order to find the area under nonlinear curves, we must either be satisfied with algebraic approximations or use the techniques of integral calculus. The development of calculus is what made classical physics possible.

Problem-Solving Strategy 16.1

If a cycle proceeds clockwise on a *P-V* diagram, then the system is a heat engine doing positive work. If the cycle proceeds counterclockwise, then work is being done on the system (as in a refrigerator).

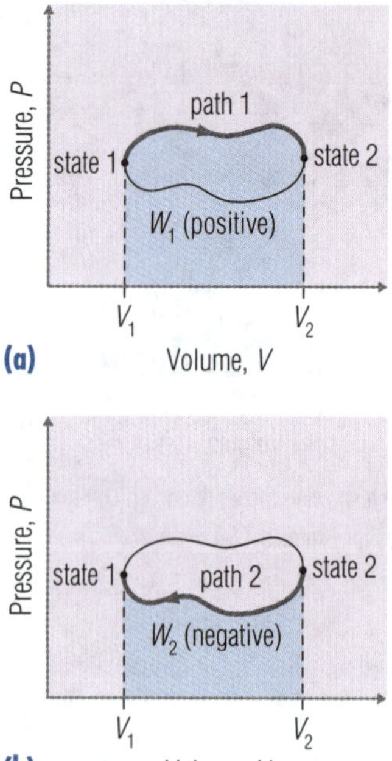

16-10 (a) The expanding gas does positive work while expanding from V_1 to V_2. (b) The gas does negative work as it is compressed from V_2 to V_1.

Problem-Solving Strategy 16.2

If a process moves from left to right along any path on a *P-V* diagram, then the system is doing work. If the process moves from right to left along any path, then work is being done on the system.

One state variable that you are familiar with is the balance of your bank account. It doesn't really matter to you what form of money (coins, bills, checks, ATM transfers, etc.) adds value to the account balance, as long as your balance increases. The state of your account is the cash balance that it has at any given moment.

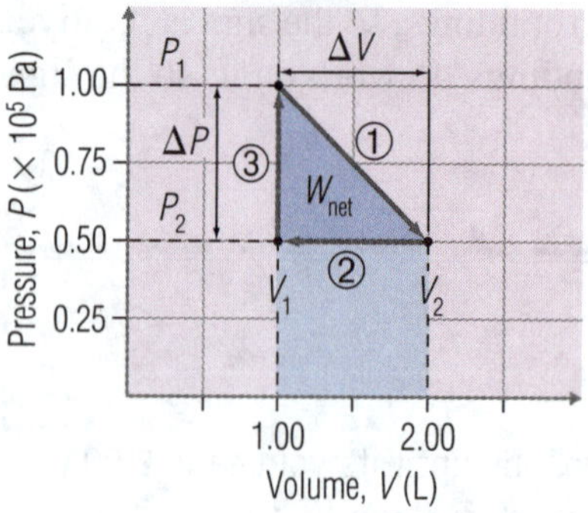

16-9 Work done by a gas that expands in a cycle

16.6 Expansion Cycles

If a gas is to be useful as a machine, it must be able to expand repeatedly, following a cycle. When the gas from the previous example returns to its original state, as Figure 16-9 shows, it is ready to expand again. For the first step in the cycle, ①, the gas expands as it did in the example, and the same amount of work is done *by* the gas on its surroundings:

$$W_1 = 7.5 \times 10^4 \text{ Pa·L} = +75 \text{ J}$$

For the second step, where the volume is reduced at constant pressure, the area under curve ② is

$$A_2 = (1.00 \text{ L})(5.0 \times 10^4 \text{ Pa})$$

$$A_2 = 5.0 \times 10^4 \text{ Pa·L}$$

$$A_2 = 50. \text{ J.}$$

Since the gas is compressed (contracts), the work *by* the gas is negative.

$$W_2 = -50. \text{ J}$$

For the third step, ③, the volume is constant; therefore, no work is done.

$$W_3 = 0 \text{ J}$$

The total work for the cycle is

$$W_{total} = W_1 + W_2 + W_3$$

$$W_{total} = (+75 \text{ J}) + (-50 \text{ J}) + 0 \text{ J}$$

$$W_{total} = +25 \text{ J.}$$

Notice that the net work by the gas is the same as the area enclosed by the path of the cycle (the triangular area in Example 16-1). This is a general principle: *For a cycle, the absolute value of the total work done is equal to the area enclosed by the path of the cycle on a P-V diagram.* A cycle that has a clockwise path on the diagram does positive work. Such a system is a heat engine. A cycle that has a counterclockwise path on the diagram does negative work. An example of a system that follows such a path is a refrigerator.

The work done by a gas, unlike work done against gravity, *depends on the path of the process on a P-V diagram.* The thermal energy added to the gas also depends on the expansion path. Experiments show, however, that for systems whose total mechanical energy is constant, the quantity ($Q - W$) is the same regardless of the thermodynamic path between two states. This quantity is the change in the internal energy of the gas, ΔU, according to the first law of thermodynamics. Since the change of internal energy does not depend on the way the energy is added (the path between two thermodynamic states), the internal energy is called path-independent. Path-independent quantities are called **state variables** since they depend only on the thermodynamic state of the gas at a given time and not on how it got there. Internal energy, temperature, pressure, and volume are some state variables. They define the state of a gas.

16.7 Thermodynamic Systems

We have mentioned in our discussion an entity called a **thermodynamic system**. As you know, a system is a piece of the universe isolated, at least mentally, for study. Other examples of possible thermodynamic systems are a calorimeter containing hot water, an ice cube, and an internal combustion engine. The boundaries of the system are set by the person studying the system. For an expanding gas, the cylinder walls and the piston are the boundaries of the system. Anything that is not a part of a system is a part of its surroundings.

Thermodynamically, neither the piston nor the cylinder walls are parts of the gas system.

Systems are classified into three categories. An **open system**, such as an ice cube resting on a kitchen counter, can exchange both energy and matter with its surroundings. A **closed system** can exchange energy but not matter with its surroundings. An expanding gas in a thermally conducting cylinder with a gas-tight piston is a closed system. An **isolated system**, such as a liquid in a perfectly insulated vacuum flask, cannot exchange either matter or energy with its surroundings. In practice, there are no truly isolated systems except for the universe itself.

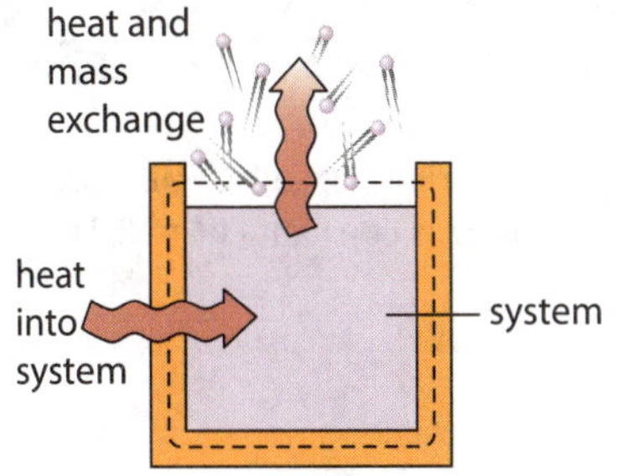

16-11a An open system

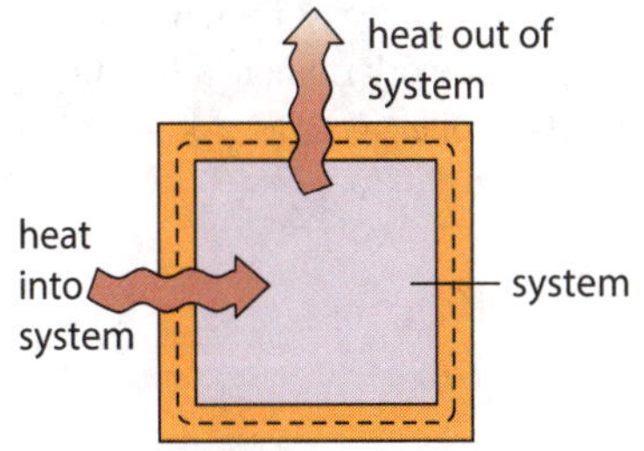

16-11b A closed system

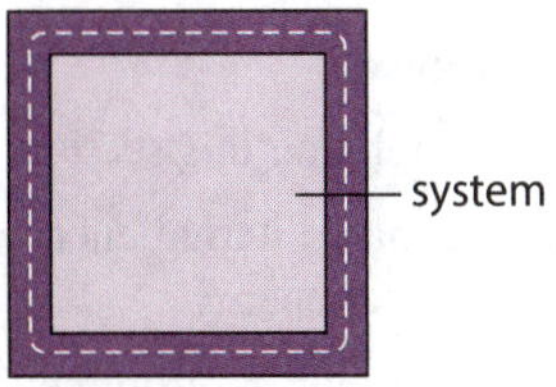

16-11c An isolated system

In an isolated system, energy is conserved. That is, the amount of energy in an isolated system is constant. Energy may be converted from one form to another, but no energy enters or leaves the system, nor does it appear or disappear. Since the universe is an isolated system, its total amount of energy is constant. When energy is exchanged between an open or closed system and its surroundings, all the energy lost by one must be gained by the other, since the system and its surroundings ultimately include the entire universe. This fact is the reason for the following equation in Chapter 15:

$$Q_{\text{system}} + Q_{\text{surroundings}} = 0\ \text{J}$$

Although no practical system is isolated, a system enclosed in a calorimeter or other vacuum-insulated container approximates an isolated system for short periods of time. The first law of thermodynamics is one of the great conservation laws of science. It states that in a isolated system, the total quantity of energy (in the forms of thermal energy and mechanical work) is constant, neither being created nor destroyed. This law is consistent with numerous biblical references proclaiming the finished work of God at the end of the Creation week.

The first law of thermodynamics is one of the most thoroughly tested and confirmed laws in physical science. If a scientist claims that energy or mass (a different form of energy) spontaneously has appeared in the universe, you know that his idea violates this law of science.

16.8 Thermodynamic Processes

A *thermodynamic process* is a change in the thermodynamic state of a system. Although it is possible for all state variables to change at the same time, analysis and calculations are easier if at least one variable is held constant throughout the process. Processes are named to indicate which variables are held constant.

An *adiabatic process* exchanges no thermal energy between the system and its surroundings. For adiabatic processes, Q is zero; therefore, the first law becomes

In *adiabatic* processes, thermal energy is constant ($Q = 0$ J).

$$0\ \text{J} = \Delta U + W, \text{ or}$$

$$\Delta U = -W.$$

Recall that W is work *on the surroundings* by the system and is the negative of the work *on* the system ($W = -W_{\text{ncf}}$).

Since internal energy of a gas is directly proportional to temperature, a change in internal energy is indicated by a change in temperature. If the gas expands, doing positive work, then internal energy decreases, and the gas temperature drops; if the gas is compressed so that it does negative work, temperature and internal energy increase.

For an isothermal process, the temperature of the system is constant. If you assume that the system is not in the process of melting, freezing, boiling, or

In isothermal processes, temperature is constant ($\Delta U = 0$ J).

This statement is not a violation of the second law of thermodynamics because it represents only a step in a cycle and not the entire cycle (see Section 16B).

In an isochoric (EYE so KOR ik) process, volume is constant ($W = 0$ J).

In an isobaric process, pressure is constant, and work done is a function of volume alone.

condensing (phase changes involving the exchange of latent heats at constant temperatures), the internal energy is also constant ($\Delta U = 0$ J). The first law becomes

$$Q = 0\text{ J} + W, \text{ or}$$
$$Q = W.$$

This means that all of the heat transferred is converted to work.

During an isochoric process, the volume is constant. Remember, work is done on or by a gas only when its volume changes. Therefore,

$$Q = \Delta U + 0\text{ J}, \text{ or}$$
$$Q = \Delta U.$$

All the thermal energy added to the system increases the internal energy of the system.

An isobaric process is a process for which the pressure is constant. Isobaric processes give the simplest equation for work,

$$W = P\Delta V.$$

The first law of thermodynamics for this process is

$$Q = \Delta U + P\Delta V.$$

The calculations for work and energy are much simpler when it is possible to use the ideal gas relationships in the general gas law. At low pressures, real gases behave much like ideal gases, and the approximation is valid. A process that allows the use of ideal gas relationships is known as an **ideal gas process**.

16A Objectives

After completing this section, I can

- ✓ differentiate internal energy from mechanical energy.
- ✓ explain why the zeroth law of thermodynamics defines thermal equilibrium.
- ✓ state the first law of thermodynamics and discuss how it pertains to the principle of conservation.
- ✓ describe the general characteristics of a heat engine.
- ✓ explain why the concept of quasi-static processes is useful for describing the operation of a heat engine.
- ✓ graph and interpret P-V diagrams.
- ✓ explain the significance of the expansion cycle of a heat engine in doing useful work.
- ✓ describe the properties of the three thermodynamic systems and give an example of each.
- ✓ differentiate among the five thermodynamic processes discussed in the text.

16A Section Review

1. What is a barrier that conducts no thermal energy called?
2. On a cold winter morning, if the air outside a car is in thermal equilibrium with the glass in the car's window and the window glass is in thermal equilibrium with the air inside the car, what can be said about the temperatures of the inside and outside air? What is your authority for your answer?
3. How is the work done by the surroundings on a thermodynamic system related to the work the system does on its surroundings? What principle of mechanics allows you to answer this question?
4. To what is the area under a P-V curve equivalent?
5. What kind of thermodynamic engine follows a counterclockwise path on a P-V diagram?
6. What kind of thermodynamic system is a sealed plastic container full of steaming hot soup in a refrigerator?
7. For a closed gas thermodynamic system, what kind of process exchanges thermal energy for change in internal energy?
8. A gas cylinder system similar to that described in this section originally has a volume of 1.80×10^{-3} m^3 at 1.01×10^5 Pa. If the volume and pressure of the gas change along a linear path on the P-V diagram to 1.22×10^{-3} m^3 and 2.02×10^5 Pa, what is the work done by the gas in joules?

16B THE SECOND AND THIRD LAWS

16.9 Heat Engines and the Second Law

All thermodynamic processes except adiabatic processes involve exchanges of energy between a system and its surroundings. The surroundings must therefore contain either a source of thermal energy, a sink (or receiver) for thermal energy, or both. An easy way to visualize a thermal energy source or sink is to imagine a **heat reservoir** at a specific temperature. The reservoir is so large that no addition or subtraction of energy can change its temperature significantly. A reservoir at a higher temperature than the system, called a *hot reservoir*, is a source of thermal energy for the system. A reservoir at a lower temperature than the system, called a *cold reservoir*, is a thermal energy sink for the system. Both a hot reservoir and a cold reservoir are used to operate a heat engine.

16-12 The sun, a vast heat reservoir, occasionally releases huge amounts of energy in the form of solar flares or prominences, such as the ones shown here.

A source of thermal energy for a system must be hotter than the system. The reason is that thermal energy flows from a hotter body to a colder body. This principle agrees with human experience. If you place a glass of hot water against a glass of cold water, the cold water warms, while the hot water cools. What would you think if, after a few minutes of contact, the cold water froze and the hot water boiled? One form of the **second law of thermodynamics** is the principle that energy flows from an area of higher concentration to an area of lower concentration.

A typical heat engine requires a hot reservoir, a cold reservoir, and a working fluid (a liquid or gas). Thermal energy absorbed from the hot reservoir causes the fluid to expand against a piston or some other movable part. The expansion forces the part to move, performing mechanical work. Then, the fluid gives up thermal energy to the cold reservoir and contracts. The fluid is then ready to be heated and expand again.

There are several ways to state the second law of thermodynamics. One statement of the second law is that energy flows from an area of higher concentration to an area of lower concentration. You observe the second law in action when thermal energy flows from hotter locations to colder ones.

16.10 Early Steam Engines

A steam engine is a heat engine using steam as the working fluid. For centuries men have known that steam can do work. The first steam engine, the *aeolipile*, was described by Hero of Alexandria, who lived about the time of Christ. The aeolipile is a sphere or cylinder with bent outlet pipes (Figure 16-13) that is connected by pipes to a sealed container of boiling water. The connecting pipes are attached in a

Hero of Alexandria (first century AD) was a mathematician and engineer whose writings were an early contribution to mechanics and related sciences.

16-13 The aeolipile

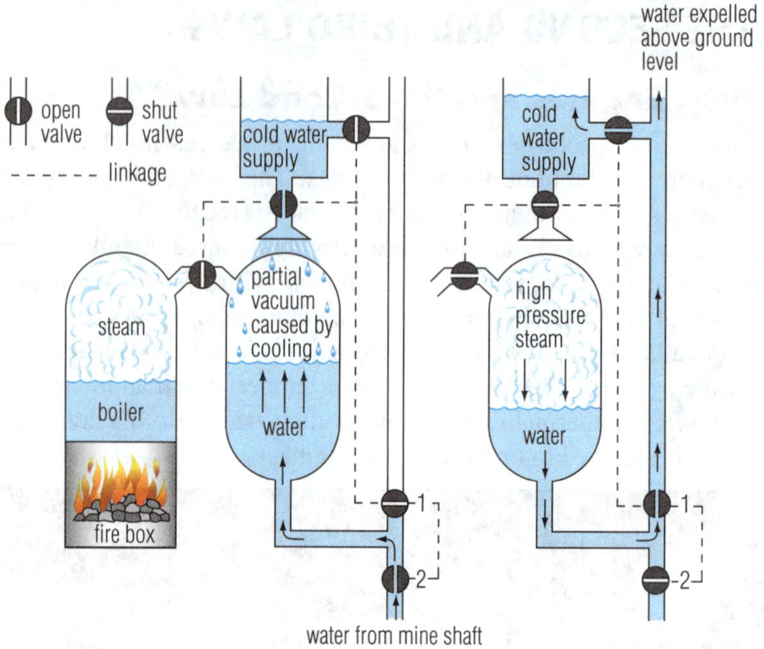

16-14 Schematic of a Savery engine (1698)

Thomas Savery (1650–1715) was an English engineer and inventor.

Thomas Newcomen (1663–1729) was an English iron magnate and inventor. His was the first practical engine to use a piston in a cylinder.

James Watt (1736–1819) was a Scottish engineer who devoted most of his life to designing and building improved steam engines.

way that allows the sphere to rotate. When the water in the container boils, steam is forced through the connecting pipes into the sphere and finally out the outlet pipes. The escaping steam exerts a reaction force on the nozzles that produces a torque, causing the sphere to rotate. The aeolipile is not a cyclic engine; the engine uses only a single irreversible thermodynamic process between two states until its water boils away. All the thermal energy (and the water) that enters the process is lost.

The first practical steam engine was a water pump invented by Thomas Savery. A better steam engine, which could be used for pumping or for raising loads, was invented by Thomas Newcomen. Newcomen's engine dominated the market until James Watt invented an even more efficient design. Watt's engine was the first to use separate chambers to heat and cool the steam to avoid the wasteful process of repeatedly heating and then cooling the cylinder walls. Watt's engine was able to power factories and transportation as well as pump water from deep mines. The Industrial Revolution began with Watt's steam engine.

The basic idea behind the steam engine is that steam expands when it is heated and contracts when it is cooled. In the Savery engine (Figure 16-14), the steam expands to push water out of a chamber and through the outlet pipe (right picture). Valve 1 is open; valve 2 is closed. When all the water is out of the chamber, valve 1 is closed and valve 2 is opened. The steam contracts, and the decrease in pressure draws water up through the inlet pipe into the chamber (left picture).

The Newcomen engine, Figure 16-15a, has a movable piston that goes up when the steam expands and down when the steam condenses. It uses a first-class lever to lift the load, so the condensing stroke does the lifting work. The condensed steam is drained from the cylinder before the next cycle begins. One problem that existed with this engine was that the cylinder was made of cast iron. The rough interior surface made it difficult to obtain a good seal with the piston. Another problem was that the cylinder was cooled during the condensation step; the steam entering at the start of the next cycle condensed prematurely as it reheated the cylinder, before any work was done.

The Watt engine, Figure 16-15b, is similar to the Newcomen engine in that it uses a piston. However, the Watt engine relied on the difference in pressure between low-pressure steam and the partial vacuum created by condensing steam

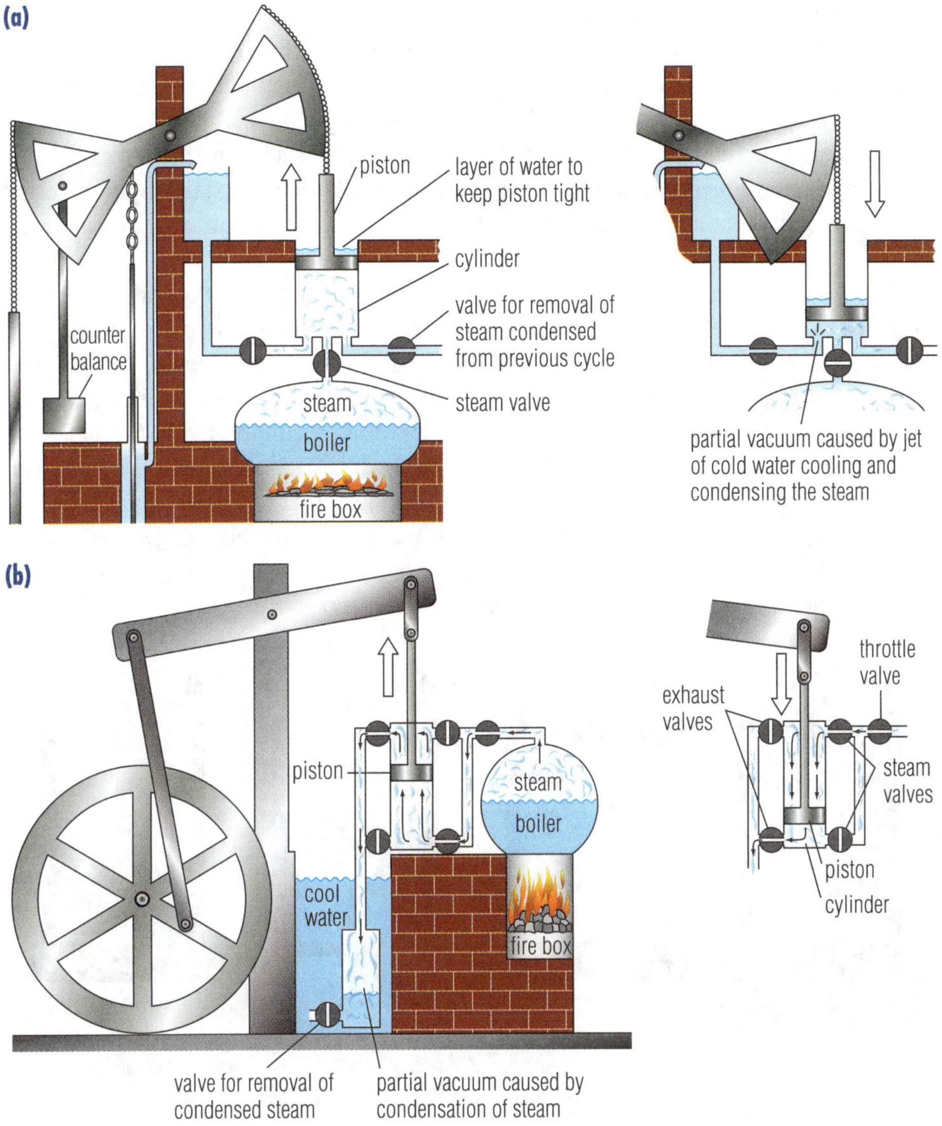

16-15 Schematics of (a) Newcomen's engine (1712) and (b) Watt's double-action rotary engine (1782)

for its power stroke. Also, in later Watt engines, the piston is *double-acting*—force is exerted on both the forward and reverse strokes of the piston through the automatic operation of appropriate steam supply and exhaust valves. The exhausted steam is directed to a second chamber (the condenser), cooled by water, to condense the steam, forming a partial vacuum that increases the net pressure applied by the steam on the piston. Watt's addition of a separate condenser greatly improved the efficiency of his engine compared to earlier designs. The Watt engine also had the advantage of a machine-bored cylinder. The extra-smooth walls allowed a good seal with the piston, preventing the waste of steam.

The valves indicated in the figures are all mechanically linked to the piston in order to ensure they operate at the proper times in the cycle.

16.11 An Ideal Heat Engine Cycle—The Carnot Cycle

A **reversible process** is a quasi-static process that leaves the system in exactly the same state after occurring twice, once normally and once in reverse. The processes that you study here can be assumed to be reversible unless the text states otherwise. However, no real processes are completely reversible due to the presence of nonconservative effects, although some are more nearly reversible than others.

A thermodynamic process is a thermodynamic change between two states. A cycle is made up of several processes that enclose an area on a *P-V* diagram. The starting and ending points of a cycle are identical, though the paths between the points may be different.

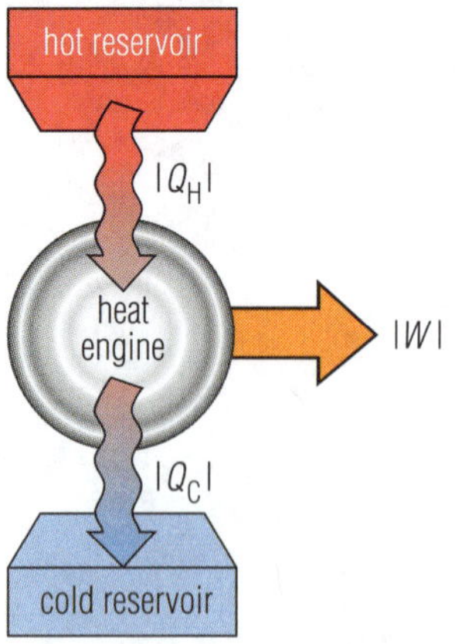

16-16 A conceptual diagram showing how a heat engine works

The **Carnot cycle** is the most efficient ideal heat engine cycle.

A reversible *cycle* leaves the system in the same state as it was before the entire process occurred. The pressure, volume, and temperature of an ideal gas are the same at the end of a reversible cycle as they were at the beginning. Reversible cycles are the most efficient means of converting thermal energy to mechanical work. Cyclic processes are practical because they are very efficient and they can repeat many times without the operating conditions of the engine having to be reset. The engine doesn't continually gain heat until it melts or lose heat until it freezes. There are several idealized cycles that convert thermal energy to work, with varying efficiency.

The most efficient cycle that can operate between two temperatures is the **Carnot cycle**, a four-step reversible cycle (see Figure 16-17). Step 1 in the Carnot cycle, (a), is an isothermal expansion from V_1 to V_2 at temperature T_H. The system absorbs thermal energy Q_H from the hot reservoir. Step 2, shown in (b), is an adiabatic expansion from V_2 to V_3. The temperature changes from T_H, the temperature of the hot reservoir, to T_C, the temperature of the cold reservoir. The system absorbs no heat in this step. In the third step, (c), an isothermal compression from V_3 to V_4 occurs at T_C. The system gives up thermal energy Q_C to the cold reservoir. Step 4, (d), is an adiabatic compression to the original conditions, T_H and V_1. The system absorbs no thermal energy in this step.

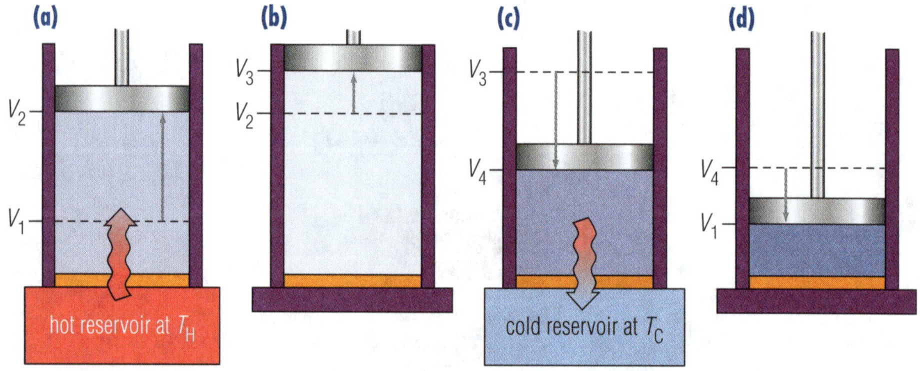

16-17 A Carnot engine completes one cycle. (a) Step 1—isothermal expansion (b) Step 2—adiabatic expansion (c) Step 3—isothermal compression (d) Step 4—adiabatic compression to original conditions

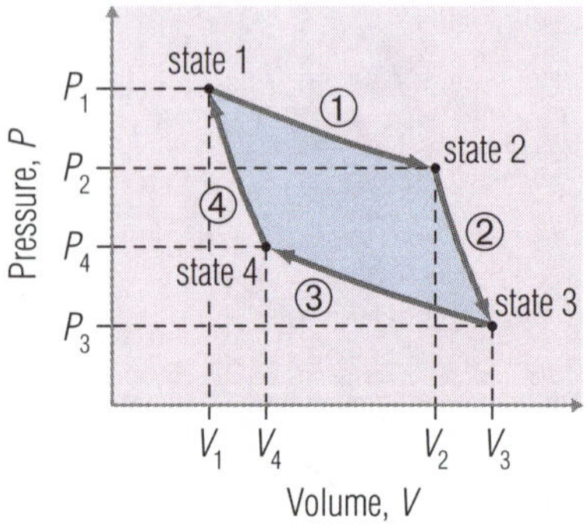

16-18 *P-V* diagram for the Carnot cycle

The work done by the system during the cycle can be found from the first law of thermodynamics:

$$Q_{cycle} = \Delta U_{cycle} + W_{cycle}$$

For a cycle, ΔU_{cycle} is zero.

$$Q_{cycle} = 0 \text{ J} + W_{cycle}$$

$$Q_{cycle} = W_{cycle}$$

Since the gas does work on its surroundings, W is positive. The sum total of the thermal energy exchanged is Q_{cycle}:

$$|Q_{cycle}| = |Q_H| - |Q_C|$$

The absolute value signs are useful, but not necessary, in order to avoid confusion with signs. By assuming that the gas heat engine does positive work, we assume that Q_H added to the system from the hot reservoir is positive and Q_C lost from the system to the cold reservoir is negative.

Substituting work for the net heat gained by the cycle,

$$|W_{cycle}| = |Q_H| - |Q_C|.$$

The curves that represent the isothermal steps 1 and 3 follow *P-V* curves called *isotherms* on the graph. There are potentially an infinite number of isotherms a process may follow. The isotherms followed by an ideal Carnot cycle are the temperatures of the hot and cold reservoirs.

The portion of the heat absorbed by the heat engine from the hot reservoir that is converted into useful work is a measure of the efficiency of the engine. The **thermal efficiency** (ε) of a Carnot engine is defined as

$$\varepsilon = \frac{|W_{\text{cycle}}|}{|Q_{\text{H}}|} \times 100\%, \text{ or}$$

$$\varepsilon = \frac{|Q_{\text{H}}| - |Q_{\text{C}}|}{|Q_{\text{H}}|} \times 100\%. \qquad (16.6)$$

The thermal efficiency (ε) of a Carnot engine is

$$\varepsilon = \frac{|Q_{\text{H}}| - |Q_{\text{C}}|}{|Q_{\text{H}}|} \times 100\%.$$

This result is true for any heat engine cycle that operates between two heat reservoirs. For the portions of the cycle involving isothermal processes at T_{H} and T_{C},

$$\frac{|Q_{\text{H}}| - |Q_{\text{C}}|}{|Q_{\text{H}}|} = \frac{T_{\text{H}} - T_{\text{C}}}{T_{\text{H}}}, \text{ or}$$

$$\varepsilon = \frac{T_{\text{H}} - T_{\text{C}}}{T_{\text{H}}} \times 100\%. \qquad (16.7)$$

You can increase the efficiency of a Carnot engine by raising the temperature of the hot reservoir, lowering the temperature of the cold reservoir, or doing both. However, even in theory, the engine cannot be 100% efficient. The efficiency of a real engine will be even lower than that of a theoretical cycle because of friction and other losses.

16.12 Heat Pumps

A heat engine that runs in the counterclockwise direction around the *P-V* diagram, with work added to move thermal energy from the cold reservoir to the hot reservoir, is called a **refrigeration engine** or **heat pump**. The Carnot cycle can be reversed to work as a heat pump. A Carnot heat pump would first adiabatically expand so that the gas's temperature drops below the cold reservoir temperature, then isothermally expand as heat is transferred from the cold reservoir to the heat pump, then adiabatically contract, raising the gas's temperature above the hot reservoir temperature, and finally contract isothermally while it discharges heat to the hot reservoir, thus returning to its original conditions. A refrigerator is a heat pump used to cool a volume. It uses work to move thermal energy from a cold place (inside the refrigerator) to a warm place (the room). An air conditioner is a kind of refrigerator. It cools a room by pumping thermal energy outside. That is the reason that some air conditioners are installed in windows. Heat pumps that function as air conditioners in the summer and as heaters in the winter (by pumping thermal energy from the outside into the room) are also available.

16-19 Domestic heat pumps transfer heat both in summer and winter from the cooler space to the warmer space.

The corresponding method for determining the efficiency of a refrigeration engine or heat pump is to determine its **performance coefficient (*K*)**. This coefficient is defined as

$$K = \frac{|Q_{\text{C}}|}{|W_{\text{cycle}}|},$$

where Q_{C} is the heat removed from the cold reservoir and W_{cycle} is the work done on the engine to move the heat in the unnatural direction from the colder temperature to the warmer. *K* is typically expressed as small numerical values (e.g., 3.00) rather than as a percentage.

16.13 Various Statements of the Second Law and the Third Law

It was stated earlier in this chapter that thermal energy naturally flows from hot bodies to cold bodies. This is one statement of the second law of thermodynamics, a law governing the direction of natural processes. Another statement of the second law is that thermal energy cannot be completely converted to work in a cyclic process. A cycle always loses some energy to a cold reservoir. No exceptions to the second law have been observed.

According to the second law, no engine can be completely efficient. Efficiency for an ideal heat engine is expressed by Equation 16.6,

$$\varepsilon = \frac{|Q_{\text{H}}| - |Q_{\text{C}}|}{|Q_{\text{H}}|} \times 100\%.$$

Statements of the second law:

- Heat flows from a place of higher temperature to one of lower temperature.
- Thermal energy cannot be completely converted to work in a cyclic process.
- No heat engine is 100% efficient.

If no energy were lost to the cold reservoir, Q_C would be zero, and efficiency would be 100%. Complete efficiency is ruled out by the second statement of the second law because some energy is always lost to the cold reservoir.

The ocean is a vast reservoir of thermal energy. Why can't we use this energy to drive a ship across the ocean? The ship could extract energy from the ocean to drive its engine, leaving the ocean a trifle colder than before. Unfortunately, the second law of thermodynamics says that thermal energy does not flow from a cold reservoir—like the ocean—to a hot reservoir—like the ship's engine—unless work is done to force this flow. The work done to extract energy from the ocean would be more than the work the energy could do, so the ship would go nowhere.

The first law says that when converting energy into work you can never get out of a transaction more than you put into it. The second law says that you cannot get as much work out of a machine as you put into it. Both laws forbid *perpetual motion machines*. The first law forbids perpetual motion machines of the first kind, which produce more energy than they consume. The second law forbids perpetual motion machines of the second kind, which are 100% efficient. Many inventors have tried to invent a perpetual motion machine; no one has succeeded.

A heat engine could be 100% efficient only if Q_C were zero:

$$\varepsilon = 100\% = \frac{|Q_H| - 0\text{ J}}{|Q_H|} \times 100\% \qquad (16.8)$$

This would mean that all the energy extracted from the hot reservoir was converted to work and none was discharged to the cold reservoir. Since the final legs of the cycle occur at T_C, this temperature would have to be the coldest possible temperature to be able to extract all possible work from the heat. The coldest possible temperature, absolute zero (0 K), has never been attained in the laboratory, though physicists have been able to approach it within a few billionths of a degree under extremely controlled artificial conditions. The coldest natural environment exists in distant interstellar space where the cosmic microwave background radiation indicates a blackbody temperature of 2.725 K.

Scientists are convinced that absolute zero cannot be reached in a particular system because all heat in it would have to be discharged to an even colder reservoir. This is the basis for the **third law of thermodynamics**, which states that absolute zero is unattainable.

The **third law of thermodynamics** states that absolute zero is unattainable. Therefore, cold reservoirs at $T_C = 0$ K are not possible.

16.14 Using Thermodynamic Processes to Solve Problems

If human health and commerce rely on cooler temperatures and low humidity, then removing heat and humidity from a location where they have adverse effects to a place where they don't would help solve the problem. Air conditioning is the obvious solution. But what can be done to make this process most efficient?

Throughout most of human history, passive methods for cooling buildings have been employed using natural air convection, pools of water, blocks of natural ice, and massive structures to provide thermal inertia. These techniques made living tolerable, but seasonal conditions still ultimately determined whether people actually felt cool. Methods for artificially creating ice began to be developed in the late 1700s. The development of true air chilling, however, began in the 1840s. The first person to effectively experiment with air chilling was a prominent medical doctor in Apalachicola, Florida, named **John Gorrie**. He noted that yellow fever arrived with hot weather and departed with cool, and he believed, as many people did at that time, that diseases originated from bad or foul air. So he concluded that cool hospital rooms would help patients recover from the disease.

John Gorrie (1802–55) was an American physician, inventor, and public servant. Born in Scotland, he spent most of his life serving the Apalachicola, Florida, community in many capacities. He was motivated to investigate air conditioning out of a humanitarian desire to help his patients.

In 1842, he succeeded in building a machine that operated on principles very similar to the modern refrigerator. Unfortunately, because of the rudimentary state of technology at that time, his machine could not produce the quantity of ice needed to cool entire rooms in a hospital.

Gorrie's invention set the stage for numerous innovations by many inventors across the world. Instead of compressed air as was used in Gorrie's unit, other fluids with greater heat capacities such as ammonia or carbon dioxide were used with increasing success and efficiency. Large public buildings such as theaters were the earliest beneficiaries of this emerging technology. Then, factory owners realized that productivity could be increased by keeping their workers comfortable.

Some businesses that relied on controlling humidity, such as the photographic film and printing industries, gained significant benefit from using true air conditioning units that controlled relative humidity as well as temperature. In the early 1900s, the American engineer Willis Carrier developed more advanced air conditioning systems and became a successful businessman. He focused on controlling the humidity of air. He sold his inventions to a variety of business clients. Though he was very successful in his work because of his personal expertise, he realized that for the air conditioning industry to become viable, it needed to be based on principles and formulas that could predict the function of a system before it was built. After extensive research and experimentation, Carrier presented a paper in 1911 in which he developed mathematical models to describe the relationship between temperature and humidity, ushering in the science of psychrometrics. Carrier continued to improve his inventions, replacing the piston gas compressors with centrifugal units, and he was the first to use chlorofluorocarbons like DuPont's Freon as the refrigerant. (Freon is a registered trademark of E.I. du Pont de Nemours and Company.) These innovations led to the development of affordable whole-house and room air conditioning units. By the late 1930s, the first car air conditioners were being installed.

16-20 A replica of John Gorrie's ice machine

Psychrometrics or psychrometry is the science of the thermodynamics of gas/liquid mixtures.

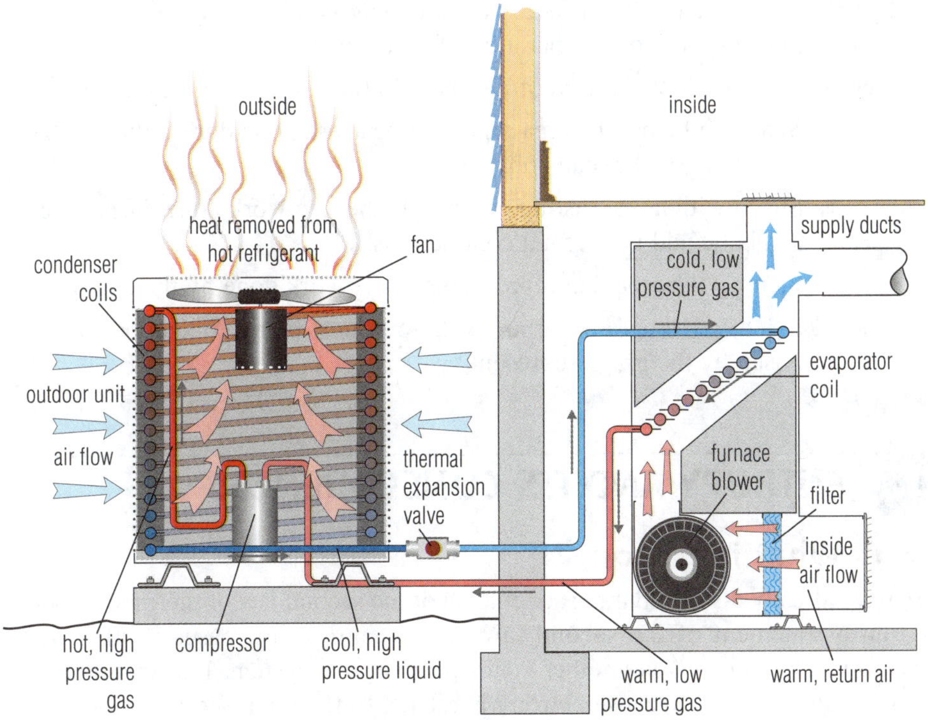

16-21 Modern air conditioning units compress refrigerant gas to a liquid, cool the liquid, then allow it to expand to a cool gas, which circulates through coils. Fans blow air through the cooled coils, cooling the air.

American society and culture were subtly and gradually revolutionized by this invention. Sections of the American South and West that were sparsely inhabited saw rapid population growth as the direct result of air conditioning. Today, nearly all cars sold in the United States are air conditioned, and in many places, air conditioning is considered a necessity rather than the luxury it once was.

16B Section Review

1. What are the essential parts of a thermodynamic heat engine?
2. a. What did the Savery engine lack that the Newcomen and Watt engines had?
   b. What was the major improvement of the Watt engine over the Newcomen engine?
3. What two basic reversible thermodynamic processes occur at various points in a Carnot cycle?
4. A steamship's propulsion plant uses the ocean water to absorb its waste heat according to the Carnot cycle. What happens to the propulsion plant's efficiency if the seawater temperature becomes warmer (as in the tropics)?
5. a. What condition would have to exist for an ideal heat engine to be 100% efficient?
   b. What principle states that this condition is not possible?
6. a. Why does your car engine have a radiator?
   b. What principle predicts that a gasoline engine would need one?
7. DS When a gas is compressed (adiabatically) inside a cylinder by a piston, what happens to its thermal energy and temperature? If the high-pressure, compressed gas is cooled, what will happen to its temperature when it is allowed to expand? What can you do with such a process?
8. DS Describe the chilling process within an air conditioning unit. How does this correspond to the operation of a heat pump?
9. ✪ a. Sketch a *P-V* diagram of a typical Carnot cycle for a heat engine.
   b. Shade and label the area representing work done during the adiabatic and isothermal expansions.
   c. Shade with a contrasting color and label the work done during the adiabatic and isothermal contractions.
   d. Identify the area that represents the net work done by the engine.
10. ✪ Sketch a *P-V* diagram Carnot cycle for a heat pump. What does the area enclosed by the graph represent?

16B Objectives

After completing this section, I can

- ✓ describe how the function of heat engines relies on the second law of thermodynamics.
- ✓ briefly describe the first successful variants of the steam engine.
- ✓ trace the path of a heat engine through a single pass of the Carnot cycle.
- ✓ compute the thermal efficiency of a heat engine.
- ✓ contrast the difference in operation between a heat engine and a heat pump.
- ✓ state the second law of thermodynamics in at least three ways.
- ✓ state the third law of thermodynamics and explain why it must be true.
- ✓ discuss the initial motivation for developing a room- or building-sized air conditioning machine.
- ✓ describe several ways that the air conditioner has changed American culture.

16C ENTROPY AND ITS CONSEQUENCES

16.15 What Is Entropy?

We have already discussed three statements of the second law of thermodynamics. A fourth statement of the second law is that entropy increases in all natural processes. **Entropy** (*S*) is another state variable of a system, like volume, temperature, or pressure. Because entropy is related to the microscopic properties of the particles of a system, it is similar in some ways to internal energy. Just as with internal energy, the *change* of entropy (ΔS) is more important and more easily

The second law of thermodynamics also states that entropy increases in all natural processes measured than the system's entropy at any particular state of thermodynamic equilibrium.

Some define entropy as a measurement of the randomness, or disorder, of the particles in a specific part of the universe. The second law says that nearly all *natural* processes make the universe more disorderly. Disorder implies unusable energy, such as the thermal energy rejected by a heat engine. The energy still exists, but it can no longer do useful work in a particular process. Mathematically, entropy for a reversible process is defined as

$$\Delta S \equiv \frac{\Delta Q}{T}, \tag{16.9}$$

where ΔQ is the change of the system's thermal energy at isothermal conditions and T is the absolute temperature of the system. The units for entropy are J/K. For quasi-static processes, the changing quantities in Equation 16.9 experience infinitesimally small changes at each step.

In reversible processes, the entropy of the universe remains constant. The entropy change in the system is exactly balanced by the entropy change in the surroundings:

$$\Delta S_{\text{system}} = -\Delta S_{\text{surroundings}}$$

In *natural* processes—irreversible processes—entropy increases (ΔS is positive). There may be local decreases in entropy for the system, but these are more than offset by an increase in entropy of the surroundings. Therefore, the entropy of the universe increases.

The entropy of a system is a state variable because the entropy change depends only on the initial and final states of the system. Therefore, the entropy change for a reversible process between two states is the same as the entropy change for an irreversible process between the same two states. Entropy calculations for a free expansion of a gas use this principle.

In a free expansion, an ideal gas is allowed to expand adiabatically into a vacuum. The gas originally occupies a volume of V at a temperature T. It expands into a vacuum also of volume V, for a final volume of $2V$. This is an irreversible process. Since the gas is adiabatically isolated, there is no heat exchanged with its surroundings. Because the gas is expanding into a vacuum, it does no work on its surroundings, so $W = 0$ J as well. Therefore, according to the first law, internal energy does not change either. The temperature remains constant at T throughout the process. A direct entropy calculation for this process using Equation 16.9 is not possible, because that equation applies only to reversible processes. However, you know that entropy has increased during the process because the particles of the gas are farther apart (less ordered), even if they don't have more energy.

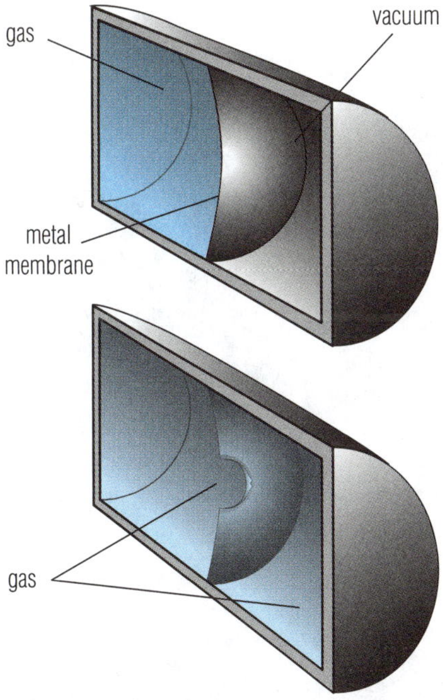

16-22 A gas trapped by a membrane expands rapidly into the evacuated volume of the vessel when the membrane ruptures.

The entropy change for the system in the free expansion is the same as that for the system during the isothermal (reversible) process. However, since the free expansion is adiabatic, the system has no contact with its surroundings, so the entropy of the surroundings does not change. The total entropy change of the gas is greater than zero.

In the kinetic theory, entropy is a measure of the likelihood that a system will be in a particular state. Because of the second law, a statistically more probable state for a system has a high entropy, and a statistically unlikely state has low entropy. The most likely state of a system is one of disorder. An example of disorder is diffusion of a dye throughout a beaker of distilled water. Even if the system is maintained at isothermal conditions, the random motion of the particles will eventually distribute the dye throughout the water. The dye particle distribution is

Statistically likely states of a system subject to natural processes have high entropies, while statistically improbable states have low entropies.

more spread out and random with time compared to the original arrangement of particles in a solid crystal of the dye. This randomness is the natural consequence of diffusion, so the entropy of the dispersed dye is greater than its entropy as a crystal. The orderliness of an isolated system tends to naturally deteriorate to maximum disorder. Another way of stating this result is to say that the entropy of an isolated system is at its maximum.

Entropy can also be related to the encoding of information. Information with high entropy contains meaningless pieces, which in communications are called *static*. Information with low entropy is exact, with few meaningless parts (e.g., when a message is communicated clearly from one place to another). Experts try to minimize the static and keep the message's entropy low. Another example of highly ordered information is the biological code stored in our genes. Entropy in this genetic information is a measure of its randomness. Because this information is highly ordered, it has low entropy. A random change, known as a mutation, destroys some information and harms a body. Such a change increases the organism's genetic entropy. When an organism dies, the cellular processes that maintain the DNA code in good, meaningful condition no longer function, and the natural random destruction of the molecules eventually erases the genetic information. Organized infusion of energy is required to maintain low entropy.

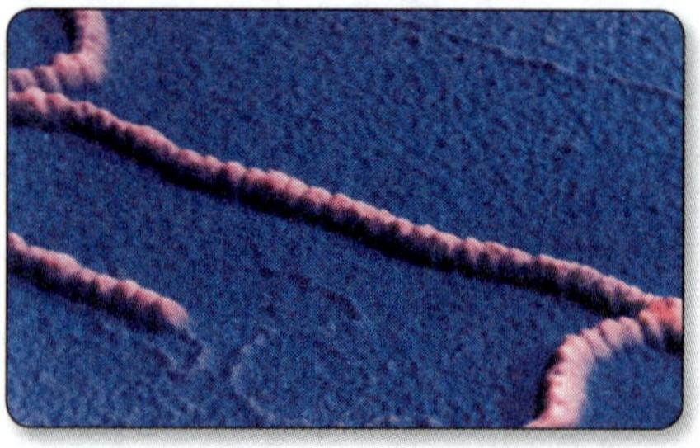

16-23 The immense amount of information stored in DNA is maintained by complex cellular mechanisms that require energy.

16.16 Conservation and Degeneration in Nature

You may wonder why the universe still has energy available for work if the second law is true. The answer is that God created the universe with an immense supply of usable energy. Obviously, the laws of thermodynamics were not valid during Creation. The first law is a result of the fact that only God can create or destroy. The only exceptions to this law are divine actions. Since the universe began with relatively little entropy, it will be a long time before it reaches a state of maximum entropy.

The first law says that energy and mass are conserved. If something is to be conserved, it must exist. Some scientists believe that the physical laws we know now have *always* been in effect. This belief is consistent with uniformitarianism. Uniformitarian scientists must believe either that matter does not now exist or that it has always existed. Many uniformitarians believe that matter is eternal. Some believe that before this universe was formed, matter existed in the form of subatomic particles. These particles eventually became compacted and then exploded (the so-called big bang). With the energy from the explosion, the particles united to form atoms. Eventually, the early universe developed to its present state. The big bang is only one of the theories of naturalistic evolutionary cosmology.

The first law, by itself, neither supports nor refutes this theory of evolution. However, assuming that the primordial universe, highly compacted and ordered, exploded and expanded in an irreversible process, the universe's entropy would have had to rapidly increase during the early period of its supposed expansion. Furthermore, in order for galaxies, stars, and planets to form, a *decrease* of entropy is required. Scientists have never been able to identify any natural ordering process that would tend to gather particles of dust together into these celestial structures. Such an ordering process could be accomplished only through an intelligent infusion of energy. Therefore, naturalistic evolutionary cosmology is thermodynamically impossible.

Evolutionists point out that, according to kinetic theory, the second law is based on probability. Although unlikely, they say, a violation of the second law is possible. Given enough time—ten billion years or so—a violation is almost inevitable. It is true that the kinetic theory suggests that an entropy decrease is possible. However, there are two reasons that evolutionists are in error when they say that the second law allows evolution. First, no violation of the second law has ever been observed. The second law is based on observations. Until a violation of it is observed, the idea that such a violation can occur is only an idea. Second, the theory of evolution requires not *one* decrease in entropy but almost *continual* decreases in entropy, whose effects are preserved by the system. If the second law can be ignored as many times as evolution requires, it is not an accurate reflection of nature. Evolution implies that there is a tendency for a system to become more ordered, but the second law says that the tendency is for a system to become more disordered. The theory of evolution directly contradicts the second law.

The second law states that energy is becoming less available for use. Eventually, if the universe is left to itself, all energy will be unavailable for use. The universe will be at a uniform low temperature. Nothing will be able to live. This effect is called the **heat death** of the universe. This hypothetical heat death is far in the future because there is still much energy available for use. Actually, the heat death will never happen, because the universe will not be left to itself. The Lord will return and replace this universe with a new one. With Him ruling and sustaining the new universe, you will not have to worry about exhausting the available energy.

16C Section Review

1. What property of a system does entropy describe?
2. According to the second law of thermodynamics, what happens to the entropy of a system subject to natural processes?
3. A mixture of ice and water is contained in a near-perfect adiabatic container. As the entropy of the melting ice increases, what must happen to the entropy of the water surrounding it?
4. Briefly discuss how the free expansion of a gas in an adiabatic container increases the entropy of the gas.
5. a. What would be the likelihood of all the oxygen in your classroom clumping together in a 1 L volume in one corner?
   b. If it did, would the entropy of the gas be higher or lower than is normally the case?
6. The belief that matter is eternal is often a tenet of what system of philosophy?
7. As one follows the progress of total entropy in the expanding universe after the supposed big bang, what direct violation of the second law of thermodynamics must have occurred for the universe to have acquired its present form?
8. What is the flaw in some evolutionists' logic when they appeal to the statistical nature of entropy over immense periods of time as a justification for evolution?

16C Objectives

After completing this section, I can

✓ define entropy in several ways.

✓ explain how entropy changes for natural irreversible processes.

✓ describe how entropy applies to the probability of certain events occurring and to information transmission.

✓ discuss how the conservation laws and entropy are strong evidences against naturalistic evolutionary hypotheses.

FACETS of PHYSICS

Growth and the Second Law

The universe is filled with countless galaxies containing billions of stars each.

The concept of "heat death"—a cold, lifeless universe—is both frightening and hard to accept. Man is naturally repelled by the second law of thermodynamics, a law that implies that all things tend to break down and that the present universe is forever losing usable energy that can never be restored. Common sense seems to indicate that the law is not true. After all, with the coming of each winter, the warm breezes of spring are sure to follow. Also, human culture is not becoming less complex but is becoming increasingly more complex; each year humans invent new and more complicated technologies. Every day in nature, photosynthesis converts sunlight into complex organic matter that animals transform into more complex tissues such as nerves and brain cells.

However, even before 1824 when Sadi Carnot first described the principles on which steam engines operate, scientists instinctively knew that systems tended to become disorganized and energy seemed to flow in preferred directions. Even the Bible teaches the idea that things and life itself deteriorate with time. "The earth shall wax old like a garment, and they that dwell therein shall die in like manner" (Isa. 51:6). "For the creature was made subject to vanity. . . . The whole creation groaneth and travaileth in pain together until now" (Rom. 8:20, 22). "All go unto one place; all are of the dust, and all turn to dust again" (Eccles. 3:20). Such impressions of the natural state of things were first formulated into the second law of thermodynamics by the German physicist Rudolf Clausius in 1850. The law's implications were controversial because they contradicted the caloric theory of heat, to which many scientists still held. But the work of contemporary scientists such as Joule and Kelvin eventually affirmed the validity of this model of nature.

The most apparent contradiction of the law of entropy is the growth of plants and animals from single cells into complex life forms. For example, acorn kernels expand into mighty one-hundred-foot oak trees. Even more amazing than oak trees is a living human being. One microscopic cell subdivides and specializes until in nine months there appears the most complex collection of matter in the universe. Within a few years, this mass of cells can think, speak, and commune intelligently with God. Social Darwinists use human progress to prove that man is evolving into a higher state and becoming perfect. But how does the second law refute their arguments, and how does it explain the growth of complex organisms?

When you consider the growth of cells, particularly of human cells, you should realize that adult organisms are no more complex than the initial fertilized egg. Every major cell function takes place in the first cell, and the first cell's DNA contains all the information needed by future cells. A baby's DNA is only as complex as his parents' DNA, and the parents' DNA ultimately came from Adam. Again, we see that God intervened, in this case forming a complex body "of the dust of the ground" (Gen. 2:7).

Human "progress" after Adam strongly supports the second law of thermodynamics. Adam's healthy body lasted 930 years before it returned to dust, but the longevity of his descendants decreased markedly after the Flood, until Moses said of his generation, "The days of our years are threescore years and ten" (Ps. 90:10). Adam's mind needed only a portion of one day to register and name every kind of beast of the field and fowl of the air.

Thermodynamics considers the result of natural processes. What is a man's life? "It is even a vapour, that appeareth for a little time, and then vanisheth away" (James 4:14). The human body shows no evidence of evolving; it merely battles to preserve the life that it was given for an appointed time. During that brief period, the body consumes massive amounts of energy to maintain itself; yet eventually it dies and decays. This process fits perfectly into the laws of thermodynamics.

Fortunately, the second law of thermodynamics is not eternal, so "heat death" will never occur. Even though God cursed the earth and the human body after Adam's sin, He promised that a Messiah would come Who would one day restore the human body and replace this universe with a new heaven and a new Earth. The creation, being "made subject to vanity," was given hope that it would "be delivered from the bondage of corruption into the glorious liberty of the children of God" (Rom. 8:20, 21). The earth shall again be a place where "there shall be no more death, neither sorrow, nor crying, neither shall there be any more pain: for the former things are passed away" (Rev. 21:4).

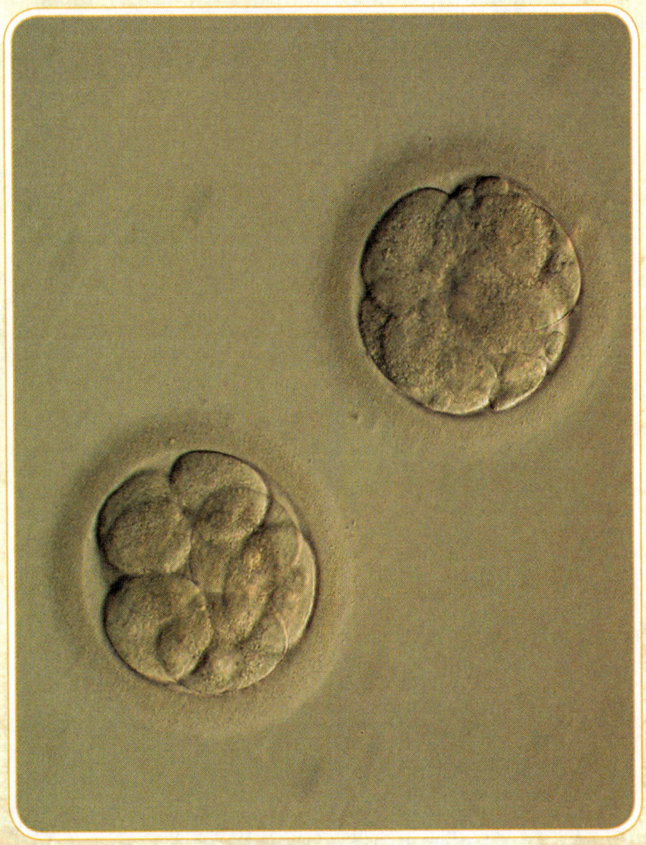

Even human embryos obey the second law of thermodynamics.

Chapter Review

In Terms of Physics

internal energy (U)	360
diathermic	360
thermal equilibrium	360
zeroth law of thermodynamics	360
first law of thermodynamics	361
heat engine	361
P-V diagram	362
pascal (Pa)	362
state variable	364
thermodynamic system	365
open system	365
closed system	365
isolated system	365
isothermal	365
isochoric	366
isobaric	366
ideal gas process	366
heat reservoir	367
second law of thermodynamics	367
reversible process	369
Carnot cycle	370
thermal efficiency (ε)	371
refrigeration engine	371
heat pump	371
performance coefficient (K)	371
third law of thermodynamics	372
entropy (S)	374
heat death	377

Problem-Solving Strategies

16.1 (page 364) If a cycle proceeds clockwise on a *P-V* diagram, then the system is a heat engine doing positive work. If the cycle proceeds counterclockwise, then work is being done on the system (as in a refrigerator).

16.2 (page 364) If a process moves from left to right along any path on a *P-V* diagram, then the system is doing work. If the process moves from right to left along any path, then work is being done on the system.

Review Questions

1. What thermodynamic state variable is dependent on the temperature of the system?
2. What is the zeroth law of thermodynamics?
3. Discuss the two ways that energy can be exchanged between a system and its surroundings.
4. State the first law of thermodynamics. What principle of mechanics does it extend?
5. Define *heat engine*. Give an example.
6. Describe the three kinds of thermodynamic systems and give an example of each.
7. Classify each of the following systems as open, closed, or isolated.
   - **a.** flour in a sifter
   - **b.** ice cube in an adiabatic calorimeter
   - **c.** a living human being
   - **d.** hot coffee in a mug in a cold room
   - **e.** a leakproof steam engine
8. Why does a heat engine require a fluid to operate?
9. What advantage did the Newcomen engine have over the Savery engine?
10. As a heat engine completes a cycle, how does its final internal energy compare to its starting internal energy?
11. Discuss two ways to thermodynamically (rather than mechanically) improve the efficiency of a Carnot heat engine.
12. What law gives the reason that a heat engine cannot be 100% efficient?
13. Which of the following is *not* a statement of the second law?
    - **a.** Thermal energy can flow from cold to hot if work is done.
    - **b.** Thermal energy cannot be converted entirely to work in a cyclic process.
    - **c.** The entropy of an isolated system tends to decrease.
    - **d.** Every natural process directly or indirectly makes the universe more disorderly.
14. How does the theory of evolution contradict the second law of thermodynamics?
15. Discuss several examples which illustrate that it is necessary to do work on a system in order to prevent an increase of entropy.
16. Discuss why the heat death of the universe will not occur.

True or False (17–32)

17. Mechanical kinetic and potential energies are properties of the macroscopic system, but internal energy is a property arising from the microscopic particles of the system.
18. The zeroth law of thermodynamics establishes the "transitive property of thermal equilibrium"—if A and B are both in thermal equilibrium with C, then they are in equilibrium with each other.
19. Fiberglass insulation approximates a good diathermic material.
20. In order to simplify the statement of the first law of thermodynamics, it is assumed that the total mechanical energy of the system does not change during a thermodynamic process.
21. A quasi-static process proceeds smoothly without any steps or stops.

22. Thermodynamics concerns the work done on the system more than the work the system does on its surroundings.

23. The work done during a thermodynamic process graphed on a *P-V* diagram does not depend on the path the process takes.

24. A steaming cup of coffee on a tabletop is an open system.

25. Pushing down on the plunger of a tire air pump causes the air in the pump to undergo an isochoric process.

26. Engines that heat the working fluid only once and then discharge it have low efficiencies compared to those that heat the working fluid over and over again.

27. James Watt invented the first practical steam-powered engine.

28. A Carnot engine is more efficient if the temperature difference between the hot and cold reservoirs is increased.

29. A heat engine would be 100% efficient if it discharged no heat to the cold reservoir.

30. Entropy is a form of energy.

31. As with some other state variables, it is simpler to determine the change of entropy than to measure it at a given state.

32. The fact that your bedroom gets messier with time is an example of the second law of thermodynamics in action.

33. A pressure of 1.52×10^5 Pa forces a gas to compress from a volume of 5.00×10^{-3} m^3 to a volume of 2.50×10^{-3} m^3. How much work is done on the gas, in N·m?

34. A gas in a cylinder with a cross-sectional area of 60.0 cm^2 holds the piston at a height of 5.00 cm when a force of 18.0 N is applied. Then, under the same force, the gas expands in a quasi-static process to a height of 7.50 cm.
  a. What is the gas's original volume, in m^3?
  b. What is the original pressure on the gas in N/m^2 (Pa)?
  c. What is the final volume of the gas, in m^3?
  d. How much work does the gas do, in J?

35. A gas expands as Figure (a) shows.
  a. How much work does the system do in step 1?
  b. How much work does it do in step 2?
  c. How much work does it do overall?

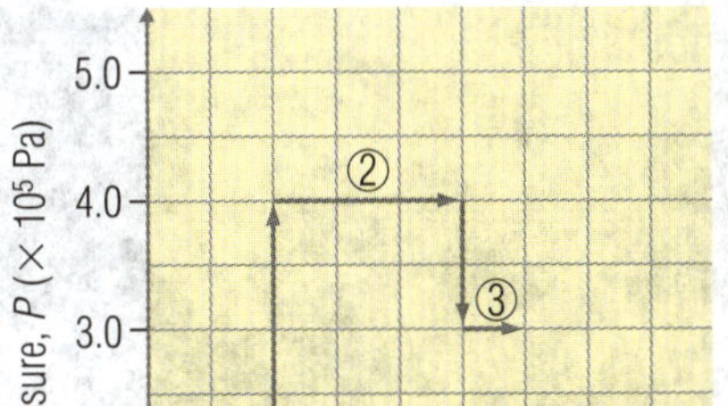

(a)

36. Find the work done by a gas following one cycle as shown in Figure (b) for
  a. each step.
  b. the entire cycle.

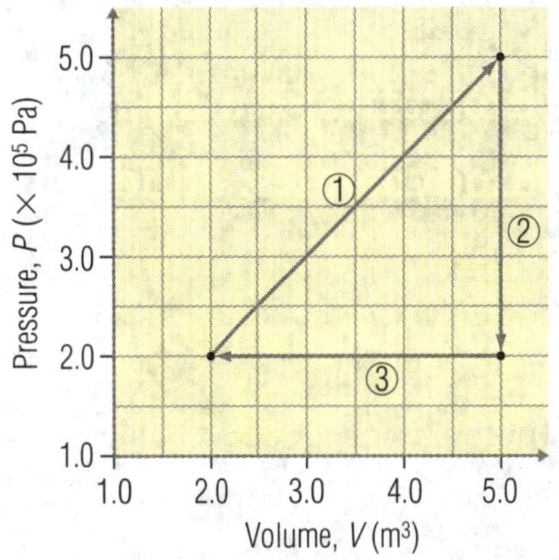

(b)

37. A Carnot engine has a hot reservoir at a temperature of 490 K and a cold reservoir at a temperature of 290 K. What is its efficiency?

38. If the hot reservoir of a Carnot engine is at room temperature (24 °C), what must be the temperature of the cold reservoir in Celsius for the engine to have an efficiency of 50%?

39. If a ship's steam propulsion boilers produce 1.20 MW of thermal energy at full power, and the propulsion plant is 15% efficient, how much power is discharged to the sea via the seawater cooling system?

40. A gas expands and contracts in the cycle shown in Figure (c). How much work does each part of the cycle do? ($A_{\bigcirc} = \pi r^2$)

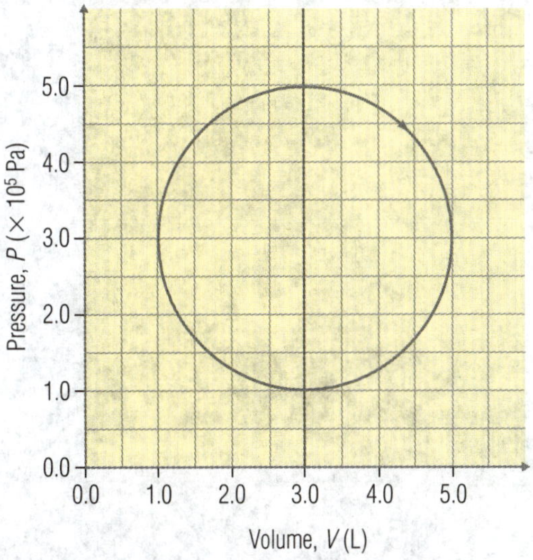

(c)

DS **41.** Do research and write a brief paper discussing how air conditioning has changed American culture. Include in your discussion an assessment of the net benefits as positive or negative. How important is or should be anticipating the cultural consequences of new technology during the development of that technology?

Water in motion can be put to useful work in hydroelectric dam generators. This view is of a penstock discharge at Hoover Dam.

Fluid Mechanics 17

Dominion Science Problem

Homespun Energy

Nearly 70% of the domestic power generation needs of America are met using fossil fuels. Almost half of the fuel is coal, and most of the rest is natural gas. Though both energy sources are fairly economical, natural gas is slightly cheaper than coal and is now preferred for new power plant construction and upgrades. Very little electrical power is supplied by oil, which has never been cheap and has experienced significant jumps in price recently. Nuclear energy meets nearly 20% of the nation's energy demands, but its cost to the consumer is two to three times as much as that of natural gas. People still oppose nuclear power plants because they seem inherently dangerous and produce long-lived radioactive wastes. Hydroelectric power spans coal and nuclear power in cost, but its availability is limited by geography. Also, it is uncertain whether any new large power dams will be built in the United States. What safe, plentiful, and renewable alternative energy source could supplement or even replace the costly, nonrenewable ones in use today?

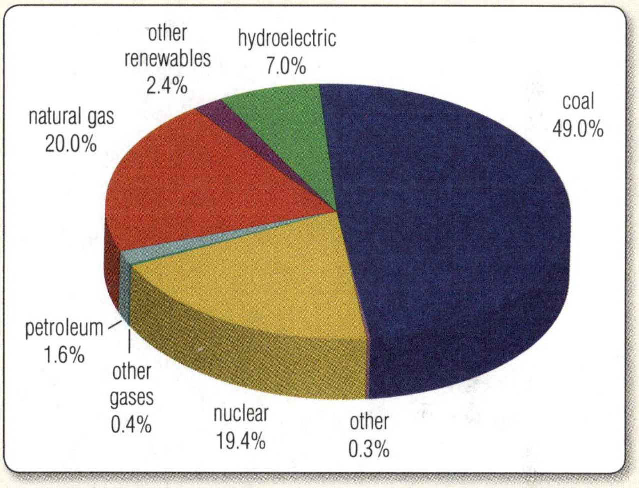

17-1 Relative contribution of power generation sources in the United States for 2006

Source: Energy Information Administration. *Electric Power Annual 2006.*

17A HYDROSTATICS: FLUIDS AT REST

17.1 The Science of Fluids

Objects can be immersed in fluids, and fluids can pass through objects. Fluids can flow around objects and structures, as well as flow through pipes and orifices. For these reasons, fluids exhibit some interesting mechanical properties that have great significance to the proper functioning of life processes as well as to strictly physical phenomena. In earlier chapters we discussed nonfluid mechanics—the mechanics of rigid objects. In this chapter you will study how Newtonian principles apply to fluids, a field called **fluid mechanics**. This area of mechanics can be broadly subdivided into **hydrostatics**—the study of stationary fluids in which all forces are in equilibrium—and **hydrodynamics**—the study of fluids in motion.

An orifice is a small opening in a barrier between two volumes through which a fluid may flow.

Recall from earlier chapters that fluids were defined as any matter that flows. Such substances include liquids, gases, and even some solid-looking matter such as the ice in glaciers.

17.2 Density

A major difference between the study of nonfluids and the study of fluids is that fluids are continuous. If you pour twenty cupfuls of water into a bucket, the water will assume one unified structure, but if you place twenty wooden blocks in a bucket, they will retain their own identities. Therefore, it is impractical to talk about the water and the wood in the same way. One consequence of this difference is that the mass of a fluid sample is less useful than the mass of a solid. The **density (ρ)** (mass per unit volume) of a fluid is usually more important in a physics context. Table 17-1 lists the densities of some common materials.

The Greek letter rho (ρ) represents **density**.

TABLE 17-1

Representative Densities (Liquids in Color)

Substance	Density (g/cm^3)
Hydrogen	0.00009
Nitrogen	0.00126
Air (dry)	0.00129
Oxygen	0.00143
Wood (balsa)	0.12
Wood (cork)	0.24
Wood (pine)	0.50
Gasoline	0.68
Wood (oak)	0.72
Ethyl alcohol	0.79
Oil (crude)	0.88
Ice	0.92
Oil (olive)	0.92
Water	1.00
Seawater	1.03
Blood (whole)	1.05
Table salt	2.17
Sand	2.32
Glass	2.40–2.80
Aluminum	2.70
Bromine	3.12
Iron/Steel	7.86
Copper	8.92
Silver	10.50
Lead	11.30
Mercury	13.60
Gold	19.30
Platinum	21.45

The densities in Table 17-1 are given in grams per cubic centimeter (g/cm^3), the common unit for density. The SI unit of density is kg/m^3. Water's density is 1000 kg/m^3. Since water's density is not numerically 1 in the SI, the densities of other substances in these units do not indicate their relative densities directly as they do in the common unit. However, if the densities of other substances are divided by the density of water, then we obtain a relative density called the **specific gravity (s.g.)**. The specific gravity of a substance is a dimensionless number that is numerically equal to the density of the substance in g/cm^3.

The **specific gravity (s.g.)** of a substance is the density of the substance divided by the density of water.

17.3 Units of Pressure

We have mentioned the concept of pressure in several contexts in earlier chapters. Fluid pressure is a distinctive property of liquids and gases. Perhaps you have experienced a change of pressure in your sinuses during a rapid descent or ascent on a mountain road or in an airplane. Skin and scuba divers do not have to go very deep before they experience significant pressure effects. How can we characterize pressure in a fluid?

Pressure is defined as the force exerted perpendicular to a unit area. If you select a small hypothetical planar area dividing two regions of fluid anywhere within the volume of the fluid (see Figure 17-2), the force exerted by the fluid on one side of the planar area is equal in magnitude but opposite to the force exerted by the fluid on the other side of the area. The two forces acting in opposite directions on different parts of the fluid form an action-reaction force pair and cancel each other according to Newton's third law. Consequently, at every point in a sample of fluid at rest the pressure at that point is equal in all directions. At the boundaries of the fluid, the container exerts a pressure on the fluid identical to the pressure the fluid exerts on the container.

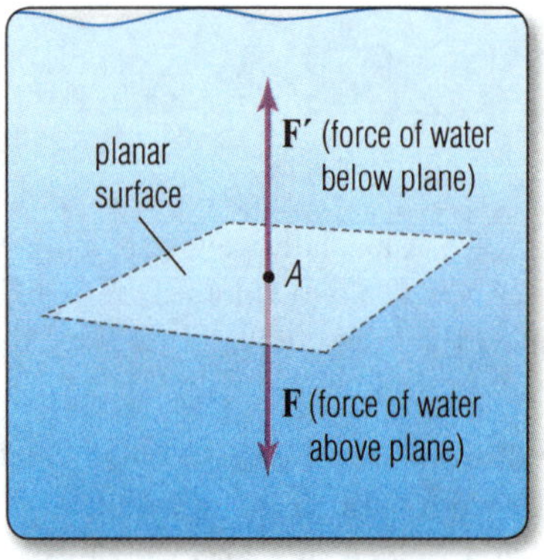

17-2 The pressure on one side of a hypothetical plane surface is exactly equal to the pressure on the opposite side when the fluid is in mechanical equilibrium.

The SI unit of pressure is the pascal (Pa). Some other units of pressure are commonly used by different professions. The **atmosphere (atm)** was probably the earliest reference pressure. It is the average pressure of the atmosphere at sea level. In pascals,

$$1 \text{ atm} = 1.013 \times 10^5 \text{ Pa}.$$

While this is the accepted average value for atmospheric pressure, the actual value at a given location will vary due to factors such as elevation, weather, and temperature.

Table 17-2

Densities and Specific Gravities for Some Selected Substances

Substance	Density g/cm³	kg/m³	Specific Gravity
Air	0.001 29	1.29	0.001 29
Ethyl alcohol	0.79	791	0.791
Water	1.00	1000	1.00
Iron	7.86	7860	7.86
Mercury	13.60	13 600	13.60
Gold	19.30	19 300	19.30

Chemists and other scientists often use the **torr**, named after Evangelista Torricelli, the inventor of the mercury barometer. This non-SI pressure unit is defined as $1/760$ of the pressure exerted by a column of mercury 760 mm high at sea level. This value was chosen because one atmosphere of pressure will support a 760 mm column of mercury in an evacuated glass tube, which was the basis for the first barometer. It should not be surprising that pressure may also be measured in mm of mercury or even inches of mercury. Atmospheric pressure is often reported by the popular media in these units.

Meteorologists are more likely to report atmospheric pressure in non-SI units called **bars** or **millibars (mb)**. A bar is 10^5 Pa, so

$$1 \text{ atm} = 1.013 \text{ bar} = 1013 \text{ mb}.$$

Millibars permit the convenient reporting of atmospheric pressures in whole numbers (or with only a single decimal place). A list of pressures equivalent to 1 atm, reported in various SI and non-SI units, is provided in Appendix C.

Engineers often report pressures in piping systems in **gauge pressure** (P_g). Mechanical pressure gauges are designed so that one side of the pressure-detecting mechanism senses atmospheric pressure, and the other side senses system pressure. When the gauge indicates zero, it still experiences 1 atm. Therefore, the actual pressure, or **absolute pressure** (P), in the system is the system gauge pressure plus 1 atm.

Some news services and other sources of weather reports use inches of mercury for atmospheric pressure. 1 atm = 29.92 in. Hg.

The torr has no unit abbreviation. Some references give the unit for the torr as 1 torr = 1 mm Hg.

Evangelista Torricelli (1608–47) was an Italian mathematician and physicist. His many works included deriving principles in plane and solid geometry, describing ballistic motion (with Galileo), and demonstrating that a vacuum could be formed.

Most problems in this textbook will involve absolute pressure. However, in everyday life you will more likely read gauge pressure. The common system pressure unit in the United States is pounds per square inch (gauge), or psig.

17.4 Pressure in Incompressible Fluids

We have seen that pressure at a point in a fluid is the same in all directions. But we also know that pressure varies with the vertical position in the fluid. How does this occur? Let's first look at how pressure changes with depth in a liquid.

The particles in liquids are very close together, so they cannot be forced much closer together than they already are. For this reason, most liquids are essentially incompressible to a first-order approximation, so their densities can be assumed to be constant throughout the bulk of the liquid.

Let's choose a vertical coordinate system with the origin at the surface of the liquid so that $y_1 = 0$ m and positive is upward. Consider a small volume of the liquid, a thin rectangular slab with a square horizontal face of area A and thickness Δd. The top of the slab is at depth $y = d_1$ (a negative value) beneath the surface of the liquid. The liquid has a density of ρ, which is constant throughout the liquid. The pressure at the surface of the liquid is P_{ref}. The pressure at depth d_1 is P_{d_1}. The pressure of the liquid at the bottom of the slab is P_{d_2}, where the depth is

$$y = d_2 = d_1 + \Delta d.$$

(Note that Δd is a change of depth in the $-y$ direction.)

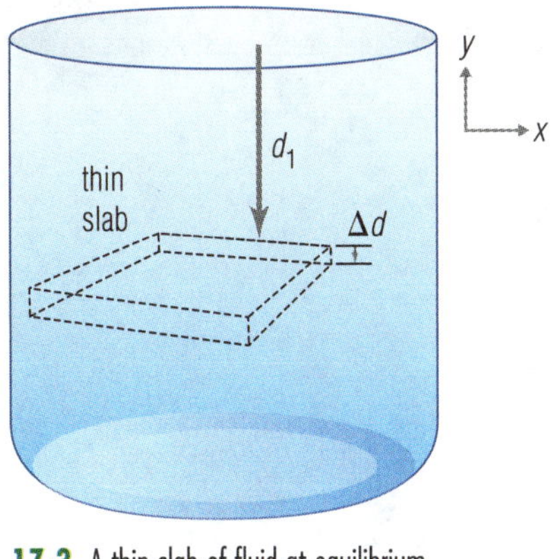

17-3 A thin slab of fluid at equilibrium

Assuming that the liquid is not in motion, then the slab is not moving, which means that the net force on the slab of liquid (the system) is zero. According to Newton's first law,

$$\Sigma \mathbf{F} = 0 \text{ N}.$$

Recall that

$$P = \frac{F}{A}, \text{ so}$$

$$F = PA.$$

The forces on the equal areas of the opposing sides of the slab are opposite and equal to each other, so they cancel and do not need to be considered further. The vertical forces acting on the slab of liquid must be in equilibrium as well. The downward force on the upper surface at depth d_1 is

$$\mathbf{F}_{d_1} = -P_{d_1}A, \tag{17.1}$$

and the upward force on the lower surface of the slab at depth d is

$$\mathbf{F}_{d_2} = +P_{d_2}A. \tag{17.2}$$

There is one more vertical force to consider—the weight of the liquid in the slab itself,

$$\mathbf{F}_{\text{w}} = m\mathbf{g}.$$

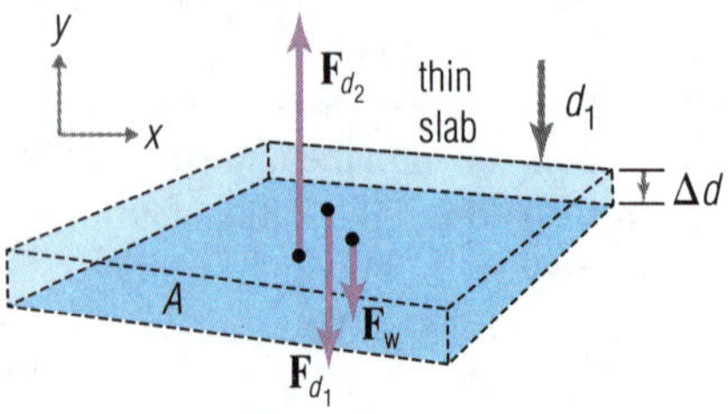

17-4 Forces on the hypothetical slab of liquid

The mass of the liquid is determined by the product of the liquid's density and its volume:

$$m = \rho V$$

But the volume of the slab is the product of the area and the magnitude of the change in depth (Δd is negative):

$$V = A|\Delta d| = A(-\Delta d)$$

So, the weight of the mass of liquid is the expression

$$\mathbf{F}_{\text{w}} = \rho A\mathbf{g}(-\Delta d) = -\rho A\mathbf{g}\Delta d, \tag{17.3}$$

which is a downward force on the slab of liquid. Referring the forces to the vertical coordinate system where upward is positive, we have the force sum in the vertical direction.

$$\Sigma \mathbf{F}_y = \mathbf{F}_{d_1} + \mathbf{F}_{d_2} + \mathbf{F}_{\text{w}} = 0 \text{ N} \tag{17.4}$$

Substituting Equations 17.1, 17.2, and 17.3 for each force in Equation 17.4, we have

$$(-P_{d_1}A)_y + (P_{d_2}A)_y + (-\rho Ag\Delta d)_y = 0 \text{ N}.$$

This is a one-dimensional problem, so we can dispense with the vector component notation. Setting the pressure terms equal to the weight term yields

$$-P_{d_1}A + P_{d_2}A = \rho Ag\Delta d.$$

Dividing both sides by the area A and rearranging the left side yields

$$P_{d_2} - P_{d_1} = \rho g\Delta d. \tag{17.5}$$

Equation 17.5 indicates that the change of pressure from the top to the bottom of any slab of a liquid is a function of only the change of the depth, since liquid density and gravitational acceleration are assumed to be constant throughout the liquid (a reasonable assumption). Therefore, we can calculate the pressure at any depth d ($\Delta d = d - 0$) in a liquid using the following equation:

$$P_d = P_{\text{ref}} + \rho g d, \qquad (17.6)$$

where d is expressed in *negative* scalar distances below the reference level at the surface of the liquid, and $g = -9.81 \text{ m/s}^2$. As you can see, pressure increases with fluid depth. If the container with the liquid is open to the atmosphere, the reference level d_1 is the liquid's surface, and the pressure $P_{\text{ref}} = P$ (the atmospheric pressure). For a closed container, P_{ref} may be greater or less than atmospheric pressure.

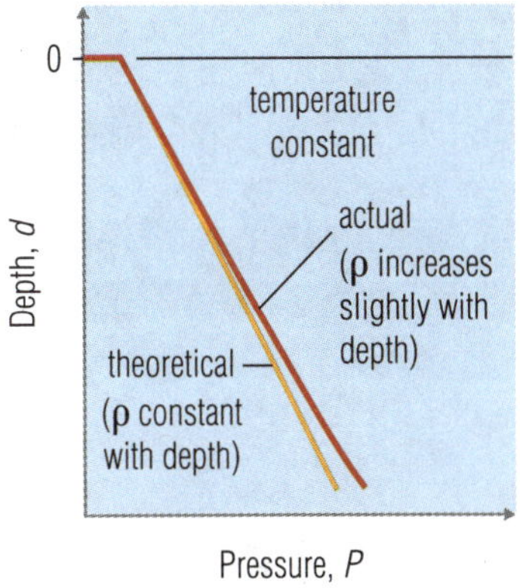

17-5 Pressure varies as an essentially linear function with depth. Compression of liquids at great depths causes the graph to depart from a truly linear relationship.

EXAMPLE 17-1

Ear Squeeze: A Practical Pressure-at-Depth Problem

Most people's eardrums can withstand a difference of up to 3.2×10^4 Pa between the external pressure and the middle ear pressure. This limitation can be a factor when free diving in a body of water. If a freshwater lake has a density of $1.00 \times 10^3 \text{ kg/m}^3$, how deep can an average person dive without fear of rupturing his eardrums? Assume that middle ear pressure remains at atmospheric pressure during the dive.

Solution:

We know that the maximum difference in pressure allowed (ΔP_{max}) is 3.2×10^4 Pa (1 Pa = 1 N/m²). Let $P_{\text{ref}} = P$ = atmospheric pressure. Equation 17.6 will permit us to calculate d_{max}.

$$\Delta P_{\text{max}} = P_d - P_{\text{ref}}$$

$$\Delta P_{\text{max}} = (P + \rho g d_{\text{max}}) - P$$

$$\Delta P_{\text{max}} = \rho g d_{\text{max}}$$

$$d_{\text{max}} = \frac{\Delta P_{\text{max}}}{\rho g} \qquad (1)$$

Substitute the known values into Equation (1) to solve for the maximum depth:

$$d_{\text{max}} = \frac{3.2 \times 10^4 \text{ N/m}^2}{(1.00 \times 10^3 \text{ kg/m}^3)(-9.81 \text{ m/s}^2)}$$

$$d_{\text{max}} \cong -3.\underline{2}6 \frac{\text{N/m}^2}{(\text{kg/m}^3)(\text{m/s}^2)} = -3.\underline{2}6 \left(\frac{\text{kg}\cdot\text{m}}{\text{m}^2\cdot\text{s}^2}\right)\left(\frac{\text{m}^3}{\text{kg}}\right)\left(\frac{\text{s}^2}{\text{m}}\right)$$

$$d_{\text{max}} \cong -3.\underline{2}6 \text{ m} \cong -3.3 \text{ m}$$

Change newtons to base SI units in order to complete the cancellation. Careful cancellation yields the expected unit for depth.

The average person can free dive to a depth of about 3.3 m, or about 10 feet, without equalizing pressure in his ears and not rupture his eardrums. As a diver approaches this depth, a painful condition called *ear squeeze* occurs. Divers avoid this condition by forcing air into their middle ears via their eustachian tubes in order to equalize the pressure across their eardrums.

The shape and volume of a container has no bearing on the pressure at a given depth in a liquid. The apparatus in Figure 17-6 demonstrates this principle. The variously shaped containers have the same external pressure—the pressure of the surrounding air. When the liquid levels are the same in each container, the pressure is the same at the same depth.

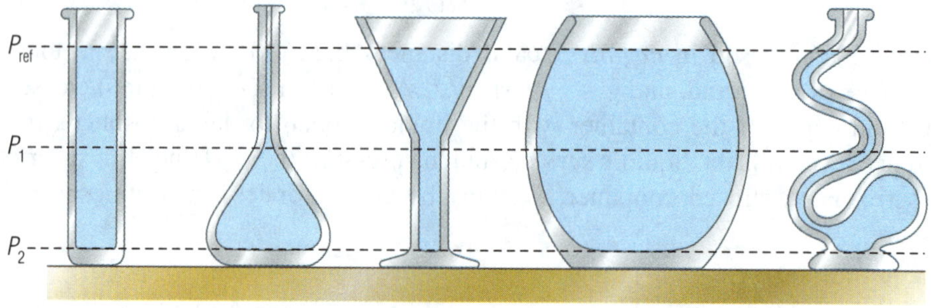

17-6 Pascal's vases, illustrating the principle that fluids at the same depth have the same pressure

17.5 Pressure in Compressible Fluids

We saw in the previous discussion that the pressure at a given depth in an incompressible fluid is proportional to the depth and the density of the liquid above that depth. This is true for pressure in a gaseous fluid as well. However, the density is not constant with height (or depth, depending on the location of the reference point) in a gas because of a gas's inherent compressibility. In fact, gas density decreases approximately exponentially with height, so gas pressure varies the same way as long as other factors remain constant. (Recall from Chapter 14 that the density of an unconfined gas is also dependent on temperature as well as pressure.) In equation form, this relationship is

In Equation 17.7, *e* is the base of the natural logarithm. Your scientific calculator should provide a function that facilitates this calculation.

$$P = P_{ref}e^{-\frac{\rho_{ref}}{P_{ref}}|g|h}, \tag{17.7}$$

where P is the gas pressure at height h above the reference height, P_{ref} is the pressure at the reference height, and ρ_{ref} is the gas density at the reference height. Note that only the magnitude of g is used in the exponent.

EXAMPLE 17-2

Mountain Sickness: Pressure in a Compressible Fluid

What is atmospheric pressure at the top of Mt. Everest (8850 m)? Assume that pressure at sea level is 1.013×10^5 Pa, and atmospheric density does not depend on temperature (*not* a realistic assumption).

Solution:

You already have the information necessary to solve Equation 17.7 for pressure at 8850 m, or you can look it up.

$$P = P_{ref}e^{-\frac{\rho_{ref}}{P_{ref}}|g|h}$$

$$P = (1.013 \times 10^5\ \text{Pa})\exp\left[-\frac{1.29\ \text{kg/m}^3}{1.013 \times 10^5\ \text{N/m}^2}(9.81\ \text{m/s}^2)(8850\ \text{m})\right]$$

$$P = (1.013 \times 10^5\ \text{Pa})\exp\left[-1.105\ \left(\frac{\text{kg}}{\text{m}^3}\right)\left(\frac{\text{m}^2\cdot\text{s}^2}{\text{kg}\cdot\text{m}}\right)\left(\frac{\text{m}}{\text{s}^2}\right)\text{m}\right]$$

$$P = (101\,\underline{3}00\ \text{Pa})e^{-1.105}$$

$$P \cong 33\,\underline{5}50\ \text{Pa} \cong 0.336 \times 10^5\ \text{Pa}$$

Problem-Solving Strategy 17.1

Complex exponential expressions are common in physics. In order to prevent confusion, the term "exp" is inserted before the expression in order to alert the reader that the following parenthetical statement is the exponent of the natural exponential base *e*. All units in the exponential expression must cancel.

Ignoring the effect of temperature on the density of the atmosphere, the pressure at the top of Mt. Everest is about one-third of sea level pressure. This lack of air pressure at high altitudes causes *anoxia*, lack of oxygen, in mountain climbers who climb without supplemental oxygen tanks. It is also a problem for athletes who have trained at elevations near sea level but must compete at high-altitude locations.

Observe that all units in the complicated exponent in the exponential expression cancel out—exponents are dimensionless numbers.

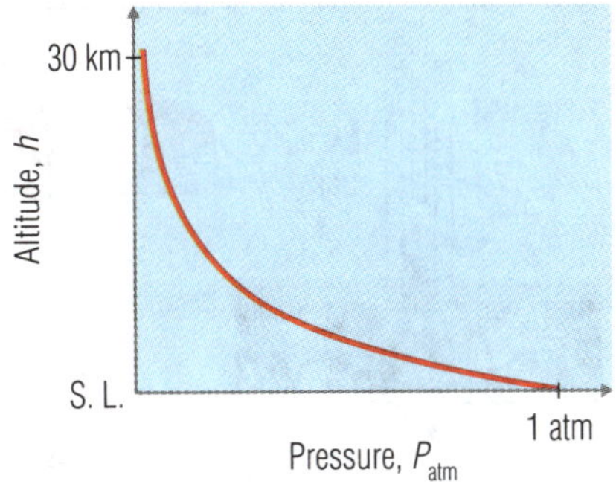

17-7 Atmospheric pressure varies exponentially with altitude. Blaise Pascal predicted that a vacuum existed above the atmosphere after he measured a decrease of atmospheric pressure with a barometer as he ascended a mountain.

17.6 Hydraulic Devices

Fill a sports bottle completely to the top. After tightening the cap with the pop top open, squeeze the bottle. What happens? The stream of water squirting from the cap indicates that the force of your hand was somehow transmitted directly to the mass of the water being ejected from the bottle.

In the seventeenth century, Blaise Pascal studied the effect of pressure on incompressible fluids (liquids) that completely filled their containers. His experiments led him to the principle that is now called **Pascal's principle**: *the external pressure applied to a completely enclosed incompressible fluid is distributed in all directions throughout the fluid.* Consider Equation 17.6 again. If we shut the cap of the sports bottle and then squeeze on the bottle to change P_d, the reference pressure P_{ref} must increase by the same amount in order to maintain the equality.

Blaise Pascal (1623–62) was a French mathematician, physicist, and philosopher known for his theory of probability and his work on pressure. He also invented one of the first calculating machines. The SI unit of pressure and a third-generation computer language are named after him.

17-8 In a famous experiment, Blaise Pascal attached a long, narrow pipe to a tightly sealed cask. After he filled the cask with water, the relatively small amount of water needed to fill the tube provided enough pressure to break the cask.

Pascal's principle provides the theoretical basis for **hydraulic devices**, which are machines that transmit forces via enclosed liquids. Figure 17-9 shows how a small input force can generate a large output force. Since we are considering the *change* in pressure caused by an applied force, we can ignore the static pressure due to liquid depth throughout the device. Static pressure is the same at equal depths throughout the device and does not affect the change of output pressure and the associated force.

The pressure created in the fluid by the force on the small piston is transmitted to the large piston. Assume that the cross-sectional area of the small piston is A and that of the large piston is nA. The pressure at the small piston is

$$P = \frac{F_{in}}{A}. \qquad (17.8)$$

Since the large piston is at the same height as the small piston, there is no effect on pressure due to height differences. Therefore, the pressure at the large piston is also P. The force on the large piston is

$$F_{out} = PnA.$$

Substituting Equation 17.8 for P, the areas cancel, giving the equation

$$F_{out} = nF_{in}. \qquad (17.9)$$

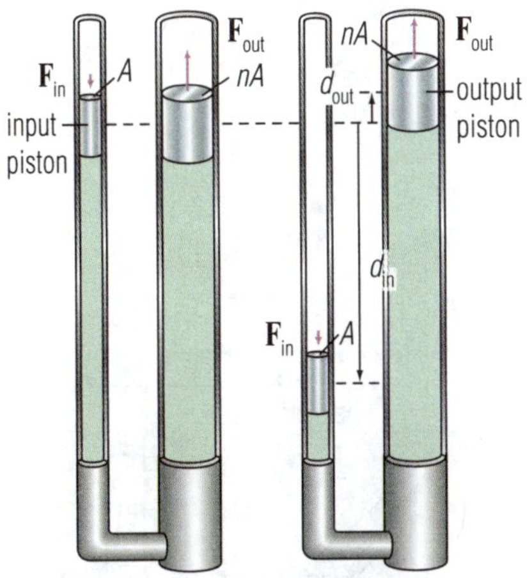

17-9 The principle behind hydraulic devices

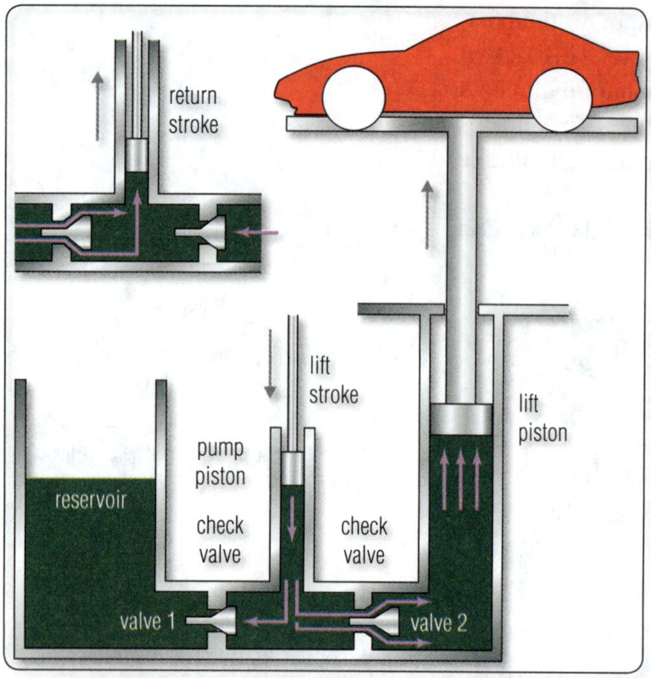

17-10 A hydraulic lift

A small force on the small piston can lift a heavy load on the larger piston if the larger piston is big enough. For example, a small force on a hydraulic brake pedal can stop the wheels of a speeding car. Similarly, a child can lift a car by using a hydraulic jack.

It seems, at first glance, that a hydraulic device produces more energy than it consumes. In reality, the device simply transmits work. Mechanical work is the scalar product of force and displacement. The small piston experiences a smaller force, but it travels a greater distance than the larger piston. This is the distance principle you observed in simple mechanical machines. The work done on the large piston is equal to or smaller than the work done on the smaller piston. In real hydraulic devices, some work is used to slightly compress the fluid and overcome friction in the device, so less work is produced than is consumed.

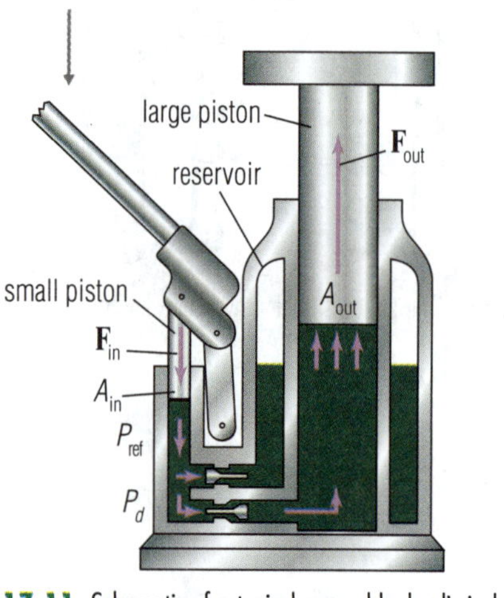

17-11 Schematic of a typical manual hydraulic jack

EXAMPLE 17-3

Multiplying Force: A Hydraulic Jack

One kind of hydraulic jack uses water as its working fluid. The face of its small input piston is 10.0 cm above the bottom of the water reservoir in the jack. The area of the input piston is $A_{in} = 2.00\ cm^2$. The output piston's area is $A_{out} = 80.0\ cm^2$. (a) If the magnitude of the net force applied to the input piston is 25.0 N, what is the pressure at the *bottom* of the water reservoir? (b) What is the magnitude of the force exerted on the output piston?

Solution:

The pressure exerted by the input piston is computed from Equation 17.8:

$$P = \frac{F_{in}}{A_{in}} = \frac{25.0\ \text{N}}{2.00\ \cancel{\text{cm}^2}} \left| \frac{10^4\ \cancel{\text{cm}^2}}{1\ \text{m}^2} \right.$$

$$P = 1.25 \times 10^5\ \text{N/m}^2 \quad (1)$$

a. Taking result (1) as P_{ref}, calculate the pressure at the bottom of the reservoir using Equation 17.6:

$$P_d = P_{ref} + \rho g d$$

$$P_{d_{-10\ cm}} = 1.25 \times 10^5\ \text{N/m}^2 + (1.00 \times 10^3\ \tfrac{\text{kg}}{\text{m}^3})(-9.81\ \tfrac{\text{m}}{\text{s}^2})(-0.100\ \text{m})$$

$$P_{d_{-10\ cm}} \cong 1.2\underline{5}9 \times 10^5\ \text{N/m}^2 \cong 1.26 \times 10^5\ \text{N/m}^2 \quad (2)$$

b. Pressure at the same height in a closed container is equal everywhere. Therefore, no pressure adjustment for height is required. The ratio of the output piston's surface area to the input piston's is

$$n = \frac{A_{out}}{A_{in}} = \frac{80.0\ \cancel{\text{cm}^2}}{2.00\ \cancel{\text{cm}^2}}$$

$$n = 40.0.$$

Use Equation 17.9 to calculate the output force (assuming equal height of the piston faces):

$$F_{out} = nF_{in} = (40.0)(25.0\text{ N})$$

$$F_{out} = 1.00 \times 10^3\text{ N}$$

The output force exerted by the jack is 40 times the input force.

Problem-Solving Strategy 17.2

Under normal circumstances, the difference of pressure due to height differences between input and output pistons in hydraulic devices is negligible compared to the force transmitted by the hydraulic system.

17.7 Liquid-Medium Pressure Indicators

It is often important to know a fluid's pressure within a system, especially in engineering applications. For example, the crankcase of a diesel engine is a sealed chamber at the bottom of the engine, containing the crankshaft and the engine's lubricating oil. Operators need to know if a significant pressure is building up in the crankcase, which would indicate that fuel and exhaust vapors are accumulating and could cause an explosion. This pressure buildup can be caused by worn piston rings. A device called a **manometer** can be connected to the crankcase to indicate a positive or negative pressure compared to atmospheric pressure.

A manometer is a U-shaped transparent tube that contains a dense liquid such as mercury or some other nonvolatile liquid. One end is attached to the container of the fluid whose pressure is unknown, and the other end is open to the atmosphere. A flexible tube connects the two ends of the instrument. Before the fluid pressure is applied, the movable end of the manometer is adjusted until the level of liquid in both ends is the same. Then, system pressure is applied to the closed end of the manometer. Assuming that the system pressure is greater than atmospheric, the liquid level will drop in the closed end and rise in the open end of the manometer. The difference in the levels of the liquid on the two sides of the tube is proportional to the difference between the pressure in the system and that in the atmosphere. This difference of levels is read as a pressure on an attached scale. Obviously, if the monitored system's pressure is less than atmospheric pressure, then the liquid will rise in the closed end and drop in the open end of the manometer.

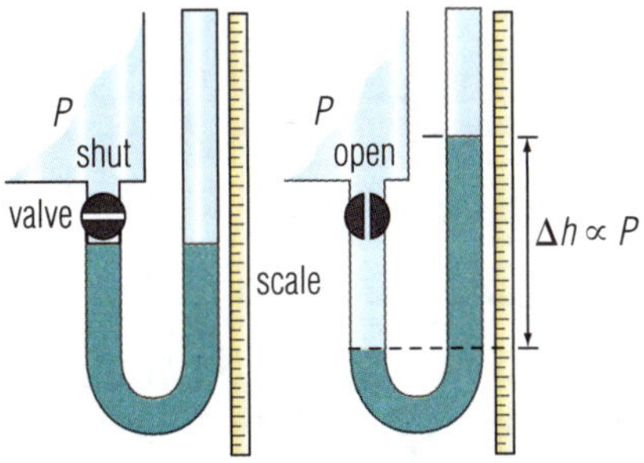

17-12 A manometer measures relative pressure.

The first instrument to accurately measure atmospheric pressure was the mercury **barometer**. This is essentially a long, straight glass tube sealed at one end, filled with mercury, and carefully overturned into an open reservoir of mercury so that no air enters the tube. If the tube is sufficiently long, the mercury meniscus at the top of the tube will drop until the atmospheric pressure exerted on the reservoir just balances the weight of the column of mercury. Greater atmospheric pressure supports a higher column of mercury. Standard atmospheric pressure, 101 325 Pa, supports a 760 mm column of mercury. As mentioned earlier in this chapter, these relationships provide various systems of units for reporting atmospheric pressure.

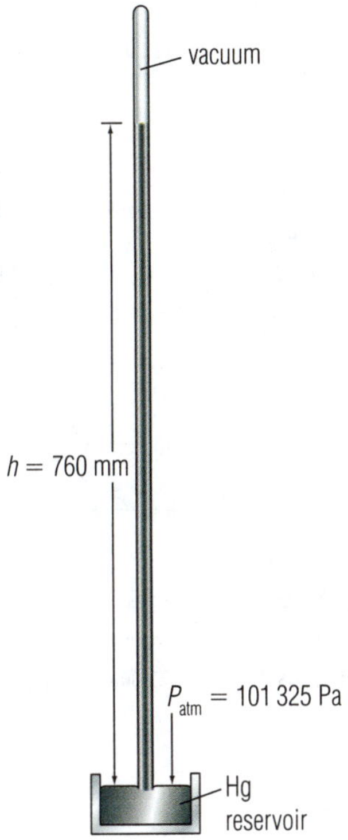

17-13 A mercury barometer

17.8 Buoyancy

A story goes that, more than two hundred years before the birth of Christ, the scientist and mathematician Archimedes was given an assignment from his king, Hieron II. The king had ordered a goldsmith to make him a crown from the gold the king provided for him. The goldsmith made the crown and delivered it to the king. Then the king began to doubt the craftsman's honesty. Had he used some silver in the crown and kept some of the gold for himself? The king immediately weighed the crown and found that it weighed the same as the gold he had given the goldsmith. But what if the craftsman had replaced the gold with an equal weight of silver? Was there any way to find out? This was the problem the king took to Archimedes.

Archimedes thought hard about the gold crown. In what way do silver and gold differ, other than in value? Of course, their densities are different, but Archimedes did not have instruments to measure density accurately. Then one day the scientist was relaxing at the local bath. As he lowered himself into the full bath, some water spilled over the edge. Suddenly Archimedes had the answer to the crown problem. If he placed the crown in a full container of water, the amount of water that spilled out would depend on the crown's volume. A given weight of gold would force out less water than would a gold-silver mixture of the same weight. When Archimedes tested the crown, he found that it indeed was a gold-silver mixture. Thinking about the solution led Archimedes to the principle that the buoyant force of water on an object equals the weight of the water it displaces.

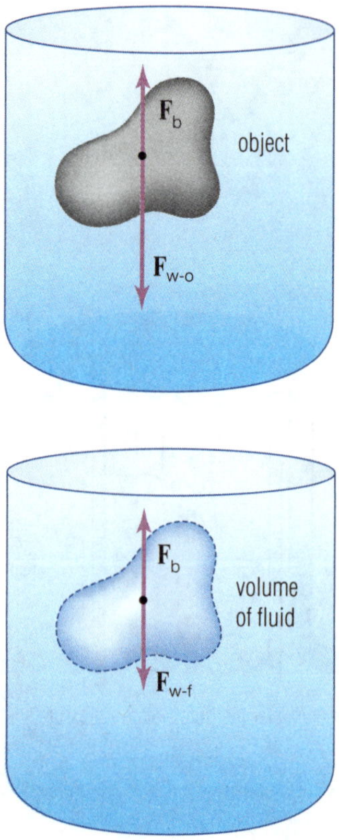

17-14 Buoyancy is the force a fluid exerts on an immersed object.

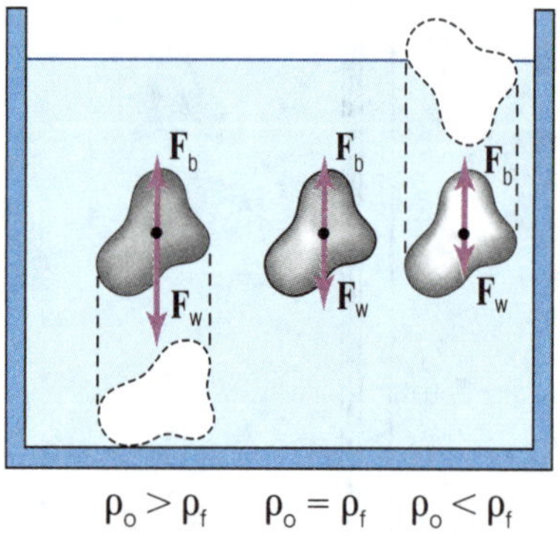

17-15 Knowing the comparative densities of the immersed object and the fluid will determine whether the object sinks, floats, or is neutrally buoyant.

17.9 Submerged Objects

What happens to an object placed in a fluid? Let's consider a volume of fluid to be the system. Its shape doesn't matter as long as it is in equilibrium with the surrounding fluid. The forces acting on the fluid-system are its weight ($\mathbf{F}_{\text{w-f}}$) and its equilibrant force, which we shall call the **buoyant force** ($\mathbf{F}_{\text{b}}$). The buoyant force results from the fluid pressure on the system. Because the fluid system is in equilibrium, these two forces are equal in magnitude but opposite to each other.

If the fluid system is now replaced by an object that is exactly the same shape, the gravitational force on the new object-system ($\mathbf{F}_{\text{w-o}}$) will be different. However, the buoyant force that is dependent on the pressure of the fluid on the *volume* of the object remains the same. The object experiences an upward force that is equal to the weight of the fluid it displaces. The magnitude of the buoyant force is determined from the equation

$$F_{\text{b}} = \rho|g|V, \tag{17.10}$$

where ρ is the density of the *displaced fluid*. The fluid-system experienced the same force, but it was balanced by its own weight. Equation 17.10 is the mathematical expression for **Archimedes' principle**. This principle states that *any system that is submerged or floats in a fluid is acted on by an upward buoyant force equal in magnitude to the weight of the fluid it displaces.*

Archimedes' principle holds for a system of any shape. The buoyant force may be greater than a system's weight, less than its weight, or equal to its weight. If it is equal to the system's weight, the system behaves exactly like the fluid it displaces. It is not accelerated at all because the forces acting on it are balanced. This implies that the overall density of the system must equal the fluid it displaces, because its volume and weight (and mass) are the same.

Similarly, if the weight of a system is greater than that of the displaced fluid, its density is greater than the fluid's. Its downward weight will exceed the upward buoyant force, and the object will begin to sink. However, the net force accelerating the system downward is the difference of the system's weight and the buoyant force. Consequently, the system appears to weigh less in the fluid. You can verify this fact by lowering a heavy stone into a pond. As it submerges, the stone seems to become lighter.

If the weight of an immersed system is less than that of the displaced fluid, its density is less than the fluid's. An object less dense than the fluid will not have enough weight to balance the buoyant force. The net force will accelerate the system upward in the fluid. If the fluid is a liquid, the system will eventually break through the surface. When this happens, the buoyant force on the system rapidly decreases as the volume of liquid that is displaced decreases. Eventually, the buoyant force equals the weight of the system, and the system stops rising with part above the liquid's surface and part below.

17-16 Because the density of icebergs ($\rho_{ice} = 0.92\ g/cm^3$) is less than that of seawater ($\rho_{sw} = 1.03\ g/cm^3$), icebergs float on the ocean. The visible portions are often huge compared to oceangoing vessels, yet only 11 percent of their entire volume is above the ocean's surface.

EXAMPLE 17-4

Floating Like a Brick: Buoyancy

A $1.00 \times 10^3\ cm^3$ brick ($\rho = 1.74\ g/cm^3$) is placed in mercury ($\rho = 13.6\ g/cm^3$). (a) What is the buoyant force on the brick? (b) Will it float?

Solution:

a. Determine the magnitude of the buoyant force on the brick fully submerged in the mercury:

$$F_b = \rho|g|V$$

$$F_b = (13.6\ g/cm^3)(9.81\ m/s^2)(1.00 \times 10^3\ cm^3)$$

$$F_b \cong 13\underline{3}\,400\ g{\cdot}m/s^2$$

Convert grams to kilograms in order to obtain newtons:

$$F_b \cong 133\ kg{\cdot}m/s^2 = 133\ N$$

b. The magnitude of the brick's weight is as follows:

$$F_w = m|g| = (\rho V)|g|$$

$$F_w = (1.74\ g/cm^3)(1.00 \times 10^3\ cm^3)(9.81\ m/s^2)$$

$$F_w \cong 17\,\underline{0}60\ g{\cdot}m/s^2 \cong 17.1\ N$$

The brick's weight is less than the maximum possible buoyant force of 133 N, so the brick will float in the mercury.

Problem-Solving Strategy 17.3

If you need to know if an object will float in a liquid, compare the densities of the object and the liquid. If the object's density is less than the liquid's density, it will float. Otherwise, it will submerge or even sink.

If the fluid is a gas and the system can rise without restriction, such as a balloon in the atmosphere, the gas density decreases with altitude, and thus the buoyant force decreases as well. Eventually, an altitude is reached at which the densities of the balloon and the displaced gas are equal, and the balloon stops rising.

17.10 Buoyancy of Real Objects

We have assumed that the submerged systems we have considered are rigid and have constant densities. These ideal systems do not respond to changes in fluid pressure as their depths or heights change. Real systems, such as helium balloons and submarines, *do* respond to pressure changes. For example, a submarine's crew can adjust the sub's weight at a specific depth so that its weight exactly equals its buoyant force—it is *neutrally buoyant.* If the submarine then moves to a shallower depth, the sea pressure becomes less and its hull expands slightly as the metal cylinder of the hull is less compressed. The submarine's volume then increases but its weight remains the same, so its density decreases. As a result, the buoyant force exceeds the weight of the submarine, and the net force acting on the ship is upward—it is *positively buoyant.* The opposite effect occurs if the ship goes deeper—it becomes *negatively buoyant.* Balloons and "rigid"

There are other fluid factors that can affect system buoyancy: variations in fluid density with depth and height, temperature changes, and changes of fluid composition. For example, seawater temperature and salinity significantly affect the buoyancy of a ship or submarine.

airships experience similar effects due to the variation in atmospheric pressure with altitude.

17.11 Center of Buoyancy

Every object submerged in a fluid has both a center of gravity (center of mass) and a center of buoyancy. The **center of buoyancy** is the center of mass of the fluid that would occupy the submerged space that the object occupies. If the object has uniform density and is completely submerged, its center of mass and center of buoyancy are the same. The same is true if the object's mass is distributed symmetrically. However, if an object's mass is not symmetrically distributed horizontally, its center of gravity will not coincide with the center of buoyancy. Therefore, the buoyant force and the gravitational force will not be on the same vertical line. A torque will exist on the object, and the object will rotate as it sinks, rises, or stays in place, until its center of gravity is on a vertical line with its center of buoyancy.

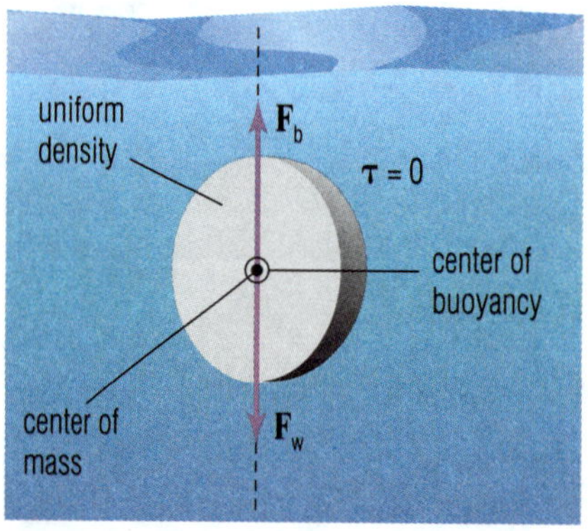

17-17a In a submerged object of uniform density and shape, the centers of mass and buoyancy are located at the same point.

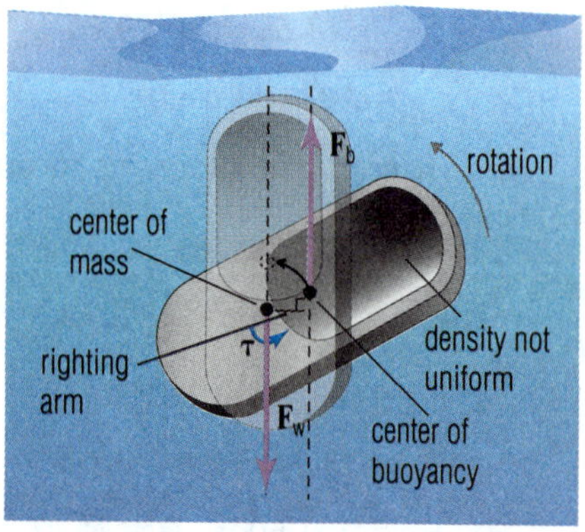

17-17b When the mass is not distributed in a horizontally symmetrical position, the resulting torque will rotate the object until its centers of mass and buoyancy fall on the same vertical line.

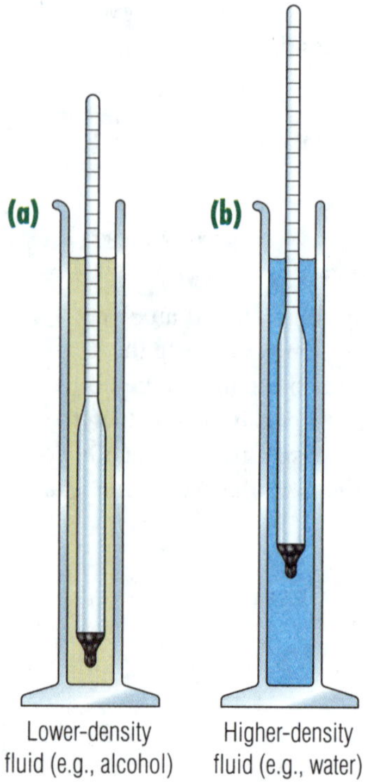

17-18 Identical hydrometers in (a) alcohol and (b) water

Do not confuse *hydrometers* with *hygrometers*. The latter instrument is used to determine the relative humidity of the atmosphere.

17.12 The Hydrometer

The relationship between density and buoyancy provides a method for measuring fluid density. One instrument used to measure density is a **hydrometer**. The hydrometer is a cylinder with a weighted end. When it is placed in a low-density liquid, the hydrometer sinks more than it does in a high-density liquid. The volume of the liquid that the hydrometer must displace to float is inversely proportional to the density of the liquid. Each hydrometer is calibrated for use with a specific liquid. The scale is placed so that the surface of the liquid intersects it at the value corresponding to the liquid's density or its specific gravity. Hydrometers are commonly used to test the amount of antifreeze in a radiator and the concentration of electrolytes in a battery. Scientists use hydrometers to measure the amount of suspended sediment in lake and river water as well.

FACETS of PHYSICS

Maple Syrup Hydrometers

Maple syrup can cost nearly forty dollars per gallon. Since the syrup is valuable, sensible manufacturers are careful to preserve it properly. An important part of this preservation is making syrup the desired density.

To make the syrup, producers gather maple sap and boil it. Some of the water in the sap evaporates, leaving a concentrated sugary solution that is much denser than before. The syrup is then filtered to remove impurities, poured into containers, and sealed.

If the syrup is not dense enough, it will spoil easily; but if it is too dense, the sugar will precipitate out of the solution and crystallize. The proper density for hot syrup is about 1.33 g/cm^3.

Vermont, which normally produces the most maple syrup in the United States, requires that each gallon labeled "pure Vermont maple syrup" have the right density. A maple syrup hydrometer determines the density of the syrup. This instrument is an ordinary hydrometer with a colored mark at 1.33 g/cm^3. A syrup producer pours boiling syrup into a tall metal cup and then places the hydrometer in the cup. If the mark is level with the surface of the syrup, the syrup has boiled enough. If the mark is below the surface, the syrup is not dense enough and must be boiled longer. If the mark is above the surface, the syrup is too dense and must be diluted with fresh sap and boiled again.

The sugar industry (also including corn, cane, and sugar beet syrup, as well as candy producers) uses a variety of traditional density units such as degrees brix, degrees baumé, degrees twaddle, and other whimsical names. Each relates a "degree" of density to a specific mass of sugar per unit volume. Maple sugar producers must ensure a minimum density of 66.0 degrees brix at 68 °F.

In recent decades, the maple syrup industry has been producing up to five different grades of syrup based primarily on color. The highest quality, fancy grade syrups (lightest in color) are often produced near the minimum of 1.33 s.g. Fuller-bodied, darker syrups are considered to be lower in grade and are usually somewhat higher in density, since the sugar is more concentrated.

17A Objectives

After completing this section, I can

- ✓ discuss the difference between fluids and nonfluids.
- ✓ define the dimensions of density and pressure and state the units used to measure them.
- ✓ discuss how pressure is generated within incompressible and compressible fluids.
- ✓ describe how Pascal's principle explains the operation of simple hydraulic machines.
- ✓ describe how manometers and barometers operate.
- ✓ use Archimedes' principle to compute the buoyant force acting on a submerged object.
- ✓ describe the stability of a submerged object by using the relative positions of its centers of mass and buoyancy.
- ✓ explain the function of a hydrometer.

17A Section Review

1. **a.** Name and briefly describe the two subsections of fluid mechanics.
   b. In what fundamental way does fluid mechanics differ from the mechanics you have studied prior to this chapter?
2. **a.** Using qualitative terms, how does fluid pressure vary horizontally through a fluid?
   b. How does it vary vertically?
3. What three factors determine the static pressure in any fluid at a given depth (or height)?
4. What determines the force a fluid exerts against a surface?
5. What two factors determine the buoyant force exerted on an object submerged in a fluid?
6. ⊙ How deep (in meters and feet) must a diver go (in fresh water) in order to experience two atmospheres of pressure? (Recall that a diver experiences 1 atm at the water's surface.) State your answer to two significant digits.
7. ⊙ Assuming that atmospheric pressure at sea level is 1.00 atm, what would be atmospheric pressure at the level of the Dead Sea (altitude: $-400.$ m)? Ignore difference in pressure due to variation of temperature.
8. ⊙ A block of solid oak (30.0 cm × 40.0 cm × 50.0 cm) floats in fresh water. What percentage of the block's volume is *not* under water?

17B HYDRODYNAMICS: FLUIDS IN MOTION

17.13 Ideal Fluids

The concepts that you have been studying in hydrostatics are fairly simple. Hydrodynamics, in contrast, is one of the most complicated areas of physics. To simplify the study, we must establish the concept of the **ideal fluid**. In the case of an ideal fluid, it is assumed that

- the fluid flows smoothly;
- the velocity of the fluid does not change with time at a fixed location in the fluid path;
- the density of the fluid is constant (the fluid, whatever its density, is incompressible);
- friction has no effect on fluid flow.

These assumptions are only approximately true for real fluids, and mainly for liquids.

The first assumption, the idea that ideal fluids flow smoothly, means that every particle passing through a given point follows the same path. Such a path is called a **streamline**. Streamlines have no physical reality, but they are useful in visualizing fluid flow. If the streamlines are continuous, then the flow is smooth, or **laminar**. If the streamlines have discontinuities (bumps, twists, or turns), then the flow is **turbulent** and thus requires much more complex modeling. Several streamlines can define a **flow tube**. A flow tube may be an actual cylindrical pipe, or it may be a small volume within a larger volume of flowing fluid. The fluid particles do not cross the boundaries of a flow tube. For the purposes of hydrodynamic studies, a flow tube defines the boundaries of a physical system.

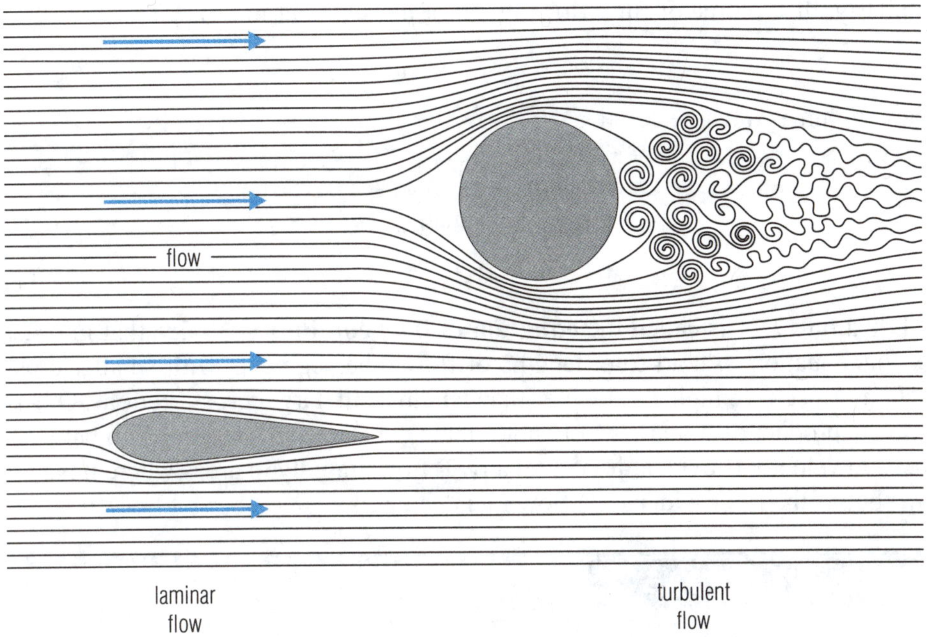

17-19 Streamlines around various objects showing laminar and turbulent flow

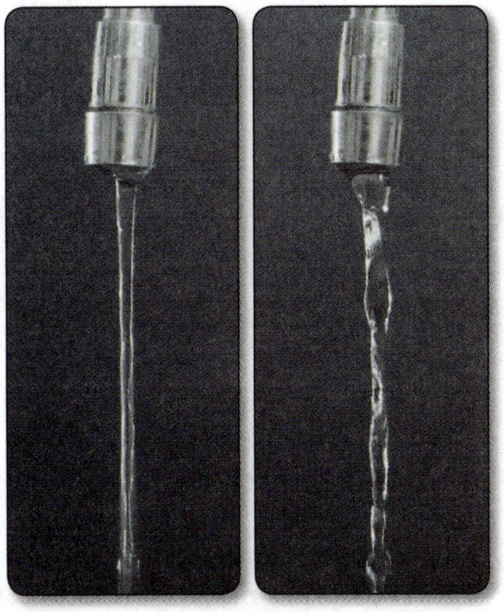

17-20 Laminar and turbulent flow in a real fluid

The second assumption establishes that the velocity of the fluid at any one location is constant with time. However, it does *not* mean that fluid velocity cannot be different at various points within a body of flowing fluid at the same time.

The third assumption is that fluids are incompressible. This, combined with the law of conservation of mass, requires that the rate of volume and mass flow into a segment of a flow tube equals the rate of volume and mass flow out of the flow tube segment. That is, if 1 L of fluid flows into one end of a flow tube in 1 s, 1 L of fluid must flow out the other end in the same second. If this were not true, then either some mass would disappear in the tube or some mass would be created in the tube.

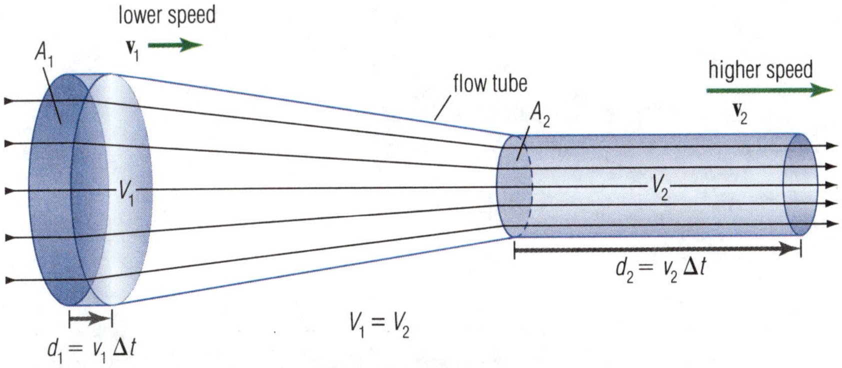

17-21 A flow tube showing how a constriction affects flow rate

17.14 Flow Continuity

The volume of fluid flowing into a flow tube segment is shaped roughly like a cylinder. The base of the cylinder is the cross section of the flow tube at the beginning of the segment, which has area A_1. The "height" of the cylinder is the length traveled by a particle of the fluid in time Δt. This distance is equal to the velocity in that segment of the flow tube multiplied by the time interval. The volume flowing into the flow tube segment is

$$V_1 = A_1 v_1 \Delta t.$$

Similarly, the volume of fluid flowing out of the segment in time Δt is

$$V_2 = A_2 v_2 \Delta t.$$

These volumes must be equal.

$$V_1 = V_2$$

$$A_1 v_1 \Delta t = A_2 v_2 \Delta t$$

$$A_1 v_1 = A_2 v_2 \quad (17.11)$$

Equation 17.11 is called the **equation of flow continuity**. Notice that in order for this equation to hold true, lengths of flow tube segments with smaller cross sections must have higher velocities than lengths with larger cross sections. Otherwise the product of A and v would not remain constant. This relationship between area and velocity is evident in nature: a river flows rapidly in areas where it is narrow; but as the river widens, its flow rate decreases.

EXAMPLE 17-5

Slow with the Flow: Flow Continuity

A fluid flows through a level pipe at a speed of 4.00 m/s at a point where the pipe's cross-sectional area is 10.0 cm^2. How fast will it flow where the pipe's area is 20.0 cm^2?

Solution:

$$A_1 v_1 = A_2 v_2$$

$$v_2 = \frac{A_1 v_1}{A_2} = \frac{(10.0\ \cancel{cm^2})(4.00\ m/s)}{20.0\ \cancel{cm^2}}$$

$$v_2 = 2.00\ m/s$$

Problem-Solving Strategy 17.4

It is permissible to use different metric systems in one calculation as long as the units cancel. This often occurs in simple proportional calculations.

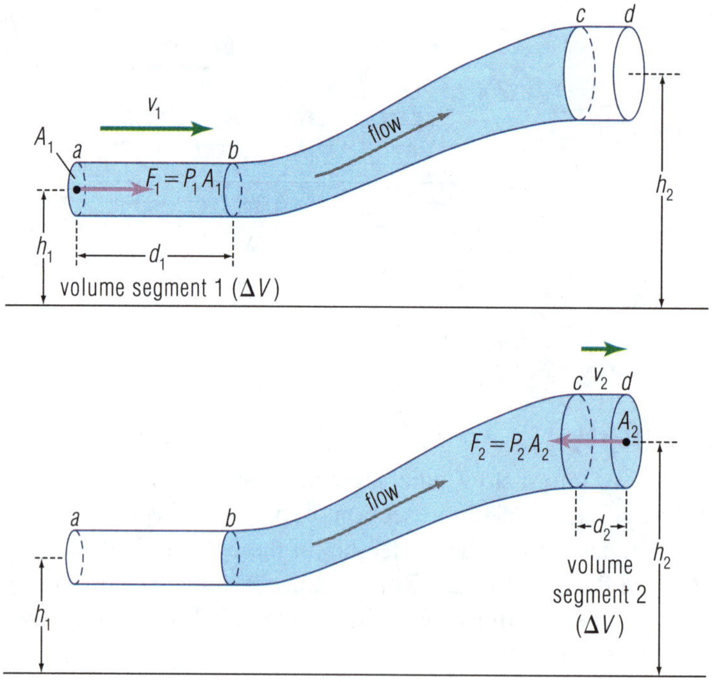

17-22 A fluid gains potential energy by moving against gravity.

17.15 Bernoulli's Principle

Consider the energy and work involved in moving a segment of fluid between *a* and *c* (the colored portion) to the volume between *b* and *d* in Figure 17-22. This is equivalent to transferring a volume of fluid, ΔV, from a height h_1, velocity v_1, and area A_1 to a height h_2, velocity v_2, and area A_2. The change in kinetic energy is

$$\Delta K = \frac{1}{2}mv_2^2 - \frac{1}{2}mv_1^2.$$

Since the density of the fluid is the mass divided by the volume of the fluid, $\rho = m/\Delta V$, then $m = \rho\Delta V$ and

$$\Delta K = \frac{1}{2}\rho\Delta V v_2^2 - \frac{1}{2}\rho\Delta V v_1^2. \quad (17.12)$$

The change in mechanical potential energy is due in this case to the change in height of the fluid sample above an arbitrary reference height,

$$\Delta U = m|g|h_2 - m|g|h_1, \text{ and}$$
$$\Delta U = \rho\Delta V|g|h_2 - \rho\Delta V|g|h_1. \quad (17.13)$$

Since we assumed that the flow is not turbulent and there are no friction or viscosity effects in or on the fluid, all the work done on the segment of fluid by all other nonconservative forces just changes the total mechanical energy of the fluid. In other words, according to the work-energy theorem,

$$W_{ncf} = \Delta K + \Delta U. \quad (17.14)$$

The total work done *on* the fluid is equal to the work done by the fluid force on the left side of the first volume segment ($\mathbf{F}_1$) plus the work done by the fluid force on the right side of the second volume segment ($\mathbf{F}_2$). The work of each force is calculated from the product of the force and its displacement through each segment of fluid. So the work of the first force is

$$W_1 = F_1\Delta d_1,$$

since the force and displacement of the fluid are oriented in the same direction. The fluid in the second segment is still moving to the right, but the external fluid force exerted on it is oriented to the left. Therefore, the work on the second fluid segment is

$$W_2 = -F_2\Delta d_2.$$

From the definition of pressure, we know that

$$F_1 = P_1A_1, \text{ and}$$
$$F_2 = P_2A_2.$$

Substituting these expressions for force into the respective work equations gives us

$$W_1 = P_1A_1\Delta d_1, \text{ and}$$
$$W_2 = -P_2A_2\Delta d_2.$$

The arbitrarily small fixed volume ΔV is the product of the area of each fluid segment and the arbitrarily small fixed displacement of the fluid in the flow tube at each segment.

$$\Delta V = A_1\Delta d_1 = A_2\Delta d_2$$

The variable ΔV is used for this derivation of Bernoulli's equation because it is intended to represent an arbitrarily small fixed volume of fluid rather than a definite volume change.

In the context of work and potential energy, ΔU means the change of mechanical potential energy as in Chapter 9 rather than the change of internal energy as in Chapter 16.

In this discussion, the symbol Δd is used to represent an arbitrarily small fixed displacement rather than a specific distance.

We can now substitute ΔV for $A_1\Delta d_1$ and $A_2\Delta d_2$ in the work equations. The work done by each of the fluid forces is then as follows:

$$W_1 = P_1\Delta V$$
$$W_2 = -P_2\Delta V$$

The work done by nonconservative forces is

$$W_{ncf} = W_1 + W_2, \text{ and}$$
$$W_{ncf} = P_1\Delta V - P_2\Delta V. \quad (17.15)$$

Substituting Equations 17.15, 17.12, and 17.13 for W_{ncf}, ΔK, and ΔU in the work-energy theorem (Equation 17.14), respectively, gives the following:

$$W_{ncf} = \Delta K + \Delta U$$
$$P_1\Delta V - P_2\Delta V = (\tfrac{1}{2}\rho\Delta V v_2^2 - \tfrac{1}{2}\rho\Delta V v_1^2) + (\rho\Delta V|g|h_2 - \rho\Delta V|g|h_1)$$

Dividing each term by ΔV gives

$$P_1 - P_2 = \tfrac{1}{2}\rho v_2^2 - \tfrac{1}{2}\rho v_1^2 + \rho|g|h_2 - \rho|g|h_1.$$

Rearranging to group terms associated with each fluid segment, we obtain

$$P_1 + \tfrac{1}{2}\rho v_1^2 + \rho|g|h_1 = P_2 + \tfrac{1}{2}\rho v_2^2 + \rho|g|h_2. \quad (17.16)$$

Equation 17.16 is called **Bernoulli's equation**.

Daniel Bernoulli (1700–82) was a Swiss mathematician who studied many different scientific topics. He is best known for his work in hydrodynamics, vibrating strings, and probability.

17.16 Bernoulli's Principle and Constant Velocity

There are two special cases of Bernoulli's equation. One is the case where the cross section of the pipe, and therefore the velocity, does not change. Then

$$v_1 = v_2 = v.$$

Substituting this into Bernoulli's equation:

$$P_1 + \cancel{\tfrac{1}{2}\rho v^2} + \rho|g|h_1 = P_2 + \cancel{\tfrac{1}{2}\rho v^2} + \rho|g|h_2$$
$$P_1 + \rho|g|h_1 = P_2 + \rho|g|h_2$$

Problem-Solving Strategy 17.5

In an ideal flow tube, if the flow cross section remains constant, then fluid pressure depends only on the height of the fluid above the reference height, according to Bernoulli's equation.

The difference in pressure ($\Delta P = P_2 - P_1$) depends only on the difference in depth in the fluid ($\Delta h = h_2 - h_1$). This is the same relationship that determines the change of hydrostatic pressure.

17.17 Bernoulli's Principle and Constant Height

The other special case is one in which the fluid remains at the same elevation. Then,

$$h_1 = h_2 = h.$$

Substituting this into Equation 17.16:

$$P_1 + \tfrac{1}{2}\rho v_1^2 + \cancel{\rho|g|h} = P_2 + \tfrac{1}{2}\rho v_2^2 + \cancel{\rho|g|h}$$
$$P_1 + \tfrac{1}{2}\rho v_1^2 = P_2 + \tfrac{1}{2}\rho v_2^2$$

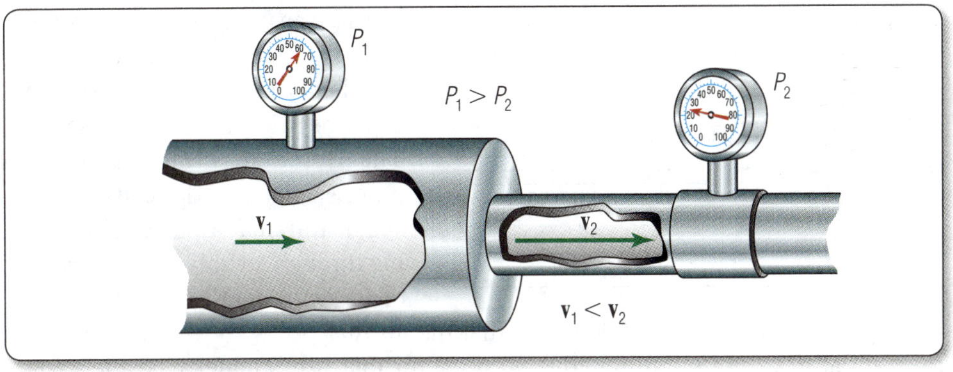

17-23 The faster flow has lower pressure.

This equation shows that in order to maintain the equality, if the velocity of the fluid increases in a segment of flow tube, the corresponding pressure must decrease, and vice versa.

You can tell which parts of a fluid are moving faster by studying the streamlines. Faster portions have smaller areas; therefore, the streamlines are diagrammatically closer together. Since faster portions have lower pressure, closely spaced streamlines also indicate lower pressure.

Problem-Solving Strategy 17.6

In an ideal flow tube, if the height remains constant, fluid pressure will increase when fluid speed decreases and vice versa, according to Bernoulli's equation.

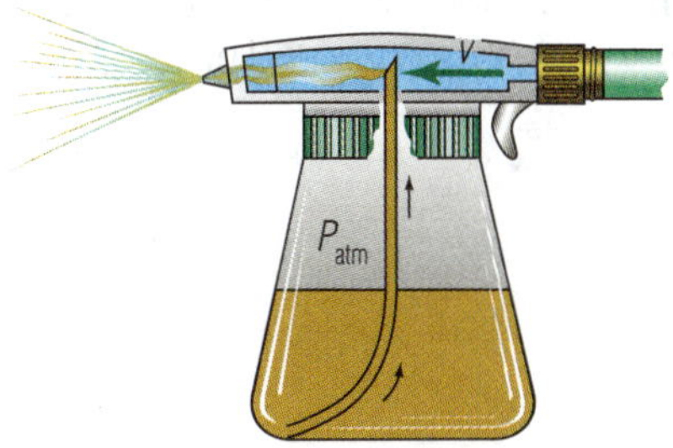

17-24 This sprayer works because air moving through a restriction lowers the pressure, drawing the liquid up the tube. This phenomenon, called the *Venturi effect*, is the consequence of the Bernoulli principle and flow continuity.

EXAMPLE 17-6

Pressure and Speed: Bernoulli's Principle

Water flowing through a level pipe has a pressure of 2.50×10^4 Pa when its speed is 1.00 m/s. The water flows through a constriction into a smaller pipe. What is its pressure when its speed increases to 3.00 m/s?

Solution:

$$P_1 + \tfrac{1}{2}\rho v_1^2 = P_2 + \tfrac{1}{2}\rho v_2^2$$

$$P_2 = P_1 + \tfrac{1}{2}\rho v_1^2 - \tfrac{1}{2}\rho v_2^2 = P_1 + \tfrac{1}{2}\rho(v_1^2 - v_2^2)$$

$$P_2 = 2.50 \times 10^4 \text{ Pa} + \tfrac{1}{2}(1.00 \times 10^3 \text{ kg/m}^3)\,[(1.00 \text{ m/s})^2 - (3.00 \text{ m/s})^2]$$

$$P_2 = 2.50 \times 10^4 \text{ Pa} - 40\underline{0}0 \text{ Pa}$$

$$P_2 = 2.10 \times 10^4 \text{ Pa}$$

$$P_2 < P_1$$

As expected, fluid pressure decreases with an increase of speed.

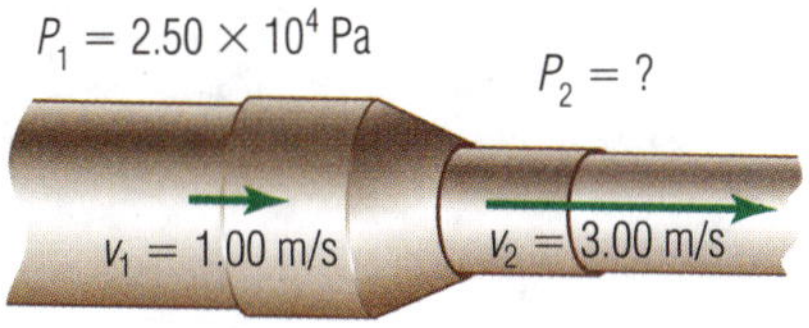

17-25 Pipe constrictions often occur in real plumbing systems.

17.18 Lift

Flowing fluids can generate dynamic forces on the surfaces they contact and on objects immersed within them. The Creator's design of fish and birds, which move easily through fluids, utilizes this principle. Long ago, humans discovered the ability of wind to generate useful forces, like those utilized by windmills and sailing ships. More recently, in attempts to emulate bird flight, people produced contraptions that create a lifting force. Any device that generates **lift** as air flows along its surface is an **airfoil**. Airfoils are typically long, narrow, and flattened solid objects that are streamlined, having a teardrop-shaped cross section. An airplane wing is one kind of artificial airfoil. Similar objects that create lift in liquid are **hydrofoils**.

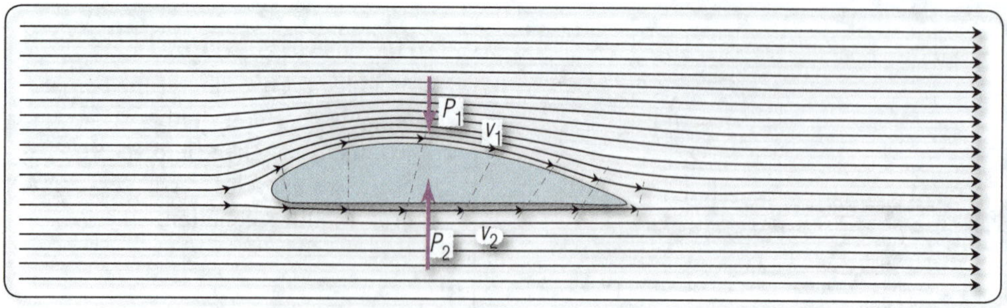

17-26 Streamlines over an airfoil illustrate the theory of the Bernoulli principle of lift.

Lift can be modeled extremely well by modern supercomputers, but there are various theories that describe what actually causes lift. The Bernoulli principle is commonly taught as the cause of lift, suggesting that a plane wing moving through air splits the air in front of it. According to this theory, the principle of flow continuity demands that the air that flows above the wing must meet up at the wing's trailing edge with the air that flows under it. If the wing's upper surface is more curved than its lower surface, the air on top will move faster, so its pressure will be lower than that on the underside of the wing. This pressure difference causes a net lifting force, as predicted by the Bernoulli principle. See Figure 17-26.

But there are several problems with this approach. First, nothing actually requires the two paths of air to reunite at the wing's trailing edge. In fact, when observed, the two flows usually do *not* reunite. Second, a wing with a steeply curved upper surface would not work upside down. High performance aircraft have wings with symmetrical cross sections, yet they still manage to generate lift. Third, airfoils divert great masses of air in the opposite direction to the direction of the lifting force (see Figure 17-28), but this is not predicted by the Bernoulli principle. Fourth, objects that present only a single surface to flowing air experience significant lift without a corresponding flow adjacent to the opposite surface (e.g., a sheet of plywood lifted off the ground on a gusty day).

Henri Marie Coandă (1886–1972) was a Romanian aeronautical engineer, aircraft designer, and inventor. He was instrumental in early aerodynamics research and built a rudimentary jet engine in 1910.

A better explanation of lift involves the Coandă effect and Newton's third law. The **Coandă effect** states that fluids moving by a gently curved surface tend to follow the curvature of the surface due to forces of adhesion and *fluid viscosity*. Fluid molecules immediately next to a moving surface adhere to it and are thus motionless. Molecules slightly farther away move slowly relative to the surface, and those at a greater distance move rapidly. These position-dependent movements compose the *boundary layer effect*. Thus, a surface, such as a wing, that moves through fluid drags that fluid along in a path similar to the surface's curvature.

Because an aircraft wing generating lift is tilted upward slightly, it compresses the air below it, causing some of the air in front of the wing to rise up and flow over it. The Coandă effect causes this air to follow the wing's surface, with air molecules spreading out above the wing due to inertia and low air viscosity. This spreading reduces the density of the air and thus the air pressure above the wing. Air even farther above the wing then rushes in to fill this low-pressure region.

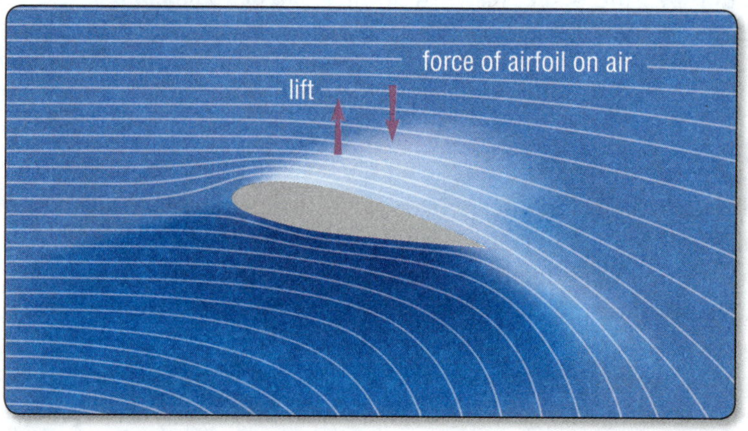

17-27 Streamlines showing the Coandă effect. The diagram exaggerates the shape of the streamlines to illustrate the effect.

The Coandă effect thus accelerates the air *above* the wing *downward* toward the top of the wing. According to Newton's third law, the surrounding air, in reaction to the force the wing exerts on the air, must exert an opposite upward force on the wing, which is lift. Notice that most of the lifting force is generated above the wing rather than below it. The downward moving air then spills behind the moving wing, as seen in the adjacent figure.

Unlike the Bernoulli principle, the Coandă effect accounts for the upside-down flight of aerobatic and military aircraft, as well as the functioning of wings streamlined on both surfaces. It also better explains why sails, propellers, and fluid switches work. Many physicists suggest that aerodynamic lift is actually a combination of the Coandă effect, the Bernoulli principle, and other complex physical principles of fluid mechanics.

You can verify the Coandă effect by a simple experiment. Hold two sheets of paper about an inch apart, as in Figure 17-30. Blow steadily between the sheets. You might expect them to fly out from the force of your blowing, but instead they draw together. The moving air between them exerts less pressure on them than the still air around them, so they are pushed together. The two photographs on the right in Figure 17-30 demonstrate how lift is created by the Coandă effect.

17-28 As predicted by the Coandă model of lift, this aircraft diverts huge amounts of air downward as it creates lift.

17-29 Helicopters generate lift using a combination of Bernoulli's principle, the Coandă effect, and Newton's laws of motion.

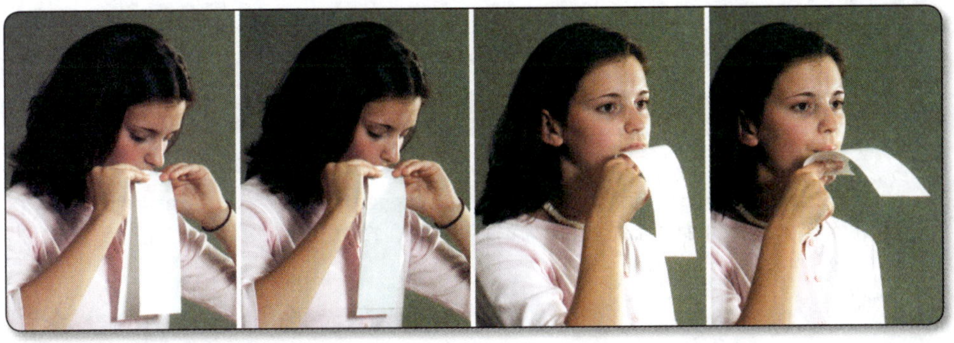

17-30 These two demonstrations show how unbalanced forces across a fluid boundary created by the Coandă effect can produce work.

17.19 Real Fluids

In real-world aerodynamics and hydrodynamics, designers and engineers must take into account friction, fluid viscosity, and turbulence, all of which greatly complicate the analysis of fluid flow around immersed objects.

The cohesive forces between particles of a fluid produce a type of internal friction called **viscosity**. Viscosity determines how freely a fluid flows. The coefficient of viscosity (η) indicates the resistance of a fluid to flow. Lower viscosity coefficients mean that the fluids flow more easily than those with higher viscosity coefficients. For instance, water has a lower viscosity than molasses. Viscosity is familiarly known as the "thickness" of a fluid.

The adhesive forces between a flowing fluid and its confining walls (such as a pipe) cause the outermost particles of fluid to be essentially stationary. Particles farther into the fluid have small velocities that increase with distance from the wall of the pipe.

17-31 Viscosity determines the ease with which objects move through a fluid. Both marbles were dropped at the same instant. The left liquid is a thick syrup; the right liquid is water.

The coefficient of viscosity (η) is given the SI unit of pascal-second (Pa·s). An older unit of viscosity is the poise. Viscosity coefficients are tabulated at a specified temperature, since temperature significantly affects cohesive forces.

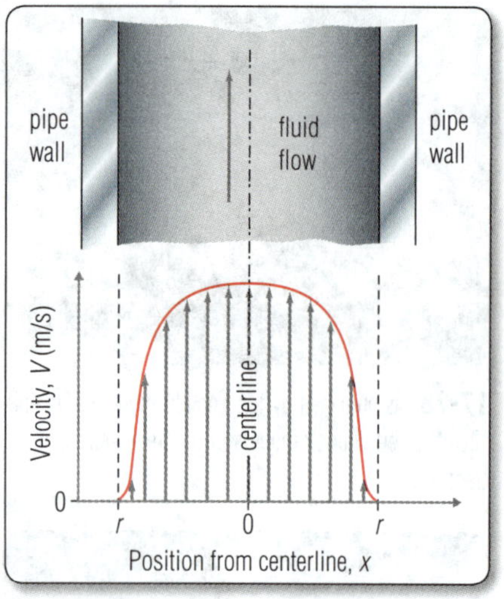

17-32 Viscosity and friction with the wall surface reduce fluid velocity near the margins of a pipe.

The fluid particles at the center have the highest velocities. The farther a particle is from the wall of the pipe, the faster it goes. Therefore, the flow of fluid at the center of a large pipe is faster than the fluid at the center of a smaller pipe at the same supply pressure. This effect is a combination of friction with the pipe and fluid viscosity. According to the principles of energy conservation, the conversion of the mechanical energy of fluid particles to thermal energy (internal energy) through friction and viscous flow results in a decrease in fluid pressure and kinetic energy (flow velocity) along the length of a pipe. This is why the shower in a one-story house farthest from the water supply pipe has the slowest flow (least pressure).

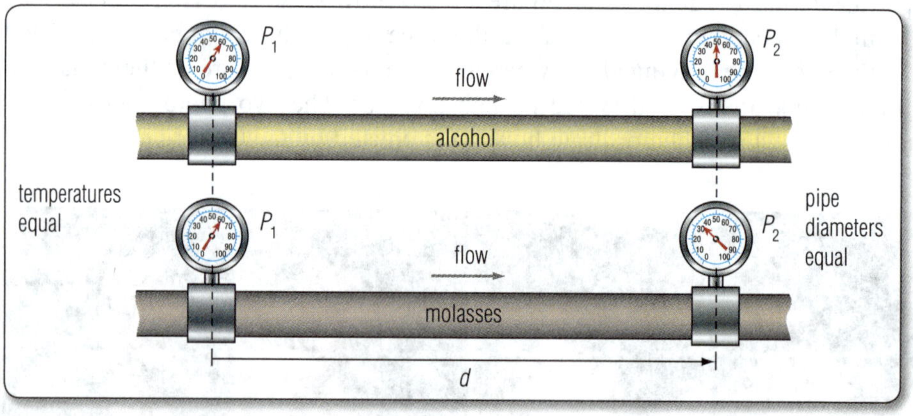

17-33 Viscosity lowers pressure with distance as a fluid moves through a pipe.

17.20 Using Fluid Mechanics to Solve Problems

Factors that drive the investigation of new electrical energy sources are cost, accessibility, safety, impact on the environment, and renewability. One emerging technology, which really isn't all that new, is wind power generation. How can wind be most effectively used to generate electricity?

People have been using wind power for centuries. For example, the Dutch used windmills to pump water from their low-lying lands, and many countries lacking extensive river systems used wind to grind grain. In an attempt to harness wind energy, engineers have used computers to develop aerodynamic *wind turbines*. To maximize efficiency and reduce structural stresses, wind turbines are elevated well above the turbulent ground breezes to expose their blades to smoother winds. The largest wind turbines have a hub height nearly three times that of the Statue of Liberty!

Wind turbines have a rotor that mounts two or three long, narrow blades. As air flows over the blades, it creates lift similar to that on an airplane's wings. The rotary force exerted on the blades spins the turbine, which turns an electrical generator that converts mechanical energy to electrical energy. Wind turbines may be used to supply part or all of the electrical needs of a single dwelling, or they may provide large-scale power generation, which requires hundreds of wind turbines in facilities called *wind farms*.

To have the greatest economical benefit, wind turbines need to be efficient. However, the maximum theoretical efficiency for flow machines such as wind turbines is only about 59%. That is, 59% of the kinetic energy available in the

17-34 The world's largest wind turbines are capable of generating more than 5 MW of electrical power.

wind is converted into mechanical energy. Even more loss occurs as mechanical energy is converted to electrical energy. Advanced technology wind turbine installations can achieve efficiencies of 40–50%.The Department of Energy estimates that the United States could potentially supply up to 20% of its energy needs with wind power by the year 2030. This would provide electricity at only a fraction of the cost of coal, gas, and nuclear energy without the accompanying environmental and public hazards. However, many political and economic challenges must be solved before this goal can be reached.

Decisions, Decisions

In spite of wind power's apparent economical benefits, many people oppose wind farms because of the noise, visual clutter, and effects on bird populations. What may people presuppose when making decisions about the benefits and liabilities of wind power?

FACETS of PHYSICS

Aerodynamically Designed Cars

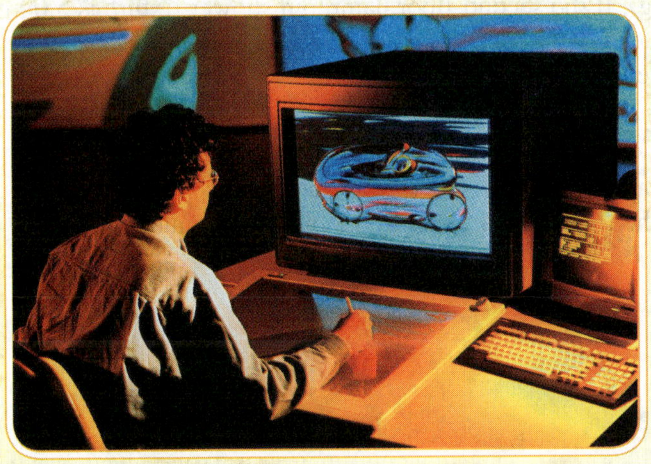

In order to conserve energy and reduce emissions, automobile designers are doing their best to make cars more fuel-efficient. Although increasing engine efficiency and making cars lighter are important, much of the fuel savings comes from designing the body to move smoothly through the air.

Actually, the idea of an aerodynamically efficient car is not entirely new. As early as 1934, Chrysler produced the Airflow, a sleek vehicle with smooth, rounded curves and fenders. The idea did not become popular, however, until the fuel shortage crises of the 1970s forced manufacturers and consumers to explore every avenue of cutting fuel consumption.

At highway speeds, a car uses more than half its power to overcome air resistance. Therefore, reducing air resistance reduces fuel consumption. The main goal is to make the flow of air around the car as nearly laminar as possible. Models of new designs are first tested using sophisticated computer modeling. Fractional-scale and full-scale testing is done in wind tunnels, where smoke streams around the models, revealing turbulence and potential drag.

If you compare newer passenger cars (including SUVs and energy hybrids) with older models, you will see that the latest are designed to let air flow smoothly over, under, and around them. Headlights are molded into the fenders, front hood lines flow smoothly into the front bumper structure, windows are flush with the body, and the overall profile promotes the laminar flow of air over the vehicle. Even the underside of the car is engineered to remove as many obstructions to the flow of air as possible. Most recreational and crossover vehicles, despite their boxy profiles, are designed to minimize wind resistance.

These efforts, combined with more efficient engines, hybrid designs, and fuel cell-powered electric motors, have dramatically improved the fuel efficiency of passenger vehicles.

17B Objectives

After completing this section, I can

- ✓ define an ideal fluid.
- ✓ describe the various types of flow in a fluid.
- ✓ use the concept of a flow tube to establish the principle of flow continuity.
- ✓ identify each term in Bernoulli's equation and state the assumptions for its use.
- ✓ use Bernoulli's equation to predict flow rate and liquid pressure under varying conditions.
- ✓ explain lift according to the model of the Coandă effect.
- ✓ explain how real fluids act when flowing through pipes and the consequences of fluid friction and viscosity.
- ✓ describe a wind turbine and state wind power's potential for contributing to the nation's electrical generation capacity.
- ✓ evaluate the challenges of implementing wind power on a large scale.

17B Section Review

1. Describe the properties of an ideal fluid. What kinds of fluids approach the description of an ideal fluid most closely?
2. When you are taking a shower, where is the water moving fastest—in the ½-inch diameter pipe connected to the shower nozzle or the ¾-inch diameter pipe connected to the house water meter? Explain your answer.
3. Explain why most municipal water supply tanks are located on top of nearby hills or on tall structural steel supports.
4. The Trans-Alaska Pipeline System connecting the Alaskan North Slope oil fields to the oil terminals on the coast extends 800 miles over land. Why does the pipeline have pumping stations placed at regular intervals along its length?

DS 5. Explain how a wind turbine converts wind energy into electricity.

DS 6. Why are tall wind turbine towers preferred for wind power generation?

⊙7. If water flows at 1.00 m/s in a 2.54 cm diameter garden hose, approximately how fast is it moving when it exits the 0.25 cm diameter orifice at the end of the hose nozzle?

⊙8. Assume that water pressure supplied to your house is 4.14×10^5 Pa (about 60 psig). If the second-floor bathroom shower fixture is 5.00 m above the water supply connection to the house, what is the (static) water pressure (in Pa) at the shower head?

⊙9. Assume that your parents' shower is on the opposite side of the wall from the shower you use. Using Bernoulli's equation, explain what happens to your shower pressure if your father turns on his shower and doubles the flow rate of water in the pipe supplying both shower fixtures.

Chapter Review

In Terms of Physics

fluid mechanics	383
hydrostatics	383
hydrodynamics	383
density (ρ)	384
specific gravity (s.g.)	384
atmosphere (atm)	385
torr	385
bar	385
millibar (mb)	385
gauge pressure (P_g)	385
absolute pressure (P)	385
Pascal's principle	389
hydraulic device	389
manometer	391
barometer	391
buoyant force ($\mathbf{F}_b$)	392
Archimedes' principle	392
center of buoyancy	394
hydrometer	394
ideal fluid	396
streamline	396
laminar	396
turbulent	396
flow tube	396
equation of flow continuity	398
Bernoulli's equation	400
lift	401
airfoil	401
hydrofoil	401
Coandă effect	402
viscosity	403
coefficient of viscosity (η)	403

Problem-Solving Strategies

17.1 (page 388) Complex exponential expressions are common in physics. In order to prevent confusion, the term "exp" is inserted before the expression in order to alert the reader that the following parenthetical statement is the exponent of the natural exponential base *e*. All units in the exponential expression must cancel.

17.2 (page 391) Under normal circumstances, the difference of pressure due to height differences between input and output pistons in hydraulic devices is negligible compared to the force transmitted by the hydraulic system.

17.3 (page 393) If you need to know if an object will float in a liquid, compare the densities of the object and the liquid. If the object's density is less than the liquid's density, it will float. Otherwise, it will submerge or even sink.

17.4 (page 398) It is permissible to use different metric systems in one calculation as long as the units cancel. This often occurs in simple proportional calculations.

17.5 (page 400) In an ideal flow tube, if the flow cross section remains constant, then fluid pressure depends only on the height of the fluid above the reference height, according to Bernoulli's equation.

17.6 (page 401) In an ideal flow tube, if the height remains constant, fluid pressure will increase when fluid speed decreases and vice versa, according to Bernoulli's equation.

Review Questions

1. What is specific gravity?

2. What is the SI unit of pressure?

3. The containers in the figure contain the same liquid at the same height. Which has greater pressure at its base?

4. The pressure of a liquid at the middle of a sealed container, 10 cm below its surface, is 1.0 atm. What pressure does the liquid exert on the sides of its container at 10 cm below its surface?

5. Explain why atmospheric pressure varies with height in a different way from the way liquid pressure varies with depth.

6. Two spheres are lowered into a pool of water to a depth of 3.0 m. One sphere has twice the surface area of the other. How does the pressure on the larger sphere compare to the pressure on the smaller sphere?

7. Which of the following instruments are designed to measure pressure?
a. thermometer
b. barometer
c. hydrometer
d. manometer

8. State whether the following will float in mercury, in water, in both, or in neither.
a. gold
b. cement ($\rho = 2.8$ g/cm^3)
c. iron
d. wood (oak)
e. lead

9. How can a ship made of steel float on water?

10. Many bony fish have an organ called a swim bladder that assists them in maintaining proper buoyancy. The volume of this gas-filled bag within the fish's organ cavity is changed as necessary. If the fish is negatively buoyant, how must the swim bladder change in volume in order to make the fish neutrally buoyant?

11. Discuss how streamlines are used in a flow diagram to indicate the movement of fluid (include ideas such as spacing, direction, etc.)

12. In what section of the pipe will water run fastest: *A*, *B*, or *C*?

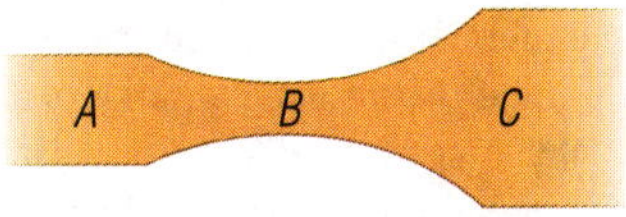

13. What assumptions are made when deriving Bernoulli's equation?

14. Using the Coandă effect, explain why a shower curtain tends to billow into the shower when the water is running.

15. Some backyard swimming pool owners remove the water from their pools using devices called aspirators attached to their garden hoses. Explain how such a device might work.

16. What effect does the cohesive attraction between particles in fluids have on fluid flow?

True or False (17–31)

17. A fluid is a liquid.

18. Specific gravity is the ratio of the density of a substance to the density of water.

19. One torr is $\frac{1}{760}$ of normal atmospheric pressure.

20. Liquid pressure increases with depth from the reference height.

21. The pressure in a fluid is dependent only on the depth (or height) in the fluid and the pressure at the reference height.

22. Atmospheric pressure decreases at a nearly constant rate with height.

23. A hydraulic device is effective because a small force applied to a large-diameter piston transmits the pressure via hydraulic fluid and produces a large force on a small-diameter piston.

24. The buoyant force on an object is proportional to the density of the fluid and inversely proportional to the volume of fluid displaced.

25. As a helium balloon rises in the atmosphere, it becomes heavier and eventually stops rising.

26. A floating object is stable if the torque produced by the misalignment of the centers of gravity and buoyancy tends to move them back into the same vertical line rather than away from it.

27. It is valid to assume that fluids are incompressible when discussing liquids, but it is not valid when discussing gases.

28. The equation of flow continuity shows that the rate at which a mass of fluid enters a flow tube must equal the rate at which it exits.

29. Doubling the speed of a fluid reduces its pressure to half the original value, according to Bernoulli's principle.

30. An airfoil produces lift because air exerts a drag on the lower surface of the foil as it moves through the air.

31. Larger pipes experience a higher flow rate for a given inlet pressure than smaller pipes.

✪32. The density of sea water is about 1030 kg/m^3. The pressure at the top of the ocean is usually 101 325 Pa. The Marianas Trench is 10 924 m below sea level. What is the pressure there?

✪33. A mercury-containing glass tube sealed at its upper end is inverted with its open end in a dish of mercury. Above the mercury in the tube is a vacuum ($P = 0$ Pa). The mercury in the tube settles 76.0 cm above the surface of the mercury in the dish. What is the pressure at the surface of the mercury in the dish? ($\rho_{Hg} = 13.6$ g/cm^3)

✪34. A hydraulic jack has a small piston with an area of 3.50 cm^2 and a large piston with an area of 31.5 cm^2.

a. What force is needed on the small piston to lift a 250. kg engine with the large piston?

b. If a man can exert a force of 175 N on the small piston, how much lifting force can he generate using the jack?

c. Suppose the jack is an ideal hydraulic device. (An ideal hydraulic device is one in which the fluid does not compress and there are no frictional losses.) How far must the small piston move to move a load 1.00 m?

d. What is the ideal mechanical advantage of the jack? (IMA = A_{out}/A_{in})

✪35. The figure below shows an open-tube manometer connected to a gas sample. Atmospheric pressure (P_{atm}) is 1.01×10^5 Pa. The mercury in the open end of the manometer stands 0.450 m above position *A*.

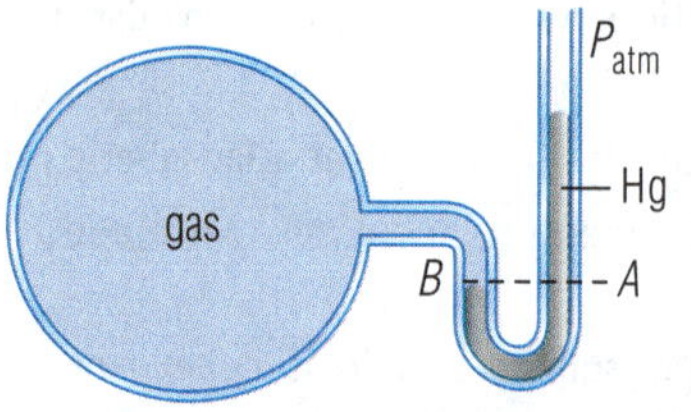

a. What is the pressure in the column of the mercury at point *A*?

b. What is the pressure of the gas at point *B*?

✪36. A rock with a mass of 1.53 kg weighs 13.5 N when it is immersed in water.

a. What is the buoyant force on the rock?

b. What is the rock's volume?

c. What is the rock's density?

✪37. When fully immersed in a liquid, a 20.0 g mass exerts a force of 1.00×10^{-1} N on a supporting spring scale. The mass has a volume of 2.75 cm^3.

a. What is the buoyant force on the mass?

b. What volume of liquid does the mass displace?

c. What is the liquid's density?

✪38. A level pipe narrows from a cross-sectional area of 11.5 cm^2 to a cross-sectional area of 5.00 cm^2. If a liquid moves at a speed of 0.50 m/s through the wider section, how fast will it go through the narrow section?

✪39. Water runs through a hose with a cross-sectional area of 11.5 cm^2 at a speed of 1.50 m/s. A child playing with the hose places a hand over its end, leaving a 0.200 cm^2 opening for the water. At what speed does the water pass the child's hand?

DS✪40. Inspect the following map of the locations and quality of wind resources in the United States. Which locations seem the best suited for wind power? Why? What kinds of challenges would their residents face? Consider the geographical, environmental, and political aspects of these decisions.

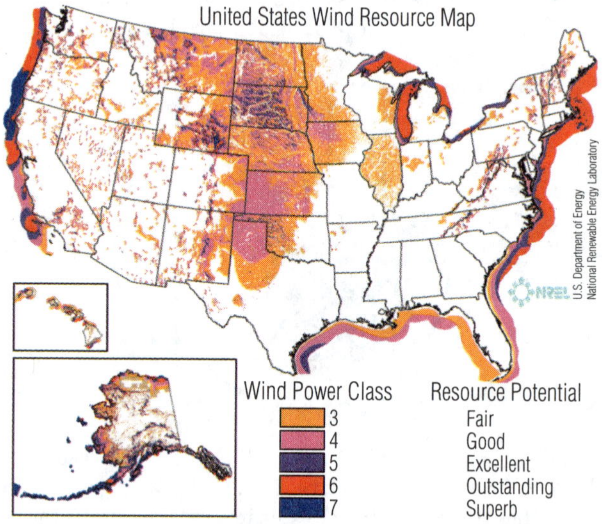

✪41. The force exerted by air pressure can be dramatically demonstrated using a pair of simple laboratory devices called *Magdeburg hemispheres*, named after the town in Germany where their inventor, Otto von Guericke, was mayor. He demonstrated his air pump invention to Emperor Ferdinand III in 1654. He constructed two hollow metal hemispheres with a gasket between them that were then evacuated using his air pump. Two teams of 8 horses each could not pull them apart. Assuming that the hemispheres were 50.0 cm in diameter, that they were completely evacuated, and that standard atmospheric pressure existed on that day,

a. what was the magnitude of the force exerted on each hemisphere by air pressure?

b. what was the minimum force each horse would have had to exert in order to pull the hemispheres apart?

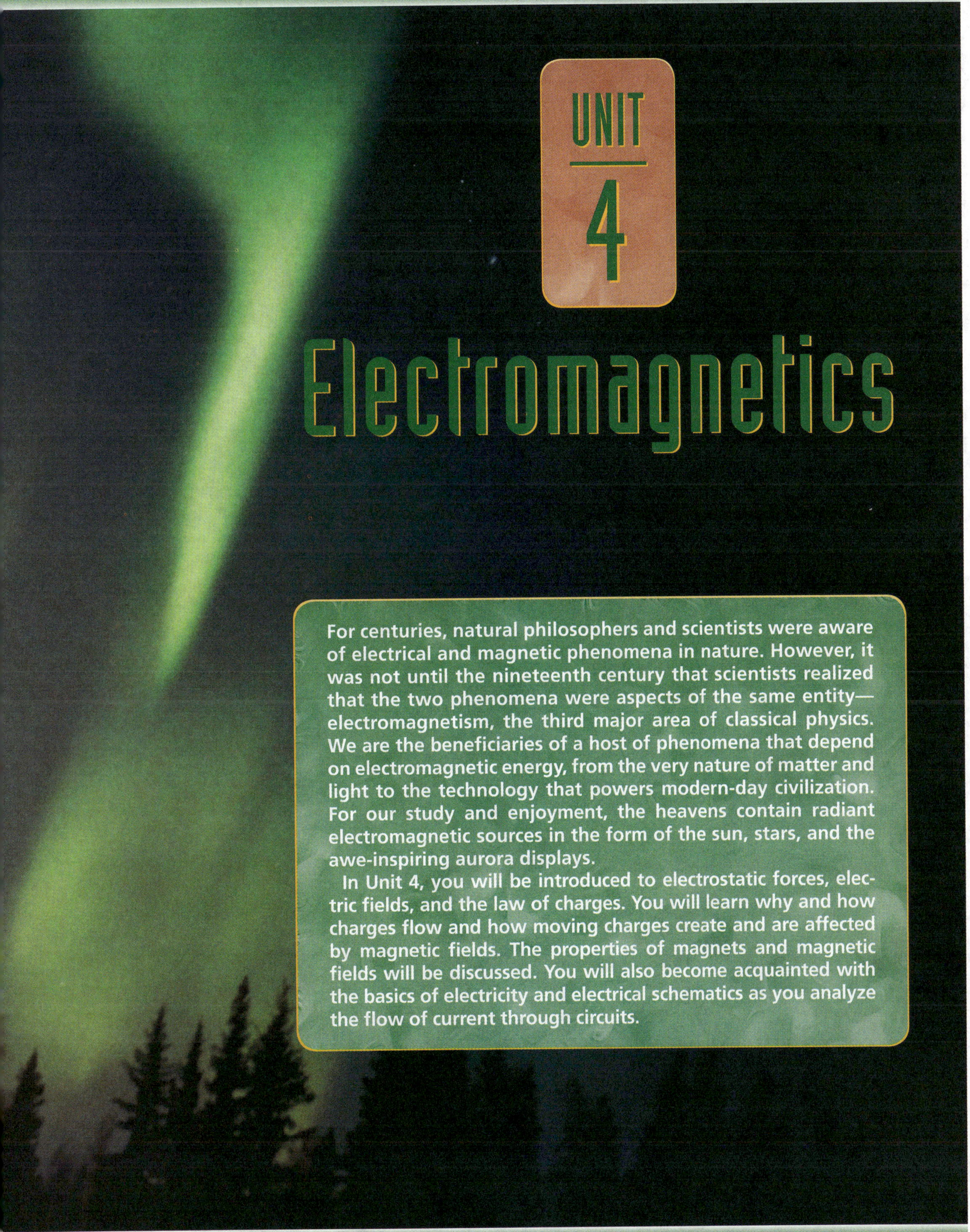

UNIT 4

Electromagnetics

For centuries, natural philosophers and scientists were aware of electrical and magnetic phenomena in nature. However, it was not until the nineteenth century that scientists realized that the two phenomena were aspects of the same entity—electromagnetism, the third major area of classical physics. We are the beneficiaries of a host of phenomena that depend on electromagnetic energy, from the very nature of matter and light to the technology that powers modern-day civilization. For our study and enjoyment, the heavens contain radiant electromagnetic sources in the form of the sun, stars, and the awe-inspiring aurora displays.

In Unit 4, you will be introduced to electrostatic forces, electric fields, and the law of charges. You will learn why and how charges flow and how moving charges create and are affected by magnetic fields. The properties of magnets and magnetic fields will be discussed. You will also become acquainted with the basics of electricity and electrical schematics as you analyze the flow of current through circuits.

Lightning is nothing more than the discharge of static electricity as charges move between the ground and the clouds.

Electric Charge 18

Dominion Modeling

Fundamental Electricity

What causes lightning? What makes your hair stand up on your head when you rub it with a balloon? Why does a doorknob spark when you touch it with your hand after shuffling across a carpeted floor? All of these are common examples of *static electricity*. When objects are rubbed, they become charged by gaining or losing electrons. Since electrons are individual charged particles, the measure of the object's charge must show a corresponding ratio to an individual unit of charge. In other words, charge is quantized, not continuous. So there must be only certain values of charge that an object can have. These values of charge are related to the charge of a single electron, but what is the charge of an electron?

18-1 A spark generated by a laboratory static electricity demonstration device called a *Wimshurst machine*

18A ELECTRIFICATION

18.1 Introduction

What is electricity? You probably use electricity every day, but what do you know about it? Most Americans realize that electricity is carried by wires; that it allows stoves to heat, refrigerators to cool, and computers to function; that it costs money; and that a modern city is crippled without it. Beyond such simple ideas, many know little about electricity.

18-2 Electricity is essential to the existence of a modern city.

The word "static" in "static electricity" means fixed or motionless. In static electricity, charges are usually fixed but are capable of moving certain distances for brief moments (even though those distances might be great, such as with lightning). In contrast, the charges in current electricity are continuously moving.

18-3 A beautiful sample of amber containing a trapped insect

Amber is fossilized resin that oozed naturally from the bark of trees thousands of years ago. Because the majority of amber samples found today are obtained from buried deposits, it is likely that these trees were inundated in the Genesis Flood. Amber has traditionally been used for jewelry.

18.2 History

Six hundred years before Christ, an unknown Greek philosopher recorded that a piece of amber rubbed with fur would attract dust and bits of dry leaves. The Greek word for amber, *elektron*, became associated with this attraction, and today it is called **static electricity**.

An Elizabethan physician, William Gilbert, was among the first to experiment extensively with *electrification*. He discovered that glass and other materials besides amber would attract small objects after being rubbed with fur or cloth. Gilbert called a material that can be electrified an *electrick*.

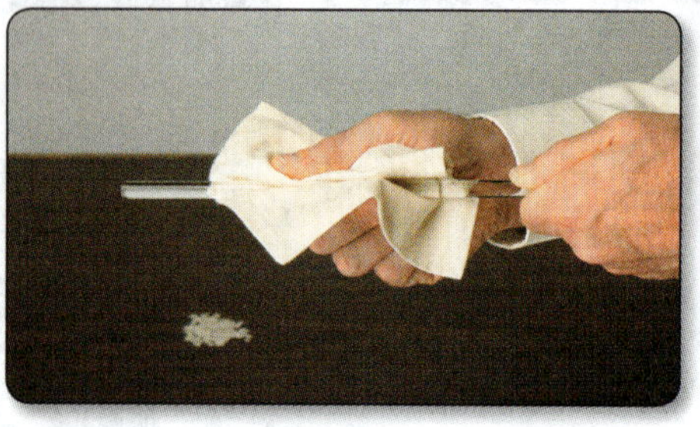

18-4a Charging an "electrick"

18-4b The charged "electrick" attracts bits of paper.

Many different materials are "electricks." As scientists experimented with these materials, they discovered that they can affect other similar materials as well as small objects. For example, an electrified amber rod repelled another electrified amber rod, and an electrified glass rod repelled another electrified glass rod. Also, an electrified glass rod attracted an electrified amber rod. Scientists divided electricks into two categories: materials that acted like amber were called **resinous** electricks (amber is fossilized resin), and materials that acted like glass were called **vitreous** electricks (*vitro* is Latin for "glass").

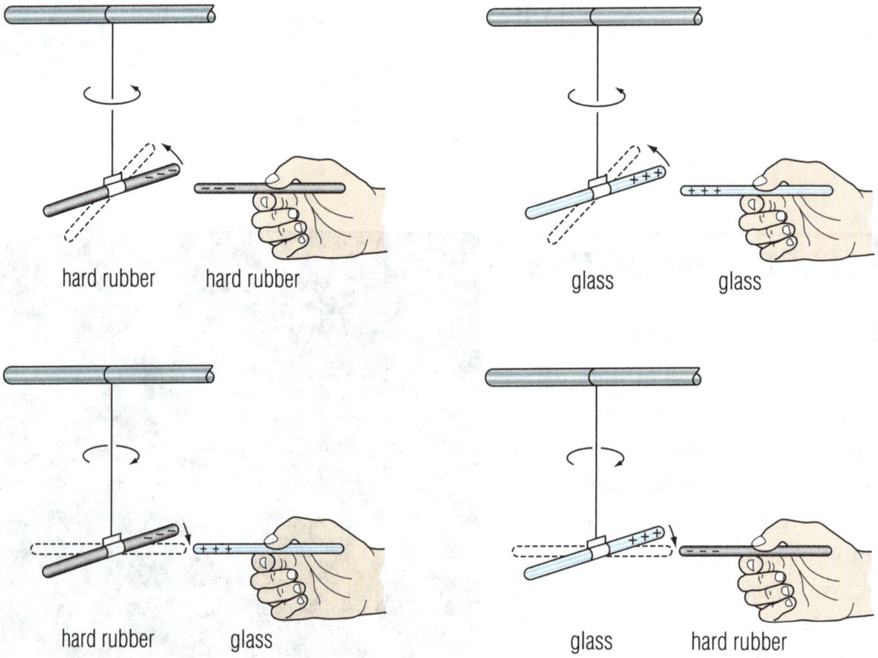

18-5 Charged rods of different materials demonstrate the law of charges. No charge attracts or repels *both* of the known kinds of charge. Therefore, it can be concluded that there are only two kinds of charge.

After Gilbert, Charles Du Fay proposed an explanation for the existence of resinous and vitreous electricks. Du Fay thought that each electrick contained an electrical fluid. Since there were two kinds of electricks, Du Fay decided that there must be two kinds of fluids: resinous fluids, contained by resinous electricks, and vitreous fluids, contained by vitreous electricks.

Benjamin Franklin, on the other hand, thought that only one kind of electrical fluid existed. He believed that vitreous electricks had an excess of fluid, while resinous electricks had a deficiency of fluid. He called vitreous electrification *positive* and resinous electrification *negative*. (This designation will become important in the discussion of electric current in later chapters.)

Franklin and Du Fay agreed about the cause of the electrification. They said that when an electrick was rubbed, some fluid was transferred between the electrick and the cloth. The resulting imbalance of fluid produced an electric *charge* on the electrick. The electric charge enabled the electrick to act according to the law of charges: unlike charges attract, and like charges repel.

Charles-François de Cisternay Du Fay (1698–1739) was an extremely gifted and intelligent French academician. Originally an army officer, he devoted his life to investigating many areas of physical and biological science, including electrostatics and the properties of crystalline solids.

Benjamin Franklin (1706–90) was an American printer, philosopher, statesman, scientist, and inventor.

The **law of charges** states that opposite charges attract each other, whereas like charges repel each other.

FACETS of PHYSICS William Gilbert (1544–1603)

From ancient times, people have been fascinated by mysterious phenomena such as lightning, electric eels, and the sparkling glow on ship's spars and mastheads (called St. Elmo's fire). Two materials received special attention: natural magnets from the fields of Magnesia (called lodestones) and fossilized tree resin (called amber). When amber was rubbed vigorously, it would attract particles of straw and wheat. The Greeks attributed this force to *sympathy*; in other words, amber was being friendly. The lodestone, on the other hand, attracted heavy metals such as iron tools. When people noticed that a lodestone needle (a compass) always pointed north, they proposed many bizarre theories to account for this magnetic effect. None of the theories were investigated scientifically until the time of William Gilbert.

Gilbert was a trained physician who served in the court of Queen Elizabeth I. In his spare time, he experimented with the effects of amber and lodestone. He felt contempt for earlier thinkers who studied these effects only by reading what ancient authorities had said. He himself would accept only those theories that he could support with experiments.

Gilbert noticed that a lodestone always has two poles and that the poles are not points but regions. He distrusted the popular explanations for the behavior of the compass needle. He offered a dramatic new theory that the earth itself is a lodestone with two poles, one north and one south. To support his idea, he made a huge spherical lodestone and placed a tiny magnetized needle on it. The needle behaved exactly like a compass would on the earth.

While studying amber, he discovered that many other common materials can become charged by friction. He called this force that attracted objects *electrick*. His experiments showed for the first time how electricity is different from magnetism: magnetism attracts few materials to two poles and is unaffected by intervening materials, whereas the electrostatic force attracts unlimited materials to one center and is greatly affected by intervening materials. Gilbert published his findings in *De Magnete* (1600), which established him as one of the foremost English scientists of his day.

Gilbert's book is important, not only because it summarizes everything known in his day about magnetism, but also because it encouraged others to test their ideas with experiments. The science of electromagnetism eventually rose from this beginning.

For some reason, early scientists were enamored with the idea of immaterial fluids (fluids that could not be sensed by normal means) as a way of explaining phenomena that involved the "flow" of some property. In the nineteenth century, scientists assumed that a fluid-like *luminiferous ether* was responsible for the transport of light. Many scientists believed that the fluid *caloric* was heat, and others proposed that *electrick* fluid gave characteristic charges to certain materials.

Sir Joseph John "J.J." Thomson (1856–1940) was an English physicist who participated in much of the early research into the structure of the atom and the nature of electrical conduction. He won the Nobel Prize in Physics in 1906.

Robert Millikan (1868–1953) was an American physicist who worked chiefly in the fields of electricity, optics, and molecular physics. He won the Nobel Prize in Physics in 1923.

18.3 Discovery of the Electron

The idea of an electric fluid implied that electric charge was continuous (i.e., there could be any amount of electric charge). In 1897 the English scientist J. J. Thomson disproved this idea when he demonstrated with cathode rays that all matter contains tiny negatively charged particles with the same charge-to-mass ratio. He initially called these particle-like entities *corpuscles*, which was merely the scientific term for unidentified particles. Previously, in 1891, the Irish physicist George Stoney suggested calling the basic electrical charge the **electron**, which became the accepted name for this charged particle. As a consequence, an object could only have a charge that was an integer multiple of the charge of an electron. The magnitude of this charge remained unknown for more than two decades.

18.4 Modeling the Fundamental Charge

In 1913 Robert Millikan reported on his experiment to find the charge of an electron. He assumed that tiny oil drops would reach terminal velocity in air when falling due to gravity and, if charged, they could also reach terminal velocity in air when falling *upward* in an electrical field. He suggested that the force exerted by the field would produce upward velocities proportional to the number of charges on the drops, and these charges would thus be multiples of the fundamental charge.

The apparatus Millikan used is shown in Figure 18-6. He sprayed oil drops into the upper chamber. They then fell through the hole in the upper electrostatic plate. X-rays ionized the air molecules between the plates, and some of the free electrons attached themselves to the oil drops. Focusing on a single oil drop with a calibrated microscope, Millikan measured its terminal velocity while allowing it to fall influenced only by gravity and air resistance. He then applied a high voltage to the upper plate, creating an electric field that caused the drop to rise. He measured the slow upward terminal velocity of the charged oil drop. He observed each drop fall and rise multiple times over the course of an hour or more. His report documents observations for 58 drops taken over a period of 60 days.

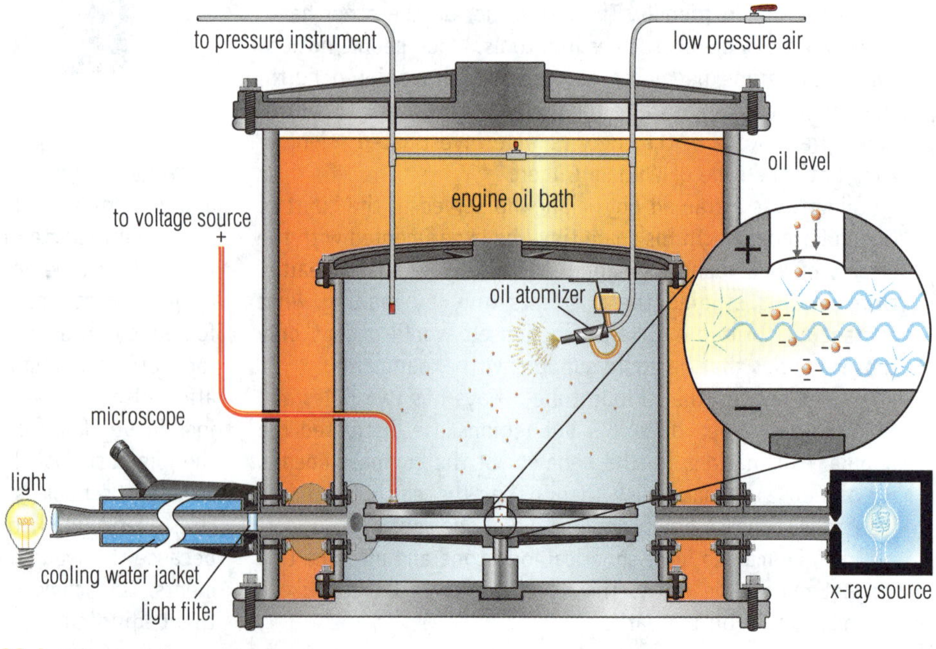

18-6 Millikan's oil-drop experiment apparatus

Millikan understood that when a drop of oil was moving at terminal velocity, all forces acting on the drop were balanced (see Chapter 10). For a falling drop, these forces are the weight of the drop and the resistance of air due to its viscosity. For a rising drop in an electric field, the forces are the electrostatic force, the drop's weight, and the air resistance. He chose to measure these two velocities to determine the charge on a single drop because it was easier to accomplish than trying to balance the weight of a drop and the electrostatic force on a motionless drop. However, it is conceptually easier to determine the charge on the drop when it is held stationary in the electric field. For this technique, only the falling terminal velocity needs to be measured in order to find the weight of the drop. Accordingly, the total charge on an oil drop in coulombs could be calculated using the following equation:

$$q_n = 1.591\,22 \times 10^{-8}(1/V)\sqrt{v_T^3/\rho},$$

where

- q_n is the total charge on a drop of oil, in coulombs (C),
- V is the potential difference between the plates, in volts (V),
- v_T is the falling terminal velocity of the drop, in m/s, and
- ρ is the net density of the oil drop in air, in kg/m^3.

The coulomb (C) is the SI unit of electrical charge. See Subsection 18.11 for a discussion of electrical charge units.

Below is a table that lists some of Millikan's actual data from his oil-drop experiment. The data is reported here in SI units, instead of the CGS units that he used (see Subsection 18.11 for a definition of the CGS system). After calculating the total charge on each drop, he plotted the data on a graph to determine the **fundamental charge (*e*)**, which would be the charge on an oil drop that had only one excess electron. You will use these values in a similar way to determine your own value of *e*.

The net density is used to determine the weight of the oil drop, because air exerts a buoyant force on the drop. It is the difference of the densities of the oil used for the experiment and air at 23 °C.

TABLE 18-1

Selected Oil Drop Experiment Data (Millikan, 1913)

Drop	Velocity (v_T)	Voltage (V)	Net Density (ρ)	Drop	Velocity (v_T)	Voltage (V)	Net Density (ρ)
no.	m/s	V	kg/m^3	no.	m/s	V	kg/m^3
1	2.357×10^{-3}	5168	918.71	31	2.649×10^{-4}	5024	919.33
3	1.032×10^{-3}	5100	918.75	33	5.601×10^{-4}	5018	919.53
5	9.575×10^{-4}	5072	918.75	35	1.103×10^{-3}	4582	919.65
7	8.368×10^{-4}	5090	918.73	37	8.496×10^{-4}	4236	919.64
9	7.375×10^{-4}	5139	918.72	39	5.115×10^{-4}	4246	919.59
11	6.720×10^{-4}	5066	918.74	41	4.254×10^{-4}	4236	919.60
13	5.463×10^{-4}	5024	918.71	43	7.340×10^{-4}	3370	919.69
15	5.274×10^{-4}	5078	918.72	45	3.425×10^{-4}	3317	919.63
17	5.545×10^{-4}	5060	918.75	47	7.921×10^{-4}	3254	919.75
19	3.570×10^{-4}	5033	918.71	49	6.815×10^{-4}	2550	919.74
21	2.893×10^{-4}	5018	918.74	51	9.346×10^{-4}	2546	919.77
23	2.554×10^{-4}	5098	918.72	53	3.055×10^{-4}	2539	919.70
25	6.737×10^{-4}	5066	919.29	55	4.206×10^{-4}	2451	919.74
27	1.085×10^{-3}	4570	919.50	57	6.599×10^{-4}	2321	919.78
29	4.393×10^{-4}	5073	917.44				

Source: R. A. Millikan, "On the Elementary Charge and the Avogadro Constant," ed. F. Bedell, *The Physical Review*, 2nd ser., 2 (1913): 109–143.

The proton was discovered as a result of a series of elegant experiments, beginning in 1886 when Eugen Goldstein reported the discovery of positively charged *canal rays*. In 1902, Wilhelm Wien, another German scientist, determined the charge-to-mass ratio of the positively charged particles in the canal rays. Ernest Rutherford, after his associates discovered the nucleus in gold atoms, may have coined the name *proton*.

Conductors are usually materials composed of the *metal* elements, principally from the middle and left side of the periodic table (see Appendix H), and their alloys. **Insulators** are usually composed of *nonmetal* elements from the upper right-hand corner of the table and their compounds. **Semiconductors** are the elements, called *semimetals* or *metalloids*, that are located between the other two kinds on the periodic table.

You experience more static cling in your clothes and static shocks from doorknobs during the winter because the relative humidity of the air is so much lower during that season. During the humid summer months, the water vapor in the air carries away the static charges before they can build up.

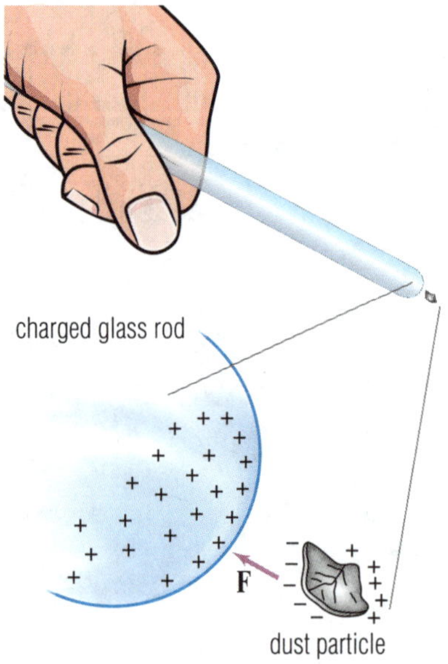

18-7 A charged rod induces a charge on a particle of dust.

Millikan rightly assumed *e* to be the charge on a single electron. In the many years since Millikan's experiment, his results have been validated and refined, but not fundamentally changed. All electrical charges on atoms, molecules, and ions are multiples of *e*; none of these has been found to be a fraction of *e*.

Experiments have shown that there are only two kinds of charge, so in this respect modern scientists agree with Du Fay. However, Franklin's terms were used to identify the charges, although any terms identifying opposites could have been used. The charge on an electron was termed negative ($-e$) because in Thomson's experiment the electrons were attracted to the electrode that was positively charged (according to Franklin's designation).

The proton was discovered several years after the electron and was found to have a positive charge. Further experimentation determined that it possessed a single fundamental charge ($+e$). In solids, the positively charged protons in an atom do not move appreciably because the atoms themselves are in fixed positions. The outer electrons in the atoms of some materials are loosely held and can be easily removed. Electrons moving from one region to another in or on the object cause the electrical charge on a solid. When electrons leave an area, the remaining protons produce a net positive charge. In the areas to which the electrons move, the extra electrons produce a net negative charge. Franklin's theory was close to modern theories for solids, but modern theories substitute electrons for his electrical fluid.

18.5 Classification of Substances by Electrical Properties

Not all solids have atoms with mobile electrons. Those solids that *do* have free electrons are called **conductors**; metals are a good example. Materials that hold their electrons tightly are called **insulators**; examples of good insulators are porcelain, rubber, and glass. Insulators can be used to confine electricity, as with the insulation around a wire. **Semiconductors** are materials that usually act like insulators but can act like conductors under certain conditions.

Gilbert's electricks were insulators that gained or lost electrons when rubbed. When a neutral object, such as a speck of dust, comes near such a charged material, its electrons are attracted or repelled by the charge. The side of the dust speck nearest the charged insulator then has a charge opposite to the insulator's charge, and the dust speck is attracted to the insulator. When the dust speck touches the insulator, electrons move to give the speck some of the insulator's excess charge. The speck then has the same charge as the insulator and is therefore repelled. The same process happens with water molecules present in the air. As a result, the charge on the insulator does not last long but leaks away into the air. A conductor also gains or loses electrons when rubbed, but because its electrons are much more mobile, it conducts its charge to the air almost immediately.

18A Section Review

1. What was the substance that was first reported to have electrical properties? How is its name preserved and used today?
2. Briefly summarize the scientists and discoveries leading to the current understanding of electrical charges.
3. State the law of charges. Explain why, based on this law, scientists believe that there are only two kinds of electrical charge.
4. Why was the electron assigned the negative charge?

5. Of what category of electrical materials would an effective lightning rod system be constructed?

6. The "electricks" (test rods) that are used in school labs are made of what category of electrical material?

DM✪7. List the variables that Millikan had to control in his famous oil-drop experiment.

DM✪8. Millikan used the terminal velocities of the drops to determine their masses, and he used this information to calculate the charge on the drops and, ultimately, the fundamental charge.

a. Calculate the charge on each oil drop listed in Table 18-1 on page 417. Use the formula given in Subsection 18.4. It is best to use a spreadsheet program. Your teacher may provide you the raw data already placed in a spreadsheet.

b. Millikan believed that each charged oil drop carried one or more excess electrons, so its charge (q_n) was an integer multiple (n) of the fundamental charge (e). This idea could be expressed mathematically as $q_n = en$. To create a simple graph of q_n vs. n, first order your charge data from the smallest charge to the largest. Create a scatter plot with q_n on the y-axis and integers (1, 2, 3, . . .) on the x-axis (most spreadsheet programs can do this automatically). Describe how the data deviates from a straight line. What might account for this?

c. Assign a best-fit linear trendline through the data points. Include the equation and r^2 value of the fit, if possible. If you can, force the line to pass through $y = 0$. What is the significance of the slope of this line?

18A Objectives

After completing this section, I can

- ✓ summarize the discovery of electrical charge.
- ✓ explain the naming of the two kinds of charge.
- ✓ state the law of charges.
- ✓ identify who discovered the electron.
- ✓ determine the value of the fundamental electrical charge (e) from Robert Millikan's data.
- ✓ classify elements, compounds, and materials by their electrical characteristics.

18B DETECTING ELECTRIC CHARGE

18.6 An Early Detection Device

To study electric charge, you must be able to detect it. The ability to measure charge is also important. Early scientists used lightweight balls made from cork or pith (material from inside plant stems) to detect charges. Two balls were hung side-by-side from silk threads, close to each other but not touching. If both balls were charged, they would either attract or repel each other. The scientists tested to see if a material had a charge by touching it to both balls. The charge would pass to the balls, and, since it would be the same kind of charge, the balls would separate.

18.7 The Electroscope

An **electroscope** is an improved instrument for detecting charge. An electroscope is usually constructed from a glass flask sealed with a rubber stopper. A straight or *L*-shaped rod extends through the stopper, and a metal spherical knob is fastened on the rod's upper end. Two thin foil leaves are attached to the lower end of the rod and hang freely inside the flask.

When you bring a charged test rod *near* the electrode of the electroscope, the leaves separate. Why? Suppose the test charge is negative. The electrons in the electrode are repelled by the negative test charge and travel down the rod to the leaves. Because both leaves receive excess electrons, both have net negative charges. Since the leaves can move, they repel each other. If the charged test rod is positive, it attracts electrons from the leaves to the top of the electrode, leaving both leaves with a net positive charge. Again, both leaves have the same charge

18-8 A basic classroom electroscope

and thus repel each other. When you move the charged test rod away from the electroscope, the leaves return to their original positions.

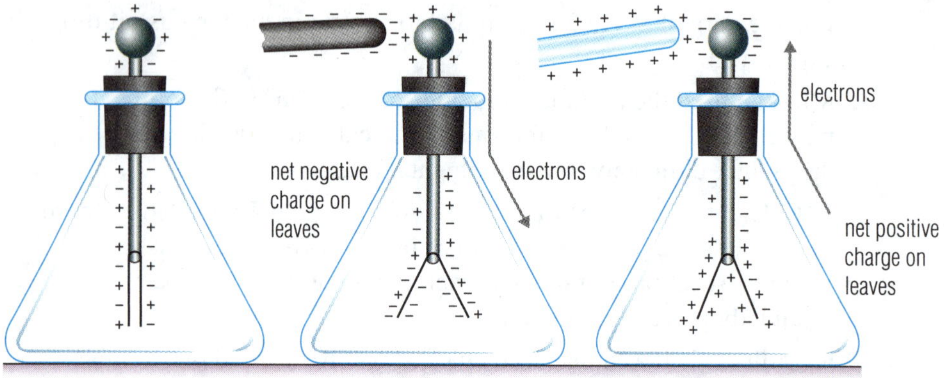

18-9 Testing for charge with an electroscope

FACETS of PHYSICS

Static Electricity

To most people, static electricity is a minor annoyance. An occasional shock or static cling does not greatly bother the average person. However, sensitive electronic devices, such as those used in computers, can be damaged or destroyed by a discharge of static electricity. Therefore, those who work with computers and other sensitive circuits have developed ways to protect the devices from static electricity.

A computer technician normally wears what is called a "grounding strap" attached to his wrist that provides a conductive path between his body and the chassis of the computer. Then, if a charge is picked up by the worker, it quickly bleeds away to the grounded chassis, preventing any possibility of the charge jumping to the computer components from the technician's hands.

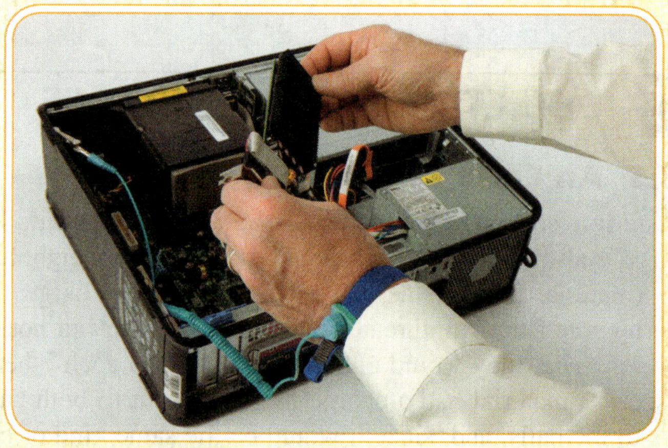

Using a grounding strap

Another method is to keep easily charged materials away from the circuit. Surrounding the sensitive circuit with a conductor insures that a charge will not build up around it.

Antistatic materials are used to prevent electrical charge buildup. Plastics have been developed that quickly drain off charges. Sensitive circuits are often wrapped in such plastics during shipping and may be surrounded by an antistatic plastic after being installed.

Some materials can be used in the vicinity of computers to dissipate charge buildup. Special antistatic carpets, sprays, and furniture materials can absorb water. Because the water conducts the charges away, charges cannot accumulate on these materials. Consumers can also buy antistatic chemicals to minimize annoying static shocks and static cling.

Suppose you *touch* an electroscope with a positively charged test rod. What happens? The electrons in the electroscope are attracted to the test rod. If the test rod is removed, the electrons go with it, leaving the electroscope positively charged. If you move another positively charged test rod near the charged electroscope, the rod will attract more electrons from the leaves. The leaves become even more charged, and they separate more. However, if you bring a negatively charged rod near the electroscope, the electrons in the test rod will force the electrode's electrons down to the leaves, reducing their positive charge. The leaves will therefore fall closer together. With a charged electroscope, you can not only detect charge but also distinguish between positive and negative charges, if you know what kind of charge is initially present.

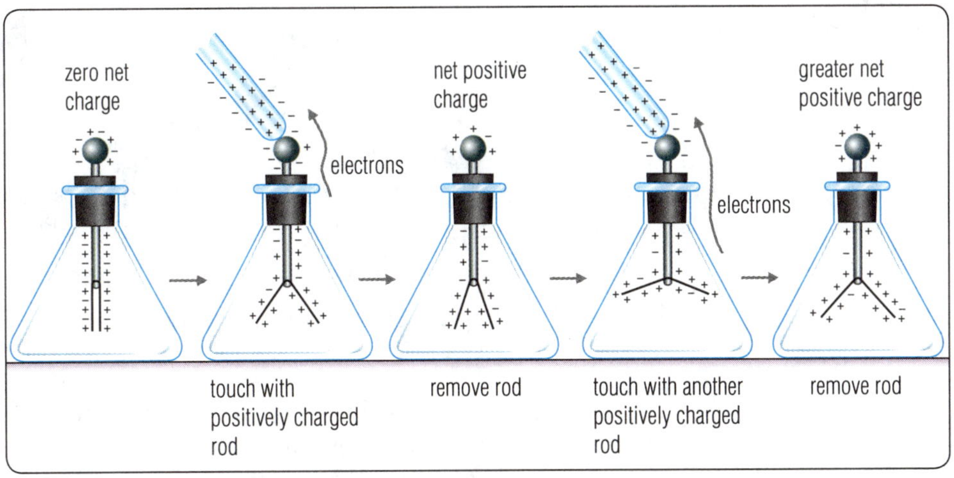

18-10 Sequentially adding charge to an electroscope

18.8 Conduction Tester

You can use a pair of electroscopes to see how well materials conduct. Begin with one neutral electroscope and one that is charged negatively. Connect the electroscopes by bridging their electrodes with the material that you want to test. If the material is a good conductor, like a metal, electrons will flow through the conductor from the charged electroscope to the neutral electroscope, charging the neutral electroscope. Its leaves will separate almost immediately. If you connect the electroscopes with an insulator, electrons cannot easily travel to the neutral electroscope. Its leaves will separate slowly, or not at all.

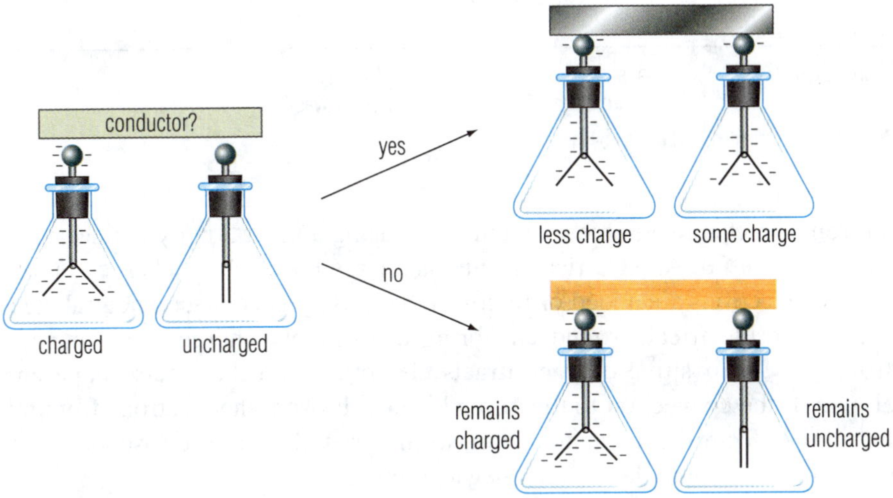

18-11 Testing for conduction

18.9 Methods of Establishing a Charge

Electroscopes illustrate two ways to charge a body: by **contact**—when a charged object touches an uncharged object and electrons are transferred from one to the other—and by **induction**—when the electrons on the surface of an uncharged object are redistributed by a charged object that is brought near.

Static charges can be produced by two methods: **contact** and **induction**.

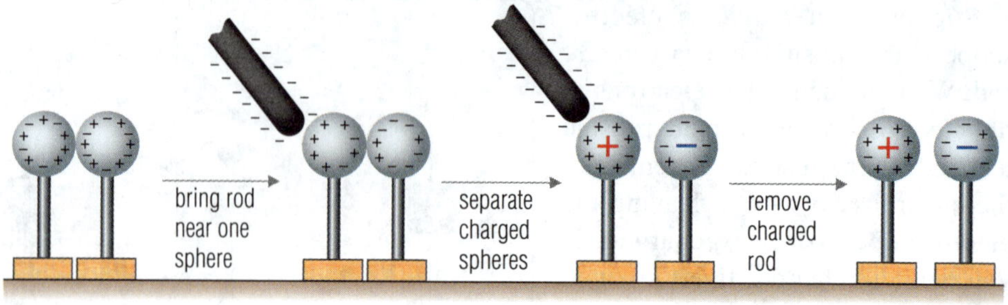

18-12 Charging by induction and separation of objects

It is possible to induce a permanent charge on a neutral body. Consider two neutral conducting spheres that are mounted on insulating stands and positioned so that they touch. When a negatively charged test rod is brought near one of the spheres, the electrons of both spheres migrate away from the rod, leaving the near sphere with a net positive charge and the far sphere with a net negative charge. If the test rod is removed, the electrons will return to their places and the spheres will again be neutral. However, if the spheres are separated while the test rod is near, they will remain charged. The rod's removal will have no effect on this charge since the spheres can no longer exchange electrons. This is charging by induction.

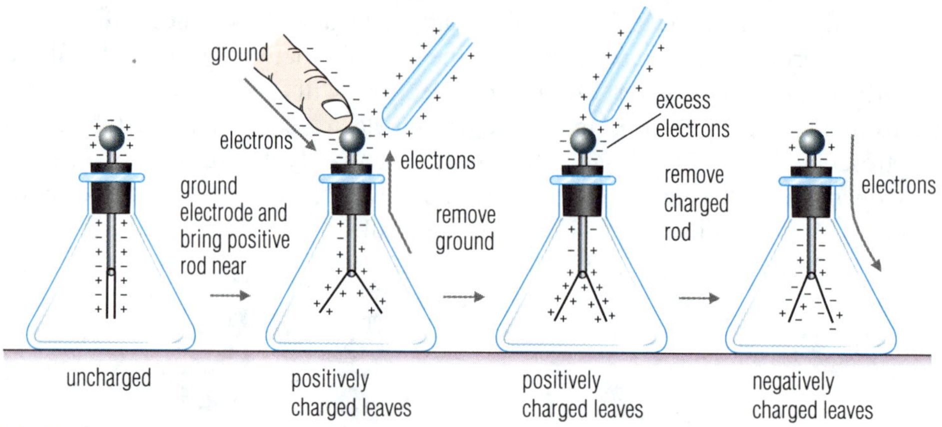

18-13 Charging a grounded electroscope by induction

An **electrical ground** is an electron reservoir similar to the heat reservoir discussed in Chapter 16. The earth is the largest electrical ground available. Your body can act as a ground for small static charges.

Electrostatic force is the force developed between two electrically charged particles or objects.

Charles-Augustin de Coulomb (1736–1806) was a prominent French physicist who discovered the electrostatic force. The SI unit of charge is named in his honor.

You can also charge an object by induction using a neutral body connected to an **electrical ground**. An electrical ground is a large body whose charge is unaffected when electrons are added or removed. Suppose you connect a neutral electroscope to an electrical ground and bring a positively charged rod near the electroscope. The positive charge attracts electrons from the electroscope and the ground. If the connection to the ground is now broken, the electrons from the ground in the electroscope cannot return to the ground. The electroscope is left with extra electrons and, therefore, a negative charge.

18.10 Coulomb's Law

Although good for detecting charges, the electroscope is not very useful for measuring electric force. The instrument that *can* measure **electrostatic force** accurately is very similar to the one that was first used to accurately measure gravitational force—the torsion balance. Charles Coulomb developed it in 1777 for his electrostatic experiments. The instrument for measuring electrostatic force consists of a silk fiber supporting a rod with an insulated metal sphere on each end. A fixed sphere is mounted near one of the hanging spheres. The fixed sphere and the hanging sphere nearest it are usually given known charges so that the fixed sphere repels the hanging sphere.

When the hanging sphere moves, it produces a torque on the fiber through the rod. As the fiber twists, a torsional force develops in the fiber that resists twisting, much like a spring resists compression. The amount of torsional force the fiber exerts against the torque exerted by the charged spheres is related to the angle through which it has twisted. Eventually, the torsion in the fiber balances the mechanical torque produced by the spheres. The torsion in the fiber can be calculated from the angle through which the fiber has twisted. From this value, the force on the movable sphere can be determined.

Coulomb studied the electrostatic force as it relates to the distance between the spheres and discovered that when he doubled the distance between the spheres, the force between them was one-fourth the original. When he tripled the distance, the force was one-ninth. Thus, he concluded that electrostatic force is proportional to the reciprocal of the square of the distance between the charges, or

$$F_{\text{elect}} \propto 1/r^2,$$

where r is the distance between the spheres. He then tried doubling the charge on one sphere, which doubled the force. Doubling the charge on both spheres quadrupled the force. Coulomb concluded that the electrostatic force is directly proportional to the charge on each sphere. He summarized his results in Coulomb's law:

$$F_{\text{elect}} = k\frac{|q_1||q_2|}{r^2}, \qquad (18.1)$$

where the constant k is a proportionality constant whose value depends on the units used. The symbol q represents a variable quantity of charge that has no physical dimension (a *point charge*).

The electrostatic force on each charged object is equal and opposite to the force exerted on the other object, according to Newton's third law of motion. This law is true even if the charges are unequal:

$$\mathbf{F}_{q_1 \to q_2} = -\mathbf{F}_{q_2 \to q_1}$$

Note that the form of Coulomb's law is identical to Newton's law of universal gravitation that predicts the force of two masses M and m on each other:

$$F_{\text{grav}} = G\frac{Mm}{r^2}$$

Two key differences exist between the gravitational force and electrostatic force. (1) There are two kinds of charge that produce the electrostatic force, but there is only one kind of mass that produces the gravitational force. Even antiparticles have "positive" mass. (2) Electrostatic forces may be attractive or repulsive, but gravity is only attractive.

You will discover additional similarities between gravitational and electrostatic force in later chapters. You should expect this, since both of these forces are central forces—they act along a line connecting the centers of the objects producing the force. They also are both conservative forces in that the work done by the force between two positions is independent of the path taken.

Interestingly, the principle of Coulomb's torsion balance was used by Henry Cavendish a few years later to determine the density of the earth. His experiment eventually enabled others to establish the value of the constant G in Newton's law of universal gravitation.

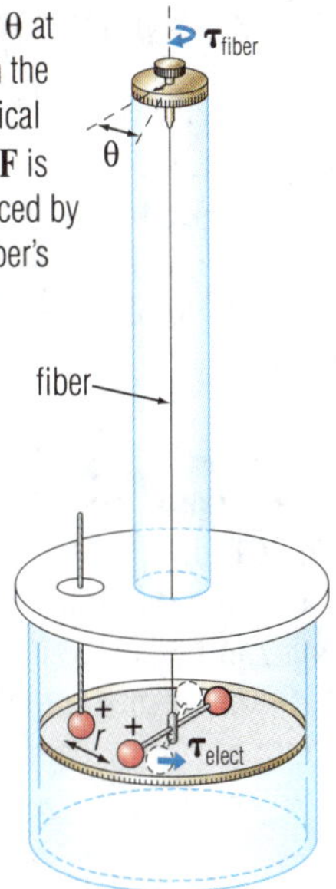

18-14 An electrostatic torsion balance

You must always use the magnitudes of the charges in the Coulomb's law equation. From this you obtain the magnitude of the electrostatic force. The direction of the force, attractive or repulsive, must be assigned based on the signs of the charges.

Coulomb's law specifically pertains to forces between point charges. As the dimensions of the charged objects become physically significant, the electrical charge spreads over their surfaces. In this case, the proportionality of force to $1/r^2$ in Coulomb's law is no longer exactly followed.

Coulomb's law is another example of an *inverse square law*, in which a property varies with the inverse or reciprocal of the square of its distance from its source. Other inverse square laws apply to gravitational force as well as light and sound intensity.

The CGS (centimeter-gram-second) system of units was suggested in 1873 by a committee of scientists in order to provide a coherent metric system in which the density of water was unity (1 g/cm^3). In 1888, Heinrich Hertz showed that magnetism and electricity, each having their own systems of units, could be combined in the CGS system by a factor equal to the speed of light, c. He called this system of units the Gaussian system. In the early 20th century, most scientific units were converted to the MKS (meter-kilogram-second) system, which was eventually replaced by the SI metric system.

Problem-Solving Strategy 18.1

When using the Coulomb's law equation, be sure that the units used for charge, force, and distance match those in the Coulomb constant. The SI units are coulombs, newtons, and meters, respectively.

18.11 Units of Charge

To use Equation 18.1, you must specify the unit of charge. One experimental unit of charge is the **electrostatic unit (esu)**. In the Gaussian or CGS system of units, one esu of charge on each of two dimensionless objects exactly 1 cm apart exerts a force of exactly 1 dyne on each object (1 dyne = 1×10^{-5} N):

$$1 \text{ dyne} = k_{esu}\frac{(1 \text{ esu})(1 \text{ esu})}{(1 \text{ cm})^2}$$

The constant k_{esu} is defined by

$$k_{esu} \equiv 1\frac{\text{dyne}\cdot\text{cm}^2}{\text{esu}^2}.$$

The esu was chosen because it allows the magnitude of the constant to be 1.

The SI unit of charge is the **coulomb (C)**, which for now will be defined as the magnitude of the charge on 6.24×10^{18} electrons. For this unit, the constant k_C is defined as

$$k_C = 8.99 \times 10^9 \frac{\text{N}\cdot\text{m}^2}{\text{C}^2}, \text{ so}$$

$$F_{elect} = (8.99 \times 10^9 \frac{\text{N}\cdot\text{m}^2}{\text{C}^2})\frac{|q_1||q_2|}{r^2}.$$

In this textbook, we will normally evaluate electrical problems using the coulomb for the unit of charge.

EXAMPLE 18-1

Distance and Charge: Coulomb's Law

Two positive point charges of 1.00×10^{-6} C each are separated by a distance of 0.100 m. (a) What is the force on each charge? (b) If the distance is 0.0100 m, what is the force?

Solution:

$$F_{elect} = k\frac{q_1q_2}{r^2}$$

a. Force at 0.100 m apart:

$$F_{0.1m} = 8.99 \times 10^9 \frac{\text{N}\cdot\text{m}^2}{\text{C}^2} \cdot \frac{(1.00 \times 10^{-6} \text{ C})(1.00 \times 10^{-6} \text{ C})}{(0.100 \text{ m})^2}$$

$$F_{0.1m} = (8.99 \times 10^9 \text{ N})(1.00 \times 10^{-10})$$

$$F_{0.1m} = 8.99 \times 10^{-1} \text{ N}$$

$$F_{0.1m} = 0.899 \text{ N}$$

b. Force at 0.0100 m apart:

$$F_{0.01m} = 8.99 \times 10^9 \frac{\text{N}\cdot\text{m}^2}{\text{C}^2} \cdot \frac{(1.00 \times 10^{-6} \text{ C})(1.00 \times 10^{-6} \text{ C})}{(0.0100 \text{ m})^2}$$

$$F_{0.01m} = (8.99 \times 10^9 \text{ N})(1.00 \times 10^{-8})$$

$$F_{0.01m} = 8.99 \times 10^1 \text{ N}$$

$$F_{0.01m} = 89.9 \text{ N}$$

As expected, the smaller distance between charges results in a much larger force between them.

18B Section Review

1. Describe how one or more pith balls hanging from a support can be used to determine the kind of charge, if any, on an object.
2. a. Describe how an electroscope works.
   b. Why should the foil leaves be as thin and light as possible?
   c. Why must the stopper supporting the electrode be made of an insulator?
3. What is the consequence of touching a neutral electroscope electrode with
   a. a positively charged test rod?
   b. a negatively charged test rod?

   Include in your discussion the location and movement of charges.
4. To what quantities is the electrostatic force between charged objects proportional?
5. ✪ Sketch and label a diagram of the force vectors on negative charge q_1 and positive charge q_2 if $|q_1| = 2|q_2|$.
6. ✪ a. What is the electrostatic force between two point charges $q_1 = +1.00$ C and $q_2 = -1.00$ C if the distance between them is 1.00 m?
   b. What can you conclude about the ability of opposing electrostatic charges to do work?

18B Objectives

After completing this section, I can

✓ describe the construction and operation of an electroscope.

✓ describe how to use two electroscopes to determine whether a material is a conductor or not.

✓ differentiate between charging an object by induction and by contact.

✓ state Coulomb's law and identify the quantities on which it depends.

✓ explain why Coulomb's law is an inverse square law.

✓ compare and contrast the two units of charge.

✓ use Coulomb's law to compute the force exerted between point charges.

Chapter Review

In Terms of Physics

static electricity	414
resinous	414
vitreous	414
law of charges	415
electron	416
fundamental charge (*e*)	417
conductor	418
insulator	418
semiconductor	418
electroscope	419
contact (charging)	421
induction (charging)	421
electrical ground	422
electrostatic force	422
Coulomb's law	423
electrostatic unit (esu)	424
coulomb (C)	424

Problem-Solving Strategies

18.1 (page 424) When using the Coulomb's law equation, be sure that the units used for charge, force, and distance match those in the Coulomb constant. The SI units are coulombs, newtons, and meters, respectively.

Review Questions

1. What did William Gilbert mean by the word *electrick*?
2. Why does amber attract dust and dry leaves after it has been rubbed with fur?
3. What happens when an electrified amber test rod is placed
   a. near an electrified glass rod?
   b. near an electrified amber rod?
4. How did the two different "electrical fluid" theories explain the differences between electrified amber and electrified glass?
5. What happens to a neutral electroscope when you hold a negatively charged object near it? Why?
6. When you bring a charged rod near a negatively charged electroscope, its leaves separate farther. Is the charge on the rod negative or positive?

7. You have two neutral conducting spheres. How can you give one a negative charge and the other an equal positive charge without touching either sphere with a charged object?

8. Explain what will happen when particles of paper picked up by a negatively charged test rod gain electrons from the rod.

9. Explain why a charged electroscope will discharge over time if left to itself.

10. What, if anything, happens when you touch the electrode of a charged electroscope with your finger?

11. Why is it advisable on a cold, dry winter day to touch the metal case of your computer before starting it or touching any other part of the system?

12. Many appliances and hand tools have at least three wires in their electrical cords. The center prong of the plug is connected to the ground wire. Where do you suppose this wire leads when it is plugged into an electrical outlet?

13. To what other kind of force is the electrostatic force similar? How is it different?

14. Is a coulomb of electrons a large charge or a small charge? In other words, is a charge of this size dangerous—can it do a lot of damaging work in a short time? (*Hint*: Review the results of Question 6 in the 18B Section Review.)

True or False (15–24)

15. An "electrick" is any substance that can easily conduct electricity.

16. Glass, amber, rubber, and Lucite are "electricks."

17. If two objects repel each other, you know that they are both negative charges.

18. The magnitude of the fundamental charge is the same for positive and negative charges.

19. When a charge is established on an object, the charge exists because electrons have moved to or from the object.

20. On humid days, charged objects lose their charges more quickly because the presence of water vapor makes the air more conductive.

21. Uncharged electroscopes can be used to indicate the presence, sign, and magnitude of electrostatic charge on an object.

22. Induction of an electrostatic charge requires only the presence of a charged object, not actual contact, to establish a charge in another object.

23. If the distance between two charged particles doubles, the force exerted between them will be halved.

24. If two charges exert an electrostatic force on each other and one charge is twice the other, then the larger charge will exert twice the force of the smaller charge, and the smaller charge will exert half the force of the larger.

DM✪25. Compare your value of the fundamental charge obtained in 18A Section Review Question 8 to the accepted value of 1.602×10^{-19} C. What are some potential sources of error? (If possible, visit the BJU Press textbook resources website to access Millikan's original paper. Reviewing his procedure will help you in this evaluation.)

✪26. Two charges separated by a distance *r* exert a force of 1.00 N on each other. If one charge is tripled without anything else being changed, what force will the charges exert on each other?

✪27. One thousand grams of water (about 1 L) contains 55.5 moles of water molecules. (One mole is 6.022×10^{23} particles.) One electron is removed from every hundredth water molecule and placed in a container 100. m from the water.
   - **a.** How many water molecules are there in the water container?
   - **b.** How many electrons have been removed from the sample of water?
   - **c.** What is the charge, in coulombs, of the electrons that were removed?
   - **d.** What is the force of attraction between the container of water and the container of electrons?

✪28. Two conducting spheres, when given a charge of 2.50×10^{-6} C each, repel each other with a force of 2.50×10^{-4} N at a certain distance apart. How far apart are the spheres?

✪29. The charged tips of two test rods are suspended 10.0 cm from each other. One rod has a charge of 2.50×10^{-5} C. The rods attract each other with a force of 5.00 N. What is the charge on the other rod?

✪30. Three conducting spheres hang in a straight line as shown in the figure below. Sphere A has a charge of -2.00×10^{-6} C. Sphere B, 5.00 cm to the right, has a charge of $+1.00 \times 10^{-6}$ C. Sphere C, 10.00 cm to the right of sphere B, has a charge of $+3.00 \times 10^{-6}$ C.

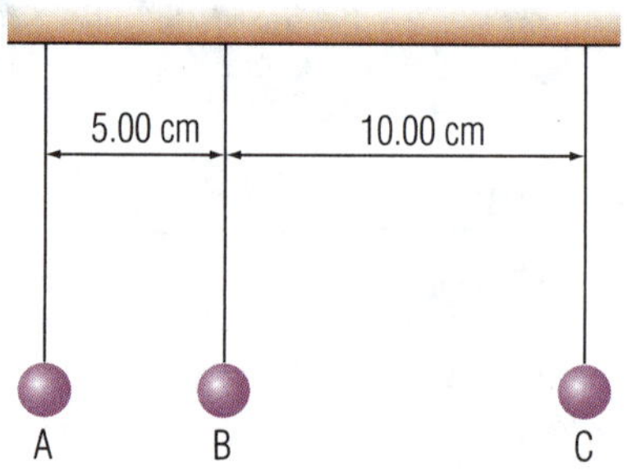

   - **a.** What is the force between spheres A and B?
   - **b.** What is the force between spheres B and C?
   - **c.** What is the total force on sphere B?

✪31. Write a brief biographical report on Charles Du Fay or Benjamin Franklin.

✪32. Research and give an oral report on the origins and properties of natural lightning.

✪33. Build and demonstrate a mechanical static electricity generator, such as a Wimshurst machine. A good reference for this exercise is *Homemade Lightning* by R. A. Ford (TAB Books, The McGraw-Hill Companies, 2001).

FACETS of PHYSICS

Michael Faraday (1791–1867)

Some scientists have special talent for devising theories to explain the observations of others; other scientists are unusually good at experimenting to test and develop theories. Michael Faraday was good at both. He is undoubtedly one of the greatest experimental geniuses in history.

Faraday was born into a poor London family. After a scanty education, he was apprenticed to a bookbinder. Faraday supplemented his education by reading the science books that came into the shop. When a customer gave him tickets to a series of lectures by Sir Humphry Davy, Faraday attended, listened attentively, and took careful notes. Later, he sent the notes with illustrations to Davy as evidence of his eagerness to "enter into the service of science." Flattered, Davy hired him as a laboratory assistant.

With Davy's help, Faraday soon became a skilled experimenter. He started with gases and successfully liquefied chlorine and isolated benzene. When Hans Christian Oersted showed that electric currents affect magnets, Faraday became interested in electromagnetism. In his first great experiment in electricity, he transformed electrical energy directly into mechanical energy by causing a current-carrying wire to revolve around a magnet. These successes brought Faraday widespread fame. He was appointed superintendent and later lab director at the Royal Institution. He was also elected a fellow of the Royal Society.

Next, Faraday tried to produce an electrical current from a magnet. He was convinced that there is a divine unity in the forces of nature. Undismayed by ten years of fruitless experiments, he was finally rewarded for his efforts when he discovered that changing magnetism induces an electrical current. The principles behind Faraday's generator were essential to the development of our modern electric power industry.

To explain his magnetic generator, Faraday proposed that a field made up of magnetic lines of force surrounds a magnet. When a conductor cuts this field, the magnetic field induces a current in the conductor. He later extended his idea to electricity: electric field lines surround electric charges. His field theory formed the basis of Maxwell's classic field theory, and it provided a strong foundation for Einstein's theory of relativity.

While experimenting with chemical solutions that conduct electricity, he arrived at the basic laws of electrolysis, now called Faraday's laws. Unfortunately, the sustained effort led to a breakdown from which Faraday never fully recovered. He returned to the laboratory long enough to discover that magnetism can affect light, but his failing mind could no longer solve any more of the scientific riddles that fascinated him.

Faraday was well respected for his speaking ability as well as his genius. His captivating lectures drew large crowds to the Institution every Friday evening. He was popular among children as well; he started a series of Christmas lectures for them that became a tradition which continues to this day.

One reason for Faraday's success was undoubtedly his love for God. A devout Christian, he believed that science is simply a search for knowledge about God's creation. In all of his experiments, he was dedicated to finding the truth. He continually checked his work for errors and abandoned any theories that did not fall in line with the facts. Faraday's life was a testimony to his fellow scientists, who realized that his strength of mind came from his worship of God. His life is evidence that a good Christian can be a good scientist.

Tesla coils generate extremely intense electric fields that permit the discharge of powerful electrical sparks.

Electric Fields 19

Dominion Science Problem

Nanovision

Have you noticed that computers, MP3 players, and cellular phones are getting thinner and smaller as technology progresses? The increased efficiency of their digital components, like microprocessors and optical elements, allows this shrinkage. To the unaided eye, microchips are rather unimpressive. But at the microscopic level, microprocessors, which contain millions or even billions of transistors (tiny electronic switches) and other electronic components, consist of dozens of layers of semiconductor and insulator compounds complexly arranged in three dimensions. In addition, specialized optical systems require extremely thin coatings of metal and other materials applied in uniform layers only a few atoms thick. Such precise technology requires special methods of inspection to ensure that the devices will work. How can technicians peer down to the atomic level when looking for defects?

19-1 Technologies like these rely on custom-built crystalline compounds.

19A MODELING THE ELECTRIC FIELD

19.1 Lines of Force

Michael Faraday, a nineteenth-century English scientist, had little formal education. He gained most of his knowledge through personal study and experimentation. As a result, Faraday's mathematical skills were somewhat limited. However, this did not keep him from taking a non-mathematical theoretical approach to explain certain phenomena that he had observed. Faraday proposed a model of forces that has proved fruitful in many areas of physics: the **field model**.

The **field model** is a theoretical approach to accurately explaining such three-dimensional phenomena as magnetic, electric, and gravitational forces, as well as sound propagation and subatomic particle fluxes.

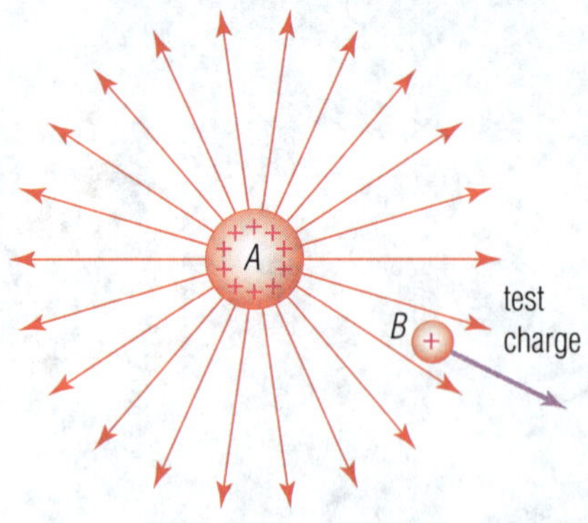

19-2 The electric field of a positively charged object

Faraday considered each charged object to be surrounded by an **electric field**. Through this field, the charged object exerts a force on any other charged object that ventures into the field. Faraday represented an electric field by **lines of force**, which give the direction of the force that the charged object exerts on a small positive charge. For example, Figure 19-2 shows the lines of force that represent the electric field of a positively charged object (A). A small positive charge (B) will experience a force in a direction directly away from the positively charged body.

Lines of force can also show the relative magnitude of the force at each point. Closely spaced lines represent a strong force. Widely spaced lines represent a weak force. For example, in Figure 19-3 the lines are close together near the charged object and far apart away from the object. This shows that the object exerts a greater force on a charge near it than on a more distant charge.

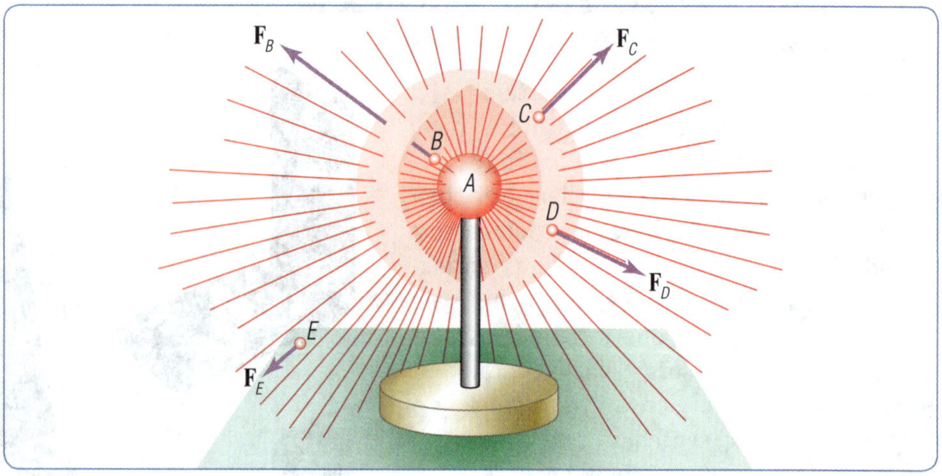

19-3 In this three-dimensional image, an electrostatic field surrounds a charged object. Note that the length of the force vector on each charge, which is related to the density of the lines of force at that point, is proportional to the repulsive force exerted by the central charge.

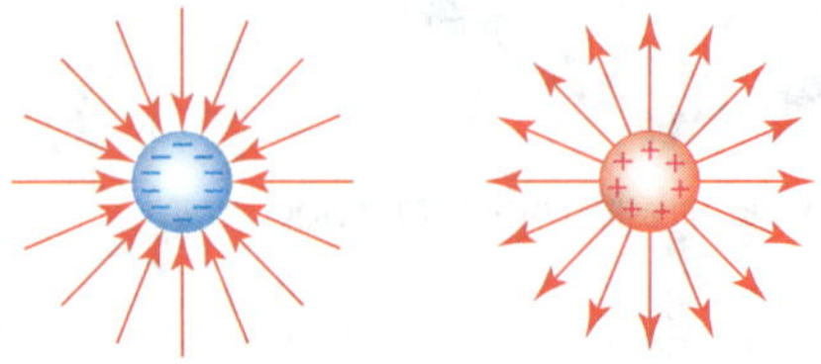

19-4 What is the shape of the electric field between two charged objects?

A **test charge** is a *positive* charge small enough that it exerts negligible force on other charges in the problem.

Equation 19.2 permits the calculation of the *magnitude* of the field strength due to a single charge Q. The orientation of the vector **E** depends on the sign of the charge. The sign associated with a positive Q is positive, and **E** points radially away from Q. If Q is negative, **E** points radially toward Q.

19.2 The Test Charge and Field Strength

Suppose there is more than one charged body, as in Figure 19-4. What is the electric field like between the objects? How can you measure its magnitude and direction? There are two ways that you can find the lines of force. First, you can use Coulomb's law to find the force that a positive **test charge (q)** would experience between the bodies. In this calculation, it is helpful to define the **electric field strength (E)**. The electric field strength is the force per unit of positive charge at each point in the electric field. The magnitude of **E** generated by a charge Q on a test charge q is defined as

$$E \equiv \frac{F}{q}. \qquad (19.1)$$

From Coulomb's law,

$$F = k\frac{Qq}{r^2},$$

where r is the distance between Q and q. Therefore,

$$E = \frac{kQ\cancel{q}}{r^2} \cdot \frac{1}{\cancel{q}}, \text{ and}$$

$$E = \frac{kQ}{r^2}. \qquad (19.2)$$

The direction of **E** at any location in the field is tangent to the electric field line through that point. The vector points radially *away* from Q if the charge is positive and radially *toward* Q if it is negative. (The direction of **E** is the same direction as the arrows representing the lines of force.)

To calculate the electric field strength between the charged objects in Figure 19-4, you must use vector addition to combine the electric field strength of the negatively charged object with the electric field strength of the positively charged object at each point. Figure 19-5 shows the result. This method is accurate but difficult and tedious to use.

19-5 Calculating the electric field strength between two charged objects. Figure (a) shows the vector addition method for determining the electric field at a single point in the electric field. This method must be applied to each point in the electric field. Figure (b) shows the electric field lines superimposed upon the individual field strengths throughout the grid. Figure (c) shows that the direction of the electric field strength vector is tangent to the field lines.

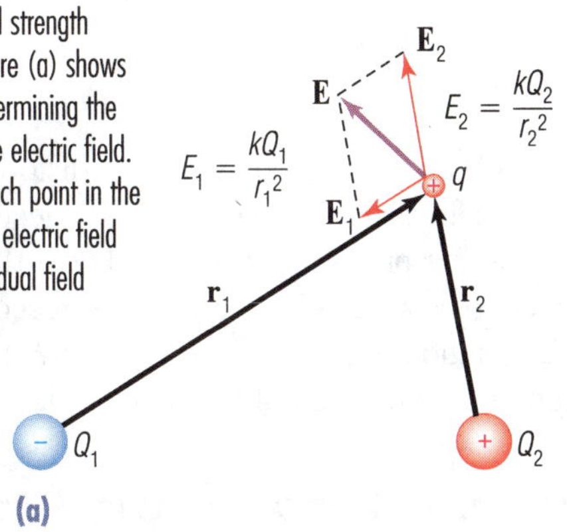

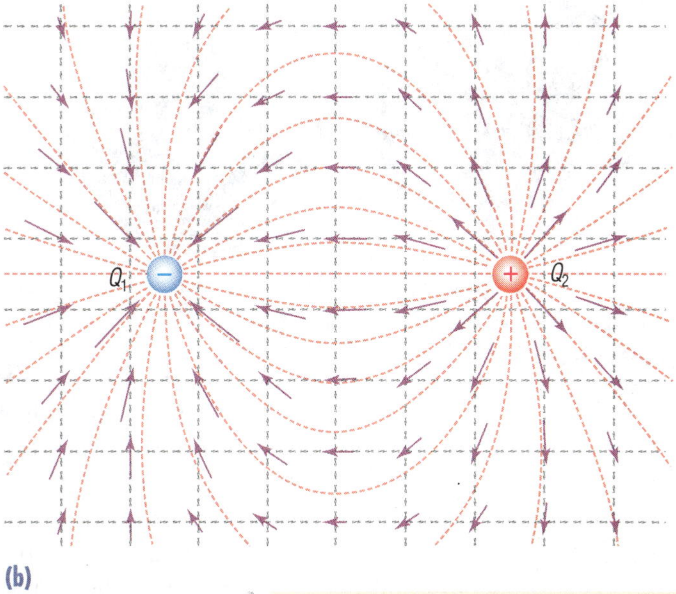

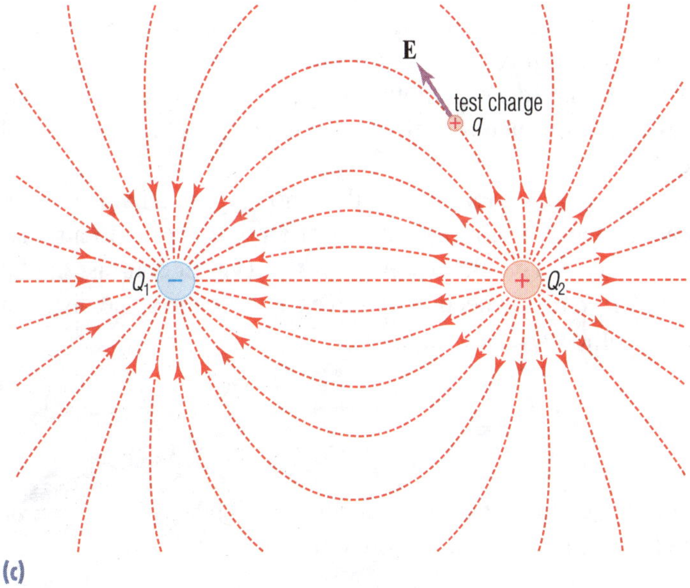

Diagrams of electric fields have the following properties:

- Lines lie parallel to the electric field strength vector **E**.
- Lines point from positive charge to negative.
- A stronger field is shown by lines that are closer together, and a weaker field is shown by lines that are farther apart.
- Field lines do not cross other field lines.

19.3 Electric Field Orientation

Notice in Figure 19-5c that the lines of force between the objects begin on the positive object and end on the negative object. That is because a test charge (which is positive) would move from the positive object to the negative object along a field line. All field lines are assumed to begin on a positive charge and end on a negative charge. However, in the figure, some field lines begin on the positive object and appear to go nowhere, whereas others appear to come from nowhere and end on the negative object. We assume that they are too far from the objects to be shown in the diagram, but they still originate at some point that has a different charge from the ones shown.

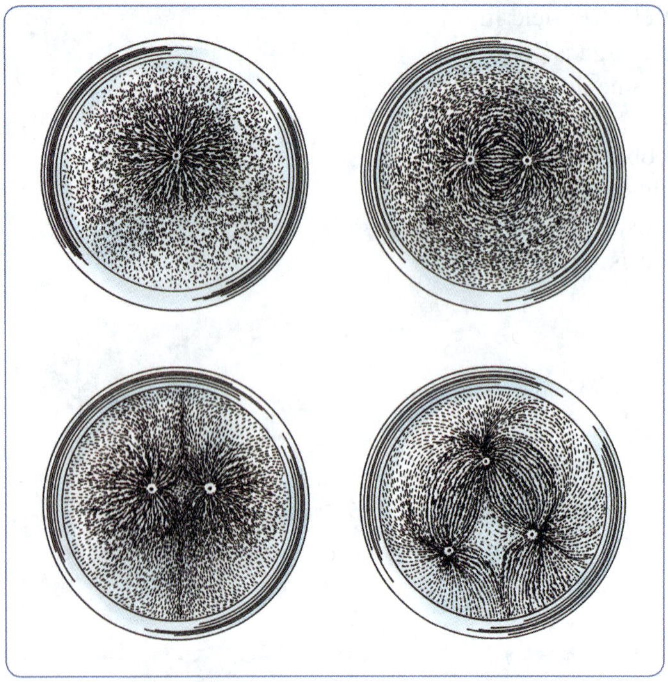

19-6 Electric field lines revealed by the positions of grass seeds in petri dishes

The second way to show the electric field is to set up an experiment. Set two charged objects in a dish containing grass seeds suspended in a liquid. Grass seeds are oblong in shape. The electrons in the grass seeds will migrate to one end of each seed so that the seed will align itself with the electric field. The seeds thus form a visible picture of the electric field. The illustrations in Figure 19-6 show the electric fields of several combinations of charged probes inserted into petri dishes containing grass seed suspended in oil.

If you know the electric field strength, E, and the magnitude of the test charge, q, but not F, you can find F by rearranging Equation 19.1:

$$F = Eq$$

If the test charge is too large, it can distort the field so that E is no longer due to the original charged object(s) alone. However, for electrostatic problems in this textbook, the magnitude of q is assumed to be so small that it has no effect on the electric field. The electrostatic force on the test charge (F_e) is due only to the field strength derived from the charged object(s) producing the field.

19.4 Electrical Potential and Electrical Potential Energy

All forces, including electrical forces, can do work. When an "electrified" piece of amber attracts a dry leaf, it does work on the leaf. Pushing like charges together or pulling unlike charges apart requires work, just as lifting a mass against gravity requires work. A lifted mass has gravitational potential energy, since it can fall. Similarly, a positive charge pushed near another positively charged object has potential energy with respect to the electrostatic force, since it will be repelled and move away, if allowed.

Electrical potential energy (U_e) is a measure of the amount of mechanical work that is necessary to move a charge from the location of a point charge to a radial location **r** from the point charge. The **electrical potential (*V*)**, in contrast, is the amount of work *per unit charge* that is needed to move a charge the same distance r from the point charge. In other words,

$$V \equiv \frac{U_e}{q}.$$

Electrical potential is analogous to gravitational potential, which equals U_g/m, or $|g|h$. Electrostatic force has the same form as gravitational force:

$$F_g = G\frac{Mm}{r^2}$$

$$F_e = k\frac{Qq}{r^2}$$

Similarly, electrical potential energy has essentially the same form as gravitational potential energy (the difference in sign is due to the repulsion of like charges):

$$U_g = -G\frac{Mm}{r}$$

$$U_e = k\frac{Qq}{r}$$

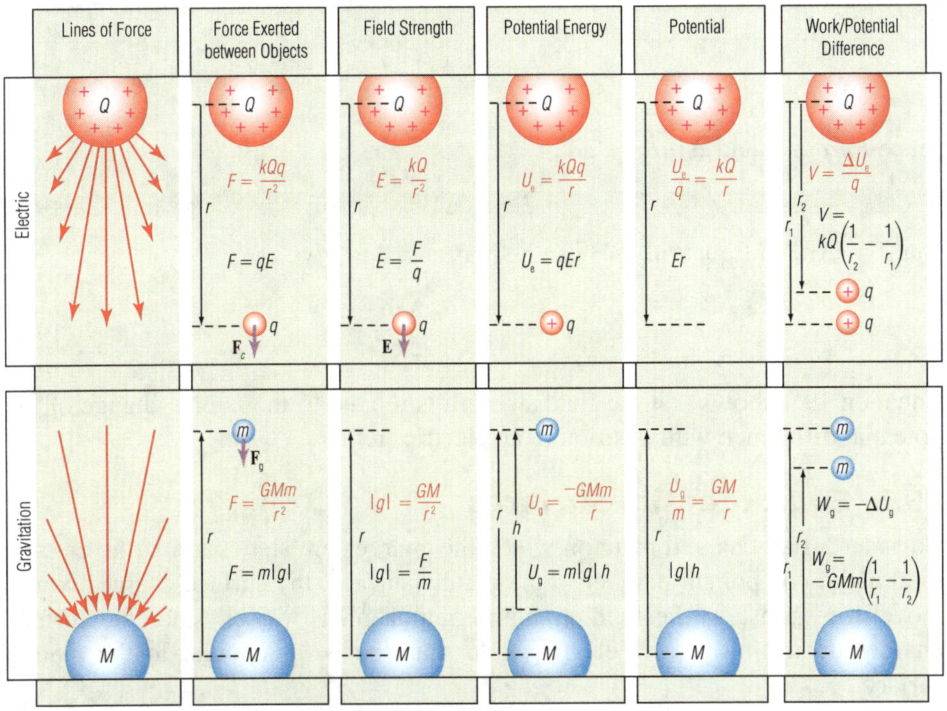

19-7 Comparison of the electrostatic force and the gravitational force

The electrical potential of a test charge q is U_e/q, so

$$V = k\frac{Q}{r},$$

where Q is a point charge and r is the radial distance of the position of the test charge from the point charge.

Like gravitational potential energy, electrical potential energy's zero point is usually taken to be an infinite separation between the charge source and the test charge. However, as we have seen elsewhere, only the difference in electrical potential energy, the **potential difference** (ΔV), is important. The potential difference between any two positions in an electrical field is the electrical work needed to move a unit charge between the positions, in the same way that mechanical work is the change in potential energy of a system in a gravitational field.

Since potential energy of all kinds is measured in joules, and charge is usually measured in coulombs, electrical potential and potential difference (ΔV) are both measured in joules per coulomb:

$$\Delta V = \frac{\Delta U_e}{q} \Rightarrow \frac{\text{J}}{\text{C}}$$

This unit of potential difference is called a **volt (V)**:

$$1 \text{ volt} \equiv \frac{1 \text{ J}}{1 \text{ C}}$$

The magnitude of the electric field strength vector (E) is related to the potential difference in the same way that force is related to the difference in gravitational potential energy. That is, for the simple case where force and displacement are in the same direction,

$$F_g \times d = \Delta U_g,$$

it follows that

$$F_e \times d = \Delta U_e.$$

Problem-Solving Strategy 19.1

Do not confuse electrical potential and electrical potential energy. Electrical potential is a property per unit charge of the *position* of any size charge relative to an electric field. It is measured in joules per coulomb (J/C) or volts (V). Its scalar value can be positive, negative, or zero, depending on the sign of the charge producing the field.

Electrical potential energy is a property of both the *size* of a charge and its *position* relative to an electric field. This quantity is expressed in joules (J).

The unit of electrical potential difference is the **volt (V)**.

Problem-Solving Strategy 19.2

Do not confuse the derived unit volt (V) with the electrical potential formula symbol, V.

Problem-Solving Strategy 19.3

The direction of the electric field is always from the more positive (less negative) values of electrical potential toward lower (more negative) values of electrical potential.

Therefore,

$$\frac{F_e}{q}d = \frac{\Delta U_e}{q}.$$

Since $F_e/q = E$ and $\Delta U_e/q = \Delta V$,

$$Ed = \Delta V. \quad (19.3)$$

But $d = |\Delta \mathbf{r}|$, so Equation 19.3 can be rearranged to give

$$E = \frac{\Delta V}{d} = \frac{\Delta V}{|\Delta \mathbf{r}|}. \quad (19.4)$$

Equation 19.4 shows that the field strength is related to the rate of change of the potential difference with position in an electric field.

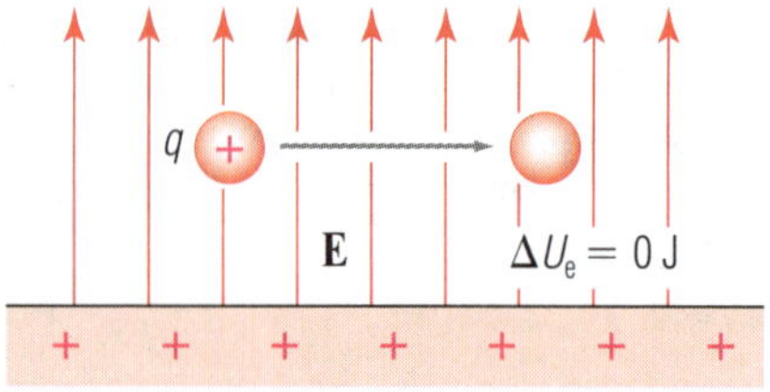

19-8 As charge q moves parallel to the charged surface, no work is done on the charge.

19.5 Equipotential Surfaces

Consider a flat, charged plate on which the charges are stationary. If the electric field had a component parallel to the plate's surface, the surface charges would move in response to the field strength component. However, since the surface charges do not move, the electric field must be perpendicular to the plate's surface.

Suppose you move a test charge parallel to the plate's surface (perpendicular to the lines of force). A force does no work on a system that moves at right angles to the line of action of the force, so you do no work with respect to the electrical potential by moving the charge parallel to the surface of the plate. Since no work is done on the test charge, its potential energy remains the same as it moves along the plate's surface. Therefore, every point on the surface of the plate must have the same electrical potential. For this reason, the surface of the plate is called an **equipotential surface**.

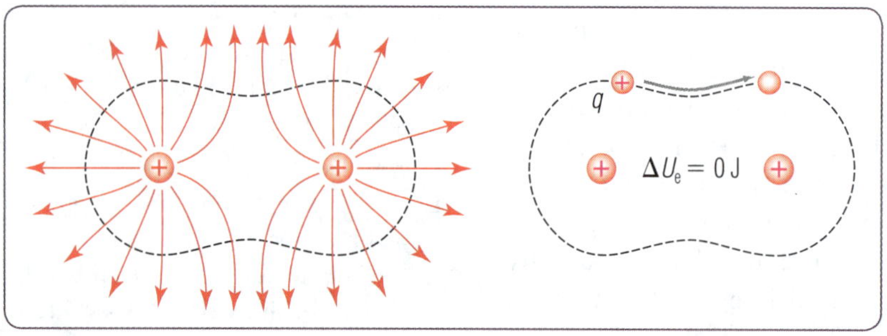

19-9 An equipotential line representing the cross section of a three-dimensional equipotential surface surrounding two charges

You can construct imaginary equipotential surfaces in any electric field by positioning the surfaces so that they are perpendicular to the electric field lines at every point on the surface. For example, the dotted line in Figure 19-9 is a two-dimensional cross section of a three-dimensional surface that is perpendicular at every point to the electric field surrounding the two charges. Therefore, it is an equipotential line. It is imaginary because there is nothing physically where the dotted line is. However, a test charge will have the same electrical potential everywhere along the line.

Every charged object is surrounded by an infinite series of three-dimensional equipotential surfaces. Since it is difficult to make three-dimensional drawings, we will represent these surfaces as two-dimensional equipotential lines. Figure 19-10 shows equipotential lines (dotted lines) around a spherical, positively charged object.

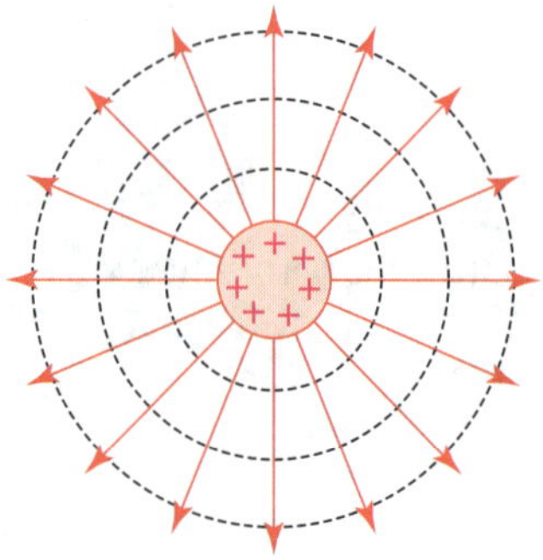

19-10 A charged sphere

The shape of the charged object affects the shape of its electric field. You have already seen the shape of a spherical object's field. What about the field of an irregular object? Figure 19-11 shows the electric field of an irregularly shaped object. The dotted lines are equipotential lines around the object. The equipotential surfaces around the object are drawn so that they are always perpendicular to the field lines representing the orientation of **E**.

Although the technique of drawing an electric field perpendicular to a conductor's surface is useful, it has limitations. For example, consider a charged cube, which Figure 19-12 represents as a square. It is easy to find a perpendicular for the sides, but what about the corners? What direction is perpendicular to a corner? In real objects, no corner is a geometrically perfect right angle. At the scale of atoms, such corners and points are broadly rounded. Interestingly, small-radii curved surfaces become locations for high concentrations of charge. Field lines are closely spaced, thus electric fields can produce high electrical potentials at these locations. This is why lightning rods are sharply pointed. Under the right conditions, the potential within the electric field surrounding the tip of a lightning rod can break down the air's insulating ability, initiating an electric discharge at the lightning rod rather than through the protected structure.

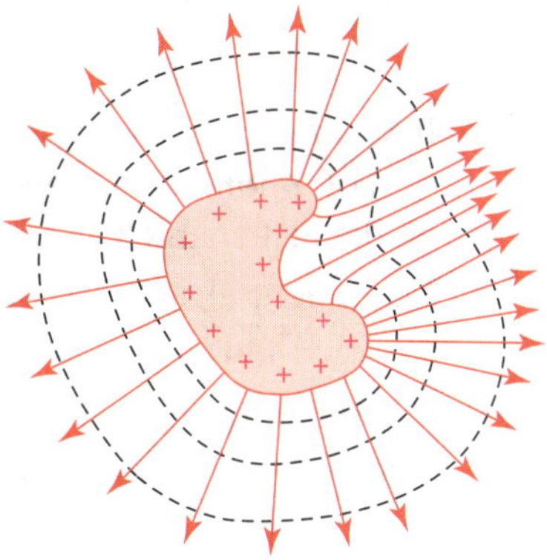

19-11 A charged object having an irregular shape

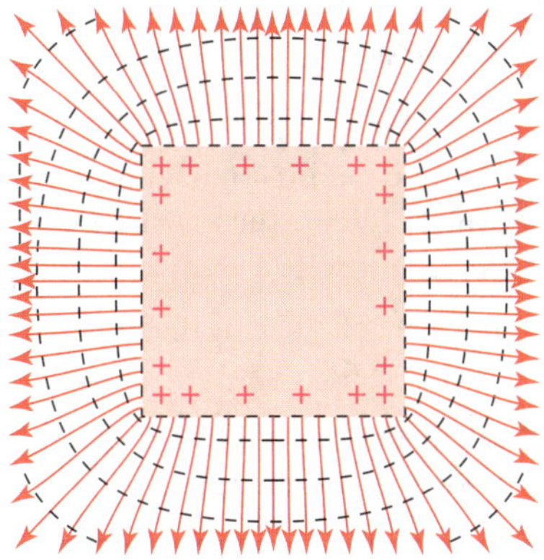

19-12 A charged cube

19.6 Using Electrical Potential to Solve Problems

So how can materials engineers use electrical potential to inspect crystalline structures at the atomic level? In 1981, two IBM researchers, Heinrich Rohrer and Gerd Binnig, found the answer by inventing the *scanning tunneling electron microscope* (STM). In 1986, they received the Nobel Prize in Physics for their invention. But how does it work?

Scanning tunneling microscopes use an extremely sharp needle with a point that is, ideally, a single atom wide. Operating within a vacuum, this needle is given a small positive charge and brought within a few nanometers of the surface to be scanned. Though the electrons on this surface are attracted to the positive needle because of their negative charge, they can't easily jump to the needle because the vacuum's insulating effect presents an electrical potential barrier. But a phenomenon called *tunneling* makes it possible for electrons to pass through this barrier. The electron, because of its wave-like nature, has a finite statistical probability of passing through the gap as a wave rather than as a particle. The stream of electrons that jump the gap forms a *tunneling current*. Though all objects have both a wave and a particle nature (see *wave-particle duality* in Chapter 28), the wave-like nature of large objects is negligible compared to a very small object like an electron. The tunneling of an electron through its potential barrier is similar to your being faced with an incredibly high fence, and, instead of jumping the barrier, you walk through it like a radio wave!

The scanning tunneling microscope moves its needle back and forth over the scanning surface by using piezoelectric actuators. As the surface changes height, the needle's distance from the surface is adjusted to maintain a constant tunneling current. The small variations of the tunneling current caused by the presence of atoms are then interpreted by a computer as atomic images (see Figure 19-13).

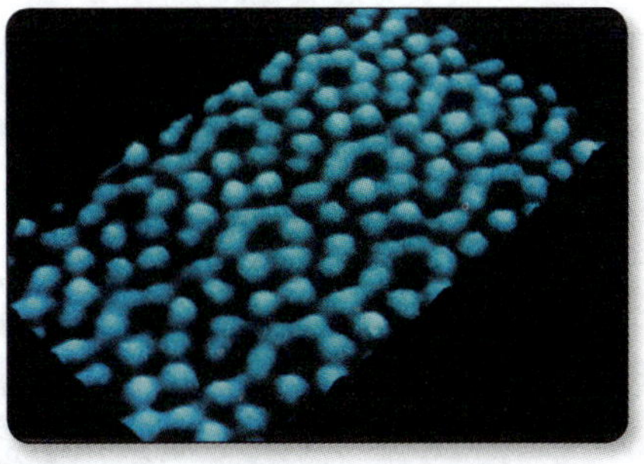

19-13 Silicon atoms as viewed with a scanning tunneling microscope (STM)

19A Section Review

1. a. What model accounts for the spatial distribution and orientation of the electrostatic force associated with a charge?
   b. What graphical aid is used to represent this force in a diagram?
2. Why must the test charge that allows you to define the electric field strength vector be as small as possible?

3. According to the accepted model of electric fields, where do lines of force begin and end?
4. If you move a test charge from position $\mathbf{r}_1$ to position $\mathbf{r}_2$ on the opposite side of a point charge Q, and $r_1 = r_2$, how much total work did you do on the test charge? Explain your answer.
5. The rate of change of what quantity determines the field strength in an electric field?
DS6. Explain why an electron can't normally jump to the needle on an STM when it is at a distance of more than a few nanometers.
DS7. Describe piezoelectric materials and how scanning tunneling electron microscopes use them.
✪8. Draw three pairs of small circles on your paper. The circles represent point charges. Position them approximately 3 cm apart. Pair (a) are both positive charges. Pair (b) are a positive and a negative charge. Pair (c) are both negative charges.
   a. For each pair, sketch enough field lines to define the behavior of the electric field between and around the two charges. Make sure to show the direction of the field.
   b. For each pair, draw at least three lines representing equipotential surfaces surrounding the charges.

19A Objectives

After completing this section, I can
- ✓ describe the field model for noncontact forces, such as the electrostatic force.
- ✓ use lines of force to illustrate an electric field.
- ✓ calculate electric field strength on a positive test charge.
- ✓ describe the shape of an electric field between charges.
- ✓ compare and contrast electrical potential and electrical potential energy.
- ✓ relate the properties of an electrostatic field to those of a gravitational field.
- ✓ define an equipotential surface and illustrate equipotential surfaces that surround various shapes.
- ✓ describe the principle on which a scanning tunneling electron microscope functions.

19B CAPACITORS

19.7 Storing Charge

One hindrance to early investigators of electricity was that they could not store charge. They could accumulate charges by using friction, but these charges soon leaked away. The investigators did discover that it is easier to store charge using two conductors separated by an insulator—positive charge on one and negative on the other—than on an isolated conductor.

In the mid-eighteenth century, two men, a Prussian cleric and a Dutch physicist, independently developed an efficient charge storage device. It was called the **Leyden jar** after Leiden, Netherlands, the hometown of the physicist. A Leyden jar is a glass jar with a lead coating on both the inside and the outside. A conducting rod inserted through an insulating stopper is connected to the inner metal surface of the jar by a metal chain. The outside of the jar was grounded. When a charged object touched the conducting rod, it charged the inner lead surface. The charges within the glass particles then aligned themselves so that the inner surface of the glass had a net charge opposite to the inner lead coating, and the outer glass surface had the same charge as the inner lead. In this way, the insulating glass maintained a neutral charge overall. The net charge of the outer glass surface induced a charge on the outer lead coating by forcing electrons to or from the ground. When the charged object was removed, the Leyden jar was charged and stayed charged for a reasonable time.

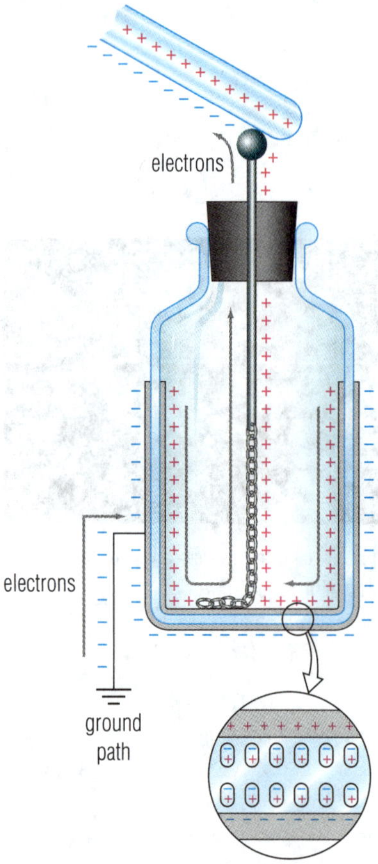

19-14 Schematic of a Leyden jar

19.8 Structure of a Capacitor

Modern descendants of the Leyden jar are called **capacitors**. A capacitor consists of two conductors, called **plates**, separated by a **dielectric**. The dielectric may be a vacuum, air, glass, or one of many insulating materials. Depending on the construction of the capacitor, the sheets of metal foil that form the plates may be

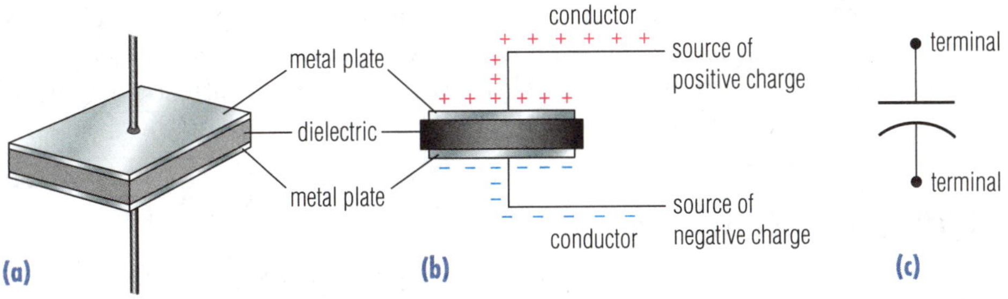

19-15 (a) shows the internal structure of a capacitor; (b) shows how a charge is established on a capacitor; and (c) is the electrical schematic symbol for a capacitor.

layered with each other in a flat, parallel configuration or rolled into a cylinder. Each conductor's surface is at a constant potential, but when the capacitor is charged, the potential is opposite on each plate. Therefore, there is a potential difference between the plates of a capacitor. For the sake of simplicity, we will limit our discussion to *parallel-plate capacitors*, capacitors containing two flat plates arranged parallel to each other.

If a conductor is connected between one plate of a capacitor and the other, the excess electrons on the negative plate travel to the positive plate, and the charge is neutralized. This process is called *discharging* the capacitor. The potential difference between the plates seems to produce a force that moves the electrons, although this is not a truly accurate explanation based on the present model of electron flow through a conductor. The more charge that there is on each plate, the greater the potential difference is between the plates, and the greater the number of electrons that can move between the plates.

The **dielectric constant** is represented by the Greek letter kappa (κ). It is a unitless, positive real number (≥ 1) that is a measure of the ease with which electrostatic fields can permeate a volume of space or matter. The larger the number, the more attenuation of the electrostatic field.

TABLE 19-1

Selected Dielectric Constants

Substance	κ
Vacuum	1
Air	1.0005
Waxed paper	2.2
Mylar	3.1
Paper	3.7
Pyrex glass	5.6
Bakelite	~5
Mica	5.4
Porcelain	6.5
Water	80

19.9 Dielectric Constant

The field lines that pass between the conducting plates of a parallel-plate capacitor with a *vacuum* dielectric have a field strength that is proportional to the charge on the plates and inversely proportional to the square of the distance between them. Through experimentation, it can be shown that when other substances are inserted between the plates of the capacitor, the field strength is reduced. This reduction of the vacuum field strength (E_0) is defined by a quantity called the **dielectric constant (κ)**. The magnitude of the field strength in the dielectric (E_{die}) is found by the equation

$$E_{die} = \frac{E_0}{\kappa}. \qquad (19.5)$$

The dielectric constant for vacuum is $\kappa_0 = 1$, so in a parallel-plate, vacuum-dielectric capacitor, Equation 19.5 reduces to an identity. The dielectric constant for paper is about 4, so the field strength in a capacitor with a paper dielectric is one-fourth that of a vacuum. Therefore, for any dielectric, the vacuum field strength is reduced by a factor of κ in a capacitor of identical dimensions. The dielectric constant for each material is found by experiment.

Consequently, the dielectric reduces the potential difference between the plates compared to a vacuum capacitor. From Equation 19.3, we know that

$$V = Ed.$$

Substituting Equation 19.5 for the field strength in the presence of a dielectric, we have

$$V_{die} = \left(\frac{E_0}{\kappa}\right)d = \frac{E_0 d}{\kappa} = \frac{V_0}{\kappa}.$$

The symbol ΔV could be used here for potential difference, but it is assumed that one of the plates of the capacitor is at zero relative potential. Therefore, it is usual to represent potential difference in such a case as just V. You saw this same principle in Chapter 9, where change in relative height determined the change in potential energy.

Thus, a charged capacitor with a paper dielectric will have approximately one-fourth the potential difference between its plates compared to a vacuum capacitor with the same amount of charge. Next, you will see how this affects the performance of a capacitor.

19.10 Capacitance and the Farad

How much charge can a capacitor store? That depends on the potential difference between the plates. In this discussion, Q represents the entire charge on a conductor. The ratio Q/V is experimentally found to be a constant for each capacitor. This ratio is called its **capacitance (*C*)**:

$$C = \frac{|Q|}{|V|} \tag{19.6}$$

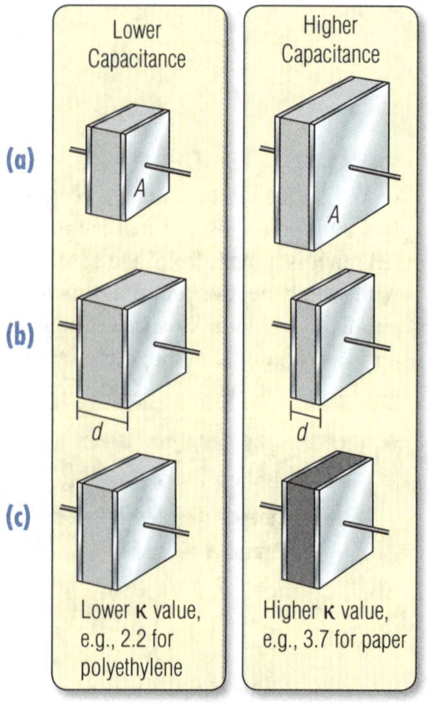

19-16 Capacitance depends upon (a) the area of the plates, (b) the separation of the plates, and (c) the dielectric constant (κ) of the medium between the plates.

A larger dielectric constant effectively increases the capacitance of a capacitor by reducing the potential difference required to store a given amount of charge. The capacitance of a nonvacuum dielectric capacitor is

$$C = \frac{|Q|}{|V_0/\kappa|} = \frac{\kappa|Q|}{|V_0|} = \kappa C_0,$$

where C_0 is the capacitance of a parallel-plate capacitor with a vacuum dielectric.

Because the capacitance of a capacitor is always positive, regardless of the kind of charge and the sign of the potential difference, the absolute value signs in Equation 19.6 are required. Since the capacitance for any given capacitor is a constant, the values of Q and V cannot vary independently of one another. That is because C is also dependent on the geometry of the capacitor. The formula for capacitance using geometric quantities is

$$C = \frac{\kappa\varepsilon_0 A}{d}, \tag{19.7}$$

where A is the area of each plate, ε_0 is called the **permittivity of free space**, and d is the distance between the plates. The constant ε_0 is defined as

$$\varepsilon_0 \equiv \frac{1}{4\pi k} = 8.854 \times 10^{-12}\ \mathrm{C^2/(N{\cdot}m^2)}.$$

The constant k in the denominator is the Coulomb's law constant, $8.988 \times 10^9\ \mathrm{N{\cdot}m^2/C^2}$. The product $\kappa\varepsilon_0$ is related to the dielectric and is called the material's **permittivity** (ε).

The farad (F) is the SI derived unit of capacitance. It is defined as one coulomb per volt:

$$1\ \mathrm{F} \equiv \frac{1\ \mathrm{C}}{1\ \mathrm{V}}$$

The farad is a very large unit, so the microfarad (μF = 10^{-6} F), the nanofarad (nF = 10^{-9} F), and the picofarad (pF = 10^{-12} F) are usually used in actual electronic circuits involving discrete electrical components.

The farad was named after Michael Faraday. A farad in SI base units is defined by

$$1\ \mathrm{F} = \frac{\mathrm{A^2 \cdot s^4}}{\mathrm{kg \cdot m^2}}.$$

An electrical component is any device that is used in an electrical circuit. Examples of electrical components include capacitors, resistors, switches, batteries, lamps, inductors, transistors, and a host of other devices.

EXAMPLE 19-1

Capacitance and Charge

A parallel-plate capacitor consists of two 1.00×10^{-4} m² plates separated by a distance of 1.00×10^{-3} m. It has air for a dielectric ($\kappa = 1.00$). (a) What is its capacitance? (b) If there is a potential difference of 1.00 V between the plates, what charge is stored on them?

Solution:

a. Find the capacitance using Equation 19.7:

$$C = \frac{\kappa \varepsilon_0 A}{d}$$

$$C = \frac{(1.00)(8.854 \times 10^{-12}\ \mathrm{C^2/N{\cdot}\cancel{m^2}})(1.00 \times 10^{-4}\ \cancel{\mathrm{m^2}})}{1.00 \times 10^{-3}\ \mathrm{m}}$$

$$C = 8.8\underline{5}4 \times 10^{-13}\ \mathrm{C^2/N{\cdot}m}$$

$$C = 8.8\underline{5}4 \times 10^{-13}\ \mathrm{C^2/J}$$

$$C = 8.8\underline{5}4 \times 10^{-13}\ \mathrm{C/V}\quad (1\ \mathrm{V} = 1\ \mathrm{J/C})$$

$$C \cong 0.885\ \mathrm{pF}$$

b. Find the charge using Equation 19.6:

$$C = \frac{|Q|}{|V|}$$

$$|Q| = C|V| = (8.8\underline{5}4 \times 10^{-13}\ \mathrm{F})|1.00\ \mathrm{V}|$$

$$|Q| = 8.8\underline{5}4 \times 10^{-13}\ (\mathrm{C}/\cancel{\mathrm{V}})(\cancel{\mathrm{V}})$$

$$Q \cong 8.85 \times 10^{-13}\ \mathrm{C}$$

19.11 Connected Capacitors

Capacitors are electrical components that store charges in an electrical circuit. In order to begin understanding some of the basic electrical properties of capacitors, we will examine how they function when connected to other capacitors. They may be connected in either of two ways. The first is a *parallel* connection. Two or more capacitors are arranged so that the wires from the plates at one end of each capacitor are all attached together (with solder or an electrical terminal), and the wires to the other plates are all attached together (see Figure 19-17). In this way, the corresponding plates of parallel-connected capacitors are at the same potential. Therefore, the potential difference between the plates of all capacitors connected in parallel is the same.

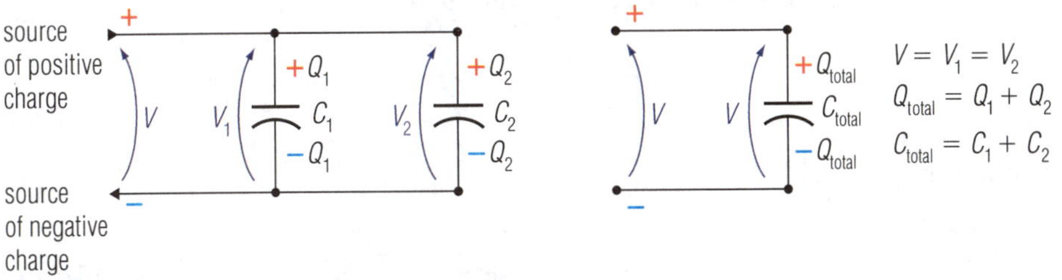

19-17 Capacitors in parallel. Each capacitor is identified by its capacitance, *C*.

Problem-Solving Strategy 19.4

The term "total capacitance" can also be called "equivalent capacitance," since you can replace all of the capacitors connected together with a single one of an equivalent value.

The total capacitance of two or more capacitors connected in parallel is the simple sum of the individual capacitances.

$$C_{\text{total}} = C_1 + C_2 + \ldots + C_n$$

An electrical **conductor** is any device or substance whose principal function is to provide a path for the transfer of electrical charges from one location to another. Conductors include wires, metal bars, neurons in biological systems, and even plasmas in space.

For two or more capacitors connected in series, the reciprocal of the total capacitance is the sum of the reciprocals of the individual capacitances.

$$\frac{1}{C_{\text{total}}} = \frac{1}{C_1} + \frac{1}{C_2} + \ldots + \frac{1}{C_n}$$

Problem-Solving Strategy 19.5

Review the process for summing the reciprocals of quantities. Recall that

$$\frac{1}{C_{\text{total}}} = \frac{1}{C_1} + \frac{1}{C_2} \neq \frac{1}{C_1 + C_2}.$$

Once you have found the sum of the reciprocal capacitances, you must take the reciprocal of *that* sum to find the total capacitance.

Problem-Solving Strategy 19.6

In the case of just two series-connected capacitors, a much simpler formula to use is

$$C_{\text{total}} = \frac{C_1 C_2}{C_1 + C_2}.$$

The total charge stored in the circuit segment is the sum of the charges on the capacitors:

$$Q_{\text{total}} = Q_1 + Q_2 \tag{19.8}$$

Since $Q = CV$, and the potential difference between the plates is the same for each capacitor, Equation 19.8 can be rewritten as

$$C_{\text{total}}V = C_1V + C_2V.$$

Canceling the Vs gives us

$$C_{\text{total}} = C_1 + C_2. \tag{19.9}$$

The total capacitance of parallel connected capacitors is just the sum of the individual capacitances.

The other way to connect capacitors is to connect one plate of one capacitor to one plate of another capacitor in a chain, as Figure 19-18 shows. The plates of the end capacitors are attached to the source of potential difference. This arrangement is called a *series* connection. The potential difference between plates A and D induces equal and opposite charges on these plates. Plate A then induces a charge of $-Q$ on plate B, and plate D induces a charge of $+Q$ on plate C, since that is how capacitors work. Note that the net charge along any continuous **conductor** containing series capacitors is zero. The total potential difference in the circuit segment is the sum of the potential differences across the capacitors.

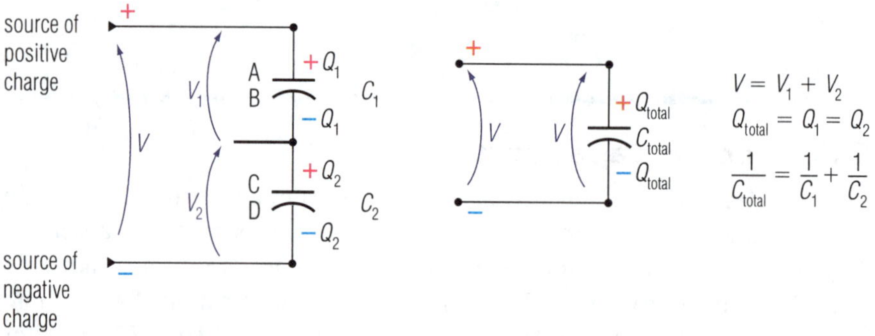

19-18 Capacitors in series

For series-connected capacitors, you can see that

$$V_{\text{total}} = V_1 + V_2. \tag{19.10}$$

Since $C = Q/V$, then $V = Q/C$, and Equation 19.10 can be rewritten as

$$\frac{Q}{C_{\text{total}}} = \frac{Q}{C_1} + \frac{Q}{C_2}.$$

Canceling the Qs yields

$$\frac{1}{C_{\text{total}}} = \frac{1}{C_1} + \frac{1}{C_2}. \tag{19.11}$$

EXAMPLE 19-2

Total Capacitance of Connected Capacitors

What is the total capacitance of a 2.00 μF capacitor and a 4.00 μF capacitor if they are (a) connected in parallel? (b) connected in series?

Solution:

a. Total capacitance in parallel using Equation 19.9:

$$C_{\text{total}} = C_1 + C_2$$

$$C_{\text{total}} = 2.00\ \mu\text{F} + 4.00\ \mu\text{F}$$

$$C_{\text{total}} = 6.00\ \mu\text{F}$$

b. Total capacitance in series using the shortcut given on the previous page:

$$C_{\text{total}} = \frac{C_1 C_2}{C_1 + C_2} = \frac{(2.00\ \mu\text{F})(4.00\ \mu\text{F})}{2.00\ \mu\text{F} + 4.00\ \mu\text{F}}$$

$$C_{\text{total}} = \frac{8.00\ (\mu\text{F})^{\cancel{2}}}{6.00\ \cancel{\mu\text{F}}}$$

$$C_{\text{total}} \cong 1.3\underline{33}\ \mu\text{F}\ (\cong 1.33\ \mu\text{F})$$

Note that the total capacitance for parallel-connected capacitors will always be higher than for the same capacitors connected in series.

If the potential difference between a capacitor's plates becomes too high, even a good dielectric will break down and conduct charge between the plates. The damage is usually permanent, and the capacitor's function is degraded or destroyed. Some dielectrics will even explode if they are exposed to a potential difference that is too high. For this reason, most capacitors are marked with a potential difference rating as well as a capacitance value. The capacitor will probably cease to function if the potential difference rating is exceeded.

Some capacitors are marked with a positive or negative sign on one terminal. These components are called *polarized capacitors* because they are designed to be connected in a circuit only one way—with the positive plate at the higher potential. If improperly connected, even normal voltages can cause the dielectric to break down, resulting in failure.

19B Section Review

1. Describe a Leyden jar and compare its structure to that of a modern capacitor.
2. a. How does electric field strength between capacitor plates in a vacuum compare to field strength when an insulating material replaces the vacuum?
   b. What electrical quantity accounts for this fact?
3. In an isolated charged capacitor of given dimensions and charge, how does the potential difference between the plates with a vacuum for a dielectric compare to the potential difference when some other insulating material is present?
4. What factors determine the capacitance of a capacitor?
5. ✪ What is the capacitance of a capacitor that has a plate area of $2.3 \times 10^{-4}\ \text{m}^2$ and a distance of 1.8×10^{-3} m between the plates? The dielectric is a Mylar-like material.

19B Objectives

After completing this section, I can

- ✓ describe the structure and operation of a Leyden jar capacitor.
- ✓ identify the parts and function of a plate capacitor.
- ✓ define the dielectric constant (κ) of a capacitor and explain its effect on the capacitor's function.
- ✓ define the capacitance (C) of a capacitor and identify the factors on which it depends.
- ✓ calculate the capacitance of a particular capacitor.
- ✓ identify the two ways to connect two or more capacitors.
- ✓ compute the equivalent capacitance of any configuration of capacitors.

6. If the capacitor in Question 5 is connected to the terminals of a 9.0 V battery, how much charge can be stored in the capacitor?

7. Describe the arrangement of capacitor connection in the adjacent figure. Determine the total capacitance of these capacitors.

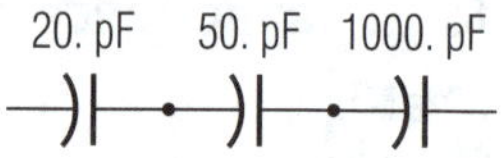

8. Describe the arrangement of capacitor connection in the adjacent figure. Determine the total capacitance of these capacitors.

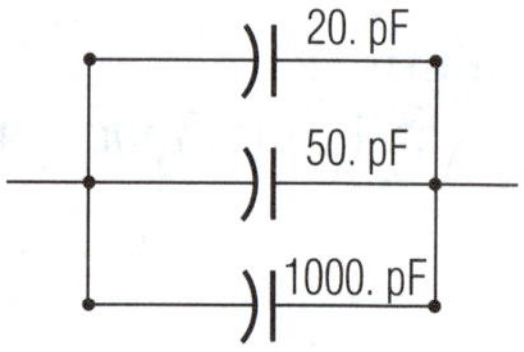

FACETS of PHYSICS

Kinds of Capacitors

Many different kinds of capacitors do many different jobs. Air capacitors, whose dielectric is air, do not need to be enclosed. However, the plates must be far enough apart that they will not touch by accident. Vacuum capacitors can tolerate a much higher potential difference without allowing a current between their plates. Of course, vacuum capacitors must be enclosed in an airtight container.

Many capacitors have solid dielectrics. Ceramic capacitors have high capacitance for small sizes, but they also become less reliable with age. Mica dielectrics are especially useful in powerful radio transmitters. Paper dielectrics are inexpensive and work well in applications where they are unlikely to get wet.

Liquid dielectrics are also common. Electrolytes (conducting solutions) form capacitors by depositing an insulator on one conducting plate. Since they operate properly when the current goes in one direction but not when the current reverses, electrolytic capacitors are *polarized* capacitors. They must be connected in the proper direction or they will not work. Modern capacitors are often polarized.

Capacitors need not have a fixed capacitance. Variable air capacitors usually consist of many semicircular plates that may be rotated to mesh. The more the plates are meshed, the greater the capacitance is. These capacitors are essential for some radio-frequency tuning circuits.

Chapter Review

In Terms of Physics

field model	429
electric field	430
line of force	430
test charge	430
electric field strength (**E**)	430
electrical potential (V)	432
potential difference (ΔV)	433
volt (V)	433
equipotential surface	434
Leyden jar	436
capacitor	436
plate	436
dielectric	436
dielectric constant (κ)	437
capacitance (C)	438
permittivity of free space (ε_0)	438
permittivity of material (ε)	438
farad (F)	438
electrical component	438
conductor	440

Problem-Solving Strategies

19.1 (page 433) Do not confuse electrical potential and electrical potential energy. Electrical potential is a property per unit charge of the *position* of any size charge relative to an electric field. It is measured in joules per coulomb (J/C) or volts (V). Its scalar value can be positive, negative, or zero, depending on the sign of the charge producing the field.

Electrical potential energy is a property of both the *size* of a charge and its *position* relative to an electric field. This quantity is expressed in joules (J).

19.2 (page 433) Do not confuse the derived unit volt (V) with the electrical potential formula symbol, V.

19.3 (page 434) The direction of the electric field is always from the more positive (less negative) values of electrical potential toward lower (more negative) values of electrical potential.

19.4 (page 440) The term "total capacitance" can also be called "equivalent capacitance," since you can replace all of the capacitors connected together with a single one of an equivalent value.

19.5 (page 440) Review the process for summing the reciprocals of quantities. Recall that

$$\frac{1}{C_{\text{total}}} = \frac{1}{C_1} + \frac{1}{C_2} \neq \frac{1}{C_1 + C_2}.$$

Once you have found the sum of the reciprocal capacitances, you must take the reciprocal of *that* sum to find the total capacitance.

19.6 (page 440) In the case of just two series-connected capacitors, a much simpler formula to use is

$$C_{\text{total}} = \frac{C_1 C_2}{C_1 + C_2}.$$

Review Questions

1. What are lines of force?

2. In the figure below, in which region is the electrical force the strongest?

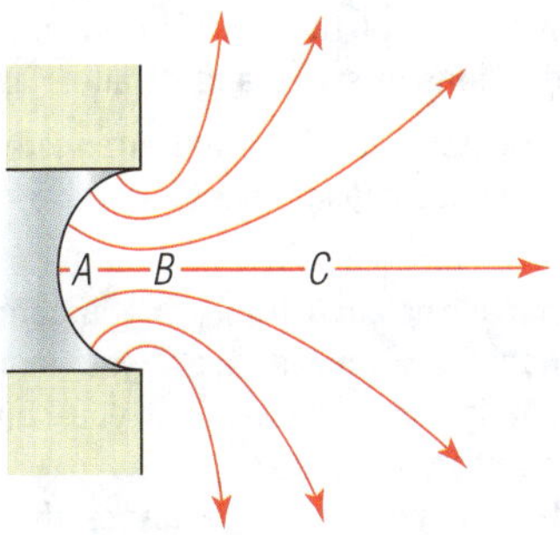

3. Draw the lines of force surrounding a spherical, negatively charged body.

4. Describe a test charge.

5. What is electric field strength?

6. An object with a charge of 2.00 mC rests far from other charges. Write an equation that gives electric field strength at a given distance r from the charge.

7. Where does an electric field line end? Why?

8. What is electrical potential?

9. The dotted lines in the figure below are equipotential lines. If the potential difference between them at A is 3 V, what is the potential difference between them at B? Why?

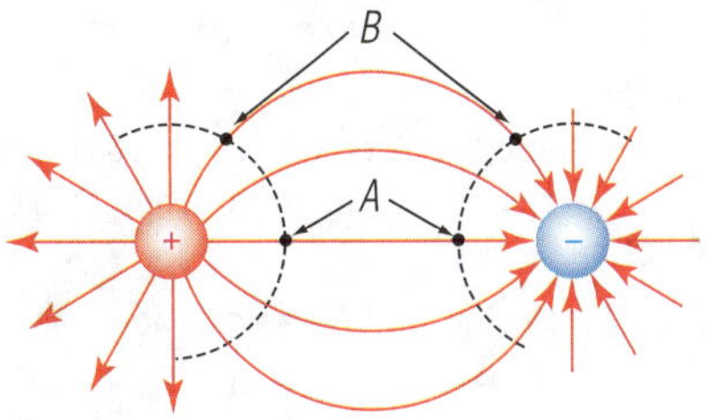

10. What does a capacitor do?

11. Name the predecessor of the capacitor.

12. List the parts of a capacitor.

13. Name three ways to increase a capacitor's capacitance.

14. During a lab period, you and your lab partner charge a capacitor with a 6 V battery. You say that the capacitor has a charge of $|Q|$. Your lab partner says that there is zero charge on the capacitor. You are both right. Explain why this is so.

15. How does charge vary with potential difference applied to a capacitor? Sketch a graph of charge versus potential difference for a capacitor with charge on the vertical axis and potential difference on the horizontal.

16. For a group of series-connected capacitors, which capacitor will have the greatest effect determining the total capacitance—the one with the largest capacitance or the one with the smallest? Explain your answer from either a mathematical or an electrical viewpoint.

True or False (17–30)

17. Lines of force can actually cross each other where electrical fields are exceptionally strong.

18. Field lines point in the direction that a small, positively charged object would move in response to the electric field.

19. The electric field strength vector **E** is tangent to the field's lines of force and points toward the positive charge.

20. Every electric field line originates at a positive charge and terminates at a negative charge, even if the line is not visible beyond the edge of a diagram.

21. Electrical potential is measured with reference to an infinite position from a charge.

22. Field strength is the rate of change of electrical potential with distance in an electric field.

23. The sharper the curvature of an equipotential surface, the more concentrated the lines of force.

24. The thicker the lead coating in a Leyden jar, the greater the amount of charge that could be stored, assuming all other dimensions and materials were held the same.

25. Within limits, the greater the potential difference supplied to a capacitor, the greater the amount of charge that can be stored.

26. A large dielectric constant implies that more charge can be stored for a given potential difference between the plates.

27. The capacitance of a vacuum dielectric capacitor is greater than for a capacitor of the same dimensions having a glass dielectric.

28. Doubling the plate area of a capacitor doubles its capacitance.

29. Capacitance in SI base units is C^2/J.

30. Doubling the number of identical capacitors connected in series doubles the amount of charge that can be stored by the entire circuit.

✪31. A large charged object exerts a force of 10^{-3} N on a 1 μC charge. What is the electrical field strength at the small charge's position?

✪32. A 3.00 μC charge has an electrical potential energy of 9.00×10^{-5} J. What is the electrical potential of its position?

✪33. A tiny spherical object with a charge of +1.00 mC rests far from all other charges.
   a. What is the electrical potential 0.0100 m from the object?
   b. What is the electrical potential 1.100 m from the object?
   c. What is the potential difference between these two positions?

✪34. Draw an equipotential line for the electric field in the figure below.

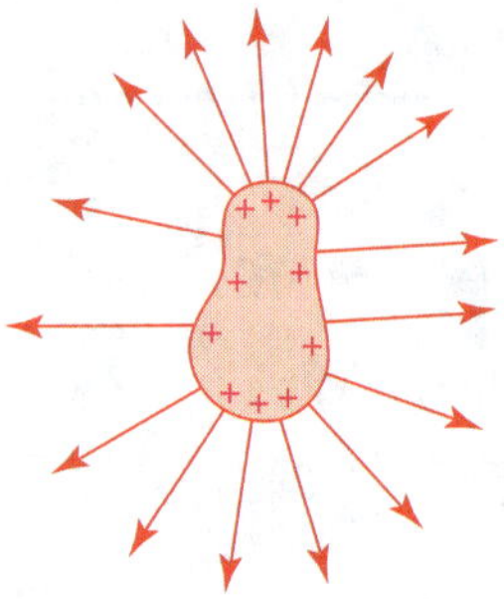

✪35. A parallel-plate, vacuum-dielectric capacitor has a plate area of 3.00×10^{-4} m² and has a distance of 3.00×10^{-2} m between the plates. What is the capacitance?

✪36. A round capacitor with a radius of 1.50×10^{-2} m has a vacuum dielectric. The distance between the parallel plates can be manually varied. The maximum distance apart is 1.00×10^{-2} m and the closest the plates can be is 5.00×10^{-4} m.
   a. What is the area of each plate?
   b. What is the capacitance at maximum plate separation?
   c. What is the capacitance at minimum plate separation?
   d. When the charge on the capacitor is 2.00×10^{-11} C, what is the potential difference across the capacitor at maximum separation?
   e. When the charge on the capacitor is 2.00×10^{-11} C, what is the potential difference across the capacitor at minimum separation?

✪37. A capacitor with square plates 5.00 cm on each side and a Pyrex glass dielectric has a separation distance of 5.00×10^{-3} m. What is its capacitance?

✪38. When a capacitor has a charge of 6.00×10^{-11} C, the potential difference between its plates is 20.0 V. What is its capacitance?

✪39. Two capacitors, one with C_1 = 2.00 μF and one with C_2 = 3.00 μF, are connected in parallel. What is the equivalent capacitance of the arrangement?

✪40. Suppose you have a box full of 1.50 pF capacitors and you need a capacitance of 1.00 pF. How could you connect capacitors to get the desired capacitance? (*Hint*: Try various combinations of series and parallel capacitors.)

✪41. Two 6.00 μF capacitors are arranged in series. What is the arrangement's capacitance?

✪42. You have three capacitors: two 4.00 pF and one 2.00 pF.
   a. What is the capacitance if you connect them in parallel?
   b. What is the capacitance if you connect them in series?
   c. What is the capacitance if you connect the two 4.00 pF capacitors in series?
   d. What is the total capacitance if you connect the series arrangement of the 4.00 pF capacitors in parallel with the 2.00 pF capacitor? (*Hint*: treat the two 4.00 pF capacitors as a unit.)

DS✪**43.** Research some applications of scanning tunneling electron microscopes and discuss your findings with the class. Explain how these applications are ways of fulfilling the Creation Mandate.

✪**44.** Write a brief report, including diagrams, of the construction and usage of 1+ farad capacitors.

✪**45.** Investigate and report on what happens when too high a voltage is applied to a capacitor. How can this be quantified?

✪**46.** Lightning is an example of breakdown of a dielectric. Prepare a report on lightning and its associated upper atmospheric effects (such as lightning sprites).

An electron-micrograph of a modern integrated circuit chip. Chemically embedded within such chips are millions or even billions of resistors, capacitors, switches, and other electronic components.

Electrodynamics 20

Dominion Science Problem

Transcontinental Telephone

It was 1907, and America's main telephone company, AT&T, formed by Alexander Graham Bell, was battling intense competition as his telephone patents began to expire. The company decided to introduce a new service to overshadow competing companies—transcontinental telephone. But there was a problem. Telephone signals would have to be significantly amplified to travel such great distances. How could engineers find a way to reliably and efficiently amplify telephone signals?

20-1 An early twentieth-century telephone sold by AT&T

20A CURRENT, VOLTAGE, AND RESISTANCE

20.1 Current

The last two chapters dealt with stationary charged objects (electrostatics). However, charges do move when they are attracted or repelled by other charges. They may set up a **current (*I*)**, which is defined as a continuous flow of charge. The study of the causes and effects of current electricity is called **electrodynamics**.

The motion of electrons creates a current in metals. This electrical conduction is similar to thermal conduction, which is the transmission of kinetic energy by collisions of free electrons. Electrical conduction is the transmission of electrical potential energy by repulsions and attractions of free electrons. Free electrons, remember, are electrons that are loosely held by the metal atoms' nuclei.

Electrical conduction in metals begins with a potential difference in the metal. For instance, if a positively charged object approaches one end of a wire, that end of the wire has a higher electrical potential than the other end. The wire's free electrons move toward the positively charged object. When the electrons move, they leave a space with a net positive charge. Therefore, new free electrons are attracted to the region that the other free electrons vacated.

Free electrons do not move far from their original positions, yet an impulse of motion spreads throughout the metal. To illustrate how this occurs, consider a line

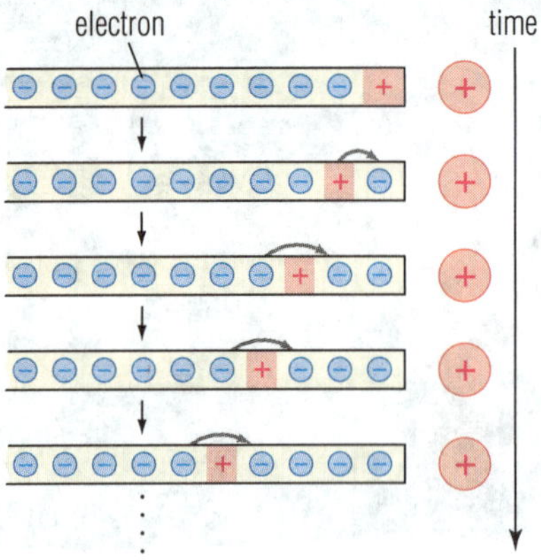

20-2 Electrons moving in a metal-like solid

This equation is the classical empirical definition of the ampere. The present-day SI definition is given in Appendix A.

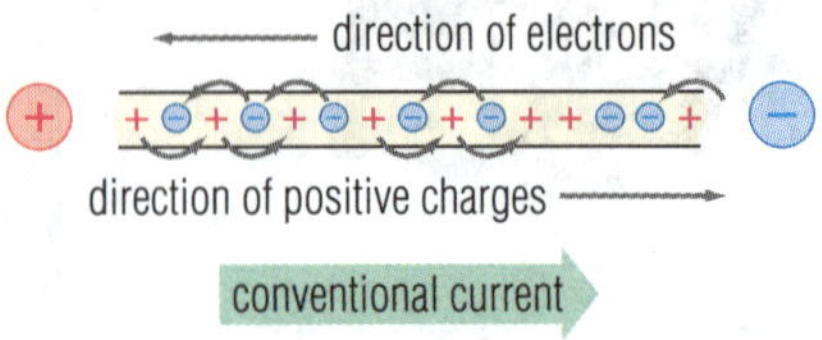

20-3 The direction of conventional current flow

The flow of current is in the direction that tends to cancel the potential difference causing the current. This textbook adopts the convention that current is the flow of positive charges from higher (positive) potentials to lower potentials.

The potential difference between a point at 10 V and a point at 5 V is 10 V − 5 V = +5 V. The potential difference between a point at 5 V and a point at 10 V is 5 V − 10 V = −5 V. The second number is always taken as the reference potential.

Polarity markings (+ and −) are used on sources of potential difference (as well as devices connected to these sources) in order to identify the location of higher and lower electrical potential.

of dominoes with many dominoes standing on end less than one domino-length apart. If you push the first domino over, the last one will eventually fall. The dominoes have not moved far from their original positions, but the impulse of falling down has spread through all the dominoes.

The velocity of the electrons, called the *drift velocity*, is slow—on the order of millimeters per second. The free electrons in a conductor tend to move in random directions, even when an electrical potential is applied, but more electrons move toward the higher potential than in any other direction. This net movement with time is the drift velocity. Each electron needs to move only far enough to repel another electron. As soon as the electron moves even a small distance, the free electrons nearest it are affected. The impulse moves much faster than the electrons do—it can approach the speed of light.

In solid conductors, electrons usually carry the current. In *electrolytic* (conducting) solutions, though, electrons are not the sole current-carriers. *Electrolytes* are materials that are able to conduct electricity when they are dissolved. They separate into positive ions, or *cations* (atoms or molecules that have a deficiency of electrons), and negative ions, or *anions* (atoms or molecules that have an excess of electrons). In a solution, the current consists of anions and cations traveling in opposite directions.

Since current is the flow of charge, it is natural to define the unit of current, the **ampere (A)**, as 1 coulomb of charge per second:

$$1\ \text{A} \equiv \frac{1\ \text{C}}{1\ \text{s}}$$

20.2 Current Direction

These facts raise a question: What is the current direction? The choice of current direction is arbitrary. The first scientists to study electricity assumed that current was the flow of positive charge (after Ben Franklin's naming convention). Therefore, they decided that current flows from a higher potential (positive charge) to a lower potential (negative charge). This text will follow this convention. When considering current in a solid, such as a wire, you need to be careful to avoid confusion. Although both positive and negative charges move through solutions, in solids only the electrons move.

Since the *conventional current* direction is the direction that positive charges would move, this direction is opposite the direction of the actual electron flow. It makes little difference, since the flow of positive charge in one direction is equivalent to the flow of negative charge in the opposite direction. In either case, the charges move to cancel the potential difference. The motion of charges to cancel a *constant-direction* potential difference is called **direct current**, or **DC**.

20.3 Potential Difference and Voltage

Potential difference is a difference in electrical potential between two positions. The potential difference between point A and point B is $V_A - V_B$, while the potential difference between point B and point A is $V_B - V_A$. The second number is always assumed to be the reference potential. Obviously, one value will be the negative of the other:

$$V_A - V_B = -(V_B - V_A)$$

Consequently, it is important to identify how potential difference is measured. Potential difference can be positive, negative, or zero. In order to identify the location of higher and lower potentials on a source of potential difference, **polarity markings** are used—positive (+) for the higher potential and negative (−) for the lower. These points do not necessarily have to be at positive or negative potentials, respectively. It just means that the point with the + is at a higher (more positive) potential than the point with the −.

Any device that creates a difference in potential between two points is a source of potential difference. If the potential difference is created by a self-contained device, especially by a chemical reaction such as in a battery, it is sometimes called *electromotive force (emf)*. Emf is a term scientists gave to potential difference before they realized that it is not a force. Even though emf is not a force as we normally use the word in physics, it is sometimes useful to resort to this term for brevity's sake.

Differences in electrical potential can be produced in several ways. Most processes involve a change from another form of energy to electrical potential energy. A *battery* changes chemical energy to electrical potential; a *photovoltaic cell* changes light energy to electrical potential; a *Van de Graaff generator* (static electricity generator) and *piezoelectric crystals* convert mechanical energy to electrical potential; and a *thermocouple* converts thermal energy to electrical potential.

Potential difference is measured in volts (V), or joules of energy per coulomb of charge:

$$1\ \mathrm{V} \equiv \frac{1\ \mathrm{J}}{1\ \mathrm{C}}$$

One coulomb (C) is defined as the total charge of a specific number of fundamental charges:

$$1\ \mathrm{C} = 6.24 \times 10^{18}\ e,$$

where e is the absolute value of the charge of an electron. Since a specific number of a large quantity of fundamental charges is impossible to measure with present technology, scientists resort to the classical definition of the ampere and define the coulomb in SI units as an ampere-second:

$$1\ \mathrm{C} \equiv 1\ \mathrm{A{\cdot}s}$$

20-4 A Van de Graaff generator can safely produce large static electrical potentials. A static charge can be the beginning of a bad-hair day!

Robert Jemison Van de Graaff (1901–67) was an American physicist who conducted research in high-energy physics. He invented the static electrical generator that bears his name.

20.4 The Voltaic Cell

Probably the most convenient source of steady potential difference for laboratory use is one based on the **voltaic cell**, an electrochemical device that spontaneously changes chemical energy to electrical energy. The conversion of chemical energy to electrical energy was first observed by Luigi Galvani as he was studying the anatomy of a frog. He discovered that if he touched a nerve in a frog's leg with a metal scalpel while the frog was on a metal tray, the touch established an electric current, which contracted the frog's leg muscles. Galvani touched the nerve with different metals and discovered that some metals caused a stronger contraction than others. He thought that the frog's tissues were the source of the current.

Considering Galvani's observations, Alessandro Volta correctly concluded that the source of the current is the contact of two different metals (the tray and the scalpel) in a conducting medium (the frog's body fluids). The difference in chemical activities of the two metals causes electrons and positive ions to flow between the metals if both metals contact an electrolytic solution. Volta made a cell from pieces of copper and zinc separated by seawater-soaked cardboard. This cell is called a *voltaic pile*, after Volta.

The metal conductors in voltaic cells are called **electrodes**. One electrode (the *cathode*) is made of a metal that has a stronger attraction for electrons (on the *external* circuit) than the other electrode (the *anode*). The cathode pulls electrons from the anode through the wire. The accumulation of electrons on the cathode attracts positive ions (cations) from the electrolytic solution that accept the electrons to form a neutral plating on the cathode. The anode, which lost electrons to

Luigi Galvani (1737–98) was an Italian physiologist and anatomist.

Voltaic cells are also called **galvanic cells** after Galvani. Both names refer to a device that uses spontaneous electrochemical reactions to produce a potential difference between the anode and cathode of the cell.

Count Alessandro Volta (1745–1827) was an Italian physicist known for his pioneering work in electricity. He invented the voltaic pile, which was the first electric battery.

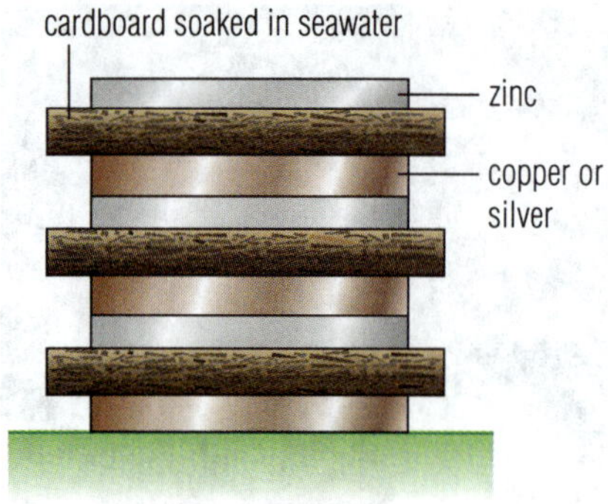

20-5 A voltaic pile

the cathode via the wire, is a site of positive charges from the solution. It attracts electrons from the negative ions (anions) in the solution. This process continues as long as electrolytic ions remain in the solution and the metal electrodes remain in contact with the solution.

Although the electrons flow from the anode to the cathode, their charge is negative, so the external *conventional current* from a voltaic cell is, by our convention, directed from the cathode to the anode. With the wire in place, the current has a complete path: positive ions go from the anode to the cathode in the *solution*, and positive charges flow back to the anode through the *wire*. The closed path of electrical current thus described is called a **circuit**.

Problem-Solving Strategy 20.1

It is not normally necessary to keep track of which direction the moving charges are going. Just remember that current flows from the highest potential to the lowest potential in the external circuit.

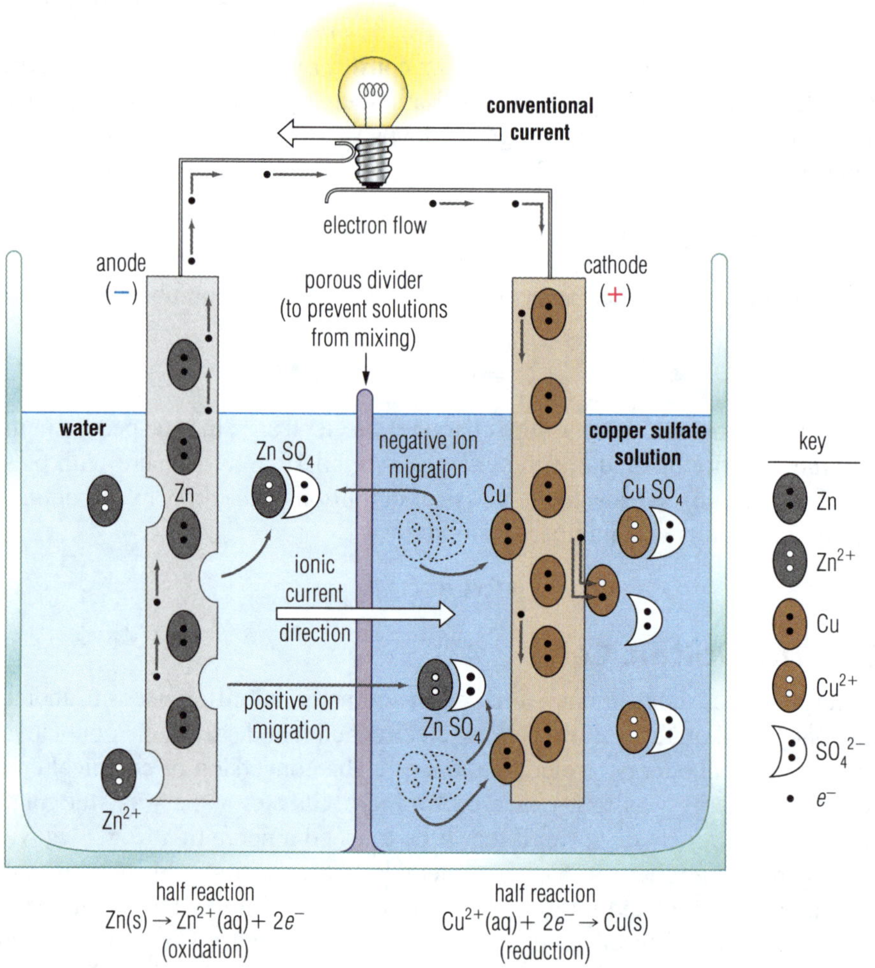

20-6 A voltaic or electrochemical cell

20.5 Types of Electrochemical Cells and Batteries

Modern voltaic cells function in a similar way to Volta's cell. There are several different types of cells. **Primary cells** simply convert chemical energy to electrical energy. A primary cell depends on irreversible (one-way) chemical reactions, so it cannot be recharged. The standard battery cells you use in a flashlight are primary cells.

Unlike primary cells, **storage cells** rely on reversible (two-way) chemical reactions. Therefore, storage cells can be recharged by the addition of electrical energy, which is converted back to chemical energy. Probably the most familiar storage cell is found in an automobile battery. These batteries usually consist of a group of connected storage cells. (Technically, a battery is a series of cells. Flashlight "batteries" are really single voltaic cells.) Rechargeable storage cells that use

An electrochemical cell consists of a single anode and cathode immersed in an electrolyte. A battery is two or more cells connected together.

lithium-ion or nickel-metal hydride technology are found in the batteries used in cellular phones, digital cameras, and laptop computers.

Another term you may hear applied to voltaic cells is "dry cell." A **dry cell** is a primary or storage cell that has an electrolytic paste instead of a liquid solution. Since there is no chance of spilling the solution, a dry cell is easier and safer to use than a "wet" cell. The common D-cell is a dry primary cell, whereas a car battery consists of wet cells.

20.6 Resistance

Every material tends to impede the flow of charge. In wires, the free electrons collide with the electric fields of the inner atoms when these atoms are in the electrons' path. Such collisions convert the kinetic energy of the free electrons to kinetic energy of the inner atoms. At higher temperatures, the inner atoms have more kinetic energy and collide more forcefully with the conducting electrons. Therefore, both the material and the temperature of an object affect how well it conducts electricity.

The tendency to impede current is represented by resistivity (ρ). Resistivity (a different quantity from resistance) is a property of a material that varies with temperature but is not dependent on its size and shape. Good conductors such as copper have lower resistivities than poor conductors such as glass. Only **superconductors**—certain materials at low temperatures (<138 K) — have zero resistivity.

Resistivity is a temperature-dependent property of a material and is the measure of the way it impedes current flow. The symbol for resistivity is the Greek letter rho (ρ). The unit of resistivity is the ohm-meter ($\Omega \cdot m$).

Table 20-1

Resistivities at Room Temperature

Material	ρ ($\Omega \cdot m$)
Conductors	
Silver	1.6×10^{-8}
Copper	1.7×10^{-8}
Aluminum	2.6×10^{-8}
Tungsten	5.5×10^{-8}
Nickel	7.8×10^{-8}
Lead	21×10^{-8}
Nichrome	100×10^{-8}
Semiconductors	
Carbon	3.5×10^{-5}
Germanium	0.5
Silicon	2.3×10^{3}
Insulators	
Glass	$10^{10} - 10^{14}$
Hard Rubber	10^{16}
Quartz	7.5×10^{17}

Any circuit component that is designed to convert electrical potential energy to thermal energy (and produce a potential difference in the process) is called a **resistor**. Electrons in a long, thin wire suffer more collisions than those in a short, thick wire, since they are forced into a very narrow space and they must travel past more inner atoms. Thus a long, thin wire is a better resistor than a short, heavy wire. The geometry of a circuit component affects how well it conducts or resists.

Resistance (*R*) is a quantity that takes into account both the resistivity of the material and the geometry of the resistor. Objects with long current paths resist more than short objects of the same material, and thin objects resist more than wide objects of the same material. The following equation represents this relationship:

$$R = \rho \frac{L}{A}, \tag{20.1}$$

where L is the length of the object's current path and A is the cross-sectional area of the current path. A good conductor has a low resistance, and a poor conductor has a high resistance. The symbol -/\/\/\- represents a resistor in a circuit diagram. The unit for resistance is the ohm (Ω), which is further defined below.

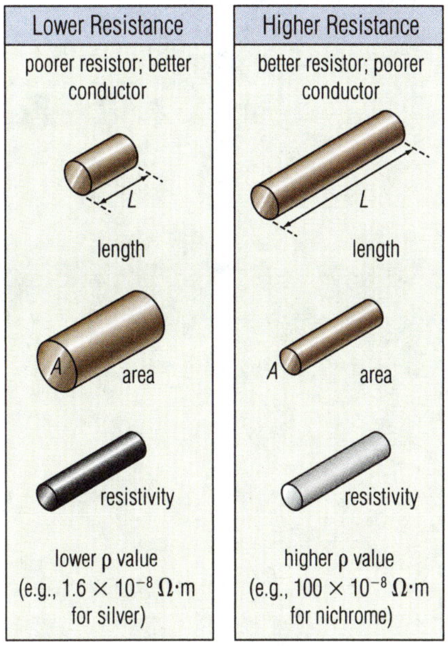

20-7 Factors influencing magnitude of resistance

20-8 Georg Simon Ohm (1789–1854) was a German physicist.

Problem-Solving Strategy 20.2

Ohm's law is true for a single circuit component, a segment of a circuit, or an entire circuit.

20.7 Ohm's Law

Every electrical component converts some electrical potential energy to thermal energy; each circuit component causes a decrease of potential in the direction of current flow. How much potential energy is lost? In 1827, Georg Ohm found the answer to this question. He began experimenting by measuring the current in a circuit made from uniform wire while he changed the potential difference, and he found that current is directly proportional to the potential difference. Then, keeping the potential difference the same, Ohm measured the current while he changed the length of the current path by adding or removing identical wires. He found that current is inversely proportional to the length of wire (the resistance). Ohm expressed his findings in an equation known as **Ohm's law**, which says that the change of potential across any circuit component is

$$V = IR, \tag{20.2}$$

where V is the potential difference across the component, I is the current through the component, and R is the resistance of the component.

The unit of resistance, the **ohm (Ω)**, is defined according to Ohm's law: one volt of potential difference across one ohm of resistance produces one ampere of current through the resistance:

$$1\ \Omega \equiv \frac{1\ \text{V}}{1\ \text{A}}$$

20.8 Electrical Work and Power

The purpose of most electrical components is to convert electrical energy to some other form of energy. That is, electrical components are intended to do work. Since potential difference is measured in volts, or joules per coulomb, and work is expressed in joules,

$$\text{potential difference} = \frac{\text{work}}{\text{charge}}.$$

From this equation you can see that

$$\text{work} = \text{potential difference} \times \text{charge, or}$$

$$W = V \times Q. \tag{20.3}$$

Equation 20.3 is true for a circuit as well as for individual components in the circuit. The rate at which work is done (work per unit time) is the definition of power:

$$P = \frac{W}{\Delta t}$$

The electrical power consumed by a circuit is

$$P = \frac{VQ}{\Delta t} = V\frac{Q}{\Delta t}.$$

But $Q/\Delta t$ is the flow of charges per unit time, or the current, I. Therefore, the electrical power used is

$$P = VI. \tag{20.4}$$

The electrical power used in the component is the product of the potential difference and the current.

Each resistor absorbs a specific amount of electric power when it changes the electrical energy of the free electrons going through it to thermal energy. Joule discovered that the thermal energy produced by a resistor each second (thermal power) is related to electrical power by the expression

$$P = I^2R. \tag{20.5}$$

Equation 20.5, known as **Joule's law**, looks different from the first equation for power. But is it? Ohm's law says that

$$V = IR.$$

Joule's law, expanded a bit, is

$$P = IR \times I.$$

So Joule's law can be written

$$P = VI,$$

which is Equation 20.4.

The electrical power used in a circuit or a circuit component can be expressed in three ways:

$$P = VI$$
$$P = I^2R$$
$$P = \frac{V^2}{R} \quad (20.6)$$

You can prove that Equation 20.6 is equivalent to the other two using Ohm's law.

Since power is energy or work per unit time, it is expressed in joules per second, or watts (W). The kilowatt (kW) is 1000 W. The **kilowatt-hour (kWh)**, a unit of energy, is equal to

$$1\ \text{kWh} = 1000\ \text{W} \times 3600\ \text{s, or}$$
$$1\ \text{kWh} = 3\,600\,000\ \text{J}.$$

Table 20-2 sums up the relationships between the various electrical variables and their units.

Electrical "power" companies sell electrical *energy*, not power. They care for only how much energy you use, not how fast you use it.

Table 20-2

Some Basic Electrical Dimensions

Dimension	Unit Symbol	Definition	Formula Symbol
current	ampere (A)	1 C/s	I
potential	volt (V)	1 J/C	V
resistance	ohm (Ω)	1 V/A	R
charge	coulomb (C)	1 A·s	q or Q
power	watt (W)	1 J/s	P

Problem-Solving Strategy 20.3

Remember that the unit symbol indicates the kind of dimension of a property, while the variable (formula) symbol represents its numerical value.

20A Section Review

1. Describe how electrical conduction occurs in a conductor.
2. Discuss the main difference between conduction in a solid conductor and in an electrolytic solution.
3. How can you tell which electrical connection on a device is supposed to be at the higher potential?
4. List at least three sources of electrical potential.
5. Which wire would impede the flow of current less, copper or aluminum? Explain your answer.
6. What is the resistance of a silver wire 10.0 m long and 0.050 cm in diameter?
7. How much power is used in a coil of copper wire 100. m long with a wire diameter 0.010 cm if there is 1.00 A of current flowing through it?
8. a. How many kilowatt-hours do you waste if you forget to turn off your 100. W bedroom lamp in the morning before school and it remains on for 9.0 hours?
   b. How many joules of energy does this quantity represent?
9. How many grams of water could you heat from 0.0 °C to 100.0 °C with the thermal equivalent of the energy used in Question 8?

20A Objectives

After completing this section, I can

- ✓ describe electric current and identify the charge carriers in various conductors.
- ✓ differentiate between electron current and conventional current.
- ✓ describe the conventions for identifying potential difference in electrical circuits.
- ✓ describe the structure and general operation of voltaic cells and batteries.
- ✓ differentiate between resistivity and resistance.
- ✓ discuss the principle behind Ohm's law and work problems using the law.
- ✓ derive the three formulas for electrical power in a circuit by using Ohm's law.

20B ELECTRICAL CIRCUITS

20.9 Terminology and Symbols

Recall from Chapter 19 that there are two ways of connecting multiple electrical components together—series and parallel. A **series** circuit has a single path for current. The components are linked together like a train or chain. The current, denoted I, passes through each circuit component in turn. A **parallel** circuit has more than one path for current. The current therefore divides up, and only a portion of the current goes through each path. The direction of current in a circuit diagram is represented by an arrow, and a value may be written by the arrow if it is known. A source of potential difference (V) is represented by a generic voltage source symbol or by a symbol of one or more voltaic cells. A curved arrow points from the lower potential to the higher potential alongside the symbol.

Problem-Solving Strategy 20.4

The symbol for a potential difference source has parallel lines of different lengths. The shorter line is conventionally the negative end of the source. Think of it as a minus sign.

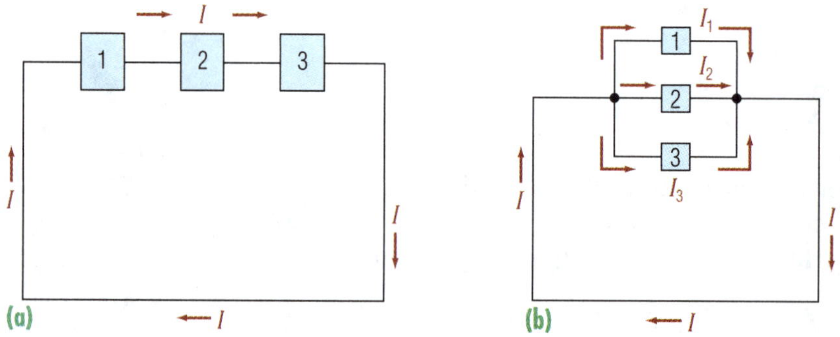

20-9 (a) A series circuit; (b) A parallel circuit

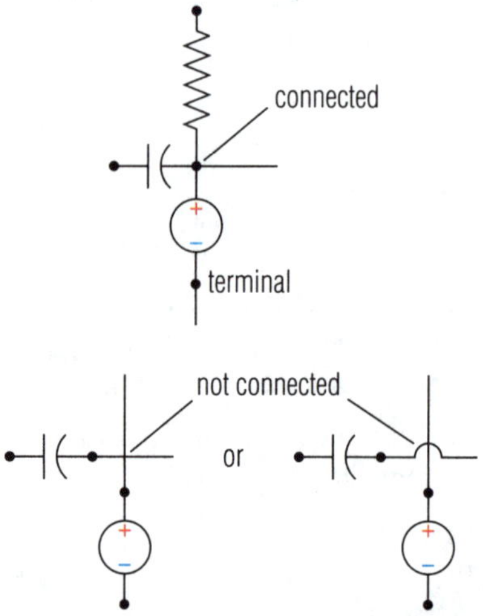

20-10 The points where components are joined to other components or to conductors are indicated on an electrical circuit schematic as a black dot. If lines on a schematic cross without connecting, this fact is indicated by the lack of a dot or an arch in the crossing line.

Electrical components must be connected to each other in order to create a circuit. Their conductors are either permanently connected by soldering (using a type of low-melting-point metal), or they are clamped together using bolts, screws, spring clamps, or some other method.

In a series circuit, the current flows through the circuit components sequentially. Since the circuit does not produce or consume electrons, the current must remain constant throughout the circuit. (That is, electrical current does not speed up or slow down in different parts of a series circuit.) The potential difference changes throughout a series circuit because each circuit component converts some electrical potential energy to kinetic energy of the conductor atoms (thermal energy). Therefore, the electrical potential is higher just before the component than just after the component. This decrease in potential is known as a **voltage drop**. Calling a potential difference "voltage" is similar to talking about the "acreage" of a piece of land, meaning its area. The magnitude of the potential difference across the voltage source is equal to the magnitude of the sum of the voltage drops in the circuit:

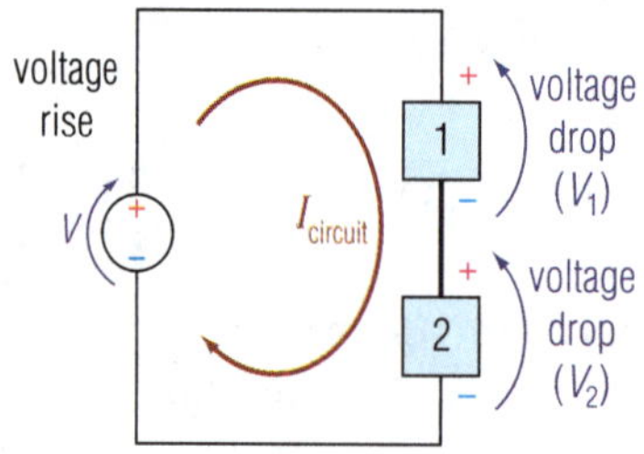

20-11 Voltage drops in a series circuit

$$|V_{\text{circuit}}| = |V_1 + V_2 + \ldots + V_n|$$

Voltage rises in the direction of current flow through a voltage *source*. Voltage drops as current flows through components connected in a circuit.

This is similar to gravitational potential energy. Imagine a forklift that raises a box to a platform 2 m high. The box must fall down two steps in order to return to floor level, where the forklift can raise it again. The sum of the drops through which the box falls is equal to the height at which the box was originally. Similarly, the sum of the voltage drops in a series circuit is equal to the potential difference of the voltage source.

In a parallel circuit, there is more than one path for the current to follow. Since there are no sources or sinks to create or consume electrons, the sum of the currents of the parallel branches is equal to the current entering or leaving the parallel portion of the circuit. In Figure 20-13,

$$I_{circuit} = I_1 + I_2.$$

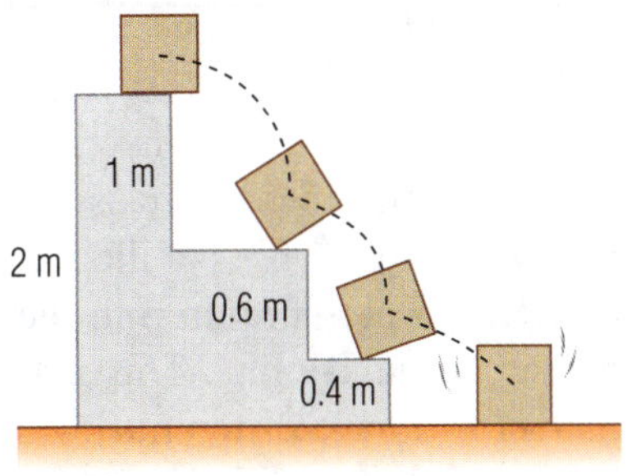

20-12 Stepwise drops in gravitational potential energy are analogous to voltage drops in a series circuit.

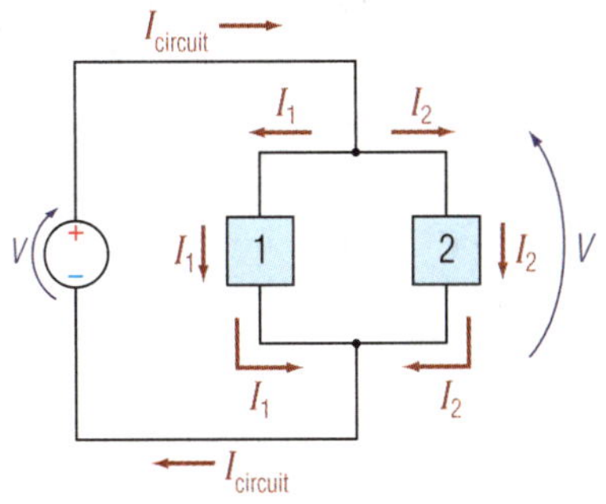

20-13 Current paths in a parallel circuit

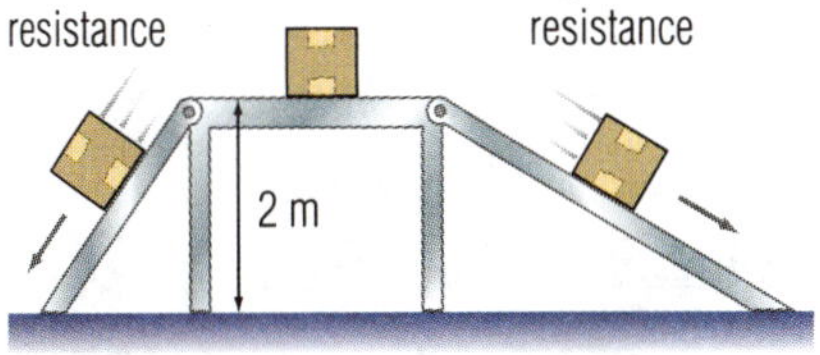

20-14 Boxes sliding down different inclined ramps from the same height are analogous to current falling through the same voltage drop via different paths.

The potential difference is the same across all parallel paths. Returning to the forklift situation, imagine that the forklift raises several boxes 2 m to a platform. There are two ramps from the platform to the floor. Some boxes slide down the steep ramp, and the other boxes slide down the shallow ramp. Both ramps span a 2 m height between the platform and the floor. Therefore, the two ramps have the same gravitational potential. Similarly, all parallel circuit paths have the same potential difference at the points where they are connected together.

20.10 Plumbing Analogy

If the principles governing electrical circuits are unfamiliar, an analogy may make them clearer. Electrons flowing through a circuit behave somewhat like water flowing through connected pipes. (This analogy is not perfect, since electrons are not identical to water in pipes, but the similarities can clarify the new ideas.) Current, the flow of charges, is similar to the flow of water. A pump acts as a source of potential difference, and pressure is analogous to voltage. Pressure is highest at the outlet of the pump and lowest at its inlet. Resistance is like the fluid friction that occurs when a constriction is encountered in the pipe. Electrical switches are like valves in a pipe. They open and close the circuit to control current flow.

Consider a "series" water system. The pump raises the pressure of the water at the pump outlet pipe. As the water encounters constrictions, friction causes drops in pressure (see Chapter 17). Since no water is added or removed, all the water must be moving at the same volumetric rate (volume per unit time). The pressure drops are similar to voltage drops. The pipes decrease the water pressure by friction just as wires decrease potential difference by resistance.

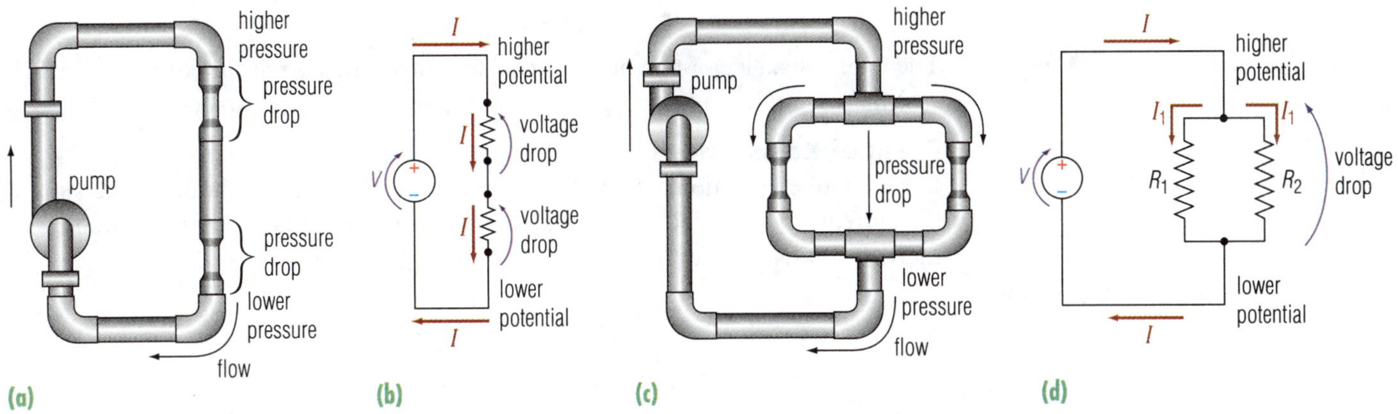

20-15 Similarities between water circuits and electrical circuits. Figures (a) and (b) show series circuits; Figures (c) and (d) compare parallel circuits.

A "parallel" water system is also possible. In this system, T-joints allow the water flow to split into several paths. Some water flows in each pipe. The total volumetric flow rate remains the same, so the water must have the same rate of flow after the pipes rejoin as it had before the pipes divided. Therefore, the sum of the flow rates in the parallel pipes is equal to the rate of flow coming out of the pump. The pressure drop in the two arms is the same, since both pipes are connected at the T-joints. This is true even if the parallel pipes are of different diameters and have different flow rates. As was noted before, this is how electricity behaves in a parallel circuit.

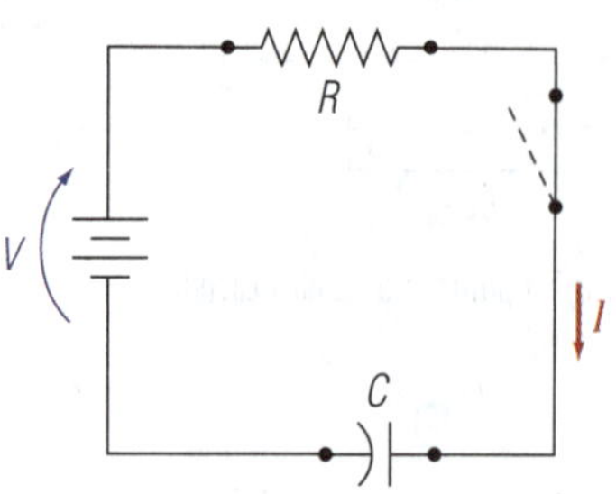

20-16 A simple resistor-capacitor (R-C) circuit

20.11 Analyzing Circuits

Figure 20-16 is a diagram of a simple circuit. Each circuit component has a symbol that represents it on a diagram. Direct-current potential difference is V or V; current is I; resistance is ; and capacitance is . A switch, , is used to start and stop current flow by opening and closing the circuit. Circuit schematic symbols can represent properties as well as discrete components. The values of components' properties are often shown adjacent to the components on the diagram.

Symbol	Meaning
	DC voltage source
	battery
	current
	resistor
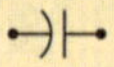	capacitor
	switch

The total resistance of series-connected resistors is the simple sum of all individual resistances: $R_{\text{total}} = R_1 + R_2 + \ldots + R_n$

20.12 Equivalent Resistance

Series Resistances

For series connections, current is uniform and the sum of the voltage drops across each component equals the total voltage drop across all of the series components. Resistance, from Ohm's law, is

$$R = \frac{V}{I}.$$

For resistors in series, the total voltage drop is

$$V_{\text{total}} = V_1 + V_2 + \ldots + V_n.$$

The total resistance is therefore:

$$R_{\text{total}} = \frac{V_{\text{total}}}{I}$$

$$R_{\text{total}} = \frac{V_1 + V_2 + \ldots + V_n}{I}$$

$$R_{\text{total}} = \frac{V_1}{I} + \frac{V_2}{I} + \ldots + \frac{V_n}{I}$$

But V_1/I is R_1, V_2/I is R_2, and so forth, so

$$R_{\text{total}} = R_1 + R_2 + \ldots + R_n. \qquad (20.7)$$

The total resistance of two or more resistors in a series is the sum of the individual resistances. Adding resistors in series increases the total resistance.

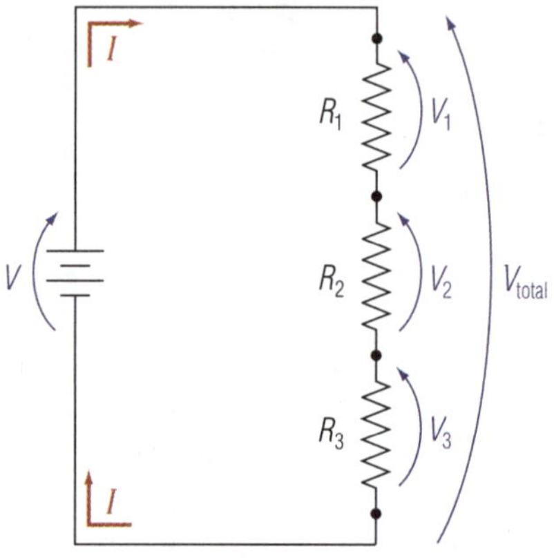

$V_{\text{total}} = V_1 + V_2 + V_3$

20-17 Resistors in series

Parallel Resistances

For parallel connections, the voltage drop is the same in each branch. The sum of the currents in the branches is equal to the current entering the divided circuit segment. From Ohm's law,

$$R_1 = \frac{V}{I_1};\ R_2 = \frac{V}{I_2};\ \ldots;\ R_n = \frac{V}{I_n}.$$

The total current flowing through the parallel branches is

$$I_{\text{total}} = I_1 + I_2 + \ldots + I_n.$$

Therefore, the combined resistance of the resistors is

$$R_{\text{total}} = \frac{V}{I_{\text{total}}}, \text{ or}$$

$$R_{\text{total}} = \frac{V}{I_1 + I_2 + \ldots + I_n}.$$

Taking the reciprocal of each side gives

$$\frac{1}{R_{\text{total}}} = \frac{I_1 + I_2 + \ldots + I_n}{V}, \text{ or}$$

$$\frac{1}{R_{\text{total}}} = \frac{I_1}{V} + \frac{I_2}{V} + \ldots + \frac{I_n}{V}.$$

The terms I_1/V, I_2/V, and I_n/V are the reciprocals of R_1, R_2, and R_n, respectively:

$$\frac{1}{R_{\text{total}}} = \frac{1}{R_1} + \frac{1}{R_2} + \ldots + \frac{1}{R_n} \qquad (20.8)$$

The total resistance of two or more resistors in parallel is the reciprocal of the sum of the reciprocals of the individual resistances. Resistors connected in parallel have less total resistance than any of the individual resistors. Adding more resistors in parallel decreases the total resistance.

A technique used to simplify circuits for analysis is that of replacing several resistors with an equivalent resistor (R_{eq}). For example, circuit (a) in Figure 20-19 looks complicated, but it can theoretically be replaced by circuit (h), which has only one equivalent resistor. The replacement must be done in steps. It uses only the two resistance rules just discussed.

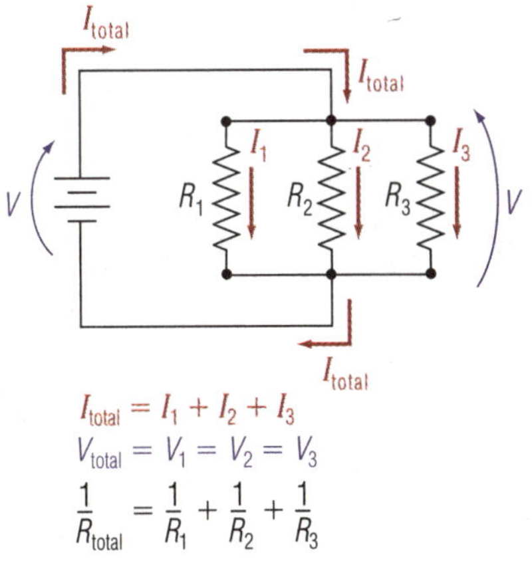

20-18 Resistors in parallel

The total resistance of a group of resistances connected in parallel is the reciprocal of the sum of the reciprocals of the individual resistances:

$$\frac{1}{R_{\text{total}}} = \frac{1}{R_1} + \frac{1}{R_2} + \ldots + \frac{1}{R_n}$$

Problem-Solving Strategy 20.5

After summing the reciprocals of the parallel resistances, remember to take the reciprocal of the sum to find the total resistance.

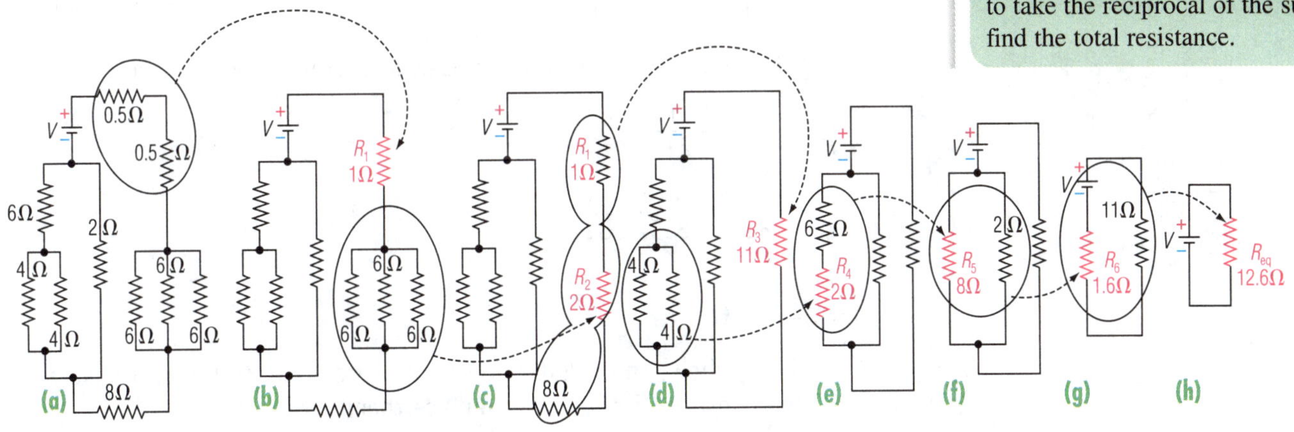

20-19 Simplifying a resistance circuit

EXAMPLE 20-1

Uncomplicate Your Life: Equivalent Resistance

Find the resistance of the equivalent resistor for circuit (a) in Figure 20-19. (The circuits in Figure 20-19 appear again as Figure 20-20 on the next page.)

Solution:

Begin at the positive terminal of the battery, and proceed in the direction of the current. You first come to two resistors in series. Replace them with an equivalent resistor of resistance R_1:

$$R_1 = 0.5\Omega + 0.5\Omega = 1\Omega$$

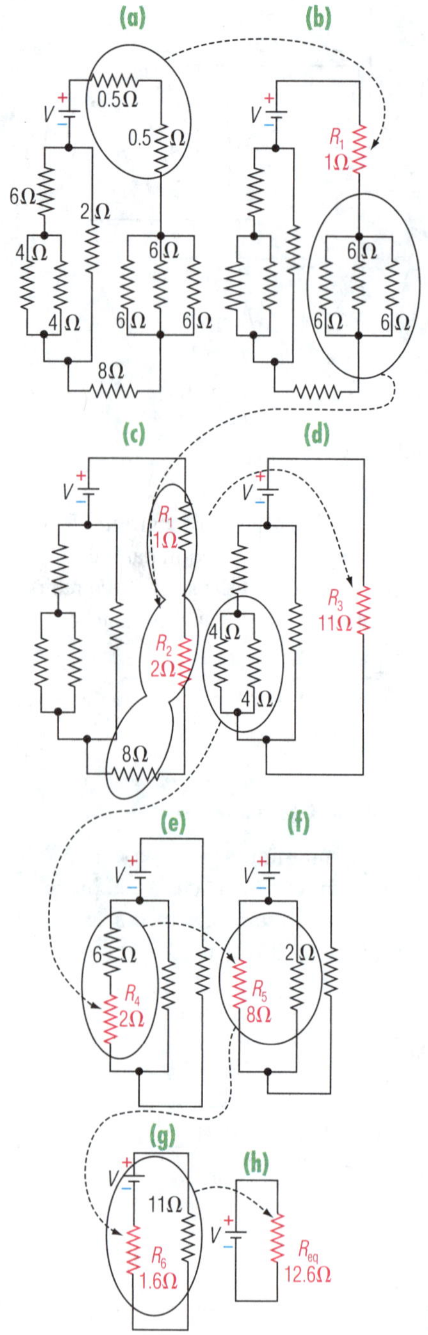

20-20 Simplifying a resistance circuit, from Figure 20-19

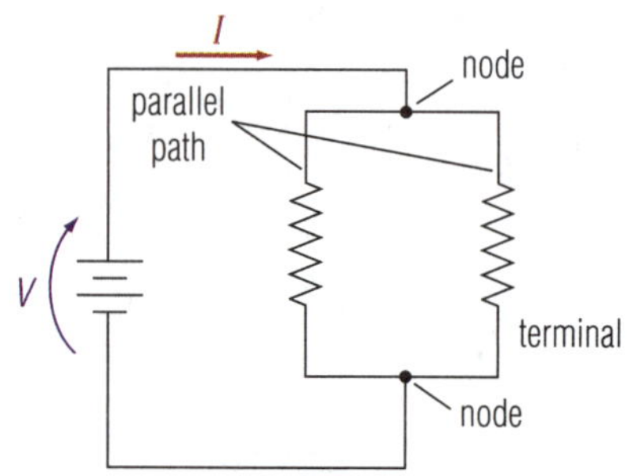

20-21 A node is a junction in a circuit where current divides or converges.

Circuit (b) is the result. Now notice that there are three resistors in parallel. Replace them with R_2:

$$\frac{1}{R_2} = \frac{1}{6\,\Omega} + \frac{1}{6\,\Omega} + \frac{1}{6\,\Omega}$$

$$\frac{1}{R_2} = \frac{3}{6\,\Omega} = \frac{1}{2\,\Omega}$$

$$R_2 = 2\,\Omega$$

Circuit (c) is the result. Notice the three resistors in series. Replace them with R_3:

$$R_3 = 1\,\Omega + 2\,\Omega + 8\,\Omega$$

$$R_3 = 11\,\Omega$$

Now you have circuit (d). You have reached the complicated parallel arrangement. Simplify the left branch first. The parallel resistors can be replaced with R_4:

$$\frac{1}{R_4} = \frac{1}{4\,\Omega} + \frac{1}{4\,\Omega}$$

$$\frac{1}{R_4} = \frac{2}{4\,\Omega} = \frac{1}{2\,\Omega}$$

$$R_4 = 2\,\Omega$$

The branch now has two series resistors. Replace them with R_5:

$$R_5 = 2\,\Omega + 6\,\Omega = 8\,\Omega$$

The parallel circuit segment is now two parallel resistors. Replace them with R_6:

$$\frac{1}{R_6} = \frac{1}{8\,\Omega} + \frac{1}{2\,\Omega}$$

$$\frac{1}{R_6} = \frac{1}{8\,\Omega} + \frac{4}{8\,\Omega} = \frac{5}{8\,\Omega}$$

$$R_6 = 1.6\,\Omega$$

The circuit (g) shows the result of these operations. Since the two remaining resistors are in series, you can replace them with R_{eq}:

$$R_{eq} = 11\,\Omega + 1.6\,\Omega$$

$$R_{eq} = 12.6\,\Omega$$

The circuit (h) contains one equivalent resistor. Circuit simplification may take a long time, but it is not hard if you consistently apply the proper rules.

In complicated circuits, it is sometimes difficult to tell where the parallel part of a circuit begins. The place where a current divides or converges is called a **node** and is represented on a circuit diagram by a connection dot. Parallel paths begin and end at a node, as in Figure 20-21.

20.13 Kirchhoff's Rules

The rules for current and potential difference in series and parallel circuits can be summed up by two general rules called **Kirchhoff's rules**:

1. The sum of voltage drops in a simple *closed* path equals the sum of voltage rises in the path. A simple closed path does not contain any cross-connecting paths. In other words, the algebraic sum of the voltage drops and the voltage rises in the path is zero.
2. The sum of currents entering a node equals the sum of currents leaving the node. In other words, the algebraic sum of the currents entering and leaving a node is zero.

For these laws to be valid, the rules for positive and negative signs must be consistent. To be consistent, you should begin the analysis of any circuit diagram by assigning positive and negative sides to each circuit component (Figure 20-22). Start with the positive side of the voltage source (the battery in this instance), and proceed in the direction of the current. Mark the first side of any circuit component you reach positive, and mark the other side negative. In this way current always flows from positive to negative.

Notice in Figure 20-22 that the current crosses all voltage drops (the resistors) from positive to negative. For all components in a circuit other than the source, the potential is higher at the current inlet of the component and lower at the outlet. This is understandable since current flows down the potential "hill." The potential difference, ΔV, for these components is therefore negative:

$$\Delta V = V_{out} - V_{in}$$

However, the current crosses the source of voltage (the battery) from negative to positive. This is consistent with what you know already. The purpose of voltage sources is to supply a high electrical potential. The potential difference across a source is therefore positive in the direction of current flow.

Gustav Kirchhoff (1824–87) was a Prussian physicist known for his work on electrical currents and spectral analysis.

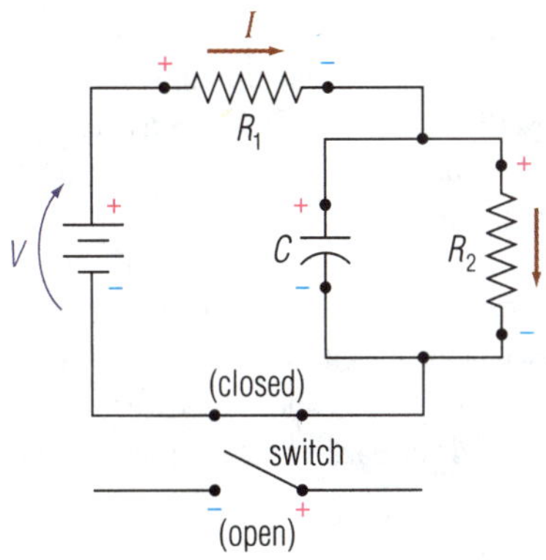

20-22 Assigning positive and negative polarities in a circuit

Problem-Solving Strategy 20.6

When applying Kirchhoff's voltage rule, remember that voltage *rises* across voltage sources (ΔV is positive) and *drops* across other circuit components (ΔV is negative) for conventional current flow. The opposite is true for negative (electron) current flow.

EXAMPLE 20-2

Using Kirchhoff's Rules

Figure 20-23 shows a circuit segment. If all the resistors have a resistance of 10 Ω, and the total current (I) is 1 A, find the current through each resistor.

Solution:

Use Kirchhoff's current rule to find the relationships among the currents. At node 1,

$$I = I_1 + I_4$$
$$1\ \text{A} = I_1 + I_4.$$

At node 2,

$$I_2 + I_3 = I$$
$$I_2 + I_3 = 1\ \text{A}.$$

You obtained no new information from node 2. That will be the case as long as there are no nodes between nodes 1 and 2. You can get more specific information from Kirchhoff's voltage rule. The only *closed* path in the figure is the path that goes through R_1, R_2, R_3, and R_4 in that order or the reverse order. Kirchhoff's voltage rule gives

$$V_1 + V_2 + V_3 + V_4 = 0\ \text{V},$$

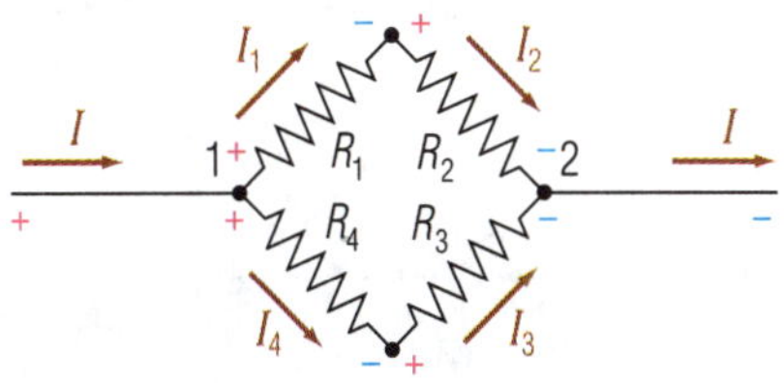

20-23 Current leaving a node must equal the current entering a node.

since there is no voltage source along the path. Note that V_1 and V_2 are negative and the other two voltages are positive along the specified path, according to the convention we are using. Apply Ohm's law to find these voltage drops.

$$(-I_1R_1) + (-I_2R_2) + I_3R_3 + I_4R_4 = 0 \text{ V}. \quad (1)$$

Since R_1 and R_2 are in series, and R_3 and R_4 are in series, then the currents through the respective pairs of resistors are equal. Therefore,

$$I_1 = I_2 = I_{1,\,2} \text{ and } I_3 = I_4 = I_{3,\,4}.$$

Substitute for currents and rearrange terms in Equation (1):

$$I_{3,\,4}(R_3 + R_4) = I_{1,\,2}(R_1 + R_2)$$

$$I_{3,\,4}(10\,\Omega + 10\,\Omega) = I_{1,\,2}(10\,\Omega + 10\,\Omega)$$

$$I_{3,\,4} = I_{1,\,2}$$

Then, $I_1 = I_2 = I_3 = I_4$.

To find the values of the currents, use the results of the current rule at node 1.

$$1 \text{ A} = I_1 + I_4 = I_{1,\,2} + I_{3,\,4}$$

$$1 \text{ A} = I_{1,\,2} + I_{1,\,2} = 2\,I_{1,\,2}$$

$$I_{1,\,2} = 0.5 \text{ A}$$

$$I_1 = I_2 = I_3 = I_4 = 0.5 \text{ A}$$

The current through every resistor is 0.5 A.

Problem-Solving Strategy 20.7

Kirchhoff's rules are used to generate as many current equations as there are unknown currents in the simple closed paths in the circuit. Solving these equations simultaneously using the techniques that you learned in algebra classes permits you to find the individual currents. After the current values are known, you can find the voltage drops and the power absorbed for each resistance.

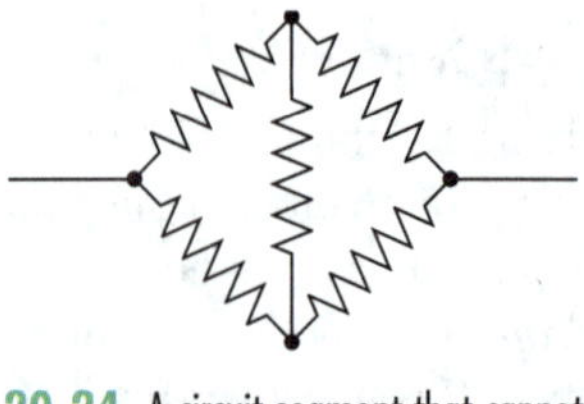

20-24 A circuit segment that cannot be simplified into an equivalent resistance

This example could have been done as easily by using equivalent resistances to calculate the voltage drop across the segment and then using that value in Ohm's law to find the current in each branch. However, some circuits cannot be simplified using equivalent resistances because there are multiple paths that the current can follow. For example, the circuit segment in Figure 20-24 would have to be analyzed using Kirchhoff's rules.

20.14 Electrical Instruments

To calculate any quantity associated with a circuit, you must have some information. You will usually obtain this information by using instruments to detect the basic quantities. An **ammeter** measures current. A **galvanometer** is a sensitive ammeter for detecting very small currents. Devices to measure current must be connected to the energized circuit in series with the current being measured.

An **ohmmeter** measures resistance. It is not connected in a circuit because it contains its own source of potential difference. An ohmmeter can be connected to a single component, a circuit segment, or an entire circuit, as long as there is no outside source of voltage present that could damage the instrument.

A **voltmeter** measures the voltage drop or rise across an energized circuit component. The instrument is connected to the circuit in parallel with the component for which the voltage is being measured.

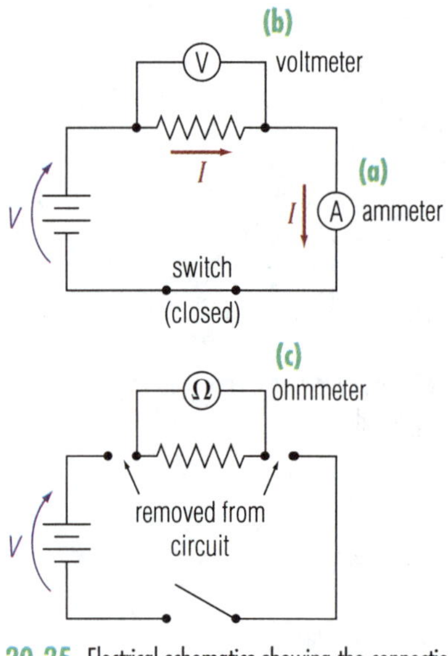

20-25 Electrical schematics showing the connections and symbols for using (a) an ammeter, (b) a voltmeter, and (c) an ohmmeter

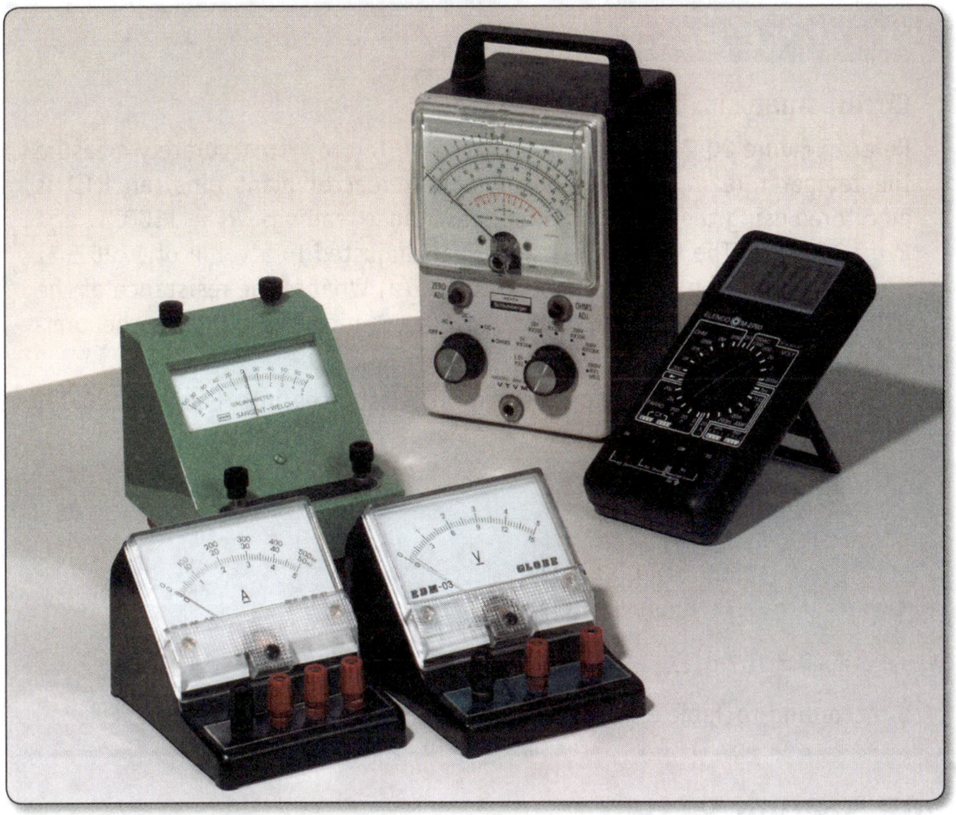

20-26 An ammeter, a voltmeter, and a galvanometer; the analog and digital multimeters (in back and right) can act as an ohmmeter as well as an ammeter and a voltmeter.

A more detailed discussion of the various electrical meters and their uses is provided in Appendix E of the Lab Manual.

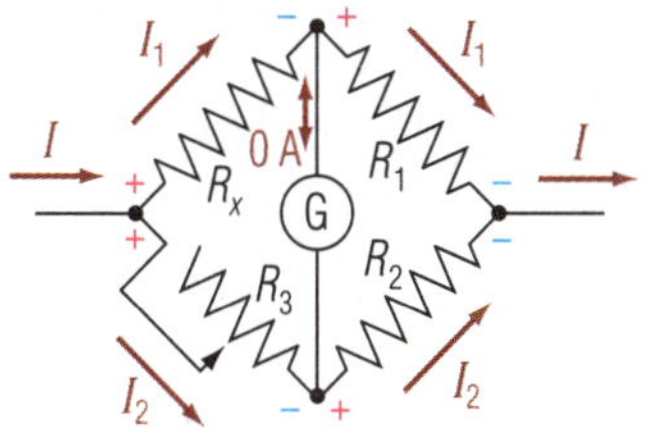

20-27a A resistance bridge circuit segment

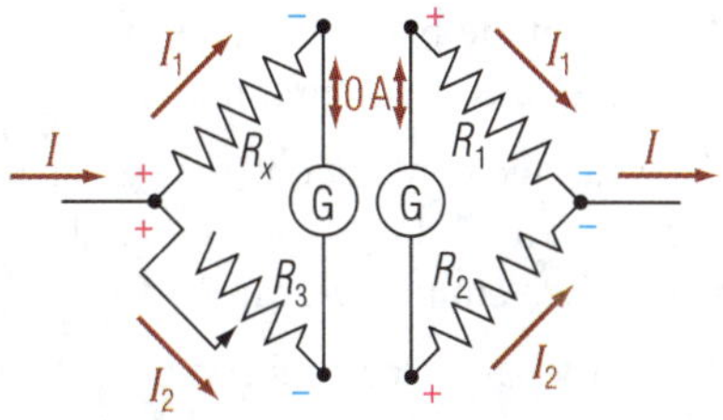

20-27b A resistance bridge broken down into two equivalent closed paths for analysis using Kirchhoff's current rule

20.15 Resistance Bridges

Figure 20-27a demonstrates the analysis of a somewhat complicated resistance circuit. Such a circuit is called a **bridge circuit** and is commonly used to accurately measure temperatures using a *resistance temperature detector (RTD)*. This circuit segment consists of two standard resistors, one precise variable resistor, and a galvanometer. The unknown resistor, labeled R_x (e.g., the RTD), is connected as Figure 20-27a shows. The *variable resistor* is adjusted until the galvanometer detects no current. Now the circuit is analyzed using Kirchhoff's rules. In Figure 20-27b, the potential difference rule for the left-hand path gives

$$-I_1R_x + I_2R_3 = 0 \text{ V}$$

$$I_1R_x = I_2R_3. \qquad (20.9)$$

For the right-hand path in Figure 20-27b, the potential difference rule gives

$$-I_1R_1 + I_2R_2 = 0 \text{ V}$$

$$I_1R_1 = I_2R_2. \qquad (20.10)$$

Now divide Equation 20.9 by Equation 20.10:

$$\frac{I_1R_x}{I_1R_1} = \frac{I_2R_3}{I_2R_2}$$

$$\frac{R_x}{R_1} = \frac{R_3}{R_2}$$

$$R_x = \frac{R_1R_3}{R_2} \qquad (20.11)$$

Equation 20.11 can give the unknown resistance precisely, and it requires only simple calculations.

A **variable resistor**, is capable of providing any resistance between two preset values by changing the length of the current path through the resistor. Typically, a variable resistor is constructed of a very long length of wire coiled around a support. A metal "wiper" is connected to one end of the circuit, and one end of the resistor's wire coil is connected to the other end of the circuit. As the current flows through the wiper resting on the wire coil, the current path length varies depending how far the wiper is from the connected end of the coil. Such a variable resistor is called a **rheostat**.

TABLE 20-3

RTD Resistance Data

Temperature (°C)	RTD (Ω)[†]
0	1000.0
10	1039.0
20	1077.9
30	1116.7
40	1155.4
50	1194.0
60	1232.4
70	1270.8
80	1309.0
90	1347.1
100	1385.1

[†]URL: http://www.weedinstrument.com/pdf/rvt.PDF

EXAMPLE 20-3

Circuit Analysis: The RTD Bridge

Refer to Figure 20-27a on the previous page. In order to accurately measure the temperature in a cooled-down nuclear reactor plant pipe, an RTD is monitored using the following bridge resistance readings: $R_1 = 1500.\ \Omega$ and $R_2 = 2500.\ \Omega$. The variable resistor, R_3, is adjusted to a value of 1796.5 Ω in order to zero the galvanometer reading. (a) What is the resistance of the RTD (R_{RTD}) for the temperature in the reactor pipe? (b) What is the temperature in the pipe?

a. Substituting the values of the known resistances into Equation 20.11, you have

$$R_{RTD} = \frac{R_1 R_3}{R_2}$$

$$R_{RTD} = \frac{(1500.\ \cancel{\Omega})(1796.5\ \Omega)}{2500.\ \cancel{\Omega}}$$

$$R_{RTD} \cong 1078\ \Omega.$$

b. According to Table 20-3, the piping temperature is 20 °C.

20B Objectives

After completing this section, I can

- ✓ differentiate between series and parallel electrical circuits.
- ✓ discuss the conventions for assigning potential drops and potential rises in a simple DC circuit.
- ✓ identify basic electrical components in an electrical circuit schematic.
- ✓ determine the equivalent resistance for various arrangements of resistors in a simple DC circuit.
- ✓ analyze simple DC circuits using Kirchhoff's rules.
- ✓ describe the purpose of basic electrical instruments and how to make connections when using them.
- ✓ find an unknown resistance using a resistance bridge circuit.

20B Section Review

1. Is the switch for your kitchen ceiling light arranged in series or parallel with the light fixture? Explain your answer.
2. Which point is at the higher potential in the figure? Explain.

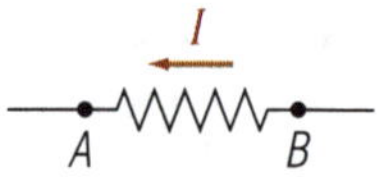

3. State Kirchhoff's voltage rule and his current rule in your own words.
4. List the three basic electrical instruments, how they are connected when in use, and the condition of the circuit/component when a measurement is being taken (i.e., energized/de-energized, connected/disconnected to a circuit, etc.)
5. What kind of circuit is used to find the value of an unknown resistance in the circuit by balancing the currents flowing through its two branches?
6. What is the equivalent resistance of the resistors 1 MΩ, 100. Ω, and 1.00 Ω
   a. in series? (Assume all resistors have a 5% tolerance for rounding purposes. See Appendix A in the Lab Manual for a discussion of manufacturer's tolerances.)
   b. in parallel?
7. For each case in Question 6, is the order of magnitude of the equivalent resistance closer to the largest or the smallest resistance?
8. Using Kirchhoff's rules, determine the current through and the voltage drop across each resistor in the accompanying figure.

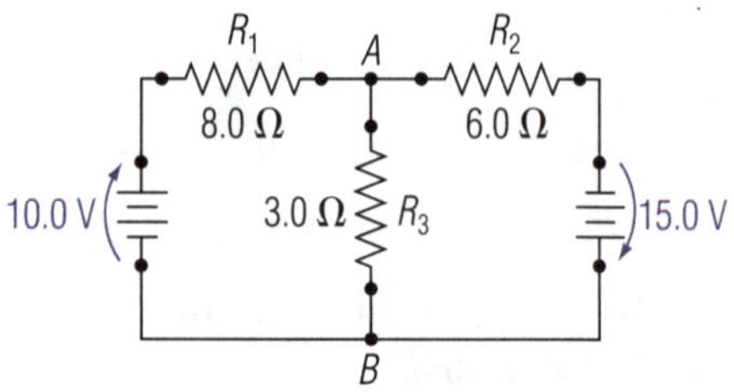

20C SEMICONDUCTORS AND TRANSISTORS

20.16 Vacuum Tubes

The fastest growing area of technology today is semiconductor electronics. Scientists and engineers are striving to make devices that are smaller, faster, more energy efficient, and more powerful. Perhaps the most surprising fact about electronics is its youthfulness. Only a little more than a century ago this discipline was unknown. Now, we would feel helpless without our calculators, computers, cell phones, GPS navigation, wireless data devices, and digital video systems.

The electronic age began with the invention of the light bulb. In 1883, in an attempt to reduce blackening on the inside of the bulb, Thomas Edison introduced a loop of wire into the bulb. He discovered that if he made the loop positive with respect to the filament, a current began to flow. If he made the loop negative with respect to the filament, no current flowed. This behavior is called the **Edison effect**. The current was surprising, considering that the bulb contained a vacuum. The charge apparently traveled through the space between the filament and the loop *without requiring a physical conductor*. A few years later, in 1897, J. J. Thomson discovered the electron, explaining the Edison effect. You will learn more about J. J. Thomson's experiment in Chapter 21.

Thomas Edison (1847–1931) was a prolific American inventor who is often credited with establishing the first industrial-scale research laboratory.

20-28 Edison's light bulb

The electrical resistance of the filament causes its temperature to rise until it emits electrons, similar to evaporation in a boiling liquid. This escape is called *thermionic emission*. The free electrons were drawn to the positively charged loop along electric field lines, producing a current. A negatively charged loop repelled the free electrons, so no electrons could reach the loop, and no current flowed.

In 1904 John Fleming placed a filament and a metal plate, connected to exterior wires, in an evacuated, sealed glass tube similar to Edison's. This device, and any other arrangement of conductors in an evacuated envelope, is called a **vacuum tube**. Because his vacuum tube, containing a filament and a plate, allowed electricity to pass in only one direction, Fleming called it a *valve*, since its action was like a check valve in a water pipe. Such devices came to be called **diodes** because they contained two electrodes.

John Ambrose Fleming (1849–1945) was an English physicist and electrical engineer who was a devout Christian and creationary scientist. He was one of the founders of the Evolution Protest Movement that eventually became the modern Creation Science Movement. He published several books about Creationism. Having no children, he left most of his estate to Christian charities.

In 1906 the American inventor Lee De Forest developed the **triode**, a vacuum tube that could be used to detect and amplify weak electrical currents, particularly signals from sound and radio waves. The triode is a vacuum tube containing three elements: a filament, a plate, and a control grid between the filament and the plate. The control grid is usually given a negative charge so that few electrons from the filament will be able to reach the plate. Electrons will be repelled by even a slightly negative control grid. If the control grid is attached to a varying current, such as an audio signal, it will vary in charge. Even a slight change in control grid charge can allow the free electrons to reach the plate or block their path, depending on the direction of the change. Thus, a small change in the grid current produces a large change in the plate current. Controlling a large current with a small one is called *amplification*. De Forest's triode vacuum tube is recognized as the first step in the development of modern electronics.

AT&T bought the patent to De Forest's triode and improved it for use in transcontinental telephone service. It was the basis of a successful expansion of telephone service nationwide. But vacuum tubes proved to be expensive to operate, generated lots of heat, and tended to be short-lived. AT&T researchers began to look for other technologies to replace them. This search eventually led to investigations involving semiconductor devices.

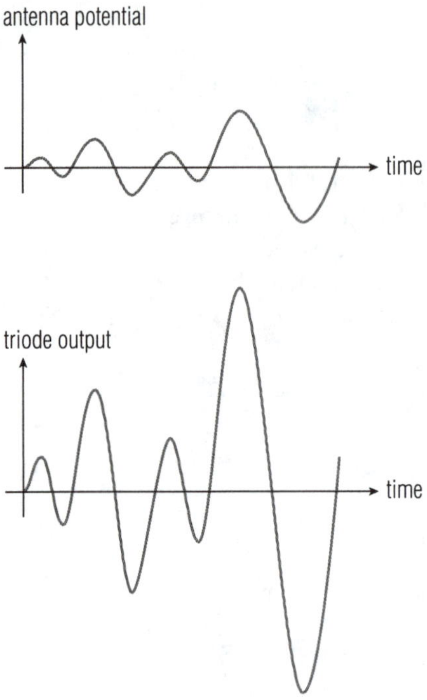

20-30 Signal amplification in a triode tube

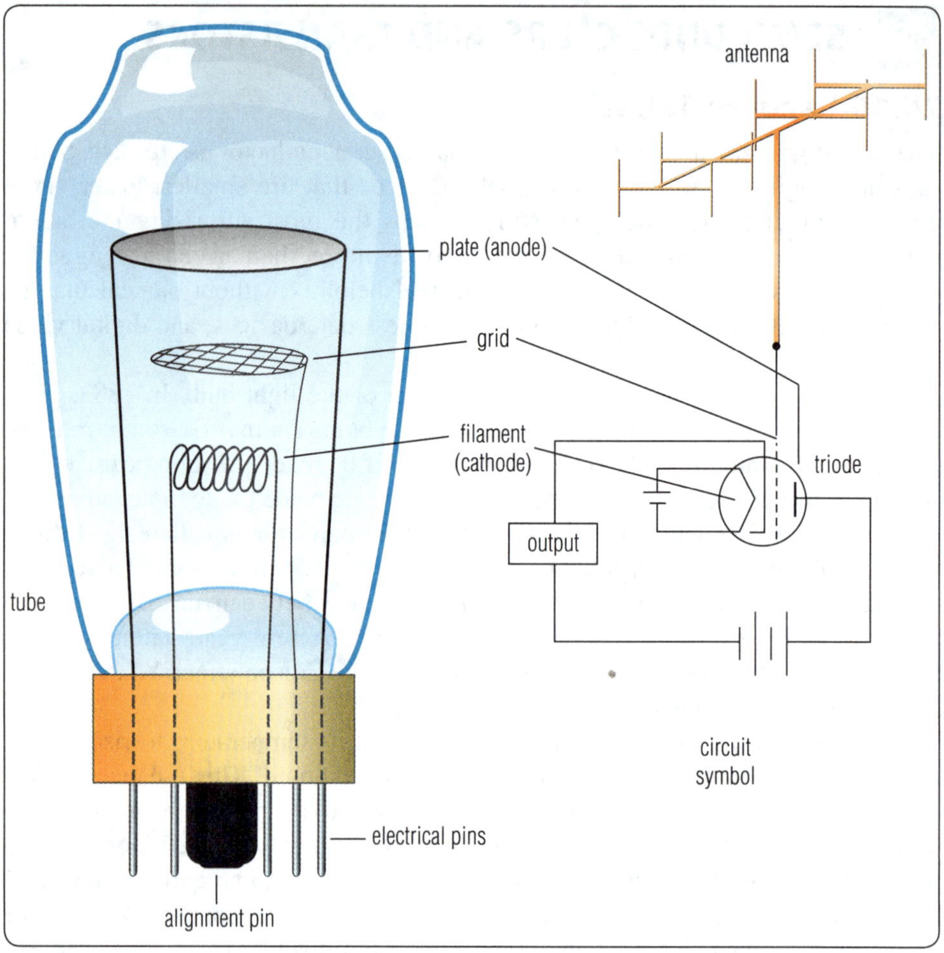

20-29 An electronic triode

20.17 Semiconductor Theory

Electronic devices that have largely replaced vacuum tubes are made of semiconductor materials. The differences in electrical behavior of conductors, insulators, and semiconductors come from the kinds and strengths of the bonds that hold their atoms together.

Conductors usually are metals. Their atoms are bonded together by metallic bonds, in which the valence electrons are loosely held and can be shared by all the nuclei in the conductor. These electrons can move freely throughout the conductor's crystal structure. Therefore, even very small potential differences applied across the conductor can cause the electrons to move.

Insulator compounds, on the other hand, usually contain strong covalent bonds between their atoms. The electrons are shared between pairs of atoms and do not move at all through the material. A relatively large amount of energy is required to remove an electron from its bond. Therefore, even a large potential difference applied across an insulator produces little current.

Semiconductor compounds are formed from the metalloid elements that lie on the stair-step region of the periodic table (see Appendix H). These elements have lower *electronegativities* than the nonmetal elements that form insulator compounds, but higher electronegativities than metals. Electrical semiconductor compounds are composed mostly of one element, such as silicon. Silicon, like carbon, has four valence electrons and makes a rigid crystal lattice similar to diamond by forming four covalent bonds with four neighboring atoms. Consequently, silicon by itself is an excellent insulator, since all of its valence electrons are tied up in covalent bonds.

The **electronegativity** of an element is a measure of how tightly an atom holds onto valence electrons when bonded to another element. Relatively high electronegativities are characteristic of nonmetal elements, and low ones are characteristic of metals.

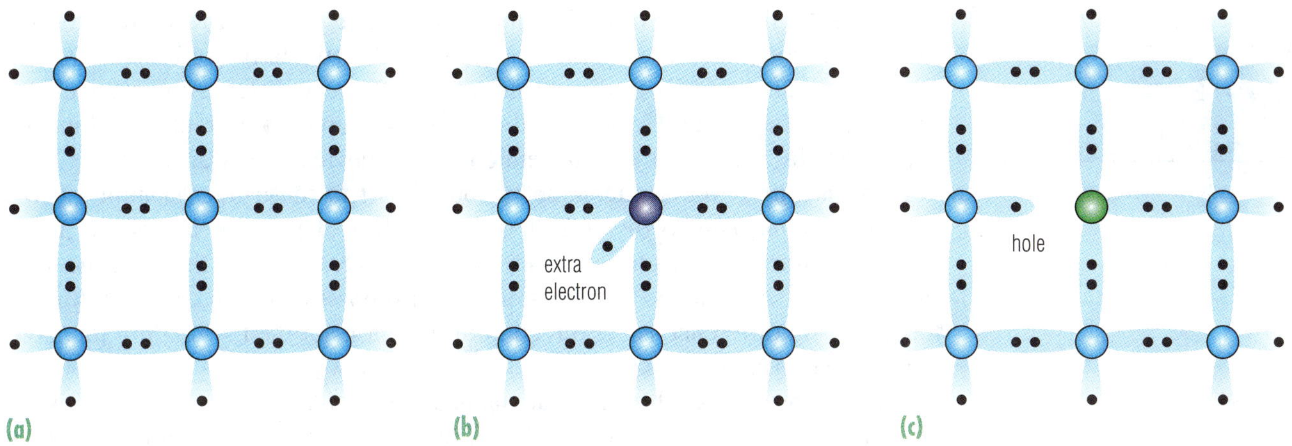

20-31 A pure silicon crystal lattice (a), an *n*-doped crystal lattice (b), and a *p*-doped crystal lattice (c)

Elements that provide excess electrons in a semiconductor lattice are called *n-dopants*, and the material is *n-doped*. If the elements create holes in the semiconductor lattice, they are called *p-dopants*, and the material is *p-doped*.

The silicon crystal lattice's conductivity can be increased in two ways. During manufacture, a small percentage of the silicon atoms can be replaced in the lattice with other atoms that have a different number of valence electrons. This process, called **doping**, creates points in the silicon lattice that have extra or missing electrons ("holes"). For example, if an element such as phosphorus, arsenic, or antimony, which all have five valence electrons, is used as the *dopant*, this produces points in the lattice where, after four covalent bonds form with adjacent silicon atoms, there is an unbonded electron left over, leaving a negatively charged point in the lattice. This kind of semiconductor compound is called an ***n*-type semiconductor**. If an element with only three valence electrons, such as boron, aluminum, or gallium, is used as the dopant, there are holes in the lattice with missing bonds, leaving atoms with a positive charge. This kind of material is called a ***p*-type semiconductor**.

When a potential difference is applied to an *n*-doped material, the unbonded electrons are pulled away from their atoms and drift down the electrical potential. With *p*-doped material, unbonded electrons jump from one hole to the next, leaving holes behind that appear to move in the opposite direction. By themselves, *n*-type and *p*-type semiconductors do conduct, although not very well. They make better resistors than conductors. However, unlike resistors, as a semiconductor's temperature increases, its resistance *decreases*.

20.18 Semiconductor Devices

By themselves, *n*- and *p*-type semiconductor materials are not particularly useful, but when they are joined together, something interesting happens. The contact surface between the two materials is called a ***p-n* junction**. Some of the excess electrons from the *n*-doped material drift across the junction and fill the holes in the *p*-doped material. This leaves a neutral region between the two materials called the *depletion zone*. A device consisting of a single *p-n* junction between two doped semiconductor materials is called a diode, and it performs the same function as the vacuum tube diode. Let's see how it works.

If a potential difference is applied to the diode so that the *n*-doped material is at the higher potential, the excess electrons in the *n*-doped material are attracted to the positive charge, and the holes in the *p*-doped material are attracted to the negative charge. The depletion region *increases* in size, and no current flows through the diode. The diode is said to be **reverse-biased**. However, if the polarity is reversed so that the *n*-material is at the lower potential and the *p*-material is at the higher potential, the electrons and holes are repelled toward the depletion zone until it disappears and charge carriers can move into the opposite materials. In this condition the diode conducts current and it is said to be **forward-biased**.

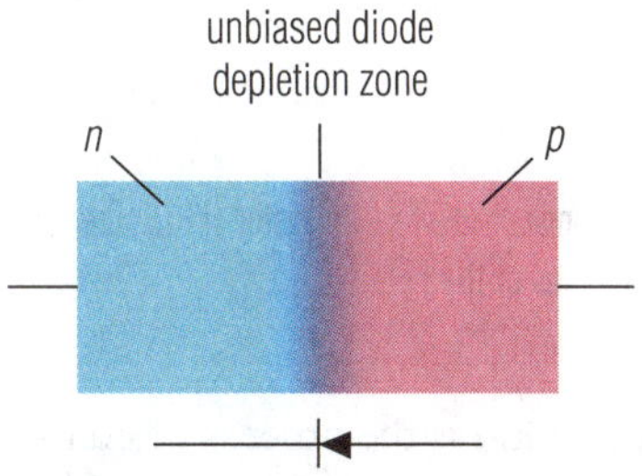

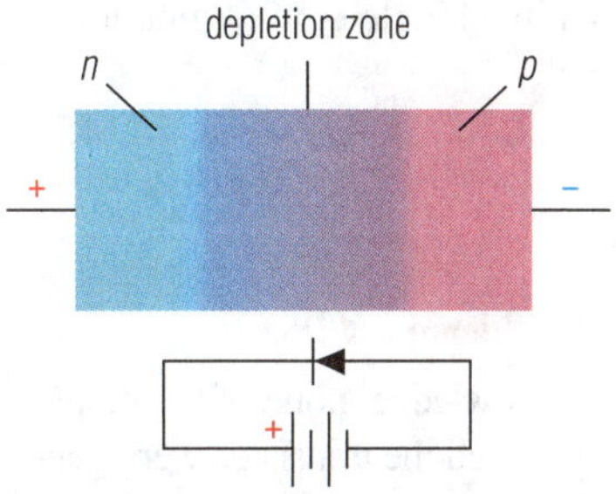

20-32 A reverse-biased diode

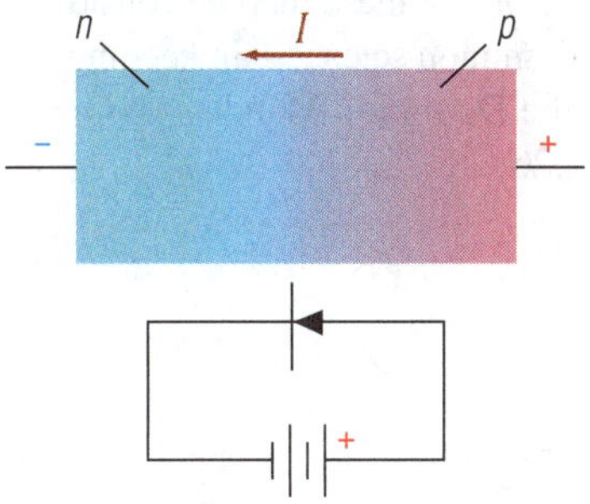

20-33 A forward-biased diode

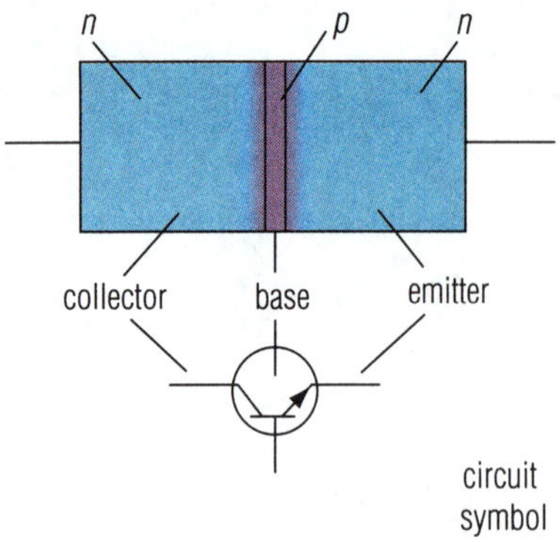

20-34 A bipolar junction transistor with an NPN structure

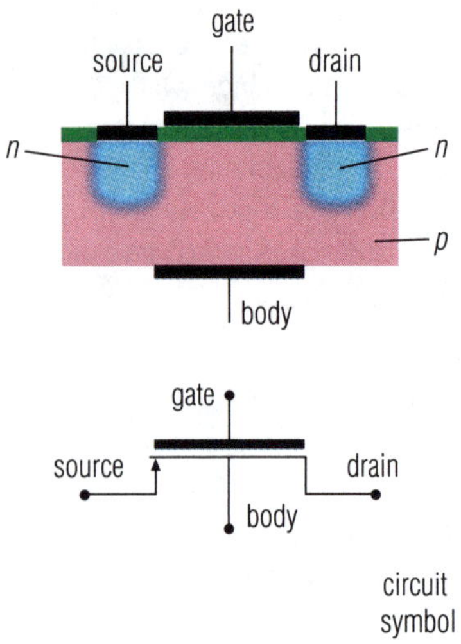

20-35 A basic field-effect transistor (FET). There are many variations of the FET.

Multiplexing is the process of combining and then separating multiple signals in a single communications channel.

Dominion Science and Non-Christians

To our knowledge, none of the men who designed the transistor were genuine Christians. Shockley himself was an avowed atheist. How can the creative efforts of these men be considered dominion science—in keeping with the Dominion Mandate of Genesis 1:26–28?

Diodes act like check valves in plumbing systems. They allow current to flow in one direction but block it in the other. They are particularly useful in converting alternating current (AC) to direct current (DC). You will learn more about this function in Chapter 22. However, a diode cannot amplify a current. This function is provided by a different kind of semiconductor device—the transistor.

A **transistor** is the semiconductor counterpart to the triode vacuum tube. A *bipolar junction transistor (BJT)* has three layers. The outer layers are doped one way (either *n*- or *p*-doped), and the inner layer is doped the opposite way. Figure 20-34 shows an NPN transistor. It is also possible to create a PNP transistor. The main difference is the direction of the biases of the internal *p*-*n* junctions within the transistor.

The regions of a BJT transistor are named as shown in Figure 20-34. If only the emitter and collector are connected to a potential difference in either direction, there will be no current flowing through the transistor. No matter which end of the transistor is connected to the higher potential, the current will face a reverse-biased *p*-*n* junction at the base. However, if a small positive potential is applied to the narrow base region, holes are forced into the *n*-regions, which then begin to conduct. If the emitter contains more dopant atoms than the collector, the resulting current through them can be much larger than the small current through the base. The small base current controls the large collector-emitter current, allowing the transistor to act as an amplifier. This is particularly useful if a varying current signal from a microphone or a radio antenna is applied to the base. The weak signals can be greatly amplified.

Another kind of transistor, called the *field-effect transistor (FET)*, utilizes *p*-*n* junctions differently to accomplish switching functions and amplification. The structure of a typical FET is shown in Figure 20-35. Current through the gate controls the size of the depletion region in the current channel between the source and the drain connections. Depending on the arrangement of the *n*- and *p*-doped materials used in the transistor, changes in the gate current cause the depletion region to expand, blocking current flow, or to shrink, allowing current to pass. FETs tend to be used as switches for external circuits. Some forms can control currents moving in either direction, which is useful for *multiplexing* circuits. The FET has largely replaced the BJT in modern electronics because of its better power efficiency.

20.19 Using Semiconductors to Solve Problems

The theoretical design for a field-effect transistor-like device was first developed by the Ukrainian emigrant physicist Julius Lilienfeld in 1925. Another field-effect device was described in a patent by the German electrical engineer Oskar Heil in 1935. However, no actual working models of these devices were built by these men. In 1945, a group of AT&T scientists at Bell Labs were assigned the task of developing a tubeless electronic amplifier—a solid-state device. In several stages of research and experimentation, plagued by interpersonal conflict, Bill Shockley, Walter Brattain, and John Bardeen developed several experimental models of a semiconductor amplifier. The most efficient design, developed by Shockley, eventually became the basis for the bipolar junction transistor. The name

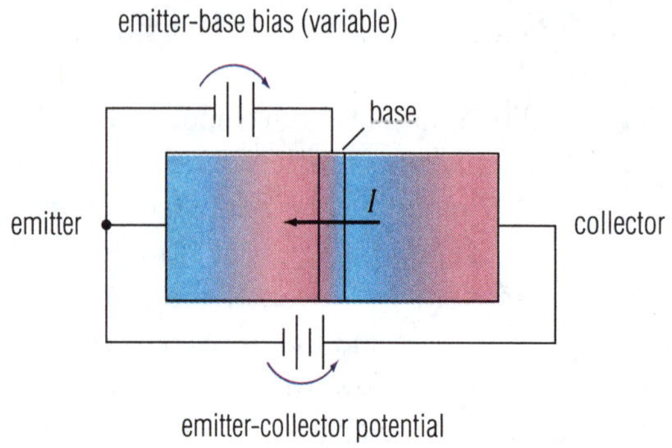

20-36 An NPN transistor in a circuit

"transistor" was adopted by Bell Labs as a combination of words describing the variable resistance properties of the device (*trans*resistance). The *–istor* ending, used in many other contemporary device names, completed the word. Despite great personal resentment between Shockley and the other two scientists, the three briefly reunited in 1956 to receive the Nobel Prize in Physics. Their invention initiated the revolution in communications and information technology.

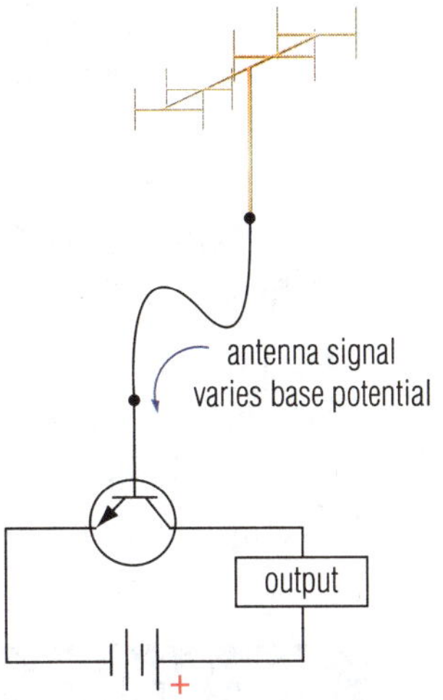

20-37 A transistor used as an amplifier

Bell Labs almost immediately began licensing the new technology for manufacturing. The objective was to miniaturize the bulky, power-hungry, and fragile telephone and radio tube amplifiers. Because transistors could act like switches, the huge, clacking mechanical telephone switching centers that filled many rooms were rapidly replaced by solid-state transistors fitting into the space of a cabinet. Telephone engineers had finally achieved their dream of replacing vacuum tubes with cooler, rugged amplifiers. Transcontinental and transoceanic telephone services became much more affordable, and the number of calls that could be made at one time jumped exponentially.

In 1952 a bright young engineer named Jack Kilby attended one of Bell Lab's transistor symposiums. When he went to work for Texas Instruments a few years later, he was working on a problem that all electronics manufacturers were trying to solve—how to pack many transistors, wires, and other components like capacitors and resistors into the smallest possible space. Miniaturization, it seemed, had come up against an insurmountable obstacle. Kilby, working alone while the rest of the company was on an extended holiday, came up with the idea that all of the major circuit components could be made from semiconductors. Capacitors, resistors, switches, and amplifiers could all be constructed from appropriate combinations of *p-n* junctions and field-effect structures. His leap of inspiration was the realization that all of these devices could be built on a single base of semiconductor material and interconnected by properly doped semiconductor compounds. The company built a successful prototype of what came to be called an **integrated circuit (IC)** on a "chip" of germanium. In modern ICs, it is possible to concentrate tens of billions of circuit elements on a single chip. The IC is the basis for all microprocessors in use today. One of the benefits of this technology is instantaneous global telephone communications, of which the AT&T engineers back in 1907 could not have even dreamed.

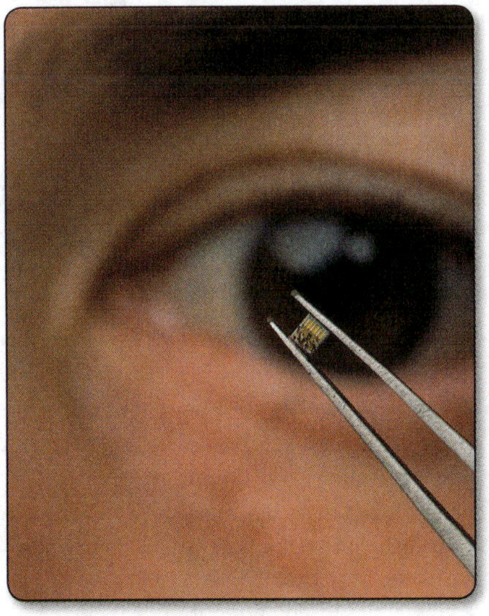

20-38 A modern integrated circuit

Jack St. Clair Kilby (1923–2005) was an American engineer who flunked the entrance exam into MIT but went on to invent the integrated circuit, for which he received the Nobel Prize in Physics in 2000. He also received patents for inventing the handheld calculator and thermal printer.

Unknown to Kilby, Robert Noyce (1927–90) of Fairchild Semiconductor was also working on an integrated circuit. Although Kilby created the first working IC, Noyce's design solved many practical problems, making it possible to mass-produce the device. Today, Kilby and Noyce are considered co-inventors of the IC.

20C Objectives

After completing this section, I can

- ✓ explain how simple vacuum tubes work.
- ✓ compare and contrast conductors, semiconductors, and insulators.
- ✓ describe how semiconductor doping establishes the properties of the material.
- ✓ explain how semiconductor junctions can be forward- or reverse-biased.
- ✓ describe the structure of simple semiconductor devices and identify their symbols in a circuit diagram.
- ✓ explain how a transistor amplifies a signal.
- ✓ summarize the historical invention of the transistor and integrated circuit.
- ✓ discuss the advantages that transistors provided to the telecommunications industry.

20C Section Review

1. What material carries the current in a vacuum tube?
2. Describe thermionic emission.
3. How does a vacuum tube diode work?
4. How does a triode amplify a current signal?

5. What process produces the different kinds of semiconductor materials used in electronic devices? What kinds of semiconductor materials are there?
6. What are the two kinds of current carriers in semiconductor materials?
7. Identify the parts of a bipolar junction transistor (BJT) and describe the function of each.
8. For what use is a field-effect transistor (FET) ideally suited?
9. What is an integrated circuit?

DS 10. What immediate advantages did AT&T gain by using transistors in their long-distance telephone service?

Chapter Review

In Terms of Physics

current (I)	447	parallel	454
electrodynamics	447	voltage drop	454
ampere (A)	448	node	458
direct current (DC)	448	Kirchhoff's rules	459
polarity markings	448	ammeter	460
voltaic cell	449	galvanometer	460
galvanic cell	449	ohmmeter	460
electrode	449	voltmeter	460
circuit	450	bridge circuit	461
primary cell	450	rheostat	461
storage cell	450	Edison effect	463
battery	450	vacuum tube	463
dry cell	451	diode	463
resistivity (ρ)	451	triode	463
superconductor	451	doping	465
resistor	451	*n*-type semiconductor	465
resistance (R)	451	*p*-type semiconductor	465
Ohm's law	452	*p-n* junction	465
ohm (Ω)	452	reverse-biased	465
Joule's law	453	forward-biased	465
kilowatt-hour (kWh)	453	transistor	466
series	454	integrated circuit (IC)	467

Problem-Solving Strategies

20.1 (page 450) It is not normally necessary to keep track of which direction the moving charges are going. Just remember that current flows from the highest potential to the lowest potential in the external circuit.

20.2 (page 452) Ohm's law is true for a single circuit component, a segment of a circuit, or an entire circuit.

20.3 (page 453) Remember that the unit symbol indicates the kind of dimension of a property, while the variable (formula) symbol represents its numerical value.

20.4 (page 454) The symbol for a potential difference source has parallel lines of different lengths. The shorter line is conventionally the negative end of the source. Think of it as a minus sign.

20.5 (page 457) After summing the reciprocals of the parallel resistances, remember to take the reciprocal of the sum to find the total resistance.

20.6 (page 459) When applying Kirchhoff's voltage rule, remember that voltage *rises* across voltage sources (ΔV is positive) and *drops* across other circuit components (ΔV is negative) for conventional current flow. The opposite is true for negative (electron) current flow.

20.7 (page 460) Kirchhoff's rules are used to generate as many current equations as there are unknown currents in the simple closed paths in the circuit. Solving these equations simultaneously using the techniques that you learned in algebra classes permits you to find the individual currents. After the current values are known, you can find the voltage drops and the power absorbed for each resistance.

Review Questions

1. What is an electrical current?
2. What carries current in metals?
3. What is the difference between conventional current and the physical movement of charges in a conductor such as a metal?
4. In what kind of current do the charges move in only one direction?
5. When installing a battery in your calculator, how do you know which end has the electrode with the higher potential?
6. List the kinds of devices that convert the following forms of energy to electrical potential:
   a. light
   b. chemical
   c. thermal
   d. mechanical
7. What carries the current in electrolytic solutions?
8. List three sources of potential difference.
9. What is the difference between a primary cell and a storage cell?
10. Discuss the difference between resistance and resistivity.
11. a. What class of materials lacks electrical resistance?
    b. What environmental condition limits the usefulness of this property?
12. Why do you think that scientists chose to define the ampere in the SI on the basis of a force between two conductors rather than using the classical definition that involved the rate of the flow of charge? (Review the Chapter 2 discussion of the SI.)
13. This series circuit has a current I_A at point A. At point B, three circuit components farther along, is I_B greater, less than, or equal to I_A?

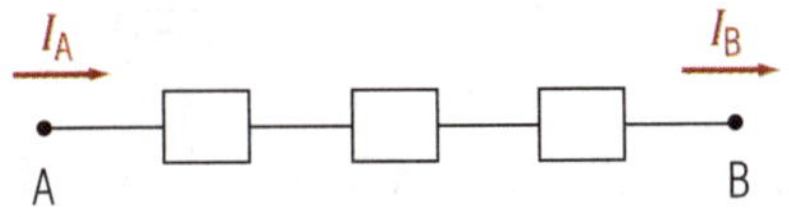

14. A circuit has a current I_A at point A just before the circuit branches. There is one circuit element in each branch. Is the current in each branch greater than, less than, or equal to I_A?
15. Using the plumbing analogy of electricity, which electrical properties or components correspond to the following plumbing concepts?
    a. piping
    b. piping constriction
    c. pump
    d. pressure drop
    e. volumetric flow rate
    f. in-line flow meter
    g. pressure gauge
    h. shutoff valve
16. In Kirchhoff's voltage rule for circuit analysis, what restriction applies when summing the voltage drops and rises along a path?
17. Name the instrument that you would use to measure a circuit's current and describe how you would connect it.
18. Name the instrument that you would use to measure a resistor's resistance and describe how you would connect it.
19. What does a resistance bridge circuit measure?
20. What happens to an electrical conductor when it is heated to incandescence? For which applications did early electronics engineers use this effect?
21. Which electrical property of metalloid elements makes them particularly suitable for use in semiconductor applications?
22. If a circuit component were made entirely of an *n*-doped or a *p*-doped semiconductor material, for what function would that component be most useful?
23. What does it mean to reverse-bias a *p-n* junction?
24. Describe the major difference in operation between a bipolar junction transistor (BJT) and a field-effect transistor (FET).

DS 25. It is commonly reported that AT&T's Bill Shockley believed that the contributions of the other two scientists working with him were of far less importance to the invention of the transistor than his, and so he felt that their names should not appear on the patent. Understandably, they strongly disagreed with him, and great animosity resulted. How does this not-uncommon problem in scientific work relate to the Creation-Fall-Redemption model of dominion science?

True or False (26–35)

26. Charged particles carrying a current in a wire move at speeds near to that of light.
27. The work done on an electron can be calculated from the equation $W = (emf)(d)$, where *emf* is the electromotive force and d is the distance that the electron moved.
28. At room temperature every material has some resistance.
29. If you applied a certain voltage across the 1 cm thickness of a 1 m^2 copper plate, you would discover that it had the same resistance as a 1 cm long copper wire.
30. Current in parallel branches must be equal in order for Kirchhoff's rules to be valid at the two nodes connected by the branches.
31. Current is constant throughout a series circuit.
32. Conventional current flowing through a resistor travels from the positive end to the negative end.
33. Voltage measured from the positive end of a resistor to the negative end is negative.
34. The *p-n* junction in the following diagram is forward-biased.

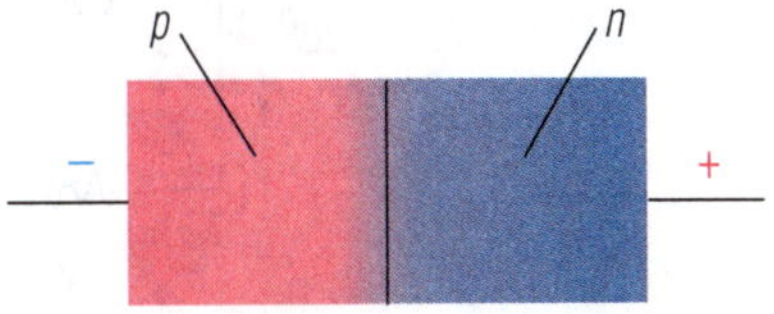

DS 35. The integrated circuit was invented by Bell Labs scientists in order to continue the trend of miniaturization of electronic circuits in their telephone system.

36. What is the resistance of a 1.00 m long copper wire with a cross-sectional area of 1.00×10^{-6} m^2?
37. A certain kind of glass has a resistivity of 1.54×10^{11} Ω·m. What is the resistance of a 1.00 m long strand with a cross-sectional area of 1.00×10^{-6} m^2?
38. A sample of silver has a volume of 2.70×10^{-5} m^3. Find the resistance along the longest dimension if
    a. the sample has the shape of a cube.
    b. the sample is a bar with a cross-sectional area of 3.00×10^{-4} m^2.
    c. the sample is a wire with a cross-sectional area of 1.00×10^{-6} m^2.
    (Remember: $V = Bh$ or $V = Al$.)

39. A current of 1.50×10^{-2} A flows through a 30.0 Ω resistor. What is the voltage drop across the resistor?

40. A 1.50 V cell is connected to a 1.50 Ω light bulb in a simple circuit. How much current flows through the bulb?

41. A 12.0 V battery produces a current of 1.00 mA when connected in a circuit to resistor R. What is R's resistance?

42. A 55.0 W bulb is connected in a circuit to a voltage source of $V_{max} = 110.$ V.
a. What current does it draw?
b. What is its resistance?

43. A 40.0 Ω resistor in a circuit draws a current of 0.100 A.
a. What is the voltage drop across the resistor?
b. How much electrical power does it absorb?

44. Find the resistance of two 1.50×10^3 Ω resistors
a. in series.
b. in parallel.

45. Find the resistance of the circuit segment shown below.

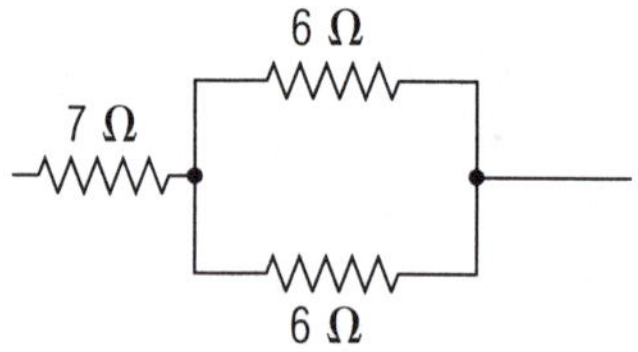

46. In the figure below, $R_1 = R_2 = 20.0$ Ω and $V = 10.0$ V. Find I.

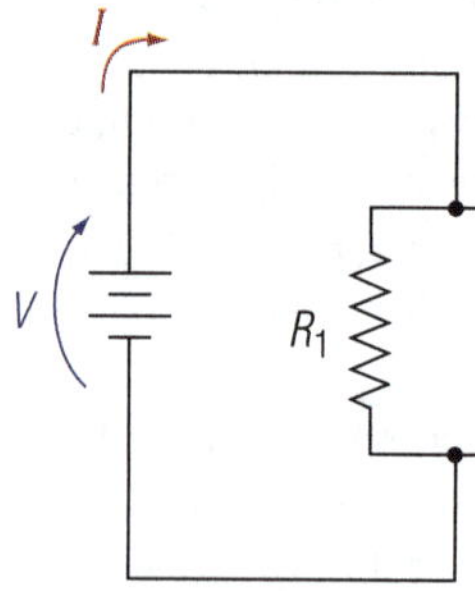

47. What is I in the figure below?

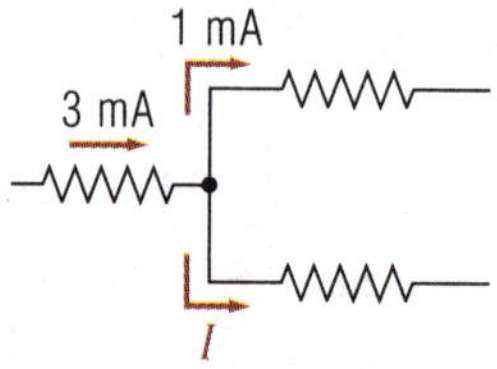

48. What is V in the figure below?

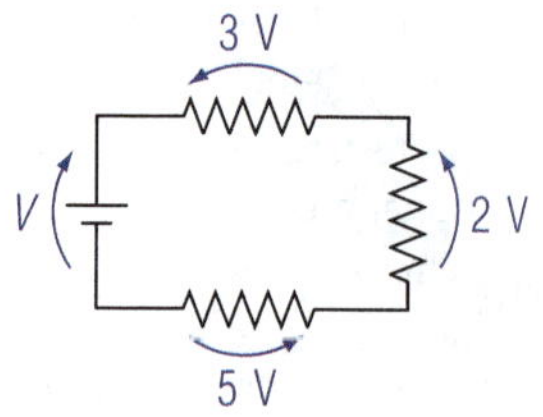

49. **a.** Find the equivalent resistance of the circuit below.

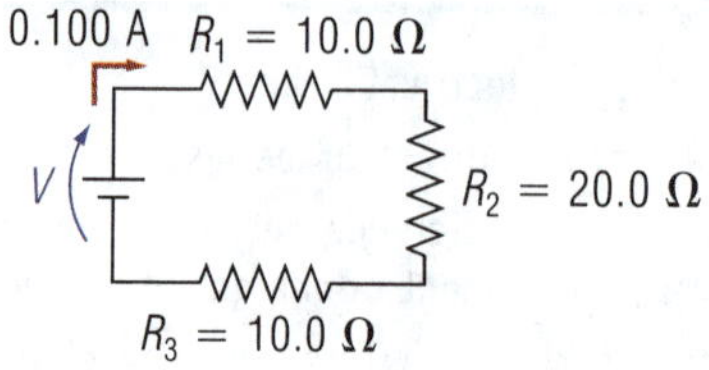

b. Find the voltage drop across R_1.
c. Find the voltage drop across R_2.
d. Find the voltage drop across R_3.
e. What is V?

50. The voltage drop across R_1 is 2.00 V. $R_1 = 20.0$ Ω; $R_2 = 30.0$ Ω; $V = 6.00$ V.

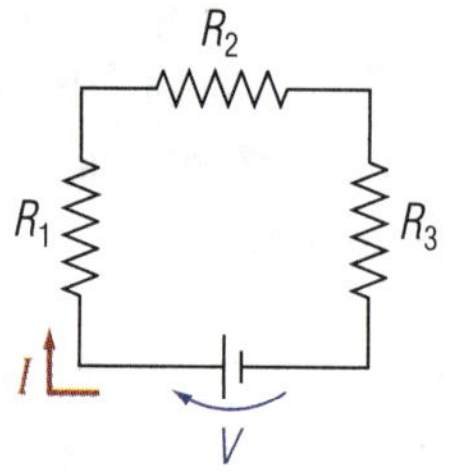

a. What is the current through R_1?
b. What is the current through R_3?
c. What is the voltage drop across R_2?
d. What is the total voltage drop in the circuit external to the voltage source?
e. What is the voltage drop across R_3?
f. What is R_3?

51. $V = 6.00$ V; $R_1 = 30.0$ Ω; $R_2 = 60.0$ Ω.

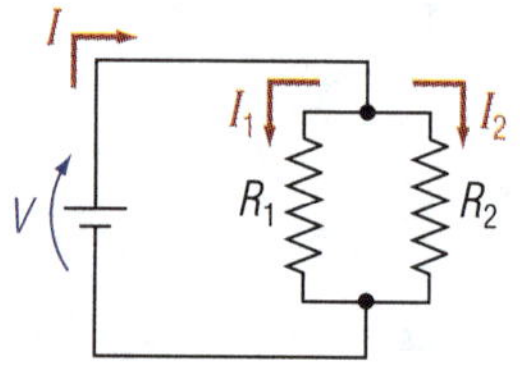

a. Find the equivalent resistance of the circuit.
b. Find the total current I through the circuit.
c. What is the voltage drop across R_1?
d. What is the voltage drop across R_2?
e. What is the current through R_1?
f. What is the current through R_2?

52. $V = 10.0$ V; $I_1 = 0.100$ A; $I_2 = 0.0500$ A.

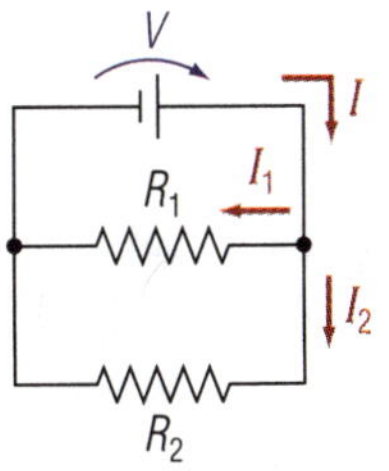

a. What is the voltage drop across R_1?
b. What is the voltage drop across R_2?
c. Calculate R_1.
d. Calculate R_2.
e. Calculate the circuit's equivalent resistance.
f. Calculate the current through the voltage source.
g. Compare R_1/R_2 with I_1/I_2 .

53. The resistor bridge in the figure below contains fixed resistors $R_1 = 10.0\ \Omega$, $R_2 = 100.0\ \Omega$, variable resistor R_3, and unknown resistance R_x. Find R_x when the galvanometer (G) reads 0 A, and $R_3 = 50.0\ \Omega$.

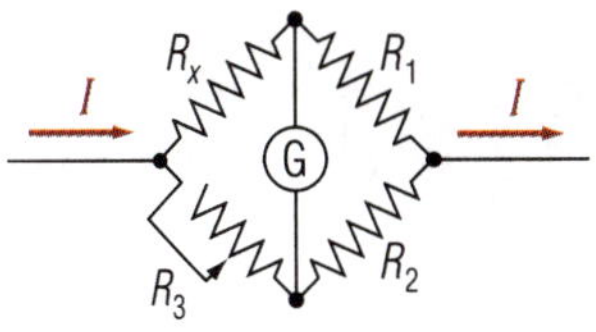

FACETS of PHYSICS

Fuel Cells

In the history of the United States, most electricity has been generated by mechanically rotating a coil of wire (called an armature) in a varying magnetic field. The mechanical power used to rotate the coil has come from a variety of sources including water wheels, windmills, steam engines, hydraulic turbines, steam turbines, gas turbines, diesel engines, gasoline engines, and Stirling engines. The fuels used in the engines (both internal and external combustion) have included charcoal, wood, coal, crude oil, natural gas, gasoline, kerosene, diesel fuel, and atomic fuel. But all of this will probably change soon because of a device called a *fuel cell*.

Fuel cells generate electricity without the use of mechanical power and rotating parts. There are five main types of fuel cells:

The *alkaline fuel cell* was the first type to be used extensively, largely because it was developed and ready for use before the others. It was used in the Apollo Command Module and the space shuttle, but it required pure hydrogen and oxygen and was too expensive for commercial use.

Three other types, using *phosphoric acid*, *solid oxide*, and *molten carbonate*, have high operating temperatures and may be suitable only for stationary power plants and marine applications.

The *proton exchange membrane fuel cell* shows the most promise for consumer applications. It consists of three parts: (1) a fuel chamber, part of which also functions as an anode; (2) a proton exchange membrane (PEM), which may have a catalytic film on the cathode side; (3) an air chamber (part of which also functions as the cathode), and an exhaust duct (see the schematic on the opposite page).

The fuel, hydrogen, enters the fuel chamber under pressure. As each atom of hydrogen reaches the PEM, its electron is stripped from the nucleus but the nucleus (a proton) passes through the membrane. The anode collects the electrons and delivers them as 0.7 volts of direct current. By connecting several cells in series, higher voltages are obtained. In the air chamber, the nuclei of the hydrogen atoms combine with oxygen to form pure water.

The PEM fuel cell operates at 70 °C and requires no cooling. It generates no pollutants, requires no lubricants, pumps, or filters, and has no moving parts to wear out. The only byproduct is pure water, suitable for drinking, cooking, washing, and irrigating.

The major problem with fuel cells is the fuel, hydrogen. Because the hydrogen atom is so small, it is difficult to make a tank that will hold hydrogen under pressure. The atoms leak out between the atoms or molecules of the tank material. Hydrogen is extremely flammable, so leaks create fire and explosion hazards. It is hazardous to transfer liquid or gaseous hydrogen from one vessel to another. And there are few filling stations (yet) where you can pull in and purchase ten gallons for your fuel tank. In spite of these problems, nearly all major

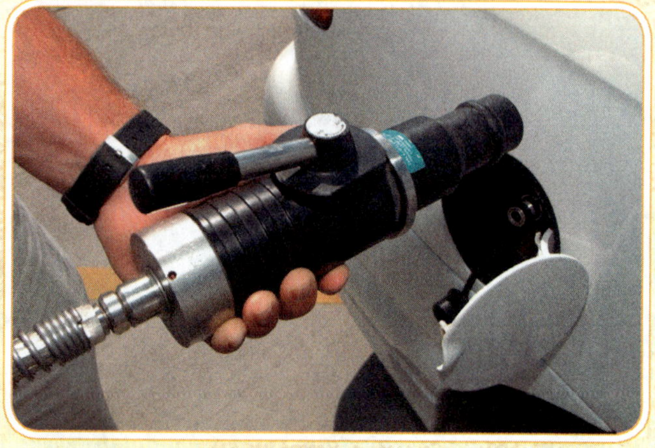

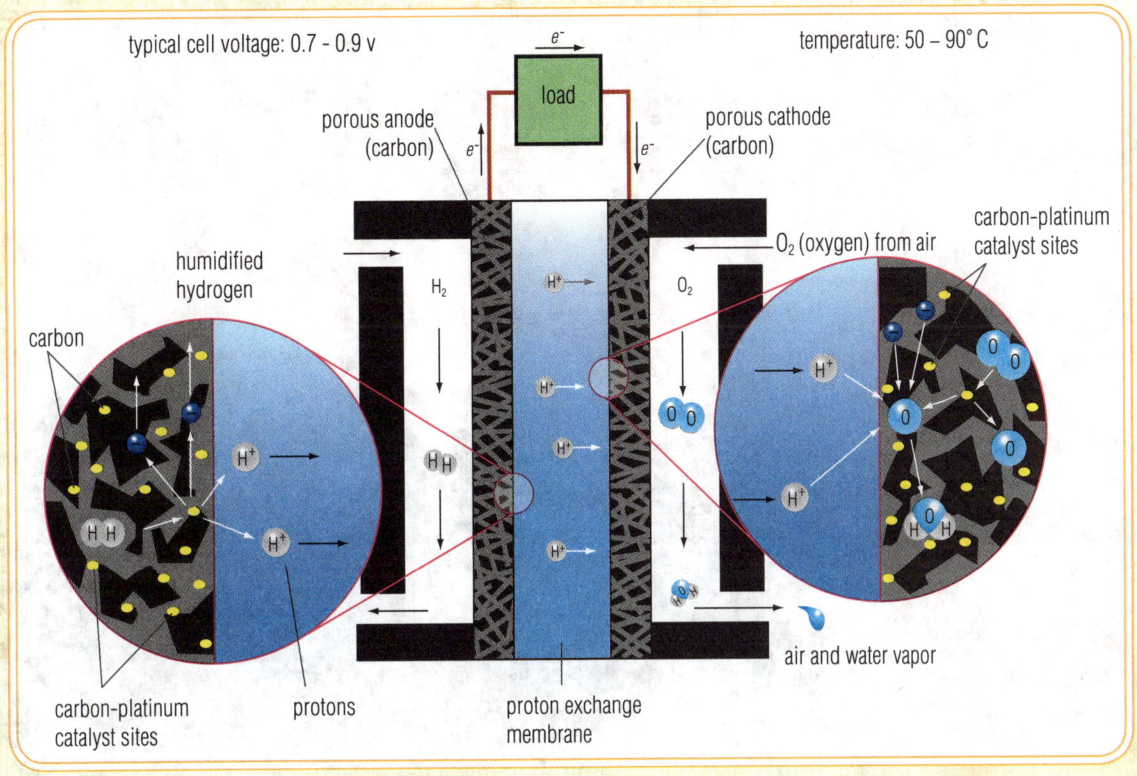

auto manufacturers around the world are building prototype vehicles that include PEM fuel cells, either in combination with a small gasoline engine generator (i.e., a *hybrid* energy vehicle) or entirely powered by electricity from fuel cells. In 2008, one company began field-testing a fleet of 200 fuel-cell cars in Southern California. These customers lease their cars and fuel up at several dozen hydrogen fuel stations that exist or are planned in the state. Public buses have also been successfully powered by PEM fuel cells in the United States, England, Japan, China, Brazil, and many countries in the European Union. These buses are nonpolluting and have done much to reduce smog and noxious emissions in cities. These vehicles are fueled at their central bus stations.

A device called a *reformer* offers one possible solution to the problems associated with the hydrogen fuel. The reformer can be tailored to work on any hydrocarbon fuel, such as gasoline, kerosene, propane, ethane, methane, butane, or natural gas. It strips the hydrogen from the fuel and supplies it on demand to the PEM fuel cell.

Another possible solution uses a compound with sodium borohydride. Drivers of PEM fuel-cell–powered vehicles will be able to pull in to a filling/reprocessing station, buy a container of this fuel, and drive away. When the hydrogen is used up, they bring the container back and exchange it for a recharged one.

Other possible solutions are being explored, but it is not yet clear which concept will be adopted. One thing is clear: the use of fuel cells has the potential to reduce or eliminate dependence on an increasingly expensive oil resource.

Earth's magnetic field is magnificent evidence of God's protection of life on the earth.

Magnetism 21

Dominion Modeling

Understanding the Electron

In the 1800s, many physicists thought the atom was indivisible. Many also thought that a fluid or "ether" was responsible for carrying and transferring charge. English physicist Joseph John ("J.J.") Thomson was convinced otherwise. He suspected that the charge carriers were particles. In the 1890s, Thomson did a series of three experiments in which he clearly identified the presence of charged subatomic particles, electrons. In a bold move, he proved both that the atom could be divided and that these particles were charge carriers. He revolutionized the existing atomic model with this discovery, receiving the Nobel Prize in Physics in 1906. But how do the size and charge of electrons compare to other particles?

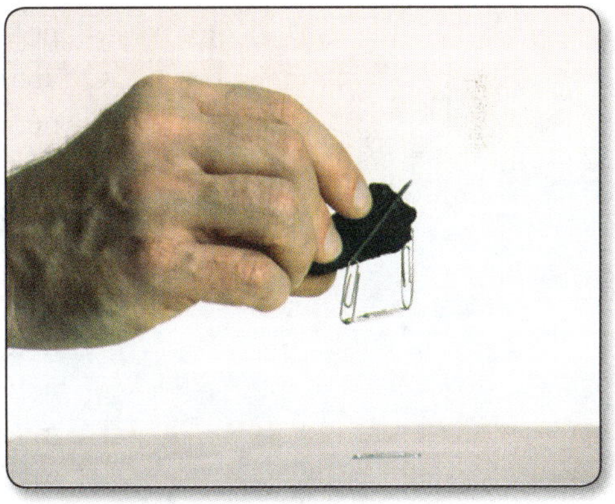

21-1 A sample of lodestone

21A DESCRIBING MAGNETISM

21.1 Historical Background

Before the time of Christ, people knew that certain stones would attract other similar stones and some metals. The Romans called such stones *magnets* after the city of Magnesia, where legend says that natural magnets were first found. During the fourth century BC, the Chinese discovered that a natural magnet suspended by a string would rotate until it aligned in a more-or-less north-south orientation. As early as the tenth century AD, Chinese sailors were navigating using stone magnets, which the Europeans called **lodestones** ("leading stones"). By the end of the twelfth century, the navigational compass, equipped with a magnetized metal needle, was in common use.

In the thirteenth century, Petrus Peregrinus de Maricourt performed several experiments to learn more about lodestones. He discovered that two regions at the opposite ends of a lodestone, which he called **poles**, had greater attraction for metals than the surface between them. He also discovered that there were two kinds of poles, north and south. Poles of the same kind repelled each other, while poles of different kinds attracted each other. This principle is called the **law of magnetic poles**. This property is very similar to the law of charges. There are two kinds of electrical charge, positive and negative. Like charges repel, whereas opposite charges attract. Early natural philosophers believed that static electricity and magnetism had much in common.

The ancient city of Magnesia was located in the coastal region of what is now Thessaly in Greece on the western shore of the Aegean Sea.

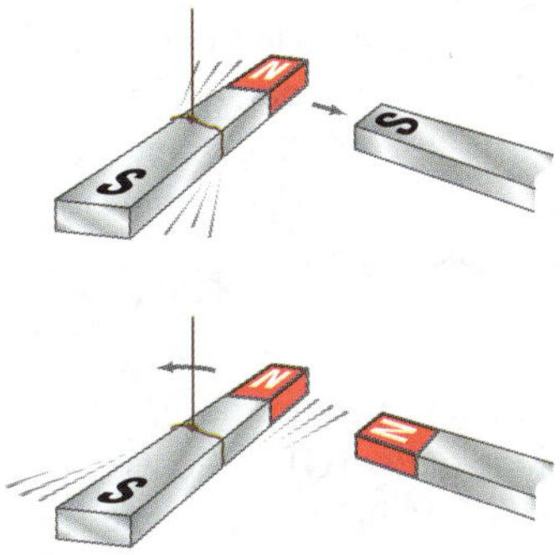

21-2 Opposite poles attract, whereas like poles repel.

Petrus Peregrinus de Maricourt (thirteenth century) was a French scholar and scientist.

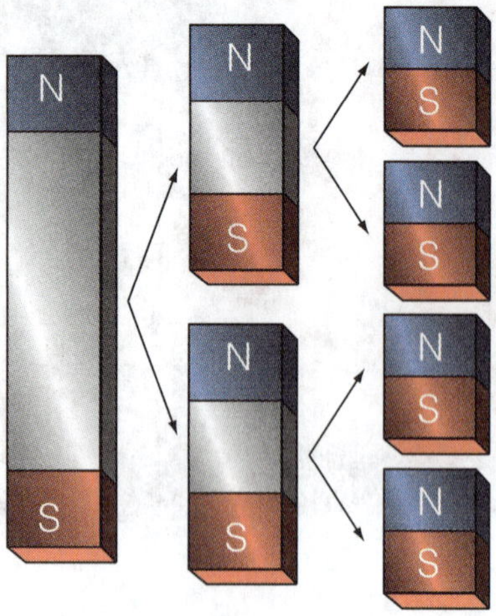

21-3 Breaking a magnet in half repeatedly just produces smaller dipole magnets.

21.2 Description of the Magnetic Field

Magnetic poles come in north-south pairs. Realizing that long magnets have little magnetic force at the center of their lengths, early experimenters tried to isolate a magnetic pole by breaking a magnet in half. They invariably got shorter, weaker magnets, each with a north and a south pole. No matter how small the magnetic pieces are made, two poles are always present. No example of a single, isolated magnetic pole has ever been observed. This is a significant difference from electrical charges.

A magnet with two poles is called a *dipole*, while one with a single magnetic pole would be called a *monopole*. While no monopole magnet has been found, electric monopoles are common. The negative and positive unit charges (e) on electrons and protons produce monopole charges.

The effect of a magnet on another magnet, metals, and moving electric charges is best described by a *magnetic field*. Like an electric field, a magnetic field is a convenient description of how the space surrounding a magnet is affected by it. A magnetic field line gives the direction that a (hypothetical) magnetic north monopole would travel if released along the line. Consequently, magnetic lines of force are curved arrows pointing toward south magnetic poles and away from north magnetic poles. If the imaginary magnetic monopole is replaced by a tiny dipole magnet that is suspended in a large magnetic field, the dipole will align itself so that it lies along the field with its north pole pointing in the direction of the arrows.

You can find the general shape of the magnetic field of a magnet easily. Simply sprinkle some iron filings onto a piece of paper and hold the paper over the magnet. The iron filings will become magnetized and align themselves along the magnetic field. Another way to map a magnetic field is to place a compass in it. The magnetized compass needle will point in the direction of the magnetic field. These methods are limited because they are able to portray only a two-dimensional cross section of the field surrounding a magnet. Like electric and gravitational fields, magnetic fields are three-dimensional.

Notice in Figure 21-4 that the magnetic field lines are closer together at a magnet's poles than anywhere else. The number of field lines passing through an area of space represents the strength of the field at a point. More lines indicate a stronger field. Both the strength and the direction of the magnetic field at a point are described by the **magnetic field vector (B)**, whose magnitude is called the *magnetic flux density*. The direction of **B** at any point in space is tangent to the magnetic field line through that point. The strength of the field, and therefore the magnitude of **B**, is measured in **teslas (T)**, the SI unit named after Nikola Tesla. One tesla is equal to one volt-second per square meter:

$$1\ \text{T} = 1\ \text{V}\cdot\text{s/m}^2$$

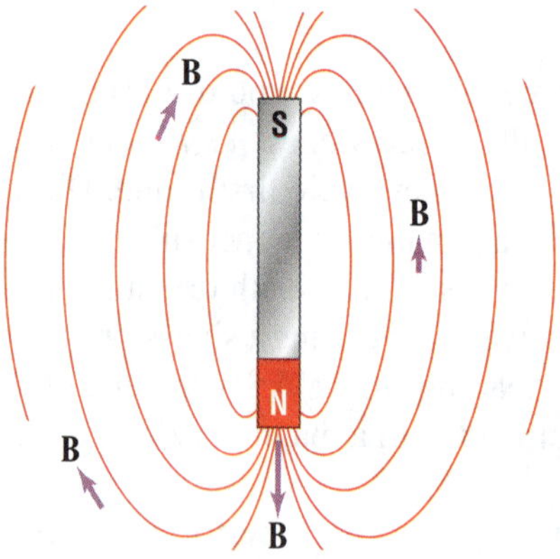

21-4 Orientation of the magnetic field vector in a magnetic field

Nikola Tesla (1856–1943) was a brilliant Serbian-American electrical engineer and inventor. He invented the methods for the practical production and use of alternating current that made the modern commercial power industry possible.

Another convenient unit of magnetism is the **gauss (G)**, often used in research involving small magnetic fields as well as in geophysics.

$$10^4\ \text{gauss} = 1\ \text{T}$$

The earth's total magnetic field strength at its surface varies between 0.3 G and 0.6 G, depending on the observer's location.

21.3 Magnetic Properties of Matter

When a material is placed within a magnetic field, the field will alter the material's magnetic character in a process called **magnetization**. Most materials lose this magnetic character when the field is removed, while others may retain some or even most of the magnetizing effect.

Materials differ in their responses to magnetic fields. A magnetic field passing through a material is usually not the same as in a vacuum. We saw this with electrical fields and capacitors. The dielectric constant κ is the ratio of the material's permittivity to the permittivity of free space (vacuum) (ε_0). The permittivity of a material is its ability to store energy in the electric field when an electrical potential difference is applied. The dielectric constant is a measure of how a material tends to concentrate electric field lines. In a similar way, scientists describe the way a material affects a magnetic field that is within it by a property called the material's *magnetic permeability* (μ). A more easily interpreted quantity is the material's **relative permeability (μ_r)**, which is defined as the ratio of the permeability of the material and the permeability of free space (μ_0):

$$\mu_r = \frac{\mu}{\mu_0} \qquad (21.1)$$

Relative permeability can be thought of as the ratio of the strength of a magnetic field within a material to the strength of the same field in a vacuum, though this is not strictly accurate for all materials under all conditions.

Materials can be classified by their magnetic properties according to the magnitude of their relative permeabilities. One group of materials is readily attracted to magnets. Their relative permeabilities are much greater than 1. Therefore, a magnetic field is much more concentrated within one of these materials than in a vacuum. These materials can be induced to become permanent magnets. A material with these properties, such as iron, nickel, cobalt, and certain alloys, is called **ferromagnetic**.

Another group of materials, **paramagnetic** materials, is not noticeably attracted to small magnets. The relative permeability of these materials is slightly greater than 1. The magnetic field in a paramagnetic material is only slightly more concentrated than it would be in a vacuum. Aluminum, platinum, and sodium are examples of paramagnetic materials.

Table 21-1

Relative Permeabilities

Material	μ_r
Ferromagnetic Materials	
Iron (Fe)	200–5000
78 Permalloy (22% Fe, 78% Ni)	5000–100 000
Supermalloy (5% Mo, 79% Ni, 16% Fe)	100 000–1 000 000
Paramagnetic Materials	
Aluminum (Al)	$1 + 2.2 \times 10^{-5}$
Sodium (Na)	$1 + 7.2 \times 10^{-6}$
Platinum (Pt)	$1 + 2.8 \times 10^{-4}$
Uranium (U)	$1 + 4.0 \times 10^{-4}$
Diamagnetic Materials	
Copper (Cu)	$1 - 9.8 \times 10^{-6}$
Lead (Pb)	$1 - 1.8 \times 10^{-5}$
Mercury (Hg)	$1 - 3.0 \times 10^{-5}$
Silver (Ag)	$1 - 2.6 \times 10^{-5}$

Materials in the third group are those that actually diminish a magnetic field. Their relative permeabilities are slightly less than 1. They are repelled by strong magnets because the external magnetic field causes the atoms of the material to align in a way that forms an opposing magnetic field. This property is called **diamagnetism**. Nearly all elements and compounds are weakly diamagnetic. However, many materials have some stronger magnetic characteristic that masks their diamagnetism. Elements that show dominant diamagnetism include most of the Group 3A–8A elements and most of the Group 1B and 2B metals on the periodic

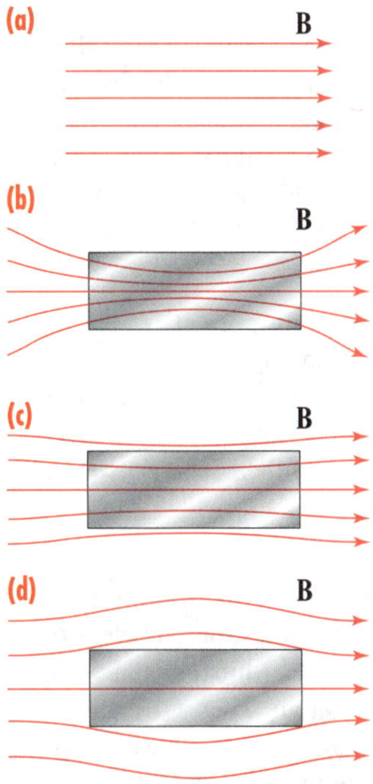

21-5 Magnetic field line density within (a) a vacuum, (b) a ferromagnetic metal, (c) a paramagnetic material, (d) a diamagnetic material

The magnetic character of materials can be described by their relative permeabilities:

Ferromagnetic	$\mu_r \gg 1$
Paramagnetic	$\mu_r > 1$
Diamagnetic	$\mu_r < 1$

21-6 When diamagnetic materials, such as those in living tissues, are exposed to a strong magnetic field, the diamagnetic force can be strong enough to levitate a small creature. The frog was not injured in this demonstration.

Hans Christian Oersted (1777–1851) was a Danish physicist who first demonstrated the relationship between electricity and magnetism—by accident—during a classroom demonstration.

André Marie Ampère (1775–1836) was a French physicist and mathematician. He was largely self-taught and eventually became the Inspector General of the French system of universities.

The quantum-mechanical model of the atom indicates that the electron exists as a three-dimensional wave within the atom. Therefore, the idea that such an entity can have "spin" reveals the inadequacy of our language to describe the properties of such a strange thing as an electron.

Magnetic Fields and God's Wisdom

The discovery of magnetism shows that God created the world with properties that humans can exploit for the benefit of others. In the future, scientists may find other useful properties built into creation.

table. Note that there are significant exceptions, such as oxygen, which is strongly paramagnetic.

A fourth category of magnetic materials consists of mixtures of different magnetic compounds that exhibit opposing magnetic orientations. The relative concentrations of the materials determine the strength and orientation of the magnetic field in an object containing the materials. These materials are **ferrimagnetic**. Natural lodestones, the most common ferrimagnetic materials, consist of two oxides of iron in a mineral called *magnetite*.

21.4 Causes of Magnetism

What makes a magnet magnetic? In 1820, Hans Christian Oersted discovered that a wire carrying a current influenced a nearby compass needle. From that discovery, André Marie Ampère concluded that all magnetic fields result from currents. Today, most scientists agree with Ampère. Yet permanent magnets need not be connected to a source of potential difference. Where then does the current come from?

Remember, an atom consists of a positively charged nucleus orbited by negatively charged electrons. The electrons are moving charges. Therefore, each atom contains a tiny current. The magnetic effect of an atomic current is represented by its **magnetic dipole moment ($\boldsymbol{\mu}$)** vector. (Do not mistake this vector quantity for the scalar magnetic permeability, μ.) The magnetic dipole moment of an atom is a vector; its direction is the direction of the atom's magnetic field, and its magnitude is related to the strength of the magnetic field. The electrons and protons in an atom also seem to "spin." Each of these subatomic particles therefore has its own magnetic moment.

Pairs of electrons "spin" in opposite directions, so their magnetic moments cancel each other within the atom. Only those substances that have unpaired electrons in certain atomic energy levels may be magnetic. Even then, in most materials the magnetic moments of their atoms point in random directions. On the average, the random magnetic moments sum to cancel each other. That is why most materials have no net magnetic field. In ferromagnetic materials, however, the atomic magnetic dipole moments are strong enough to influence each other. The magnetic moments of all the atoms within tiny crystalline regions of the material are aligned in one direction. These regions are called **domains**.

Usually, the alignment of the domains varies randomly, so there is no overall reinforcement of the individual domain dipole moments. In the presence of a strong external magnetic field, however, the domains tend to align with the field, and the material becomes magnetized. This magnetization becomes permanent if the individual atomic dipole moments are strong enough to overcome the tendency for the moments to become disoriented due to the thermal motion of their atoms (ferromagnetic materials). Otherwise, the magnetism lasts only until the field is removed (paramagnetic materials).

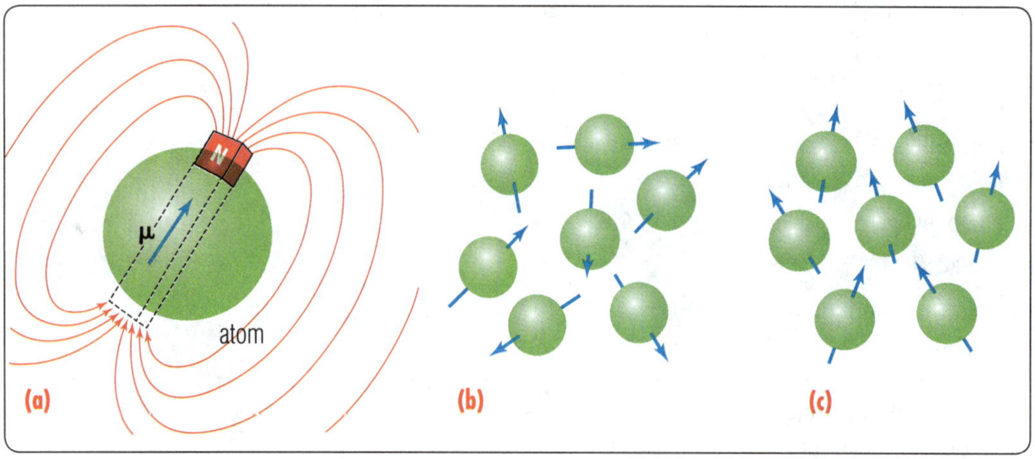

21-7 The atoms of some materials act as tiny bar magnets (a). If they are randomly oriented, the material is completely unmagnetized (b). In ferromagnetic materials, atoms within individual crystals of the material have nearly the same orientation, producing domains of magnetism (c).

Diamagnetic materials not exposed to an external magnetic field exhibit no net magnetic moment. When an external magnetic field is applied, a magnetic dipole moment is induced in the atoms similar to the way that electrostatic charge is induced in a neutral object. North is attracted to south. The resulting induced field opposes the external field and diminishes its strength within the material. In strong fields, a diamagnetic object can actually be repelled by a magnet or even levitated (see Figure 21-6).

The ordered state of magnetic domains is greatest very close to absolute zero, when there is little thermal motion. At higher temperatures, atomic thermal motion interferes with the alignment of magnetic moments. When it exceeds a characteristic temperature, called its **Curie temperature (T_C)**, a ferromagnetic material loses its ferromagnetic properties and becomes paramagnetic.

Pierre Curie (1859–1906) was a French physicist who discovered the piezoelectric effect and showed that magnetic properties of a substance change at a given temperature. He and his wife, Marie, discovered the elements radium and polonium and shared the 1903 Nobel Prize in Physics with Henri Becquerel for their studies in radiation.

21-8 A magnetic compass that a hiker might use

21.5 Terrestrial Magnetism

Why does the north pole of a compass point north? A magnetic compass is a small, light magnet mounted on a pivot so that it can rotate freely in a horizontal plane. The poles of a magnet are called the north magnetic pole and the south magnetic pole. A magnet's *north magnetic pole* points toward the earth's magnetic north pole (which is actually a *south* magnetic pole). This confusing state of affairs exists because opposite poles attract, and a north magnetic pole is defined as one that points toward the magnetic north pole.

William Gilbert reasoned that magnets are attracted by magnets. Therefore, the earth must be a magnet. To test his conclusions, Gilbert milled a large piece of lodestone into a sphere. When he placed a compass at various positions on the surface of the lodestone, the compass reacted in the same way as it would when

The earth's magnetic north pole is continually drifting, and its location is not precisely known from year to year. A computer model places the pole at approximately 85°N, 130°W in 2010. This is well to the northwest of Ellesmere Island in northern Canada in the Arctic Ocean. This location is about 560 km (345 mi) south of the geographic north pole. The pole has drifted more than 1700 km since it was first located in the nineteenth century, and it is currently moving about 50 km per year. The drift speed is variable and is expected to slow in coming years.

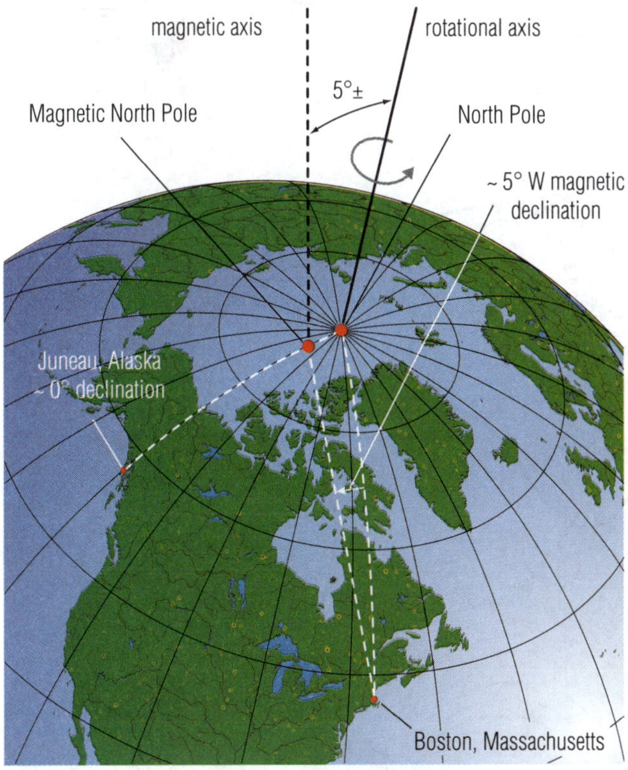

21-9 The poles of the earth's magnetic field are not aligned with its geographic poles. This is the main reason for magnetic declination.

21-10 In 1905, the Department of Terrestrial Magnetism of the Carnegie Institute, Washington, D.C., commissioned the first detailed magnetic survey of the Pacific Ocean using a nearly nonmagnetic wooden sailing vessel, the *Galilee*.

responding to the earth's magnetic field. Though some describe the earth as if it had a large bar magnet at its center, this analogy is not completely accurate. The magnetic north and south poles do not lie on a line that passes through the geographic center of the globe, nor do they move at the same rates. These observations indicate that there are highly complex factors deep inside the earth that affect the source of the earth's magnetic field.

The earth's magnetic axis is not in the same place as its rotational axis. The magnetic north pole is inclined about 5° from the rotational axis. A compass points to the magnetic pole. Assuming that the earth's magnetic field is symmetrically distributed around the line passing through the magnetic north and south poles, a compass would not point to geographic north unless it were aligned with a magnetic field line that coincides with the meridian containing both the geographic north pole and the magnetic north pole. When the magnetic field is not aligned with this meridian, it points a certain angular distance away from the geographic pole. The angular difference between true north and magnetic north is called **magnetic declination**. It is usually given in degrees east (or +) or west (or −).

Determining magnetic declination is far more complicated than a mere exercise in spherical geometry. The distribution and motions of the material in the earth's core that produces the terrestrial magnetic field are not uniform. In addition, the crust's thickness and local conditions also affect the orientation of the earth's field at its surface. As noted above, the locations of the magnetic poles continuously change as well. Therefore, magnetic field data must be periodically taken and plotted on maps such as that shown in Figure 21-11.

Problem-Solving Strategy 21.1

The direction of magnetic declination at your location can be determined by imagining the meridian line connecting your position and the geographic north pole. If the magnetic north pole is to the left (west) of this line, then the declination is west; if it is to the right (east) of the line, then declination is east.

The rate of the annual change of direction of the earth's magnetic field vector at any given point is fairly constant over a span of several years. This rate is called the **annual variation**, and it is usually printed in the legend on better-quality maps. It is given in units of degrees per year east or west, depending on the direction of change of local declination.

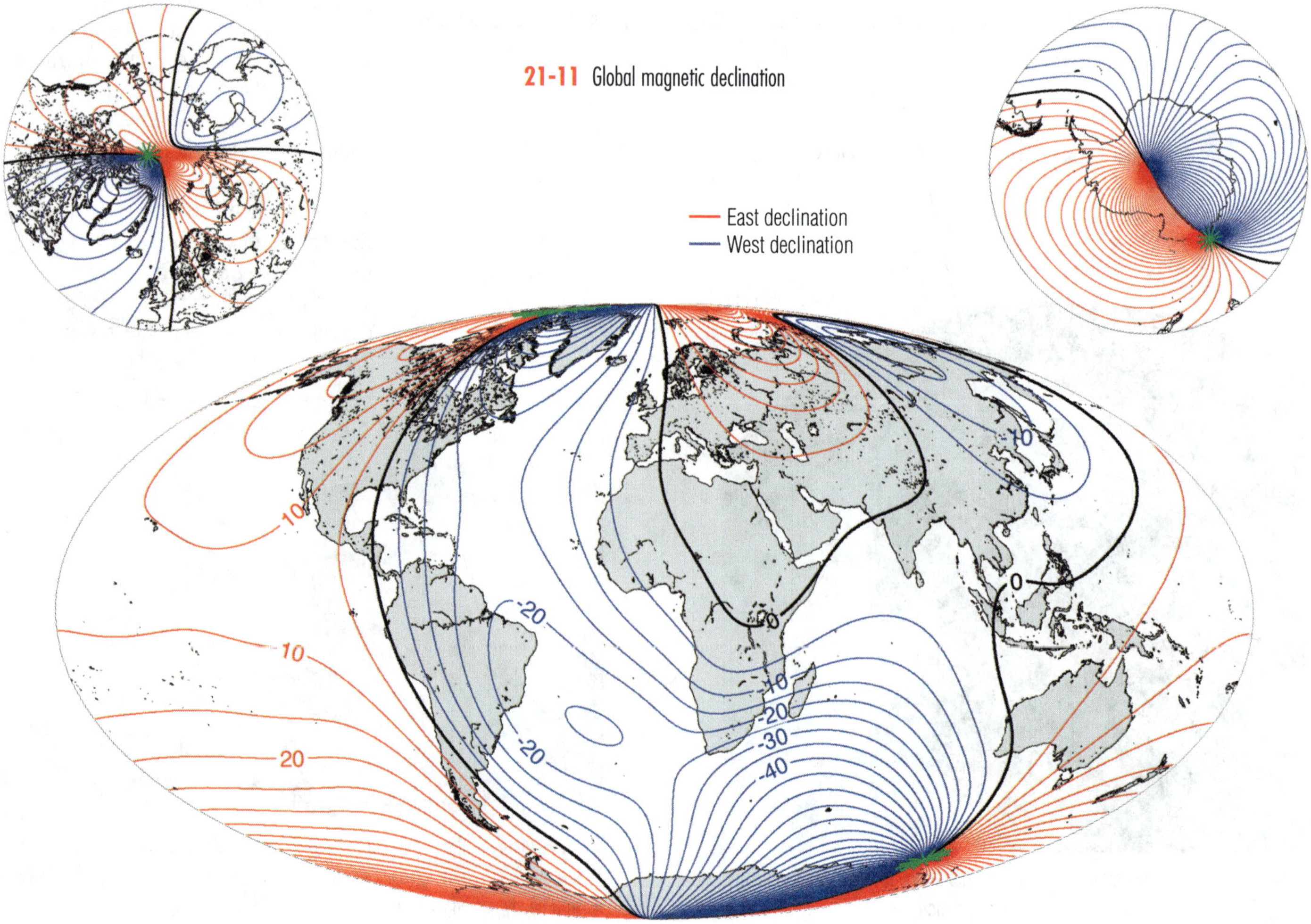

21-11 Global magnetic declination

21.6 The Magnetosphere

The volume of space containing the earth's effective magnetic field is called the **magnetosphere**. The shape of the magnetosphere is not exactly like the symmetrical field surrounding a laboratory bar magnet. It is distorted by the **solar wind**, the continuous stream of charged particles emitted by the sun. These charged particles form a high-energy fluid (plasma) and are deflected by the magnetic field. The field, in turn, is carried along with the plasma flow. Therefore, the magnetosphere facing the solar wind is not as thick as the side facing away from the sun, which tends to be drawn out like a magnetic tail (Figure 21-12).

21-12 The earth's magnetic field, or magnetosphere, distorted by the solar wind

James Alfred Van Allen (1914–2006) was an American physicist who worked in the field of geomagnetism. After the success of the Soviet *Sputnik*, Van Allen began designing a satellite that would transmit data about the earth's magnetosphere. He is most famous for the discovery of the Van Allen belts—doughnut-shaped bands of high-intensity radiation surrounding the earth.

Some particles are trapped in two areas of the magnetosphere. These belts of charged particles are called the **Van Allen belts** after James Van Allen, who discovered them in 1958 during the flight of America's first satellite, *Explorer I*. If the earth's magnetic field did not trap or deflect the radiation of the solar wind, it would reach the earth and damage all living things on the earth's surface. The existence of the strong magnetosphere is essential to preserve life on the earth and is evidence of God's protection of His creation.

The particles trapped in the Van Allen belts spiral back and forth from one end of the earth to the other in a few seconds. Sometimes, when the solar wind is especially strong, a high electrical potential (>10 000 V) is established between the lower ends of the belts and the ionosphere. High-energy electrons rain down and collide with the atmosphere's molecules about 100 km above the earth's surface. The energy of the collisions is dissipated by the emission of light. The light as observed from the ground (and space) is known as an **aurora** (from the Latin word for *dawn*). The aurora in northern latitudes is called the *aurora borealis* or the *northern lights*. The southern aurora is called the *aurora australis* or the *southern lights*.

"And all the days of Methuselah were nine hundred sixty and nine years: and he died" (Gen. 5:27). Why did people live so much longer before the Flood? Nobody knows for sure, but one theory is that the earth's magnetic field could have been much stronger, giving better protection from the solar wind and cosmic radiation. With the weakening magnetic field, more mutation-inducing radiation has penetrated to the earth's surface, destroying the natural vitality and longevity that God originally created in living things. Other explanations related to the Flood itself are also possible.

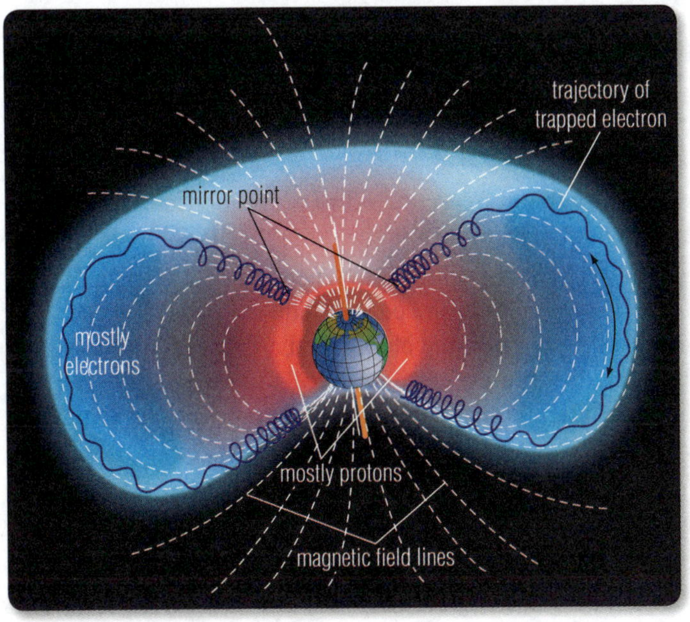

(a)

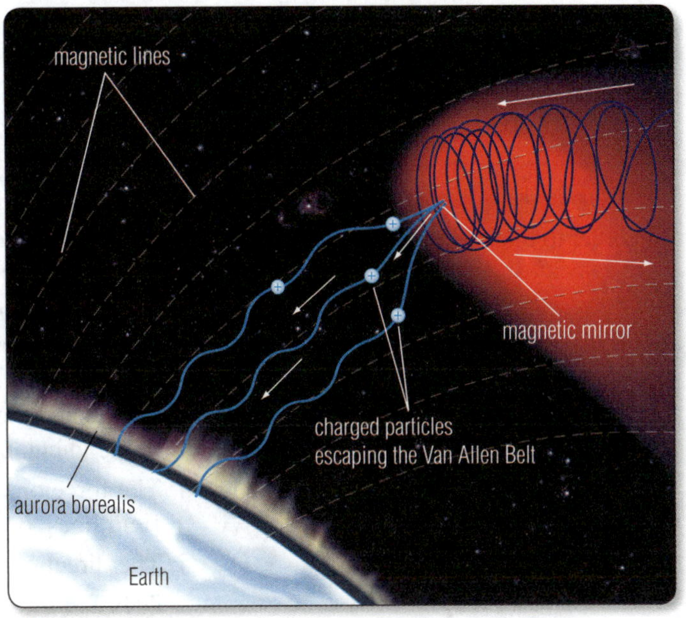

(b)

21-13 (a) The Van Allen belts and (b) the cause of auroras

21A Objectives

After completing this section, I can

- ✓ summarize the discovery of and early investigations into magnetism.
- ✓ describe the shape, properties, and units associated with a magnetic field.
- ✓ differentiate between the three kinds of magnetic responses a material may exhibit to an external magnetic field.
- ✓ describe how magnetism is produced at the particle level of a material.
- ✓ define the Curie temperature of a ferromagnetic material.
- ✓ differentiate between the earth's geographic and magnetic north poles and explain why magnetic declination and variation exist.
- ✓ describe the magnetosphere and how it is evidence for a good design of the earth.

21A Section Review

1. a. What is a magnetic pole?
   b. How many kinds of poles are there?
   c. State the law of magnetic poles.
2. How is the magnitude and direction of the magnetic field represented *at a point* in a diagram?
3. What is the standard unit of magnetism and its SI equivalent?
4. a. What is the process that makes magnets out of unmagnetized metals?
   b. What kind of metals retains some magnetism after this process?
5. Ultimately, what two factors result in magnetism in a given material?
6. What condition can negate ferromagnetism in even the strongest magnets?
7. For a point on the surface of the earth, what is the horizontal angle between the directions to magnetic and true north called?
8. Describe the safety mechanism God provided to protect life on the earth from the effects of the solar wind.

21B ELECTROMAGNETISM AND CHARGES

21.7 Magnetic Force on a Moving Charge

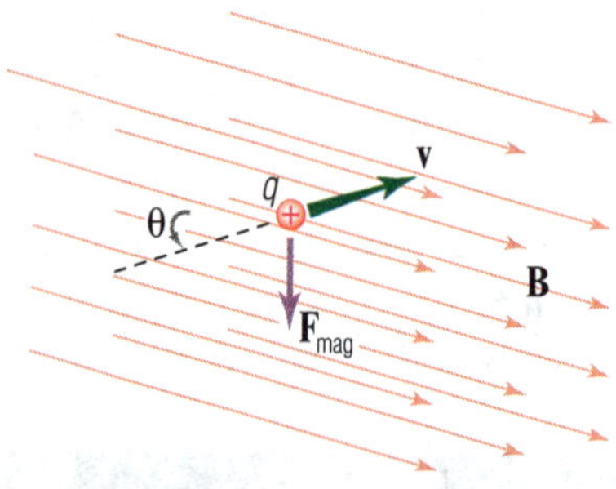

21-14 A point charge moving through a magnetic field

A current-carrying wire can produce a magnetic force that can attract or repel a magnet. When Oersted discovered this fact, scientists wondered, "Do magnets also affect electrical charges?" The answer is yes—if the charges are moving. A magnet has no effect on a stationary charge, and a stationary charge has no effect on a magnetic field. Scientists have made the following observations regarding the interplay of magnetic fields and electric charges:

1. A point charge of either polarity at rest in a magnetic field experiences zero force.
2. A charge of either polarity that moves parallel to magnetic field lines in either direction experiences zero force.
3. If any point charge moves so that it crosses magnetic field lines at an angle, a *magnetic force* (F_{mag}) is exerted on the charge. The magnitude of the force is proportional to the speed of the charge and a function of the angle with which it crosses the field lines. For a given relative speed between the charge and the field, the force is maximum when the charge's velocity is perpendicular to the magnetic field lines.
4. The force on a moving charge is proportional to the density of the magnetic field lines, or the magnitude of flux density (B).
5. The direction of the force vector on a charge depends on the polarity of the charge.

Problem-Solving Strategy 21.2

In order to find the angle between the velocity and the magnetic field vectors, place their vectors tail-to-tail. The angle θ is the smallest angle between the vectors.

FACETS of PHYSICS

Paleomagnetism and the Age of the Earth

Most scientists agree that the earth's magnetic field is the result of some kind of current in the earth's core, but they disagree on the source of the current. The core is far too hot for a permanent magnet to be the cause of the earth's magnetic field. There are many "dynamo" theories that say the earth contains some kind of self-perpetuating electric generator. None of these theories satisfactorily explains the earth's magnetic field. A theory more consistent with observations is that the current was started at some time in the past, and it has not yet been overcome by flow resistance. This theory would predict that the current, and therefore the earth's magnetic field, is decreasing. The earth's magnetic field has decreased by a significant amount since it was first measured in 1839 by Carl Friedrich Gauss.

The idea that the earth's magnetic field is consistently decreasing bothers those who believe that the earth is billions of years old. The measurements of the earth's magnetic field show that its strength is decreasing exponentially. The field's half-life—the time in which the magnetic field energy halves—is about 1465 y. Based on these facts, if the earth's magnetic field has been decreasing in the same way since it started, then one million years ago it would have had sufficient energy to melt the crust! Some evolutionists believe that the recorded magnetic field data indicates that the field is decreasing in a linear fashion. If this were the case, then only twenty thousand years ago the magnetic field would have been stronger than most man-made magnetic fields. Such a strong field would require a current so large that it would produce enough heat to damage the earth's core. If the field continues decreasing at the same rate, it will disappear in just a few thousand years, and life on Earth will be seriously endangered.

The rapidly decreasing magnetic field creates a significant problem for evolutionists. Most of those who believe that the earth is billions of years old believe that its magnetic field must be about as old because the solar wind would destroy or damage life on the earth if the magnetic field did not deflect it. Therefore, since a constantly decreasing field would already have disappeared, they have to theorize that the earth's magnetic field oscillates. According to this theory, we are observing only the decrease since the last maximum. For evidence, they cite rocks containing magnetic materials that show magnetism with a polarity opposite to the earth's present magnetism.

When a hot ferromagnetic material cools, it often becomes magnetized by aligning with an external magnetic field. When a volcano erupts, magnetite crystals in the lava usually orient themselves along the earth's magnetic field. Geologists have noticed that igneous rocks bordering the mid-Atlantic ridge contain bands of materials magnetized in opposite directions, as the figure on the next page shows. Assuming that (according to the uniformitarian plate-tectonics model) as the continental plates slowly spread apart and new magma wells up into the crack, the rocks "record" the orientation of the earth's magnetic field at the time of their formation, geologists conclude that the earth's magnetic north and south poles have exchanged places several times in the earth's history.

Scientists working within a Creation-Flood model believe that the magnetic banding detectable near the mid-ocean ridges was caused by the devastating breakup of the earth's crust during and immediately following the Genesis Flood. Except for this short disruption, the earth's overall magnetic field seems to have been decaying continuously in an exponential way since Creation, about 6000 years ago.

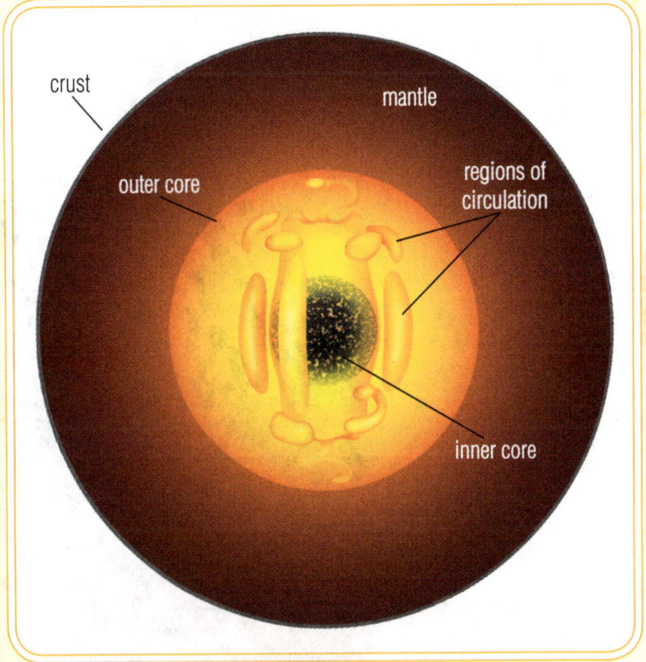

A model of the magnetically active core of the earth. Geophysicists believe that the earth's magnetic field is produced by complex circulatory flows of liquid metals in the outer core.

Some scientists, especially those who believe that the earth is relatively young, do not find the evidence for changing magnetic fields convincing for several reasons. One reason is that the magnetization in rocks is so small that it is difficult to measure. Some measuring devices can even change the magnetic field they measure. In addition, the evolutionary model assumes that the decay and recovery of the earth's magnetic dipole field is accomplished by storing energy in the weaker 4-pole and 8-pole fields that are components of the much stronger dipole field. However, recent data shows that the slight gain in "poly-pole" field intensity does not account for the loss of energy in the main dipole field during the past 40 years. The theory also assumes that the energy transfer between the fields is 100% efficient. Nowhere else in the universe is energy conversion 100% efficient. In short, the data "proving" that the earth's magnetic field has reversed is not very reliable.

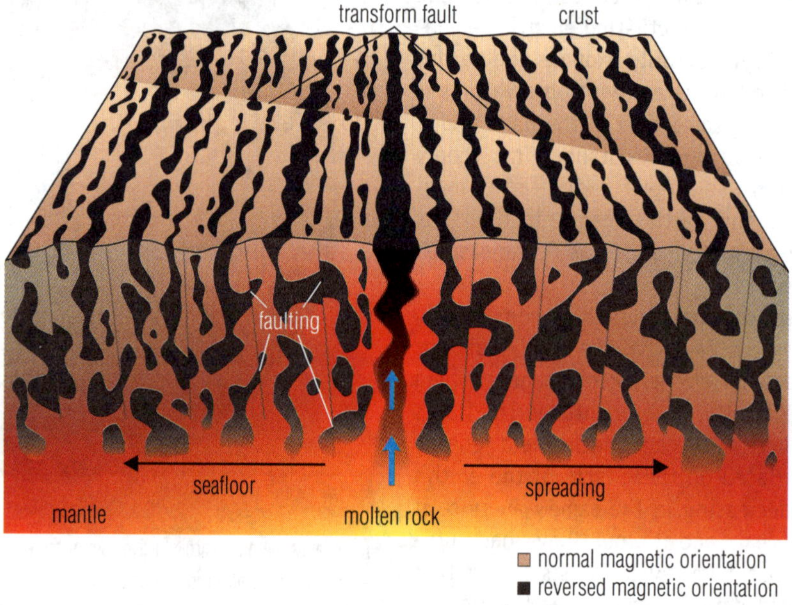

Recent research has confirmed that the magnetic banding along midocean ridges exists. However, drill core samples from the bands reveal that the magnetic orientation in rocks beneath the surface is highly variable with depth, showing no coordinated pattern. Rather than being evidence for a slow, gradual spreading of the seafloor according to uniformitarian geology, the data supports a catastrophic model of rock formation that occurred quickly, along with rapid local changes of magnetic field orientation, during the time of the Genesis Flood.

The velocity in Equation 21.2 is the *relative* velocity between the charge and the field. This means that the charge alone, the field alone, or both the charge and field may be moving in an external reference frame.

If Equation 21.2 is solved for B, an analysis of the resulting units gives

$$\frac{\text{N}}{\text{C}\cdot\text{m/s}} = \frac{\text{N}\cdot\text{s}}{\text{C}\cdot\text{m}}.$$

This ratio in SI units is the tesla (T).

The magnitude of the magnetic force on a point charge is calculated by the formula

$$F_{\text{mag}} = |q|vB\sin\theta\ , \qquad (21.2)$$

where θ is the smallest angle ($\leq 180°$) between the velocity vector and the magnetic induction vector (**B**). The absolute value sign is required because the charge q may be either positive or negative.

Surprisingly, the force of a magnetic field on a moving charge (or a moving field on a stationary charge) is *not* in the direction of the charge's relative motion, nor is it in the direction of the field; it is perpendicular to both the magnetic induction vector and the velocity vector of the charge. Because the magnetic force on the charge has no component in the direction of the charge's motion, it does no mechanical work on the charge. This interaction is similar to the force of gravity on an orbiting object. Earth's gravity keeps the moon in its orbit. However, the force is perpendicular to the moon's motion, so it does no work on the moon.

Similarly, a magnetic field changes the direction of a charged particle's motion but does no work on the particle. You can see that the magnetic force on a charge is a conservative force just like the electrostatic and gravitational forces.

When the charge's velocity vector is perpendicular to the magnetic field, Equation 21.2 reduces to

$$F_{mag} = |q|vB,$$

since the sine of 90° is 1. The direction of **F** for such a case is shown in Figure 21-15. The symbol ⊙ stands for a vector pointing up out of the page. Think of it as the view of the tip of an archery arrow pointing toward you. The symbol ⊗ denotes a vector pointing down into the page. It is the view you would have when looking at the feathers on an arrow pointing away from you. These symbols will be used throughout our discussion of magnetic fields to represent vectors in the third dimension.

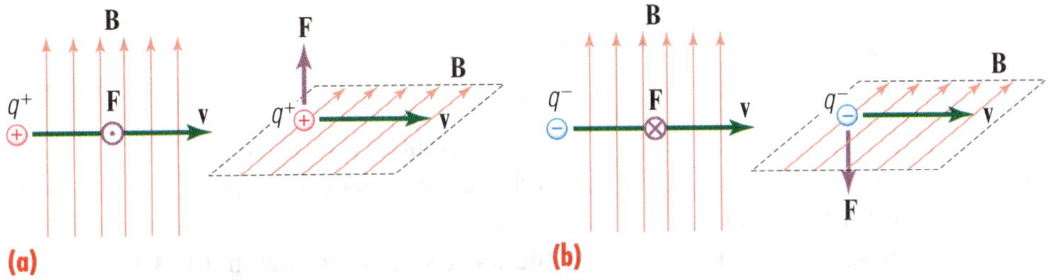

21-15 Magnetic force on a moving charge. The charge in (a) is positive; the charge in (b) is negative.

21.8 Right-Hand Rule for Magnetic Force on a Positive Charge

The force vector ($\mathbf{F}_{mag}$) is always perpendicular to the plane defined by **B** and **v**. You can find the direction of $\mathbf{F}_{mag}$ on a *positive* charge by another right-hand rule. Refer to Figure 21-16. Find the component of **v** that is perpendicular to **B**, as in figure (a). Extend your right index finger in the direction of the magnetic field. Extend your thumb in the direction of the perpendicular velocity component. Your middle finger extended at a right angle to the index finger and thumb points in the direction of the magnetic force vector ($\mathbf{F}_{mag}$). You may have to adjust the position of your hand in order to align your fingers in the required directions. Figure (c) shows how to position your hand to show the direction of the magnetic forces in figure (a). The direction of the force on a negative charge is opposite to that on a positive charge.

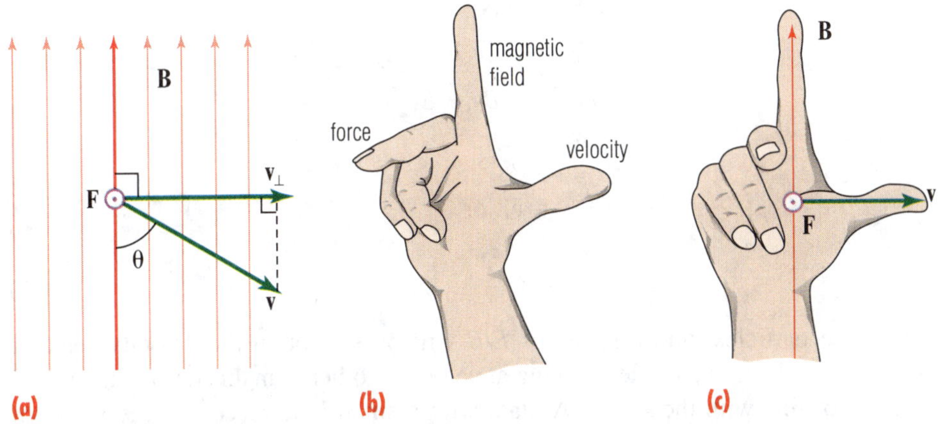

21-16 The right-hand rule for magnetic force on a moving positive charge

21.9 Velocity Selectors and Mass Spectrometers

An electric field displaces a charged particle parallel to an electric field line. A magnetic field deflects a moving charged particle perpendicular to a magnetic field line. What happens when both an electric field and a magnetic field influence a charged particle? The net force on the particle is the vector sum of the electrical force and the magnetic force:

$$\Sigma \mathbf{F} = \mathbf{F}_e + \mathbf{F}_{mag} \tag{21.3}$$

If the electric field, the magnetic field, and the velocity vector of the particle are all perpendicular to one another, the magnetic force will be either in the same direction as the electric force or in the opposite direction, depending on the sign of the charge. Figure 21-17a shows a case in which the magnetic and electrical forces are in opposite directions. Using the components of the vectors, Equation 21.3 becomes

$$\Sigma F_y = F_{e\,y} + F_{mag\,y}.$$

In Chapter 19 you learned that

$$\mathbf{F}_e = |q|\mathbf{E}, \text{ so}$$

$$F_{e\,y} = -|q|E_y. \tag{21.4}$$

Equation 21.4 is true if the electric field is oriented downward compared to the positive y-axis. If the charge q is positive, the force is downward. If the charge is negative, the force is upward.

Also, you know that for **v** (along the x-axis) perpendicular to **B** (along the z-axis),

$$F_{mag\,y} = |q|v_x B_z. \tag{21.5}$$

Equation 21.5 is true for any possible combination of charge polarity (positive or negative), velocity component (right or left), and magnetic field direction (into or out of the page).

Therefore,

$$\Sigma F_y = -|q|E_y + |q|v_x B_z$$

$$\Sigma F_y = |q|(v_x B_z - E_y).$$

One analytical use of the arrangement of fields and forces in Figure 21-17 is as a **velocity selector**. In a velocity selector, the fields and the velocity are adjusted until the charged particles pass through the fields undeflected. In order for the particles to pass through undeflected (i.e., unaccelerated), the net force on them must be zero according to Newton's first law of motion:

$$\Sigma F_y = 0\ \text{N}$$

$$0\ \text{N} = |q|(v_x B_z - E_y)$$

$$0\ \text{N} = v_x B_z - E_y$$

$$E_y = v_x B_z$$

$$\frac{E_y}{B_z} = v_x, \text{ or}$$

$$\frac{E}{B} = v \tag{21.6}$$

Only those particles with a speed of E/B will pass through the velocity selector undeflected. Charged particles moving at velocities other than this ratio will be scattered out of line with the x-axis. A plate with a small hole placed on the x-axis of the velocity selector can prevent scattered particles from emerging from the selector. Only undeflected particles with a specific speed pass through the hole.

The velocity selector is a component of a mass spectrograph. A charged particle emerges from the velocity selector into a region with a magnetic field of precisely known strength oriented perpendicular to the particle's velocity. The

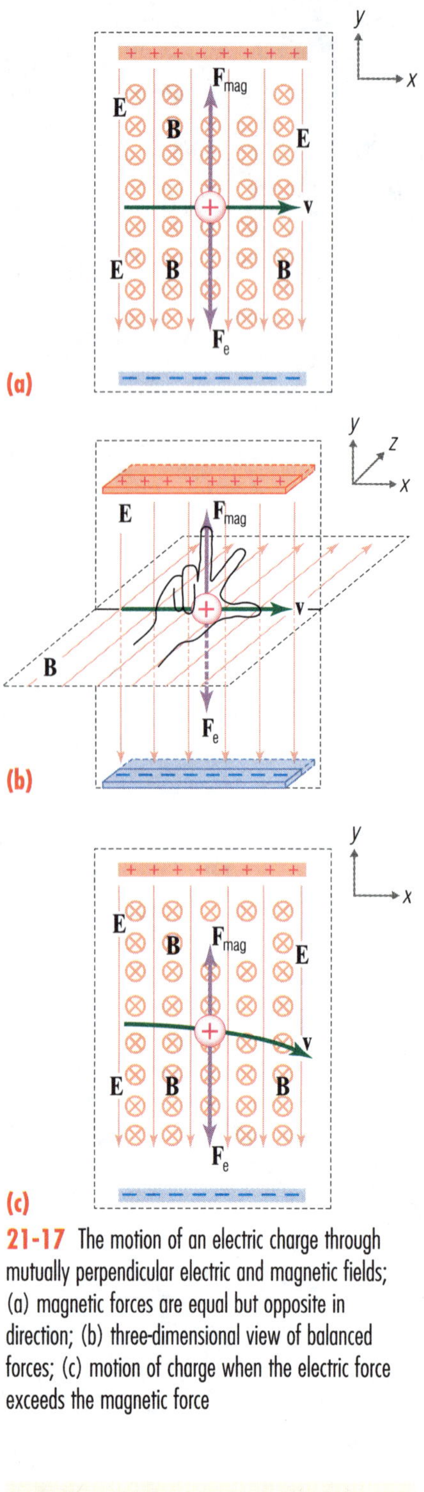

21-17 The motion of an electric charge through mutually perpendicular electric and magnetic fields; (a) magnetic forces are equal but opposite in direction; (b) three-dimensional view of balanced forces; (c) motion of charge when the electric force exceeds the magnetic force

A mass spectrograph is an instrument that analyzes and records the composition of mixtures or pure materials by determining the atomic or molecular masses of their constituent substances. Such devices are used in atmosphere quality monitors, for manufacturing quality assurance, and in forensic and archaeological research.

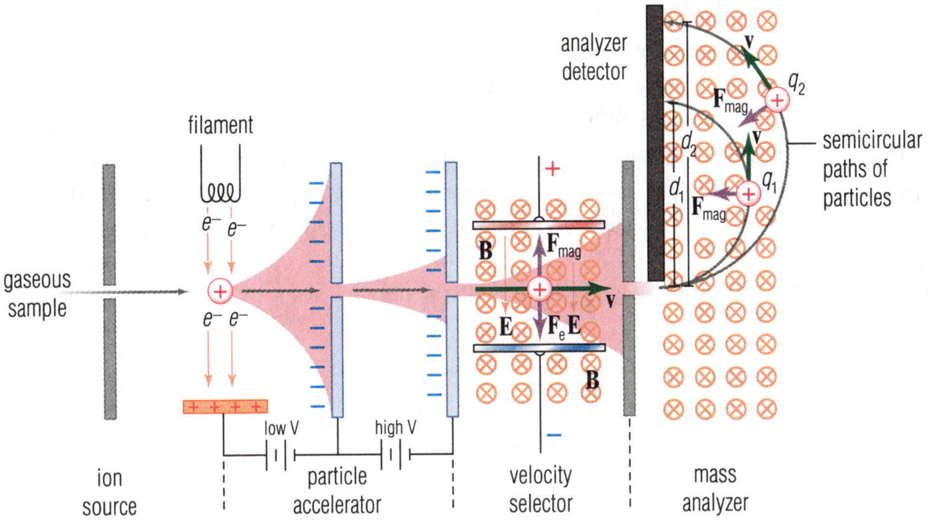

21-18 Functional diagram of a simple mass spectrograph

magnetic field influences the particle to follow a semicircular path onto a photographic film or into an electronic detector. The radius of this path is directly proportional to the particle's mass.

To find the exact relationship between path radius and the mass, remember the formula for centripetal acceleration:

$$a_c = \frac{v_t^2}{r}$$

The magnetic force to cause the acceleration must be

$$F_{mag} = ma_c$$

$$F_{mag} = m\frac{v_t^2}{r}.$$

However, when the particle exits the velocity selector and enters the spectrometer magnetic field, its velocity vector is perpendicular to the magnetic field. You know that when the particle's velocity and the magnetic field are perpendicular,

$$F_{mag} = |q|vB.$$

The particle's tangential speed and its velocity selector exit speed are the same. Therefore,

$$m\frac{v_t^2}{r} = m\frac{v^2}{r} = |q|vB$$

$$mv^2 = |q|vBr$$

$$m = \frac{|q|Br}{v}. \quad (21.7)$$

The distance from the exit of the velocity selector to where the particle of interest enters the detector or is imaged on film is the diameter of its semicircular path, *d*. The radius (*r*) is half the diameter. If you know the charge on a particle, the magnetic and electric fields in the velocity selector, and the magnetic field in the spectrograph, you can find the mass of the particle as a function of its deflection radius. In order to get a detectable reading with this technique, many particles of the same mass must be present.

21.10 Discovery of the Electron

In one of his many experiments with cathode rays, J. J. Thomson used an arrangement of magnetic and electric fields similar to a velocity selector to confirm that

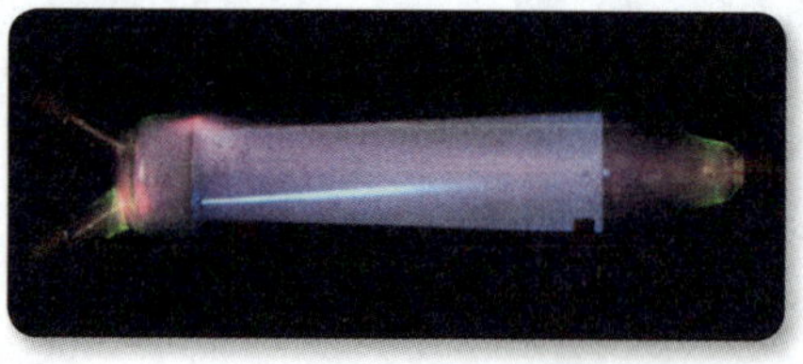

21-19 A cathode-ray tube contains an inert gas to make the beam of electrons visible.

Thomson concluded that cathode rays were actually particles having mass because their deflection in electric and magnetic fields demonstrated that they had inertia and otherwise obeyed Newton's laws of motion.

cathode rays were particles and to determine their mass-to-charge ratio. He used several different **cathode-ray tubes (CRT)** to do this. Though Thomson varied the configuration of this important experimental device for his purposes, cathode-ray tubes contain a minimum of three essential parts—a cathode, an anode, and a sealed glass tube to separate and insulate them from each other. The cathode is typically a solid plate of metal attached to a source of negative electrical potential, and the anode is a separate plate, often with a small hole or slot in it, connected to a potential that is positive relative to the cathode. The electrical connections to the external voltage sources pass through airtight penetrations in the glass envelope.

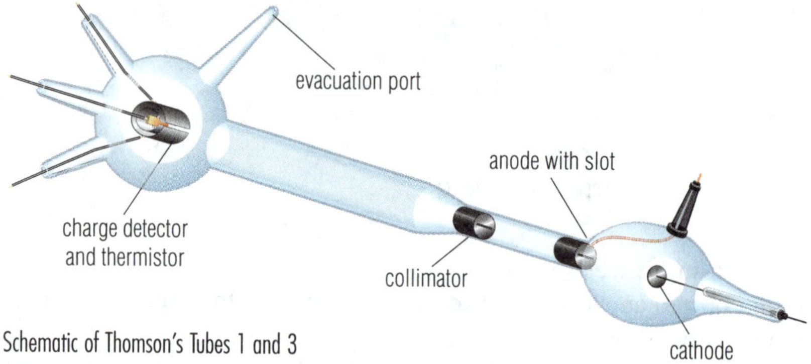

21-20 Schematic of Thomson's Tubes 1 and 3

When a high potential difference existed between the cathode and the anode, cathode rays streamed from the cathode toward the anode. Because the anode had a hole in it, many of the rays passed through the anode to the end of the tube. Thomson showed that magnetic and electric fields could divert the path of the cathode rays, especially when the tube was almost completely evacuated. Through a series of experiments, Thomson was able to prove that the cathode rays were actually negatively charged bits of matter and that they constituted a more basic particle of matter than atoms or atomic ions. He called the particles *corpuscles*.

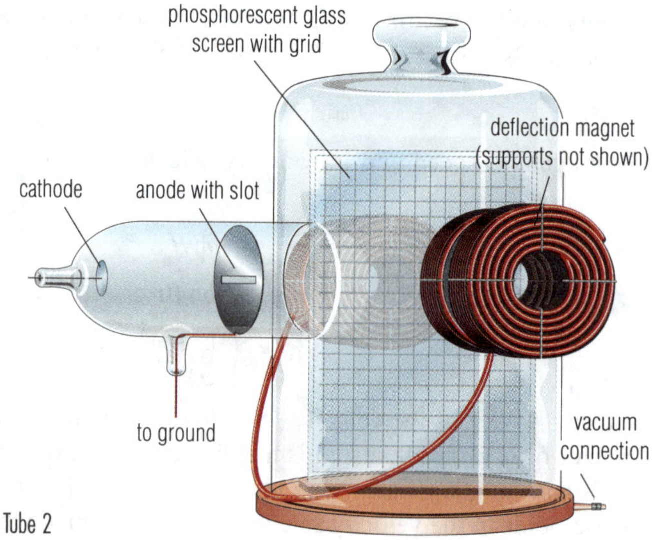

21-21 Schematic of Tube 2

21.11 Modeling the Electron's Mass-to-Charge Ratio ($m/|q|$)

In his research, Thomson suggested that the kinetic energy (E) of the corpuscles in a discrete quantity of cathode rays was given by

$$E = \tfrac{1}{2}Nmv^2, \qquad (21.8)$$

where E was given in the CGS unit of energy, the erg, N was the total number of corpuscles, m was the mass of each corpuscle in *eleventh-grammes (gXI)*, and v was their average velocity in cm/s. Their total charge, Q, could be expressed as

$$Q = Nq, \quad (21.9)$$

where q was their individual charge in electrostatic units (esu).

If r is the radius of curvature of the deflected cathode rays under the influence of a stable magnetic field, B, then from Equation 21.7,

$$Br = \frac{mv}{|q|}. \quad (21.10)$$

Thomson labeled the quantity Br with the symbol I in his equations to simplify them. This quantity had the unit of gauss-centimeters (G·cm). Combining and rearranging Equations 21.8–10, he could determine the mass-to-charge ratio of a corpuscle by measuring the total charge deposited by the corpuscles in the tube, the energy deposited, the magnetic field used to deflect the corpuscles, and the radius of deflection:

$$\frac{m}{|q|} = \frac{I^2|Q|}{2E} \quad (21.11)$$

In Thomson's 1897 paper, the quantity m/q had the units eleventh-grammes/esu (gXI/esu).

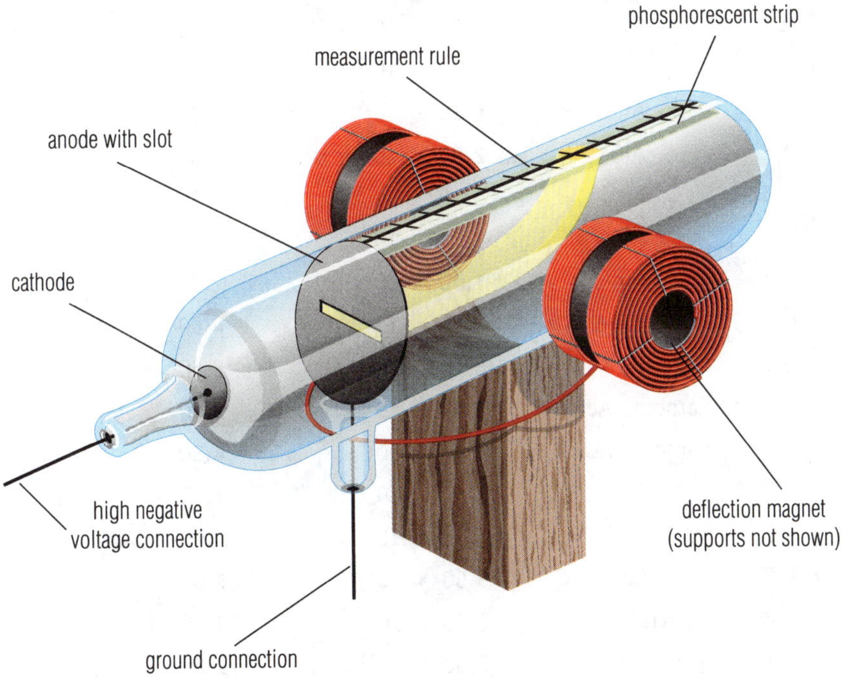

21-22 CRT used to measure the magnetic deflection radius

Thomson measured the total charge deposited by the cathode rays with an electrometer connected to collector plates inside the CRT. Using a thermistor to measure the increase of temperature at the cathode rays' point of impact, he determined the total kinetic energy deposited by the particles. He calculated the magnetic field according to the known characteristics of electromagnetism between the two current-carrying coils surrounding the CRT. The curvature of the deflected path the corpuscles took could be observed by the glow they imparted to a phosphorescent material applied to the inside of the CRT's glass envelope. He calculated the radius of the path of the rays using simple geometry. This data

1 **erg** = 10^{-7} J

In England around 1900, the British Association (BA) introduced a complex system of units that supposedly simplified calculations for electromagnetic investigations. The **eleventh-gramme (gXI)** was an extremely small unit of mass equal to 1×10^{-11} g, which seems to be the mass unit used by Thomson.

Thomson's intent in his experiments with cathode rays was twofold: first, to prove that they were matter rather than immaterial rays, and second, that they were substantially less massive than an atom. Scientists had already estimated the mass-to-charge (m/q) ratio of positive hydrogen ions (protons) in a chemical solution. Thomson's task was to show that a cathode ray's m/q ratio was much smaller than a hydrogen ion's.

A **thermistor** is an electrical device consisting of a conductive material whose resistance varies proportionately with temperature.

was obtained using three different kinds of CRTs with a variety of cathode metals and filling gases. You will use Thomson's data to model his results in order to determine the mass-to-charge ratio of an electron.

Thomson discovered that the $m/|q|$ ratio was essentially constant no matter what material the cathode was made of or the kind of gas filling the tube. He concluded that all materials contain small, negatively charged particles with a mass-to-charge ratio of

$$\frac{m}{|q|} \cong 2.1 \times 10^{-12} \text{ kg/C}$$

in SI units. Considering the relatively crude methods Thomson employed, this result compares well with the modern accepted $m/|q|$ value for the electron of 5.686×10^{-12} kg/C. As a matter of comparison, the m/q ratio of a hydrogen ion (a proton), the smallest possible atomic ion, is 1.044×10^{-8} kg/C, so Thomson thus concluded that corpuscles were distinct particles found within atoms.

TABLE 21-2

Selected Cathode Ray Experiment Data (Thomson, 1897)

Tube	CRT Gas	E/Q (erg/esu)	I (G·cm)
1	Air	4.60×10^{11}	2.30
1	Air	1.80×10^{12}	3.50
1	Air	6.10×10^{11}	2.30
1	Air	2.50×10^{12}	4.00
1	Air	5.50×10^{11}	2.30
1	Air	1.00×10^{12}	2.85
1	Air	1.00×10^{12}	2.85
1	Hydrogen	2.10×10^{12}	4.60
1	Carbonic Acid*	8.40×10^{11}	2.60
1	Carbonic Acid*	1.47×10^{12}	3.40
1	Carbonic Acid*	3.00×10^{12}	4.80
2	Air	2.80×10^{11}	1.75
2	Air	4.40×10^{11}	1.95
2	Air	3.50×10^{11}	1.81
2	Hydrogen	2.80×10^{11}	1.75
2	Air	2.50×10^{11}	1.60
2	Carbonic Acid*	2.00×10^{11}	1.48
2	Air	1.80×10^{11}	1.51
2	Hydrogen	2.80×10^{11}	1.75
2	Hydrogen	4.40×10^{11}	2.01
2	Air	2.50×10^{11}	1.76
2	Air	4.20×10^{11}	2.00
3	Air	2.50×10^{11}	2.20
3	Air	3.50×10^{11}	2.25
3	Hydrogen	3.00×10^{11}	2.50

* Former name for carbon dioxide

21B Section Review

1. What factors determine the magnitude and direction of the magnetic force on an electrical charge?
2. Why is it necessary to use the absolute value of the charge in Equation 21.2 when finding the magnitude of the magnetic force?
3. How do you determine the direction of the magnetic force on a charge moving through a magnetic field?
4. When a charged particle is propelled along a path that is perpendicular to mutually perpendicular magnetic and electric fields, what condition must exist for the particle to maintain a straight path?
5. Briefly discuss the principle of operation of a mass spectrograph.

DM 6. a. What type of device was used by J. J. Thomson to determine the mass-to-charge ratio of an electron?
   b. Briefly describe its construction.

DM 7. After completing his experiments, what fact led Thomson to conclude that all matter contains the small, negatively charged particles that he called corpuscles?

DM✪ 8. What atomic theory of matter had to be discarded with the discovery of the electron?

DM✪ 9. Equation 21.11 can be rearranged in the following way:

$$\frac{|Q|}{E} = \frac{m}{|q|}\frac{2}{I^2}.$$

   a. Using Thomson's data in Table 21-2, make a scatter plot of $|Q|/E$ versus $2/I^2$.
   b. Apply a best-fit trendline to your data. If the best fit seems to be linear, force the y-intercept to zero, if possible. Report the equation and r^2 value of the fit.
   c. What is the relationship of the $m/|q|$ factor in the above equation to your scatter plot trendline?

21B Objectives

After completing this section, I can

✓ state the factors that affect the magnetic force exerted on a charge moving through a magnetic field.

✓ calculate the magnetic force acting on a moving charge.

✓ use the right-hand rule for magnetic force to determine the direction of the force on a charge moving through a magnetic field.

✓ describe how a velocity selector works and its use in mass spectrometers.

✓ describe J. J. Thomson's experiments with cathode-ray tubes that led to the discovery of the electron.

✓ determine the mass-to-charge ratio of an electron using Thomson's original data.

✓ explain the significance of the value of the mass-to-charge ratio Thomson obtained from his experiment.

21C ELECTROMAGNETISM AND CONDUCTORS

21.12 Magnetic Force on Currents

A current is simply a collection of moving charges. What, then, is the magnitude of the magnetic force on a current-carrying wire segment placed in a magnetic field? In Section 21B you learned that the magnitude of the magnetic force on a moving charge is

$$F_{mag} = |q|vB \sin \theta.$$

The product qv for a single charge has the units C·m/s. This is analogous to the product Il for a current of I in a wire of length l (units: (C/s)·m). The latter product includes the speed and polarity of all the charges in the wire. The units are the same:

$$|q|v \Rightarrow \text{C·m/s} \Rightarrow Il$$

In fact, experiments have shown that the force a magnetic field **B** exerts on a wire segment of length l carrying a current I is

$$F_{mag} = IlB \sin \theta, \qquad (21.12)$$

where θ is the smallest angle (≤ 180°) between the current direction and **B**.

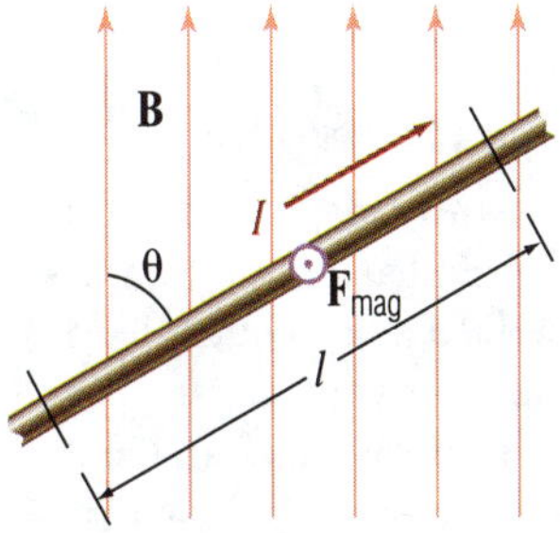

21-23 Magnetic force on a current-carrying conductor

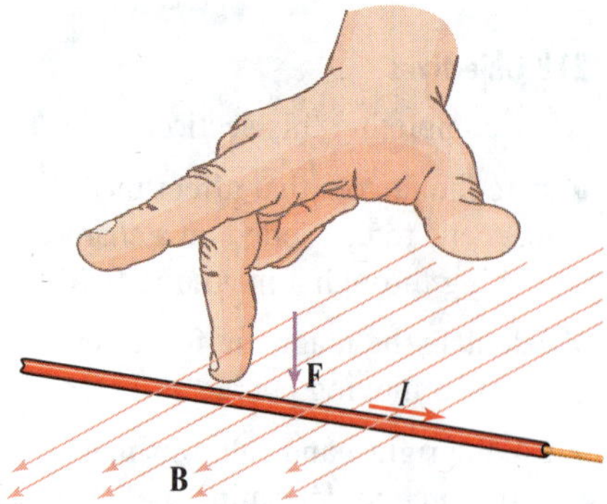

21-24 The right-hand magnetic rule for magnetic force on a conductor

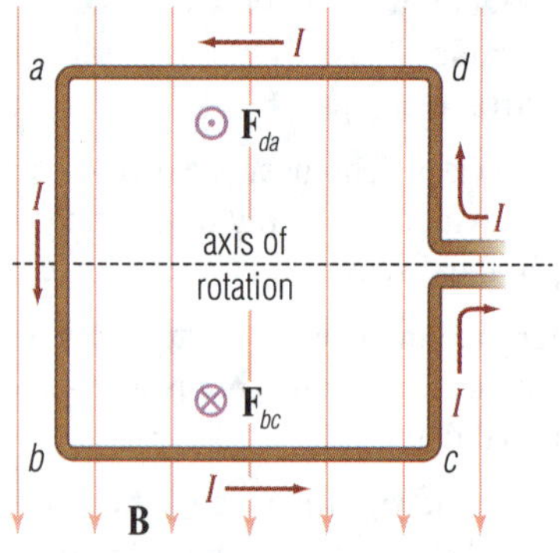

21-25 Magnetic force on a current loop

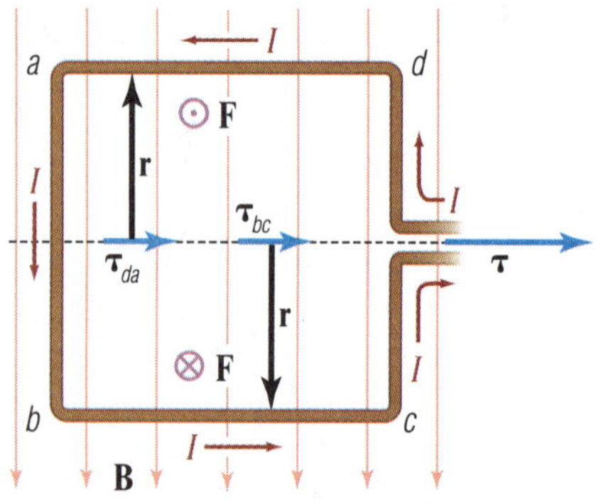

21-26 Torque on a current loop

Problem-Solving Strategy 21.3

The right-hand rule for torques: Hold your right hand flat with the thumb extended at a right angle to the fingers. Point your fingers away from the axis of rotation parallel to the position vector **r**. Curl your fingers toward the direction of the force vector where it is applied to the moment arm. Your thumb points in the direction of the torque vector.

21.13 Right-Hand Rule for Magnetic Force on a Conductor

The direction of the force on a conductor is found in the same way as the direction of the force on an individual moving positive charge. Point your right thumb in the direction of conventional current flow (or its component) that is perpendicular to the magnetic field. Point your index finger at a right angle to your thumb in the direction of the magnetic field. Hold the middle finger perpendicular to the other two fingers to indicate the direction of the magnetic force on the conductor.

21.14 Rotating Current Loops

One problem with Equation 21.12 is that it applies only to a segment of a conductor. In practice, continuous currents exist only in complete circuits. Therefore, in order to make use of the magnetic force, conductors in the shape of a loop must somehow move within a magnetic field. In order to find the net force exerted on a loop conductor, you must sum the forces acting along its length.

Consider a square conductor loop supplied with a continuous current (I), such as in Figure 21-25. The loop is positioned in a uniform magnetic field (**B**) as shown. The currents in segments *ab* and *cd* are in the same direction as **B** and −**B**, respectively. Therefore, sides *ab* and *cd* experience no magnetic force (sin 0° = sin 180° = 0).

The force on side *bc* is

$$F_{bc} = IlB \sin 90°$$

$$F_{bc} = IlB.$$

The direction of the force is into the page, as shown in the figure.

The force on side *da* is

$$F_{da} = IlB \sin 90°$$

$$F_{da} = IlB$$

$$F_{da} = F_{bc} = F.$$

However, the direction of $\mathbf{F}_{da}$ is out of the page and opposite to the direction of $\mathbf{F}_{bc}$. The sum of these forces is zero, so the square loop will not be translated from its position.

Although there is no net translational force on the loop, it will rotate if free to do so. Therefore, there must be a net torque on the circuit. Recall from Chapter 7 that the torque (**τ**) on an object results when a force is applied to a moment arm at a distance r from the axis of rotation. The formula for the magnitude of torque is

$$\tau = rF \sin \theta,$$

where θ is the smallest angle between **F** and **r** when the two vectors are placed tail-to-tail. In Figure 21-26 the torque on side *bc* is

$$\tau_{bc} = rF \sin 90°$$

$$\tau_{bc} = rF.$$

The direction of the torque is found by using the *right-hand rule for torques.* The torque on side *da* is

$$\tau_{da} = rF \sin 90°$$

$$\tau_{da} = rF.$$

The magnitudes of the torques on segments *bc* and *da* are equal and in the same direction. Therefore, the total torque is

$$\tau = \tau_{bc} + \tau_{da}$$

$$\tau = rF + rF$$

$$\tau = 2rF.$$

In Figure 21-28, each segment of the square current loop is perpendicular to **B**. Therefore, all the sides experience forces of the same magnitude,

$$F = IlB.$$

The net torque is now zero, since all the forces are parallel to their respective moment arms (**r**) and the current loop does not rotate. The current loop in Figure 21-25 will tend to rotate until its plane is perpendicular to **B**. At that point it has no net torque, so it continues to rotate at a constant speed. As the loop moves through the perpendicular position and is no longer perpendicular to **B**, the torque generated by the magnetic forces on the current loop tends to return it to a perpendicular position. Because this is a restoring force, the current loop will oscillate about its equilibrium position, perpendicular to **B**.

Circular current loops have no straight segments. You can get an idea of the direction of the force or torque on a circular circuit by approximating it with a polygon. The currents in segments *ab* and *ef* in Figure 21-29 are parallel to **B**, so there is no force on these segments. The direction of the forces on the other segments can be found from the right-hand rule and Equation 21.12. They are indicated in the figure. As you can see, the circular loop will also tend to rotate.

The torque vector is parallel to the axis of rotation. For counterclockwise motion in a horizontal plane, the torque vector points up. For clockwise motion, the torque vector points down.

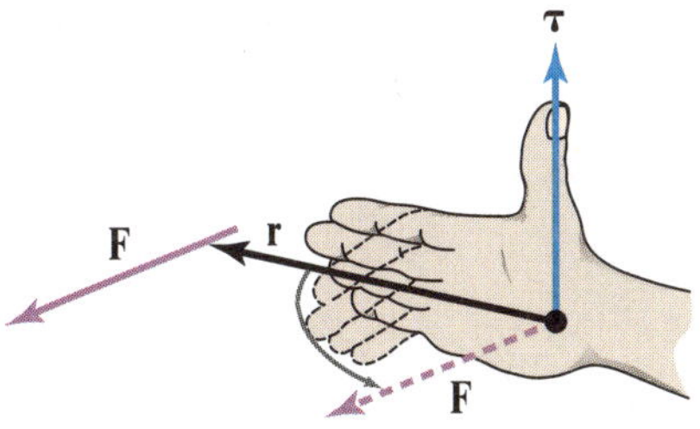

21-27 Right-hand rule for torques: for counterclockwise rotation, the torque vector points up; for clockwise rotation, the torque vector points down.

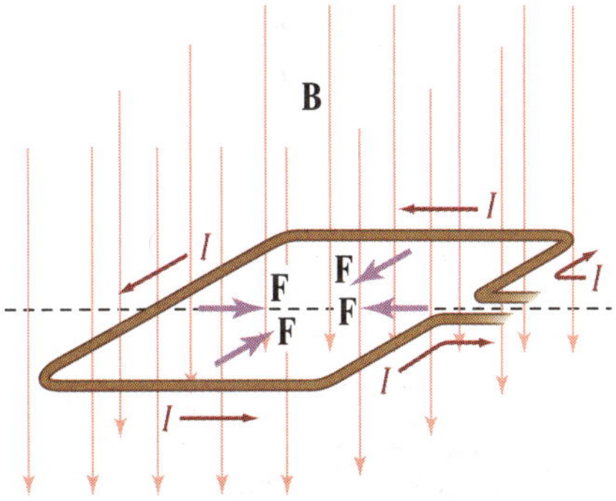

21-28 Magnetic forces on a current loop oriented perpendicular to the magnetic field

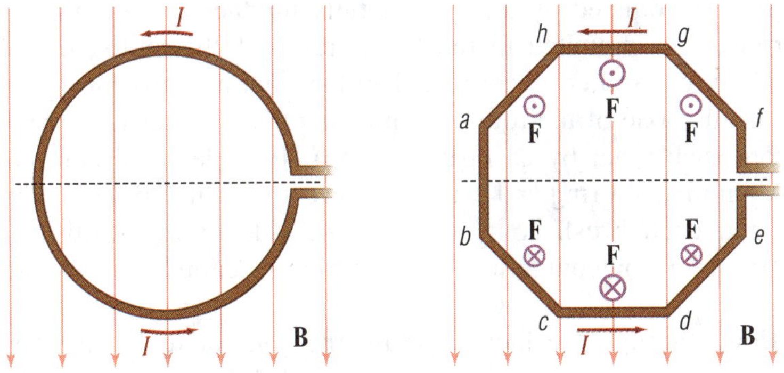

21-29 Analyzing the magnetic force on a circular current loop

Using the methods of calculus, the circular loop is broken down into innumerable segments. The forces on each segment are summed to find the total torque on the loop.

21.15 Rotating Coils

Coils of wire behave like stacks of current loops. The wire may be wound around a solid core, or it may be wound without a core. Consider a coil of *n* wraps or windings. The force of the magnetic field on this coil is the same as the force of the field on *n* current loops with the same width and height. That is, for a coil, the force on a side is

$$F_{mag} = nIlB \sin \theta \cdot \tag{21.13}$$

21-30 A coil of wire acts like a stack of individual current loops.

A *galvanometer* uses the rotation of a coil in a magnetic field to measure current. Torque is proportional to force, which is proportional to current. Therefore,

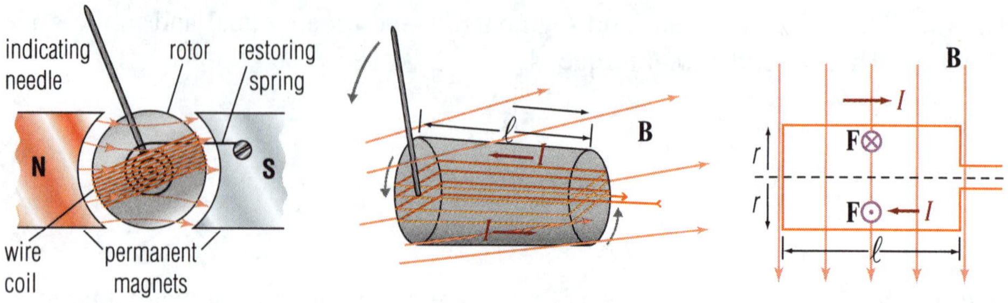

21-31 Functional diagram of a galvanometer

rotational torque can be used to measure current. The galvanometer consists of a coil wound on a cylindrical core, a spring mounted so that it opposes the turning of the coil, a needle to show the amount of turning, a scale, and a magnet with concave poles to produce a uniform magnetic field. When a current flows in the wire, a torque on the coil rotates it in the magnetic field. The rotation of the coil stretches the spring and turns the needle. The needle stops moving when the torque of the magnetic force is balanced by the opposing torque applied by the stretched spring. Greater currents cause greater deflection of the needle. Opposite currents cause deflections in opposite directions. The zero-current position is at the middle of the scale. Sensitive galvanometers contain more turns of wire in their coils in order to increase the magnetic force for a given amount of current.

21.16 DC Motors

The DC motor is, in some ways, similar to the galvanometer. The purpose of the DC motor is to convert electrical energy to mechanical work. A motor, like a galvanometer, has a coil wound on a ferromagnetic core. The core is mounted on an axle supported by bearings to reduce friction. This assembly is called the **rotor**. The coil is surrounded by a magnetic field produced by permanent magnets or electromagnets mounted in the fixed frame of the motor, called the **stator**. The DC current source is connected to the rotating coils through a split ring mounted on the axle of the rotor. (A split ring is two conducting half-rings separated from each other by an insulated gap). Two electrical contacts called **brushes** ride on the split ring and carry the current supplied to the motor. At any given instant, each brush contacts only one side of the split ring. The subassembly of the rotor containing the split ring is called a **split-ring commutator**.

When current flows through the motor circuit, the coil rotates. If the DC current were to remain constantly in one direction, the coiled winding would tend to rotate until it was perpendicular to the magnetic field, then oscillate around that position as discussed above. However, as the rotor turns, the split ring segments move with it. When the rotor has turned far enough, the split ring gap slides under each brush just as the coil becomes perpendicular to the field. Inertia carries the rotor a little farther, and the opposite commutator segments then contact the brushes. This action reverses the current flow through the rotor coils, maintaining the magnetic torque in the same direction as it was initially. The rotor continues another half turn, and the commutator again reverses the current, sustaining the torque in the correct direction. The DC motor continues to rotate, doing useful work.

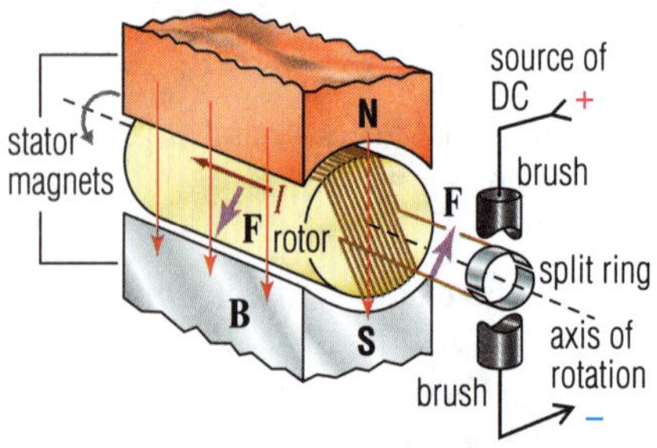

21-32 A DC motor uses a split-ring commutator to reverse the electric current in the rotating coil.

The commutator of a DC motor permits continuous rotation of the motor in one direction without having to continually change the polarity of the voltage source connected to the motor.

21C Section Review

1. What factors determine the magnitude of the magnetic force on a current moving through a straight conductor in a magnetic field?
2. Copy the following conductor loops onto your paper. Using proper symbols, indicate the direction of the magnetic force on each segment of the loop.

a.

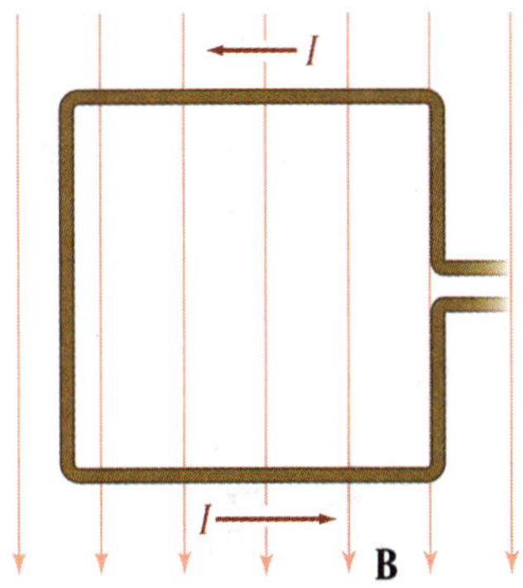

b.

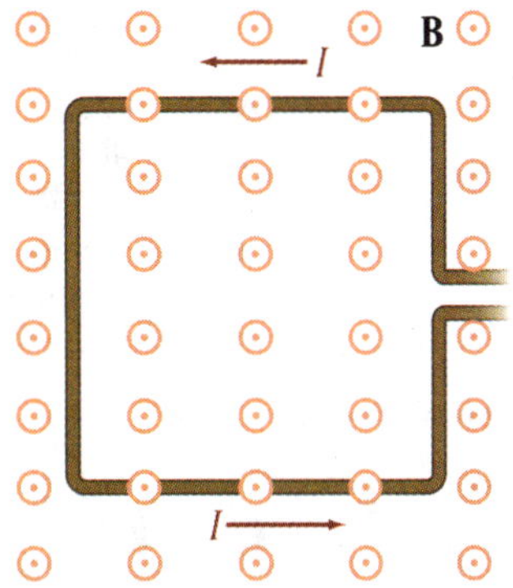

3. Which of the current loops in Question 2 will experience a torque? Explain your answer.
4. What factors determine the magnitude of the magnetic force on a coil of wire containing a current in a magnetic field?
5. The figure below shows a single current loop mounted so that it can rotate. Explain what will happen when the switch is closed.

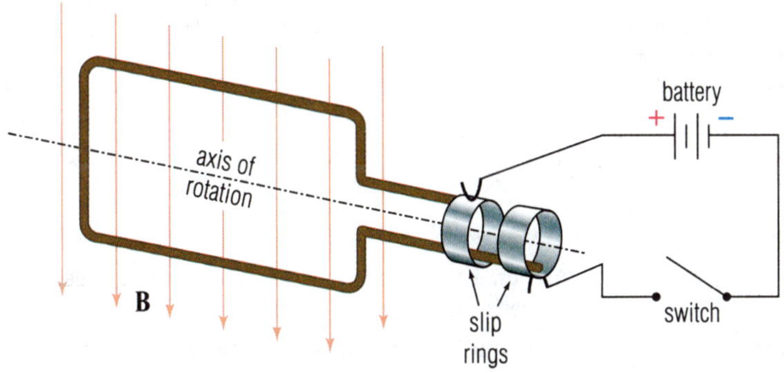

6. The figure below shows a different current loop mounted so that it can rotate. Explain what will happen when the switch is closed.

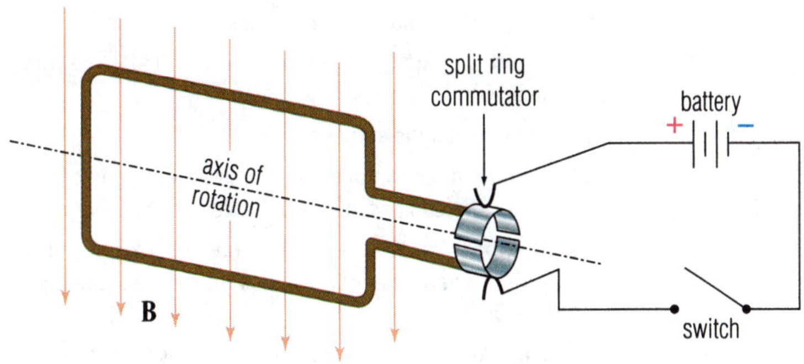

21C Objectives

After completing this section, I can

- ✓ apply the theory of magnetic force on moving charges to current-carrying conductors.
- ✓ explain how a current-carrying loop of wire in a magnetic field can generate a torque.
- ✓ use the right-hand rule for torques to predict the direction a current-carrying loop will rotate in a magnetic field.
- ✓ describe how devices containing coils of wire interacting with a magnetic field can be used to measure current.
- ✓ describe the construction and operation of a simple DC motor.

Chapter Review

In Terms of Physics

Problem-Solving Strategies

21.1 (page 480) The direction of magnetic declination at your location can be determined by imagining the meridian line connecting your position and the geographic north pole. If the magnetic north pole is to the left (west) of this line, then the declination is west; if it is to the right (east) of the line, then declination is east.

21.2 (page 482) In order to find the angle between the velocity and the magnetic field vectors, place their vectors tail-to-tail. The angle θ is the smallest angle between the vectors.

21.3 (page 492) **The right-hand rule for torques:** Hold your right hand flat with the thumb extended at a right angle to the fingers. Point your fingers away from the axis of rotation parallel to the position vector **r**. Curl your fingers toward the direction of the force vector where it is applied to the moment arm. Your thumb points in the direction of the torque vector.

Review Questions

1. Can you obtain isolated poles by breaking a long magnet in half? Explain.
2. Name one way to find the shape of a magnetic field.
3. **a.** Where is this magnet's north pole?
   b. Where is its field strongest?

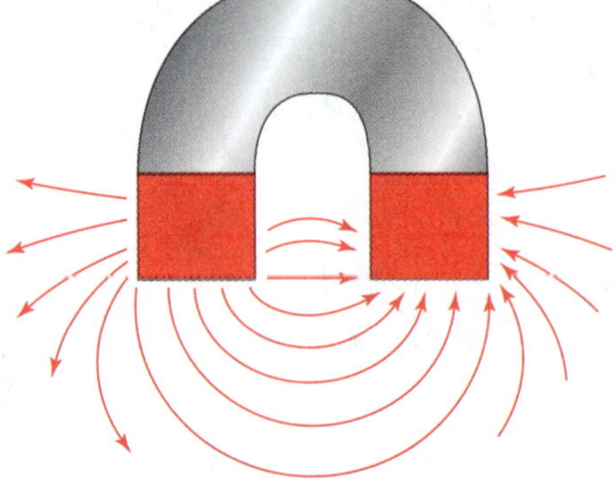

4. What vector represents a magnetic field?
5. How does a magnet attract an unmagnetized piece of ferromagnetic metal?
6. How can you distinguish ferromagnetic steel from paramagnetic aluminum?
7. Bismuth has a relative permeability of 0.99983. Is this element ferromagnetic, diamagnetic, or paramagnetic?
8. If you want a strong magnetic field, in what kind of material should you generate the field?
9. What is the best model for explaining magnetism today?
10. Why does a magnetic compass point north?
11. Are the earth's magnetic poles at its geographic poles? Explain.
12. What is magnetic declination?
13. In what direction does the north-seeking end of a compass point when the compass is placed at the north geographic pole?
14. In what direction does the north-seeking end of a compass point when the compass is placed at the north magnetic pole?
15. What keeps harmful charged particles from the sun from reaching the earth's surface?
16. What causes the auroras?

*17. If the earth's magnetic field has always been decreasing as it has been doing since men began observing it, how old might it be?

18. For a charge of a given magnitude moving at a constant velocity through a uniform magnetic field, what is the difference in the magnetic force on the charge when it is positive compared to when it is negative?

19. Study the figure below. In what direction will the charge be deflected?

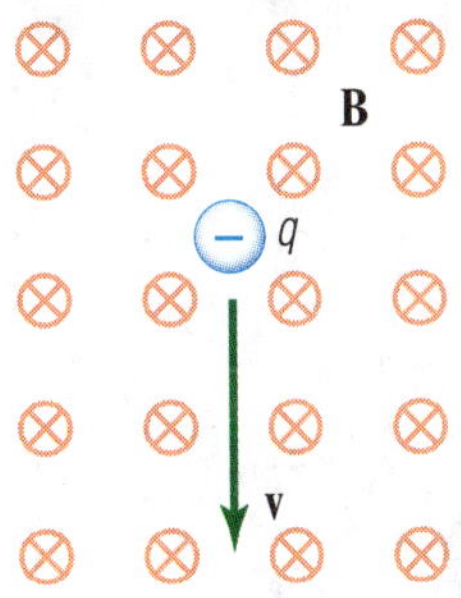

20. Describe a mass spectrometer.

21. Two kinds of particles are detected in a mass spectrometer sample. One kind of particle is detected at 9.5 cm from the entrance point. The other is detected at 13.8 cm from the entrance. Which kind of particle has greater mass?

22. A current-carrying conductor is aligned parallel to a magnetic field. In which direction is the magnetic force on the conductor?

23. A current loop lies in a magnetic field so that all of its sides are perpendicular to the field lines. Will the loop rotate? Explain.

24. A circular current loop lies in a magnetic field so that it is in a plane parallel to the magnetic field lines. Will the loop rotate? Explain.

25. Why does a DC electric motor rotate continuously rather than oscillate?

True or False (26–40)

26. All magnets have both a north and south pole.

27. One tesla is approximately the strength of the terrestrial magnetic field at the earth's surface.

28. A magnetic field is not affected in any way by the material in which it is located.

29. The strength of a magnet depends not only on the dipole moments of its atoms but also on their orientation.

30. Magnetic declination over the surface of the earth is relatively constant from one decade to the next.

31. A charge at rest in a magnetic field will be accelerated perpendicularly to the field lines.

32. A moving electric charge is deflected in the direction of a magnetic field line.

33. The closer the crossing angle between a charge's velocity and a magnetic field line is to 90°, the greater the magnetic force on the charge.

34. Magnetic force is a conservative force.

35. The right-hand rule for magnetic force is valid only for positive charges.

36. The torque vector on a current loop in a magnetic field points in the direction of rotation.

37. The formula for magnetic force in a coil of wire (Equation 21.13) assumes that all the coils have the same dimensions.

38. A galvanometer works by balancing an electromagnetic torque against a spring torque.

39. In a simple DC motor, the stator contains the electric current that drives the motor, and the rotor contains the magnetic field.

40. The purpose of a commutator is to reverse the current in the motor coil so that it can continue to turn in one direction.

DM✪41. Using the accepted mass-to-charge ratio for an electron given on page 490, determine the mass if the electron's charge is -1.602×10^{-19} C.

✪42. A $+1.00 \times 10^{-6}$ C charge moves through a magnetic field, $B = 3.00 \times 10^{-5}$ T, initially perpendicular to the field, with a constant speed of 1.33×10^{-3} m/s. What is the magnitude of the magnetic force initially exerted on the charge?

✪43. A magnetic field exerts a force of 5.50×10^{-10} N on a -1.10 mC charge initially moving perpendicularly to the field at a constant speed of 2.50×10^{-2} m/s. What is B?

✪44. A point charge moving at a constant speed of 1.50×10^{-2} m/s perpendicular to a magnetic field of $B = 3.00 \times 10^{-4}$ T experiences a force of 9.00×10^{-12} N. What is the magnitude of the point charge?

✪45. A velocity selector has a magnetic field of 3.00×10^{-4} T and an electric field of 4.50 N/C. What velocity does it select?

✪46. You want a velocity selector to allow charged particles with a speed of 2.25×10^{3} m/s to pass undeflected. The magnet is capable of producing a field with $B = 3.333 \times 10^{-4}$ T. What must the electric field strength be?

DM✪47. In the history of science, at the time when J. J. Thomson discovered that his corpuscles were less than one-thousandth the mass of the smallest atom, what impact did his discovery have on the model of the atom? What were some new scientific questions that came out of his discovery?

✪48. Prove that the unit ratio $\frac{\text{N·s}}{\text{C·m}}$ equals the tesla (V·s/m^2). This combination of units occurs when solving for the magnetic field (B) by rearranging the magnetic force equation (21.12).

✪49. If there are two kinds of particles detected in a mass spectrometer, such that one kind is twice as far from the beam entrance as the other, how do their masses compare? (Assume that both kinds of particles have the same velocity and the same charge when they enter the spectrograph.)

Alternating current transformers are essential for the transmission of electrical power over great distances.

Electromagnetism 22

Dominion Science Problem

An Electrifying Idea

On September 4, 1882, Thomas Edison flipped the switch in his new electric generating facility, providing electricity for the first time in history to customers in Manhattan. This electricity was relatively low-voltage direct current (DC), the type of electricity upon which Edison had focused his experimentation and his reputation. This type of current works well for light bulbs and simple motors, the most common use of electricity at the time. But the system had its disadvantages. Because DC voltage cannot be easily adjusted to compensate for different sizes of loads, much larger wires with heavier insulation had to be run from the power generating station to supply factories, with their large industrial motors. The electrical resistance of long transmission wires caused large voltage drops, requiring power plants every few miles. This added to the expense of maintaining and operating the DC current electrical grid. Using higher DC voltages reduced some of the problems, but changing voltage at customer locations involved large resistors or step-down motor-generators that wasted electrical energy and generated huge amounts of heat. In addition, this tangle of wires and power plants spelled danger. During the Great Blizzard of 1888, energized DC power lines in New York City collapsed, killing several people. How could electricity be economically and safely delivered to customers?

22-1 This photo shows the tangle of electrical and telephone wires brought down by the Great Blizzard of 1888.

22A CURRENTS AND MAGNETIC FIELDS

Recall that conventional current is opposite to the flow of electrons in a wire.

22.1 Magnetic Field Around a Conductor

As Oersted discovered in 1820, moving charges produce magnetic fields. The magnetic field generated by charges moving in a long, straight wire can be graphically represented by concentric circles around the wire. (Of course, each end of the long wire must be connected to a source of potential difference. If the connections are far from the middle of the wire, then the magnetic field near the wire at its middle is like that of an isolated wire.) You can find the direction of the magnetic field vector (**B**) near a wire by a new right-hand rule. Point your right thumb in the direction of the (conventional) current flowing through the wire and wrap your fingers around the wire (mentally, of course, unless the wire is insulated). Your curled fingers point in the direction that the field circles the wire. Figure 22-3 on the next page shows the direction of **B** as it relates to your fingers. This illustrates the magnetic field vector **right-hand rule for conductors**.

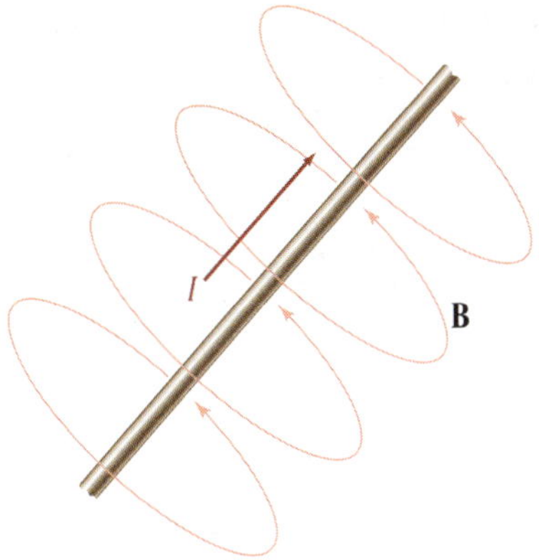

22-2 The magnetic field surrounding a current-carrying conductor

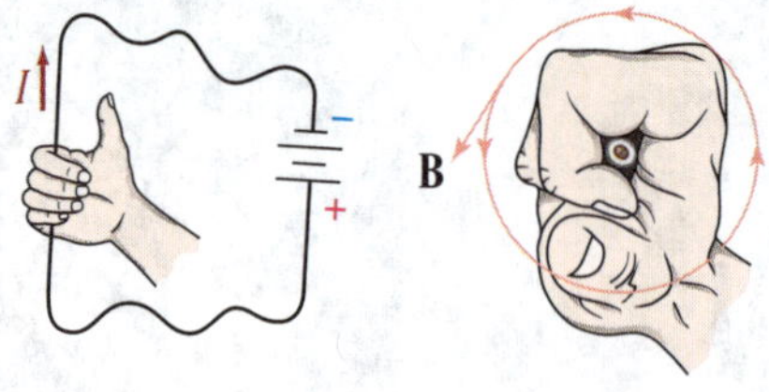

22-3 The right-hand rule when used to determine the magnetic field vector **B**

Jean-Baptiste Biot (1774–1862) was a French mathematician and scientist who investigated extensively in astronomy, cartography, geometry, thermodynamics, and magnetism. He held a number of prominent chairs in academic institutions and was even involved in a short-lived political rebellion.

Felix Savart (1791–1841) was a French professor who experimented with magnetism and acoustics. With Biot, he discovered that a magnetic field generated by a current-carrying conductor was proportional to the inverse square of the distance from the conductor.

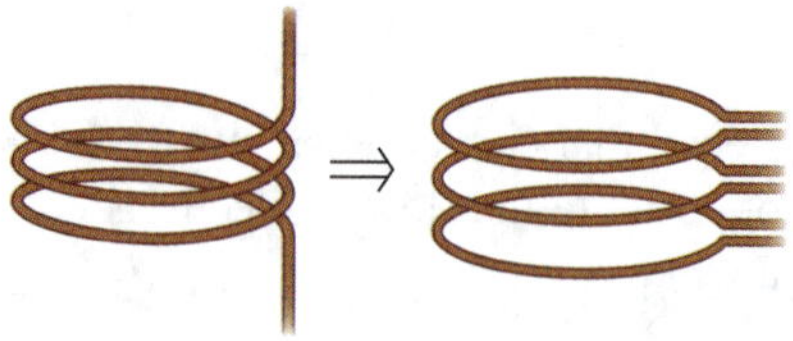

22-5 A solenoid coil is like a stack of individual current loops.

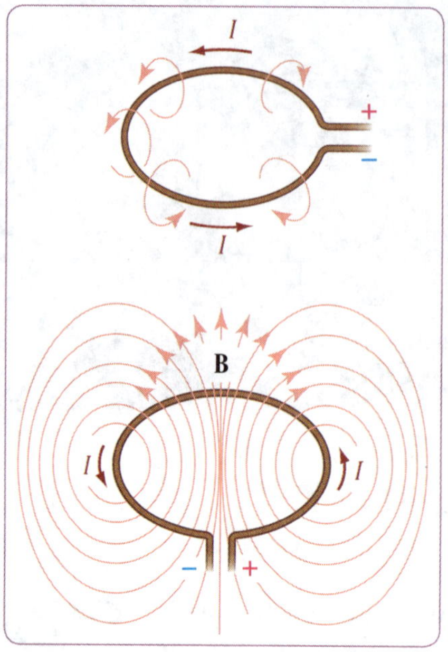

22-4 A circular current loop and its magnetic field

A circular current loop, such as the one in Figure 22-4, can be mentally divided into many small, nearly straight segments. Each segment contributes to the net magnetic field, and the field is simply the sum of the contributions from all segments. The magnetic field of each segment points straight up inside the loop and straight down on the outside. The entire magnetic field of a circular loop looks like the magnetic field of a bar magnet. Therefore, the magnetic field of a current loop is said to be a *magnetic dipole.*

The magnetic induction vector at a point in space produced by current flowing through a very short segment (Δl) of a long wire is given by the **Biot-Savart law**:

$$\Delta B = \frac{\mu_0}{4\pi} I \frac{\Delta l \sin\theta}{r^2} \qquad (22.1)$$

The **permeability of free space (μ_0)** is a measure of how effectively the magnetic field penetrates a vacuum. The angle θ is the angle between the wire segment and a line connecting the center of the segment with the point in space. The term r is the distance from the center of the segment to the point. The magnetic field vector (**B**) at the point representing the total field of the current loop is the sum of the magnetic field vectors (**ΔB**) at that point contributed by each tiny segment (Δl).

The Biot-Savart law is important because it demonstrates that the magnetic field generated by a small segment of a conductor is similar to a gravitational field generated by a small mass and to an electric field generated by a point charge.

- Each is proportional to a constant related to the kind of field.
- Each is an inverse-square law.

22.2 Electromagnets

A single turn of a current-carrying wire can act as a type of magnet called an **electromagnet**. A **solenoid** is a series of wire coils used for this purpose. If the solenoid is wound tightly, it resembles a stack of independent current loops. The magnetic field of a solenoid with n turns is n times as strong as the magnetic field of a single current loop. Like the magnetic field of a single current loop, a solenoid's magnetic field is shaped like a bar magnet's field. This fact supports the model that bar magnets are magnetized by atomic currents.

22.3 SI Definition of the Ampere

The magnetic field of one long, straight wire will interact with the magnetic field of another long, straight wire. If the currents in the wires are moving in the same direction, the wires will attract each other. If the currents are in opposite directions, the wires will repel each other. This interaction occurs because each wire is positioned in the other's magnetic field. Use the magnetic field vector right-hand rule to determine the direction of the field due to the other wire at each wire's location. Then use the first right-hand rule from Chapter 21 to determine the force on the wire. Point your thumb in the current direction and your index finger at a right angle to your thumb, parallel to the field vector, **B**, *from the other wire.* The middle finger, held at a right angle to the other two fingers, points in the direction of the force on the wire. In Figure 22-6, $\mathbf{F}_{1\to2}$, which represents the force of wire 1's magnetic field on wire 2, points toward wire 1. Similarly, the force of wire 2's magnetic field on wire 1, $\mathbf{F}_{2\to1}$, points toward wire 2. The wires are drawn together.

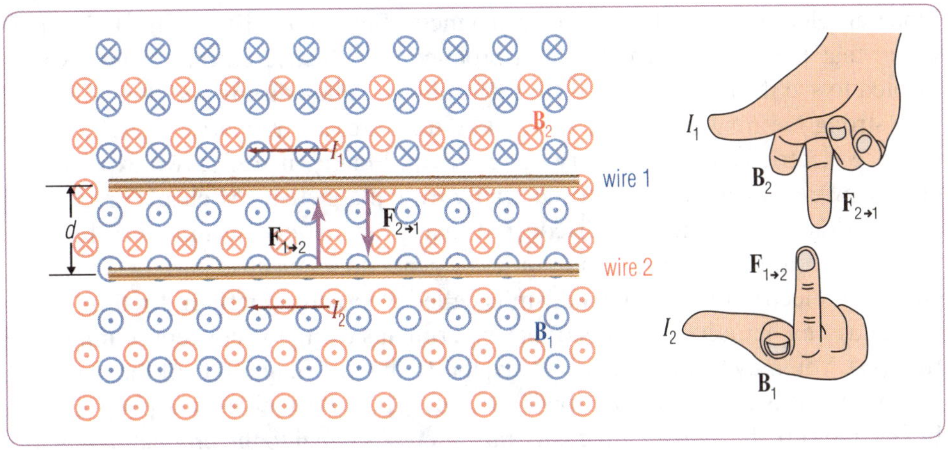

22-6 Forces between parallel current-carrying conductors

Each wire exerts on the other wire a force of

$$\frac{F_{\text{mag}}}{l} = \frac{kI_1I_2}{d},$$

where d is the distance between the wires and k is a constant related to the permeability of free space. This force gives a practical measurement of the ampere. The ampere was defined in Chapter 20 as the current in a conductor in which 1 C of charge flows past a given point in 1 s. However, a coulomb, defined as 6.24×10^{18} electron charges, is difficult to measure. To avoid this problem, the ampere has been redefined as follows: "1 A is the amount of current that, when it exists in each of two infinitely long wires separated by a distance of 1 m, causes each wire to exert on the other a force of 2×10^{-7} N for each meter of length." This is the present-day SI definition of the ampere. The forces can be measured accurately on a current balance (Figure 22-7). The coulomb has been redefined as the amount of charge that flows in 1 s past a point in a conductor carrying a current of 1 A.

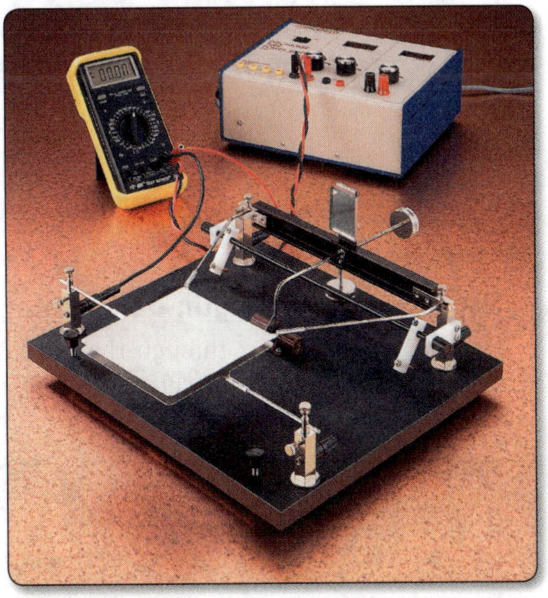

22-7 A current balance

22.4 Electromagnet Applications

Most magnets used today in industry are electromagnets, usually solenoids wound on ferromagnetic cores. Using a ferromagnetic core greatly concentrates a solenoid's magnetic field. Since electromagnets that have magnetic fields far stronger than those of permanent magnets can be constructed, they are used when strong magnetic fields are required. Electromagnets are also used when a continuous magnetic field is not desired. Doorbells, for instance, use electromagnets to ring their bells, and tiny electromagnets are used to read and write information on digital magnetic media.

22.5 Electromagnetic Induction

Since an electric current produces a magnetic field, scientists in the 1820s reasoned that perhaps a magnetic field produces an electric current. They experimented to see if they could prove this speculation. They discovered that no matter how strong the magnet or how conductive the wire, a stationary magnet will not produce electricity in a stationary current loop. Finally, in 1831, two experimenters discovered a way to generate electricity from magnetism. Michael Faraday and Joseph Henry independently discovered that a magnetic field must be changing with respect to the current loop in order to generate an electric current. This generation of electricity from magnetism is called **electromagnetic induction.**

Joseph Henry (1797–1878), an American physicist, was one of the founders of the National Academy of Sciences and the first Secretary of the Smithsonian Institution.

Faraday explained the results of his experiments using the concept of **magnetic flux (ϕ)**. You may remember from Chapter 21 that the magnitude of the magnetic field vector **B** is the *magnetic flux density* (B). Thus, **B** involves flux per unit area. In fact, the flux through an imaginary planar area is equal to the product of the area and the perpendicular component of **B** passing through the area:

$$\phi = \mathbf{B}_{\perp}A \qquad (22.2)$$

If **B** is not uniform throughout the region, the region must be divided into smaller areas for which **B** is approximately uniform. The total flux through the region is the sum of the fluxes through the smaller areas. For a coil, the **total magnetic flux (Φ)** is equal to the flux (ϕ) through each loop multiplied by the number of loops (n).

Magnetic flux is measured in webers (Wb). From Equation 22.2 we see that

$$1 \text{ Wb} = 1 \text{ T}\cdot\text{m}^2, \text{ but}$$

$$1 \text{ T} = 1\frac{\text{V}\cdot\text{s}}{\text{m}^2}, \text{ so}$$

$$1 \text{ Wb} = 1\frac{\text{V}\cdot\text{s}}{\cancel{\text{m}^2}}\cdot\cancel{\text{m}^2} = 1 \text{ V}\cdot\text{s}.$$

The unit of magnetic flux is the weber (Wb):

$$1 \text{ Wb} = 1 \text{ V}\cdot\text{s}$$

Wilhelm Weber (1804–91) was a German physicist whose work helped Maxwell develop the electromagnetic theory of light.

EXAMPLE 22-1

Magnetic Field Strength: Determining Magnetic Flux

Find the magnetic flux through the current loop in Figure 22-8. The magnetic field (**B**) is perpendicular to the loop, and $B = 5.00 \times 10^{-2}$ T. The loop has a radius of $r = 0.100$ m.

Solution:

Use Equation 22.2:

$$\phi = B_{\perp}A$$

The vector **B** is perpendicular to the loop, so

$$B_{\perp} = B = 5.00 \times 10^{-2} \text{ T}.$$

The area of the loop is

$$A = \pi r^2 = \pi(0.100)^2$$

$$A \cong 0.031\underline{4}1 \text{ m}^2.$$

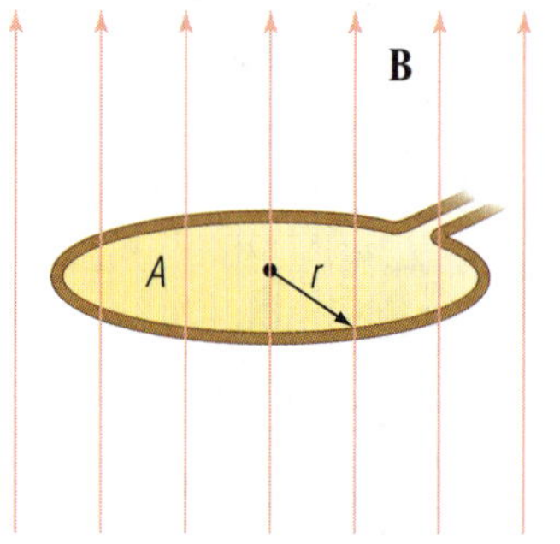

22-8 Magnetic flux depends on the strength of the field through a certain area of space.

Therefore, after substituting into Equation 22.2, ϕ is

$$\phi = (5.00 \times 10^{-2}\ \text{T})(0.031\underline{4}1\ \text{m}^2)$$

$$\phi \cong 1.5\underline{7}0 \times 10^{-3}\ \text{V·s}.$$

Since 1 V·s = 1 Wb, the flux is

$$\phi \cong 1.57 \times 10^{-3}\ \text{Wb}.$$

22.6 Faraday's Law

Faraday's experiments showed that a difference in electrical potential is induced in any current loop through which the flux changes. This result is called **Faraday's law**. Mathematically,

$$V_i = -\frac{\Delta\phi}{\Delta t}, \qquad (22.3)$$

where V_i is the induced potential difference and $\Delta\phi$ is the change in flux during a time Δt. The term *induced potential difference* is sometimes replaced by *induced emf* in order to avoid confusion with the voltage drop across a circuit or circuit component.

Problem-Solving Strategy 22.1

Recall that *emf* is not a true Newtonian force. Emf is used for describing potential differences that are induced in an electrical circuit by non-electrical sources, such as magnetic fields.

What causes the induced current? A conductor has free electrons. When the conductor is moving through a nonuniform magnetic field, its free electrons are affected just as individual moving negative charges would be. For instance, the electron in Figure 22-9 is deflected as the figure shows. As it moves, it repels free electrons in the left side of the loop. The other electrons in the left side are deflected in the same direction as the electron shown. This deflection and repulsion establishes a conventional current (of positive charges) in the direction opposite to the electron flow, according to the right-hand rule for magnetic force.

In Figure 22-9, the conductor is moving to the right, which is the original motion of the charge in the magnetic field. Only after the magnetic force thus generated acts on the charge and deflects it according to the right-hand rule does a current become established in the conductor.

One way the magnetic flux can be changed is by motion. A changing magnetic flux may be caused by moving the current loop through a nonuniform magnetic field, as in Figure 22-9, or by moving a magnetic field through a current loop. Both methods have the same deflecting effect. Only relative motion matters.

A second way the magnetic flux can be changed is by changing the current. A changing current produces a changing magnetic field, and this changing magnetic field has the same effect as motion. A current loop connected to a source of changing potential difference can induce a potential difference in another current loop. Figure 22-10 shows a possible arrangement for one current loop to induce current in another. At the instant the switch is closed, the magnetic field of the first loop expands and "moves out" through the second loop. This change of the magnetic field generates a current in the second loop. When the current in the first loop reaches its steady-state value, the magnetic field stops changing. Then the current in the second loop decays to zero. At the instant that the switch is opened, the magnetic field of the first loop collapses, moving past the second loop. This change generates another current in the second loop, but in the opposite direction to the earlier current.

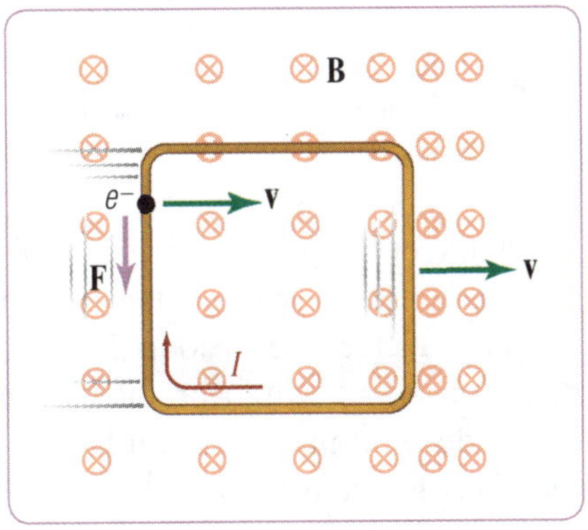

22-9 An electron in a conductor is deflected as it moves through a magnetic field.

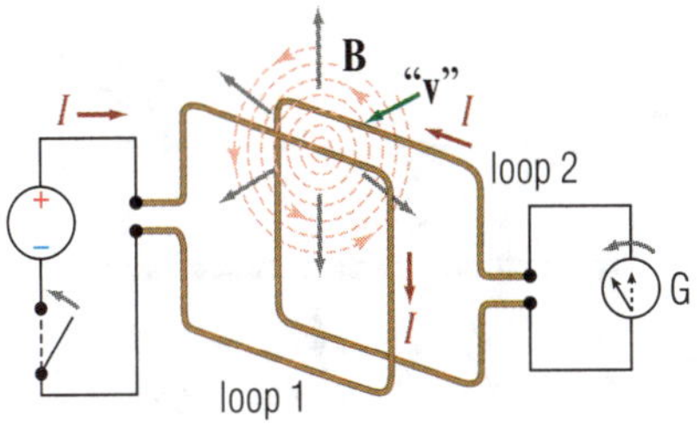

22-10 One current loop can induce a current in another loop.

Heinrich F. E. Lenz (1804–65) was a Russian physicist who discovered the relationships between induced magnetic fields, voltage, and current when a conductor passes through a magnetic field.

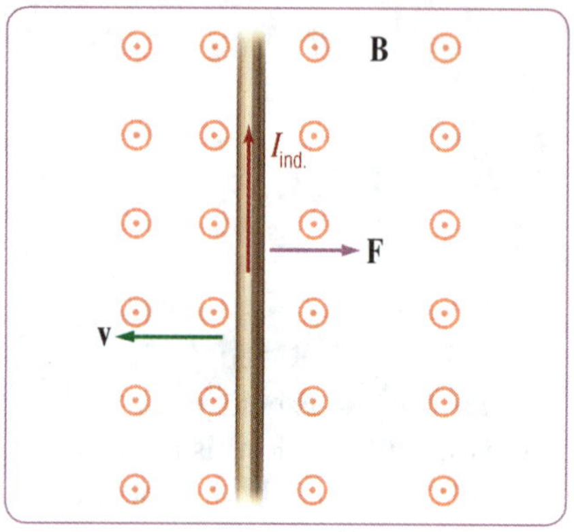

22-11 Lenz's law as it applies to a simple, straight conductor

Use the magnetic field vector right-hand rule with Faraday's law to determine the direction of the induced current in loop 2. Use the right-hand rule for conductors to determine the direction of the magnetic field around loop 2 and its inductive effect on the other loop.

A fairly continuous alternating current could be induced in the second loop by rapidly opening and closing the switch in the first loop, causing a continuously changing magnetic field.

22.7 Lenz's Law

The direction of an induced potential difference, or emf, is described by **Lenz's law**: the emf always opposes the change in the magnetic field that caused it. Consider a wire segment moving through a nonuniform magnetic field. The magnetic flux around the wire changes with time. The change in magnetic flux induces a potential difference across the segment. The direction of the induced potential difference is such that the magnetic field exerts a force on the segment opposite to the direction of the segment's motion. The magnetic force on the induced current will halt the segment unless the segment is pushed or pulled along.

Another example of Lenz's law is the two-current-loop system we studied earlier. Figure 22-12 is a detail of the two current loops from Figure 22-10. In figure (a), the switch supplying loop 1 has just closed, so the current (I_1) and therefore the magnetic field in loop 1 rapidly increase to their steady-state value. The increasing magnetic field ($\Delta\mathbf{B}_1$) induces a potential difference in loop 2. The potential difference produces a current (I_2) opposite to the current in loop 1, which generates a changing magnetic field ($\Delta\mathbf{B}_2$). This changing magnetic field tends to induce a current in loop 1 in the direction opposite to I_1, which opposes the change in the magnetic field ($\Delta\mathbf{B}_1$) that caused I_2 in the first place. Therefore, the temporary induced magnetic field in loop 2 opposes the buildup of the magnetic field in loop 1.

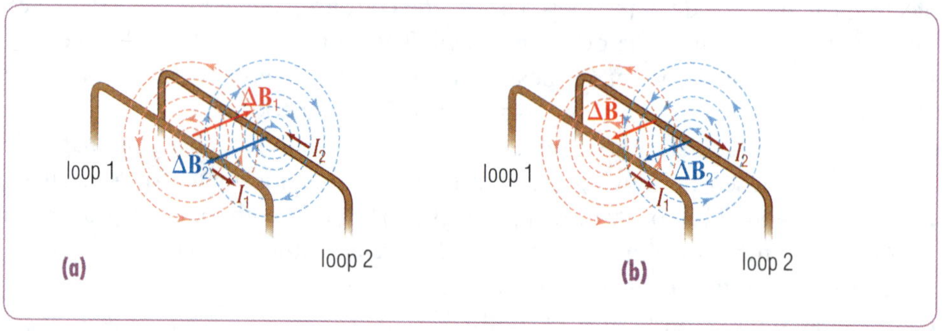

22-12 Lenz's law for induction in two loop circuits

On the other hand, in figure (b) the switch has just opened. The current (I_1) and therefore the magnetic field of loop 1 decrease. This change ($\Delta\mathbf{B}_1$) induces a potential difference in loop 2, which causes a changing current (I_2) in the *same* direction as I_1, which produces a changing magnetic field ($\Delta\mathbf{B}_2$). The induced magnetic field in loop 2 reinforces the dying current in loop 1 and thus opposes the change of the field from loop 1 that caused I_2; in other words, it opposes the *decrease* in the original magnetic field. Therefore, the temporary induced magnetic field in loop 2 opposes the decay of the field around loop 1.

Lenz's law is a natural consequence of the conservation of energy. Study Figure 22-13: a wire segment connected by long wires to a galvanometer moves through a nonuniform magnetic field. If the induced current were in the direction shown in figure (a), the force of the magnetic field on the current would be in the direction of the segment's motion. The force would continue to move the segment even if it were not pulled along. The continuing motion would continue to generate

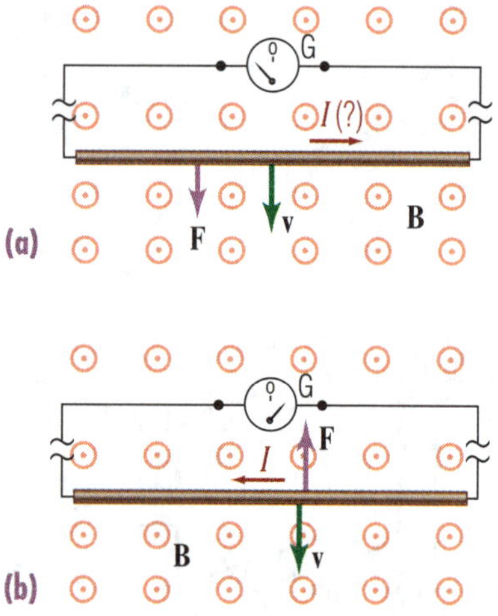

22-13 Lenz's law is a consequence of the conservation of energy.

electricity and would produce a perpetual motion generator (an impossibility). The conservation of energy requires that Lenz's law be obeyed. The actual direction of the current and the resulting force on the wire are shown in figure (b). Energy must be supplied to keep the segment moving and the induced current flowing.

22.8 Eddy Currents

When a conductor moves through a magnetic field, a potential difference is induced. If the conductor is a wire, the current flowing through it can travel in only one path since the wire has only one significant dimension and is small in all others. In a conductor other than a wire, such as a metal plate or bar, the current has many paths available to it. The current will divide into **eddy currents**. This name is derived from the current path's resemblance to water eddies around a floating object. Figure 22-14 shows eddy currents in a conducting plate within a changing magnetic field.

The magnetic field exerts a force on eddy currents within the material to oppose the conductor's motion. Therefore, eddy currents have a braking effect on a solid conductor's motion. A rotating conductive disk can be brought to a standstill by impressing a magnetic field on it, which induces eddy currents. The eddy currents, like friction, dissipate the kinetic energy of the plate as thermal energy.

Eddy currents are usually undesirable in electrical equipment. The strategy used to minimize eddy currents is to break up the area or volume of the conductor by interspersing conductors and insulators. In disks or plates, slits provide these breaks. Air is not a good conductor, and so the slits restrict the eddy currents. In solenoids and other coils, eddy currents in the core are a problem. Therefore, cores are laminated in a direction that is perpendicular to potential eddy currents. Thin strips of iron or other ferromagnetic (and conducting) materials are alternated with thin insulating strips to restrict the paths of eddy currents.

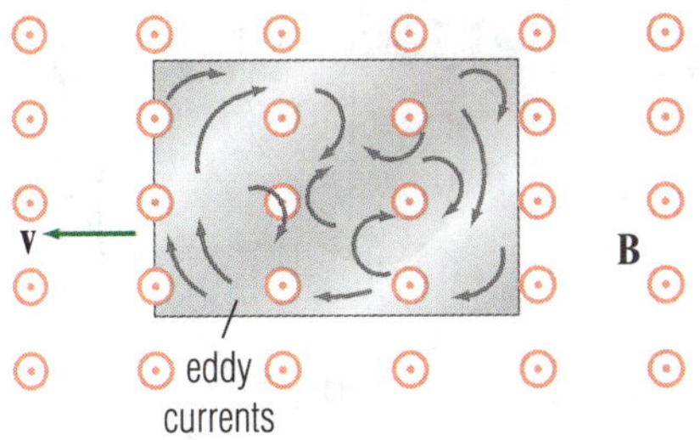

22-14 Eddy currents generated in a metal plate

22-15 Transformers use laminations of thin strips of lacquered metal to minimize eddy current formation.

22A Section Review

1. Name and describe the rule for determining the orientation of the magnetic field around a straight conductor carrying a continuous current.
2. What is the significance of the Biot-Savart law?
3. a. To what could you compare the shape of the magnetic field produced by a current-carrying loop or coil of wire?
   b. What is the name given to such a magnetic field?
4. a. What happens to the charges within a conductor placed in a magnetic field that changes strength with time?
   b. What law describes this observation?
5. According to Lenz's law, what happens when a magnetic field changes around a conductor?
6. How can you change the magnetic field surrounding a conductor (i.e., produce a $\Delta\phi/\Delta t$)?
7. a. What occurs in large, thick conducting materials when a magnetic field passing through them changes?
   b. How can this effect be minimized?
8. What diameter of a circular current loop would be required in order to produce an average magnetic flux of 2.50×10^{-5} Wb if the average field generated perpendicular to the loop is 6.30×10^{-2} T?

22A Objectives

After completing this section, I can

✓ use the right-hand rule for conductors to determine the direction of the magnetic field around a conductor.

✓ show that magnetic field strength obeys an inverse-square law.

✓ explain how the ampere is defined as a function of the interaction of two magnetic fields in a current balance.

✓ describe electromagnetic induction according to Faraday's and Lenz's laws.

✓ describe how eddy currents form and why they are usually detrimental.

✓ illustrate a way to minimize eddy currents in conductors.

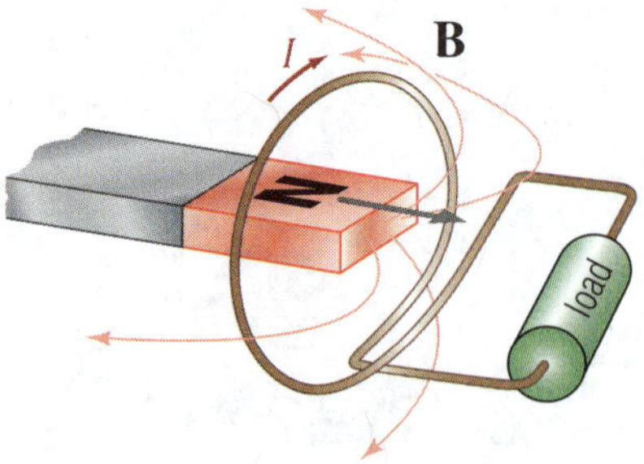

22-16 Generating a current with a magnet

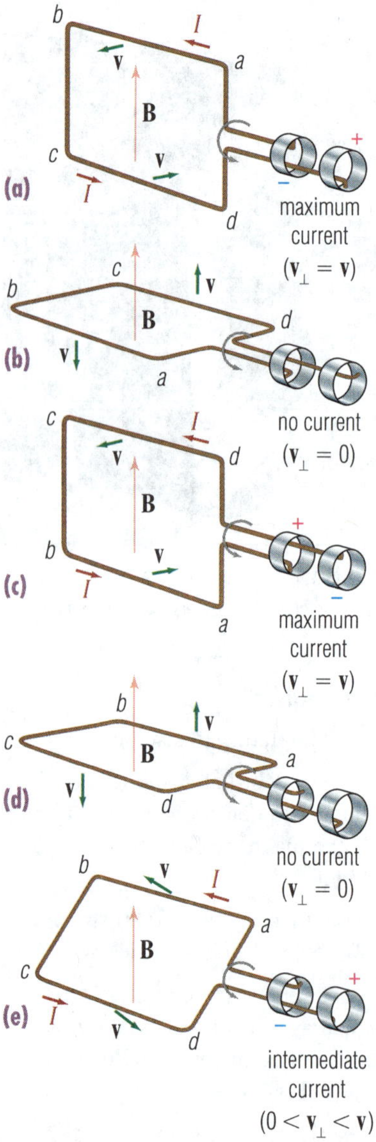

22-18 Steps in the production of alternating current

22B ALTERNATING CURRENT

22.9 Generating Alternating Current

There are many ways to generate electricity using electromagnetic induction. A simple demonstration of a generator is a magnet and a loop of wire. The magnet is pushed in and out of the loop. This changing magnetic field induces a current in the loop. The current in this generator changes directions (alternates) during each motion. It tends to oppose the magnet's field when the magnet is approaching the loop and to reinforce the magnet's field when the magnet recedes.

The most common type of generator works by rotating a wire loop (really multiple coils) in a magnetic field (Figure 22-17). A rotating coil produces alternating current. **Alternating current (AC)** is electricity that is characterized by the periodic change of voltage polarity and current direction within the circuit. Figure 22-18 shows why the current alternates. When the wire loop rotates through the magnetic field in the direction of **v**, a clockwise current is induced in the loop. As the loop rotates, its orientation with respect to the magnetic field changes. In figure (a) the induced current flows through the loop from *a* to *b* to *c* to *d*. When the loop has rotated 180°, as figure (c) shows, the current (which is still clockwise) moves from *d* to *c* to *b* to *a*. Since each end of the current loop is always connected to the same side of the load circuit through the slip rings, the current in the load circuit alternates.

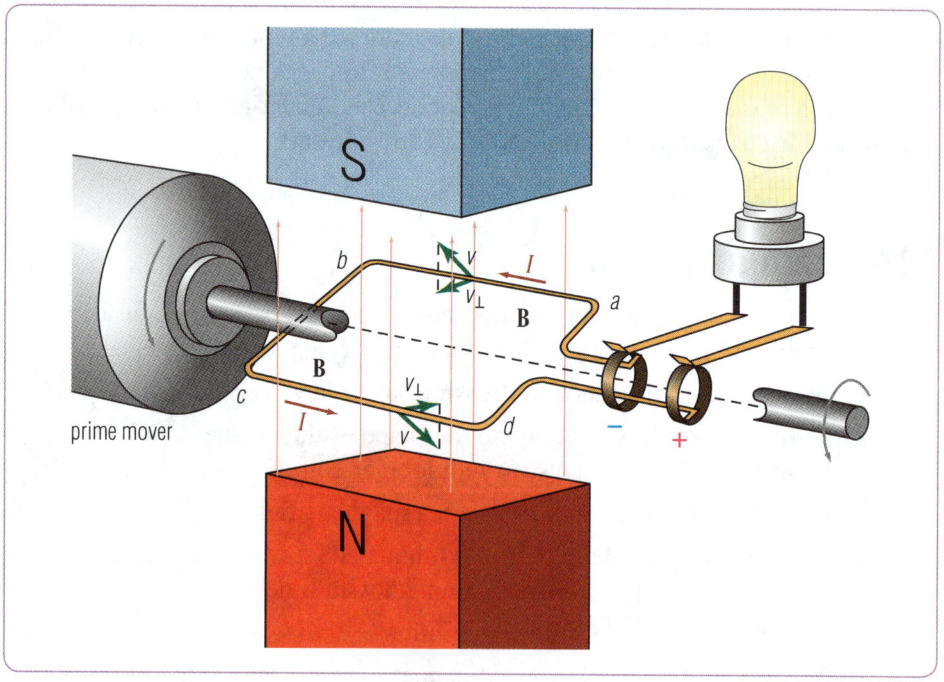

22-17 An AC generator

The rotating coil-core assembly of the generator is called the **armature**. The rotation of the armature coil in a generator is the result of an input of mechanical energy, which may come from one of several sources. Commercial generators often use steam turbines to turn the coils. Coal, natural gas, and nuclear reactions are used to produce steam to turn the turbines. Hydroelectric generators use falling water instead of steam, and wind turbines connected to generators use the wind to turn the coil. Highly efficient gas turbines are being used more for new power plants. Emergency electrical generators are often powered by diesel and gasoline engines.

The potential difference and current produced by any generator change in magnitude as well as in direction. The current is caused by the magnetic force on a moving charge. From Chapter 21, you learned that the magnitude of this force is

$$F_{mag} = |q|vB \sin \theta.$$

As the loop rotates, θ changes and F_{mag} has the form of a sine wave. Similarly, the current and potential difference produced by F_{mag} have sine-shaped graphs. Figure 22-19 is a graph of alternating current. The labeled points correspond to the steps in Figure 22-18.

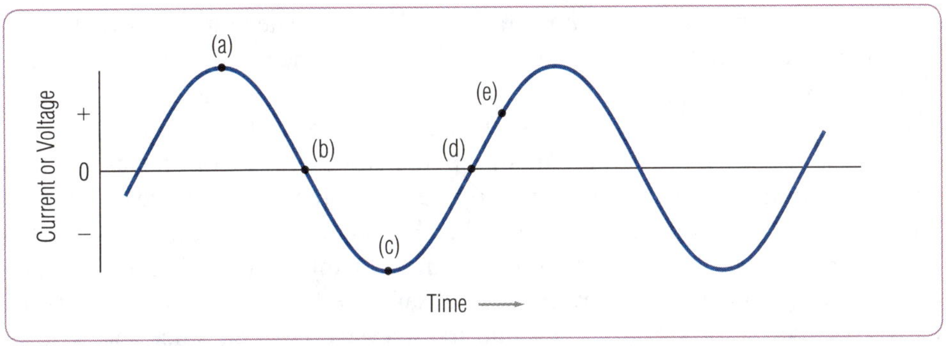

22-19 Generated alternating voltage and current versus time

Since the potential difference varies, it is helpful to know the *effective* or *rms* potential difference. The rms potential difference is the equivalent DC (nonvarying) potential difference that would produce the same power that the AC potential difference does. For a potential difference that varies as a sine function of time, the rms potential difference is

$$V_{rms} = V_{max}/\sqrt{2}, \quad (22.4)$$

where V_{rms} is the rms potential difference and V_{max} is the peak (maximum) AC potential difference. Similarly, the effective or rms current is

$$I_{rms} = I_{max}/\sqrt{2}. \quad (22.5)$$

It is somewhat difficult to compute the energy, work, and power associated with any physical quantity that varies with time according to a sine or cosine function. A more useful quantity is the average (or mean) value of the varying quantity. Scientists and engineers therefore often determine the **root mean square (rms)** values of alternating current or voltage by squaring the maximum value (a square is always positive), dividing the squared quantity by 2 (finding the square's mean), and then taking the square root to find the root mean square.

A single sine waveform is not always the best form for delivering electricity, especially for large electrical motors and other loads. Three-phase generators, which have armatures containing three separate coils, deliver a given amount of electrical power in a more compact way. Three wires, each of which carries an alternating current that is 120° "out of phase" with the currents in the other wires, connect the generator to the loads. Peak values of current are produced whenever a rotor coil is aligned in the same direction with the magnetic field. In a three-phase generator, this alignment occurs three times in every rotation (every 120°). The induced emf for the entire generator is three out-of-phase sine waves that have been superimposed (Figure 22-21).

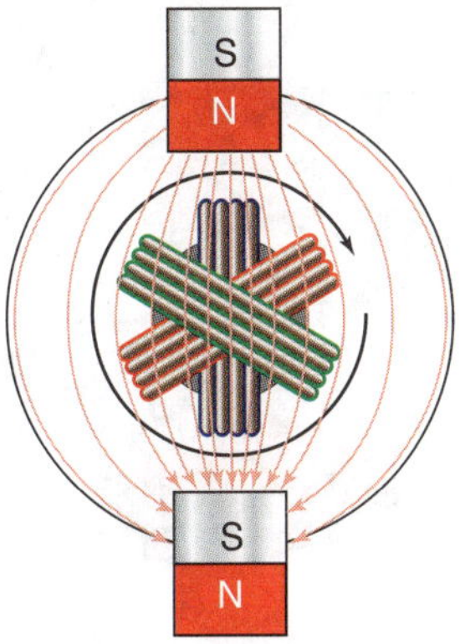

22-20 A three-phase AC generator

The circuit symbol for a source of DC potential difference consists of a generic voltage source or one or more voltaic cells. Because alternating current is usually produced by a rotating generator, AC voltage sources are given the symbol -⊙- in electrical schematics.

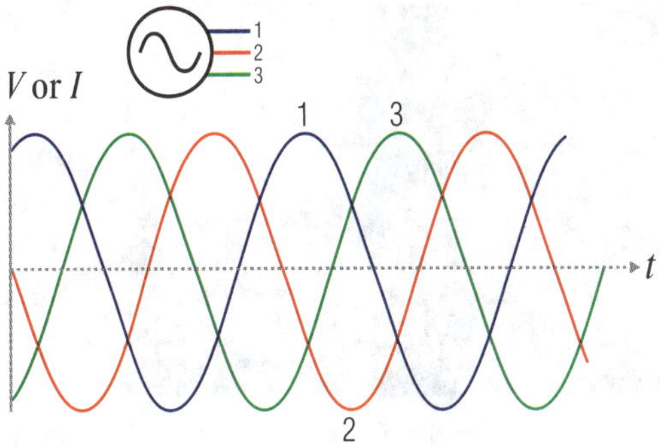

22-21 The voltage or current output of a three-phase AC generator

22.10 Direct Current Generation

The current generated in a rotating coil is always alternating current. In spite of this fact, it is possible to make a DC generator. For a DC motor, DC was turned into AC within the rotating coil by the use of a split-ring commutator. The same device can convert the AC in the generator's coil to

There are special electrical circuits that can "condition" the pulsating DC output to establish a nearly constant DC voltage. The easiest way to generate DC is to use an AC generator and electronically convert the AC output to a nearly steady DC voltage. The alternator in your car is probably such a generator.

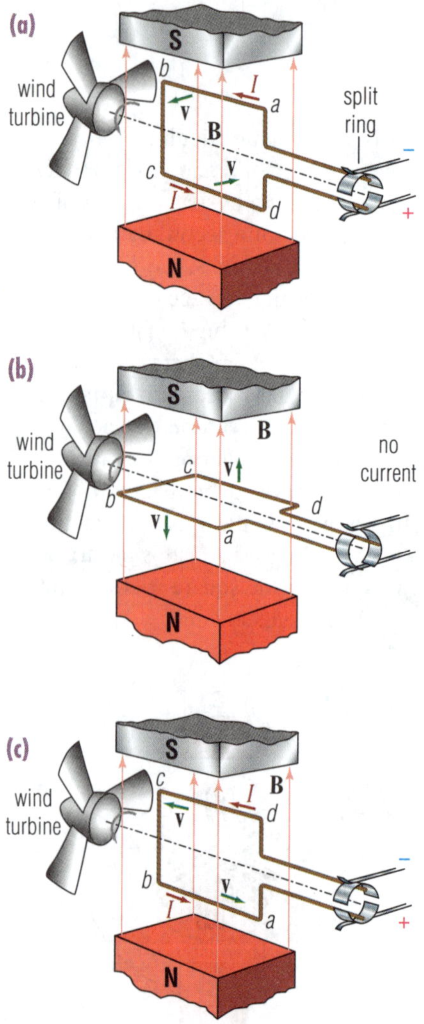

22-22 Steps in the production of pulsating direct current

22-24 Transformers like this supply the 110/220 V AC to residences in the United States.

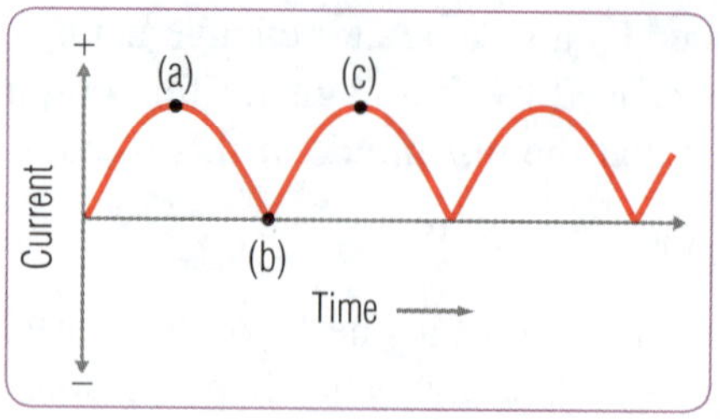

22-23 Pulsating direct current versus time

DC at its output terminals. The current coming from the commutator will no longer change its direction, but it will still change its magnitude. The split ring is connected to the coil, and the output terminals are connected to the commutator by brushes. When the coil has turned far enough so that its current is reversed in the magnetic field, the brushes are in contact with the opposite sides of the split ring. The current therefore continues in the same direction. The output current, which is plotted against time in Figure 22-23, is called **pulsating DC**.

A motor, a device that converts electrical energy to mechanical work, is fundamentally a generator run in reverse. Instead of the armature being rotated to generate electricity, it is supplied electricity that causes it to rotate. Since the armature of a motor rotates in a magnetic field, a potential difference, or emf, is induced across the armature. This induced emf opposes the potential difference supplied to the motor (Lenz's law); therefore, it is called **back emf (ε)**. The current generated by the back emf reduces the total current in the armature by opposing the supplied current. This "coincidence" provides several safety features. Back emf reduces the danger of the motor's overheating due to excessive current, since the wire in the armature winding presents little resistance to current flow. It also prevents the motor from increasing in angular speed unchecked until it flings itself apart.

22.11 Using Alternating Current to Solve Problems

In 1884, Edison hired a brilliant young Serbian engineer named Nikola Tesla. Previously, Tesla had worked in Hungary and France, experimenting with rotating magnetic fields and electromagnetic induction. Edison put him right to work, eventually assigning him the task of improving the Edison DC generator. Tesla succeeded, producing several new patents for Edison's company, but he had a vision for a better way to deliver electricity—alternating current. This vision eventually alienated Edison and Tesla because Edison had staked his professional reputation on the DC electrification of whole cities.

Tesla continued his work at his own expense, even digging ditches to fund his research. He eventually invented the first AC motor in 1887. In 1888, Tesla paired up with George Westinghouse at the Westinghouse Electric and Manufacturing Company. Westinghouse saw Tesla's ideas as a way to gain a commercial advantage in the new electrical industry. In 1893, Tesla and Westinghouse unveiled alternating current electricity at the Chicago World's Fair, using alternating current and Westinghouse-brand light bulbs to light up the exhibition. The War of the Currents had begun.

In 1896, Westinghouse Electric began supplying AC electricity to industries in Buffalo, NY, using a hydroelectric plant at Niagara Falls driving generators that bore Tesla's name. Because alternating current voltage could be easily adjusted with transformers at the customer's location to accommodate different load sizes, it could be delivered at high voltages and low currents, vastly increasing the economy of the entire electrical grid. Fewer transmission wires were needed, and power stations could be built much farther apart. AC motors were also more compact and simpler in construction than their DC counterparts. Because of these factors, AC electricity quickly overtook and eventually replaced DC electricity distribution for most commercial and domestic uses.

But AC distribution is not without its problems. It is difficult to connect separate AC utilities or individual power generators that operate at different frequencies or at the same frequencies but with different timings.

Research into DC power distribution has never died out completely, and recent developments in alternative energy sources, such as solar and wind power, may restore DC to the national electrical grid in the form of high-voltage DC (HVDC). HVDC can make seamless connections between wind farms in California and a solar generating facility in Arizona without synchronization concerns.

HVDC also promises to be more economical for long-distance electricity transmission. As opposed to the three transmission cables required for long-distance three-phase AC, HVDC requires only one cable, thus using less wire. HVDC is immune to the interactions of alternating current and magnetic fields that increase resistance in AC conductors, so HVDC experiences less energy loss per kilometer of wire compared to AC. A wire's insulation limits the maximum voltage it can handle. Since the AC peak voltage must be higher in order to achieve the equivalent power of a lower DC voltage, AC transmission lines require heavier insulation.

The return of DC to power distribution illustrates an important point about technology: it adapts to meet emerging needs. Alternating current solved the problem of distributing electricity to customers in the early days of the electrical industry, just as HVDC may help solve long-distance electrical transmission problems today. In the future, the electrical grid will probably make use of both kinds of power, which will be the best way of exercising dominion and helping people.

22.12 Inductance

An armature is not unique in experiencing a back emf. Any coil or loop with a changing current produces a changing magnetic field. The changing field causes a change in magnetic flux through the coil; the result is an induced potential difference across the coil. This induced emf opposes the change in current in the coil, and so it is a back emf.

Consider a coil in a DC circuit. There is a switch in the circuit. At the instant that the switch is closed, the current through the coil begins increasing. The increasing current causes an increasing magnetic field in the coil. The resulting back emf creates an induced current whose direction is opposite to that of the main current in the circuit. This opposition causes the current in the circuit to take a measurable time—usually much less than one second—to reach its full value. Similarly, when the switch is opened, the decreasing magnetic field induces a current in the same direction as the main current. The current will therefore not stop immediately but will fall to zero over a measurable period of time. The induction of a back emf in a circuit element by a current flowing through that element is called **self-induction**.

The magnitude of a back emf depends on the geometry of the coil as well as the change of the current in the coil. A factor that takes into account the geometry is called the self-inductance, or simply **inductance (L)**:

$$L = \frac{n\phi}{I} \tag{22.6}$$

The symbol I is the current, and n represents the number of turns in the coil. The values of $n\phi$ and I are independent of each other—they depend only on the geometry of the magnetic coil. Such a coiled component in a circuit is called an **inductor**. The dimensions of inductance are units of magnetic flux (T·m^2 or Wb) divided by current (A). Inductance is measured in **henries (H)**:

$$\text{H} \equiv \frac{\text{T}\cdot\text{m}^2}{\text{A}} = \frac{\text{Wb}}{\text{A}}$$

The magnetic flux in a given inductor is found by rewriting Equation 22.6:

$$LI = n\phi \tag{22.7}$$

22-25 Inductors are manufactured in a variety of shapes and sizes.

If the current through an inductor is varying (ΔI), then the magnetic flux will change with time as well ($n\Delta\phi$). Faraday's law indicates that an emf will be induced in the inductor:

$$\text{induced emf} = -\frac{n\Delta\phi}{\Delta t} \tag{22.8}$$

Problem-Solving Strategy 22.2

The formulas for induced emf have a negative sign because the emf opposes the applied potential difference in the circuit.

If we substitute Equation 22.7 into Equation 22.8, we have

$$\text{induced emf} = -\frac{L\Delta I}{\Delta t}. \tag{22.9}$$

The induced emf of an inductor can be found by either Equation 22.8 or 22.9.

EXAMPLE 22-2

Determining Inductance

An inductor with ten coils produces a magnetic field with a flux of $\phi = 2.50$ Wb per coil when a current of $I = 1.25$ A flows through its coils. What is its inductance?

Solution:

This relationship of current and magnetic flux is determined by the structure of the inductor. Recall that ϕ is the flux per current loop, so the total magnetic flux (Φ) for the inductor is $n\phi$.

$$L = \frac{\Phi}{I} = \frac{n\phi}{I}$$

$$L = \frac{(10\ \cancel{\text{loops}})(2.50\ \text{Wb}/\cancel{\text{loop}})}{1.25\ \text{A}}$$

$$L = 20.0\ \text{Wb/A}$$

$$L = 20.0\ \text{H}$$

When a varying magnetic field of one current loop passes through another current loop, it induces a potential difference in the other loop. This process is called **mutual induction**. The resulting potential difference depends on the size, shape, and relative orientation of the current loops as well as the current change in the active inducing loop or coil. The geometric factors are accounted for in the **mutual inductance (M)**,

$$M = \frac{n_A\phi_{B\to A}}{I_B} \quad \text{or} \quad M = \frac{n_B\phi_{A\to B}}{I_A},$$

where n_A is the number of turns in inductor A, n_B is the number of turns in inductor B, $\phi_{B\to A}$ is the flux that inductor B causes in inductor A, $\phi_{A\to B}$ is the flux that inductor A causes in inductor B, I_A is the current in inductor A, and I_B is the current in inductor B. Notice that the mutual inductance is the same no matter which inductor is the inducing component. The potential difference induced in inductor B by inductor A is given by either of the following equations:

$$\text{induced emf}_B = -\frac{n_B\Delta\phi_{A\to B}}{\Delta t}, \text{ or}$$

$$\text{induced emf}_B = -\frac{M\Delta I_A}{\Delta t}$$

The mutual inductance, like the self-inductance, is measured in henries.

22.13 Inductors

An inductor is a circuit element whose purpose is to resist changes in current. It is represented in a circuit diagram by the symbol . Inductors are usually coils of wire wound with either no core or with a ferromagnetic core. Those with ferromagnetic cores have higher inductance than air-core inductors of the same size.

Recall that, in order for an inductor to function, the current must change with time through the inductor. In a DC circuit, current does not normally change with time except when the circuit is turned on or off; therefore, an inductor appears as a straight length of wire to direct current. Inductors are important in circuits where current is expected to change with time, such as in alternating current circuits.

Inductors may be connected in series or in parallel. However, since two inductors in parallel have less inductance than either inductor alone, most circuits do not use inductors in parallel. Two inductors in series produce a total inductance equal to the sum of the individual inductances as long as the inductors are not close enough to cause mutual induction:

$$L_{\text{total}} = L_1 + L_2$$

If the inductors are close enough for mutual induction or are wound on the same core, they may enhance or detract from each other's inductance. If the inductors are placed so that their windings are in the same direction, the total inductance is more than the sum of the individual inductances:

$$L_{\text{total}} = L_1 + L_2 + 2M$$

If they are placed so that their windings are in opposite directions, the total inductance is less than the sum of the individual inductances:

$$L_{\text{total}} = L_1 + L_2 - 2M$$

Inductors store energy in magnetic fields, just as capacitors store energy in electric fields. A capacitor and an inductor in parallel can (ideally) keep a current alternating between themselves indefinitely once a current is started in the circuit. In real circuits, the wires in the circuit and in the inductor have resistance that will dissipate the electrical energy as heat.

Parallel inductors follow the same rules as parallel resistors. For two parallel-connected inductors,

$$L_{\text{total}} = \frac{L_1 L_2}{L_1 + L_2}.$$

The total inductance of a series of inductors is the simple sum of the inductors:

$$L_{\text{total}} = L_1 + L_2 + \ldots + L_n$$

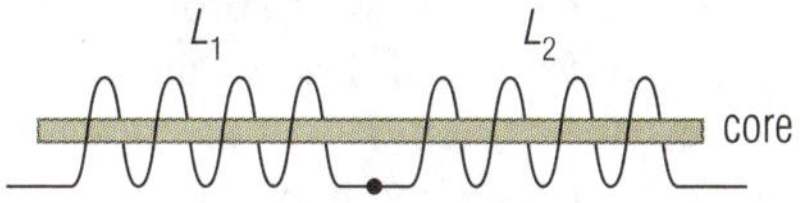

22-26 Inductors in series that reinforce each other

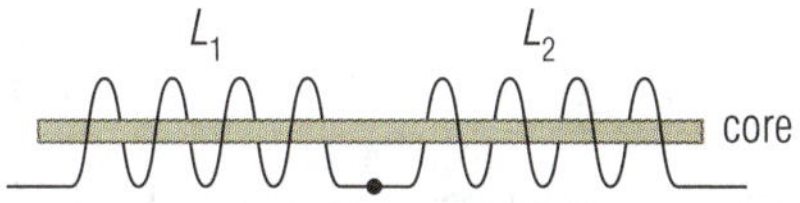

22-27 Inductors in series that oppose each other

22.14 Transformers

Transformers are important induction devices. A **transformer** consists of two coils of insulated wire wound on a common ferromagnetic core. One coil, the *primary winding*, is attached to an AC source. The other coil, the *secondary winding*, is attached to an electrical load. The primary induces a potential difference across the secondary. The potential difference across each turn in the winding is the same. If the voltage across each turn is V_{turn} and the primary has n_p turns, the potential difference across the primary is

$$V_p = n_p V_{\text{turn}}.$$

The secondary has n_s turns. Therefore,

$$V_s = n_s V_{\text{turn}}.$$

Dividing the secondary voltage by the primary voltage, you obtain the following:

$$\frac{V_s}{V_p} = \frac{n_s \cancel{V_{\text{turn}}}}{n_p \cancel{V_{\text{turn}}}}$$

$$V_s = V_p\left(\frac{n_s}{n_p}\right) \qquad (22.10)$$

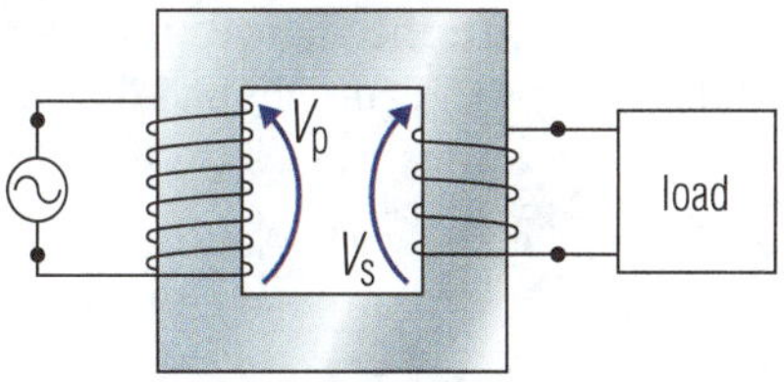

22-28 Schematic of a simple transformer with attached circuits

22-29 A transformer used for supplying current at low voltage to a household doorbell

If the secondary has more turns than the primary, the potential difference across the secondary is larger than across the primary. The transformer therefore is a **step-up transformer**. If the secondary has fewer turns than the primary, as in Figure 22-28, the secondary potential difference is less than the primary potential difference. This transformer is a **step-down transformer**.

The power in the secondary *must* be the same as that in the primary in order to maintain the conservation of energy. Recall that one of the electrical power equations is

$$P = VI.$$

Therefore, using the effective values of current and voltage,

$$V_s I_s = V_p I_p$$

$$\frac{V_s}{V_p} = \frac{I_p}{I_s}. \qquad (22.11)$$

The current in the secondary winding is *smaller* than the primary current in a step-up transformer but *greater* than the primary current in a step-down transformer. These factors allow efficient transfer and safe use of electricity. High-current transmission of power would cause too much heating (energy loss) to be practical. Therefore, electricity suppliers use a step-up transformer to raise the voltage and lower the current for transmission. High voltage would be unsafe for home use, so electricity suppliers use substations with transformers to step down the voltage. The transformers make electricity both safe and economical to use. Transformers within electronic devices are used to step down the supplied 120 V to an appropriate level for the operation of the device.

Transformer cores are ferromagnetic, and most ferromagnetic materials are conductors. The core is therefore subject to the induction of eddy currents. As stated before, unrestricted eddy currents create magnetic fields that interfere with the operation of the transformer. Therefore, transformer cores are laminated to restrict eddy currents without destroying the core's ferromagnetic properties (see Figure 22-29).

Recall that (instantaneous) power used in an electrical component is

$$P = VI.$$

High-voltage overland power transmission lines operate at voltages up to half a million volts AC. Power delivered to large industrial complexes is stepped down to around 12 800 volts. Transformers that supply residences step the voltage down to 220 volts, divided between two conductors and a ground wire. Two lines at 110 volts each are connected to the house.

22B Objectives

After completing this section, I can

- ✓ describe how alternating current is generated.
- ✓ state the difference between peak AC voltage and current, and the effective, or rms, values for these.
- ✓ differentiate between AC and DC generation.
- ✓ discuss why AC electrical power distribution systems are usually more practical than DC systems.
- ✓ explain why high-voltage direct current (HVDC) transmission may be better in certain situations.
- ✓ explain what prevents a motor winding from being an electrical short when the motor is running.
- ✓ list the factors that affect inductance in a coil of wire or solenoid.
- ✓ describe inductors and explain how they affect current in DC and AC circuits.
- ✓ describe how AC transformers function and be able to solve transformer problems.

22B Section Review

1. What kind of electricity is produced by a current loop, rotating in a magnetic field, that is connected by slip rings to an external circuit?
2. What is the rms value of an alternating quantity?
3. What prevents the coiled wire in the armature of a rotating DC motor from acting like a short circuit?
4. DS Why was DC current originally used for electricity distribution? Why was it later replaced by AC?
5. DS Despite the fact that AC has largely replaced DC, how can DC be used to solve some of the problems associated with alternating current?
6. What factors determine the induced emf generated across a circuit component containing coils of a conductor?
7. What does an inductor in a DC circuit act like?
8. ✪ The inductance of an inductor is 15.0 H. If it normally has 0.90 A flowing through it, what is the total magnetic flux in the inductor?
9. ✪ Calculate the total inductance for two inductors of 5.0 H each connected
   a. in series.
   b. in parallel.

10. A transformer has 500 primary winding turns and 750 secondary winding turns.
   a. What function does this transformer perform?
   b. If the rms primary voltage is 120 VAC, what is the secondary voltage?
   c. If the load connected to the secondary side draws 1.5 A (rms), how many amps flow through the primary windings?

22C AC CIRCUIT CHARACTERISTICS

22.15 Resistor-Capacitor (RC) Circuits

Not all circuit elements react the same to AC and DC. Purely resistive circuits, which have already been discussed, do not distinguish between AC and DC. Other circuits do. A resistor-capacitor (RC) circuit is said to block DC but pass AC. The circuit in Figure 22-30 is a DC circuit. When the switch is positioned to *A*, current begins to flow from the DC voltage source through the resistor. Positive charge builds up on one side of the capacitor, and negative charge is attracted to the other side. When the capacitor is fully charged, based on its capacitance and the potential difference across the capacitor ($C = Q/V$), current no longer flows anywhere in the circuit. The direct current flow is blocked. When the switch is positioned to *B*, the charge on the capacitor flows through the resistor, discharging the capacitor. The resistor transforms the electrical energy to thermal energy.

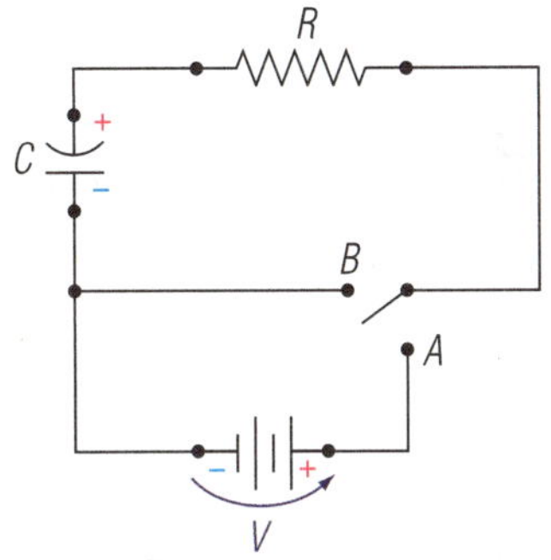

22-30 A DC-RC circuit

In an AC-RC circuit (Figure 22-31), the direction of the current changes regularly. The capacitor is charged in one direction until the AC source potential difference changes polarity. The positive side of the capacitor now faces the negative side of the AC source. The positive charges flow toward the lower potential, and the negative charges flow toward the higher potential. This process is repeated every time the current reverses. Although no current passes through the capacitor (ideally), there is a current throughout the circuit. Since a fully charged capacitor blocks the current, the ideal AC-RC circuit response is more nearly achieved when a large capacitor (which takes longer to charge completely) is used. Frequent changes in current direction also allow the current to flow more continuously by giving the capacitor less time to charge. At high current frequencies, a capacitor appears as a short (that is, as just a segment of wire).

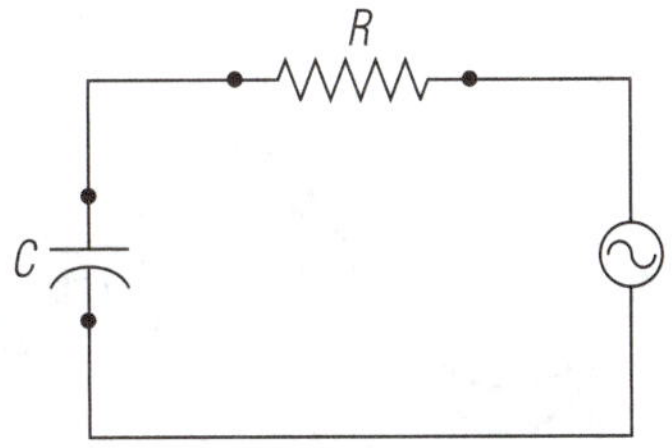

22-31 An AC-RC circuit

22.16 Resistor-Inductor (RL) Circuits

A resistor-inductor (RL) circuit impedes AC more than DC. At the instant that the switch in Figure 22-32 is positioned to *A*, the current through the inductor starts increasing. The induced back emf opposes the circuit current so that it takes time to reach its full value. When the current has reached its full value, it is no longer changing. Therefore, there is no more back emf, and circuit current is determined by the resistor. At the instant that the switch is positioned to *B*, the current in the inductor starts decreasing. The induced back emf sets up a current in the same direction as the original current. The induced current from the inductor produces thermal energy in the resistor as the electrical energy is dissipated.

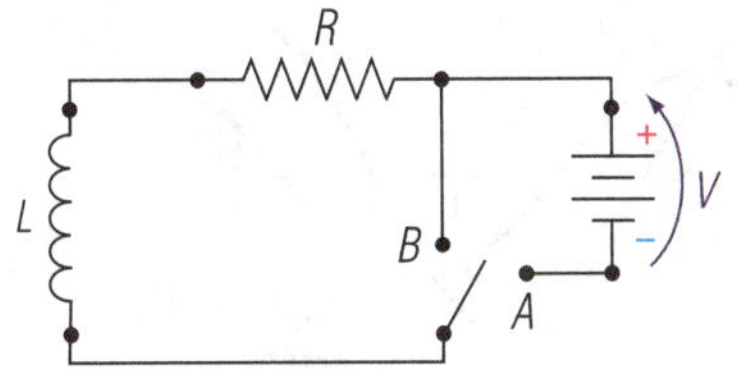

22-32 A DC-RL circuit

In an AC-RL circuit, the current is continually changing direction and magnitude. The induced back emf opposes the changes in current. This constant opposition lowers the peak current below that which would exist without the inductor. Higher inductance produces a higher back emf. Faster current changes also cause a higher back emf and, therefore, a lower total current in the circuit.

22.17 Inductor-Capacitor (LC) Circuits

An inductor-capacitor (LC) circuit is usually part of a much more complex circuit. If the capacitor is charged from another part of the circuit, its charges move around

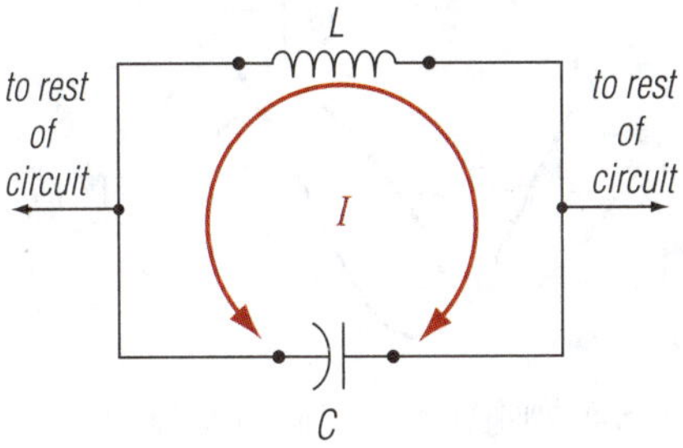

22-33 A segment of a circuit containing parallel LC components that exchange potential and current

the LC portion of the circuit. This motion of charge produces an increasing current through the inductor. The induced back emf opposes the current and slows the capacitor's discharging. Then, when the capacitor's potential difference drops below the corresponding potential difference across the inductor, the inductor sets up a current to oppose the decrease in current. This induced current charges the capacitor with the opposite polarity to that which it was originally. The capacitor then begins to discharge again, and the process repeats itself. In real LC circuits, the electrical energy in the current is converted to thermal energy by the resistance of the wires and components in the circuit. The oscillating current rapidly dies out unless the voltage across the capacitor is periodically restored by the outside circuit.

22.18 Resistor-Inductor-Capacitor (RLC) Circuits

A resistor-inductor-capacitor (RLC) circuit contains a capacitor, which blocks DC. It also contains an inductor, which carries small amounts of AC. The characteristics of a circuit containing resistors, inductors, and capacitors depend on the frequency (f) of the current changes in the circuit.

Electrical frequency is measured in Hz. The United States power grid that supplies homes, schools, and businesses operates at a frequency of 60 Hz, which is extremely stable over time.

The frequency of a current, like the frequency of an oscillating spring, is the number of cycles it undergoes per second. One cycle for the current is the change from a maximum in one direction through a maximum in the other direction and back to another maximum in the first direction. Current frequency is measured in cycles per second, or hertz (Hz).

The frequency of a current flowing through a resistor is not affected by the presence of the resistor. Ohm's law is followed exactly. The voltage drop across the resistor and the corresponding current are exactly in phase. This result does not occur with the other two kinds of components.

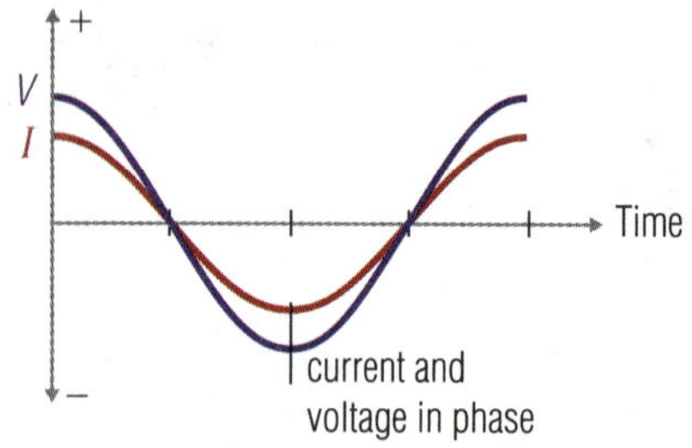

22-34 Voltage and current are in phase across a resistor.

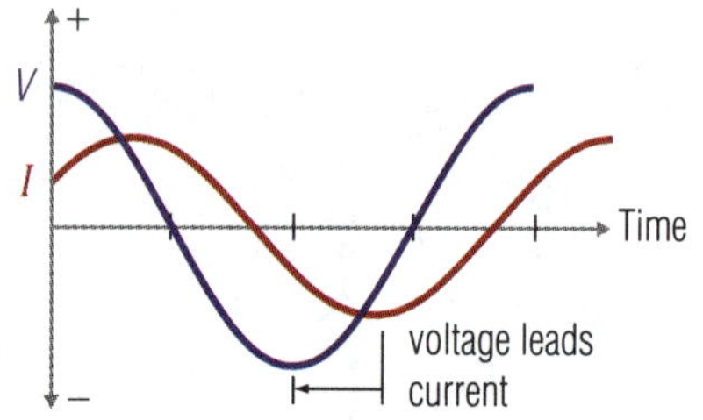

22-35 Voltage leads current in an inductor.

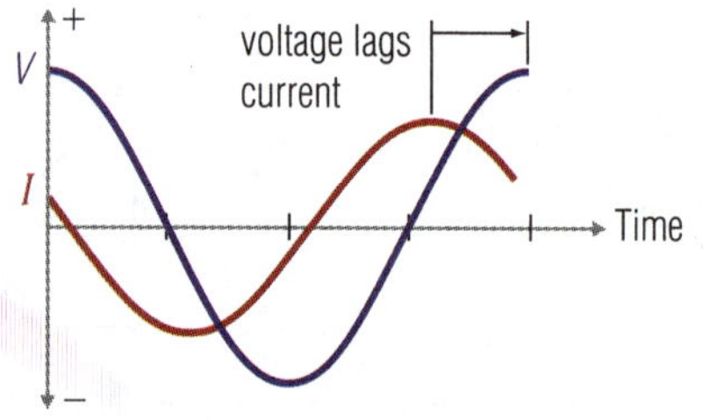

22-36 Voltage lags current in a capacitor.

As the alternating voltage changes across an inductor, its inductance tends to retard the buildup of current in response to the increasing voltage and delay the reduction of current as voltage decreases. This introduces a shift in the timing of the voltage and current wave forms (see Figure 22-35). This difference in timing is called a *phase shift*. Though the timing changes, the frequencies of both voltage and current changes stay the same. The obstacle that an inductor's back emf presents to AC is called **inductive reactance (X_L)**. The magnitude of inductive reactance is found from the formula

$$X_L = 2\pi f L. \qquad (22.12)$$

Inductive reactance increases for higher inductance and for higher current frequencies. For very high frequencies, an inductor appears to be almost like an open circuit. Since voltage must change before current can change in an inductor, we say that voltage *leads* the current in an inductor.

Capacitors also affect AC. As an alternating voltage is applied across a capacitor, current flows into and out of a capacitor in an attempt to neutralize the applied voltage. The potential difference across the capacitor plates doesn't equal circuit voltage until the current stops flowing. Therefore, capacitor voltage is said to *lag* the current. A capacitor's tendency to impede voltage changes is called **capacitive reactance (X_C)**. The magnitude of capacitive reactance is calculated with the formula

$$X_C = \frac{1}{2\pi f C}. \qquad (22.13)$$

Note that capacitive reactance decreases for greater capacitance and for higher AC frequencies. For very high frequencies, X_C approaches zero, and a capacitor appears to be just a piece of wire in the circuit.

The combined effect of both resistance and reactance in a circuit is called **impedance (Z)**. The reactance and resistance add as though they were vectors at right angles to each other. Their "vectors" are called *phasors* and are shown in Figure 22-37. The impedance is therefore

$$Z = \sqrt{R^2 + X^2},$$

where X is the vector sum of the X_C and X_L phasors.

AC circuit impedance is analogous to simple resistance in a DC circuit. Impedance opposes the flow of current in an AC circuit. For an AC circuit, Ohm's law becomes

$$V = IZ.$$

The magnitude of the total impedance of a series RLC circuit is

$$Z_{\text{total}} = \sqrt{R^2 + (X_L - X_C)^2}. \quad (22.14)$$

This equation shows that impedance is smallest when the resistance and the differences of the reactances are smallest. Although there is always some resistance in a real circuit, the total reactance can be zero. Therefore, the impedance of an RLC circuit is at a minimum when the difference of the reactances is zero.

In an AC-RLC circuit, the current is largest when the impedance is the smallest. The maximum current in such a circuit occurs when $X_L - X_C = 0$. This state in which there is zero reactance is called **resonance**. Resonance happens at a specific AC frequency for a given circuit. The *resonance frequency* (f_{res}) is found by setting the reactance term in Equation 22.14 equal to zero:

$$(X_L - X_C)^2 = 0\,\Omega$$

Taking the square root of both sides gives

$$X_L - X_C = 0\,\Omega,\text{ or}$$
$$X_L = X_C.$$

Since X_L is $2\pi f_{\text{res}}L$ and X_C is $1/(2\pi f_{\text{res}}C)$,

$$2\pi f_{\text{res}}L = \frac{1}{2\pi f_{\text{res}}C}. \quad (22.15)$$

Multiplying both sides by $f_{\text{res}}/(2\pi L)$ to isolate f_{res} on the left side gives

$$f_{\text{res}}^2 = \frac{1}{(2\pi)^2 LC}.$$

Now take the square root of both sides:

$$f_{\text{res}} = \frac{1}{2\pi\sqrt{LC}} \quad (22.16)$$

A circuit called a **tuning circuit** can be constructed so that the resistance is small. The resonance effect then makes the resonance current large compared to the current at other frequencies. If the circuit is connected to a voltage source at or near the resonance frequency, the current will suddenly become very large. If the frequency of the voltage source is much different from the resonance frequency, the current will drop to a very low level. An obvious application of tuning circuits is in radios and televisions.

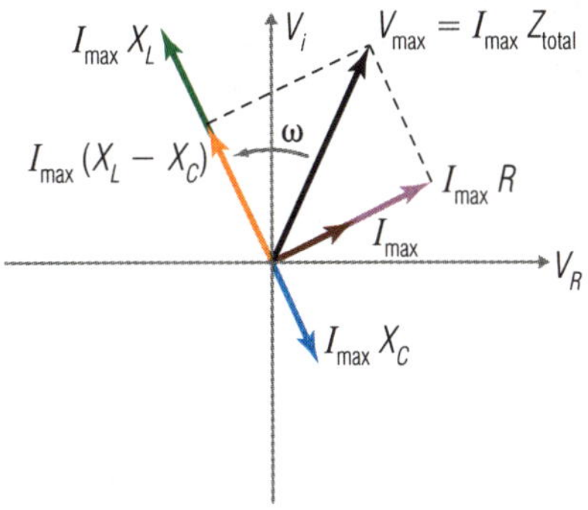

22-37 An RLC voltage phasor diagram is plotted on a coordinate system consisting of real and imaginary voltage axes. The whole plot rotates counterclockwise with time at the frequency ω. The inductive reactance leads current by 90°, and the capacitive reactance lags current by 90°. The maximum circuit voltage is the peak current multiplied by the vector sum of the total reactance and the circuit resistance, or the product of peak current and total impedance.

The magnitude of the peak voltage drop across each kind of circuit element follows the form

$$V = IZ,$$

where for a

resistor $Z_R = R$,
inductor $Z_L = jX_L = j2\pi fL$, and
capacitor $Z_C = jX_C = \dfrac{1}{j2\pi fC}$.

Do not concern yourself with the complex factors. This list is only intended to show that impedance is related to AC resistance and reactance.

22.19 Electromagnetic Waves

Faraday, like many others who have proposed a theory, did not fully develop his field model of electricity. Much of our current understanding of electric fields comes from Faraday's younger contemporary, James Clerk Maxwell, a theorist. One of Maxwell's most successful predictions from the field theory was that

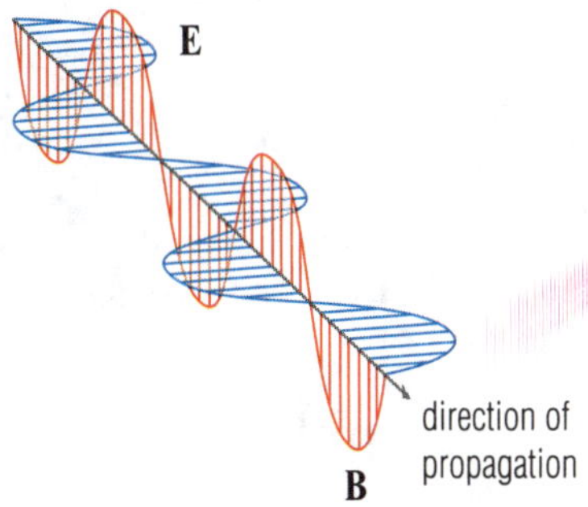

22-38 A transverse electromagnetic wave

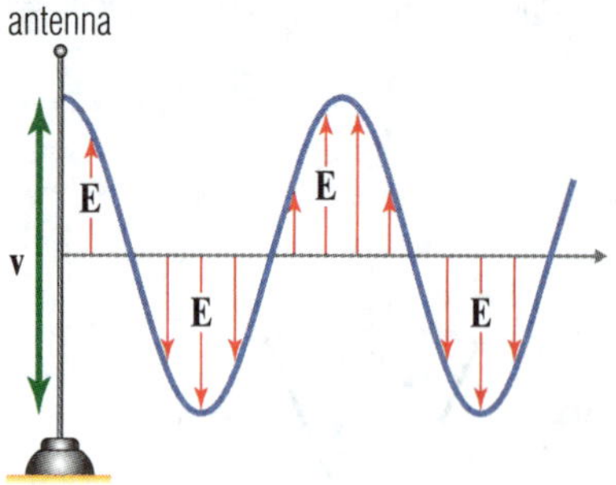

22-39 An antenna emits an EM wave.

electromagnetic waves exist. He reasoned that a changing electric field, like that of a moving charge, induces a magnetic field. A changing magnetic field induces a potential difference, which defines an electric field. If the electric field is changing nonuniformly—for instance, if an electric charge is accelerating—it induces a changing magnetic field. The changing magnetic field induces an electric field in the same direction as the original electric field, which in turn induces a magnetic field, and so on. If there is no circuit with resistance to convert the energy to thermal energy, these fields will continue to reinforce one another, forming an **electromagnetic (EM) wave**.

The mutually reinforcing fields are perpendicular to each other, and both are perpendicular to the direction in which the wave moves. The latter fact gives the wave the name *transverse electromagnetic wave.* Figure 22-38 shows a sinusoidal EM wave. This wave can be started by oscillating charges in a rod called an **antenna**. The electric field oscillates in the same way as the charges. The charges must be accelerating, since they change direction. Therefore, the nonuniformly changing electric field induces a changing magnetic field. The resulting EM wave may look like Figure 22-39. The charges on the antenna are said to *radiate* because they produce an EM wave. In fact, all accelerating charges produce **electromagnetic radiation**. The waves have the same frequency as the oscillating charges.

22.20 Speed of Light

All electromagnetic waves propagate with the same speed, c, in a vacuum:

$$c \cong 3.00 \times 10^8 \text{ m/s}$$

The speed of light (c) is approximately 3.00×10^8 m/s in a perfect vacuum.

Light is an electromagnetic wave, and it travels at the same speed as other electromagnetic waves. The speed c is therefore called the speed of light. Light travels fastest in a vacuum, so in any other medium its speed is less than c. The speed of electromagnetic waves in a medium is related to the electric permittivity (ε) and the magnetic permeability (μ) of the medium. A medium's permittivity and permeability interact to determine the speed with which any electromagnetic wave travels through the medium. This speed is

$$v_{EM} = \frac{1}{\sqrt{\mu\varepsilon}}. \quad (22.17)$$

EXAMPLE 22-3

Slowing Light: The Speed of Light in Water

A beam of orange light travels through pure water at 20°C. Water at this temperature has an electric permittivity (ε) of $1.76\varepsilon_0$ and a magnetic permeability of $1.00\mu_0$ for this color of light. (The permittivity and permeability for a vacuum are ε_0 and μ_0, respectively.) What is the speed of light through water under these conditions?

Solution:

$$v_{light} = \frac{1}{\sqrt{\mu\varepsilon}}$$

$$v_{light} = \frac{1}{\sqrt{(1.00\mu_0)(1.78\varepsilon_0)}}$$

Rearranging,

$$v_{light} = \frac{1}{\sqrt{\mu_0\varepsilon_0} \cdot \sqrt{1.78}}$$

But since $1/\sqrt{\mu_0\varepsilon_0}$ is the speed of light in a vacuum,

$$v_{light} = c \cdot \frac{1}{\sqrt{1.78}}$$

$$v_{light} = \frac{3.00 \times 10^8 \text{ m/s}}{\sqrt{1.78}}$$

$$v_{light} \cong 2.2\underline{4}8 \times 10^8 \text{ m/s} \cong 2.25 \times 10^8 \text{ m/s}.$$

The energy contained in an EM wave is reflected in the *Poynting vector*. The Poynting vector is a measure of the power passing through a unit area perpendicular to the wave's direction of motion. The Poynting vector has units of watts per square meter.

22C Section Review

1. If you wanted to increase the current flowing through the resistor of an AC-RC circuit segment, would you increase or decrease the value of the capacitor?
2. If you wanted to decrease the peak current in an AC-RL circuit segment, would you increase or decrease the value of the inductor?
3. Discuss the relationship of the factors that contribute to the inductive reactance of a circuit component.
4. Discuss the relationship of the factors that contribute to the capacitive reactance of a circuit component.
5. Discuss the characteristics of an electromagnetic wave. What kind of medium is required for the propagation of EM waves?
6. ✪ What is the total impedance of a series-connected, 60 Hz RLC circuit where $R = 100.\ \Omega$, $L = 15.0$ H, and $C = 500.$ pF?
7. ✪ In a series-connected, 400 Hz RLC tuning circuit, what must be the size of the capacitor that will be in resonance with a 5.0 H inductor? (*Hint*: See Equation 22.15.)
8. ✪ What is the resonance frequency of a tuning circuit that has a 8.80 H inductor and a 300. μF capacitor?

22C Objectives

After completing this section, I can

- ✓ describe simple series circuits containing combinations of resistors, capacitors, and inductors.
- ✓ describe the response of RC, RL, and RLC series circuits to changes in current flow.
- ✓ explain why capacitors and inductors have different responses to AC.
- ✓ differentiate between reactance and impedance, and state how each is determined for resistors, capacitors, and inductors.
- ✓ explain what occurs when an RLC circuit is at resonance and how this condition can be achieved.
- ✓ describe electromagnetic waves according to the model of magnetism discussed in this chapter.
- ✓ calculate the speed of light in a medium given the permittivity and permeability of the medium.

Chapter Review

In Terms of Physics

right-hand rule for conductors	499	electromagnetic induction	502
Biot-Savart law	500	magnetic flux (ϕ)	502
permeability of free space (μ_0)	500	total magnetic flux (Φ)	502
electromagnet	500	weber (Wb)	502
solenoid	500	Faraday's law	503

Lenz's law	504
eddy current	505
alternating current (AC)	506
armature	506
root mean square (rms)	507
pulsating DC	508
back emf (ε)	508
self-induction	509
inductance (*L*)	509
inductor	509
henry (H)	509
mutual induction	510
mutual inductance (*M*)	510
transformer	511
step-up transformer	512
step-down transformer	512
inductive reactance (X_L)	514
capacitive reactance (X_C)	514
impedance (*Z*)	514
resonance	515
tuning circuit	515
electromagnetic (EM) wave	516
antenna	516
electromagnetic radiation	516
speed of light (*c*)	516

Problem-Solving Strategies

22.1 (page 503) Recall that *emf* is not a true Newtonian force. Emf is used for describing potential differences that are induced in an electrical circuit by non-electrical sources, such as magnetic fields.

22.2 (page 510) The formulas for induced emf have a negative sign because the emf opposes the applied potential difference in the circuit.

Review Questions

1. The adjacent figure shows a current-carrying wire surrounded by its magnetic field. Is the current in the wire coming out of the page or going into the page?

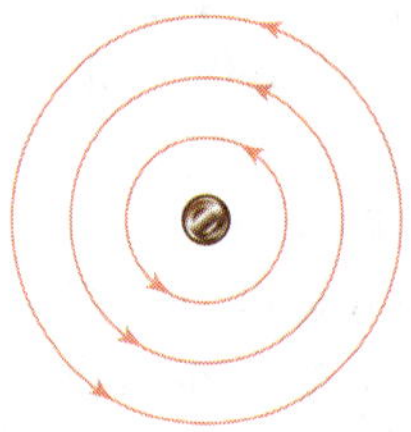

2. How can you reverse the magnetic field of a stationary current-carrying loop?
3. A moving magnetic field is needed to generate a current. Earth, which has a magnetic field, is moving rapidly through space. Why can we not generate electricity simply by constructing large coils on the ground, since we are so close to a moving magnetic field?
4. Some lever-type laboratory balances are equipped with magnets that establish a magnetic field through a plate of paramagnetic metal attached to the end of the balance arm. What is the magnet's function?
5. How can you minimize eddy currents within a conductor?
6. What is alternating current?
7. What is the principal difference between an AC generator and a DC generator?
8. What is the difference between an AC motor and a DC motor?
9. What is the rms or effective potential difference of an AC electrical supply?

DS10. Discuss at least three ways that alternating-current electrical service and devices are different from what would exist if we received our electricity as DC.

11. Describe self-induction.
12. What is the difference between self-induction and mutual induction? Can they ever occur simultaneously?
13. Why does it make a difference whether two inductors in a circuit segment are wound in the same direction on a common core or not? (A diagram may help.)
14. By what principle do transformers work?
15. Why do power companies transmit electricity over long distances at high voltages?
16. Suppose you wish to increase the reactance of an RC circuit. Would you increase or decrease its capacitance?
17. If you want to increase the reactance of an RL circuit, should you increase or decrease its inductance?
18. Identify each circuit described below as RC, RL, LC, or RLC.
    - **a.** It permits a large alternating current to pass in a narrow band of current frequencies near its resonance frequency; it allows only a small current to pass at other frequencies.
    - **b.** It blocks DC but passes AC.
    - **c.** It will, in theory, sustain an oscillating current for an unlimited time.
    - **d.** It presents a greater obstacle to AC than to DC.
19. An RLC circuit has a resonance frequency (f_{res}) under certain conditions. Will the resonance frequency increase, decrease, or remain constant in the following situations?
    - **a.** Another resistor is added in series.
    - **b.** Both *L* and *C* are increased.
    - **c.** Both *L* and *C* are decreased.
    - **d.** *L* doubles, while *C* halves.
    - **e.** *L* is increased, but *C* stays the same.
    - **f.** *C* decreases, while *L* stays the same.
20. What is an electromagnetic wave?

True or False (21–34)

21. If a straight length of wire not connected to an external circuit moves through a magnetic field perpendicular to the magnetic field lines, a current will *not* flow in the conductor.
22. The magnitude of a magnetic flux is dependent on the size of the area perpendicular to a certain density of field lines.
23. Faraday's law concerns the potential produced by a voltaic cell.
24. According to Lenz's law, the current generated by a collapsing magnetic field around a conductor produces a field that opposes the original magnetic field.

25. Alternating current is produced by a loop rotating in a magnetic field when the segments cutting the field lines are connected to the same ends of the external circuit by slip rings.
26. The effective, or rms, voltage of an AC electricity supply is the average of the minimum and maximum voltages.
27. A split-ring DC generator produces constant direct current and voltage.
28. Back emf is the potential induced in the winding coils by a changing magnetic field that opposes the potential difference (and its current) that caused the magnetic field in the first place.
29. Inductance is dependent on the dimensions of the coiled conductor and the material in the core on which the wire is coiled.
30. A transformer converts AC to DC.
31. A circuit segment that acts more and more like a straight piece of wire with increasing frequency is an RC circuit.
32. To increase the alternating current in a circuit segment at a given supply frequency, you could add a large inductor to the segment.
33. The resonance frequency of an RLC circuit segment is reduced by reducing the capacitance of the circuit.
34. The speed of light (c) is constant.
35. Using the equation of force for each of two parallel wires and the definition of the ampere, find the constant k.
36. A 1200 W toaster is connected to 120 V. The wires in the 1.00 m toaster cord are 5.00×10^{-3} m apart. Since one wire carries current to the toaster and the other carries current away from the toaster, the currents are always in opposite directions.
    - **a.** What is the effective current in the toaster circuit?
    - **b.** What is the force between the wires in the cord?
    - **c.** Will the wires attract or repel each other?
37. In the adjacent figure, a 0.10 m × 0.10 m square conduction loop is entirely within a uniform magnetic field with $B = 1.0 \times 10^{-5}$ T. The loop is traveling at a speed of 1.0 m/s.

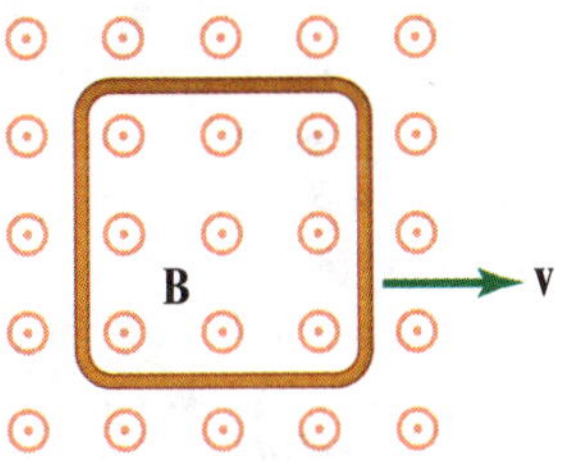

    - **a.** What is ϕ at the instant shown in the drawing?
    - **b.** One second later, the loop is still entirely within the magnetic field. What is ϕ now?
    - **c.** Calculate $\Delta\phi$ and V_i.
    - **d.** Explain your answer for the magnitude and sign of V_i by showing what happens to the free electrons.
38. In (a) a current loop with width $w = 0.100$ m and length $l = 0.300$ m has just begun to enter a uniform magnetic field $B = 1.5 \times 10^{-3}$ T. The rectangular loop has a length $l' = 0.100$ m inside the field. In (b), 1.0 s later, the entire loop has just entered the field.

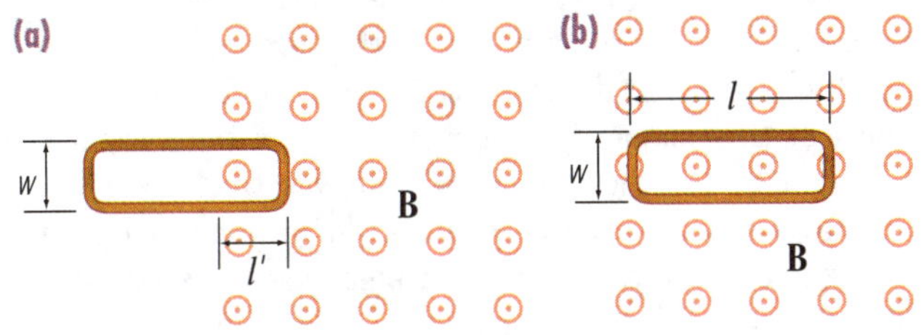

    - **a.** Find $\phi_{(a)}$.
    - **b.** Find $\phi_{(b)}$.
    - **c.** Find $\Delta\phi$ and the potential difference induced in the loop in the second between (a) and (b).
39. In residential circuits the rms potential difference is about 120 V. In the circuit pictured below,

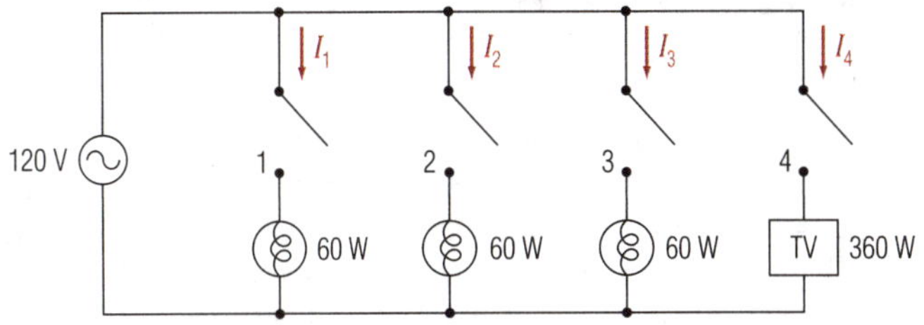

    - **a.** find the rms current through each branch. (Assume all loads are resistances and all switches are closed.)
    - **b.** calculate the maximum (peak) voltage of the AC supplied to the house.
    - **c.** find the total rms current when switches 1, 2, and 3 are closed.
    - **d.** find the total rms current when all the switches are closed.
40. What is the inductance of a coil with fifteen turns if a current of 3.0 mA produces a flux of 2.0×10^{-10} Wb per turn?
41. A coil with an inductance of 6.00 μH carries a current of 10.0 mA. Over a period of 3.00 ms, the current increases to 12.0 mA. What potential difference is induced in that time?
42. Two coils of insulated wire are wound on opposite sides of a hollow square iron core. Coil 1 has six turns and an inductance $L_1 = 5.00$ μH. Coil 2 has three turns. The current in coil 1 is 0.500 A at $t_1 = 0$ s, and 0.000 A at $t_2 = 2.00 \times 10^{-3}$ s. The mutual inductance $M = 1.00$ μH.
    - **a.** What potential difference is induced in coil 2 during that time?
    - **b.** What is the current induced in coil 2 at each time?
    - **c.** What is the potential difference in coil 1?
    - **d.** What is the back emf in coil 1?
43. A series AC-RLC circuit has an rms potential difference $V_{rms} = 120$ V at a frequency of $f = 60.0$ Hz. $R = 10.0\,\Omega$; $L = 14.0/\pi$ mH; $C = 1.0/\pi$ μF.
    - **a.** Find the capacitive reactance, X_C.
    - **b.** Find the inductive reactance, X_L.
    - **c.** Assuming that the capacitor and inductor have no resistance, find the impedance, Z, for the circuit.
    - **d.** Find the rms current in the circuit.
    - **e.** Find the circuit's resonance frequency (f_{res}).
44. What is the ratio of the speed of light in Lucite (clear plastic) to the speed of light in air? Lucite's permeability is μ_0, and its permittivity is $2.5\varepsilon_0$. Air's permeability is μ_0, and its permittivity is ε_0.
45. Prove that the units of inductance and capacitance yield frequency units in Equation 22.16.

DS 46. Research and write a paper on Nikola Tesla.

FACETS of PHYSICS

Diesel Locomotives

When you think of an electric generator, you probably imagine a huge rotating machine in a power plant. Such generators stay in the same place for most of their existence. However, there are some common portable generators. In fact, most vehicles carry their own generators. The alternator that charges your car battery is an AC generator whose output is changed to DC. In rail mass transportation, most locomotives carry diesel engines that generate electricity to turn their wheels.

Why are most rail locomotives powered by diesel-driven generators? The modern *diesel-electric* locomotive engine is the culmination of more than a century of technological evolution that began with the invention of the compression-ignition engine in 1898 by Rudolf Diesel, a German mechanical engineer. A diesel engine is an internal-combustion engine that does not require spark plugs like a gasoline engine. Diesel's original engine was too bulky and too underpowered for practical use in transportation, but within two decades, railroads were beginning to experiment with this new form of motive power.

In the early 1900s, the steam locomotive was the principal means for moving freight and passengers over long distances. Steam engines did have a few problems, however. They were expensive to build and operate. Builders designed locomotives for specific routes, so most engines were unique, relying on many non-interchangeable parts. Steam engines required frequent water and coaling stops, a big problem on routes that crossed sparsely populated plains and deserts. At best, steam locomotives were less than 30% efficient. They emitted large quantities of smoke and coal dust, eventually causing them to be banned in many large cities. They required long hours of startup preparation before a run, continual attention while operating, and an elaborate cool-down, clean-up, and lubrication process at the end of a run. A small fire had to be continually maintained in freezing weather to prevent damage to the boiler. The larger steam locomotives required a crew of three or more men.

Early in the 1900s, the national drive for electrification led railroad operators to consider all-electric rail lines. This effort required stringing a tangle of hundreds of miles of heavy wires for overhead power supply or using a dangerous third rail to power the electric trains. These solutions were effective only for urban routes because of the electrical power losses over long distances. Railroad operators began to look to the diesel engine for a solution.

After unsuccessful attempts to drive the train wheels directly with diesel engines, engineers considered the idea of carrying a diesel-powered generator that would drive the locomotive through electric motors. General Electric developed the first locomotive DC diesel-generator and special *traction motors* to turn the wheels. A controller was invented that varied both the engine speed and the motor characteristics simultaneously so that the necessary power could be delivered to the wheels. The Canadian National Railway began operating the first long-haul diesel-electric locomotives in 1929.

Diesel-electric locomotives eliminated many of the problems associated with steam. They could be built in large batches of identical units. They were far more efficient than steam, approaching 70% efficiency or more. They could be started up and shut down in just a few minutes, or be left to idle unattended for hours or even days. They required only a single operator. And compared to steam locomotives, they were far less polluting.

Most modern diesel-electric locomotives use AC diesel-generators (known as alternators). The AC is converted or *rectified* to DC, just as in a car's electrical system. This design eliminates a potentially dangerous condition that DC generators are prone to—commutator *flashover* or arcing between commutator bars. A flashover could severely damage the generator or even cause a fire in the locomotive. The latest development in diesel-electric propulsion is the variable-voltage/variable-frequency controller, which offers precision control of AC drive motors. The variable-frequency power supply adjusts the AC output frequency in response to commands from the engineer, thereby controlling the speed and torque on the drive wheels. This approach is much simpler than earlier control systems and eliminates the commutators and brushes on the motor end of the system.

Innovation continues to improve the efficiency and economy of this important mode of transportation. Though supplanted by the diesel-powered highway truck in the United States, perhaps in the near future, due to high fuel prices, we will see a return to a more widespread use of diesel-electric or hybrid railroad locomotives.

James Clerk Maxwell (1831–79)

Many advances in science and technology stem from our understanding of electromagnetic waves. Radios, televisions, cellular phones, pagers, and the GPS, for example, send or receive messages through space using electromagnetic waves. Yet little more than a century ago few people even knew that these waves existed. The first man to predict their presence was a Scottish physicist named James Clerk Maxwell.

Maxwell was reared by his father on a farm. His classmates at Edinburgh Academy made fun of his unsophisticated clothes and manners, nicknaming him "Dafty." But they were soon surprised on two counts. First, he was a swimmer, gymnast, rower, and horseman. But more than that, his brains matched his physique. He had a knack for mathematics and science: when he was only fifteen, he presented a paper on drawing ellipses to the Royal Society of Edinburgh. After studying mathematics at Cambridge, he became a professor at Marischal College and later at King's College in London. There Maxwell met Michael Faraday, whose magnetic theory fascinated him.

Maxwell's knowledge of mathematics enabled him to make major contributions in many fields, including the kinetic theory of heat, the velocity of gas molecules, human perception of color, and even the composition of Saturn's rings. In 1871 he was invited to become the first professor of experimental physics at Cambridge. There he used his genius to design and staff the famous Cavendish laboratory.

The accomplishment for which Maxwell is most remembered is his derivation and extension of four differential equations, later developed by others into a coherent model of electromagnetic energy. The equations explained Faraday's electric field effects, and, for the first time in history, mathematically demonstrated that electricity, magnetism, and light were all related. His model suggested that light was simply the small band of electromagnetic energy to which the human eye is sensitive.

His equations showed that the velocity of an electric signal is related to electric permittivity and magnetic permeability. When experimenters showed that electric signals travel at nearly the speed of light, he proved that the two velocities are theoretically equal because light results from variations within the electromagnetic field. He further predicted the existence of wave frequencies outside the range of visible light. His prediction was confirmed a few years later when Heinrich Hertz discovered radio waves.

Maxwell shared with Faraday not only a theory of field lines but also a faith in Jesus Christ as Savior. Like Faraday, Maxwell saw science as an attempt to find out more about God's creation. He strongly opposed the theory of evolution, which was becoming popular among the scientists of his time. Like Faraday, Maxwell was both a good Christian and a good scientist. Maxwell was the best theoretical physicist in his century and is ranked with Newton and Einstein as one of the three greatest physicists of all time.

UNIT 5

Geometric Optics and Light

Although light is one of the most ubiquitous phenomena in nature and provides most of our sensory information about our surroundings, it was one of the least-understood phenomena until the late nineteenth and early twentieth centuries. Newton thought that light consisted of particles, but later experimentation confirmed that it acted as a wave. Light is now known to be just a small portion of the entire electromagnetic energy spectrum, the portion that our Creator has designated as the stimulus for sight.

In Unit 5, you will first investigate the principles of geometric reflection and refraction. You will also study the interesting phenomena resulting from the wave nature of light, such as diffraction and iridescence. The properties of color and color mixing will be reviewed. Finally, the characteristics of various optical instruments, such as telescopes and microscopes, will be discussed.

The Hobby-Eberly telescope at the McDonald Observatory in the Davis Mountains of Texas. Its mirror consists of 91 hexagonal, spherically curved, 1 m segments that yield a mirror diameter of 11 m and an effective aperture of 9.2 m. Each segment is supported by three computer-controlled actuators that adaptively adjust the mirror's surface shape to compensate for atmospheric motion. The telescope is primarily used for spectroscopy.

Light and Reflection 23

Dominion Modeling

Reflecting on Albedo

How can you tell the difference between a planet and a star in the night sky? Over time, you might observe that planets change brightness and position, while the stars remain in the same positions with respect to each other. You might also notice that if you used a telescope, the planets would appear as disks, rather than pinpoints of light as stars do. One of the most obvious ways to tell planets and stars apart is the fact that the principal planets in our solar system generally appear to be brighter than most stars. Since planets are not sources of light themselves, they are visible because sunlight reflects from them, causing them to appear to shine. The fraction of the incoming light that is reflected is called **albedo**. Albedo is usually expressed as a unitless decimal between 0 and 1, where 1 is the brightest-possible reflection. How can we model the modes of reflection in order to help us understand solar system objects better?

23-1 How can you tell a planet apart from a star in the night sky?

albedo (al BEE doh): L. *albedo*—whiteness; from L. *album*—white

23A LIGHT AND THE ELECTROMAGNETIC SPECTRUM

23.1 Light as Energy

In this unit you will study a familiar form of energy—visible light. Perhaps you regard light more as an illuminator than as a form of energy. But consider the following evidences that light is energy:

1. Light powers the chemical reactions that expose photographic film.
2. Light produces an electrical current in a photovoltaic cell.
3. Light rotates a radiometer.
4. Sunlight exerts pressure on the tails of comets, causing them to point away from the sun.
5. Light from a laser produces thermal energy in the material that it strikes.

In each instance, light proves itself to be energy by doing work. On a microscopic level, light moves electrons, atoms, and dust particles. On a macroscopic level, the energy of a laser can be concentrated enough to destroy large objects.

23-2 A radiometer demonstrates that light energy can be converted into mechanical energy.

23.2 The Electromagnetic Spectrum

Visible light is a member of a family of waves called *electromagnetic waves.* Also included in the family are radio waves, microwaves, infrared waves, ultraviolet rays, x-rays, and gamma rays. All of these waves travel at the same speed in a vacuum, 3.00×10^8 m/s. Looking at Figure 23-8 on page 528, you can see that radio waves have the longest wavelengths, and gamma rays have the shortest. The boundaries between the various kinds of electromagnetic waves are generally defined by the processes that create the waves. When the entire range of a phenomenon is arranged according to the value of some property, a **spectrum** is defined. Arranging electromagnetic waves by either increasing frequency or decreasing wavelength forms the **electromagnetic spectrum**.

Waves with shorter wavelengths have higher frequencies. Because energy is proportional to frequency, waves with shorter wavelengths also have greater energies. Thus, radio waves are the least energetic, and gamma rays are the most energetic.

23-3 Radio telescopes investigate the radio-frequency portion of the electromagnetic spectrum in space.

23.3 Radio Waves

The electromagnetic waves called *radio waves* include the waves that transmit television signals and long-range radio communications. These waves are generated by electrons accelerating in an artificial or natural *antenna.* Radio frequencies extend from around 300 GHz (10^9 Hz) down to undetectably low (>0 Hz), and radio wavelengths may range from a few millimeters to many thousands of kilometers long. The longest radio waves created by man are in the ELF band (30 to 300 Hz). These radio waves were used by the military to communicate with submerged submarines. Television signals are transmitted in the VHF and UHF bands (30 MHz to 3 GHz). Eventually, more than 60% of all over-the-air television stations in North America will broadcast HDTV in the UHF band. The highest frequency waves in the radio frequency band are called *microwaves.* Microwaves, which have frequencies from 1 to 300 GHz, are used in radar, line-of-sight communications, and microwave ovens.

Radio-frequency waves are produced by accelerating charges in a conductor or conductive medium.

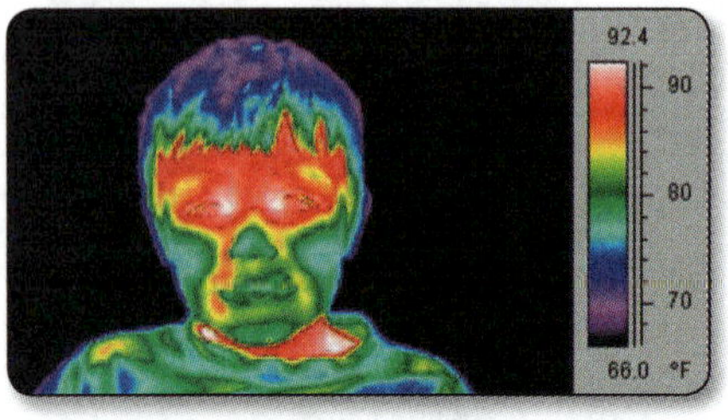

23-4 In infrared frequencies, warmer sources are brighter, while cooler spots are darker.

23.4 Infrared Waves

The next region of the electromagnetic spectrum is occupied by *infrared (IR)*, or "below red," waves, with frequencies from 10^{11} to 10^{14} Hz (λ = 1 mm–750 nm). Infrared waves, often (incorrectly) called heat rays, are produced by the thermal motion of atoms. Since all matter has thermal motion, all matter emits infrared waves. Infrared rays, having frequencies lower than those of red light, are invisible to humans, although some animals, such as pit-viper snakes, can sense near-infrared. Infrared is used for medical and industrial heat treatments, remote sensors in resource satellites and fire-detection systems, communications systems, and night-imaging sensors.

Infrared radiation is produced by the thermal motion of atoms and molecules.

23.5 Visible Light Waves

Visible light, the most familiar form of electromagnetic radiation and the subject of this chapter, includes only a narrow band of the electromagnetic spectrum, ranging from 3.9×10^{14} to 7.7×10^{14} Hz (λ = 770 nm–390 nm, corresponding to the color spectrum from deep red to deep violet). Visible light is generated when energetic electrons in atoms fall to lower energy levels, emitting light in order to conserve energy. Some infrared is also produced this way.

As with the rest of the electromagnetic spectrum, the various colors of visible light form a continuous spectrum. Contrary to what you may see in typical drawings of rainbows, the visible spectrum contains more than seven colors. Computer graphics adapters and monitors are capable of displaying millions of distinct colors. Theoretically, there are an infinite number of colors. We will discuss color and the related properties of visible light in Chapter 26.

Visible light is emitted by electrons as they drop to lower energy levels in atomic electron clouds.

23.6 Ultraviolet Waves

Ultraviolet (UV), or "beyond violet," light is also produced by electrons returning to their lowest-possible energy levels, but it involves greater energy than visible light. Its frequencies are higher than those of visible violet light, ranging from around 10^{15} to 10^{17} Hz (λ = 300 nm–3 nm).

UV light is arbitrarily subdivided into three bands. Near UV (or UV-A) is the lowest frequency band and has the lowest energy UV. It forms about 90% of the solar UV that reaches the earth's surface. It cannot cause sunburns, but sufficient exposure can injure unprotected eyes. Near UV is used medically for treating certain skin disorders and for setting some dental polymer restorations. Also called "black light," it causes phosphorescent paints to glow and can be used to identify compounds and minerals that fluoresce. Far UV (UV-B) is the middle band of UV energies. It is the predominant form of solar UV at midday and can cause sunburns. It is used industrially for quality assurance in manufacturing and for chemical conditioning. Extreme UV (UV-C) is the highest UV energy band. UV-C radiation from the sun is completely absorbed by the atmosphere and does not reach the ground. It is used medically for germicidal treatments and for food, water, and air purification. Chronic exposure to any form of UV has been implicated as a cause of skin cancers.

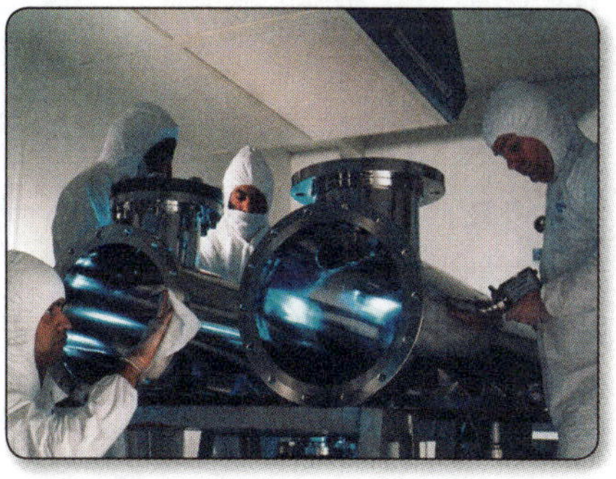

23-5 Ultraviolet radiation is used to kill bacteria in many municipal water supplies.

Ultraviolet radiation is emitted when energetic electrons within an atom fall to their lowest-possible energy levels.

23.7 X-rays

X-rays, with frequencies from 10^{17} to 10^{21} Hz (λ = 3 nm–0.1 pm), are produced when high-energy electrons strike atoms and suddenly decelerate. Such radiation is technically called *bremsstrahlung* in German, or "braking radiation." The x-rays' most impressive characteristic is their ability to penetrate solid matter. X-rays are used for medical and industrial diagnostics, though precautions are necessary because x-rays are very damaging to biological systems. Fortunately, all but the most energetic solar and cosmic x-rays are absorbed by the atmosphere before they reach the earth's surface. As a result, astronomical x-ray observatories must be placed in orbit in order for scientists to study x-rays. From above the earth's atmosphere, they allow astronomers to examine the sun's most energetic processes and to map high-energy objects in the universe.

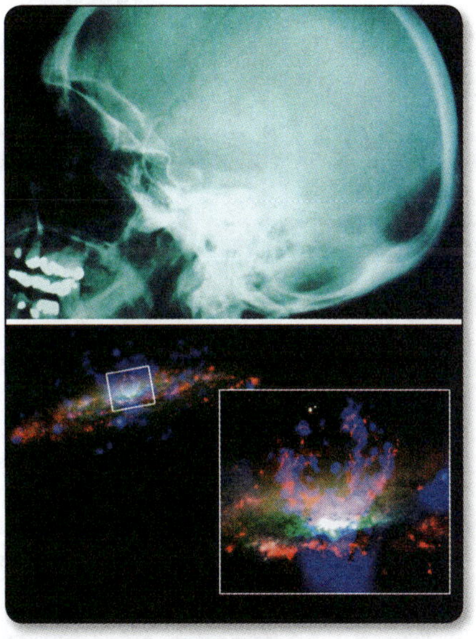

23-6 Throughout the last century, x-rays have been ranked among the most important medical diagnostic tools. Today, space probes are extending the imaging of celestial objects into the x-ray spectrum.

X-rays are generated by high-energy free electrons as they smash into the electron cloud of an atom and slow down, releasing their excess energy.

23.8 Gamma Rays

Although *gamma rays*, with frequencies of more than 10^{19} Hz (λ = 3 pm and shorter), are much like x-rays and overlap the x-ray spectrum band, their source is entirely different. Gamma rays are produced by high-energy changes in subatomic particles, such as nuclear decays and electron-positron annihilations (see Chapter 29). Gamma rays are stopped only by very thick or dense materials. The extremely high-energy extraterrestrial gamma rays that almost reach the earth's

23-7 High-energy radioactive gamma sources are used to take photos of the internals of large metal structures to find flaws in welds and materials.

Gamma rays are emitted during decays of nuclear particles (protons and neutrons) and other changes in nuclei and subatomic particles.

surface (also called "cosmic rays") can cause significant damage to tissues through which they pass. Fortunately, we are exposed to very few of them. Decaying radionuclides in the earth's crust are the major source of gamma rays at the earth's surface. They make up a significant portion of what is called the "background" radiation that we all are exposed to throughout life. Gamma rays are difficult to produce on demand. A few highly radioactive isotopes (byproducts of nuclear reactions) emit sufficient gamma rays to be useful for inspecting the internals of thick metal structures and for performing biological experimentation.

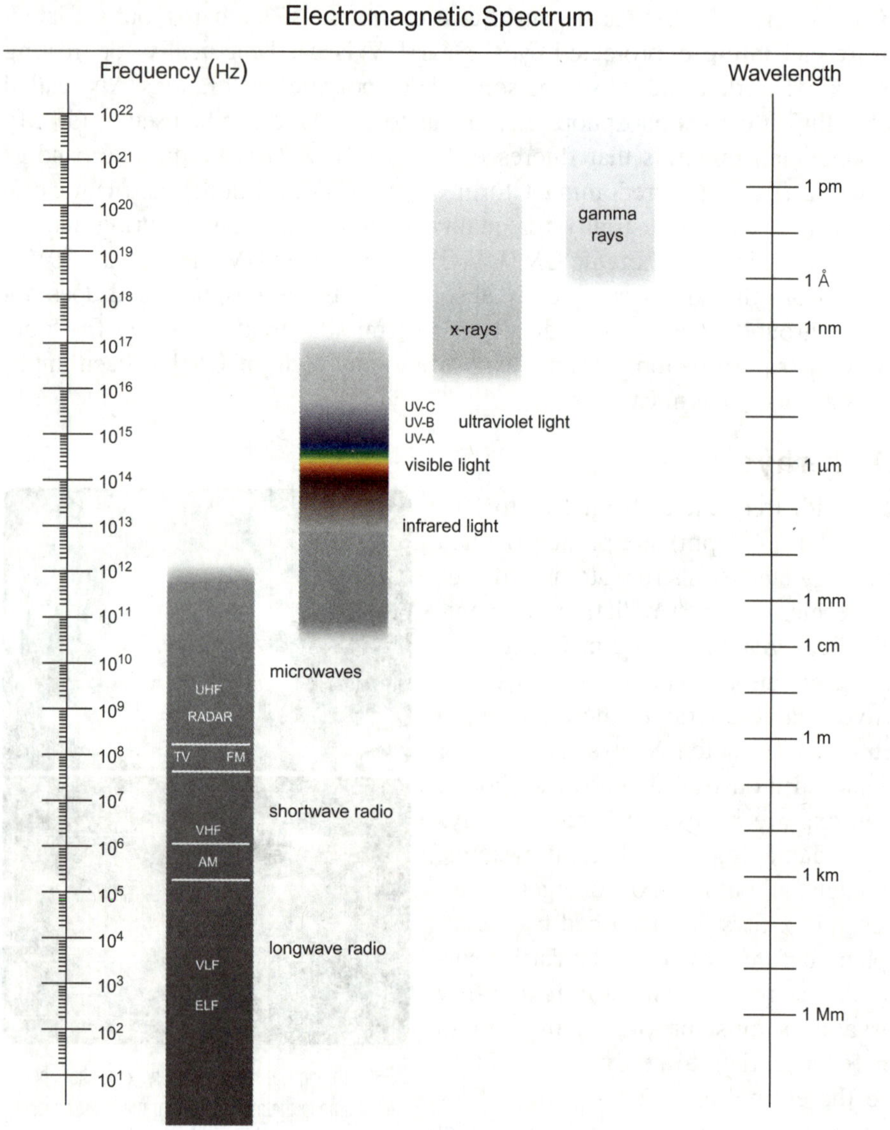

23-8 The main regions of the electromagnetic spectrum

23A Objectives

After completing this section, I can
- ✓ provide evidence that light is a form of energy.
- ✓ define what a spectrum is.
- ✓ identify the properties of light that define the electromagnetic (EM) spectrum.
- ✓ list the major bands of the EM spectrum according to increasing frequency or decreasing wavelength.
- ✓ identify the sources of each major kind of EM energy.
- ✓ suggest useful applications for each kind of EM energy.

23A Section Review

1. Discuss at least three ways in which light energy can be converted into other forms of energy.
2. Describe a spectrum.
3. a. How are radio waves produced?
   b. What man-made radio waves have the longest and shortest wavelengths?

4. a. How are infrared waves produced?
   b. What is one technology that depends on IR radiation?
5. a. How are visible light waves produced?
   b. How many colors are there?
6. a. How are ultraviolet waves produced?
   b. What is one benefit and one hazard of UV radiation?
7. a. Discuss the origin of x-rays.
   b. What is their most useful application in technology?
8. a. What is the source of gamma rays?
   b. Why are they so difficult to produce in a controlled way?

23B SOURCES AND PROPAGATION OF LIGHT

Sources of Light

You have seen that light is produced when an energetic electron loses some energy. There are many natural and artificial sources of light.

The relationship between the Kelvin temperature of a blackbody radiator and the wavelength of the radiation having maximum intensity is summarized in **Wien's displacement law**:

$$\lambda_{max}T = 2.90 \times 10^{-3} \text{ m·K}$$

A broad spectrum of radiation may be emitted by a blackbody radiator, but λ_{max} is the wavelength having the greatest intensity.

23.9 Incandescent Sources

23-9 When a material is heated sufficiently, it will begin to glow in the visible region of the electromagnetic spectrum.

Incandescent sources are objects that are heated until they glow. All bodies emit electromagnetic energy in the form of infrared rays due to the thermal motion of their atoms; however, hotter bodies emit EM radiation of higher frequencies than cooler bodies. At room temperature, a blackbody (a perfect radiator and absorber—see Chapter 15) emits radiant energy with the majority of its frequencies near 10^{13} Hz (mid-infrared). At 700 °C it glows red (visible light). As the blackbody's temperature continues to rise, it becomes orange, yellow, and so on, until at a temperature of about 1800 °C it glows yellowish-white. Incandescent lamps consist of a tungsten filament surrounded by gases such as nitrogen and argon. The filament heats rapidly, and the gas prevents chemical reactions that would destroy the filament. Incandescent sources usually emit radiant energy in all visible frequencies.

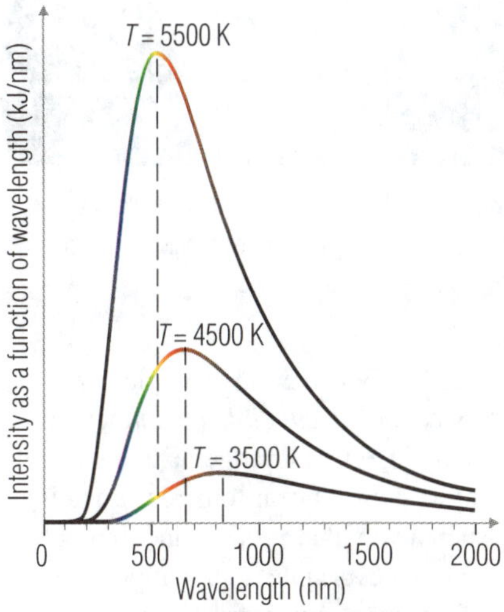

23-10 A blackbody radiator emits electromagnetic radiation in wavelengths that depend on its Kelvin temperature. A narrow band of wavelengths has the greatest intensity for a specific temperature.

23.10 Gas-Discharge Tubes

Gas-discharge tubes also produce light. Gas-discharge tubes consist of a glass envelope fitted with electrodes and containing a gas sealed within it. When a large potential difference is established across the electrodes, a current flows through the tube using the valence (highest energy) electrons of the gas as the current carrier. These electrons jump to higher energy levels, then drop back, emitting visible light in the process. Each kind of gas emits light of a unique color, because each gas has an atomic structure with a unique arrangement of electrons. For example, mercury vapor emits a bluish light, while sodium vapor emits a yellowish light. Unlike incandescent sources, gas-discharge tubes generally emit light at only a few discrete frequencies or in narrow bands of frequencies. So-called "neon lights," fluorescent bulbs, and sodium-vapor street lamps fall into this category.

A variation on the gas-discharge tube is the fluorescent light. Mercury vapor or other gas in the fluorescent tube emits ultraviolet light rather than visible light. The inside of the glass tube is coated with *phosphors*—chemicals that emit visible light when struck by high-energy EM radiation. This indirect method of producing light is more efficient than heating a filament. A 22 W fluorescent fixture emits more light than a 75 W incandescent bulb. For this reason, the Environmental Protection Agency encourages consumers to replace all incandescent bulbs with compact fluorescent bulbs.

23-11 Automobile xenon-vapor headlamps are gas-discharge tubes.

23.11 Lasers

Lasers are devices that produce coherent light—light at a single frequency (i.e., monochromatic) where all the light waves from atoms in the laser source are in phase and combine to form a single, energetic electromagnetic wave. All other light sources produce noncoherent light (i.e., light waves that are not combined in phase), and most are polychromatic. Laser light is extremely intense because of its coherent nature. Lasers are used in many industrial, medical, communications, and military applications, but are not suitable for area lighting.

23-12 Lasers are used to measure the motion of the atmosphere in order to compensate for the optical distortion that the moving air causes.

23.12 Light-Emitting Diodes (LEDs)

A **light-emitting diode (LED)** is a solid-state electronic component manufactured from a semiconductor material that emits monochromatic light when a small potential difference is established across it. LEDs emit light only when the voltage polarity is in the right direction. They are important components in digital technology, where they function as light sources for a wide variety of applications, from simple on-off indicators to full-color digital displays. Their principal advantages are very low power requirements and high efficiency. They can also be manufactured in extremely small sizes using integrated-circuit production techniques, making them suitable for miniature digital devices. LEDs were originally developed for displaying information and showing the status of electronic equipment. Recent developments in manufacturing large, powerful LEDs that produce near-white light have made LED flashlights and other area illumination applications practical.

23-13 LEDs are used for illumination and as indicators on many kinds of electronic equipment.

23.13 Cold Light

Cold-light sources generate light with minimal heat through chemical reactions. People have known for centuries that phosphorus can glow in the dark because of slow oxidation. In the past century, scientists developed several methods for producing "cold light" from chemical reactions using silicon-based substances such as siloxene and other proprietary preparations. For example, one *chemiluminescent* product produces light when a complex organic chemical is combined with an oxidizer, such as hydrogen peroxide. The light can be used directly, or it can activate a fluorescent dye to produce a wide range of colors. The ever-popular light sticks and glow-in-the-dark safety equipment use this cold-light technology.

Bioluminescence is the production of cold light by living things. Some insects, deep-sea fish, squids, jellyfish, bacteria, protozoans, and fungi produce light ranging in color from blue to yellow-green. In these living creatures, an enzyme

23-14 A firefly utilizes a bioluminescent chemical reaction to produce its flashing light.

The firefly is a good example of the concept of *irreducible complexity* in nature. Certain arrangements in living systems appear to be designed in such a way that proper functioning cannot occur unless all components of the arrangement are simultaneously present. These arrangements are evidence for design as opposed to gradual or even punctuated evolution.

combines with another chemical in a biochemical reaction to generate light. These reactions are among the most efficient in nature, converting as much as 80% of the available chemical energy into light energy.

The Speed of Light

23.14 Historical Setting

When emphasizing that something happens very quickly, people commonly say that it happens at the speed of light. How fast does light travel? Galileo, intrigued with this question, tried to measure the speed of light. One night, he and his assistant held shaded lanterns on hilltops about a mile apart. Galileo uncovered his lantern. As soon as his assistant saw its light, he uncovered his own lantern, and Galileo recorded the time that passed between uncovering his own lantern and seeing the light from his assistant's lantern. He came to the conclusion that he had measured only his assistant's reaction time and that the speed of light is too great to measure, if not infinite.

As we saw in Chapter 2, another scientist, Ole Rømer, while investigating eclipses of the moons of Jupiter, found that the speed of light is not infinite. After much study, he made a schedule of eclipses. His schedule worked perfectly when the earth was on the same side of the sun as Jupiter, but when it was on the far side of the sun, the eclipses fell behind schedule. Rømer suspected that this time difference was due to the additional time required for light to traverse the extra distance. Rømer never actually attempted to calculate the speed of light from his data. He concluded, however, that light takes twenty-two minutes to cross the earth's orbit and therefore does not move with infinite speed.

Ole Rømer (1644–1710) was a Danish astronomer who made the first measurement of the speed of light through astronomical observations possible.

In 1849, Armand Fizeau, a French physicist, improved on Galileo's experiment. Instead of having an assistant a mile (1.61 km) away, Fizeau placed a mirror 8.63 km away. He placed a bright light source behind a spinning wheel with evenly spaced notches. He theorized that light would go out through one notch, strike the mirror, and return to the wheel. If the wheel rotated at the right speed, each burst of returning light would be blocked by the next tooth on the wheel. Knowing the rotational speed of the wheel, Fizeau calculated the speed of light with only a 4% error.

Armand Fizeau (1819–96) was a French physicist who made the first experimental measurement of the speed of light on the earth. He also discovered that radiant heat from the sun behaves as wave motion.

23.15 Current Value for the Speed of Light

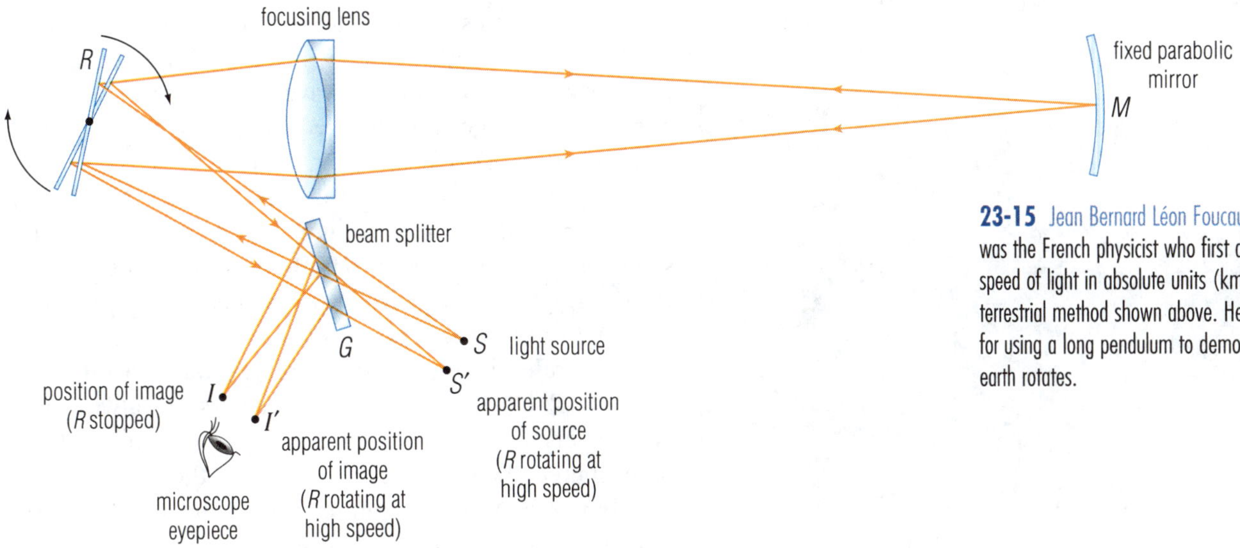

23-15 Jean Bernard Léon Foucault (1819–68) was the French physicist who first determined the speed of light in absolute units (km/s) using the terrestrial method shown above. He is also well known for using a long pendulum to demonstrate that the earth rotates.

Similar techniques involving rotating mirrors allowed experimenters to measure the speed of light with increasing accuracy. In 1862, Léon Foucault was the first to determine the speed of light in metric terms using a terrestrial method. Finally, however, directly measuring the speed of light gave way to an indirect method developed by Albert Michelson using interferometers (see Chapter 25). As you learned in Chapter 12, the speed of a wave of length λ and frequency f is found by

$$v = \lambda f.$$

The currently accepted value for the speed of light (c), 299 792 458 m/s (exact), was approximated in the 1970s by measuring a laser light's wavelength and frequency with great precision and then multiplying them together. Eventually, the speed of light was known to a much higher precision than the meter itself was defined. So, in 1983, the speed of light was defined to the nearest whole meter per second, and the meter was redefined as the distance that light travels in a vacuum in 1/299 792 458 s. The letter c is the abbreviation for the Latin *celeritus*, speed, and is usually rounded to 3.00×10^8 m/s.

As was just mentioned, the value of c is the speed of light in a vacuum. When traveling through matter, light travels more slowly than it travels through a vacuum. In most gases, light travels at nearly c; in liquids and solids, however, speed reduction is significant. Light travels through water with a speed of about $(3/4)c$, through glass with a speed of around $(2/3)c$, and through diamond with a speed of roughly $(2/5)c$.

(a) 3-dimensional waves

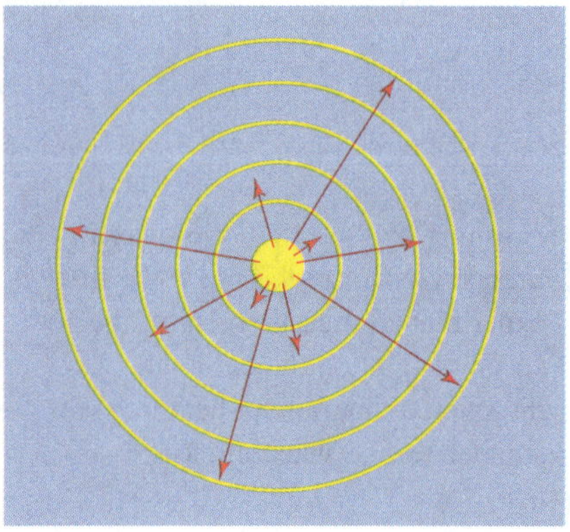

(b) 2-dimensional representation

23-16 (a) A point light source emits spherical light waves. (b) These are represented on a flat diagram as two-dimensional circles concentric with the source.

Wave Characteristics of Light

23.16 Physical Description of Light Waves

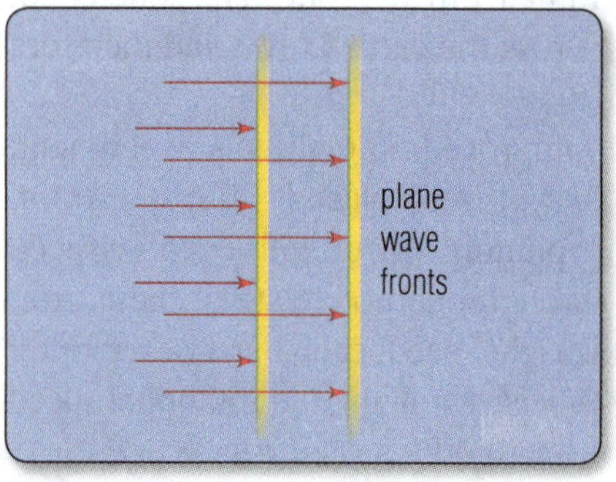

23-17 When a spherical wave front travels far enough from its source, it appears to be a nearly flat, or plane, wave front.

You learned in Chapter 22 that electromagnetic waves consist of self-supporting, changing electric and magnetic fields. The electric and magnetic field vectors are always at right angles to each other, and both vary at right angles to the direction that the wave travels. The changing electric field reinforces the magnetic field, and the changing magnetic field reinforces the electric field, perpetuating the wave.

Light travels outward in concentric *spherical waves* from a point source. Since this movement is hard to illustrate, light is usually shown in a two-dimensional diagram as moving in a circular wave (see Figure 23-16). Like a water wave, light travels at equal speeds through a uniform medium in every direction from a source. The sphere (representing a single wave cycle) expands as the light travels from its source. At great enough distances, the sphere is so large that a small portion of its surface is nearly flat (a straight line segment in a two-dimensional diagram). A straight-line wave surface is called a *plane wave*.

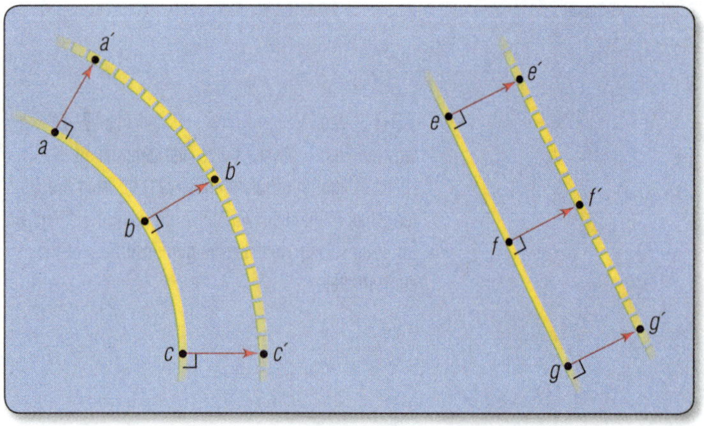

23-18 Points on a wave travel perpendicularly to the orientation of the wave front.

The circles in figure (b) represent **wave fronts**. The wave displacement and velocity vectors are always perpendicular to the wave front in a uniform medium. For a point source, the wave front is spherical; the waves of light travel radially outward from the source and are concentric. For plane waves, the wave fronts are planar, and the light waves are parallel to each other (Figure 23-17).

If you know where the wave started and how fast it is going, you can easily find where it will be at a given time. But what if you do not know where the wave

Christiaan Huygens (1629–95)

The behavior of light puzzled scientists world wide until a Dutchman named Christiaan Huygens (HOW ghens) proposed the wave theory of light. He suggested, for example, that an advancing wave front creates secondary spherical waves at each point. Huygens's principle, as it is now called, has remained an outstanding element of our modern wave theory. His ideas were the culmination of a lifetime of work in mathematics and physics.

Like many fortunate scientists of his time, Huygens had wealthy parents who wanted him to spend his life studying rather than making money. At age sixteen, Christiaan left home to study mathematics and law at the University of Leiden and the College of Breda. Four years later, he returned to the family estate, where, for the next twenty years, he made discoveries in mechanics, astronomy, optics, and chronometry. His knowledge of mathematics enabled him to consider the technical complexities of his findings.

Huygens began his studies with astronomy. He and his brother Constantijn ground their own lenses for homemade telescopes. To correct the problem of spherical aberration in the standard single-lens system, they invented the Huygens eyepiece. With this improved telescope, Huygens studied Saturn more closely than anyone had before. After many years of observation, he discovered Saturn's largest moon, Titan, and realized that Saturn's rings are not connected to the planet.

During the same period, the later 1650s, Huygens's interest in the motion of pendulums led him to design a pendulum clock that could be accurate enough to measure time in minutes and seconds. He published his studies of the pendulum nearly twenty years later in the treatise *Horologium Oscillatorium* (Oscillating Timepiece). In this work he gave a thorough mathematical description of the working of his clock, of the motions of pendulums, and of the laws of "centrifugal force."

By 1666, Huygens had enough prestige to be elected a founding member of the French Académie des Sciences in Paris, and the Académie gave him a pension and an apartment in its building. He enjoyed Paris life and made friends with many intellectual leaders, including Pascal and Leibniz (whom he tutored for three years). His stay in Paris was cut short in 1681 when illness forced him to return home to Holland. His hopes of returning to France were dashed in 1685 when Louis XIV revoked the Edict of Nantes, which had guaranteed Protestants freedom from persecution.

While he lived in Paris, Huygens (wrongly) came to the conclusion that light, like sound, is a mechanical wave moving through a medium. This medium could not be air, because light travels through a vacuum. So Huygens assumed that some substance permeates all space. He called this medium *ether.* He defined candlelight as the sum of the spherical waves from each tiny particle in the flame. Huygens's principle, that each point on a wave front generates secondary spherical waves, was a logical extension of his view of the candle flame. Huygens's theory of light enabled him to explain satisfactorily the reflection and refraction of light.

The publication of Newton's *Principia* in 1687 caused a stir in the international scientific community. Two years later Huygens broke the monotony of his stay in Holland with a lengthy trip to England. Although he admired Newton's mathematics, Huygens disagreed with his action-at-a-distance theory of gravity. Nevertheless, Huygens failed to gain a following for his own mechanical theory of gravity.

Besides studying light and pendulums, Huygens spent much time studying collisions. He discovered the conservation of momentum and kinetic energy in elastic collisions. He used the concept of the center of mass, which he had studied extensively in pendulums, to explain the behavior of colliding bodies.

Huygens's contributions to the study of physics were many and varied. By questioning his surroundings and studying until he found satisfactory answers, he was able to build on the foundations laid by Galileo. Later scientists further built on what he had learned. Although some of his ideas were mistaken, many helped to form the basis of modern science.

An **envelope** is a geometrical curve that is tangent to an associated set of curves or surfaces. Its purpose is to show graphically the net effect of the collection of curves.

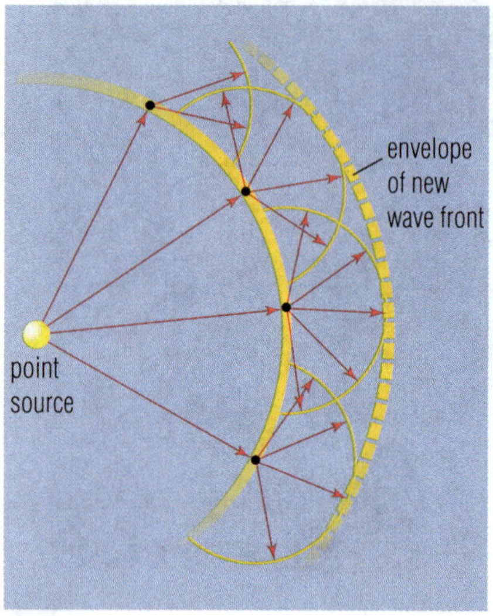

23-19 Huygens's principle of wave front propagation

started? Christiaan Huygens expressed a principle to answer this question. Huygens postulated that every point on a wave front acts like a new source of small spherical waves, called wavelets. As the wavelets expand with time, their wave fronts coalesce into a new wave front, defined by the *envelope* of the wavelets (see Figure 23-19). A spherical wave front propagates as a spherical surface, and a plane wave front propagates as a plane surface. This principle is called *Huygens's principle.*

23.17 Mathematical Description of Light Waves

An electromagnetic wave is mathematically described by time-dependent sine functions. The magnitude of the electric field strength (E) and the magnitude of the magnetic field vector (B) both act as sine waves. That is, they can be written as

$$E = E_{\text{max}} \sin \omega t \text{ and}$$
$$B = B_{\text{max}} \sin \omega t,$$

where ω is a time-dependent variable equivalent to angular speed.

The equations are similar because the electric and magnetic fields are in phase. That is, both reach their maximum strength at the same time, and both reach their minimum strength at the same time.

Most significant in the mathematical description of light waves was the work of James Clerk Maxwell (see Chapter 22), who successfully related electricity, magnetism, and light. By showing the relationship between four known equations describing the behavior and interaction of electrical and magnetic fields, and by modifying one of them, he demonstrated that light is related to electromagnetism.

The first equation, Gauss's law for electricity, states that the net electric field through a closed surface is proportional to the charge enclosed by the surface. The second equation, Gauss's law for magnetism, states that the net magnetic flux through a closed surface is zero. The third equation is Faraday's law, which states that a changing magnetic field induces an electric field. The fourth equation, Ampere's circuital law, was corrected by Maxwell to account for changing currents and electric fields and the way that they induce magnetic fields.

Maxwell's equations are as important to the study of electromagnetism as Newton's laws are to the study of mechanics, or as the laws of thermodynamics are to the study of heat. Using these equations, Maxwell predicted the existence of electromagnetic waves, deduced that light is an electromagnetic wave, and calculated the speed of light. These equations affect such diverse things as motors, generators, transformers, radios, televisions, radar units, and microwave ovens. Even the equations for E and B can be derived from Maxwell's equations.

23B Objectives

After completing this section, I can
- ✓ identify the principal sources of visible light.
- ✓ state the important properties of light produced by different sources.
- ✓ summarize the history of establishing the speed of light.
- ✓ state the speed of light, c, to three significant digits.
- ✓ discuss the geometrical ways that light propagates.
- ✓ describe how EM energy is mathematically modeled as a wave.

23B Section Review

1. **a.** Discuss how incandescent light emission occurs.
   b. List three sources of incandescent light not mentioned in the text.
2. Which two measurable dimensions does Wien's displacement law link for a blackbody radiator?
3. What is probably the most common gas-discharge tube used for illumination?
4. What are the two key properties of laser light?
5. What makes bioluminescence such an elegant testimony of design in creation?
6. The speed of light and the length of a meter ultimately depend on which measurement?

7. Where do plane electromagnetic wave fronts come from?
8. What was the significance of Maxwell's contributions to understanding electromagnetic theory?

23C REFLECTION AND MIRRORS

23.18 Ray Optics

Light can be regarded as a group of **rays**, which are represented as lines drawn in the direction that the wave is traveling, perpendicular to its wave front. Light rays travel in straight lines through any single material, but they may bend when passing from one material to another. We know that light travels in reasonably straight lines because objects cast sharp shadows. For example, a spherical Jovian moon casts a circular shadow on the planet it orbits. A pinhole viewer also demonstrates that light rays travel in straight lines. A shoebox with a pinhole in one end will display an inverted image of a light source placed outside it. Figure 23-20 shows why the image is inverted.

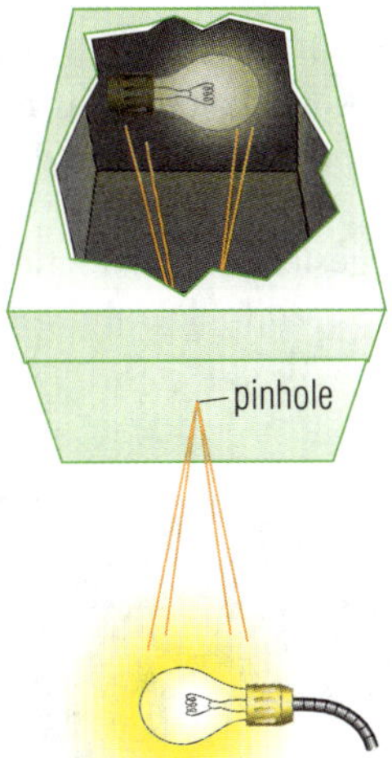

23-20 A pinhole projector produces an inverted and reversed image because rays travel in straight lines.

When light strikes an object, some of the light rays "bounce off" its surface. This change of direction is known as **reflection**. The way light rays reflect from a surface depends on the material that the surface is made of. Irregular surfaces, such as paper, cloth, and brick, reflect rays in random directions. The sizes of the irregularities are much larger than one wavelength of the incoming light. Reflection in random directions is called *diffuse reflection.* Polished opaque surfaces, such as mirrors or smooth metals, reflect rays in a predictable pattern. The dimensions of the irregularities on such surfaces are much smaller than one wavelength of light. Predictable reflection is called *regular* or *specular* reflection.

The word "specular" is derived from the Latin word *speculum*, a mirror.

Consider a ray of light striking a mirror, as Figure 23-21 shows. The imaginary line perpendicular to the mirror's surface at the point of reflection is called the normal line, or just the *normal*. You first saw normal lines in Chapter 8 when studying friction and the normal force. The angle that the incoming ray makes with the normal is called the **angle of incidence (θ_i)**. The angle that the reflected ray makes with the normal is called the **angle of reflection (θ_r)**. How is the reflected ray related to the incoming ray? The answer is the two-part **law of reflection**:

1. The incoming (incident) ray, the normal, and the reflected ray all lie in the same plane.
2. The angle of incidence equals the angle of reflection.

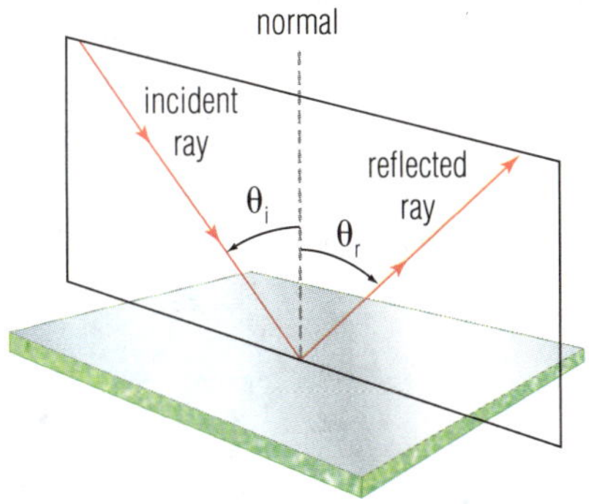

23-21 The law of reflection

23.19 Modeling Albedo

Sunlight from a planet, moon, or other nonluminous celestial object reflects diffusely. Visible-light albedo is actually a ratio of the reflected light to the incident light. If all light is reflected, then an astronomical body has an albedo of 1.00. If all light is absorbed, then an astronomical body has an albedo of 0. The reflectance of all materials in a planet's atmosphere and on its surface contributes to its albedo. Clouds are the most significant factor for planets with atmospheres. For airless planets, ices, rocks, and dust form the reflective surface. Generally, clouds and ices produce relatively high albedos, while rocks and dust are generally darker and less reflective. Some moons have very dark surfaces that absorb most of the incident light.

There are two types of albedo used by astronomers. The differences are based on the location of the light source relative to the position of observation. *Geometric albedo* is measured when the sun is directly behind the observer relative to the

Albedo and Global Warming

Many people are concerned that the increase of human-generated greenhouse gases will lower the albedo of the earth (i.e., incident solar energy will be retained rather than reflected). They believe that this will accelerate global warming, causing catastrophes such as melting ice caps, dramatic sea level increases that could submerge coastal cities, disturbed ocean currents, and temperatures so high that only the poles would be habitable. How should a Christian view such concerns?

George Phillips Bond (1825[6?]–1865) was an American astronomer and surveyor. He pioneered the photography of astronomical objects and suggested that photographs could be used to determine their magnitude (brightness).

23-22 The much greater albedo of Saturn's moon Enceladus is evident as it is pictured passing in front of the larger moon Dione. Enceladus is the brightest planetlike object in the solar system.

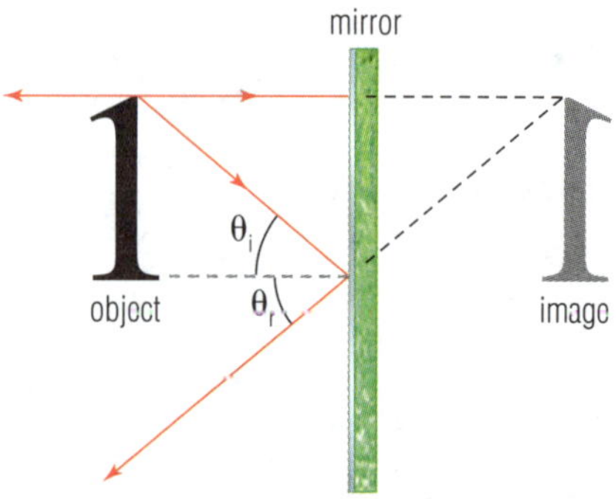

23-23 Reflection in a plane mirror and a virtual image

observed object. It is related to the absolute brightness of the object and its diameter, and compares the albedo of the body to a flat, ideal reflector of the same dimensions. *Bond albedo* is simply the ratio between the incident and reflected electromagnetic energy without regard to the position of the sun. These values are usually similar but can be significantly different. You will use the information in the table below to model a relationship between geometric and Bond albedos for selected astronomical objects in our solar system.

TABLE 23-1

Astronomical Albedo Data

Body	Geometric Albedo	Bond Albedo
Mercury	0.14	0.12
Venus	0.84	0.75
Earth	0.37	0.29
Moon	0.11	0.12
Mars	0.15	0.16
Jupiter	0.52	0.34
Saturn	0.47	0.34
Enceladus	1.40	0.99
Uranus	0.51	0.30
Neptune	0.41	0.29
Pluto	0.44	0.52

23.20 Plane Mirrors

An image in a flat, or plane, mirror is formed by many rays of light, each obeying the law of reflection. Figure 23-23 shows two rays from the top of the object being reflected. The eye and brain, detecting these rays, assume that they come from the same point in straight lines, so the object appears to be located *behind* the mirror. Since the rays do not really come from behind the mirror, the image of the object in the mirror is said to be a **virtual image**. The mind interprets the reflected rays as originating at a point that doesn't exist in reality. A virtual image cannot be projected onto a screen.

When you think of a mirror image, you probably think of an image with its left and right sides reversed. This idea is a semantic misnomer. Actually, the sides are not reversed. A mirror reflects your right side on your right and your left side on your left. Since your reflection is facing you, your mind considers your reflection another person and automatically orients the reflection accordingly. You naturally consider the image's right side (from the "image's perspective") to be a reflection of your left side and vice versa. So, you assume that the image is reversed. In fact, the left side of the image (as you face it) is a reflection of your left side, and the image's right side is a reflection of your right side. Your reflection is a "mirror image."

What happens when two plane mirrors are set at right angles? When an object is approximately equal distances from each mirror, it has *three* images. Images I_1 and I_2 are formed as they would be if the mirrors were separate. But where does the third image come from? Some light rays are reflected from both mirrors before meeting the observer's eye. These twice-reflected rays form the third image, I_3, which is reversed so that the object's left side is reflected to the object's right.

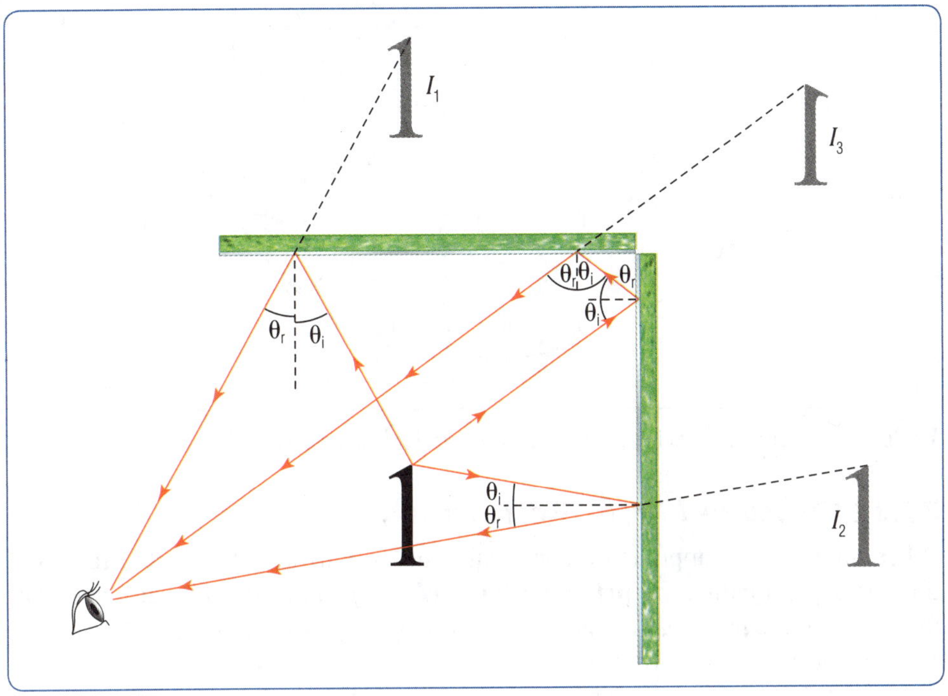

23-24 Reflection in two perpendicular mirrors

If the angle between the mirrors is reduced to 60°, five images appear. If the angle is decreased to 45°, an object will produce seven images. The number of images at a given angle is determined by the formula

$$n = \frac{360°}{\theta} - 1,$$

where θ is the angle between the mirrors in degrees. What if the mirrors face each other? Then $\theta = 0°$, and

$$n = \frac{360°}{0} - 1,$$

$$n = \text{d.n.e. } (\rightarrow \infty).$$

There will be an unlimited number of images if the mirrors are properly adjusted. (You may have seen this effect at amusement parks or in furniture stores where wall mirrors are sold.)

23.21 Curved Mirrors

Mirrors with curved surfaces are useful in many different ways. You have probably seen such mirrors used as cosmetics mirrors, side-view mirrors in cars, safety mirrors in the aisles of stores, and as flashlight reflectors. Such mirrors may be **concave**, formed from part of the inside of a curved surface, or **convex**, formed from part of the outside of a curved surface. Concave mirrors that are exactly spherical are the easiest to manufacture and are very common. However, they produce slightly blurry images, an effect called **spherical aberration**. Parabolic mirrors avoid spherical aberration. Curved mirrors, like plane mirrors, follow the law of reflection. We will discuss both kinds of curved mirrors in the following sections.

23-25 Apollo-program astronauts left several box-mirror laser reflectors on the surface of the moon. Each reflector cell is the right-angle corner of a box. Such reflectors can reflect an incident ray in the direction from which it came. The moon reflectors are used to measure the distance between the earth and the moon to a precision within a few centimeters.

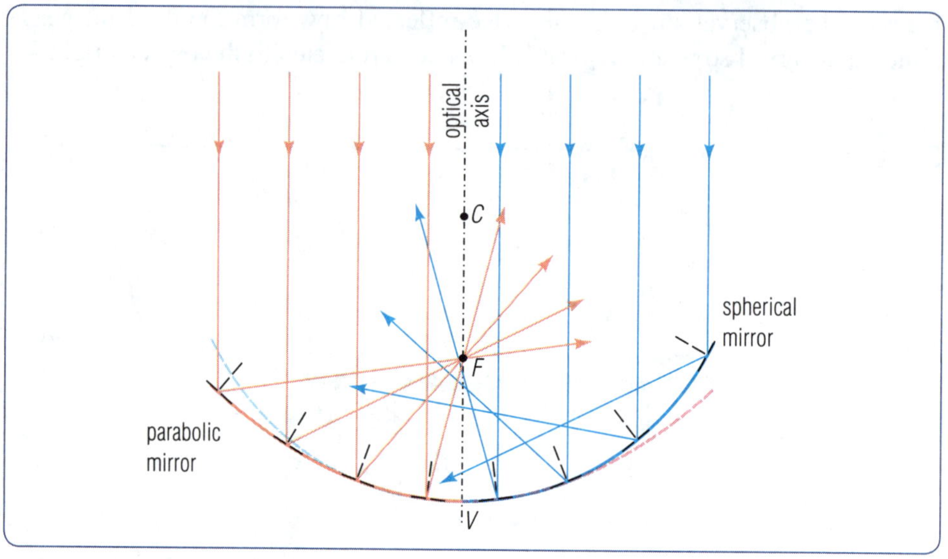

23-26 Spherical mirrors do not focus parallel light rays at a point. Parabolic mirrors do.

23.22 Concave Mirror Reflections

When parallel rays approach a spherical concave mirror, the reflected rays pass through a point called the **principal focus (*F*)** or *focal point*. The distance from the mirror to the principal focus is the **focal length (*f*)** of the mirror. The distance *R* is the *radius* of the mirror—the radius of a sphere that the mirror's surface is a part of. The point *C* is the *center* of the spherical surface. Every point on the surface of a spherical mirror is *R* units from *C*. The line through *F* and *C* intersects the mirror at its *vertex* (*V*) and is called the *principal* or **optical axis** of the mirror. The focal point is midway between *V* and *C*:

$$f = \frac{R}{2}$$

For Figure 23-27 and the following similar figures, refer to the legend below.

Ray 1 (green): Incident on mirror parallel to optical axis, then reflected through *F*

Ray 2 (yellow): Incident on mirror via *F*, then reflected parallel to optical axis

Ray 3 (purple): Incident on mirror via *C*; always reflects back through *C*

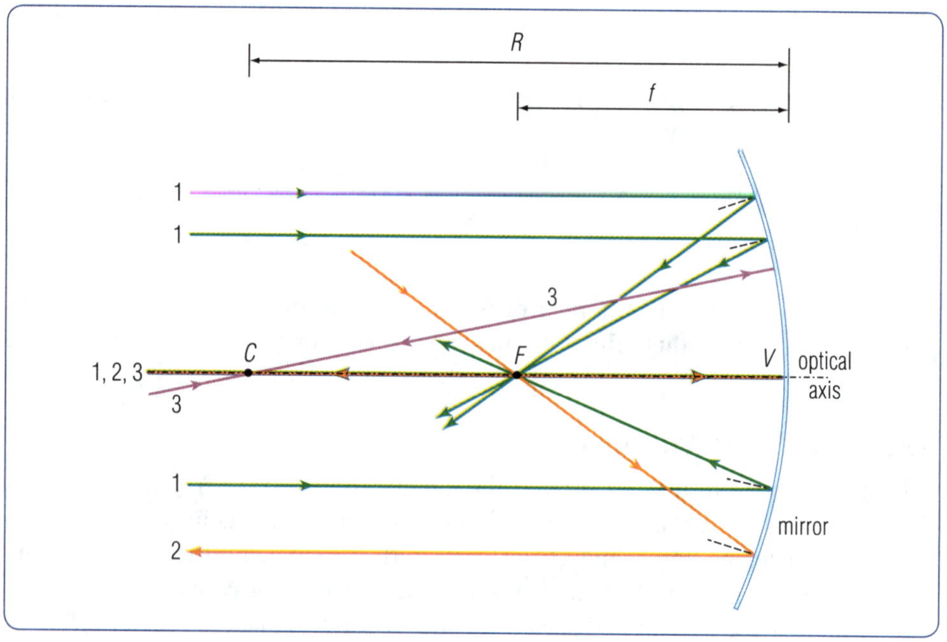

23-27 Terminology and rules for reflection in a concave spherical mirror

A visual image of a reflection is formed when certain rays from a point on an object are reflected so that they come together at a point on the retina of your eye.

In order to determine how the rays from an object are reflected, three specific rays from a point on the object are projected to the mirror's surface and then reflected according to the law of reflection. Where they intersect is the position of the image. The special rays obey the following rules (the corresponding rays are numbered in Figures 23-27 and 23-29 through 23-34):

1. Incident rays that are *parallel to the optical axis* are reflected through the principal focus (F).
2. Incident rays that *pass through the principal focus* are reflected parallel to the optical axis.
3. Incident rays that *pass through the center of curvature* (C) are reflected back through C.

Keeping these rules in mind, let us see where the image of an object will appear. The **object distance (d_O)** is the distance of the object from the mirror. The **image distance (d_I)** is the distance of the image from the mirror. For simplicity, we will assume that the object is located on the optical axis, although this discussion applies to any location where all three rays can be reflected from the mirror. There are six optically different object locations to consider.

Case 1: at "infinity" ($d_O \Rightarrow \infty$) or far enough from the mirror so that all rays from the object are nearly parallel

Case 2: beyond C ($d_O > R$)

Case 3: at C ($d_O = R$)

Case 4: between C and F ($f < d_O < R$)

Case 5: at F ($d_O = f$)

Case 6: between F and the mirror ($0 < d_O < f$)

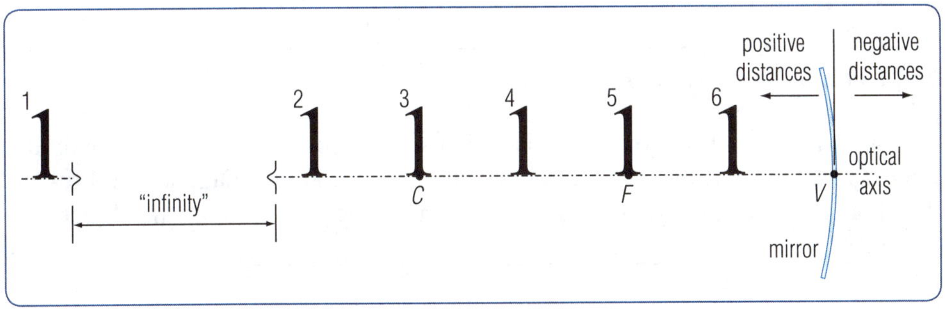

23-28 Six possible object locations that affect image location and orientation

Let us first consider Case 2, an object beyond the mirror's center of curvature, C. The object in Figure 23-29 is represented as a solid black numeral "1" at a distance d_O from the mirror, and its image is a gray "1" at a distance d_I from the mirror. By convention, all distances in front of the mirror are positive, and distances behind the mirror are negative.

To locate the object's image, trace the paths of two rays from the top of the numeral, one parallel to the optical axis and one through the focus. Where they meet, the image of the corresponding point on the object will form. As the figure shows, they intersect between C and F on the opposite side of the optical axis. As a result, the image is inverted compared to the object's orientation. Notice that the light rays converge at the point where your eye would have to be in order to see the image. Since the image location is in front of the mirror, it is called a **real image**. Real images can be focused on a screen.

The key difference between a real image and a virtual image is the place where they are (or appear to be) formed in relation to the mirror's surface. Real images are formed by the convergence of light rays from a real point in front of the mirror's surface, while virtual images appear to be formed by the divergence of light rays from a virtual point behind the mirror's surface.

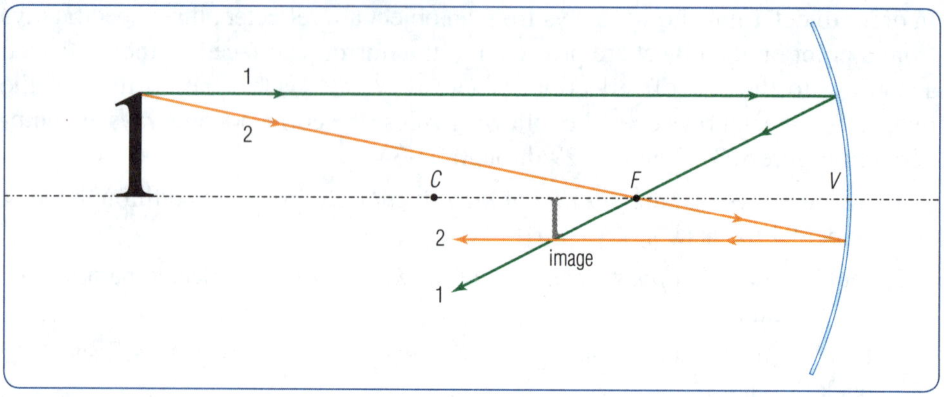

23-29 Reflection of an object when $d_O > R$ (Case 2)

Now consider Case 4, in which the object is placed between *C* and *F*, and the image appears beyond *C* ($d_I > R$), as Figure 23-30 shows. Notice that the image here is larger than the object. Whenever $d_O < d_I$, the image is larger than the object. However, when $d_O > d_I$, the image is smaller than the object. In both Case 2 and Case 4, the image is upside-down, or *inverted.*

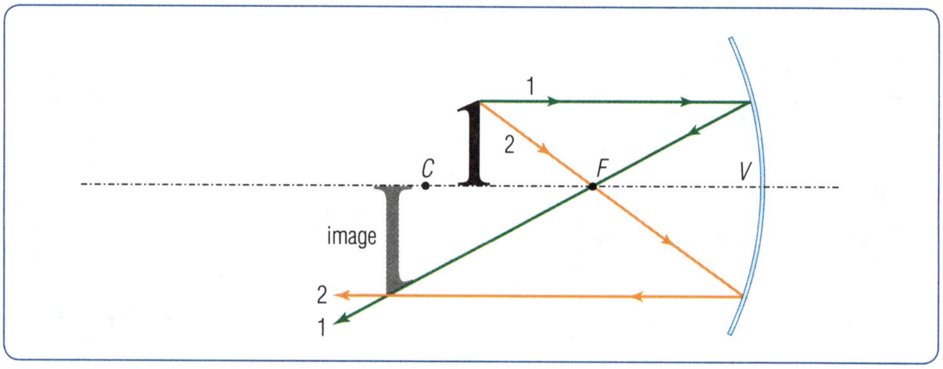

23-30 Reflection of an object when $f < d_O < R$ (Case 4)

For Case 3, when the object is placed at *C* (Figure 23-31), its image will also appear at *C* but will be inverted so that the object and its image are end-to-end. The image in this arrangement is also real and is the same size as the object.

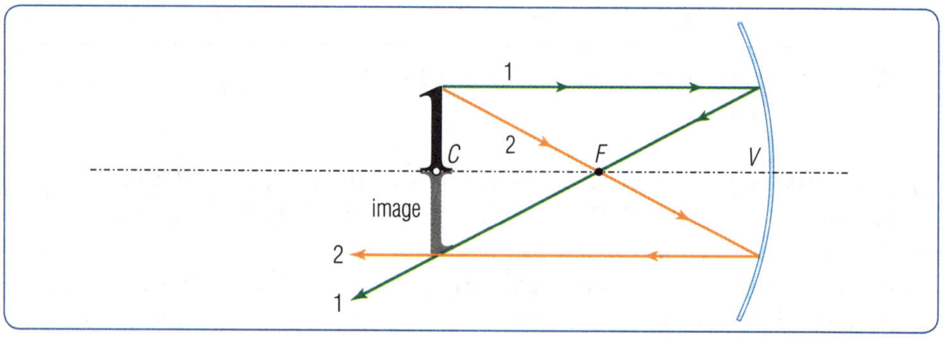

23-31 Reflection of an object when $d_O = R$ (Case 3)

Real images are formed for Cases 2 through 4, but not for the other three cases. Suppose the object is at infinity (Case 1). At that distance *all* the rays that reach the mirror from the object are essentially parallel to the optical axis. Therefore, all the rays are reflected through the focus, *F*. The image is formed where the rays meet—at the focus. So, when an object is at infinity, its image is a miniscule point at the focus.

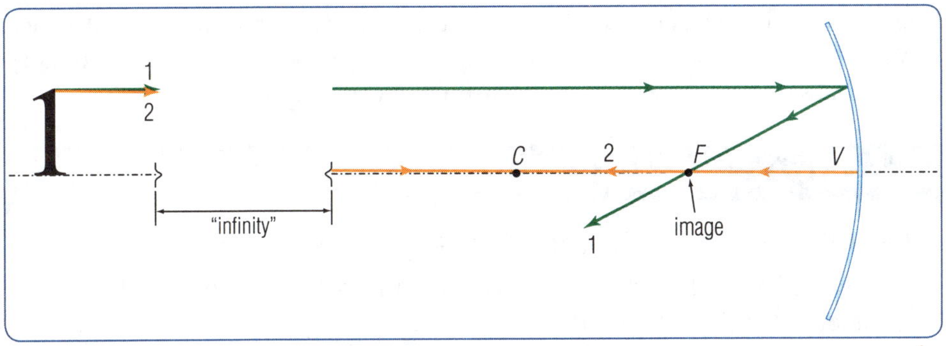

23-32 Reflection of an object located at an infinite distance from the mirror (Case 1)

Suppose, instead, that the object is at the focus (Case 5). Rays from every point of the object are reflected so that they never intersect. The image is said to be formed at infinity and is infinitely large. It cannot be seen, so it does not exist.

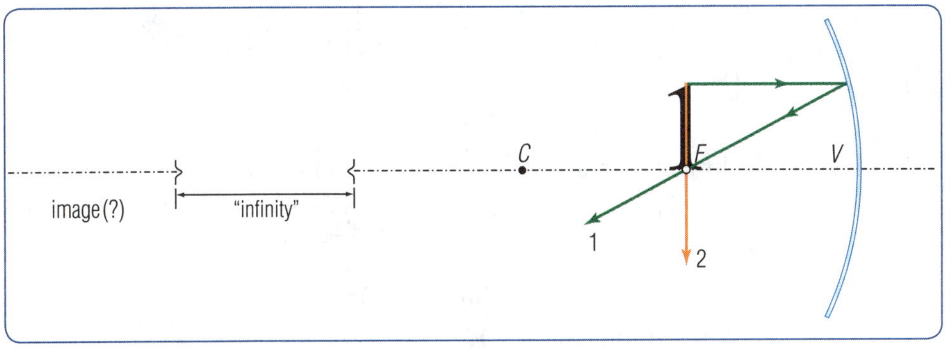

23-33 Reflection of an object when $d_O = f$ (Case 5)

Finally, consider an object between *F* and the mirror (Case 6). As Figure 23-34 shows, the rays do not converge. However, as with the plane mirror, the eye sees the rays as diverging from the same point, apparently behind the mirror. All points of the image are in the same vertical relationship as the object, so the image is upright, or *erect*. This image, like that in a plane mirror, is virtual—not real.

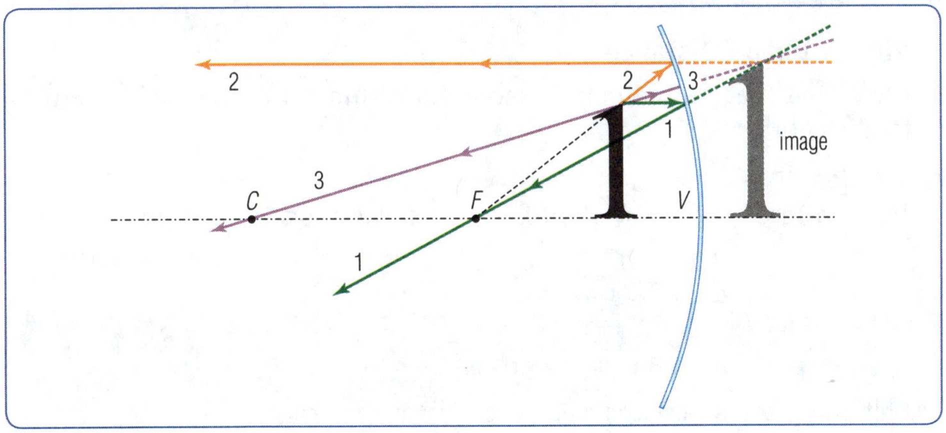

23-34 Reflection of an object when $d_O < f$ (Case 6)

23.23 Finding Image Position

Can the exact location of an image be predicted? Yes, if the mirror's focal length and the position of the object are known. The **mirror equation** is

$$\frac{1}{d_O} + \frac{1}{d_I} = \frac{1}{f}, \qquad (23.1)$$

where d_O is the distance between the object and the mirror, d_I is the distance between the image and the mirror, and f is the mirror's focal length. For laboratory mirrors, these distances are usually measured in centimeters.

Problem-Solving Strategy 23.1

Distances in front of the mirror are assumed to be positive; distances behind the mirror are negative. Upright orientations are considered positive; inverted orientations are considered negative.

EXAMPLE 23-1

Good Reflection: Finding the Image Distance

Locate the image formed by a concave mirror whose focal length is 10.0 cm if the object is 35.0 cm from the mirror.

Solution:

You know that $d_O = 35.0$ cm and $f = 10.0$ cm, and you want to find d_I. Use Equation 23.1:

$$\frac{1}{d_O} + \frac{1}{d_I} = \frac{1}{f}$$

$$\frac{1}{35.0\text{ cm}} + \frac{1}{d_I} = \frac{1}{10.0\text{ cm}}$$

$$\frac{1}{d_I} = \frac{1}{10.0\text{ cm}} - \frac{1}{35.0\text{ cm}}$$

Multiply by the lowest common denominator, $(70.0\text{ cm})d_I$.

$$\frac{(70.0\text{ cm})\cancel{d_I}}{\cancel{d_I}} = \frac{(70.0\cancel{\text{ cm}})d_I}{10.0\cancel{\text{ cm}}} - \frac{(70.0\cancel{\text{ cm}})d_I}{35.0\cancel{\text{ cm}}}$$

$$70.0\text{ cm} = 7.00d_I - 2.00d_I$$

$$70.0\text{ cm} = 5.00d_I$$

$$d_I = 14.0\text{ cm}$$

The image is 14 cm from the mirror. Since $f = 10.0$ cm and $R = 2f = 20.0$ cm, the image is between F and C (Case 4).

EXAMPLE 23-2

Virtual Image Distance

Locate the image of an object placed 4.0 cm from a concave mirror with focal length $f = 8.0$ cm.

Solution:

This is a case in which $d_O < f$ (Case 6). Use Equation 23.1 again:

$$\frac{1}{4.0\text{ cm}} + \frac{1}{d_I} = \frac{1}{8.0\text{ cm}}$$

$$\frac{1}{d_I} = \frac{1}{8.0\text{ cm}} - \frac{1}{4.0\text{ cm}}$$

Multiply by the lowest common denominator, $(8.0\text{ cm})d_I$.

$$\frac{(8.0\text{ cm})\cancel{d_I}}{\cancel{d_I}} = \frac{(\cancel{8.0\text{ cm}})d_I}{\cancel{8.0\text{ cm}}} - \frac{(8.0\cancel{\text{ cm}})d_I}{4.0\cancel{\text{ cm}}}$$

$$8.0\text{ cm} = d_I - 2.0d_I$$

$$8.0\text{ cm} = -1.0d_I$$

$$-8.0\text{ cm} = d_I$$

A negative d_I means that the image is *behind* the mirror; it is a virtual image.

EXAMPLE 23-3

Loss of Focus: Object at Infinity

Locate the image of an object at infinity (Case 1) if the concave mirror's focal length is 18 cm.

Solution:

You can solve this in one of two ways. First, if you know that the image of an object at infinity is at the focal point, you can simply say that

$$d_I = f = 18 \text{ cm}.$$

Or you can use Equation 23.1 to work it out.

$$\frac{1}{d_O} + \frac{1}{d_I} = \frac{1}{f}$$

But $d_O = \infty$, so

$$\frac{1}{\infty} + \frac{1}{d_I} = \frac{1}{f}.$$

As d_O approaches ∞, $1/\infty$ approaches 0:

$$0 + \frac{1}{d_I} = \frac{1}{f}$$

$$d_I = f = 18 \text{ cm}$$

Problem-Solving Strategy 23.2

No object is truly at an infinite distance from a mirror. However, you can assume that an object is at infinity if its incident light rays are nearly parallel.

23.24 Magnification

For all spherical mirrors, the height of the image (H_I) relates to the height of the object (H_O) by this equation:

$$\frac{H_I}{H_O} = -\frac{d_I}{d_O} \qquad (23.2)$$

The negative sign maintains the correct sign relationship. The object distance and object height are always positive for mirrors. For real images, image distance is always positive because it is in front of the mirror, and image height is always negative because it is inverted. Virtual images have negative image distances and positive heights because they are erect.

Notice in Equation 23.2 that if the image is farther from the mirror than the object is, the image must be larger than the object to maintain the proportionality. If the image is closer to the mirror than the object is, the image is smaller than the object. The **magnification** of an image is the absolute value of the image height to the object height:

$$m = \left|\frac{H_I}{H_O}\right| \qquad (23.3)$$

Problem-Solving Strategy 23.3

The magnification of an object may be greater or less than one but never negative as long as an image is formed. Magnification may be zero only when an object at an infinite distance forms a point image at the principal focus.

EXAMPLE 23-4

Supersizing: Magnification

Find the size of the image of a 15.0 cm object that is 30.0 cm from a concave mirror whose focal length is 20.0 cm. What is the magnification?

Solution:

You must first find d_I using Equation 23.1, then H_I using Equation 23.2.

$$\frac{1}{30.0\text{ cm}} + \frac{1}{d_I} = \frac{1}{20.0\text{ cm}}$$

$$\frac{1}{d_I} = \frac{1}{20.0\text{ cm}} - \frac{1}{30.0\text{ cm}}$$

Multiply by $(60.0\text{ cm})d_I$.

$$\frac{(60.0\text{ cm})\cancel{d_I}}{\cancel{d_I}} = \frac{(60.0\ \cancel{\text{cm}})d_I}{20.0\ \cancel{\text{cm}}} - \frac{(60.0\ \cancel{\text{cm}})d_I}{30.0\ \cancel{\text{cm}}}$$

$$60.0\text{ cm} = 3.00d_I - 2.00d_I$$

$$60.0\text{ cm} = d_I$$

Solve for image height.

$$\frac{H_I}{H_O} = -\frac{d_I}{d_O}$$

$$H_I = -\frac{H_O d_I}{d_O} = -\frac{(15.0\text{ cm})(60.0\ \cancel{\text{cm}})}{30.0\ \cancel{\text{cm}}}$$

$$H_I = -30.0\text{ cm}$$

The real image ($d_I > 0$) is 30.0 cm tall and inverted ($H_I < 0$).

Find the magnification using Equation 23.3.

$$m = \left|\frac{H_I}{H_O}\right| = \left|\frac{-30.0\text{ cm}}{15.0\text{ cm}}\right| = 2.00\times$$

The image is magnified to twice the object height.

The symbol "×" is often used as a unit of magnification "power." A more formal way is to refer to the number of diameters of magnification.

23.25 Spherical Mirror Aberration and Parabolic Mirrors

For a concave spherical mirror, not all incoming rays parallel to the optical axis are reflected so that they actually pass through the focal point. Although light rays aligned with the optical axis are reflected directly back to the focal point, light rays that are off-axis (but still parallel to the axis) are reflected to points slightly in front of the focal point, toward the mirror. The farther away from the vertex of the mirror the incident ray is, the farther in front of the focal point the reflected ray lies. For this reason, images formed by a highly curved spherical mirror are blurry. As we mentioned earlier, this problem is called spherical aberration, and it exists in every spherical mirror. One method of compensating for spherical aberration is to use a relatively flat spherical mirror (R is large compared to the mirror's diameter). Because the curvature is small, reflected light rays focus close to the focal point. This method minimizes spherical aberration, but it cannot eliminate it entirely because it is a built-in characteristic of spherical mirrors.

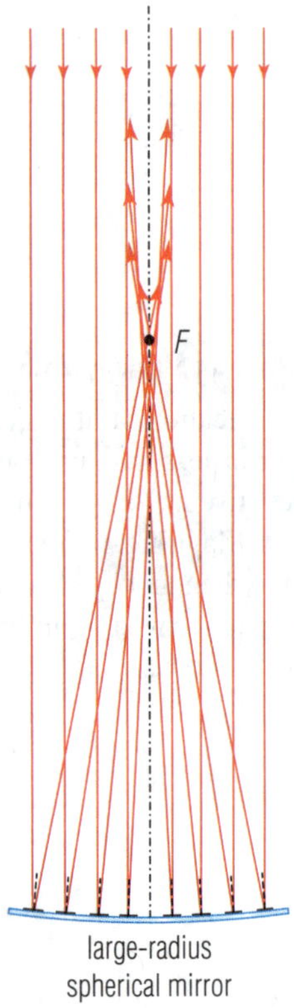

23-35 A spherical mirror with a long focal length is relatively flat. The reflected rays tend to focus very close to the geometric focal point ($R/2$).

A *paraboloidal* reflecting surface is not subject to spherical aberration. A paraboloid is formed by rotating a parabola around its axis of symmetry. Geometrically, every incident light ray parallel to the optical axis of the mirror is reflected through the principle focus, F. Remember that the key differences between spherical and parabolic mirrors are the properties of their centers of curvature and their focal points. Spherical mirrors have definite centers of curvature and indefinite focal points, while parabolic mirrors have indefinite centers of curvature and distinct focal points. So, the location of the effective center of curvature (C) for a parabolic mirror is defined in the same way as for spherical mirrors: $R = 2f$. Parabolic mirrors are the most common light-gathering element of Newtonian reflector telescopes.

Instead of focusing rays to form an image, parabolic mirrors can also project rays. When a light source is located at the principal focus, the mirror reflects parallel rays in a beam. This arrangement is used for flashlights, spotlights, searchlights, and automobile headlights.

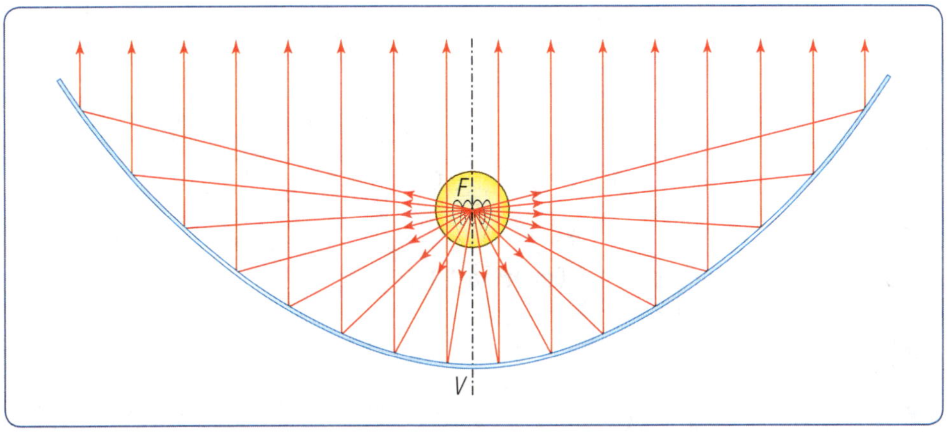

23-36 Reflection of a light source at the focus of a parabolic reflector

23.26 Convex Mirrors

Convex mirrors (mirrors that bulge out) can produce only one type of image. The image is always virtual (behind the mirror), erect, and smaller than the object. The principal focus is *behind* the mirror, again at $f = R/2$. The rules that you used for concave mirrors must be modified slightly. First, an incoming ray that is parallel to the optical axis has a reflection that, if extended through the mirror, seems to originate at the virtual focus. Second, an incoming ray that would, if extended through the mirror, pass through the focus is reflected parallel to the optical axis.

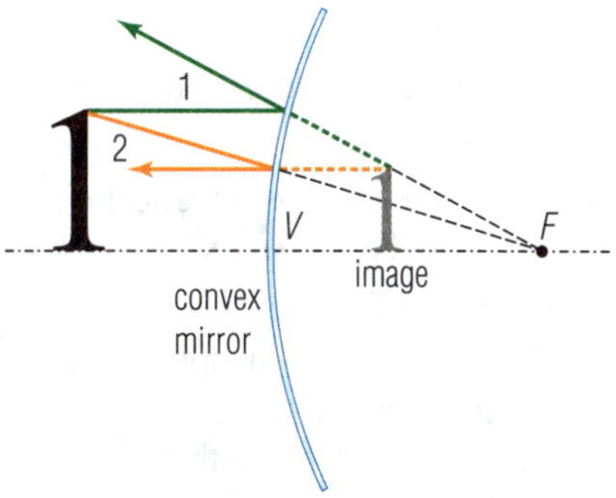

23-37 Reflection in a spherical convex mirror

Because convex mirrors reflect over a wide range of angles and their images are smaller than the objects, they give a panoramic view of their surroundings. They are therefore used as rearview mirrors, especially for trucks, as shoplifting deterrents in stores, and as safety devices to prevent forklift collisions at busy intersections in factories and warehouses. The earth's oceans also act as convex mirrors. Astronauts often see an image of the sun reflected in the oceans.

Equation 23.1 applies to convex as well as to concave mirrors. Of course, for convex mirrors f and d_I are negative numbers. Computing image size using Equation 23.2 is also valid for convex mirrors.

EXAMPLE 23-5

Through a Looking Glass: Virtual Images

Locate the image of a 2.0 cm object that is 10.0 cm from a convex mirror whose focal length is 10.0 cm. How high is the image?

Solution:

$d_O = 10.0$ cm and $f = -10.0$ cm

Use Equation 23.1 to find d_I.

$$\frac{1}{d_O} + \frac{1}{d_I} = \frac{1}{f}$$

$$\frac{1}{10.0 \text{ cm}} + \frac{1}{d_I} = \frac{1}{-10.0 \text{ cm}}$$

$$\frac{1}{d_I} = -\frac{1}{10.0 \text{ cm}} - \frac{1}{10.0 \text{ cm}}$$

$$\frac{1}{d_I} = -\frac{2}{10.0 \text{ cm}} = -\frac{1}{5.0 \text{ cm}}$$

$$d_I = -5.0 \text{ cm}$$

Solve for image height using Equation 23.2:

$$\frac{H_I}{H_O} = -\frac{d_I}{d_O}$$

$$H_I = -\frac{H_O d_I}{d_O} = -\frac{(2.0 \text{ cm})(-5.0 \text{ cm})}{10.0 \text{ cm}}$$

$$H_I = +1.0 \text{ cm}$$

The image is 5.0 cm behind the mirror and 1.0 cm high. The image is erect, which is another clue that it is a virtual image.

23C Objectives

After completing this section, I can

- ✓ explain optical reflection geometrically.
- ✓ illustrate plane mirror reflection and state the properties of a virtual image.
- ✓ given two plane mirrors that form an angle, predict the number of reflected images that they will produce.
- ✓ describe the optical properties of curved mirrors.
- ✓ identify the geometric characteristics of curved mirrors and determine the type, size, and location of images formed in them.
- ✓ illustrate the various kinds of real and virtual images formed by curved mirrors, based on the location of the reflected object.
- ✓ compute image position using the mirror equation.
- ✓ compute magnification of the image in a curved mirror.
- ✓ discuss several methods for reducing or eliminating spherical aberration in a curved mirror.

23C Section Review

1. What aspect of the way light propagates makes it relatively simple to study reflection?

DM 2. What kinds of information can scientists infer from astronomical albedo data?

DM 3. How could an astronomical object's albedo change over time?

4. a. Discuss the law of reflection.
   b. What are the two kinds of reflection?

5. a. Is your image reflected from a plane mirror really reversed? Explain your answer.
   b. What is meant by a mirror image?

6. Describe how to determine the location of an object's image after reflection by a concave spherical mirror.

7. Why is a virtual image always erect and a real image always inverted?

DM ✪ 8. Using the information in Table 23-1 and a spreadsheet program or graphing calculator, create a scatter plot of geometric albedo (p) (x values) vs. Bond albedo (A) (y values).
   a. Apply a trendline using a least-squares fit to your data. If the trendline seems to be linear, force the y-intercept to zero, if possible. Report the trendline's equation, substituting p for the x variable and A for the y variable. Also report the r^2 value of the trendline.

b. What is the relationship between these two variables?

c. Your equation should resemble the equation

$$A = qp,$$

where A is the Bond albedo, p is the geometric albedo, and q is called a *phase integral*, which computes the fraction of the total reflected light from the object as a function of the phase angle at which it is scattered. The phase angle is the angle between the light source (usually the sun) and the observer. According to your model, what is the value of q? What are you assuming about the nature of q? Evaluate your assumptions considering the variety of planetary objects in the solar system.

9. In a high-school laboratory, a spherical concave mirror 10.0 cm in diameter has a focal length of 35.0 cm. If a paper clip 4.70 cm long is mounted vertically in a stand on the mirror's optical axis 40.0 cm from the center of the mirror, where will the paper clip's image be? What will be its orientation?

10. Find the height of the paper clip's image if the paper clip in Question 9 is positioned 15.0 cm from the concave mirror. What is the magnification of the image at this point?

11. Solve Question 9 assuming that the mirror is a convex mirror.

Chapter Review

In Terms of Physics

albedo	525
spectrum	526
electromagnetic spectrum	526
Wien's displacement law	529
light-emitting diode (LED)	530
wave front	532
ray	535
reflection	535
angle of incidence (θ_i)	535
angle of reflection (θ_r)	535
law of reflection	535
virtual image	536
concave mirror	537
convex mirror	537
spherical aberration	537
principal focus (F)	538
focal length (f)	538
optical axis	538
object distance (d_O)	539
image distance (d_I)	539
real image	539
mirror equation	541
magnification ($\times$)	543

Problem-Solving Strategies

23.1 (page 542) Distances in front of the mirror are assumed to be positive; distances behind the mirror are negative. Upright orientations are considered positive; inverted orientations are considered negative.

23.2 (page 543) No object is truly at an infinite distance from a mirror. However, you can assume that an object is at infinity if its incident light rays are nearly parallel.

23.3 (page 543) The magnification of an object may be greater or less than one but never negative as long as an image is formed. Magnification may be zero only when an object at an infinite distance forms a point image at the principal focus.

Review Questions

1. What is visible light?
2. Name three types of waves in the electromagnetic spectrum.
3. **a.** For a given amplitude, which electromagnetic waves are the most energetic?
   b. Which are the least energetic?
4. Which kinds of rays are produced by objects converting thermal energy to radiant energy?
5. Which of the following is *not* a direct source of visible light?
   a. incandescence
   b. electrical current in a wire
   c. gas-discharge tubes
   d. chemical reactions
6. When it is initially formed, what is the shape of a light wave front from a point light source?
7. How did Maxwell further our understanding of the nature of light?
8. Did the scientist who described light waves mathematically devise the equations that he used?
9. What is a *normal,* in a discussion of reflection?
10. State the law of reflection.
11. **a.** Where is the image in a plane mirror reflection?
    b. Is it real or virtual?
12. What is a virtual image?
13. How is your mind tricked into thinking that the left and right sides of your reflection in a plane mirror are reversed?
14. Suppose you cut out a letter *B* and hold it up so that you read the letter correctly. Will its reflection in a plane mirror in front of you be backwards?
15. Where should you place an object in front of a concave mirror if you want its image to be larger than the object?
16. Where should you place an object if you want its image in a concave mirror to be virtual?
17. Name two uses for convex mirrors.

True or False (18–29)

18. Radio waves are always less energetic than visible light waves.
19. Red light has a higher frequency than blue light.
20. Extreme ultraviolet light from the sun is not a danger to life at the surface of the earth.
21. High-energy gamma rays measurable at the earth's surface probably originate from outside the earth.
22. Single-frequency light is obtained from gas-discharge and laser light.
23. Coherent light is obtained from cold-light sources.
24. The speed of light is a fundamental physical constant. It is approximately 3.00×10^8 m/s wherever it is measured.
25. Spherical wave fronts eventually become plane waves if allowed to propagate long enough and far enough.
26. Light rays travel in straight lines through uniform media.
27. The incident ray, the reflected ray, and the normal at the point of reflection all lie in the same plane.
28. Ideally, all of the incoming light rays parallel to the optical axis of a concave mirror reflect and pass through the same point.
29. A convex mirror cannot produce a real image.

DM✪30. From 2000 to 2001, a space probe called NEAR Shoemaker orbited the near-earth asteroid Eros and gathered data in an intense survey of the asteroid. It measured both its geometric and Bond albedo.
   a. Using the model that you developed in 23C Section Review, calculate Eros's Bond albedo if its geometric albedo was measured as 0.23.
   b. Eros's actual Bond albedo is 0.093. Calculate the percent error for your modeled value.
   c. Assess your model based on the percent error you calculated in Part b. What does this reveal about the nature of models?

✪31. Calculate the number of images produced by two plane mirrors at a right angle.

✪32. How many images are produced by two plane mirrors at a 30° angle?

✪33. An object is located 30.0 cm from a concave mirror with a focal length of 10.0 cm.
   a. Where will its image be?
   b. Will the image be larger or smaller than the object?

✪34. An object placed 20.0 cm from a concave mirror has an image of the same size at 20.0 cm from the mirror. What is the mirror's focal length?

✪35. A 6.00 cm long pencil is placed 30.0 cm from a concave mirror with a focal length of 10.0 cm. Where is the image and how tall is it?

✪36. A 2.00 cm high eraser is 20.0 cm from a convex mirror with a focal length of 5.00 cm. Where is the image and how tall is it?

DM✪37. Saturn's moon Enceladus has the highest albedo of nearly any solid object in the solar system. After researching, provide at least two factors to explain this observation.

Digital cameras still require an optical lens to form the image from which a picture is made.

Refraction 24

Dominion Modeling

Refraction and Glass Composition

How many things can you name that use glass? The lenses in your glasses, TV screens, glass beverage bottles, car and building windows, fiberglass insulation, fiber-optic cable, light bulbs, drinking glasses—all of these things are common applications of glass. The primary ingredient of glass is silicon dioxide, or sand, but the many different kinds of glass contain other elements or compounds to customize them for their specific purposes. In fact, there are about 50 000 types of glass. Some glasses even contain iron oxide (rust) or water! How do these components affect the ability of glass to refract light?

24-1 Glass is a versatile material that is used in a large variety of useful products, only a few of which are shown.

24A THEORY OF REFRACTION

24.1 Cause of Refraction

When light travels through matter, it moves more slowly than it does in a vacuum. Some media (matter through which light traverses) slow light down more than others. When light travels from one medium to another in which its speed is different, the light rays may bend. This bending is called **refraction**.

An illustration may help you to understand why light refracts. Imagine a company of soldiers marching in formation. Each column of soldiers represents a light ray, and each row of soldiers represents a wave front. The soldiers are marching on a level concrete surface at an angle to a boundary with an expanse of tall grass and mucky ground. As each soldier reaches the grass, he is forced to slow down. He cannot keep up with the adjacent soldier who is still on the paved surface. Therefore, the row bends. At the same time, the soldiers in the grass are trying to keep aligned with the soldiers in their row, so the column also bends.

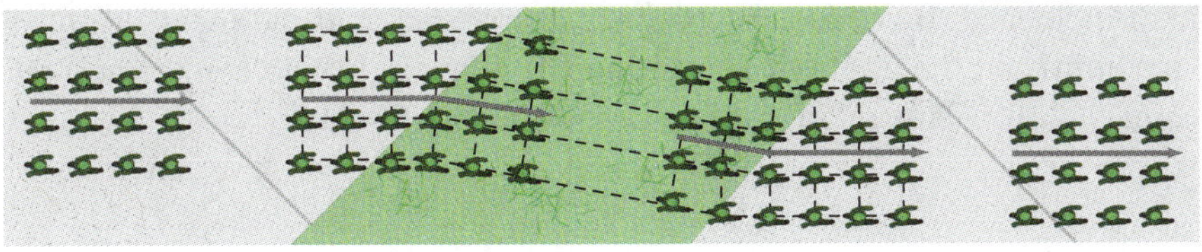

24-2 Refraction can be illustrated by a group of soldiers marching into tall grass from a smooth concrete surface while attempting to keep formation.

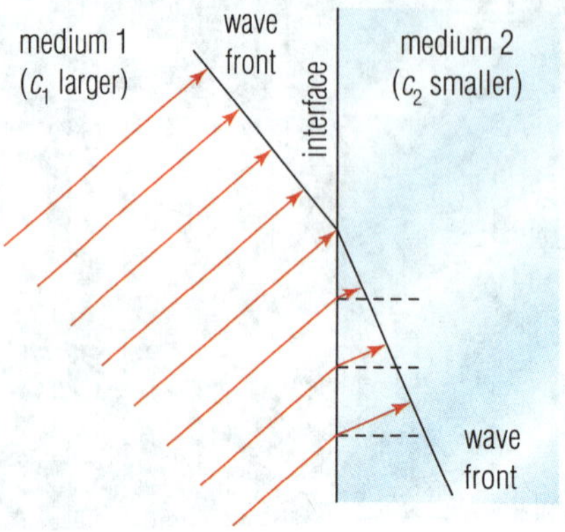

24-3 Light refracting as it passes from one medium into an optically denser one

Similarly, when light passes from one medium to another at an angle, its wave fronts bend. The light rays, which are always perpendicular to the wave front, bend too. Figure 24-3 shows the direction that a light ray bends. When light passes from a medium in which its speed is faster to a medium in which its speed is slower, it bends toward a *normal* line (a line drawn perpendicular to the boundary at the point where the ray crosses the boundary). When light goes from a medium in which it moves slower to a medium in which it moves faster, it bends away from the normal. The boundary between two media is called the **interface** between the media. Light travels slower in a medium that is optically denser. The optical density of a medium is often related to its mass density.

How much does a medium slow light? The optical density of a medium is indicated by its **index of refraction (n)**, which is the ratio of the speed of light in a vacuum to the speed of light in the medium:

$$n = \frac{c_{\text{vac}}}{c_{\text{med}}} \qquad (24.1)$$

From the medium's refractive index, you can calculate the speed of light in the medium.

The word **interface** is not strictly an optical term. It is useful in any study of something occurring at the boundary between two different substances, or even between two systems.

Problem-Solving Strategy 24.1

When the two media at an interface are named, the first contains the incident ray, and the second contains the refracted ray. In an air-glass interface, the ray is moving from the air into the glass.

The **index of refraction (n)** (or the *refractive index*) of a medium is the ratio of the speed of light in a vacuum to the speed of light in the medium. Notice that indices of refraction are dimensionless because the units cancel.

TABLE 24-1

Selected Indices of Refraction

Medium	n
hydrogen	1.0001
air	1.0003
carbon dioxide	1.0004
ice	1.309
methyl alcohol	1.33
water	1.333
ethyl alcohol	1.36
fluorite	1.43
Lucite	1.51
glass (crown)	1.52
sodium chloride	1.54
glass (flint)	1.64
diamond	2.42

EXAMPLE 24-1

Speed Zone: Speed of Light in Water

Water's refractive index is 1.33 at room temperature. What is the speed of light in water?

Solution:

$$n_{\text{water}} = 1.33$$

$$n_{\text{water}} = \frac{c_{\text{vac}}}{c_{\text{water}}}$$

$$c_{\text{water}} = \frac{c_{\text{vac}}}{n_{\text{water}}} = \frac{3.00 \times 10^8 \text{ m/s}}{1.33}$$

$$c_{\text{water}} \cong 2.2\underline{5}5 \times 10^8 \text{ m/s} \cong 2.26 \times 10^8 \text{ m/s}$$

24.2 Description of Refraction

In 1621, **Willebrord Snell** discovered a mathematical relationship between the angle of incidence (θ_i) (the angle between the normal and the incident ray in the first medium) and the **angle of refraction (θ_r)** (the angle between the normal and the refracted ray). This relationship is known as **Snell's law**:

$$n_i \sin\theta_i = n_r \sin\theta_r \qquad (24.2)$$

In this equation n_i is the index of refraction of the medium containing the incident ray, n_r is the index of refraction of the medium containing the refracted ray, θ_i is the angle of incidence, and θ_r is the angle of refraction.

Willebrord Snell (1580–1626) was a Dutch mathematician best known for showing that different materials have different indices of refraction.

Figure 24-4 shows why Snell's law is true. A light ray is being refracted as it travels from the first medium to the second medium, and a circle is drawn with its center at the point of refraction. Both $\overline{AB}$ and $\overline{BC}$ are radii of the circle; therefore, $AB = BC$. The dotted lines $\overline{AE}$ and $\overline{CD}$ are perpendicular to the normal and are called *semichords*, since they each form half a chord of the circle. Results of experiments that measure various angles of incidence and refraction at a glass-air interface and the corresponding ratios of *AE/CD* are tabulated in Table 24-2.

TABLE 24-2

θ_i (°)	θ_r (°)	*AE* (cm)	*CD* (cm)	*AE/CD*
10	6.7	3.43	2.30	1.51
20	13.3	6.84	4.60	1.49
30	19.6	10.00	6.71	1.49
40	25.2	12.83	8.52	1.51
50	30.7	15.32	10.2	1.50
60	35.1	17.32	11.5	1.51
70	36.6	18.79	12.5	1.50
80	40.6	19.69	13.02	1.51

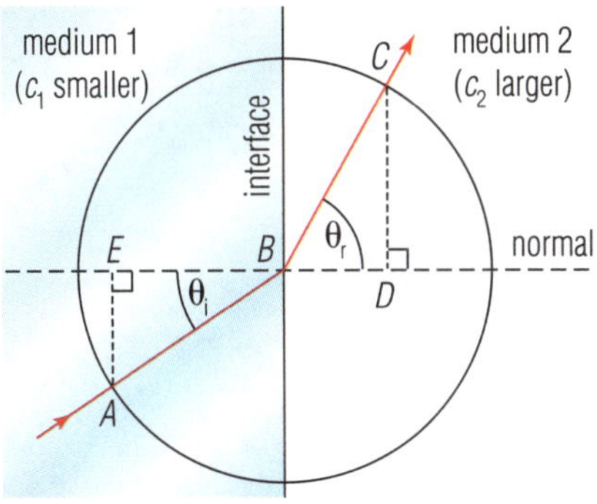

24-4 Derivation of Snell's law

The symbol $\overline{AE}$ represents the line segment between endpoints *A* and *E*. The magnitude of the line segment is indicated simply by the endpoints (*AE*).

The ratio *AE/CD* is a constant within the limits of experimental error. It is also close to the ratio of the index of refraction of glass to the index of refraction of air,

$$\frac{AE}{CD} = \frac{n_{\text{glass}}}{n_{\text{air}}}.$$

How is *AE/CD* related to the angles of incidence and refraction? Notice that ΔBDC and ΔBEA are right triangles. Recall that the angles and sides of right triangles can be related by the use of sine ratios. The sine of an acute angle in a right triangle is defined as

$$\sin\theta = \frac{\text{opposite}}{\text{hypotenuse}}.$$

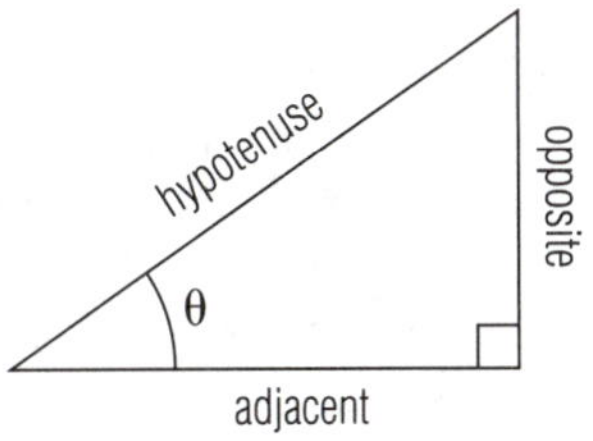

24-5 The relationship of the sides to an angle in a right triangle

In Figure 24-4,

$$\sin\theta_i = \frac{AE}{AB}, \text{ and } \sin\theta_r = \frac{CD}{BC}.$$

Since $AB = BC$, $\sin\theta_i$ can be rewritten as

$$\sin\theta_i = \frac{AE}{BC}.$$

The ratio $\sin\theta_i/\sin\theta_r$ is

$$\frac{\sin\theta_i}{\sin\theta_r} = \frac{AE/\cancel{BC}}{CD/\cancel{BC}} = \frac{AE}{CD}.$$

We know that

$$\frac{AE}{CD} = \frac{n_{\text{glass}}}{n_{\text{air}}}, \text{ so}$$

$$\frac{\sin\theta_i}{\sin\theta_r} = \frac{n_{\text{glass}}}{n_{\text{air}}}. \quad (24.3)$$

Equation 24.3 is true for any combination of transparent materials. If a light ray passes from a medium containing the incident ray with an index of refraction n_i to a second medium with an index of refraction n_r, then

$$\frac{\sin\theta_i}{\sin\theta_r} = \frac{n_r}{n_i}.$$

By cross-multiplying, you get the equation

$$n_i \sin\theta_i = n_r \sin\theta_r,$$

which is the mathematical statement of Snell's law.

EXAMPLE 24-2

Finding the Angle of Refraction, θ_r

Find the angle of refraction when an incident ray ($\theta_i = 30.0°$) passes from air into ethyl alcohol (ETOH).

Solution:

From Table 24-1 you find that $n_{ETOH} = 1.36$.

$$n_i \sin\theta_i = n_r \sin\theta_r$$

Dividing by n_r, you get the following:

$$\frac{n_i \sin\theta_i}{n_r} = \sin\theta_r$$

$n_{air} = n_i = 1.00$; $n_{ETOH} = n_r = 1.36$; $\theta_i = 30.0°$

$$\sin\theta_r = \frac{1.00 \sin 30.0°}{1.36}$$

$$\sin\theta_r \cong 0.3676$$

$$\theta_r = \sin^{-1}(\sin\theta_r) \cong 21.\underline{56}°$$

$$\theta_r \cong 21.6°$$

The angle of refraction in the ethyl alcohol is about 21.6° from the normal.

Problem-Solving Strategy 24.2

You can avoid rounding error when using Snell's law to compute angles of incidence or refraction by taking the inverse sine (arc-sine) of the right-hand side of the equation.

For example, the angle of refraction is found by

$$\theta_r = \sin^{-1}\left(\frac{n_i \sin\theta_i}{n_r}\right).$$

EXAMPLE 24-3

Finding an Unknown Index of Refraction

A ray of light passes through the wall of a glass container ($n_{glass} = 1.50$) to the liquid in the container at the glass-liquid interface. The angle of incidence is 35.0°, and the angle of refraction is 40.0°. What is the liquid's index of refraction?

Solution:

$$n_i \sin\theta_i = n_r \sin\theta_r$$

Divide by $\sin\theta_r$ to isolate n_r:

$$\frac{n_i \sin\theta_i}{\sin\theta_r} = n_r$$

$n_{glass} = n_i = 1.50$; $\theta_i = 35.0°$; $\theta_r = 40.0°$

$$n_r = \frac{1.50 \sin 35.0°}{\sin 40.0°}$$

$$n_r \cong 1.3\underline{38} \cong 1.34$$

The liquid's index of refraction is about 1.34.

24.3 Critical Angle of Incidence

Let's look at Snell's law a little closer:

$$n_i \sin \theta_i = n_r \sin \theta_r$$

If n_i is greater than n_r, then $\sin \theta_i$ must be less than $\sin \theta_r$ to maintain the equality. Therefore, light moving from one medium to another with a lower index of refraction has an angle of refraction larger than its angle of incidence. Such a boundary has interesting effects on light with large angles of incidence.

When a light ray passes from a medium of low n to one of higher n, the ray bends *toward* the normal. When the ray passes into a medium of lower n, the ray bends *away* from the normal.

EXAMPLE 24-4

Losing the Refracted Ray: Critical Angle

A light ray passes from water to air with an angle of incidence of 48.61°. What is its angle of refraction?

Solution:

$$n_i \sin \theta_i = n_r \sin \theta_r$$

Divide by n_r to isolate $\sin \theta_r$.

$$\frac{n_i \sin \theta_i}{n_r} = \sin \theta_r$$

$$\theta_r = \sin^{-1}\left(\frac{n_i \sin \theta_i}{n_r}\right)$$

$n_{water} = n_i = 1.333$; $n_{air} = n_r = 1.000$; $\theta_i = 48.61°$

$$\theta_r = \sin^{-1}\left(\frac{1.333 \sin 48.61°}{1.000}\right)$$

$$\theta_r = 90.0°$$

The refracted ray will travel along the interface boundary and be lost.

Problem-Solving Strategy 24.3

Critical angles exist only when the refraction index of the medium containing the incident ray is larger than that of the medium containing the refracted ray ($n_i > n_r$).

What if the angle of incidence in the example were greater than 48.61°? Then $\sin \theta_r > 1.000$. But no angle has a sine greater than 1. Therefore, the ray is not refracted at all but is *reflected* back into the water. This phenomenon, called **total internal reflection**, occurs at a water-air interface whenever light rays have an angle of incidence greater than 48.61°. This angle of incidence is called the **critical angle (θ_c)**. Every interface in which light passes from a medium with a higher index of refraction into a medium with a lower index of refraction has a critical angle of incidence at which light is refracted 90° along the interface surface. At angles of incidence greater than the critical angle, the light experiences total internal reflection.

Problem-Solving Strategy 24.4

Snell's law is valid as long as the critical angle at the interface of the two media is not exceeded.

EXAMPLE 24-5

Finding the Critical Angle

Find the critical angle for a glass-air boundary.

Solution:

You want to find θ_i for $\theta_r = 90.0°$.

$$n_i \sin \theta_i = n_r \sin \theta_r$$

24-6 The image of the diver reflected from the water's surface is an example of total internal reflection. Light rays from objects visible from above the water follow a path to the observer that is refracted at an angle less than the critical angle.

Divide by n_i to isolate $\sin \theta_i$:

$$\sin \theta_i = \frac{n_r \sin \theta_r}{n_i}$$

$$\theta_i = \sin^{-1}\left(\frac{n_r \sin \theta_r}{n_i}\right)$$

$$n_{glass} = n_i = 1.50;\ n_{air} = n_r = 1.00;\ \theta_r = 90.0°$$

$$\theta_i = \sin^{-1}\left(\frac{1.00 \sin 90.0°}{1.50}\right) \cong 41.\underline{8}1°$$

$$\theta_i \cong 41.8°$$

The critical angle for the glass-air interface is about 41.8°.

Total internal reflection has many uses. One application is in **fiber optics**, a technology in which a light beam is directed into a glass fiber. The light experiences total internal reflection within the fiber, even if the fiber bends, as long as it is not bent sharply. A fiber used in this way is called a *light pipe.* Bundles of fibers in the form of cables transmit high-speed digital signals across great distances. The advantages of fiber optics in digital telecommunications are the higher signal-to-noise ratio possible with the optical signal and the fact that fiber optics are impervious to *electromagnetic interference (EMI)* along the cable, which can degrade standard electrical or radio-frequency signals carried by a wire. Fiber optics is also used for optical inspection of inaccessible locations within machinery and the human body. Doctors regularly use fiber-optic devices for observing surgical procedures within small cavities in a patient's body as well as for carrying precision laser beams required for delicate surgery. These are yet more examples of using technology to love one's neighbor.

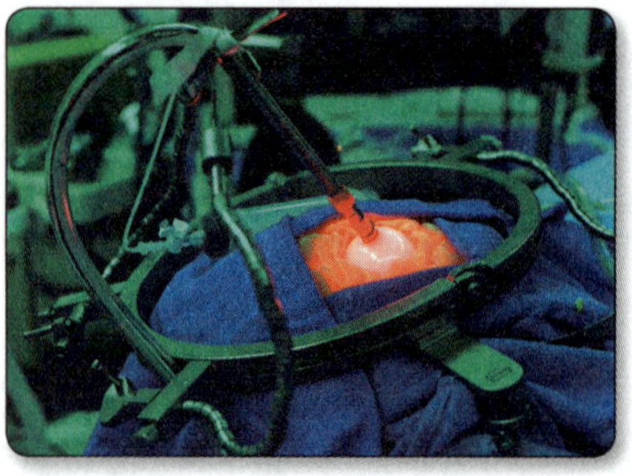

24-7 Laser light delivered by a fiber-optic cable is used for extremely precise brain surgery for which a scalpel could not be used.

In any communications system, noise always travels along with the desirable information. Engineers use the term **signal-to-noise ratio** (SNR) to express the quality of the resulting combination. Signal-to-noise ratios can express quality in a wide variety of signaling applications and mediums, including those involving electricity, sound, and light. SNR is calculated as the ratio of the desirable signal's power to the noise's power. A high SNR is desirable because it means that the signal is much stronger than the noise.

24.4 Phenomena Involving Refraction

We have been discussing the refraction of light as though all light refracts in the same way. Actually, different frequencies of light have different indices of refraction in the same materials. These differences occur because the wavelengths of light toward the blue end of the spectrum are closer to the dimensions of the particles in matter, so those colors tend to interact more strongly with matter and are slowed more than colors toward the red end of the spectrum. In our examples and tables, we have been using an average index of refraction, based on a yellow-green light. Since each frequency of light has a different index of refraction, different frequencies with the same angles of incidence have different angles of refraction. If white light—light of all frequencies—is incident on a refractive interface, the different frequencies will not emerge from the interface at the same angle. Instead, they will be spread out, or *dispersed*, displaying a spectrum of colors.

A prism is a triangular solid, usually made of glass, that causes **dispersion**. When white light enters the glass, the different frequencies are refracted at different angles.

24-8 A prism disperses light.

Light toward the blue end of the spectrum is refracted more than red light. This is the reason that prisms disperse white light into its color spectrum.

Emerging from the glass, the rays of different frequencies separate even more. (For glass with parallel edges, the second edge would reverse the dispersion, not continue it.) Since different frequencies have different colors, it is easy to see the dispersion of white light through a prism. The dispersion can be reversed incompletely with a second prism placed upside down in the dispersed beam. However, reconstruction of the original light beam is not easily done without a series of prisms and lenses.

An atmospheric rainbow results from dispersion only when specific conditions exist. Numerous water droplets must be illuminated directly by sunlight, and the observer must be between the sun and the water droplets. The colors of the sunlight are dispersed by being refracted twice and reflected once inside each raindrop (see Figure 24-9). The arching shape of the rainbow is due to the geometry of the effect. The observer sees the light rays returning from the water droplets within a certain angle of an imaginary line, called the *antisolar line*, that passes from the sun through the observer's head. Since blue light is refracted more than red, it appears about 40° all around the antisolar line. Red light, being least refracted, appears about 42° away. The other colors fall in between. If you were in an airplane observing a rainbow, you could see a complete circle of colors. Because the ground intersects the circle under most circumstances, you see only the upper arch. Secondary rainbows can sometimes be observed about 51° from the antisolar line. These rays are reflected twice inside the water drop, reversing the order of the colors in the process.

Refraction is often the cause of optical illusions. If you dip a stick into a glass of water, it looks bent and distorted in size. As Figure 24-10b shows, the ruler looks bent because the eye receives refracted rays of light from the stick. The mind is designed so that it interprets light rays as straight lines. Therefore, it extrapolates the rays back to where they would have met if they had not been refracted, and it constructs the image at that location.

Refraction can occur in the atmosphere when layers of hot air are close to layers of cool air. Air's index of refraction varies with temperature and humidity—hot air is less dense, and cold air is more dense. Index of refraction generally increases with air density, so light moving from the cool air to the hot, or vice versa, is refracted. In extreme cases this atmospheric refraction makes light rays from the sky appear to come from the ground, or rays from an object near the ground appear to come from the sky. The former is called a **mirage**, which is the refraction of

Dispersion as a Sign

The rainbow reminds us of God's promise not to destroy the earth again with a flood. "And God said, This is the token of the covenant which I make between me and you . . . I do set my bow in the cloud" (Gen. 9:12–13).

It is something of scientific folklore that a dispersed light beam can be reconstituted by using a second identical prism. You will be greatly frustrated if you attempt this procedure, because the divergent paths of the light beams cannot be made to converge back into the original beam in one optical step.

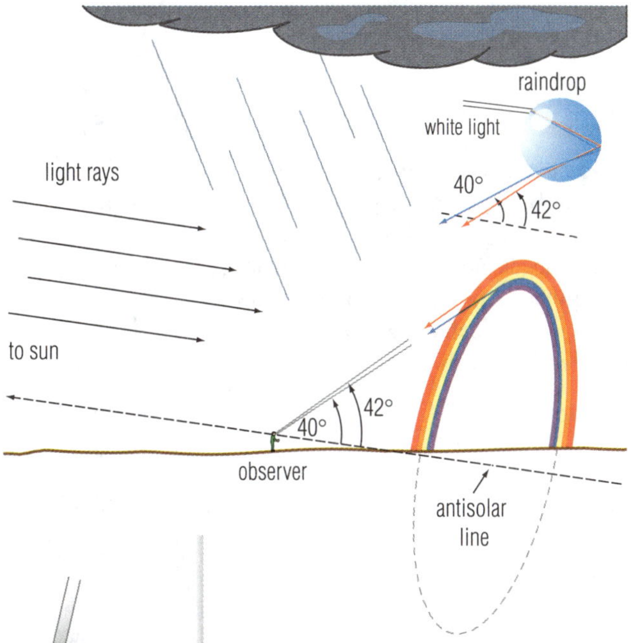

24-9 Rainbows are observed at an angle between 40° and 42° from the antisolar line.

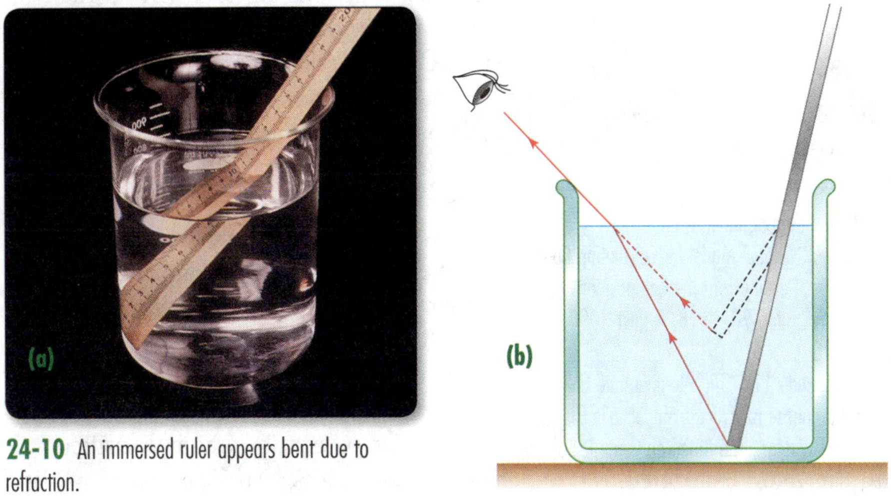

24-10 An immersed ruler appears bent due to refraction.

Snell's law assumes that a distinct interface exists between two uniform media. The word "uniform" is key. If the optical characteristics of a medium gradually change with position or time, then a ray passing through the medium is going to be refracted in curves rather than in angles between straight segments. This kind of refraction is responsible for many atmospheric visual phenomena as well as the propagation of sound through the deep ocean basins.

the sky and horizon that gives the appearance of a reflection from the surface of water in the desert. The latter, called an *inferior mirage* or *looming*, causes ships and structures to appear to float above the horizon and makes distant lighthouses and ships visible that are normally hidden by the earth's curvature.

Minor refraction in the atmosphere due to atmospheric movement makes the stars "twinkle." Beyond our atmosphere, the stars are seen to shine steadily. Similarly, on hot days distant objects viewed through the unstable air over a black road surface shimmer because of refraction.

24.5 Modeling the Index of Refraction and Glass Composition

From the data listed in Table 24-1, you can see that different types of materials have different indices of refraction. You learned that this is because their optical densities affect the speed of light in these materials. Interestingly, the optical density of a transparent material, especially glass, is related to its mass density.

Different types of glass are formulated with different elements and compounds, depending on the glass's intended purpose. Certain glasses called borosilicate glasses contain compounds like boron oxide (B_2O_3), which limit how much these glasses expand with temperature. This is important for glasses that are to be used in the oven or over a flame. Some very expensive glasses consist completely of fused silicon dioxide (SiO_2) and can withstand temperatures of almost 1200 °C for short periods of time. TV glass panels contain a higher percentage of aluminum oxide (Al_2O_3) which makes them tough and resistant to breakage and scratching. Glass used in textile fibers, bottles and jars, and windows contains a higher percentage of calcium oxide (CaO), which makes the glass easier to mold. Other properties affected by additives include a glass's viscosity, its electrical and thermal conductivities, and its optical properties. In particular, additives determine a glass's index of refraction. Not surprisingly, chemical additives alter the mass density of glass compared to pure silicon dioxide.

In order to evaluate the relationship between mass density and index of refraction, you will make a model using data obtained from experiments with container glass. Container glass is composed of various mixtures of silicon dioxide and oxides of sodium, calcium, potassium, aluminum, magnesium, lithium, titanium, iron, sulfur, and chromium. The percent by weight for silicon dioxide is never less than 63%. Table 24-3 lists

TABLE 24-3

Density and Refraction Data for Container Glass at 20°C

Density (ρ) (g/cm^3)	Index of Refraction (n)
2.40314	1.5019
2.45052	1.5106
2.46402	1.5138
2.46594	1.5150
2.46831	1.5170
2.48110	1.5266
2.48149	1.5205
2.48686	1.5239
2.49411	1.5228
2.49806	1.5261
2.50124	1.5318
2.53491	1.5336
2.53636	1.5336
2.54388	1.5351
2.56401	1.5439

Table based upon data obtained from:
Alexander Fluegel: "Global Model for Calculating Room-Temperature Glass Density from the Composition"; J. Am. Ceram. Soc., vol. 90, no. 8, August 2008, pp. 2622–2625.
Shelby J. E.; Appendix I in: "High Temperature Glass Melt Property Database for Process Modeling"; Eds. T.P. Seward III and T. Vascott; The American Ceramic Society, Westerville, Ohio, 2005, ISBN 1-57498-225-7.

Trashing God's Green Earth?

God made us stewards of His creation. But unfortunately, we generate lots of trash, much of it as waste glass. Americans alone discard more than 13 million tons of glass per year! How can we use technology to exercise better stewardship?

the mass density of different container glasses and their associated indices of refraction. The refraction data was obtained for yellow light at 589 nm.

24A Section Review

1. How does refraction occur?
2. What measurable quantities are considered in Snell's law?
3. What is the angle of refraction when the incident ray is at the critical angle for the two materials forming the interface?
4. Explain why red light is refracted the least and blue light is refracted the most.
5. Calculate the speed of light in diamond.
6. A new thermoplastic is being tested to determine its properties. A laser beam enters a sample with an incident angle of 35.0°. The beam is refracted at an angle of 21.3° to the normal. What is the plastic's refractive index?
7. What is the critical angle for a diamond-air interface?
8. Imagine that you are sitting on the bottom of a swimming pool looking upward toward the sky. What is the maximum angle from the vertical at which you can see the sky?
9. DM Using the information in Table 24-3, create a scatter plot of index of refraction (n) (y values) versus glass density (ρ) (x values) in a spreadsheet program or with your graphing calculator.
   a. Create a trendline using a least-squares fit of the data, if possible. Give the trendline's equation and r^2 value, if applicable.
   b. What kind of proportionality exists between glass density and index of refraction?
   c. Describe the limitations and applicability of your model.

24A Objectives

After completing this section, I can

- ✓ describe optical refraction.
- ✓ identify the property of a medium that causes refraction.
- ✓ use Snell's law to solve problems in optical refraction.
- ✓ define the critical angle of incidence and state the conditions for which it is applicable.
- ✓ explain how total internal reflection occurs and is usefully applied.
- ✓ discuss how refraction relates to prismatic dispersion, rainbows, optical illusions, and mirages.
- ✓ model the relationship between mass density and optical density for certain kinds of glass.

24B APPLICATION OF REFRACTION—LENSES

24.6 Common Lenses

One familiar application of refraction is in lenses—disks of transparent material with one or both sides machined into a three-dimensional curved surface. It is likely that many of you are wearing corrective lenses for your eyes, either glasses or contact lenses. Some lenses *converge*, or bring together, light rays, while other lenses *diverge*, or spread apart, light rays. Common lenses and their effects are summarized in the following table.

TABLE 24-4

Selected Lens Types

Lens Type	Shape	Effect	Lens Type	Shape	Effect	Lens Type*	Shape	Effect
Plane Convex		converging	Plane Concave		diverging	Positive Meniscus	 $R_{front} < R_{rear}$	converging
Double Convex		converging	Double Concave		diverging	Negative Meniscus	 $R_{front} > R_{rear}$	diverging

* R_{front} is the radius of the front lens surface relative to the incoming ray.
R_{rear} is the radius of the rear lens surface relative to the incoming ray.

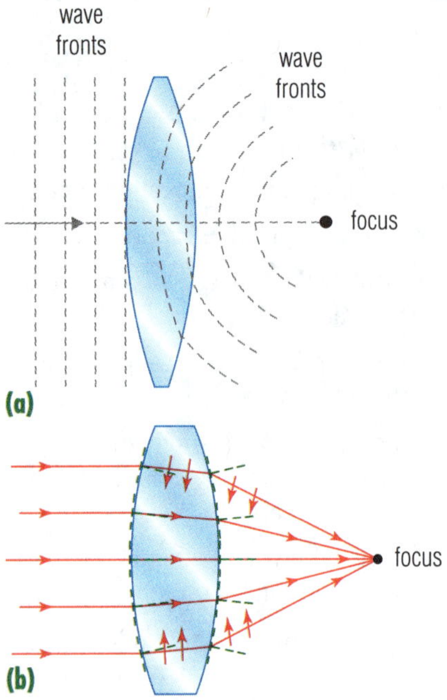

24-11 Converging lens function considering (a) wave fronts and (b) ray paths

Problem-Solving Strategy 24.5

All lens distances should be measured from the center of the lens, not from the surface.

24.7 Converging Lenses

Notice that a lens which is thicker in the middle than at the edges is a **converging lens**. Figure 24-11 shows why. Assume that the wave front is planar at first. The rays near the edges of the lens have to go through less of the lens; thus they are slowed less than the rays near the middle, which must go through more lens material. Therefore, the outer waves "get ahead" of the inner waves, and the wave-front envelope curves. The wave fronts are refracted so that after passing through the lens they meet at a point called the *focus* or *focal point* of the lens.

One ray's refraction can also show how the rays converge. The ray strikes the lens, and, since it is entering a material with a higher index of refraction than air, it bends toward the normal. When the ray leaves the lens, it enters the air, which has a lower index of refraction than the lens, so the ray bends away from the normal. Both refraction paths tend to bend the ray toward the focus.

A thin converging lens is much like a concave mirror. The same terminology is used for both. Like concave mirrors, converging lenses have a focus at a distance f from their centers on the optical axis. For symmetrical spherical lenses, f is one-half the radius (R) of the lens's curvature. One difference between mirrors and lenses is that a lens's real image is located on the opposite side of the lens from its object. We will always assign the object distance (d_O) a negative value. The image may be real or virtual at a distance d_I from the center of the lens. If the image is real, d_I will have a positive value; if the image is virtual, it will appear to be on the same side of the lens as the object and will have a negative d_I. The mirror equation must be modified to relate these quantities correctly:

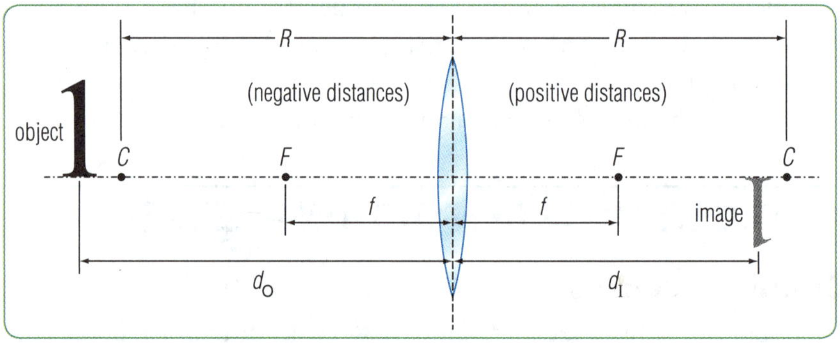

24-12 Optical parameters used in the thin-lens equation

A *thin lens* consists of two refracting surfaces having a small distance between them compared to (a) image and object distances and (b) the curvature radii of the surfaces. The diagrams of lenses in this chapter are shown relatively thick for illustrative purposes only.

$$-\frac{1}{d_O} + \frac{1}{d_I} = \frac{1}{f} \quad (24.4)$$

Since only thin lenses obey this equation, it is called the **thin-lens equation**.

The height of a thin-lens image can be calculated using an equation similar to Equation 23.2 but without the negative sign on the right. This equation is valid if the medium on both sides of the lens is the same, as in the case of a hand magnifier:

$$\frac{H_I}{H_O} = \frac{d_I}{d_O} \quad (24.5)$$

Magnification of a thin lens uses the same formula as Equation 23.3:

$$m = \left|\frac{H_I}{H_O}\right| \quad (24.6)$$

As with mirrors, lens magnification is denoted by the word "power" or the "×" symbol. A lens that achieves a ten-fold magnification is a ten-power (10×) lens.

Determining image height when the medium on each side of the lens is different is more difficult because three indices of refraction are involved. Therefore, the location and size of the object and image for each surface must be considered individually. Contact lenses are one application in which these factors must be taken into account.

EXAMPLE 24-6

Maintaining Focus: Using the Thin-Lens Equation

A double convex thin lens has a focal length of 16.0 cm. An object 6.00 cm high is placed 40.0 cm from the lens. Find (a) the image distance and (b) the height of the image.

Solution:

a. Find the image distance using Equation 24.4:

$d_O = -40.0$ cm; $f = 16.0$ cm

$$-\frac{1}{d_O} + \frac{1}{d_I} = \frac{1}{f}$$

$$\frac{1}{d_I} = \frac{1}{f} + \frac{1}{d_O} = \frac{1}{16.0 \text{ cm}} + \frac{1}{-40.0 \text{ cm}}$$

Multiply by the common denominator (80.0 cm)d_I.

$$\frac{(80.0 \text{ cm})\cancel{d_I}}{\cancel{d_I}} = \frac{(80.0 \cancel{\text{cm}})d_I}{16.0 \cancel{\text{cm}}} - \frac{(80.0 \cancel{\text{cm}})d_I}{40.0 \cancel{\text{cm}}}$$

$$80.0 \text{ cm} = 5.00d_I - 2.00d_I = 3.00d_I$$

$$d_I = \frac{80.0 \text{ cm}}{3.00}$$

$$d_I = 26.6\overline{6} \text{ cm} \cong 26.7 \text{ cm}$$

b. Calculate the height of the image:

$$\frac{H_I}{H_O} = \frac{d_I}{d_O}$$

$$H_I = H_O\frac{d_I}{d_O}$$

$H_O = 6.00$ cm; $d_O = -40.0$ cm; $d_I = 26.6\overline{6}$ cm

$$H_I = 6.00 \cancel{\text{cm}} \cdot \frac{26.6\overline{6} \text{ cm}}{-40.0 \cancel{\text{cm}}}$$

$$H_I \cong -3.9\underline{99} \text{ cm} \cong -4.00 \text{ cm}.$$

The real image is about 26.7 cm from the lens and 4.00 cm high (inverted).

Problem-Solving Strategy 24.6

Remember to include the negative sign in front of the object term in the thin-lens equation. This sign is required because the object distance is always negative for lenses.

Chapter 23 explained the six cases for the concave mirror. The same six cases may be applied to the converging lens, with the same results. Table 24-5 lists the characteristics of the image for each location of the object, and Figure 24-13 shows how each case produces an image.

Table 24-5

Properties of Geometric Refraction in a Lens

Object Location	Image Location	Image Size	Orientation	Type
$d_O = \infty$	$d_I = f$	minuscule	inverted	real
$d_O > R$	$f < d_I < R$	$H_I < H_O$	inverted	real
$d_O = R$	$d_I = R$	$H_I = H_O$	inverted	real
$f < d_O < R$	$d_I > R$	$H_I > H_O$	inverted	real
$d_O = f$	d.n.e. (∞)	d.n.e.	—	—
$d_O < f$	$d_I < 0$	$H_I > H_O$	erect	virtual

Note: "d.n.e." means "does not exist" (value is infinitely large).

Legend:

Ray 1 (green): Enters lens parallel to optical axis, then refracts through F

Ray 2 (yellow): Enters lens via F, then refracts parallel to optical axis

Ray 3 (purple): Passes through optical center of lens

C: Center of curvature for each lens surface

F: Focal point of lens $\left(f = \tfrac{1}{2}R\right)$

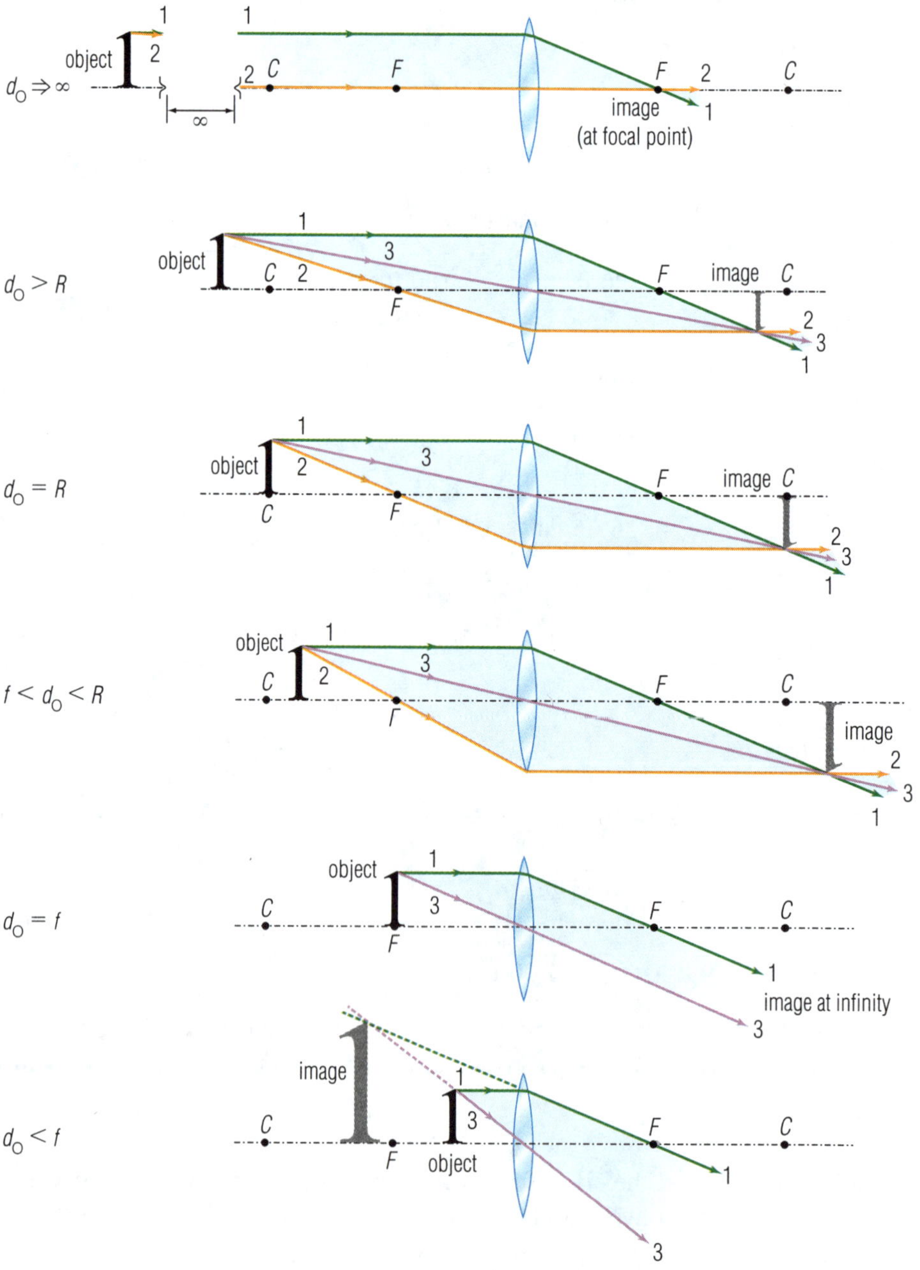

24-13 Positions and orientations of images produced by a thin convex lens for various object positions

The last case, where $d_O < f$, is most effective for magnifying. The image appears to be on the same side of the lens as the object; therefore, the image is virtual. A magnifying glass uses a converging lens to magnify an object that is less than one focal length away from the lens.

EXAMPLE 24-7

Virtually Larger: Magnification in a Thin Lens

A magnifying glass has a focal length of 5.00 cm. If an object 0.500 cm high is placed 4.50 cm from the lens, (a) where is the image, and (b) how large is it?

Solution:

a. Calculate the image position:

$$-\frac{1}{d_O} + \frac{1}{d_I} = \frac{1}{f}$$

$$\frac{1}{d_I} = \frac{1}{f} + \frac{1}{d_O} = \frac{1}{5.00 \text{ cm}} + \frac{1}{-4.50 \text{ cm}}$$

Multiply by the common denominator $(45.0 \text{ cm})d_I$.

$$\frac{(45.0 \text{ cm})\cancel{d_I}}{\cancel{d_I}} = \frac{(45.0 \cancel{\text{cm}})d_I}{5.00 \cancel{\text{cm}}} - \frac{(45.0 \cancel{\text{cm}})d_I}{4.50 \cancel{\text{cm}}}$$

$$45.0 \text{ cm} = 9.00d_I - 10.0d_I$$

$$45.0 \text{ cm} = -1.0d_I$$

$$d_I \cong -4\underline{5}.0 \text{ cm} \cong -45 \text{ cm}$$

b. Determine the image height:

$H_O = 0.500$ cm; $d_I = -4\underline{5}.0$ cm; $d_O = -4.50$ cm

$$H_I = H_O\frac{d_I}{d_O} = 0.500 \text{ cm}\cdot\frac{-4\underline{5}.0 \cancel{\text{cm}}}{-4.50 \cancel{\text{cm}}}$$

$$H_I \cong +5.\underline{00} \text{ cm} \cong +5.0 \text{ cm}$$

The image appears to be 45 cm behind the lens and is 5.0 cm high (erect). It has been magnified by a factor of 10.

24.8 Diverging Lenses

A lens with edges thicker than the middle is a **diverging lens**. Consider a planar wave front approaching a diverging lens. The light near the edges of the lens must pass through more of the lens than the light near the middle of the lens. Therefore, the waves near the center emerge ahead of the waves near the ends, and the wave-front envelope is curved.

Following a single ray through the lens also shows why the light diverges. When a ray enters the lens, it bends toward the normal, as Figure 24-14b shows. When it emerges from the lens, it bends away from the normal. You can see that both of these refractions direct the ray away from the optical axis; therefore the rays diverge.

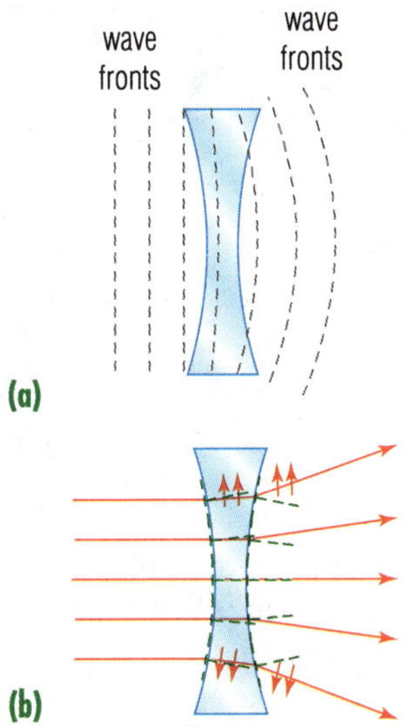

24-14 (a) Wave fronts refracted by diverging lens; (b) Light rays refracted by diverging lens

Diverging lenses produce only virtual images. For this reason, the virtual image appears to be on the same side of the lens as the object, which means that the image distance has a negative value.

The diverging lens behaves much like a convex mirror. Also, like a convex mirror, a diverging lens produces only one kind of image: virtual, erect, and smaller than the object. The thin-lens equation describes the location of the image. Because the virtual image appears to be on the same side of the lens as the object, diverging lenses are given negative focal lengths.

EXAMPLE 24-8

Divergent Point of View: Image in a Diverging Lens

An object 8.00 cm high is 30.0 cm from a lens with a focal length of −10.0 cm. Find (a) the position and (b) the height of the image.

Solution:

a. Calculate the image position:

$d_O = -30.0$ cm and $f = -10.0$ cm

$$-\frac{1}{d_O} + \frac{1}{d_I} = \frac{1}{f}$$

$$\frac{1}{d_I} = \frac{1}{f} + \frac{1}{d_O} = \frac{1}{-10.0\text{ cm}} + \frac{1}{-30.0\text{ cm}}$$

Multiply by the common denominator $(30.0\text{ cm})d_I$.

$$\frac{(30.0\text{ cm})d_I}{d_I} = -\frac{(30.0\text{ cm})d_I}{10.0\text{ cm}} - \frac{(30.0\text{ cm})d_I}{30.0\text{ cm}}$$

$$30.0\text{ cm} = -3.00d_I - 1.00d_I = -4.00d_I$$

$$d_I = -7.50\text{ cm}$$

b. Determine image height:

$H_O = 8.00$ cm; $d_I = -7.50$ cm; $d_O = -30.0$ cm

$$H_I = H_O\frac{d_I}{d_O} = 8.00\text{ cm}\cdot\frac{-7.50\text{ cm}}{-30.0\text{ cm}}$$

$$H_I = +2.00\text{ cm}$$

The image appears to be 7.50 cm behind the lens and is 2.00 cm high (erect).

Problem-Solving Strategy 24.7

Images formed by diverging lenses are always virtual and erect.

Remembering the following rules for signs will help you to use the thin-lens equation. These rules only apply to images formed by single lenses.

1. All object distances (d_O) are negative (behind the lens from the observer's perspective).
2. All real images are in front of the lens. Their image positions (d_I) are positive.
3. All virtual images are behind the lens; hence their positions are negative.
4. Focal lengths of converging lenses are positive; focal lengths of diverging lenses are negative.
5. Objects are assumed to be erect (positive heights). Real images are inverted (negative heights), while virtual images are erect (positive heights).

From his point of view, an observer is in front of the lens, and the object is behind the lens. Thus, all positions in front of the lens have positive distances, and all positions behind the lens have negative distances.

24.9 Combining Lenses

So far, we have discussed only forming an image with a single lens. How does a combination of lenses affect light? Telescopes, microscopes, binoculars, and cameras all use combinations of lenses to form an image. If one lens receives light that has passed through another lens, it reacts just as it would to any other light. A real image formed by the first lens is treated as an object by the next lens. Figure 24-15 shows a lens producing image I_1 of the object. The next lens takes I_1 as its object and produces a second image, I_2. The location and size of I_1 can be found by applying the thin-lens equation to the object's distance from the first lens. The location and size of I_2 can be found by applying the thin-lens equation to I_1 and the second lens.

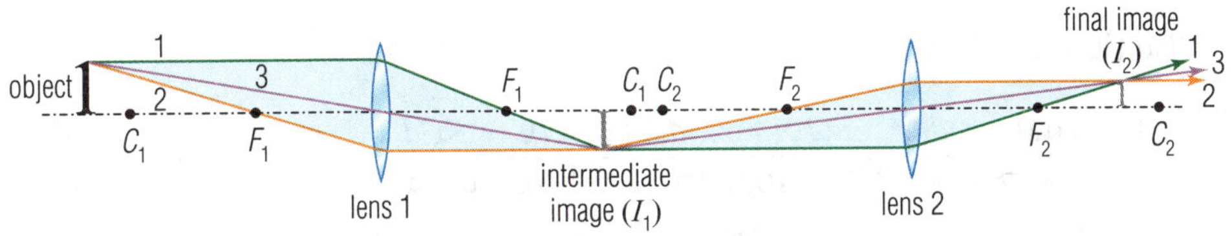

24-15 An image formed by a combination of lenses

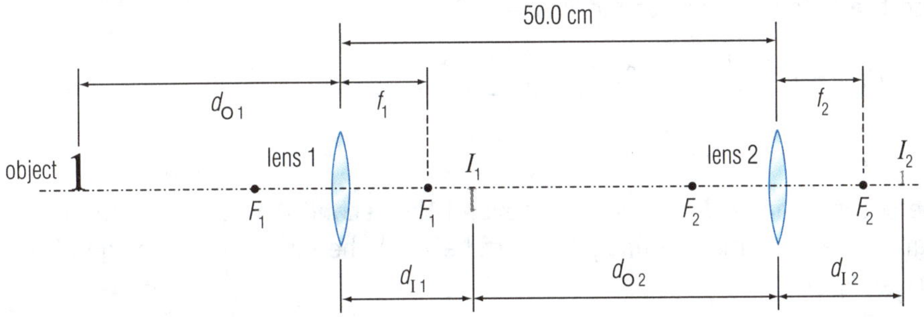

24-16 Schematic of the lenses and images for Example 24-9

EXAMPLE 24-9

Scoping It Out: Combining Lenses

A 1.00 cm object is 30.0 cm from a lens with a focal length of 10.0 cm. The first lens is 50.0 cm from a second lens that has a focal length of 10.0 cm. What are the sizes and locations of (a) the image from the first lens and (b) the image from the second lens?

Solution:

a. For I_1, $d_{O\,1} = -30.0$ cm, $f_1 = 10.0$ cm:

$$\frac{1}{d_{I\,1}} = \frac{1}{f_1} + \frac{1}{d_{O\,1}} = \frac{1}{10.0\text{ cm}} + \frac{1}{-30.0\text{ cm}}$$

Multiply by $(30.0\text{ cm})d_{I\,1}$:

$$\frac{(30.0\text{ cm})\cancel{d_{I\,1}}}{\cancel{d_{I\,1}}} = \frac{(30.0\,\cancel{\text{cm}})d_{I\,1}}{10.0\,\cancel{\text{cm}}} - \frac{(30.0\,\cancel{\text{cm}})d_{I\,1}}{30.0\,\cancel{\text{cm}}}$$

$$30.0\text{ cm} = 3.00d_{I\,1} - 1.00d_{I\,1} = 2.00d_{I\,1}$$

$$d_{I\,1} = +15.0\text{ cm}$$

Find the height of I_1:

$$H_{I1} = H_{O1}\frac{d_{I1}}{d_{O1}} = 1.00\ \text{cm}\cdot\frac{+15.0\ \cancel{\text{cm}}}{-30.0\ \cancel{\text{cm}}}$$

$$H_{I1} = -0.500\ \text{cm}$$

The real image, I_1, is 15.0 cm from lens 1 between lenses 1 and 2. Since there is 50.0 cm between the lenses, I_1 is 35.0 cm from lens 2.

b. For the second lens, $d_{O2} = -35.0$ cm, $f_2 = 10.0$ cm:

$$\frac{1}{d_{I2}} = \frac{1}{f_2} + \frac{1}{d_{O2}} = \frac{1}{10.0\ \text{cm}} + \frac{1}{-35.0\ \text{cm}}$$

Multiply by $(70.0\ \text{cm})d_{I2}$:

$$\frac{(70.0\ \text{cm})\cancel{d_{I2}}}{\cancel{d_{I2}}} = \frac{(70.0\ \cancel{\text{cm}})d_{I2}}{10.0\ \cancel{\text{cm}}} - \frac{(70.0\ \cancel{\text{cm}})d_{I2}}{35.0\ \cancel{\text{cm}}}$$

$$70.0\ \text{cm} = 7.00d_{I2} - 2.00d_{I2} = 5.00d_{I2}$$

$$d_{I2} = +14.0\ \text{cm}$$

Find the height of the second image:

$$H_{I2} = H_{I1}\frac{d_{I2}}{d_{O2}} = -0.500\ \text{cm}\cdot\frac{+14.0\ \cancel{\text{cm}}}{-35.0\ \cancel{\text{cm}}}$$

$$H_{I2} = +0.200\ \text{cm}$$

The second image, I_2, is 14.0 cm beyond the second lens and is 0.200 cm high and erect. This *real* image is erect because the intermediate image (I_1) was inverted.

Problem-Solving Strategy 24.8

In a compound lens system, it is *not* necessary for the refracted rays from the first lens to form a focused image before they enter the second lens.

24.10 Correcting Lens Aberrations

24-17 Refractor telescopes depend entirely on combinations of lenses to collect light and form a magnified image.

Spherical mirrors cannot reflect parallel light rays to a single point and therefore experience spherical aberration. Similarly, spherical lenses cannot refract rays near their edges according to the rules that you have studied. The *spherical aberration* of a lens produces blurry images. Several solutions to help prevent this blurriness are possible. First, the edges of the lenses may be covered by an iris diaphragm, an opaque ring that reduces the effective diameter of the lens, called its **aperture**. A smaller diameter reduces the total lens curvature in the light path, thus reducing the effect of this kind of aberration. However, iris diaphragms restrict the amount of light through the lens, so they cannot be used where maximum light gathering is required, as in telescopes. Another solution is to make the two halves of a double convex lens asymmetrical. The difference in curvature causes peripheral rays to focus properly. The most satisfying approach to the problem, however, is to use two lenses so that the second lens corrects the problem of the first. Lenses of different indices of refraction are often able to compensate for each other's aberrations.

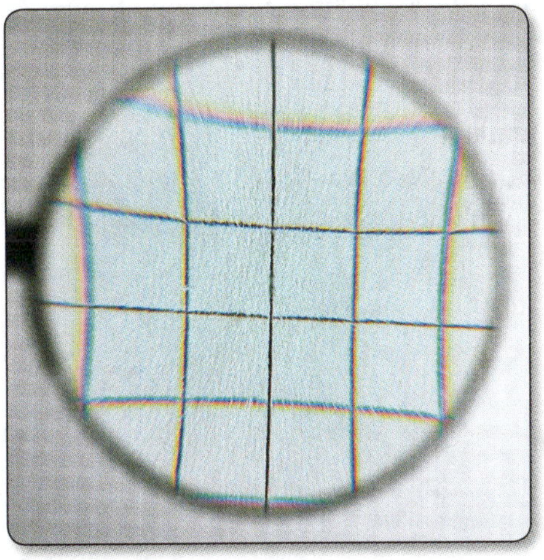

24-18 Simple spherical lenses have inherent optical aberrations that degrade the images they produce. Note the blurring and color dispersion near the edge of the lens.

24-19 Achromatic lenses come in many different combinations of elements in order to provide the correct combination of aperture and focal length.

Another kind of problem that lenses have is called **chromatic aberration**—the unwanted dispersion of colors. The solution for this problem is the same as the solution for spherical aberration. Simply combine two lenses of different materials so that the second lens can recombine the colors dispersed by the first lens. For example, a converging crown-glass ($n \approx 1.5$) lens bonded to a diverging flint-glass ($n \approx 1.6$) lens will exhibit little or no spherical or chromatic aberrations. Such a combination is called an *achromatic doublet.*

The term *achromatic* means "without color."

24.11 Special Lenses

Lenses are not restricted to the shapes discussed above. One special kind of lens is the *meniscus lens*, so called because it has a concave and a convex surface. Meniscus lenses may be either converging or diverging, depending on whether the center or the edges are thicker. If the lens's center is thicker, it is a converging lens; if its edges are thicker, it is a diverging lens. Meniscus lenses are used in eyeglasses to allow for movement of the eyelashes and to reduce distortions near the edges due to movement of the eyeball. Converging lenses are used for farsightedness, and diverging lenses correct nearsightedness.

24-20 A meniscus lens, like the kind used in eyeglasses

24B Section Review

1. Discuss the differences between the shape of a converging lens and that of a diverging lens.
2. a. When you are considering the positions of the object and the image relative to a lens, what sign is conventionally given to the object distance (d_O)?
   b. Is this position in front of or behind the lens in relation to the observer?
3. What restriction exists on the use of the image-height and magnification equations (24.5 and 24.6)?

24B Objectives

After completing this section, I can

✓ classify a lens based on its shape.

✓ describe how various shapes of lenses refract light.

✓ use the thin-lens equation to solve problems in lens refraction.

✓ calculate the magnification of simple lenses under various conditions.

✓ compare and contrast converging and diverging lens refraction and the images that they form.

✓ explain how simple multi-lens systems form images.

✓ discuss several methods for correcting common optical aberrations in lenses.

✓ discuss the shapes and characteristics of meniscus lenses.

4. Where must the object be in relation to the lens for the lens to be used as a magnifier?
5. What can a manufacturer do to correct a lens for spherical and chromatic aberrations?
6. What is the size of the image of the sun produced by a lens with a 20.0 cm focal length? (*Note*: Use the earth and sun's astronomical data found in Appendix E.)
7. In a school laboratory, a physics student observes that a 12.0 cm candle 36.0 cm from a thin spherical lens produces on a screen a sharp inverted image that measures 3.0 cm high. What is the focal length (f) and the radius of curvature (R) of the lens?
8. What is the magnification of the lens in Question 7?

Chapter Review

In Terms of Physics

refraction	551
interface	552
index of refraction (n)	552
angle of refraction (θ_r)	552
Snell's law	552
total internal reflection	555
critical angle (θ_c)	555
fiber optics	556
dispersion	556
mirage	557
converging lens	560
thin-lens equation	560
diverging lens	563
aperture	566
chromatic aberration	567

Problem-Solving Strategies

24.1 (page 552) When the two media at an interface are named, the first contains the incident ray, and the second contains the refracted ray. In an air-glass interface, the ray is moving from the air into the glass.

24.2 (page 554) You can avoid rounding error when using Snell's law to compute angles of incidence or refraction by taking the inverse sine (arc-sine) of the right-hand side of the equation. For example, the angle of refraction is found by

$$\theta_r = \sin^{-1}\left(\frac{n_i \sin\theta_i}{n_r}\right).$$

24.3 (page 555) Critical angles exist only when the refraction index of the medium containing the incident ray is larger than that of the medium containing the refracted ray ($n_i > n_r$).

24.4 (page 555) Snell's law is valid as long as the critical angle at the interface of the two media is not exceeded.

24.5 (page 560) All lens distances should be measured from the center of the lens, not from the surface.

24.6 (page 561) Remember to include the negative sign in front of the object term in the thin-lens equation. This sign is required because the object distance is always negative for lenses.

24.7 (page 564) Images formed by diverging lenses are always virtual and erect.

24.8 (page 566) In a compound lens system, it is *not* necessary for the refracted rays from the first lens to form a focused image before they enter the second lens.

Review Questions

1. What is refraction?
2. What is the surface where water and glass meet called?
3. State Snell's law.
4. **a.** What is a critical angle?
   b. Does it apply to the interface between any two transparent materials? Explain.
5. What happens to a light ray at an interface when its angle of incidence is greater than the critical angle?
6. How is total internal reflection applied in long-distance communication technology?
7. Why does a prism transform a beam of white light into a spectrum of colors?
8. If a person needs a large correction in his glasses for near or far vision, what could be one undesirable consequence of using a material with a low index of refraction for the lenses?
9. Which color is on the outside or on top of a primary rainbow?
10. Based on what you know about refraction in air, does the ocean horizon normally appear closer or farther away than it would appear if air had an index of refraction of $n = 1$? Assume that the air is coolest and densest near the ocean's surface. Explain your answer.
11. You are looking through a pair of binoculars at a distant lighthouse that appears to be floating above the water's surface at the horizon. What kind of optical phenomenon is this?
12. Classify the following lenses as converging or diverging.

a.

b.

c.

d.

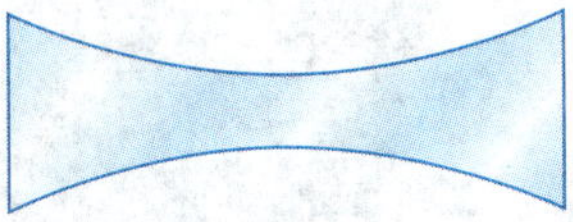

e.

13. Where are real images formed by lenses in relation to their objects?
14. Why does the thin-lens equation have a negative sign in front of the object-related term?
15. If you want to magnify an object, where should you put it with respect to the focal point of the lens?
16. Can a diverging lens be used as a magnifying glass? Explain your answer.
17. Some scuba divers who are nearsighted or farsighted own special dive masks that have corrective lenses instead of plate glass lenses to help them see more clearly under water. What must be considered when making these lenses that is not true for the divers' regular glasses or contact lenses?
18. The *Johnson Sea Link* is a deep-diving research minisubmersible. The pilot sits inside a 1.5 m diameter clear acrylic sphere whose wall is 13 cm thick. When the minisub is sitting on its support ship's deck, will the submarine's pilot appear smaller or larger to an outside observer than he really is?
19. An object is viewed through two converging lenses so that neither the object nor its image is between a lens and its focus. Is the resulting image from the second lens real or virtual, erect or inverted?
20. Can a virtual image be an object for a lens? If not, show why not with a ray diagram. If it can, give an instance from everyday life.

True or False (21–30)

21. A light ray moving through a glass-air interface will bend toward the normal.
22. A light ray from a star that passes straight down the optical axis of a refractor telescope will be refracted.
23. The path taken by a ray of light across the interface between two media is the same regardless of the direction the light is moving.
24. Dense materials have an index of refraction less than 1.
25. Reflection of sunlight by water droplets is the only process responsible for the formation of atmospheric rainbows.
26. When using the thin-lens equation, the object distance is always positive.
27. An image that is smaller than its object is indicated by a negative magnification.
28. A virtual image cannot be focused on a screen.
29. A negative image height produced by a single lens implies that the image is real.
30. When combining lenses, the refracted rays from the first lens do not *have* to form a focused image before they enter the second lens.
31. Find the speed of light in each of the following materials.
    a. crown glass
    b. ice
    c. hydrogen
    d. flint glass
32. Given the speed of light in each of the following materials, find its index of refraction.
    a. ethyl acetate ($c_{\text{ETAC}} = 2.19 \times 10^8$ m/s)
    b. corn syrup ($c_{\text{syrup}} = 2.01 \times 10^8$ m/s)
    c. iodine crystal ($c_{\text{I}} = 8.98 \times 10^7$ m/s)
    d. ruby ($c_{\text{ruby}} = 1.69 \times 10^8$ m/s)
33. A ray of light moves from water into salt. Will the angle of refraction be greater or less than the angle of incidence? Why?
34. The angle of incidence of a ray of light at an air-fluorite interface is 45.0°. At what angle will it be refracted?
35. A ray of light with an angle of incidence of 0.0° in flint glass enters crown glass. At what angle is it refracted?

36. Light from a laboratory lamp falls on the surface of a beaker of methyl alcohol at an angle of 30.0° and is refracted at an angle of 22.1° with the normal. What is the index of refraction of methyl alcohol?

37. A waterproof lamp is submerged at the bottom of a beaker containing water. One light ray approaches the water-glass interface at an angle of 45.0° to the normal and has an angle of refraction of 34.6°. What is the glass's index of refraction?

38. At what angle would a light ray have to approach a flint glass-diamond interface to have a refraction angle of 30.0°?

39. Find the critical angle for the following interfaces if the light ray passes
   a. from flint glass into crown glass.
   b. from diamond into water.
   c. from fluorite into air.

40. A 0.500 cm ant is observed through a hand magnifier ($f = 10.0$ cm). Find the location, size, orientation (erect or inverted), and type (real or virtual) of the ant's image when the ant's distance from the lens is
   a. 10 000.0 cm.
   b. 25.0 cm.
   c. 20.0 cm.
   d. 15.0 cm.
   e. 10.0 cm.
   f. 5.00 cm.

41. Which of the following will magnify an object more, placing it 9.5 cm from a lens with a focal length of 10.0 cm or placing it 5.0 cm from the lens? (*Hint*: Find d_I for each case first.)

42. A 10.0 m high tree is 100.0 m from a diverging lens with a focal length of −10.0 cm.
   a. Where is the tree's image?
   b. How high is it?
   c. Is it real or virtual?
   d. Is it erect or inverted?

DM 43. Low-expansion, borosilicate glass contains oxides of sodium, calcium, potassium, aluminum, barium, and boron. Using the trendline equation that you obtained from the modeling exercise in Section 24A, predict the refractive index of a sample of this glass, which has a mass density of 2.23 g/cm^3. Do a percent error calculation assuming that it has an actual refractive index of $n = 1.474$. Explain why it may not be valid to use your model for this example. Considering your percent error, what do you conclude?

Facets of Physics

Fresnel Lenses

Lighthouses and navigational beacons must produce strong beams of light, but they need not be the same strength in all directions simultaneously. A single beam, sweeping across the sea, is enough to warn passing ships of any danger. Beacons use less energy if their light is directed in a single beam.

How can a beacon's light be concentrated in one direction? The easiest method might seem to be a lens. However, the lens required to focus a strong beam of light is so large and heavy that it tends to be impractical for most navigational aids such as buoys and unmanned light towers.

While studying this problem in the early 1800s, Augustin-Jean Fresnel (1788–1827), a French physicist, decided to modify the normal lens. Instead of making the lens a solid portion of a sphere, Fresnel cut its curved surface into rings of uniform width. Then he laid all the rings together on a flat surface and fastened them together. This lens focuses light as well as the original solid lens, but it is much lighter and easier to make.

Today Fresnel lenses are popular wherever strong, lightweight lenses are needed. For example, Fresnel lenses work well in spotlights and in overhead projectors. Modern Fresnel lenses are usually stamped or molded from glass or plastic.

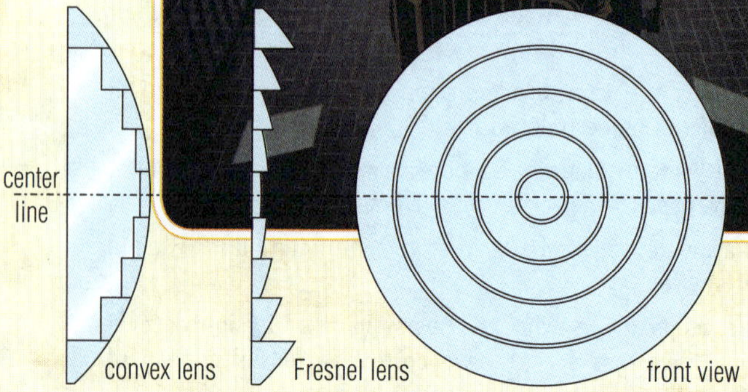

These soapy bubbles exhibit thin-film diffraction, just one of many kinds of light-wave phenomena. It lends an iridescent appearance to the bubbles, very similar to that produced in butterfly wings, bird feathers, and other materials with fine structures.

Wave Optics 25

Dominion Science Problem

Busting Counterfeiters

As long as there have been commodities in the world and currency to buy them, there has been counterfeiting. Terrorists, corrupt governments, and criminals often fund their activities with counterfeited items and monies. CDs, DVDs, music, movies, pharmaceuticals, clothing, watches, musical instruments, and even food are profitable targets. Counterfeiting threatens our economy as fakes of name brands multiply in the market, devaluing the real articles, infringing on copyrights, and, especially, placing human health at risk. Counterfeiting reduces the value of real money and increases inflation. Widespread money counterfeiting can decrease confidence in the currency, forcing merchants and consumers to shift to other mediums of exchange. Some estimate that counterfeiting costs the United States hundreds of billions of dollars annually. And it doesn't just cost in dollars. It costs in lives.

In Asia and Africa, malaria kills over 2.5 million people a year, even more than AIDS. In fact, about 40% of the population of the world is at risk of contracting malaria. One of the most effective medications against a particularly virulent strain of malaria is called artesunate. But counterfeit versions of this medication abound in Asia, often containing absolutely no active ingredients. In Southeast Asia, nearly two-thirds of the drug tested were fake. Thousands of people die, believing that they have taken the medication that is supposed to heal them. How can we solve this problem?

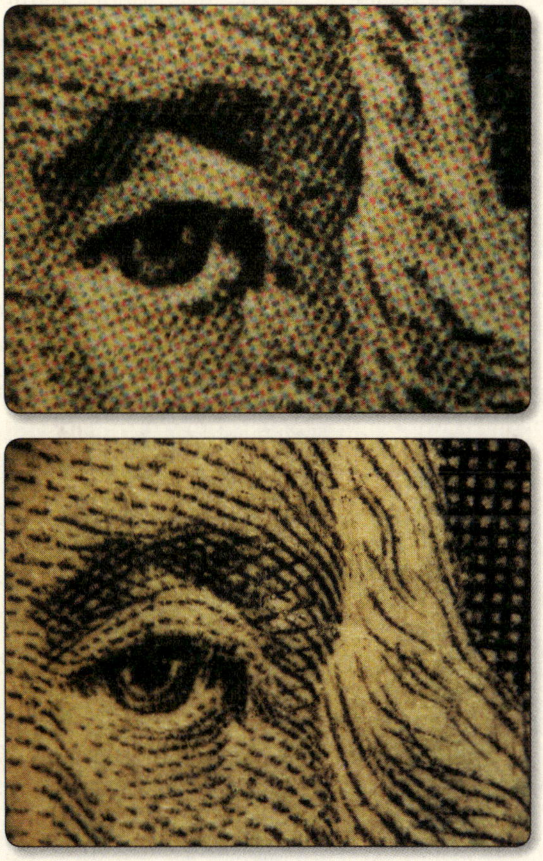

25-1 Can you tell which is real and which is counterfeit?

25A WAVE INTERFERENCE

25.1 Describing Interference

Before the nineteenth century, scientists debated whether light was made up of particles or of waves. Isaac Newton believed that, in spite of some wave-like characteristics, light could be best explained as a stream of particles. Later scientists hesitated to question Newton's ideas. In 1801, however, Thomas Young showed conclusively that light has wave-like properties that a particle model cannot explain.

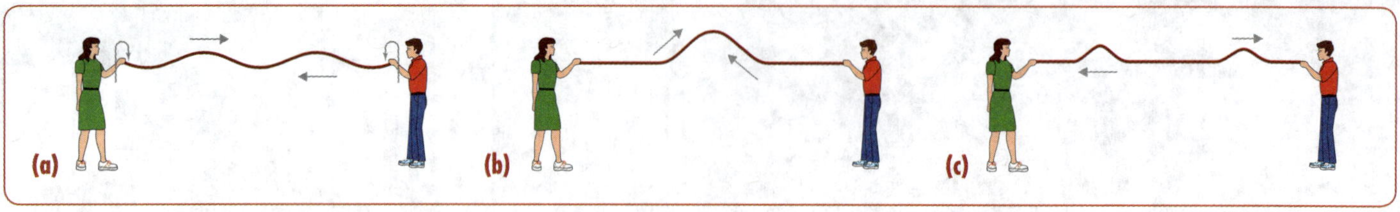

25-2 Constructive interference of pulses in a rope

A limited burst of wave activity is called a **pulse**. The duration of a pulse is short compared to the travel time of the wave.

The **superposition principle** states that the net displacement of a point in a medium at a given instant is the vector sum of the displacements of all waves passing through the point at that instant.

One unique characteristic of waves is called **interference**. Suppose two students hold a rope stretched between them. First one student and then the other snaps the rope upward, starting a *pulse* toward the rope's center. What happens when the pulses meet? For an instant, they become one displacement equal to the sum of the displacements from the two pulses. Then they separate, and each pulse continues along its original path. The increased amplitude in this case indicates **constructive interference**.

Suppose, instead, that one student snaps the rope upward, and the other student snaps the rope downward. The pulses again travel toward the rope's center. Now what happens when they meet? For an instant, if the pulses have the same amplitude, the rope will have zero displacement. The downward pulse's amplitude is *subtracted* from the upward pulse's amplitude. Although the rope has no displacement, the waves are still moving, and, after an instant, each pulse continues on its original path. This case demonstrates **destructive interference**. The fact that the displacements of two impulses add at their meeting place to give the rope's net displacement is called the **superposition principle**.

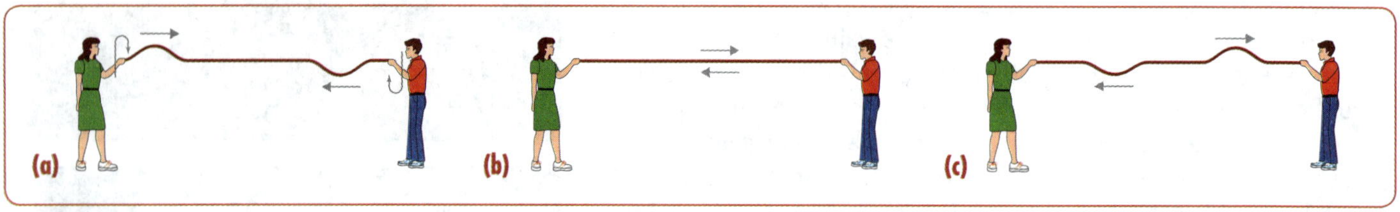

25-3 Destructive interference

The superposition principle applies to waves in any kind of medium, to electromagnetic waves, and to any number of waves. If waves are moving in phase—that is, if their troughs and crests (lowest and highest points, respectively) coincide—they will interfere constructively. If they are completely out of phase—the crests of one wave coincide with the troughs of the other—they will interfere destructively. If the waves are neither in phase nor completely out of phase, they will experience some constructive and some destructive interference. Many different effects can result from wave interference, but the most distinctive is called an *interference pattern.*

If light is a wave, it should show interference, obeying the superposition principle. When two light waves interact, they may show constructive interference, destructive interference, or, more probably, both. Areas of constructive interference appear brighter, while areas of destructive interference appear dimmer than light from either original wave. Does light really show interference? Does light plus light ever equal darkness?

25.2 Young's Double-Slit Experiment

In 1801, Thomas Young demonstrated light interference with his "double-slit" experiment. Light interference is more striking when waves come from two sources that are in phase. Since light sources are hard to synchronize, Young split a single source into two parts. He also found that the effect he was observing was much clearer if he used colored filters to restrict the transmitted light to a narrow range of frequencies.

Young's experiment involved three opaque screens. One screen was pierced by a single pinhole, which generated a point source of light for the distant second screen. The second screen had two closely spaced narrow slits. The third screen was a white screen used to observe the interference patterns. Sunlight falling on the filter and first screen passed through the pinhole. The wave front arriving at the second screen was essentially a plane wave. As the waves spread out from the slits, a pattern of alternating light and dark bands (or *fringes*) appeared on the third screen.

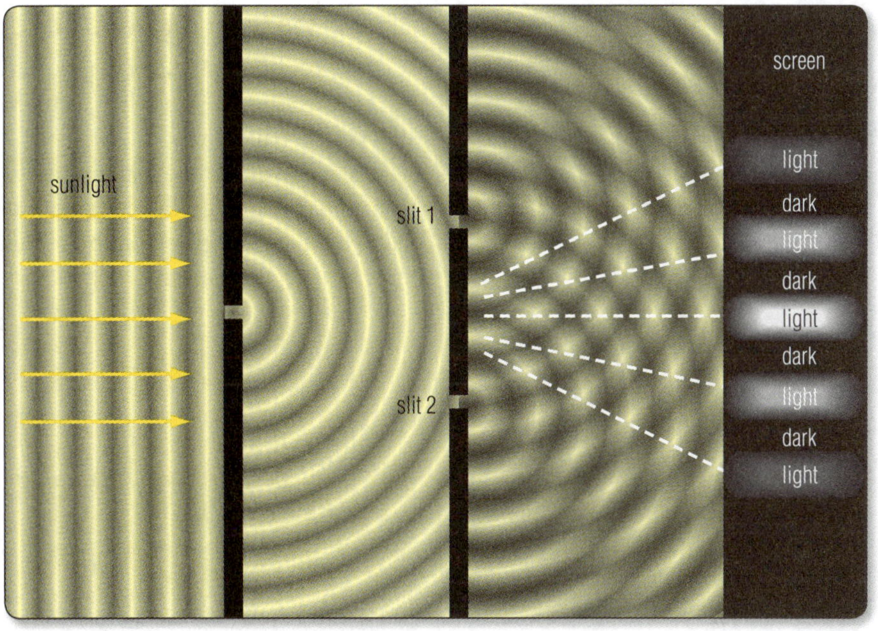

25-4 Schematic of Young's double-slit experiment

Why does the interference pattern appear? By considering light strictly as a wave, we can explain the interference fringes according to Huygens's principle. (Note that light also has a particle nature, which we shall investigate in Chapter 28.) Each slit acts as a source of wave fronts that spread out from the slit toward the projection screen. Since both slits are illuminated by a common light source, the light has the same frequency. Thus, the two wave fronts will tend to interfere constructively and destructively at different distances from their respective slits. This interference is particularly noticeable if a coherent light source (such as a laser) is used. Notice in Figure 25-5 that the bright spot in the center ($n = 0$), called the *central maximum*, is the same distance from each slit. Thus, light reaching the central maximum from the slits is in phase, creating constructive interference. At the maximum labeled $n = 1$ (known as the first order maximum), the light from the farther slit travels exactly one wavelength farther than the light from the nearer slit, so they are still in phase. The maxima mark areas where the *path difference*—the difference in the distances that the light from the two sources travels—is an integer multiple of the light's wavelength.

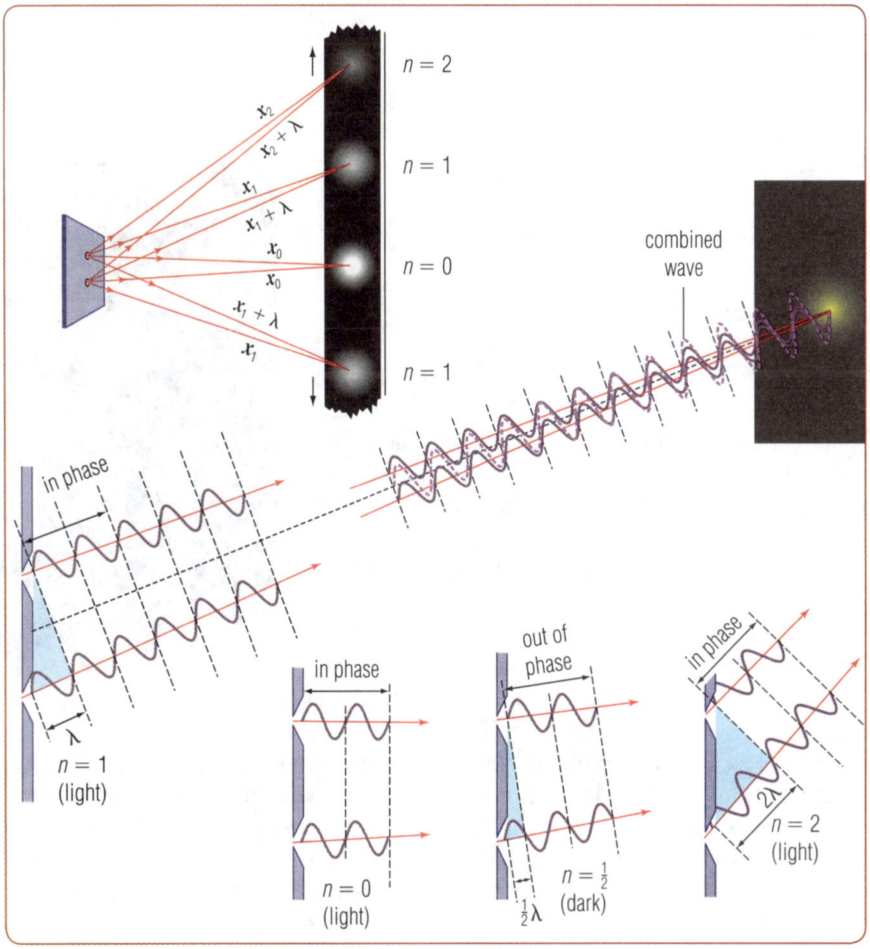

25-5 Constructive and destructive interference in Young's experiment

The interference *maxima* are separated by dark areas, or *minima*. The light reaching the minimum next to the central maximum from the farther slit travels exactly one-half wavelength farther than light from the nearer slit. The light is completely out of phase, and it destructively interferes; therefore, the area is dark. Young reasoned that physical particles cannot interact and destroy each other as these light rays seemed to be doing, so he concluded that light acts like a wave.

In an interference pattern, the bright fringes are called **maxima**, and the dark fringes are called **minima**.

25.3 Math of Wave Interference

Is it possible to calculate where maxima and minima will appear if you know the light's wavelength? Yes, it is possible, if you ensure that the distance between the slits is far smaller than the distance between the second and third screens. Figure 25-6 shows the conditions that produce an interference maximum. The rays of light from slits 1 and 2 are labeled rays 1 and 2, respectively. We have arranged the rays so that there is a maximum on the screen at their intersection (point E). Therefore, ray 2 must be longer than ray 1 by $n\lambda$, where n is an integer. If we drop a perpendicular

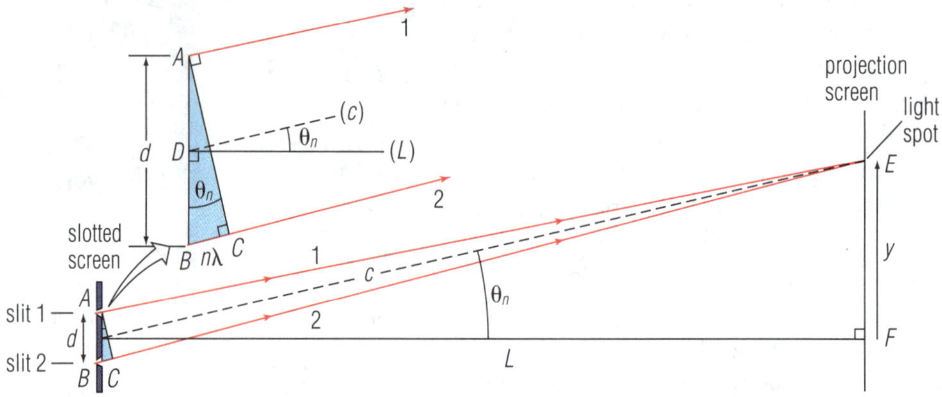

25-6 Conditions that produce a bright fringe in Young's experiment

Thomas Young (1773–1829)

English scientists after Newton believed that he was almost infallible. Because they inferred that Newton advocated the particle theory of light, so did they. When the Dutch physicist Huygens proposed the wave theory as an alternative, no Englishman would listen to this foreigner's unproven ideas. However, in the early nineteenth century, experiments forced these scientists to see that only a wave theory could explain many of light's properties.

Although trained as a doctor, Thomas Young had abilities in many fields. A child prodigy, he learned to read when he was only two, had read through the Bible twice by the time he was four, and spoke several languages before he began his medical studies. While still a student, Young formulated the theory that the lens of the eye focuses by changing thickness. The twenty-one-year-old student was elected a fellow of the Royal Society. After studying in three countries, Young received his medical degree from the university in Göttingen, Germany.

The young doctor was interested in the way human senses work. He was one of the first scientists to propose that receptors in the human eye receive only three colors—red, green, and blue. He also studied the ear's reception of sound waves.

From his studies, Young became convinced that light is a transverse wave. His double-slit experiment in 1801 proved that light undergoes interference, something that only waves can do. At first, other English scientists refused to listen to Young, who they thought was contradicting Newton. Eventually, however, experiments like Young's double-slit experiment proved Young's position.

Young's interests were not limited to the human body or light. For example, he analyzed the elastic properties of solids. With his expertise in languages, Young became a leading Egyptologist and was among the first to decipher parts of the Rosetta Stone. In fact, one of his last publications was a work about Egypt, not physics.

line from slit 1 to ray 2, ray 2 is divided into two segments—one with the same length as ray 1 and one of length $n\lambda$ (see the inset to Figure 25-6). This segment ($\overline{BC}$) forms a right triangle ABC with an angle θ_n at vertex A. The sine of θ_n is

$$\sin\theta_n = \frac{\text{opposite}}{\text{hypotenuse}} = \frac{n\lambda}{d}. \qquad (25.1)$$

The symbol n represents the *order number* of the maximum as well as the number of wavelengths in the path difference between the two rays. The central maximum is the zeroth order ($n = 0$) maximum. The first-order maxima have $n = 1$, and so on.

Since the separation distance of the screens (L) is much greater than the distance between the slits (d), rays 1 and 2 and line $\overline{DE}$ (the line segment connecting the center between the slits to the maximum) are essentially parallel. Therefore, $\overline{AC}$ is perpendicular to $\overline{DE}$, and $\overline{DF}$ is perpendicular to $\overline{AB}$. Angle EDF is equal to θ_n, and triangle DFE is similar to triangle ACB. The length y is the distance between the maximum being studied and the central maximum. The tangent of θ_n for the larger triangle is

$$\tan\theta_n = \frac{\text{opposite}}{\text{adjacent}} = \frac{y}{L}. \qquad (25.2)$$

For small angles,

$$\tan\theta_n \approx \sin\theta_n.$$

Therefore,

$$\sin\theta_n \approx \frac{y}{L}.$$

By substituting y/L for $\sin\theta_n$ in Equation 25.1, we get

$$\frac{y}{L} = \frac{n\lambda}{d} \text{ (for maxima).} \qquad (25.3)$$

This equation is especially useful for determining the distance between slits or the wavelength of light, assuming that the other parameters are known.

For analyzing the *minima* between maxima, the ray path length difference is in half-wavelengths. Therefore, you substitute $n - \frac{1}{2}$ for n, and Equation 25.3 changes to

$$\frac{y}{L} = \frac{(n-\frac{1}{2})\lambda}{d} \text{ (for minima).} \qquad (25.4)$$

In this equation, y is still the distance from the minimum to the central maximum, L is the screen separation, and d is the slit separation.

Problem-Solving Strategy 25.1

The symbol n in the interference equation is equal to two quantities: (1) the difference between path lengths (in whole wavelengths of the light) that produce the maximum and (2) the order number of the same maximum.

Problem-Solving Strategy 25.2

When analyzing the *minima*, be sure to use the order number (n) of the next-higher maximum in the interference formula.

EXAMPLE 25-1

Determining Wavelength Using Interference Patterns

Red light illuminates a screen containing two slits separated by 0.500 mm and passes through to a second screen 3.00 m away. The distance between the central maximum and the first-order maximum is 0.400 cm. Find the light's wavelength.

Solution:

Since you are given the distance to a *maximum*, use Equation 25.3:

$$\frac{y}{L} = \frac{n\lambda}{d}$$

$$\lambda = \frac{y}{L} \cdot \frac{d}{n}$$

Wavelengths of visible light are conveniently measured in nanometers (1 nm = 10^{-9} m).

Since the problem involves a *first-order maximum*, $n = 1$. Convert all distances to meters.

$$\lambda = \frac{(4.00 \times 10^{-3}\ \text{m})(5.00 \times 10^{-4}\ \cancel{\text{m}})}{(3.00\ \cancel{\text{m}})(1)}$$

$$\lambda = 6.66\overline{6} \times 10^{-7}\ \text{m}$$

$$\lambda \cong 6.67 \times 10^{-7}\ \text{m (or 667 nm)}$$

You can check your answer by noting that light with a wavelength of 667 nm is red (see Table 26-3, page 602).

EXAMPLE 25-2

Determining Slit Spacing

Monochromatic yellow light with a wavelength of 580. nm illuminates a double-slit screen. The centers of the minima adjacent to the central maximum are 0.435 mm from the center of the pattern. The projection screen is 120. cm from the slits. Calculate the distance between the slits.

Solution:

You are working with the distance of a *minimum*; therefore, use Equation 25.4. The next-greater maximum is 1. Convert all distances to meters.

$$\frac{y}{L} = \frac{(n - \frac{1}{2})\lambda}{d}$$

$$d = \frac{L(n - \frac{1}{2})\lambda}{y}$$

$$d = \frac{(1.20\ \text{m})(1 - \frac{1}{2})(580. \times 10^{-9}\ \text{m})}{4.35 \times 10^{-4}\ \text{m}}$$

$$d = 8.00 \times 10^{-4}\ \text{m}$$

The slits are 0.800 mm apart.

25.4 Thin-Film Interference

Did you ever wonder why a soap bubble or a film of spilled gasoline displays a shimmering color? The answer is interference. In a thin film, light reflects from both the top and the bottom film surfaces. Depending on the path difference and the light's wavelength, the reflected waves may interfere constructively or destructively. If white light illuminates the film, some wavelengths (colors) at each point on the film will probably be the right wavelength to interfere constructively. Other wavelengths will interfere destructively; therefore, only the constructively interfering colors will be seen. Since most films vary in thickness, different colors will be reflected by different portions of the film. Swirling bands of color will be seen in the film as air currents move the film around.

A thin convex lens resting on a flat glass plate produces interference patterns. These patterns are caused by interference between light reflected from the flat plate below the lens and light reflected from the curved lower surface of the lens. The thickness of the air below the lens increases from the center of the lens out to

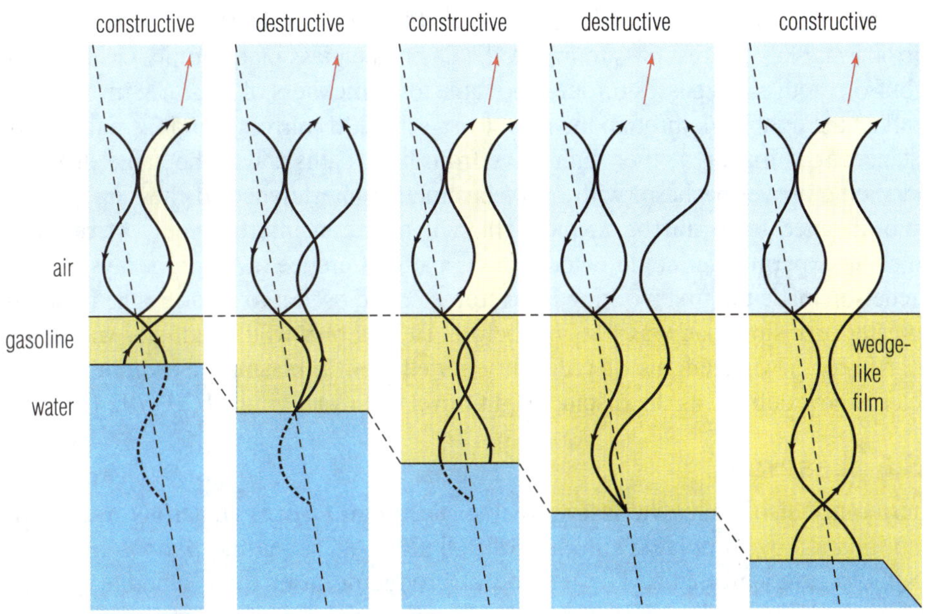

25-7 Interference of light reflected from a thin film

the edges. Thus, there are concentric "rings" of air under the lens, each ring having a different width. Where constructive interference occurs, it produces a bright ring of light. Where destructive interference occurs, a dark ring appears. These rings are called *Newton's rings* after Isaac Newton, who first observed them. Interference also occurs between two flat glass plates separated by a tiny wedge of air. The light is reflected above and below the air wedge, and the two reflected rays interfere. White light separates into colors when reflecting from this device.

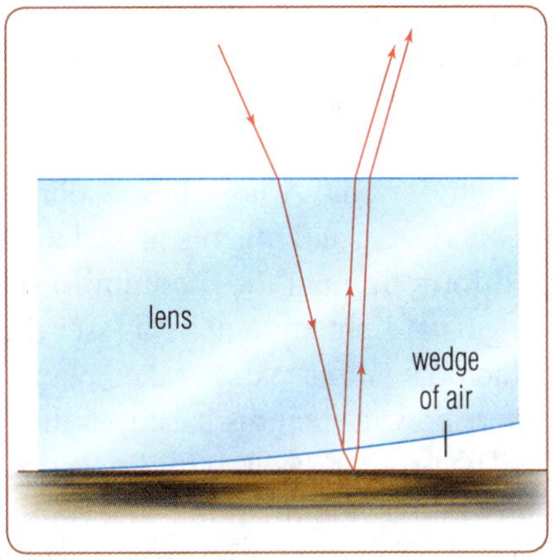

25-8 Light rays producing Newton's rings

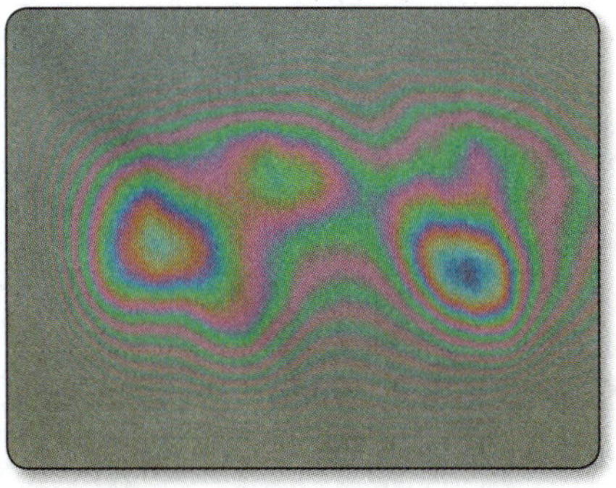
25-9 Thin-film interference in water between two sheets of glass

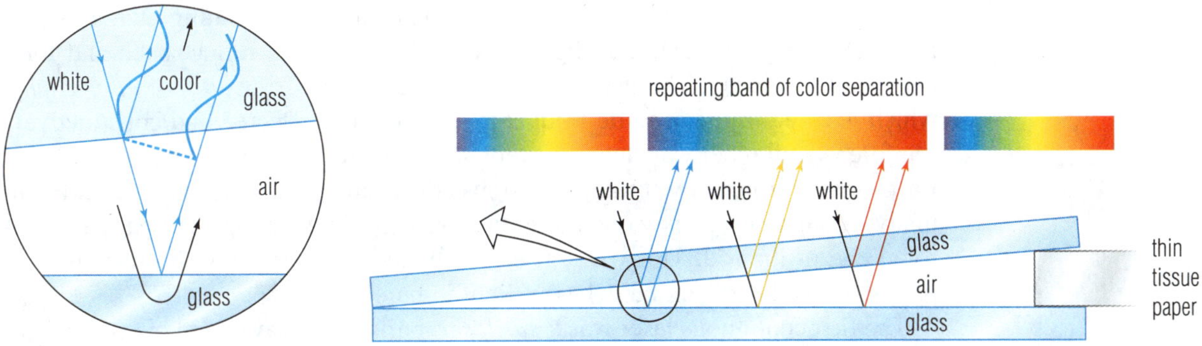

25-10 Interference of rays reflected from two glass plates separated by a thin wedge of air

25.5 Interferometers

An **interferometer** uses the interference of light to make precise measurements of distance. A monochromatic (one-color) light source directs its light toward a

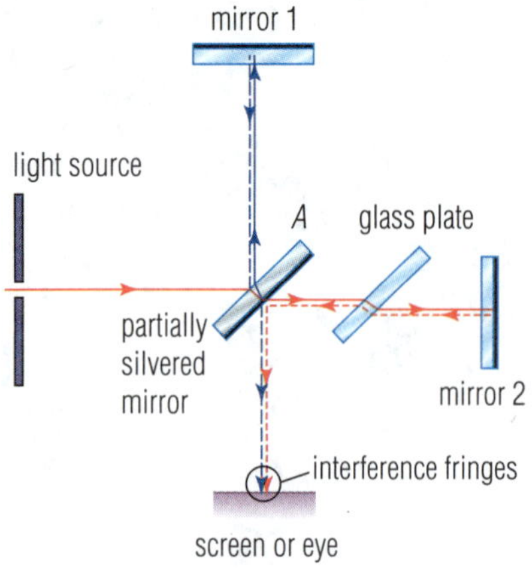

25-11 Functional schematic of an interferometer

partially silvered mirror (mirror *A*, Figure 25-11), which reflects half of the light to mirror 1 and lets the rest pass to mirror 2. (The plane glass plate simply ensures that light from both paths passes through the same total thickness of glass.) Mirrors 1 and 2 reflect the light back through the mirror to a screen. If mirrors 1 and 2 are the same distance from mirror *A*, then light rays from both paths travel the same distance and should arrive in phase, with only constructive interference. If they are not the same distance from mirror *A*, the light will make an interference pattern from which an experimenter could determine the path difference of the light rays. Albert Michelson made the first working interferometer and used it to measure the meter to a greater precision than was possible before. He determined that a meter was equal to 1553165.5 wavelengths of red light emitted from a plasma of cadmium. (The meter is now defined as the distance light travels in a vacuum in 1/299792458 s.)

25.6 Lasers

Interference also makes the laser possible. The word *laser* is an acronym for *l*ight *a*mplification by *s*timulated *e*mission of *r*adiation. A laser emits *coherent* light that is almost monochromatic. The method of producing laser light generates a very concentrated beam. Properly focused, a laser's light is intense and can retain a narrow beam over great distances. Incandescent light sources, on the other hand, emit many wavelengths, and their light disperses in all directions.

There are many kinds of lasers, but they all use the same principle. Some atoms can retain excess energy in their electron orbitals for extended periods of time. If they absorb additional energy from a wave of light, they can emit a light ray identical to the original and in phase with it. The result is amplification—getting two rays of light for one. A laser has a source of energy (a light "pump"), a gain medium to be energized, and mirrors at each end to reflect the spontaneously emitted light back and forth through the medium to stimulate the emission of more light. The wavelengths of light emitted by a laser depend mainly on the material or *gain medium* in which the amplification takes place. When stimulated, the medium will emit characteristic wavelengths based on the type of chemical bonding in the medium.

Theodore Maiman (1927–2007) was an American physicist who invented the ruby laser in 1960.

The original laser, built by Theodore Maiman in 1960, was a *crystal laser* with ruby as the gain medium. The ruby rod, about the size of a short pencil, has parallel ends. The front of the rod is partially silvered to reflect some light back into the rod but also to allow the laser to emit some light. The back of the rod is fully silvered. Surrounding the rod is a spiral flash tube like those used in photography. The flash tube "pumps" energy into the ruby to raise the average energy level of the ruby atoms. Some atoms emit light spontaneously, stimulating others to emit light. The light rays that are aligned axially with the rod reflect back and forth between the mirrors. The distance between the mirrors forms a *resonator cavity*, which is tuned to the specific frequency that will be reflected and constructively reinforced—all other wavelengths eventually die out. As the wave passes back and forth, it is reinforced by further emissions at the same frequency and phase. Eventually, enough energy is gained to emit a burst of light through the partially mirrored end of the rod. This early laser functioned by energizing the flash tube in pulses, so the laser emitted coherent light in bursts.

The first continuous laser was a *gas laser*, built by Ali Javan and his co-workers in 1960. The helium-neon gas in this laser is in a long quartz tube containing electrodes. The tube has a pair of mirrors at its ends, as in the crystal laser. When a high voltage is connected across the electrodes, the helium is ionized. Its electrons energize the neon atoms, which emit light, and thus stimulate the emission of more in-phase light rays. Since the voltage is continuous, the laser light is also continuous. Gas lasers emit more nearly monochromatic light than crystal lasers; for this reason they are used for communications and measurements, where frequency stability is important. Crystal lasers emit more energy, so they are used for drilling holes and welding metals.

Semiconductor or *diode lasers* also produce laser light directly from electrical energy. Since these lasers are tiny—often less than 1 mm in each dimension—they are frequently used for reading from and writing to optical digital information storage and retrieval mediums, especially CDs and DVDs.

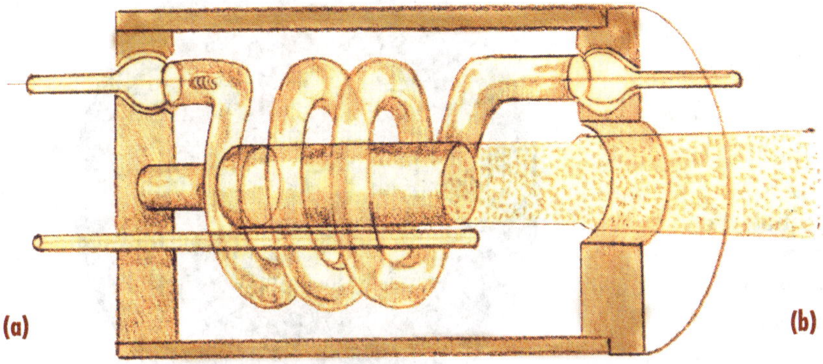

25-12 (a) A schematic of the first ruby laser (b) Ali Javan (1926–), the Iranian-American physicist who invented the gas laser in 1960

The potential uses for lasers are nearly unlimited. Lasers have found a place in precision macro- and micro-scale measurement. Lasers are used in medicine for delicate, bloodless surgery. High-speed photography, data recording and reading, communications, metallurgy, presentation pointers, sensors, barcode readers, and even directed-energy weapons all use laser technology. See the facet on the next page for more information on this subject.

25A Section Review

1. Patterns of ocean waves sometimes cross each other such that there are extraordinarily tall waves formed where individual waves intersect. What specific wave phenomenon causes these waves to form?
2. Discuss what produces the alternating bright and dark fringes in Young's experiment.
3. While conducting an experiment similar to Young's, an experimenter notes third-order maxima. If the wavelength of light used is λ, what is the length difference of the ray paths that form the maxima?
4. Why do lasers emit essentially monochromatic light?
5. ⊙ Young's experiment is performed in a school laboratory using 660. nm light and slits 5.00×10^{-6} m apart. What is the angular separation between the central maximum and the second-order maximum?
6. ⊙ Referring to Question 5, if the slits are 1.00 m from the projection screen, what is the distance of the second-order maximum from the central maximum?
7. ⊙ You are performing Young's experiment using 540. nm light and slits that are separated by 7.50×10^{-6} m. What is the distance between the central maximum and the first-order minimum if the projection screen is 1.25 m from the slits?
8. ⊙ a. Will the spacing of interference maxima be greater for longer or for shorter wavelengths of light for a given experimental setup?
   b. Work out an example to justify your answer.

25A Objectives

After completing this section, I can

- ✓ describe wave interference according to the superposition principle.
- ✓ summarize Thomas Young's experiment.
- ✓ explain the geometry of the light waves that form the interference patterns Young observed in his experiment.
- ✓ solve problems involving slit experiments.
- ✓ explain thin-film interference.
- ✓ describe the apparatus associated with interferometry and know that interferometry of light was used to define the length of a meter.
- ✓ discuss the operation of lasers and the applications of laser light.

FACETS of PHYSICS

Lasers

Men have long dreamed of producing a pure beam of light powerful enough to vaporize any matter. When Theodore Maiman introduced the first successful laser in 1960, the world realized that his invention had even more potential than they had dreamed. In fact, lasers were at first called a solution waiting for a problem. Not only did the laser concentrate tremendous energy in one direction, but it also generated a nearly monochromatic light (one frequency of "pure color") that could be pulsed with incredible speed. Within four decades, many different types of lasers were being produced for a wide variety of jobs.

Because laser light is virtually monochromatic, lasers can measure fine details from almost any distance. High-flying planes using LIDAR (*li*ght *d*etection *a*nd *r*anging) can distinguish objects less than 10 cm in size. Improved interferometers can measure wavelengths and minute linear displacements. For example, three interferometers, located 15 miles from California's San Andreas fault, detect changes in strain across the fault as small as a ten-millionth of an inch. Any detected strain indicates a build-up of stress, which alerts seismologists to a possible earthquake.

Because laser light is coherent (a single frequency in phase) and is generated by axial reflections within a tube or crystal, the beam, when properly focused, forms a nearly perfect straight line. It is ideal for surveys and large construction projects. Land engineers find the laser indispensable when they drill tunnels and lay pipelines, and jet-aircraft engineers align their 200-foot jigs (aligning tools) with accuracies up to a few hundredths of a centimeter. Lasers also monitor the production of precision tools, and they gauge super-fine wire.

The beam from a laser can travel long distances without losing its identity. Can you imagine having a "flashlight" that could illuminate something thousands of miles away? If you knew the time that the light took to reach an object and bounce back to you, you could figure out the object's distance. Scientists do just that. Erbium-doped glass lasers are used to measure the height of clouds, and lasers on airplanes measure the altitude of the plane. Some pulsed ruby lasers on earth track distant satellites to an accuracy of a few millimeters! Something even more unbelievable happened after the first men who landed on the moon, "Buzz" Aldrin and Neil Armstrong, placed multi-prism corner reflectors on the moon during the *Apollo 11* mission.

A laser-reflecting satellite (LAGEOS I)

Laser-equipped earth stations were able to pick up beams bounced off the reflectors and precisely measure, with millimeter precision, the distance from the earth to the moon. Instruments taking readings three times each day determined that the moon is gradually receding from the earth at the rate of a few centimeters per century.

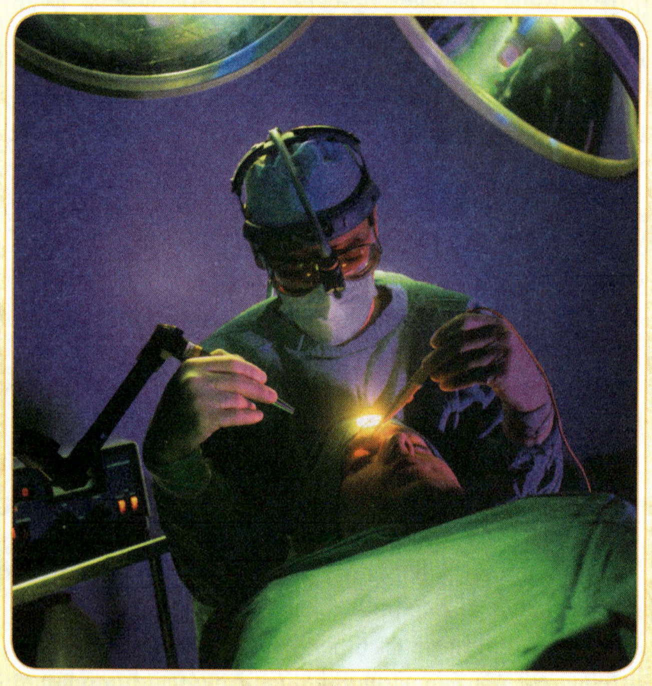

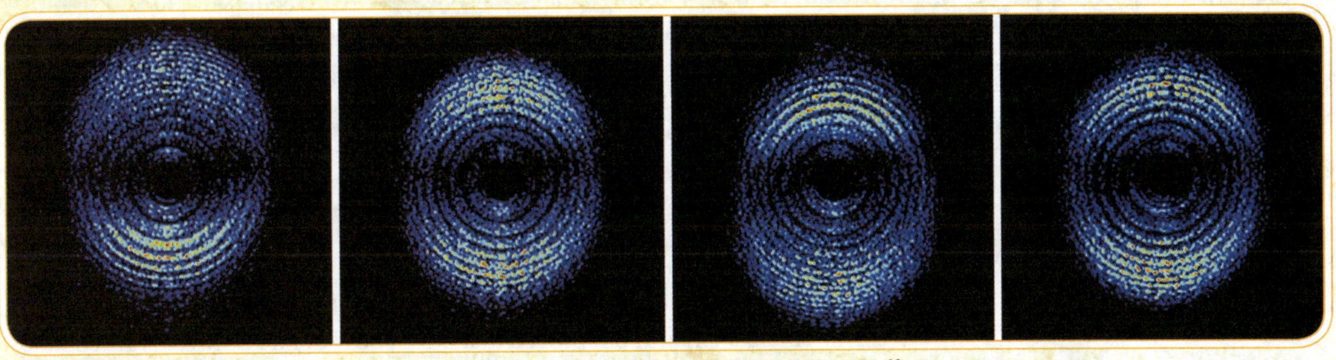

This series of images reconstructs the motions of an electron near an ionized atom by using light from an attosecond (10^{-18} s) pulsed IR laser. The target was argon atoms ionized with extreme UV pulses synchronized with the laser. The images are very similar to a stroboscopic, stop-action photo.

Industries routinely employ laser energy in drilling, cutting, and welding. The energy in one pulse of a "mode-locked" laser can produce tens of trillions of watts—more power in one instant than is being generated by all the earth's power plants combined. Pulsating ruby lasers blaze small holes through the three hardest substances on the earth: sapphires and rubies for bearings in watches as well as diamonds for wire-drawing dies. Other kinds of lasers drill holes in tiny circuit boards and cut grooves so that the boards can be broken into individual circuits. The ultimate in ink erasers, lasers remove ink without touching the paper. Clothing manufacturers cut materials quickly and precisely. Construction crews can cut 5-centimeter stainless steel at 40 centimeters per minute and weld 2-centimeter sheets at 30 centimeters per minute.

Microsurgery is making advances almost every day. Without touching the skin, lasers remove both tattoos and skin blemishes, which otherwise would remain for life. "Bloodless surgery" is becoming more practical—as lasers remove tumors and other tissue, the heat cauterizes the outlying blood vessels and keeps the incision sterile. Ophthalmologists can "spot weld" detached retinas using laser pulses as well as reshape the cornea in a procedure called *laser-assisted in situ keratomileusis* (LASIK). Many nearsighted and farsighted people who undergo successful LASIK surgery can go without glasses for decades after. Researchers can even alter the shape of chromosomes in individual cells with finely focused laser beams.

The military has taken special interest in the potential power of the laser beam. All modern tanks are equipped with laser range finders (LADAR) to determine the target's exact distance. Laser-guided shells (and "smart bombs") home in on targets that have been illuminated by ground YAG (*y*ttrium *a*luminum *g*arnet) lasers: YAG lasers are ideal because they produce an invisible infrared beam. Gas-injection lasers shine a light on the enemy to make him an easy target. Recently, air-, ground-, and sea-borne high-energy lasers are being developed for use in strategic and tactical ballistic missile defense. Such weapons were the stuff of science fiction stories for nearly a hundred years.

High-speed photography takes advantage of the rapid pulses achieved by crystal lasers. You probably cannot detect the time your own camera takes to expose each flash because it is very fast. But these high-tech cameras can expose a frame in one ten-trillionth of a second! Unlike hand-held video cameras, which record images at a rate of 30 frames per second (fps), the best laser cameras take as many as one billion fps. This breakthrough helps in studying ultra-fast, momentary phenomena, such as the formation of vapor during laser welding.

Lasers can perform tasks either difficult or impossible by other means. Research continues in all fields to find new uses for lasers. Communications researchers are investigating how to send even more information by laser light waves via fiber-optic networks. Fiber-optic data transmission based on laser sources is improving all the time. Currently, main trunk fiber-optic data transmission lines can carry 14 terabytes of data per second, which is far greater by many orders of magnitude than equivalent copper-wire data cables can manage.

Laser applications may soon involve the initiation of controlled fusion reactions, the production of metallurgical plasmas, photochemical reactions, and phase transitions of matter. In the distant future even 3D television might be possible.

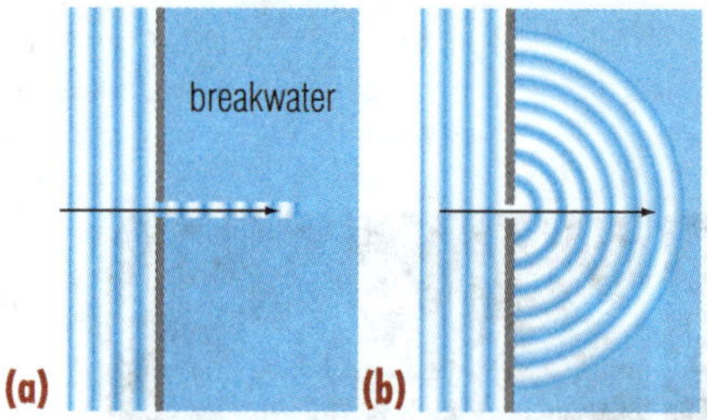

25-13 Water wave crests showing (a) no diffraction and (b) diffraction around the edges of the gap

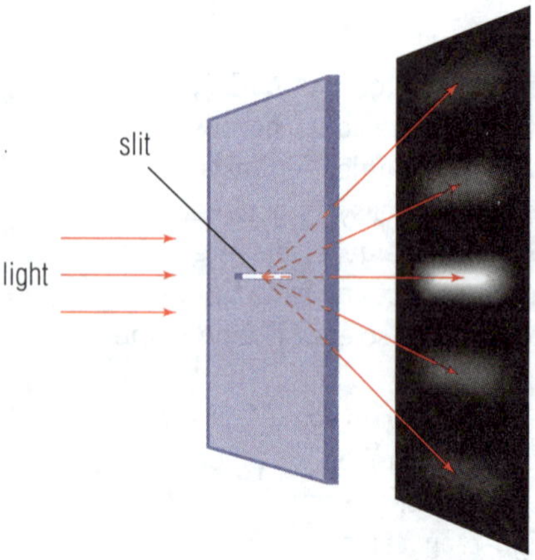

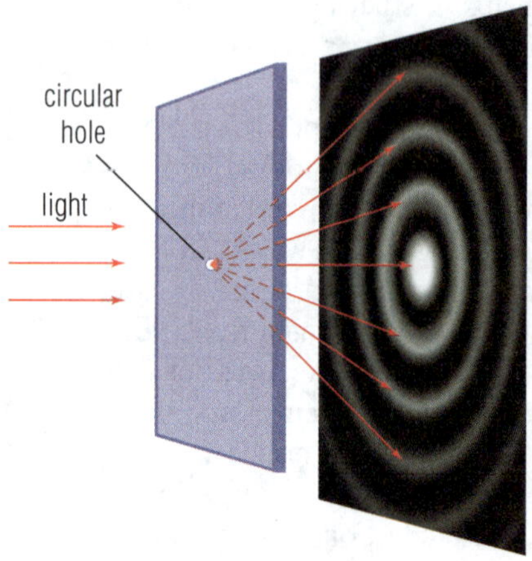

25-14 Opening diffraction

25B DIFFRACTION

25.7 Describing Diffraction

To early experimenters, light seemed to travel in straight lines and cast sharp shadows. Other waves, such as ocean and sound waves, were known to **diffract**; that is, they spread into a region behind an obstruction where, if they could only travel in straight lines, a shadow zone should exist. For example, consider water waves approaching a gap in a harbor breakwater. If water waves could only travel in straight lines, waves inside the breakwater, such as those in Figure 25-13a, would result. However, since water waves do diffract, the resulting waves are actually like those shown in figure (b).

Light's apparent lack of diffraction convinced Newton that light cannot be a wave. But Young's experiment proved that light does act like a wave. How can these statements be reconciled? The explanation lies in light's short wavelengths. Water waves diffract only around obstacles whose dimensions in the direction of wave propagation are small compared to the wavelength. With the long wavelengths (measured in meters) of most water waves, most obstacles are small compared to the wavelengths. In like manner, light diffracts only around objects that are small compared to its wavelengths. However, obstacles on the order of several hundred nanometers in size are too small to be seen without a microscope; therefore, you cannot readily see light diffracting around them. Light diffraction may be observed at the edges of distant shadows cast by bright, small light sources that are themselves distant from the object casting the shadow. Under these circumstances, the tiny diffraction fringe has an opportunity to visibly spread into the shadow zone.

Diffraction is a form of interference caused by parts of a wave front encountering an obstacle. The interference produces light and dark bands parallel to the obstacle. How is this interference different from the interference in Young's experiment? Young's double-slit experiment involved the interference of diffracted light waves from two openings. But as we have seen, simple diffraction and the resulting interference patterns can occur when a light wave interacts with just a single obstacle or opening.

Diffraction can be separated into four categories: edge, obstacle, opening, and grating diffraction. The shadow of a knife edge under monochromatic light illustrates *edge diffraction.* Instead of a clear-cut shadow, bands of light and darkness appear inside the area expected to be in shadow. The light bends around the edge into the shadowed region, creating an interference pattern.

Obstacle diffraction results when a small object is in the path of a beam of monochromatic light. Again, the object has no clear-cut shadow, but it has alternating light and dark bands inside the shadowed area. For a small spherical object, the shadow may have a bright spot in the center.

Opening diffraction occurs when light passes through a small opening. When you look through a single slit, alternating bands of dark and light appear in the opening. For a circular opening, alternating rings of dark and light appear in the opening. Figure 25-14 shows the resulting patterns. You can observe opening diffraction for yourself with no special equipment. Place the tips of your thumb and index finger about 0.5 cm apart, and look through the opening at a light source with your hand a few centimeters from your eye. Now gradually narrow the space between your fingers. As the opening closes, dark lines parallel to the surface of your finger and thumb will appear. This is opening diffraction. You can also see this type of diffraction by shining light through an opening onto a screen.

Grating diffraction occurs when light passes through a piece of transparent material scratched with evenly spaced, fine parallel lines, called *rulings.* Between the rulings the glass is still transparent, so light can easily pass between them. The rulings obstruct the path of light. The light passing through the grating forms a diffraction pattern with minima and maxima.

You can calculate the locations of bright regions for gratings the same way that you calculated bright regions for interference patterns. The fact that there are now many sources of light (unruled areas) instead of just two makes little difference. Figure 25-16 shows the light waves that constructively interfere to form a maximum. The path difference between two waves from *adjacent* rulings to the nth maximum is an integer multiple of wavelength ($n\lambda$); the path difference to the same maximum for *any* two rulings is an integer multiple of $n\lambda$. The geometry for any two lines is the same as Figure 25-16, from which the following equation is obtained:

$$\sin\theta_n = \frac{n\lambda}{d}$$

This equation is identical to Equation 25.1. The angle θ_n is the same as it was for Young's experiment with interference patterns. It is the angle between the imaginary line connecting the center of the grating to the central maximum on the projection screen and the line connecting the grating to the nth-order maximum. Equation 25.1 is especially useful for grating diffraction.

Diffraction gratings are so named from the collection of closely spaced linear scratches, or rulings, whose appearance is similar to a fireplace grating.

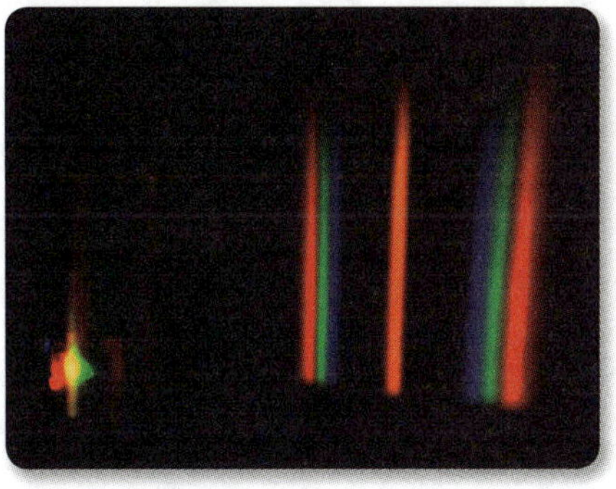

25-15 Grating diffraction

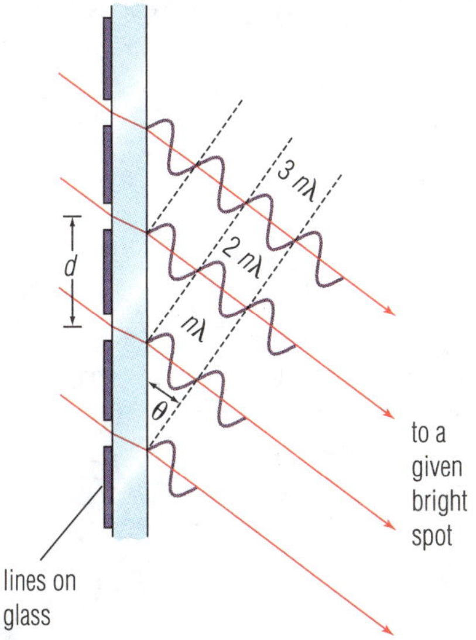

25-16 Waves in grating diffraction forming a bright fringe

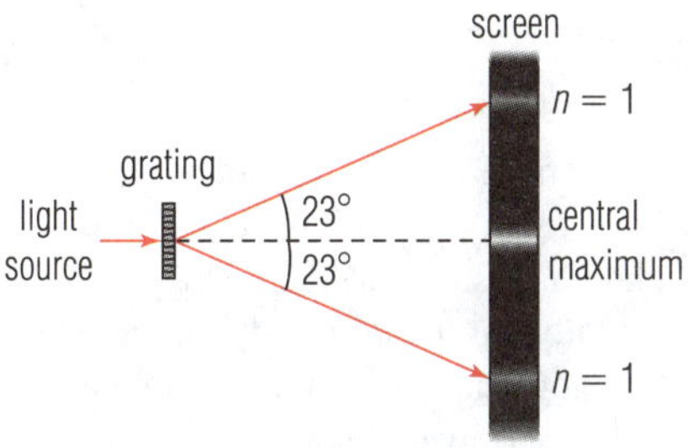

25-17 Diffraction grating information for Example 25-3

EXAMPLE 25-3

Diffraction Rules: Wavelength from Diffraction

A diffraction grating has 6000 ruled lines per centimeter. Find the wavelength of light that places the first-order maxima at $\theta_1 = 23.0°$ from the central maximum (see Figure 25-17).

Solution:

You know that $n = 1$ (the *first-order maximum*), $\theta_1 = 23.0°$, $d = 1.00$ cm/6000 lines $\cong 1.6\underline{6}6 \times 10^{-4}$ cm $\cong 1.6\underline{6}6 \times 10^{-6}$ m (assuming that the width of the ruling lines is negligible, the spacing between rulings is just the reciprocal of the line density). You need to find λ. Use Equation 25.1.

$$\sin\theta_n = \frac{n\lambda}{d}$$

$$\lambda = \frac{d\sin\theta_n}{n} = \frac{(1.6\underline{6}6 \times 10^{-6}\ \text{m})\sin 23.0°}{1}$$

$$\lambda \cong 6.5\underline{0}9 \times 10^{-7}\ \text{m} \cong 651\ \text{nm}$$

The light's wavelength is about 651 nm, so its color is in the red region of the visible spectrum.

EXAMPLE 25-4

Determining the Location of Diffraction Maxima

Light with a wavelength of 480. nm illuminates a grating with 6000 lines per centimeter. At what angle are the second-order diffraction maxima?

Solution:

You know that λ = 480. nm and n = 2. You need to find θ_2.

$$\sin\theta_n = \frac{n\lambda}{d}$$

$$\theta_n = \sin^{-1}\left(\frac{n\lambda}{d}\right)$$

$$\theta_2 = \sin^{-1}\left[\frac{2(480. \times 10^{-9}\ \text{m})}{1.6\underline{6}6 \times 10^{-6}\ \text{m}}\right]$$

$$\theta_2 \cong 35.\underline{1}8° \cong 35.2°$$

The second-order fringes will be about 35.2° from the central maximum.

Problem-Solving Strategy 25.3

When you determine SDs associated with diffraction grating ruling densities, the number of lines is *counted* and therefore does not affect SDs in calculations. Its reciprocal is a distance, however, so assume that grating spacing has three SDs.

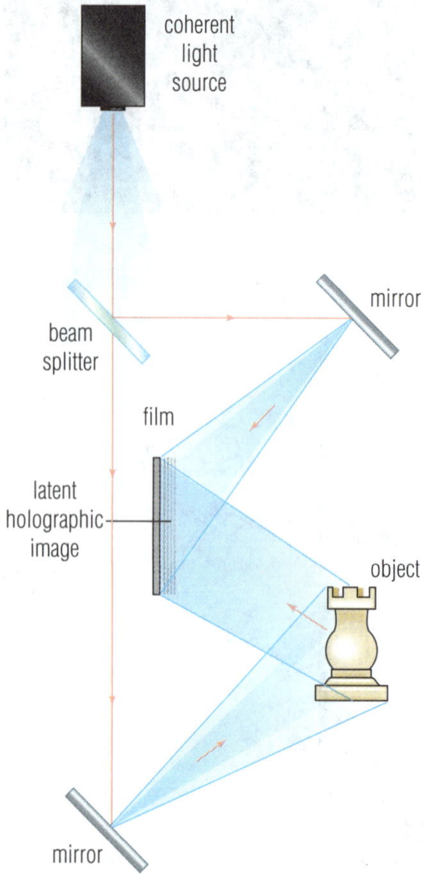

25-18 The holographic recording process

Holograms as Art

Holograms have been gaining popularity as an art medium. The Massachusetts Institute of Technology (MIT) actually has a museum of holography. But holographic art often degenerates into a portrayal of something weird, grotesque, or cultish. What should a Christian think about the use of holograms as art?

25.8 Using Interferometry and Diffraction to Solve Problems

Holography, the science of using diffraction and interferometry to reconstruct three-dimensional images, was invented by Hungarian Dennis Gabor in 1947, for which he received the Nobel Prize in Physics in 1971. However, the first *hologram*, a three-dimensional image, was not created until 1962 because Gabor lacked a coherent light source.

Interest in holography revived when pure colors from lasers became readily available. Two beams of monochromatic light are directed to a photographic film or plate. One beam reflects off the subject before it joins the other beam on the light-sensitive surface. The resulting interference pattern is meaningless until the developed film is illuminated by another beam with the same frequency of light used to create the image. A three-dimensional image forms that looks identical to the original. Looking into the hologram is like looking through a window—as the point of view moves from side to side, different views of the object can be seen.

Even if the hologram is broken into tiny pieces, the entire scene can be reconstructed from a single piece, though the image will be less distinct. The immense advantage of holograms is the depth of information that can be stored. Moving particles like pollen and fog droplets cannot be photographed more than once before they begin to move, so only a hologram can provide the detailed information needed for motion studies.

Natural-light holograms or *rainbow transmission holograms* are designed to provide a multicolored, 3D holographic image when viewed under ordinary light. These holograms are commonly found on credit cards and packaging for copyrighted materials. Holograms are filling a crucial need in the effort to reduce counterfeiting and piracy. Many paper currencies around the world, including US bills, contain holographic features to reduce the potential for counterfeiting.

25-19 A holographic trademark seal

The packaging for the anti-malarial drug artesunate contains a layered rainbow transmission holographic certification seal that attests to the authenticity of the drug. According to a report published in the 2008 International Anti-Counterfeiting Directory, this seal, while potentially useful, suffers from several defects that make it easy to counterfeit. It is small, has a simple design, and uses a technically simple process to produce the hologram. In addition, the seal has been used for over ten years without change. The international anti-counterfeiting organization has suggested that the Chinese company that manufactures the real drug immediately revise the seal, increase its size and complexity, update its holographic process, and add holographic packaging to the pill sleeve inside the box. These efforts should reduce the incidence of thousands of people dying from malaria after unknowingly taking useless counterfeit drugs.

The increasing sophistication of counterfeiters demands that legitimate food and drug manufacturers continue to improve the marking and packaging of their products. The health of people all over the world must be protected from these conscienceless criminals.

25.9 Spectrometers

One important application of diffraction is the *grating spectroscope*, which separates light into its component wavelengths. Since the diffraction angle is dependent on wavelength, each color range has a maximum at a different angle. The diffraction grating, like a prism, generates a continuous color spectrum when illuminated by white light. The light entering the spectroscope first passes through a **collimator**. A collimator adjusts the shape of the incident light beam and makes the rays parallel. Light emerging from the collimator lens consists of parallel light rays that enter the diffraction grating. From the grating, the light is dispersed into a diffraction pattern consisting of a bright first-order maximum and dim higher-order fringes. A rotatable eyepiece detects light parallel to its optical axis. At each angle of a first-order maximum, the eyepiece detects light of a different wavelength, so the angles on a fixed scale can be labeled with the wavelength that will be detected there.

Spectrometers are most useful when one is measuring the line spectra of sources that emit just a few wavelengths of light, such as sodium vapor lamps.

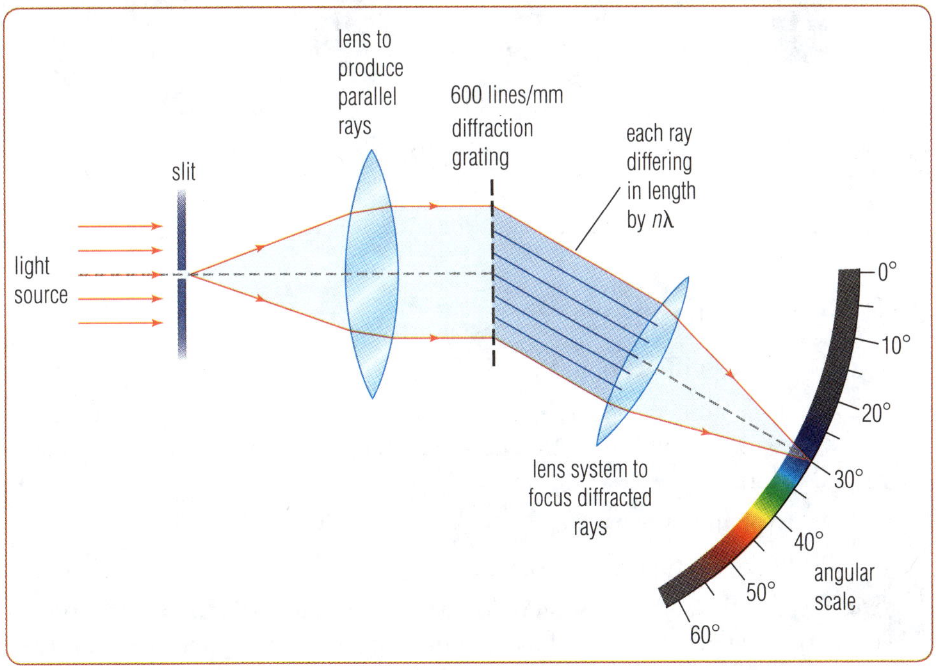

25-20 A diffraction-grating spectrometer

25.10 Optical Resolving Power

Lenses also show diffraction effects. In the ordinary use of lenses, you can assume that the image of a point is a point, but actually it is a group of concentric rings. This phenomenon makes it difficult for a lens to distinguish between two points of light that are close together. Such a problem is evident in astronomical telescopes that attempt to separate—or *resolve*—two stars optically close together. The ability to separate close objects is called the **resolving power** of the lens. Contrary to what you may think, the resolving power of a lens or system of lenses is *not* a function of the total magnification available, but rather the angular diameter of the first-order interference fringes seen surrounding the objects. These interference fringes associated with point sources of light, such as stars in telescopes and tiny sub-cellular objects in microscopes, are called **Airy discs**.

George Biddell Airy (1801–92) was a British mathematician and astronomer who served as the Astronomer Royal (chief astronomer) for the British government. He was responsible for establishing the Prime Meridian at Greenwich, England.

There is a limit to how close point-like objects can be if they are to be resolved as separate images. Lord Rayleigh, an English physicist, showed that two points must be separated by at least an angle of θ_R for a given lens to show them as distinct images. The angle θ_R is called the **Rayleigh angle** and is computed by

John William Strutt, Lord Rayleigh (1842–1919) was an English physicist who won the Nobel Prize in Physics in 1904 for his investigations of the densities of gases.

$$\theta_R = 1.22\frac{\lambda}{d}, \qquad (25.5)$$

where λ is the wavelength of light coming from the point sources, d is the aperture of the lens or mirror, and θ_R is the angle, given in radians. Any points with greater angular separation can be clearly resolved. For a given wavelength of light, it is clear that resolving power is inversely proportional to the diameter of the optical aperture. Therefore, the larger the lens (or mirror, since this relationship is applicable to both kinds of optical components), the smaller the Airy discs and the sharper and more detailed the image.

Problem-Solving Strategy 25.4

Remember that the Rayleigh angle (θ_R) is given in radians.

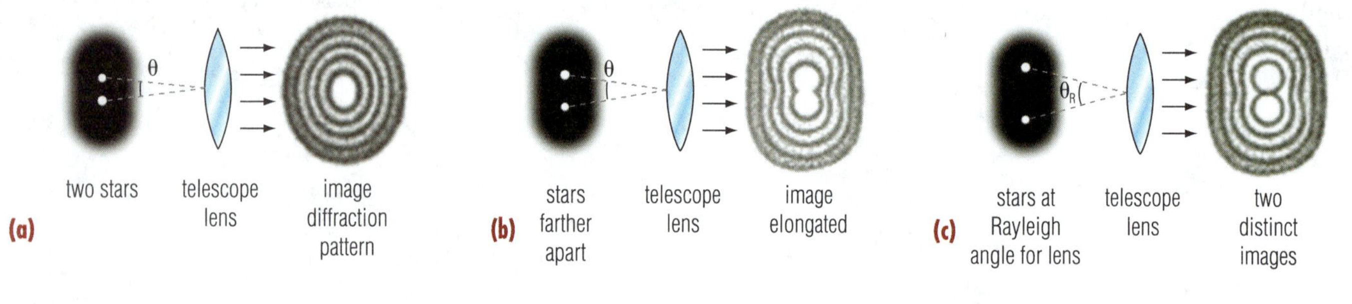

25-21 Two stars must have a minimum angular separation in order to be resolved by a given lens. The angle shown is exaggerated for clarity. A 10 cm lens can theoretically resolve two point-like objects about 1 second of an arc apart.

25B Section Review

1. How does diffraction differ from refraction?
2. a. If the number of rulings per centimeter in a diffraction grating increases, will the diffraction maxima get closer together or farther apart?
   b. Explain your answer.
3. DS Why do most quality holograms require a coherent light source?
4. DS Even if the authentication hologram for the anti-malarial medication artesunate is revised and made more counterfeit-resistant, what must be done to ensure that the millions of users in Southeast Asia know that they have purchased the real thing?
5. What is the purpose of a grating spectrometer?
6. What would happen to the image of two stars that could barely be resolved if you placed a circular mask over the end of the telescope in order to reduce the aperture to half its original diameter?

The ability to separate details in an optical image is a function of the aperture of the lens or mirror. The larger the aperture, the smaller the angular separation objects may have and still be resolved by the optical system. This is one reason astronomers seek to build larger telescopes—to separate extremely distant, close-together celestial objects in photographic images.

7. In a diffraction-grating experiment, what is the spacing of the grating rulings if the angular displacement of the first-order maximum is 31.33° from the central maximum, and you are using light with a wavelength of 650. nm?

8. What is the highest-possible order number (n) for interference maxima if the diffraction grating is ruled at 12 000 lines/cm, and you are using red light (λ = 700. nm)? (Remember that n must be a whole number.)

9. What is the theoretical Rayleigh angle for the 60.0 mm objective lens of a refractor telescope when viewing yellow stars (effective wavelength λ = 580. nm)? (Remember to compute your answer in radians.)

10. How many seconds of an arc is the Rayleigh angle in Question 9? (*Hint*: Degrees are further subdivided into minutes and seconds.)

25B Objectives

After completing this section, I can

✓ describe wave diffraction and state under which conditions it can occur.

✓ state the factors that control the appearance of diffraction-related interference patterns.

✓ solve problems involving ruled diffraction gratings.

✓ describe ways that holographic images are being used in the battle against counterfeiting.

✓ describe the function of diffraction-grating spectrometers.

✓ state the principle that limits the resolving power of optical systems.

✓ compute the Rayleigh angle for a given optical system.

25C POLARIZATION OF LIGHT

25.11 Describing Wave Polarization

You have learned that light waves can be modeled as *transverse waves*. That is, their electric (**E**) and magnetic (**B**) field vectors oscillate at right angles to the direction of travel of the light wave. In addition, the E- and B-fields oscillate in planes mutually perpendicular to each other. Since the B-field is always perpendicular and proportional to the E-field, physicists, by convention, describe the properties of a light wave by the orientation of its E-field. As a light wave propagates with time, the change in the direction and magnitude of the electric field vector **E** defines the **polarization state** of the light wave. Typically, the E-fields of light waves emitted by atoms are oriented in random directions, even if the rays themselves all travel in the same direction (see Figure 25-22). Such randomly oriented light waves are called **unpolarized**. If the E-fields in a beam of light are *not* randomly oriented, the light is **polarized**. There are two special polarization states and one general state of interest in optical physics.

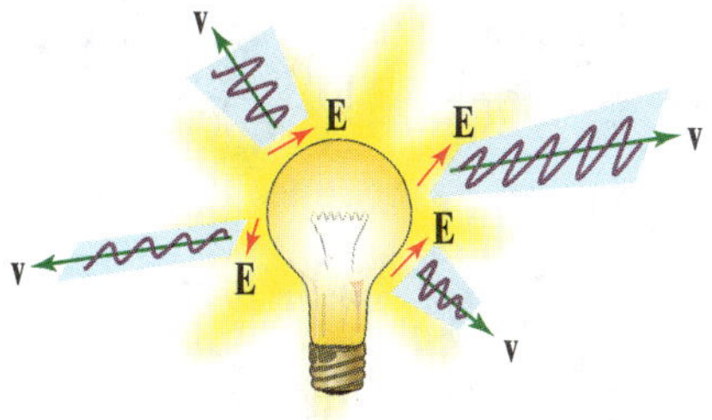

25-22 Unpolarized light consists of waves with different directions of **E**.

A light wave's sinusoidal waveform (E-field) moving through space can be resolved into perpendicular x and y components oriented to an arbitrarily chosen 3D coordinate system in space. The z-axis is aligned in the direction of the light wave propagation. If the x- and y-components' waves are in phase, the E-field vector oscillates in a single plane along its path of propagation. Such a light wave is said to have *linear* or *plane polarization*, as in Figure 25-23a. If the wave passes through an imaginary plane perpendicular to the path of propagation, the oscillating tip of the E-field vector traces a line segment on the plane.

If the two components have exactly the same amplitude but are out of phase by 90°, the E-field vector has a constant length and rotates in a helical pattern as the light wave moves through space. As the wave moves through the perpendicular reference plane, the tip of the E-field vector describes a circle, so the light wave has *circular polarization*, as in Figure 25-23b.

Both plane and circular polarizations are special cases of a more general state in which the E-wave components are not in phase and either do not have the same amplitude or are not exactly 90° out of phase, or both. The E-field vector describes an ellipse as the light wave propagates, so it is *elliptically polarized*, as in Figure 25-23c.

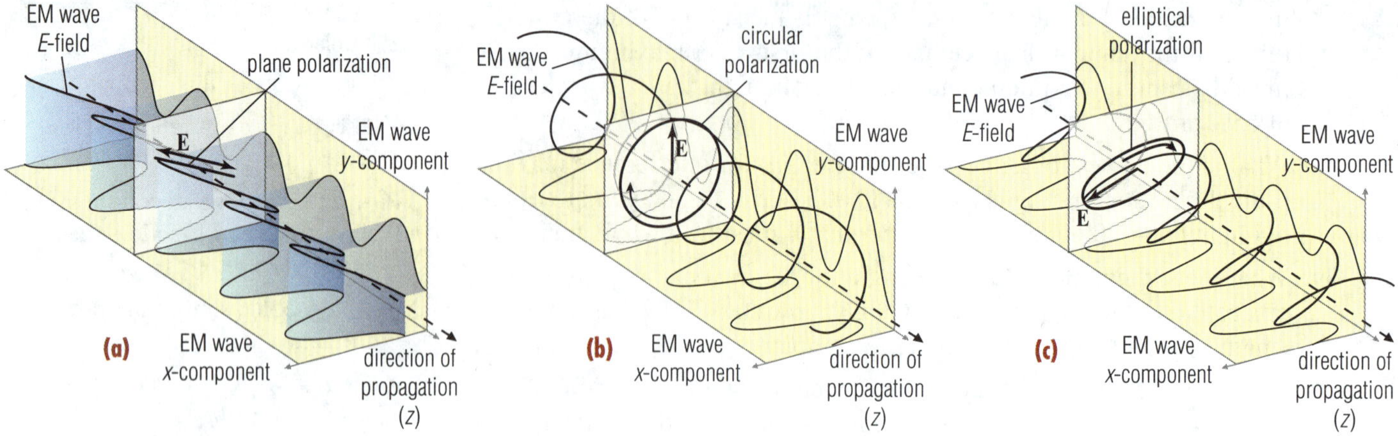

25-23 Light can experience plane polarization (a), circular polarization (b), or elliptical polarization (c).

Since light leaving a luminous source is usually unpolarized, something must act on the light to polarize it. There are four principal ways to polarize light: *reflection*, *double refraction*, *selective absorption*, and *scattering*.

25.12 Reflected Polarization—Malus's and Brewster's Laws

Étienne Louis Malus (1775–1812) was a French physicist who discovered that reflected light is polarized. He coined the word "polarize" because he accepted the particle theory of light and believed that the particles had to have poles in order to show polarization.

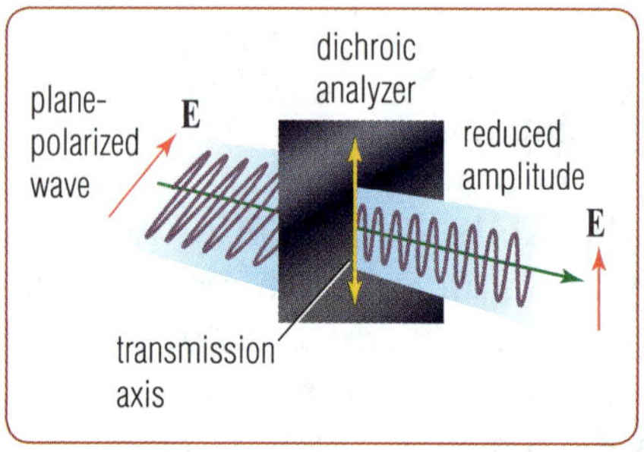

25-24 Malus's law describes the effect a dichroic analyzer has on the intensity of a beam of plane-polarized light.

In 1808, the French physicist Étienne Malus discovered that when a light beam strikes a pane of glass or the surface of water, the reflected light is plane polarized. He determined this fact after he looked through a special crystal of the mineral calcite, called *Iceland spar*, at light reflecting off building windows. As he rotated the crystal, reflections were extinguished at certain angles. Iceland spar and other materials transmit polarized incoming light in a preferred direction, called the **transmission axis**. Only light waves oscillating parallel to the transmission axis pass through. Such polarizing materials are called **dichroic**. **Malus's law** states that the intensity of a polarized light beam transmitted by a dichroic material (the *analyzer*) is a function of the angle between the wave's plane of polarization and the analyzer's transmission axis.

When unpolarized light reflects off a surface, some of the light is reflected and some is refracted into the substance. The light that is reflected tends to contain more waves that are horizontally plane polarized, but the refracted rays tend to contain more vertically plane polarized waves. The proportion of each kind depends on the angle of incidence and the materials in contact at the reflection interface.

Sir David Brewster (1781–1868) was a Scottish physician who invented the kaleidoscope and discovered the phenomenon of light reflected at specific angles.

David Brewster discovered that for any pair of materials in contact at an interface, the reflected light is completely polarized when the reflected light rays in the incident medium are 90° from the refracted light rays in the other medium. At this special angle of incidence, called **Brewster's angle (θ_B)**, the reflected beam consists entirely of horizontally polarized waves, but the refracted beam contains mostly (but not entirely) vertically polarized waves. This discovery is called **Brewster's law**. The mathematical form of Brewster's law is

Problem-Solving Strategy 25.5

Brewster's angle (θ_B) may be given in degrees or radians.

$$\tan \theta_B = \frac{n_r}{n_i}, \qquad (25.6)$$

where θ_B is the Brewster's angle of incidence that produces the greatest polarization, and n_r and n_i are the refraction indices of the media containing the refracted and the incident rays, respectively. Thus, you can easily calculate the Brewster's angle of incidence that produces the greatest polarization in the reflected light beam.

25-25 Polaroid sunglasses contain a dichroic material whose vertical transmission axis blocks the horizontally polarized light reflected from the lake.

EXAMPLE 25-5

Reflecting on Polarization: Brewster's Angle

What is Brewster's angle for an air-water interface? Use the indices of refraction from Table 24-1 on page 552.

Solution:

You find that $n_{air} = 1.0003$ and $n_{water} = 1.333$. Use Equation 25.6:

$$\tan\theta_B = \frac{n_r}{n_i} = \frac{n_{water}}{n_{air}}$$

$$\theta_B = \tan^{-1}\left(\frac{n_{water}}{n_{air}}\right) = \tan^{-1}\left(\frac{1.333}{1.0003}\right)$$

$$\theta_B \cong 53.\underline{1}1^\circ \cong 53.1^\circ$$

When the sun is about 37° above the horizon (90° − 53.1°), its rays reflecting off the surface of water are horizontally polarized.

Problem-Solving Strategy 25.6

Remember that Brewster's angle is the incident angle of the incoming ray of light. The 90° angle is measured between the reflected ray and the refracted ray.

25.13 Polarization by Double Refraction

25-26 Double refraction in Iceland spar calcite

A second method of polarizing light uses a property of many mineral crystals called *double refraction* (or *birefringence*), which was first observed by Erasmus Bartholinus in 1669. It was so named because the indices of refraction are different through the different crystal planes. Some common examples of crystals with this property are aragonite, calcite, ice, mica, quartz, selenite, sugar, and topaz. In such a crystal, a refracted ray is split into two rays—the *ordinary* (O) ray and the *extraordinary* (E) ray. The ordinary ray follows Snell's law of refraction exactly. The extraordinary ray does not refract according to Snell's law but will be refracted at an angle dependent upon the microscopic structure of the crystal. It does emerge parallel to the ordinary ray, however. Both rays are polarized so that their planes of polarization are at right angles to each other.

Erasmus Bartholinus (1625–98) was a Danish mathematician who wrote the first scientific description of the polarization of light in a crystal of Iceland spar calcite.

A *Nicol prism*, invented in 1828 by William Nicol, is made from a calcite crystal that has been cut and cemented together again with Canada balsam, an adhesive often used in optical applications. The calcite produces an O-ray and an E-ray, but the O-ray is reflected by the Canada balsam so that the E-ray continues alone. This device produces polarized light. Nicol prisms may be used to identify unknown minerals. In this application, two Nicol prisms are placed in the path of an unpolarized beam of light with a thin section of the material of interest between them. The polarized light from the first prism enters the sample and is rotated by its crystal lattice. The amount of rotation is detected by rotating the second prism

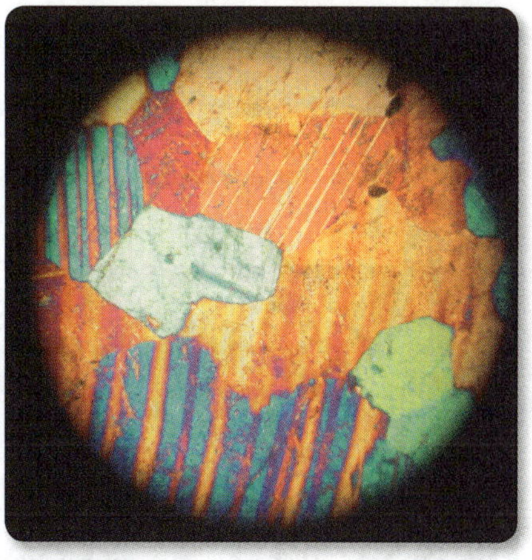

25-27 Minerals in a thin section of rock sample can be identified with a polarizing microscope by using crossed Nicol prisms.

William Nicol (1768–1851) was a Scottish physicist.

until it transmits the maximum amount of light, showing that its transmission axis is aligned with the plane of the light's polarization. The amount of rotation is characteristic for each crystalline material and is useful for identifying unknown substances.

25-28 When two sheets of selectively absorptive material are placed at right angles to each other in the optical path, they will absorb all of the originally unpolarized light.

25.14 Polarization by Selective Absorption

A few materials not only doubly refract but also absorb one of the refracted rays. These materials naturally act like a Nicol prism, although they may emit the *O*-ray rather than the *E*-ray. This behavior is called selective absorption. The transmission axes of two crystals of tourmaline, for example, turned at 90° to each other will together absorb all the light that is directed through them. The first will emit a polarized *O*-ray, and the second will absorb that ray.

Polaroid filters, invented by Edwin Land, selectively absorb a wide selection of optical wavelengths. This means that incident, unpolarized white light will be partially absorbed and partially transmitted with a preferred polarization. The absorbed light is converted to thermal energy. Certain Polaroid sunglasses are designed to reduce the overall intensity of sunlight by screening about half of the unpolarized sunlight rays. Others are designed to absorb horizontally polarized light reflected off water and glass. Such materials are found in windshields and in sunglasses used by fishermen, boaters, and seamen (see Figure 25-25).

25-29 Try this: look through two Polaroid sunglasses held at right angles to each other, and note the significant reduction of light intensity.

Edwin H. Land (1909–91) was an American researcher and industrialist who made many contributions to the science, technology, and art of optics. He held some 20 honorary doctorate degrees but never completed a formal college course of study.

25.15 Polarization by Scattering

Scattering is the final way to polarize light. The colors that you see in the sky are a result of the scattering of sunlight by air molecules and dust particles in the air. Short wavelengths are scattered more than long wavelengths; when you look at the sky, you see the greatly scattered blue color. At sunrise and sunset, sunlight must travel a much longer path through the atmosphere to get to your eye. Much more scattering of the shorter wavelengths takes place, so only the longer orange and red wavelengths remain visible.

When you look in directions that are about 90° from the sun, you see mostly scattered light. You can show that this light is polarized by using a selectively absorptive material such as a Polaroid filter or lens. Polarized light illuminating a Polaroid filter must be aligned with the filter's transmission axis in order to pass through. If the sunlight were unpolarized, then about the same amount of light would pass through the Polaroid filter no matter at what angle you held it. However, the intensity of light observed through the filter changes as the filter is rotated, showing that the scattered light must be partly polarized. For this reason, photographers often use polarizing filters to darken clear skies and make clouds stand out more by removing scattered light.

25-30 Crossed polarizers were used to reveal a flaw in this material.

Some materials become doubly refracting when they are stressed. Polarized light can be used to detect stresses. If two polarizing filters are aligned 90° to each other (they are *crossed*), no light passes through them. However, if a material placed between them changes the polarization of the light, as a doubly refracting material can, light will be transmitted through the second filter. Thus, if a material that has been stressed is placed between two crossed polarizers, the assembly will transmit light in a pattern showing the stressed areas. If the material has not been stressed, the assembly will not transmit light, and the object will appear dark. Polarized light is used to detect areas of high stress in models of complicated structures where mathematical calculations would be difficult.

Facets of Physics

Optical Testing

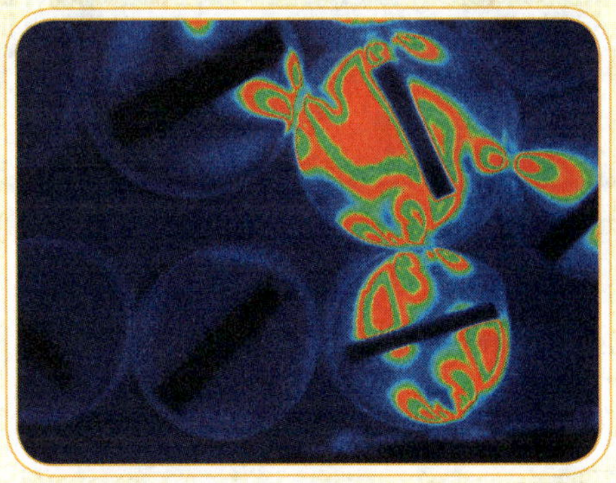

When designing buildings, bridges, aircraft, ships, and so forth, engineers employ a multitude of mathematical formulas to calculate the loads and the resulting stresses and strains on each member of their design. Analyzing loads is a difficult procedure. How can the engineer estimate the maximum weight of trucks, buses, and cars that will ever be on a bridge at any one time? To make sure that a design will do what it is supposed to do, engineers design in a "safety factor." A safety factor of three, for example, means that the design will handle loads three times as heavy as the analysis assumes as the maximum anticipated load.

With complex designs there are serial loads; that is, the load on one member may depend on the loads on other connecting members. This relationship often makes precise calculations of stresses and strains impossible. In the past, it was customary to increase the safety factor to compensate for possible impreciseness, but this also increased the construction cost. Recent advances in photoelastic plastics and holography are helping engineers to accurately analyze complex stress patterns and increase the precision of their calculations.

Photoelastic plastics are created with one special property. Under normal conditions they reflect or transmit light evenly in all directions, but under stress they doubly refract light. Polarizers can be positioned to block out all light except the second refracted beam that follows a slightly different path. The engineer places a small-scale plastic model of his design in the light path between two crossed polarizers. At first, the unstressed model will appear black. But as the engineer applies different forces on the model, the stressed areas will appear as light regions. The intensity or color of the light is proportional to the strain in the material. This information can help the engineer to verify or improve his design.

Another new aid in analyzing stress is the hologram. First, an engineer makes a hologram of the machine part or model. After the hologram is completed, he returns it to the laboratory where it was exposed. He then illuminates both the hologram and the object. When the engineer superimposes the hologram on an image of light reflected directly from the object, the image is uniform. However, when the object is stressed, an interference fringe pattern forms in areas of strain. The advantage of this testing method is that the full-sized object made out of the actual materials can be tested. The engineer can experiment with many complex and shifting stresses.

As industrial materials and designs continue to be improved, photoelasticity and holography will be great helps in testing designs and keeping construction costs to a minimum.

25C Section Review

1. Describe the difference between polarized and unpolarized light sources.
2. What happens to the different light waves that reflect off the surface of glass?
3. How can you tell if a crystalline substance causes double refraction?
4. What happens to the energy in the light waves that are *not* transmitted through a Polaroid filter?
5. At what angle should you align the transmission axis of a Polaroid filter in order to screen out the majority of direct sunlight reflected from a lake's surface?
6. What is the principal cause for the red color of sunsets?
7. What is Brewster's angle for an air-flint glass interface? The incident light ray is in air. See Table 24-1 for indices of refraction.

25C Objectives

After completing this section, I can

- ✓ describe the different polarization states of light.
- ✓ explain dichroism and polarization from Malus's law.
- ✓ explain how to determine the index of refraction of a material using Brewster's angle of incidence.
- ✓ discuss birefringence and explain how it is used to identify certain materials.
- ✓ describe how Polaroid filters produce polarized light through selective absorption.
- ✓ explain polarization by scattering.
- ✓ show how sky light is polarized.

Chapter Review

In Terms of Physics

interference	574
constructive interference	574
destructive interference	574
superposition principle	574
interferometer	579
diffraction	584
collimator	587
resolving power	588
Airy discs	588
Rayleigh angle (θ_R)	588
polarization state	589
unpolarized	589
polarized	589
transmission axis	590
dichroic	590
Malus's law	590
Brewster's angle (θ_B)	590
Brewster's law	590

Problem-Solving Strategies

25.1 (page 577) The symbol n in the interference equation is equal to two quantities: (1) the difference between path lengths (in whole wavelengths of the light) that produce the maximum and (2) the order number of the same maximum.

25.2 (page 577) When analyzing the *minima*, be sure to use the order number (n) of the next-higher maximum in the interference formula.

25.3 (page 586) When you determine SDs associated with diffraction grating ruling densities, the number of lines is *counted* and therefore does not affect SDs in calculations. Its reciprocal is a distance, however, so assume that grating spacing has three SDs.

25.4 (page 588) Remember that the Rayleigh angle (θ_R) is given in radians.

25.5 (page 590) Brewster's angle (θ_B) may be given in degrees or radians.

25.6 (page 591) Remember that Brewster's angle is the incident angle of the incoming ray of light. The 90° angle is measured between the reflected ray and the refracted ray.

Review Questions

1. What is wave interference?
2. Two pulses of equal amplitude but completely out of phase move toward each other. Will they annihilate each other? Explain.
3. The two pulses in each figure travel toward each other. Will they interfere constructively or destructively?

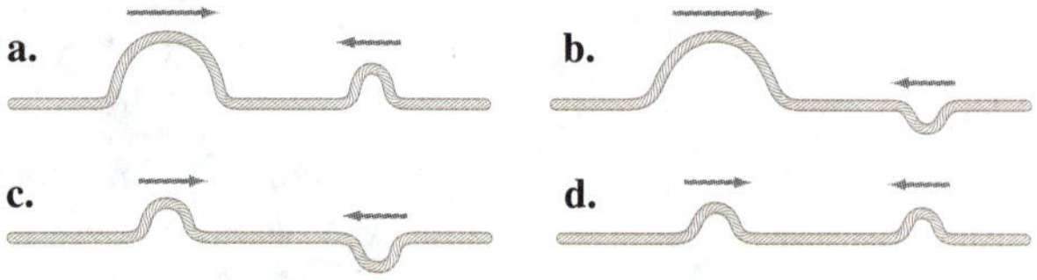

4. Why did Young's experiment demonstrate that light acts as a wave?
5. What does an interference pattern from two close slits look like?
6. The figure below shows a double-slit experiment. Is point P dark or bright? Why?

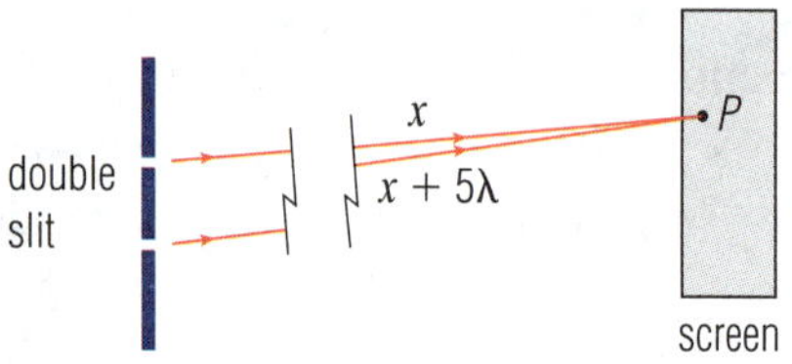

7. **a.** What factors can affect the distance between adjacent interference maxima in a double-slit experiment?
   b. Discuss the effect of each factor.
8. Describe briefly why gasoline spilled on water looks multicolored.
9. Referring to Question 8, why does a thick or opaque film not produce the same multicolored effect?
10. What are some characteristics of laser light?
11. An interferometer was instrumental in refining the precision of which physical constant?
12. What is one common use for diode lasers?
13. What is diffraction?
14. Why did Newton *not* notice light's diffraction?
15. Name three kinds of diffraction.
16. What can diffraction produce at the center of the shadow of a small spherical object?
17. Which kind of diffraction produces a distinct color spectrum?
18. What limits the ability of an optical system to resolve two visual objects that are separated by a small angle?
19. What is polarized light?
20. Name two ways to polarize light.
21. Discuss how you could determine the theoretical index of refraction of an opaque material by using reflected polarization. (For example, the index of refraction for gold is 0.47.)

True or False (22–34)

22. When two wave pulses of opposite phases are superposed, the waves are destroyed.
23. Monochromatic light produces clearer maxima in Young's experiment.
24. The interference maxima in Young's experiment can be brought closer together by decreasing the wavelength of the light used in the experiment.
25. The thinner a film is, the closer together the observed interference bands are.
26. The length of the crystal rod in a laser is the main factor that determines the wavelength of light emitted.
27. The reason you can hear someone speaking to you from behind you is mostly due to sound diffraction.
28. A grating spectrometer will produce bright, distinct lines of different colors when analyzing pure white light.
29. A telescope is said to have a high resolving power if two stars that appear to be very close to each other can be optically separated.
30. A larger Rayleigh angle would be required to resolve two blue stars than to resolve two red stars.
31. Direct, unscattered light from the sun is unpolarized.
32. Brewster's angle is the angle of incidence that produces the strongest reflection.
33. The direction of the extraordinary ray in calcite is dependent on the orientation of the mineral's crystal lattice.
34. Acting a little crazy, you wear your Polaroid sunglasses upside down. The filtering effect of the sunglasses no longer works in this position.

✪35. An opaque screen is pierced by two slits separated by a distance of 0.250 mm. A second screen, 4.00 m away, reflects the interference pattern. The light has a wavelength of 600. nm.
   a. How far is the center of the first dark minimum from the central maximum?
   b. How far is the fifth bright maximum from the central maximum?
   c. Is the edge of the screen 25.0 cm from the central maximum bright or dark?

✪36. Monochromatic light falls on two small holes that are 3.00×10^{-4} m apart in a window shade. Their interference pattern encounters a wall 1.50 m away. If the first bright fringe is 2.40 mm from the central maximum, what is the light's wavelength?

✪37. A laser light illuminates a screen with two circular holes 5.00×10^{-4} m apart and is projected onto a screen 10.0 cm in diameter that is 5.00 m away. The central maximum is at the circular screen's center. Two bright spots are at the perimeter of the screen on opposite sides. The wavelength of the light is 500. nm.
   a. What is the highest-ordered maximum on the screen? ($n = ?$)
   b. What is the highest-ordered minimum on the screen?
   c. Are the spots at half the screen's radius from the central maximum the $n^{\text{th}}/2$ maxima? (n = highest-order number of maxima on the screen)

✪38. How closely should the lines on a diffraction grating be spaced if you want the first-order maximum from green light ($\lambda = 550.$ nm) to be at 29.7°?

✪39. A light beam directed into a spectroscope with a diffraction grating ruled in 10 000 lines per centimeter has a wavelength of 500. nm. At what angle is its first-order maximum?

✪40. A spectroscope contains a diffraction grating with 7500 rulings per centimeter. What wavelengths of visible and infrared light have a first-order maximum at the following angles?
   a. 20.0°
   b. 25.0°
   c. 30.0°
   d. 35.0°
   e. 40.0°

✪41. Judging from your answers to Question 40, will a spectroscope have a linear scale? That is, will the same change in angle always indicate the same change in wavelength?

✪42. What is the index of refraction of a new plastic material if its Brewster's angle is 18.5°? The incident ray is in air.

DS✪43. Research and write a paper about some of the ways different types of holograms are being used to thwart counterfeiters.

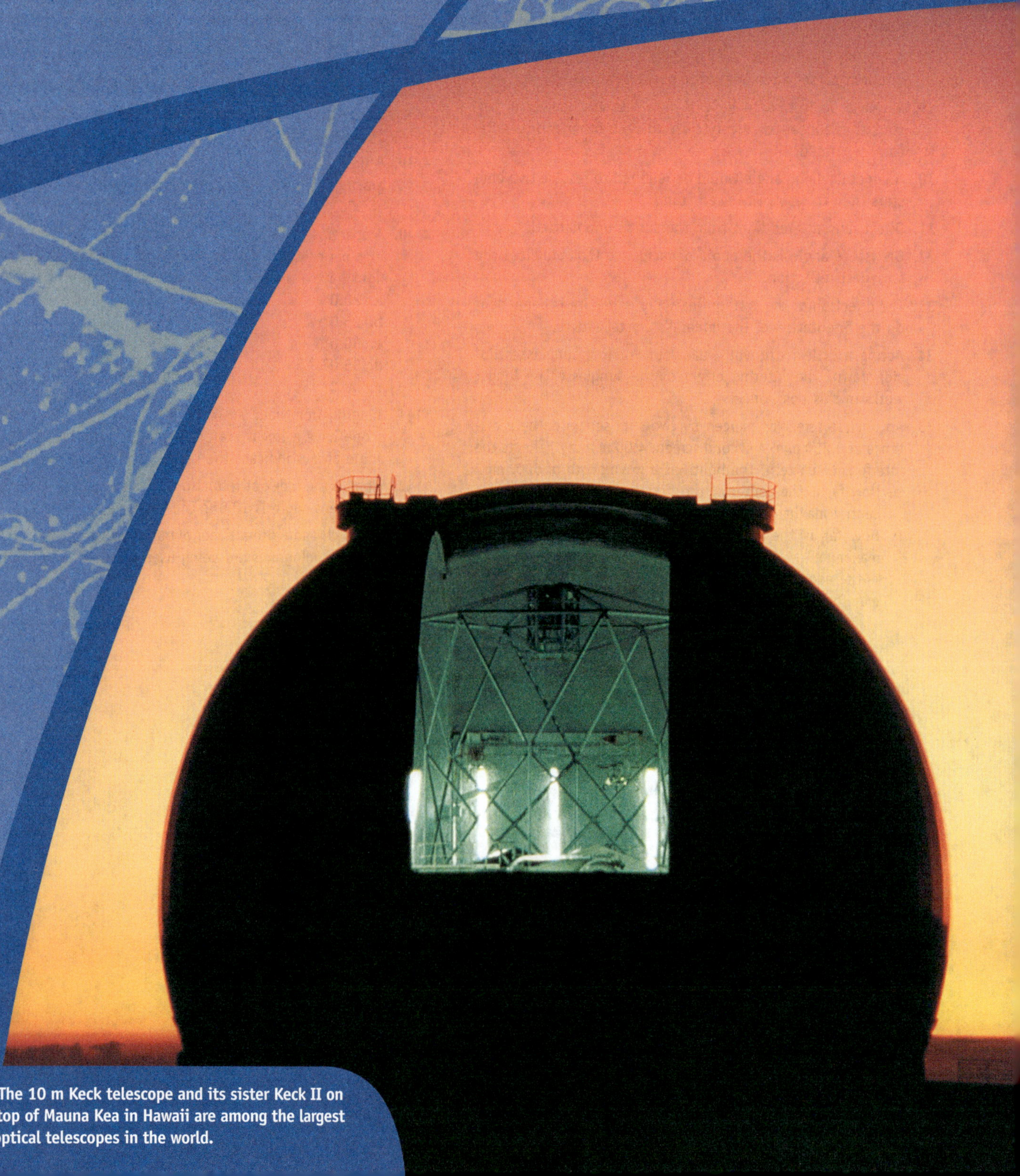

The 10 m Keck telescope and its sister Keck II on top of Mauna Kea in Hawaii are among the largest optical telescopes in the world.

Using Light 26

Dominion Science Problem

A Better Picture

A picture is truly worth a thousand words. When you look at a photograph in this textbook or elsewhere, besides studying the subject of the image, do you ever consider why the photo was included in the first place? Photographs are capable of conveying more information than any other means of communication available to our senses. Photographs are an efficient, convenient, and often essential medium for archiving historical moments, crimes, scientific events, medical information, and even family memories.

The art and science of photography is nearly two hundred years old. Until relatively recently, photographers accepted the limitations and hazards of the medium. For example, most photographic print processes require that the negative film be removed from the camera and delivered to a photo-processing facility. When using standard print film, even with 1-hour processing, you have no way of knowing immediately if the photographs that you have taken are satisfactory. In addition, most people feel that it would be wasteful to shoot just one or two pictures on a roll of film before developing it, so potentially one could go for days or weeks before an entire roll of film is exposed. In addition, the photographic chemicals that form the permanent images on film and prints are based on compounds of silver, a heavy metal hazardous to organisms in the environment. Furthermore, because of the chemical basis for film photography, images tend to fade or their colors change over time, affecting their archival value. And finally, once the photographs are obtained, they are bulky to store and inconvenient to share. What has been done to overcome the disadvantages of traditional film photography, an important medium for recording, sharing, and storing visual information?

26-1 What kind of information has been lost in this aging photograph?

26A INTENSITY AND COLOR

26.1 Light-Source Intensity—The Candela

How much light does a bulb produce? Light bulb manufacturers rate their bulbs in watts, a unit of power. Does a 100 W bulb produce four times as much light as a 25 W bulb? No. Although the amount of light a bulb emits is related to the power the bulb consumes, the relationship is not direct. The brightness of a light source is called its **luminous intensity** (**I_L**). How do scientists and engineers measure luminous intensity?

The unit steradian (sr) in this definition is a conical solid angle (an angle measured in three dimensions) with its vertex at the center of a sphere that cuts off a circular area on the surface of the sphere equal to the square of the sphere's radius. The central angle of a steradian is about 65.5°.

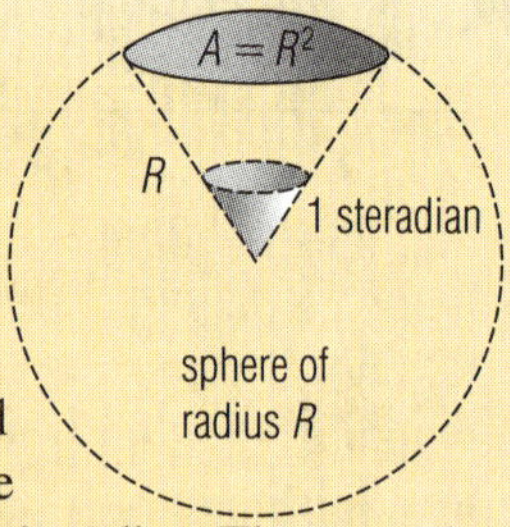

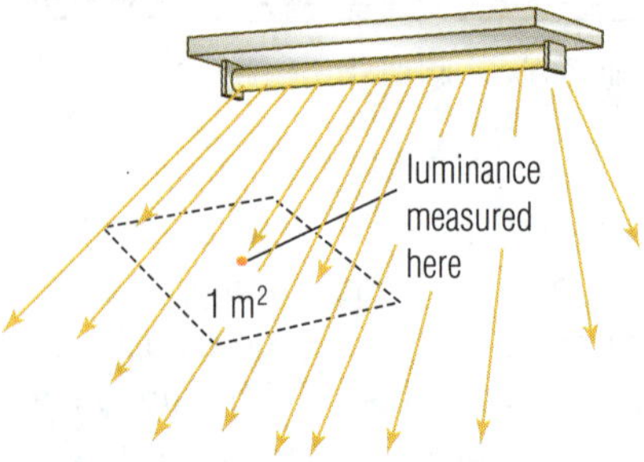

26-3 Luminance measures the rate of light energy falling on a unit surface.

Table 26-1

Selected Luminance Data

Luminance (cd/m²)	Source
0.03	white paper in moonlight
2900	surface of the Moon
3200	clear sky at midday
5000	candle flame
6000	fluorescent lamp
25 000	white paper in sunlight
10^7	tungsten filament
2×10^9	surface of the sun

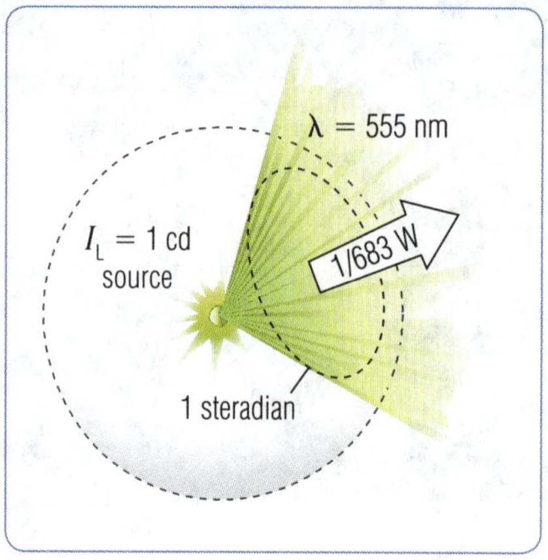

26-2 Luminous intensity is the rate of light energy emitted by a light source (its power) in a specific direction.

Throughout the nineteenth century, a light source was compared to a standard candle to determine its luminous intensity. These luminous standards varied, and no candle flame has a constant brightness. The first real attempt to standardize luminous intensity was based on carbon-arc incandescent lamps. However, even these were not stable enough to provide the reliable standard required by scientists. A theoretical standard based on a blackbody radiator at the freezing point of platinum (2045 K) was proposed in 1933. This standard was too difficult to achieve in the laboratory, so a new standard was developed in 1979 based on radiometry. The **candela (cd)** is now defined as *the luminous intensity, in a given direction, of a source that emits monochromatic radiation of frequency* 540×10^{12} *hertz and that has a radiant intensity in that direction of 1/683 watt per steradian.* Let's look at this definition in detail.

The human eye is not equally sensitive to all frequencies of visible light. Therefore, the frequency that our eyes are most sensitive to was chosen for the definition. Frequency was used rather than wavelength because wavelength varies depending on the material that light passes through. A frequency of 540×10^{12} Hz corresponds to a wavelength of 555 nm (in a vacuum), which is greenish-yellow light.

The fraction 1/683 makes the current definition compatible with the previous definition based on the freezing point of platinum.

The requirement to measure the radiated power within a solid angle called a steradian (sr) makes the candela unit compatible with other SI power units.

An incandescent light bulb gives off roughly one candela for each watt it consumes. A 25 W bulb produces 21 cd; a 50 W bulb, 55 cd; and a 100 W bulb, 125 cd. Thus, a 100 W bulb produces more than five times as much light as a 25 W bulb produces. That is, a high-wattage bulb uses its power more efficiently than a low-wattage bulb.

26.2 Extended Lighting Sources—Luminance

The definition of luminous intensity assumes a point light source. How do you specify the intensity of a real light source, such as a light bulb, an overhead fluorescent lighting unit, or even the sun? The rate of flow of light energy reaching a surface in a given direction from the source is called the **luminance** of the light source. Luminance is measured at a point on an illuminated surface and has units of power per unit of area, or cd/m².

26.3 Luminous Flux—The Lumen

We mentioned that the luminous intensity of a light source is a measure of the illumination strength in a particular direction, but how do you measure the total amount of light that a source gives off? This quantity is called the **luminous flux (Φ)** around the light source. Luminous flux is independent of the direction of measurement. It is measured in units called **lumens (lm)**, which is the product of candelas and steradians (cd·sr). In other words, 1 lumen is the luminous flux of a 1 cd light source in 1 sr. Since there are 4π sr in a sphere, a 1 cd point light source has a total luminous flux of 12.57 lm ($4\pi = 12.57$). Luminous flux is a measure of power, as is luminous intensity. Both are derived from the SI unit for power, the watt.

26.4 Illumination at a Point—Illuminance

Lighting engineers want to know how bright the light falling on a surface appears to the human eye at a particular point. The lighting required for a desk in an office differs from that required for a sidewalk at night. You know from experience that the effectiveness of a source of light varies with the distance from the source. For example, the headlights of an approaching car seem to grow brighter as the car nears. The distance of an object from a light source also affects the amount of light that the object receives, with the light intensity being greater at shorter distances from the source than it is farther away. The lighting value at a particular point is measured by the **illuminance (*E*)**. The illuminance at a distance from a light source is directly proportional to the luminous flux of the source and inversely proportional to the square of the distance from the source. In equation form,

$$E = \frac{\Phi}{r^2}, \qquad (26.1)$$

where r is the distance from the light. The units of illuminance are lumens per square meter (lm/m^2), which are called **lux (lx)**. A 1 cd source produces an illuminance of 1 lux at every point that is 1 m from the source. Illuminance is directly proportional to luminous intensity. For example, a 2 cd light gives twice as much light at a given distance as a 1 cd light gives at the same distance. Notice that Equation 26.1 is an inverse-square law. That is, an object that is $2x$ meters from a light source receives only one-fourth the light received by an object that is x meters from the source.

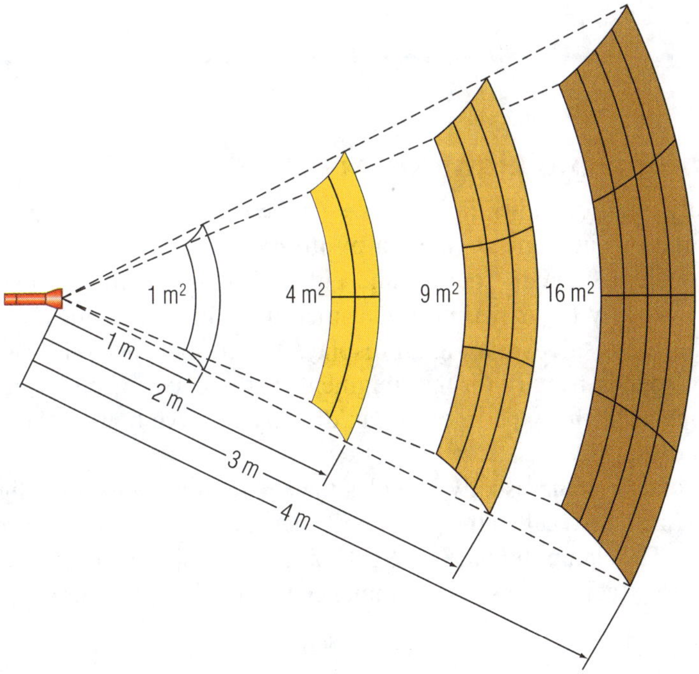

26-6 As the distance between a point light source and an illuminated surface increases, illuminance decreases because light spreads out over a greater area.

$$1 \text{ lm} = 1 \text{ cd} \times 1 \text{ sr}$$

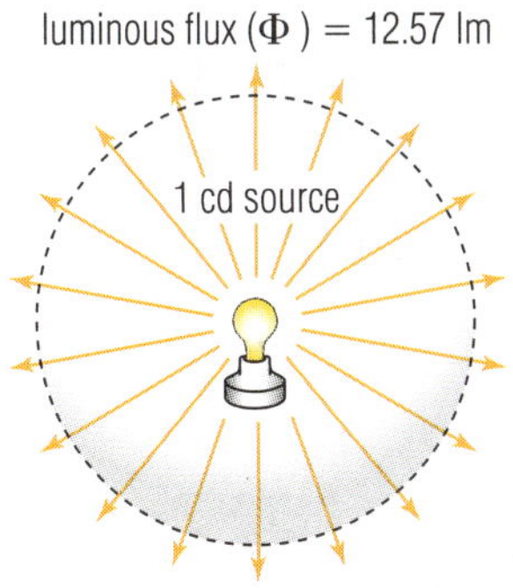

26-4 Luminous flux measures the total rate of light energy emitted by a source in all directions.

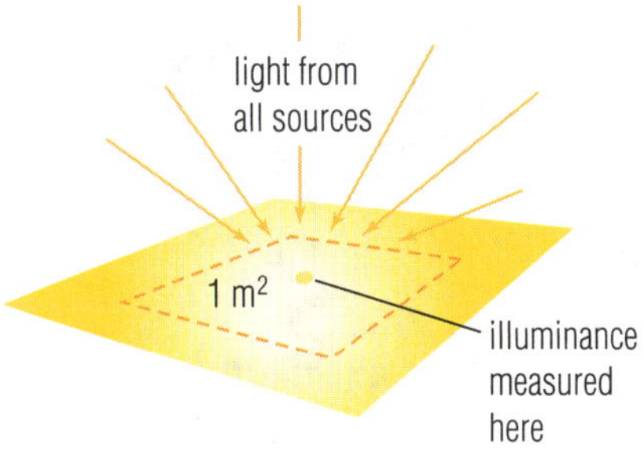

26-5 Illuminance is the total illumination at a point on a surface.

$$1 \text{ lx} = \frac{1 \text{ lm}}{1 \text{ m}^2}$$

Table 26-2

Selected Illuminance Data

Illuminance (lm/m^2)	Conditions
0.003	starlight
0.2	full moonlight
100	precision work, artificial light
1000	office near window
10 000	bright overcast day
100 000	clear sky at midday

EXAMPLE 26-1

Calculating Illuminance

What is the illuminance at a point 5.0 m from a light source emitting 600. lm?

Solution:

Use Equation 26.1:

$$E = \frac{\Phi}{r^2}$$

$$E = \frac{600.\ \text{lm}}{(5.0\ \text{m})^2}$$

$$E = 24\ \text{lm/m}^2 \text{ or } 24\ \text{lx}$$

EXAMPLE 26-2

How Bright a Bulb? Determining Intensity

How strong a light source, in lumens, is needed to give an illuminance of 80. lx at a distance of 1.5 m?

Solution:

Use Equation 26.1:

$$E = \frac{\Phi}{r^2}$$

$$\Phi = Er^2$$

$$\Phi = (80.\ \text{lm/m}^2)(1.5\ \text{m})^2$$

$$\Phi = 180\ \text{lm}$$

26-7 A high-school laboratory photometer

The photometer described here is a *comparator photometer*. Other kinds of photometers, including those called *light meters* by photographers, measure incident light either from all directions at a point or from a specific direction.

26.5 Comparing Light Sources

The intensity of an unknown light source can be determined by comparing it to a known source. An instrument called a **photometer** receives beams of light from the unknown and standard sources and either displays them side by side in an eyepiece viewed by the experimenter or measures their intensities electronically and provides the results via a digital display. The former method is less accurate because it depends heavily on the judgment of the experimenter, but it is more instructive because the experimenter can directly observe how apparent intensity varies with distance.

The photometer setup usually includes the standard and unknown light sources and a comparison box, called the photometer head, or a digital detector mounted on a meter stick. The clamps holding the light sources are adjusted until the comparison box or detector receives the same illuminance from each light source. That is,

$$E_{\text{unk}} = E_{\text{std}}.$$

Since luminous flux (Φ) is proportional to luminous intensity (I_L), we can say that

$$E \propto I_L/d^2,$$

$$E_{unk} \propto \frac{I_{L\,unk}}{d_{unk}^2} \text{ and } E_{std} \propto \frac{I_{L\,std}}{d_{std}^2}, \text{ or}$$

$$\frac{I_{L\,unk}}{d_{unk}^2} = \frac{I_{L\,std}}{d_{std}^2}.$$

Rearranging,

$$I_{L\,unk} = I_{L\,std}\frac{d_{unk}^2}{d_{std}^2}. \quad (26.2)$$

Since you can read d_{unk} and d_{std} from the meter stick, and you know $I_{L\,std}$, you can calculate $I_{L\,unk}$. Notice that you do not have to know E_{unk} to find $I_{L\,unk}$.

Problem-Solving Strategy 26.1

Use the source's luminous intensity (in candelas) rather than its luminous flux (in lumens) when using the photometer equation.

EXAMPLE 26-3

Finding an Unknown Intensity

A standard 100. cd bulb that is 32.0 cm from the photometer detector has the same luminous intensity as an unknown bulb that is 25.0 cm from the photometer head. Find the unknown bulb's luminous intensity.

Solution:

Use Equation 26.2:

$$I_{L\,unk} = I_{L\,std}\frac{d_{unk}^2}{d_{std}^2}$$

$$I_{L\,unk} = (100.\text{ cd})\frac{(25.0\text{ cm})^2}{(32.0\text{ cm})^2}$$

$$I_{L\,unk} \cong 61.\underline{0}3\text{ cd} \cong 61.0\text{ cd}$$

26.6 Transmittance

Materials differ in their ability to transmit light. *Transparent* materials, such as glass, water, air, Lucite, and cellophane, transmit light so that you can see clearly through them. Since a transparent material does reflect and absorb some light, a thick piece of transparent material is harder to see through than a thin piece of transparent material. The **transmittance** of an object is the ratio of the transmitted luminous flux to the incident flux.

A *translucent* material, like frosted glass, tissue paper, or cloth, transmits light but distorts it so that you cannot see clearly through it. An *opaque* material, such as wood, metal, or concrete, does not transmit visible light at all unless the material is extremely thin. The divisions between transparent and translucent materials and between translucent and opaque materials are not distinct, so the thickness or smoothness of an object may determine which category it belongs in.

26.7 Color

If you doubt the importance of color, look around you and try to imagine the world as if it were a black-and-white photograph. Scientists long ago determined that a

Color is not an intrinsic property of an object apart from the visual system that perceives colors. The EM waves that stimulate specialized cells to send signals to the visual regions of the brain are just a part of the larger continuum of the EM spectrum. Certain cells are sensitive to specific bands of EM energy corresponding to just a few distinct *primary colors*. It is the way that these color signals are combined in the brain that defines the colors that are seen. Thus, color vision is totally dependent on the processing of received light in the visual system (photoreceptors, neurons, and brain) of an organism.

significant portion of an average adult human's sensory information is received through his eyes. It is therefore understandable that people have developed color photography, paints, dyes, colored inks, color television, and color digital displays in order to fully employ this sense.

26-8 Without color, much visual information becomes uninteresting or may even be lost, as shown by the monochrome half of the picture.

Problem-Solving Strategy 26.2

The equation $E = hf$ uses frequency rather than wavelength because wavelength changes as light passes through various forms of matter.

What you see as color is actually different frequencies of light. You can receive this light in three ways: directly from its source, by reflection from a surface, or by transmission through a material. When light reaches you directly from its source, its color depends on the frequencies of light the source's atoms emit. The light's frequency, which determines its color, is related to the energy change of the electrons by the equation

$$E = hf, \quad (26.3)$$

where the coefficient of the frequency (f) is *Planck's constant (h)*. As you can see, larger changes in electron energy produce higher energy emissions that are observed as higher frequencies of light.

Planck's constant (h) is one of the four universal physical constants (along with k, c, and G) that define the basic character of energy and matter in the universe:

$$h = 6.626 \times 10^{-34}\ \text{J·s}.$$

Light that reflects from a surface or passes through a medium may have some of its frequencies attenuated (lessened) or removed. The energy from the absorbed light heats the material that absorbs it.

The wavelength (or frequency) of light is not the only factor that determines how the color appears to you. Colors are specified by hue, saturation, and brightness. *Hue* refers to the mixture of wavelengths in the light. The basic color type is determined by hue. *Saturation* is a measure of the amount of neutral colors present, such as gray, that tend to dull the color. The difference between a greenish-tinged gray and a brilliant kelly green is saturation. *Value, brightness*, or *lightness* is how dark or light the color appears. A nearly black color has low value, while a nearly white color has a very high value. The term can also refer to the amplitude of the light, where high-amplitude light appears brighter than low-amplitude light. Figure 26-9 provides four different models of graphing the colors seen by the human eye. Note that the eye is not equally sensitive to all colors or color ranges.

Table 26-3

Representative Colors

Color	Wavelength, λ (nm)
violet	400–450
blue	450–500
green	500–570
yellow	570–590
orange	590–610
red	610–700

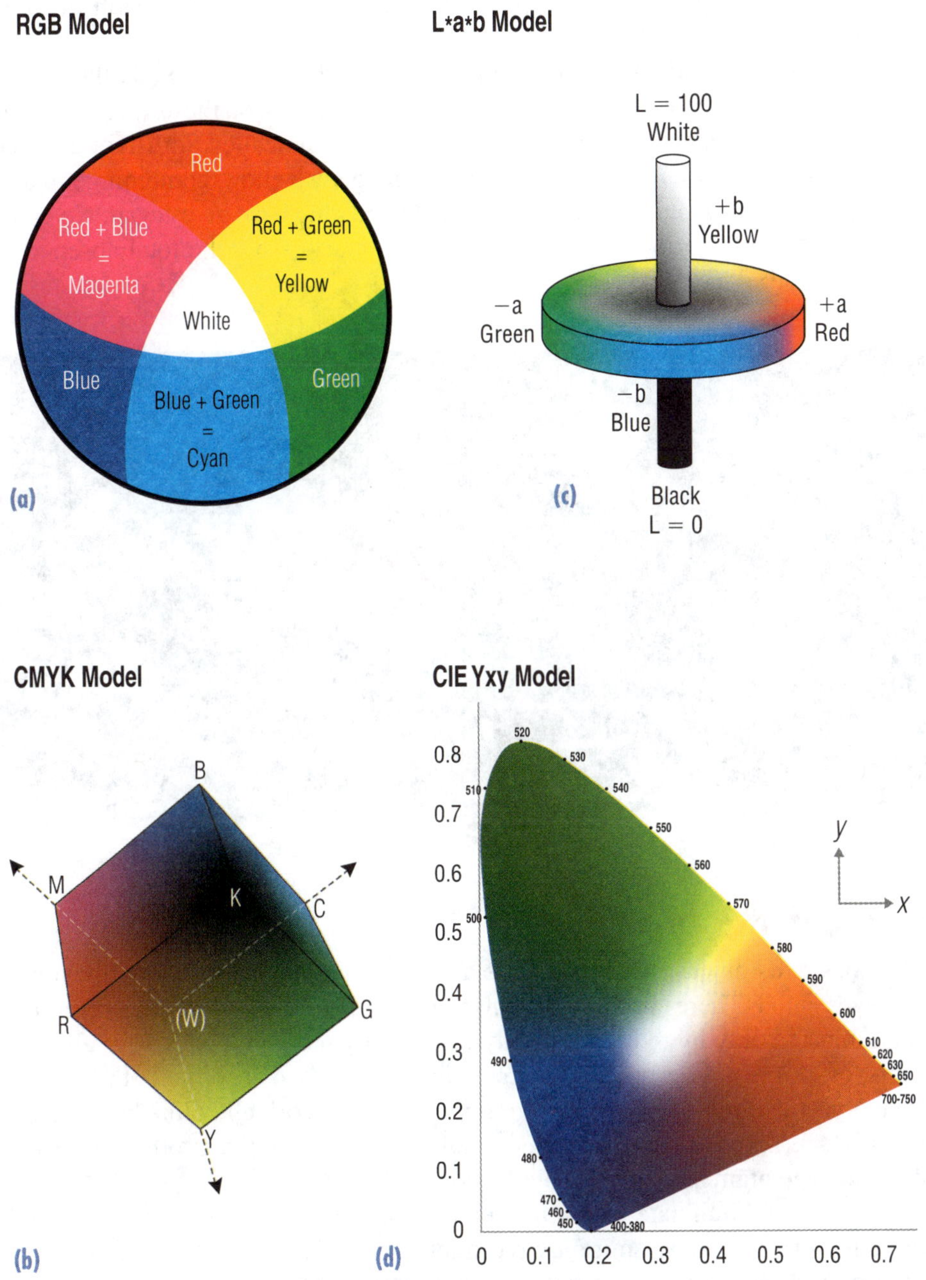

26-9 (a) The standard red-green-blue (RGB) color model used in color TVs and computer monitors; (b) The cyan-magenta-yellow-black (CMYK) color model used in print media; (c) A 3D color-wheel model approximating the relationship between hue (perimeter of the wheel), saturation (radial distance from the center), and value (vertical position); (d) One of many 2D representations showing the range of colors perceived by the human eye

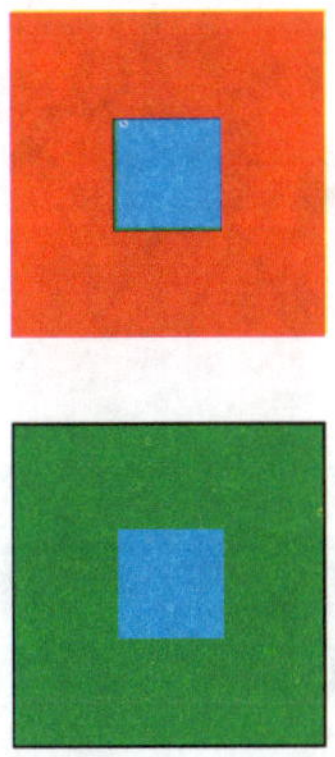

26-10 A color is perceived in relation to its surroundings.

An object's color also depends on the colors of surrounding objects. That is, you see color in contrast to its surroundings (see Figure 26-10).

26.8 Additive Color Mixing

White light is usually a mixture of all visible wavelengths, but some combinations of two or three wavelength bands produce a light that looks white to us. For example, yellow and blue lights, green and magenta (bluish-red) lights, or red and cyan (greenish-blue) lights combine to produce white light. Since these pairs appear to produce the complete spectrum (i.e., white light), they are called **complementary colors**.

The additive primary colors are red, green, and blue (also called the RGB color model).

If beams of red, green, and blue light shine on a white screen so that they overlap, the region where all overlap is white. Combinations of these three colored lights can produce any color in the spectrum, so they are called additive primary colors. Notice that combining red and green light gives yellow, combining red and blue light gives magenta, and combining green and blue light gives cyan. No combination of additive primary colors can produce black. Only the total absence of light is perceived as black.

26-11 Colored lights combine in an additive way to produce colors on a white surface.

26.9 Colored Objects

Why does a red object appear red? The answer is that the light coming from it to your eyes is red. There are three reasons that the light may be red: the object may be illuminated only by red light; it may reflect only red light and absorb all other colors; or, if it is not opaque, it may transmit only red and absorb all other colors. If the object is illuminated with white light, it must absorb all colors but red. The substance in an object that absorbs certain colors and reflects others is called **pigment**. The pigment is named by the color that it reflects.

Suppose you illuminate with red light an object containing red pigment. It will continue to reflect or transmit only red light. However, since the object is illuminated only with red light, all light reaching it will be reflected or transmitted; therefore, the object will still be red. What happens if you illuminate the object with green light? Since the object receives no red light, and it absorbs all the rest, it will not reflect or transmit light. The object will appear black.

26.10 Subtractive Color Mixing

Figure 26-12 shows white light from behind passing through a yellow filter and then a cyan filter to your eyes. When the light passes through the yellow filter, the filter absorbs all but the yellow light. Yellow light can be thought of as a combination of red light and green light. The cyan filter absorbs all but cyan light, that is, all but blue and green light. However, in this example only yellow (red and green) light reaches it. The cyan filter absorbs the red light and transmits the green light. Therefore, the light coming through the overlapping filters will be green. This process is called *subtractive* color mixing because the remaining color is the light that pigments have not absorbed (subtracted) from the original light.

Subtractive color mixing also governs the color of reflected light. For example, if an object contains both yellow and cyan pigments, it will absorb all light reaching it except green, which it will reflect. With subtractive mixing, you can produce all colors from the **subtractive primary colors**: cyan, magenta, and yellow. A mixture of yellow and magenta pigments reflects orange light; a mixture of cyan and magenta pigments reflects purple; and, as you have seen, a mixture of cyan and yellow pigments reflects green. The correct mixture of the three subtractive primary colors produces black. Note that only white pigments (those that do not absorb *any* colors) can reflect white light. No mixture of the subtractive primary pigments can produce white.

The **subtractive primary colors** are cyan, magenta, and yellow (also called the CMY model).

26-12 Subtractive color mixing. Notice that the light that has passed through both the yellow and cyan filters appears green.

In the publishing industry, the preferred color model consists of the three subtractive primary colors, cyan, magenta, and yellow, plus black (CMYK). The separate black color (K) permits obtaining true black in printed characters and increases the value range possible in color images.

26A Section Review

1. a. Compare the terms *luminous intensity*, *luminous flux*, *luminance*, and *illuminance*.
   b. Which term is important to an architect who wants to know how much light is needed in a hallway?
2. Both luminous intensity and luminous flux involve a dimension associated with energy. What is the unit of this dimension, and how does it relate to the candela and the lumen?
3. Using a photometer to compare two light sources, what can you conclude about the intensity of a lamp if it produces the same illuminance at twice the distance as the standard lamp?
4. a. Which has the greater amount of energy for a given amplitude, red light or blue light?
   b. Explain your answer.
5. What color term classifies the colors grass green or apple red?
6. a. If possible, how would you produce the color *black* on a white surface by mixing primary additive colors or their complements?
   b. Explain.
7. Which subtractive color pigments are required to produce a white-colored surface in white light?
8. ✪ What is the illuminance of an omnidirectional lamp that produces 251 lm on a flat surface that faces the lamp 3.0 m away?

26A Objectives

After completing this section, I can

✓ define the various means of measuring light intensity.
✓ describe how light intensity changes with distance from the source.
✓ calculate illuminance and the intensity of an unknown light source.
✓ explain how light energy is dependent on frequency.
✓ describe the perception of color and the means of qualitatively describing colors.
✓ compare various color-system models, including the RGB and CMY[K] systems.
✓ differentiate between additive color mixing and subtractive color mixing, and state the primary colors for each method.

26B OPTICAL INSTRUMENTS

Optical instruments use one or more lenses, mirrors, or prisms to form an image that is to be enlarged or reduced for direct observation or that is to be recorded in some way. For example, microscopes, telescopes, cameras, binoculars, projectors, and periscopes are all optical instruments. Looking at two of these should demonstrate the principles that nearly all optical instruments use.

Anton van Leeuwenhoek (LAY u ven HOOK) has been mistakenly called the "father of the microscope." Simple and compound microscopes had been in use for several decades before his birth. A contemporary of Robert Hooke, the great English microscopist, Leeuwenhoek was a true amateur who made astounding discoveries of microscopic life in drops of water and bodily fluids. He was fortunate in that he was able to share his findings with a wider scientific audience through his association with the British Royal Society. However, it was Hooke's famous book *Micrographia*, written in English, that introduced the public to the world of the microscope.

26.11 The Microscope

A **microscope** magnifies minute objects in order to allow observers to see fine details that are closer together than the resolving power of our eyes. Magnifying lenses have been used since the time of the ancient Greeks, but not until the seventeenth century were small, accurately shaped lenses like those in a modern microscope made. The first microscopes were simple, with only one converging lens. Anton van Leeuwenhoek constructed a simple microscope through which he could see blood capillaries, red corpuscles, muscle fibers, hairs, skin cells, nerve cells, and reproductive cells. In his microscope a double convex lens with a short focal length was mounted in a frame with a handle. A specimen was mounted on one side of the lens and focused with a screw adjustment that moved the specimen nearer to or farther from the lens.

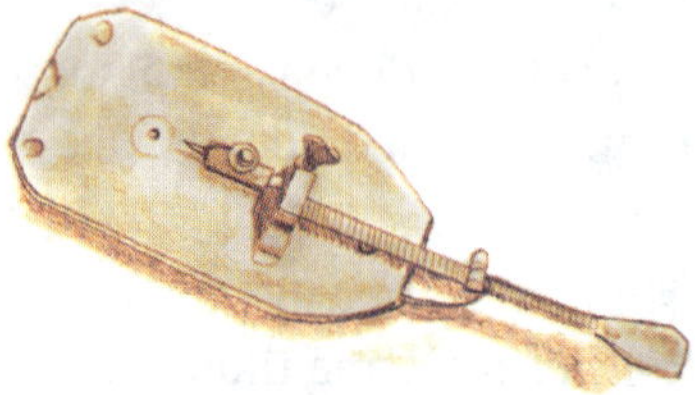

26-13 Dutch biologist Anton van Leeuwenhoek (1632–1723) and one of his microscopes

Though simple, Leeuwenhoek's microscopes were capable of producing images comparable to the high-quality compound microscopes of today, up to a magnification of 300× or more. The main limitations of his instruments were their inconvenience of use and their limited ability to vary the optical path to improve the image quality. Robert Hooke's microscope solved these problems by mounting two lenses in a light-tight barrel, which is the basic design of nearly all modern light microscopes.

Using two lenses "in series" produces an overall magnification equal to the product of the individual magnifications of the lenses. This strategy underlies the **compound microscope**, which is essentially two short-focal-length converging lenses. The compound microscope acts as two simple microscopes. The first lens, the *objective*, receives light from the object, producing a magnified, inverted real image (located beyond F of the objective lens). The second lens, the *ocular* (eyepiece), uses the real image of the objective as *its* object and magnifies it, producing a much-enlarged virtual image. The image in a standard compound microscope is inverted top-to-bottom and reversed right-to-left.

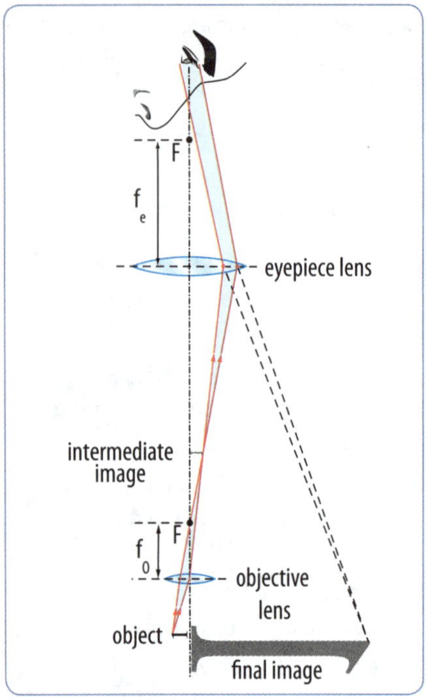

26-14 Rays of light passing through a microscope

The total magnification of a microscope is the product of the magnification powers of its objective and ocular. For example, a microscope with a 10× objective and a 5× ocular produces a total magnification of 50×.

The objective is usually a combination of lenses. The first lens has a short focal length so that it magnifies an object placed near it, and the other lenses correct the aberrations of the first. Because the ocular magnifies the image produced by the objective, the objective is the most important component in determining the overall quality of the image. Most compound microscopes that you will see in a high-school laboratory have three separate objectives mounted in a rotatable turret nosepiece so that the user can choose low, medium, or high magnification.

The ocular is usually also a combination of lenses so that there will be as little distortion as possible. Better-quality microscopes are provided with a selection of oculars to increase the choices of magnification. Some microscopes are used with two oculars *(binocular)* to reduce the strain of observing with a single eye for long periods. If it has two objectives as well as two oculars, the instrument is a *stereo binocular microscope*, or simply a **stereo microscope**. A stereo microscope permits depth perception. Such microscopes are used for dissection, geology field-sample identification, and manufacturing quality inspections where low magnifications, spatial-perception requirements, or opaque objects are involved. A stereo microscope has an additional prism system in the light path that restores proper up-down, right-left orientation to make manual work with the microscope natural.

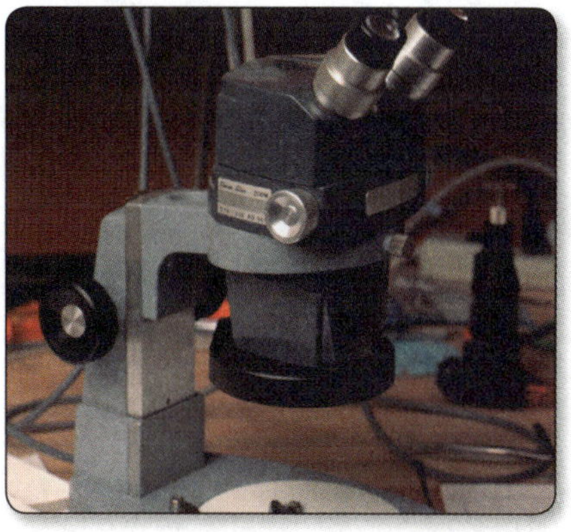

26-15 A stereo microscope

FACETS of PHYSICS

Microscopy

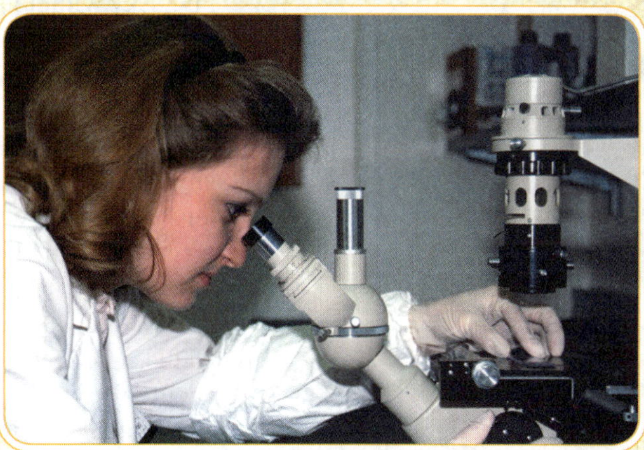

Following the invention of the electron microscope, many people expected the light microscope to become an oddity of the past—like a Ford Model T. Yet both electron and conventional light microscopes have severe limitations when used for studying live organisms. Because the inherent contrast in the cells of organisms is very low, oftentimes they must be killed and stained before they can be observed. So a biologist cannot be certain how much the organisms' structures have changed as a result of death; many structures disappear or remain invisible after death.

In 1934, the Dutch physicist Frits Zernike (Nobel Prize in Physics, 1953) solved this dilemma when he invented the *phase-contrast microscope*. By diffracting light through an organism rather than reflecting light off it, this microscope can produce high-contrast images directly from living, moving cells. Both it and the microscopes discussed below have become standard tools in all serious biological studies.

The phase-contrast microscope takes advantage of the different indexes of refraction in specimens. Light passing through a specimen arrives at the objective lens a little behind the light that passes around it. A transparent plate between the specimen and the objective lens increases the phase difference. As the different light rays interfere with each other, an image is formed. Biologists have been able to study and film dynamic events such as cell division and cell response to foreign agents such as chemicals. Interesting discoveries have also been made about hormones, and the discovery of phytochrome pigments in plants has helped us to understand how plants detect and follow the sun.

The *interference microscope* also uses the slowing of light within the optical system to form clear images of transparent objects. However, the light is divided into two beams, one that passes through or reflects off the object and another that does not. The beams later recombine to form an interference pattern that depicts the object. The interference microscope is far more complex and expensive than the phase-contrast microscope, and it fell out of favor for several decades. However, it has seen something of a revival in recent years now that digital cameras can be used to obtain images that can be recombined digitally using special software.

A third optical microscopy technique is called *differential interference contrast* (DIC) microscopy. Unlike the previous two techniques, this microscope system polarizes the incoming light before it enters the optical system. The light passes through a special dichroic filter that splits the light into a beam that contains light rays polarized at 90° to each other. These rays are focused within the object, and then they pass through the objective lens. The two systems of rays are recombined in another dichroic filter where their waves create interference fringes that enhance the details of the object. The advantages of the DIC method are a clearer image than that produced by phase-contrast microscopy and less expense and complexity than the interference microscope.

These techniques have vastly increased our understanding of living cells and have been the means of many profound discoveries, such as the function of muscle cells. Optical interferometry has allowed scientists to develop contour maps of the mass of material per unit volume in cells that have greatly aided the study of cell theory in biology. Research in metallurgy has also profited from these technologies.

Research microscopes are often built with the ability to record or transmit the images observed by the user. Such systems include built-in or add-on film or CCD cameras, or projection screens. Typically, there is a splitting prism in the optical path between the objective and the ocular. Part of the light is transmitted to the ocular and part to the camera. Some instruments have selectors that permit moving the prism completely into or out of the optical path to provide as much light as possible to the ocular or the camera.

26.12 The Telescope—Refractors

A **telescope** makes distant objects seem nearer. The inventor of the first telescope is not known. Several men in the Netherlands claimed to be the inventor, but Hans Lipperhey, a Dutch lensmaker, was the first to apply for a patent from his government in 1608 and to make the instrument well-known. Having received a description of this instrument, Galileo built his own telescope and observed the night sky with it in 1609. He thus became the first person to see the moon's mountains, Saturn's rings, Venus's phases, and the four largest moons of Jupiter. When he studied the Milky Way, Galileo saw for himself that "the host of heaven cannot be numbered" (Jer. 33:22).

Hans Lipperhey (1570?–1619) was a Dutch spectacle maker.

26-16 One of Galileo's telescopes

Galileo's first telescope had little magnifying power. Like a microscope, it had an objective lens and an eyepiece. The objective was a double convex converging lens. Its ocular was a double concave diverging lens, which produced a virtual, erect image of the objective's image. Since the ocular was placed *between* the objective's focal point and the lens, the image remained erect throughout the entire optical path. However, the telescope had an extremely narrow *field of view*. Galileo continued to improve his telescope until he achieved a magnification of about thirty power.

The field of view of a telescope, measured in degrees, is the maximum angular size of an object that can be observed.

Eventually, scientists realized that they could get greater magnification if both lenses were double convex. Today's "all-lens" telescopes, or **refracting telescopes**, use two double convex converging lenses. The objective lens is a long-focal-length lens that gathers light from the object, so it works better if its diameter is large. A telescope is described by the diameter of its objective—a 75 mm telescope has an objective whose diameter is 75 mm. The standard ocular forms a virtual image of the objective's inverted image; therefore, the final image is also inverted compared to the object.

The magnifying power of a telescope is the ratio of the angle that the image subtends, or takes up (θ_I, Figure 26-17b), to the angle that the object subtends (θ_O, Figure 26-17a). A planet's image in a telescope is much easier to see than the planet itself because it occupies a greater proportion of your field of vision. The ratio of θ_I/θ_O, the magnifying power (M), is equal to the ratio of the focal lengths of the objective (f_o) and the ocular (f_e), as expressed by Equation 26.4:

$$M = \frac{\theta_I}{\theta_O} = \frac{f_o}{f_e} \quad (26.4)$$

For example, if the objective's focal length is 100. cm and the ocular's focal length is 1.25 cm, the telescope's magnifying power is

$$M = \frac{f_o}{f_e} = \frac{100.\ \cancel{\text{cm}}}{1.25\ \cancel{\text{cm}}} = 80.0\times.$$

The largest refractor telescope in the world is the 102 cm telescope at the Yerkes Observatory in Williams Bay, Wisconsin.

As you learned in Chapter 25, the larger the lens, the greater the resolving power. This is true because the larger aperture has greater light-gathering ability. As you would expect, the best refracting telescopes have very large objective lenses. However, there is a practical limit to the size of a lens. If the lens is much more than 1 m in diameter, gravity causes the center of the lens to sag, distorting

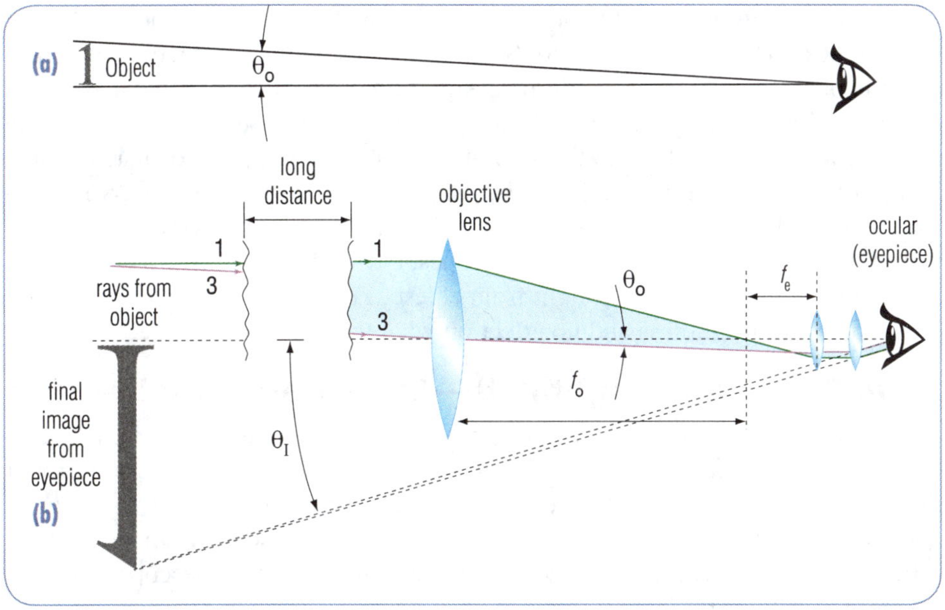

26-17 Actual angular size of object (a); Apparent angular size of object (b)

its shape and the images it produces. Thus, the largest refractor telescopes have objective lenses that are around 1 m in diameter.

26.13 Reflector Telescopes

Because a refractor's objective can be supported only at its edges, large lenses have a sagging problem. To circumvent the problem of distortion and other optical aberrations associated with large lenses, the **reflecting telescope** was developed. Reflector telescope mirrors can be supported across the entire back surface; thus, much larger and more massive light-gathering elements are possible (exceeding diameters of 8.3 m for single mirrors and 10 m for segmented mirrors).

The first reflecting telescope was designed and built by the Italian monk Niccolò Zucchi in 1616. Though he was able to make some observations of the planets, he couldn't grind the proper curve in the mirror, so he abandoned his efforts. In 1663, the Scottish mathematician James Gregory designed a reflecting telescope that was similar to many modern telescope designs. Robert Hooke built a working model of Gregory's design around ten years later. About 1670, Isaac Newton built a reflector telescope of his own design to solve the problem of chromatic aberration prevalent in the refractor telescopes of his day.

One drawback to the reflecting telescope is that the mirror forms a real image at its focal point on the same side as the object. In order to observe the image, a person has to somehow get between the object and the mirror. How, then, does the image reach a human eye without the body obstructing light from the object? For large astronomical telescopes, there is a special housing positioned at the principal focus of the main mirror that receives the reflected image. Depending on the kind of observations being made, the housing may use mirrors to reflect the light to analytical equipment through the side of the telescope framework or back through a hole in the center of the main mirror. Sometimes, instruments or even the observers themselves can be placed right in the prime-focus housing to analyze the light directly. Since each part of the mirror receives light from every point of the object, this housing reduces the total amount of light reaching the mirror, but it does not create a hole in the middle of the image. For a large telescope, this reduction of light is acceptable.

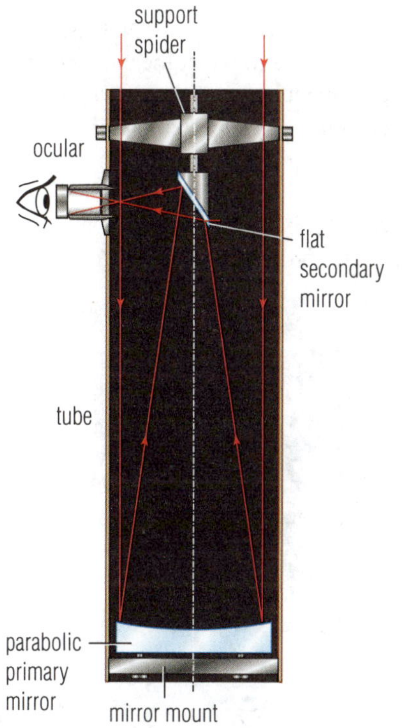

26-18 The Newtonian form of the reflector telescope uses a small, flat secondary mirror located at the main mirror focus to direct the image through a hole in the side of the tube.

At the time of this textbook's printing, the world's largest single-mirror reflector telescope is the Subaru reflector (8.3 m) operated by the National Astronomical Observatory of Japan (NAOJ) atop the Mauna Kea volcano in Hawaii. However, the reflector telescope with the largest individual mirrors is the Large Binocular Telescope (LBT) at the Mount Graham International Observatory in Arizona (each 8.4 m). The LBT is operated by the United States, Italy, and Germany. The largest general-purpose segmented-mirror reflector is the Gran Telescopio Canarias (GTC), which has a 10.4 m mirror and is located in the Spanish Canary Islands off the northwest coast of Africa.

For small telescopes, on the other hand, placing an observer at the focal point would block too much light. Newton's means of observing the image was to place a small plane mirror at an angle to the principal axis near the focal point of the main mirror. The image was reflected perpendicularly to the optical axis of the main mirror through a hole in the side of the telescope tube. An ocular mounted in the hole focused the light reflected by the plane mirror so that an observer could use the telescope. This arrangement is called the **Newtonian reflector** form of the reflecting telescope. As with larger telescopes, the secondary mirror blocks only a small portion of the incoming light from celestial objects, so it has only a minor effect on image brightness and no effect on what the image looks like.

Cassegrain (1629?–1693?) Almost nothing is known of this man—even his first name is in doubt, for he is variously known as Laurent, Jacques, Guillaume, or N. Cassegrain. His only documented accomplishment was in France, where he designed the telescope that bears his name. He didn't even build a telescope of his design—that was first accomplished by James Short (1710–68).

26.14 Folding the Light Path—Cassegrainian Telescopes

Another method of observing the image in a small reflecting telescope was devised by the French inventor Cassegrain. In this telescope, a convex secondary mirror at the main mirror's focal point reflects the image back through a hole in the main concave mirror to an eyepiece behind the mirror. The **Cassegrainian** form of the reflecting telescope is especially convenient in larger amateur telescopes since its eyepiece is behind the main mirror. With a larger Newtonian telescope, the observer must sometimes use a platform or ladder to reach the eyepiece, which is located near the open end of the telescope tube.

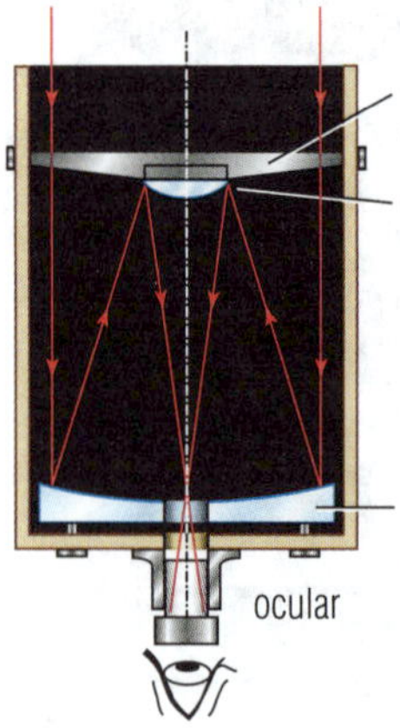

26-19 Cassegrainian focus; a convex secondary mirror near the main mirror focus directs the image through a hole in the main mirror.

26.15 The Schmidt-Cassegrainian Telescope

One problem introduced by the basic Cassegrainian design is spherical aberration from the convex secondary mirror. This problem was solved by Bernhard Schmidt. He used a spherically ground main concave mirror and installed a "correcting plate" at the open end of the telescope tube to cause the incoming rays to focus at a point when reflected by the spherical mirror, thus canceling the spherical aberration problem. The design was originally applied to the Newtonian design but was quickly adapted to the Cassegrainian form, producing what is called the **Schmidt-Cassegrainian** reflector telescope. This is a very popular telescope among amateurs because the focal length of the main mirror can be "folded up" within a shorter tube, larger apertures are possible in a more compact form, and the interior of the tube is sealed from dust and thermal currents that can affect the optical path of an open-tube telescope.

Bernhard Schmidt (1879–1935) was born in Estonia, but he was educated and lived in Germany. He was an optical technician for a prominent German observatory.

26-20 The interior of the Keck observatory showing the telescope structure. The position of each of the 36 segments of the 10.0 m primary mirror can be controlled independently by computer to compensate for the restless motion of the atmosphere.

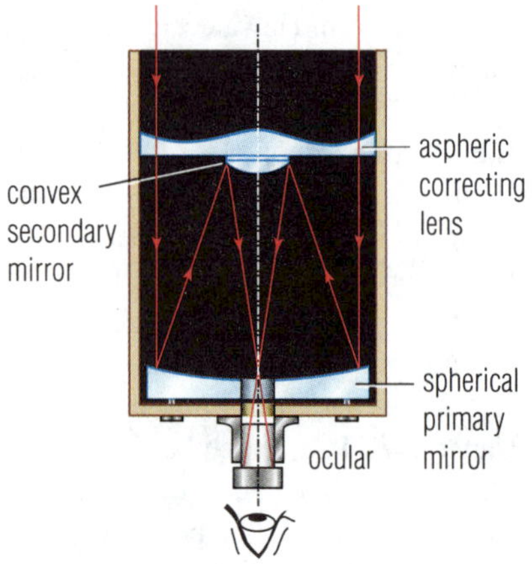

26-21 A Schmidt-Cassegrainian telescope

26.16 Advantage of Size

Why do astronomers prefer large telescopes to small ones? At least three reasons are apparent. First, a large telescope gathers more light than a small telescope, and, therefore, produces a brighter image. The light-gathering power of a telescope depends on the area of its objective (or the square of its objective's radius, since $A_{\text{circle}} = \pi r^2$). Therefore, a 200 cm telescope gathers four times as much light as a 100 cm telescope and can detect stars that are only one-fourth as bright as the faintest star a 100 cm telescope can detect.

Second, a large telescope has better resolution than a small telescope. Remember from Chapter 25 that the smallest angular distance between two objects that will allow the objects to be resolved is

$$\theta_R = 1.22\frac{\lambda}{d}. \qquad (26.5)$$

Objects with a small angular separation can be seen as separate objects with an objective or mirror of a large diameter (d). Small instruments tend to blur two close objects, making them look like a single object.

Third, a large telescope can magnify finer details better than a small telescope. Magnification is dependent on focal length, and large telescopes have longer practical focal lengths than smaller ones. Because of the lower resolving power in a smaller telescope, magnifying an indistinct image beyond a certain power produces no more detail.

26.17 Using Optics to Solve Problems

Many of the optical instruments mentioned in this section have benefited from recent advances in photographic technology that have eliminated most of the disadvantages of film photography discussed at the beginning of the chapter. In most imaging systems, photographic negative film, with its need for positive prints, has been replaced with a single electronic component—the charged-coupled device. A **charge-coupled device (CCD)** is a light-sensitive semiconductor surface designed to acquire optical images and convert them to a digital format that can be stored and processed by computers and special integrated circuits. A lens gathers light and focuses it on the photoactive layer of the CCD that converts light into an electric charge proportional to the light's intensity. These charges are distributed downward through semiconductor capacitors to a charge amplifier in the transmission layer of the CCD, where they are converted to a voltage. The numerical equivalent of the voltage for each *picture element (pixel)* is stored in a digital memory. In many digital cameras, color-sensitive pixels on the CCD (see Figure 26-22) or color filters in front of the CCD allow images to be captured in color.

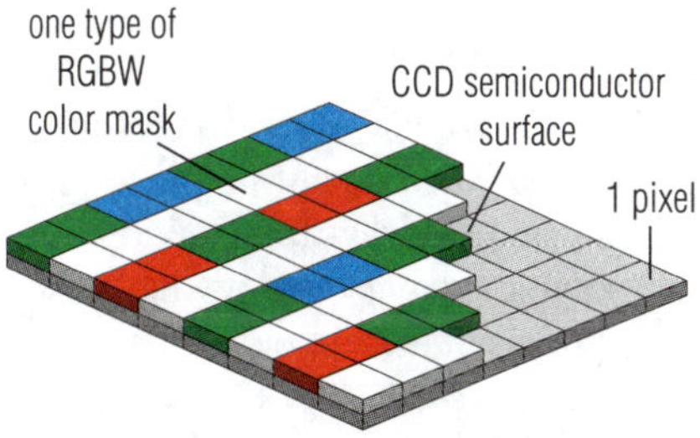

26-22 A modern CCD photographic element, showing the pixels sensitive to the primary colors. Clusters of 16 CCD pixels combine to form one digital photograph pixel.

In nearly every way, CCDs have eliminated the disadvantages of film photography. There are no chemical wastes. Feedback to the photographer on the quality or value of an image is immediate, so a poor photo can often be retaken. The digital medium is reusable, and the longevity of the digital image is as permanent as the medium on which it is stored. The need to store bulky collections of print photographs is eliminated. And digital images can be shared instantaneously across the world.

Digital photography has transformed information communication and documentation. CCDs are commonly found in consumer digital cameras and camera phones. They have replaced television cameras in space probe imaging systems. Researchers using microscopes and telescopes find CCDs to be much more compact than equivalent-quality film cameras, and they can easily switch between still- and movie-mode as needed. CCDs have reduced the amount of x-radiation

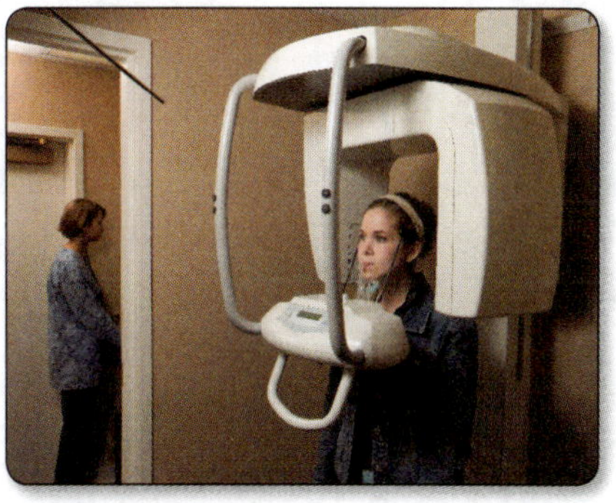

26-23 Digital medical x-ray procedures involve far less radiation exposure to the patient and technicians than earlier techniques.

required for medical x-rays, virtually eliminating the protective lead apron used during dental x-rays. And fluoroscopy, the process of using real-time video x-rays during medical procedures, such as angioplasties and pacemaker installation, has become a routine procedure because of CCD reception elements, yielding images of unparalleled clarity.

26B Objectives

After completing this section, I can

- ✓ explain how a microscope works.
- ✓ describe how the optical system of a simple compound microscope works.
- ✓ compare the application of various microscope configurations.
- ✓ describe how the optical system of a simple refractor telescope works.
- ✓ define optical magnification with respect to the characteristics of a telescope's optical components.
- ✓ compare and contrast the various forms of reflector telescopes.
- ✓ explain why large telescopes are generally better for astronomical observation.
- ✓ summarize the major disadvantages of traditional film photography and the ways that digital photographic technology has eliminated these problems.

26B Section Review

1. Discuss the purpose (the optical function) of the following instruments:
   a. camera
   b. periscope
   c. telescope
   d. microscope
2. What was the main limitation of Leeuwenhoek's microscope?
3. a. What is the orientation of the image produced by a standard compound microscope?
   b. If you move a specimen slide to the left while observing it through the microscope, which way does the image move?
4. What kind of microscope would you want to use if you were a US Mint worker verifying that the anti-counterfeiting microprinting on twenty-dollar bills was printed correctly?
5. Refractor telescopes are limited in size by the mass and diameter of their objective lenses. Explain why the need to correct spherical aberration further limits the effective aperture of the objective in a refracting telescope.
6. Does the secondary mirror in a Newtonian reflector make the image in the ocular erect? Explain your answer.

DS 7. What are the three steps that a digital camera performs to convert light to a digital image?

✪ 8. What is the magnification of a refractor telescope having a 1.25 m focal length objective and a 25 mm focal length ocular?

✪ 9. Describe any image-producing instruments that are not optical instruments within the definition given in this chapter.

DS ✪ 10. The raw video feed of the first moonwalk on July 21, 1969, was recorded on a now-obsolete one-inch magnetic tape. Unfortunately, machines that read these tapes are also obsolete, so veterans from NASA had to go digging to locate one. To further complicate matters, the original tape with the raw video is missing, leaving us with only a lower-quality television conversion of the video. What challenges lie ahead for those who archive digital information, such as images, who must preserve historically important information like the first moonwalk for posterity?

Chapter Review

In Terms of Physics

Problem-Solving Strategies

26.1 (page 601) Use the source's luminous intensity (in candelas) rather than its luminous flux (in lumens) when using the photometer equation.

26.2 (page 602) The equation $E = hf$ uses frequency rather than wavelength because wavelength changes as light passes through various forms of matter.

Review Questions

1. What is the difference between *luminous intensity* and *illuminance*?

2. **a.** Consider a standard incandescent light bulb. Is all the illumination it provides useful if the bare bulb is used primarily as a reading light?
b. What is used to improve the bulb's efficiency by directing more of its light onto the page of the book?

3. **a.** What does a photometer measure?
b. If one side of a comparator photometer screen is brighter than the other, does that necessarily mean that the lamp on that side is brighter than the lamp on the other? Explain.

4. Which kind of optical material transmits light but distorts the path of light rays so that an image cannot be formed?

5. Why is it not possible for a material to have a transmittance greater than 1?

6. What is color? Does color exist in nature in the absence of a mind to perceive it?

7. The energy in a light wave is proportional to what property of light?

8. What property of color describes how dull or brilliant a given color may be?

9. Under what circumstances may colors be additively combined?

10. Under what circumstances may colors be subtractively combined?

11. What is the minimum number of primary and complementary colors that must be added in order to produce white?

12. List the subtractive primary colors. Are they the same as the additive primary colors?

13. What is the minimum number of primary and complementary colors that you must combine in order to produce black?

14. **a.** List three optical instruments.
b. Why is a photometer not an optical instrument according to the definition given in this chapter?

15. What is the difference between a simple microscope and a compound microscope?

16. Is a telescope's image larger or smaller than the object?

17. What is a refracting telescope?

18. Why are the world's largest telescopes reflector telescopes, not refracting telescopes?

19. Where in relation to the object and the primary light-gathering element does a Newtonian reflecting telescope's image form?

20. Describe two ways to bring the image in a small reflecting telescope to the observer's eye.

21. Give one advantage that a large telescope has over a small one.

True or False (22–33)

22. The total amount of light a bulb produces is measured simply in watts.

23. Luminous intensity is the measure of the rate of light energy emitted in a particular direction.

24. The apparent or observed intensity of a light source is inversely proportional to the distance from the source.

25. Opaque materials do not transmit light.

26. Shorter wavelengths of light have less energy than longer wavelengths.

27. A dark burgundy color has less hue than a light burgundy color.

28. Combining two additive primary colors produces the third primary color.

29. Subtractive color mixing only works when light interacts with pigments—not with light by itself.

30. The image from a microscope is smaller than the object.

31. The shorter the focal length of a telescope's objective, the higher its resolution.

32. A Schmidt-Cassegrainian telescope is generally easier to use than a Newtonian telescope of equal aperture.

33. A telescope with a mirror three times the diameter of a smaller telescope can view stars one-ninth the intensity of those viewable by the smaller.

✪34. What is the illuminance on a surface that is 2.0 m from a 1600 lm source?

✪35. What luminous intensity is needed to produce the following illuminances?

a. 100. lx at 1.50 m
b. 333 lx at 3.00 m
c. 1200. lx at 5.00 m
d. 1500. lx at 0.400 m

✪36. Find the unknown luminous intensity ($I_{L\ unk}$) for the following cases:

a. An unknown bulb at $d_{unk} = 0.070$ m has the same illuminance as the standard source of $I_{L\ std} = 1.00$ cd at $d_{std} = 0.350$ m.
b. $I_{L\ std} = 1.00$ cd; $d_{std} = 0.250$ m; $d_{unk} = 0.500$ m
c. $I_{L\ std} = 1.00$ cd; $d_{std} = 0.0100$ m; $d_{unk} = 0.200$ m
d. $I_{L\ std} = 1.00$ cd; $d_{std} = 0.100$ m; $d_{unk} = 0.700$ m

✪37. What influence does the ratio of a telescope's focal length to the diameter of its objective (or mirror) have on its performance?

✪38. Consider Question 6 again. Discuss the philosophical position that some scientists take when they state that if one cannot perceive something (either directly with the senses or with the aid of instruments), then it has no reality. Is this a biblical philosophy?

DS✪39. Research some of the uses of CCDs in astrophotography.

UNIT 6

Modern Physics

The dawning of the twentieth century revealed to scientists that the universe was much more complicated than the laws of classical physics predicted. For one thing, energy didn't come in a continuous range of values—it was quantized. The atom wasn't the ultimate particle of matter, and matter itself was mostly space. Further research uncovered sources of energy that have become the most efficient commercial fuels as well as the most devastating weapons of war. Today, scientists are still struggling with Einstein's theory of relativity and are seeking the theory that will unify all the forces of nature in one grand theory.

In Unit 6, you will just scratch the surface of the various theories of relativity and coordinate system transformations. You will learn how matter and energy are quantized and are interchangeable. After reviewing the structure of the atom, you will study nuclear reactions in some detail and learn about the truly fundamental particles that make up all matter.

Some of the first spectra of distant galaxies obtained with the European Southern Observatory's (ESO) visible multi-object spectrograph (VIMOS). More than 220 distant galaxies were observed simultaneously in this first-ever image. These spectra allow astronomers to obtain the redshift, a measure of distance, as well as to assess the physical status of the gas and stars in each of these galaxies. According to the theory of relativity, the laws of physics are the same in these distant galaxies as they are here on the earth.

Relativity 27

DOMINION SCIENCE PROBLEM

"Relatively" Accurate GPS

Have you ever been lost with a map in hand? Many times, this happens because a person cannot locate his position on the map. Recent technology has provided a solution to this problem—the Global Positioning System (GPS). GPS has been incorporated into many applications, including precision surveying, car-trip planning and navigation, cell phone and auto accident emergency location, ship and aircraft distress transponders, personal positioning devices, military navigation and targeting, and scientific research.

The GPS consists of a constellation of 24–32 NAVSTAR satellites containing atomic clocks, microwave radio transmitters and receivers, and associated computer hardware, orbiting at a height of about 20 200 km (12 552 mi) above the earth at a speed of 3.87 km/s. A minimum of three satellites is needed to determine a GPS receiver's horizontal location through a process known as *trilateration*. This navigation technique involves analyzing for a unique intersection of the *spheres of position* created by the radial distances to the GPS satellites. These distances are determined by computing the difference between the time a satellite message says it was transmitted and the time when the receiver copies it. The actual distance is calculated by knowing the speed of light, including any speed corrections needed as the signal passes through the atmosphere.

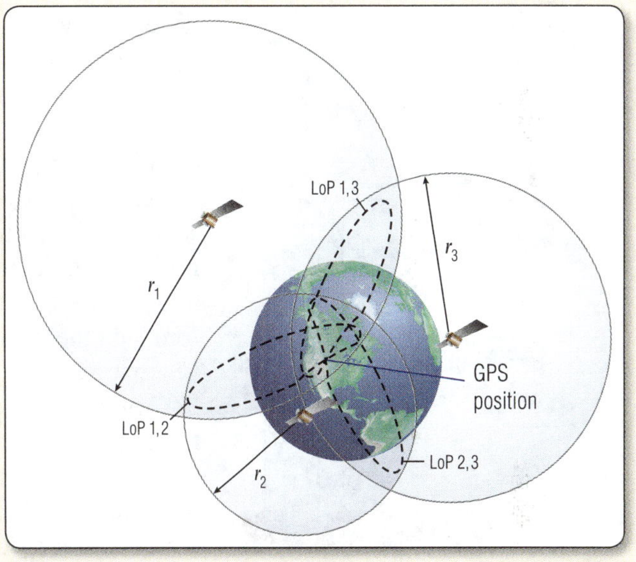

27-1 GPS positions are determined by finding the distances to at least three GPS satellites, a navigation technique called trilateration. (Differences in radii exaggerated for illustrative purposes.)

Because of the very high speed of light and the relatively short distances involved, it is crucial that information from the atomic clocks on the satellites be synchronized with the ground receiver to within 20–30 ns in order to be able to provide accurate position data. What problem affects GPS satellite clock synchronization, and how can it be corrected?

27A GALILEAN RELATIVITY

27.1 Phenomenological Relativity

An avid runner runs along, thinking of nothing but the speed he can achieve. His objective is a six-minute mile, sustained over ten miles. Breathing heavily, he keeps close track of his pedometer and his stopwatch. Finally, he is reaching his objective! He has maintained his six-minute-mile speed for nine miles—nine and one-half miles—almost there! Then, his wife walks past his treadmill to call him to supper.

The word "phenomenological" describes any event or entity that can be observed or measured.

Some people extend the idea of relativity to the sphere of morality and human experience through the philosophy known as *relativism*. Relativism states that no conceptions of truth and morality are absolute but are instead relative to each person or group. This is the "it's right for me" philosophy. However, unlike the physical world where absolute reference frames do not exist, God has established in His Word fixed points of reference for moral decisions and the standards by which humans should live.

The man was running fast relative to the belt of his treadmill but was motionless compared to his exercise room. Which standard—the treadmill or the exercise room—is correct? Neither. Only relative motion has meaning, since there is no valid standard of "motionless-ness" to which motion can be compared. This truth, called **relativity**, is emphasized in the study of objects in one reference frame (the set of points used to locate objects) as seen from another reference frame that is moving with respect to the first. The principle of no absolute rest applies to all realms of physics, including mechanics, optics, thermodynamics, and electromagnetics. The idea originated with Galileo Galilei, who applied it to mechanics.

27.2 Galileo's Observations

In Galileo's time, Copernicus's heliocentric model of the solar system was the subject of much debate. If Copernicus were right, the earth must move, and its motion must be rapid. Many scholars who disagreed with Copernicus declared that this conclusion was obviously wrong. As an example, they said, suppose that the earth were moving rapidly, as Copernicus said, and a man dropped a ball from the top of a tall building. After being released, the ball would fall straight down, but as it fell, the earth would move from under it, and the ball would eventually land a noticeable distance from the building. However, a ball dropped from a building actually lands close to the building. Therefore, they reasoned, the earth cannot be moving.

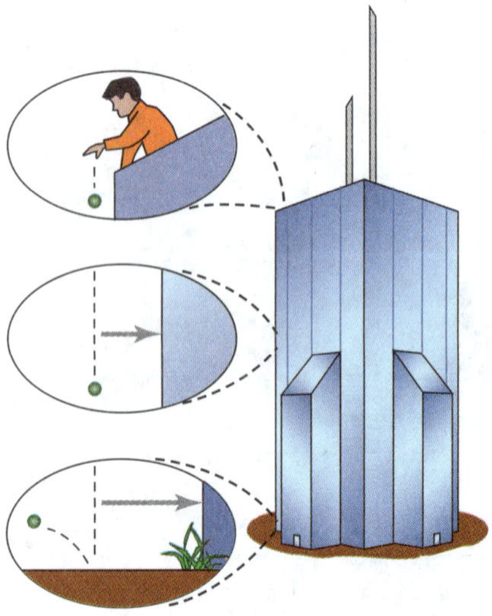

27-2 According to the anti-Copernican scientists of the sixteenth century, a ball dropped on a moving Earth could not fall straight down.

Galileo's answer to this argument was that the laws of mechanics are the same in steadily moving reference frames as they are in resting reference frames. This idea is called **Galilean relativity**. He pointed out that before the ball is released, it travels at the same velocity that the earth and the building travel. There are no horizontal forces on the falling ball, so its horizontal velocity does not change. The ball continues to travel along with the building as it falls, and it lands close to the building.

Galileo's explanation was corroborated by an experiment using a reference frame that everyone agreed was moving—a ship. As the ship was sailing with constant velocity, a man dropped a weight from the mast. The anti-Copernicans predicted that the weight would fall toward the back of the ship, but Galileo predicted that it would fall beside the mast. The weight landed next to the mast, so Galileo's argument was proved correct.

27-3 An experiment proved that on a ship moving at a constant velocity with respect to the earth, a dropped weight appears to fall straight down when observed from the deck of the ship.

A proper analysis of the weight's fall must take into account all of its velocities. To predict the path of the weight as seen from the shore, you must add the velocity of the weight as seen by a man on the ship to the velocity of the ship as seen by the man on the shore. This becomes more obvious when the velocities are in the same direction. Suppose a ship is moving away from a shore with a speed

$$v_s = 10 \text{ km/h}.$$

A man on the ship begins walking toward the front of the ship at the speed of

$$v_m = 3 \text{ km/h}.$$

How fast is the man on the ship moving with respect to the observer on the shore? After a little thought, the answer should be obvious.

$$v = v_s + v_m$$
$$v = 10 \text{ km/h} + 3 \text{ km/h}$$
$$v = 13 \text{ km/h}$$

This is a prediction of Galilean relativity.

27-4 The crewman's total velocity is $\mathbf{v}_s + \mathbf{v}_m$.

27.3 Galilean Coordinate Transformations

How is Galilean relativity useful? Let's use a contrived scenario to illustrate its usefulness. A physics teacher, exhausted after a long but fruitful year of teaching, takes a vacation at the beach. He relaxes in his floating lounge chair just off shore in the gentle swell, reading Newton's *Principia*. Completely absorbed in his book, he is unaware that the current has carried him well out to sea. His absence is noticed, a rescue is initiated, and the Coast Guard officer, standing where the teacher was last seen on the beach, directs a helicopter to fly directly away from that point, parallel to the current, at 50 m altitude.

27-5 The officer's coordinate system

The officer's coordinate system consists of three dimensions and time—x, y, z, and t. He chooses to align his x-axis horizontal and parallel to the direction of the current. His y-axis is horizontal and perpendicular to the x-axis, and his z-axis is vertical. The origin for his system is located at his position.

The helicopter's coordinate system (using primed symbols) is parallel in all respects to the officer's, but the helicopter's origin is located at its (changing) position.

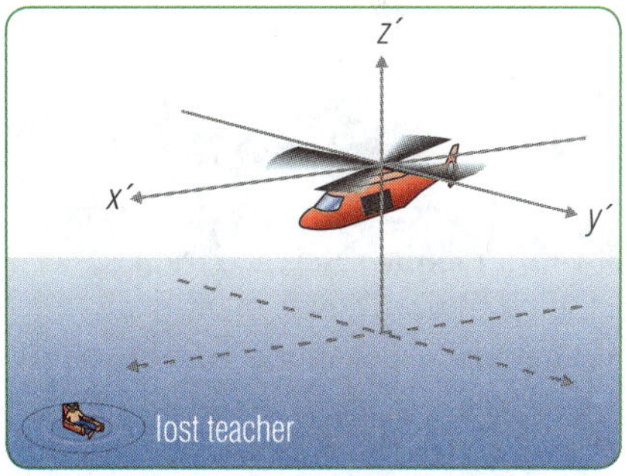

27-6 The helicopter's coordinate system

Eventually, the helicopter sights the physics teacher in his lounge chair 200 m ahead ($x' = +200$ m), 50 m to the left ($y' = +50$ m), and 50 m below ($z' = -50$ m). The pilot reports the teacher's location as (200 m, 50 m, −50 m). The officer looks at the point $x = 200$ m, $y = 50$ m and is perplexed, wondering how the teacher can be at $z = -50$ m (below sea level) from the officer's position on the shore. Obviously, he has assumed that the helicopter's coordinate system is identical to his own—that

$$x = x', y = y', \text{ and } z = z'.$$

But the coordinates are actually different.

How can the officer convert the helicopter's coordinate system into his own? After studying Figure 27-7, you can see that both the officer and helicopter pilot will agree that the teacher's position is 50 m to the left. Therefore, $y = y'$. The helicopter is flying at a constant altitude of $z = 50$ m, so

$$z = z' + 50 \text{ m}.$$

If the helicopter flew over the officer at time $t = t' = 0$, then at that time x and x' were equal. The distance between the origins of the two systems, Δx, is just the distance the helicopter travels in time Δt:

$$\Delta x = v\Delta t$$

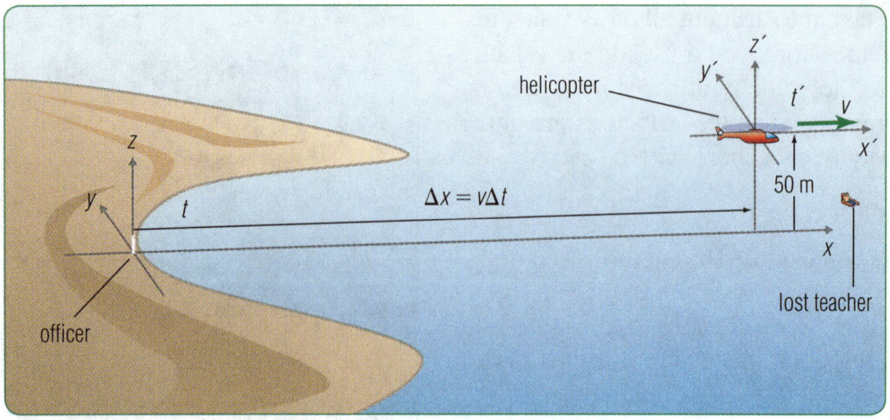

27-7 Locating a position in two different reference frames

Therefore, according to the officer on the shore,

$$x = x' + v\Delta t.$$

The officer can translate the coordinates from the helicopter to his own system using this **Galilean transformation**:

$$x = x' + v\Delta t,$$
$$y = y',$$
$$z = z' + 50 \text{ m, and}$$
$$t = t'$$

Similarly, if the officer on the shore gives coordinates to the helicopter, the pilot must translate them. His transformation will be exactly like the officer's transformation, except that he sees the shore 50 m *below* him *receding* with a velocity of $-v$. His translation is

$$x' = x + (-v)\Delta t,$$
$$y' = y,$$
$$z' = z - 50 \text{ m, and}$$
$$t' = t.$$

Problem-Solving Strategy 27.1

Time in Galilean transformations is always assumed to be the same in every reference frame:

$$t = t'$$

27.4 Inertial Reference Frames

The laws of mechanics—those known to Galileo and those discovered since his time—have the same mathematical form in every reference frame *that does not accelerate*. Such a reference frame is called an **inertial reference frame**. As long as the helicopter in the previous illustration moves with a constant velocity, mechanical experiments carried out in the aircraft will give the same results as they would have if carried out on the "stationary" shore. Without looking out the window, the teacher (still avidly reading his book) would be unable to tell whether the helicopter was moving at a constant speed or hovering. If he dropped the book, it would appear to him to fall straight down in either case. Any other mechanical experiments will also give the same results whether the helicopter is moving in a straight line at constant speed or not moving at all. There is, in fact, no mechanical experiment that can determine absolute motion. Only relative motion has meaning in mechanics.

Problem-Solving Strategy 27.2

Note that accelerations may occur within inertial reference frames, but the reference frames themselves cannot be accelerating.

27.5 Problems Arising from Galilean Relativity

Scientists of the nineteenth century understood that mechanics cannot determine absolute motion, but they thought that optics could. All the waves that they had studied had been disturbances in a medium, as sound is a disturbance in the air. They assumed that light was the same. However, light travels in an apparent vacuum in space. To explain this phenomenon, scientists theorized that a colorless, odorless, weightless substance called the **luminiferous** (light-bearing) **ether** filled all space and all transparent objects. The ether was assumed to be at rest, with the earth and other bodies moving through it, so it should provide a reference frame that was absolutely at rest. For example, if a light were shined from the back of a moving ship to the front, it would have to travel through the stationary ether at 3.0×10^8 m/s. The light's speed relative to the ship would be less—it would equal the speed of light minus the ship's speed. Therefore, it would take longer for the light to reach the front on a moving ship than on a stationary ship.

Like every other theory, the ether theory needed to be tested. Every theory makes predictions, but the test of the theory is whether its predictions are borne out by experiment. If they are not, the theory must be modified or discarded. From the ether theory, Albert Michelson predicted that it would take light longer to travel a path parallel to the earth's motion through the ether than to travel a path of the same length perpendicular to that motion.

Michelson used an interferometer, shown in Figure 27-8, to test his prediction. A light source, *L*, at the left directs light toward a partially silvered mirror, which reflects some light to *R* and allows the rest of the light to continue, via a folded path created by reflecting the light off of two sets of mirrors, to *T*. Mirrors at *R* and *T* reflect the light back to *C*, which directs the light to the detector telescope, *D*. Michelson reasoned that if the instrument were oriented so that the light rays from *L* to *T* were traveling in the same direction as the motion of the earth through the ether, the round-trip journey for the light would take a maximum time. The light going by way of *R* would take less time. When the light reached the telescope, the light from the two paths would be out of phase, and the telescope would display an interference pattern. He also reasoned that the relative phase of the light from the two paths would change if the apparatus were rotated, since the path toward *T* would no longer be parallel to the earth's motion through the ether. The diffraction pattern would therefore change. Michelson's interferometer was sensitive enough to detect changes much smaller than the calculated change. Detecting this change would corroborate the ether theory.

Albert A. Michelson (1852–1931) was an American physicist born in Poland. He is best remembered for his work in determining the speed of light. Awarded the Nobel Prize in Physics in 1907, he was the first American to receive a Nobel Prize in the sciences.

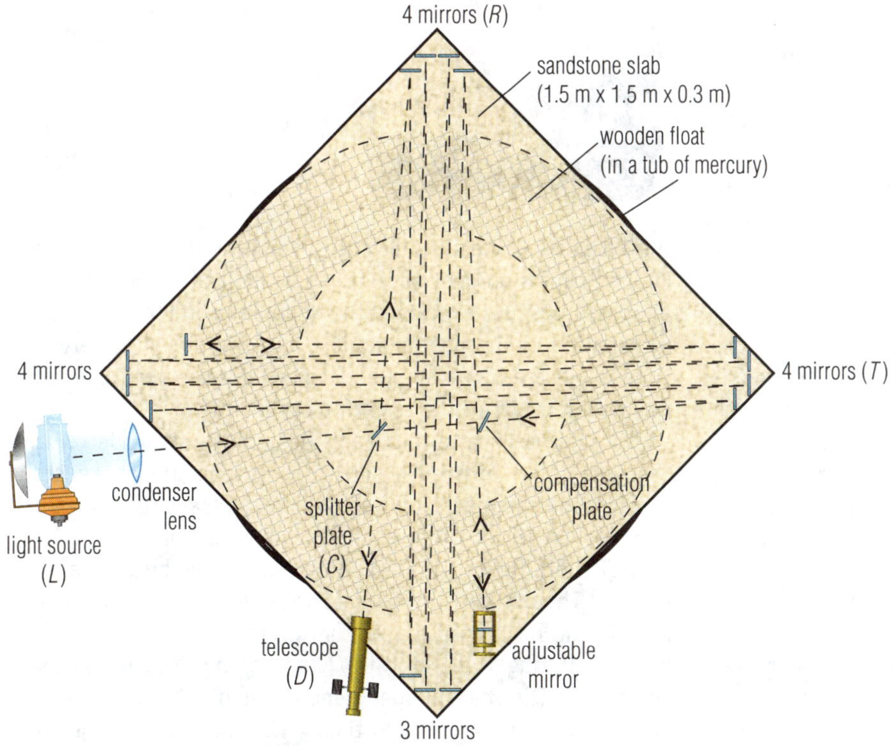

27-8 Michelson and Morley's interferometer

Michelson and his partner, Edward Morley, tried repeatedly to find the predicted change. According to the ether theory, if either path were aligned along the direction of the earth's motion, the interference pattern would change. To achieve this alignment, the experimenters rotated the instrument slowly through a full circle as they recorded their data, and they performed the experiment again three months later, when the earth's motion through the hypothetical ether was in a different

Edward W. Morley (1838–1923) was an American chemist.

The inconclusive results of the Michelson-Morley experiment is an excellent illustration of the fact that an experiment does not have to prove a hypothesis true to be considered successful. This experiment was a success in that it provided the corroborating evidence supporting some of the predictions of Einstein's theories of relativity.

direction. In every case, they obtained the same result: the interference pattern remained essentially constant. The prediction of the ether theory was not borne out by this experiment nor by any others over the next 50 years. However, it wasn't until after 1920 that physicists came to the conclusion that the experiment actually supported the idea that *there was no ether*.

27A Section Review

1. Discuss what is meant by *relativity*.
2. What is the main premise of Galilean relativity?
3. What did the anti-Copernicans fail to take into account when describing the path of a falling object?

Albert Einstein (1879–1955)

The best-known modern physicist is Albert Einstein. When you hear his name, you probably picture a genius scribbling incomprehensible equations on a chalkboard. At one time, it was commonly reported that only a dozen people could understand Einstein's higher mathematics, but today his theories have become a fundamental part of every college course in theoretical physics.

Einstein's Jewish father and uncle ran a small engineering plant in Munich, Germany. The young boy did not always perform well at school because he disliked the rigid discipline of his German teachers. It was his uncle who stimulated his passion for mathematics and science. When his family moved to Switzerland, Albert managed to pass the entrance exam to the Swiss Federal Polytechnic School on his second try. After graduation, he found a quiet job at the patent office in Berne, where he spent his spare time puzzling over the nature of the universe.

In 1905, the 26-year-old clerk published four papers that forever changed our understanding of space, time, mass, and the universe. The first paper earned him a doctorate from the University of Zurich. The next three papers won him world renown. One gave a theoretical explanation for Brownian motion, a mystery since it had been discovered in 1827. The next paper suggested that light consists of both particles (photons) and waves; his quantum theory of light explained why light striking a metal plate stimulates it to emit electrons (called the photoelectric effect). The last paper introduced his special theory of relativity that included the famous equation $E = mc^2$.

Soon, this unknown clerk was in demand as an instructor at several universities. He changed professorships almost once a year until he finally settled in Berlin. The position offered by the Prussian Academy of Sciences freed him to research and publish papers on general relativity, heat, and other topics in physics. In 1921, he received the Nobel Prize in Physics for his explanation of the photoelectric effect. The Royal Society of London elected him a foreign member the same year. Einstein was now a leader in the scientific community.

But the rise of the Nazi Party in Germany coincided with the ebb of Dr. Einstein's scientific output. He spoke out against the prospects of war and Nazi persecution of the Jews. When Hitler became chancellor of Germany in 1933, Einstein left for the United States and later became a US citizen. Though he continued to receive honors throughout his life, he no longer was at the forefront of discoveries in physics. During the last thirty years of his life, he parted company with the younger generation of physicists, who believed that a unified field theory was impossible. Einstein remained convinced that the electromagnetic field and gravity are two forms of the same phenomenon. He never found the secret formula, which has eluded scientists to this day. Nevertheless, Einstein did not believe that he had wasted his time searching for this unification. He had, he declared, showed future scientists where *not* to look.

4. a. What mathematical process must be performed to convert positions in one coordinate system to another coordinate system?
   b. Do the coordinate systems need to be stationary to accomplish this conversion?
5. a. What are unaccelerated coordinate systems called?
   b. Give an example of an accelerated coordinate system as well as an unaccelerated one.
6. a. During the late nineteenth century, what was considered the absolute reference frame for all motion?
   b. What experiment inadvertently disproved this theory?
7. At time t_1, a car is traveling west at a constant 95 km/h on flat terrain. Its coordinate system is defined by $x_2 = x_1 - (26.4\text{ m/s})\Delta t$, y, z, and t. Another car starting 5.0 km to the east of the first car at time t_1 heads directly north at 60. km/h. Determine the transformation equations to describe the position of the first car within the reference frame of the second car. In other words, $x = ?$, $y = ?$, $z = ?$, and $t = ?$

27A Objectives

After completing this section, I can

- ✓ justify the need to understand phenomenological relativity.
- ✓ discuss the essential features of Galilean relativity.
- ✓ perform Galilean coordinate transformations between two inertial reference frames.
- ✓ discuss the historical significance of the concept of a luminiferous ether.
- ✓ describe the Michelson-Morley experiment and explain its inconclusive results.

27B SPECIAL RELATIVITY

27.6 Assumptions of Special Relativity

The ether theory presupposed that a fixed, universal reference frame existed. The theory's failure was readily addressed by another model, the **special theory of relativity**. Albert Einstein, the theory's originator, stated that there could be no preferred reference frame and therefore no ether. His theory was corroborated by the Michelson-Morley experiment's inconclusive results. The theory consists of two assumptions, or postulates:

1. Each law of physics has the same mathematical form in all *inertial* reference frames.
2. The speed of light in a *vacuum* is the same in all reference frames.

The first postulate is simply an extension of Galilean relativity, which said that the laws of mechanics are the same in every inertial reference frame. Einstein extended the rule to include thermodynamics, electromagnetism, and optics. He assumed that *no* physics experiment can distinguish between a reference frame moving at a constant velocity and a stationary one.

In contrast, the second postulate is contrary to all our experience. If a car is approaching you, and a person in the car throws a ball at you, the ball will reach you sooner than it would if the car were stationary. However, Einstein said that if the person in the car turned on the headlights instead of throwing a ball, the light would reach you at the same time that it would if the car had been stationary. Relative motion of the source of the light and the observer has no effect on the observer's measurement of the speed of light.

27.7 The Lorentz Transformation

The special theory of relativity does not agree with the Galilean transformation, so a new transformation is needed. However, since normal occurrences follow the

Hendrik A. Lorentz (1853–1928) was a Dutch physicist who won the Nobel Prize in Physics in 1902.

Galilean transformation, the new transformation must reduce to the Galilean transformation at normal speeds. The new transformation, called the **Lorentz transformation**, is as follows:

$$x = \frac{x' + v\Delta t'}{\sqrt{1-(v/c)^2}}$$
$$y = y'$$
$$z = z'$$
$$t = \frac{t' + vx'/c^2}{\sqrt{1-(v/c)^2}} \quad (27.1)$$

In these equations, v is the speed of the object measured within the observer's frame of reference in the x direction.

At normal speeds, v is much smaller than the speed of light in a vacuum (c), so the terms $(v/c)^2$ and v/c^2 become very small compared to the other quantities. For everyday situations, the transformation reduces to

$$x = x' + v\Delta t,$$
$$y = y',$$
$$z = z', \text{ and}$$
$$t = t',$$

which is the Galilean transformation. "Much smaller than c" includes speeds under $0.1c$. No observable terrestrial object outside of particles in a particle accelerator has ever come close to $0.1c$, so everyday experience does not include Lorentz transformations. Subatomic particles that approach the speed of light follow the Lorentz transformation. A particle whose speed approaches the speed of light is said to have a **relativistic** speed.

27.8 Predictions of Special Relativity

To test the special theory of relativity, simply use it to make predictions. Some of the predictions are contrary to experience, or "common sense," but the circumstances described by the theory are outside normal experience. Experiments have so far confirmed the predictions. It should not be surprising that not everything in the universe conforms to man's common sense. The universe was designed by God, not by man. Einstein's theory is presently man's most workable model for describing what God has made. Only God knows how nearly correct it is. Our job as stewards of God's creation is to study and model what we find to the best of our abilities as we exercise good and wise dominion.

The special theory of relativity leads to four interesting predictions:

1. Simultaneity—If two observers are moving at a relativistic speed with respect to each other, two events that appear to be simultaneous to one observer will not appear to be simultaneous to the other.
2. Time dilation—A clock moving at a relativistic speed with respect to an observer will appear to be running slow.
3. Length contraction—When a stationary observer sees an object moving at a relativistic speed, the object's length parallel to the direction of motion appears to be shortened.
4. Mass increase—The mass of an object moving at a relativistic speed appears to increase with increasing speed.

In this discussion of special relativity, the terms "stationary" and "moving" are relative to the observer. The observer is normally considered to be stationary relative to all other objects moving within his frame of reference.

27.9 Simultaneity

One prediction of the special theory of relativity is that if one observer says that two events in two places are simultaneous, another observer moving with respect to the first will not see the events as simultaneous.

Consider the following example. A rocket boat is traveling parallel to the shore at nearly the speed of light. A man on the shore, O, stands at equal distances from the seaside corners (A and B) of a pier. A woman in the boat, O', sits at the midpoint of the boat. There are strobe lights at A and B. When A' is opposite A, the strobe at A goes off. When B' is opposite B, the strobe at B goes off. The observers judge when the strobes went off by when they see the light.

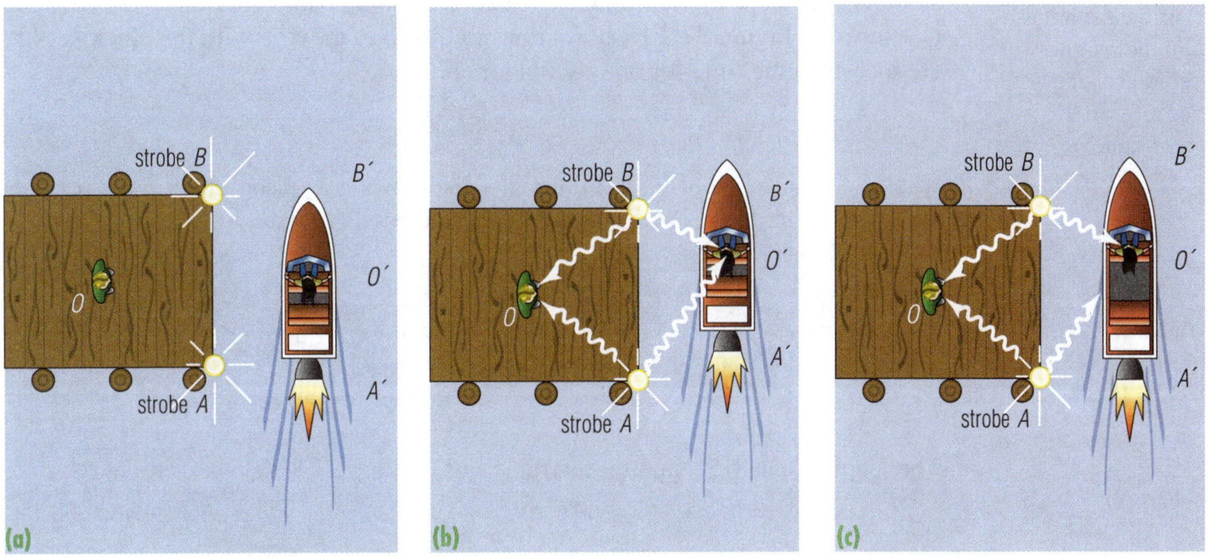

27-9 These observers disagree about simultaneity.

Suppose O receives the light from both strobes at the same time. He decides that the strobes went off simultaneously and that, therefore, the boat is as long as the width of the pier. The boat, which is moving with a speed near that of light with respect to O, moves between the time the strobes go off and the time the light reaches O (Figure 27-9b). From the perspective of O, the light from B has less distance to travel to O' than to O, so B's light reaches O' first. The light from A has farther to travel to O' than to O, so A's light reaches O first. Therefore, from the perspective of O, the light from B reaches O' before the light from A.

The observer O' agrees that she received light from B before she received light from A. Since she knows that the light had the same distance to travel from both points, she assumes that the strobe at B went off before A. She realizes that B' must have reached B before A' reached A, so she concludes that her boat is longer than the width of the pier. She disagrees with O in two observations: O says the flashes were simultaneous; O' says B flashed first. Also, O says that the boat length and the pier width are the same; O' says that the boat is longer (Figure 27-9c).

The essence of this discussion on simultaneity is this: if two events are simultaneous in a given frame of reference, they will *not* be simultaneous in a frame of reference moving with respect to the first.

27.10 Time

The problem of simultaneity brings up two more predictions of special relativity: time dilation and length contraction. Time **dilation** means that an observer perceives a clock moving with respect to him as running slow. Consider a moving "light clock."

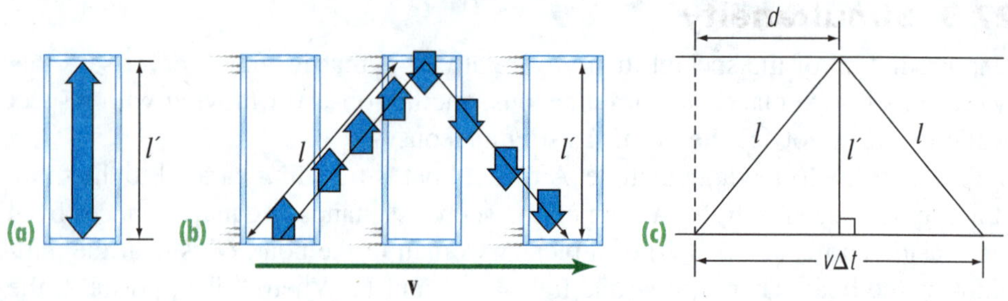

27-10 A moving light clock seen (a) from within its own reference frame and (b) from the reference frame in which it is moving. Figure (c) gives the geometry for determining l.

In this discussion, primed symbols refer to quantities in the moving clock's frame of reference, and unprimed quantities refer to the stationary observer's frame of reference.

Time is measured by the number of oscillations a beam of light makes between two mirrors. Figure 27-10a shows how an observer moving with the clock sees it. He measures the time for one oscillation as

$$\Delta t' = \frac{2l'}{c}, \tag{27.2}$$

where c is the speed of light. Figure 27-10b shows how a stationary observer sees one oscillation. He measures the time for one oscillation as

$$\Delta t = \frac{2l}{c}. \tag{27.3}$$

The distance l relative to l' can be found using trigonometry. The distance d is half of the distance traveled by the clock in time Δt:

$$d = \tfrac{1}{2}v\Delta t \tag{27.4}$$

The length l' can be found by rearranging Equation 27.2:

$$l' = \tfrac{1}{2}\Delta t'c \tag{27.5}$$

The length l, from the Pythagorean theorem, is

$$l = \sqrt{d^2 + l'^2}. \tag{27.6}$$

Substituting the values d and l' from Equations 27.4 and 27.5 results in

$$l = \sqrt{(\tfrac{1}{2}v\Delta t)^2 + (\tfrac{1}{2}\Delta t'c)^2}.$$

Since both terms contain $(\tfrac{1}{2})^2$, it can be factored out:

$$l = \tfrac{1}{2}\sqrt{(v\Delta t)^2 + (c\Delta t')^2}$$

The time elapsed, according to the stationary observer, is equal to the distance the light traveled, $2l$, divided by its speed:

$$\Delta t = \frac{2l}{c} = \frac{2\left[\tfrac{1}{2}\sqrt{(v\Delta t)^2 + (c\Delta t')^2}\right]}{c}$$

Squaring both sides yields

$$(\Delta t)^2 = \frac{(v\Delta t)^2 + (c\Delta t')^2}{c^2} = \frac{(v\Delta t)^2 + c^2(\Delta t')^2}{c^2}$$

$$(\Delta t)^2 = \frac{(v\Delta t)^2}{c^2} + (\Delta t')^2.$$

Rearranging,

$$(\Delta t')^2 = (\Delta t)^2 - \frac{(v\Delta t)^2}{c^2} = (\Delta t)^2 - (\Delta t)^2 \cdot \frac{v^2}{c^2}.$$

Factor out $(\Delta t)^2$:

$$(\Delta t')^2 = (\Delta t)^2\left(1-\frac{v^2}{c^2}\right) = (\Delta t)^2\left[1-\left(\frac{v}{c}\right)^2\right]$$

Solving for $(\Delta t)^2$ gives

$$(\Delta t)^2 = \frac{(\Delta t')^2}{1-(v/c)^2}.$$

Taking the square root of each side results in

$$\Delta t = \frac{\Delta t'}{\sqrt{1-(v/c)^2}}. \qquad (27.7)$$

The factor

$$\frac{1}{\sqrt{1-(v/c)^2}} \qquad (27.8)$$

occurs so often in the study of relativity that it is called the **relativistic factor (γ)**. Therefore, Equation 27.7 becomes

$$\Delta t = \gamma \Delta t'. \qquad (27.9)$$

The factor

$$\gamma = \frac{1}{\sqrt{1-(v/c)^2}}$$

appears in every Lorentz transformation in special relativity.

Problem-Solving Strategy 27.3

Remember that γ is always greater than 1.

In Equation 27.9, the factor γ is always greater than 1, which means that the relativistic clock's time interval $\Delta t'$ is observed to be a longer time interval than Δt when measured by the stationary observer's own clock. A clock moving with respect to an observer always seems to him to show less elapsed time than, say, his wristwatch. That is, a moving clock always appears to run slow to an observer in the stationary reference frame (the one that he is observing from). A person in the reference frame of the moving clock will observe the clock of the stationary observer, which appears to him to be moving, as running slow also.

27.11 Length

Special relativity also predicts that the observed dimensions of a moving object are affected by its relativistic velocity. *Length contraction* is related to time dilation. If an object is moving with respect to an observer, its length along the direction of motion appears shorter. The length of an object as measured by an observer moving with the object (i.e., in the object's reference frame) is its **proper length (l_0)**. Its length as measured by an observer in a stationary reference frame is l, which is less than l_0. The exact relationship is

$$l = l_0\sqrt{1-(v/c)^2}, \text{ or}$$

$$l = \frac{l_0}{\gamma}. \qquad (27.10)$$

Notice that only the dimension parallel to the direction of motion is affected, and the effect is noticeable only to an observer outside the moving system.

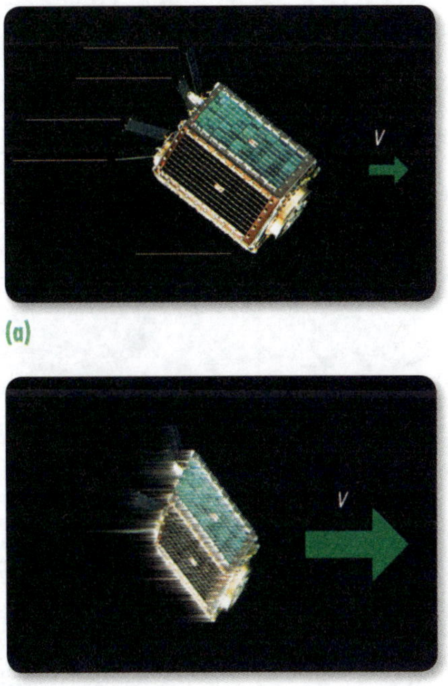

(a)

(b)

27-11 Length contraction at relativistic speeds as seen by a stationary observer. Figure (a) is observed moving at speeds $<<0.1c$; Figure (b) is observed moving at speeds $>>0.1c$.

27.12 Mass

Another prediction of special relativity is that apparent mass increases with speed. This requirement is necessary for the conservation of momentum. The mass of an object as measured by one who is moving with the object is its **rest mass (m_0)**. The mass of the same object measured by one who is moving with respect to the object *appears* to be

$$m = \frac{m_0}{\sqrt{1-(v/c)^2}} = \gamma m_0. \qquad (27.11)$$

Note that mass is an intrinsic property of an object and cannot change. Therefore, the only mass that needs to be considered is the rest mass (m_0). The apparent increase of mass is related to an object's momentum at relativistic speeds in an observer's frame of reference.

27.13 The Practical Effect of γ

For relativistic time, length, and mass, the velocity-dependent factor that describes the changes observed by the hypothetical stationary observer is γ. To describe your observations of systems moving at "normal" speeds, this factor must be nearly 1. Consider two reference frames with a relative velocity of 10^4 m/s (10 km/s, or nearly the escape speed from the earth). The denominator of γ becomes

$$\begin{aligned}\sqrt{1-(v/c)^2} &= \sqrt{1-(10^4\text{ m/s}/3\times10^8\text{ m/s})^2}\\ &= \sqrt{1-(10^8\text{ m/s}/9\times10^{16}\text{ m/s})}\\ &= \sqrt{1-1.\bar{1}\times10^{-9}}\\ &\cong 0.999\,999\,999\,4.\end{aligned}$$

The factor γ is then

$$\begin{aligned}\gamma &= \frac{1}{\sqrt{1-(v/c)^2}} \cong \frac{1}{0.999\,999\,999\,4}\\ &\cong 1.000\,000\,001.\end{aligned}$$

Most measurements are not made to a precision of nine decimal places, so an observer of a rocket moving at 10^4 m/s, for example, will not notice any measurable relativistic effects.

27.14 The Theoretical Consequences of γ

The predictions of the special theory of relativity about objects traveling at the speed of light are interesting. Consider two observers, one in a spaceship moving at the speed of light and the other in a stationary space station. Even though only one observer is being propelled, each views the other as moving at the speed of light in his own reference frame. Each observer has a large clock that can be seen by the other observer. The moving observer notes that 1 s elapses on his clock ($\Delta t' = 1$ s). How much time does he see elapse on the other's clock?

$$\Delta t = \frac{\Delta t'}{\sqrt{1-(c/c)^2}}$$

$$\Delta t = \frac{\Delta t'}{\sqrt{1-(1)^2}}$$

$$\Delta t = \frac{\Delta t'}{0} \Rightarrow \text{d.n.e. (infinitely large time interval).}$$

According to the observer in the spaceship, time in the space station is standing still. The observer on the space station also says that while 1 s passed on his clock, the moving observer's time seems to be standing still.

Each observer holds a meter stick parallel to the direction of relative motion. The observer on the spaceship measures the other observer's stick as

$$l = (1\text{ m})\sqrt{1-(c/c)^2}$$

$$l = 0\text{ m}.$$

Not only does the stationary observer's meter stick have zero length, but the entire space station appears as a vanishingly thin line perpendicular to the direction of relative motion! (Don't ask how you can read the meter stick—this is all hypothetical anyway.) The stationary observer also thinks that the moving observer's stick (and the observer himself) has no length in the direction of motion.

Finally, both observers bounce a 1 kg ball on their respective floors and measure the mass of the ball based on the conservation of momentum of the ball's collision with the floor. Each finds that

$$m = \frac{m_0}{\sqrt{1-(c/c)^2}} = \frac{m_0}{0}$$

$$m \Rightarrow \text{d.n.e. (mass seems infinitely large).}$$

Each observer says that the other's ball has infinite mass. This observation prohibits any material object from accelerating at, or even reaching, the speed of light. Since an apparent infinite mass requires an infinite outside force to accelerate it, and since there are no natural infinite forces, nothing can exceed the speed of light *in a vacuum.* (However, some particles can go faster than the speed of light in a medium; for example, some subatomic particles can go through water faster than light can.)

Recall that momentum is the product of mass and velocity:

$$p = mv$$

27.15 Velocity Addition in Special Relativity

Another prediction of special relativity concerns velocity addition. Consider two reference frames with observer O' moving at a relative velocity of $\mathbf{v}$ to observer O. Observer O' launches a rocket with a speed of $\mathbf{v}_r'$ with respect to O', parallel to the relative velocity $\mathbf{v}$. What is the rocket's velocity with respect to the observer O in the other reference frame? The speed of anything is simply the change in distance divided by the change in time. For the rocket,

$$v_r = \frac{\Delta x}{\Delta t}. \tag{27.12}$$

From the Lorentz transformations (Equations 27.1):

$$\Delta x = \frac{\Delta x' + v\Delta t'}{\sqrt{1-(v/c)^2}}$$

$$\Delta t = \frac{\Delta t' + v\Delta x'/c^2}{\sqrt{1-(v/c)^2}} \tag{27.13}$$

Therefore, dividing Δx by Δt gives the speed of the rocket (v_r) in observer O's frame of reference:

$$\frac{\Delta x}{\Delta t} = \frac{\Delta x' + v\Delta t'}{\cancel{\sqrt{1-(v/c)^2}}} \cdot \frac{\cancel{\sqrt{1-(v/c)^2}}}{\Delta t' + v\Delta x'/c^2}$$

$$v_r = \frac{\Delta x' + v\Delta t'}{\Delta t' + v\Delta x'/c^2} \tag{27.14}$$

This can be simplified somewhat by dividing the numerator and the denominator by $\Delta t'$, giving

$$v_r = \frac{\Delta x'/\Delta t' + v}{1 + v\Delta x'/c^2\Delta t'}. \tag{27.15}$$

But $\Delta x'/\Delta t'$ is simply v_r' (in observer O''s reference frame):

$$v_r = \frac{v_r' + v}{1 + (vv_r'/c^2)} \tag{27.16}$$

As long as v_r' and v are less than c (which must be the case, since nothing can exceed the speed of light in a vacuum), v_r (the speed of the rocket as observed by O) is less than c. At low speeds, the denominator is so close to 1 that the equation reduces to

$$v_r = v_r' + v,$$

which is the familiar Galilean transformation (assuming that all speeds are in the same direction).

Problem-Solving Strategy 27.4

Relativistic velocity additions always yield velocities that are less than or equal to the speed of light. If you obtain a total summed speed greater than c, you can be sure that you have made a mistake.

27.16 Mass-Energy Equivalence

The most familiar prediction of special relativity is expressed by the equation

$$E = mc^2, \qquad (27.17)$$

where E is a particle's total energy and m is its relativistic mass (the mass it appears to have when moving at a high speed). The kinetic energy of the moving particle is

$$E_k = mc^2 - m_0c^2. \qquad (27.18)$$

The rest-mass energy (E_0) is sometimes just called rest energy.

When the particle is at rest, therefore, it has a rest-mass energy (E_0) of

$$E_0 = m_0c^2. \qquad (27.19)$$

Equation 27.19 shows a most useful prediction of special relativity: mass and energy are equivalent entities. Mass can be converted to various kinds of energy, and energy can be converted to mass. Any addition of energy to a particle results in an increase in its apparent mass.

27.17 Experimental Evidence for Special Relativity

Experiments have supported the predictions of special relativity. One of these experiments involves *muons*, particles somewhat smaller than protons. Muons produced in laboratories (with low relativistic speeds) have a half-life of 2.2×10^{-6} s. That is, after 2.2×10^{-6} s, half of the muons have decayed into other subatomic particles. However, muons that are produced in the atmosphere from cosmic ray interactions may fall toward the earth with speeds of $0.9c$ or more.

At $0.9c$, a muon travels almost 600 m in one half-life. Therefore, you would expect to find half as many muons reaching the earth's surface as there are 600 m above the earth's surface. Actually, much more than half reach the earth. The explanation is that, to an observer on the earth, the muons' time is running slow. When we say that 2.2×10^{-6} s has passed, only 9.6×10^{-7} s has passed for the muons as measured by the stationary observer (their time appears to move slower). Since the muons' time determines their half-life, they last longer than we would expect.

27-12 Particle accelerators can generate huge relativistic velocities and masses for subatomic particles.

From the muons' perspective, the muons "see" the earth rushing toward them at $0.9c$. Why does the earth travel 600 m in only 9.6×10^{-7} s? Actually, it does not, since the muons also "see" a length or distance contraction. The 600 m to them is around 260 m. The measured proportion of muons reaching the earth agrees well with these calculations.

Another specific prediction of the theory is that the mass of a particle emerging from a particle accelerator is more than its rest mass. This prediction can be tested by a mass spectrometer (see Chapter 21). A larger mass requires a stronger magnetic field to deflect it. By using a spectrometer, scientists have determined that particles moving near the speed of light may have masses that are hundreds or thousands of times larger than their rest masses.

27-13 An atomic clock is the most accurate time-measuring tool available. Four of these clocks were used to test the clock paradox.

Special relativity also predicts what is known as the **clock paradox**. Two observers, each with a clock, begin at rest with respect to each other. The clocks are synchronized. One observer, O', decides to travel. He moves at a constant speed to his destination and then returns to the other observer, O, at the same speed. The observer O has not moved. When O' returns, they compare their clocks. They discover that the traveler's clock registered less time for the journey than the stationary observer's clock.

The paradox is that it seems we have defined a preferred frame of reference. We can tell which observer traveled. Why can we not view the trip as O' staying still and O traveling in the opposite direction? The solution to the paradox is that O stayed in one inertial reference frame, but O' was in *three* different reference frames. He began in O's reference frame. On his trip to his destination, O' was in a reference frame moving in one direction with respect to O. On his return trip, O' was in a reference frame moving in the opposite direction with respect to O. Finally, O' returned to O's reference frame. This changing of reference frames makes O''s clock slower than O's clock.

The clock paradox was corroborated in an experiment using atomic clocks on jet aircraft. Four clocks were compared to stationary atomic clocks. Then, the four went around the world on jets. They began in a reference frame at rest with respect to the earth, changed to a reference frame moving with respect to the earth, and finally returned to rest with respect to the earth. When the traveling clocks were again compared to the clocks that stayed in place, the traveling clocks were a fraction of a second slower than the resting clocks. The results of the experiment agreed with the prediction of special relativity.

You may have concluded from the discussion of the Lorentz transformations earlier in this section that the differences in time, length, velocity, and mass between reference frames are virtual—illusions—due to differences in observer reference frames. This conclusion is not accurate. If it were possible to actually measure each of those parameters in the other reference frame, they would show the change predicted by the transformation equations. The measured dimension would be as "real" as any other measurement in the observer's frame of reference.

27.18 Applications of Special Relativity

The predictions and phenomena of special relativity can be effectively used in dominion science to glorify God and to benefit other people. For example, mass-energy equivalence is probably the most useful result of the theory. Since mass is equivalent to energy, it is not surprising that mass is sometimes converted to energy in certain processes.

For instance, the mass of an atom's nucleus is less than the sum of the rest masses of the particles that compose the nucleus. The "lost" mass is called the *mass defect* due to nuclear *binding energy*. When the particles are joined to form the nucleus, they give up some of their mass as energy. If two smaller nuclei merge, each gives up some of its mass. This kind of reaction, called *fusion*, produces large quantities of energy that have potential use. On the other hand, before a particle can leave the nucleus, it must gain enough energy to restore its full mass. If the binding energy is large, it is difficult for a particle to gain enough energy to leave the nucleus. Therefore, nuclei with large binding energies per nucleon are more stable than nuclei with small binding energies. We will study nuclear stability further in Chapter 29.

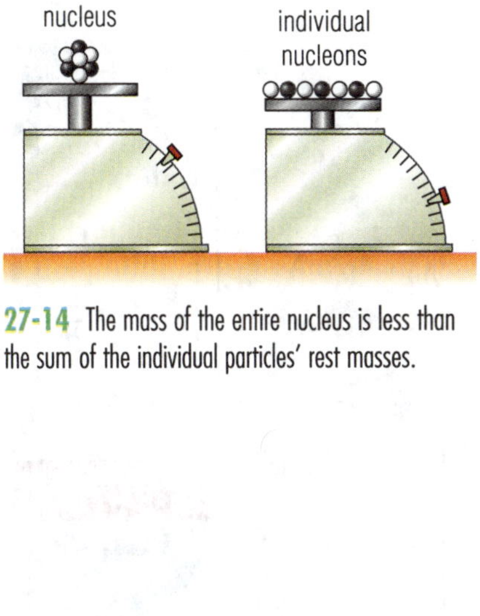

27-14 The mass of the entire nucleus is less than the sum of the individual particles' rest masses.

A technological application of nuclear binding energy is nuclear power. Nuclear power encompasses both its peaceful use in electrical power generation, medicine, and propulsion, and its use in weapons of war. Matter is a potential source of immense quantities of energy. One gram of matter theoretically contains 9×10^{13} J of energy. If mass could be totally transformed into electrical energy and sold at one cent per kilowatt-hour, a 1 g paper clip would produce energy worth \$2.5 million. Actually, only a small fraction of mass is converted to energy in nuclear reactions. But even that fraction contains a great potential for doing work, as demonstrated by the energy production of nuclear power plants and the destructiveness of nuclear weapons. Power plants employ a controlled conversion of matter to energy over a long time span, while nuclear weapons are designed to use an uncontrolled reaction to convert matter to energy in an instant.

27B Objectives

After completing this section, I can

✓ state the two postulates of Einstein's special theory of relativity.

✓ explain why the Lorentz coordinate transformations are necessary, according to special relativity.

✓ define *relativistic speed.*

✓ state four predictions of special relativity and briefly describe their consequences.

✓ describe the relativistic factor and explain its significance at typical and relativistic speeds.

✓ perform velocity additions at relativistic speeds.

✓ discuss the conditions for which the equation $E = mc^2$ is true.

✓ briefly describe two experimental evidences for the validity of special relativity.

✓ discuss how special relativity applies to the use of nuclear energy.

27B Section Review

1. What are the two postulates of the special theory of relativity?
2. If two spaceships are approaching each other at $0.5c$, and one ship shines a laser beam at the other, how fast is the light in the laser beam moving when measured by the second ship?
3. Describe four predictions made by the special theory of relativity.
4. Explain why it is impossible for physical objects to attain the speed of light in a vacuum.
5. The *New Horizons* space probe is the fastest-moving man-made object, traveling through space at about 17.69 km/s relative to the sun.
   a. At what fraction of c is this spacecraft moving?
   b. Does *New Horizons* experience significant relativistic effects?
6. A meteoroid with a rest mass of 1.00 g is moving at $0.8c$ in an observer's reference frame. What is the mass of the meteoroid according to the observer?
7. In a test of Einstein's special theory of relativity, a space probe traveling at $0.65c$ relative to the earth fires a projectile at a speed of $0.70c$ relative to the probe in the same direction as the probe's movement. What is the projectile's speed as measured from the earth?
8. a. If the 5.5 g mass of a US quarter were converted completely into energy, how many joules would be produced?
   b. How many meters could this energy move a railway car against a 10 000 N rolling friction force if all the energy were converted into mechanical work?

27C GENERAL RELATIVITY

27.19 The Limitation of Special Relativity

Special relativity is *special* because it deals only with inertial reference frames. Actually, no known reference frame is completely free from acceleration. The velocity of the earth continually changes as it revolves around the sun. Every point on the earth's surface is accelerated as it rotates around the earth's axis. Inertial reference frames are usually defined as those that do not accelerate with respect to the earth. It is difficult, if not impossible, to define a truly inertial reference frame. Fortunately, Einstein was able to extend his special theory to all reference frames, accelerated as well as inertial.

27.20 Gravity = Acceleration

The **general theory of relativity** mainly states that acceleration and gravity are indistinguishable. For example, neither a voyager in a spaceship far from any significant mass nor an astronaut in free fall in orbit feels any gravitational force. The space traveler is not in a gravitational field. The astronaut is experiencing unrestrained acceleration (no normal force). It is not possible for a person to tell which is his condition if he cannot observe his surroundings.

As another example, consider the distant spaceship accelerating at a rate of $9.81\ m/s^2$. The astronaut standing on the deck experiences the same weight that he would if he were standing on the earth's surface. If he were to release a ball inside the ship, it would appear to accelerate toward the deck. No mechanical experiment

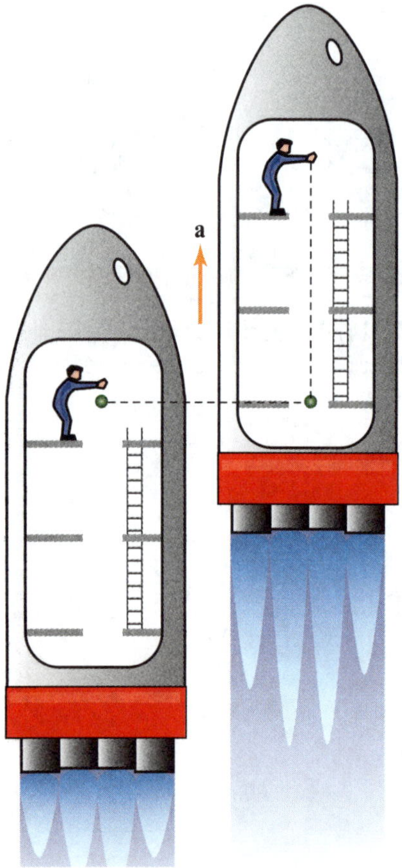

27-15 Acceleration makes a ball appear to fall.

that he could perform inside the spaceship would unambiguously reveal whether he was accelerating in free space or at rest on the earth.

An interesting outgrowth of this example is gravity's effect on light. If the space traveler shines a flashlight on the far wall of his accelerating spaceship, the light beam will seem to be deflected. The light beam initially has an upward speed equal to the spaceship's speed as the light leaves the flashlight. The ship's speed increases before the light reaches the opposite wall, so the ship travels farther upward than does the light in that time. Therefore, the light strikes the wall at a lower level than the place from which it started. This effect is also evidence for the particle theory of light. It would be harder to explain if light consisted of trains of light waves.

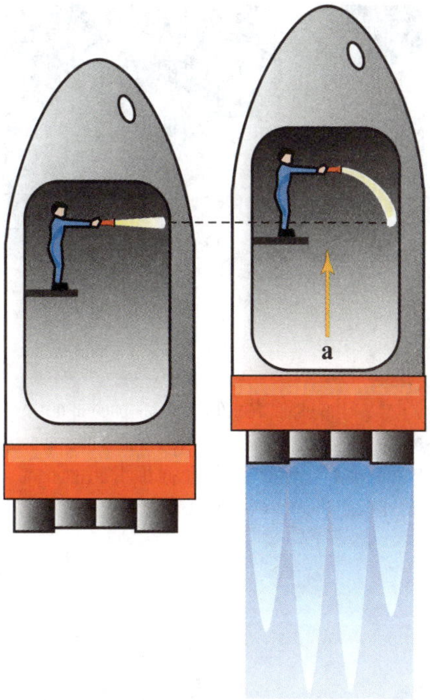

27-16 Acceleration makes a light beam bend.

27.21 Gravitational Lensing

General relativity also predicts that gravity will attract light. The earth's gravity is too small and too restricted in space to have a noticeable effect on the fast-moving light waves. However, the effect of the sun's gravity is observable on starlight. Starlight coming past the sun is deflected; therefore, stars near the limb of the sun appear to be in the wrong places. Most of the time you cannot see light from stars near the sun because the sun's light is too intense, but during a solar eclipse these stars become visible. Observations of stars during eclipses have confirmed this prediction.

Further proof of this prediction of the general theory was the discovery, through deep-space photography using the Hubble Space Telescope, of the phenomenon called **gravitational lensing**. Massive distant galaxies contain enough matter to deflect the light of even more distant objects beyond them that would not be directly visible from the earth. Distorted images of the farther object appear at small angular separations from the intervening galaxy, as light rays on opposite sides of the galaxy are bent toward the earth. Figure 27-17 shows an example of gravitational lensing.

27-17 Hubble photograph of gravitational lensing by a galaxy of an even more distant galaxy (arrow)

27.22 Black Holes

A strong gravitational field may even prevent light from escaping it. A body with such a strong gravitational field is called a **black hole**. Black holes are thought to be the remnants of super-massive stars that collapse due to exhaustion of hydrogen and the cessation of the normal fusion processes that power a star. It is theorized that as the star violently collapses, its matter is crushed into a mass of subatomic particles. Astronomers believe that the shock wave of rebounding gases from this hyper-dense core blasts away the outer layers of the star in a supernova. The remaining matter is so dense and massive that not even light can escape. Since no light ever leaves a black hole, it cannot be observed directly. However, their intense gravitational fields accelerate ionized dust and gases for great distances, generating tell-tale electromagnetic signatures that are observable. Detection of such electromagnetic evidence, as well as the wobble of companion stars, has all but confirmed the existence of these strange objects. Some astronomers believe that super-massive black holes are the central gravitational anchors that hold galaxies together.

27-18 X-ray photograph of the torus dust ring surrounding a suspected super-massive black hole at the center of the spiral galaxy M77 (NGC 1068) in the constellation Cetus

27.23 Other Evidence for the General Theory

Other observations have confirmed the validity of the general theory of relativity. The planet Mercury is the planet closest to the sun in our solar system. The

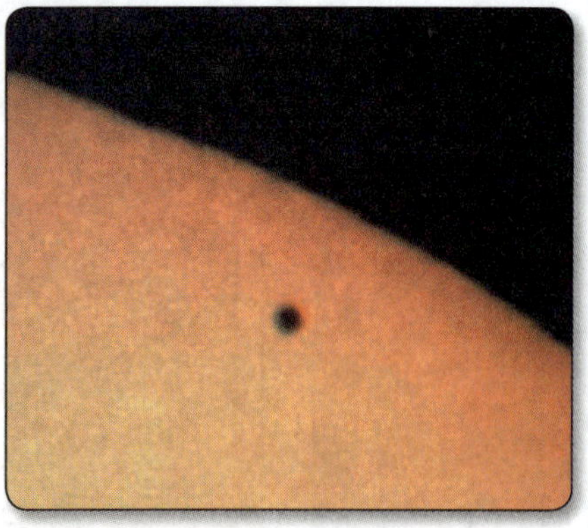

27-19 The planet Mercury photographed in a transit of the sun. Precession of Mercury's orbit is due, in part, to the sun's immense gravitational field, according to the general theory of relativity.

alignment of the major axis of its elliptical orbit changes (precesses) about 1/6 of a degree per century. The majority of this precession is due to the gravitational effects of the other inner planets, but a small amount of the angular change could not be explained without resorting to the general theory of relativity.

The general theory predicts that clocks in stronger gravitational fields run slower than in weaker fields. Therefore, your watch will be fast after completing a transcontinental trip in an airplane compared to what it would have been if you had taken a car. For the same reason, clocks in the basement of a skyscraper run slower than clocks on the roof. Atomic clocks have verified this prediction.

Gravity causes wavelengths of light to shift depending on whether the light is leaving a strong gravitational body or falling toward it. Outgoing light experiences a **gravitational red shift** as its wavelengths are stretched by gravity, while inbound light shows a *gravitational blue shift* as its wavelengths are shortened. Gravitational red shift was first observed in 1925 and later confirmed by experiment in 1959.

27.24 Using Relativity to Solve Problems

Review the dominion science introduction to this chapter. The atomic clocks in GPS satellites orbit at a distance of $1\frac{1}{2}$ earth diameters from the earth's surface, while the clocks in GPS receivers are at or very near the surface of the earth. Because the earth's gravity is very different at these two locations, general relativity predicts that these clocks will "tick" at different rates. The principle of trilateration positioning requires that the GPS receiver accurately compute the distance (r) to a satellite by using the transit time for the radio signal traveling at the speed of light ($r = c\Delta t$). The transit time is determined by subtracting the transmission time from the receipt time of the coded signal. According to general relativity, GPS satellite clocks should gain 38 microseconds, or 38 000 nanoseconds, per day compared to GPS clocks on the ground. That error is far too great to enable positions to be determined with the requisite accuracy. An accumulation of errors like this would quickly render GPS positions absolutely useless for navigation. What can be done?

27-20 Surveyors using signal-corrected GPS instruments can determine survey points to within a few millimeters of the actual position.

Physicists designed the GPS satellites so that their atomic clocks would essentially match the atomic clocks on the earth's surface. Before launch, the satellites' clocks are programmed to "tick" slower to counter the effects of general relativity once they are in orbit. In addition, satellite clocks receive frequent program updates so that they remain synchronized with coordinated universal time (UTC) at the earth's surface. GPS receivers also have built-in microcomputers that perform relativistic corrections to the time signals. Consequently, users can have accurate real-time position data with accuracies of 5–10 m for basic receivers, and millimeter accuracy for precision surveying.

27C Objectives

After completing this section, I can

- ✓ state the limitations of special relativity and the need for a general theory of relativity.
- ✓ show the equivalence of gravity and accelerated reference frames.
- ✓ describe several phenomena that exhibit the effects of general relativity.
- ✓ discuss how general relativity must be accounted for in the operation of the global positioning system (GPS).

27C Section Review

1. What is the main restriction on the special theory of relativity?
2. Discuss the main point of Einstein's general theory of relativity.
3. If you were completely enclosed in a windowless cubicle, what unambiguous experiment could you perform to prove that you were at rest on the earth rather than accelerating at g through space?
4. Discuss three predictions of the general theory of relativity.
5. **a.** What is the primary observational property of a black hole?
   b. How do scientists believe black holes can be detected?
6. **a.** What is a gravitational red shift?
   b. How was this prediction of the general theory of relativity confirmed?

DS7. How does the operation of the GPS demonstrate the workability of the general theory of relativity?

DS8. Considering that clock adjustments are necessary for GPS satellites (located between the earth and the moon) to synchronize with earth GPS receivers for positional accuracy, what implications does this have for precision communications between the earth and probes to the moon, Mars, Mercury, or interstellar space? Would such corrections be needed for space probes?

FACETS of PHYSICS

Gravitational Red Shift

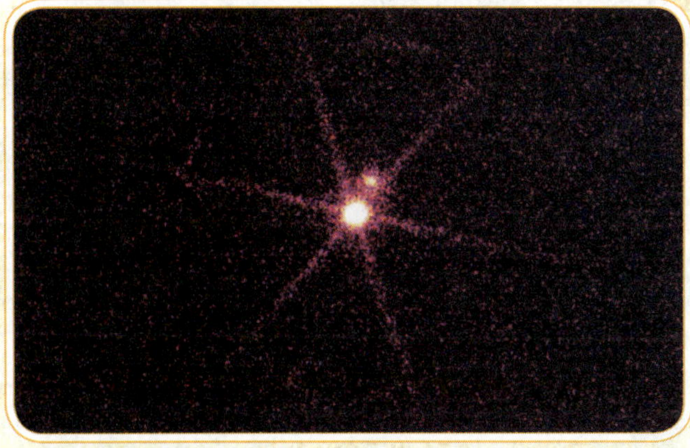

One conclusion reached from Einstein's general theory of relativity is that gravity affects light. As light passes through a gravitational field, its path changes slightly, and some of its energy is expended in escaping the field. Therefore, the light emerges from the field with less energy.

As you learned in the last chapter, light's energy is related to its frequency by the equation

$$E = hf,$$

where h is Planck's constant. Light with high energy has a high frequency, and light with low energy has a low frequency. Therefore, light's frequency decreases after it passes through a gravitational field. Because the lowest frequency in which light is visible is in the red band, this change, predicted by Einstein, is called Einsteinian red shift to distinguish it from Doppler shift.

Only a very strong gravitational field removes enough energy to cause a noticeable red shift. Observers first detected this shift in light traveling past Sirius B, an extremely dense star known as a white dwarf.

Gravitational red shift is presently used to detect features of the universe with unusually strong gravitational fields. Astronomers believe that they have identified black holes and other exotic objects in space by the way they affect light.

Chapter Review

In Terms of Physics

relativity	620
Galilean relativity	620
Galilean transformation	622
inertial reference frame	622
luminiferous ether	622
special theory of relativity	625
Lorentz transformation	626
relativistic	626
dilation	627
relativistic factor (γ)	629
proper length (l_0)	629
rest mass (m_0)	629
rest-mass energy (E_0)	632
clock paradox	632
general theory of relativity	634
gravitational lensing	635
black hole	635
gravitational red shift	636

Problem-Solving Strategies

27.1 (page 622) Time in Galilean transformations is always assumed to be the same in every reference frame:

$$t = t'$$

27.2 (page 622) Note that accelerations may occur within inertial reference frames, but the reference frames themselves cannot be accelerating.

27.3 (page 629) Remember that γ is always greater than 1.

27.4 (page 631) Relativistic velocity additions always yield velocities that are less than or equal to the speed of light. If you obtain a total summed speed greater than c, you can be sure that you have made a mistake.

Review Questions

1. What is relative when one is considering relativity?
2. What does a Galilean transformation do?
3. Describe an inertial reference frame.
4. **a.** What theory did the Michelson-Morley experiment test?
   b. Did it confirm the theory?
5. What are the assumptions of special relativity?
6. Suppose you and a friend are in identical spacecraft traveling at $0.9c$ with respect to each other.
   a. If you can see your friend's clock, will you say that it runs faster, slower, or the same as your clock?
   b. Will your friend say that your clock is faster, slower, or the same as his clock?
   c. You measure your friend's spacecraft and find it about half as long as yours. Will your friend agree with this measurement?
7. What mathematical factor accounts for the differences in mass, length, velocity, and elapsed time of an object traveling at a relativistic speed when observed from a stationary frame of reference?
8. Give two examples of an experiment that supports a prediction of the special theory of relativity.
9. What properties of a system are observed to be affected by relativistic speeds?
10. What is "special" about special relativity?
11. What is the main postulate of general relativity?
12. Discuss the effects of gravitational red shift on light.
13. Give two examples where light is affected by extreme gravity.

True or False (14–25)

14. The concept of relativity, that physical laws can be the same in different reference frames moving with respect to each other, was invented by Einstein.
15. If you throw a baseball while riding a bicycle, the ball's speed relative to the ground is the sum of the bicycle's speed and the speed at which you threw the ball.
16. Relativity means that all motion can be described relative to one absolute frame of reference.
17. Matter and energy are interchangeable.
18. Two events that occur simultaneously in one inertial reference frame will be observed to be simultaneous in all inertial reference frames.
19. You could get a few extra minutes of sleep each night if you could charter a very fast vehicle with a sleeping compartment that would return you to your house in time to go to school.
20. You and a friend are traveling in the same very fast spacecraft. You face each other so that one of you faces the front and the other faces the back. Each of you would appear to be like flattened photographs to the other.
21. If all of the subatomic particles in the universe were separated, the universe would have more mass than it does now.
22. There is no experimental difference between occupying a position in a gravitational field and experiencing an equal acceleration in a place with negligible gravity.
23. Since light bends under a high acceleration or gravity, it must experience inertia. Objects that exhibit inertia have mass. Therefore, light must have mass.
24. Black holes cannot be directly observed, but their presence is believed to produce observable effects.
25. The light we see coming from a very massive star is more blue than the light emitted at the star's surface.

✪26. You are riding north in a minivan that is crossing a bridge over a railroad track. A freight train engine is passing underneath in a perpendicular direction from right to left at the same moment. (You wave at the engineer.) The van is moving at 10.0 m/s and the train is moving at 7.0 m/s. If the bridge is 12 m above the track, write the Galilean transformation equations of the minivan's position coordinates with respect to the train's frame of reference (let the train's coordinates be primed).

✪27. A rocket traveling at $0.75c$ with respect to the earth launches a probe at a speed of $0.75c$ in the same direction with respect to the rocket. What is the probe's speed with respect to the earth?

✪28. You are in a spacecraft moving at $0.90c$. The deck you are standing on is perpendicular to the direction of motion. How tall would you appear to an observer in a stationary space station? Be sure to include your rest length (height) as well as your relativistic length in your answer.

✪29. A spacecraft with a rest length of 20.0 m shoots by you at $0.250c$.
   a. How long does the vessel appear to you?
   b. According to your stopwatch, how long does it take to go by?

✪30. A linear particle accelerator is a long, tube-like device that contributes to the study of nuclear matter by accelerating subatomic particles to high velocities and smashing them into target nuclei. One such linear accelerator is 3.0 km long. How long does it appear to an electron moving at $0.999\,999c$?

DS✪31. Write an in-depth report about how the GPS works.

A "stadium" of iron atoms on a substrate of copper atoms. The structure was created by moving individual atoms into position with a scanning-tunneling electron microscope (STM). The image was obtained using the same instrument.

Quantum Physics

28

Dominion Science Problem

Information Security

When was the last time you made an Internet purchase? Did you even consider how your personal information is kept secure? Your credit card number is kept safe by digital *encryption*, which is related to the principles of cryptography. *Cryptography* is the ancient art of communicating between a sender and a recipient in a way that cannot be easily deciphered by an unauthorized third party. The message text (the *plaintext*) is altered in some way, according to a formula or algorithm known only to the sender and recipient, that makes the actual meaning of the message unintelligible to others (the *ciphertext*). The basis for scrambling the message is known as the *cipher key*. The effectiveness of the encryption is completely dependent on the complexity of the encryption algorithm, the length of the cipher key, and the difficulty in figuring out the key. Banks, businesses, governments, and particularly the military, have expended immense resources to develop encryption techniques for secure communications.

In this age of computers, however, even the most complex algorithms and secure cipher keys potentially can be unraveled by supercomputers that test millions or billions of trial keys per second to see if anything intelligible can be gleaned from the intercepted information. The war of cryptography has seemingly become an endless race between those who develop more sophisticated algorithms to encrypt messages and those who use computers to try to break the encryption. All personal information stored or transmitted digitally over publicly accessible data circuits is vulnerable. How can we make information more secure?

“Ì5_YÛ d © ØaP´‡□,³2ÉõYn’□ie†ö
B-w-ÝVWj°µB3e+•¯»ûûÝî½¿™Àmjÿ
èø¹□ß½ç÷:ßß9¿ó»wtÄ□cÇ1h□â\Û¤
Üüè¥ßÛXÔÊ¾, hüÁ&Š” ˜Ö’çœÏ®û
FÔ□ØìJ/øpÖ§ã]MÊ Ò…:Æ26»€±“0°|
Ìeb¿•_ª~´@èp?¥WâÜ˜ •Ë□qÚrþG
±Ã)3›xÄÌ¦Öœ ÊòZa>ZPnÆð»,çvlÖY
Ã~K•““öÞøÊÈ>s□¡Ø•çÌ “□ü•ß

28-1 This is an example of information from a portion of an image that has been encrypted in ordinary text.

28A QUANTUM THEORY

28.1 Blackbody Radiation

As we mentioned in the previous chapter, most physicists near the end of the nineteenth century believed that few major discoveries remained to be made in physics. The accepted theories in mechanics, thermodynamics, and electromagnetics, now called *classical physics*, appeared to be able to explain all experimental facts. Physics, it seemed, was complete. Then, all at once, some facts came to light that challenged many basic assumptions of physics. Classical physics, which had seemed nearly perfect, failed. One area that classical physics failed to explain was blackbody radiation.

Every object whose temperature is above absolute zero emits thermal radiation. Some bodies are more efficient radiators than others. You will remember that a perfect emitter of thermal radiation is called a *blackbody*. Although perfect emitters do not exist, *cavities*—voids in opaque materials that have small openings to the outside—are a close approximation. The energy emitted by a cavity is called **blackbody radiation**. The rate at which energy is emitted follows **Stefan's law**,

$$S = \sigma T^4, \qquad (28.1)$$

where S is the rate of radiated energy emitted per square meter, σ is the Stefan-Boltzmann constant, and T is the cavity's absolute temperature.

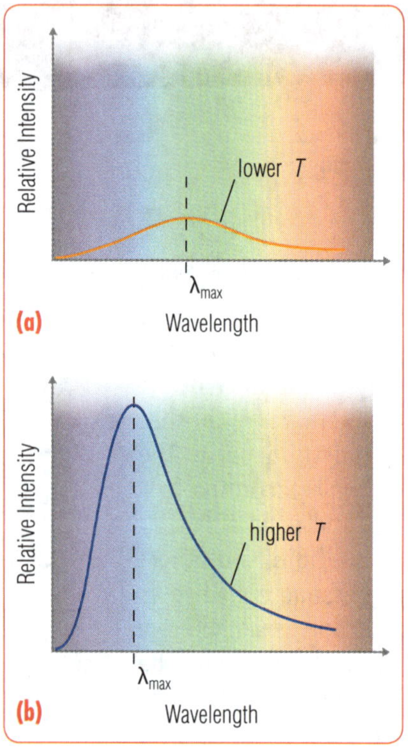

28-2 Energy emitted by a blackbody vs. wavelengths emitted at (a) low temperature; (b) high temperature

28.2 Wien's Displacement Law

A blackbody cavity emits a continuous spectrum of radiation that includes radiation at all frequencies. At higher temperatures the cavity emits more of its radiation at higher frequencies with shorter wavelengths. Figure 28-2 shows the relative intensity of energy emitted at each wavelength. Figure (a) is for a lower absolute temperature than figure (b). The symbol λ_{max} represents the wavelengths at which the most energy is emitted.

In some ways the predictions of classical physics were correct. For instance, *Wien's displacement law* says that the most intense radiant energy emitted by a cavity progresses to shorter wavelengths as the cavity's temperature rises:

$$\lambda_{max} T = 2.8978 \times 10^{-3}\ \text{m·K} \qquad (28.2)$$

Since frequency increases as wavelength decreases, Wien's displacement law predicts the experimental finding that Figure 28-2 shows.

28.3 The Rayleigh-Jeans Function

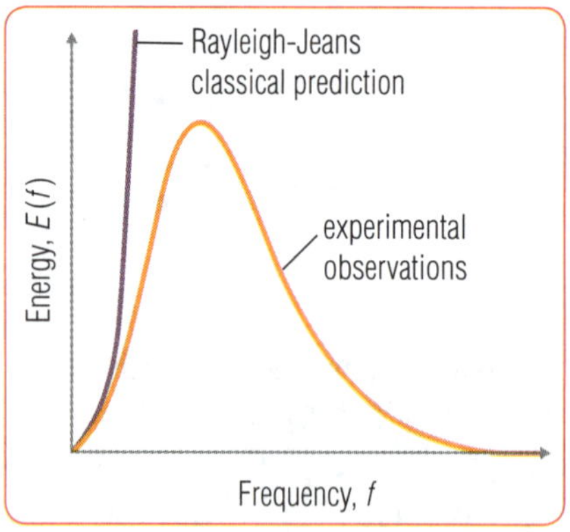

28-3 At higher frequencies, the Rayleigh-Jeans function does not agree with the observed energies and frequencies emitted from a blackbody at a given temperature.

Another formula based on classical physics, developed by Lord Rayleigh and James Jeans, was much less successful. The function below is their attempt to find the relationship between frequency and the energy emitted at that frequency from a blackbody radiator at a given temperature. If $E(f)$ is the emitted energy as a function of frequency, the Rayleigh-Jeans function is

$$E(f) = \frac{8\pi f^2 kT}{c^3}. \qquad (28.3)$$

The symbol k is *Boltzmann's constant*, and c is the speed of light. All other quantities in Equation 28.3 are constant at a given

Wilhelm Wien (1864–1928) was a German physicist who won the Nobel Prize in Physics in 1911.

According to Wien's displacement law, the wavelength of the most intense radiant energy emitted by a blackbody radiator is inversely proportional to the radiator's Kelvin temperature.

Sir James H. Jeans (1877–1946) was an English astronomer, physicist, and mathematician.

The constant k in Equation 28.3 is called Boltzmann's constant, where $k = 1.381 \times 10^{-23}$ J/K. You were first introduced to this constant in Chapter 14 in connection with the universal gas constant (R). Ludwig Boltzmann, who developed the statistical basis for thermodynamics in the nineteenth century, showed that the probability of a molecule being in any of two states was a function of its energy, which was proportional to its absolute temperature. The proportionality constant in that function is Boltzmann's constant. Boltzmann's constant is considered to be one of the four fundamental constants of physics, along with G, c, and h.

Max Planck (1858–1947)

Modern theories about the atom's structure stem from the quantization of energy. Though just a little more than one hundred years old, this idea has come to dominate all branches of modern physics. Many investigators have added to our knowledge of quanta, but most of the credit must go to the man who first introduced the idea, Max Planck.

Planck's specialty was thermodynamics. After studying physics and mathematics in Munich and Berlin, he presented his doctoral thesis on thermodynamics. He later became a professor of theoretical physics in Berlin, where he served for thirty-seven years.

Planck's interest was drawn to the problem of blackbody radiation, the thermal radiation that a completely black cavity emits. He tried to solve it with the classic principles of thermodynamics. However, the only equation that agreed with his experiments required a radical new theory of physics. On October 19, 1900, he announced that the energy of radiation is proportional to its frequency ($E = hf$, where h is a constant named after Planck). To explain his theory, he proposed that blackbodies emit radiation in "packets" of energy, called quanta.

He wondered whether his equation was an accurate reflection of nature or merely a product of artificial manipulations. However, physicists began to solve many other problems with the quantum theory. Einstein explained the photoelectric effect and even suggested a light quantum particle, later called the *photon*. Niels Bohr, using quantum theory, gave structure to the electrons surrounding Rutherford's atomic nucleus. Planck was awarded the Nobel Prize in Physics in 1918 for his introduction of quantum theory.

Although his discovery revolutionized physics, Planck himself remained a conservative physicist. Throughout his life he was reluctant to accept his own concept of quanta in blackbodies, and he spoke out against its application in other fields. He belonged to an older generation of physicists who supported classical theories and feared that their young followers were chasing unsound new ideas.

Kelvin temperature. Figure 28-3 shows that the function is only partly successful in its attempt to describe the energy profile for a blackbody radiator's spectrum. The function provides a close fit to the observed curve for low frequencies. However, as the figure shows, Equation 28.3 predicts that the amount of energy emitted at higher frequencies increases without limit. This prediction is not only untrue but also preposterous. The failure of this function, which is based on classical electromagnetics and thermodynamics, is called the **ultraviolet catastrophe**.

28.4 Planck and Quantized Energy

Because classical physics could not explain the results of experiments with blackbody radiation, Max Planck proposed "out of desperation" (in his words) a nonclassical equation that agreed with experimentation:

$$E(f) = \frac{8\pi f^2}{c^3} \cdot \frac{1}{e^{hf/kT} - 1} \tag{28.4}$$

Equation 28.4 is called *Planck's blackbody spectrum.* It produces a curve extremely close to experimental observations. As noted in previous chapters, the symbol h is *Planck's constant*,

$$h = 6.626 \times 10^{-34}\ \text{J·s},$$

The kind of desperation expressed by Planck as he developed a workable model of energy versus frequency that agreed with empirical data is a classic example of how scientific revolutions are disruptive to the existing scientific paradigm (see the Thomas Kuhn facet in Chapter 1 on page 8). Kuhn also notes that older scientists, used to the former paradigm, are almost never comfortable working within the new one, and some never make the switch.

and k is Boltzmann's constant. To derive Equation 28.4, Planck assumed that energy is emitted as *quanta*, or certain discrete values proportional to the frequency of the radiation. These allowed values are given by the equation

$$E = nhf, \text{ where } n = 1, 2, 3, \ldots \qquad (28.5)$$

The idea that energy is not continuous but can take on only certain values seemed unreasonable to classical physicists. Waves, in classical theory, could have any energy. To assume that radiant energy is **quantized** (emitted as quanta) implies that it originates not as waves but as particles. Planck himself had trouble accepting the quantum hypothesis. However, Planck's equation agreed well with experiments. It was therefore eventually accepted that energy is quantized.

28.5 Photons and the Photoelectric Effect

Planck intended his idea of quanta to apply only to the *emission* of energy, but Albert Einstein expanded the idea to energy in transit. Light, Einstein said, travels in packets of definite amounts of energy, which he called light quanta, but which later were called **photons**, each with an energy of hf. This extension of Planck's quantum theory addressed another of the failures of classical physics—the photoelectric effect.

Light, especially ultraviolet light, falling on a polished metal surface can dislodge electrons from the metal. This is called the **photoelectric effect**, and the electrons so dislodged are called *photoelectrons*. According to classical theory, a light wave falls on the metal and distributes its energy over the surface of the metal. When an electron has absorbed enough energy to overcome the metal's bonds, it leaves the metal. If it has absorbed more energy than it needs to break the metal's bonds, the electron acquires kinetic energy. From the classical theory you would expect that

1. any frequency of light can liberate electrons if it is intense enough;
2. there is a measurable time between the instant that the light strikes the metal and the first electrons' ejections;
3. a more intense light gives electrons more energy.

28.6 Photoelectric Work Function

The quantum theory of Planck and Einstein, on the other hand, explains the photoelectric effect in terms of photons. As light strikes a metal target, a few photons will probably strike electrons. However, it is unlikely that more than one photon will strike the same electron at the same time. Therefore, each electron receives an energy of hf. If this energy is greater than the **work function** that an electron must gain to escape the metal, the electron leaves the metal and acquires some kinetic energy. If the energy is less than the electron's work function, the electron remains in the metal and soon dissipates its extra energy. From this theory, you would expect that

1. the frequency of light must be such that hf is greater than the work function for any electrons to be liberated;
2. there may be no time between the instant the light strikes the metal and the first electrons' ejections;
3. a more intense light ejects more electrons at the same maximum energy.

28.7 Photoelectric Cutoff Frequency

To investigate the photoelectric effect, experimenters used cathode-ray tubes. Electrons are emitted when light falls on a polished cathode. The number of

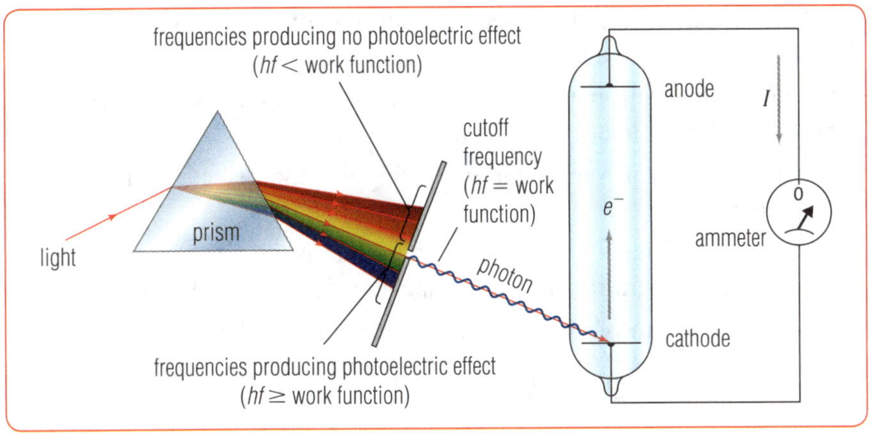

28-4 Light at or above the cutoff frequency removes electrons from a polished metal surface.

photoelectrons is measured by the current through the tube. Experiments showed that the current is zero for light of frequencies below a **cutoff frequency**, which depends only on the cathode material. Light below this frequency, regardless of its intensity, will not liberate any electrons. Light above the cutoff frequency will liberate at least a few electrons immediately ($< 10^{-10}$ s) after striking the cathode, no matter how weak the light. These findings supported the quantum theory.

28.8 Experimental Support for the Quantum Theory

A cathode-ray tube can also be used to measure the kinetic energy of the electrons. When the tube is connected as Figure 28-5 shows, the external voltage opposes the motion of the photoelectrons. The voltage is increased until the current is zero. Then, even the most energetic electrons cannot reach the anode. A voltage V_0 will obstruct the motion of electrons with energies up to

$$E_0 = eV_0, \tag{28.6}$$

where e is the magnitude of the charge on an electron (the fundamental charge). If V_0 stops the current, then E_0 is the kinetic energy of the most energetic electrons. Experiments showed that E_0 depends only on the light's frequency, not its intensity. In other words, classical predictions were wrong, but the quantum theory agreed with experiments.

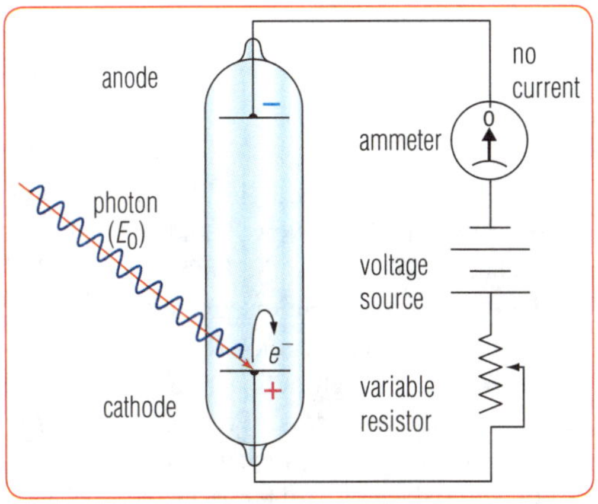

28-5 Because of the opposing voltage, none of the dislodged electrons reaches the anode.

28.9 The Compton Effect

Another experiment corroborated the quantum theory. Arthur Compton discovered some unexpected lines in an x-ray diffraction spectrum. He found that he could completely explain the lines by assuming that x-ray photons, with a momentum of E/c, collided with electrons in the diffracting crystal. Such a collision would transfer some of the x-ray's energy to an electron and change the x-ray's direction. With less energy after diffraction, the x-ray would have a lower frequency. The changes of the x-ray's frequency and direction are called the **Compton effect**. Compton's interpretation of his discovery strengthened the quantum theory.

Arthur Holly Compton (1892–1962) was an American physicist. He is considered a pioneer of high-energy physics and won the Nobel Prize in Physics in 1927.

The effective momentum of a photon with energy $E = hf$ can be found from Einstein's special relativity equation

$$E = mc^2.$$

If $E = hf = mc^2$, then the photon's momentum ($p = mv = mc$) is

$$mc = E/c.$$

28A Section Review

1. In a blackbody radiator, how is the wavelength of the most intense radiant energy related to the blackbody's Kelvin temperature?
2. What was the *ultraviolet catastrophe*?
3. Why did Max Planck doubt the validity of his spectrum equation?

28A Objectives

After completing this section, I can

✓ describe the properties of a blackbody radiator.

✓ summarize the historical development of the quantum theory by Max Planck.

✓ describe various experimental evidences for the quantum theory of energy.

4. What was one result of Planck's quantum theory that was inconsistent with the classical wave theory of electromagnetic radiation?
5. Discuss how electromagnetic radiation incident on a metal surface produces the photoelectric effect.
6. What is the critical variable in producing the photoelectric effect in a given metal?
7. According to Wien's displacement law, what is the most intense wavelength of light radiated by a blackbody at 1.80×10^3 K?
8. Assuming the sun is a blackbody radiator, what is the temperature of its photosphere if λ_{max} = 550. nm (yellow light)?

28B QUANTUM MECHANICS AND THE ATOM

28.10 Naming the New Physics

The radical new discoveries of the first three decades of the twentieth century revealed a whole new class of phenomena completely unanticipated by classical physics. Planck's model of blackbody radiation, the discovery of Compton scattering, and the unveiling of atomic structure forced scientists to develop a new model that could describe and predict the behavior of very small particles. This model came to be known as **quantum mechanics**. Over time, physicists realized that quantum mechanics could be applied to matter at all scales and at all velocities and energies, while classical mechanics was limited to dealing with matter at scales much larger than those of the building blocks of matter. Quantum mechanics thus accurately modeled phenomena that had previously baffled classical physicists.

Ernest Rutherford (1871–1937) was a physicist from New Zealand and is considered to be the Father of Nuclear Physics. He discovered that some elements become different elements through radioactive decay. He won the Nobel Prize in Chemistry in 1908 for his study of radioactive decay and the chemistry of radioactive substances.

28.11 The Electrical Atom

Before the 1890s most physicists believed that atoms were hard spheres that could not be divided. However, in 1897, J. J. Thomson discovered that the charge carriers in different conductors have the same mass-to-charge ratio (see Chapter 21). This evidence suggested that all atoms contain the same charge carriers, now called *electrons*. Since atoms are electrically neutral, Thomson suggested that an atom is a ball of positive charge containing possibly thousands of electrons scattered evenly throughout. Another British physicist, Lord Kelvin, suggested that Thomson's atomic model looked a lot like an English plum pudding filled with raisins. Thomson's model came to be known as the **plum-pudding model**.

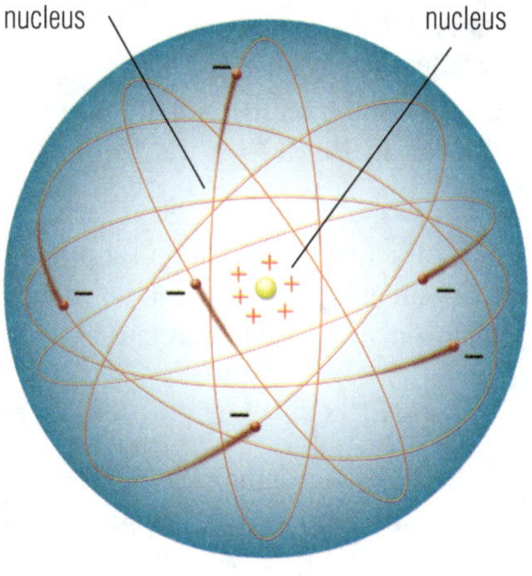

28-6 Rutherford's model of the atom

28.12 The Nuclear Atom

Thomson's atomic model was soon challenged. Ernest Rutherford studied atoms by directing alpha particles, heavy nuclear particles emitted naturally by certain atoms, at gold metal foils that were just a few atoms thick. Alpha particles are more than 7300 times as massive as electrons and have a positive charge twice as large as the electron's negative charge. According to Thomson's model, an alpha particle should be slightly deflected by the positive charge throughout the atom and perhaps by a collision with an electron. Rutherford and his research team found, however, that most of the particles were not deflected at all. Of those that were deflected, some were deflected slightly, some at large angles, and a few even bounced directly back the way they had come. These observations were inconsistent with Thomson's model, so it had to be modified.

Rutherford proposed a new atomic model based on the results of his experiment. Since most alpha particles were undeflected, he suggested that an atom is mostly empty space. The positive charge, he said, is concentrated in a small but massive **nucleus**. The electrons orbit the nucleus as planets orbit the sun. The masses of most nuclei are at least as large as the mass of an alpha particle. In fact, Rutherford later showed that an alpha particle is a helium nucleus. It is not surprising that the few alpha particles striking a massive gold nucleus bounced off like BBs hitting a bowling ball.

Rutherford's experiments indicated that the nucleus is on the order of 10^{-5} the diameter of the atom itself.

28.13 Isotopes

The nucleus is made up of two major kinds of particles, collectively called **nucleons**. The total number of nucleons in a given nucleus is its **mass number (*A*)**. One kind of nucleon is the **proton (p)**, a positively charged particle with 1836 times the mass of an electron but with the same magnitude of charge. The number of protons in a nucleus, called the **atomic number (*Z*)**, determines what element the atom is. For instance, all oxygen atoms contain eight protons in their nuclei, while gold atoms have seventy-nine protons. The other kind of nucleon is the **neutron (n)**, a neutral particle that is slightly more massive than a proton. In general, elements with small atomic numbers have almost equal numbers of neutrons and protons. In the nuclei of the heaviest elements, the ratio can be as high as about 1.5 neutrons per proton.

Atoms of the same element but with different mass numbers (different numbers of neutrons) are called **isotopes** of the element. The *isotopic symbol* for an element displays the mass number at the top of the element symbol and may include its atomic number at the bottom as well (see Figure 28-7). For instance, carbon (C) has six protons ($Z = 6$). One isotopic form of carbon has six neutrons, so its mass number is twelve ($A = 12$), and the isotope is called carbon-12. Another carbon isotope has eight neutrons ($A = 14$), so it is called carbon-14.

Another term for describing different combinations of protons and neutrons in a nucleus is **nuclide**. While *isotope* and *nuclide* are often used interchangeably, we will use *isotope* when discussing different atomic species of the same element and *nuclide* when discussing isotopes of different elements. For example, oxygen-18, carbon-14, and nitrogen-16 are all radioactive nuclides.

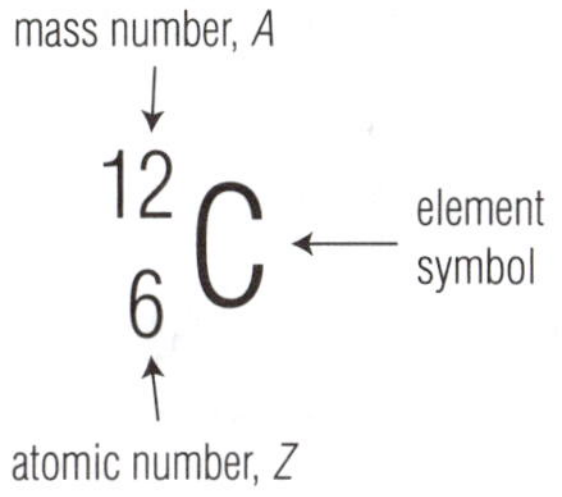

28-7 Isotopic notation used for distinguishing different nuclear forms of an element

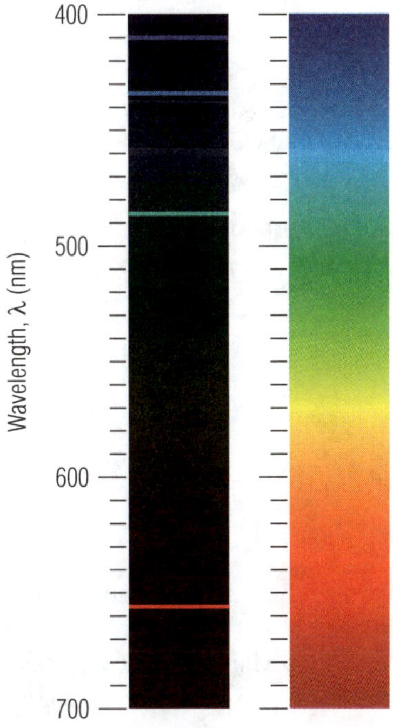

28-8 Line spectrum of hydrogen gas compared to a continuous spectrum

28.14 The Bohr Atomic Model

Although Rutherford's atomic model was useful, it did have a serious fault. Electrons (imagined to be tiny BBs) orbiting a nucleus are continuously accelerating because their velocities are continually changing direction. In classical electromagnetic theory, Maxwell demonstrated that accelerating charges radiate electromagnetic energy. As electrons radiate electromagnetic energy, they lose total energy, which is a combination of their kinetic energy of motion and potential energy relative to the nucleus. If they continuously lose energy as they orbit, they will slow down and "lose altitude," eventually spiraling into the nucleus. Thus, the atom collapses. However, atoms do *not* collapse. Therefore, Rutherford's theory needed to be modified.

Another problem with Rutherford's theory was its failure to account for the energy observed to be emitted by a low-pressure, ionized gas at high temperature. Unlike other glowing materials, the gas does not emit a continuous spectrum of light. Instead, each gas emits certain characteristic frequencies of EM energy. Since these frequencies appear in a spectrograph as lines, a gas is said to emit a **line spectrum**. According to Rutherford's model, an electron may emit radiation at any frequency. His model does not explain why low-pressure, glowing gases emit only a few characteristic frequencies.

The lifetime of an atom with a continuously spiraling electron would be about 10^{-11} s.

28.15 Energy Levels

Both of these problems—that electrons do not continuously radiate and the existence of line spectra (implying that atoms *do* radiate specific energies under certain circumstances)—led Niels Bohr to propose a new atomic model. Bohr postulated that electrons are restricted to certain circular orbits characteristic of an atom of a given element. The radii of these orbits were functions of an integer number. In other words, the orbit radii were quantized. He further postulated that when electrons remain in these allowed orbits, they do not radiate. The energy in each of these orbits was also quantized by the same integer numbers.

The energy of an electron is measured in **electron volts (eV)**—the energy required to move a fundamental charge through a potential difference of 1 volt. The ground-state energy of hydrogen's single energy level is 213.6 eV.

The allowed electron orbits in Bohr's model are properly called **energy levels** and represent levels of energy of an electron. When all electrons in an atom are at their lowest energy states, the atom is considered to be in its **ground state**. Usually, atoms have energies greater than this condition and are in an **excited state**. Electrons in the energy level nearest the nucleus have the least energy (E_1). (Similar to gravitational potential, E_1 is a relatively large negative number.) Electrons in energy levels farther away from the nucleus have more energy (smaller negative numbers). The energy levels are designated by integer numbers from the ground state, $n = 1$, to a theoretical orbit so far from the nucleus that electrons have zero potential energy with respect to the nucleus ($n = \infty$). The energy levels become closer together as n increases. If an electron's energy is positive with respect to a nucleus, it is not bound to that nucleus.

Recall that the potential energies of electron energy levels are negative and $E_{n-x} < E_n$. Equation 28.8 will always yield a positive number for energy levels associated with the same nucleus.

Each energy level is associated with an energy E_n. When an electron moves from a level with greater energy (E_n) to a level of less energy (E_{n-x}), it emits a single photon with an energy of

$$E_{\text{phot}} = hf = -(E_{n-x} - E_n). \tag{28.7}$$

The frequency of the emitted photon can be found by solving Equation 28.7 for f:

$$f = -\frac{E_{n-x} - E_n}{h} \tag{28.8}$$

For a photon emission,

$$E_2 - E_1 = E_{\text{phot}}.$$

For an absorption,

$$E_1 + E_{\text{phot}} = E_2.$$

Now consider an electron moving from an energy of E_n to a higher energy of E_{n+x}. To accomplish this move, the electron must gain an energy of

$$\Delta E = E_{n+x} - E_n. \tag{28.9}$$

Equation 28.9 implies that the electron must absorb a photon of $hf = \Delta E$ in order to move to the higher energy level.

28.16 Evidence that Supports the Bohr Model

Hydrogen-like atoms include atomic hydrogen, singly ionized helium, and doubly ionized lithium. Each of these atoms has only a single outer electron.

Bohr's atomic model explained the line spectra of hydrogen-like gases (gases containing atoms with only one electron). By extension, he proposed that each atom in a gas has energy levels in which electrons have different energies. To absorb or radiate, an electron must move from one level to another. The difference in energy between these levels determines the frequency of the radiation absorbed or emitted. Since all atoms of a gas have the same energy levels, each gas can radiate only certain frequencies that are related to the differences between those levels. At low gas pressures, the photons from the different atoms of a given gas do not interfere to form other frequencies, so only the frequencies that would be emitted by any single atom are detected. Different elements have energy levels with different energies and thus give off light of different frequencies.

There are several lines in a spectrum because several transitions are possible. For instance, an electron in the fourth level may fall to the third level, the second level, or the first level. Photons produced by any transition *to* a given level have frequencies of the same order of magnitude. A group of frequencies caused by a change to the same level is called a **spectral series**. For instance, in hydrogen-like atoms, transitions to the ground state produce frequencies in the ultraviolet portion of the spectrum and are part of the **Lyman series**. Transitions from $n > 2$ to the $n = 2$ energy level produce visible light and are part of the **Balmer series**.

Theodore Lyman (1874–1954) was an American physicist.

Johann Jakob Balmer (1825–98) was a Swiss mathematician and physicist.

For an electron to move to a higher energy level, it must absorb a specific amount of energy. The highest energy associated with a given nucleus occurs when $n = \infty$. This energy level has been assigned a value of 0 eV. With greater energy, the electron is free from any atom. The bound electrons have lower, and therefore negative, energy. The energy required to remove an electron from the nth energy level of an atom is

$$\Delta E = E_\infty - E_n, \text{ so}$$

$$\Delta E = -E_n. \qquad (28.10)$$

If this electron is the one in the highest energy level (a *valence* electron) of a neutral, ground-state atom, then $-E_n$ is called the **first ionization energy** of the atom—the energy needed to remove the most loosely held electron from an atom in its ground state.

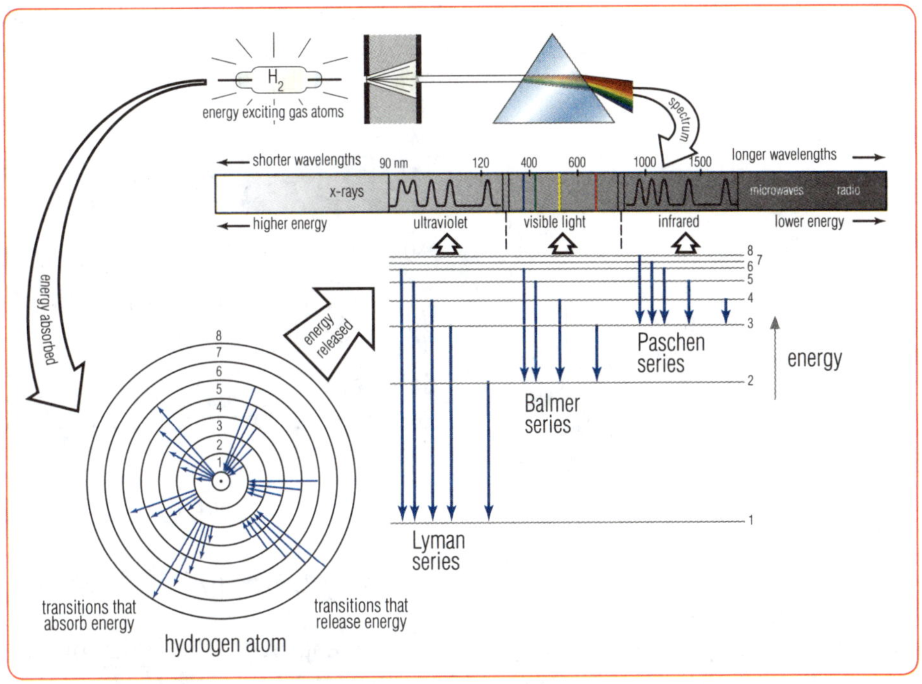

28-9 Producing a line spectrum that includes lines from the Lyman, Balmer, and Paschen series

Bohr's model says that atomic electron energy is quantized, just as Planck and Einstein said that electromagnetic energy is quantized. However, some predictions of the classical theory (which assumes that energy is continuous) were also correct. How could both theories make correct predictions? The *correspondence principle* states that new theories must give the same results as old theories where the old theories were supported by experiments. Bohr's model and classical theories give the same results for the outermost energy levels of large atoms. In the Bohr atomic model, the energy levels get closer together as n increases. For large n the levels are so close together that the energy appears to be continuous. Thus, when n is large, Bohr's theory and the classical theory predict the same results.

28.17 X-ray Spectra

When atoms, especially ones with large atomic numbers, are placed in the path of high-speed electrons, high-energy, high-frequency radiation in the x-ray portion of the spectrum is emitted. Figure 28-10 shows two types of x-rays. The lower, smooth, continuous curve represents photons emitted as the colliding electrons slow down within the cloud of electrons surrounding the atom as they pass through the atom. This radiation is called *bremsstrahlung*. Occasionally, the colliding electrons remove ground-state electrons from the atom. Since electrons in an atom tend to seek the lowest possible energy level, an electron in a higher level ($n \geq 2$) will promptly fill the vacancy in the ground state. In doing so, the electron emits an extra-high-energy x-ray. The frequencies of the x-ray line spectra are characteristic of the bombarded, or target, atom. These characteristic lines are superimposed on the normal continuous spectrum.

Bremsstrahlung is a compound German word meaning *braking* and *radiation*. The *braking* refers to the slowing down of the electrons as they fly through an atom.

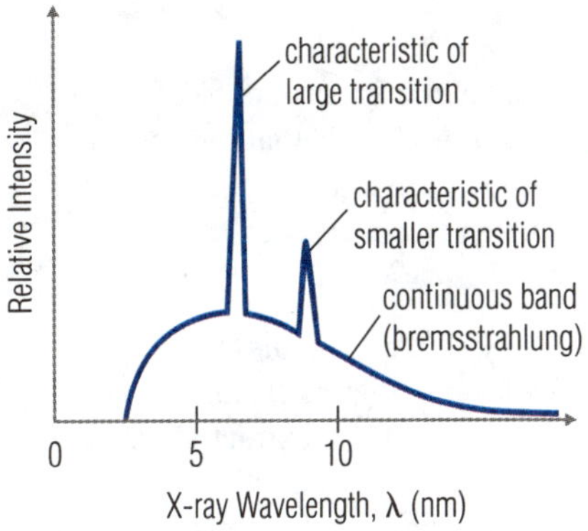

28-10 An x-ray spectrum resulting from a specific energy of colliding electrons

Bremsstrahlung radiation is the opposite process to the photoelectric effect. In the former, energy is added to an atom by an energetic electron, and it emits a photon. In the latter, a photon is absorbed by an atom, which then emits an electron.

28.18 Particles as Waves

If waves (like light) can be considered to be particles (like photons) under some circumstances, why can't particles act like waves at times? This question caused Louis de Broglie to calculate what a *matter wave* would be like. He predicted that the wavelength of a matter wave would be

$$\lambda = \frac{h}{mv}. \qquad (28.11)$$

Louis de Broglie (1892–1987) was a French physicist who won the Nobel Prize in Physics in 1929 for his discovery of the wave nature of electrons.

This wavelength is so small for ordinary-sized objects that it cannot be detected by diffraction effects. Diffraction appears only when a wave passes an obstacle whose width is nearly the wavelength of the wave. The smallest obstacles that can be used are atoms, which have radii on the order of 10^{-10} m. Matter wavelengths are much smaller than this.

EXAMPLE 28-1

Wave Runner: Calculating a Matter Wavelength

Calculate the wavelength of a 75 kg athlete running at a speed of 5.0 m/s.

Solution:

You are given the runner's speed and his mass. Use Equation 28.11:

$$\lambda = \frac{h}{mv} = \frac{6.626 \times 10^{-34}\ \text{J}\cdot\text{s}}{(75\ \text{kg})(5.0\ \text{m/s})}$$

$$\lambda \cong 1.76 \times 10^{-36}\ \frac{(\text{kg}\cdot\text{m}^2/\text{s}^2)\cdot\text{s}}{\text{kg}\cdot\text{m}/\text{s}} \cong 1.8 \times 10^{-36}\ \text{m}$$

A wavelength on the order of 10^{-36} m is undetectable by any known instrument.

An undetectable wave is not very useful. However, matter waves are undetectable only for ordinary-sized objects. The waves associated with electrons, for instance, are detectable. An electron traveling at 3 000 000 m/s ($10^{-2}c$) has a wavelength of about 2×10^{-10} m. This is near x-ray wavelength, so, like x-rays, electron waves should be diffracted by a crystal. Davisson and Germer tested this idea by directing high-speed electrons at a nickel crystal. They found a pattern identical to that caused by a wavelength in the x-ray range.

Clinton J. Davisson (1881–1958) was an American physicist who discovered the diffraction of electrons by crystals and shared the Nobel Prize in Physics in 1937 with George P. Thomson.

Lester Halbert Germer (1896–1971) was an American physicist who assisted Clinton Davisson in confirming Louis de Broglie's hypothesis about the wave nature of electrons. He also studied thermionics and surface chemistry.

28.19 Schrödinger Wave Equation

Because waves are sets of time-dependent positions rather than a single point, they are described by functions. Erwin Schrödinger developed an equation that uses the *wave function*, ψ, of an electron's wave. For one-dimensional components of a wave, this equation is

$$-\frac{\hbar^2}{2m}\cdot\frac{\partial^2\psi}{\partial x^2} + U\psi = i\hbar\frac{\partial\psi}{\partial t}, \qquad (28.12)$$

where $\hbar = h/2\pi$, m is the electron's mass, U is the electron's potential energy function, i is the complex root ($\sqrt{-1}$), and the entire right side is the total energy function. The term $\partial^2\psi/\partial x^2$ is used in calculus to describe change in the wave's motion with respect to the x-axis. Similarly, the term $\partial\psi/\partial t$ describes the rate of change of some aspect of the wave with time. This equation is a *differential*

The term $h/2\pi$ occurs so often in quantum physics that the special symbol $\hbar$ is used instead. In spoken words, it is called "h-bar."

Many of the symbols in Equation 28.12 are what mathematicians call *operators*—they extract values from the associated quantities, just like the sine operator extracts a dimensionless number from an angle.

equation. That means its solution is itself a function rather than a point or two. Equations such as this one require such advanced mathematics to solve for even simple cases that most scientists use computers to solve them. However, *Schrödinger's wave equation* gives some useful information about an electron whose total and potential energies are known.

The function ψ for an electron's wave gives the probable location of the particle. The electron can be thought of as being somewhere along the waveform. The electron is more likely to be where the wave has a high amplitude than where it has a low amplitude. It will not be where the wave's amplitude is zero. The waveform is therefore called a *probability wave.*

The relationship between the wave and the electron's location is used to conceptually explain the existence of allowed orbits in atoms, though the 3D spatial model is far more complicated. Using the Bohr model, if an orbit's circumference is a whole number of wavelengths, the wave will reinforce itself around the orbit. The electron is likely to be in that orbit. If, however, the orbit's circumference is not a whole number of wavelengths, the wave will destructively interfere and cancel itself. The electron is unlikely to be there. Thus, Bohr orbits are those orbits that can contain a whole-number multiple of the electron's wave function.

According to modern theory, the electrons' paths around a nucleus are not as clear-cut as in Bohr's model. There are three-dimensional regions of greater probability of finding an electron at the distance of the Bohr orbits. However, it is possible to find the electrons outside of Bohr orbits—even near the nucleus or completely free from the atom. Electron locations are thought to be better described as **orbitals** than as clear-cut orbits. An orbital is a subregion of an energy level in which a single electron or pair of electrons may be found. The region of space occupied by individual orbitals may have a variety of shapes, depending on their electrons' energies within the energy level. The sum total of all possible electron orbitals in an atom produces an indistinct region around the atom called the *electron cloud.*

Erwin Schrödinger (1887–1961) was an Austrian physicist who developed the equation that describes wave-like behavior of subatomic particles, the basic equation of quantum physics. With Paul Dirac, he shared the Nobel Prize in Physics in 1933.

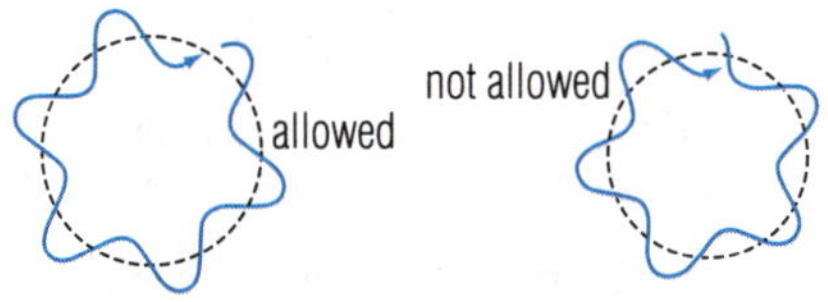

28-11 These diagrams illustrate the idea that an electron orbit at a certain distance (and energy) from the nucleus can accommodate only certain electron wave functions. If the wave functions cannot fit in at a particular energy level, the corresponding orbit cannot exist.

28.20 Heisenberg Uncertainty Principle

Physicists are forced to deal with probability waves because they cannot locate an electron and know its momentum at the same time. Werner Heisenberg showed that this limitation is a result of making an observation of subnuclear objects rather than due to imperfections of instruments. To observe an object, at least one photon must be reflected or diffracted by it. Many photons may be reflected by ordinary objects without noticeably changing the object or its motion. However, an electron's motion is changed unpredictably by an interaction with a single photon. If you set up an experiment to find an electron's location, the interaction with a photon will change its momentum so that you cannot predict where the electron will be later. Heisenberg showed that even under ideal conditions, the product of the uncertainties in measuring the electron's position and momentum is greater than $\hbar/2$. As the uncertainty in position decreases (the measurement becomes more precise), the uncertainty in momentum increases. This concept is called the **Heisenberg uncertainty principle**.

Werner Karl Heisenberg (1901–76) was a German physicist who won the Nobel Prize in Physics in 1932 for the development of quantum physics and the principle of indeterminacy.

28.21 Wave-Particle Duality

The uncertainty principle illustrates the principle of **complementarity**: the concepts of waves and particles are both needed to describe reality. In the measurement of an electron's position, the electron acts as a wave until a photon strikes it. Then, like a particle, it is moved by the collision. Both concepts are required to explain observations.

28B Objectives

After completing this section, I can

- ✓ summarize the historical development of the atomic model.
- ✓ describe the properties of the atomic model at different points in history.
- ✓ identify the key nuclear particles and state their properties.
- ✓ define isotopes of elements and identify all parts of the isotopic notation.
- ✓ discuss the relationship between line spectra and the Bohr model of the atom.
- ✓ describe electron energy levels and explain how their energies compare with each other.
- ✓ discuss evidences for the Bohr model of the atom.
- ✓ describe the wave-particle nature of all matter.
- ✓ explain how the wave-particle nature of matter contributed to the development of the quantum model of the atom.

28B Section Review

1. What experiment conclusively proved that atoms are *not* a solid matrix of positive charge with embedded negative charges?
2. Approximately how large is an atom's nucleus compared to the atom itself?
3. What are isotopes of an element?
4. What key prediction of classical electromagnetism was violated in Rutherford's planetary nuclear model of the atom?
5. Why do atoms radiate specific wavelengths of EM radiation under certain situations?
6. a. What energy transitions produce the Lyman series of line spectra in hydrogen-like atoms?
   b. In what portion of the electromagnetic spectrum do these emissions fall?
7. What causes bremsstrahlung radiation?
8. What did the combination of de Broglie's and Schrödinger's theories do to the classical particle theory of matter?
9. ✪ What is the matter wavelength of the *New Horizons* spacecraft (m = 479 kg) traveling at 17.69 km/s?

28C MODERN ATOMIC MODELS

28.22 The Quantum Numbers

Just as the equation $y = 3x + 2$ has ordered pairs of numbers as unique solutions, such as (2, 8), so the Schrödinger wave equation can be solved to give four **quantum numbers** that uniquely specify each electron's energy in an atom. The **principal quantum number (*n*)** (also called the *energy level number*) stands for the probable distance of the electron from the nucleus. As you learned before,

$$n = 1, 2, 3, 4, \ldots$$

and it identifies the principal energy level that the electron occupies.

The **azimuthal** (or sublevel) **quantum number (*l*)** refers to the angular momentum of the electron. Its possible values depend on n:

$$l = 0, 1, 2, \ldots, n - 1$$

In other words, for every energy level n, there are n azimuthal numbers.

The **magnetic** (or orbital) **quantum number (*m*)** relates to the electron's orientation as detected by a magnetic field. The possible values are

$$m = -l, -(l - 1), \ldots, 0, \ldots, l - 1, l.$$

The **spin quantum number (m_s)** stands for the direction of the electron's spin. It has two possible values:

$$m_s = +\tfrac{1}{2} \text{ or } -\tfrac{1}{2}$$

An analogy may help you to understand these numbers. The n is like the distance at which a planet orbits a sun. The number l is analogous to a planet's angular momentum, mvr, as it orbits (which is related to its distance from the sun and its energy). The m-number shows the orientation of the plane of the planet's orbit.

Each quantum number has a maximum number of allowable values:

n: no limit

l: n (for each n)

m: $2l + 1$ (for each sublevel l)

m_s: 2 (for each m)

In nuclear physics parlance, positive spin is "up," and negative spin is "down." Try to remember that these words do not have their strictly physical meanings in this context.

For $m = 0$, the orbital plane contains the sun's equator. For other values of m, the planet's orbit may be symmetrical about a position axis not centered on the sun, or in some other configuration (this picture does not fit the planetary analogy very well). Finally, m_s gives the direction in which the planet rotates about its own axis—its spin. For the closest orbit, the planet must orbit at one speed, and it has only one possible orbital plane. Each farther orbit has more freedom. This analogy is not perfect, but it may help you to understand quantum numbers.

EXAMPLE 28-2

Using Quantum Numbers

What values of l are allowed if $n = 5$?

Solution:

The allowable numbers are $l = 0, 1, 2, \ldots, n - 1$. Therefore,

$$l = 0, 1, 2, 3, 4 \text{ (five numbers)}.$$

EXAMPLE 28-3

By the Numbers: Sets of Quantum Numbers

List all possible quantum numbers for an electron in the second energy level ($n = 2$) and the second sublevel ($l = 1$).

Solution:

$n = 2$

$l = 1$

For $l = 1$, $m = -1$, 0, or $+1$

$m_s = +\frac{1}{2}$ or $-\frac{1}{2}$

Sets:

$2, 1, -1, +\frac{1}{2}$ $\quad$ $2, 1, -1, -\frac{1}{2}$

$2, 1, 0, +\frac{1}{2}$ $\quad$ $2, 1, 0, -\frac{1}{2}$

$2, 1, +1, +\frac{1}{2}$ $\quad$ $2, 1, +1, -\frac{1}{2}$

This means that there are six possible electron energies in the second sublevel of the second energy level.

Problem-Solving Strategy 28.1

Follow these steps to determine all of the possible quantum numbers for a given energy level:

1. List all the azimuthal numbers for the associated principal quantum number. The number of possible sublevels equals the numerical value of the energy level.
2. For each azimuthal number, determine the possible orbital numbers. There are $2l + 1$ orbitals for each l number.
3. For every combination of n, l, and m, there are two possible spin numbers.

28.23 Electron Ordering Using Quantum Numbers

The idea that an electron responds to magnetic fields as though it were spinning about its own axis was demonstrated by the Stern-Gerlach experiment in 1922. In this experiment, neutral silver atoms were propelled through a narrow rectangular collimating window into an asymmetrical magnetic field and then deposited on a metal screen. Classical electromagnetic theory predicted that a single spot would form. Instead, the

(a) (b)

28-12 Target deposition pattern in the Stern-Gerlach experiment that demonstrated the existence of electron spin. (a) Pattern predicted by classical theory; (b) Observed pattern

Otto Stern (1888–1969) was a German physicist whose work demonstrated the wave nature of atoms and molecules. He won the Nobel Prize in Physics in 1943.

Walther Gerlach (1889–1979) was a German physicist.

FACETS of PHYSICS

Electron Microscope

When Leeuwenhoek used refracted light to magnify unseen "animalcules," the scientific world was astounded. For the next two centuries glass lenses were continually refined in an effort to increase the resolving power of the light microscope. The improved lenses focused more light on the specimen and spread the refracted light farther, thus increasing the magnification. Smaller and smaller structures were discovered. But by the turn of the twentieth century, one weakness in the light microscope prevented scientists from making further improvements: a wave of light cannot resolve anything smaller than its own wavelength of about 10^{-7} m.

Theoretically, electrons, which have wavelengths as short as 10^{-10} m, could replace light because electrons can also be refracted. But no one was able to create a practical lens until 1933, when Ernst Ruska developed the first successful electron microscope. Ruska used a magnetic "lens" rather than a glass lens to focus and spread the stream of electrons. The refracted electrons could not be viewed directly because, unlike light, they are invisible to the human eye. The image must be photographed, projected onto a fluorescent viewing screen, or reproduced electronically.

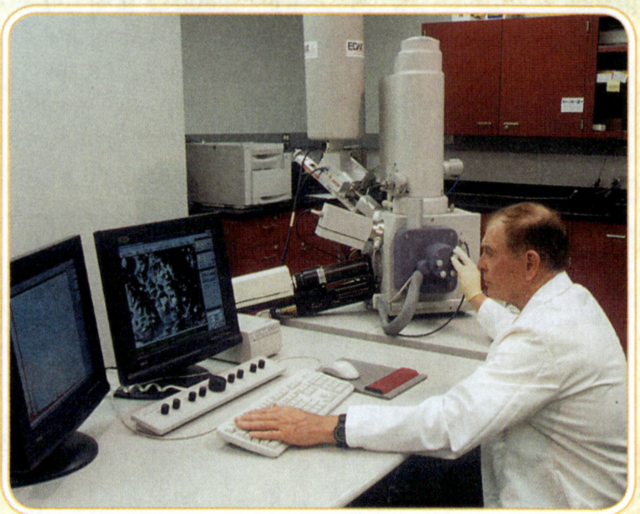

An electron microscope

Electron microscopes provide a wealth of information about ultra-small structures. The *transmission microscope*, which passes electrons through the specimen, can view large molecules such as DNA and even large atoms. Metallurgists gain valuable knowledge about crystals and metal surfaces, and medical laboratories examine viruses, bacteria, and pollen grains. But live specimens must be killed, treated, and sliced thinner than 1 micron (10^{-6} m) before the transmission microscope will work.

experimenters observed the pattern shown in Figure 28-12b, indicating that the atoms had two different magnetic moments. After considering the evidence and performing additional experiments, scientists realized that the *electrons* in the atoms had two different magnetic moments, which were analogous to spin in opposite directions.

Electrons seek the lowest possible energy level. However, even non-energetic atoms have electrons in energy levels above the lowest possible energy level. This phenomenon is explained by the **Pauli exclusion principle**, which states that no two electrons in the same atom can have identical sets of quantum numbers. For an atom with a single energy level in its ground state, $n = 1$, so the remaining electron quantum numbers are either $l = 0$, $m = 0$, and $m_s = +\frac{1}{2}$ or $l = 0$, $m = 0$, and $m_s = -\frac{1}{2}$. There are only two possible sets of quantum numbers for the ground-state electrons. Therefore, the lowest energy level in the ground state can hold only two electrons.

Wolfgang Ernst Pauli (1900–58) was an Austrian physicist who made significant contributions to quantum theory. He was first to predict the existence of the neutrino. He won the Nobel Prize in Physics in 1945 for his exclusion principle.

Electrons occupy the sublevel with the lowest energy that is open to them. In the planet analogy, a planet in a smaller orbit with a greater angular momentum may have more energy than a planet in a larger orbit with less angular momentum. Similarly, an electron in the fifth energy level with a high l-number may have more energy than an electron in the sixth energy level with a low l-number. The azimuthal numbers are given the name *sublevels*. For convenience, the sublevels are labeled *s*, *p*, *d*, *f*, *g*, *h*, and so on for $l = 0, 1, 2, 3, 4, 5, \ldots$, respectively. Each principal energy level has "room" for the number of sublevels allowed by the number of l-values for that energy level number (n). For example, the fifth energy

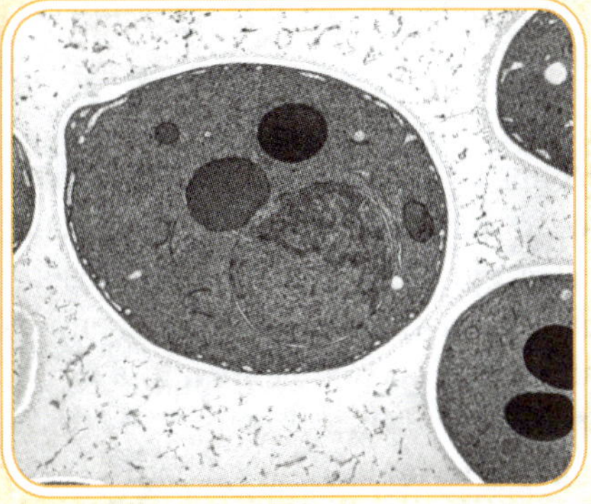

A transmission electron micrograph of a yeast cell

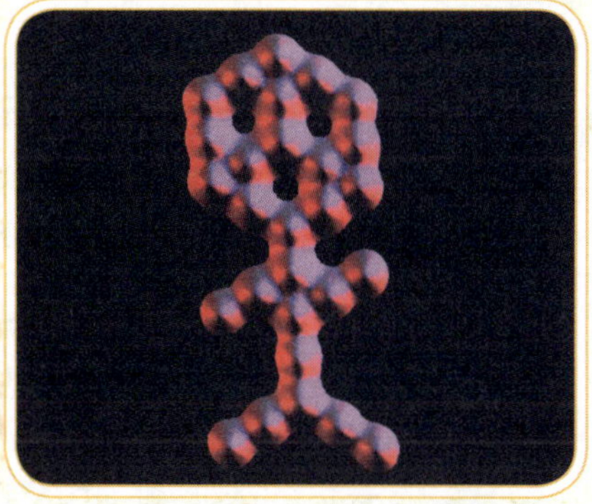

"Carbon monoxide man." An STM image of carbon monoxide molecules on a platinum substrate

A helpful supplement to this type of electron microscope is the *scanning electron microscope (SEM)*, which reflects an electron beam off the surface of the specimen. Specimens are much easier to prepare, and they do not need to be sliced up. Minuscule structures do not have to be removed from large specimens, such as flies, mosquitoes, and butterflies. Some studies simply show the beauty and economy of God's creation, while others reveal practical details about the human body that could not be observed before. The inner ear, taste buds, and the color cones inside the eye are only three examples.

Even newer kinds of microscopes use the ability of electrons to "tunnel" across a vacuum gap between two charged surfaces. No image is formed directly by the electrons. Instead, a charged needle is slowly moved over the prepared surface of a specimen at extremely small distances. As the distance varies between the specimen and the needle, the current of electrons that "tunnel" across the gap varies in proportion to the distance. The current signal is stored by the digital memory of the instrument along with the position of the needle. An image of the surface is reconstructed by plotting the tunneling distance versus position. With such a *scanning tunneling electron microscope (STM)*, the surfaces of small atoms have been indirectly imaged.

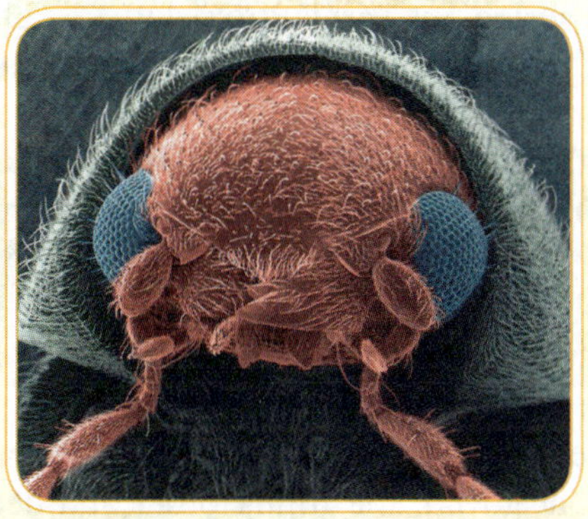

A scanning electron micrograph of the head of a cigar beetle

level ($n = 5$) has 5 allowable azimuthal (l) numbers. Therefore, sublevels s, p, d, f, and g may exist in the fifth energy level if the element's atomic number is large enough.

Sublevels are identified by their principal quantum number and the sublevel letter. The fourth sublevel (f) of the fifth energy level (5) is denoted 5f. Similarly, the first sublevel of the second energy level is denoted 2s. The number of electrons actually present in the sublevel is denoted by a superscript on the sublevel letter. For instance, the presence of two electrons in 5f is denoted $5f^2$ (spoken, "five-eff-two"). Figure 28-13 provides a guide that orders the energies of the sublevels from least energy to greatest. The energies of electrons occupying the sublevels of atoms in their ground state can be generally ordered using this guide, though there are some exceptions.

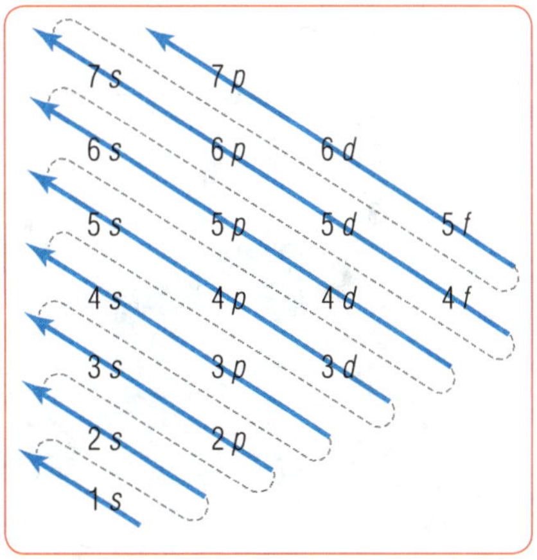

28-13 The arrows go from lower-energy to higher-energy sublevels.

Each type of sublevel can hold a definite number of electrons. The s sublevel ($l = 0$), for instance, can hold two; $l = 0$, $m = 0$, $m_s = \frac{1}{2}$ and $l = 0$, $m = 0$, $m_s = -\frac{1}{2}$. The total number of electrons that each sublevel can contain is

$$2(2l + 1)\,e^-.$$

The locations of electrons for a specific element are given by a string of shorthand notations, in order of energy. For instance, the eight electrons in ground-state oxygen atoms are represented by $1s^2 2s^2 2p^4$.

28.24 The Periodic Table

The arrangement of electrons in an atom determines its chemical and many of its physical properties. Particularly important are the electrons in the highest occupied energy level, the valence electrons. These electrons are the most easily removed and added. Therefore, they are the ones usually involved in chemical bonding. More than 100 years ago, scientists organized the elements into a useful table that relates repeating (periodic) similarities in both their electron structures and chemical properties. This table is called the **periodic table** (Appendix H). It represents one of the most impressive pieces of evidence that the basic building blocks of matter are designed.

The periodic table is arranged so that element atomic numbers increase in rows, called *periods* or *series*, from left to right. Interestingly, the elements in each vertical column, or chemical *group*, usually have the same number of valence electrons. For this reason, the elements in a group typically have similar chemical properties. The periods of the table are numbered according to the highest occupied energy level (when the atom is in its ground state).

Figure 28-14 is a reproduction of the periodic table, omitting the element symbols. Instead, the blocks indicate the sublevel where each electron is added with increasing atomic number. Referring back to Figure 28-13, you can see that sublevels are not filled from lowest to highest in order. Rather, certain sublevels are filled in higher energy levels before other sublevels in lower energy levels. The rules that govern the filling order and determine valence electron structure are normally introduced in high-school chemistry courses. You should note that in each period, the valence electrons normally occupy the *s* and *p* sublevels.

The term *filling* refers to the way electrons are added to the electronic structure of atoms of elements arranged in a tabular fashion, such as in the periodic table (see Appendix H). Such an order assumes that the atoms are in their lowest energy state. In an actual atom, individual electrons will occupy allowed sublevels according to their excited energies. For a highly energetic atom, lower sublevels may be briefly vacated as electrons move from one energy level to another.

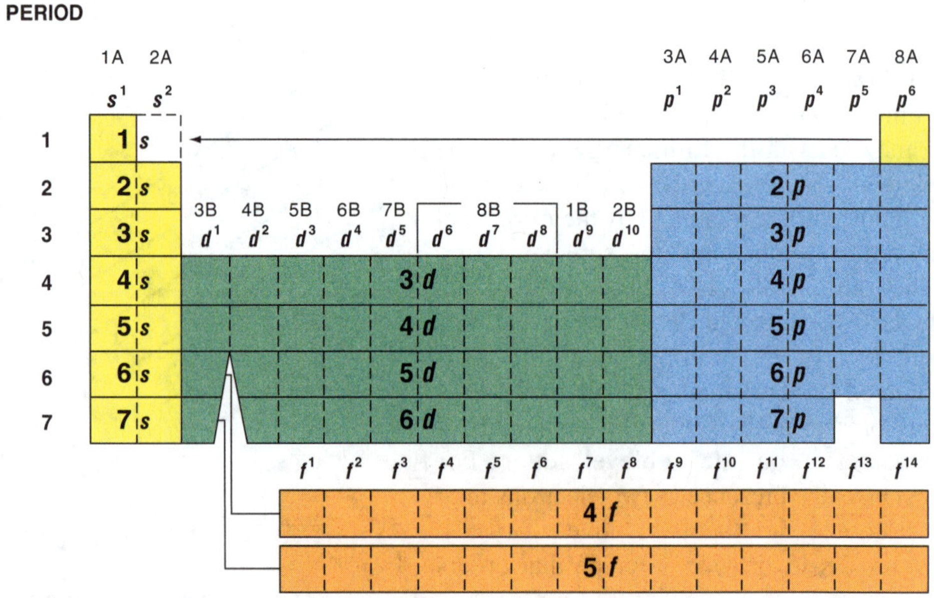

28-14 The periodic table reflects the arrangements of electrons within atoms by their position on the table.

28.25 Applications of Quantum Physics

Masers and Lasers

The quantization of energy provides a way to get *coherent*—single phase and frequency—radiant energy. Most radiant energy sources emit photons of different frequencies. If one or more electrons in each atom in a material could be raised to the same energy level and then stimulated to emit photons at the same time, the result would be a coherent emission. This condition does not readily occur.

In the middle of the twentieth century, Charles Townes discovered a practical way to produce nearly coherent emissions. A photon passing an excited (unusually energetic) electron stimulates the electron to emit photons in phase with the original photon. However, keeping the electrons excited until a photon passes is difficult because excited electrons tend to radiate almost immediately. It is necessary to find a relatively stable excited energy level, called a *metastable state.* Not every material is capable of establishing metastable states. Those that do tend to be crystalline or molecular materials that have bonding orbitals that hold electrons for relatively long times. The electrons are "pumped" to an energy slightly greater than the metastable energy. They immediately fall to the metastable state, where they remain briefly until a photon passes. Then, they emit a photon in phase with the original photon. If the process of pumping more electrons continues, and the photon beam is allowed to bounce back and forth through the material, two things happen: more electrons radiate photons, and the photons constructively interfere to produce a single, high-intensity beam of one (i.e., monochromatic) wavelength. The result is strong, coherent radiant energy. Townes called his device, which emitted microwaves, a **maser** (*m*icrowave *a*mplification by *s*timulated *e*mission of *r*adiation). An optical maser, emitting light, is called a *laser* (see the facet on lasers in Chapter 25).

Charles Hard Townes (1915–) is a brilliant American physicist who pioneered the studies of the stimulated emission of coherent electromagnetic energy. He received the Nobel Prize in Physics in 1964 for his groundbreaking work.

Masers are often used as amplifiers for radio astronomy—astronomy that analyzes celestial low-frequency radiation, such as radio waves. Hydrogen masers are used as extremely stable short-term time standards for scientific research and for the calibration of national time standards. Masers occur naturally in the gas nebulae surrounding certain stars and galaxies.

Semiconductors

Quantum mechanics provides the best explanation to date for conduction in solid materials. As you know, we can classify solids into conductors, insulators, and semiconductors, according to their conductive properties. Electrons fill orbitals in atoms according to their energies. Each atomic orbital occupies a discrete allowable energy according to quantum mechanics. When atoms bond in solids, their valence electron orbitals merge with the orbitals of the other atoms. No two bonding orbitals can have the same energy. For very large numbers of bonded atoms, the energies of these bonding orbitals are so close together that they form a *band* of energies. The valence electrons occupy the highest energy bands, collectively called the *valence band.* Some spans of energies are unoccupied by electrons at the ground state of the material's atoms. These regions are called *band gaps.* If these band gaps cannot be occupied even at elevated material energies, they are called *forbidden gaps.* In some materials under certain conditions, valence electrons can acquire enough energy to jump from the bonding orbitals and move freely among the atoms. These *unlocalized electrons* have kinetic energies that fall in the *conduction energy band* of the material, if it has one.

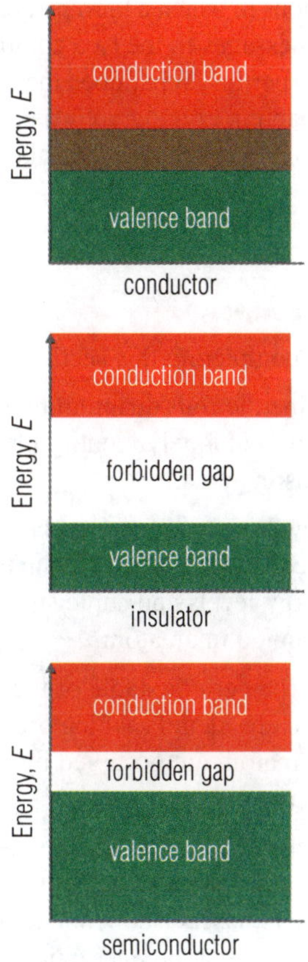

28-15 The band theory of conduction

In conductors, the valence energy band overlaps the conduction band. Electrons can easily move from one orbital to another and from one atom to another if they are given a little energy. There is no forbidden band between the valence energy band and the conduction band. In an insulator, the valence energy band is full of electrons. The combined interaction of the electron waves reduces the allowable energy states significantly. Therefore, a large forbidden gap exists between the valence band and the conduction band. For an individual electron to escape, it must gain a very large amount of energy. The amount of energy necessary to cause an insulator to readily conduct electricity is usually sufficient to melt or destroy the material first.

In a semiconductor, as in an insulator, the electrons nearly fill the valence energy band. Unlike an insulator, however, the forbidden gap is smaller. It is not too difficult to give electrons enough energy to reach the next conduction energy band. Therefore, semiconductors can conduct with a moderate addition of energy.

28.26 Using Quantum Mechanics to Solve Problems

Quantum mechanics, through a relatively new field called quantum cryptography, may hold part of the solution to the increasing problem of information security. *Quantum cryptography*, the brainchild of an American named Stephen Wiesner, uses quantum mechanics to make communication absolutely secure. One prominent quantum encryption method is called *quantum key distribution (QKD)*. The communicating parties create a cipher key by transmitting a bit stream (1s and 0s) of photons through a *quantum communications channel*, such as a fiber-optic cable. Both the quantum state and the polarization of a photon are established by using a random number generator. These two photon properties are subject to quantum effects. The data stream is received, and a similar random process is used to read both the states and polarizations of the photons. If a photon's state and polarization are the same as the state of the receiver, the photon is read. If the state, the polarization, or both are different, the photon is rejected. The recipient compares his matches against what the sender actually transmitted. The key results from the unique random sequence of bits that match. The key is then used to encrypt the actual information to be transmitted over a public circuit, such as radio or the Internet. The communications are essentially unbreakable, since the key is used only once—a "one-time pad" in cryptologic parlance.

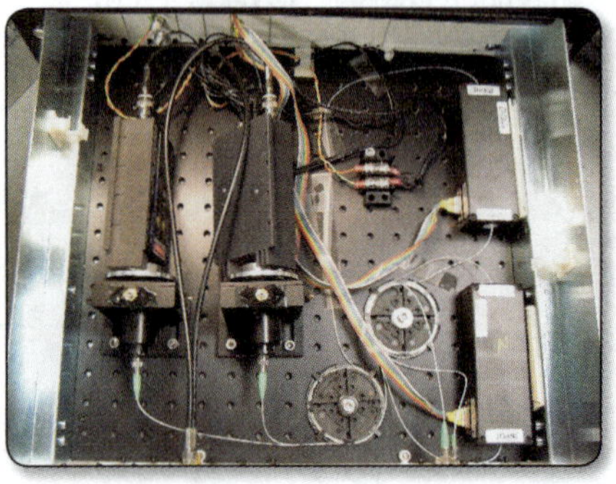

28-16 A prototype built by the National Institute of Standards and Technology (NIST) of a high-speed quantum key distribution (QKD) system for encrypting transmitted data

Cryptography and the Nature of Man

Despite its strengths, quantum cryptography is still susceptible to hacking by "man-in-the-middle" attacks, situations in which a third party sends fake messages to the two communicating parties. What does the need for quantum cryptography and cryptography in general tell us about the nature of man?

Quantum cryptography has some very attractive features for businesses and governments. One of the most important and unique features of quantum cryptography is its ability to detect eavesdroppers. If a third party attempts to intercept or read the key, one of the two communicating parties will detect errors in the decoding or encoding because of *quantum indeterminacy*, a concept that includes the Heisenberg uncertainty principle. When the third party measures the photon, its polarization and quantum state are altered from what was sent. His intrusion alerts the communicating parties to the interception and also gives them an idea as to how much of the key was read. Since no actual information was transmitted, this step guarantees the channel's security in addition to providing a key to encrypt the data.

This form of cryptography is currently still in development, though it is being used by a handful of businesses. In fact, in 2004, the mayor of Vienna, Austria, used quantum cryptography to make a crucial bank transfer that needed guaranteed security—the first time in history that quantum cryptography was used for this purpose. This development has the potential to revolutionize information exchange when absolute security is essential.

28C Objectives

After completing this section, I can

- ✓ define the four quantum numbers used to uniquely identify electrons in an atom.
- ✓ show how quantum numbers determine the order in which energy levels and sublevels are occupied in an atom.
- ✓ use the periodic table of the elements to identify which orbitals in an atom are occupied by electrons.
- ✓ discuss various applications of quantum physics, including masers, lasers, semiconductors, and information encryption.

28C Section Review

1. **a.** Describe the four quantum numbers that are solutions to the Schrödinger wave equation.
   b. What do these numbers uniquely identify?
2. Why can't all of the electrons in an atom occupy the lowest energy level?
3. Discuss why the second energy level of an atom in the second period of the periodic table might not contain all 8 electrons allowed by the quantum number rules for $n = 2$.

4. What is the outermost energy level being filled by atoms in the fourth period?
5. Which energy levels are incompletely filled in the element iron?
6. a. What is an optical maser called?
   b. What are the principal uses of masers?
7. How does quantum mechanics explain the mechanism of conduction and semiconductors?

DS8. How does quantum cryptography depend on the consequences of the Heisenberg uncertainty principle?

Chapter Review

In Terms of Physics

blackbody radiation	642
Stefan's law	642
ultraviolet catastrophe	643
quantized	644
photon	644
photoelectric effect	644
work function	644
cutoff frequency	645
Compton effect	645
quantum mechanics	646
plum-pudding model	646
nucleus	647
nucleon	647
mass number (*A*)	647
proton (p)	647
atomic number (*Z*)	647
neutron (n)	647
isotope	647
nuclide	647
line spectrum	647
energy level	648
electron volt (eV)	648
ground state	648
excited state	648
spectral series	649
Lyman series	649
Balmer series	649
first ionization energy	649
orbital	651
Heisenberg uncertainty principle	651
complementarity	651
quantum number	652
principal quantum number (n)	652
azimuthal quantum number (l)	652
magnetic quantum number (m)	652
spin quantum number (m_s)	652
Pauli exclusion principle	654
periodic table	656
maser	657

Problem-Solving Strategies

28.1 (page 653) Follow these steps to determine all of the possible quantum numbers for a given energy level:

1. List all the azimuthal numbers for the associated principal quantum number. The number of possible sublevels equals the numerical value of the energy level.
2. For each azimuthal number, determine the possible orbital numbers. There are $2l + 1$ orbitals for each l number.
3. For every combination of n, l, and m, there are two possible spin numbers.

Review Questions

1. What was the ultraviolet catastrophe?
2. What does the term *quantized* mean?
3. a. What is a photon?
   b. Why was the photon not predicted under classical physics?
4. Explain the photoelectric effect.
5. Describe Thomson's *plum-pudding* model of the atom.
6. Describe Rutherford's model of the atom.
7. Name the major subatomic particles—or nucleons—in a nucleus.
8. A nitrogen atom ($Z = 7$) has seven neutrons. What would you call this isotope?
9. What were Bohr's postulates about electrons?

10. Why do we not notice the wave nature of ordinary objects?

11. Why can scientists not locate an electron precisely and know its momentum precisely at the same time?

12. What are valence electrons?

13. Must all sublevels of an energy level be filled before any in the next level are? Explain.

True or False (14–26)

14. The power of radiant energy from a blackbody emitter is proportional to the fourth power of the emitter's Kelvin temperature.

15. According to Wien's displacement law, as the temperature of a blackbody rises, the peak frequency of the radiant spectrum is lower.

16. Planck's blackbody spectrum equation was predicted by classical physics.

17. Photoelectrons from a specific metal come only in certain energies.

18. Rutherford's experiment demonstrated that the atom is mostly empty space.

19. Elements ^{16}N and ^{16}O are both nuclides of the same isotope.

20. Line spectra are characteristic of low-pressure, glowing gases.

21. An emission line spectrum series consists of all possible wavelengths produced by electrons dropping *to* a specified energy level.

22. It is impossible to see the wave characteristic of normal forms of matter such as baseballs because their wavelengths are too long to be noticed.

23. An electron with a principal quantum number of 3 could have any of 14 combinations of quantum numbers.

24. Valence electrons are the highest-energy electrons in the atom, regardless of the energy level in which they are found.

25. Groups of elements on the periodic table have the same number of energy levels in the ground state of their atoms.

26. Coherent waves have the same amplitude and the same frequency.

✪27. What is the peak wavelength of radiation emitted from a snowball (t = 0.0 °C) after it is placed in a dark black box (i.e., no reflected light)?

✪28. Find the de Broglie wavelength of a 0.149 kg baseball traveling at 40.0 m/s.

✪29. How many electrons in one atom can have the quantum numbers $n = 1$, $l = 0$, $m = 0$, $m_s = +$?

✪30. How many electrons in one atom can have the quantum numbers $n = 2$, $l = 1$?

✪31. After conducting some research, write a paper or give an oral report on one of the following topics:
   a. the significance of Planck's constant to the order of nature
   b. the properties of a wave function
   c. a biography of one of the scientists mentioned in this chapter

DS✪32. The process for using quantum key encryption is complicated, and the equipment required is expensive. How do you think this will affect its future applications, especially for protecting personal information?

Niels Bohr (1885–1962)

At the turn of the twentieth century, the world's leading researchers concentrated on the atom and its structure. Models were constantly being proposed, accepted, proved inadequate, and reformulated. By 1913, a twenty-eight-year-old Dane named Niels Bohr proposed the model on which our present-day concept of the atom is based.

Bohr was born in Copenhagen, Denmark. His father, a professor at the University of Copenhagen, encouraged him to pursue his scientific interests and abilities. Bohr's genius became apparent when he attended the university. His doctoral thesis has become a classic on the electron theory of metals. Upon graduation, Bohr journeyed to England to work under the famous researcher J. J. Thomson, but he was soon attracted by the experiments of Ernest Rutherford.

While working in Rutherford's laboratory, Bohr became interested in the radiation emitted by hydrogen atoms. To explain what he observed, Bohr applied the recently proposed quantum theory to Rutherford's nuclear model of the atom. His radical new theory argued that atoms exist in stable energy states and that they emit or absorb radiation only during a transition between two of these states. Every scientist before that time had assumed that atoms emit energy continuously rather than in quanta. But only Bohr's new theory could explain both the spectra of light emitted by hydrogen atoms and the activity of an atom's valence electrons.

He returned to his old university to direct the newly created Institute for Theoretical Physics. In 1922, he received the Nobel Prize in Physics for his atomic model. He was already a member of the Danish Academy of Sciences and was soon elected a foreign member of the Royal Society of London. However, Bohr did not rest on his past achievements. He continued to study the atom and proposed modifications to his model, concluding that no model could ever predict the exact location of an electron at any one time. This belief sparked a lifelong dialogue between Bohr and other leading scientists such as Albert Einstein, who strongly disagreed.

When Germany overran Denmark during World War II, Bohr was forced to leave Copenhagen and seek refuge in England. He then journeyed to the United States to assist the scientists at Los Alamos to develop the atom bomb. After the war, he returned to his homeland, where, like many of the Los Alamos scientists, he became a lifelong opponent of nuclear war. Bohr is held by many to be one of the greatest thinkers in the history of science.

The interior of a commercial power plant nuclear reactor during refueling. The reactor core is visible at the bottom of the photo. Note the faint blue glow of Cherenkov radiation surrounding a spent fuel module as it is removed from the reactor.

Nuclear Physics 29

Dominion Modeling

The Sands of Time

A treasure was awaiting discovery in the Judean wilderness. In 1947, a Bedouin shepherd looking for a lost goat threw a rock into a cave near the Dead Sea—and he heard something shatter. Climbing up to investigate, he stumbled upon clay jars containing ancient scrolls and manuscript fragments. The shepherd didn't know it, but some of those scrolls were a special treasure—early copies of the Hebrew Scriptures (the Old Testament) more than a thousand years older than any other known manuscripts. Soon, more ancient scrolls were found in a whole network of caves dotting the surrounding cliffs. Many of these "Dead Sea scrolls" were sold and passed from one antiquities dealer to another, until, at immense effort and cost on the part of several individuals, the state of Israel was able to acquire them. Today, the greatest find, the Great Isaiah Scroll, is on display in Israel's Shrine of the Book.

Many scholars believe that a cultish Jewish sect called the Essenes, who occupied a local settlement during the time of Jesus, stored a library of documents in these caves above the Dead Sea. Other scholars believe that important documents were hidden by the Jews to protect them from the Roman occupiers. No one can know for sure. Whatever the reason for their presence in the caves, they were forgotten and remained hidden for eighteen centuries. Their finding coincided exactly with the rise of the state of Israel in 1948, making for an exciting, romantic story.

The Dead Sea Scrolls were analyzed by archaeological scholars, who assigned them historical dates going back as far as several centuries before the time of Christ. But how could these dates be confirmed for one of the most remarkable archaeological discoveries of the twentieth century?

29-1 One of the Qumran caves in which the Dead Sea Scrolls were discovered.

29A RADIATION AND RADIOACTIVITY

29.1 Discovery of Radiation

Many important scientific discoveries were made unintentionally. For instance, Wilhelm Roentgen was studying electricity with a cathode-ray tube when he noticed that a nearby screen coated with barium platinocyanide was **fluorescing**,

Wilhelm Roentgen (1845–1923) was a German physicist who, in 1901, won the first Nobel Prize ever awarded in physics.

or emitting visible light. Materials fluoresce only after they have absorbed energy from light or other sources. The barium platinocyanide fluoresced even in a darkened room, as long as the cathode-ray tube was operating. Roentgen concluded that the cathode-ray tube emitted some previously unknown rays that excited the materials to fluoresce. He called the rays *x-rays.*

Antoine Henri Becquerel (1852–1908) was a French physicist who won the Nobel Prize in Physics in 1903 for his discovery of spontaneous radiation in uranium.

Henri Becquerel reasoned that if x-rays *cause* fluorescence, they may also be *emitted* in fluorescence. X-rays, like visible light, can expose film. Becquerel decided to wrap a photographic plate in black paper so that no light could enter. He intended to place a fluorescent material on the wrapped plate and set the experiment in the sunshine. The sun's ultraviolet light would cause the material to fluoresce, and any x-rays produced would expose the plate through its protective wrapper.

29-2 Marie Sklodowska Curie (1867–1934) was a French physicist. She was the first female professor of physics at the Sorbonne. She won two Nobel Prizes: in 1903 she shared the prize with two others (one was her husband) for her work on the spontaneous radiation discovered by Becquerel; in 1911 she won the Nobel Prize in Chemistry for her discovery of and investigations into the chemistry of radium and polonium.

Becquerel chose uranium potassium sulfate as a fluorescent material. The day he was to set the equipment in the sunshine was cloudy, so he put the uranium and plate into a drawer until a sunny day. The next several days were also cloudy. Finally, Becquerel decided to develop the plate without putting it in the sunshine. He expected at most a weak exposure, but he found a surprisingly strong image. Investigating further, Becquerel found that uranium will emit plate-exposing radiation even if it has not been exposed to an external energy source for months or years.

The terms associated with nuclear decays are often incorrectly used. *Radiation* consists of the particles and rays emitted by an unstable nucleus. *Radioactivity* is the act of emitting such particles and rays and is measured as the time rate of radiation emission. *Radioactive* is the property describing a material that emits radiation.

Other scientists soon began studying this strange form of energy. Pierre and Marie Curie discovered two new elements in uranium ore that produced even stronger photographic exposures. They named these *polonium* and *radium*—the former after Marie's home country of Poland. They also coined the term **radioactivity** for the emission of these mysterious rays.

The short-term rates of the emission of radiation have been found to be essentially stable under normal conditions, remaining the same regardless of changes in the temperature, pressure, or chemical form of the radiator. Emission rates are found to depend only on the kind of atoms and the number present. Therefore, scientists concluded that radioactivity comes from the nucleus of an atom. A radioactive nucleus emits a particle, or electromagnetic energy, or both, and may become another element. Radioactive nuclei are therefore unstable—they are likely to change into other kinds of nuclei.

The change of one nucleus into another, which would effectively change the element, was the age-old aspiration of alchemists. Their goal was to *transmute* a "base" metal, such as lead, into precious metal, such as gold.

29.2 Cause of Radioactivity

What makes a nucleus unstable? A better question is, "What makes a nucleus stable?" You learned in earlier chapters that nuclei are composed of positively charged protons and neutral neutrons. One positive charge repels another, and the electrostatic force is one of the strongest forces in nature. So it is astounding that up to ninety-two protons can be found in naturally occurring nuclei. What holds protons together in a nucleus?

Since protons do manage to remain together in a nucleus, physicists postulated a strong nuclear interaction force that binds protons and neutrons together. This force must apply to all nucleons, regardless of charge. It must be stronger than the electrostatic repulsion of adjacent protons. Experiments have shown that this force has a range of 10^{-15} m. This distance is smaller than the size of some nuclei.

In large nuclei, the electrical repulsion between protons on opposite sides of the nucleus is stronger than the strong nuclear interaction force between the same protons. Therefore, those protons tend to repel each other. Neutrons surrounding each proton help to stabilize the nucleus by contributing to the strong nuclear interaction force but not to electrical repulsion. Figure 29-3 is a chart of all known nuclides plotted according to the numbers of protons (Z) and neutrons (n) found in their nuclei. The cells are colored by the nuclide's half-life, if it is unstable. Notice that for lighter nuclei, there are about the same number of neutrons as protons. For heavier nuclei, more neutrons are required per proton for stability. For nuclei with more than eighty-three protons, no number of neutrons has been found to provide stability. These unstable nuclei are radioactive.

The element technetium, Tc ($Z = 43$), is unusual in that, even though it has relatively few protons, it has no known stable isotopes. It decays so quickly that it is not found in nature, but is known only as a product of artificial nuclear reactions.

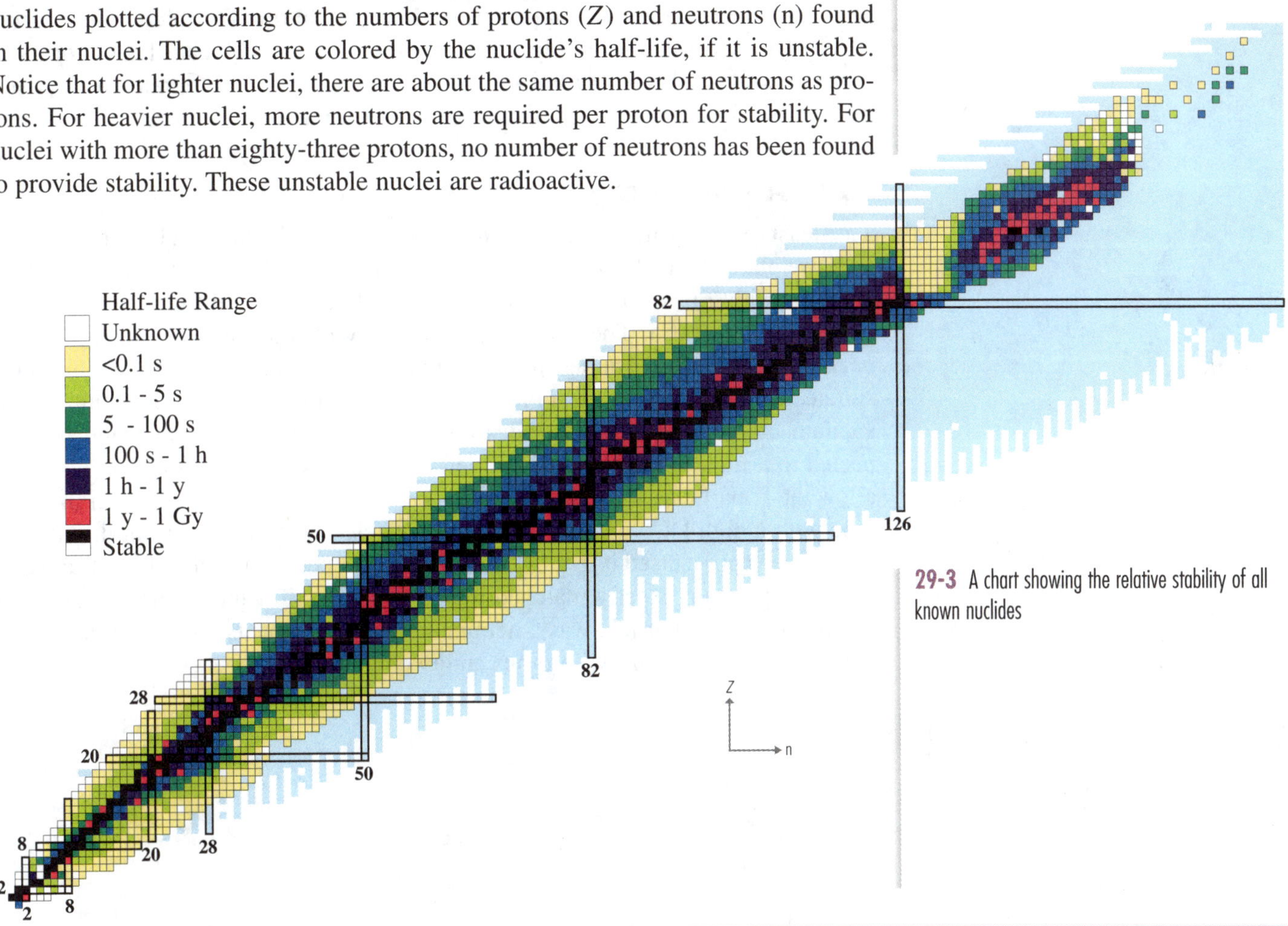

29-3 A chart showing the relative stability of all known nuclides

29.3 Kinds of Radiation

Not all emissions from naturally radioactive materials are the same, as a simple experiment can demonstrate. Figure 29-4 shows a schematic of such an experiment set up in a vacuum. A lead block has a hole drilled in it, and a mixture of various radioactive materials is placed in the hole. Not far from the hole is a series of three sheets of photographic film. A lead plate with a small collimating hole is placed between the radioactive source and the first sheet of film. Between the collimator and the film is an electric field (a magnetic

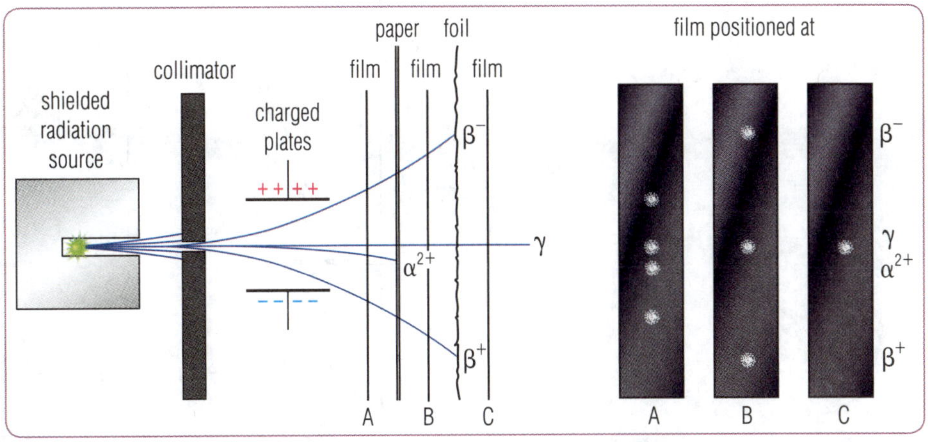

29-4 Alpha, beta, and gamma radiation, demonstrating charge and penetrating ability

field would also work as well). The sources are allowed to radiate, after which the films are developed. Film A is exposed in four areas, apparently by four types of radiation, some of which are deflected by the electric field. One kind of radiation is greatly deflected toward the positive plate. A second kind of radiation is deflected toward the negative plate an equal distance as the first. A third is slightly deflected toward the negative plate. The fourth kind of radiation is undeflected.

The four types of radiation also differ in penetrating power. If a piece of heavy paper is placed between film A and film B, the radiation that formed the slightly deflected spot on film A does not appear on film B, while the others do. This "missing" radiation is called **alpha radiation (α)**. On film C, which follows a thin sheet of metal foil, the highly deflected spots also disappear. These forms of radiation are called **beta radiation (β)**. The radiation that penetrated both coverings, forming the undeflected spot, is called **gamma radiation (γ)**.

29.4 Alpha Decay

The least penetrating radiation, α-radiation, is deflected slightly toward the negative plate of the electric field. It was found to be made up of positively charged particles with a mass-to-charge ratio of about four thousand times the electron's mass-to-charge ratio. These massive, positively charged particles are called *α-particles.* Rutherford studied them by placing a source of α-particles, uranium sulfate, in a thin-walled tube, or *ampule.* He sealed this alpha source within a glass vacuum tube containing electrodes (similar to Figure 29-5) that could be connected to a high-voltage source. He left the tube unenergized for several days to allow the α-particles to accumulate as a low-pressure gas. Then, he energized the tube with a high DC voltage between the plates. The electric current through the α-particle vapor caused it to glow, emitting a line spectrum. Rutherford discovered that this spectrum exactly matched the line spectrum for the element helium. He decided that an α-particle was a helium atom stripped of its two electrons, or just a particle consisting of two protons and two neutrons.

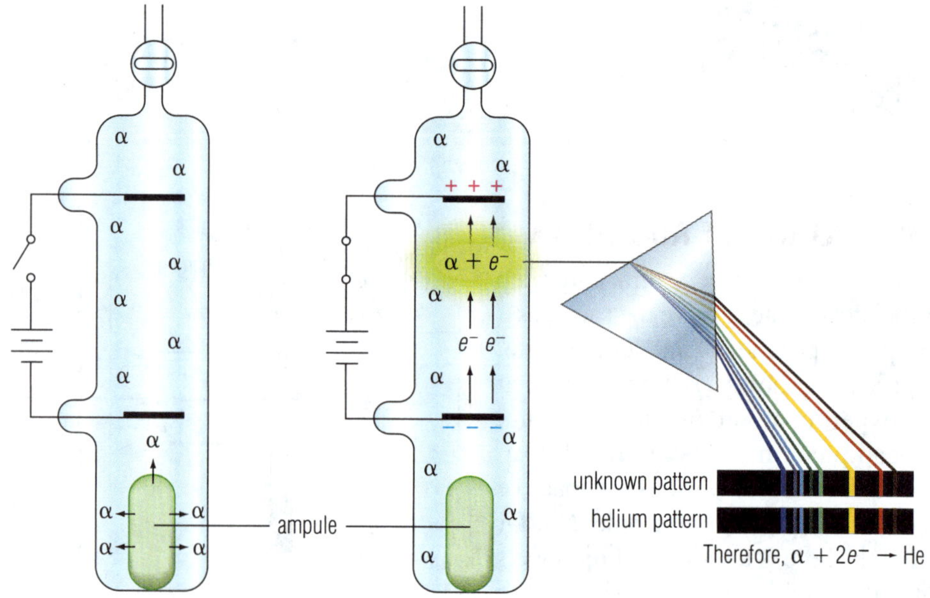

29-5 Excited alpha particles emit helium's line spectrum.

It should not be surprising that a nucleus is changed when it emits an α-particle. For instance, when a uranium nucleus loses the two protons in the α-particle, it is no longer the element uranium. This α-decay is expressed by the equation

$$^{238}_{92}\text{U} \rightarrow {}^{234}_{90}\text{Th} + {}^{4}_{2}\text{He},$$

in which $^{4}_{2}$He is the isotopic symbol for an α-particle. By this decay, uranium-238 changes to thorium-234. Notice that both mass number and nuclear charge must be conserved during a nuclear decay.

In some textbooks, this reaction is written simply as

$$^{238}_{92}\text{U} \rightarrow {}^{234}_{90}\text{Th} + \alpha.$$

In the above form, the conservation of mass and charge is not as obvious but is implied by what the α symbol represents.

29.5 Beta Decay

The next radiation, β-radiation, was found to be made up of particles with the same mass-to-charge ratio as electrons. However, the particles seemed to be split between those with the normal negative charge of an electron and a species with a positive charge. All the evidence accumulated about β-radiation indicated that it consisted of negative and positive electrons. However, these electrons come from the nucleus rather than the electron cloud. The negative nuclear electrons were named β^--particles, and the positive electrons were named **positrons (β^+)**. Collectively, electrons and positrons are called *beta particles*. The evidence about β-decay shows that when a β^--particle is emitted, a neutron disappears and a proton is left in the nucleus. If a positron is emitted, a proton disappears and a neutron is left in its place. Beta decay, like alpha decay, changes the atomic number of the nuclide, and thus the element. For instance, carbon-14 decays to nitrogen-14 by the equation

$$^{14}_{6}\text{C} \rightarrow {}^{14}_{7}\text{N} + \beta^-, \text{ or}$$
$$^{14}_{6}\text{C} \rightarrow {}^{14}_{7}\text{N} + {}^{0}_{-1}e,$$

where $^{0}_{-1}e$ is a convenient symbol for an electron to demonstrate the conservation of mass number and charge in a nuclear reaction.

The symbol for a *positron* in a nuclear equation is $^{0}_{+1}e$.

29.6 Gamma Decay

The most penetrating radiation, γ-radiation, resembles x-rays. For instance, our hypothetical experiment shows that γ-rays, like x-rays, are not affected by electric fields. Also, both x-rays and γ-rays can be diffracted by optically opaque crystalline materials. Not surprisingly, gamma rays are high-frequency electromagnetic radiation. The dividing line between x-rays and γ-rays is arbitrary in terms of energy and frequency. They are differentiated primarily by their source. X-rays originate in the electron cloud of large atoms when energetic free electrons collide with them. Gamma rays are generated by nucleons releasing excess energy as they transition from very high energy levels to lower energy levels. Gamma rays usually accompany α- or β-emissions, but they may be emitted by themselves as well.

Since no nucleons are changed during a gamma emission, the element's isotope remains the same. Therefore, an atom may undergo gamma decays repeatedly without changing its identity.

29.7 Natural Fission

A fourth possible, although naturally rare, form of radioactivity is **fission**, the splitting of a nucleus into two nuclei, called *fission fragments*, and a few individual particles. Spontaneous fission occurs at statistically random intervals only in nuclei with very large mass numbers. Fission may also occur when energy is added to a large nucleus by the absorption of a gamma ray or from the impact of speeding particles. Because susceptible nuclei are so far apart in natural ores, and the density of free neutrons and gamma rays (which themselves are produced from other fissions) is so low, fission rarely occurs in nature.

In 1972, mineralogists discovered the evidence of extensive natural fission within a uranium ore deposit in the small country of Gabon, located on the west coast of Africa on the equator.

Fission can be produced artificially by bombarding certain nuclei with neutrons. We will discuss artificial fission used for power generation later in this chapter.

29.8 Detection of Radiation

In order to study radioactivity, you must have some way of detecting radiation. Photographic film can detect large numbers of particles, but this method is cumbersome and time-consuming. More modern imaging systems use fluorescing screens, but the equipment is hardly portable. Several instruments developed early in the twentieth century can detect even single α- or β-particles and individual γ-rays. All of these instruments depend on the principle that energetic radiation particles and rays can remove electrons from atoms as they collide with them. In fact, nuclear radiation is classified as **ionizing radiation** to differentiate it from nonionizing radiation, such as microwaves.

Nuclear radiation is classified as **ionizing radiation** in order to differentiate it from nonionizing radiation, such as microwaves, infrared, and visible light.

The basic alpha-beta-gamma detector is often called the *Geiger counter*, first invented in 1911 by Hans Geiger. The radiation detection component of the instrument consists of a specially designed electronic tube called a *Geiger-Mueller (G-M) tube.* It is a sealed, thin-walled glass chamber containing a metallic cylinder and a central wire conductor insulated from the cylinder. If it is designed to detect alpha particles, the end of the chamber will be sealed with a very thin mylar film instead of glass. The tube is filled with an inert gas. The detector is connected to a source of several hundred volts DC and to measuring circuits. In operation, when a particle or ray passes through the gas, some of the atoms are ionized. The ions and electrons are accelerated to the electrodes by the electrical potential, colliding with other gas atoms in the process and producing additional charges in a cascade effect. The large momentary current produces a pulse in the circuit. The pulses are counted over time and displayed as a rate on a meter. Usually an audio speaker installed in the instrument produces the characteristic clicking sound associated with Geiger counters. More rapid clicks indicate more ionizing events in the G-M tube. Figure 29-6 shows several modern portable radiation detectors based on the Geiger principle.

Hans Geiger (1882–1945) was a German physicist whose experiments led to Rutherford's identification of the α-particle as a helium nucleus.

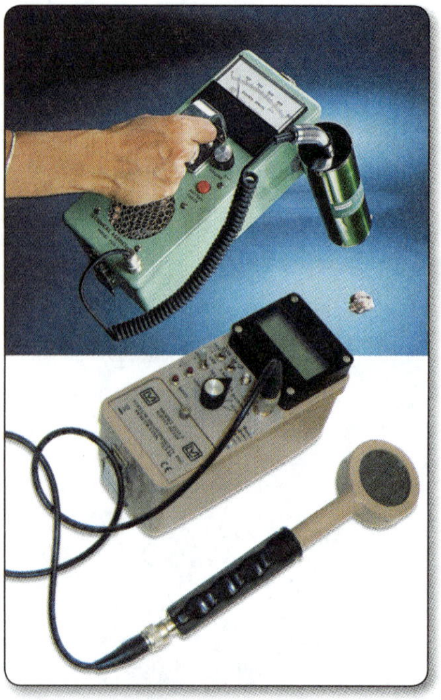

29-6 Portable radiation detection instruments. (top) This instrument can detect only beta and gamma radiation. (bottom) This instrument has a thin mylar window for detecting alpha particles in addition to beta and gamma radiation.

Other portable and fixed radiation detectors have been designed for specific purposes. Some instruments have been designed to detect fast and slow neutrons. Such instruments are used to detect power levels in nuclear reactors. Portable neutron detectors must be shielded from other forms of radiation, so their detectors are usually contained in heavy cylindrical "pigs" of lead and are not easily carried.

Lead is used for radiation shielding because it is economical and consists of heavy nuclei that easily stop beta and gamma radiation. High-energy neutrons are not easily captured by massive nuclei and can pass through the lead shielding to the detector.

Workers in areas containing radioactive sources must be able to conveniently monitor their exposure in real time. They carry instruments called pocket *dosimeters* that they can view as they work to see how much radiation they have been exposed to. For longer-term medical monitoring, workers also carry personal dosimeters that must be read by a machine.

In nuclear research, several types of instruments, in addition to ionizing detectors, have been developed to detect nuclear particles. The purpose of research detectors is to display the nuclear particle in such a way that its energy and momentum can be calculated. These detectors are often surrounded by magnetic and electric fields that produce spiral tracks (in the case of charged particles) for this purpose. A *cloud chamber* contains a cold gas near its condensation point. Gases tend to condense around ions, so the gas will condense first along a particle's ion trail. A *bubble chamber* contains a liquid near its vaporization point. The liquid will vaporize first around ions, causing the particle to leave a trail of bubbles in the liquid. A *spark chamber* contains closely spaced conducting plates maintained at a high potential difference with respect to one another. A spark easily jumps

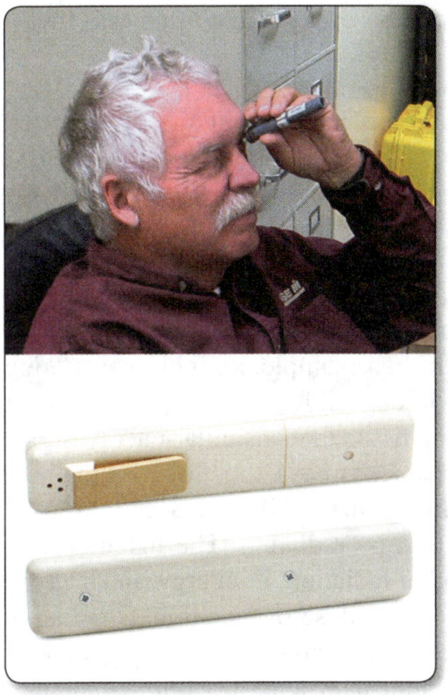

29-7 Two types of personal radiation-detection instruments. (top) A worker checking a pocket dosimeter; (bottom) A personal radiation dosimeter that emits audible chirps at a rate proportional to the intensity of ionizing radiation detected

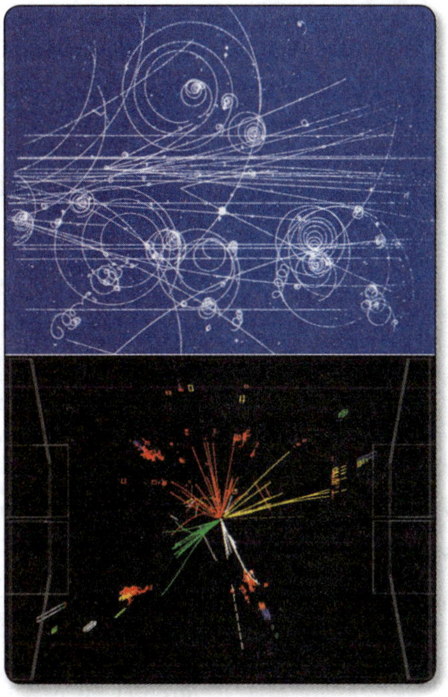

29-8 (top) Bubble-chamber particle tracks; (bottom) Modern high-energy particle-collider display

along a trail of ions such as those that a particle leaves. The trails of particles in cloud, bubble, and spark chambers can be photographed for future study. These trails are much larger than the particles, just as the contrail—the cloudy trail that follows a jet—is often seen when the jet is too small to be visible. The newest detector instruments are far more complex, such as those built into the Large Hadron Collider (LHC) located at the CERN facility that spans the border between France and Switzerland near Geneva. These detectors are constructed as concentric shells of ionizing chambers, subdivided in such a way that a computer can reconstruct the paths of thousands of individual particles as they are flung from the point of collision. Each detector is enclosed by powerful magnets that curve the paths of the particles. From the trajectory data, the charge and momentums of the particles can be determined, assisting in their identification. The tracks can be analyzed and displayed in 3D using the latest in distributed computer processing technology.

29.9 Units of Radiation and Radioactivity

Radioactivity and radiation can be measured in a seemingly bewildering array of dimensional units, depending on the property of concern. The most elementary measurement is the number of particles or rays produced per unit time. This *decay rate* is called the **activity (*A*)** of the sample. The older unit for activity is the **curie (Ci)**, the activity of about 1 g of radium, or 3.70×10^{10} disintegrations per second. The current SI unit for activity is the **becquerel (Bq)**, where

In nuclear physics, a single ionizing emission is considered a disintegration or decay of the nucleus, even if the emission is just a gamma ray.

$$1 \text{ Bq} = 1 \text{ disintegration per second.}$$

Decay rates say nothing about how much energy is deposited in material exposed to radiation. In order to quantify this, a new SI unit called the **gray (Gy)** was established in 1975. One gray is equal to one joule of energy absorbed by one kilogram of material. The number of grays quantifies the **dose (*D*)** of radiation received by a material.

Louis Harold Gray (1905–65) was a British physicist whose later career focused on medical physics and radiobiology.

The older unit of absorbed energy was the *rad (rad)*:

$$1 \text{ Gy} = 100 \text{ rad}$$

The Gy is equivalent to J/kg, or m^2/s^2 in base SI units.

The ionizing effect of electromagnetic radiation on a sample of air was measured with a unit called the **roentgen (R)**:

$$1 \text{ R} = 25\,800 \text{ coulombs of charge per kilogram of air}$$

The roentgen is no longer an important unit. Scientists and health physicists are more interested in the effects of ionizing radiation on biological tissues. The ionization of biological molecules in cells damages or destroys cellular function. Some kinds of radiation cause more damage than others in living tissue because they travel short distances and deposit more energy per unit path length. For example, a highly penetrating γ-ray of a certain energy will deposit less energy per centimeter of path length through living tissue than a similar-energy, short-range α-particle. Health physicists have established a method, called the **radiation weighting factor (W_R)**, to account for the differences in damage caused by radiation. The magnitude of W_R is dependent on the kind of radiation, its energy, and the type of biological tissue receiving the dose of radiation. β-radiation and γ-radiation are assigned a W_R of around 1 since they deposit relatively little energy per unit path length, while α-radiation has a W_R of about 20.

Table 29-1

Radiation Weighting Factors (W_R)

Radiation Type	Energy Range	W_R
x-rays, γ-rays	All	1
electrons, positrons, muons	All	1
neutrons	< 10 keV	5
	10 keV–100 keV	10
	> 100 keV–2 MeV	20
	> 2 MeV–20 MeV	10
	> 20 MeV	5
protons	> 2 MeV	2–5
α-particles, fission fragments	All	20

A measure of the comparative amount of damage is the **equivalent dose (H_T)** of the radiation. The H_T is the product of the radiation dose and the weighting factor of the radiation:

$$H_T = D \times W_R$$

Equivalent dose is measured in SI units called **sieverts (Sv)**, named after a Swedish radiologist who extensively studied the effects of radiation on living organisms.

The older dose-equivalent unit was called the *rem* (*r*oentgen *e*quivalent *m*an): 1 Sv = 100 rem

The Sv is equivalent to J/kg, or m^2/s^2 in base SI units.

The effects of radiation on organisms, especially people, are difficult to assess for a variety of reasons. One measurable effect that has been identified is the slight increase of cancer rates in large populations of workers exposed over periods of time to low levels of ionizing radiation. For this reason, workers in nuclear power plants, medical-radiation personnel, crew members in nuclear-powered ships, nuclear-weapons personnel, and nuclear materials processing personnel are closely monitored, and their exposure records are retained indefinitely in order to assess the long-term effects of radiation exposure.

Table 29-2

Typical Equivalents Doses for Various Radiation Exposures

Exposure	H_T
Natural background radiation	2.5–3.0 mSv/y
(soil, cosmic, radon, etc.)	(250–300 mrem/y)
chest x-rays	0.4 mSv (40 mrem)
dental x-rays	0.2 mSv (20 mrem)

The International Commission on Radiation Protection recommends a whole-body equivalent dose of no more than 5 mSv (0.500 rem) per year for the general public and 50 mSv (5.000 rem) per year for radiation workers.

Bone is considered the tissue most sensitive to radiation damage because, compared to other kinds of tissues, it has the least ability to replace damaged cells.

29A Section Review

1. Why would it be incorrect to say, "This mineral is emitting strong radioactivity"?
2. Discuss why radioactive decay is a natural way of transmuting one element into another.

3. What seems to be the main factor in determining the stability of an atom's nucleus?
4. Discuss how beta decay changes the nucleus.
5. What is the main difference between gamma rays and x-rays?
6. a. Why is nuclear radiation classified as ionizing radiation?
   b. Describe the standard instrument used for detecting all types of ionizing radiation.
7. a. What is the SI unit for radioactivity?
   b. What is the SI unit for measuring the equivalent dose to biological tissues?
8. Why could breathing in a tiny particle of an alpha-emitting isotope that lodges in your lung cause more biological damage than receiving the same dose of gamma radiation over your whole body?
9. How many chest x-rays would it take to exceed the recommended annual dose of ionizing radiation for the general public?

29A Objectives

After completing this section, I can

- ✓summarize the discovery of radiation.
- ✓describe the nuclear conditions that lead to instability and the production of nuclear radiation.
- ✓compare and contrast the four principal types of nuclear radiation and their properties.
- ✓describe how atomic nuclei change when they emit nuclear radiation.
- ✓explain how radiation is detected.
- ✓list various units of radioactivity and biological dosage.
- ✓explain why different kinds of radiation have different biological effects.

29B RADIOACTIVE DECAY

29.10 Radioactive Decay Law

Naturally radioactive nuclei emit radiation at constant rates under nearly all circumstances. For instance, 1 g of organic carbon in living tissues typically has an activity of 0.227 Bq from the decay of carbon-14. About every four seconds, on the average, it emits a β-particle. If the carbon-14 continues to emit radiation for a long time, its atoms will eventually change to mostly nitrogen-14 mixed in with the stable carbon-12 atoms of the sample.

As time progresses, the sample's activity decreases when there are fewer carbon-14 atoms to decay. This means that the decay rate of the sample is proportional to the amount of radioactive material present, or

$$\frac{\Delta N}{\Delta t} \propto N, \qquad (29.1)$$

where N is the number of a certain kind of decaying nuclei, and $\Delta N/\Delta t$ is the rate of decay of the nuclei in disintegrations per unit time. A proportionality constant can be added to give us an equation:

$$\frac{\Delta N}{\Delta t} = -\lambda N \qquad (29.2)$$

The factor λ is called the decay constant. The units of the decay constant are inverse time, such as s^{-1} or y^{-1}. The negative sign before the right-hand side indicates that the decay rate results in a decreasing number of decaying atoms with time. Note that the activity of the radioactive nucleus (A) is $|\Delta N/\Delta t|$ because activity is reported simply as disintegrations per unit time.

The **decay constant (λ)** is defined by the equation

$$\lambda = \frac{1}{\tau},$$

where τ is the average lifetime of a radioactive particle. The decay constant is characteristic of a specific isotope and *decay mode*. An isotope may decay via alpha-, beta-, or gamma-emissions, each of which has a different decay constant.

Radioactive decay is a *statistical* process. While it is fairly straightforward to predict how much of a sample of billions of nuclide atoms will change to another nuclide with time, there is no way to determine exactly *when* a particular nucleus will experience a decay. This aspect of nuclear decay seems to be completely random.

Through calculus, it can be shown that the number of nuclei remaining at a certain time after an initial number of nuclei begin to decay can be found from the formula

$$N = N_0 e^{-\lambda \Delta t}, \qquad (29.3)$$

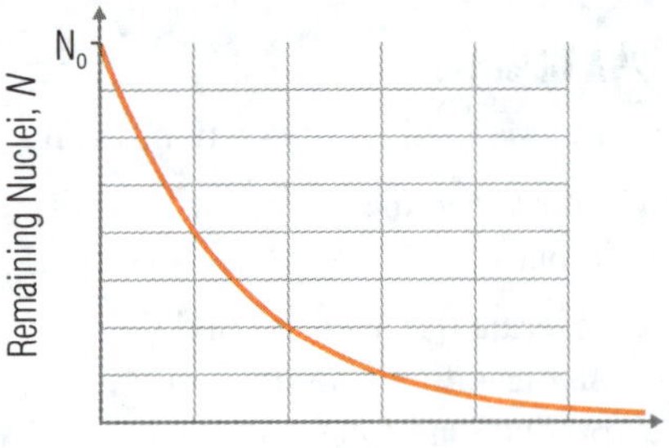

29-9 A radioactive decay curve showing the number of undecayed nuclei remaining with time

Problem-Solving Strategy 29.1

Here is a brief review of the laws of logarithms:

If $n^y = x$, then $\ln_n x = y$

$\ln e = 1$

$\ln 1 = 0$

$\ln x{\cdot}y = \ln x + \ln y$

$\ln x/y = \ln x - \ln y$

$\ln x^y = y \ln x$

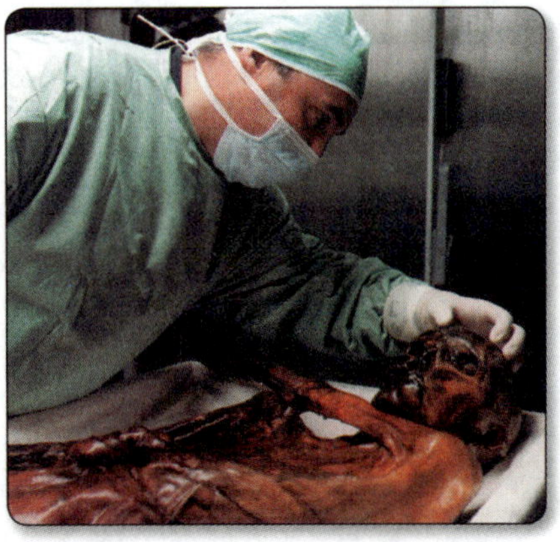

29-10 The age of the mummified remains of the "Ice Man" was determined using radiocarbon dating.

where N_0 is the original number of decaying atoms, and N is the number of atoms after time interval Δt, which is the total elapsed time since the original sample N_0 began to decay. As Equation 29.3 shows, the number of undecayed nuclei remaining with time decreases exponentially. Figure 29-9 is a typical decay curve of a sample of radioactive nuclei.

The time that it takes to reduce the original number of radioactive atoms by half is called the isotope's **half-life ($t_{1/2}$)**. Clearly, this value is more useful than the decay constant for indicating how rapidly a particular isotope will decay away. An isotope's half-life can be determined from Equation 29.3 by calculating the time that it takes for N to equal $N_0/2$:

$$N = N_0 e^{-\lambda \Delta t}$$

$$\frac{N_0}{2} = N_0 e^{-\lambda t_{1/2}}, \text{ or}$$

$$\frac{1}{2} = e^{-\lambda t_{1/2}} \tag{29.4}$$

Taking the natural logarithm of both sides of Equation 29.4 gives

$$\ln 1 - \ln 2 = -\lambda t_{1/2}.$$

Since $\ln 1 = 0$, we can rearrange the remaining terms to solve for $t_{1/2}$:

$$t_{1/2} = \frac{\ln 2}{\lambda}$$

$$t_{1/2} = \frac{0.693}{\lambda} \tag{29.5}$$

Equation 29.5 relates the half-life of a radioactive nuclide to the decay constant. The half-life is constant for a given nuclide.

Equation 29.3 is valid not only for the numbers of nuclides left over time but also for the decay rate or activity of the nuclides. Since activity is easier to measure than the actual number of nuclide atoms present in a sample, the equation's more useful form is

$$\frac{\Delta N}{\Delta t} = \left(\frac{\Delta N}{\Delta t}\right)_0 e^{-\lambda \Delta t}, \text{ or}$$

$$A = A_o e^{-\lambda \Delta t}, \tag{29.6}$$

where A is the activity in becquerels.

29.11 Radioactive Dating

Knowing the number of half-lives that have passed for a given sample of radioactive nuclide atoms can quickly indicate how much radioactivity is left. After one half-life there is half, $(1/2)^1$, the original activity. After two half-lives there is $(1/2)^2$, or $1/4$ the original activity. After three half-lives there is $(1/2)^3$, or $1/8$, and so on. Conversely, scientists claim that if the original activity of a sample is known or can be reliably estimated, the number of half-lives that elapsed to produce the activity observed in the present can be determined. The number of half-lives then determines the age of the sample since it was formed. This concept is the basis for **radioactive dating**.

For instance, it takes the activity of a sample of carbon-14 5730 years to decrease by half ($t_{1/2} = 5730$ y). Therefore, if a sample begins with an activity of 0.5 Bq, after 5730 years its activity will be only 0.25 Bq. After another 5730 years the activity will halve again to 0.125 Bq.

Because no physical or chemical change, except under extraordinary conditions, is known to affect a nuclide's half-life, radioactive nuclides (or *radionuclides*) are used to date objects. If a scientist assumes the activity of the nuclide at the "beginning" of the sample's existence (A_0), measures the activity present now (A), and knows the radionuclide's half-life ($t_{1/2}$), he can calculate the time that has passed since the "beginning" (Δt) by using Equation 29.6. Solving the equation is straightforward. Measuring the present activity of a radionuclide can be done with a variety of instruments, and the decay constants are known for nearly all possible radionuclides. The problem comes in determining the original activity of the radionuclide in a sample.

For most samples, the amount of the original radionuclide, the **parent nuclide**, is calculated from the amount of the radionuclide remaining in the sample today plus the amount of the **daughter nuclide**, the element that the radionuclide becomes after decaying. For instance, in 12.0000 g of an organic carbon sample (e.g., from tree wood), you measure 11.8668 g of C-12, 0.1332 g of C-13, and about 0.6×10^{-11} g of C-14. Since an identical modern carbon sample should have about 1.2×10^{-11} g of C-14, you could infer that the sample was the age of about one half-life of C-14, or 5730 years old. But this conclusion *could* be wrong.

The activity of C-14 in living tissue is 0.227 Bq/g. This results in about 15 decays per minute per gram of natural carbon.

The problem of determining how much of the daughter nuclide was in the original sample is a serious one for those who use radioactive dating techniques. The normal way of resolving the problem is to assume that the elements have always had isotopes in the same proportion that they do today. For example, the element strontium (Sr) has isotopes 84, 87, 86, and 88 in a 1:13:18:147 ratio today. Suppose a geologist finds a sample that includes 1 g of rubidium-87 (Rb-87), which decays into strontium-87. In the same sample are 0.01 g of Sr-84, 0.63 g of Sr-87, 0.18 g of Sr-86, and 1.47 g of Sr-88. The other isotopes of strontium are in their normal ratio, but there should be only 0.13 g of Sr-87. Therefore, the geologist (or his colleague, a radioactivity expert) concludes that the "extra" 0.5 g of Sr-87 is **radiogenic** (i.e., produced by radioactive decay). The radioactivity expert knows how much Rb-87 decays into 0.50 g of Sr-87; therefore, he can calculate the original amount of Rb-87 and the time that it has taken to decay.

In calculating the original amount of a radionuclide in a sample, it is necessary to make several assumptions:

1. The proportions of isotopes for each element have remained constant throughout the life of the sample.
2. The sample has not been contaminated with either the parent or the daughter nuclide(s) originating from outside the sample (e.g., by diffusion or through transport by ground water).
3. None of the parent or daughter nuclides has escaped from the sample (e.g., diffusion or leaching by ground water).

The second and third assumptions are often checked using the first assumption; that is, a person assumes that if any element entered or left the sample, it did so in the normal proportions.

29.12 Problems with Radioactive Dating

Keeping these assumptions in mind, let us examine carbon-14 dating. Carbon-14 is continually being produced in the atmosphere by cosmic rays interacting with atmospheric nitrogen-14, causing positron emissions. Radioactive carbon, like ordinary C-12, forms carbon dioxide in the atmosphere and is incorporated into all living tissues directly or indirectly via photosynthesis. While the tissue is alive, it exchanges carbon with the atmosphere and so has the same proportion of C-14 as the atmosphere has. When an organism dies, it stops exchanging carbon with the

The nuclear reaction equation for the production of C-14:

$${}^{14}_{7}\text{N} + {}^{0}_{0}\gamma \rightarrow {}^{14}_{6}\text{C} + {}^{0}_{+1}e$$

FACETS of PHYSICS

Radioactive Decay Halos

Radionuclide halos in mineral crystals are spherical regions of discoloration surrounding a minute radioactive source called an *inclusion*. Alpha decays of the radioactive nuclei in the inclusion send out alpha particles with several MeV (mega-electron volts) of kinetic energy. As the alpha particles plow through their surroundings, they knock hundreds of thousands of electrons loose and cause damage to the crystalline structure. Since the alpha particles will eventually lose all of their kinetic energy, their range depends on their starting energy. Different decays produce rings of different radii called *halos*. The photo shows the halos formed by the decay of polonium-218 and its daughter isotopes.

The decay chain of Po-218 includes three isotopes that emit alpha particles—Po-218, Po-214, and Po-210. Other nuclides in the chain undergo beta decay, but beta particles are not usually energetic enough to cause halos. This decay chain ends with Pb-206, which is a stable nucleus with a long half-life. Thus, there will be one alpha particle from the Po-218, one from the Po-214, and one from the Po-210. Experiments have determined the energies of alpha particles, which travel known distances before losing all of their energy. Hence, the largest of the three rings shown in the picture goes with a 7.69 MeV alpha, the middle ring with a 6.00 MeV alpha, and the small ring with a 5.30 MeV alpha.

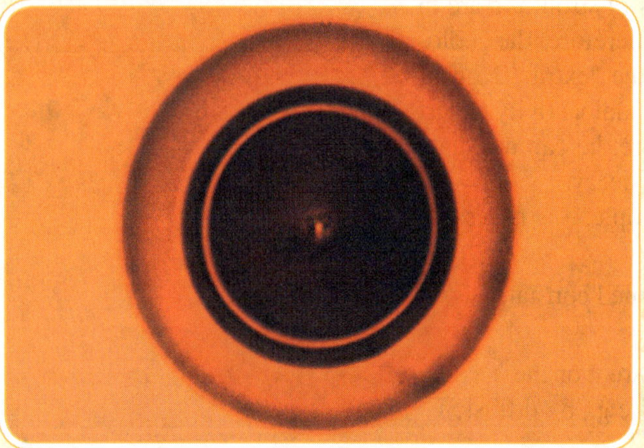

A Po-218 halo magnified 1450×. A radioactive inclusion in biotite mica (from Sweden) produces spherical regions of discoloration, which appear as circles in this photograph.

Robert V. Gentry, the scientist with the most experience in studying these halos, worked at Oak Ridge National Laboratory for over a decade. Gentry found that, because of the short half-lives of Po-218 (3.10 minutes) and Po-214 (163.7 microseconds), the existence of these halos is very difficult to explain using evolutionary timeframes. Halos form only in crystallized, nonmolten rock. In order for all three halos to form, the rock must have already been crystallized when the Po-218 was placed there. If there had been any significant cooling period, virtually no Po-218 or Po-214, because of their short half-lives, would have been left to generate any halo. The only reasonable explanation for this phenomenon is that the crystallized rock, along with the polonium inclusions, was created instantaneously rather than by uniformitarian processes.

Some scientists believe that the polonium halos are really generated by uranium inclusions since polonium is also in the decay chain of uranium-238. According to this theory, the daughter Po-218 migrated and formed halos at its new position. However, Gentry found no alpha particle pits and fission tracks (from very large fission particles) in the crystals surrounding the polonium halos, which would have been expected if migration were true. Thus, Gentry showed that the polonium could not have originated from the migration of uranium atoms into the mineral crystal. Instead, he favors a Creation model in which the polonium was created in place in the rocks.

Reference: Robert V. Gentry, *Creation's Tiny Mystery.* (Knoxville, TN: Earth Science Associates, 1992).

atmosphere. Its own C-14 continues to decay. Theoretically, it is possible to determine when an organism died by its present C-14 activity.

To use carbon-14 dating, you must know the atmosphere's C-14/C-12 ratio when the organism died. The present-day ratio is approximately 1.20×10^{-12}. However, the ratio has not been constant throughout the earth's history. It is possible to calibrate the dating technique for periods up to five thousand years ago by using human archaeological artifacts. However, most uniformitarian scientists believe that other techniques, including counting tree rings in ancient trees, growth layers in ancient corals, and stratigraphic data from microscopic shelled organisms called foraminifera, can provide absolute C-14 calibration as far back as 21 500 years or more. To do this, you must determine the age of an organic object by linking it to a specific date in history. For instance, you determine the age of a tree at the time it was felled or died in a historical context (building construction, battle, fire, etc.) by counting its rings (assuming that one ring is formed per year—not necessarily a good assumption). Other organic objects that can be linked to a historical date, such as textile fabrics, wooden implements, and mummies are used to calibrate the carbon ratio throughout post-Flood human history. Then, the activity of C-14 is measured, and from this value the number of C-14 atoms present in the sample is found. The ratio of C-14/C-12 can then be calculated. Any deviation from the current ratio must be considered when determining ages in materials showing corresponding activities per gram.

This procedure is the reason why meaningful C-14/C-12 ratios can be determined only for the last five-thousand-or-so years. Within that timeframe the carbon-14 dating method can give somewhat reliable ages. However, before that time there is no way to determine age from the C-14/C-12 ratio of the atmosphere without making unsupportable assumptions, so carbon-14 dating does not give reliable dates prior to historical times. Of course, for all C-14 dates, an investigator must be careful to avoid contaminated samples and samples that do not represent equilibrium conditions with atmospheric levels of C-14 and C-12. For example, seawater contains dissolved calcium carbonate with almost no C-14. Thus, animals living in this water are surrounded by an environment with a C-14/C-12 ratio very different from that found in the atmosphere. Standard radiocarbon dating techniques have assigned living mollusks ages of thousands of years!

The need for calibrated C-14/C-12 ratios has driven researchers to develop a *calibration curve*, based on human artifacts and other methods mentioned, for as far back in time as is technically feasible.

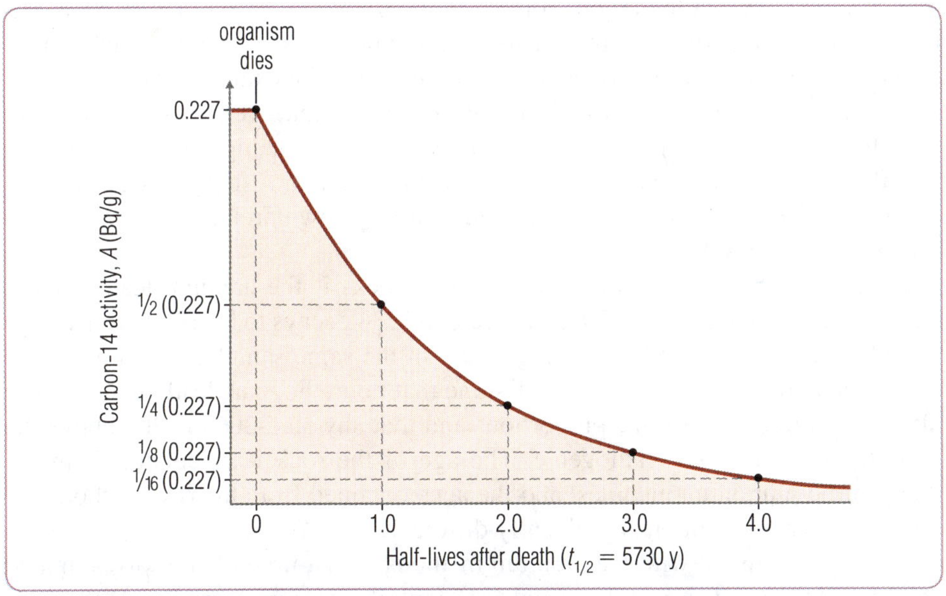

29-11 When an organism dies, the carbon-14 present in its tissues begins to decay with a half-life of 5730 y.

Many factors can affect the C-14/C-12 ratios in organic materials. Plants prefer C-12 to C-14, so plant material seems older than it actually is. Assumptions must be made about the constancy over time of the cosmic ray flux that produces C-14 from N-14 high in the atmosphere. The concentration of N-14 in the atmosphere may vary with time. Other factors, such as volcanic activity that introduces C-12 into the air and the burial of organic material that removes C-12 from the organic life cycle, must be considered as well. Both of these latter effects were significant at the time of the Genesis Flood. The Industrial Revolution was

By mentioning the various methods of *geochronology*, we are not giving credence to any of them. These are methods utilized by the scientific community based on the prevailing old-earth evolutionary model. *Index fossil* dating assumes that certain fossils are representative of specific biological epochs in the geologic column. *Stratigraphic* dating depends on the uniformitarian assumption that identifiable rock layers (strata) were laid down in a specific order keyed to the geologic column. *Radiometric* dating uses ratios of parent and daughter radionuclides to obtain absolute ages for rocks.

responsible for the introduction of additional carbon dioxide into the atmosphere. Atomic-weapons testing in the twentieth century increased the concentration of C-14 above natural levels. Recent evidence suggests that the C-14/C-12 ratio in the atmosphere is not in a state of equilibrium due to the above factors, so the activity of C-14 that we see in organic materials today is not representative of equilibrium conditions. As you can see, attempting to use C-14 as a dating technique, especially for samples believed to be much older than recorded history, is dependent on many unprovable assumptions based on one's presuppositions about the history of the world.

29.13 Geochronology: Determining the Age of the Earth

29-12 Lava rocks from eruptions of Mt. Ngauruhoe in New Zealand within the past 60 years have been dated as old as 3.5 million years using the K-Ar method.

Carbon dating is valid only for formerly living things. Scientists resort to other dating methods when dating inorganic rocks and, ultimately, determining the age of the earth. Dating the earth's rocks by any method is called **geochronology**. Geologists may use index fossil, radiometric, and/or stratigraphic methods for obtaining relative or absolute ages for rock. One radiometric method uses the decay of potassium-40 to argon-40 (K-Ar method). This decay has a half-life of 1.3 billion years.

One problematic aspect of the potassium-argon method of dating is that the daughter nuclide, Ar-40, is a monatomic noble gas that does not chemically react with other substances and is therefore free to move around. Scientists assume that molten magma ejects gas atoms and molecules, or *degasifies*, as it reaches shallow depths or the surface, so there is little argon in the rock at the time that it solidifies. They assume, then, that most of the argon found in the rock is radiogenic because they also assume that the argon atoms become trapped in the mineral crystal lattices after solidification. Neither of these assumptions is supported in laboratory experiments or field observations of recent eruptions. For example, the potassium-argon method has been used to date volcanic rocks that were observed to have been formed within the past two hundred years. Due to the high concentrations of Ar-40 found, the age obtained was several million years—much older than the known age. These observations are an indication of how unreliable the potassium-argon dating method is.

Perhaps the best-known method for dating rocks is the uranium-lead method. Uranium-235 decays to lead-207, and uranium-238 decays to lead-206. Since both nuclides of uranium and lead usually occur in the same sample, they are used to check each other. Again, geologists assume that nonradiogenic lead nuclides have always occurred in the same proportions and that any lead-206 or -207 above the expected proportions is radiogenic. The age of the rock is calculated from the decay of both uranium nuclides, and the ages obtained from the two methods are compared. They are often significantly different.

A serious problem with the uranium methods is that uranium dissolves in acid solutions, but lead does not. This quality is exploited to purify uranium in the mining process. Ground water is often slightly acidic, and uranium samples are not isolated from ground water. There is a possibility that significant amounts of uranium are carried away from the sample in ground water.

As you can see, there are significant technical difficulties with obtaining valid ages from radionuclide dating techniques. Other issues surrounding the process include a lack of scientific honesty when interpreting the dates. Many cases have been identified by both creationary and non-creationary scientists in which valid age data was discarded by researchers because it did not agree with their preconceived notions of what the date *should* be. If all the data were considered, the computed age of a rock sample might be a lot younger, a lot older, or indeterminate.

Nowhere else in naturalistic science do presuppositions intentionally displace objectivity more than in geochronology—dating the age of the earth. Biological evolution presumes billions of years for the first glimmer of cellular activity to be recorded in the geologic column. Accordingly, geologists (supported by cosmologists) provide the data that supports a 4.5 billion-year age of the earth.

The only scientific conclusion one can make is that radioactive dating of the earth's rocks will not produce meaningful ages unless the underlying assumptions can be assured in some way and that the data is honestly interpreted.

29.14 Modeling Using Radiocarbon

In the 1990s, a group of experts took samples from the Dead Sea Scrolls to analyze and date using radiocarbon dating. Small samples were removed, studied under a microscope, and treated, if they appeared to be contaminated, before analysis with an accelerator mass spectrometer (AMS) located in Arizona. The table below shows some of the documents sampled, their historical age, and their C-14 age obtained from radiocarbon dating. C-14 age is reported in years before present (BP), which, in this case, is defined as years before AD 1950.

TABLE 29-3

Radiocarbon Dating of the Dead Sea Scrolls

Document Identity	Document Historical Date	C-14 Age (y BP)
Papyrus	135 AD	1758
Papyrus	130 AD	1799
Papyrus	128 AD	1827
Damascus document	25 BC	2094
Commentary on Habakkuk	30 BC	2054
Paleo Exodus	63 BC	2044
Damascus document	75 BC	1954
Community Rule	88 BC	2041
Messianic Apocalypse	90 BC	1984
Community Rule	100 BC	1823
Community Rule	100 BC	1964
Book of Isaiah	138 BC	2141
Astronomical Enoch	200 BC	2095

29B Section Review

1. (True or False): The activity of a specific isotope is independent of the amount of the isotope.
2. How does the amount of a radioactive isotope in a sample vary with time?
3. How does isotope half-life vary with the magnitude of the decay constant?
4. Discuss the assumptions that are made when using a radioactive dating technique.
5. Discuss the factors that can make age determination of an object using a C-14 radiometric method unreliable.
6. DM Why were the Dead Sea Scrolls good candidates for using the C-14 dating method? What valuable information can archaeologists obtain from dating these scrolls that can advance the radiocarbon dating technique?
7. ✪ Write the nuclear decay equations for the alpha decay of U-238, followed by beta decays to U-234.

29B Objectives

After completing this section, I can

✓ describe the modeling of radioactive decay using the exponential decay law.

✓ state the significance of a nuclide's decay constant and explain how it is computed.

✓ relate a nuclide's half-life to its decay constant.

✓ discuss the assumptions and limitations of radioactive dating.

✓ describe how C-14 is produced and how it may be employed in radioactive dating.

✓ use actual radiocarbon data to determine the ages of historical objects.

8. How many years would it take for a sample of highly radioactive cobalt-60 to decay to 1/1000 of its initial activity? The half-life of Co-60 is 5.3 y.

9. Assuming that the C-14/C-12 ratio has not varied over recorded history, about how old is an axe handle that emits 1.6×10^{-1} Bq/g of carbon? Refer to Figure 29-11.

DM 10. a. Copy the data listed in Table 29-3 into a spreadsheet. Using the historical dates of the Dead Sea Scrolls given in Table 29-3, calculate the historical ages (or *paleographic age*) of the scrolls BP (AD 1950).

b. Create a scatter plot in the spreadsheet, displaying the document's paleographic age on the horizontal axis versus the corresponding C-14 age. The scatter plot should trend from the oldest to the most recent ages. This may require reversing the horizontal scale in your spreadsheet program.

c. Discuss the pattern of data points in your scatter plot. What general relationships do you observe? Include in your discussion what you would expect to see if the paleographic and C-14 ages were identical.

DM 11. a. If a smooth curve were drawn through the Dead Sea Scrolls data points in the scatter plot created in Question 10, for what purpose could an archeologist use the curve?

b. Convert the scatter plot to a smooth curve by altering the data series formatting in your spreadsheet plot.

c. Using the curve that you created in Part b, determine the paleographic age of a wooden bowl found in one of the jars containing the Dead Sea Scrolls if it has a C-14 age of 1900 years BP.

29C NUCLEAR REACTIONS

29.15 Nuclear Energy

It is well known that seemingly limitless energy can be extracted from the nucleus of the atom. But just exactly how is that done? As with any process that releases energy, potential energy within the system is converted to other forms of energy. In the case of the nucleus, we must examine the energies of the individual nucleons (protons and neutrons) that make up the nuclear system. The strong nuclear interaction force between nucleons is the source of nucleon potential energy. Recall that the gravitational potential energy of a ball falling due to the earth's gravity is highest at the height from which the ball is dropped and lowest when the ball hits the ground. Similarly, the nuclear potential energy between nucleons is greatest at a distance where the strong nuclear interaction force is just able to begin to attract other nucleons (a very short distance apart) and is least when the nucleons are bound together. If a hypothetical nucleus is assembled from protons and neutrons, the potential energy lost from the system is released as heat, light, and kinetic excitation of the nucleons. This released energy is called the **binding energy** of the nucleus. Binding energy is the difference between the sum of the kinetic and potential energies of all the nucleons before binding due to the strong nuclear interaction force and the total energy of the assembled nucleus in its minimum energy state (ground state). It is also equal to the mechanical energy needed to disassemble and separate the nucleons far enough from each other so that the strong nuclear interaction force between them is negligible.

Nuclear stability is a function of the amount of nuclear binding energy *per nucleon*. In other words, the lower the nuclear potential energy (and the greater the amount of binding energy released) for each nucleon, the more stable the nucleus. This makes sense because we know that a physical object is most stable mechanically when its gravitational potential energy is lowest. Figure 29-13 is a graph of binding energy per nucleon versus mass number, A. Notice that nuclei with mass numbers around $A = 60$ have the highest binding energy per nucleon and are thus the most stable.

Early in their study of the nuclear atom, physicists discovered that the mass of a nucleus is less than the sum of the masses of the individual nucleons that compose the nucleus. The difference in mass, called the **mass defect**, is attributed to binding energy released during assembly of the nucleus, according to Einstein's special relativity equation, $E = mc^2$. Thus, nuclei that have larger binding energies per nucleon have smaller masses per nucleon. This mass-energy equivalence is the basis for the practical application of nuclear energy. In a nuclear reaction, binding energy and mass can be converted to other forms of energy, or vice versa, but total mass-energy is conserved.

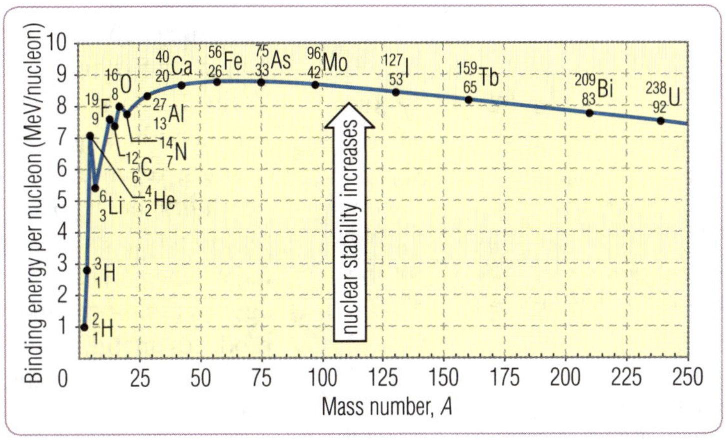

29-13 Graph of binding energy per nucleon. Note that the maximum energy per nucleon (greatest mass defect) occurs when $A \approx 60$. This means that atoms in this range have the greatest nuclear stability.

Many nuclear reactions can be artificially forced to occur. Usually, the purpose of a nuclear reaction is to produce useful energy. In other words, the combination of all nuclear changes during the reaction results in the liberation of energy. Such reactions are **exoergic**, which is just another way of saying *exothermic*. Since energy and mass are equivalent, and mass-energy must be conserved, a release of energy must mean a loss of total mass in the products of such a reaction. On the other hand, some nuclear reactions are studied where the products have more mass than the reacting particles. In such cases, the reactions absorb energy and are called **endoergic**.

How can you predict whether a reaction is exoergic or endoergic? Look at Figure 29-14. Find the reactants and the products on the graph. If the products are *uphill* from the reactants, the reaction is *exoergic*. This means that the products have less mass per nucleon than the original nucleus due to the higher binding energy per nucleon. The mass difference was released as binding energy in the reaction. If the products are *downhill* from the reactants, the reaction is *endoergic*.

There are two types of exoergic reactions, fission and fusion. In fission, a large nucleus splits into smaller nuclei. This reaction is exoergic as long as both the reactant and the product nuclei have mass numbers greater than about 60. Fission reactions start with reactants far to the right side of Figure 29-14, and the products are generally to the right of the peak of the graph. In **fusion**, two or more small nuclei merge to form a larger nucleus. The reactants are at the far left end of the curve in the figure,

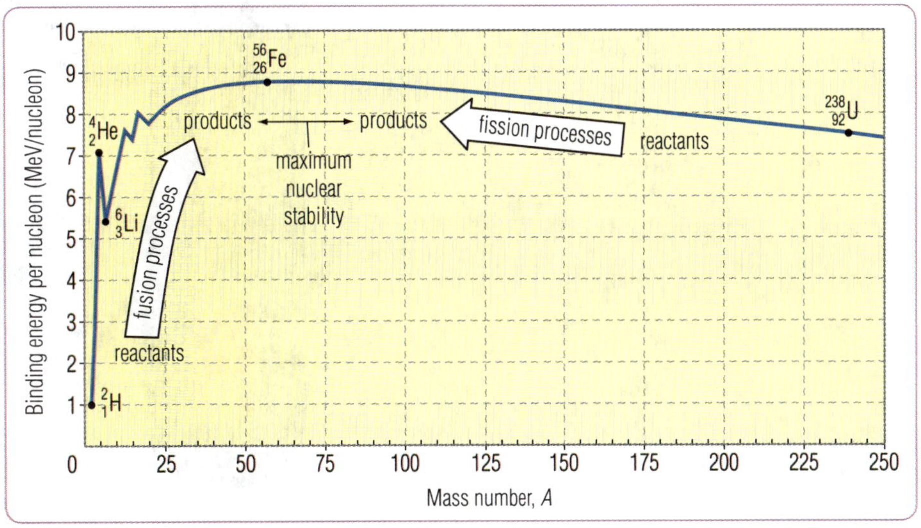

29-14 Fission and fusion products are uphill from reactants on a binding-energy-per-nucleon graph.

and the products are to the left of the curve's peak. This reaction is exoergic as long as the reactants and the product have mass numbers less than about 60.

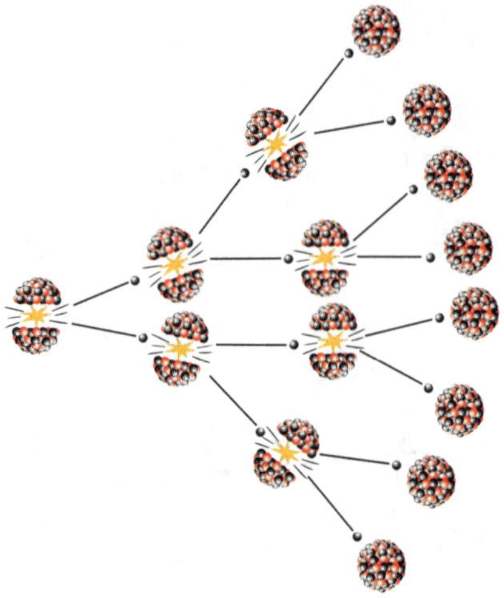

29-15 Under the right conditions, a single uranium fission can produce many fissions in a supercritical chain reaction during a brief interval of time.

The time delay between the production of a neutron from fission until the next fission occurs is on the order of micro- to nanoseconds. Therefore, reaction rates can increase very rapidly (or conversely, die out rapidly) if appropriate conditions exist in the fissionable material.

29.16 Fission Reactions

At our present state of technology, it is easier to force a fission reaction than a fusion reaction. Some materials—notably U-235 and Pu-239—undergo fission when they are bombarded with neutrons. The nucleus absorbs a neutron, and then splits. The fission produces more neutrons. If there are enough fissionable nuclei, it is likely that at least one of the new neutrons will strike another nucleus and cause it to split. A series of sequential fissions caused by fission neutrons is called a **chain reaction**. If the rate of neutron production from fissions and the concentration of fissionable nuclei are both optimal, a constant rate of fissions can be attained. The necessary concentration of nuclei for a continuous chain reaction is contained in a quantity of material called the **critical mass**. A constant rate of fissions, regardless of the number of fissions per second, is a condition called *criticality*, or the overall nuclear process is said to be *critical*. Since each U-235 fission produces an average of 2.47 neutrons in addition to the other fission fragments, the potential exists for the number of fissions to increase with each neutron "generation"—the interval between the production of a neutron and a fission that produces another neutron (see Figure 29-15). A neutron generation can be as short as a few micro- or nanoseconds, so the increase of the rate of neutron production can be very rapid. A reaction where the rate of fissioning is slowly or rapidly increasing with time is *supercritical*. The reaction is *subcritical* if the rate of fissions is decreasing with time.

Power Reactors

Most US nuclear power plants operate by the fissioning of uranium-235. This reaction begins when a U-235 nucleus captures a neutron and undergoes fission, producing more neutrons. Natural uranium contains about 0.7% U-235, and the remaining 99.3% is U-238. The uranium used as fuel for commercial power plants is a mixture of about 3% U-235 and 97% U-238. The percentage of U-235 is artificially increased through a very expensive chemical and mechanical process called *enrichment*. The U-238 has a low probability of undergoing fission, but it does absorb neutrons. U-238 is more likely to absorb fast neutrons than U-235, while U-235 is more likely to absorb slow, or *thermal*, neutrons. Most of the neutrons produced by fission are fast. In order to slow the neutrons down so that they are more likely to cause a fission in U-235, a **moderator** material is interspersed among the uranium to slow the neutrons. In most power reactors, the moderator is simply water, although some older designs use carbon or some other material.

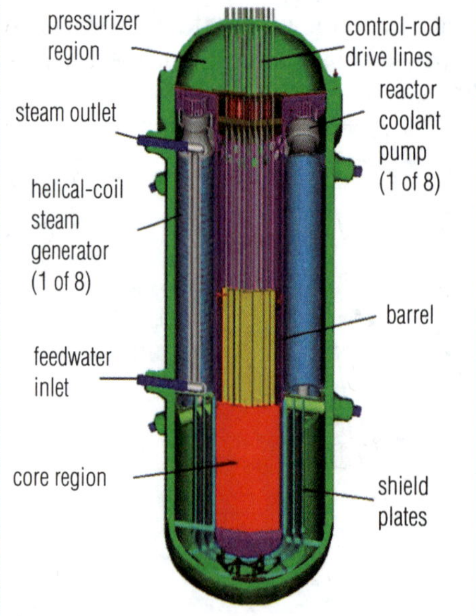

29-16 A cutaway of a nuclear reactor showing the location of the fuel and control rods, reactor vessel parts, and cooling water flow

With water as a moderator, a critical chain reaction can be established and easily sustained. The goal at constant power is to get exactly the same number of fissions from each generation of fissions. A few more per generation, and the reaction rapidly runs out of control. A few less, and the reaction quickly dies out. The mass of uranium fuel exposed to the neutron flux in a reactor is the critical mass, just enough to sustain the critical reaction. The water moderator contains hydrogen atoms that quickly slow down the fast fission neutrons to thermal energies (see the margin box). Some neutrons leak out of the reactor, but enough are reflected back into the fuel to sustain the reaction. If too many fissions occur, producing higher temperatures in the water, the water density decreases, allowing more neutrons to leak out and causing the reaction to go subcritical. Then, as the water cools, its density increases slightly, reflecting more neutrons and increasing neutron level back to criticality. This "feedback" process is very rapid and makes reactors stable.

A mechanical device that contains an appropriate arrangement of nuclear fuel and moderator to sustain a nuclear reaction is called a **nuclear reactor**. In order to establish criticality in a reactor and to compensate for fuel-atom burnout as the atoms are depleted by fission, **control rods** are used. They are made of an alloy of hafnium, cobalt, or boron that has a high affinity for thermal neutrons but will not fission itself. Neutrons removed by these nuclear "poisons" cannot cause another fission. The rods are remotely operated and can be manually raised or lowered. They can also be automatically and rapidly inserted to shut down the reactor in an abnormal situation.

Another kind of fission reactor uses plutonium-239. Pu-239 is mixed with U-238. Both Pu-239 and U-238 absorb fast neutrons, so there is less need of a moderator. When Pu-239 absorbs a neutron, it splits, producing energy and more neutrons. Some neutrons strike other Pu-239 atoms, while others are absorbed by U-238. The resulting U-239 soon undergoes β-decay to form Pu-239. If the reactor is designed to produce more Pu-239 than it consumes, it is called a **breeder reactor**. Breeder reactors were used in the United States to produce nuclear-weapon-grade plutonium. Since nuclear weapons programs have been curtailed in the United States, breeder reactors have been deactivated in this country. In other countries, such as France, breeder reactors are used to produce plutonium for power reactors because fissionable uranium is scarce. Interestingly, most power reactors breed plutonium fuel during their operating lifetimes and, toward the end of each operating cycle before refueling, can generate from ⅓ to ½ of their energy from fissions in fuel produced from breeding. The next generation of nuclear reactors being considered by several nations will work this feature into their designs to improve efficiency. With the development of on-site fuel reprocessing at the nuclear facility, the generation of long-lived nuclear waste by the industry will become negligible.

When a fast neutron elastically collides with a hydrogen nucleus in a water molecule, classical mechanics shows us that the neutron's rebound velocity is reduced to as little as half of what it was before the collision. This occurs because the neutron's original momentum is divided between two particles with similar masses (a lone proton has nearly the same mass as a neutron). Elastic collisions with heavier nuclei leave the neutron with more of its original velocity, which means many more collisions are required to slow it to thermal energies. In this case, the neutron has a greater potential to "leak out" of the fuel without causing a fission.

Fission Bombs

An *uncontrolled* fission reaction is the basis of the nuclear bomb. Where a nuclear reactor uses just the critical mass of a U-235/U-238 mixture and sustains a critical reaction, a bomb requires a *supercritical* mass of nearly pure (>99% enriched) U-235 or Pu-239. To avoid a spontaneous nuclear reaction, the fissionable material is separated into two or more subcritical masses. When the bomb is to be detonated, high explosives force the masses together. The rapidly expanding chain reaction instantaneously produces far more fast neutrons than are needed to sustain an ordinary critical reaction. This *prompt-critical* reaction releases immense amounts of energy in a brief moment.

29-17 The first successful test of a fission bomb at Alamogordo, New Mexico, in 1945 was code-named "Trinity."

29.17 Fusion Reactions

Fusion reactions usually involve hydrogen or helium as the reactant. In one fusion process, four hydrogen nuclei go through several steps to emerge as a helium

29-18 A segment of the vacuum magnetic vessel of the proposed ITER fusion reactor during manufacture. The magnetic "bottle" will lie within the torus ring formed by the completed assembly of these segments. Note the size of the man in relation to the vessel.

nucleus. Other reactions produce lithium, another light nucleus. This reaction releases huge amounts of energy. Because of the energy involved and the abundance of hydrogen and helium in the sun, most scientists believe that this type of fusion reaction produces the sun's energy. Hydrogen-helium fusion converts much more mass into energy than fission reactions do. However, scientists have not yet found a way to produce usable power from the reaction.

The "energy of the future" is thought to be fusion reactions. Researchers are trying to produce a self-sustaining fusion reaction that lasts for more than a few milliseconds and that produces usable energy. The main obstacle is that to induce fusion, the light nuclei must be heated to millions of kelvins. No known material has a melting point of more than a few thousand kelvins; therefore, the superheated plasmas cannot be held in a material container. Researchers are using two methods to initiate a fusion reaction. One is creating a high-temperature plasma of hydrogen or helium and containing the ions in a magnetic "bottle." A **tokamak** is a fusion reactor built on this principle. The other method involves forcing tiny pellets together with such force that fusion takes place. These *inertial fusion* reactors use lasers to drive the nuclei together so rapidly that they cannot escape before the reaction occurs. No reactor has succeeded in generating more energy than it consumes to initiate a reaction, although scientists believe that they are close to attaining this important milestone. A consortium of nations has joined to design and build a fusion reactor prototype in France based on the tokamak plan. It is called the ITER project. Construction has begun and is expected to be completed in 2016, when the first thermonuclear plasma will be generated. The ITER is not intended to generate electricity on a commercial scale, but rather to evaluate and develop the technologies for doing so. The advantages of developing practical fusion energy are numerous. Eventually, we will be able to reduce or eliminate the use of coal and other fossil fuels for electricity generation, which will dramatically reduce atmospheric pollution and the mining operations associated with coal. Fusion energy utilizes hydrogen, which is readily available from water, as fuel. And fusion reactor operations produce far less radioactive waste than fission reactors, mostly low-level waste. Fusion energy is truly the energy of the future.

Fusion Energy and Dominion

How will the development of widespread fusion energy affect the people around the world who have been involved in coal mining for generations? How does this problem relate to the proper exercise of dominion?

ITER originally was the acronym for "International Thermonuclear Experimental Reactor," but this usage has been discontinued. The Latin meaning for the word *iter* ("the way") is preferred by the project partners.

29-19 The first thermonuclear bomb tested by the United States was 500 times more powerful than the "Trinity" test shown in Figure 29-17.

Fusion Bombs

Like fusion reactors, fusion bombs work on the principle of creating heavier nuclei out of lighter. These *thermonuclear* devices employ a complicated process of first creating deuterium and tritium nuclei (^2H and ^3H, respectively) powered by the detonation of a miniature fission bomb within the warhead and then driving these nuclei together under extremely high temperatures and pressures generated by x-rays emanating from other materials in the warhead. A fusion detonation then occurs. The whole reaction takes place in less than a few billionths of a second, producing hundreds of times more energy than a fission bomb of equivalent mass.

29.18 Nuclear Medical Tools

Medicine, as well as energy production, has benefited from the discovery of radioactivity. Gamma therapy for cancer uses γ-rays emitted by cobalt-60 sources to kill cancer cells. A machine directs the γ-rays at the cancer. Nearby healthy tissue is shielded with lead, or the beam is collimated to deliver radiation precisely to the tumor. The radiation doses are usually spread out over time to kill the cancer without causing radiation sickness.

Since radionuclides have the same chemical properties as stable nuclides, they are absorbed by the body in the same way. For instance, iodine is absorbed mostly by the thyroid. If a doctor suspects that the thyroid is working improperly, he may give the patient a dose of radioactive iodine to evaluate the thyroid's function. The doctor then traces the iodine with a radiation detector. If the proper proportion does not arrive at the thyroid, the doctor's suspicion is confirmed. Radionuclides used in this way are called **tracers**. In a similar way, radioactive "bullets" can be created that target specific organs or tissues where cancers occur. These bullets are specially formulated compounds containing intensely radioactive isotopes that bind to specific sites on cancer cells. The radiation kills the sensitive cancer cells while doing little damage to healthy cells.

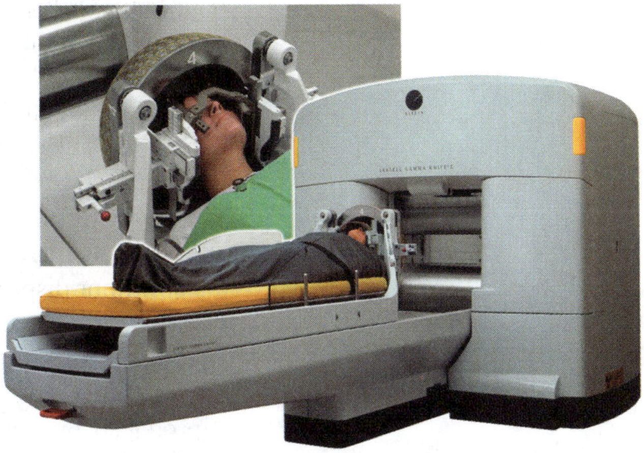

29-20 A gamma knife machine uses multiple beams of gamma rays aimed at a precisely located point in a tumor within the patient's head.

29C Objectives

After completing this section, I can

- ✓ define and relate nuclear binding energy and mass defect.
- ✓ contrast exoergic and endoergic nuclear reactions.
- ✓ compare and contrast fission and fusion reactions.
- ✓ predict whether a particular nuclear reaction will be exoergic or endoergic.
- ✓ summarize the process of nuclear fission and the factors that affect the likelihood of nuclear fission.
- ✓ contrast critical, supercritical, subcritical, and prompt-critical chain reactions.
- ✓ contrast the design and function of a fission reactor versus a fission bomb.
- ✓ contrast the design and function of a fusion reactor versus a fusion bomb.
- ✓ identify the most important technological hurdle impeding commercial fusion energy.
- ✓ name several ways radionuclides are used in medicine

29C Section Review

1. Explain why isotopes with the greatest binding energy per nucleon have the most stable nuclei.
2. Do exoergic reactions yield product nuclei with a total mass more than or less than the original reactant nuclei? Explain your answer.
3. Why would an unrestrained supercritical chain reaction be hazardous?
4. What is the purpose of a moderator in a reactor that uses U-235 for fuel?
5. If U-238 doesn't fission when it absorbs a neutron, why is the majority of the fuel U-238 in a uranium-fueled commercial power reactor?
6. a. Discuss the differences between fission and fusion reactions.
   b. Which produces more energy per reaction?
7. What is the greatest technological hurdle in developing fusion power?
8. If a radionuclide having a half-life of 18.0 h were injected into a patient for medical therapy, how long would the doctor have to retain the patient under observation in order to reduce the radionuclide activity remaining in the patient's body to 1/100th of its initial level? (This is done to minimize exposure to family members and others who might come into close proximity with the patient.)

29D SUBATOMIC PARTICLES

29.19 Early Evidence of More Subatomic Particles

In some nuclear reactions, it appeared at first that energy was not conserved. For instance, a β-particle does not have enough kinetic energy from a β-decay to account for the difference in mass between a neutron and the resulting proton and electron β-particle. (The proton remains fixed in the nucleus, but both particles have kinetic energy.) Because the conservation of energy had not been violated before, Pauli suggested in 1931 that there may be another particle emitted in a β-decay. He proposed that this neutral particle would have no rest mass and would account for the "missing" energy. The following year, Enrico Fermi named this mysterious particle the *neutrino* ("little neutral one"). How could experimenters confirm or disprove Pauli's suggestion? Unknown to researchers, neutrinos could travel through several light-years of matter before being absorbed. They left no tracks in the standard cloud, bubble, and spark chambers available at the time. Eventually, researchers designed an experiment that detected neutrinos produced by a nuclear reactor. The first neutrino interactions were detected in 1956. In this way, twenty-five years after it was first predicted, the existence of the neutrino was confirmed.

29.20 Nuclear Building Blocks

It is interesting to note that there are far more subatomic and elementary particles forming the nucleons of the atom than there are elements forming all of matter!

Nuclear reactions and particle accelerators have shown that more than 300 **subatomic** and **elementary particles**, particles smaller than atoms, may exist. These particles include the familiar proton, neutron, and electron, and some of these particles are themselves made of even more basic particles. Just as with whole nuclei, subatomic and elementary particles enter into reactions with each other and decay as well.

Physicists use two basic rules when studying these particles. First, every reaction will happen at some time unless it is prohibited by a conservation law. Even if a particular reaction has never been observed, physicists presume that if no known conservation law prohibits it, a reaction can and does occur. Thus, physicists have found that the classical conservation laws governing energy, momentum, angular momentum, and electric charge apply to the subatomic particles too.

The second rule of subatomic particles is that every particle has an **antiparticle**. That is, for every particle there is another particle of the same mass and same spin, if applicable, but its charge or, for neutral particles, its quark composition, and magnetic moment are opposite. For example, physicists studying nuclear decays discovered the positron, a particle with the same mass as an electron but with opposite charge. The positron could also be called the *anti-electron.* The symbol for an antiparticle is the particle's symbol under a bar, an opposite sign (if the particle is charged), or both. For example, the symbol for an antiproton is $\overline{\text{p}}$.

When a particle collides with its antiparticle, both completely disappear in a burst of electromagnetic energy equivalent to their rest masses ($E = m_0c^2$) plus whatever kinetic energy they possessed. This process is called *pair annihilation.*

The opposite process to pair annihilation is *pair production.* This occurs when a high-energy x-ray photon passes close to a nucleus and disappears, producing an electron and a positron pair.

It is theoretically possible for antimatter—atoms consisting of antiprotons, antineutrons, and positrons—to exist. Anti-hydrogen was even synthesized at the CERN particle physics laboratory in 1995. However, if antimatter meets ordinary matter, both will be annihilated. Naturally occurring antimatter has not been observed anywhere in the universe—only "normal" matter. This poses a serious problem for supporters of the big-bang origin of the universe. Presuming that all matter in existence originated with the condensation of particles from the intense energy "soup" shortly after the big bang occurred, there should have been an equal number of particles and antiparticles formed. Since these particles were uniformly mixed together, many (one would think, most) particles should have been quickly eliminated in pair-annihilation events, leaving nothing but high-energy photons in a dark universe. This obviously is not the case. Physicists have suggested that this lack of symmetry in the big-bang theory is due to an arcane abnormality of symmetry in particle-antiparticle processes called the *CP violation*. This concept claims that conservation laws hold true for all processes and conservative forces except for those associated with the weak nuclear interaction force. Physicists concluded from detailed statistical reviews of particle-collider experiments that subatomic particle interactions are not symmetrical for some particles and their antiparticles. Thus, they suggest that the CP violation, or something similar to it, is responsible very early in the big bang for the predominance of normal matter in the universe. Most physicists dismiss this idea, since it can't account for the fact that there are *no* observable quantities of antimatter in the universe. Therefore, the only reasonable explanation is that the universe was created with only "normal" matter.

29.21 Elementary Particle Zoo

With such a large number of subatomic particles discovered, physicists were compelled to classify them in a manner consistent with the **Standard Model** of the theory of matter. While subatomic particles can be categorized in several ways, one common method is to classify them by their property of spin. Recall that *spin* in particle physics does not mean rotational motion, but rather it is just a name for a distinguishing property.

The first group of subatomic particles, called **fermions**, have spins of $\frac{1}{2}$. These consist of the *leptons* and *quarks*, plus their antiparticles, 24 particles in all. **Leptons** include the electron (e), muon (μ), tauon (τ), and their neutrinos (ν). The lepton antiparticles are identical to the normal particles in nearly all respects except for opposite charges. Quarks will be discussed in the next subsection. Fermions also include composite subatomic particles with spins of $\frac{1}{2}$ or $\frac{3}{2}$.

Bosons make up the other major group of subatomic particles. These are particles with integer spins of 0, 1, or 2. They are generally associated with mediating the force interactions identified as the four fundamental forces, but they also include composite particles with integer spins (see Subsection 29.23). Four of these particles, including the photon, have been confirmed, while seven remain hypothetical. It is hoped that the Large Hadron Collider will provide evidence for one or more of these particles, including the elusive Higgs boson, which is believed to be a particle that could be the key to completing the Grand Unification Theory of the fundamental forces.

The **Grand Unification Theory** suggests, among other things, that the four fundamental force interactions and, in particular, the gravitational force and its hypothetical exchange particle, the graviton, are all related within a single model of the universe.

Elementary fermions and bosons can combine to form larger composite subatomic particles and their antiparticles in a supergroup called the **hadrons**. Composite fermions are generally heavy particles, classified as **baryons**, that include the familiar proton and neutron. Composite bosons form the **mesons**. There are many extremely short-lived particles that fall within these categories, as well as a host of hypothetical particles believed to exist because of predictions of various theories within the Standard Model. Table 29-4 summarizes some of the properties of the major elementary and composite subatomic particles.

Table 29-4

Characteristics of Selected Subatomic Particles*

Group	Family	Particle	Particle Symbol	Antiparticle Symbol	Charge (×*e*)	Spin	Rest Mass (MeV/c^2)
Fermions (Elementary) (½ spin)	Leptons	electron	e^-	e^+	∓1	½	0.5
		muon	μ^-	μ^+	∓1	½	106
		tauon	τ^-	τ^+	∓1	½	1777
		electron neutrino	ν_e	$\bar{\nu}_e$	0	½	<2 eV/c^2
		muon neutrino	ν_μ	$\bar{\nu}_\mu$	0	½	<0.2
		tauon neutrino	ν_τ	$\bar{\nu}_\tau$	0	½	<16
	Quarks	up	u	$\bar{u}$	±⅔	½	~2.5
		down	d	$\bar{d}$	∓⅓	½	~4.8
		charm	c	$\bar{c}$	±⅔	½	~1250
		strange	s	$\bar{s}$	∓⅓	½	~100
		top	t	$\bar{t}$	±⅔	½	~170 000
		bottom	b	$\bar{b}$	∓⅓	½	~4250
Bosons (Elementary) (Integer spin)	—	photon	γ	self	0	1	0
		W boson	W^-	W^+	∓1	1	80 400
		Z boson	Z	self?	0	1	91 200
		gluon	g	self?	0	1	0
		graviton(?)	G	self	0	2	0
Hadrons (Composite)	Baryons (Leptons)	proton	p	$\bar{p}$	±1	½	938
		neutron	n	$\bar{n}$	0	½	940.
		Lambda	Λ	$\bar{\Lambda}$	0	½	1116
		Sigma	Σ^+	$\bar{\Sigma}^-$	±1	½	1189
			Σ^0	$\bar{\Sigma}^0$	0	½	1193
			Σ^-	$\bar{\Sigma}^+$	∓1	½	1197
		Xi	Ξ^0	$\bar{\Xi}^0$	0	½	1315
			Ξ^-	$\bar{\Xi}^+$	∓1	½	1321
		Omega	Ω^-	Ω^+	+1	3/2	1672
	Mesons (Bosons)	Pion	π^+	π^-	±1	0	135
			π^0	self	0	0	140.
		Kaon	K^+	K^-	±1	1	494
			K^0	$\bar{K}^0$	0	1	497
		D meson	D^0	$\bar{D}^0$	0	0	1865
			D^+	D^-	±1	1	1870.

**Note*: Many variants exist of the composite particles that are not listed in this table.

Only a few of the subatomic particles are stable. Even the neutron, a major part of the atom, has a half-life of only 14.8 minutes when it is not combined with protons. Protons and electrons have such extremely long half-lives that they are considered stable. That is, their half-lives are many orders of magnitude longer than the known age of the universe by anyone's reckoning. Why are protons and electrons stable? No one really knows, although scientists have postulated conservation principles to account for this puzzling fact. However, if protons and electrons were not stable, atoms would not be stable, and the universe would soon disintegrate. God has designed atoms so that the universe will continue as long as He wants it to.

29.22 Quarks and Strange Flavors

After scientists discovered the existence of subatomic particles, they believed that they had discovered the basic building blocks of the universe. (Does this sound familiar?) Initially, the categories of particles included some of the leptons and several hadrons, discovered as discrete particles. By the 1950s and 1960s, many hadrons of both the baryon and meson group had been discovered. In fact, the explosive growth in the discovery of new particles became a challenging problem as researchers scrambled to bring some order to their rapidly expanding collection of particles. Physicists noted that regular patterns in these particles' rest masses and other properties suggested that they might be composites of even more elementary subatomic particles.

During the 1950s, J. Robert Oppenheimer, the American physicist called the "father of the atom bomb," quipped that he would give a Nobel Prize to the person who DIDN'T discover a new subatomic particle that year!

Then, in 1963, physicists Murray Gell-Mann and George Zweig suggested that the baryon subatomic particles could be subdivided into the fundamental entities that are now called **quarks**. There are six kinds, or *flavors*, of quarks having definite properties. In order to distinguish the flavors, scientists chose some whimsical names for their distinctiveness rather than for their descriptiveness of any given property. The six flavors in order of increasing mass are *up*, *down*, *strange*, *charm*, *top* (or *truth*), and *bottom* (or *beauty*). The leptons do not appear to have any internal structure, which qualifies them as truly fundamental particles along with quarks. The names of leptons are now called flavors as well. Table 29-4 lists the quark flavors and their charges. Note that every quark has its associated antiquark with an opposite fractional charge of the same magnitude as the quark.

Murray Gell-Mann (1929–) is an American physicist who won the Nobel Prize in Physics in 1969 for his work on elementary particles.

George Zweig (1937–) is an American physicist and neurobiologist.

The up, down, and strange quarks were discovered first. When combined, they form the hadron particles. Scientists recognized that since some of the baryons were charged, the quarks must have fractional charges in order to combine in such a way that they produce unit elementary charges in the larger particles. Figure 29-21 shows how several of the hadrons are formed by various quarks.

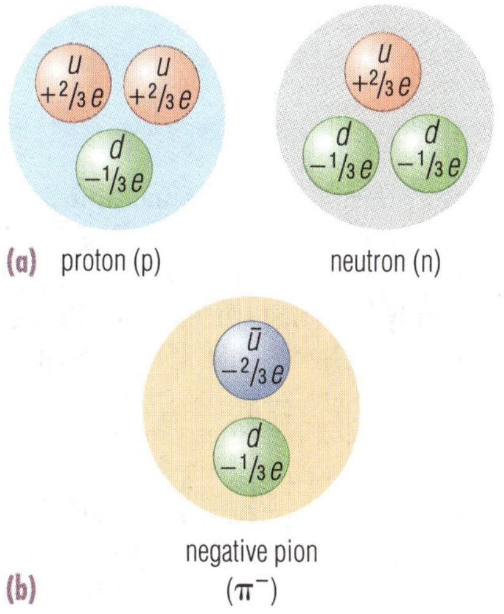

29-21 (a) Three quarks are required to form the baryons, such as the proton and neutron. (b) A quark and antiquark are required to form the mesons, such as the negative pion.

The individual masses of quarks are determined by complicated calculations that take into account their kinetic energies in particle-photon interactions. Even though they cannot be observed directly, their existence has been experimentally verified repeatedly through the application of conservation laws in high-energy particle experiments.

Quarks can exist only inside of hadrons. They cannot exist independently like electrons and the other leptons can, no matter how much energy is expended to tear the hadron particle apart. This theory is called **quark confinement**. The forces that hold quarks tightly within the hadron particles are described by a property called *color charge*, or just *color*. The color charges are analogous to electric charges between the quarks, and there are three of them: red, blue, and green. Color charges are related to the strong nuclear interaction force as described in the area of quantum mechanics called *chromodynamics*. Note that "color charge" and the three colors are whimsical names that physicists chose for this property of quarks. They have nothing to do with the visual appearance of quarks.

Are there even more basic particles than quarks? Scientists today will not make any predictions one way or the other. Christians can be sure that, ultimately, it is God Who sustains the universe through His providential control manifested in these subatomic particles.

29.23 Exchange Particles

All of the forces in nature have been placed in four categories: gravitational force, weak nuclear interaction force, electromagnetic force, and strong nuclear interaction force, in order of increasing strength. Each kind of force can be understood in terms of an **exchange particle**. If one object transfers an exchange particle to another object, there is a force or interaction between the two objects.

The weakest force is the *gravitational force*. If the strong nuclear interaction force has a strength of 1, gravity would have a strength of 10^{-45}. However, gravity's range is unlimited. The predicted exchange particle for gravity is the massless, uncharged *graviton*, which has not yet been observed. The force of gravity is the only force that affects all particles. In spite of its relative weakness, however, gravity is most important in long-distance interactions. For example, gravity holds planets, solar systems, and galaxies together.

The *weak nuclear interaction force* is much stronger than gravity but still much weaker than the strong nuclear interaction force. If the strong nuclear interaction force is 1, the weak nuclear interaction force would be 10^{-8}. The range of the weak nuclear interaction force is the shortest range—only 10^{-17} m, or less than 1/1000 the diameter of the nucleus. The exchange particle for the weak nuclear interaction force was predicted to be the heavy *W-particle*. The W-particle was first detected in the 1980s. This particle is more than 100 times more massive than the proton. It comes in three varieties: W^+, W^-, and Z^0, with charges of $+e$, $-e$, and 0, respectively. The weak nuclear interaction force is involved in the transmuting of particles that exist within the nucleus. For instance, β-decay in neutrons and protons is mediated by the weak nuclear interaction force among protons, neutrons, and electrons/positrons.

The *electromagnetic force* is nearly as strong as the strong nuclear interaction force. If the strong nuclear interaction force is 1, the electromagnetic force would be 10^{-3}. Its range is unlimited. The exchange particle for the electromagnetic force, according to the theory, is the photon, which has been detected. Electromagnetic forces affect all charged particles. For example, they hold electrons within orbitals around an atom's nucleus.

The strongest force is the *strong nuclear interaction force.* Its range is short—just 10^{-15} m. The predicted exchange particle, the *pion* (π meson), has been detected. This particle is about 270 times more massive than an electron. Like the W-particle, a pion may have a charge of $+e$, $-e$, or 0. The strong nuclear interaction force affects baryons and mesons. For example, it holds the nucleus of an atom together and relates to the interactions between quarks.

The strengths of the four fundamental forces that have been identified are related in the exact ratio that our present universe requires. If the ratios of the forces were even one order of magnitude different—for example, if the relative strength of gravity were 10^{-46} or 10^{-44} instead of 10^{-45}—the universe would be radically different from the one we know, and life could not exist. Atheistic scientists are amazed that the universe is exactly right for us but deny any anthropic principle. Bible believers, however, realize that God has created the universe harmoniously in a way that preserves His creation and brings honor to Himself.

The **anthropic principle** states, in essence, that the universe as it exists with all of its laws and features is the only kind that can support life—specifically human life. If it were different, it could not. The concept is controversial, because some believe that it leaves the door open to an intelligent Designer, which is an untestable theory and therefore, they say, has no part in naturalistic science.

29D Section Review

1. a. What was the first elementary subatomic particle discovered?
   b. Of what class of elementary subatomic particles is it a member?
2. a. Discuss the characteristics of an antiparticle.
   b. Are particles and their antiparticles compatible?
3. a. What seems to be the relative proportion of antimatter to "normal" matter in the universe?
   b. What does this proportion suggest about the origin of the universe?
4. What group of particles are the building blocks of the hadron composite subatomic particles?
5. What three quarks would have to be combined in order to produce an antiproton?
6. a. What is an exchange particle?
   b. Suggest a reason why scientists believe that the graviton exists.

29D Objectives

After completing this section, I can

✓ summarize the discoveries of elementary subatomic particles.

✓ discuss the general properties of subatomic particles and their antiparticles.

✓ differentiate between the supergroups, groups, and main classes of elementary and composite subatomic particles.

✓ describe how quarks combine to produce the composite subatomic particles.

✓ identify the principal exchange particles and explain how they are responsible for the four fundamental forces and interactions.

Chapter Review

In Terms of Physics

Problem-Solving Strategies

29.1 (page 672) Here is a brief review of the laws of logarithms:

If $n^y = x$, then $\ln_n x = y$

$\ln e = 1$

$\ln 1 = 0$

$\ln x \cdot y = \ln x + \ln y$

$\ln x/y = \ln x - \ln y$

$\ln x^y = y \ln x$

Review Questions

1. Describe the three kinds of nuclear radiation.
2. a. What happens to a nucleus's atomic number when it emits a β-particle?
   b. What happens to its mass number?
3. Name the unidentified product in the reaction: $^{22}_{11}\text{Na} \rightarrow ^{22}_{10}\text{Na} + ?$
4. How does the energy of γ-rays generally compare to that of x-rays?
5. Name two detectors of radiation.
6. What kind of instrument provides real-time indication of a radiation worker's dose in millirem or sieverts?
7. What kind of mathematical curve does radioactive decay follow when you plot activity versus time?
8. What is the half-life of a radionuclide?
9. What can carbon dating, when used properly, reveal about the life of an organism?
10. Why is carbon dating more reliable than, for instance, potassium-argon dating?
11. Give one weakness of
    a. the K-Ar dating method.
    b. the U-Pb dating method.
12. Thermodynamically speaking, what kind of reactions are fission and fusion?
13. What is the difference between fission and fusion?
14. If *A* produces *B* and *C*, will the reaction be endoergic or exoergic?

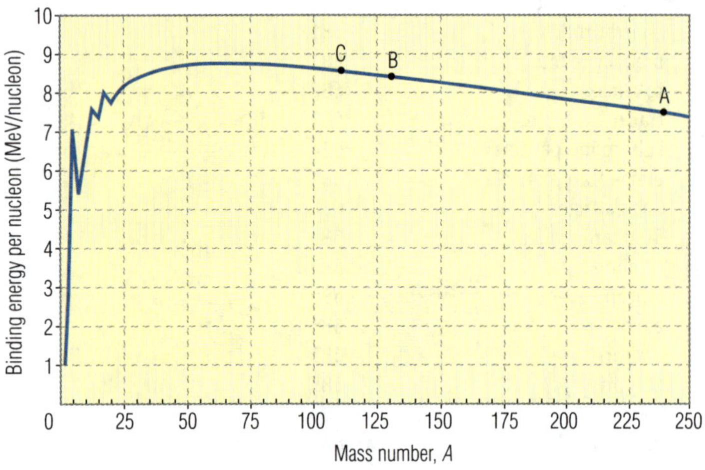

15. What is a chain reaction?
16. Why does a U-235 fission reactor need a moderator?
17. What do reactor control rods do?
18. What undergoes fission in a breeder reactor?
19. Describe two methods of initiating fusion in a controlled fashion.
20. What is the main obstacle to practical fusion power?
21. Which early discovered elementary subatomic particle hardly interacts with matter at all?
22. Explain why pair annihilation argues against the plausibility of the big-bang origin of the universe.
23. What is the significance of the names used for the six flavors of quarks?
24. Make a table of the four fundamental forces, listing them from weakest to strongest. In the table, provide columns for relative strength (strong nuclear interaction force = 1), effective distance, associated exchange particle, and whether the exchange particle has been detected.

True or False (25–40)

25. Both x-rays and nuclear radiation were unintentional discoveries.
26. The total number of neutrons determines the stability of the nucleus.
27. A more penetrating form of radiation will have a higher radiation weighting factor than a form that is stopped at the surface of the skin.
28. A person can receive essentially zero sieverts of radiation per year by living as far away from a nuclear reactor as possible.
29. Nothing seems to change the radioactivity of a given amount of a nuclide.
30. The larger the decay constant, the shorter the half-life of an isotope.
31. As a young-earth creationist, you would expect the oldest possible corrected radiocarbon date to be around one C-14 half-life.
32. The steadiness of cosmic ray intensity over thousands of years has little to do with carbon-14 dating techniques.
33. Fission reactions are exoergic but fusion reactions are endoergic.
34. A nuclear chain reaction does not *have* to be a supercritical reaction.
35. Control rods function by removing neutrons from the reactor so that they cannot cause more fissions.
36. One function of a breeder reactor is to produce more U-235.
37. The neutrino was the first lepton elementary subatomic particle discovered.

38. There are only a few known subatomic particles.

39. The antiproton has a charge of $-e$.

40. Force-exchange particles are included in the collection of elementary subatomic particles.

✪**41.** A bismuth-212 nucleus emits an α-particle.

a. What is the daughter product's mass number, *A*?

b. What is the daughter product's atomic number, *Z*?

c. Use the periodic table in Appendix H to find the name of the daughter product. Write it as where X is the product's proper abbreviation.

✪**42.** If a certain nuclide has a half-life of 5.4 y, what is its decay constant?

✪**43.** Berillium-8 has a decay constant of $1.03 \times 10^{16}\ s^{-1}$. What is its half-life?

✪**44.** Construct a positive pion using any two of up, down, or strange quarks.

DM✪**45.** In 29B Section Review Question 11, you were asked to create what is essentially a C-14 calibration curve from the Dead Sea Scrolls data provided in Table 29-3 on page 677. Other calibration curves have been created from various organic materials that have chronological properties, such as ancient tree-ring data (dendrochronology). One such recent calibration curve, called the IntCal04 curve, is a composite C-14 calibration curve developed from a variety of sources, including tree rings, corals, and foraminifera/varve data. Scientists claim that this curve extends valid C-14 dates back to 26 000 y BP.

a. How does this view align with a biblical view of the earth's history?

b. For this step, your teacher will have to provide you a portion of the IntCal04 data set pertaining to the span of time encompassing the Dead Sea Scrolls. Add this curve to the graph that you created in 29B Section Review Question 11. Create trendlines for both curves.

c. Discuss how well the two calibration curves match up, and suggest reasons for the differences between them.

✪**46.** After doing some research in other references, write the reactions of one possible decay chain leading from U-238 to Pb-206. (There are at least six.)

✪**47.** Where did the term "quark" originate?

✪**48.** Write a paper discussing the experiment and describe the data that confirmed the existence of the neutrino.

✪**49.** Give a report on the latest research in refining the quark model, neutrino research, or the Grand Unification Theory.

Facets of Physics

Cosmic Rays

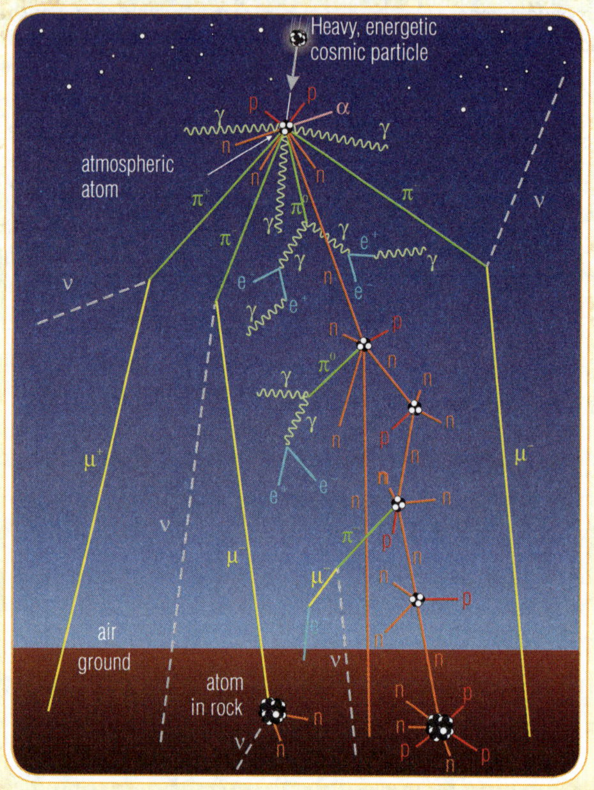

What is the main source of the high-energy particles found on the earth? It is not a particle accelerator, as you might think. Long before artificial accelerators had been developed, astronomers discovered that a steady stream of high-energy radiation from outer space bombards the earth day and night. At first, they thought it was gamma radiation, so they called the radiation cosmic *rays.*

But they were wrong. Instead of being rays, the radiation was a mixture of about 10% electrons and 90% atomic nuclei (83% of these were protons, 16% helium, and 1% from all other elements on the periodic table). What first intrigued astrophysicists about this super-energetic radiation was the variety of subatomic particles that it produced after colliding with atmospheric nuclei. In a sense they were witnessing the opposite of a nuclear explosion—massive energy was being replaced by tiny particles of matter! These particles (positrons, muons, pions, and neutrinos) had never been seen before because they require such concentrated energy and are unstable (neutral pions last about 10^{-16} s).

Most *primary cosmic rays* (radiation from outer space) never make it to the earth's surface because they run into atmospheric nuclei first. Even at high altitudes, only one in a thousand survive the trip. Therefore, most observatories study the rays indirectly by examining the particles produced during atmospheric collisions. When a ray strikes a nucleus, its kinetic energy is dispersed in one of two ways: either the energy turns into pions with kinetic energy, or the ray splits the nucleus (a fission) and imparts kinetic energy to the fragments. The pions and fission fragments are called *secondary cosmic rays.* They, in turn, decay or sometimes collide with other nuclei to form more secondary cosmic rays, some of which will eventually reach the earth.

The energy in some of the primary rays is incredible. A typical particle carries billions of electron volts (10^9 eV or GeV). By contrast, the particles at room temperature have a kinetic energy of 1/20 eV, and particles from a nuclear explosion carry 10^7 eV (or 10 MeV). A few electrons and positrons exceed 10^{15} eV, which cannot be directly measured on the earth. The highest-energy particle measured so far had 10^{20} eV. That is enough energy, in one nuclear particle, to raise a 1 kg mass 1 m! Such high-energy particles begin a chain reaction when they enter our atmosphere, producing an electron shower, or *cascade shower,* similar to meteor showers. In some instances, one particle has produced 10 billion new particles that spread over a 1000 m range before reaching the earth.

After the development of artificial accelerators, astronomers stopped looking at subatomic physics in space and began inquiring about the source of cosmic rays and how the rays are accelerated. The issue remains a mystery of astronomy, and many hypotheses have been proposed to account for the rays.

Only one thing is certain: the radiation is deadly to life on the earth. Without our atmosphere's cushion, we would all die from radiation poisoning. Because astronauts lack this protection, their cabin and space suits must be made of heavy, thick material to impede radiation. Currently, the protection is enough to block most rays, and the flights are so short that the limited exposure is not threatening. But new designs and materials will have to be developed for lengthy trips to places such as Mars.

APPENDIX A

Base and Derived* Units of the SI†

Dimension	Name	Symbol	Definition
Base Units			
length	meter	m	The meter is the length of the path traveled by light in a vacuum during a time interval of 1/299 792 458 of a second.
mass	kilogram	kg	The kilogram is the unit of mass; it is equal to the mass of the international prototype of the kilogram.
time	second	s	The second is the duration of 9 192 631 770 periods of the radiation corresponding to the transition between the two hyperfine levels of the ground state of the cesium-133 atom.
electric current	ampere	A	The ampere is that constant current which, if maintained in two straight, parallel conductors of infinite length, of negligible circular cross section, and placed 1 meter apart in a vacuum, would produce between these conductors a force equal to 2×10^{-7} newtons per meter of length.
thermodynamic temperature	kelvin	K	The kelvin, the unit of thermodynamic temperature, is the fraction 1/273.16 of the thermodynamic temperature of the triple point of water.
amount of substance	mole	mol	1. The mole is the amount of substance of a system that contains as many elementary entities as there are atoms in 0.012 kilogram of carbon-12. 2. When the mole is used, the elementary entities must be specified and may be atoms, molecules, ions, electrons, other particles, or specified groups of such particles.
luminous intensity	candela	cd	The candela is the luminous intensity, in a given direction, of a source that emits monochromatic radiation of frequency 540×10^{12} hertz and that has a radiant intensity in that direction of 1/683 watt per steradian (see definition below).
The Radian and Steradian			
plane angle	radian	rad	The radian is the plane angle between two radii of a circle that cuts off on the circumference an arc equal in length to the radius.
solid angle	steradian	sr	The steradian is the solid angle that, having its vertex in the center of a sphere, cuts off an area of the surface of the sphere equal to the radius of the sphere squared.
Derived Units			
absorbed dose	gray	Gy	$J/kg = m^2 \cdot s^{-2}$
activity (radionuclide)	becquerel	Bq	s^{-1}
capacitance	farad	F	$C/V = A^2 \cdot s^4 \cdot kg^{-1} \cdot m^{-2}$
Celsius temperature	degree	°C	K
electric charge	coulomb	C	$A \cdot s$
electric potential difference, emf	volt	V	$J/C = kg \cdot m^2 \cdot A^{-1} \cdot s^{-3}$
electric resistance	ohm	Ω	$V/A = kg \cdot m^2 \cdot A^{-2} \cdot s^{-3}$
energy, work, quantity of heat	joule	J	$N \cdot m = kg \cdot m^2 \cdot s^{-2}$
equivalent dose	sievert	Sv	$J/kg = m^2 \cdot s^{-2}$
force	newton	N	$kg \cdot m \cdot s^{-2}$
frequency	hertz	Hz	s^{-1}
illuminance	lux	lx	$lm \cdot m^{-2} = cd \cdot sr \cdot m^{-2}$
inductance	henry	H	$Wb/A = kg \cdot m^2 \cdot A^{-2} \cdot s^{-2}$
luminous flux	lumen	lm	$cd \cdot sr$
magnetic flux	weber	Wb	$V \cdot s = J/A = kg \cdot m^2 \cdot A^{-1} \cdot s^{-2}$

Base and Derived* Units of the SI† *[continued]*

Dimension	Name	Symbol	Definition
Derived Units (continued)			
magnetic flux density	tesla	T	$Wb/m^2 = kg \cdot A^{-1} \cdot s^{-2}$
plane angle	radian	rad	$m \cdot m^{-1} = 1$
power, radiant flux	watt	W	$J/s = kg \cdot m^2 \cdot s^{-3}$
pressure, stress	pascal	Pa	$N/m^2 = kg \cdot m^{-1} \cdot s^{-2}$
solid angle	steradian	sr	$m^2 \cdot m^{-2} = 1$
Experimentally Derived Units			
particle energy	electron volt	eV	The electron volt is the kinetic energy acquired by an electron in passing through a potential difference of 1 V in a vacuum; 1 eV = 1.602 176 53 × 10^{-19} J‡
particle mass	unified atomic mass unit	u	The unified atomic mass unit is equal to 1/12 of the mass of an atom of the nuclide ^{12}C; 1 u = 1.660 538 782 × 10^{-27} kg.
Other Units Used with the SI			
pressure	bar	bar	1 bar = 0.1 MPa = 100 kPa = 10^5 Pa
level of power	bel	B	$L_B = \log_{10}(P/P_0)$, where L_B is the relative power in bels, P_0 is the reference power, and P is the measured power. A power level of 1 B exists when $P = 10P_0$.
	decibel	dB	$L_{dB} = 10\log_{10}(P/P_0)$, where L_{dB} is the relative power in decibels, P_0 is the reference power, and P is the measured power. A power level of 1 dB exists when $P = 10^{0.1}P_0$ (or approximately $1.259P_0$).

*Derived and Other units listed are those used in this textbook.

†NIST (URL: http://physics.nist.gov/Pubs/SP811/cover.html)

‡Peter J. Mohr and Barry N. Taylor (January 2005). CODATA recommended values of the fundamental physical constants: 2002; Reviews of Modern Physics 77: 1–107. doi:10.1103/RevModPhys.77.1, p. 62

APPENDIX B

Commonly Used Unit Abbreviations

Unit	Abbreviation	Unit	Abbreviation
ampere	A	kilowatt-hour	kWh
atmosphere	atm	meter	m
atomic mass unit (unified)	u	millibar	mb
coulomb	C	minute	min
day	d	mole	mol
decibel	dB	newton	N
degree (angle)	°	ohm	Ω
degree Celsius	°C	pascal	Pa
degree Fahrenheit	°F	radian	rad
electron volt	eV	revolution	rev
farad	F	second	s
gram	g	tesla	T
henry	H	torr	torr
hertz	Hz	volt	V
hour	h	watt	W
joule	J	weber	Wb
kelvin	K	year	y
kilogram	kg		

APPENDIX C

Unit Conversions

The conversion factor required to convert a dimension in the left vertical column to the dimension in the top row is found in the table where the two dimensions cross. The factor may be used in dimensional analysis.

Length	cm	m	km	in.	ft	mi
1 cm =	1	10^{-2}	10^{-5}	0.3937	3.281×10^{-2}	6.214×10^{-6}
1 m =	100	1	10^{-3}	39.37	3.281	6.214×10^{-4}
1 km =	10^5	1000	1	3.937×10^4	3281	0.6214
1 in. =	2.540	2.54×10^{-2}	2.540×10^{-5}	1	8.333×10^{-2}	1.578×10^{-5}
1 ft =	30.48	0.3048	3.048×10^{-4}	12	1	1.894×10^{-4}
1 mi =	1.609×10^5	1609	1.609	6.336×10^4	5280	1

1 angstrom (Å) = 1.000×10^{-10} m

For astronomical distances, see Appendix E.

1 nautical mile (nm) = 6076 ft = 1852 m

Mass	g	kg	u	oz	lbm	t
1 g =	1	0.001	6.022×10^{23}	3.527×10^{-2}	2.205×10^{-3}	1.102×10^{-6}
1 kg =	1000	1	6.022×10^{26}	35.27	2.205	1.102×10^{-3}
1 u =	1.660×10^{-24}	1.660×10^{-27}	1	5.855×10^{-26}	3.660×10^{-27}	1.829×10^{-30}
1 oz =	28.35	2.835×10^{-2}	1.708×10^{25}	1	6.250×10^{-2}	3.125×10^{-5}
1 lbm =	453.6	0.4536	2.732×10^{26}	16	1	5.000×10^{-4}
1 t =	9.072×10^5	907.2	5.465×10^{29}	3.2×10^4	2000	1

Time	s	min	h	d	y
1 s =	1	1.667×10^{-2}	2.778×10^{-4}	1.157×10^{-5}	3.169×10^{-8}
1 min =	60	1	1.667×10^{-2}	6.944×10^{-4}	1.901×10^{-6}
1 h =	3600	60	1	4.167×10^{-2}	1.141×10^{-4}
1 d =	8.640×10^4	1440	24	1	2.738×10^{-3}
1 y =	3.156×10^7	5.260×10^5	8.766×10^3	365.25	1

Pressure	atm	$in._{Hg}$	torr	Pa	mb	lbf/in.^{2}
1 atm =	1	29.92	760	1.013×10^5	1013	14.70
1 $in._{Hg}$ =	3.342×10^{-2}	1	25.4	3386	33.86	0.4912
1 torr =	1.316×10^{-3}	3.937×10^{-2}	1	133.3	1.333	0.01934
1 Pa =	9.869×10^{-6}	2.953×10^{-4}	7.501×10^{-3}	1	0.01	1.450×10^{-4}
1 mb	9.869×10^{-4}	2.953×10^{-2}	0.7501	100	1	1.450×10^{-2}
1 lbf/in.^{2} =	6.805×10^{-2}	2.036	51.71	6.895×10^3	68.95	1

Volume	cm^3	m^3	L	in.^{3}	ft^3
1 cm^3 =	1	1.0×10^{-6}	1.000×10^{-3}	6.102×10^{-2}	3.531×10^{-5}
1 m^3 =	10^6	1	1000	6.102×10^4	35.31
1 L =	1000	1.000×10^{-3}	1	61.02	3.531×10^{-2}
1 in.^{3} =	16.39	1.639×10^{-5}	1.639×10^{-2}	1	5.787×10^{-4}
1 ft^3 =	2.832×10^4	2.832×10^{-2}	28.32	1728	1

1 US fl. gal. = 4 US fl. qt. = 8 US pt. = 128 US fl. oz. = 231 in.^{3}

1 British imperial gal. = 277.4 in.^{3}

APPENDIX D

Physical Constants

Quantity	Symbol	Value
Fundamental		
Atomic mass unit (unified)	u	1.661×10^{-27} kg
Avogadro's number	N_A	6.022×10^{23} mol^{-1}
Boltzmann's constant	k	1.381×10^{-23} J/K
Electron mass	m_e	9.109×10^{-31} kg
Fundamental charge	e	1.602×10^{-19} C
Gas constant (SI units)	R	8.314 J/(mol·K)
Gravitational constant	G	6.674×10^{-11} N·m^2/kg^2
Neutron mass	m_n	1.675×10^{-27} kg
Permeability of free space	μ_0	$4\pi \times 10^{-7}$ T·m/A
Permittivity of free space	ε_0	8.854×10^{-12} C^2/(N·m^2)
Planck's constant	h	6.626×10^{-34} J·s
	$\hbar \equiv h/2\pi$	1.055×10^{-34} J·s
Proton mass	m_p	1.673×10^{-27} kg
Speed of light in vacuum	c	2.998×10^{8} m/s
Stefan-Boltzmann constant	σ	5.670×10^{-8} W/(m^2·K^4)
Wien displacement law constant	$\lambda_{max}T$	2.898×10^{-3} mK
Other Useful Constants		
Absolute zero	0 K	−273.15 °C
Atmospheric pressure, standard	1 atm	1.013×10^{5} Pa
Coulomb constant	k	8.988×10^{9} N·m^2/C^2
Density of air	—	1.20 kg/m^3 (20 °C and 1 atm)
Density of water	—	1.000×10^{3} kg/m^3 (20 °C and 1 atm)
Gravitational acceleration, standard	$\lvert g \rvert$	9.81 m/s^2 (average, near the earth's surface)
Latent heat of fusion of water	L_f	333.5 J/g
Latent heat of vaporization of water	L_v	2256 J/g
Rest energy of 1 u	—	931.5 MeV
Speed of sound in air	—	343 m/s (20 °C and 1 atm)
Triple point of water	T_{tpw}	273.16 K (at 0.006 atm)

APPENDIX E

Planetary and Astronomical Data†

Object	Radius (km)	Mass (kg)	Density (g/cm^3)	Orbital Radius* ($\times 10^8$ km)	Rotational Period (d)	Orbital Period (y/d)
Sun	696 000	1.989×10^{30}	1.42	—	25.6	—
Mercury	2440	3.302×10^{23}	5.43	0.5791	58.66	0.2408 / 87.97
Venus	6052	4.869×10^{24}	5.24	1.082	243.01	0.6152 / 224.70
Earth	6378	5.974×10^{24}	5.515	1.496	0.9973	1.0000 / 365.26
(Moon)	1738	7.348×10^{22}	3.34	0.00384	29.53	0.074 80 / 27.32
Mars	3397	6.419×10^{23}	3.93	2.279	1.026	1.8809 / 686.98
Jupiter	71 492	1.899×10^{27}	1.33	7.783	0.4135	11.862 / 4332.7
Saturn	60 268	5.685×10^{26}	0.69	14.294	0.445	29.458 / 10 759.5
Uranus	25 559	8.663×10^{25}	1.27	28.710	0.72	84.01 / 30 685
Neptune	24 764	1.028×10^{26}	1.64	45.043	0.673	164.79 / 60 190
(Pluto)**	(1195)	(1.85×10^{22})	(1.75)	(59.064)	(6.405)	(248.68 / 90 465)

†Reference: NASA Planetary Fact Sheets: http://nssdc.gsfc.nasa.gov/planetary/planetfact.html

*Mean orbital radius (semi-major axis)

**Since 2006, the International Astronomical Union (IAU) has designated Pluto as a "dwarf planet." Information about Pluto is included here with the other planets for completeness.

Other Data

Astronomical unit: 1 ua = 1.496×10^8 km

Lunar gravitational acceleration (g_{Moon}) = 1.62 m/s^2

Light year: 1 ly = 9.461×10^{12} km

Parsec: 1 pc = 3.086×10^{13} km

(A parsec is the astronomical distance unit defined by the distance from the earth an object must be in order to exhibit 1 arc-second of optical parallax caused by the earth's revolution about the sun.)

APPENDIX F

Moments of Inertia for Selected Regular Objects

Object	Description	Formula	Notes
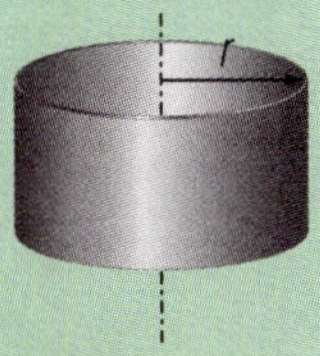	Thin ring Thin cylindrical shell	$I = mr^2$	Thickness of ring << r
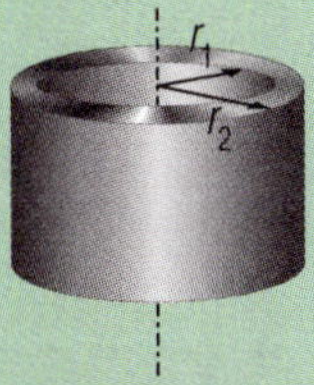	Thick ring Thick cylindrical shell	$I = \frac{1}{2}m(r_1^2 + r_2^2)$	r_1 and r_2 are radii of inner and outer wall respectively
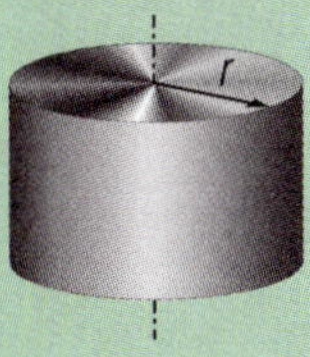	Solid disk Solid cylinder } center axis	$I = \frac{1}{2}mr^2$	
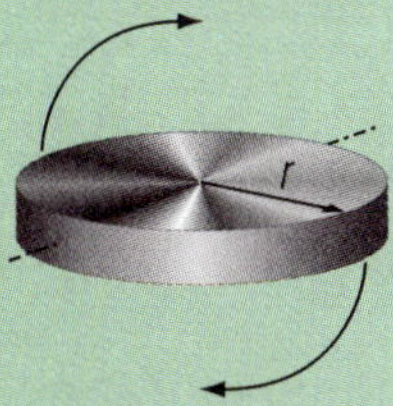	Thin solid disk; transverse axis	$I = \frac{1}{4}mr^2$	
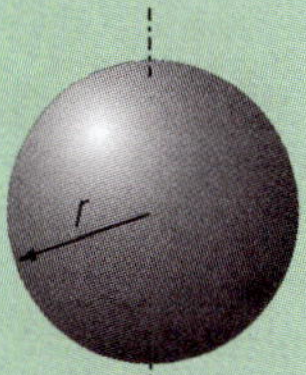	Solid sphere	$I = \frac{2}{5}mr^2$	
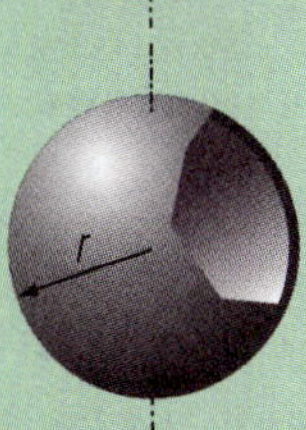	Thin spherical shell	$I = \frac{2}{3}mr^2$	Thickness of wall << r
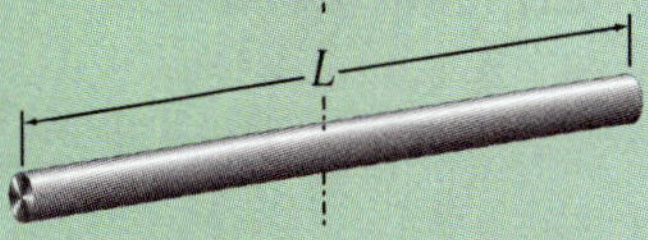	Thin solid rod—axis at middle	$I = \frac{1}{12}mL^2$	L is length of rod
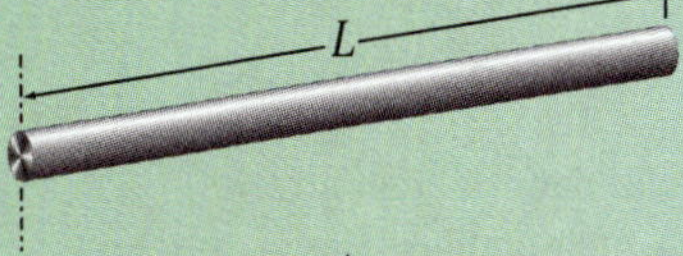	Thin solid rod—axis at end	$I = \frac{1}{3}mL^2$	L is length of rod
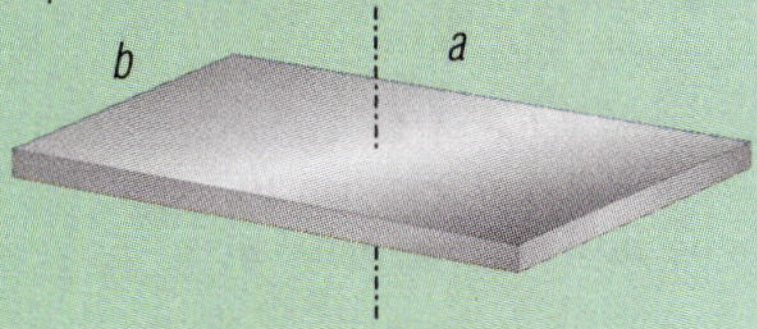	Rectangular plate	$I = \frac{1}{2}m(a^2 + b^2)$	a and b are the length and width; axis is perpendicular through center of face ab

APPENDIX G ◆ MATHEMATICAL REFERENCE

This appendix is provided as a ready reference for most of the mathematical principles required for this textbook. It is not intended to be exhaustive.

A. Mathematical Symbols

1. *Relational Symbols*

Symbol	Definition
$\equiv$	is defined as
$=$	equals exactly; assigns a value to
$\neq$	does not equal
$\cong$	a rounded or cut-off calculator result (not rounded to correct significant digits)
$\cong$	answer rounded to correct significant digits
$\sim$	approximately equal to; result is representative of actual value
$<>$	less than; greater than
$\leq\geq$	less than or equal to; greater than or equal to
$\ll\gg$	much less than; much greater than
$\Leftrightarrow$	if and only if
$\Rightarrow$	implies; "if *a*, then *b*" is $a \Rightarrow b$
$\propto$	proportional to

2. *Other Symbols*

$\pm$	plus or minus—value following may be positive or negative
$\lvert x \rvert$	absolute value of x; magnitude of x
Δx	change in x; final value minus initial value, or $\Delta x = x_2 - x_1$
Σx	summation of x; the sum of all values represented by x
$x \to 0$	x approaches zero; the magnitude of x becomes as small as desired
∞	infinity; the value is as large as desired

B. Rounding Rules

1. Determine the position of the digit to which the number must be rounded. This digit will be referred to as the least significant digit (the "least SD"). The rules for determining significant digits are discussed in Chapter 2.
2. Note the digit to the right of the least SD and round according to the following rules:
   a. If the number is 0 through 4, drop all digits to the right of the least SD. If necessary, substitute the dropped numbers with zeros as placeholders to locate the decimal point, but do not write in the decimal point.
   b. If the number is 5 through 9, add 1 to the least SD and drop the remaining digits, replacing them with zero placeholders, if required.
   c. Do not consider any digits other than the one immediately to the right of the least SD. Specifically, do *not* begin rounding from the right-most digit. This often results in a final digit that has been rounded up in error.
3. Examples:

   Round 37 547 m to the nearest 1000 m. *Answer*: 38 000 m

   Round 1.448 cm to the nearest 0.1 cm. *Answer*: 1.4 cm

C. Scientific and Engineering Notation

1. Scientists and engineers must often deal with extremely large or small numbers that involve many zeros for locating the decimal point. To make working with such numbers easier, a shorthand method of writing them is used called *scientific* or *engineering notation.*
2. In *scientific notation*, the number is written in the form $M \times 10^n$. The value of M can be positive or negative, but its magnitude must be greater than or equal to 1 and less than 10 ($1 \leq \lvert M \rvert < 10$). The exponent, n, is an integer (not a fractional or decimal number) and can be positive, negative, or zero.

3. In order to convert a number to scientific notation, move the decimal point left or right as necessary until only one nonzero digit appears to the left of the decimal point. The number of places that the decimal point has moved is the exponent (n). If you moved the decimal point to the left, the exponent is positive. If you moved the decimal point to the right, the exponent is negative. Retain in M all significant digits present in the original number. All nonsignificant trailing zeros should be dropped. For example:

Decimal Form	*Improper*	*Proper*
0.000057	0.57×10^{-4}	5.7×10^{-5}
34500000	345×10^{5}	3.45×10^{7}
176.00	1.76×10^{2}	1.7600×10^{2}

As you can see in the third example, scientific notation can be used to indicate that the trailing zeros in a number are significant.

4. If all the trailing zeros in a small number are significant, indicate this fact by a decimal point following the final zero.

5400 cm	2 SDs	50 m	1 SD
5.40×10^{3} cm	3 SDs	50. m	2 SDs

5. *Engineering notation* is similar to scientific notation in that it has both a decimal and an exponential part. However, it is often formatted in one or more of the following ways for convenience and clarity:

   a. The exponential part is often expressed in powers divisible by 3. In other words, the position of the decimal point in the M term is moved so that the power of 10 is divisible by 3.

   Example: 3.94×10^4 m = 39.4×10^3 m

   b. Similar to Part a, but the exponential term is replaced by the appropriate metric unit prefix.

   Example: 3.94×10^4 m = 39.4×10^3 m = 39.4 km

   c. The factor "$\times 10^n$" is often replaced by the symbol "E*n*."

   Example: 3.94×10^4 m = 3.94E4 m

   (*Note*: This notation is used by many spreadsheet programs for entering and displaying scientific/engineering data.)

   Engineering notation is also useful for comparing the orders of magnitude of a group of numbers. The decimal part, M, may be any value as long as the power of 10 is the same for each number.

   Example: Compare the masses of the planets Mercury, Earth, and Jupiter in scientific notation.

   Mercury, 3.302×10^{23} kg; Earth, 5.974×10^{24} kg; Jupiter, 1.899×10^{27} kg

   In engineering notation:

   Mercury, 0.3302×10^{24} kg; Earth, 5.974×10^{24} kg; Jupiter, 1899×10^{24} kg

   The relative masses of the planets are made much more evident by using engineering notation.

   When using engineering notation, you should be aware that it does not always preserve significant digits, so some loss of precision may occur.

D. Logarithms and Exponential Operations

The concept of relating a geometrical progression of numbers to an arithmetic one by using logarithms seems to have been first considered by the German mathematician Michael Stifel (1486–1567) in his book *Arithmetica integra* ("The Complete Arithmetic"). However, logarithms were not generally accepted as a mathematical tool until after 1614, when the Scottish politician and mathematician John Napier (1550–1617) published a paper titled *Mirifici logarithmorum canonis descriptio* ("A Description of the Wonderful Law of Logarithms"). His purpose was to simplify the process of computation by reducing multiplication and division to addition and subtraction. While his original effort was associated with trigonometric functions, he and a collaborator, Henry Briggs (1561–1631), generalized the concept by developing logarithms using the base 10, which is highly compatible with the common base-10 mathematics we use today.

Logarithms and Exponential Operations (continued)

1. Any positive real number, *n*, can be expressed as a base number, *a*, raised to some exponent, *b*. In an equation, this can be written as

$$n = a^b.$$

2. The number *b* is called the *logarithm* of *n* to the base *a*, so a logarithm is nothing more than an exponent. This expression is written as:

$$\log_a n = b$$

3. Any positive number may be the base of a logarithm, but you are likely to come across only certain numbers used as bases in physics. Logarithms of base 10 are called *common logarithms*, while those of the base *e* (2.71828. . .) are called *natural logarithms*. The common log of *n* is expressed as

$$\log n = b,$$

where the base (*a*) is understood to be 10. The natural log of *n* is expressed as

$$\ln n = b,$$

where the base of ln is understood to be *e*.

4. The logarithms of numbers greater than 1 are positive decimal values. The logarithms of numbers between 0 and 1 are negative decimals. The logarithm of 1 is zero. The logarithm of zero does not exist. For a negative number, you may obtain only the logarithm of its magnitude.

5. Examples using a calculator:
   a. $\log 25 \cong 1.3979 \ldots$ (that is, $10^{1.3979} \sim 25$)
   b. $\log 0.4 \cong -0.3979 \ldots$ (that is, $10^{-0.3979} \sim 0.4$)
   c. $\ln 25 \cong 3.2188 \ldots$ (that is, $e^{3.2188} \sim 25$)
   d. $\ln 0.4 \cong -0.9162 \ldots$ (that is, $e^{-0.9162} \sim 0.4$)

6. *Operations with Logarithms.* The following list of operations is valid for both common and natural logarithms:
   a. $\log 10 = 1; \ln e = 1$
   b. $\log 1 = 0; \ln 1 = 0$
   c. $\log ab = \log a + \log b$
   d. $\log a/b = \log a - \log b$
   e. $\log a^b = b \log a$
   f. $\log \sqrt[b]{a} = (\log a) \div b$

7. *Rules of Exponents.* These rules look similar to the laws of logarithms but are more general because they can apply to *any* base, including variable symbols.
   a. $x^1 = x$
   b. $x^0 = 1$
   c. $1/x^a = x^{-a}$
   d. $x^a \cdot x^b = x^{a+b}$ (bases must be the same)
   e. $x^a/x^b = x^{a-b}$ (bases must be the same)
   f. $\sqrt[a]{x} = x^{1/a}$

E. Angular Measure and Converting Angles

1. Angular measure provides the means to quantify direction and circular motion.

2. You are most familiar with the unit of angular degree, whose symbol is °. There are 360° in a full circle. Because of the way a degree is defined, the kinds of mathematical operations that can be performed on angular measurements using this unit are limited.

3. The derived SI unit for angular measure is the *radian (rad)*. The radian is preferred in physics because the unit may be used in any legitimate mathematical operation involving angular measure, for it is actually *dimensionless* (i.e., has no SI units associated with it). A "dimensionless unit" may seem to be contradictory, so let's see how it is derived.

4. A radian is equal to the central angle of a circle formed by two radii that subtend (cut off) an arc of the circle equal in length to its radius (see Figure G-1). Since 1 rad = *r* meters/*r* meters = 1, radians have no SI base units and are dimensionless.

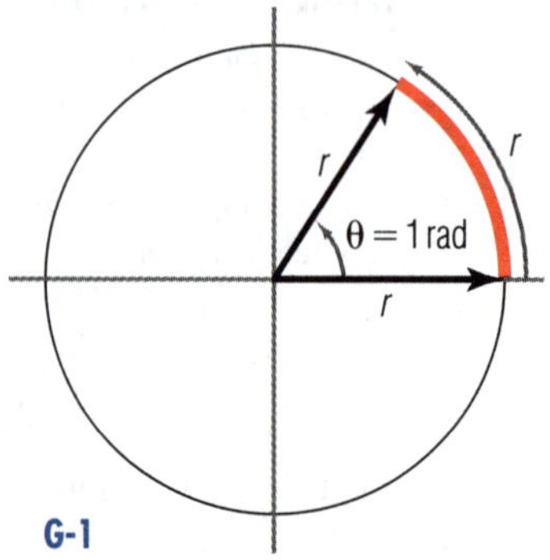

G-1

5. You know that the circumference of a circle with radius r is equal to $2\pi r$. Accordingly, there are 2π radians in a full circle, or $360° = 2\pi$ rad. This relationship can be used to develop a formula for converting degrees to radians and radians to degrees using the following conversion factors:

$$360° = 2\pi \text{ (rad)}$$

$$180° = \pi \text{ (rad)}$$

$$1 = \frac{\pi \text{ (rad)}}{180°} \qquad \frac{180°}{\pi \text{ (rad)}} = 1$$

$$\theta_{rad} = \frac{\pi \text{ (rad)}}{180°}\theta_{deg} \qquad \theta_{rad}\frac{180°}{\pi \text{ (rad)}} = \theta_{deg}$$

6. Examples:

Convert 38° to radians:

$$38° \cdot \frac{\pi \text{ (rad)}}{180°} \cong 0.6632 \text{ (rad)}$$

Convert 1.5π (rad) to degrees:

$$1.5\cancel{\pi} \text{ (rad)} \cdot \frac{180°}{\cancel{\pi} \text{ (rad)}} = 270°$$

F. Geometric Figures

This section summarizes some basic geometric figures that are referred to in this textbook and the formulas that are associated with them.

Circle of radius r	area $= \pi r^2$	perimeter $= 2\pi r$
Sphere of radius r	surface area $= 4\pi r^2$	volume $= {}^4\!/_3\pi r^3$
Triangle	area $= {}^1\!/_2 bh$	
Rectangle	area $= wl$	perimeter $= 2(w + l)$
Rectangular prism ($B = wl$) of height h	lateral area $= 2h(w + l)$ total area $= 2(B + h(w + l))$	volume $= wlh = Bh$
Right circular cylinder of radius r and height h	lateral area $= 2\pi rh$ total area $= 2(\pi r^2 + \pi rh)$	volume $= \pi r^2 h$
Right circular cone of radius r and height h	lateral area $= \pi r\sqrt{r^2 + h^2}$ total area $= \pi r^2 + \pi r\sqrt{r^2 + h^2}$	volume $= {}^1\!/_3\pi r^2 h$

G. Algebra and Trigonometry

The rules and operations of algebra and trigonometry are extensive, and discussing them in any detail is beyond the scope of this appendix. If you need to brush up on some aspect of solving algebraic equations or trigonometric identities, you should review the applicable sections of a mathematics textbook such as *Algebra 2* or *Precalculus*, both published by BJU Press.

H. Functions

One characteristic of God's creation that man discovered is that many phenomena can be described by simple equations that relate a physical quantity to one or more others. For example, if you know the speed of a car and how long it has been traveling at that speed, you can tell how far the car traveled during the time interval. Such relationships are called *functions* because a set of input numbers (called the *independent* variable set or *domain*) produces a unique set of output numbers (called the *dependent* variable set or *range*). You will study a variety of functions in physics. In this course they will be limited to a few simple types. We will discuss their characteristics and graphs in this section.

1. *Linear functions* are the simplest type of function. They represent a directly proportional relationship between the independent and dependent variables and produce a straight-line graph. Linear functions may be written in the form

$$y = mx + b,$$

where m is the *slope* of the line and b is the coordinate where the line crosses the y axis. If the relationship between the independent and dependent variables is *direct*, then the line passes through the origin of the graph ($b = 0$). Figures G-2a and G-2b show a linear function and a direct function, respectively. One example of a linear function is the expansion of a thin metal rod with temperature as it is heated.

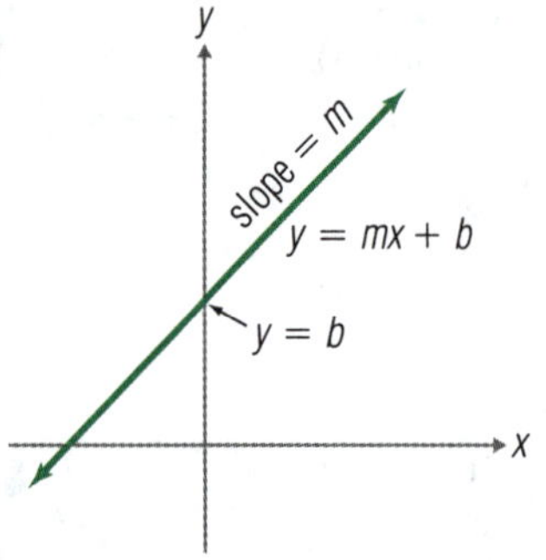

G-2 (a)

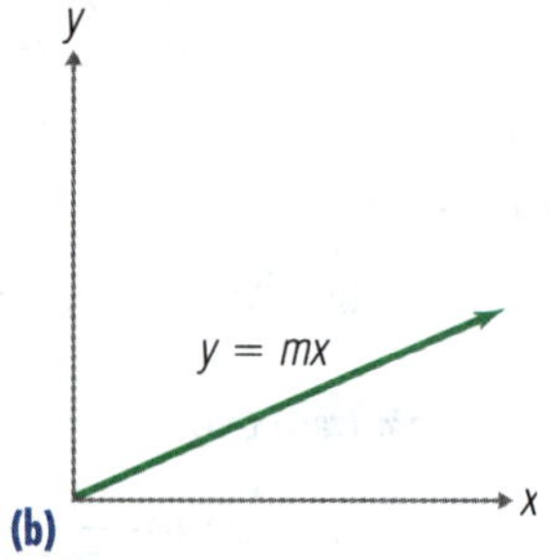

(b)

2. *Quadratic functions* are relationships in which the dependent variable is a function of the square of the independent variable. Graphs of quadratic functions are parabolas. Quadratic functions are written in the form

$$y = ax^2 + bx + c,$$

where a, b, and c are constants. An example of a quadratic function is the height of a ballistic projectile with time. Figure G-3 shows the graph of a quadratic function.

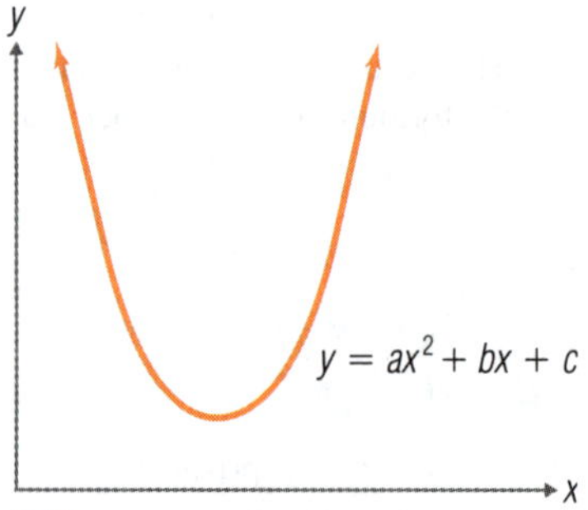

G-3

3. *Inverse functions* are relationships in which as the independent variable increases, the dependent variable decreases in a reciprocal fashion. Such relationships are called inverse proportions. Graphs of inverse functions are half-hyperbolas. These functions may be written in the form of either

$$y = \frac{k}{x} \text{ or } yx = k,$$

where k is a proportionality constant. An example of an inverse function is Boyle's law. Figure G-4 is the graph of an inverse function. Note that neither variable ever reaches zero nor crosses its axis in the inverse function of a physical phenomenon.

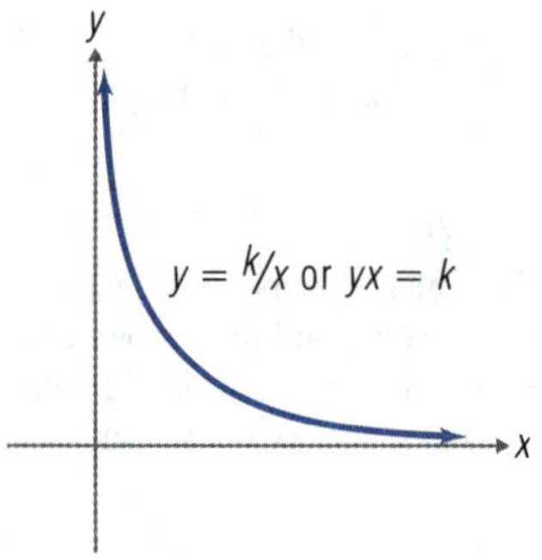

G-4

4. *Exponential functions* are very common in natural processes. In a basic exponential function, the dependent variable is the result of some base number raised to the power equal to the independent variable. If the values of the independent variable are not restricted, then the dependent variable may take any positive real number value. These functions can be written in the form of either

$$y = n^x \text{ or } y = n^{-x},$$

where n is a positive real number. The magnitude of n determines how steep or how shallow the curves are. For natural processes, n is usually the base of the natural logarithms, e. The graphs of the exponential functions form sweeping rising or falling curves (see Figure G-5). Note that when $x = 0$, both curves pass through $y = 1$.

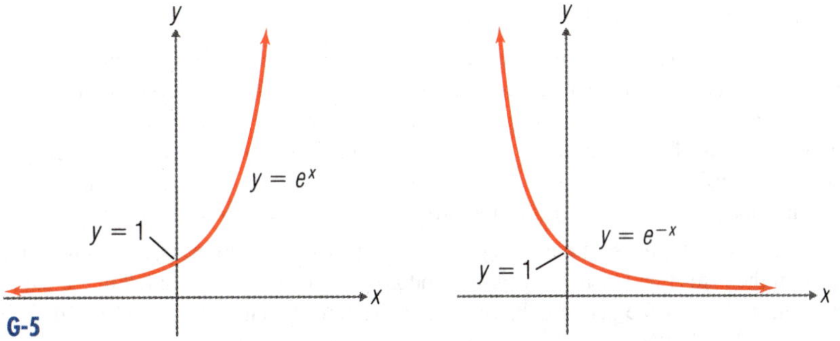

G-5

5. *Periodic functions* are functions that are repetitive or show an oscillation of the dependent variable over the domain of the independent variable. Such functions include the sine, cosine, and tangent functions. The equation form of these functions varies depending on the operator and argument but generally follows the form

$$y = A\ op\ (\theta + k) + c,$$

where A, k, and c are all constants and the operator op ( ) may be any periodic function operator, such as a trigonometric ratio. The quantity within the parentheses may be an angle or an angular frequency that is dependent on time. An example of a periodic physical phenomenon is the displacement of a pendulum from its vertical hanging position. Figure G-6 shows two graphs of typical periodic functions.

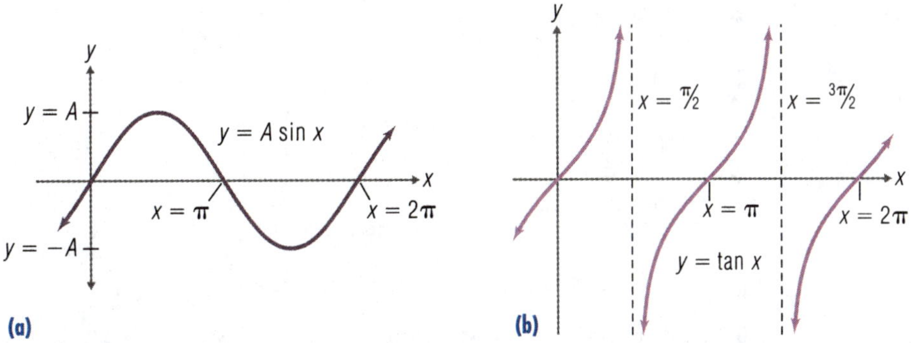

(a) (b)

G-6 (a) A sine function; (b) A tangent function

Both exponential and periodic functions are called *transcendental functions*.

I. Graphical Analysis

The dominion modeling exercises presented in this textbook involve analyzing graphical plots of real-world data. The concept of dominion modeling is discussed in Subsection 1.19 of Chapter 1. By plotting the data and creating mathematical models, you will better understand the related topics. Since science is essentially devoted to developing workable models of the real world, these exercises allow you to assume the role of a scientist.

Creating mathematical models is probably unfamiliar to you. This section provides some additional background information to help you understand the tools that you will use to create the models. Graphical analysis can be done with a computer spreadsheet program, which is preferred, but you can also complete these assignments by using a graphing calculator. For guidance on using a graphing calculator to analyze data, see Appendix D in *ALGEBRA 2*, 2nd ed. or Appendix D in *PRECALCULUS*, both published by BJU Press. The following are descriptions of the modeling tools and the procedures for using them.

Scatter plots

A scatter plot is a graph of individual data points consisting of related independent and dependent variable values (i.e., x and y). One variable is plotted against the horizontal axis and the other against the vertical axis. The resulting plot is a set of points scattered across the graph, hence the name. You will be given guidance in the dominion modeling review questions as to which axis should be used for each variable.

To create a scatter plot with a spreadsheet program, you will normally enter the data for the independent variable, in order of increasing values, into one spreadsheet column, and then enter the associated dependent data into the adjacent column. Typically, you will then block-select the two columns of data, omitting any column headings, and create a scatter plot chart using the program's charting function. Just select the simple scatter plot chart choice (also called an XY chart in some programs), and follow the program prompts to complete the chart.

When scientists measure data, they include in the measurement an estimation of the error, based on the instrument or measuring methodology. This is often included in the data tables as a ± value following the actual data value. For example, a length might be listed as 2.965 cm ±0.020 cm. The first value, 2.965 cm, is the measurement, and the 0.020 cm is the potential error of the measurement. This information can be represented on a scatter plot by *error bars*, which are vertical bars that extend above and below the data point according to the magnitude of the associated error. Figure G-7 is an example of a scatter plot that includes error bars. Horizontal error bars can be applied to the data points as well if the independent variable is a measured quantity.

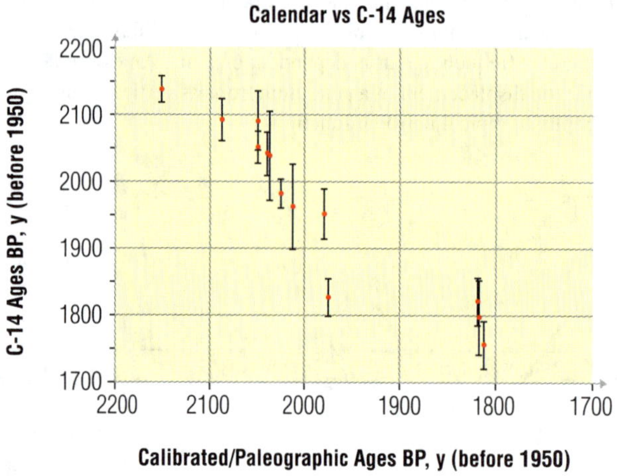

G-7 A scatter plot that includes error bars for *y* values

Trendlines

The scatter plot graphs the raw data, but due to the uneven nature of real data, any governing mathematical relationship may not be readily apparent. To create a model of the data, you must develop a mathematical function that accounts for the data and from which *extrapolations* beyond the data set may be made. The first step in evaluating the data is to determine the kind of relationship the graph suggests. Do the variables show a linear (directly proportional), parabolic (quadratic), hyperbolic (inversely proportional), or exponential relationship? You can estimate the kind of relationship by comparing the general shape of the scatter plot against the function graphs displayed in Part H of this appendix.

To build an actual model, however, you must find the mathematical function that best describes your data. This is accomplished by performing a process known as *curve-fitting*, which applies a *best-fit trendline* to the data. This line is a smooth curve, based on one of the types of functions, that passes through the plot of data points without necessarily connecting every point. The equation for the best-fit trendline becomes your mathematical model. The entire process of fitting the curve and generating its equation is called *regression analysis*.

Your spreadsheet program or calculator can automatically apply a best-fit curve of your choosing and display the curve's equation. Refer to your program or calculator manual for instructions. The program creates the curve by performing a statistical analysis called a *least-squares fit*. When a smooth curve is plotted through a scatter plot, there are usually differences between the y-values of the curve at the data's x-values and the actual y-values for those data points. These differences are called *residuals*. A least-squares fit creates a curve that minimizes the sum of the squares of these residual values. This is why a program or calculator is needed.

How do you know which type of function is the best one to represent your data? You make this decision by determining which function type generates a curve that most closely fits the actual data. The closer the curve corresponds to the data, the better the fit and thus, it may be assumed, the better the model. Statisticians use a tool called the *coefficient of determination* (r^2) to evaluate how well a trendline fits the data. An r^2 value of 1.00 indicates a perfect fit between the trendline and the data. Fractional values indicate poorer fits. Your spreadsheet program or calculator can compute this quantity. You choose the best function type by trying to maximize the r^2 value. For highly scattered or sparse data, it may not be possible to achieve a good fit with any function type.

Another technique that helps define the model, especially for direct relationships, is to make the trendline pass through the origin of your plot. This is called "forcing the y-intercept to zero." This feature is available in most spreadsheet programs. Forcing a zero intercept helps the equation of the trendline resemble the expected behavior of the system more closely and can affect the overall shape of the curve.

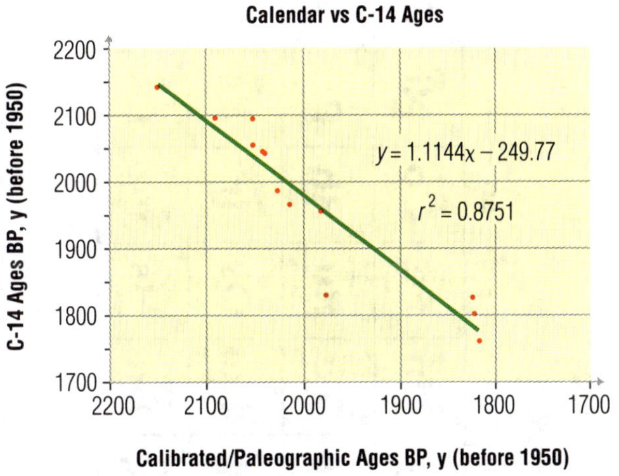

G-8 A least-squares regression fit with the equation and coefficient of determination displayed.

Using Your Model to Make Inferences

A model becomes truly valuable when it is used to explain observations or make predictions. Sometimes, the predictions a model makes are used to test its accuracy. Some dominion modeling exercises ask you to use a trendline equation to extrapolate data results beyond the data range given. Others will lead you through the steps that scientists have used in the past to arrive at the scientific formulas we use today. As you create and study models built from real-world data, you will come to recognize both the usefulness and the limitations of model-making in science.

APPENDIX H ◆ PERIODIC TABLE OF THE ELEMENTS

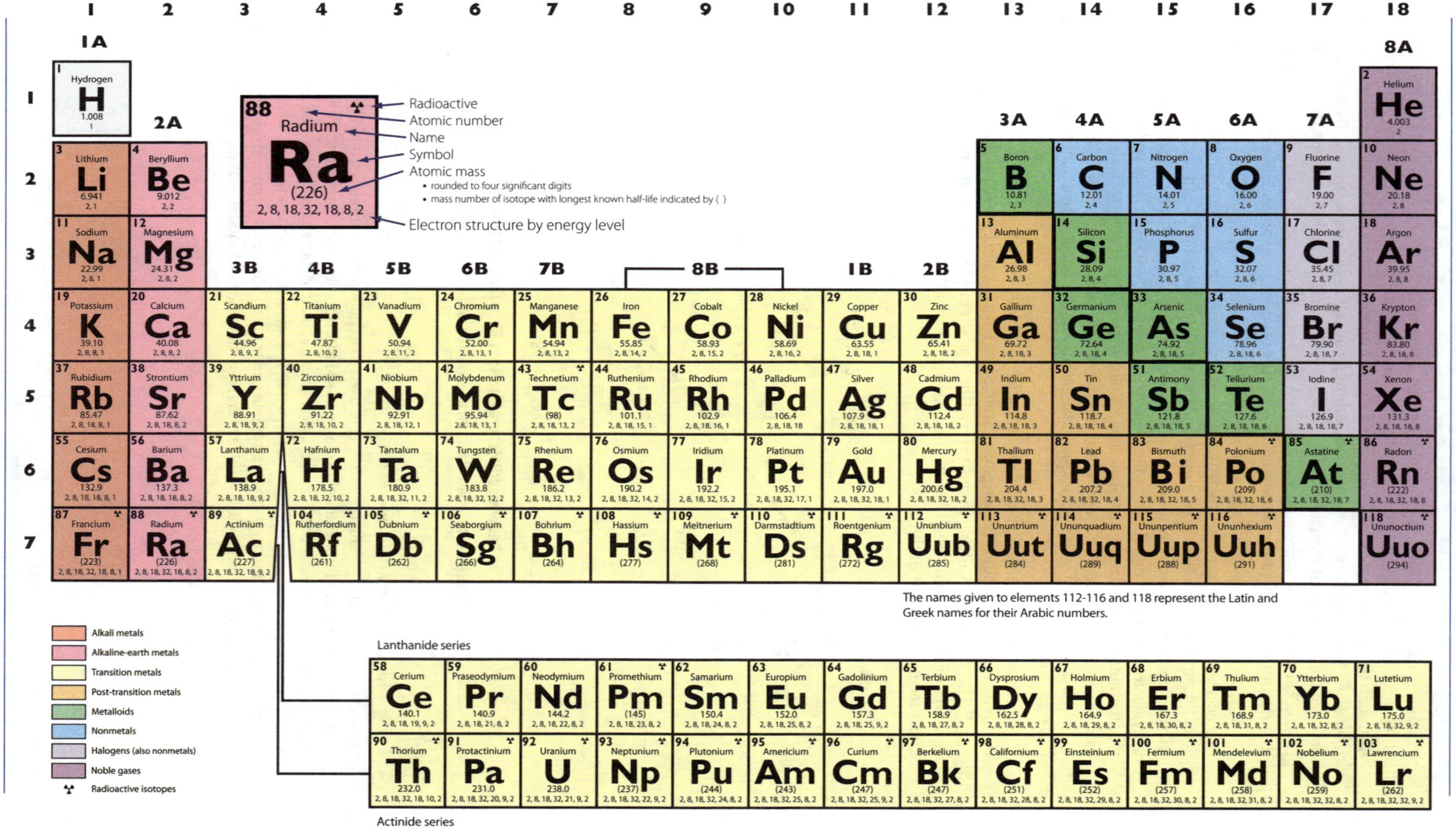

APPENDIX I

Prefixes for Powers of 10

Prefix	Symbol	Factor	Example
yotta-	Y	$\times 10^{24}$	yottagram (Yg)
zetta-	Z	$\times 10^{21}$	—
exa-	E	$\times 10^{18}$	—
peta-	P	$\times 10^{15}$	—
tera-	T	$\times 10^{12}$	terawatt (TW)
giga-	G	$\times 10^{9}$	gigabyte (GB)
mega-	M	$\times 10^{6}$	megajoule (MJ)
kilo-	k	$\times 10^{3}$	kilometer (km)
hecto-	h	$\times 10^{2}$	(rarely used in physics)
deka-	da	$\times 10^{1}$	(rarely used in physics)
(base)	—	$\times 10^{0}$	—
deci-	d	$\times 10^{-1}$	decibel (dB)
centi-	c	$\times 10^{-2}$	centimeter (cm)
milli-	m	$\times 10^{-3}$	millivolt (mV)
micro-	μ	$\times 10^{-6}$	micropascal (μPa)
nano-	n	$\times 10^{-9}$	nanometer (nm)
pico-	p	$\times 10^{-12}$	picofarad (pF)
femto-	f	$\times 10^{-15}$	—
atto-	a	$\times 10^{-18}$	—
zepto-	z	$\times 10^{-21}$	zeptometer (zm)
yocto-	y	$\times 10^{-24}$	—

APPENDIX J

Greek Alphabet

Letter	Lower-case	Upper-case	Letter	Lower-case	Upper-case
alpha	α	A	nu	ν	N
beta	β	B	xi	ξ	Ξ
gamma	γ	Γ	omicron	ο	O
delta	δ	Δ	pi	π	Π
epsilon	ε or ϵ	E	rho	ρ	P
zeta	ζ	Z	sigma	σ or ς	Σ
eta	η	H	tau	τ	T
theta	θ	Θ	upsilon	υ	Υ
iota	ι	I	phi	ϕ	Φ
kappa	κ	K	chi	χ	X
lambda	λ	Λ	psi	ψ	Ψ
mu	μ	M	omega	ω	Ω

GLOSSARY

A

absolute pressure (*P*) (385) Fluid pressure measured with respect to a perfect vacuum; equals gauge pressure plus 1 atm

absolute zero (0 K) (322) The minimum theoretical temperature; the temperature at which all thermal motion ceases

acceleration (a) (59) The time rate of change of velocity (**v**); $\mathbf{a} = \Delta\mathbf{v}/\Delta t$, and $a = \Delta v/\Delta t$.

accuracy (31) An assessment of how close a measurement is to the accepted or known value

action-reaction force pair (128) The two equal but opposite forces that exist when two systems act on each other

activity (*A*) (669) The rate of emission of particles or energy from a radioactive nucleus; measured in bequerels (Bq)

actual mechanical advantage (AMA) (228) A method for indicating the mechanical advantage of a simple machine that accounts for all mechanical losses sustained through using the machine; calculated from the ratio of a machine's output force to its input force, or of actual work output to the input energy or work

additive primary color (604) One of the three colors of light that can additively combine to form any color of light sensible in the human visual spectrum; red, green, or blue light

adhesion (298) Attraction between the molecules of different substances

adiabatic (345) A property or process thermally isolated from its surroundings so that no heat is exchanged between systems

airfoil (401) The cross-sectional shape of an object immersed in a gas, such as a wing or rudder, that produces lift, principally through the Coandă effect

Airy discs (588) The disc, rather than a point, of light that is the image normally formed of a point light source by an optical system of lenses and that is caused by diffraction within the system; The size of the Airy disc determines the pratical resolution limit of the optical system.

albedo (525) The fraction of the incident electromagnetic energy that is reflected diffusely at a particular observation angle from a celestial object

alpha radiation (α) (666) Alpha particles emitted by radioactive nuclei

alternating current (AC) (506) An electric current that regularly changes direction

ammeter (460) A device that measures and displays the current flowing through a circuit; An ammeter is connected in series with the circuit during measurement.

amorphous solid (299) A rigid solid without any definite or repetitive crystalline structure

ampere (A) (448) The SI unit of electrical current; A current of 1 A in each of two conductors of infinite length and negligible cross section separated by 1 m in a vacuum will produce a force of 2×10^{-7} N per meter of length on each wire.

amplitude (*A*) (267) The maximum displacement from the equilibrium position of a system experiencing periodic motion; applies to the medium or field that carries periodic waves

angle of incidence (θ_i) (535) The angle between the incident ray of light and the normal line at the point of incidence

angle of reflection (θ_r) (535) The angle between the reflected ray of light and the normal line at the point of incidence

angle of refraction (θ_r) (552) The angle between the refracted ray of light and the normal line at the interface between two media

angular acceleration (α) (142) The time rate of change of angular velocity (**ω**); $\boldsymbol{\alpha} = \Delta\boldsymbol{\omega}/\Delta t$, and $\alpha = \Delta\omega /\Delta t$.

angular momentum (L) (259) The momentum of a rotating system that is dependent on the angular velocity and mass distribution of the system

angular speed (ω) (139) The time rate of change of angular position (θ); $\omega = \Delta\theta/\Delta t$.

angular velocity (ω) (141) The time rate of change of angular position in a specified direction

annual variation (480) The annual change of direction of the earth's magnetic flux vector at a specified location

antenna (516) A device in which charged particles accelerate either to form an electromagnetic wave or in response to an electromagnetic wave

anthropic principle (688) The concept that the universe in all of its aspects and details is finely tuned to support life as we know it; If even a few of its many features, universal constants, and fundamental structures were different, then life would not be possible; leads to the controversial conclusion that the universe is designed rather than a product of chance

antiparticle (684) A subatomic particle whose properties except for charge and magnetic moment are identical in all respects to the associated particle (e.g., proton p and antiproton $\overline{p}$)

aperture (566) The opening through which light is admitted into an optical instrument

apparent weight (174) The weight of an object measured while the object is under acceleration

Archimedes' principle (392) The buoyant force of a fluid on an immersed object is equal to the weight of the fluid that the object actually displaces.

armature (506) The component that generates the primary rotating magnetic field in an electric motor

astronomical unit (ua) (153) The unit of astronomical distance equal to the mean orbital radius of the earth from the sun; 1.496×10^8 km

atmosphere (atm) (385) The average pressure of the atmosphere at sea level; 1 atm = 1.013×10^5 Pa.

atom (295) A particle of matter composed of protons, electrons, and usually neutrons; the basic building block of matter and the smallest representative particle of an element

atomic mass unit (amu) (296) A convenient unit for measuring the masses of atom-sized particles, originally equal to 1/16 the mass of an oxygen-16 atom, but now generically referring to the relative mass unit for atom-sized particles; See *unified atomic mass unit*.

atomic number (*Z*) (647) The number of protons in the nucleus of an atom; Elements on the periodic table are ordered by their atomic numbers.

aurora (481) A polar light display resulting from the interaction between high-energy charged particles trapped in the earth's magnetic field and upper atmospheric gas molecules

average acceleration ($\bar{\mathbf{a}}$) (96) The average rate of change of a system's velocity during an interval of time

Avogadro's constant (N_A) (296) The number of atoms in exactly 12 g of carbon-12; the number of identical atoms whose mass in grams is numerically equal to the weight of one atom in atomic mass units; $N_A = 6.022 \times 10^{23}$ particles.

azimuthal quantum number (l) (652) The second quantum number, which identifies the orbital that an electron occupies

B

back emf (ε) (508) A potential difference induced across an inductor by a changing magnetic field that opposes the change in current through the inductor

balanced forces (117) Concurrent forces that have a zero vector sum

ballistic (98) The portion of a projectile's trajectory in a gravitational field during which only external forces act on the projectile

ballistic pendulum (252) An inertial device used to measure the energy of ballistic projectiles

Balmer series (649) A spectral series of emission lines produced when an electron in a hydrogen-like atom falls from the third or higher energy levels to the second energy level

bar (385) A non-SI unit for measuring atmospheric pressure; 1 atm = 1.013 bar.

barometer (391) An instrument that measures and displays atmospheric pressure

baryon (685) A class of composite fermion subatomic particles; Baryons are among the most massive subatomic particles and include the proton and neutron.

battery (450) A source of electrical potential difference consisting of two or more voltaic cells connected in series

becquerel (Bq) (669) The derived SI unit of radioactivity; 1 Bq = 1 ray or particle emission per second.

Bernoulli's equation (400) The equation that illustrates the conservation of energy associated with a flowing, incompressible fluid's pressure, speed, and elevation

beta radiation (β) (666) Electrons (e^-) and positrons (e^+) emitted by radioactive nuclei

bimetallic strip (316) Two metals welded together forming a long, narrow strip, each metal having a different coefficient of expansion that causes the strip to bend in response to temperature changes

binding energy (678) The amount of energy that a particle must give up in order to join a nucleus; energy equivalent to the nuclear mass defect according to Einstein's equation $E = mc^2$

Biot-Savart law (500) Defines the relationship between the current in an infinitesimal portion of a conductor and the magnetic field induced by it

blackbody (354) A perfect emitter and absorber of electromagnetic radiation

blackbody radiation (642) The electromagnetic radiation emitted in a characteristic spectrum by a blackbody emitter

black hole (635) An astronomical body so dense and massive that even light cannot escape its gravitational field

block and tackle system (231) An arrangement of fixed and movable pulleys connected by strings, ropes, or cables that provides a mechanical advantage for exerting a force

boiling (305) The process of liquid vaporization at the boiling point associated with the prevailing atmospheric pressure; rapid vaporization within the bulk of a liquid

Boltzmann's constant (k) (329) A proportionality constant for expressing the kinetic energy of a particle of an ideal gas per kelvin; $k = 1.3807 \times 10^{-23}$ J/K.

boson (685) Elementary and composite subatomic particles that carry an integer spin of 0, 1, or 2; Elementary bosons include the exchange particles that form the fundamental forces according to the Standard Model of matter.

Boyle's law (326) The pressure of a certain mass of an ideal gas is inversely proportional to its volume if the temperature is held constant; $PV = k$.

breeder reactor (681) A fission reactor designed to use Pu-239 as a nuclear fuel; causes the transmutation of U-238 to Pu-239 and thus produces more fuel than it consumes

Brewster's angle (θ_B) (590) The special angle of incidence at the boundary between two media in contact where the reflected light rays in the incident medium are exactly 90° from the refracted rays in the other medium; The reflected beam will be entirely horizontally polarized, and the refracted beam will be mostly vertically polarized.

Brewster's law (590) The angle of incidence that best polarizes reflected light is θ_B in the formula $\tan \theta_B = n_i$, where n_i is the index of refraction of the reflecting medium.

bridge circuit (461) An electrical sub-circuit composed of a series-parallel arrangement of fixed and variable resistors; used to determine the unknown resistance of a component in the circuit

Brownian motion (340) The random, chaotic movements of microscopic particles suspended in a fluid medium, caused by the collision of the fluid particles with the suspended particles

brush (494) One of a set of electrical contacts in a motor or generator that connects the external electrical circuit to the rotating portion of the device (the rotor); The brushes rest on a split-ring commutator in DC machines or on continuous slip rings in some AC machines.

buoyant force (F_b) (392) An upward force exerted by a fluid, according to Archimedes' principle, on an object immersed in the fluid

C

caloric theory (341) The now-obsolete theory of heat in which heat is a material fluid that flows from hot objects to cold ones

calorie (cal) (342) An obsolescent unit of heat; the amount of heat gained or lost that changes the temperature of 1 g of water by 1 °C

calorimeter (345) A laboratory apparatus that establishes an approximately adiabatic boundary within which energy exchange processes can be quantitatively evaluated with minimal influence from the surroundings

candela (cd) (598) The SI unit of light intensity; 1 cd is the luminous intensity, in a given direction, of a source that emits monochromatic radiation of frequency 540×10^{12} hertz and that has a radiant intensity in that direction of 1/683 watt per steradian.

capacitance (C) (438) The amount of charge per volt of potential difference that a capacitor can store; measured in farads

capacitive reactance (X_C) (514) The effective opposition of a capacitor to alternating current

capacitor (436) A simple electrical device that is used to store charge, containing conducting plates separated by an insulating material, or dielectric

capillarity (305) The movement of a liquid through a narrow passage or tube caused by the adhesion of the molecules in the liquid to the surface with which it is in contact

Carnot cycle (370) A theoretically ideal cycle of a heat engine that operates at the greatest possible efficiency between two given temperatures

Cassegrainian reflector telescope (610) A reflecting telescope where the light incident on the main mirror is reflected by a secondary mirror through a hole in the center of the main mirror to the eyepiece located behind the main mirror

cathode-ray tube (CRT) (488) An evacuated glass envelope containing at least two oppositely charged plates at high electrical potential; Electrons are emitted from the negative plate (the cathode) and propelled by the voltage difference toward the anode.

center of buoyancy (394) The geometric center of the volume of the fluid occupied by the immersed object; equivalently, the point in the object where the buoyant force can be assumed to be acting

center of mass (257) The point within a system where the entire mass of the system can be assumed to be concentrated when one is evaluating the behavior of the system according to Newton's laws of motion; also called the *center of gravity*

central force (221) A force that always acts along the line connecting the centers of two objects; a conservative force

centrifugal force (146) When considering forces acting on a revolving system, a pseudoforce that seems to pull the system away from the axis of rotation due to the inertia of the system resisting the acceleration of the centripetal force

centripetal acceleration ($\mathbf{a}_c$) (140) The center-seeking acceleration that results when a central force holds a revolving system in a circular path

centripetal force ($\mathbf{F}_c$) (146) A central force responsible for accelerating a system in a circular path

chain reaction (680) A nuclear fission process in which neutrons produced by fissions trigger more fissions in a continuous chain of events

charge-coupled device (CCD) (611) A semiconductor photoactive surface that converts incident light into voltages, which are then usually digitally coded as picture elements (*pixels*); the heart of digital photographic and video systems

Charles's law (325) The volume of a certain mass of an ideal gas is directly proportional to its absolute temperature when its pressure is held constant; $V = kT$.

chemical compound (295) A combination of two or more different elements in definite proportions that cannot be separated by ordinary physical means

Christian worldview (7) A perspective on life that views all things through the biblical teachings on Creation, Fall, and Redemption

chromatic aberration (567) Blurred images resulting from color fringes produced by the dispersion of white light by simple lenses because they refract different frequencies of light by different amounts

circuit (450) A complete conducting path for an electrical current

circular motion (138) Two-dimensional motion in a circular path around a central axis of motion; includes both revolving and spinning motions

classical physics (11) Physics that is based upon the three major areas of mechanics, electrodynamics, and thermodynamics, all of which were well developed prior to the establishment of quantum physics in the twentieth century

clock paradox (632) The relativistic principle that when one clock moves away from and then back toward another clock, it runs slow compared to the second clock; also known as the *twin paradox*

closed system (365) A system that can exchange energy but not matter with its surroundings

Coandă effect (402) A fluid flowing along a gently curving surface will tend to follow the surface due to adhesion to the surface and cohesion within the fluid; believed to be the principal phenomenon in generating lift by air and hydrofoils

coefficient of friction (μ) (177) A dimensionless proportionality constant that relates the friction exerted by two surfaces in contact with each other to the normal force between them; $\mu = f/N$.

coefficient of linear expansion (α) (315) A dimensionless proportionality constant that relates the amount a material expands in one direction to the change in temperature

coefficient of viscosity (η) (403) A measure of a fluid's resistance to flow under a constant force and laminar conditions

coefficient of volume expansion (β) (316) A dimensionless proportionality constant that relates the amount a material expands in three dimensions to the change in temperature; $\beta = 3\alpha$.

cohesion (298) Attraction between atoms or molecules of the same type within the volume or bulk of a substance

collimator (587) A component within an instrument that adjusts the shape of an incident beam of light or particles to make their paths parallel

combined gas law (328) A law that relates the pressure, volume, and Kelvin temperature of a certain mass of gas according to Boyle's, Charles's, and Gay-Lussac's laws; $PV/T = k$.

complementarity (651) The quantum mechanics principle that all matter can be viewed as being particles or waves, but never both at the same time; attributed to Niels Bohr

complementary color (604) One of a pair of colors that, when combined in the correct proportion produce white light if additive or a very dark gray if subtractive; In a three-color model, the complement of a primary color is formed by the combination of the other two primary colors.

component (electrical) (438) In electrical circuits, any distinct device in the circuit connected to other components by conductors (wires)

component (of a vector) (81) See *component vector*.

component vector (81) One of the two (or three) vectors parallel to the designated coordinate axes that can be summed to produce the vector being analyzed in two (or three) dimensions

compound microscope (606) A microscope that uses two or more lenses or groups of lenses to magnify an object

compression (281) In general, the reduction of volume a certain amount of matter occupies (i.e., an increase in density); in a mechanical longitudinal wave, the region of higher density analogous to the crest of a water wave

compressive force (122) One of a pair of opposing forces that act in line so as to compress the system they are acting on

Compton effect (645) The loss of energy and increase of wavelength when photons of x- or gamma rays interact with electrons; one of the key evidences for the particle nature of electromagnetic energy

concave mirror (537) A mirror surface that is shaped like the inner surface of a sphere

conceptual model (6) A scientific model that cannot be represented by a physical object

condensation (307) The change of state from a gas to a liquid due to the loss of thermal energy to the surroundings

conduction (electrical) (418, 657) The flow of electricity through a current-carrying path, caused by a change of electrical potential along the path

conduction (heat) (352) The flow of thermal energy from one object to another through contact

conductor (418, 440) A material consisting of particles with loosely held electrons that allows the easy transmission of thermal energy or electricity

conservation law (219) One of many similar principles that state that a given dimensional quantity is conserved (remains constant) throughout a process, even if it changes forms

conservation of mechanical energy (212) Kinetic and potential energies of a system are conserved unless it is acted on by external nonconservative forces (e.g., friction).

conservation of momentum (244) The total momentum of a system, including a multipart system, is conserved if the net external force acting on the system is zero.

conservative force (204) A force that can convert kinetic energy to potential energy and vice versa without loss

constructive interference (574) The superposition of wave forms that produces larger wave amplitudes through reinforcement

contact (charging) (421) One of two ways to charge an object; Electrons are transferred from one object to the other when a charged object touches an uncharged object.

contact force (122) A force that can be exerted only through the physical contact of two systems

continuum (8) An unbroken span or sequence of values within a measurable dimension

control rod (681) An elongated rod-like device, usually with flat blades mounted along the shaft, that is made of a material that can absorb neutrons produced by a nuclear chain reaction; The more fission neutrons absorbed, the fewer that are available for causing fissions. The nuclear reaction rate depends on the positions of the control rods within the reactor core.

convection (353) The transfer of thermal energy from one place to another by the physical translation of particles between locations

converging lens (560) A lens that focuses incident light rays at a point

conversion factor (29) A multiplication factor equal to 1 that converts the given units of a quantity to other units of the same dimension while preserving the size of the quantity

convex mirror (537) A mirror surface that is shaped like the exterior surface of a sphere

cosine θ (80) In a right triangle, the ratio of the length of the side adjacent to the acute angle θ to the length of the hypotenuse

coulomb (C) (424) The derived SI unit of electrical charge; $1\ \text{C} = 1\ \text{A}\cdot\text{s}$.

Coulomb's law (423) The law relating the electrical force between two objects to the magnitude of their charges and the distance between them;

$$F_{\text{elect}} = k\frac{|q_1||q_2|}{r^2}.$$

Creation Mandate (2) God's command to humans to exercise wise and good dominion over His creation to the glory of God and for the benefit of their fellow humans (Gen. 1:26, 28)

critical angle (θ_c) (555) For light traveling from a medium with a higher index of refraction to one with a lower, it is the angle of incidence that produces an angle of refraction of 90° (the light ray is lost parallel to the interface surface); Greater angles of incidence produce total internal reflection.

critically damped oscillation (278) A truncated oscillation that barely overshoots the equilibrium point due to damping

critical mass (680) The smallest mass of a fissionable material that can sustain a nuclear chain reaction

crystalline solid (299) A solid in which the atoms or molecules are arranged in a regular, repeating pattern of geometrically identical subunits

curie (Ci) (669) A non-SI unit for the decay rate or radioactivity; $1\ \text{Ci} = 3.70 \times 10^{10}\ \text{Bq}$.

Curie temperature (T_C) (479) The temperature at which a permanent magnet loses its magnetic field

current (I) (447) The flow of electrical charges through a conductor or conductive space; measured in amperes

cutoff frequency (645) The lowest frequency of EM radiation that will eject photoelectrons from a material

cycle (267) One complete path or oscillation of a system; usually repeated periodically

D

damped harmonic oscillator (277) A system that oscillates periodically with decreasing amplitude due to an opposing force acting on the system

damping (266) Reducing the amplitude of periodic motion through friction or another nonconservative force

data (17) All scientific observations that are subject to interpretation according to the governing scientific paradigm; may be objective quantitative data or subjective qualitative data

daughter nuclide (673) The nuclide resulting from a nuclear decay involving the emission of a particle

decay constant (λ) (671) The probability of decay for a given isotope during a specified time period

decibel (dB) (284) A non-SI unit of relative intensity; In audible acoustics, 1 dB is the smallest difference in loudness that the average human ear can detect.
density (ρ) (384) A measure of the concentration of matter; It is expressed as the ratio of an object's mass to its volume.
deposition (308) Solidification from a gas directly to a solid, without passing through the liquid phase, at temperatures below the freezing point of the solid; formerly referred to as one of the *sublimation* phase changes
derived unit (28) In the SI, any unit that may be expressed as a combination of fundamental or other derived SI units
destructive interference (574) The superposition of wave forms that produces smaller wave amplitudes or cancellation of the wave forms at a point
diamagnetism (477) The property of materials that are not attracted by magnets; having a magnetic permeability (μ) less than 1
diathermic (360) The ability to conduct heat; a thermal conductor
dichroic (590) A property of some materials to polarize light passing through them, producing one or two separate rays
dielectric (436) The insulator between the plates of a capacitor
dielectric constant (κ) (437) The ratio of a given electric field strength in a vacuum to the same field passing through a particular material
difference vector (59) The resultant of a vector subtraction operation; The difference of two velocities would be represented by $\Delta\mathbf{v}$.
diffraction (584) The bending of waves around the edges of openings, objects, or gratings that have dimensions similar to the wavelength of the wave
diffusion (340) The process whereby two systems of particles mix and spread into each other due to thermal particle motion
dilation (time) (627) The apparent slowing of the clocks of a system moving at a relativistic speed as seen by an observer assumed to be at rest; a prediction of the special theory of relativity
dimension (24) A measureable property of an object or system (e.g., length, width, height, mass, force, time, etc.)
dimensional (10) Of or having to do with a dimension of a system; See *dimension.*
diode (463) An electrical component that permits the passage of current in only one direction
direct current (DC) (448) An electric current that flows in only one direction and at a constant magnitude
dispersion (556) The refraction by transparent materials of different frequencies of light at different angles; The dispersion of a beam of white light produces a rainbow effect.
displacement (d) (50) The distance and direction between an object's position at one time and its position at a later time; in oscillatory motion, the distance and direction of the system from its equilibrium position
disturbance (280) A pulse or oscillation in a medium
diverging lens (563) A lens that spreads incident light rays apart
domain (478) A microscopic magnetic region of a material in which the magnetic moments of the particles are more-or-less aligned in the same direction; The overall magnetic character of the material is the net effect of all the domains in the material.
dominion modeling (6) Creating descriptive and predictive models of observable phenomena using real-world data for the purpose of exercising dominion in the world according to the Creation Mandate
dominion science (10) Empirical science that is done in response to God's commandment to exercise dominion over the earth; See *Creation Mandate.*
doping (465) The process of adding trace amounts of elements to a substrate semiconductor material during manufacturing in order to alter its electrical characteristics
Doppler effect (285) The apparent change of frequency compared to the emitted frequency of a source as received by an observer moving relative to the source
dose (*D*) (669) The amount of energy absorbed per unit of mass by a substance exposed to nuclear radiation; measured in grays (Gy)
driven oscillation (279) Oscillations resulting from a periodic addition of energy at the natural frequency of the system
dry cell (451) A voltaic cell with electrodes immersed in an electrolytic paste instead of a liquid electrolyte
dynamics (50) The branch of mechanics that mathematically describes the causes of motion

E

eddy current (505) A circulating current induced by an oscillating magnetic field in a two- or three-dimensional conductor
Edison effect (463) The emission of electrons from a heated cathode in a vacuum tube; discovered by Thomas Edison; also called *thermionic emission*
efficiency (η) (233) The ratio of useful work output to work input for a machine; Other methods of measuring efficiency involve the ratio of the effective output to input of the relevant parameter.
elastic collision (247) A collision in which kinetic energy is conserved
elastic modulus (*E*) (300) The measure of the elasticity of a uniform material; the ratio of the normal stress to the linear strain; also called *Young's modulus*;

$$E = \frac{\sigma}{\varepsilon}.$$

elastic potential energy (210) Energy possessed by a system due to its position or condition with respect to an elastic force
electrical ground (422) A source or sink of charge large enough that adding or subtracting a finite amount of charge will not change its charge; The earth is the ultimate electrical ground.
electrical potential (*V*) (432) The amount of work per unit charge that is needed to move a charge a certain distance *r* from a point charge; $V = U_e/q$.
electric field (430) The region surrounding a charged object in which it can exert an electrical force
electric field strength (E) (430) The direction and magnitude of the electric field at a point in space
electrode (449) A conductor that allows an electrical current to enter or leave an electrolytic cell (or other apparatus)
electrodynamics (447) The study of the causes and effects of current electricity
electromagnet (500) A solenoid with a core of ferromagnetic material
electromagnetic (EM) wave (516) A self-propagating wave consisting of mutually perpendicular electric and magnetic component fields

electromagnetic force (121) One of the four fundamental forces in nature; manifested in the forces exerted between atoms; The photon is the exchange particle for electromagnetic forces.

electromagnetic induction (502) The generation of an electrical potential in a conductor by a changing magnetic field; may be created by the movement of the conductor through a magnetic field or by the change of the field surrounding the conductor

electromagnetic radiation (516) Energy transferred by electromagnetic waves

electromagnetic spectrum (526) The entire continuum of electromagnetic energy from radio waves to gamma rays

electromagnetism (11) The field of study involving any aspect of electricity and magnetism; also, the production of electricity from magnetism or vice versa

electron (416) A negatively charged particle-wave entity that occupies the space immediately surrounding the nucleus of an atom and which is involved in the formation of bonds between atoms; a lepton subatomic elementary particle whose antiparticle is the positron

electron volt (eV) (648) The kinetic energy acquired by an electron accelerated through a potential difference of one volt

electroscope (419) An instrument that can detect and measure electrostatic charge

electrostatic force (422) The attraction or repulsion force between two electrical charges

electrostatic unit (esu) (424) A non-SI unit of charge; Within the CGS (or Gaussian) system of units, it permits the constant k in the Coulomb's law equation to be 1.

element (295) A pure substance that consists of atoms with the same atomic number

elementary particle (684) A basic building block of atoms; may exist as a unique entity, such as a quark or lepton, or in combinations with other elementary particles to form composite entities (i.e., hadrons)

empirical science (10) Science conducted through direct, present-day observation of phenomena or that utilizes credible documentation of observations in the past; sometimes referred to as *experimental science*

endoergic (679) A nuclear reaction that converts other forms of energy to mass; any endothermic process

energy (196) A property of a system that gives it the ability to do work

energy level (648) In quantum mechanics, one of the allowable atomic energy states occupied by orbiting electrons having the same principal quantum number

entropy (*S*) (374) The measure of randomness or disorder in a specified portion of the universe; a measure of the likelihood that a system will be in a particular state; the measure of the amount of unusable energy in the system

equation of flow continuity (398) In fluid dynamics, the equation that states that the product of an ideal fluid's speed and the cross-sectional area of its flow tube is constant

equilibrant force (119) A force that yields a zero force sum when added to the resultant of all other concurrent forces acting on a system

equilibrium position (199) The rest position of a system

equipotential surface (434) An imaginary surface in an electrical field on which the electrical potential is the same at all points

equivalent dose (H_T) (670) A measure of the damage done by exposure to nuclear radiation; The type of radiation and the kind of material being irradiated are used to qualify the dose.

error (32) The difference between a measured value and the actual or accepted value

escape speed (223) The minimum initial radial speed that an object must possess in order to permanently escape the gravitational field of a planet

evaporation (306) The process by which energetic liquid-phase molecules spontaneously enter the gaseous phase at temperatures below the boiling point and above the freezing point of the liquid

exchange particle (688) In the Standard Model of matter, a particle that when passed from one entity to another results in a force exerted between the entities

excited state (648) The condition of an electron that is in an energy level higher than its lowest possible energy state (i.e., the ground state)

exoergic (679) A nuclear reaction that converts mass to other forms of energy; any exothermic process

F

farad (F) (438) The derived SI unit of capacitance; 1 F = 1 C/V.

Faraday's law (503) The induced voltage, or emf, in a circuit is proportional to the rate of change of the magnetic flux through the circuit.

fermion (685) Elementary or composite subatomic particles identified by a fractional spin; Fermions include the elementary leptons and quarks, and the composite baryons.

ferrimagnetic (478) Describes the net magnetic moment of a material that contains two sources of opposite and unequal magnetic moments; Magnetite is an example of a natural ferrimagnetic material.

ferromagnetic (477) Having a strong magnetic moment produced by all domains being aligned essentially in the same direction; having a magnetic permeability (μ) much greater than 1; the property normally associated with permanent magnets

fiber optics (556) The use of glass fibers to transmit light signals by total internal reflection

field (208) A physical property of space that can be quantified at every point in a volume by a physical law relating such quantities as distance, force, flux, or flux density

field model (429) A theoretical approach to accurately explaining such three-dimensional phenomena as magnetic, electric, and gravitational forces, as well as sound propagation and subatomic particle fluxes

field theory (121) The theoretical model that explains gravitational attraction and the attraction or repulsion of magnetic or charged objects as properties imposed on the space surrounding the objects (the field) by the matter or kind of matter of the objects

first-class lever (229) A lever-type simple machine having the fulcrum between the load force and the effort force

first commandment (2) "Thou shalt love the Lord thy God with all thy heart, and with all thy soul, and with all thy mind, and with all thy strength" (Mark 12:30); the most important moral command in Scripture; When we put this together with the Creation Mandate, we learn that we exercise dominion out of love for God and our desire to glorify Him.

first equation of motion (61) Final velocity is directly proportional to the uniform acceleration calculated from an initial velocity; $v_{2x} = v_{1x} + a_x\Delta t$.

first ionization energy (649) The energy required to remove the highest-energy (valence) electron in a neutral atom

first law of thermodynamics (361) Matter and energy can be neither created nor destroyed in any process; One practical application is the observation that heat transferred to or from a system is equal to the sum of the change of the system's internal energy and the work that the system does on its surroundings.

fission (667) The nuclear process of splitting a massive atomic nucleus, which results in the production of two smaller nuclei and several neutrons, as well as the release of a large amount of energy (thermal, light, and kinetic)

flow tube (396) An imaginary volume of moving fluid bounded by streamlines defining a fluid dynamics system

fluid mechanics (383) The application of Newtonian principles to fluids; subdivided into hydrostatics (the study of the mechanics of relatively stationary fluids) and hydrodynamics (the study of moving fluids)

fluoresce (663) To release energy, usually visible light, after exposure to ultraviolet rays or higher-energy radiation

focal length (f) (538) The distance from the center of a lens or mirror to its principal focus

force (117) A push or a pull on a system; according to the Standard Model of quantum mechanics, the effect or interaction resulting from the exchange of meson elementary particles

force table (123) A circular laboratory apparatus designed to demonstrate the summation of concurrent forces

formula unit (295) The smallest chemical formula that completely describes the composition of a compound; a single molecule of a molecular chemical compound; the smallest ratio of elements in an ionic compound

forward-biased (465) The application of a potential difference across a *p-n* junction such that the higher potential is on the *p* side and the lower is on the *n* side, causing the *depletion zone* at the junction to diminish or be eliminated, allowing current to flow through the junction

frame of reference (94) A coordinate system or point of view that an observer chooses to consider stationary and from which all motions are observed

free-body diagram (165) A functional diagram used in mechanics that isolates systems to analyze the forces on each system

free fall (66) The state of falling toward the earth or other planetary object under the influence of gravity only

freezing (307) The change of physical state from a liquid to a solid at the substance's freezing point

frequency (f) (267) Number of cycles completed per unit time; measured in hertz (Hz) or s^{-1}

friction (**f**) (122, 176) The contact force between two surfaces sliding against each other that opposes their relative motion

fundamental charge (*e*) (417) The magnitude of and unit for the electrostatic charge on a single electron or proton; $1\,e = 1.602 \times 10^{-19}\,C$.

fundamental (frequency) (284) The dominant and lowest frequency in a complex sound, such as a single tone produced by a musical instrument; Harmonic frequencies are whole-number multiples of the fundamental frequency.

fundamental force (120) One of four fundamental forces or interactions in nature: gravity, electromagnetic force, strong nuclear interaction force, and weak nuclear interaction force

fusion (679) The nuclear process of combining two or more smaller nuclei into one larger nucleus, which results in the release of thermal, electromagnetic, and kinetic energies; generally more exoergic than fission per event

G

Galilean relativity (620) The laws of mechanics are the same in all nonaccelerating reference frames; Velocity is not absolute but must be measured relative to a selected frame of reference.

Galilean transformation (622) A change in coordinate systems between inertial reference frames that are moving at nonrelativistic speeds with respect to each other

galvanic cell (449) A chemical source of electrical potential difference created by dissimilar metals immersed in an electrolyte

galvanometer (460) A device that measures and displays very small electric currents

gamma radiation (γ) (666) Gamma rays emitted by radioactive nuclei

gauge pressure (P_g) (385) The pressure of a fluid system with reference to ambient atmospheric pressure; Absolute pressure equals gauge pressure plus 1 atm.

gauss (G) (476) The unit of magnetic flux density (**B**) in the CGS system of units; used for weak magnetic fields; 10 000 gauss = 1 tesla.

Gay-Lussac's law (327) The pressure of a confined ideal gas is proportional to its absolute temperature, provided its volume is held constant; $P = kT$.

general theory of relativity (634) Acceleration and gravity are indistinguishable; The laws of mechanics are valid in all reference frames (inertial and accelerated) with appropriate transformation rules.

geocentric theory (151) The earth is the unmoving center of the universe; All celestial objects are in orbit around the earth or move with respect to the earth; an early manifestation of the anthropic principle attributable to a simplisitc reading of the Christian Bible

geochronology (676) Naturalistic methods used to estimate the age of individual rocks and of the earth itself

gravitational force ($\mathbf{F}_g$) (120) The force between two systems originating from their masses, exerted over the distance between their centers of mass; always attracts; the weakest of the four fundamental forces

gravitational lensing (635) The bending of light from more distant celestial objects around closer massive objects, such as galaxy clusters, that produces telescopically distorted images of the more distant objects

gravitational potential ($U_g(r)$) (208) Gravitational potential energy per unit mass at a specific height or distance from the source of gravity; $U_g(r) = U/m(r) = |g|r = |g|h$, where the notation "$(r)$" means "a function of r."
gravitational red shift (636) The apparent stretching of light wavelengths due to the proximity of an extremely strong source of gravity; A consequence of the general theory of relativity, it results in a shifting of the received frequency toward the red end of the spectrum.
gray (Gy) (669) The derived SI unit of radiation dose to organic tissues; 1 Gy = 1 J absorbed/kg tissue.
ground state (648) The lowest possible energy state of an atom, where all of its electrons are in their lowest energy levels

H

hadron (685) The supergroup of composite subatomic particles that consists of the baryons (composite fermions) and the mesons (composite bosons); In the Standard Model of quantum mechanics, hadrons are distinguished by being "colorless" (i.e., they carry no *color charge*).
half-life ($t_{1/2}$) (672) The amount of time required for one-half of a given number of a particular radioactive nuclei in a sample to undergo a certain kind of decay; $t_{1/2} = 0.693/\lambda$, where λ is the decay constant of the nuclide.
harmonic (284) A mechanical wave whose frequency is an integer multiple of the principal or fundamental frequency
heat (Q) (342) Thermal energy transferred from one system to another
heat capacity (C) (342) The amount of heat needed to be gained or lost by a system to change its temperature 1 degree
heat death (377) The uniformitarian theory that, given enough time, all of the energy in the universe will become unavailable for work, and the temperature will be uniform at a few kelvins
heat engine (361) A device that converts thermal energy to mechanical work
heat pump (371) A device that moves thermal energy from a cold reservoir to a hot reservoir by expending work on a heat-carrying fluid
heat reservoir (367) A theoretical body that is assumed to remain at a constant temperature no matter how much thermal energy is added to it or taken from it
Heisenberg uncertainty principle (651) It is impossible to know the momentum and exact position of a quantum mechanical object at the same time; Measuring one property of such an object disturbs the other, so no two properties can be known simultaneously.
heliocentric theory (151) The model of the solar system whereby the sun is its gravitational center and all other periodic objects are in orbit about the sun
henry (H) (509) The derived SI unit of inductance; 1 H = 1 Wb/A.
hertz (Hz) (267) The derived SI unit of frequency; 1 Hz = 1 s^{-1}.
heterogeneous solid (299) A solid material that consists of a mixture of crystalline and/or amorphous substances, or one that exhibits some large-scale, nonuniform structure not related to a crystal lattice (e.g., wood)
historical science (10) The body of scientific knowledge aquired by making inferences about events and processes in the past, based on observations of evidence in the present; sometimes called *origins science*
Hooke's law (199) Within the elastic limit of a material, the displacement of the system from its equilibrium position is directly proportional to the applied force.
Hooke's-law force (199) Any varying force exerted on or by a system that obeys Hooke's law
horizontal projection (99) The motion and trajectory of a ballistic projectile launched in a horizontal direction; assumed to have a constant horizontal velocity and zero initial vertical velocity
hydraulic device (389) A simple machine that transmits, according to Pascal's principle, a force via an enclosed liquid
hydrodynamics (383) The study of the mechanics of fluids in motion
hydrofoil (401) An immersed wing-like surface that generates lift as it moves relative to a liquid medium
hydrometer (394) A weighted, immersible instrument with a calibrated scale that indicates by the depth that it floats the density of a liquid
hydrostatics (383) The study of the mechanics of fluids at rest
hypothesis (9) An educated guess at the solution to a scientific problem; a tentative explanation for a phenomenon before the explanation has been tested

I

ideal fluid (396) A hypothetical incompressible fluid that is assumed to flow smoothly (without turbulence), without friction, and with a constant velocity at every point in the fluid
ideal gas (324) A hypothetical gas that behaves exactly according to the ideal gas law; Ideal gas particles are infinitesimally small, collide elastically, do not interact, and move randomly.
ideal gas law (330) Relates pressure, volume, temperature, and amount of an ideal gas under moderate conditions; $PV = nRT$.
ideal gas process (366) A process that allows the use of ideal gas relationships or formulas
ideal mechanical advantage (IMA) (226) The factor that indicates the reduction in effort when doing work using a simple machine; can be computed by various ratios such as effort distance to resistance distance, or effort arm to resistance arm
ideal pulley (168) A grooved, frictionless wheel and axle used to change the direction of a tension force on a string, rope, cable, or chain
ideal spring (199) A hypothetical spring that has no internal resistance, has no mass, and perfectly obeys Hooke's law
ideal string (166) A hypothetical string that has no mass and does not stretch, perfectly transmitting tension forces
illuminance (E) (599) The amount of light an object receives from a light source; proportional to the intensity of the source and inversely proportional to the square of its distance from the object
image distance (d_I) (539) The distance of an image from the center of the mirror or lens producing it
image of God (2) The sum total of all the created human attributes that reflect, in a limited way, the perfections of God; Included in the image of God are all aspects of the human nature such as our abilities to love, hate, create, reason, plan, and so on.

impedance (Z) (514) The resistance to current flow in an AC component or circuit; depends on the resistance and/or reactance of the component or circuit

impulse (I) (241) The change in momentum due to the application of a force; $\mathbf{I} = \Delta\mathbf{p} = \Sigma\mathbf{F}\Delta t$.

inclined plane (228) A plane surface that connects two different heights relative to a reference height; a simple machine that is incorporated into wedges, screws, and ramps

index of refraction (n) (552) The ratio of the speed of light in a vacuum to the speed of light in a given medium; $n = v_{medium}/c$; a measure of a medium's optical density

inductance (L) (509) A measure of an AC circuit component's induced potential difference in the presence of a changing current; measured in henries (H)

induction (charging) (421) Creating a temporary electrical dipolar charged condition in a neutral object by causing a segregation of charges with a nearby source of charge

inductive reactance (X_L) (514) A measure of how strongly an inductive AC component opposes the change of alternating current through it

inductor (509) An AC circuit component that produces an induced potential difference that opposes a change in the current flowing through it

inelastic collision (247) A collision in which kinetic energy is not conserved; Some or all of the initial total kinetic energy is consumed in deforming the colliding objects.

inertial reference frame (622) An unaccelerated coordinate system in which Newton's laws hold exactly

inference (7) A method of reasoning whereby one arrives at a conclusion based on premises known or assumed to be true

insulator (418) A material having few mobile electrons so that it does not easily conduct thermal energy or electricity

integrated circuit (IC) (467) A complete, microscopic electronic circuit consisting of dozens to billions of semiconductor electronic components formed on the surface of a single wafer or chip

intensity (I_s) (284) The strength or power of a wave phenomenon; proportional to the square of a mechanical wave's amplitude

interface (552) The boundary between two media in contact

interference (574) The superposition of two or more waves that exist in the same place at the same time

interferometer (579) An instrument that uses the interference of waves to determine the characteristics of the waves or to measure distance

internal energy (U) (360) The sum total of energy consisting of the potential and kinetic energies associated with the particles of a substance; includes the thermal energy of the substance

ion (295) An atom or group of atoms that has either fewer or more electrons than the number of protons in the particle, resulting in a positive or negative net charge on the particle, respectively

ionizing radiation (668) Any form of radiation energetic enough to remove electrons from atoms, thereby creating ions; The principal forms of nuclear radiation (α, β, and γ) are ionizing radiation.

isobaric (366) A process or condition at constant pressure

isochoric (366) A process or condition at constant volume

isolated system (365) A system that does not exchange energy or matter with its surroundings

isothermal (365) A process or condition at constant temperature

isotope (647) Any atom of a given element having a different number of neutrons compared to other atoms of the same element

J

joule (J) (196) The derived SI unit of energy, work, and heat; $1\ \text{J} = 1\ \text{N·m}$.

Joule's law (453) The power dissipated in a resistor is equal to the product of the resistance and the square of the current through the resistor; $P = I^2R$.

K

Kepler's constant (K) (153) For a given planetary gravitational system, the ratio of the cube of the object's orbital semi-major axis and the square of its orbital period is a constant. $K = R^3/T^2$.

Kepler's laws (152) Mathematical rules that geometrically describe the motions of objects orbiting a star or planet; (1) orbits are ellipses with the orbited body located at one focus, (2) planetary radial lines sweep out equal areas in equal times, (3) $R^3 = KT^2$.

kilogram (kg) (27) The SI unit of mass; 1 kg is defined as the mass of a manufactured Pt-Ir alloy standard maintained at the International Bureau of Weights and Measures in Sèvres, France.

kilowatt-hour (kWh) (453) The unit of electrical energy used by utility companies to measure electrical energy consumption

kinematics (49) The branch of mechanics that describes how systems move, without addressing what causes the motion

kinetic energy (K) (202) The mechanical energy possessed by a moving system; $K = \frac{1}{2}mv^2$.

kinetic friction (f_k) (177) The friction force that opposes the motion of an object that is moving; $f_k = \mu_k N$.

kinetic theory of matter (298) Matter consists of tiny particles in constant motion; The properties of matter are consequences of that motion.

Kirchhoff's rules (459) Electrical conservation rules; Voltage Rule: All voltage drops in a circuit are equal to the voltage rises in the circuit; Current Rule: All currents entering an electrical node must equal all the currents leaving the node.

L

laminar (396) A term used to describe fluid flow; smooth, layer-like, with continuous streamlines

latent heat of fusion (L_f) (348) The amount of absorbed thermal energy required to change a certain mass of a solid at its melting point to a liquid at the same temperature; the same amount of heat liberated to freeze the same mass of the liquid

latent heat of vaporization (L_v) (348) The amount of thermal energy required to change a certain mass of a liquid at its boiling point to a gas at the same temperature

law (10) A statement (usually mathematical) describing a consistent relationship among physical quantities in nature; Explanation of phenomena is not an inherent property of scientific laws.

law of charges (415) Like charges repel; opposite charges attract.

law of magnetic poles (475) Like magnetic poles repel; opposite poles attract.

law of moments (149) For simple lever systems, rotational equilibrium exists when the opposing moments applied to both ends of the lever are equal in magnitude; $F_1\ell_1 = F_2\ell_2$.

law of reflection (535) In ray theory, the angle of incidence to the normal equals the angle of reflection.

Lenz's law (504) A magnetically induced potential difference in a conductor opposes the change in the current that caused it.

lepton (685) The class of elementary subatomic particles that possess a spin of 1/2; includes the electron, neutrino, muon, and their antiparticles

lever (229) A device that can pivot about a fulcrum, transmitting a force applied at one location to move a load at another

Leyden jar (436) An early capacitor-like device designed to store electrical charge; consists of a lead-coated and lead-lined glass jar plugged by an insulating stopper and penetrated by a metal probe with an attached chain that rests on the interior metal lining

lift (401) An aerodynamic or hydrodynamic force exerted more-or-less perpendicular to a surface by a fluid flowing past the surface, primarily in response to the Coandă effect

light-emitting diode (LED) (530) A semiconductor device that emits light when a minimum specified potential difference is established across its terminals

line of force (430) One of a number of imaginary lines defining the direction and intensity of the force exerted by the field at all points surrounding an electric charge or magnet

line spectrum (647) A spectrographic display consisting of lines representing electromagnetic emissions in discrete frequencies or wavelengths

lodestone (475) A natural magnet of an iron oxide ore

longitudinal wave (281) A wave in which the medium oscillates parallel to the direction of the wave's propagation

Lorentz transformation (626) A change of system coordinates from one inertial reference frame to another that is valid at any relative speed between them; accounts for the effects of special relativity at relativistic speeds

lumen (lm) (598) The derived SI unit of luminous flux; equal to a 1 cd light source in 1 sr; 1 lm = 1 cd·sr.

luminance (598) The rate of light energy reaching a surface in a given direction; Unit is cd/m^2.

luminiferous ether (622) A hypothetical fluid, believed by nineteenth-century scientists to be the physical medium necessary for the propagation of light waves

luminous flux (Φ) (598) The measure of the total power produced by a light source; units of lumens (lm); weighted power of a light source adjusted to the sensitivity of the human eye for all wavelengths emitted

luminous intensity (I_L) (597) The intrinsic brightness of a light source; the power of light energy emitted in a specific direction; Unit is the candela (cd).

lux (lx) (599) The derived SI unit for illuminance; 1 lx = 1 lm/m^2, or the illuminance of a 1 cd light source at 1 m distance.

Lyman series (649) A spectral series of emission lines produced when an electron in a hydrogen-like atom falls from the second or higher energy levels to the first energy level

M

machine (simple) (226) A device that changes the magnitude or direction of a force, or both

Mach speed (M) (265) The speed of sound in air for the existing conditions of temperature and air pressure; measured in mach units (M)

magnetic declination (480) For a given location on the earth's surface, the angle between the directions to true north and magnetic north

magnetic dipole moment (μ) (478) The vector representing the magnetic effect of the electron current in an atom or molecule

magnetic field vector (B) (476) The strength and direction of the magnetic field at a specified point in space; Unit is the tesla (T).

magnetic flux (Φ) (502) The strength or amount of a magnetic field in a given volume of space; Unit is the weber (Wb).

magnetic quantum number (*m*) (652) The third quantum number, which defines the shape and orientation of an electron's orbital

magnetization (476) The process of aligning the magnetic dipoles within a ferromagnetic or paramagnetic material by the imposition of an external magnetic field on the material

magnetosphere (481) The earth's magnetic field that extends into the space surrounding the planet; the principal shield protecting the earth against bombardment by energetic particles from the solar wind

magnification (×) (543) The ratio of the height of an image to the height of the object from which it is formed

magnitude (50) The absolute value of the measure of a quantity; the size of a measurement without regard to its direction or sign

Malus's law (590) The intensity of a polarized light beam transmitted by a dichroic material (the analyzer) is a function of the angle between the wave's plane of polarization and the analyzer's transmission axis.

manometer (391) An apparatus consisting of a U- or J-tube that measures the pressure in a fluid system by supporting a liquid column against atmospheric pressure; The pressure in the system is measured by comparing the meniscus of the liquid against a scale.

maser (657) A device that uses amplification through stimulated emission to produce a coherent beam of microwave radiation

mass (26) The quantity of matter in an object; the measure of an object's inertia; Unit is the kilogram (kg).

mass defect (679) The difference between the mass of a nucleus and the sum of the masses of the individual particles from which it was formed; equivalent to the binding energy of the nucleus

mass number (*A*) (647) An integer number equal to the number of protons and neutrons in the nucleus of an atom

mass spectrograph (486) A device that determines the masses of particles in a sample using the principle of the velocity selector

matter (24) Any created substance that occupies volume and has mass; one of the three fundamental dimensions of Creation

mechanical equilibrium (124) The state of a system where the vector sum of all external forces on the system is zero

mechanics (11, 49) The study of the motion of macroscopic objects; the study of how (kinematics) and why (dynamics) things move

medium (280) In wave mechanics, the substance that contains a mechanical wave; more generally, a substance that contains any

kind of wave, whether the substance is necessary for the wave's propagation or not
melting point (302) The temperature at which a solid substance enters its liquid phase for a given set of physical conditions (e.g., atmospheric pressure)
meson (685) The family of subatomic particles consisting of composite bosons, or particles with integer spins; includes particles with a wide range of masses, such as the pion, kaon, and D-meson, and their antiparticles
meter (m) (25) The SI unit of length; 1 m is defined as the distance traversed by light in a vacuum in exactly 1/299 792 458 second.
microscope (606) An optical instrument designed to magnify extremely small objects
millibar (mb) (385) A non-SI unit of pressure used for measuring and reporting atmospheric pressure; 1/1000 of a bar; 1 mb = 100 Pa.
mirage (557) Any atmospheric optical illusion caused by the refraction of light near the horizon that alters the shape of a distant object or its location with respect to the horizon
mirror equation (541) A formula that relates object distance, image distance, and mirror focal length;

$$\frac{1}{d_O} + \frac{1}{d_I} = \frac{1}{f}$$

model (6) Any theoretical or physical construct that provides a workable description or explanation of a concept or phenomenon; A model does not have to be exact in every detail, but must be useful for the purpose for which it is intended.
moderator (680) A material included in a U-235 fission reactor consisting of light particles (such as water) with which neutrons can collide to slow down; A moderator both slows down neutrons to thermal energies, which is necessary for U-235 fission, and reflects them back into the core to improve fission efficiency.
modern physics (13) As distinguished from classical physics, all revolutionary discoveries in physics made since approximately 1900; deals primarily with quantum mechanics, high-energy physics, and relativistic astrophysics
molar mass (344) The mass, in grams, of a pure substance equal in number to its atomic mass stated in unified amu (u); The molar mass of a substance contains Avogadro's number of atoms, molecules, or formula units of the substance.
mole (mol) (296) The SI unit for numbering atoms, molecules, or other similar particles; One mole contains Avogadro's number of particles.
molecule (295) A separate, distinct, independent unit consisting of two or more covalently bonded atoms
moment arm (ℓ) (148) The perpendicular distance between the pivot point and the line of the applied force when computing torque; $\ell = r\cos\theta$, where θ is the angle between the force and the position vector of the point of application of the force.
moment of inertia (I) (275) The inertia characteristic of a rotating system that is defined by the distribution and quantity of mass and its distance from the rotational axis of the system
momentum (p) (239) Newton's "quantity of motion" characteristic of any moving system, directly proportional to both the system's mass and its velocity; $\mathbf{p} = m\mathbf{v}$.
mutual inductance (M) (510) A measure of a circuit element's contribution to the induced potential difference in another circuit element
mutual induction (510) Induction of a potential difference across an inductor-like circuit element as a result of a changing magnetic field associated with another circuit element

N

natural oscillation frequency (f_0) (278) The frequency at which a system oscillates if it is given a single pulse of energy
neutron (n) (647) A neutral particle with slightly more mass than a proton; normally found in the nucleus of an atom
Newtonian reflector telescope (610) A reflecting telescope that uses a flat secondary mirror to project the image from the main mirror to an eyepiece mounted on the side of the telescope tube
Newton's law of universal gravitation (156) The law that describes the force of attraction between all matter;

$$F_g = G\frac{Mm}{r^2}.$$

Newton's laws of motion (123) Principles describing the effects of forces on "normal"-sized objects moving at nonrelativistic speeds
node (458) The junction of two or more conducting paths in an electrical circuit; the location of no displacement in a standing wave.
nonconservative force (205) Any force that when exerted on an object over a closed path produces non-zero work; Similarly, the amount of work done on an object with respect to such a force depends on the path taken.
noncontact force (121) A force that exists between systems that are not touching; a force exerted according to a field theory
nonuniform circular motion (142) Circular motion that involves a change in angular speed during a time interval
normal force (N) (171) The force exerted by a supporting surface on a system and that is perpendicular to the supporting surface
normal science (9) According to Thomas Kuhn, the regular task of scientists working within the accepted paradigm of science; Normal science primarily involves "puzzle solving," which, according to Kuhn, consists of the day-to-day "mop-up work" that affirms and extends the paradigm.
***n*-type semiconductor** (465) A manufactured material consisting of a semiconductor doped with another element that carries an extra valence electron compared to the semiconductor, which thus introduces extra negative charge carriers into the crystal lattice
nuclear reactor (681) The container for a controlled fission reaction; a massive, complex device containing an arrangement of nuclear fuel, a moderator, and control mechanisms; The heat liberated is used to boil water for use in a connected steam plant.
nucleon (647) A proton or neutron bound in the nucleus of an atom
nucleus (647) The dense central part of an atom made up of protons and neutrons; The nucleus contains virtually all of the atom's mass but occupies only a tiny portion of the atomic volume.
nuclide (647) A specific isotope of an element; often used when comparing isotopes of different elements

O

object distance (d_O) (539) The distance of the object from the center of a mirror or a lens
observation (17) A deliberate application of the senses to a phenomenon, with or without the aid of instruments; In science,

an observation implies the creation of a permanent record of the observation for future reference.

ohm (Ω) (452) The derived SI unit of electrical resistance; $1\ \Omega = 1\ \text{V/A}$.

ohmmeter (460) An instrument that measures and displays electrical resistance

Ohm's law (452) The potential difference across a circuit or circuit component is equal to the product of the current through it and its resistance; $V = IR$.

open system (365) A system that allows the exchange of both energy and matter with its surroundings

operational definition (13) Defining something by specifying certain tests that an object or phenomenon must meet rather than by using simpler or more fundamental terms; usually needed for fundamental concepts in science

operational science (10) The body of scientific knowledge aquired by methods of investigation that involve real-time observations of present-day phenomenon or records of first-person observations of events in the past

optical axis (538) The imaginary line that passes through the optical center of a lens or mirror and its center of curvature

orbital (651) The probable location and energy of an electron defined by its first three quantum numbers

orbital motion (138) Circular motion about an axis of revolution that does not pass through the moving object itself

order of magnitude (43) The approximate value of a numerical quantity expressed to the nearest power of 10; a ratio of two values expressed to the nearest power of 10

overdamped oscillation (278) An incomplete oscillation where damping is so strong that the system returns directly to its equilibrium position with no overshoot

P

paradigm (8) A set of assumptions, concepts, values, beliefs, and practices that controls the way a person or a group of persons perceives some aspect of reality

parallel (circuit) (454) The type of electrical circuit that has more than one simultaneous path for the current flowing from the voltage source

paramagnetic (477) Describes a material that is slightly attracted to magnets; having a magnetic permeability (μ) slightly greater than 1; Paramagnetic materials do not retain a permanent magnetic field when an external magnetic field is withdrawn.

parent nuclide (673) Generally, the heaviest nuclide in a radioactive *decay chain* that emits alpha or beta particles to become a daughter nuclide

partially elastic collision (247) A collision in which the colliding objects exchange some of their kinetic energy for other forms of energy but do not stick together

pascal (Pa) (362) The derived SI unit of pressure; $1\ \text{Pa} = 1\ \text{N/m}$.

Pascal's principle (389) An enclosed fluid exerts the same pressure in all directions at the same depth, and a change of pressure in the contained fluid is transmitted undiminished to all parts of the fluid.

Pauli exclusion principle (654) No two electrons in the same atom can have identical sets of quantum numbers.

peer review (18) A review of a scientific paper by qualified scientists working in the area of the reported work in order to offer critical comments, suggest corrections, and to ensure that the highest standards of scientific research are followed

pendulum (270) A mass suspended at some distance from a pivot point and that is free to swing in periodic motion

pendulum arm (*L*) (270) The distance of the center of mass of a pendulum from the system pivot point

percent error (32) The measurement error given as a percentage of the known or accepted value;

$$\text{percent error} = \frac{\text{measured value}}{\text{accepted value}} \times 100\%.$$

performance coefficient (*K*) (371) A numerical rating of the efficiency of a refrigeration engine or heat pump

period (*T*) (153) The time required to complete one cycle in periodic motion

period (chemistry) (656) A horizontal row in the periodic table of the elements

periodic motion (266) Oscillatory or vibratory motion that continuously repeats itself with a constant, or nearly constant, period

periodic table (656) A table of the elements, listed in order of increasing atomic number and arranged to relate their electron structures and chemical similarities

permeability of free space (μ_0) (500) A measure of how effectively a magnetic field penetrates a vacuum

permittivity of free space (ε_0) (438) A measure of how effectively an electric field penetrates a vacuum

permittivity of material (ε) (438) A measure of the ability of a dielectric material between the plates of a capacitor to resist the formation of an electric field within it

phase diagram (308) A graph showing the relationship between the phases of a substance and controlling factors such as temperature and pressure

phenomenon (6) An observable event or process

photoelectric effect (644) Electrons are ejected from a polished metal surface when it is struck by light photons having a frequency whose energy exceeds the work function of the electrons in the metal.

photometer (600) An instrument that compares the luminous intensity of an unknown light source to a standard light source

photon (644) An electromagnetic "particle" with energy hf and zero rest mass; According to the Standard Model of quantum mechanics, the photon is the exchange particle for the electromagnetic force.

physical pendulum (275) A real pendulum, where some mass is distributed along the pendulum arm

pigment (604) A substance that selectively absorbs certain wavelengths of light; Its color is the result of the wavelengths it reflects.

pitch (mechanical) (229) The distance between two adjacent threads on a screw

pitch (sound) (284) The perceived quality of highness or lowness of a tone that depends upon its fundamental frequency

plate (436) One of the conductors in a capacitor; the anode in a vacuum tube

plum-pudding model (646) The atomic model suggested by J. J. Thomson that views an atom as a sphere of positive charge in which numerous negative electrons are embedded

***p-n* junction** (465) In a semiconductor device, the boundary between *p*- and *n*-doped materials, characterized by a depletion zone where the adjacent charges neutralize each other

polarity marking (448) With an electrical device, the "+" sign indicates a point of higher (more positive) potential, and the "−" sign indicates a point of lower potential.

polarization state (589) The specific manner in which a light beam is polarized; States include plane, circular, and elliptical polarization.

polarized (589) The condition in which the *E*-waves of the photons in a beam of EM energy oscillate in a preferred orientation or change in a predictable way (i.e., not randomly oriented)

pole (475) A region or location on an object where magnetic flux or electric charge is concentrated

position-time graph (52) A graph of an object's position versus time

position vector (r) (84) A vector with its tail at the origin of a coordinate system and its head at the position of an object at a given time

Positivism (8) The philosophical view that the only valid source of knowledge is personal experience and that such knowledge can only be affirmed through the scientific method

positron (β^+) (667) A particle with the mass of an electron but a positive charge; an antielectron

potential difference (ΔV) (433) The difference of electrical potential between two locations in a circuit or in space

potential energy (U) (204) Mechanical energy that an object possesses because of its position or condition with respect to a conservative force

pound-force (lbf) (131) The unit of force in the English Engineering System; the gravitational force on 1 lbm

pound-mass (lbm) (131) The unit of mass in the English Engineering System

power (P) (201) The time rate at which work is performed; $P = W/\Delta t$.

precision (32) The number of significant decimal places a measuring instrument can yield; the repeatability of a certain measurement within specified measuring tolerances

presupposition (11) An assumption one derives from his worldview when making a decision about the validity of a stated fact; A paradigm includes a collection of presuppositions.

primary cell (450) A voltaic cell that uses an irreversible chemical reaction to produce an electrical potential

principal focus (F) (538) For a mirror or lens, the point at which incident light rays parallel to the principal axis converge after reflection or refraction

principal quantum number (n) (652) The first quantum number, which identifies the main energy level of an electron

principle of inertia (115) An object in motion will continue in its original state of motion unless or until an outside agent acts on it; attributed to Galileo Galilei

projectile (98) Any flying object that is given an initial velocity and whose trajectory is determined by the influence of external forces only, such as gravity

proper length (ℓ_0) (629) The length of an object as measured by an observer who is at rest with respect to the object

proton (p) (647) A positively charged particle in the nucleus of an atom with 1836 times more mass than an electron; a baryon consisting of two up-quarks and one down-quark

***p*-type semiconductor** (465) A manufactured material consisting of a semiconductor doped with another element that carries one less valence electron compared to the semiconductor, which thus introduces missing bonds into the crystal lattice that appear as positive holes or positive charge carriers

pulsating DC (508) A current that periodically oscillates between zero and a maximum but does not drop below the zero level

***P-V* diagram** (362) A graph on which the pressure of a working gas is plotted as a function of the gas's volume

Q

quality (284) The ear's interpretation of the mixture of fundamental and harmonic tones in a sound wave

quantized (644) Produced or transmitted as quanta; Values change in discrete steps as opposed to a continuum of values.

quantum mechanics (646) The study of the mechanics of matter at the atomic and subatomic level, which emphasizes the quantized nature of their energies and momentums

quantum number (652) One of four numbers that describe the location and energy of any electron in an atom at its ground state; the arguments of the Schrödinger wave equations

quark (687) One of a family of elementary subatomic particles within the fermion supergroup; There are six kinds (or "flavors") of quarks, along with their corresponding antiquarks; Quarks have 1/2 spin and a fractional charge ($\pm 2/3$ or $\pm 1/3\ e$).

quark confinement (687) The theory that quarks can exist only inside hadrons, and are held together by gluons exerting a property called *color charge*

R

radian (rad) (39) The derived SI unit of angular measure; in a circle, the central angle that subtends an arc on the circle equal in length to the radius of the circle

radiation (energy) (354) The transportation of energy or matter without the use of or dependence on an intervening medium

radiation (nuclear) (664) The particles and rays emitted by an unstable nucleus

radiation weighting factor (W_R) (670) A radiation dose correction factor based on the kind of radiation (α, β, or γ) and the kind of material exposed; used in calculating the equivalent dose (H_T)

radioactive dating (672) Estimating the age of a material by comparing the concentration of a radioactive nuclide and its daughter products, and then calculating the amount of time necessary for the measured proportion to occur; requires making a variety of assumptions, many of which are usually not scientifically verifiable

radioactivity (664) The emission of rays and particles from an unstable nucleus; the time rate of such emissions

radiogenic (673) A substance derived from radioactive decay

range (R) (99) The horizontal distance covered by a projectile

rarefaction (281) In a mechanical longitudinal wave, the region where the medium particles are farther apart than average; the region of lower density analogous to the trough of a water wave

ray (535) An imaginary line representing the direction of travel of a light wave; It is perpendicular to the wave front.

Rayleigh angle (θ_R) (588) For a given lens, the smallest angular distance between two objects at which the lens is able to resolve them into two distinct images; directly proportional to the wavelength of light and inversely proportional to the lens diameter

real image (539) An optical image formed by convergent light rays; Real images are inverted with respect to the object and can be projected on a screen.

recoil (244) The reaction force of a projectile on the device that launched it

reference angle (α) (82) The positive acute angle between a vector and its horizontal component vector

reference direction (74) The direction from which a vector angle is measured; usually the direction of the positive x-axis or the direction to the North Pole, depending on the situation

reference frame (94) See *frame of reference*.

reflecting telescope (609) A telescope that uses a mirror as the principal light-gathering element

reflection (535) The bouncing of a wave off a surface or boundary between media

refracting telescope (608) A telescope that uses a lens as the principal light-gathering element

refraction (551) The bending of waves as they pass from one medium to another with different transmission characteristics (optical density, temperature, salinity, water depth, etc.)

refrigeration engine (371) A heat engine that moves thermal energy from a cold reservoir to a hot reservoir

regelation (303) The melting of ice by the application of extreme pressure; When the pressure is removed, the liquid water instantly refreezes.

relative permeability (μ_r) (477) The ratio of the magnetic permeability (μ) of the material and the permeability of free space (μ_0)

relativistic (626) A relative speed of about $0.1c$ or greater between systems within a single frame of reference

relativistic factor (γ) (629) A mathematical factor that transforms properties of a system moving at a relativistic speed with respect to a stationary observer in order to match his observations of those properties; accounts for the effects of special relativity on the system

relativity (theory of) (620) If the appropriate transformations are made, it is impossible to distinguish between moving systems and those at rest by using the laws of classical mechanics.

resinous (414) Electrical behavior like that of amber or other resinous materials, which acquire a negative charge when rubbed with fur

resistance (*R*) (451) An electrical property that quantifies an entire object's opposition to current flow; measured in ohms

resistivity (ρ) (451) A measure of the degree to which a given material impedes current flow at a given temperature; Unit is the ohm-meter.

resistor (451) Any electrical component that depends primarily on its material, rather than its shape or construction, to impede the flow of current through it; a discrete circuit component whose purpose is either to establish a specific potential difference or to reduce the current in a portion of a circuit

resolving power (588) The ability of an optical system to resolve two distinct objects close together in angular separation into two separate images

resonance (electronics) (515) The state of an AC oscillator circuit exhibiting nearly zero impedance (for series-connected components) or very large to theoretically infinite impedance (for parallel-connected components)

resonance (periodic motion) (279) The state of an oscillating system driven at its natural frequency

rest mass (m_0) (629) The mass of an object as measured by an observer and that is at rest with respect to the observer

rest-mass energy (E_0) (632) The total energy of an object at rest according to Einstein's special theory of relativity; $E_0 = m_0c^2$.

restoring force (F_r) (266) A force that acts parallel to a system's displacement and that tends to return it to its equilibrium position

resultant (76) The vector sum or difference of two or more vectors

reverse-biased (465) The state of a *p-n* semiconductor junction in which the potential difference across the junction increases the size of the depletion zone and thus prevents conduction of current through the junction

reversible process (369) A controlled, quasi-static process that operates in either direction between two distinct states; a cyclic process that can be operated in reverse (e.g., the Carnot cycle)

rheostat (461) A manually controlled variable resistor

right-hand rule for circular motion (141) A rule for determining the orientation of the coordinate system associated with rotating or spinning systems; When the right hand is positioned such that the thumb extends along the rotational axis so that the fingers curl in the direction of rotation, the thumb points in the positive direction for indicating angular velocity and acceleration.

right-hand rule for conductors (499) A rule for determining the direction of the magnetic field around a current-carrying conductor; When grasping a conductor with the right hand and with the thumb pointing in the direction of conventional current flow, the fingers wrap in the direction of the magnetic field.

right-hand rule for magnetic force (485) A rule for determining the force exerted on a *positive* charge moving through a magnetic field; When the thumb, index finger, and middle finger are extended perpendicular to each other so that the index finger is parallel to the magnetic flux vector **B** and the thumb is parallel to the component of the charge's velocity vector perpendicular to the field ($\mathbf{v}_\perp$), then the middle finger points in the direction of the imposed magnetic force on the moving charge. The force on a negative charge is in the opposite direction.

right-hand rule for solenoids (500) A rule for determining the orientation of the magnetic dipole formed by a current-carrying solenoid coil; Grasping a solenoid coil so that the fingers of the right hand wrap in the direction of the conventional current flow, the extended thumb points in the direction of the north end of the magnetic dipole thus formed.

right-hand rule for torque (492) A rule for determining the orientation of the torque vector resulting from a moment on a rotating system; With the right hand extended such that the thumb is extended along the rotational axis and the fingers are gently curled in the direction of the applied moment, the thumb points in the direction of the torque vector.

right triangle (79) A triangle containing a 90° angle

roentgen (R) (670) A non-SI unit of the ionizing effect of nuclear radiation on air; 1 R = 2.58×10^4 C/kg of air.

rolling friction (f_r) (181) A nonconservative force that opposes the rolling motion of a system or the rolling components of a system

root mean square (rms) (507) The effective (rather than the peak) value of an alternating quantity; the peak value divided by $\sqrt{2}$

rotational equilibrium (149) The condition of a system when the sum of all torques is zero; The system will either continue to rotate at a constant speed or will remain stationary.

rotor (494) That part of an electric motor, generator, or other electromechanical device that rotates under the influence of electromagnetic forces

S

scalar (50) A quantity that can be completely described with only one piece of numerical information; a number that can be positive, negative, or zero

scalar component (82) The scalar value of a component vector

Schmidt-Cassegrainian telescope (610) A reflecting telescope having a spherical primary mirror, a convex secondary mirror, a rear-mounted ocular lens, and a front correcting plate that eliminates spherical aberration; The viewing position is thus behind the main mirror.

science (5) The collection of observations, inferences, and models produced through a systematic study of nature for the purpose of enabling humans to exercise good and wise dominion over God's world; The word science is also used to refer to the systematic methods that produce the observations, inferences, and models.

scientific method (16) The commonly used term for describing what scientists do; See *scientific methodology*.

scientific methodology (16) A systematic process for solving a scientific problem or presenting an explanation for a phenomenon; It involves a logical approach to the problem, suggesting a hypothesis, gathering data, testing the hypothesis, and using deductive and inductive reasoning to evaluate the validity and workability of the explanation. No scientific methodology would be complete without a peer review of the findings followed by publishing the work to give others access to the findings.

screw (228) A fastener consisting of a cylindrical shaft on which a thin, helically wrapped wedge is attached; The screw is a member of the inclined-plane family of simple machines.

second (s) (26) The SI unit of time; 1 s is defined as exactly 9 192 631 770 cycles of vibrating gaseous cesium-133 atoms.

second-class lever (229) A lever-type simple machine having the load force between the fulcrum and the effort force

second commandment (2) "And the second is like unto it, Thou shalt love thy neighbour as thyself" (Matt. 22:39); the second-most important moral command in Scripture; When we put this together with the Creation Mandate, we learn that we exercise dominion out of love for others.

second equation of motion (64) Displacement or position is dependent on the initial velocity, uniform acceleration, and the interval of time under consideration; $d_x = v_{1x}\Delta t + \frac{1}{2}a(\Delta t)^2$ or $x_2 = x_1 + v_{1x}\Delta t + \frac{1}{2}a(\Delta t)^2$.

second law of thermodynamics (367) During any energy transformation, some energy becomes unusable; includes the principle that heat flows from higher to lower temperatures; Every natural process tends toward minimum usable energy and maximum entropy.

self-induction (509) The production of back emf within an inductor-like circuit component by current changes in that component; Self-induction opposes the change in current that generates the inductor field in the first place.

semiconductor (418) A solid substance that has electrical properties intermediate between conductors and insulators

semi-major axis (*R*) (153) Half of the average distance from a planet's perihelion to its aphelion

series (chemical) (656) A subset of a horizontal row or period of the periodic table such as the Lanthanide Series; sometimes used to refer to an entire period of the table

series (electrical) (454) A circuit with a single current path

shear force (122) An asymmetrical force pair that tends to cause strain within a material along planes parallel to the applied force; By illustration, a deck of cards is distorted by a shear force as the cards slip past one another.

shear modulus (*G*) (301) The ratio of the shear stress to the shear strain;

$$G = \frac{\text{shear stress}}{\text{shear strain}}.$$

SI (24) Abbreviation for *Système International d'Unités* (International System of Units); the system of standardized metric units used by scientists worldwide

sievert (Sv) (670) The derived SI unit of equivalent dose from ionizing radiation exposure

significant digit (SD) (36) In a measurement, one of the digits that is known for certain based on the instrument's scale markings; Significant digit also applies to the one estimated digit included in a scientific measurement, which is a decimal fraction of the smallest scale increment.

similar triangle (79) A triangle whose three angles are congruent to the corresponding angles of another triangle

simple harmonic motion (SHM) (266) Periodic motion having a constant period perpetuated in an ideal system by a restoring force

sine θ (80) In a right triangle, the ratio of the length of the side opposite the acute angle θ to the length of the hypotenuse

Snell's law (552) A law governing the angles of incidence and refraction of light across the interface between two optical media; The ratio of the sines of the angles of incidence and refraction is equal to the ratio of the indices of refraction for the two media; $n_i \sin \theta_i = n_r \sin \theta_r$.

solar wind (481) The stream of high-energy charged particles emitted by the sun

solenoid (500) An electromagnetic device consisting of a tightly coiled conductor for the purpose of establishing a dipole magnetic field using electricity; If a solenoid is wrapped around a ferromagnetic core, an electromagnet is formed.

space (24) One of the fundamental dimensions of Creation; a three-dimensional volume in which matter and energy may be found and in which events take place

special theory of relativity (625) The model framework establishing that (1) all of the laws of physics have the same form in all nonaccelerated frames of reference, and (2) the speed of light in a vacuum is the same in all reference frames

specific gravity (s.g.) (384) The ratio of a sample's density to the density of water, yielding a dimensionless quantity that provides a convenient relative density

specific heat (c_{sp}) (343) The amount of thermal energy (in joules) that must be gained or lost in order to change the temperature of 1 g of a pure substance 1 °C; sometimes given in terms of joules per kilogram-kelvins

spectral series (649) A series of distinct wavelengths of electromagnetic radiation emitted by hydrogen-like atoms when their electrons fall from any higher energy level to a specified energy level

spectrum (526) The ordered distribution of some periodic aspect of a physical system; an arrangement of electromagnetic or acoustic emissions ordered by wavelength or frequency

speed of light (c) (516) The speed of propagation of all forms of electromagnetic radiation in a vacuum; defined as 299 792 458 m/s (exactly)

spherical aberration (537) The distortion of an image formed by a simple spherical mirror or lens due to the geometric impossibility of focusing parallel rays at a single point

spin (138) Rotation of an object about an axis that passes through the object

spin quantum number (m_s) (652) The fourth quantum number, which denotes the spin state of an electron

split-ring commutator (494) A device mounted on the shaft of a DC motor or generator consisting of conducting segments separated by insulating material, which are alternately connected to opposite ends of the windings mounted on the rotor; Electrical brushes touching the segments carry current to or from the coils.

spring constant (k) (199) The proportionality constant in Hooke's law for springs that allows equating the force exerted by a stretched or compressed spring to the displacement from its rest position

Standard Model (685) The theory of particle physics that accounts for the interactions of three of the four fundamental forces and the elementary particles that make up all matter in the universe; Only gravity cannot be accounted for within the Standard Model of matter. Unifying all four fundamental forces is the goal of the Grand Unification Theory (GUT).

standard temperature and pressure (STP) (330) A set of agreed-upon conditions for using the ideal gas laws; defined as 273.15 K and 1 atm

state variable (364) A variable physical quantity that defines the state of a material or system; a path-independent variable

static electricity (414) Related to or the production of relatively stationary electrostatic charges

static friction (f_s) (178) Friction acting between two surfaces in contact with no relative motion between them; Static friction opposes any force that may cause relative motion parallel to the surfaces.

stator (494) The stationary part of an electric motor, generator, or other electromechanical device that contains the magnets or coils that surround the rotor

Stefan-Boltzmann constant (σ) (354) A proportionality constant relating radiant energy emitted by a body to its absolute temperature

Stefan's law (354, 642) The rate of thermal emissions radiated by an object; (S) is proportional to the fourth power of its absolute temperature; $S = \sigma T^4$.

step-down transformer (512) A transformer designed such that the input coil has more turns than the output coil; Therefore, output AC voltage is lower than the input voltage.

step-up transformer (512) A transformer designed such that the input coil has fewer turns than the output coil; Therefore, output AC voltage is higher than the input voltage.

steradian (sr) (598) The conical solid angle, with its vertex at the center of a sphere, that cuts off a circular area on the surface of the sphere equal to the radius of the sphere squared

stereo microscope (607) A microscope that has two objectives and two eyepieces, and thus produces a stereoscopic image

storage cell (450) A voltaic cell that uses a reversible chemical reaction to produce an electrical potential

strain (ε) (299) The ratio of the amount of change in a dimension of an object caused by the stress to the object's original dimension; $\varepsilon = \Delta\ell/\ell$.

streamline (396) In fluid mechanics, an imaginary line that shows the path of fluid particles within a flow path

stress (σ) (299) The ratio of the magnitude of the force applied perpendicularly to a specified surface area and the area itself, having units of pressure; also called *normal stress*

strong nuclear interaction force (121) The short-range interaction between nucleons and within other subatomic particles; responsible for holding the nucleus together and for the assembly of many composite subatomic particles; strongest of the four fundamental forces

subatomic particle (684) Any particle simpler than a hydrogen atom and which is one of the major constituents of an atom (e.g., protons, neutrons, and electrons); a simple or composite elementary particle

sublimation (305) A change of phase directly between the solid and gaseous states, bypassing the liquid phase; The opposite phase change is called *deposition*.

subtractive primary color (605) One of the set of pigments that, when added in various combinations, produce all colors possible within that color system by the absorption of incident or transmitted white light; The most common subtractive primary colors used in the publishing industry are cyan, magenta, yellow, and black (CMYK). Other combinations are possible.

superconductor (451) A material that when cooled to a certain temperature offers no detectable resistance to current

superposition principle (574) In wave theory, interacting wave trains within a given medium will generate constructive or destructive interference in proportion to how much they are in or out of phase, their amplitudes, and how closely they match in frequency; applies to any kind of wave or medium

surface tension (304) The tendency for the surface of a liquid to form a barrier to penetration of its particles due to unbalanced cohesive forces on the surface particles

surroundings (124) Everything outside the designated boundaries of a system; The surroundings may be ignored when one is performing a physics analysis unless the surroundings somehow affect the system of interest. See *system*.

system (123) A part of the universe isolated from its surroundings by a physical or conceptual barrier for the purpose of simplifying the analysis of forces on the system

T

tangent θ (80) In a right triangle, the ratio of the length of the side opposite the acute angle θ to the side adjacent to the angle

tangential acceleration (a_t) (143) The time rate of change of the tangential velocity

tangential velocity (v_t) (139) The velocity vector of a point in a rotating or spinning system whose magnitude is equal to the tangential speed and whose orientation is perpendicular to the rotating position vector

telescope (608) An optical instrument that provides high-resolution images of distant objects using lenses or mirrors to collect and focus visible light so that the objects may be examined in greater detail; may also apply to any instrument that can increase the image resolution of an object using any portion of the electromagnetic spectrum (e.g., a radio or infrared telescope)

tensile force (122) A contact force that tends to stretch or pull an object apart

tension (T) (165) A tensile force exerted via a string-like connection between systems; acts in concert with an equal but opposite force at the opposite end of the string

terminal velocity (220) The velocity of a falling object at which air resistance and its weight balance; the limiting velocity at which an object can fall through a fluid

tesla (T) (476) The derived SI unit of magnetic flux density; $1\ \text{T} = 1\ \text{Wb/m}^2$.

test charge (430) A positive charge small enough that it does not distort an electric field and that is used to determine the direction of the electric field vector

theory (9) A general or specific model that attempts to account for a related set of observations or phenomena; It explains many different observations and inferences within the context of the applicable scientific paradigm.

thermal efficiency (ε) (371) The ratio of the absolute value of the net work done by a heat engine to the thermal energy absorbed by the engine from the hot reservoir;

$$\varepsilon = |W_{\text{cycle}}| / |Q_{\text{H}}|.$$

thermal energy (339) The sum total of the kinetic energies of the particles composing an object or substance; measured in joules

thermal equilibrium (360) Two systems are in thermal equilibrium when they are separated by a diathermic boundary, yet they exchange no thermal energy (i.e., they must be at the same temperature).

thermodynamics (12) The study of the sources and utilization of heat, and its conversion to other forms of energy

thermodynamic system (365) A system isolated for the purpose of studying its thermodynamic properties and in which its total mechanical energy is assumed to be constant

thermometer (319) A device that uses a thermometric property of a material to measure and display temperature

thermometric property (320) Any property of a material that varies proportionately with temperature in a way that can be employed to measure or display temperature; Examples include volumetric expansion and electrical resistivity.

thin-lens equation (560) A formula, valid only for thin lenses, that relates object distance, image distance, and lens focal length;

$$\frac{1}{d_{\text{O}}} + \frac{1}{d_{\text{I}}} = \frac{1}{f}.$$

third-class lever (230) A lever-type simple machine having the effort force applied between the fulcrum and the load

third equation of motion (64) The one-dimensional displacement or position of a system is proportional to the difference of the squares of its final and initial velocities during an interval and is inversely proportional to its uniform acceleration;

$$d_x = \frac{v_{2x}^2 - v_{1x}^2}{2\overline{a}_x} \text{ or } x_2 = x_1 + \frac{v_{2x}^2 - v_{1x}^2}{2\overline{a}_x}.$$

third law of thermodynamics (372) The principle that absolute zero is unattainable

thrust (244) The impulsive force on a rocket- or jet-propelled vehicle accompanying the expulsion of propellant exhaust that accelerates the vehicle

time (24) A nonphysical continuum that orders the sequence of events and phenomena

tokamak (682) A type of nuclear fusion reactor using magnetic fields to contain high-temperature hydrogen plasma within an evacuated torus (doughnut-shaped) chamber

torque (τ) (148) The effect of a force applied at a distance from an axis of rotation; the vector product of a force and the moment arm; also called a *mechanical moment*

torr (385) A non-SI unit of pressure; 1 torr of atmospheric pressure will support a column of mercury in an evacuated tube 1 mm high.

torsion (122) A force pair applied to an object that tends to twist it; a mechanical moment or torque

torsion balance (156) A sensitive instrument for measuring an unknown force by balancing it against a known torsional force

total acceleration (a_{total}) (271) The vector sum of tangential and centripetal accelerations acting on an object moving around an axis of revolution

total internal reflection (555) Reflection of the light ray at the interface between two media in which the incident angle in the optically denser medium exceeds the critical angle for the two media

total magnetic flux (Φ) (502) The sum of the flux produced by each loop in a solenoid or inductor coil; net magnetic flux in the core of an inductor

total mechanical energy (*E*) (211) The sum of the potential and kinetic energies of a system
tracer (683) A small amount of a radioactive isotope used to detect the movement of an element through a biological system
traction (f_t) (176) The frictional force that opposes the slipping motion of a wheel, shoe, or other similar device against the ground; the force that permits motion over ground
trajectory (98) The path of a projectile through air or space
transformer (511) An AC electrical device, consisting of two coils of wire wound on a common ferromagnetic core, that employs mutual induction to produce an AC output voltage or current separate and often different from the input values; also used to electrically separate an AC circuit from its source for safety reasons
transistor (466) A semiconductor device consisting of two or more *p-n* junctions; utilizes a small current to control the flow of a larger main current through the device; may be designed for use as an electronic switch or amplifier
transmission axis (590) The orientation that transmits the *E*-field of light undiminished within the crystal lattice of an optically transparent dichroic material; defines the polarization orientation of plane-polarized light transmitted by such a material
transmittance (601) A measure of transparency; the ratio of transmitted luminous flux to the incident flux
transverse wave (281) A wave in which the oscillating component vibrates at right angles to the direction of the wave's propagation
triode (463) A vacuum tube that includes the normal cathode and anode as well as a control grid between them that uses an externally imposed voltage to control the flow of current through the tube; may be employed as an electronic switch or amplifier
triple point (308) The combination of pressure and temperature for many substances in which their solid, liquid, and gaseous phases exist simultaneously in equilibrium
tuning circuit (515) A resonating AC circuit that amplifies only in response to a narrow band of input frequencies
turbulent (396) Fluid flow characterized by discontinuous or irregular streamlines

U

ultraviolet catastrophe (643) The prediction, according to classical physics, that a blackbody will radiate an unlimited amount of energy at high frequencies; This production has been contradicted by actual observations.
unbalanced forces (117) Concurrent forces that sum to a nonzero resultant force; the unequal forces on an object that cause the object to accelerate
unified atomic mass unit (u) (296) The relative mass unit used when dealing with atoms and molecules; equal to 1/12 of the mass of a carbon-12 atom; See *atomic mass unit*.
uniform circular motion (139) Circular motion in which the angular speed is constant
uniformitarianism (7) The belief that physical processes have always followed the same laws—that is, the present is the key to the past; allows for no miracles or extraordinary physical events in the earth's history other than isolated, naturally caused catastrophes
uniformly accelerated motion (60) Linear motion in which acceleration during the time interval of interest is constant; motion in which the average acceleration equals the instantaneous acceleration
unit analysis (30) As a problem-solving method, the cancellation of units in factors on one side of an equation so that the remaining units can be compared to the expected units of the solution
universal gas constant (*R*) (330) The product of Boltzmann's and Avogadro's constants; In SI units, $R = 8.314\ \text{J/(K·mol)}$.
universal gravitation constant (*G*) (156) The proportionality constant in Newton's law of universal gravitation; $G = 6.674 \times 10^{-11}\ \text{N·m}^2/\text{kg}^2$.
unpolarized (589) The condition in which the *E*-waves of the photons in a beam of EM energy oscillate in randomly oriented planes

V

vacuum tube (463) An evacuated and sealed glass tube containing at least an anode and cathode, and usually other control grids for the purpose of controlling and amplifying current through the tube; utilized in early electronic equipment and mostly replaced today by transistors
Van Allen belt (481) One of two toroidal regions encircling the earth high above the atmosphere, consisting of charged particles from the solar wind trapped by the earth's magnetic field
Van der Waals gas (331) A gas that obeys the *Van der Waals equation*, which predicts the relationship between pressure, volume, and temperature at extremes of temperature and/or pressure where the ideal gas law fails
vaporization (305) The general term for the process whereby particles of a solid or liquid enter the gaseous phase; includes evaporation, boiling, and sublimation
vapor pressure (306) The equilibrium pressure exerted by the gaseous phase of a substance in contact with its liquid or solid phase in a closed container under specified conditions of pressure and temperature
vector (50) A physical quantity that requires two pieces of information in order to be completely described—a scalar value and a direction relative to a specified reference frame
vector angle (θ) (74) The direction of a vector measured from a specified reference direction; expressed in degrees or radians
velocity (v) (58) The time rate of change in position in a specified direction
velocity selector (486) An analytical device that uses superimposed electric (*E*) and magnetic (*B*) fields to deflect all particles in a collimated beam except those with a speed of *E*/*B*
velocity-time graph (59) A graphical plot of velocity (usually one component) versus time; may be used to evaluate displacement and accelerations during selected time intervals
vernier scale (35) An auxiliary scale on a measuring instrument used to increase the precision of measurements; provides a precise means for obtaining an additional significant digit in the measurement
virtual image (536) An image formed by light rays that seem to diverge from points behind a mirror or lens; A virtual image cannot be projected without the use of additional optical elements and is always erect.

virtual particle (121) In the exchange-particle theory of forces, a nonmaterial particle that is believed to carry energy and create interactions that we perceive as forces between other subatomic particles; Confirmed virtual exchange particles are photons, gluons (pions), and the *W*/*Z*-particles.

viscosity (η) (403) A measure of the resistance of a fluid to flow; the "thickness" of a fluid

vitreous (414) Electrical behavior like that of glass, which acquires a positive charge when rubbed with silk

volt (V) (433) The derived SI unit used to measure electrical potential and potential difference; 1 V = 1 J/C.

voltage drop (454) The change in electrical potential across a circuit component due to current flow interacting with the component's resistance; The electrical energy is expended flowing through the resistance of the component.

voltaic cell (449) See *galvanic cell*.

voltmeter (460) A device that measures and displays the potential difference between two points in a circuit; A voltmeter is connected in parallel to the circuit during measurement.

W

watt (W) (201) The derived SI unit of power; 1 W = 1 J/s.

wave (280) For mechanical waves, an oscillatory disturbance traveling through a medium; for electromagnetic waves, a self-propagating interaction between oscillating electric and magnetic fields that does not require a physical medium

wave front (532) A three-dimensional model of wave propagation that represents the position of a corresponding point in the wave, such as a wave crest, throughout the medium; a geometrical surface formed by a wave that is perpendicular at all points to the direction of the wave's travel

wavelength (λ) (282) The distance between corresponding points on adjacent waveforms

wave speed (*v*) (282) The speed with which a wave travels through a medium; a property of the medium, not due to the origin or shape of the wave

weak nuclear interaction force (121) One of the four fundamental forces in nature; extremely short-ranged—about 1/1000 the diameter of the nucleus; extremely weak field strength—about 10^{-13} of that of the strong nuclear interaction force; Most significant effect is its involvement in beta decay.

weber (Wb) (502) The derived SI unit of magnetic flux; 1 Wb = 1 V·s.

weight ($\mathbf{F}_w$) (119) The force of gravity acting on an object's mass at the earth or other planet's surface

Wien's displacement law (529) In a blackbody radiator, the wavelength of maximum radiant intensity is inversely proportional to the blackbody's kelvin temperature; $\lambda_{max}T = 2.90 \times 10^{-3}$ m·K.

work (*W*) (196) The energy transferred to or from a system as a force acts on it through a displacement; $W = Fd \cos \theta$.

workability (6) Effectiveness of a model to account for an observed phenomenon as well as to make accurate predictions; makes no claim as to ultimate accuracy or conformance to absolute truth

work-energy theorem (202) The work done on a system by all external forces is equal to the change in the system's kinetic energy.

work function (644) The minimum electromagnetic energy (*hf*) needed to eject the most loosely bound electrons from a metal surface; a feature of the photoelectric effect

worldview (6) The framework or point of view from which a person perceives and interprets all aspects of the world; a collection of presuppositions enabling a person to observe and understand the world

Y

Young's modulus (300) See *elastic modulus*.

Z

zeroth law of thermodynamics (360) Two systems in thermal equilibrium with a common third system are in thermal equilibrium with each other.

zero vector (0) (94) Any vector with zero magnitude; useful for determining difference vectors originating at the origin of the coordinate system

INDEX

D

E

F

G

M

Q

R

S

T

U

V

W

Acknowledgments

Careful effort has been made to trace the copyright ownership of selections included in this textbook. Errors or omissions are inadvertent and will be corrected in subsequent editions, provided written notification is made. The publisher gratefully acknowledges the following individuals and publishers for copyrighted material:

Chapter 2

"History of the Speed of Light Experiments" by Barry Setterfield. Used by permission.

Chapter 3

Aerospaceweb.org: "Mach vs. Altitude Tables" © Aerospaceweb.org. By permission of Aerospaceweb.org.

Chapter 12

Brooks/Cole Publishing Co.: "Comparison of Sound Levels" From REESE. *University Physics (with Info Trac)*, 1E. © 2000 Brooks/Cole, a part of Cengage Learning, Inc. Reproduced by permission. www.cengage.com/permissions.

USA TODAY: "Standard Atmosphere Tables" by Jack Williams, From USATODAY.com. Copyright © 2005 Jack Williams. From USA TODAY, a division of Gannett Co., Inc. Reprinted with Permission.

Chapter 15

Taylor & Francis Group, LLC: "Table 12-196 and 197" Copyright © 2005-2006 From *CRC Handbook of Chemistry & Physics/86th* edition by David R. Lide, Editor-In-Chief. Reproduced by permission of Taylor and Francis Group, LLC, a division of Informa plc.

Chapter 29

Dept. of Geosciences, University of Arizona: "Radiocarbon Dating of Scrolls and Linen Fragments from the Judean Desert" by A.J. Timothy Jull, et al., From *Radiocarbon* Vol. 37, No. 1. Copyright © 1995 A.J. Timothy Jull. From the Department of Geosciences, The University of Arizona and The University of Arizona Libraries.

Earth Science Associates: "Creation's Tiny Mystery", Copyright © 1992 by Robert V. Gentry. Reprinted by permission of Earth Science Associates.

Photograph Credits

The following agencies and individuals have furnished materials to meet the photographic needs of this textbook. We wish to express our gratitude to them for their important contribution.

A2 WIND TUNNEL, LLC
Adam Hart-Davis/DHD Multimedia Gallery
Advanced Micro Devices (AMD)
AFP/Getty Images
AirTeamImages
Alamy
Altizer, Suzanne R.
American Educational Products, LLC
American Physical Society
Answers in Genesis
Aquionics, Inc.
Arbor Scientific
Arrow-Technologies, Inc.
Artemis Images
Associated Press (AP)
AstronomyOnline.org
Boeing
Bjerk, John
The Bridgeman Art Library
Britten, William/lighthousegetaway.com
Carrier Corporation
Carvalho, Tina/University of Hawaii
Cassini Imaging Central Laboratory for Operations (CICLOPS)
CelebrateBoston.com
Celestron
The College of Wooster
Commissariat à l'Énergie Atomique (CEA)
Commonwealth Scientific and Research Organization (CSIRO)
Conseil Européen pour la Recherche Nucléaire (CERN, Geneva, Switzerland)
Corbis
Coss, Dr. Tom
Creation Ministries, Inc.
Creation Ministries International
Crosby, Benjamin
Curtis, Jan
Department of Defense (DoD)
Department of Energy (DoE)
Dept. of Terrestrial Magnetism, Carnegie Institution, Washington
DHD Multimedia Gallery
Dieman, Paul
Digital Vision
Dilyard, Matt
Dorling Kindersley
Drama in Nature/G. Arnell Williams
Edgar Fahs Smith Collection, University of Pennsylvania Library
Edmund Industrial Optics
Edwards AFB History Office
Elekta, Inc.
Ely Diocesan Association of Church Bell Ringers
Energy Northwest
Espenak, Fred/www.MrEclipse.com
European Southern Observatory (ESO)
European Space Agency (ESA)
Fisher, Tammy
Fitzgerald, Jim
Flammarion, Camille
Florida Park Service
Fluke Biomedical
Fotolia.com
French, Peter
Gentry, Dr. Robert V.
Getty Images
GOMACO Corporation
Goshen College
Hansen, Brenda
Harris, Marty/McDonald Observatory
Hemera Technologies, Inc.
High Field Magnet Laboratory/Radboud University/Nijmegen, the Netherlands
Hinton, Peter
Howell, Daniel and R.P. Behringer/Duke University
HowStuffWorks.com
IBM Research/Almaden Research Center
Instituto e Museo di Storia della Scienza (IMSS)
International Business Machines Corporation (IBM)
International Thermonuclear Experimental Reactor (ITER)
iStockphoto.com
Jagiellonian Library
Jenkins, John/www.sparkmuseum.com
Johns Hopkins University
Johnson, Thomas E.
JupiterImages Corporation
Koh, Tracey/ButterflyUtopia.com
Landis, Joyce
Lawrence Berkeley National Laboratory
Library of Congress
Lockheed Martin
Los Alamos National Laboratory
Ludlum Measurements, Inc.
Mansion House Maple Syrup
Maus, S./University of Colorado
Mitchell, Rita
Mitutoyo America Corporation
NASA Ames Research Center
NASA and Space Telescope Science Institute (STSci)
NASA Jet Propulsion Laboratory (NASA-JPL)
NASA/JPL/Caltech
NASA/JPL/CICLOPS
NASA Marshall Space Flight Center (NASA-MSFC)
National Academy of Sciences, USA
National Aeronautics and Space Administration (NASA)
National Center for Atmospheric Research (NCAR)
National Institute of Standards and Technology (NIST)
National Maritime Museum
National Oceanic and Atmospheric Administration (NOAA)
National Parks Department
National Renewable Energy Laboratory (NREL)
National Science Foundation (NSF)
Nature Magazine
Neiconi, Emanuel
Northrop Grumman
Oak Ridge National Laboratory
Ocean Remote Sensing Group/Johns Hopkins University Applied Physics Laboratory
Ohaus Corporation
Parsons, David
Perry, Susan
Pflug, Kathy
PhotoDisc/Getty Images
Photo Researchers, Inc.

photovault.com
Repower Systems AG
Reuters
Robert Morris Photography
Rodon Products, Inc., Huntington Beach, CA
Rorah, Robert A.
Rowin, Stanley
Science Kit and Boreal Labs
Science Magazine
Science Photo Library
Shimadzu Corporation
Six Flags
Smithsonian Institution
Solar and Heliospheric Observatory (SOHO)
STEM Labs, Inc.
Strainoptic Technologies, Inc.
Super Stock, Inc.
Texas Instruments
United States Air Force
United States Bureau of Reclamation
United States Department of Transportation
United States Geological Survey (USGS)
United States Marine Corps
United States Navy
University of Utah Andrology Laboratory
Unusual Films
Visuals Unlimited, Inc.
VWR International
Ward's Natural Science Laboratory, Inc.
Washington State Historical Society
Wikimedia Commons
Wikipedia.com
W. M. Keck Observatory
Woodfill, Jerry
World Figure Skating Museum and Hall of Fame
www.arttoday.com
Zero G
Zurbuchen, Thomas

Front Matter

NASA iv; Zero G xi

Unit Openers

Courtesy CERN, Geneva, Switzerland (all unit openers); Original black-and-white image by Camille Flammarion, 1888; Courtesy History of Science Collections, University of Oklahoma Libraries/Colorized by John Bjerk xiv–xv; PhotoDisc/Getty Images 46–47, 522–523 (background); © 1997 by the Ocean Remote Sensing Group, Johns Hopkins University Applied Physics Department 290–91 (satellite image); Image courtesy AIRS Science Team, NASA/JPL 290 (background); Jan Curtis 410–411 (aurora); ©Tracey Koh/ButterflyUtopia.com 522–523 (butterfly); AFP/Getty Images 616–617 (man in Tore Supra); Published with permission of ITER 616–617 (inset)

Chapter 1

Reuters xvi; AP Photo/John Rucosky 1 (top); Lockheed Martin 2 (middle); John McColgan/Wikipedia.org 2 (bottom); NASA 3 (top), 12 (middle, bottom), 13 (middle), 14 (top, middle left); AP Photo/Wang Jiaowen, ColorChinaPhoto 3 (bottom); Getty/The Image Bank/WIN-Initiative 4 (top); © David R. Frazier Photolibrary, Inc./Alamy 4 (middle); AP Photo/Gene J. Puskar 4 (bottom); Kathy Pflug 5; A2 WIND TUNNEL, LLC, www.A2WT.com 6 (top); Unusual Films/© 2008 JupiterImages Corporation 6 (bottom); Jeremy Walker/Science Photo Library 7 (bottom); Copyright © 2008 Stanley Rowin 8; © California Association for Research in Astronomy/Photo Researchers, Inc. 9; © Geoff Tompkinson/Photo Researchers, Inc. 10; Copyright © CERN/Photo Researchers, Inc. 11 (middle); Wikimedia.org 11 (bottom left); Portrait of Sir Isaac Newton after an original painting by Sir Godfrey Kneller (oil on canvas) by Goldschmit, 1847(detail), Academie des Sciences, Paris, France/Archives Charmet/The Bridgeman Art Library 11 (bottom right); © 2008 www.arttoday.com 12 (top), 13 (bottom), 14 (bottom left); Simon Frasere/Photo Researchers, Inc. 13 (top); © Clark Brennan/Alamy 14 (middle right); © 2003 Hemera Technologies, Inc. All rights reserved 14 (bottom right); Courtesy Northrop Grumman 15; NOAA 16; edhiker/Wikimedia.org 17 (top); NASA, ESA/Wikipedia.org 17 (bottom); Used by permission from Benjamin Crosby 18 (top); PhotoDisc/Getty Images 18 (bottom); © National Academy of Sciences, USA/© 2008 Nature, by permission from Macmillan Publishers Ltd/Science 2008, reprinted with permission from AAAS/photo by Unusual Films 19 (top); Answers Research Journal used by permission of Answers in Genesis, www.answersingenesis.org/CSRQ used by permission of Creation Ministries International/Journal of Creation used by permission of Creation Ministries, Inc/photo by Unusual Films 19 (bottom)

Chapter 2

NASA/JPL/CICLOPS 22; NASA and STSci 24; National Institute of Standards and Technology 25, 26 (top); Peter Hinton, Ely Diocesan Association of Church Bell Ringers 26 (bottom); NASA 27 (top); © National Maritime Museum 27 (bottom); © Copyright CSIRO Australia, 2008 28 (top); STEM Labs, Inc. 28 (bottom); David R. Frazier Photolibrary, Inc./Photo Researchers, Inc. 31 (top); U.S. Geological Survey 31 (middle); Unusual Films 31 (bottom both), 33 (top), 35 (middle left, bottom), 36, 42; From Christiaan Hugyens' Treatise On Light, translated by Silvanus P. Thompson 33 (bottom); Ohaus Corporation 35 (top); Mitutoyo America Corporation 35 (middle right); Texas Instruments images used with permission/Rita Mitchell 37; © 2008 www.arttoday.com 41

Chapter 3

U.S. Navy photo by Photographer's Mate First Class William R. Goodwin 48; NASA 49 (top left), 56, 57; NCAR/NSF 49 (top right); PhotoDisc/Getty Images 49 (bottom left), 54, 66; Dr. Tom Coss 49 (bottom right); Robert Morris Photography 58; Edwards AFB History Office 62; Emanuel Neiconi 65

Chapter 4

PhotoDisc/Getty Images 72, 75 (bottom); AP Photo/Stan Honda 73 (top); DoD photo by Cpl. Branden P. O'Brien, U.S. Marine Corps 74; U.S. Geological Survey 75 (top); National Parks Department in Washington, DC 84; NOAA 88

Chapter 5

AP Photo/Vincent Thian 92; AP Photo/Jockel Finck 93; Suzanne R. Altizer 96, 99; U.S. Geological Survey 98; U.S. Air Force/Wikimedia.org 101; Corbis 104; U.S. Air Force photo by Senior Airman Ricky J. Best 106; PhotoDisc/Getty Images 107

Chapter 6

NASA 112, 114, 128; U.S. Navy photo by Senior Chief Mass Communications Specialist Spike Call 113; Wikimedia.org 116 (left); Principia Cover/Brenda Hansen 116 (right); SOHO (ESA & NASA) 117; © Humberto Olarte C./Alamy 118; © 2003 Hemera Technologies, Inc. All rights reserved 121; Ohaus Corporation 122 (top), 127 (bottom), 132 (bottom all); © 2009 JupiterImages Corporation 122 (bottom); Courtesy of Science Kit and Boreal Labs, www.sciencekit.com 123; U.S. Navy photo by Martin S. Fuentes 126 (top); © iStockphoto.com/Vlado Janžekovi 126 (bottom); Corbis 127 (top); U.S. Navy/MC2 Robert C. Long 129 (top); U.S. Navy/MCSN Brandon Morris 129 (bottom); Copyright © Boeing. All Rights Reserved 130 (top); U.S. Air Force 130 (bottom); © iStockphoto.com/Jill Fromer 132 (top); Eric Gaba/Wikimedia.org 135

Chapter 7

NASA/JPL 136, 137 (top); Ricky Leon Murphy, MSc: AstronomyOnline.org 137 (bottom); Unusual Films 142 (top), 142 (bottom); Brenda Hansen 145; Jagiellonian Library 152 (top); © 2008 www.arttoday.com 152 (bottom); © iStockphoto.com/Drazen Vukelic 154 (top); NASA 154 (bottom); Courtesy of VWR International 157; NASA/JPL/Space Science Institute 158; Courtesy of Six Flags 162

Chapter 8

© 2008 Hemera Technologies, Inc. All rights reserved. 164; © iStockphoto.com/Tammy Bryngelson 165; Artemis Images 172, 176 (bottom); NASA 175, 191; PhotoDisc/Getty Images 176 (top); Brenda Hansen 181; Wikimedia.org 187; NASA Ames Research Center 192

Chapter 9

PhotoDisc/Getty Images 194, 203 (bottom), 209; Brad Sauter © Fotolia 195 (right); © iStockphoto.com/landbysea 195 (left); Edward Kinsman/Photo Researchers, Inc. 203 (top); AP Photos/Du Huaju 213

Chapter 10

NASA Marshall Space Flight Center (NASA-MSFC) 218; Paolo Donati © Dorling Kindersley 219; NASA 220 (all); Carrier Corporation 225; © 2008 JupiterImages

Corporation 228a; © iStockphoto.com/Alasdair Thomson 228 (bottom); Steve Carroll © Fotolia 228 (center); PhotoDisc/Getty Images 231; Library of Congress, Prints and Photographs Division, Historic American Buildings Survey or Historic American Engineering Record, Reproduction Number HAER NY, 31-NEYO, 81-16 232; Image provided courtesy of HowStuffWorks.com/BJUP 233

Chapter 11

NASA 238; © Enigma/Alamy 239; © 2008 Getty Images 240; Galinka © Fotolia 242; Photo of Air Track device used with permission of American Educational Products, LLC, Fort Collins, CO 246; Steven Belanger © Fotolia 247; © iStockphoto.com/A. C. Maritz 248; Courtesy of VWR International 252; Wikipedia.org 254 (top); © Trip/Alamy 254 (bottom); Cover is from the collection of Jerry Woodfill, used by his permission 255; NASA Jet Propulsion Laboratory (NASA-JPL) 256; Drama in Nature—G. Arnell Williams 257; © 2008 www.arttoday.com 259 (top); Joyce Landis 259 (bottom); World Figure Skating Museum and Hall of Fame 260

Chapter 12

NASA 264; U.S. Navy 265; Copyright © The Bridgeman Art Library 270; National Museum of American History, Smithsonian Institution 274; AP Photo/Dima Gavrysh 276; Unusual Films 277; © iStockphoto.com/Patricia Nelson 278 (bottom); Goshen College 279; Washington State Historical Society 280 (both)

Chapter 13

NOAA 292; Alain LE BOT/Getty Images 293 (top); © 2003 Hemera Technologies, Inc. All rights reserved 293 (bottom), 294 (bottom), 295 (top); NASA 294 (top); © CERN 295 (bottom); Unusual Films 296, 303, 305 (all), 307; © Glow Images/Super Stock, Inc. 304

Chapter 14

PhotoDisc/Getty Images 312; © PHOTOTAKE Inc./Alamy 313; U.S. Department of Transportation 314 (top); GOMACO Corporation 314 (bottom); Unusual Films 316, 320 (bottom), 325 (all), 326 (all); Joyce Landis 320 (top); Dell Foutz/Visuals Unlimited, Inc. 332 (top); John Birdsall/Visuals Unlimited, Inc. 332 (bottom)

Chapter 15

© 2003 Hemera Technologies, Inc. All rights reserved 338; © 2008 JupiterImages Corporation 339 (left, middle); Nuclear RegulatoryCommission/Wikimedia.org 339 (right); Unusual Films 340 (all), 345 (top); Courtesy of Shimadzu Corporation 345 (bottom); Suzanne R. Altizer 348

Chapter 16

Image By: Wernher Krutein/photovault.com 358; © JupiterImages/Brand X/Alamy 359; SOHO (ESA & NASA) 367; NASA 355; Carrier Corporation 371; Photo Courtesy of the Florida Park Service 373; Science VU/IBMRL/Visuals Unlimited, Inc. 376; NASA, R. Williams and The Hubble Deep Field Team (STSci) 378; University of Utah Andrology Laboratory 379

Chapter 17

U.S. Bureau of Reclamation 382; Paul Dieman 395 (left); Jim Fitzgerald/Courtesy of Mansion House Maple Syrup 395 (right); Unusual Films 397, 403 (top left), 403 (bottom left); Steve Morris/AirTeamImages 403 (top right); © iStockphoto.com/Dieter Spears 403 (bottom right); Repower Systems AG 404; Getty Images/Stone/Michael Rosenfeld 405 (left); Stefan Emmanuel © Fotolia 405 (right); DoE/NREL 408

Chapter 18

PhotoDisc/Getty Images 412; © Mauro Rinaldi/Alamy 413 (top); © 2003 Hemera Technologies, Inc. All rights reserved 413 (bottom); Ward's Natural Science Laboratory, Inc. 414 (left); Unusual Films 414 (middle, right), 419, 420 (both)

Chapter 19

PhotoDisc/Getty Images 428; © 2008 JupiterImages Corporation 429 (top left, bottom left); Fatman73 © Fotolia.com 429 (right); Science VU/IBMRL/Visuals Unlimited, Inc. 435; Unusual Films 442

Chapter 20

Reprint Courtesy of International Business Machines Corporation, copyright © International Business Machines Corporation 446, 467; © 2008 JupiterImages Corporation 447; Matt Dilyard/Courtesy of The College of Wooster 449; Edgard Fahs Smith Collection, University of Pennsylvania Library 452; Unusual Films 461; © The London Art Archive/Alamy 463; NASA 472 (left); AP Wide World Photo/Steve Yeater 472 (right); David Parsons 473 (bottom)

Chapter 21

NASA 474, 481 (top); Unusual Films 475, 493; High Field Magnet Laboratory, Radboud University, Nijmegen, the Netherlands 477; Joyce Landis 479 (top); Dept. of Terrestrial Magnetism, Carnegie Institution, Washington 479 (bottom); Reproduced with permission of S. Maus, University of Colorado 480; John Jenkins/www.sparkmuseum.com 488

Chapter 22

U.S. Department of Energy 498; Courtesy of CelebrateBoston.com 499; Courtesy of VWR International 501; Unusual Films 505, 511; Robert A. Rorah 508; Courtesy Rodon Products, Inc., Huntington Beach, CA 509; Thomas E. Johnson 520

Chapter 23

Marty Harris/McDonald Observatory 524; NASA/Jimmy Westlake of Yampa, Colorado 525 (top); Image Courtesy of Arbor Scientific 525 (bottom); David Parker, 1997/ Science Photo Library 526 (top); Courtesy of NASA/JPL/Caltech 526 (bottom); Compliment of Aquionics, Inc. 527 (top); © 2003 Hemera Technologies, Inc. All rights reserved. 527 (middle); NASA 527 (bottom), 528, 537; PhotoDisc/Getty Images 529; Sergfoto © Fotolia 530 (top); NASA, Goddard Space Flight Center 530 (middle); © iStockphoto.com/BrooksElliott 530 (bottom left); Susan Perry 530 (bottom right); NASA/JPL/Space Science Institute 536

Chapter 24

Unusual Films 550, 557, 567 (top left, bottom); © 2009 JupiterImages Corporation 551 (top left, center, top right, bottom); Tammy Fisher 556 (top); NASA 556 (center); Adam Hart-Davis/DHD Multimedia Gallery 556 (bottom); Telescope photo provided by Celestron, a manufacturer of telescopes and related accessories, binoculars, spotting scopes and microscopes. Product information available at www.celestron.com 566; Photo Courtesy of Edmund Industrial Optics 567 (top right); Photo by William Britten/lighthousegetaway.com 570

Chapter 25

Lawrence Lawry/SPL/Photo Researchers, Inc. 572; Rita Mitchell 573 (top, bottom); Unusual Films 579, 585, 591 (center, bottom), 592 (top); NASA 582 (top); PhotoDisc/ Getty Images 582 (bottom); American Physical Society/Johan Mauritsson and Marko Swoboda 583; Image courtesy of AMD 586; Joyce Landis 591 (top); Strainoptic Technologies, Inc. 592 (bottom); Daniel Howell and R.P. Behringer, Duke University 593

Chapter 26

Peter French 596; © iStockphoto.com/Nic Taylor 597; Courtesy of VRW International 600; Digital Vision 602; Rijksmuseum/Wikimedia.org 606; PhotoDisc/Getty Images 607 (top, bottom); Instituto e Museo di Storia della Scienza 608; W. M. Keck Observatory 610; Unusual Films 611

Chapter 27

European Southern Observatory 618; NASA 629 (top, bottom), 635 (bottom), 637; © CERN 632 (left); National Institute of Standards and Technology 332 (right); NASA, ESA, J. Blakeslee and H. Ford (Johns Hopkins University) 635 (top); © Fred Espenak, www.MrEclipse.com 636 (top); © iStockphoto.com/Henryk Sadura 636 (bottom)

Chapter 28

Courtesy: IBM Research, Almaden Research Center. Unauthorized use not permitted. 640, 655 (right); Inga Spence/Visuals Unlimited, Inc. 654; Lawrence Berkeley National Laboratory 655 (left); Tina Carvalho/University of Hawaii 655 (bottom); NIST 658

Chapter 29

Courtesy of Energy Northwest 662; Grauesel/Wikipedia.org 663; Edgar Fahs Smith Collection, University of Pennsylvania Library 664; Richard Firestone/Lawrence Berkeley National Laboratory 665 (top); Courtesy of VWR International 668 (top); Photo courtesy of Ludlum Measurements, Inc. 668 (bottom); Arrow-Technologies, Inc. 669 (top left); Photo courtesy of Fluke Biomedical 669 (bottom left); Courtesy CERN, Geneva, Switzerland 669 (top right); Courtesy DELPHI experiment, CERN, Geneva, Switzerland 669 (bottom right); AP/Wide World Photos 672; Used by permission of Dr. Robert V. Gentry 674; NOAA 676; Courtesy of Oak Ridge National Laboratory, managed by UT-Battelle, LLC, for the U.S. Dept. of Energy 680; Los Alamos National Laboratory 681; Published with permission of ITER 682 (top); U.S. Department of Energy 682 (bottom); Elekta, Inc. 683